Nature Conservation 5:

Conservation in Production Environments: Managing the Matrix

Nature Conservation 5:

Conservation in Production Environments: Managing the Matrix

Edited by

John L. Craig, Neil Mitchell and Denis A. Saunders

Published by

Surrey Beatty & Sons Pty Limited

The National Library of Australia
Cataloguing-in-Publication entry:

Conservation in Production Environments: Managing the Matrix.

Includes index.

ISBN 0 949324 88 4.

1. Nature conservation — Congresses. 2. Agricultural ecology — Congresses. I. Craig, John L. II. Saunders, Denis A. III. Mitchell, Neil, 1950– . (Series: Nature conservation; 5)

333.72.

Released December 2000

PRINTED IN AUSTRALIA BY
SURREY BEATTY & SONS PTY LIMITED
43 Rickard Road, Chipping Norton, NSW 2170

PREFACE:

NATURE CONSERVATION SERIES

The first volume in Surrey Beatty & Sons' Nature Conservation Series, *Nature Conservation: the Role of Remnants of Native Vegetation* (1987) was based on an international conference held at Busselton, Western Australia in 1985. It examined the conservation and management of remnants of native vegetation. During this conference the issue of landscape linkages or connectivity as a method of fostering the movement of native biota was raised. Accordingly, the role of linear strips of vegetation or corridors in landscape function was discussed at an international conference at Busselton in 1989. This material was published as *Nature Conservation 2: the Role of Corridors* (1991).

The issue of reconstruction of fragmented ecosystems was explored at a workshop held at Tammin, Western Australia in 1991. Ecologists from Australia, Canada, Finland, Norway, Sweden, UK and USA met to discuss ways in which their ecological ideas could be put to practical application towards landscape reconstruction. This workshop formed the basis of *Nature Conservation 3: Reconstruction of Fragmented Ecosystems — Global and Regional Perspectives* (1993).

At all three meetings, the importance of people in achieving conservation, and the issues of how networks of people can be formed, maintained and involved in nature conservation were raised. So an international conference on this topic was held at Geraldton, Western Australia in May 1994. The results of this meeting constituted *Nature Conservation 4: the Role of Networks* (1995).

The growing importance of integrating nature conservation in production land and seascapes provided the impetus for this volume, the fifth in the series *Nature Conservation in Production Environments — Managing the Matrix*. This volume is based on an international conference held in Taupo, New Zealand in 1997.

The Nature Conservation Series has developed along the lines that the emerging discipline of Conservation Biology has followed. That is, each volume examines a range of issues that are aimed at relating ecology and conservation biology to the practical applications necessary for reconstructing our rapidly degrading world. There is a need to develop the science of Conservation Biology and see that this science is applied. To this end, the Nature Conservation Series will continue to publish work that fits this mould.

The Executive Editor and Publisher are interested in receiving proposals for future volumes in this series. Any proposals should be based on the proceedings of an international conference or workshop exploring a theme of international significance within the topic of nature conservation.

Denis A. Saunders
Executive Editor

Ivor E. Beatty
Surrey Beatty & Sons

PREFACE

John Craig,[1] Denis Saunders[2] and Neil Mitchell[1]

The Nature Conservation Series has grown from the initial conference/workshop in 1985 on the role of remnants that was attended mainly by academics and conservation managers, to the networks conference at Geraldton in 1994, that was attended by academics, conservation managers, students from secondary and tertiary institutions, members of non-government organizations concerned with nature conservation, farmers and individuals. In the lead up to the Geraldton conference, Denis Saunders and John Craig discussed whether there would be a follow up to *Nature Conservation 4: The Role of Networks*. It became apparent to John and Denis during the conference that there was considerable pressure for a conference following-up on some of the issues raised at Geraldton.

As the 1990s progressed, there was an increasing realization that if conservation of biodiversity was to achieve what most people wanted, then relying on a system of reserves run solely by government agencies would be inadequate. Not only did conservation have to become part of private and commercial enterprise, conservation needed to be integrated across the total matrix of land and water. Retaining the territorial mindsets that had generated reserves and non-reserves with different environmental, economic and social rules on each side of the boundary appeared little more than an extension of the zoological park attitude. Many of the talks at the Networks Conference in Geraldton expressed the need for a broader view on conservation and so the theme for the conference on which this volume is based emerged by the final discussion session of the conference.

The conference structure adopted for the Nature Conservation Series has been to meet in a town where few participants are locals, provide excellent food plus ample opportunities for social interactions to encourage networking among participants, and to provide the opportunity for as many delegates as possible to present their ideas. This structure has been extremely successful. In addition, the condition of acceptance of an offer to present a paper at each conference has been conditional on the presenter providing the organizers at the conclusion of the conference with a manuscript suitable for refereeing for the proceedings. This conference structure and conditions associated with presentation has resulted in a book series that has become an important reference set for conservation. New Zealand, offering to host the next conference, provided the chance to make the series more international while retaining an Australasian focus.

A conference organizing committee of Denis Saunders, John Craig, Neil Mitchell and Simon Towle of World Wide Fund for Nature New Zealand was formed. Taupo, on New Zealand's North Island, was chosen as an ideal venue. Not only did it have excellent conference facilities, it had stunning scenery, a full range of accommodation and restaurants as well as access to a wide diversity of production and conservation landscapes. The organizers met with potential sponsors as well as government agencies and advertised the conference internationally. All papers given at the conference were offered and no attempt was made to present a particular view using an invited address. Moreover, we sought participation from a wide set of industries including those under attack from non-government agencies involved in conservation. Our attitude was that if we are to manage biodiversity across the full matrix, then there can be no boundary and everyone has to be welcomed into discussions. A single incident of graffiti on a poster of a major sponsor of the conference showed that a minority of participants held a narrower view of the way towards effective dialogue and collaborative management.

The committee sought the assistance of Mel Galbraith to involve senior secondary students from both Auckland and Taupo. This approach to listening to the views of some of "the next generation" had been a marked success at Geraldton. As expected, the students rapidly delivered a message of impatience and hope. They despaired, like those at Geraldton had, at the way conservation professionals and others talked past each other and used obtuse communication. Given money to reward the better speakers, they chose David Lindenmayer, Wendy Lynch and Jon Elder.

Two hundred and eight people from eight countries attended all or part of the conference. Ninety-three spoken papers and 16 poster papers were delivered at the conference. All speakers were required to submit a manuscript for refereeing and were given until early 1998 to revise their manuscript in light of what they had heard at the

[1]School of Environmental and Marine Sciences, University of Auckland, Private Bag 92019, Auckland, New Zealand.
[2]CSIRO Wildlife and Ecology, G.P.O. Box 284, Canberra, Australian Capital Territory, Australia 2601.

conference. Most did so. Authors of posters were given the opportunity to present a short communication on their work and eight used this opportunity. All manuscripts were sent to two referees for review and the three of us relied heavily on these comments in determining acceptance. Seventeen manuscipts were withdrawn or rejected through this process.

The chapters in the volume are set out in groupings of similar subject matter with a sequence similar to their order at the conference. The final chapter represents the editors' overview of the material presented at the conference and our recommendations for future directions.

ACKNOWLEDGEMENTS

The success of the Taupo conference and the production of these proceedings were due to the commitment and enthusiasm of many people. We are extremely grateful to the following individuals and organizations for their assistance in the lead up to and the conduct of the conference, and in the preparation of these proceedings.

Bill Dyck and Carter Holt Forests plus Kit Richards and Timberlands West Coast provided generous financial assistance that allowed eight high school students to attend free of charge and 22 students to participate at a much reduced rate. Carol Ellis of Worldwide Fund for Nature New Zealand organized conference carry bags. Telecom provided free national phone and fax facilities for conference delegates, and Taupo Native Plant Nursery provided potted native trees to decorate the venue.

Timi Te HueHue of Ngati Tuwharatoa welcomed participants and opened the conference. The Hon Dr Nick Smith, Minister of Conservation, delivered the opening address. The Mayor of Taupo District Council, Mrs Joan Williamson welcomed delegates to the conference. The Manager and staff of the Great Lakes Centre ensured that the conference ran smoothly at their superb venue. Chris Johnston and staff of Replete Gourmet provided excellent food throughout the conference. Carter Holt Harvey and Harry Keys and Paul Green of the Tongariro-Taupo Conservancy of the Department of Conservation organized field trips for conference participants. The Taupo Tourist Information Bureau organized accommodation for conference delegates.

Anne Brown and Margaret Lichtwark from Auckland University acted as conference secretaries. Monique Devondens, Rob Jessop, Graham Ussher, V Saunders and Katherine Short assisted with organizing the venue, registration and projection.

Conference sessions were chaired by: Margaret Brock, John Craig, Alison Davis, Jean Edwards, Mel Galbraith, David Goldney, Toni Hayman, Richard Hobbs, Andrew Hull, Robert Lambeck, David Lindenmayer, Sue McIntyre, Craig Miller, Neil Mitchell, Ray Nias, Emma Pharo, Jamie Pittock, Kelly Price, Helen Ritchie, Denis Saunders and Denise True.

Refereeing of manuscripts submitted for publication in these proceedings was carried out by: Mike Austin, David Barker, Mark Bellingham, Carl Binning, Paul Blaschke, Ann Boggiano, Dick Braithwaite, Lynn Brake, Ann Brandenburg, Clive Briffett, Sue Briggs, Margaret Brock, Stuart Bunn, Peter Cale, Lisa Chambers, Bruce Clarkson, Rochelle Constantine, Don Cowley, John Craig, Jenny Davis, Peter Davis, John Dell, Kath Dickenson, Alison Doley, Grant Dumbell, Rachel Ebbett, Jean Edwards, Pip Forer, Richard Forman, Mel Galbraith, David Goldney, Romy Greiner, Ron Hacker, Rod Hay, Richard Hobbs, Stuart Harrop, Pierre Horwitz, Penny Hussey, John Ive, Sue Jennings, Rob Jessop, Bronwyn Keighery, Richard Kingsford, Jamie Kirkpatrick, Robert Lambeck, David Lamont, Margaret Lawton, Gillian Lee, Wendy Lynch, Tim McBride, Janet McDonald, Mike McFarlane, Sue McIntyre, Allan Mark, Louise Marra, Colin Meurk, Joanne Millar, Craig Miller, Jim Miller, Neil Mitchell, Sue Moore, Steve Morton, Owen Nichols, Jim Noble, David Norton, Donna Nunan, Margaret O'Brien, Bob Pressey, Patrick Prevett, Kit Richards, Helen Richie, Ralph Riley, Mere Roberts, Paul Ryan, Rod Safstrom, Guy Salmon, Denis Saunders, Melody Serena, Basil Sharp, Greg Siepen, Simon Smalley, Ian Spellerberg, Willie Smith, Graeme Smith, Anne Stewart, Laura Stocker, Lesley Stone, Ian Swingland, David Towns, Denise True, Graham Ussher, Marjorie van Roon, Ken Wallace, John Watson, Garry Whisson, John Williams, Colin Yates and Euan Young.

We are especially grateful to Ivor E. Beatty and staff of Surrey Beatty & Sons for their encouragement and support of the conference including flyers and programmes, these proceedings and the Nature Conservation Series.

OPENING ADDRESS

by

Hon. Dr Nick Smith, the Minister of Conservation

The history of Taupo is very important in New Zealand's story of conservation. It is the birhplace of New Zealand's national parks. In 1887, Ngati Tuwharetoa chief, Te Heuheu Tukino donated the land surrounding Tongariro, to become the basis of our first national park.

It is timely as we approach the millennium to review our guardianship, or Kaitiakitanga of this planet over the last century. It does not look a pretty picture. The century has been one of indulgence and abuse, that if we repeat through the next century, the very survival of mankind will hang in the balance.

In the last century we have extensively damaged the globe's forests. World-wide only 20% of our ancient forests are still standing. Seventy-six countries have cut down all their ancient forests. We have plundered the earth's fishing stocks, to the point where some fisheries have completely collapsed. We have poisoned our atmosphere with billions of tonnes of carbon and sulphur, and used chemicals that have put the ozone layer at risk. Man induced erosion, desertification and chemical contamination has made vast tracts of land unusable. With the loss of forests and wetlands we have caused the extinction of 20% of the world's species of birds. Over that same century our population has increased six fold. Only a fool would pretend we could take the same destructive practices with us from the 20th century into the 21st.

This conference is about providing hope and inspiration. It is about going beyond the cliches and into some of the serious grunty issues that we need to confront in the new century. The theme of conserving nature in productive environments is one of the more difficult, but relevant, challenges of our future. The policy questions of managing an area set aside for conservation are relatively easy. Here in New Zealand we have set aside about 30% of our land area for conservation, and that is one of the highest proportions of any country in the world. But if we are serious about the challenges of the new century we need to do more than provide sanctuaries for nature, we need to see how we can bring together the outcomes for both conservation and production on the far larger land mass of unprotected areas. There are those in the Green movement who take the view that the choice is between the environment or economic growth. It is either one or the other. I hear the same sort of observations from some of our backward looking business leaders. I openly acknowledge that a pro-environment, pro-growth strategy is far more intellectually challenging. It requires us to work harder and smarter but it is the only way forward. We have got to earn a living as a country. We want to see improvements in our education and health care systems — and that comes at a cost. It is immoral to derive such growth by mining our environmental balance sheet. The challenge at the core of the conference agenda is how to marry together those conservation and economic objectives to create a win-win situation.

I want to talk about some of the sectors in which this challenge is brought into stark focus. I want to talk about the revolution that has occurred in the forestry sector over the last two decades. I want to talk about the conflicts within our fishing industry and the risks we are running with the high incidence of sea-bird by-catch and accidental killings of marine mammals, as well as the issues of over-fishing. I also want to talk about some issues in the farming sector, and questions over South Island High Country leases. Tourism too has environmental impacts to be managed and I want to touch on those as well. I want to conclude in talking about some key strategic issues for the New Zealand Department of Conservation with respect to bio-diversity and environmental indicators.

As I outlined earlier, I want to start with a discussion concerning some areas in New Zealand, where we are trying to manage the matrix between conservation and environment.

FORESTRY

To talk about forestry, I am going to start talking about fishing for trout in this beautiful area of Taupo. This fishery is considered one of the best in the world – thanks to some forward-thinking in the 1970s. Then, the Lake Taupo forest was established, along with a scheme to plant the margins around the edge of the lake. This scheme saw land adjacent to streams, or prone to erosion, retired from grazing and fenced out. In some instances these were replanted with native or timber species but usually were just left to revegetate naturally, forming a buffer along the stream margins. Not only has the scheme had many conservation benefits but it has also provided financial benefits. It is conservatively estimated that 5% more land than that retired would have been lost to pastoral farming from erosion if the scheme was not in place.

On the eastern side of the lake, Lake Taupo Forest provides an excellent example of a development which contributes significant conservation benefits and also a major return for the owners. In 1969 a lease was drawn up between Ngati Tuwharetoa, as the landowners of 31 000 hectares of land, and the Crown, to establish this project. The primary objectives were to prevent soil erosion, conserve and protect fish and wildlife habitat and to safeguard historic and sacred sites and areas of natural beauty and unique vegetation. Only in a manner consistent with these objectives was a fourth objective to maximize the financial return from the forest.

Roading was planned and implemented in consultation with the fishery managers so that it did not act as a source of sediment into the streams. Fish passes were also installed in the culverts so that passage was not blocked for migrating fish species. The forest was developed leaving extensive riparian margins along all the water courses, up to 400 m wide in places and sites of historic significance or with special features were left untouched. Several important rivers pass through the forest but the riparian strips have proven very successful in filtering off sediment loss and nutrients. The forest is now at an age for harvesting and the combination of extensive riparian margins, established roading network and carefully managed harvesting operations means that the logging is having no measurable impact on water quality and the owners are receiving their long awaited return. And the fishing in Lake Taupo is brilliant.

So as you think about issues of conservation in production environments, do not forget to look out the window. Think about the real world, and apply your ideas to solving actual problems. Perhaps this conference will be the genesis of an idea or concept, of the calibre of those 20 years ago, which promoted the Lake Taupo Forest project.

FISHING

At sea, in the fishing industry, there are classic problems in nature conservation, arising from production. One example of this is around the sub-antarctic islands. In these southern waters, Hooker sea lions are being caught accidentally by squid fishing. In the 1997 season, the squid fishery was closed after the limit of 73 sea lions were accidentally killed.

This sent a very strong message to the industry that excessive by-catch of sea lions will not be tolerated by government. After the closing of the fishery, the Industry has moved to seriously look at ways of reducing the by-catch of sea lions. It is supporting the development of escape devices in trawlers, as well as research into sea lions themselves.

Seabird by-catch is another issue in the same waters. This year we had better observer coverage than in the past and that led to better information on what was happening on boats. Still, hundreds of seabirds were drowned while diving for bait, principally intended to catch tuna. Sea-bird by-catch has a serious impact on the survival of a species, like Albatross, Petrels, Mollymawks. This is a message that the New Zealand fishing industry has picked up on and is taking seriously. Currently two setting devices are being trialed, that avoid the capture of sea-birds. Australia and New Zealand are funding the development of these devices, in recognition that this is an international issue.

For instance, Japan has seen that sea-bird by-catch may cause a major threat to the tuna fishing industry and is starting to act. I might add though, somewhat tardily.

For both sea lions and sea-bird by-catch, fishing industries are waking up to the fact that the impact on birds and mammals can potentially close down their industry. This means the industry jeopardizes its sustainability and if it wants to survive, then it needs to do something about it, now.

But it is no good just being saints domestically — this is an international issue. By-catch issues are on the agenda of the next meeting for the Food and Agriculture organization of the United Nations, in March in Tokyo. Other international recognition of this issue includes the Australian government proposing to list albatross on its endangered list for the convention on migratory species. This would protect albatross throughout their range and have a positive effect on the survival of the species.

Another fishing issue is the overcatch of Patagonia toothfish. These fish are widely distributed in polar and sub-polar regions, and are mostly caught on longlines. There is an international agreement on the number of Patagonia toothfish to be caught. But over and above the number, it is estimated that between 60 and 80 thousand tonnes of toothfish have been caught in the past year illegally.

Some countries, having agreed to limit their toothfish catch, have then registered their vessels under different flags, so as to avoid abiding by international treaties that their country has signed up to.

This is a classical problem for managing industrial fisheries. It is totally unacceptable that those nations from Europe, and in particular Spain and Norway, are raping resources from future New Zealanders. Just because they have destroyed their own fishing stocks, gives them no right to come south and destroy ours. All that the illegal catching of fish will do is lead to the collapse of stocks that much quicker.

1998 is international year of the oceans. If we are seriously committed to making the year more than tokenism, then a large effort needs to go into protecting our marine bio-diversity. New Zealand has the fourth largest exclusive economic zone of any country in the world, and the protection of this environment will be an issue that, as Minister of Conservation, I am focusing on.

FARMING

Farming has been the engine room of the New Zealand economy for well over a century. Compared to the other pioneering industries like whaling, sealing, mining and indigenous logging, it has proved to be the most sustainable. However, we still have some way to go if we are to claim that it is a truly sustainable industry.

The sustainability of pastoral farming has been most sorely tested in the South Island High Country. This environment is highly fragile and has been through dramatic boom/bust cycles from poor management.

We are about to embark on a major period of restructuring of high country farming through the process of tenure review. There are 2.8 million hectares of land in perpetual lease. Some of it is imminently suitable for farming as freehold title. Some land can be farmed with suitable covenants, and some is fragile mountain ecosystems that is totally unsuitable for grazing. The process of taking 200 plus leases through the land tenure review, is one of the most important environmental steps that New Zealand needs to take in the decade ahead. It will see our conservation lands expanded by millions of hectares.

Improving the sustainability on freehold farmland is a real political minefield. Through the *Resource Management Act,* councils up and down New Zealand are identifying areas of particular conservation value and are developing suitable rules to try and protect them. In some areas of New Zealand, this process has been managed very poorly and has landed councils in the political pooh. On Wednesday the Minister for the Environment, Hon Simon Upton, and myself are travelling to the Far North to try and find a way through the political quagmire. It is not the only local authority that has got into trouble. There is delicate balance between the rights of property owners and the wider community's conservation goals, that needs to be found. The *Resource Management Act* has triggered a rash of enthusiasm for regulatory controls, when a greater use of voluntary mechanisms would achieve better conservation outcomes. If we are to win the biodiversity battle, and have farmers with us, and not against us, we are going to have to get far better at communicating what we are on about.

TOURISM

When we talk about production in the environment, we need to think about the emergence of industries, like tourism. Tourism now earns more for New Zealand than the dairy industry — but we are only just starting to recognize the impacts on our environment. The Parliamentary Commissioner for the Environment's report on the impacts of tourism indicates that: One of the issues in tourism that we must consider is a totally unseen one. It is the issue of noise in our recreation spaces. Aircraft movements in most of the more scenic part of the South Island can spoil an outdoor wilderness experience. We need to consider why people go into the great outdoors and if a plane or helicopter overhead is going to make a major impact on the experience they enjoy.

If we do not want to end up like Yosemite or the Grand Canyon, then we need to think about this point. The most obvious options are controlling aircraft movements, having quieter planes, and observing height limits on aircraft.

Net conservation benefits come into this argument. Where is the point to draw a line in the sand, or perhaps I should say, the glacier, and decide enough is enough.

The challenge, and I suggest one focus of the week, is to find a way of working through these issues. This is just one example of a whole range of clashes developing in the tourism industry and I look forward to reading the Parliamentary Commissioner for the Environment's report on this very issue.

BIODIVERSITY STRATEGY

A Resource Economist, Dr Murray Paterson, from Massey University, has recently valued New Zealand's biodiversity at $39 billion. The total economic value of New Zealand's marine biodiversity considerably adds to this. Dr Paterson estimates that $9 billion per year is of a direct use value — that is the value of all goods and services derived from the direct use of biodiversity. He estimates another $30 billion per year is an indirect value. That is the value derived from biodiversity supporting or protecting direct use activities. Like erosion control, waste treatment, pollution, solid formulation, etc.

Add to these, also passive value. Very simply put, this is what we are prepared to pay to preserve tuatara, or knowing that offshore island programmes to save our birds exist, and will be there for future generations. The study also found that New Zealand was rich in biodiversity, having one and a half times the world average for terrestrial biodiversity and 22 times the world average value for marine biodiversity. This says so much about not just our national, but also our international obligations.

What is unique about planet earth is our huge diversity of biological organisms. A key focus of the Earth Summit in Rio was in getting Nation States to accept their responsibilities in maintaining that diversity. The United Nations Convention on Biological Diversity that we signed at the Earth Summit commits us to a work programme of species protection and ecosystem management. That will cost money to implement and spark some hot debates between competing interests.

This is not just an issue about protecting our native species.

Protecting our biodiversity goes hand in hand with the long-term security of our key primary industries like fishing, forestry, agriculture and horticulture. The modern techniques of plant propagation and artificial breeding are narrowing the genetic base of those animals and plants which we are dependent on for our economic prosperity.

An example in New Zealand would be our pip fruit industry where millions of cultivars of Braeburn are being planted here and overseas. If we allow the genetic base of our pip fruit stock to become too narrow we limit the potential of the apple industry in the future. For this reason we need to protect the genetic stock of apple trees.

A key project for the Department is in developing a comprehensive biodiversity strategy for New Zealand, and this is due out early next year. The government considers this work to be very important. You people here today have a great deal to contribute to that strategy, with respect to both our natural and productive biodiversity. However, we lag considerably behind Australia in this area. I hope you take this opportunity to learn from your Australian colleagues' experience, so that when we ask for your submissions this year they are imaginative and constructive.

ENVIRONMENTAL INDICATORS

For all the positive that has taken place, and there is much, the most basic worry for me, is the lack of reliable information on which quality decisions can be made. At the beginning of October this year, my colleague, the Minister for the Environment released New Zealand's first State of the Environment report. This is as a direct consequence of the Earth Summit held in Rio de Janiero in Brazil in 1992.

This report outlined the state of our environmental resources, some of the pressures on these resources and how we might respond. It identified how we can improve the information on which we base our decisions for the future and how we can better manage our natural resources.

It is on this issue of environmental data that I think we face one of the greatest challenges. As we attempt to make sound policy decisions the lack of good quality information is quite scary. If we look at air quality the information is questionable. On biodiversity, again we only know some of the picture. In the energy field we have good quality information, but it is the exception. With fisheries we have some but there are huge unknowns. And when it comes to greenhouse gases the information is scant. We have some idea of the picture of 7 000 hazardous waste sites but the extent of the pollution of each site is unknown. With ozone depletion we are getting good quality information but there is a glaring hole in our information databases on soil quality and transport impacts on the environment.

The reason I highlight this problem of poor information is that it goes to the heart of the environmental challenge we face next century and it is one in which we can make the greatest contributions. My frustration is that when it comes to the economic implications of our decisions we are able to quantify them down to the last dollar but when it comes to the environmental cost our information is so poor that the trade-offs become quite vague. Far too often decisions go in favour of economics over the environment only because the economic outcomes are more measurable.

Think for a moment of the national resources that go into measuring economic outcomes. There is a whole profession of bean counters in the public and private sectors dedicated to monitoring and recording every dollar spent. If somebody robs the till, there is a public outcry. As a government, we are very focused on budgets, deficits, our national debt and our national assets.

But what of the less tangible environmental and conservation assets. How do we track gains or losses in their value? Do we react to those who rob us of our natural assets as those who rob us financially?

If we are to balance the scales we should be putting the same energies into quantifying the environmental impacts as we do with economic impacts. The way forward will be with concepts like resource accounting and net conservation gains. These will provide a far more robust decision making process. I emphasize again though that these will not work unless we have a quantum leap in the quality of our environmental databases. This is one of the greatest challenges the Department of Conservation faces in putting together our Strategic Business plan for the next five years.

CONCLUSION

I have presented many areas that need some creative solutions, and also some areas that New Zealand is doing well in. The real challenge over this week is to take the conservation ethic into those areas that are outside the protected estate.

As you work through the range of technical issues covering agriculture, forestry and fishing, you need to be searching for policy options that meet both ecological and human needs. Your work is about building a bridge between the production and conservation sectors. Succeed and the fruits will be awesome, but if we fail, the consequences are too awful to contemplate. The templates you help develop will define the sort of planet we pass on to the next generation. Can I conclude with a quote from Edwin Teale that beautifully sums up the focus for this conference.

"The long fight to save wild beauty represents democracy at its best. It requires citizens to practise the hardest of virtues, self-restraint."

It is my great pleasure to now declare the Nature Conservation in Production Environment Conference open. May I wish you the best of self-restraint, the best of debate and the best of discovery and inspiration.

CONTENTS

CHAPTER 1

Nature conservation in agricultural landscapes: real progress or moving deckchairs?

Richard Hobbs[1] and Denis Saunders[2]

After 13 years of research on, and involvement in, nature conservation issues in fragmented systems within an agricultural landscape in Western Australia, we take stock of what the achievements have been and what has yet to be done. From a programme which initially focussed on the dynamics of biota within conservation reserves, we have moved into an interdisciplinary framework which examines the broad context within which nature conservation operates.

During the period of the research programme, significant areas continued to be cleared in the Western Australian wheatbelt until recently. Activities such as fencing of remnant vegetation have increased dramatically, as have the number of individuals and groups involved in conservation activities. This has not, however, been matched by the large increase in revegetation activities thought to be necessary to reverse land degradation and loss of biodiversity. Current barriers to better land management are as much economic, political and social as they are technological. While there have been major shifts in management attitudes over the course of our research, these have been translated into action in only a marginal way. There is a need to concentrate on changing behaviours rather than attitudes. We believe that we need to have the combination of a clear collective vision of what we would like our region (and the world) to be like 20–30 years from now, sound strategies for achieving this vision, and a set of tactics to put the strategy into operation. The challenge for the future is to build on current good will, to continue building and maintaining networks, and to ensure that our science is embedded in a sound grasp of social and political realities. At the same time, application-oriented research must still maintain a strong strategic and innovative component to meet the challenges of the next century.

INTRODUCTION

After four conferences on issues relating to nature conservation in production landscapes over the past 12 years, what remains to be said? Four volumes of proceedings (Saunders *et al.* 1987; Saunders and Hobbs 1991; Saunders *et al.* 1993b; Saunders *et al.* 1995) and numerous other publications provide information on, and proposed solutions to, many of the problems to be tackled, together with exhortations for scientists to get their acts together, integrate their research with that of others, communicate more and more effectively, and ensure that their work is relevant in the bigger picture. Has progress been made, or have these meetings simply been good opportunities to meet up with old friends, make new ones, and generally have a great time? We take this opportunity to assess where we (collectively) have come from, where we are up to, and where we should perhaps go next.

On a more personal level, we also take the opportunity to assess our own research efforts within the context of changes that have taken place over the past decade. We started working together in 1984 in a field which was then relatively poorly articulated and about which there was relatively little interest "on the ground". What followed were 13 years of ups and downs, progress and frustration, and many dramatic changes. In June 1997, Denis

[1]CSIRO Wildlife and Ecology, Private Bag, P.O. Wembley, Western Australia, Australia 6014.
[2]CSIRO Wildlife and Ecology, P.O. Box 284, Canberra, Australian Capital Territory, Australia 2601.
Pages 1–12 *in* NATURE CONSERVATION 5: NATURE CONSERVATION IN PRODUCTION ENVIRONMENTS: MANAGING THE MATRIX ed by J. L. Craig, N. Mitchell and D. A. Saunders. Surrey Beatty & Sons, 2000.

left Western Australia for the barren wastes of Canberra, marking an endpoint to our day to day collaboration. As with any point of major change, this has been a period of reflection and taking stock. However, instead of being purely retrospective and wallowing in nostalgia and past glories and/or failures, we also look to the future.

CONTEXT

While we will discuss mainly nature conservation issues in the agricultural areas of Australia, with particular emphasis on the wheatbelt of Western Australia, it is useful to remember the overall context of these discussions. We have to keep in mind the global trends of human population growth and resource use, and the effects that these have in terms of land transformation, global biogeochemistry and climate change and loss of biological diversity. Vitousek *et al.* (1997) draw evidence from a variety of sources which suggests that world-wide, 39—50% of the land surface has been transformed or degraded by human activity (with the corollary that an equivalent proportion of the land's biological production is co-opted for human use). Humans currently utilize more than half of the fresh and reasonably accessible runoff water. In addition, human activities since the start of the industrial revolution have caused a 30% increase in carbon dioxide in the atmosphere, with burning of fossil fuel adding 5.5 ± 0.5 billion metric tons of CO_2-C to the atmosphere annually. More nitrogen is now fixed by human activities than by all natural terrestrial sources combined. Humans have also significantly impacted Earth's biota, with, for instance, about 25% of bird species going extinct over the past 2 000 years, many as a result of human activities. Sixty-six per cent of all known fish stocks are also thought to be either already depleted, overexploited or fully exploited. Vitousek *et al.* (1997) conclude from all this that "By these and other standards, it is clear that we live on a human-dominated planet".

Set against these human-induced transformations is the need to feed and provide material comfort for ever increasing numbers of humans without terminally degrading the resource base needed to do this. At the same time, it is increasingly seen as a good idea to prevent further losses of biological diversity, both for its own sake and because of the services collectively provided by Earth's non-human species (Baskin 1997; Chapin *et al.* 1997; Daily 1997).

THE WESTERN AUSTRALIAN WHEATBELT

The rapid and pervasive transformation of the Western Australian wheatbelt from a landscape dominated by a mosaic of woody perennial vegetation communities to one consisting of mixed annual crops and pastures with scattered remnants of native vegetation has been documented elsewhere (Saunders *et al.* 1985; Hobbs *et al.* 1993). The region now produces a significant proportion of Australia's agricultural output, while retaining, albeit in mostly small patches, an incredibly diverse and fascinating biota. The development of the region has come at a cost, both in terms of loss of biota, and in terms of increasing land degradation (McFarlane *et al.* 1993; Nulsen 1993; George *et al.* 1995). Here we examine recent trends in clearing of native vegetation, protection of remaining vegetation, revegetation, and community involvement in Landcare and other activities.

Saunders *et al.* (1985) documented the phases of agricultural development, with periods of rapid clearing occurring this century both between the world wars and in the decades following World War II. Over 50% of the area was cleared in this

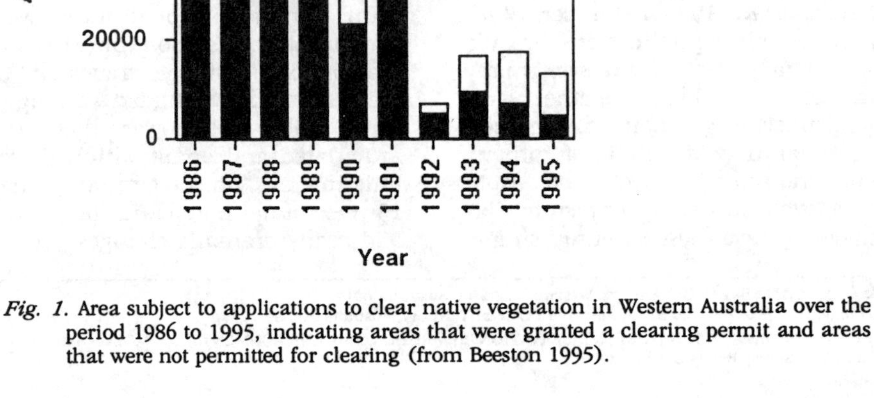

Fig. 1. Area subject to applications to clear native vegetation in Western Australia over the period 1986 to 1995, indicating areas that were granted a clearing permit and areas that were not permitted for clearing (from Beeston 1995).

post-war period, when the catch-cry was to release one million acres of uncleared land a year for development into wheat-sheep farms. In the mid-1980s, when our research group commenced work on the fragmented systems of the wheatbelt, the general perception was that the rate of clearing had slowed significantly, and that land clearance *per se* was not a major threat to nature conservation. Data gathered subsequently indicate, on the contrary, that clearance of native vegetation has continued at appreciable rates over the past decade (Fig. 1). Shur (1990) and Beeston (1995) estimated that between 1986 and 1995, 213 000 ha of land had been cleared in the wheatbelt, indicating a rate of clearance still in excess of 20 000 ha yr^{-1}. This matches figures published recently in the federal State of the Environment Report (Commonwealth of Australia 1996), which indicated a rate of 26 000 ha yr^{-1} in the Western Australian wheatbelt over the period 1983 to 1993.

There is an ongoing rate of clearance nationally of an estimated 664 000 ha yr^{-1}, with extensive clearing continuing in most states, especially in New South Wales and Queensland (Sivertsen 1994; Alexandra 1995; Commonwealth of Australia 1996; Catterall *et al.* 1997). Suggestions that the clearing of native vegetation is gradually being balanced by replanting of trees misses the vital point that revegetation efforts cannot hope to replicate the habitat lost through clearing within any sensible timeframe. In addition to the impact in terms of continued habitat loss, land clearance also contributes significantly to Australia's greenhouse gas emissions. In fact, estimates of the extent of this contribution vary from 7 to 45% of total emissions, with an average estimate of 27% (Department of Environment 1994; Commonwealth of Australia 1996). Clearly, the recent debate over targets for cuts in greenhouse gas emissions failed to take this potentially large source into account or to recognize reducing rates of land clearing as an opportunity to deal with two critical environmental issues at the same time.

Closer inspection of the annual figures for Western Australia indicate a reduction in areas cleared over the period 1992 to 1996, but no reduction in the numbers of clearing applications being received (Government of Western Australia 1997). Comparisons of clearing activity in four shires over the periods 1976 to 1986 and 1986 to 1996 also clearly indicate a marked reduction in clearing activity in the past decade (Table 1; Coates 1987; Jenkins 1996). In 1995 regulations relating to the *Soil and Land Conservation Act* were modified to incorporate nature conservation values into the assessment of clearing applications, rather than just land and water conservation as had previously been the case (Jenkins 1996). In addition, clearing would be discouraged in shires with less than 20% cover of native vegetation, essentially making clearing difficult over much of the wheatbelt.

Remnant protection

Given that so little remnant vegetation remains over large areas of the wheatbelt, a corollary of reducing rates of clearing is protecting remaining patches of bush. Given the recognized impacts of stock grazing on native vegetation (e.g., Scougall *et al.* 1993; Norton *et al.* 1995; Petit *et al.* 1995), an essential first step in remnant protection is exclusion of domestic livestock. Hence, fencing of remnant bush has been seen as a priority exercise. Data from four shires indicates a clear increase in the proportion of bush on farms that has been fenced over the past decade (Table 1). Additionally, government initiatives have provided funding to cover part of the costs of fencing remnants. The Western Australian Remnant Vegetation Protection Scheme resulted in over 38 000 ha of bush being fenced over the period 1989 to 1995 (Beeston 1995), although the areas fenced and number of grants awarded dropped over that period (Fig. 2). Other schemes, including the federal Save the Bush scheme, also injected significant funds into protection of remnant vegetation. These levels of funding are to be maintained or increased under the National Heritage Trust Bushcare arrangements (Hill 1997).

In addition to government-assisted protection, there is undoubtedly an increase in unfunded activity carried out by individual landholders or community groups, as indicated in Table 1. Such activities seem very difficult to quantify effectively, since we found no good data on their extent.

The emphasis on remnant protection is perhaps understandable given the desire to prevent further degradation to the remaining native vegetation in the agricultural areas. However, it has been

Table 1. Activities relating to vegetation removal, retention and replacement in five shires in Western Australia. Source: Coates (1987); Jenkins (1996). Data from 1996 survey unless otherwise indicated (Data not available for Kellerberrin in 1986). The 1986 and 1996 surveys did not survey the same set of landholders: hence comparative figures within shires are indicative rather than absolute.

Shire	% remnant vegetation total (private)	% farmers who cleared land 76–86	86–96	% private remnant veg. that is fenced 1986	1996	% farmers who revegetated	% of farm revegetated	Area per farm revegetated (ha)
Pingelly	9.5 (3.7)	45	8	11	37	79	1.3	19
Lake Grace	76.5 (14.2)	70	13	12	45	81	1.0	31
Dumbleyung	12.7 (6.8)	33	4	24	70	83	1.2	23
Kellerberrin	8.1 (7.1)	NA	0	NA	39	85	0.9	19
Tammin	3.2 (1.6)	7	0	24	21	100	1.5	32

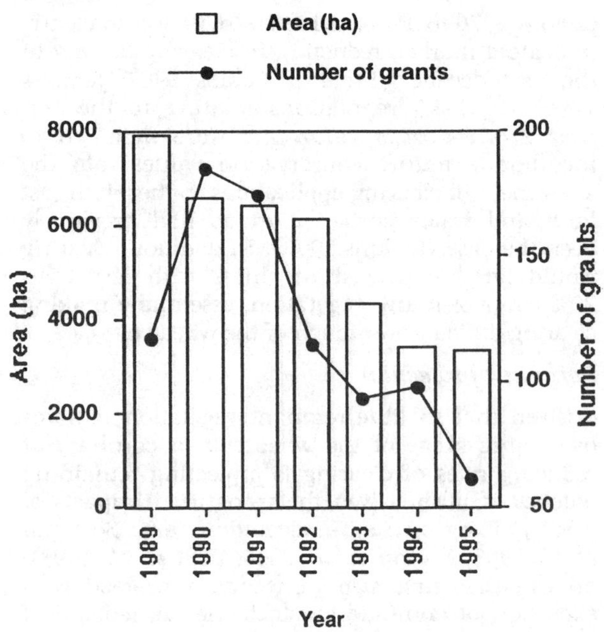

Fig. 2. Number of grants allocated under the Western Australian Remnant Vegetation Protection Scheme, and area of remnant vegetation fenced over the period 1989 to 1995. Source: Beeston (1995).

accompanied by the implicit assumption that a remnant is secure once it is fenced. This does not take account of the need for active management of remnants, for instance to ensure that regeneration processes are maintained (Yates and Hobbs 1997a,b). It also does not take account of the external factors arising from the agricultural matrix which are likely to impact on the remnant (Saunders *et al.* 1991), particularly hydrological changes (Hobbs 1994; George *et al.* 1995). While the latter issue is increasingly recognized (e.g., Agriculture Western Australia *et al.* 1996a), the need for active management of remnants is only slowly being accepted, despite the availability of information on the issues involved (Hussey and Wallace 1993). Wallace (1995b) recently concluded that, although "management of remnant vegetation is undoubtedly more effective and better resourced than it was in 1983 . . . management is neither sufficiently effective nor well enough resourced to conserve the existing biodiversity at a district or regional level".

Revegetation

There are clear indications that revegetation will form a major part of the solution to the dual problems of land degradation and loss of conservation values in the Western Australian wheatbelt (Lefroy *et al.* 1993). Clearly, since many of the problems of rising water tables, salinization and other forms of degradation have been caused by the extensive removal of deep rooted perennial vegetation, it makes sense that replacing such vegetation will help reverse current trends. There is still active debate on the types and extent of revegetation that will be both effective and acceptable, and on the relative merits of other options (such as perennial pastures and phase cropping). The extent of revegetation needed is likely to vary greatly from place to place, depending on rainfall and recharge rates and on local geology. Options also vary regionally, with much greater potential for plantation forestry and agroforestry solutions in higher rainfall areas (Agriculture Western Australia *et al.* 1996a,b). Nevertheless, current strategies indicate a push for extensive revegetation across the agricultural area. For instance, an aim of the Western Australian Salinity Action Plan is to plant three million ha of trees and shrubs across the area, although the time-frame for doing this is not specified (Agriculture Western Australia *et al.* 1996b).

Given this degree of interest in revegetation, the actual amount that has taken place remains remarkably small. Data from five shires indicate that revegetation to date occupies only 1–2% of farm properties (Jenkins 1996). This matches data collected nationally that indicates that the mean area of revegetation per farm is 9.7 ha (Nicholl and Dobbie 1996). A mean farm size of 1 375 ha implies that revegetation takes up less than 1% of farmland nationally. More detailed data for the South Tammin catchment, collated by Brandenburg and Majer (1995), indicate increasing participation in revegetation over the period 1982 to 1992 (Fig. 3). However, the data also indicate that a real increase in revegetation effort occurred only following the inception of a sponsorship arrangement between the catchment group and the mining company Alcoa in 1989. Even with this major injection of funds, revegetation covered only 3.5% of the catchment in 1992.

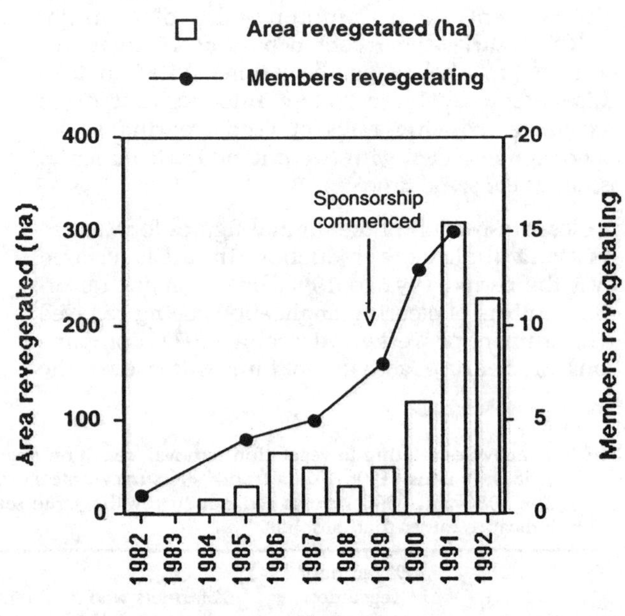

Fig. 3. Number of members of the South Tammin Catchment Group who engaged in revegetation activities over the period 1982 to 1992, and the area revegetated over the same period. Sponsorship by Alcoa commenced in 1989. Data from Brandenburg and Majer (1995).

Community groups

In Western Australia, as in Australia as a whole, the past decade has seen a huge increase in the number of community groups involved in land and nature conservation activities in agricultural areas. The most noticeable manifestation has perhaps been the Landcare movement (Campbell 1994), which has seen a dramatic and ongoing rise in the number of Landcare groups across the agricultural regions of Australia (Fig. 4). Similar trends can be found in other indicators of community involvement. For instance, numbers of properties registered in the Land for Wildlife Scheme in Victoria increased one hundred fold over the period 1981 to 1994 (Platt and Ahern 1995). Accurate estimates of current numbers of community groups such as Landcare groups are difficult to obtain. Groups come into and out of existence in a very fluid way and what constitutes a landcare group varies enormously from state to state — ranging from "six farmers and a dog" to very large groups (David Mussared, pers. comm.). At the same time, some Landcare groups do not actually register as such, and there are a variety of other possible group designations.

In addition to groups linked to national or state initiatives, a wide range of independent groups have also been formed which focus on elements of biodiversity conservation. In Western Australia, notable examples include several discussed in the Networks volume, such as the Waddi Forest Group (Doley 1995), the Malleefowl Preservation Group (Orsini *et al.* 1995; Dennings 1997) and the Fitzgerald Biosphere Reserve (Watson *et al.* 1995),

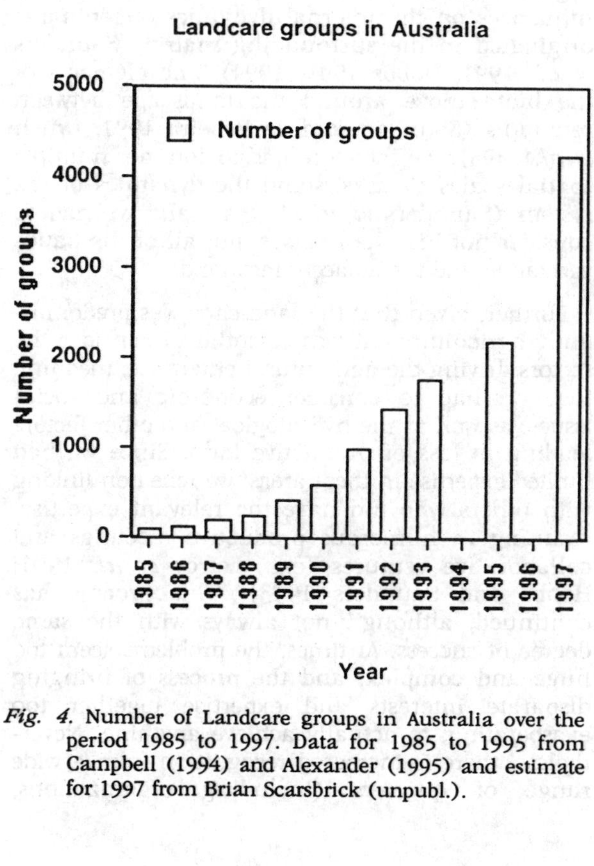

Fig. 4. Number of Landcare groups in Australia over the period 1985 to 1997. Data for 1985 to 1995 from Campbell (1994) and Alexander (1995) and estimate for 1997 from Brian Scarsbrick (unpubl.).

and others such as the Dongolocking Corridors Project (Lloyd 1995). The majority of these groups have formed within the last 10–15 years, and provide significant amounts of time and resources for conservation-related activities.

ATTITUDES *VS* ACTION

The approach we and others adopted throughout our research and communication activities was to take every opportunity to talk to people individually and in groups, and in that way increase awareness and understanding of the problems and some of their solutions. Groups addressed included local Landcare groups running field days on their properties, community information meetings, and more formal seminars and conferences/workshops ranging from local to international. We also used every opportunity to involve others in our research, including students and international volunteers (Gilmour and Saunders 1995). Our mental model assumed that through increased awareness and changing attitudes, changes in actions would ensue. From the information presented above, it is clear that there have been some very encouraging shifts in management activities and policies. However, it is also clear that the magnitude of the response still falls far short of the magnitude of the problems. Why is this?

Although environmental awareness has undoubtedly increased globally over the past few decades, there is still considerable confusion about, and lack of understanding of, basic issues relating to the environment. A growing literature, loosely categorized as "environmental optimism", casts doubts on the validity of claims of environmental damage and the need for concern (e.g., Bailey 1995; Easterbrook 1995; Simon 1995). While much of this is based on incorrect or misleading argument (Ehrlich and Ehrlich 1996), it plays to an audience which has not received clear messages from scientists on environmental issues, which does not understand the fundamental issues, and which probably does not readily want to accept that adverse environmental impacts are a likely result of current lifestyles.

Locally in Western Australia, a recent study (P. Mazzucato, unpubl.) of over 800 people randomly selected from the community in Perth assessed the level of knowledge of environmental issues by asking a range of questions on these issues. Of the 11 questions asked, only two were answered correctly by a large proportion of the sample. The investigator concluded that the answers to these could be deduced logically without any knowledge of the issues. The conclusion was that the level of environmental knowledge was relatively low. The same study indicated a relatively high level of awareness and concern over environmental issues, but relatively little willingness to alter behaviours such as motor vehicle use, energy conservation, recycling and so on.

This study highlights two fundamental problems which stand in the way of effective action on environmental problems such as those found in the Western Australian wheatbelt. The first is the lack of clear understanding of the causes and consequences of environmental degradation, reflecting perhaps the often complex nature of the processes involved. Secondly, there is not a clear link between awareness, attitudes and behaviour. There is increasing recognition of this phenomenon in the agricultural context. In particular, several authors have noted that landholders and community groups may possess a strong conservation or stewardship ethic, but that this is not necessarily translated into action (Binning 1997; Curtis 1997).

Several authors have pointed out that this apparent contradiction arises from an overly simplistic model of the relationship between attitudes and behaviour (Brasden 1994; Cosgrove *et al.* 1994; Vanclay and Lawrence 1995). A simple model (Fig. 5a) suggests a straightforward link between perception of the environment leading to the formulation of attitudes and a resultant response (in terms of changed behaviour) (Vanclay and Lawrence 1995). On the other hand, a more realistic model (Fig. 5b) seems to suggest that perceptions and attitudes are filtered on the basis of available information and in response to the current context, especially in relation to available resource, financial situation and so on.

The messages from this are that attitudes are not a good predictor of behaviour and that an emphasis on changing attitudes may be attacking the wrong part of the equation, or at least only part of it. While attitudinal change may be a prerequisite

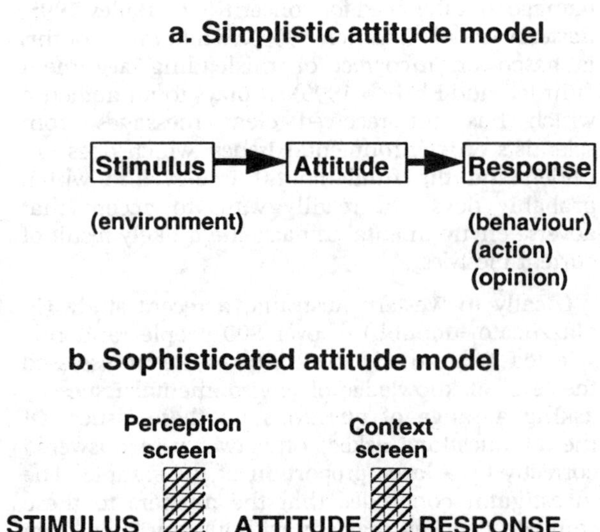

Fig. 5. Simple and sophisticated models of the relationship between information, attitudes and behaviour, redrawn from Vanclay and Lawrence (1995).

for behavioural change, it is not sufficient. As Binning (1997) has suggested, we now need to put considerably more emphasis on finding out how to change behaviours rather than attitudes. This requires, amongst other things, a much more detailed analysis of the socioeconomic setting within which we operate, and the options available for encouraging behaviour which promotes the conservation of biodiversity in production landscapes (Price 1995; Young *et al.* 1996).

THE ROLE OF SCIENCE

Science has increasingly been under scrutiny for what it can and cannot offer in today's rapidly changing world. We explored elsewhere the problems relating to increased fragmentation and specialization in science and the ability to tackle complex environmental problems (Hobbs and Saunders 1995). Suzuki (1990) summed up the situation thus: "It is clear that while science provides insights into the complexity of the world around us, those insights . . . present a fractured mosaic rather than a seamless whole".

Our scientific path has been one of building up from examining one piece of the mosaic to attempting to get at the whole picture. When our programme started in 1984, the objective was to examine the ecology and management of conservation reserves and their associated fauna. Accordingly, we started examining the dynamics of species and communities within those reserves. Quickly, however, it became clear that this approach was providing only part of the picture, since the reserves were only a subset of the remnants of native vegetation, and in addition, most of the major influences on the internal dynamics of remnants originated in the surrounding matrix (Saunders *et al.* 1991; Hobbs 1993, 1994), and elements of the biota move around the landscape between remnants (Saunders and de Rebeira 1991; Lynch *et al.* 1995). We therefore had to consider multiple spatial scales to understand the dynamics of the system (Saunders *et al.* 1993a), and we had to consider not just the reserves, but all of the native vegetation and the adjacent farmland.

Further, given that the landscape was predominantly agricultural, it was essential to consider the factors driving the agricultural enterprise, meaning that we had to consider economic and social issues, as well as the hydrological and other factors leading to loss of productive land. Since we had limited expertise in these areas, we relied on linking with others who did have the relevant expertise, resulting in numerous productive meetings and collaborative ventures (e.g., Lefroy *et al.* 1991; Hobbs and Saunders 1993). This process has continued, although not always with the same degree of success. At times, the problems seem too huge and complex, and the process of bringing disparate interests and expertise together too exasperating, to actually achieve anything. Nevertheless, there is increased recognition from a wide range of quarters (including organizations, individuals and funding agencies) that the approach is both valid and necessary.

Our approach has increasingly been one of trying to fit individual bits of research into a broader perspective so that the whole adds up to more than just the sum of the parts. We have tried to do this both within our own ecological research (Hobbs 1995; Saunders 1996a) and by fitting that research into the broader picture which includes socio-economic and agricultural perspectives (Fig. 6; Hobbs and Saunders 1991, 1993). The importance of these activities from a manager's perspective was emphasized by Wallace (1995a) who suggested that "... ideas and 'bits' of information must be synthesized into theory or a 'big picture' if they are to provide maximum benefit". Unfortunately, it is still often difficult to achieve this for the reasons discussed by Hobbs and Saunders (1995), although there are encouraging signs that funding bodies and research organizations now place more value on synthetic and interdisciplinary approaches.

THE WAY AHEAD

At the 1994 Networks conference, David Goldney presented a paper entitled "Winning battles but losing the war", in which he questioned whether the encouraging signs in remnant protection and management represented real progress or whether they simply masked an ongoing and more or less unchecked decline in conservation values in remnant vegetation (Goldney *et al.* 1995). In this chapter, we have asked a similar question using a different metaphor — i.e., are we making a real difference or is what we are doing similar to rearranging the deckchairs on the Titanic. We feel that the answer to our question is that we are doing more than rearranging deckchairs, but that we are certainly sailing in troubled waters full of icebergs. We therefore need careful planning and navigation to avoid sinking completely.

To return to David Goldney's military analogy, we need to think not only about the tactics of winning battles, but also the strategy to win the war. Both strategy and tactics are essential to the conduct of warfare. Broadly stated, strategy is the planning, co-ordination, and general direction of military operations to meet overall political and military objectives. Tactics implement strategy by short-term decisions on the movement of troops and employment of weapons on the field of battle. The great military theorist Carl von Clausewitz put it another way: "Tactics is the art of using troops in battle; strategy is the art of using battles to win the war."

We also argue that there is another essential element, which can be broadly defined as "vision". The dictionary definition of vision is "statesmanlike foresight", and implies the ability to see a goal towards which efforts should be directed. The broad vision thus answers the question "Where do we want to go?". The strategy answers the question "How do we want to get there?", and tactics provide the answer to "What actions do we need to take?". We suggest that all three elements are essential for success, and part of the failure to date can be attributed to the lack of one of these elements. We have argued elsewhere that a lack of vision has hampered our attempts to develop strategies for biodiversity conservation in agricultural landscapes (Saunders 1996b). Similarly, strategies are useless unless they are backed up with adequately resourced and planned tactical measures.

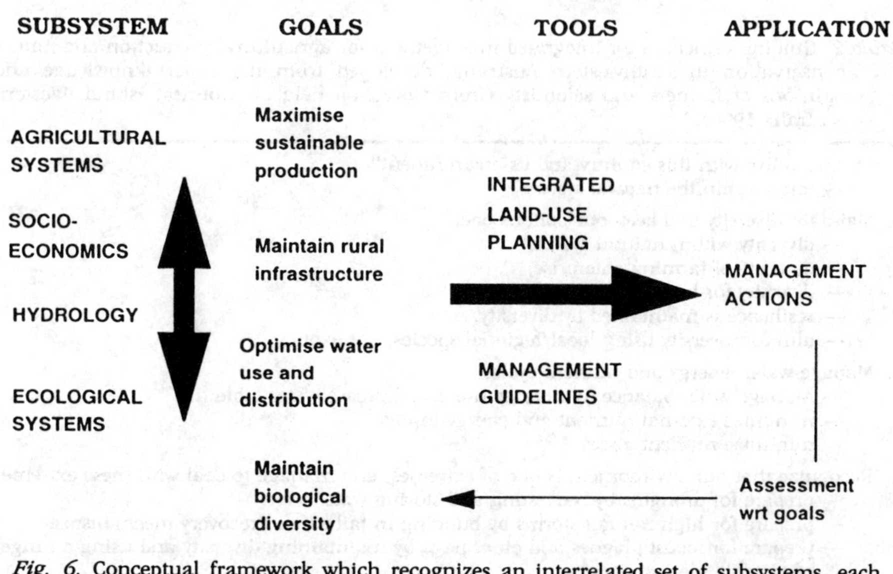

Fig. 6. Conceptual framework which recognizes an interrelated set of subsystems, each with its own goals (of which only examples are shown), which need to be translated into management actions via a toolkit of planning, decision and integration tools. Subsystems include: (1) **sustainable agriculture,** covering all agricultural activities including growing of deep-rooted perennial crop species, (2) **hydrology,** primarily in relation to rising water tables and salinization, (3) maintenance and re-establishment of **biodiversity,** (4) **socio-economic** aspects of farming and nature conservation in the Agricultural Region (sustainable communities).

We believe that there is a need to develop a collective vision of where we, collectively, would like to be in 20–30 years time — a 2020 vision — and identify a set of strategies and associated tactics in order to get us there. Perhaps a vision for 100 years time would be more appropriate, given the timeframes needed for restoration. However, we feel that a 20–30 year timeframe is more tractable and at the same time challenging enough in its scope. Such a vision needs to be challenging and yet open to change as opportunities arise. For instance, who in 1898 would have developed a vision for the world which included mass air flight, personal computers and world-wide telecommunications?

We started the process of developing a vision for the Western Australian wheatbelt with a group of farmers, managers and scientists at a workshop on Rottnest Island in 1996. It was relatively easy to arrive at a common vision of where we wanted to be, but more difficult to decide on the strategy and tactics for getting there. The workshop produced a set of guiding principles which could be used as a blueprint for developing strategies (Table 2), and recognized that there may be a multiplicity of ways of achieving the broad vision. Given that agricultural landscapes are populated by individual landholders with widely varying attitudes and values, we should perhaps concentrate on the strategic dimensions of the problem and leave the tactical decisions up to the individual.

This presupposes that adequate incentives exist to encourage individuals to contribute to an overall strategic response, but it also presupposes that these individuals can share a common vision in the first place. It also involves the provision of adequate and accessible information on which tactical decisions can be made. Fulfilling these three criteria are likely to be the major challenges ahead of us in the coming years.

Vision

A vision for 2020 might include elements which focus on the development of sustainable agricultural systems which reverse current trends of land degradation and rising water tables, while at the same time preventing further loss of biodiversity. The exact definitions of these elements need careful thought and some degree of consensus needs to be reached on these.

In order for a shared vision such as this to be generally accepted and acted on, the general level of awareness of the problems and their potential solutions needs to be high. The value of what might be lost in the absence of action needs to be better enunciated. People in general are only likely to take action if there is a good reason for doing so. This therefore involves a continuation of the type of "awareness-raising" activities that were discussed previously — across a range of audiences including managers, policy makers and children.

Strategy

Numerous strategies already exist which deal with elements of the problems discussed here. For instance, there are strategies on biodiversity, weeds, salinity and others either at the federal or state level in Australia. Strategies relating to biodiversity need, as essential elements, pathways to the cessation of vegetation clearance, and effective guidelines for remnant vegetation management and strategic revegetation in relation to specified conservation goals.

Table 2. Guiding principles for integrated management for agricultural production and nature conservation in southwestern Australia, developed from the expert knowledge and opinions of farmers and scientists (from workshop held on Rottnest Island, Western Australia 1996).

1. "Learn to live with this country and its environment"
   — farm within the natural system
2. Maintain diversity and keep our options open
   — diversity within natural system
   — diversity of farming enterprise
   — diversity for biological control
   — resilience is maintained by diversity
   — aim for diversity using local/regional species, not exotics
3. Manage water, energy and nutrient cycles
   — manage water balance to increase use and decrease water table rise
   — minimize external nutrient and energy inputs
   — minimize nutrient losses
4. Recognize that our environment is one of extremes, and manage to deal with these extremes
   — prepare for droughts by harvesting and storing water
   — prepare for high rainfall storms by building in fail-safe or recovery mechanisms
   — prepare for locust plagues and other pests by maintaining diversity and using a range of suitable species
   — aim for fail-safe and recovery mechanisms (resilience)
5. Manage according to natural boundaries
   — farming to soil type
   — individual farms managed within the context of catchment planning
   — bioregional planning as a basis for setting priorities and guiding local actions
6. Aim for social and economic systems which are congruent with achieving these goals.

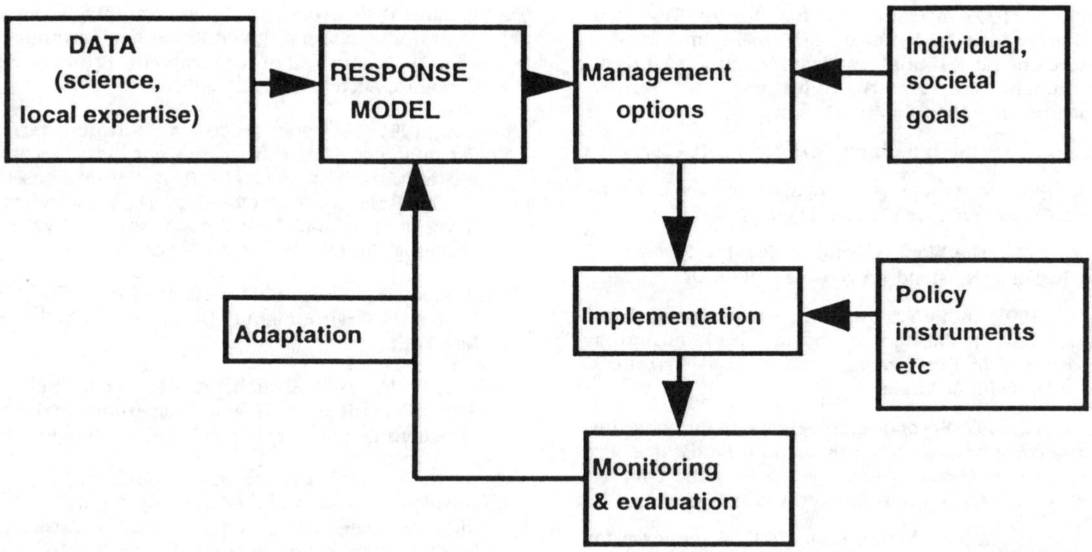

Fig. 7. An adaptive management framework, in which data from both scientific investigations and local expertise are used to generate ecological response models, from which a range of management options can be derived. The options selected will depend on the goals of both individual landholders and society as a whole. Implementation of the selected options will be affected by a range of factors including the set of policy instruments (incentives etc.) in place and broader socioeconomic issues. Implementation is accompanied by monitoring and evaluation of progress relative to the recognized goals, checked against the response models and the management options are modified as necessary. At the same time the response models are modified as necessary in the light of new data.

Strategic research is also needed which "asks questions for tomorrow" rather than just "answering questions for today". Of particular interest here is the development of the ability to partition regions on the basis of landscape characteristics, so that management guidelines can be produced for particular landscape types and applied wherever that type exists. This in turn requires a clearer understanding of the functional linkage between landscape patterns and processes. In addition, strategic research will need increasingly to be interdisciplinary. However, this will involve not just bringing different disciplines together, but actively seeking out the opportunities for innovative approaches found in the murky zones between disciplines.

Tactics

Most activity to date has been at this level, and hence there already exists a battery of tools which can be used in a tactical way. These include land use planning models and packages, management guidelines, revegetation guidelines and databases and so on. However, there is still a clear need for the more effective transfer of scientific information to managers in accessible ways. Hence there is an undiminished requirement for more effective management tool kits, coupled with a battery of policy instruments aimed at achieving greater on-ground implementation rates.

A further important tactic will be adaptive management, which has taken a variety of guises (Holling 1978; Walters and Hilborn 1978; Walters and Holling 1990), but is taken here to mean an iterative process of using a set of response models to provide a set of management options which are implemented and modified as necessary (Fig. 7). This melding of experimental research with real-life management operations seems essential if we are to make progress with broad-scale issues.

CONCLUSION

Most of the approaches discussed here are already underway or at least being considered. We are aware of Bert Main's admonition that if you really want to see change happening in the way things are done, you have to be prepared for a long haul (Main 1995). We are also aware of the urgency with which action is required. This urgency both at a local and global level is probably greater now than it was when we started over 10 years ago. The icebergs are closer, but at least we can see them coming now and are accumulating the tools to navigate through them. Let us hope we have the collective wisdom to do so.

REFERENCES

Agriculture Western Australia, Department of Conservation and Land Management, Department of Environmental Protection, and Water and Rivers Commission, 1996a. Salinity: A Situation Statement for Western Australia. Government of Western Australia: Perth, Western Australia.

Agriculture Western Australia, Department of Conservation and Land Management, Department of Environmental Protection, and Water and Rivers Commission, 1996b. Western Australian Salinity Action Plan. Government of Western Australia: Perth, Western Australia.

Alexander, H., 1995. A framework for change: The State of the Community Landcare Movement in Australia. Report on the National Landcare Facilitator Project. Department of Primary Industries and Energy: Canberra, Australian Capital Territory.

Alexandra, J., 1995. Bush bashing. *Habitat* **23:** 19–26.

Bailey, R., 1995. Seven doomsday myths about the environment. *Global Climate Change Digest* **8:** 5.

Baskin, Y., 1997. The Work of Nature: How the Diversity of Life Sustains Us. Island Press: Washington, DC.

Beeston, G., 1995. Initiatives to protect native vegetation on private land in Western Australia. South Australian Department of Environment and Natural Resources: Adelaide, South Australia.

Binning, C. E., 1997. Beyond reserves: options for achieving nature conservation objectives in rural landscapes. Pp. 155–68 *in* Frontiers in Ecology: Building the Links ed by N. Klomp and I. Lunt. Elsevier: Oxford.

Brandenburg, S. A. and Majer, J. D., 1995. A database for revegetated areas in the Tammin region of Western Australia: implications for land owners, managers and researchers. Pp. 258–70 *in* Nature Conservation 4: The Role of Networks ed by D. A. Saunders, J. L. Craig and E. M. Mattiske. Surrey Beatty & Sons: Chipping Norton, New South Wales.

Brasden, J., 1994. Alternatives for achieving sustainable land use. Pp. 172–92 *in* Restoring the Land: Environmental Values, Knowledge and Action ed by L. Cosgrove, D. G. Evans and D. Yenken. Melbourne University Press: Melbourne, Victoria.

Campbell, A., 1994. Landcare. Communities Shaping the Land and the Future. Allen & Unwin: St Leonards, New South Wales.

Catterall, C. P., Storey, R. J. and Kingston, M. B., 1997. Reality versus rhetoric: a case study monitoring regional deforestation. Pp. 367–77 *in* Conservation Outside Nature Reserves ed by P. Hale and D. Lamb. Centre for Conservation Biology, University of Queensland: Brisbane, Queensland.

Chapin, F. S., Walker, B. H., Hobbs, R. J., Hooper, D. U., Lawton, J. H., Sala, O. E. and Tilman, D., 1997. Biotic control over the functioning of ecosystems. *Science* **277:** 500–04.

Coates, A. M., 1987. Management of Native Vegetation on Farmland in the Wheatbelt of Western Australia. Department of Agriculture: Perth, Western Australia.

Commonwealth of Australia, 1996. Australia: State of the Environment 1996. CSIRO Publishing: Collingwood, Victoria.

Cosgrove, L., Evans, D. G. and Yenken, D., 1994. Introduction. Pp. 1–6 *in* Restoring the Land: Environmental Values, Knowledge and Action ed by L. Cosgrove, D. G. Evans and D. Yenken. Melbourne University Press: Melbourne, Victoria.

Curtis, A. L., 1997. Landcare, stewardship and biodiversity conservation. Pp. 143–53 *in* Frontiers in Ecology: Building the Links ed by N. Klomp and I. Lunt. Elsevier: Oxford.

Daily, G. C. (ed), 1997. Nature's Services: Societal Dependence on Natural Ecosystems. Island Press: Washington, DC.

Dennings, S., 1997. It's gnow or never: malleefowl conservation in Western Australia. Pp. 214–16 *in* Conservation Outside Nature Reserves ed by P. Hale and D. Lamb. Centre for Conservation Biology, University of Queensland: Brisbane, Queensland.

Department of Environment, Sports and Territories, 1994. Australia's national greenhouse gas inventory 1988 and 1990. Australian Government Printing Service: Canberra, Australian Capital Territory.

Doley, A., 1995. Networks needed by individual farmers: a personal perspective from the northern wheatbelt of Western Australia. Pp. 241–50 *in* Nature Conservation 4: The Role of Networks ed by D. A. Saunders, J. L. Craig and E. M. Mattiske. Surrey Beatty & Sons: Chipping Norton, New South Wales.

Easterbrook, G., 1995. A Moment on Earth. The Coming of Age of Environmental Optimism. Penguin Books: New York.

Ehrlich, P. R. and Ehrlich, A. H., 1996. Betrayal of Science and Reason: How Anti-environmental Rhetoric Threatens our Future. Island Press: Washington, DC.

George, R. J., McFarlane, D. J. and Speed, R. J., 1995. The consequences of a changing hydrologic environment for native vegetation in south Western Australia. Pp. 9–22 *in* Nature Conservation 4: The Role of Networks ed by D. A. Saunders, J. L. Craig and E. M. Mattiske. Surrey Beatty & Sons: Chipping Norton, New South Wales.

Gilmour, J. and Saunders, D. A., 1995. EARTHWATCH: an international network in support of research on nature conservation. Pp. 627–33 *in* Nature Conservation 4: The Role of Networks ed by D. A. Saunders, J. L. Craig and E. M. Mattiske. Surrey Beatty & Sons: Chipping Norton, New South Wales.

Goldney, D., Bauer, J., Bryant, H., Hodgkins, D. and Watson, G., 1995. Winning battles but losing the war: the education marketing imperative. Pp. 547–88 *in* Nature Conservation 4: The Role of Networks ed by D. A. Saunders, J. L. Craig and E. M. Mattiske. Surrey Beatty & Sons: Chipping Norton, New South Wales.

Government of Western Australia, 1997. 1997 State of the Environment Reference Group Draft Working Papers. Section 8. Tracking Progress Towards Ecologically Sustainable Development. Department of Environmental Protection: Perth, Western Australia.

Hill, Senator, The Honorable, R., 1997. Investing in our natural heritage: the Commonwealth's Environmental Expenditure 1997–98. Australian Government Publishing Service: Canberra, Australian Capital Territory.

Hobbs, R. J., 1993. Effects of landscape fragmentation on ecosystem processes in the Western Australian wheatbelt. *Biol. Conserv.* **64:** 193–201.

Hobbs, R. J., 1994. Fragmentation in the wheatbelt of Western Australia: landscape scale problems and solutions. Pp. 3–20 *in* Fragmentation in Agricultural Landscapes ed by J. Dover. IALE: Garstang, UK.

Hobbs, R. J., 1995. Research into remnant management issues: nuts and bolts and the big picture. Pp. 21–23 *in* Remnant Native Vegetation Ten Years On: A Decade of Research and Management ed by K. Wallace. Department of Conservation and Land Management: Perth, Western Australia.

Hobbs, R. J. and Saunders, D. A., 1991. Reintegrating fragmented landscapes — a preliminary framework for the Western Australian wheatbelt. *J. Env. Manag.* **33:** 161–67.

Hobbs, R. J. and Saunders, D. A. (eds), 1993. Reintegrating Fragmented Landscapes. Towards Sustainable Production and Conservation. Springer-Verlag: New York.

Hobbs, R. J. and Saunders, D. A., 1995. Conversing with aliens: do scientists communicate with each other well enough to solve complex environmental problems? Pp. 195–98 *in* Nature Conservation 4: The Role of Networks ed by D. A. Saunders, J. L. Craig and E. M. Mattiske. Surrey Beatty & Sons: Chipping Norton, New South Wales.

Hobbs, R. J., Saunders, D. A., Lobry de Bruyn, L. A. and Main, A. R., 1993. Changes in biota. Pp. 65–106 *in* Reintegrating Fragmented Landscapes. Towards Sustainable Production and Nature Conservation ed by R. J. Hobbs and D. A. Saunders. Springer-Verlag: New York.

Holling, C. S. (ed), 1978. Adaptive Environmental Assessment and Management. Wiley and Son: Chichester.

Hussey, B. M. J. and Wallace, K. J., 1993. Managing Your Bushland. Department of Conservation and Land Management: Perth, Western Australia.

Jenkins, S., 1996. Native vegetation on farms survey 1996. Agriculture Western Australia: Perth, Western Australia.

Lefroy, E. C., Hobbs, R. J. and Atkins, L., 1991. Revegetation Guide to the Central Wheatbelt. Western Australian Department of Agriculture, CSIRO and Greening Australia: Perth, Western Australia.

Lefroy, E. C., Hobbs, R. J. and Scheltema, M., 1993. Reconciling agriculture and nature conservation: toward a restoration strategy for the Western Australian wheatbelt. Pp. 243–27 *in* Nature Conservation 3: Reconstruction of Fragmented Ecosystmems, Global and Regional Perspectives ed by D. A. Saunders, R. J. Hobbs and P. R. Ehrlich. Surrey Beatty & Sons: Chipping Norton, New South Wales.

Lloyd, T., 1995. The Dongolocking corridors project. Pp. 65–66 *in* Remnant Native Vegetation Ten Years On: A Decade of Research and Management ed by K. Wallace. Department of Conservation and Land Management: Perth, Western Australia.

Lynch, J. F., Carmen, W. J., Saunders, D. A. and Cale, P., 1995. Use of vegetated road verges and habitat patches by four bird species in the central wheatbelt of Western Australia. Pp. 34–42 *in* Nature Conservation 4: The Role of Corridors ed by D. A. Saunders, J. L. Craig and E. M. Mattiske. Surrey Beatty & Sons: Chipping Norton, New South Wales.

Main, A. R., 1995. Decision makers and ecologists: networks and difficulties. Pp. 141–47 *in* Nature Conservation 4: The Role of Networks ed by D. A. Saunders, J. L. Craig and E. M. Mattiske. Surrey Beatty & Sons: Chipping Norton, New South Wales.

McFarlane, D. J., George, R. J. and Farrington, P., 1993. Changes in the hydrologic cycle. Pp. 146–86 *in* Reintegrating Fragmented Landscapes. Towards Sustainable Production and Nature Conservation ed by R. J. Hobbs and D. A. Saunders. Springer-Verlag: New York.

Nicholl, C. L. and Dobbie, M. J., 1996. The CSIRO National Farm Tree Survey: Landholders' Observations of Trees Planted for Land Degradation Control. Division of Water Resources Technical Memorandum 96.26. CSIRO Division of Water Resources: Canberra, Australian Capital Territory.

Norton, D. N., Hobbs, R. J. and Atkins, L., 1995. Fragmentation, disturbance and plant distribution: mistletoes in woodland remnants in the Western Australian wheatbelt. *Conserv. Biol.* **9**: 426–38.

Nulsen, R. A., 1993. Changes in soil properties. Pp. 107–45 *in* Reintegrating Fragmented Landscapes: Towards Sustainable Production and Nature Conservation ed by R. J. Hobbs and D. A. Saunders. Springer-Verlag: New York.

Orsini, J.-P., Hall, G. and Members of the Malleefowl Preservation Group, 1995. The Malleefowl Preservation Group in Western Australia: a case study in community participation. Pp. 517–22 *in* Nature Conservation 4: The Role of Networks ed by D. A. Saunders, J. L. Craig and E. M. Mattiske. Surrey Beatty & Sons: Chipping Norton, New South Wales.

Petit, N. E., Froend, R. H. and Ladd, P. G., 1995. Grazing in remnant woodland vegetation: changes in species composition and life form groups. *J. Veg. Sci.* **6**: 121–30.

Platt, S. J. and Ahern, L. D., 1995. Nature conservation on private land in Victoria, Australia — the role of Land for Wildlife. Pp. 300–11 *in* Nature Conservation 4: The Role of Networks ed by D. A. Saunders, J. L. Craig and E. M. Mattiske. Surrey Beatty & Sons: Chipping Norton, New South Wales.

Price, P. (ed), 1995. Socio-economic aspects of maintaining native vegetation on agricultural land. Land and Water Research and Development Corportation: Canberra, Australian Capital Territory.

Saunders, D. A., 1996a. Habitat networks in the fragmented landscape of the Western Australian wheatbelt: preliminary results, involvement with landcare groups, and experience in implementation. Pp. 69–80 *in* Species Survival in Fragmented Landscapes ed by J. Settele, C. R. Margules, P. Poschlod and K. Henle. Kluwer Academic Publishers: The Netherlands.

Saunders, D. A., 1996b. Does our lack of vision threaten the viability of the reconstruction of disturbed ecosystems? *Pac. Conserv. Biol.* **2**: 321–26.

Saunders, D. A., Arnold, G. W., Burbidge, A. A. and Hopkins, A. J. M. (ed), 1987. Nature Conservation: The Role of Remnants of Native Vegetation. Surrey Beatty & Sons: Chipping Norton, New South Wales.

Saunders, D. A., Craig, J. L. and Mattiske, E. M. (ed), 1995. Nature Conservation 4: The Role of Networks. Surrey Beatty & Sons: Chipping Norton, New South Wales.

Saunders, D. A., Hobbs, R. J. and Margules, C. R., 1991. Biological consequences of ecosystem fragmentation. *Conserv. Biol.* **5**: 18–32.

Saunders, D. A. and de Rebeira, C. P., 1991. Values of corridors to avian populations in a fragmented landscape. Pp. 221–40 *in* Nature Conservation 2: The Role of Corridors ed by D. A. Saunders and R. J. Hobbs. Surrey Beatty & Sons: Chipping Norton, New South Wales.

Saunders, D. A. and Hobbs, R. J. (eds), 1991. Nature Conservation 2: The Role of Corridors. Surrey Beatty & Sons: Chipping Norton, New South Wales.

Saunders, D. A., Hobbs, R. J. and Arnold, G. W., 1993a. The Kellerberrin project on fragmented landscapes: A review of current information. *Biol. Conserv.* **64**: 185–92.

Saunders, D. A., Hobbs, R. J. and Ehrlich, P. R. (eds), 1993b. Nature Conservation 3: Reconstruction of Fragmented Ecosystems, Global and Regional Perspectives. Surrey Beatty & Sons: Chipping Norton, New South Wales.

Saunders, D. A., Rowley, I. and Smith, G. T., 1985. The effects of clearing for agriculture on the distibution of cockatoos in the southwest of Western Australia. Pp. 309–21 *in* Birds of Eucalypt Forests and Woodlands: Ecology, Conservation, Management ed by A. Keast, H. F. Recher, H. Ford and D. Saunders. Surrey Beatty & Sons: Chipping Norton, New South Wales.

Scougall, S. A., Majer, J. D. and Hobbs, R. J., 1993. Edge effects in grazed and ungrazed Western Australian wheatbelt remnants in relation to ecosystem reconstruction. Pp. 163–78 *in* Nature Conservation 3: Reconstruction of Fragmented Ecosystems, Global and Regional Perspectives ed by D. A. Saunders, R. J. Hobbs and P. R. Ehrlich. Surrey Beatty & Sons: Chipping Norton, New South Wales.

Shur, B., 1990. W.A.'s biggest nature conservation problem: land clearing in the south west. *Land and Water Research Notes* **5**: 6–8.

Simon, J. L. (ed), 1995. The State of Humanity. Blackwell: Oxford.

Sivertsen, D., 1994. The native vegetation crisis in the wheatbelt of NSW. *Search* **25**: 5–8.

Suzuki, D., 1990. Inventing the Future: Reflections on Science, Technology and Nature. Allen and Unwin: London.

Vanclay, F. and Lawrence, G., 1995. The Environmental Imperative: Eco-social Concerns for Australian Agriculture. Central Queensland University Press: Rockhampton, Queensland.

Vitousek, P. M., Mooney, H. A., Lubchenco, J. and Melillo, J., 1997. Human domination of Earth's ecosystems. *Science* **277**: 494–99.

Wallace, K. J., 1995a. Conclusions. Pp. 89–95 *in* Remnant Native Vegetation Ten Years On: A Decade of Research and Management ed by K. Wallace. Department of Conservation and Land Management: Perth, Western Australia.

Wallace, K. J., 1995b. Setting the scene — remnant management during the past decade. Pp. 51–54 *in* Remnant Native Vegetation Ten Years On: A Decade of Research and Management ed by K. Wallace. Department of Conservation and Land Management: Perth, Western Australia.

Walters, C. J. and Hilborn, R., 1978. Ecological optimization and adaptive management. *Ann. Rev. Ecol. Syst.* **9**: 157–88.

Walters, C. J. and Holling, C. S., 1990. Large-scale management experiments and learning by doing. *Ecology* **71**: 2060–068.

Watson, J., Lullfitz, W., Sanders, A. and McQuoid, N., 1995. Networks and the Fitzgerald River National Park Biosphere Reserve, Western Australia. Pp. 482–87 *in* Nature Conservation 4: The Role of Networks ed by D. A. Saunders, J. L. Craig and E. M. Mattiske. Surrey Beatty & Sons: Chipping Norton, New South Wales.

Yates, C. J. and Hobbs, R. J., 1997a. Temperate eucalypt woodlands: A review of their status, processes threatening their persistence and techniques for restoration. *Aust. J. Bot.* **45**: 949–73.

Yates, C. J. and Hobbs, R. J., 1997b. Woodland restoration in the Western Australian wheatbelt: a conceptual framework using a state and transition model. *Restoration Ecology* **5**: 28–35.

Young, M. D., Gunningham, N., Elix, J., Lambert, J., Howard, B., Grabosky, P. and McCrone, E., 1996. Reimbursing the future: an evaluation of motivational, voluntary, price-based, property-right, and regulatory incentives for the conservation of biodiversity. Biodiversity Unit, Biodiversity Series Paper No. 9. Department of the Environment, Sport and Territories: Canberra, Australian Capital Territory.

CHAPTER 2

Managing unreserved forest land for biodiversity conservation: the importance of the matrix

David B. Lindenmayer[1] and Jerry F. Franklin[2]

Matrix or off-reserve forests where such commercial activities as the production of timber and pulp takes place will dominate most forest landscapes around the world. As a result, if managed appropriately, the majority of forest biodiversity will occur in the matrix; a fact that has particular importance because of the inadequacy of existing reserve systems and limitations on their future expansion. Management strategies in the matrix will govern overall levels of forest biodiversity that are present there through influencing: (1) the size, viability and spatial dispersion of populations of many forest-dependent organisms, (2) facilitating or obstructing the movement of organisms, and, (3) buffering sensitive areas and reserves. Appropriate management of the matrix also will be critical for underpinning the sustained production of resources like wood fibre and water.

Strategies for conserving biodiversity in managed matrix forests include: (1) longer rotation times, (2) the retention of natural vegetation throughout the matrix (including cutover sites), particularly structural and floristic components that provide habitat for wildlife such as large living and dead standing trees, logs on the forest floor, and intact thickets of understorey vegetation and other plants, (3) reservation from cutting of sensitive areas such as stands on unstable soils and landforms, and, (4) establishment of systems of retained habitat such as wildlife corridors and riparian stream buffers. The application of all of these approaches for matrix management is an important risk-spreading strategy that improves chances of providing suitable conditions for the array of different forest-dependent taxa characterized by different resource and habitat requirements.

Despite the importance of matrix-based strategies, quantitative data to guide on-the-ground prescriptions are limited. For example, information is lacking on the number and spacing of trees needed to facilitate both recolonization of logged areas and dispersal through managed forests. Carefully designed management-by-experiment studies are urgently required to assist the acquisition of data needed to guide effective matrix-based management. Long-term monitoring and the adoption of adaptive management principles will be integral to the success of such experiments in managed forests. In addition, fields of research like conservation biology need to place greater emphasis on topics such as the response of sets of species to matrix conditions, how animals use and move through the matrix, and the rate at which modifications in the matrix alter edge effects. Conservation biologists, forest managers and conservationists also need to become more aware of the important role the matrix plays in the conservation of forest biodiversity. In the absence of such attention, large losses of forest biodiversity are inevitable.

---

[1]Centre for Resource and Environmental Studies and Department of Geography, The Australian National University, Canberra, Australian Capital Territory, Australia 0200.
[2]Box 352100, University of Washington, Seattle, Washington, USA, 98195-2100.

Pages 13–25 *in* NATURE CONSERVATION 5: NATURE CONSERVATION IN PRODUCTION ENVIRONMENTS: MANAGING THE MATRIX ed by J. L. Craig, N. Mitchell and D. A. Saunders. Surrey Beatty & Sons, 2000.

INTRODUCTION

STRATEGIES for conserving forest biodiversity have traditionally focused on networks of dedicated reserves or biodiversity priority areas (Noss and Cooperrider 1994; Anon. 1996). The establishment of reserves is important where particular forest types or age classes are being rapidly modified or eliminated (McNeely 1994a). Examples include Brigalow dominated-vegetation communities in eastern Australia (Nix 1994; Fensham 1996), old growth forests in Scandinavia (Virkkala *et al.* 1994) and late-successional forests in northwestern America (Thomas *et al.* 1990; Kohm and Franklin 1997). However, most forest landscapes are going to be comprised primarily (or even exclusively) of managed forests. This is particularly true in temperate regions where most productive forested lands have already been extensively modified by human activities (Franklin 1988; Virkkala *et al.* 1994). Even in countries and regions where policies call for the allocation of up to 15% of the forested land for conservation (e.g., Australia; Anon. 1996), most biodiversity will have to exist outside reserves and, often, in areas managed primarily for commercial activities such as the production of wood fibre (McNeely *et al.* 1990; Pimental *et al.* 1992, 1997; Lindenmayer and Franklin 1997a). Hence, managed forests, which we refer to as the "matrix", are critical for the maintenance of forest biodiversity (Wilcove 1989; Franklin 1993).

How the matrix is managed will influence the size and viability of populations of many (if not most) forest taxa and thus biological diversity *per se*. Conditions in the matrix also will greatly influence the movement of organisms and connectivity between reserves. They will also control the effectiveness of reserves and other areas by buffering them. Furthermore, the matrix provides many goods such as wood as well as services like the regulation of hydrological regimes.

Despite its importance, the matrix has received limited attention from conservation biologists and environmentalists (Lindenmayer and Franklin 1997a). In this chapter, we outline the major roles of matrix lands in the conservation of biological diversity. We also discuss management approaches which can be employed to enhance the conservation value of the matrix. Our objective is to provide an increased understanding of the importance of the matrix for the conservation of forest biodiversity.

WHAT IS THE MATRIX?

The term "matrix" has several definitions and usages. In the technical language of landscape ecology, the matrix is defined as the dominant and most extensive "patch type" (Forman 1995). Within this context, several criteria can be used to identify the matrix; total area is the most obvious and straight-forward, but connectivity and a controlling influence over the landscape are also important (Forman 1995). The term matrix is often used in the conservation biology and forest planning literature to refer to areas not reserved primarily for nature conservation and which are available for commodity production and other human uses. This usage has emerged primarily in developed temperate regions where the majority of the landscape is unreserved. In these cases, the definitions of the matrix are consistent from both a landscape ecology and conservation biology perspective. Conversely, in undeveloped regions, the matrix *sensu* landscape ecology, may not be equivalent to the matrix *sensu* conservation biology because the majority of the forested land remains in a "natural" condition. For the purposes of this chapter, we have defined the matrix as forest not reserved for conservation regardless of its current condition (i.e., natural or developed).

THE IMPORTANCE OF THE MATRIX

The importance of the matrix for biodiversity conservation stems primarily from its sheer size and scale. The matrix currently dominates most tropical and temperate forest landscapes. Since the amount of land reserved in temperate forest regions around the world is unlikely to exceed 10 to 15% (even in developed countries like the United States of America, Canada and Australia), the majority of forest biodiversity will either exist in the matrix outside systems of protected areas — or those species for which reserves do not provide sufficient suitable habitat may cease to exist (Franklin 1993; Lindenmayer and Franklin 1997a). Indeed, McNeely *et al.* (1990) noted that during future decades, the preservation of biodiversity will occur in areas with primary uses other than nature conservation.

The importance of the matrix is further magnified by problems and limitations of the existing reserve system. Some of these include:

Reserves have often been set aside for reasons other than nature conservation (Hall 1988; Noss and Cooperrider 1994) such as the protection of places of scenic beauty or because such areas were unsuitable for other uses (e.g., agriculture or forestry).

Reserves are typically an unrepresentative sample of the range of environments, plant communities or rare taxa that occur in a region (e.g., Johnson 1992; Specht *et al.* 1995; Khan *et al.* 1997). Reserves also are biased toward places characterized by steep terrain and/or low productivity (Pressey 1995; Pressey *et al.* 1996) — areas that can be unsuitable for the conservation of some species (Braithwaite 1984).

Populations confined to reserves may not be viable in the long term (Newmark 1985; Wilcove 1989; Grumbine 1990). Even small species, such as some invertebrates, may only be viable in large numbers (Thomas 1990; Tscharntke 1992) or need large areas for long-term persistence (Okland 1996).

The units used as a basis for reserve selection (e.g., forest types, plant communities or ecological vegetation units [Commonwealth of Australia and Department of Natural Resources and Environment 1997]) may not be a good surrogate for other forms of biodiversity (Burgman and Lindenmayer 1998).

For example, some taxa like invertebrates and small mammals are closely associated with understorey plants (Gullan and Robinson 1980; Woinarski and Cullen 1984) rather than dominant tree species.

Taxa with fine scale or restricted distributions such as species-rich invertebrate groups (e.g., Onycophorans [New 1995; Reed 1996 in Commonwealth of Australia and Department of Natural Resources and Environment 1997] may not be captured with reserve systems.

Forest ecosystems can be dynamic with concomitant changes in species composition (e.g., Woinarski *et al.* 1992). As a result, species conserved in reserves now may not remain there, either temporarily or permanently (Margules *et al.* 1994; Witting and Loeschcke 1995).

Some mobile species require resources that vary in their temporal and spatial availability and occur within and outside reserves (Redford and de Fonseca 1986; Woinarski *et al.* 1992; Date *et al.* 1996).

Opportunities to overcomes these problems and expand reserves to make them more viable, representative, and improve their geographic coverage, are very limited in most temperate forest regions.

CRITICAL ROLES OF THE MATRIX

Several critical roles can be identified for the matrix and these are discussed briefly below.

Supporting populations of species

The matrix is the dominant patch type in most forest ecosystems and if it is managed appropriately, it will support well-distributed populations of species, as well as higher overall population levels of taxa. Larger populations have a lower risk of extinction through demographic and genetic stochasicity (Pimm *et al.* 1988; Lacy 1993) and environmental variability (Thomas 1990; Tschartntke 1992). The maintenance of larger, well-distributed populations also minimizes the risks that an entire population could be extirpated in a single catastrophic event (Gilpin 1987) such as a wildfire (Lindenmayer and Possingham 1995a).

The importance of the matrix for maintaining populations of forest organisms has led, in some cases, to the development of conservation strategies based entirely on the management of the matrix, rather than the establishment of reserves. A current example is the California Spotted Owl; a declining sub-species which viability analysis has indicated may be vulnerable to extinction (La Haye *et al.* 1994). The conservation strategy recommended for the California Spotted Owl and adopted as an interim strategy by the United States Forest Service, provides for management of the entire forest matrix to maintain well-distributed populations (Verner *et al.* 1992). This approach was selected in preference to the establishment of a system of reserves which would predictably result in a significant reduction in the total number of owls. A specific provision of the strategy is the retention of all trees exceeding 85 cm in diameter on harvested lands (Verner *et al.* 1992).

The maintenance of populations in the matrix has other important functions such as supplementing populations in reserves. Diamond *et al.* (1987) and Laurance (1991) demonstrated that species that persist in the matrix will also be those likely to reside in reserves. For example, evidence of rapid species turnover within protected areas (e.g., Margules *et al.* 1994) suggests that dispersalists from populations residing in the matrix may be required to reverse localized extinctions within reserves (Thomas *et al.* 1992; Hanski *et al.* 1995).

Maintaining connectivity and facilitating the movement of organisms

The matrix controls connectivity (*sensu* Bennett 1998) in forest landscapes. Therefore, movement of organisms can be facilitated or obstructed depending on the conditions in the matrix (Rosenberg *et al.* 1997). A connectivity function is important because habitat fragmentation and the subsequent potential risks associated with the creation of smaller, isolated and/or sub-divided populations is widely considered to be a major process threatening wildlife populations in temperate forests around the world (Harris 1984; Wilcove *et al.* 1986). Facilitating the movement of organisms across the matrix may prevent populations residing in reserves from becoming isolated and fragmented. The maintenance of connectivity between populations may increase their demographic and genetic size (Lacy 1993) and enhance their long-term probability of persistence (Scotts 1994).

Conservation management strategies applied throughout the matrix also will be particularly important for those species which disperse randomly (e.g., Cabbage Butterfly [Fahrig and Paloheimo 1988]) rather than using corridors to move between areas of potentially suitable habitat. Indeed, if conditions in the landscape matrix are suitable, then only a relatively small amount of movement may be required to rescue populations in reserves or habitat remnants (Stacey and Taper 1992). In Australia, matrix suitability also may allow migratory and nomadic species such as Pigeons and Honeyeaters access to resources that fluctuate in temporal and spatial availability (Date *et al.* 1996).

Buffering sensitive areas and reserves

In most regions, small or medium-sized reserves will be embedded within the matrix. Therefore, conditions in the matrix can influence their effectiveness (Schoenwald-Cox 1988; Kelly and Rotenberry 1993). Hostile conditions in the matrix can have negative biotic and abiotic influences on reserves and sensitive areas (e.g., Lovejoy *et al.* 1986). For example, changed wind, light, temperature and moisture conditions in edge environments are well known (reviewed by Saunders *et al.* 1991; Forman 1996; Parry 1997). Biotic edge effects such as increased tree mortality, weed invasion, increased nest predation and nest parasitism also have been well documented (e.g., Wace 1977; Temple and Cary 1988; Chen *et al.* 1992; Luck 1996). Indeed, Nelson (1991) argued that the effectiveness of many reserves and other protected areas in Canada was related to conditions in the surrounding areas. Janzen (1983) pointed out that matrix management assumed increasing importance with decreasing reserve size.

The intensity of the edge interactions between two patches is directly related to the level of structural contrast between them. Hence, management of the matrix to reduce the contrast in conditions has a major influence on the intensity and depth of edge effects. By managing the matrix to buffer sensitive areas, edge effects can be reduced and the area available for species requiring forest-interior habitats expanded — often dramatically (Harris 1984). For example, Burgman and Ferguson (1995) and Parry (1997) stressed the importance of buffer systems in the matrix to protect rainforest areas in Victoria from edge effects.

Providing for the production of commodities and services

Management of the matrix will influence the extent to which the production of goods and services can be sustained in the long term (Chapin *et al.* 1997). Humans derive a wide array of goods and services from forest ecosystems (Constanza *et al.* 1997). The production of wood fibre is one of the major goods derived from forest ecosystems. Key services include the regulation of streamflow, soil protection, nutrient cycling, and the provision of habitat for forest-dependent organisms. The conservation of biodiversity is critical for maintaining forest productivity as well as the continued provision of ecosystem services (Pimental *et al.* 1992, 1997). Losses of forest biodiversity or large changes in species abundance may impair important ecosystem functions such as pollination, seed dispersal and the formation of mycorrhizal associations. These changes could influence the long-term floristic composition and stand architecture of forests and will have major negative ramifications for sustained production of commodities. For example, fungi which form mycorrhizae are critical to the regeneration and growth of trees in the Douglas-fir forests of northwestern United States of America (Perry 1994) — as well as most other forest regions. The retention of vegetation within cutover sites in the Pacific Northwest in the United States of America provides a reservoir of mycorrhizae and soil microbes to inoculate the adjacent harvested area (Perry 1994).

Failure to maintain key ecosystem processes (and in turn, goods and services) from the matrix will have major social and economic costs (Pimental *et al.* 1992, 1997). Constanza *et al.* (1997) estimated that the value of goods and services from forest ecosystems approached five trillion dollars annually. Most of these goods and services will be derived from matrix forests. As an example of just one service, populations of natural parasites and predators have been estimated to undertake the equivalent of $100–200 billion worth of pest control — compared to $20 billion expended on artificial control measures such as spraying (Pimental *et al.* 1992). Loss of natural populations of parasites and predators has massive financial implications for timber and pulp production from native and plantation forests (Pimental *et al.* 1992).

MANAGING THE MATRIX TO CONSERVE BIOLOGICAL DIVERSITY

A range of strategies for matrix management can be used to promote the conservation of biodiversity in production forests landscapes. These include the protection of sensitive areas, the maintenance of key habitat attributes on logged sites, and modification of the spatial and temporal extent of timber harvesting. These strategies are discussed briefly below.

Protect sensitive areas

The protection of sensitive and ecologically important areas such as wetlands, riparian zones and lake margins is a key component of any approach to managing the matrix for improved conservation of biological diversity. For example, riparian areas typically contain rich faunal assemblages (e.g., Loyn *et al.* 1980; Recher *et al.* 1980; Naiman *et al.* 1993). Harvesting operations close to or within them could have negative effects on an array of aquatic taxa (Michaelis 1984; Doeg and Koehn 1990) by altering environmental conditions (e.g., sediment and temperature regimes; Silsbee and Larson 1983; Graynoth 1989; Metzeling *et al.* 1995). In the State of Victoria, increased sedimentation of streams and rivers is specifically listed as a process threatening nature conservation under the *Flora and Fauna Guarantee Act* (Wilson and Clark 1995). Loss of tree cover from riparian buffer strips can change microclimatic conditions within streams (such as light and temperature regimes) and impair habitat suitability for freshwater fish (Barton *et al.* 1985; Budd *et al.* 1987; Doeg and Koehn 1990). Streamside logging also eliminates the source of large woody debris which are critical structural elements of streams and rivers (Harmon *et al.* 1986; Maser *et al.* 1988).

Maintaining or facilitating the movement of organisms

Management of the matrix for the conservation of forest biodiversity can facilitate the movement and dispersal of organisms between suitable habitat patches. Modifications to enhance matrix management can include the following:

1. Creation of networks of wildlife corridors (Bennett 1998). Linking patches of suitable habitat with wildlife corridors can be valuable for those species that avoid dispersing through open areas (such as some birds; Martin and Karr 1986; Desrochers and Hannon 1997) or taxa where habitat suitability is an important factor

influencing dispersal (e.g., Garrett and Franklin 1988; Baur and Baur 1992; Nelson 1993).

2. "Softening" the matrix (Franklin 1993). This may enhance the movement of some animals through the matrix, particularly those taxa that employ random dispersal strategies and do not move along systems of more consolidated networks of retained habitat such as wildlife corridors (Thomas *et al.* 1990; Murphy and Noon 1992). Softening the matrix can be achieved by permanently retaining parts of original stands during harvesting including key habitat components such as large living and dead trees as well as large logs on the forest floor. For example, the retention of logs provides important protected travel routes for microtine rodents in Douglas-fir forests and allows animals to disperse into, and through, disturbed areas (Maser *et al.* 1977). A matrix-based retention strategy was considered to be the most effective strategy to facilitate the movement of the Northern Spotted Owl in the Pacific Northwest of the United States of America through the matrix and thus between stands of old-growth forest (Thomas *et al.* 1990). This approach was embraced because the species is believed to disperse randomly, and the use of a corridor-based conservation strategy was considered inappropriate (Murphy and Noon 1992). In this case, 50% of the matrix was retained as trees averaging 11 inches (27.5 cm) in diameter and with 40% canopy cover (Thomas *et al.* 1990; Tuchmann *et al.* 1996).

3. Limiting barriers to dispersal. This will facilitate the movement of organisms through matrix landscapes. For example, roading networks may need careful planning because they can be dispersal barriers for forest-dependent organisms such as small mammals (Barnett *et al.* 1978; Burnett 1992; Richardson 1992). In other cases, harvesting prescriptions may need to prevent the development of barriers across particular topographic features such as streamlines or ridge-tops that appear to be used by some species as dispersal routes (Taylor *et al.* 1985) (e.g., Black-tailed Prairie Dogs [Garrett and Franklin 1988]).

Maintain essential habitats

Careful management of matrix forests allows some species to remain on logged areas or to rapidly recolonize harvested stands. This can be accomplished by retaining structural elements that fulfil habitat requirements for various organisms. On this basis, suitable habitat for forest-dependent organisms may need to be retained across spatial scales ranging from a few metres (e.g., a log on the forest floor) to hundreds of metres (e.g., a wetland buffer).

Retention of intact patches of forest in managed landscapes in Malaysia allowed some species to persist and act as population sources to recolonize the surrounding regenerating stands (Johns 1996). In another case, populations of small mammals in southeastern Australia can persist within otherwise unsuitable areas of exotic softwood (Radiata Pine) plantation by using windrows of felled dead trees from the previously cleared stand of native Eucalyptus stand (Friend 1982). A number of studies in logged Australian forests have highlighted the potential value of retained trees in promoting the recolonization of logged and regenerated forests by birds (e.g., Recher *et al.* 1980; Smith 1985; Kavanagh and Turner 1994). Remnant trees in harvested areas allow epiphytic lichens to persist which would otherwise be lost from harvested areas (Peck and McCune 1997).

Another potential value of retaining selected structures as part of a harvesting operation is that they can provide key resources required by non-autotrophic organisms. The retention of freshly cut logs within harvested Scandinavian forests facilitates the persistence of diverse groups of fungi that might otherwise be lost from production landscapes (Niemela *et al.* 1995). Dead trees left in these environments are important for saproxylic beetles (Kaila *et al.* 1997) that contribute to key ecosystem processes like decomposition and nutrient cycling (Niemela *et al.* 1995; Kaila *et al.* 1997). Similar relationships among beetle populations in harvested forests have been recorded in the boreal region of North America (Niemela *et al.* 1993).

Modify the method of harvesting

Many of the strategies outlined above for protecting biodiversity on matrix lands require modification of traditional silvicultural practices. Hunter (1994, P. 273) asserted that the conservation of biodiversity in production landscapes requires management to be as consistent as possible with natural ecological processes. In this context, Attiwill (1994), Bunnell (1995) and a range of other authors have hypothesized that logging operations will have minimal effects on biodiversity if they are within the bounds of natural disturbance regimes. This requires strong congruence between natural disturbances and disturbances created by logging operations at a range of spatial scales including the stand level (Fries *et al.* 1997) and the landscape level (Welsh and Healy 1993; McNeely 1994b).

Maintaining structural features and plant species composition in logged forests that are typical of forests recovering from natural disturbances (e.g., wildfire) is important at the stand level (Franklin *et al.* 1997; Lindenmayer and Franklin 1997a). Fries *et al.* (1997) provides examples in Swedish boreal forests based on the modification of existing clearfelling techniques. Franklin *et al.* (1997) have formalized the concept as the "variable retention harvest system".

Silvicultural approaches in the Mountain Ash forests of the Central Highlands of Victoria provide another useful example. Large old living and dead trees with hollows are characteristic structural attributes of burnt stands of Mountain Ash forest. These structures are key habitat components for many vertebrate and invertebrate taxa (Gibbons and Lindenmayer 1997). However, large old living and dead trees with hollows can be scarce in intensively logged areas such as those subject to even-aged (clearfelling) silvicultural systems (Lindenmayer *et al.* 1991). By modifying silvicultural prescriptions, multi-aged stands can be created over at least a portion of the harvested areas, thereby providing key components of forest structure. Since field and simulation studies have shown that 10—30% of stands of Mountain Ash forest are multi-aged, this is probably an appropriate target for the percentage of logged areas which should be maintained in a structurally diverse condition (McCarthy and Lindenmayer 1998).

Modifying harvesting operations to achieve biodiversity objectives in logged areas requires changes in the traditional philosophy of silviculture. Objectives must be broadened from the singular objective of cutting and regenerating a stand of "crop trees" to the long-term maintenance of key elements of stand structure and plant species composition (e.g., large old hollow-bearing trees and intact thickets of understorey vegetation [Commonwealth of Australia and Department of Natural Resources and Environment 1997]). Forest managers must accept the need for the modification of timber production approaches to incorporate other forest values such as the conservation of biodiversity (Lindenmayer and Franklin 1997b).

The retention and perpetuation of key components of stand structure and floristic composition within logged stands will require careful planning. It also will complicate on-the-ground logging operations. However, increased complexity in forest management should be regarded as normal in production forests where the maintenance of other values such as biodiversity is important (Lindenmayer and Franklin 1997b).

Although silvicultural prescriptions and logging operations can better mimic conditions created by natural disturbance (e.g., see Fries *et al.* 1997) there is never complete congruence. For example, a selective harvesting operation does not result in the same structural features as a partial or low-intensity burn. Similarly, at a landscape scale, there will be differences between logged and fire-disturbed regions in the mean age and spatial characteristics of patches (McCarthy and Burgman 1995). The lack of a complete match between natural and human disturbances emphasizes the need for some reserves where natural disturbance regimes are allowed to occur (Baker 1992) and human activities are excluded.

Modify the rate of harvesting

Some species and ecological processes are sensitive to the proportion of the landscape that is disturbed (e.g., Milledge *et al.* 1991; Lamberson *et al.* 1994). One approach to managing the matrix to enhance the conservation of sensitive taxa is to change the rate and spatial pattern (see below) of timber harvesting. The rate of timber harvesting can be modified by lengthening rotations (Curtis 1997; Thiollay 1997).

Longer rotation times may have a number of benefits for nature conservation and these include:

1. Increasing the time available for organisms to re-invade cut-over areas. However, their ability to do this can be influenced by the amount, type and spacing of vegetation retained on cutover areas (Franklin *et al.* 1997; Lindenmayer and Franklin 1997a).
2. Expanding the array of age classes that occur within a landscape, thereby contributing to landscape heterogeneity. This will reduce the total area of forest in very young cohorts which may be unsuitable for taxa sensitive to the effects of logging operations (Curtis 1997).
3. Increasing the opportunity for the development of key structural and floristic attributes of stands that are important habitat components for wildlife but take a long time to recruit (e.g., large old cavity-bearing trees or logs).
4. Expanding the range of silvicultural options on a site to encompass thinning operations (Curtis 1997) and/or modified post-logging site treatment prescriptions. For example, young Mountain Ash trees in southeastern Australia are highly susceptible to the effects of fire. However, old living trees frequently show signs of fire-damage (Lindenmayer *et al.* 1991; see also Banks 1993). This is an indication that larger stems are more fire resistant. Fire is used in the regeneration of logged Mountain Ash forests (Squire *et al.* 1991) and retained trees within cutover areas might better survive if they are older larger individuals left as part of extended rotation times.
5. Maintaining a large array of future forest management options by delaying some aspects of decision making (Curtis 1997). This has the additional advantage that more informed decisions can be made in the future on the basis of new and better data gathered on forest ecosystems.

Curtis (1997) noted that extended rotation times can generate a range of other benefits such as minimizing the visual impacts of harvesting, increasing long-term timber yields (particularly if thinning takes place), improving the quality of forest products derived from harvested areas, and facilitating the quality and quantity of other forest commodities such as water (see Langford *et al.* 1982). However, although extended rotation times may have some important advantages, they are not panaceas. Even prolonged rotation times may still be too short to allow the development of some key features of old growth stands (Stohlgren 1992; Lindenmayer and Franklin 1997a). In some Eucalyptus forests in eastern Australia, the development of trees with cavities on logged sites would require increasing the period between clearcutting operations from the current 50 to 80 year period to beyond 200 years (Lindenmayer and Possingham 1995b). However, even with 200 year rotations, only 25% of harvested stands would be suitable for cavity-using animals at any one time (forest stands >150 years old). In addition, while the number of cavities increases in eucalypt stems as they mature (Ambrose 1982; Lindenmayer *et al.* 1993a) this process reduces their value as sawlogs.

Modify spatial patterns of harvesting

Many studies have shown that some forest wildlife populations respond not only to the amount of habitat but also the spatial arrangement of habitat (e.g., Lamberson *et al.* 1994). An important consequence of this is that spatially isolated areas of suitable habitat may remain unoccupied. Another consequence may be that remnant forest patches are too small to provide suitable habitat (Lindenmayer *et al.* 1999a).

These types of relationships between the size and spatial location of habitat patches and their rates of occupancy are a well known phenomenon in ecology (reviewed by Hanski 1994; Forman 1995). Hence, even if production landscapes are covered with trees, suitable habitat may be unavailable or too fragmented for some species. A theoretical study by Franklin and Forman (1987) showed that large contiguous areas of old forest were lost rapidly under a dispersed "chequerboard" approach to harvesting coupes. Such approaches to the scheduling of logging operations have negative implications for species that require extensive, relatively intact, areas of suitable habitat such as the Northern Spotted Owl in northwestern North America (Thomas *et al.* 1990; Lamberson *et al.* 1994) and the Yellow-bellied Glider and Sooty Owl in the Mountain Ash forests of southeastern Australia (Milledge *et al.* 1991; Incoll 1995; Lindenmayer *et al.* 1999a). The conservation of these types of "area-sensitive" species requires careful consideration of the appropriate size and spatial juxtaposition of patches of suitable habitat. In these cases, planning may reveal a need to aggregate harvested areas to provide large contiguous areas of suitable habitat (Franklin and Forman 1987). Parker (1997) provides an interesting example of such an approach where logged areas were aggregated (up to 500 ha in size) to better create pre-European conditions appropriate for pine barren ecosystems in central United States of America.

AN INTEGRATED APPROACH TO FOREST BIODIVERSITY CONSERVATION

Matrix-based strategies are essential for the conservation of biological diversity in forest ecosystems. As outlined above, some matrix-based strategies will include: (1) lengthening rotation times (Curtis 1997; Thiollay 1997), (2) retention of structural elements characteristic of natural forests within cutover sites (Thomas *et al.* 1990; Franklin 1993) such as large living and dead standing trees (Loyn 1985; Gibbons and Lindenmayer 1997; Rose and Muir 1997), logs on the forest floor (Maser *et al.* 1977), and intact thickets of understorey vegetation and other plants (Ough and Ross 1992; Mueck *et al.* 1996), (3) protection of geologically and edaphically unstable areas such as stands on steep terrain (e.g., Department of Nature Resources and Environment 1996); and, (4) establishment of systems of retained habitat such as wildlife corridors (Wardell-Johnson and Nicholls 1991; Scotts 1994) and riparian stream buffers (Naiman *et al.* 1993; Barling and Moore 1994).

Although each strategy makes an important contribution to biological conservation, any single approach is not effective for all species. For example, a network of wildlife corridors will not protect all taxa potentially at risk from the effects of logging (Murphy and Noon 1992; Lindenmayer 1994). Corridors also will not overcome problems created by continued habitat loss in the remainder of the matrix (Rosenberg *et al.* 1997). Streamside or riparian reserves are unsuitable for taxa requiring forest habitat on ridges or uplands (McComb and McGarigal 1992; Claridge and Lindenmayer 1994). Finally, areas on steep terrain appear to be unsuitable for some species (Taylor *et al.* 1972; Lindenmayer *et al.* 1997a).

The diverse needs of taxa operating at markedly different spatial scales necessitates the implementation of an array of differently-scaled options ranging from large reserves to strategies at the landscape-scale (such as wildlife corridors) and stand level (such as the maintenance of structural attributes and vegetation complexity). A diverse array of strategies increases the likelihood that suitable habitat conditions will exist for most species in at least some part of a production landscape (i.e., the matrix).

Another benefit of strategies implemented at different spatial scales is the potential for synergism. Field data from forests in southeastern Australia suggest that smaller coupe sizes and/or increased amounts of structural retention (e.g., large trees) on cutover areas may reduce rates of windthrow in wildlife corridors (Lindenmayer *et al.* 1997b). This will enhance their long-term effectiveness as habitat for many species.

Multi-scale conservation strategies should produce a heterogeneous landscape containing the spatially dispersed array of resources needed by some species. Baudry (1984) describes such a response in the field/hedgerow landscapes of France. In addition, factors at multiple scales influence the distribution and abundance of many species (e.g., Forman 1964) and multi-scale conservation strategies will be critical. Leadbeater's Possum in the forests of Victoria is a good example. Reserves, wildlife corridors, and stand-retention strategies have been found to be, on their own, insufficient. This is because large reserves are susceptible to recurrent wildfires that could eliminate resident populations of Leadbeater's Possum (Lindenmayer and Possingham 1995a). Wildlife corridors do not appear to provide suitable habitat for the species (Lindenmayer *et al.* 1993b). Current silvicultural methods in logged areas do not conserve habitat at the stand scale (Lindenmayer *et al.* 1997b). An over-arching approach that encompasses an array of different strategies implemented at multiple scales is required for the conservation of Leadbeater's Possum (Lindenmayer and Franklin 1997a).

Adaptive management and gathering new knowledge to guide matrix management

Despite the importance of the matrix, quantitative data on the long-term effectiveness of matrix-based strategies for forest biodiversity conservation is virtually non-existent. For example, there is limited information on the effectiveness of wildlife corridors within wood production landscapes (Simberloff *et al.* 1992; Hobbs 1992, 1997; Lindenmayer 1994). Similarly, the number, type, size and spatial pattern of patches of retained vegetation on cutover areas to achieve specific conservation objectives is largely unknown (Gibbons and Lindenmayer 1997; Franklin *et al.* 1997).

Given such limitations, new information is required to assess and improve the effectiveness of conservation prescriptions for matrix lands. This is best done within an adaptive management framework (Walters and Holling 1990). Under such an approach, disturbances resulting from logging would be treated as experiments within which biodiversity responses would be carefully monitored (see also Margules and Lindenmayer 1996; Lindenmayer and Franklin 1997a). In these new experiments, disturbance regimes in the matrix would be varied; hence the range of prescriptions set in place for biodiversity conservation would also be varied (including sites with no cutting and sites subject to very intensive cutting). This would provide a range of conditions for formal hypothesis testing. Information on target species or key ecosystem processes gathered in such monitoring "experiments" would then be used to assess the efficacy of logging prescriptions (e.g., the extent of canopy retention needed across the landscape, the width of wildlife corridors, or the number of hollow trees retained within cutover sites). Data collected in this way would be used to inform forest managers about the efficacy of conservation strategies. Prescriptions and codes of practice for timber harvesting would be adjusted accordingly. New sites where modified prescriptions and strategies are applied would then be added to an expanded logging and monitoring experiment.

Margules and Lindenmayer (1996) further outline some of the important principles underpinning the adaptive forest management-by-experiment-and-monitoring approach like the need for replication of sites, the selection of "controls" (uncut sites), and the spatial independence of replicate areas. Several other factors are integral to the success of such studies. These include: (1) long-term financial and logistic commitments to the instigation and of monitoring studies; (2) a commitment by the scientific community to provide intellectual support for monitoring programmes so as to ensure that the results are statistically robust and ecologically valid; (3) extensive dialogue between scientists and forest managers to ensure that the necessary range of stand treatment conditions are provided; and, (4) a willingness on the part of resource management agencies and governments to embrace the findings from research and monitoring programmes.

A number of adaptive management forest experiments are under way in production landscapes around the world (e.g., Carey 1994; United States Department of Agriculture Forest Service Pacific Northwest Research Station 1995; Schmiegelow and Hannon 1993; Arnott *et al.* 1995; Phillips 1996; Pickett 1996). Excellent examples of those that have explored different timber harvesting regimes include the Demonstration of Ecosystem Management Options in the Pacific Northwest of the United States of America and the Montane Alternative Silvicultural Systems experiment on Vancouver Island in Canada (Phillips 1996). Others previously begun (e.g., Squire *et al.* 1987) have been hampered or lost as a result of cuts in long-term funding. There is an urgent need for more studies which will provide a basis for better management of matrix lands for biodiversity conservation.

Other research

Monitoring-by-experiment studies will take time to generate results that are meaningful for applied forest management. However, considerable useful information could be gleaned relatively quickly from retrospective studies designed using some of the criteria listed above and which aim to examine sites logged in the past. The value of such retrospective studies is dependent on the availability of good data on the history of past disturbance in managed stands. In these cases, sites cut with new types of silvicultural practices may need to be added to monitoring experiments to expand the range of conditions being tested. Indeed, Dunning *et al.* (1995) highlighted the value of "quasi experiments" in which a rigorous experimental design is employed to take advantage of existing landscape characteristics. This approach has recently been applied in studies of spatial context effects on vertebrates in forest landscapes in southeastern Australia (Lindenmayer *et al.* 1999b). Haila (1988) described the value of "analytical descriptions" in which ecological phenomena are systematically described and related to key underpinning ecological theory and assumptions.

CONCLUSIONS

The matrix will be the dominant patch type in most forest landscapes making appropriate management strategies in the matrix essential for the conservation of forest biodiversity. Some general approaches are available that have the potential to enhance conservation in matrix lands such as the retention of structural components throughout logged areas, the establishment of wildlife corridors and riparian reserves, and changes in the temporal and spatial pattern of harvesting. However, the effectiveness of these approaches remains untested. This lack of knowledge is partly because much of conservation biology has focused on a narrow set of topics such as reserve design and allocation, the effects of habitat fragmentation, and the use of wildlife corridors. Conservation of biodiversity in the matrix has not been a prominent part of conservation biology or other emerging sub-disciplines such as landscape ecology. For example, the contribution of the matrix to the effectiveness of wildlife corridors is rarely considered (Simberloff *et al.* 1992; Lindenmayer 1994; Rosenberg *et al.* 1997), many studies of fragmentation effects ignore species occurrence in the adjacent production environments (Lindenmayer *et al.* 1999b), and reserve design algorithms do not account for matrix conditions (Burgman and Lindenmayer 1998).

Given its importance, the matrix must be given more attention by conservation biologists. Work is needed on topics such as the response of sets of species to matrix conditions (e.g., Thiollay 1992), how animals use and move through production landscapes (Gustafson and Gardner 1996), and the rate at which edge effects change as the matrix changes (Parry 1997).

In summary, researchers, conservationists and forest managers need to be aware of the important role that the matrix plays in the conservation of forest biodiversity. In the absence of such attention large losses of biodiversity will occur.

ACKNOWLEDGEMENTS

We thank Dr D. Saunders and Professor J. Craig for asking us to complete this review. Comments by Dr K. Viggers, Dr K. Kohm and an anonymous reviewer improved earlier versions of the manuscript. David B. Lindenmayer would like to thank a range of colleagues for collaborative studies from which some of the themes addressed in this chapter have been based. These include Mr R. Cunningham, Mr P. Gibbons, Mr M. Pope, Ms C. Donnelly, Associate-Professor M. Burgman, Dr S. Dovers, Professor H. Nix, Dr R. Lacy, Dr M. McCarthy, Dr C. Margules, Professor H. Possingham, and Professor H. Recher.

REFERENCES

Anon., 1996. National forest conservation reserves. Commonwealth proposed criteria. Commonwealth of Australia: Canberra, Australian Capital Territory.

Arnott, J. T., Beese, W. J., Mitchell, A. K. and Peterson, J., 1995. Montane alternative silvicultural systems (MASS). FRDA Report No. 238. Forest Resource Development Agreement.

Attiwill, P. M., 1994. Ecological disturbance and the conservative management of eucalypt forests in Australia. *For. Ecol. Manage.* **63**: 301–46.

Baker, W. L., 1992. The landscape ecology of large disturbances in the design and management of nature reserves. *Landscape Ecol.* **7**: 181–94.

Banks, J. C., 1993. Tree-ring analysis of two Mountain Ash trees *Eucalyptus regnans* F. Muell from the Watts and O'Shannassy Catchments, Central Highlands, Victoria. A Report to the Central Highlands Old Growth Forest Project. August 1993. Department of Conservation and Natural Resources: Melbourne, Victoria.

Barling, R. D. and Moore, I. D., 1994. The role of buffer strips in the management of waterway pollution: a review. *Environ. Manage.* **18**: 543–58.

Barnett, J. L., How, R. A. and Humphreys, W. F., 1978. The use of habitat components by small mammals in eastern Australia. *Aust. J. Ecol.* **3**: 277–85.

Barton, D. R., Taylor, W. D. and Biette, R. M., 1985. Dimensions of riparian buffer strips required to maintain habitat in southern Ontario streams. *Nth Am. J. Fish. Manage.* **5**: 364–78.

Baudry, J., 1984. Effects of landscape structure on biological communities: the case of hedgerow network landscapes. Pp. 55–65 *in* Proceedings of the First International Seminar on Methodology in Landscape Ecological Research and Planning ed by J. Brandt and P. Agger. Volume 1. Roskilde Universitetsforlag Geo, Roskilde: Denmark.

Baur, A. and Baur, B., 1992. Effect of corridor width on animal dispersal: a simulation study. *Global Ecol. Biogeogr. Letters* **2**: 52–56.

Bennett, A. F., 1998. Linkages in the landscape. The role of corridors and connectivity in wildlife conservation. IUCN, Gland: Switzerland.

Braithwaite, L. W., 1984. The identification of conservation areas for possums and gliders in the Eden woodpulp concession district. Pp. 501–08 *in* Possums and Gliders ed by A. P. Smith and I. D. Hume. Surrey Beatty & Sons: Chipping Norton, New South Wales.

Budd, W. W., Cohen, P. L., Saunders, P. R. and Steiner, F. R., 1987. Stream corridor management in the Pacific northwest. I. Determination of stream-corridor widths. *Environ. Manage.* **1**: 587–97.

Bunnell, F., 1995. Forest-dwelling fauna and natural fire regimes in British Columbia: patterns and implications for conservation. *Conserv. Biol.* **9**: 636–44.

Burgman, M. A. and Ferguson, I., 1995. Rainforest in Victoria — a review of the scientific basis of current and proposed protection measures. DCNR Forest Service Technical Report 95–4. Department of Conservation and Natural Resources: Melbourne, Victoria.

Burgman, M. A. and Lindenmayer, D. B., 1998. Conservation Biology for the Australian Environment. Surrey Beatty & Sons: Chipping Norton, New South Wales.

Burnett, S., 1992. Effects of a rainforest road on movements of small mammals: mechanisms and implications. *Wildl. Res.* **19**: 95–104.

Carey, A., 1994. Forest ecosystem study: summary, background, and rationale. Olympia Forest Science Laboratory. Northwest Research Station. USDA, Olympia: Washington.

Chapin, F. S., Walker, B., Hobbs, R. J., Hooper, D. U., Lawton, J. H., Sala, O. E. and Tilman, D., 1997. Biotic control over the functioning of ecosystems. *Science* **277**: 500–04.

Chen, J., Franklin, J. F. and Spies, T. A., 1992. Vegetation responses to edge environments in old-growth Douglas-fir forests. *Ecol. Appl.* **2**: 387–96.

Claridge, A. W. and Lindenmayer, D. B., 1994. The need for a more sophisticated approach toward wildlife corridor design in the multiple-use forests of southeastern Australia: the case for mammals. *Pac. Conserv. Biol.* **1**: 301–07.

The Commonwealth of Australia and Department of Natural Resources and Environment, 1997. Comprehensive Regional Assessment — Biodiversity. Central Highlands of Victoria. The Commonwealth of Australia and Department of Natural Resources and Environment: Canberra, Australian Capital Territory.

Constanza, R., d'Arge, R., de Groot, R., Farber, S., Grasso, M., Hannon, B., Limburg, K., Nacem, S., O'Neill, R. V., Paruelo, J., Raskin, R. G., Sutton, P. and van den Belt., 1997. *Nature* **387:** 253–60.

Curtis, R. O., 1997. The role of extended rotations. Pp. 165–70 *in* Creating a Forestry for the 21st Century ed by K. A. Kohm and J. F. Franklin. Island Press: Covelo.

Date, E. M., Recher, H. F., Ford, H. A. and Stewart, D. A., 1996. Conservation biology of rainforest pigeons in northern New South Wales. *Pac. Conserv. Biol.* **2:** 299–308.

Department of Natural Resources and Environment, 1996. Code of Practice. Code of Forest Practices for timber production. Revision No. 2, November 1996. Department of Natural Resources and Environment: Melbourne, Victoria.

Doeg, T. J. and Koehn, J. D., 1990. A review of Australian studies on the effects of forestry practices on aquatic values. Silvicultural Systems Project Technical Report No. 5. Fisheries Division, Department of Conservation and Environment: Melbourne, Victoria.

Dunning, J. B., Borgella, R., Clements, K. and Meffe, G. K., 1995. Patch isolation, corridor effects, and colonisation by a resident sparrow in a managed pine woodland. *Conserv. Biol.* **9:** 542–50.

Fahrig, L. and Paloheimo, J., 1988. Effect of spatial arrangement of habitat patches on local population size. *Ecol.* **69:** 468–75.

Fensham, R. J., 1996. Land clearance and conservation of inland dry rainforest in north Queensland, Australia. *Biol. Conserv.* **75:** 289–98.

Forman, R. T., 1964. Growth under controlled conditions to explain the hierarchical distributions of a moss, Tetraphis pellucida. *Ecol. Monogr.* **34:** 1–25.

Forman, R. T. T., 1995. Land mosaics. The ecology of landscapes and regions. Cambridge University Press: New York.

Franklin, J. F., 1988. Structural and functional diversity in temperate forests. Pp. 166–75 *in* Biodiversity ed by E. O. Wilson. National Academy Press: Washington DC.

Franklin, J. F., 1993. Preserving biodiversity: species, ecosystems or landscapes? *Ecol. Appl.* **3:** 202–05.

Franklin, J. F. and Forman, R. T., 1987. Creating landscape patterns by forest cutting: ecological consequences and principles. *Landscape Ecol.* **1:** 5–18.

Franklin, J. F., Berg, D. E., Thornburgh, D. A. and Tappeiner, J. C., 1997. Alternative silvicultural approaches to timber harvest: variable retention harvest systems. Pp. 111–39 *in* Creating a Forestry for the 21st Century ed by K. A. Kohm and J. F. Franklin. Island Press: Covelo.

Friend, G. R., 1982. Mammal populations in exotic pine plantations and indigenous eucalypt forests in Gippsland, Victoria. *Aust. For.* **45:** 3–18.

Fries, C., Johansson, O., Petterson, B. and Simonsson, P., 1997. Silvicultural models to maintain and restore natural stand structures in Swedish boreal forests. *For. Ecol. Manage.* **94:** 89–103.

Garrett, M. G. and Franklin, W. L., 1988. Behavioral ecology of dispersal in the black-tailed prairie dog. *J. Mamm.* **69:** 236–50.

Gibbons, P. and Lindenmayer, D. B., 1997. Conserving hollow-dependent fauna in timber-production forests NSW National Parks and Wildlife Service Environmental Heritage Monograph Series 3, 1–110.

Gilpin, M. E., 1987. Spatial structure and population vulnerability. Pp. 126–39 *in* Viable Populations for Conservation ed by M. E. Soulé. Cambridge University Press: New York.

Graynoth, E., 1989. Effects of logging on stream environments and faunas in Nelson. *N.Z. Mar. Freshw. Res.* **13:** 79–109.

Grumbine, R. E., 1990. Viable populations, reserve design, and Federal lands management: a critique. *Cons. Biol.* **4:** 127–34.

Gullan, P. K. and Robinson, A. C., 1980. Vegetation and small mammals of a Victorian forest. *Aust. Mammal.* **3:** 87–96.

Gustafson, E. J. and Gardner, R. H., 1996. The effect of landscape heterogeneity on the probability of patch colonisation. *Ecol.* **77:** 94–107.

Haila, Y., 1988. The multiple faces of ecological data and theory. *Oikos* **53:** 408–11.

Hall, C. M., 1988. The "worthless lands hypothesis" and Australia's national parks and reserves. Pp. 441–59 *in* Australia's Ever Changing Forests ed by K. J. Frawley and N. M. Semple. Australian Defence Force Academy: Canberra, Australian Capital Territory.

Hanski, I., 1994. Patch occupancy dynamics in fragmented landscapes. *TREE* **9:** 131–34.

Hanski, I., Pakkala, T., Kuussaari, M. and Lei, G., 1995. Metapopulation structure of an endangered butterfly in a fragmented landscape. *Oikos* **72:** 21–28.

Harmon, M., Franklin, J. F., Swanson, F., Sollins, P., Gregory, S. V., Lattin, J. D., Anderson, N. H., Cline, S. P., Aumen, N. G., Sedell, J. R., Lienkaemper, G. W., Cromack, K. and Cummins, K., 1986. Ecology of coarse woody debris in temperate ecosystems. *Adv. in Ecol. Res.* **15:** 133–302.

Hobbs, R. J., 1992. The role of corridors in conservation: solution or bandwagon? *TREE* **7:** 389–92.

Hobbs, R. J., 1997. The right path or the roads to nowhere? *Nature Australia, Autumn* **1997:** 57–63.

Holling, C. S. (ed.), 1978. Adaptive Environmental Assessment and Management. International Series on Applied Systems Analysis 3, International Institute for Applied Systems Analysis. John Wiley and Sons: Toronto.

Hunter, M. L., 1994. Fundamentals of conservation biology. Blackwell: Melbourne, Victoria.

Incoll, R. D., 1995. Landscape ecology of glider populations in old-growth forest patches. B.Sc. Honours thesis, Department of Zoology, University of Melbourne: Melbourne, Victoria.

Janzen, D. H., 1983. No park is an island: Increase in interference from outside as park size decreases. *Oikos* **41:** 402–10.

Johns, A. G., 1996. Bird population persistence in Sabahan logging concessions. *Biol. Cons.* **75:** 3–10.

Johnson, S., 1992. Protected areas. Pp. 447–78 *in* Global Diversity: Status of the Earth's Living Resources. World Conservation Monitoring Centre. Chapman and Hall: London.

Kaila, L., Martikainen, P. and Puntilla, P., 1997 Dead trees left in clearcuts benefit Coleoptera adapted to natural disturbances in boreal forests. *Biodivers. and Conserv.* **6:** 1–8.

Kavanagh, R. P. and Turner, R. J., 1994. Birds in eucalypt plantations: the likely role of retained habitat trees. *Aust. Birds* **28:** 32–41.

Kelly, P. A. and Rotenberry, J. T., 1993. Buffer zones and ecological reserves in California: replacing guesswork with science. *In* Interface Between Ecology and Land Development in California ed by J. E. Keeley. Southern California Academy of Sciences: Los Angeles.

Khan, M. L., Menon, S. and Bawa, K. S., 1997. Effectiveness of the protected area network in biodiversity conservation: a case study of Meghalaya state. *Biodivers. and Conserv.* **6**: 853–68.

Kohm, K. A. and Franklin, J. F. (eds), 1997. Creating a Forestry for the 21st Century. Island Press: Covelo.

Lacy, R. C., 1993. Impacts of inbreeding in natural and captive populations of vertebrates: implications for conservation. *Persp. Biol. and Med.* **36**: 480–96.

La Haye, W. S., Gutierrez, R. J. and Akçakaya, H. R., 1994. Spotted owl metapopulation dynamics in southern California. *J. Anim. Ecol.* **63**: 775–85.

Lamberson, R. H., Noon, B. R., Voss, C. and McKelvey, R., 1994. Reserve design for territorial species: the effects of patch size and spacing on the viability of the Northern Spotted Owl. *Conserv. Biol.* **8**: 185–95.

Langford, K. J., Moran, R. J. and O'Shaughnessy, P. J., 1982. The Coranderk Experiment — the effects of roading and timber harvesting in mature Mountain Ash forest on streamflow and quality. *In* The First National Symposium on Forest Hydrology ed by E. M. O'Loughlin and L. J. Bren. Institution of Engineers: Canberra, Australian Capital Territory.

Laurance, W. F., 1991. Ecological correlates of extinction proneness in Australian tropical rain forest mammals. *Conserv. Biol.* **5**: 79–89.

Lindenmayer, D. B., 1994. Wildlife corridors and the mitigation of logging impacts on forest fauna in south-eastern Australia. *Wildl. Res.* **21**: 323–40.

Lindenmayer, D. B., Cunningham, R. B., Nix, H. A., Tanton, M. T. and Smith, A. P., 1991. Predicting the abundance of hollow-bearing trees in montane ash forests of south-eastern Australia. *Aust. J. Ecol.* **16**: 91–98.

Lindenmayer, D. B., Cunningham, R. B., Donnelly, C. F., Tanton, M. T. and Nix, H. A., 1993a. The abundance and development of cavities in montane ash-type eucalypt trees in the montane forests of the Central Highlands of Victoria, south-eastern Australia. *For. Ecol. Manage.* **60**: 77–104.

Lindenmayer, D. B., Cunningham, R. B. and Donnelly, C. F., 1993b. The conservation of arboreal marsupials in the montane ash forests of the Central Highlands of Victoria, south-east Australia. IV. The distribution and abundance of arboreal marsupials in retained linear strips (wildlife corridors) in timber production forests. *Biol. Conserv.* **66**: 207–21.

Lindenmayer, D. B. and Possingham, H. P., 1995a. The conservation of arboreal marsupials in the montane ash forests of the Central Highlands of Victoria, south-east Australia. VII. The contribution of changed wood production practices to metapopulation persistence. *Biol. Conserv.* **73**: 239–57.

Lindenmayer, D. B. and Possingham, H. P., 1995b. Modelling the impacts of wildfire on metapopulation behaviour of the Australian arboreal marsupial, Leadbeater's Possum, *Gymnobelideus leadbeateri*. *For. Ecol. Manage.* **74**: 197–222.

Lindenmayer, D. B., Cunningham, R. B., Pope, M., Donnelly, C. F., Nix, H. A. and Incoll, R. D., 1997a. The Tumut fragmentation experiment in south-eastern Australia: the effects of landscape context and fragmentation on arboreal marsupials. CRES Working Paper, 1997/4.

Lindenmayer, D. B., Cunningham, R. B. and Donnelly, C. F., 1997b. Tree decline and collapse in Australian forests: implications for arboreal marsupials. *Ecol. Appl.* **7**: 625–41.

Lindenmayer, D. B. and Franklin, J. F., 1997a. Forest structure and sustainable temperate forestry: A case study from Australia. *Conserv. Biol.* **11**: 1053–068.

Lindenmayer, D. B. and Franklin, J. F., 1997b. Re-inventing forestry as a discipline — a forest ecology perspective. *Aust. For.* **60**: 53–55.

Lindenmayer, D. B., Cunningham, R. B. and McCarthy, M., 1999a. Landscape analysis of the occurrence of arboreal marsupials in the montane ash forests of the Central Highlands of Victoria, south-eastern Australia. *Biol. Conserv.*

Lindenmayer, D. B., McCarthy, M. and Pope, M. L., 1999b. A test of Hanski's inclidence function metapopulation model for patch occupancy in a fragmented ecosystem. *Oikos*.

Lovejoy, T. E., Bierregaard, R. O., Rylands, A. B., Malcolm, J. R., Quintela, C. E., Harper, L. H., Brown, K. S., Powell, A. H., Schubert, H. O. and Hays, M. B., 1986. Edge and other effects of isolation on Amazon forest fragments. Pp. 257–85 *in* Conservation Biology. The Science of Scarcity and Diversity ed by M. E. Soulé. Sinauer Associates: Sunderland.

Loyn, R. H., 1985. Strategies for conserving wildlife in commercially productive eucalypt forest. *Aust. For.* **48**: 95–101.

Loyn, R. H., Chesterfield, E. A. and Macfarlane, M. A., 1980. Forest utilization and the flora and fauna in Boola Boola State Forest in south-eastern Victoria. Forest Commission of Victoria Bulletin 28.

Luck, G., 1996. Bird population responses and artificial nest predation at inherent and induced edges in the Murray Mallee, South Australia. Honours thesis, Department of Environmental Science and Management. University of Adelaide, Roseworthy, South Australia.

McCarthy, M. A. and Burgman, M. A., 1995. Coping with uncertainty in forest wildlife planning. *For. Ecol. Manage.* **74**: 23–36.

McCarthy, M. A. and Lindenmayer, D. B., 1998. Development of multi-aged Mountain Ash *(Eucalyptus regnans)* forest under natural disturbance regimes: implications for wildlife conservation and logging practices. *For. Ecol. Manage.* **104**: 43–56.

McNeely, J. A., 1994a. Protected areas for the 21st century: working to provide benefits to society. *Biodiv. and Conserv.* **3**: 390–405.

McNeely, J. A., 1994b. Lessons from the past: forests and biodiversity. *Biodiv. and Conserv.* **3**: 3–20.

McNeely, J. A., Miller, K. R., Reid, W. V., Mittermeier, R. A. and Werner, T. B., 1990. Conserving the world's biodiversity. International Union for the Conservation of Nature and Natural Resources, Gland: Switzerland.

Margules, C. R., Nicholls, A. O. and Usher, M. B., 1994. Apparent species turnover, probability of extinction and the selection of nature reserves: a case study of the Ingleborough limestone pavements. *Conserv. Biol.* **8**: 398–409.

Margules, C. R. and Lindenmayer, D. B., 1996. Landscape level concepts and indicators for the conservation of forest biodiversity and sustainable forest management. Pp. 65–83 *in* Proceedings of the UBC-UPM Conference on the Ecological, Social and Political Issues of the Certification of Forest Management, Malaysia.

Martin, T. E. and Karr, J. R., 1986. Patch utilisation by migrating birds: resource orientated? *Ornis Scandinavia* **17**: 165–74.

Maser, C., Trappe, J. M. and Ure, D. C., 1977. Implications of small mammal mycophagy to the management of western coniferous forests. Trans. *Nth Am. Wildl. Nat. Res. Conf.* 78–88.

Maser, C., Tarrant, R. F., Trappe, J. M. and Franklin, J. F., 1988. From the forest to the sea: a story of fallen trees. U.S. Department of Agriculture Forest Service, Pacific Northwest Research Station. General Technical Report PNW-GTR-229.

Metzeling, L., Doeg, T. and O'Conner, W., 1995. The impact of salinisation and sedimentation on aquatic biota. Pp. 126–36 *in* Conserving Biodiversity: Threats and Solutions ed by R. A. Bradstock, T. A. Auld, D. A. Keith, R. T. Kingsford, D. Lunney and D. P. Sivertsen. Surrey Beatty & Sons: Chipping Norton, New South Wales.

Michaelis, F. B., 1984. Possible effects of forestry on inland waters of Tasmania: a review. *Environ. Conserv.* **11**: 331–43.

Milledge, D. R., Palmer, C. L. and Nelson, J. L., 1991. "Barometers of change": the distribution of large owls and gliders in Mountain Ash forests of the Victorian Central Highlands and their potential as management indicators. Pp. 55–65 *in* Conservation of Australia's Forest Fauna ed by D. Lunney. Royal Zoological Society of New South Wales: Mosman, New South Wales.

Mueck, S. G., Ough, K. and Banks, J. C., 1996. How old are wet forest understories? *Aust. J. Ecol.* **21**: 345–48.

Murphy, D. D. and Noon, B. R., 1992. Integrating scientific methods with habitat conservation planning: reserve design for Northern Spotted Owls. *Ecol. Appl.* **2**: 3–17.

Naiman, R. J., Decamps, H. and Pollock, H., 1993. The role of riparian corridors in maintaining regional biodiversity. *Ecol. Appl.* **3**: 209–12.

Nelson, J., 1991. Beyond parks and protected areas: from public and private stewardship to landscape planning and management. *Environments* **21**: 23–34.

Nelson, M. E., 1993. Natal dispersal and gene flow in white-tailed deer in northeastern Minnesota. *J. Mamm.* **74**: 316–22.

New, T. R., 1995. Onychophora in invertebrate conservation: priorities, practice and prospects. *Zool. J. Linn. Soc.* **14**: 77–89.

Newmark, W. D., 1985. Legal and biotic boundaries of western North American national parks: A problem of congruence. *Biol. Conserv.* **33**: 197–208.

Niemela, J., Langor, D. and Spence, J. R., 1993. Effects of clear-cut harvesting on boreal ground-beetle assemblages (Coleoptera: Carabidae) in western Canada. *Conserv. Biol.* **7**: 551–61.

Niemela, T., Renvall, P. and Penttila, R., 1995. Interactions of fungi at late stages of wood decomposition. *Ann. Botanicol Fennici* **32**: 141–52.

Nix, H. A., 1994. The Brigalow. Pp. 198–233 *in* Australian Environmental History ed by S. Dovers. Oxford University Press: Melbourne, Victoria.

Noss, R. F. and Cooperrider, A. Y., 1994. Saving Nature's Legacy. Protecting and Restoring Biodiversity. Island Press: Covelo, California.

Okland, B., 1996. Unlogged forests: important sites for preserving the diversity of mycetophilids (Diptera: Sciaroidea). *Biol. Conserv.* **76**: 297–310.

Parker, L., 1997. Restaging an evolutionary drama: thinking big on the Chequamegon and Nicolet National Forests. Pp. 218–19 *in* Creating a Forestry for the 21st Century ed by K. A. Kohm and J. F. Franklin. Island Press: Covelo.

Parry, B. B., 1997. Abiotic edge effects in wet slcerophyll forest in the central highlands of Victoria. Msc. thesis, School of Botany, University of Melbourne: Melbourne, Victoria.

Peck, J. E. and McCune, B., 1997. Remnant trees and canopy lichen communities in western Oregon: a retrospective approach. *Ecol. Appl.* **7**: 1181–187.

Phillips, E. J., 1996. Comparing silvicultural systems in a coastal montane forest: productivity and cost of harvesting operations. FERIC Special Report, No. SR-109.

Pickett, S. T., 1996. Sustainable forestry in Chilean Tierra del Fuego. *TREE* **11**: 450–51.

Pimm, S. L., Jones, H. L. and Diamond, J., 1988. On the risk of extinction. *Am. Nat.* **132**: 757–85.

Pimental, D., Stachow, U., Takacs, D. A., Brubaker, H. W., Dumas, A. R., Meaney, J. J., O'Neill, J. A., Onsi, D. E., Corzilius, D. B., 1992. Conserving biological diversity in agricultural/forestry systems. *BioScience* **42**: 354–62.

Pimental, D., Wilson, C., McCullum, C., Huang, R., Dwen, P., Flack, J., Tran, Q., Saltman, T. and Cliff, B., 1997. Economic and environmental benefits of biodiversity. *BioScience* **47**: 747–57.

Pressey, R. L., 1995. Conservation reserves in N.S.W. Crown jewels or leftovers? *Search* **26**: 47–51.

Pressey, R. L., Ferrier, S., Hager, T. C., Woods, C. A., Tully, S. L. and Weinman, K. M., 1996. How well protected are the forests of north-eastern New South Wales? — Analyses of forest environments in relation to formal protection meaures, land tenure, and vulnerability to clearing. *For. Ecol. Manage.* **85**: 311–33.

Recher, H. F., Rohan-Jones, W. and Smith, P., 1980. Effects of the Eden woodchip industry on terrestrial vertebrates with recommendations for management. Forestry Commission of New South Wales, Research Note No. 42.

Rose, C. R. and Muir, P. S., 1997. Green tree retention: consequences for timber production in forests of the Western Cascades, Oregon. *Ecol. Appl.* **7**: 209–17.

Rosenberg, D. K., Noon, B. R. and Meslow, E. C., 1997. Biological corridors: form, function and efficacy. *BioScience* **47**: 677–87.

Saunders, D. A., Hobbs, R. J. and Margules, C. R., 1991. Biological consequences of ecosystem fragmentation: a review. *Conserv. Biol.* **5**: 18–32.

Schmiegelow, F. K. and Hannon, S. J., 1993. Adaptive management, adaptive science and the effects of forest fragmentation on boreal birds in northern Alberta. *Trans. Nth Am. Wildl. Nat. Res. Conf.* **58**: 584–98.

Schoenwald-Cox, C. M., 1988. Boundaries in the protection of nature reserves. *BioScience* **38**: 480–86.

Scotts, D. J., 1994. Sustaining sensitive wildlife within temperate forest landscapes: regional systems of retained habitat as a planning framework. Pp. 85–106 *in* Ecology and Sustainability of Southern Temperate Ecosystems ed by T. Norton and S. Dovers. CSIRO: Melbourne, Victoria.

Simberloff, D. A., Farr, J. A., Cox, J. and Mehlman, D. W., 1992. Movement corridors: conservation bargains or poor investments? *Conserv. Biol.* **6**: 493–504.

Silsbee, D. G. and Larson, G. L., 1983. A comparison of streams in logged and unlogged areas of the Great Smoky Mountains National Park. *Hydrobiol.* **102:** 99–111.

Smith, P., 1985. Effects of intensive logging on birds in eucalypt forest near Bega, New South Wales. *Emu* **85:** 15–21.

Specht, R. L., Specht, A., Whelan, M. B. and Hegarty, E. E., 1995. Conservation atlas of plant communities in Australia. Centre for Coastal Management and Southern Cross University Press: Lismore, New South Wales.

Squire, R. O., Dexter, B., Smith, R., Manderson, A. and Flinn, D., 1987. Evaluation of alternative systems for Victoria's commerically important mountain forests — project brief. Lands and Forests Division. Department of Conservation, Forests and Lands: Melbourne, Victoria.

Squire, R. O., Campbell, R. G., Wareing, K. J. and Featherston, G. R., 1991. The Mountain Ash forests of Victoria: ecology, silviculture and management for wood production. Pp. 38–57 *in* Forest Management in Australia ed by F. H. McKinnell, E. R. Hopkins and J. E. D. Fox. Surrey Beatty & Sons: Chipping Norton, New South Wales.

Stacey, P. B. and Taper, M., 1992. Environmental variation and the persistence of small populations. *Ecol. Applica.* **2:** 18–29.

Stohlgren, T. J., 1992. Resilience of a heavily logged grove of Giant Sequoia (Sequoiadendron giganteum) in Kangs Canyon National Park, California. *For. Ecol. Manage.* **54:** 115–40.

Taylor, C. R., Caldwell, S. L. and Rowntree, V. J., 1972. Running up and down hills: some consequences of size. *Science* **178:** 1096–097.

Taylor, R. J., Bryant, S. L., Pemberton, D. and Norton, T. W., 1985. Mammals of the Upper Henty River region, Western Tasmania. *Pap. Proceed. Roy. Soc. Tas.* **119:** 7–15.

Temple, S. A. and Cary, J. R., 1988. Modelling dynamics of habitat interior bird populations in fragmented landscapes. *Cons. Biol.* **2:** 340–47.

Thiollay, J., 1992. The influence of selective logging on bird species diversity in a Guianian rain forest. *Cons. Biol.* **6:** 47–63.

Thiollay, J., 1997. Disturbance, selective logging and bird diversity: a Neotropical forest study. *Biodivers. and Conserv.* **6:** 1155–173.

Thomas, C. D., 1990. What do real population dynamics tell us about minimum viable population sizes. *Conserv. Biol.* **4:** 324–27.

Thomas, J. W., Forsmann, E. D., Lint, J. B., Meslow, E. C., Noon, B. R. and Verner, J., 1990. A conservation strategy for the Northern Spotted Owl. U.S. Government Printing Office: Portland.

Tscharntke, T., 1992. Fragmentation of Phragmites habitats, minimum viable population size, habitat suitability, and local extinction of moths, midges, flies, aphids and birds. *Conserv. Biol.* **6:** 530–36.

Tuchmann, E. T., Connaughton, K. P., Freedman, L. E. and Moriwaki, C. B., 1996. The Northwest Forest Plan. A Report to the President and Congress. U.S.D.A. Forest Service, PNW Research Station, Portland: Oregon.

Verner, J., McKelvey, K. S., Noon, B. R., Gutierrez, R. J., Gordon, G. I. and Beck, T. W., 1992. The California Spotted Owl: a technical assessment of its current status. USDA Forest Service General Technical Report PSW-GTR-133.

Virkkala, R., Rajasarrkka, A., Vaisanen, R. A., Vickholm, M. and Virolainen, E., 1994. The significance of protected areas for the land birds of southern Finland. *Conserv. Biol.* **8:** 532–44.

Wace, N., 1977. Assessment of the dispersal of plant species — the car borne flora of Canberra. *Proc. Ecol. Soc. Aust.* **10:** 166–86.

Walters, C. J. and Holling, C. S., 1990. Large scale management experiments and learning by doing. *Ecol.* **71:** 2060–068.

Wardell-Johnson, C. and Nichols, O., 1991. Forest wildlife and habitat management in southwestern Australia: knowledge, research and direction. Pp. 161–92 *in* Conservation of Australia's Forest Fauna ed by D. Lunney. Surrey Beatty & Sons: Chipping Norton, New South Wales.

Welsh, C. J. and Healy, W. M., 1993. Effects of even-aged timber management on bird diversity and composition in northern hardwoods of New Hampshire. *Wildl. Soc. Bull.* **21:** 143–54.

Witting, L. and Loeschcke, V., 1995. The optimization of biodiversity conservation. *Biol. Conserv.* **71:** 205–07.

Wilcove, D. S., 1989. Protecting biodiversity in multiple-use lands: lessons from the US Forest Service. *TREE* **4:** 385–88.

Wilcove, D. S., McLellen, C. H. and Dobson, A. P., 1986. Habitat fragmentation in the temperate zone. Pp. 237–56 *in* Conservation Biology: the Science of Scarcity and Diversity ed by M. E. Soulé. Sinauer: Sunderland.

Wilson, B. A. and Clark, T. W., 1995. The Victorian Flora and Fauna Guarantee Act 1988: a five-year review of its implementation. Pp. 87–103 *in* People and Nature Conservation. Perspectives on Private Land Use and Endangered Species Recovery ed by A. Bennett, G. Backhouse and T. Clark. Transactions of the Royal Zoological Society of New South Wales. Surrey Beatty & Sons: Chipping Norton, New South Wales.

Woinarski, J. M. and Cullen, J. C., 1984. Distribution of invertebrates on foliage in forests in south-eastern Australia. *Aust. J. Ecol.* **9:** 207–32.

Woinarski, J. C., Whitehead, P. J., Bowman, D. and Russell-Smith, J., 1992. Conservation of mobile species in a variable environment: the problem of reserve design in the Northern territory, Australia. *Global Ecol. and Biogeogr. Letters* **2:** 1–10.

CHAPTER 3

Managing the matrix: realigning paradigms toward sustainability

Neil D. Mitchell[1] and John L. Craig[1]

The ecosphere, inhabited by people and all other biodiversity, has been arbitrarily divided into conservation and production areas. Not only are the areas spatially separate, they are managed with distinct approaches within legislation, economic models and social values. To achieve sustainable living, there is a need to integrate conservation and production into the same set of paradigms. Ten examples of existing mindsets (paradigms) that currently hinder this integration are offered along with replacement values that are more likely to achieve sustainability. Managing the ecosphere matrix in ways that maintain and recognize ecosystem services is the underlying base of the revised mindset.

Key words: Conservation, Production environments, Sustainability, Ecological paradigms.

INTRODUCTION

A WORLD-WIDE environmental problem, addressed by this conference, is the treatment of conservation and production as two separate and distinct activities. The reality is that both "sectors" have much in common, they are both attempting to manage assets important to both present and future generations, yet dialogue between the two is often acrimonious and mutually distrustful. Historically the distinction was seen as rational but unfortunately the issue is not as simple as just re-establishing a dialogue. Through a century or more of divergence the sectors have philosophically and legislatively separated and have developed their own paradigms.

In the past when human populations were small relative to the dimensions of the natural world, aggressive use of natural assets as a source of resources was seen as a way to enhance human welfare. Unsustainable use tended to be localized and was readily addressed by local people. From the time of the industrial revolution, the size of the human population and their effects on the environment increased at an exponential rate (e.g., Ehrlich 1992). One response was to argue for "protected" areas to be set aside from economic development, to balance the extremes of developed land. This became the first attempt toward sustainability.

This almost complete separation of conservation and production as an artefact of our society, has meant that people have become isolated from their environment and the essential services it provides (Daly 1997). Meanwhile, producers making use of natural services for their production (clean water, clean air, natural waste assimilation etc. see Costanza *et al.* 1997 and Daly 1997) seem to have similarly become distanced from the underlying ecological services they utilize. To ensure the long-term existence of both the natural environment and the production environment, there has developed an awareness that sustainable use of the environment involves more than managing a dichotomy; this has been sparked by the expansion from local environmental degradation to global environmental impacts. Hence, we believe that to actually achieve progress toward environmental sustainability requires integration of "production" and "conservation".

In many societies the separation of "conservation" and "production" has become codified in legislation designed to advance the interests of the two sectors. Theoretically, it would be possible to establish legislation that removes this distinction; but unless the underlying philosophical issues have been resolved, it would only achieve resentment. We propose that there should be a re-evaluation of the paradigms, under which the two sectors operate, in particular how people view the natural environment. We offer a set of common paradigms, that can bring the two sectors together. If this can be achieved then there could be an integrated and sustainable

[1]School of Environmental and Marine Sciences, The University of Auckland, Private Bag 92019, Auckland, New Zealand.

Pages 26–34 *in* NATURE CONSERVATION 5: NATURE CONSERVATION IN PRODUCTION ENVIRONMENTS: MANAGING THE MATRIX ed by J. L. Craig, N. Mitchell and D. A. Saunders. Surrey Beatty & Sons, 2000.

management of our biological and physical environment.

PHILOSOPHICAL ISSUES

Many mainstream ethical beliefs have promoted a separation of people from nature (Clair *et al.* 1995). Modern "western" tradition is underpinned with the Judeo-Christian imperatives of the Old Testament. "Let them [humans (Man)] have dominion over . . . all the earth and over every creeping thing" (Genesis 1.26), ". . . be fruitful, and multiply, replenish the earth, and subdue it and have dominion over the fish of the sea, and the fowl of the air, and over every living thing that moveth upon the earth" (Genesis 1.28). Doctrines such as this typify the underlying mindset used to foster the frontier mentality that accompanied colonial domination by Europe. Commonly humans were and still are seen as superior and the controllers of nature. This approach of newly colonizing peoples should of course be contrasted with the beliefs of many long established indigenous peoples, in the unity of humans and nature (e.g., Maori and Aboriginal societies) (see Roberts 1995; Walsh 1995; Roberts *et al.* 1997; Taipa *et al.* 1997). The industrialization and associated urbanization of societies has been achieved through a tacit acceptance of the ghettoization of people, as a species. This separation of people into urban deserts has reinforced the superficial independence of people from the natural world (Clair *et al.* 1995).

Legislatively nature is treated quite separately to other activities in the environment, i.e., management of "nature" is independent of people's lives and commercial activity. For example in New Zealand: the *Conservation Act (1987)* with all its subsidiary Acts such as the *Reserves Act, Wildlife Act* along with international treaties such as Ramsar and Convention on International Trade in Endangered Species are based on separation as a way of achieving "protection and preservation". Protected areas are exempt from commercial laws such as the *Commerce Act* and even sustainability laws such as the *Resource Management Act (1991)* that control production activities. When these distinctions are challenged (e.g., Hartley 1997), conservation authorities and non-governmental organization groups argue loudly for the ongoing separation of activities and law (see Craig 1998). Australia and other countries have similar legislative distinctions.

The overall consequence is that the industrial/urban society experiences nature vicariously in zoos, on television, in books, in reserves or at some point physically distant from where they live. Even in reserves, extreme lengths are often taken to distance people from "their" environment. Signs typically tell "visitors" what they can not do. So it is not surprising that people and societies have forgotten that our very lives depend upon functioning ecosystems. Moreover, reassured that natural systems are "protected" elsewhere, production environments use and abuse the very life supporting fabric of our environment.

We are strongly of the view that:

> *societies need to re-establish their links with nature in order to understand the inter-dependence of all things;*

and that

> *conservation and production environments be managed together so that there can be a better understanding of their inter-dependence.*

To achieve this integration we perceive that societies re-define their core paradigms, if they are to move more rapidly toward sustainable systems. Essentially production and conservation need to be managed from a shared set of paradigms; the future of human life depends on this (Costanza *et al.* 1997).

INTEGRATION THROUGH SHARED PARADIGMS

Reserve boundaries separate not only ecology from economy but also the management of nature and production. World-wide, conservation has become the responsibility of government at both national and state levels. Sitting outside commercial life, conservation has become dependent on a "welfare funding" approach administered through government monopolies (Craig 1997; Hartley 1997). Business on the other hand has used this situation to justify arguments that "environmental welfare", including conservation, is solely the responsibility of government. One outcome of this is that most effects of production systems on the environment are regarded as externalities not dealt with in normal business accounting.

With business free of many of the financial implications of their environmental effects and government conservation activities free of restrictions on monopolistic and non-competitive commercial activity (Hartley 1997), conservation and production are indeed poles apart. Until both commerce and conservation are regulated by the same goal of sustainability the world's biodiversity is indeed in peril. This situation will persist until the nature-destroying activities of commerce are part of accounting rather than externalities and that government conservation activities cease trading as a protected monopoly subsidized by government welfare. Until then, conservation will always be a stop gap measure.

Revisiting entrenched but separate paradigms, as a way of establishing new shared paradigms, is an important avenue forward.

PARADIGMS TO BE CHANGED AND NEW PARADIGMS TO BE DEVELOPED

1: A goal of business is sustainability

Present paradigm to be addressed:

> *Maximizing annual returns on shareholder investment is a key performance indicator.*

Production is a business, and an important measure of success is the ability to make profits — measured solely as money. To date, business has concentrated on the short-term returns from human made capital investment, but has mostly ignored the natural capital that forms the basis for production. Adding value has been confounded with appropriating value from nature. The importance of profit as a monetary gain also stems from the perspective that the principal responsibility of a company is to its shareholders. This is despite an underlying reliance on ecological services, whose continuation are seen as "the role of nature — our birthright". Effects of business on external ecological services and the need to maintain them are rarely considered a part of core business. At its worst this can be expressed as: if you can pour it down the river then it is someone else's problem.

Changed paradigm:

Sustainable production involves investment in the environment.

(The natural environment is integral to sustainable production/business)

The average longevity of supposedly highly successful corporates (e.g., Fortune 500 companies) is around 40 years (de Geus 1997) suggesting that sustainability has for many not been more than a short-term goal. Some companies have survived for hundreds of years suggesting that long-term wealth creation through sustainability is possible. Part of the change needed to achieve this, requires approaches that guarantee more than financial stability. For example, substituting cyclical processes for linear processes as well as interacting with the full range of stakeholders will enhance the future as well as rationalize costs. In general cyclical processes are more sustainable than the older "take, make, waste" approaches. With cyclical processes (driven by the Value Cycle rather than the Value Chain) fewer resources are used but they are used more intensively. Hence, assuming that company longevity matters, survival into the long term requires that resources should be used both efficiently and sustainably. Investment in natural and human capital is as important as technological and made capital.

Companies do not live in isolation of the natural environment or society. Hence, as awareness grows of the interactive nature of all human activities, then there is a constituency to be satisfied that is not just the shareholders, but consists of all stakeholders (customers, staff, neighbours and shareholders) who will evaluate company performance. Ultimately, if a company ignores the wider stakeholders, people will not buy its products, use its services or allow it to operate (Hawken 1993). Many companies, e.g., Interface and a range of Swedish and German companies, are building sustainability models, such as The Natural Step, into their decision making as a way of enhancing long-term sustainability (Natrass and Altomare 1999).

2: Equate natural and production values

Present paradigm to be addressed:

Human based productivity is always more important than natural productivity.

At present, there are many societal subsidies on production that add to profits. This can work through local land taxes, e.g., in New Zealand, Australia and elsewhere. Local land taxes are based on the notional market value of the land, not on the services provided to or received from the land. A classic example is flood control in the lowlands being paid for by those whose land is flooded or by central government, not by those who caused it or who could control it, i.e., it is unrelated to hill country forest clearance. Similarly, for water quality, where water is abstracted from a river or groundwater by urban dwellers, it is usually the consumers who pay to clean the water, not those whose practices resulted in it requiring cleaning.

An example related to current awareness of global warming, is the inability of current taxation or rating schemes to differentiate companies and individuals that are high producers of greenhouse gasses and other pollutants, from those who are low producers. The perversity of current incentives is that those who retain forest on their land, acting as both a carbon sink and an oxygen producer, typically pay the same rates (land taxes) as those who removed such life supporting services. Indeed conserving natural resources becomes a cost; whereas destroying natural systems to maximize human based production is rewarded (see also Davis and Cocklin, this volume).

Changed paradigm:

Balancing natural and human productivity is essential for sustainability.

If instead, the services and products of the natural environment were properly costed, then product and service charges would more closely represent true costs and benefits. Land taxes and other government charges should include environmental services used or in need of cleaning or repairing after use, but the value of services provided should also be costed, e.g., oxygen production, carbon fixation, biodiversity conservation, etc. The money earned by ensuring an area provides ecosystem services would then act as an incentive for people to conserve natural production. A mix of market mechanisms and regulation can then determine the mix of production and conservation needed to ensure sustainability. Accurate costings are less important than the changed mindset initially. Indeed getting the valuations right is likely to be an ongoing process influenced by values and market mechanisms.

3: Conservation is a business

Present paradigm to be addressed:

Protection and preservation achieve conservation.

Conservation is often perceived in simplistic terms, in particular that protection and preservation equates to sustainable land management. This is often represented as the greater the area under protection the greater the conservation achieved and hence the more "sustainable" the environment. This has been reinforced by Agenda 21, the biodiversity convention and state of the environment reporting, with the result that national reporting concentrates on area under "protection" and species numbers. Such national statistics never comment on habitat quality or area restored to a functioning state! These would be outcomes worth reporting on, rather than simply describing actions taken during the past reporting period.

The typical initial response to the destruction of biodiversity associated with the development of production landscapes, has been one of preservation and protection of natural areas. Indeed during the phase of nature colonization, protection was a necessary process. Once sustainability laws are invoked, preservation based conservation can then be counter productive as it provides no incentives for commerce to consider conservation as part of its business. Moreover, the extent of human induced disturbance world-wide results in conservation lands requiring ongoing and often intensive management (Craig 1998). Management with a mindset of protection and associated welfare will tend to preclude sustainability.

Changed paradigm:

> *Conservation is a business with all of society as shareholders.*

Conservation should seek diversified management and outcomes. Thus it should involve and report on, not just government activity but also that by individuals and companies. There are numerous chapters in this volume (e.g., Webber and Fowke), which exemplify the variety of management approaches that can occur and the variety of organizations actively participating in "conservation" activities. In aggregate, this is being achieved through effective and outcome directed management.

As an active part of this process "shareholders" should be able to experience a variety of "conservation" experiences (Hall and MacArthur 1997). Organizations and governments should compare degraded habitats with restored and/or functioning systems; protection environments with production environments. At present this occurs by default, but all sectors of society should understand the consequences of ignoring non-sustainable environmental activities. All those interested in investing in conservation business should receive normal commercial rewards.

4: Nature is dynamic not static

Present paradigm to be addressed:

> *Preserve biodiversity and ecosystems as representatives of pre-human pasts.*

Commonly held beliefs by conservation practitioners include the view that (a) there is some original natural state that is stable and represents the ideal and (b) we have an ability to return to this original natural state. This is based on early ecological concepts of climax communities postulated by Tansley (1936) but subsequently challenged by many (e.g., Gleason 1926). In New Zealand this is enshrined in the *Reserves Act (1977):*

> *"Ensuring as far as possible, the survival of all indigenous species of flora and fauna, both rare and commonplace, in their natural communities and habitats, and the <u>preservation of representative samples</u> of all classes of natural ecosystem and landscapes which in the aggregate <u>originally</u> gave New Zealand its own recognizable character"* (our emphases)

and recently reiterated in the draft New Zealand Biodiversity Strategy (Department of Conservation and Ministry for the Environment 1999).

Effectively this is a key paradigm to be overturned since it implies that nature is static and that there is a single natural state. It is often expressed as an ecological goal of seeking "representativeness". As if by finding and protecting, in the words of New Zealand legislation, "representative samples"; we will retain our natural heritage. If this goal was ever achieved, then it would forever doom our natural environments to the status of a museum rather than being a dynamic and integral part of the environment. It is worth posing the question — in how many other sectors would we accept such a static approach? What would our world look like if we had "preserved" representative examples of every old business, every past human settlement including medical, legal and religious codes?

Very commonly there is an associated aim of re-creating or restoring to some original "pre-human" state. Nowhere on the face of the earth are there ecosystems untouched or unmodified by humans. Over much of the earth, humans have been part of the local ecosystems for thousands, tens of thousands and probably hundreds of thousands of years, and permanently "altered" them. As such, it is impossible to return to some pristine or pre-human state. At its worst expression, native peoples have been forcibly removed from natural areas in the belief that a pristine state is being created, e.g., Yellowstone and Little Barrier Island.

Changed paradigm:

> *Retain native biodiversity in functioning ecosystems.*

Instead, recognize a dynamic goal such that natural systems must be able to survive the future. Otherwise, all our management will be for naught if our natural systems do not survive inevitable future changes. The future will be environmentally different to today and certainly to the past, so if we keep trying to return systems to some poorly described previous state, we are doing those systems and ourselves a disservice.

The key, is to put in place a change from management through "benign" neglect to one of seeking internal "homeostasis" such that natural systems are largely self-supporting. At the same time, recognize the reality that it is necessary to manage most "conservation" alongside "production". Both functions are needed since natural areas often provide services for production areas, and clearly human activities including production affected natural areas.

5: We can choose the matrix

Present paradigm to be addressed:

What naturally is, is best!

Another ecological myth that needs to be resolved is that each place has a characteristic and unique ecosystem. Firmly set within plant ecology is the belief that there are unique plant associations. Earlier this century there was a long, acrimonious and partly unresolved debate, based on the ideas of Clements (1936) and Gleason (1926) as to "the nature of the plant community". While the two views expounded by these authors were antagonistic to each other, they come to the same conclusion: that there is a characteristic plant community for a given place.

For animal ecologists, the equivalent perception was the belief in fixed habitat requirements for species. Thus there is a fixed niche within which species function and if only that niche could be fully defined, then management would be simplified. Unfortunately, this often becomes translated into the idea that: what is, is best. In other words, where we find a species must be its core niche to which it is best adapted.

There are numerous examples from within New Zealand and elsewhere that show this is simply not true. The classic New Zealand example is the takahe (*Notornis mantelli*), a large flightless gallinule (Gray and Craig 1991; Craig 1994). It was reduced to a population of 200, in alpine regions of the South Island — this was assumed to be its core niche. When *in situ* management failed to reverse its decline, individuals were relocated to islands. Considerable success is being achieved on most of these islands including Tiritiri Matangi Island, 1 000 km to the north, where the birds favour pasture grasses and there is now a thriving population. Takahe now share pastures with sheep! The prevarication in management up until its relocation had stalled its conservation and redirected many resources such that management of this and other species was hindered (Craig 1997).

Changed paradigm:

Diversified and functioning landscapes can be designed.

There is no doubt that we can choose the nature of the matrix, our landscape management is testament to this. However, we suggest that irrespective of human land management, we should recognize the possibility that there may be multiple naturally stable states at any one place, the difference will solely relate to the level of management. A further issue associated with modern landscapes is that many "modern" environments are outside the evolutionary history of most species, so we cannot rely on observations in historically constructed natural environments to tell us how a species can survive. In part we do not know enough about most species to be sure that we have identified all their limits or potential. For example, in the case of most translocated species (either accidentally or intentionally) unresolved and/or unidentified genetic potential commonly becomes expressed.

It is surprising the level of rigidity that is ascribed to native species compared to the way societies approach the introduction of highly selected domesticated animals and plants. Also palaeoecological examination of past species associations usually shows changing associations through time, the mix at any one time relating to current environmental conditions e.g., Davis 1981. Humans happen to be a current major environmental force, especially in the lowlands. It is to be expected that species associations resulting from our effects will be as unique as associations formed in response to past environmental fluctuations.

A distinction is commonly made between "production" and "natural" environments. Certainly, there may be gross biological differences in species composition; although the wise land manager may try to reduce the difference in functional type between the expected natural and production species association. Very commonly, "natural" and "production" environments co-exist alongside each other, i.e., our landscapes are already a matrix, and in many parts of the world a matrix has been in existence for centuries. The key to this maintenance is management, i.e., we choose the nature of the matrix. Management is the dominant ecological force in human dominated landscapes. The growing movement worldwide to introduce farming of native species instead of introduced species, e.g., kangaroos instead of sheep and cattle in Australia (Grigg 1987; Lunney 1995), or bison instead of cattle in North America shows the changing mindsets of some conservationists.

However, to achieve better environmental management we need to design an integrated matrix rather than the separation into "production" or "natural" reserves.

6: Integrate biodiversity and socio-economics

Present paradigm to be addressed:

Conservation is best achieved away from people.

One of the problems that prevent better integration of nature conservation into society as a whole, is the belief that nature conservation requires unique and special consideration. For example, in New Zealand, government legislated conservation mostly functions outside the main planning and commercial laws (see above, Hartley 1997), such as the *Resource Management Act* and the *Commerce Act;* the public conservation estate also does not pay local government rates (land taxes) (Craig 1997). Consequently such agencies operate by different rules and in a commercially protected market that largely precludes others participating.

Moreover, people are commonly seen as the problem and much "conservation" activity centres around the exclusion of people from either entry or participation in the management of conservation activities. In a sense, this is because it is seen that "no use" values override "use" values. A clear example is seen in the New Zealand Department of Conservation Visitor Management Strategy (Department of Conservation 1997) where the dominant theme is avoiding potential threats rather than identifying both opportunities and threats. The drawback to this approach is that few people have the opportunity to live sustainable lifestyles within functioning indigenous ecosystems and because conservation takes place away from people, society tends to ignore conservation activities in urban and other production environments.

Changed paradigm:

Conservation is part of social and business responsibilities.

The fundamental change in this approach is to recognize that humans are part of every ecosystem and that ecosystem services are imperative for human existence – oxygen, carbon fixation, water management, mineral nutrition, etc. In this context we are using all environments, whether "production" or "natural". Because of this, all environments need to be included in the economic system. Costanza *et al.* (1997) have begun that process and illustrate one approach to the integration of "natural" environments into the global economy.

If this approach was adopted (see also Masood and Garwin 1998), then nature conservation could benefit from "use" and "abuse" values being placed upon all human activities and integrated into economic and social frameworks. In simple terms this comes down to charging users and paying providers for environmental services!

7: Manage the landscape sustainably

Present paradigm to be addressed:

Conservation and production happen in different places.

At present, the landscape is divided arbitrarily into protection and production zones. A line on a map delineates that ecology dominates on one side and economy on the other. Even though both are important everywhere as are social issues, the regulatory and accounting mindsets of society reinforce the non-sustainable division. Because "natural" areas currently have minimal economic value (New Zealand Government year book reports conservation estate at $149 per hectare) and because the ecosystem services in production areas are appropriated at no cost, it becomes difficult to integrate the two. The overall consequence is that we do not have sustainable landscapes either naturally or economically.

Changed paradigm

Integrate production and conservation.

Therefore to manage the landscape in its totality for sustainable processes it is necessary to recognize that "production" and "natural" environments are a matrix and not as usually portrayed at extreme ends of a continuum. It is better to perceive the landscape as a jigsaw of inter-locking features.

Natural processes can be used to sustainably manage production. In some areas, this already happens, e.g., in sewage management and storm-water control. There are also some production values already in natural systems, e.g., harvesting, tourism, etc. However, such production in conservation environments requires that these are managed sustainably — an impossibility where "free use" rather than "free access" is offered as a birthright (Craig 1997). If conservation plus production is used in a portfolio of management, there will be a wide range of uses within both current extremes. For example some less disturbed environments may receive minimal direct use but will be important for the provision of ecosystem services. At the other extreme restored natural areas such as riparian zones may be integrated into production landscapes to avoid abuse costs. These would not only provide biodiversity reserves, they would be used for all the services they provide as a counter to nearby production activities (e.g., carbon sinks, water cleaning) and in many cases could be used for harvesting or non-extractive uses.

8: Functioning ecosystems not just species

Present paradigm to be addressed:

Management of rare or keystone species saves ecosystems (is conservation).

Species-centric conservation dominates management and legislation. From the earliest legislation about nature, politics and legislation focuses predominantly on species and threatened species. Most of this type of legislation was aimed at protecting species that were heading for extinction or endangerment (Simberloff 1998). With all due respect to Worldwide Fund for Nature and other non-government organizations, fund raising and political action have become focused on such creatures as the panda, koala and kakapo. In particular the largely exclusive emphasis on "charismatic megafauna" emphasizes an incomplete message to the public, politicians and financiers. It also attacks symptoms rather than causes (Craig and Stewart 1994). There has also been a focus on so called "keystone", "indicator" and "umbrella" species, leading to the management approach of the type – "if species x is present then it can be safely assumed the ecosystem is in good health". Additionally, protection of such species is also assumed to automatically result in the appropriate protection of the rest of the ecosystem (Simberloff 1998). Except for the few situations where it is known how the "indicator" species reflects the multitude of ecosystem interactions, it is likely that that protection of such species will still only protect a subset of the overall ecosystem (Simberloff 1998). We suggest that the exclusive application of such species-centric approaches to ecosystem management needs to be balanced with the concept of ecosystem services.

This would also apply to restoration schemes, e.g., on Tiritiri Matangi Island, invertebrates, herbs, ferns and bryophytes were never considered (Craig *et al.* 1995). This restoration scheme (with which we are closely associated) like so many others, still emphasizes megaflora and megafauna, with the tacit assumption that the "less apparent" species will somehow arrive there by themselves or are already there. Moreover, there is a tacit assumption in species based approaches that common species are not at risk and that restoring rare species is the most urgent problem. Barrett *et al.* (1994), Saunders *et al.* (1987) and Craig (1997) address this issue with respect to birds.

The species-centric approach also creates problems in the production environment. The typical request to ecological consultants assessing or assisting a development is: "Are there any rare species present at place a or b". Typically, if anything regarded as rare is found (irrespective of how it became rare or listed), then activity will be prevented or extreme mitigation imposed. On the other hand, whole ecosystems and the services they provide can be destroyed, if they are not considered to contain anything rare. Valuing single species independently of other ecosystem services is an anathema to sustainability.

Changed paradigm:

> *Manage ecosystem services including rare species.*

Recognition of processes at differing scales from landscape to catchment to ecosystem to species to genes allows management to consider a hierarchy of effects that all require attention. Together they relate to integration of biogeochemical cycles and energy flows. If this approach is taken, it will enable functioning systems and ensure provision of ecological services. If there are functioning ecosystems, then for those species that may still require special management because of their rarity, it should be much easier and not at the expense of other species.

This general approach has also been criticized as it is still unable to identify the best configuration of landscape features to ensure the survival of biota in a region (Lambeck 1997). We suggest that a mixed approach including the focal species approach of Lambeck (1997) is more appropriate. However, in multi-use landscapes, management of *both* biota and processes is required. Both processes and biota require management because the existing land cover has been changed, usually from some previously indigenous character. It is generally assumed that for a given region, indigenous ecosystems will provide the best "tools" to manage ecosystem processes. In production landscapes therefore, the specific location of natural ecosystems may become crucial with respect to the provision of ecosystem services. However, for that managed landscape to retain functionality of its indigenous ecosystems, many of the existing (mostly species-centric) approaches then need to be employed, in addition to wider ecosystem focused methodologies.

9: Oceans start at the mountain tops

Present paradigm to be addressed:

> *Land, air and water require separate and special management.*

One of the great failures of 20th century "environmental management" is the separation of land and water management, and terrestrial and marine management through independent legislation and regulation. Undeniably, what happens on the land affects what happens in the water and sea. This inconsistent legislation has been further reinforced by the physical separation of terrestrial, freshwater and marine reserves. For example, in New Zealand, marine and terrestrial reserves are covered by separate legislation and are, with few exceptions, not physically linked. The best known marine reserve in New Zealand adjoins intensively farmed and modified landscapes!

Further, this legislative and philosophical separation means that those most basic of ecosystem functions — biogeochemical cycles are being ignored.

Changed paradigm:

> *Land, air and water are an integrated management unit — the ecosphere.*

Instead, if we develop an approach whereby the values required for functioning marine and freshwater ecosystems are adopted, then land management will have to adapt to achieve these standards. This can be achieved if management reflects energy flows, and water and nutrient cycles.

Indigenous peoples, commonly recognize the connectivity within their local environment. Quite often their environmental practices have been put in place as a consequence of their own previous poor "management" (Flannery 1994). If during the last few centuries, colonial land developers had listened to indigenous peoples, then we suspect that environmental and resource management would have been in a much better state than it is today. We would not have been condemned to making some of the same mistakes that new colonists inevitably make. We suggest that legislative structures would still benefit from the inclusion of indigenous approaches; in New Zealand, we are seeing the beginnings of this in the *Resource Management Act*.

Given the proper integration of land, air and water, then conservation and production benefits are possible and expected by all sectors of society.

10: Sustainability needs everybody

Present paradigm to be addressed:

Conservation is for conservationists.

Management by conservation professionals has not always been appropriate. Indeed most decisions are made without reference to society or economy. For example, it is commonly stated e.g., New Zealand *Conservation Act 1987,* that management of a place or species is "for future generations and/or intrinsic values". Unfortunately, the clients are not here, or cannot speak and so can never complain or judge (Craig and Stewart 1994). Decisions are rarely made for people alive today, so that appropriate feedback is rarely obtained and management can appear non-accountable (Craig 1997; Hartley 1997). It should also be pointed out that business makes decisions without involving conservation. Finally, because there are no clear clients or consumers conservation is usually set up as government welfare to be funded on a minimalist and survival basis, rather than as an active and essential part of society. Unfortunately, this tends to absolve and exclude the production sector from conservation. This is exacerbated by the highly political nature of conservation groupings. For example, when the New Zealand Business Roundtable attempted to evaluate conservation, they were given no official help and when their report was published (Hartley 1997), the Department of Conservation and the Minister of Conservation totally rejected all findings (Craig 1998). In contrast, some conservation groups have automatic representation on the community board that oversees the actions and policies of the Department of Conservation. Such rejection of representatives of the production sector along with the outright acceptance of groups of like philosophy, does little to advance conservation or sustainability.

Changed paradigm:

Conservation is part of sustainability and involves everybody.

Sustainability requires negotiated outcomes and joint action. It is important to recognize that biodiversity conservation does and must occur away from government conservation lands (there are many examples reported in this volume). This suggests that business and the community can contribute to significant conservation outcomes and often do in spite of considerable disincentives.

Local communities can also make an important contribution, since the environment is the responsibility of all of us and we cannot, and should not assume, that central government can or will manage it effectively on our behalf. It is quite clear that given proper recognition, nature conservation through individual involvement and responsibility can achieve more than lock up protection (see Saunders *et al.* 1995). In the long term such local action and local responsibility, can help achieve sustainable and interesting environments.

CONCLUSIONS

Through these 10 examples of existing beliefs, we have shown that there are considerable societal difficulties to managing a matrix of conservation and production. If society continues to manage the environment through versions of these existing paradigms, then it will not be possible to achieve a sustainable environment. We will witness a continued degradation of "natural values" and ecological services including biological diversity. Changes needed to ensure progress toward sustainable use of the ecosphere will require a considerable shift in mindset, especially from legislators and existing land and water managers, conservation agencies and the production sector. Open societal debate is urgently needed to achieve change.

However, given adoption of revised paradigms in line with those suggested above, we believe that: the re-connection of people, production and "nature" can be achieved; sustainable environments will result from the integration of conservation and production; and, integrated legislation will effect conservation plus sustainable production.

REFERENCES

Barrett, G. W., Ford, H. A. and Recher, H. F., 1994. Conservation of woodland birds in a fragmented rural landscape. *Pac. Cons. Biol.* **1**: 245–56.

Costanza, R., d'Arge, R., de Groot, R., Farber, S., Grasso, M., Hannon, B., Limburg, K., Naeem, S., O'Neil, R., Paruelo, J., Raskin, R., Sutton, P. and van den Belt, M., 1997. The value of the world's ecosystem services and natural capital. *Nature* **387**: 253–60.

Craig, J. L., 1997. Managing bird populations: for whom and at what cost. *Pac. Cons. Biol.* **3**: 172–82.

Craig, J. L., 1998. An economic analysis of New Zealand's conservation strategies. *Agenda* **5**: 311–22.

Craig, J. L., Stewart, A. M., 1994. Conservation: a starfish without a central disk? *Pac. Cons. Biol.* **1**: 163–68.

Craig, J. L., Mitchell, N. D., Walter, B., Walter, R., Galbreith, M. and Chalmers, G., 1995. Communities restoring natures networks: Tiritiri Matangi Island. Pp. 534–41 *in* Nature Conservation 4: the Role of Networks ed by D. A. Saunders, J. L. Craig and E. M. Mattiske. Surrey Beatty & Sons: Chipping Norton, New South Wales.

Daly, G., 1997. Nature's Services: Societal Dependence on Natural Ecosystems. Island Press: New York.

Davis, M. B., 1981. Quaternary history and the stability of forest communities. Pp. 132–53 *in* Forest Succession ed by D. C. West, H. H. Shugart and D. B. Botkin. Springer-Verlag: New York.

de Geus, A., 1997. The Living Company: Habits for Survival in a Turbulent Business Environment. Harvard Business School: Boston.

Ehrlich, P. R., 1993. The scale of human enterprise. Pp. 3–8 *in* Nature Conservation 3: The Reconstruction of Fragmented Ecosystems ed by D. A. Saunders, R. J. Hobbs and P. R. Ehrlich. Surrey Beatty & Sons: Chipping Norton, New South Wales.

Department of Conservation and Ministry for the Environment 1998. Draft of New Zealand Biodiversity Strategy: our chance to turn the tide. Department of Conservation, Ministry for the Environment: Wellington, New Zealand.

Flannery, T. T., 1994. The Future Eaters. Reed New Holland: Sydney, New South Wales.

Gleason, H. A., 1926. The individualistic concept of the plant association. *Bull. Torrey Bot. Club* **53:** 7–26.

Government Printer, 1999. The New Zealand Biodiversity Strategy, draft. Government Printer: Wellington, New Zealand.

Gray, R. D. and Craig, J. L., 1991. Theory really matters: hidden assumptions behind the concept of habitat requirements. *Acta XX Congressus Internationalis Ornithologici* **4:** 2554–560.

Grigg, G. C., 1987. Kangaroos — a better economic base for our marginal grazing lands. *Aust. Zool.* **24:** 204–05.

Hall, C. M. and McArthur, S., 1996. Heritage management in Australia and New Zealand: the human dimension. Oxford University Press: Melbourne, Victoria.

Hartley, P., 1997. Conservation Strategies for New Zealand. New Zealand Business Roundtable: Wellington, New Zealand.

Hawken, P., 1993. The Ecology of Commerce. Harper Collins: New York.

Lambeck, R. J., 1997. Focal species: a multi-species umbrella for nature conservation. *Cons. Biol.* **11:** 849–56.

Lunney, D., 1995. Kangaroo harvesting in the context of ecologically sustainable development (ESD) and biodiversity conservation. Pp. 166–75 *in* Conservation Through Sustainable Use of Wildlife ed by G. C. Grigg, P. T. Hale and D. Lunney. Centre for Conservation Biology: University of Queensland, Queensland.

Masood, E. and Garwin, L., 1998. Costing the earth: when ecology meets economics. *Nature* **395:** 426–30.

Natrass, B. and Altomare, M., 1999. The Natural Step for Business. New Society: Gabriola Is., B.C.

Roberts, R. M., 1995. Biculturalism in Conservation in Aotearoa. Pp. 132–40 *in* Nature Conservation 4: The Role of Networks ed by D. A. Saunders, J. L. Craig and E. Mattiske. Surrey Beatty & Sons: Chipping Norton, New South Wales.

Saunders, D. A., Arnold, G. W., Burbidge, A. A. and Hopkins, A. J. M., 1987. The role of remnants of native vegetation in nature conservation: future directions. Pp. 387–92 *in* Nature Conservation: the Role of Remnants of Native Vegetation ed by D. A. Saunders, G. W. Arnold, A. A. Burbidge and A. J. M. Hopkins. Surrey Beatty & Sons: Chipping Norton, New South Wales.

Saunders, D. A., Craig, J. L., Mattiske, E. M., 1995. Nature Conservation 4: the Role of Networks. Surrey Beatty & Sons: Chipping Norton, New South Wales.

Simberloff, D., 1998. Flagships, umbrellas, and keystones: is single-species management passé in the landscape era? *Biol. Cons.* **83:** 247–57.

Taiepa, T., Lyver, P., Horsley, P., Davis, J., Bragg, M. and Moller, H., 1997. Co-management of New Zealand's Conservation Estate by Maori and Pakeha: A Review. *Environ. Cons.* **24:** 236–50.

CHAPTER 4

Nature conservation in New Zealand plantation forestry

W. J. Dyck[1]

Plantation forestry is an increasingly important component of New Zealand's economy currently contributing up to 5.3% of Gross Domestic Product and 12% of export earnings. It does this by producing and processing 17 million m³ of wood from a land base of 1.5 million hectares. The country's indigenous forests are almost totally reserved for conservation values. Wood production from these forests in 1996 was less than 100 000 m³.

In 1991 New Zealand's forest industry and the major environmental groups signed the New Zealand Forest Accord. This agreement, considered to be unique in the world, makes a clear distinction between natural indigenous forests and commercial plantation forests. It recognizes the important heritage value of the remaining natural forest and the value of commercial plantation forests as an essential source of renewable fibre. It also acknowledges that the existing area of natural indigenous forest should be maintained and enhanced and that plantation forests should not replace natural forest and other natural areas.

Despite the recognition that plantation forests are grown primarily for commercial purposes, they are also managed for many conservation values including soil and water protection, and to some extent, biodiversity. This chapter reviews the contribution that plantation forestry makes to nature conservation in New Zealand.

Key words: Plantation forestry, Sustainability, Nature conservation, Aquatic habitats, Land use, Exotic forest, Pasture, Production.

INTRODUCTION

PLANTATION forestry is important to New Zealand. The forest industry is currently the second largest export earner (12%) in New Zealand and contributes 5.3% of the country's Gross Domestic Product and directly employs 32 000 people (New Zealand Forest Owners Association 1997). Plantation harvest is currently 16.8 million m³ yr⁻¹ and will double by 2010. The plantation base currently occupies 6% (1.5 million ha) of the country's land area and is expanding at approximately 70 000 ha/yr, primarily as small landowners convert marginal farmland to productive forest.

Because of the size of the industry and the increasing importance to the economy, it is important to understand the real and potential impacts of plantation forestry on the environment. The environmental impacts of plantation forestry have been a major focus of the forest industry for many years and especially since the *Resource Management Act* was introduced in 1991. This "effects-based" legislation was welcomed by landowners, and particularly the forestry sector, which saw the new law as a huge improvement over the previous *Town and Country Planning Act* that openly discriminated against forestry as a land use in favour of pastoral farming.

In addition to legislative controls, the plantation forest industry has a number of significant agreements with the country's major conservation groups that influence how both natural and plantation forests should be managed. In contrast to many countries, plantation forestry is regarded as making positive contributions to the New Zealand environment. This chapter reviews some of these contributions and particularly focuses on the role of plantation forest management in nature conservation.

[1]Carter Holt Harvey Forests, Private Bag 92–106, Auckland, New Zealand.

Pages 35–43 *in* NATURE CONSERVATION 5: NATURE CONSERVATION IN PRODUCTION ENVIRONMENTS: MANAGING THE MATRIX ed by J. L. Craig, N. Mitchell and D. A. Saunders. Surrey Beatty & Sons, 2000.

The role of plantation forestry in conserving indigenous forest

Our ancestors began moving out of the "hunter/gatherer" mode some ten thousand years ago (Lewin 1997) and within a few thousand years civilization became dependent on intensive agricultural production. Today relatively few people rely on hunting animals and gathering wild fruit for their daily sustenance. Forestry, for the most part, has yet to make this shift to intensive production, mainly because wood has always been plentiful relative to food supply. However, with declining natural forest resources and increasing environmental pressures, coupled with a rapidly expanding population increasing the demand for wood products (Fig. 1), the supply of wood from natural forests can no longer be taken for granted.

Few countries have made the distinction between forests for conservation and forests for commercial wood production purposes as most countries still manage their forests for multiple use objectives. Foresters world-wide, have generally been trained in a multiple use ethic and forests are often regarded as special entities that can produce many different values to satisfy the needs of society, and even as having mystical qualities. Management intensity is low in most forests as natural regeneration methods are often used following harvesting and the application of herbicides and fertilizers is limited. However, as the world population continues to explode and the demand for wood products rises exponentially, the ability of natural forests to continue to satisfy demand will decrease. Multiple use forestry, as practised in most managed natural forests, is inefficient at achieving high rates of wood production in much the same way that agriculture would be inefficient if farm managers had to provide equal emphasis on other values, such as recreation, wildlife, and biodiversity, while attempting to maximize food production (Table 1). Faced with this dilemma, forest managers have little choice but to apply more efficient strategies to maximize wood production from a declining land base.

While we generally accept that agricultural production should replace natural biodiversity with

Table 1. Comparison of wood production from natural versus plantation forestry.

Natural	Plantation
Expensive	Cost effective
Inefficient	Highly efficient
Low productivity	High productivity
Leads to non-renewable substitutes	Wood = CO_2 + H_2O + sun
Leads to high energy substitutes	Promotes wood

a totally artificial system, society is generally far less willing to accept that forestry should do the same. As a species we depend on a relatively small number of plants and animals for sustenance. In fact, 70% of our calorie intake comes from just nine food items (potato, wheat, maize, rice, barley, cassava, soy bean, sugar and oats) (Sutton 1995). It is quite appropriate for agriculture to frequently apply herbicides, insecticides and fertilizers to raise productivity, and in fact it is expected. Furthermore, the question of monocultures is never raised in an agricultural context where even clonal production, as in the case of grapes, has been used for thousands of years. Forestry, it seems, can only escape close scrutiny when trees are grown on short rotations, such as for energy or specialized fibre production, and in these cases the operation is no longer considered to be "forestry" but rather an agricultural tree cropping operation.

Plantation forestry is a 20th century phenomena and New Zealand was one of the first countries to realize the benefits of "farming" trees. Today plantation forests provide approximately 10% of the world's wood harvest (Kanowski 1997) and the figure is increasing as the area of natural forests available for harvesting declines. The most important reason for pursuing plantation forestry versus managing natural forests for wood production is the considerable increase in productivity available through intensive plantation forest management, including selection of species and genetics. As a comparison, softwood growth rates in New Zealand plantations are in the order of 25 m³/ha/year compared to much less than 5 m³/ha/year in traditional forestry regions such

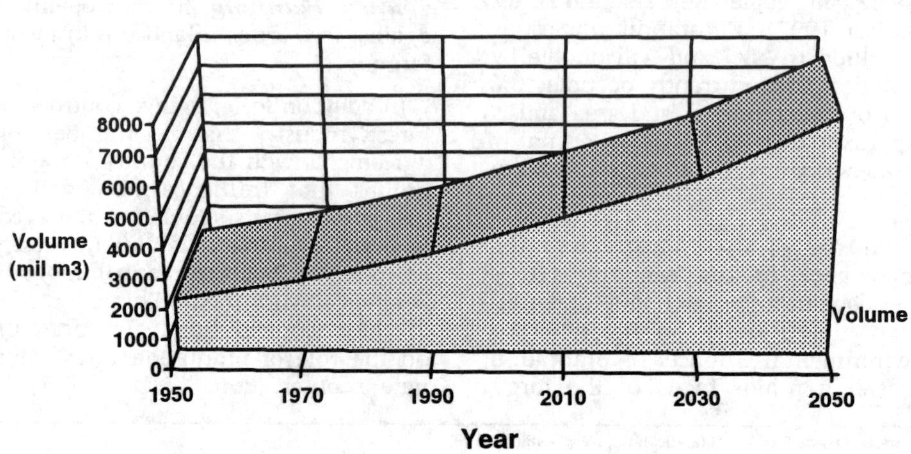

Fig. 1. Projected world wood demand assuming population growth to 12 billion by 2050.

as Canada, Russia, and Scandinavia (New Zealand Forest Owners Association 1997).

In view of the rapidly expanding world population together with concerns about climate change and the need to reduce atmospheric CO_2 levels, there is considerable incentive to increase the world's area of plantation forest. One of the major environmental benefits of increasing the plantation estate is that the pressure to harvest natural forests is substantially reduced, primarily because intensively managed plantations are often an order of magnitude more productive than natural forests.

Forestry in New Zealand

When Maori arrived in New Zealand approximately 1 000 years ago they found a land heavily covered in forest. The forest vegetation was dominated by tall podocarps (*Podocarpaceae*) in lowlands and beech (*Nothofagus* spp.) at higher elevations, except in the north where kauri (*Agathis australis*) forest prevailed. The forests must have appeared very unfamiliar to Maori as 95% of the tree species are unique to New Zealand and vastly different to those of their homeland. Prior to Maori, and later European settlement, which began in earnest in the early part of the 19th century, 75% of the land area was forested with only small areas below treeline in tussock grassland. However, today this figure has been reduced to 24%, mainly as a consequence of land clearing for agriculture (Fig. 2).

In 1918, predictions of a timber famine together with concern that New Zealand timber should be conserved for New Zealanders, prompted Sir Francis Bell, the Commissioner of State Forests, to outline, for the first time, a national forest policy. Timber export restrictions were imposed for the purpose of conserving timber and also to control domestic timber prices.

McIntosh Ellis, a young Canadian forester, was appointed as the first Director of Forestry for New Zealand. One of Ellis' first tasks was to assess the current situation in New Zealand forestry in terms of resource and timber consumption. In his 1920 report he basically concluded that New Zealand was running out of timber and, that amongst other recommendations, he stated that the country required a simple Forest Act, a Forest Service, and a fund for forest development and demarcation. While initially Ellis was of the view that the conservation and re-establishment of indigenous forests would meet the future needs of New Zealand, the results of the national forest inventory in 1923 indicated otherwise. The official assessment was that in 80 years three-quarters of the virgin timber would be gone and that the plantation establishment effort must be stepped up. In 1925 there were 5 300 hectares (13 000 acres) of State plantations and Ellis recommended that this be increased to 125 000 hectares (300 000 acres) by 1935. Ellis became so enthusiastic about the potential of plantation forestry for New Zealand that in 1927 he suggested that the plantation estate be extended to 2 million hectares (5 million acres), "a breath-taking suggestion" (Roche 1990).

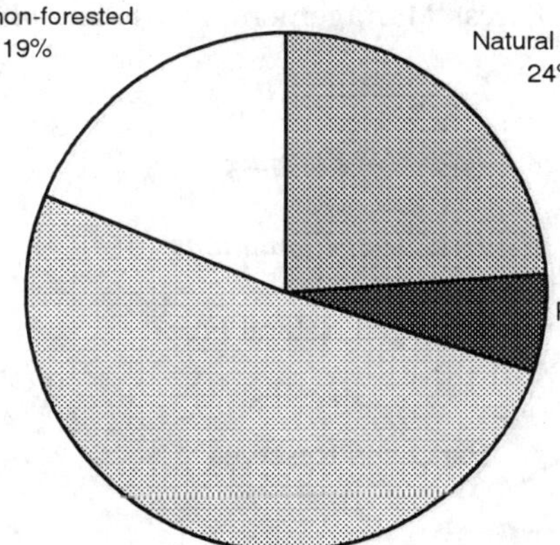

Fig. 2. New Zealand land use (New Zealand Forest Owners Association 1997).

Today there are approximately 1.5 million ha of plantation forest and opportunity for considerable expansion. Radiata pine *Pinus radiata* is very much the preferred species for plantation forestry in New Zealand, not only because of its fast growth and good wood properties, but also because it will grow on a wide range of sites and responds well to silvicultural treatment (e.g., pruning and thinning). Very little wood is harvested from the indigenous estate and this relatively unmanaged, and hence unproductive (at least from a timber perspective) resource is primarily managed for conservation purposes (Table 2).

Table 2. New Zealand Forestry Statistics.

	Plantation forestry	Natural forestry
Area (million ha)	1.5	6.4
Expansion (000 ha)	70 000	Similar[1]
Harvest (million m^3)	17	0.1
Productivity ($m^3\ ha^{-1}yr^{-1}$)	25	3
Contribution to		
Gross Domestic Product	5.4%	nm[2]
Export Earnings	12%	nm

[1]Indigenous estate unofficially @ 9 million ha in 1997 (Ministry of Forestry, unpubl.).
[2]Not measured.

One of the biggest challenges facing the New Zealand forest industry is getting the message across to the rest of the world that our industry is sustainable and that many of the initiatives taking place in other countries happened here several decades ago. In many ways we are in danger of being so far out in front of the pack that we have lapped it and are about to join in again. We also face a similar challenge in New Zealand as many of our current stakeholders sit outside of mainstream forestry and are not aware of some of the more significant developments that have taken place in the sector. In particular, we need to ensure that staff from regional councils and other territorial authorities, government departments, and also concerned environmental groups, are well informed about forestry, and that their understanding of the environmental effects of forestry is based on scientific evidence and not folklore.

New Zealand is fortunate that the country's major environmental groups have an excellent working relationship with the major players in the forest industry. However, it has not always been the case as up until the late 1980s there was considerable acrimony between forestry companies and the government forestry department, both which had a stake in commercial production from the natural forest, and the environmental groups who were concerned that native forest logging was destroying the country's natural heritage.

As Ken Shirley (formerly Executive Director of the New Zealand Forest Owners Association) described it, "peace broke out" when the New Zealand Forest Accord was signed in 1991.

The New Zealand Forest Accord (1991)

The objectives of the New Zealand Forest Accord (1991) are to:

- define those areas where it is inappropriate to establish plantation forestry
- recognize the important heritage values of New Zealand's remaining natural indigenous forests and the need for their protection and conservation
- acknowledge that the existing area of natural indigenous forest in New Zealand should be maintained and enhanced
- recognize that commercial plantation forests of either introduced or indigenous species are an essential source of perpetually renewable fibre and energy offering an alternative to the depletion of natural forests
- acknowledge the mutual benefits emanating from an accord between New Zealand commercial forestry enterprises and conservation groups and the example that this unique accord can provide for the international community.

The main constraints that the Accord imposes on the forestry signatories is to greatly restrict the harvest of native forests and the conversion of indigenous forest and scrub land to plantation forestry. New Zealand appears to be unique in having a signed agreement between the main forestry companies and the major conservation groups that aims to both protect the nation's remaining natural forest from logging and promotes plantation forestry as a means of producing wood products on a sustainable basis.

Principles for commercial plantation forest management in New Zealand (1995)

More recently, the New Zealand Forest Owners Association, together with the Farm Forestry Association and the country's major environmental groups, took another step towards ensuring environmental sustainability by agreeing to a set of "Principles for Commercial Plantation Forest Management". These were signed in December 1995 with the objective of "promoting understanding between the signatory parties with a view to New Zealand achieving environmental excellence in plantation forest management and participating as an effective advocate internationally for the sustainable management of plantation forests and the protection, preservation, and sustainable management of natural forests".

PLANTATION FORESTS, ENVIRONMENTAL IMPACTS AND NATURE CONSERVATION

Forest management will have an impact on the environment but whether or not it is an acceptable impact will depend on the type of forestry and society's views at the time. The environmental values associated with parks and reserves clearly must be greater than those associated with industrial plantations, where the primary goal is wood production (Fig. 3). Similarly, the intensity of vegetation management will be considerably greater in an industrial plantation than in a less intensively managed natural forest type.

Commercial production and environmental values do not need to be in conflict, although clearly it is impossible to have both intensive farming and high environmental values, such as natural bio-diversity, on one site. The Principles for Commercial Plantation Forest Management recognize what is

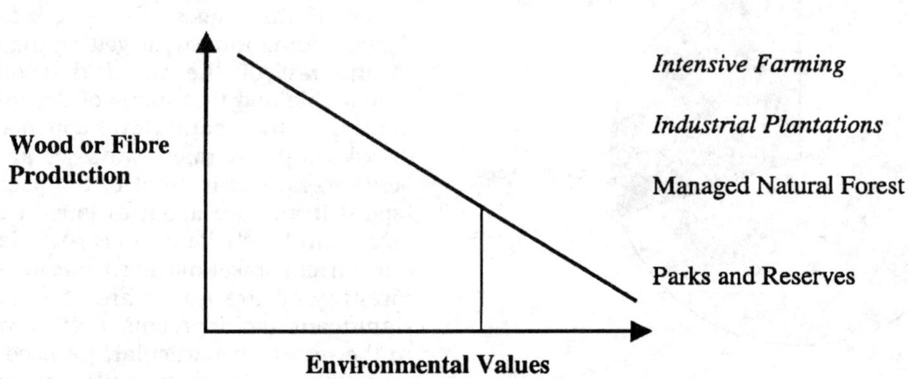

Fig. 3. Relative importance of wood or fibre production and environmental values for various land uses.

possible in a production plantation forest system. Some of the Principles most relevant to nature conservation are discussed in the following sections.

Protection and rehabilitation of aquatic ecosystems

New Zealand Forest Owners Association members are committed to the Principle that:

> "Forestry operations shall be conducted in such a way that safeguards stream margins and water bodies with the objective of achieving healthy aquatic ecosystems."

Compared to other land uses forests have a reputation for producing high quality water (Dyck and Cooke 1981; National Council Aim and Stream Improvement 1994) while agricultural drainage water has been reported to be of poor quality (Smith *et al.* 1993), and for some parameters inferior to urban drainage water (Omernik 1977). Forestry operations certainly have the potential to degrade water quality. However, many of the sedimentation problems associated with forestry operations have been reduced by the application of Best Management Practices. The underlying principles of these guidelines are to minimize disturbances in the vicinity of stream channels, reduce the erosive power of water on bare surfaces, and to protect the infiltration capacity of forest soils.

More recently the industry has been more proactive in commissioning studies to determine the effect of radiata pine forestry on water quality and fish populations. One recent study conducted by the National Institute of Water and Atmospheric Research Ltd. (Rowe 1996, unpubl.) has provided evidence that "there is no difference in native fish communities between streams in mature exotic forests and streams in native forests. However, fish communities in pasture streams differed significantly from those in forested streams." (Fig. 4). Research is continuing to determine the effects of various management practices on native fish populations in respect to the effects of other land uses, as well as the influence of geology.

The influence of land use on freshwater aquatic ecology has received less emphasis than simple measures of streamwater physical or chemical parameters. However, in many situations it is the stream ecology, and particularly the indigenous biodiversity that is of greatest importance. Measures of benthic invertebrate communities (e.g., stream dwelling insects) integrate the effects of alterations in water quality characteristics and provide a relatively sensitive indicator of ecological response to land use changes. A recent study in New Zealand compared benthic invertebrate communities between major land uses: native forest; pasture (cleared from native forest 60 years earlier); and 15-year-old radiata pine planted on to pasture (after 45 years in pasture) (Quinn *et al.* 1994). Sixteen catchments were studied including a 15-year-old pine catchment that had 21% of its area in native riparian vegetation. Despite only 15 years of radiata pine development the results of the study showed that afforestation of pasture had restored many aspects of the streams towards native forest conditions. Where 21% native riparian vegetation had been retained in the pine catchment there was an even greater similarity to the pure native catchments, despite 45 years of pasture production prior to afforestation with radiata pine.

A recent study of fish species richness and diversity in Bay of Plenty and Coromandel streams showed that native species diversity was higher in streams draining exotic forest compared to native forest or pasture (Table 3) (Rowe 1996, unpubl.). The available data, while still preliminary, indicate that the type of forest vegetation (e.g., native *vs* exotic) may not be important for some fish species and many native fish will readily occur within catchments of mature pine forest.

Further research is being conducted to determine the effects of harvesting on fish populations and

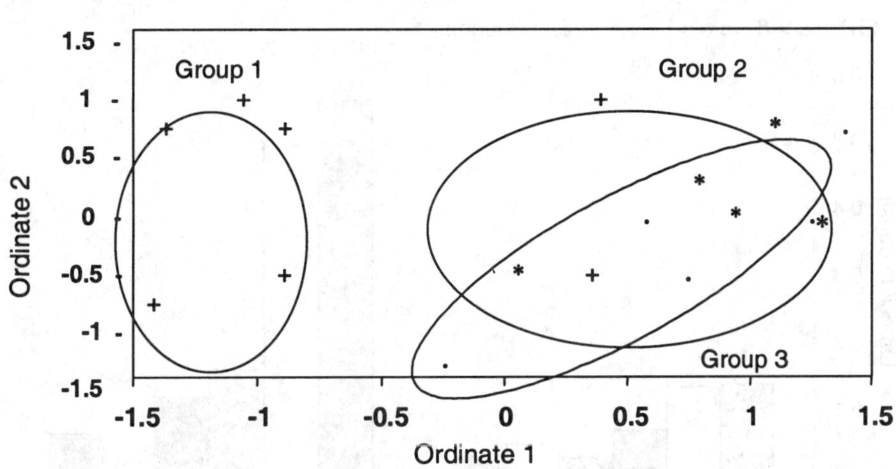

Fig. 4. Results from stream sampling and analysis of fish populations in streams of varying land uses showing similarities between fish populations in exotic forest streams (Group 2) and native forest streams (Group 3) in contrast to pasture streams (Group 1) (Rowe 1996, unpubl. report).

the influence of riparian vegetation on aquatic habitats. Preliminary evidence from a study in Carter Holt Harvey forests on the Coromandel Peninsula indicates that aquatic habitat is only slightly impaired as a consequence of harvesting and streams appear to recover quickly (Halliday and Quinn 1997; unpubl. National Institute of Water and Atmospheric Research Ltd. report).

Table 3. Effect of land use on native fish species richness (number of species) and diversity (complement, Simpsons D) in Bay of Plenty and Coromandel streams (mean +/− S.E.) (Rowe 1996, unpubl. National Institute of Water and Atmospheric Research Ltd. report).

Variable	Native forest	Exotic forest	Pasture
Species richness	3.8 ± 0.5	4.7 ± 0.7	4.6 ± 0.6
Species diversity	0.46 ± 0.03	0.60 ± 0.05	0.45 ± 0.09

This table does not indicate the change in species from one land use type to another.

Restoration of natural soil processes

The Principle dealing with soil quality states:

"Plantation managers shall maintain or enhance soil quality and minimize soil erosion for the purpose of maintaining site productivity and water quality."

Plantation forest managers today, at least in New Zealand but also overseas, realize the importance of soil quality to the current and future productivity of their forests. However, this has not always been the case as earlier forestry practices in New Zealand and elsewhere tended to disregard the importance of soil quality to long-term productivity (see Dyck *et al.* 1994). The importance of planting forests to reduce erosion and prevent sedimentation of streams and lakes is also well known and several New Zealand radiata pine forests were established primarily for soil stabilization.

Soil stabilization

An excellent documented example of the benefits of pine plantations in reducing erosion comes from the East Coast of the North Island where Marden and Rowan (1993) examined the effect of different vegetation types on controlling erosion in a very unstable geology (Fig. 5). Prior to Cyclone Bola, the landslide density for pine forest greater than 6-years-old was half that of pasture and decreased even further as the forest aged. This significant benefit of planting pine trees for erosion control became even more obvious during Cyclone Bola when pine forest greater than 8-years-old was found to be just as effective as indigenous forest at preventing landslides. A subsequent study showed that pasture had 0.68 landslides per hectare compared to 0.06 for pine forest greater than 8-years-old (Phillips *et al.* 1990). In a recent review of the available literature on soil erosion Maclaren (1996) concluded that:

"There is indisputable evidence, from a substantial number of studies, that at least several major types of erosion are far less severe under a vegetation cover of pines than of pasture, regardless of the farm management regime."

Soil quality improvement

There has been considerable discussion over the last century on the effects of conifer forests on soil quality with early evidence from Germany suggesting that spruce forests acidified the soil (for a review see Will and Ballard 1976). Although originally thought to be caused by the trees

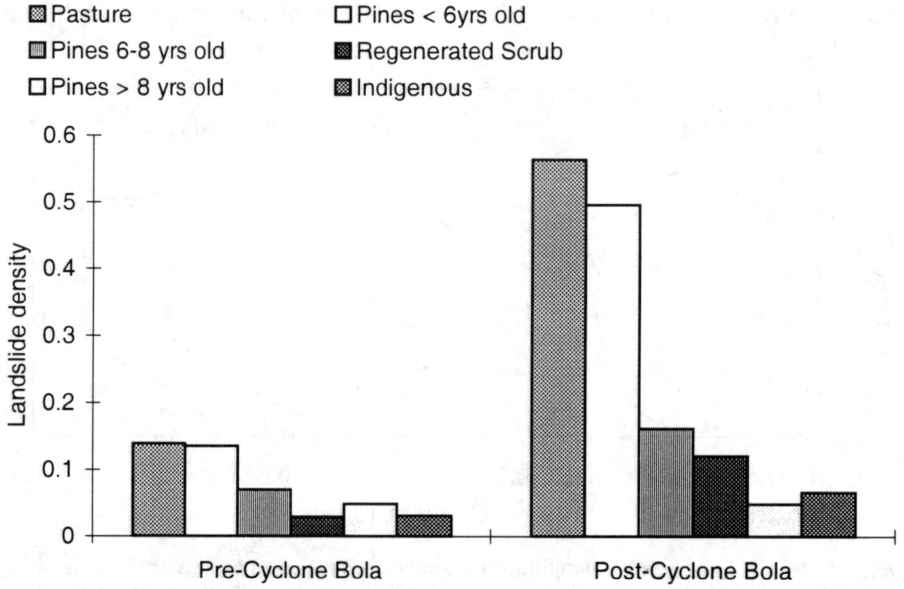

Fig. 5. Landslide densities for different vegetation types, East Coast, North Island (Marden and Rowan 1993).

themselves, it was later concluded that second-rotation productivity decline was caused by nutrient removal by peasants regularly collecting forest needle litter for animal bedding.

Professor Earl Stone, an eminent American soil scientist reviewed the available evidence in 1973 in a Report of the President's Advisory Panel on Timber and the Environment and concluded that:

> *"much of the famed European experience with rapid soil deterioration under conifers has little real meaning in modern Europe and still less in the United States" (cited in Will and Ballard 1976)*

In fact, rather than a soil degrader, radiata pine is actually a soil improver, a fact that has been known since the days of Leonard Cockayne, one of New Zealand's earliest and very prominent botanist. In 1914 he wrote:

> "there is widespread belief that Pinus radiata is a great soil-exhauster, and that ground once occupied by this tree becomes very infertile . . . in practice, however, it is found that when the pine-trees are cut down and removed the soil is admirably suited for the growing of all kinds of crops . . . it is evident that the growing of Pinus radiata, instead of weakening the soil fertility has the very opposite effect (see Will and Ballard 1976).

Radiata pine has been found to improve soil quality by increasing porosity (Ballard 1972) and improving soil nutrient and organic matter levels (Davis and Lang 1991; Dyck *et al.* 1991; Smith *et al.* 1994). Additionally studies comparing the effects of exotic conifers to adjacent grassland has shown that the trees can improve the levels of available nutrients (Table 4), often improving the soil for both tree and grass growth (Davis and Lang 1991).

Table 4. Effect of vegetation on available nutrients in topsoil under grassland and exotic conifers (Mean and S.E.) (Davis and Lang 1991).

Chemical property	Grassland	Conifer
Mineralizable N (mg kg^{-1})	8.0 (3.33)	9.0 (4.25)
Sulphate-S (mg kg^{-1})	4.9 (0.98)	13.5 (4.57)
Inorganic P (mg 100 g^{-1})	34.0 (5.20)	42.0 (6.10)
Bray-2 K(mg kg^{-1})	31.0 (13.40)	62.0 (24.10)
Bray-2 K (me 100 g^{-1})	3.2 (0.70)	3.6 (0.58)
Bray-2 Ca (me 100 g^{-1})	3.9 (0.71)	3.6 (0.80)
Bray-2 Mg (me 100 g^{-1})	0.90 (0.15)	0.83 (0.15)
Exchangeable Al (me 100 g^{-1})	0.9 (0.30)	1.64 (0.29)

Many soils are so impoverished that they require fertilizer additions to get trees to grow, however, long-term research results have shown that phosphorus fertilizer applied early in one rotation still has a carry-over effect into the next rotation, thus reducing the need for additional fertilizer applications (Table 5) (Zhou and Skinner, unpubl.).

Perhaps a more obvious indicator that radiata pine has not caused soil deterioration in New Zealand is the fact that many soils are now supporting a third crop of radiata pine with improved productivity with subsequent rotations.

Table 5. Inorganic P (Bray extraction) in Riverhead Museum plot topsoil 40 years after treatment (Zhou and Skinner, unpubl.).

Treatment	Inorganic P (ug g^{-1})
Control	1.144
Fertilized (255 kg P ha^{-1})	1.689

Increased biodiversity vs *alternative productive land uses*

There are several principles that deal with biodiversity, including:

> "Indigenous biodiversity will be protected primarily in natural areas."
>
> "The protection of indigenous biodiversity in plantation forests is not the primary objective but should be recognized and provided for where appropriate."
>
> "Where threatened species are known to occur within plantation forests and their presence is considered significant by the Department of Conservation, plantation managers shall consult with the Department of Conservation on management practices with the objective of conserving the population."
>
> "Plantation forests shall not replace natural forest and other natural areas, as agreed under the instruments of the New Zealand Forest Accord."
>
> "Plantation managers shall take all practical steps to protect indigenous vegetation along the margins of water bodies where appropriate."
>
> "Plantation managers shall recognize, and where appropriate, facilitate the restoration of depleted indigenous habitat on critical areas under their management."

The intention of these Principles is to recognize that it is the indigenous forest that is the most important reserve of New Zealand's unique biodiversity and this resource should be conserved. Because plantation species were exotic and intensively managed, it was considered impossible to provide for indigenous ecosystems on the same piece of ground that was supporting exotic pine trees or other plantation species. However, it was also recognized that indigenous biodiversity could be maintained, and even enhanced, in areas within plantation estates including indigenous vegetation reserves, along stream margins, and also within aquatic systems within the plantation forests. Terms such as "practical" and "where appropriate" were specifically included in these statements to recognize that it would not always be possible to avoid damage to streamside vegetation and reserves, particularly during harvesting operations, and that factors of worker safety and economics also needed to be considered.

Threatened species are not common in plantation forests but the commitment is there to protect

Table 6. Number of shrub and fern species in a radiata pine chronosequence, Kinleith Forest (from Ogden *et al*. 1997).

Category	Stand age (years)								
	6	11	20	24	29	32	35	40	67
Shrubs									
total	6	10	9	13	6	9	7	9	12
Native	*3*	*6*	*7*	*12*	*6*	*9*	*6*	*9*	*11*
Ferns	5	15	12	21	19	17	12	13	22
Total	11	25	21	34	25	26	19	22	34

significant populations where they do occur. Kiwis, for example, are known to inhabit Carter Holt Harvey forests on the Coromandel Peninsula and Hochstetter frogs have been found in forests in Northland. In the case of kiwis, the most important steps that managers can take are to control predator populations and restrict dog access.

Perhaps surprisingly to some people, radiata pine plantations are not "biological deserts" but in fact contain a large variety of understorey plant species (Allen *et al.* 1995a,b; Ogden *et al.* 1997). The composition of the understorey will vary with geographic location, geology, previous land use, and silvicultural regime. For example, first rotation forests established on pastureland may have understoreys dominated by native ferns, such as parts of Mahurangi Forest, or exotic "weeds" if planted on to gorse *Ulex europaeus* infested hillsides. In a recent study of nine radiata pine stands in Carter Holt Harvey's Kinleith Forest located on volcanic soils in the central North Island, Ogden *et al.* (1997) found considerable species diversity in the forest understorey (Table 6). The understorey in the older stands was comparable to that found in native podocarp and kauri forests. Allen *et al.* (1995a) found similar results in Kinleith Forest and stated that the level of species richness was greater than for many natural New Zealand forests (Allen *et al.* 1995b).

Current annual afforestation rates are in the order of 70 000 to 80 000 ha. As well as increasing the productivity of what is generally considered to be marginal pastoral land and providing a greater economic return to the landowner and the nation, afforestation will also be providing many environmental benefits, including increased indigenous biodiversity. Countries, such as Britain, provide subsidies to encourage greater biodiversity and make biodiversity management economically attractive. For example, in Britain landowners have grants available to allow up to 20% of their land to be left unplanted and managed as open ground for biodiversity values (Hodge *et al.*, unpubl. Forestry Commission report). Despite the absence of subsidies in New Zealand, New Zealand forestry companies are committed to managing for biodiversity as per the Principles and many have extensive areas of their land in indigenous vegetation. Carter Holt Harvey, for example, has 20% of its estate in indigenous cover, much of it in reserves.

CONCLUSION

The New Zealand forest industry has readily acknowledged that environmental damage may occur as a consequence of inappropriate management practices and has developed Codes of Practice and Guidelines to avoid such problems. It has also worked very closely with the country's environmental groups to develop both the New Zealand Forest Accord (1991), which clearly sets out to maintain and enhance the indigenous forest estate, while recognizing the importance of plantation forests for wood production, and more recently agreed to the Principles for Commercial Plantation Forest Management in New Zealand (1995), which promote environmental excellence in plantation management. The industry has been much less active in publicising the environmental benefits of plantation forestry to New Zealand.

Plantation forestry in New Zealand plays a major role in conserving nature in a number of ways:

Providing a source of renewable wood

Conserving remaining natural forest

Increasing indigenous biodiversity by converting farmland to forest

Protecting and enhancing aquatic biodiversity

Improving soil properties

It also plays a more subtle role by acting as a carbon sink and reducing methane and nitrous oxide production, both greenhouse gases, when pasture land is converted to forest.

REFERENCES

Allen, R. B., Platt, K. H. and Coker, R. E. J., 1995a. Understorey species composition patters in a *Pinus radiata* D. Don plantation on the central North Island Volcanic Plateau, New Zealand. *NZ J. For. Sci.* **25:** 301–17.

Allen, R., Platt, K. and Wiser, S., 1995b. Biodiversity in New Zealand plantations. *NZ For.* **39:** 26–29.

Ballard, R., 1972. Influence of heavy phosphate dressing and subsequent radiata pine response on the properties of a Riverhead clay soil. *NZ J. For. Sci.* **2:** 202–16.

Davis, M. R. and Lang, M. H., 1991. Increased nutrient availability in topsoils under conifers in the South Island high country. *NZ J. For. Sci.* **21:** 165–19.

Dyck, W. J., Cole, D. W. and Comerford, N. B. (eds), 1994. Impacts of Forest Harvesting on Long-term Site Productivity. Chapman and Hall: London.

Dyck, W. J. and Cooke, J. G., 1981. Exotic forestry and its effect on water quality. Pp. 63–84 *in* Waters of the Waikato. University of Waikato: Hamilton, New Zealand.

Dyck, W. J., Hodgkiss, P. D., Oliver, G. R. and Mees, C. A., 1991. Harvesting sand-dune forests: Impacts on second-rotation productivity. Pp. 163–76 *in* Long-term Field Trials to Assess Environmental Impacts of Harvesting ed by W. J. Dyck, C. A. Mees. Halliday and Quinn Forest Research Institute Bulletin No. 161, Forest Research Institute: Rotorua.

Kanowski, P. J., 1997. Plantation forestry for the 21st century. XI World Forestry Congress, 13–22 October 1997. 10 Pp.

Lewin, R., 1997. Ancestral echoes. New Scientist, 5 July 1997. Pp. 32–37.

Maclaren, J. P., 1996. The Environmental Effects of Planted Forests in New Zealand — the implications of continued afforestation of pasture. New Zealand Forest Research Institute, Forest Research Institute Bulletin No. 198. 180 Pp.

Marden, M. and Rowan, D., 1993. Protective value of vegetation on tertiary terrain before and during Cyclone Bola, East Coast, North Island, New Zealand. *NZ J. For. Sci.* **23:** 255–63.

National Council Aim and Stream Improvement, 1994. Forests as nonpoint sources of pollution, and effectiveness of best management practices. National Council of the Paper Industry for Air and Stream Improvement, Inc. Technical Bulletin No. 672. 57 Pp.

New Zealand Forest Owners Association, 1997. NZ Forestry Facts and Figures 97. New Zealand Forest Owners Association Inc., Wellington, New Zealand. 22 Pp.

Ogden, J., Braggins, J., Stretton, K. and Anderson, S., 1997. Plant species richness under *Pinus radiata* stands on the central North Island volcanic plateau, New Zealand. *NZ J. Ecology* **21:** 17–29.

Omernik, J. M., 1977. Nonpoint source — stream nutrient level relationships: a nation-wide study. EPA-600/3-77-105. Corvallis, Oregon.

Phillips, C. J., Marden, M. and Pearce, A. J., 1990. Effectiveness of reforestation and control of landsliding during large cyclonic storms. Pp. 340–50 *in* Proc. of XIX World IUFRO Congress, Montreal, Canada, August 1990. Div. 1, Vol. 1.

Quinn, J. M., Cooper, A. B., Davies-Colley, R. J., Rutherford, J. C. and Williamson, R. B., 1994. Land-use effects on New Zealand hill country streams and implications for riparian management. Pp. 19–27 *in* The Ecology and Management of Aquatic-Terrestrial Ecotones. G. Lee Link (Compiler). Proceedings of the International Workshop, University of Washington, Seattle, USA, February 1994.

Roche, M., 1990. History of Forestry. New Zealand Forestry Corporation Ltd. GP Books. 466 Pp.

Rowe, D. K., 1996. Interim report on studies related to the effects of land-use changes on native fish. National Institute of Water and Atmospheric Research Ltd. unpubl. report, 10 Pp.

Smith, C. M., Wilcock, R. J., Vant, W. N., Smith, D. G. and Cooper, A. B., 1993. Towards sustainable agriculture: freshwater quality in New Zealand and the influence of agriculture. Ministry of Agriculture and Fisheries Policy Technical Paper 93/10. Ministry of Agriculture and Fisheries, Wellington. 208 Pp.

Smith, C. T., Lowe, A. T., Beets, P. N. and Dyck, W. J., 1994. Nutrient accumulation in second-rotation *Pinus radiata* after harvest residue management and fertiliser treatment of coastal sand dunes. *NZ J. For. Sci.* **24:** 362–89.

Sutton, W. R. J., 1995. The need for planted forests and the example of radiata pine. Paper presented at the symposium "Planted Forests — Contributions to Sustainable Societies", June 1995, Portland, Oregon. 14 Pp.

Will, G. M. and Ballard, R., 1976. Radiata pine — soil degrader or improver? *NZ J. For.* **21:** 248–52.

CHAPTER 5

The design of exotic softwood plantations to enhance wildlife conservation: preliminary lessons from the Tumut fragmentation experiment, southeastern Australia

D. B. Lindenmayer[1] and M. L. Pope[1]

Extensive areas of land have been planted with exotic softwood (Radiata Pine) plantation in southeastern Australia over the past 50 years. In the past, large areas of native *Eucalyptus* forest were cleared to establish Pine forests. Present expansions of plantations in states like New South Wales are continuing primarily on grazing lands that support remnant forest and woodland. Preliminary results from a large-scale fragmentation experiment taking place within a softwood plantation near Tumut, southern New South Wales, indicate that although softwood establishment is high wood-fibre-yielding enterprise, new plantations can be planned to increase their value for wildlife conservation. Design strategies include the retention of existing native vegetation remnants within the boundaries of softwood plantations (particularly larger patches), restoration of native vegetation along gully-lines, and targeting strategic areas for re-planting with indigenous species to link retained patches of native vegetation within and across catchments. Harvesting practices within long-established softwood plantations also could be modified to help conserve biodiversity such as staggering clearfall schedules in stands of pine trees that surround large patches of native forest and ensuring that thickets of important indigenous understorey vegetation are not extensively damaged during cutting operations.

INTRODUCTION

EXTENSIVE areas of exotic softwood plantations have been established in Australia over the past few decades and they are expanding at a rate of more than 30 000 ha per year (Resource Assessment Commission 1992; State of the Environment Australia 1996). Much of the softwood plantation estate in southeastern Australia is dominated by stands of Radiata Pine *Pinus radiata*. Timber extracted from exotic softwood plantations now constitutes an increasing proportion of the total sawn timber consumed in the various Australian states (Clark 1992). Thus, the plantation industry makes an important contribution to the Australian economy. However, a number of authors have suggested that there may be negative effects of plantation establishment on biodiversity (e.g., Pawsey 1966; Disney and Stokes 1975; Suckling *et al.* 1976; Suckling and Heislers 1978; Smith 1982; see review by Ray *et al.* 1983). This is because: (1) in the past, large areas of native *Eucalyptus* forest were cleared to establish softwood plantations (Tyndale-Biscoe and Smith 1969; Routley and Routley 1975), and, (2) stands of pine forest often lack key habitat attributes and food sources such as tree hollows, flowers, seeds and fruit that are needed by Australian forest-dependent animals (Smith 1982).

In Australian states like New South Wales, much of the present expansion of the softwood plantation estate is occurring on predominantly cleared agricultural and grazing lands that often support

[1]Centre for Resource and Environmental Studies, The Australian National University, Canberra, Australian Capital Territory, Australia 0200.

Pages 44–49 *in* NATURE CONSERVATION 5: NATURE CONSERVATION IN PRODUCTION ENVIRONMENTS: MANAGING THE MATRIX ed by J. L. Craig, N. Mitchell and D. A. Saunders. Surrey Beatty & Sons, 2000.

remnant patches of the original woodland and forest cover. Although the primary goal of plantation forestry is clearly the efficient production of large quantities of timber and pulp, important opportunities exist to design new softwood plantations to enhance their value for the wildlife. Similar opportunities exist as part of harvesting existing softwood plantations and the next rotation is being established. These initiatives are important as the softwood plantation estate is presently large; more than one million hectares (Resource Assessment Commission 1992; State of the Environment 1996) and is set to increase substantially in the coming years. Indeed, although the removal of native vegetation in some Australian states has declined in more recent times, clearing is set to increase again (e.g., in Victoria) as a consequence of proposals to establish plantations, particularly exotic softwoods (Commonwealth of Australian and Department of Natural Resources and Environment 1997).

In this chapter, we briefly outline some of the possible approaches to the design and management of exotic softwood plantations to promote their value for wildlife conservation. The strategies we discuss are preliminary ones based on the outcomes of a large-scale landscape quasi-experiment (*sensu* Dunning *et al.* 1995) at Tumut in southern New South Wales (Lindenmayer *et al.* 1996, 1997a, 1997b, 1999a, 1999b) together with the findings of a radio-tracking investigation of patch use by one species; the Greater Glider *Petauroides volans* (Pope and Lindenmayer 1997).

THE TUMUT FRAGMENTATION EXPERIMENT

The large-scale fragmentation experiment at Tumut has focused on 192 patches of remnant eucalypt forest embedded within a 45 000 ha softwood (Radiata Pine) plantation in the Buccleuch State Forest. These patches vary in size (0.2–124 ha), shape, dominant tree species, time since isolation (5–65 years) and a host of other variables (see Lindenmayer *et al.* 1996, 1997a). Extensive areas of native forest abut the plantation, including those located in Kosciuszko National Park, Bondo State Forest, and Bungongo State Forest. The main objective of the experiment is to compare the presence and abundance of vertebrates (small mammals, arboreal marsupials and birds) in three broad groups of sites: (1) sites in large contiguous areas of native eucalypt forest adjacent to the softwood plantation (40 sites), (2) sites within remnant patches of native eucalypt forest surrounded by stands of Radiata Pine (86 remnants), and, (3) sites dominated by Radiata Pine trees that comprise the softwood plantation in the study area (40 sites) (Lindenmayer *et al.* 1997a). A rigorous approach to experimental design has underpinned the study and this has involved: (i) A stratified, randomized and replicated patch selection procedure to guide the selection of remnants in four different size classes, two different shape classes, five different forest types, and two different isolation time categories. (ii) Cross-matching between groups of remnants and sites in large contiguous native forest on the basis of climate, geology and vegetation types. And, (iii) The measurement of a suite of covariates at each site such as stand structure and plant species composition as well as isolation distance and other landscape indices (for the 86 remnants only) (see Lindenmayer *et al.* 1997a, 1999a).

The landscape experiment at Tumut has been underway since mid-1995 and preliminary results are available for arboreal marsupials (Lindenmayer *et al.* 1997a, 1999a; Pope and Lindenmayer 1997) and small mammals (Lindenmayer *et al.* 1999b). Some analyses have been completed for birds (Lindenmayer *et al.* 1996; Cunningham *et al.* 1999) but a detailed exploration of data gathered for this group has only recently commenced. The following discussion of principles for plantation design is based predominantly on the preliminary findings for arboreal marsupials and small mammals. We expect that these design principles would be expanded or modified on the basis of new information, particularly for birds. This is because birds are by far the most species-rich group of vertebrates at Tumut range and they exhibit a very wide range of responses to landscape context and fragmentation effects.

NEW SOFTWOOD PLANTINGS

The value of existing remnant native vegetation

As outlined above, the current expansion of the softwood plantation estate in New South Wales is occurring in areas that have a history of agriculture and/or grazing. The native vegetation in these areas has been largely cleared and only remnants of the original forest or woodland remain. The results of the Tumut fragmentation experiment clearly indicate that remnant patches of eucalypt forest surrounded by softwood plantations have value as habitat for native mammals and that they should not be cleared as part of plantation establishment. Studies of arboreal marsupials and small mammals have indicated that remnant patches, including those that have been isolated for prolonged periods (up to 65 years), are occupied by a wide range of taxa including the Greater Glider *Petauroides volans*, Common Ringtail Possum *Pseudocheirus peregrinus*, Sugar Glider *Petaurus breviceps*, Feathertail Glider *Acrobates pygmaeus*, Common Brushtail Possum *Trichosurus vulpecula*, Mountain Brushtail Possum *Trichosurus caninus*, Brown Antechinus *Antechinus stuartii*, Dusky Antechinus *Antechinus swainsonii*, Bush Rat *Rattus fuscipes*, Echidna *Tachyglossus aculeatus*, Common Wombat *Vombatus ursinus*, Eastern Grey Kangaroo *Macropus giganteus* Red-necked Wallaby *Macropus rufogriseus* and Black Wallaby *Wallabia bicolor*. Thus, while some species do not appear to be able to persist in vegetation remnants — Yellow-bellied Glider *Petaurus australis* and Squirrel Glider *Petaurus norfolcensis* were absent from these areas; the patches did provide suitable habitat for many taxa.

The importance of large patches

Larger patches of native vegetation on semi-cleared grazing lands should have priority for exemption from clearing plantation establishment. Analyses of data gathered at Tumut have revealed positive statistical relationships between patch size and the presence and abundance of several species of mammals (Lindenmayer *et al.* 1997a). Figure 1 shows the relationship derived between the probability of occurrence of any species of arboreal marsupial and the area of remnant patches surveyed at Tumut. The probability of occurrence of an arboreal marsupial was >50% in remnants 3 ha or larger and >80% in those exceeding 20 ha, with the lower bound of the 95% confidence interval exceeding 50% in remnants measuring about 8 ha (Fig. 1). Small mammals such as the Brown Antechinus and the Bush Rat show similar responses to patch size (Lindenmayer *et al.* 1999b). On the basis of these results, we recommend that remnants 3 ha or larger be exempt from clearing.

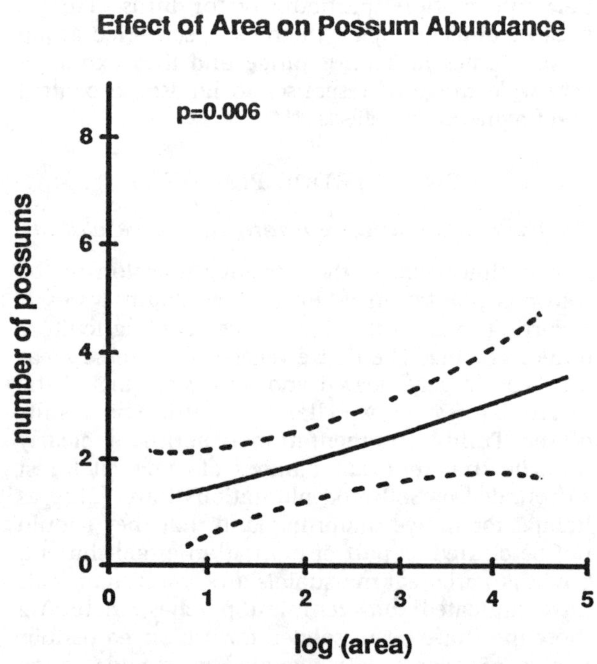

Fig. 1. Relationships between the abundance of arboreal marsupials and the area of remnant patches of eucalypt forest at Tumut which are surrounded by extensive stands of Radiata Pine.

Many of the agricultural landscapes presently targeted for plantation establishment in the Tumut region and other areas in southeastern New South Wales are characterized by paddocks supporting tiny remnants comprised of one or two isolated eucalypt trees. The results of our field experiment indicate that tiny remnants of this size embedded within stands of Radiata Pine appear to have limited value as habitat for native mammals at Tumut. For example, the Common Ringtail Possum was the only species of arboreal marsupial detected in extensive surveys of remnants measuring one ha. However, even it was rare in these sites and was detected at only three of the 41 tiny remnants examined (Lindenmayer *et al.* 1997a). Notably, the Common Ringtail Possum also was the only species observed more than once during surveys of more than 40 sites in stands of Radiata Pine (Lindenmayer *et al.* 1997a). These findings further support the conclusion that plans for plantation establishment should focus on the retention of larger, more consolidated remnants rather than preservation of isolated paddock trees. Finally, isolated trees as well as remnants comprised of only a few stems often suffer from severe dieback (*sensu* Landsberg and Wylie 1983). Structural and floristic measurements completed throughout the array of sites examined in the Tumut experiment have indicated that the bigger remnants are typically the only ones characterized by younger regrowth stems and hence the ones more likely to maintain integrity in the long term.

EXISTING SOFTWOOD-DOMINATED LANDSCAPES

Softwood plantations have existed in the Tumut region for many decades and the first plantings were established in the early 1930s. The results of our studies at Tumut have indicated some modifications to current practices within existing softwood plantations could enhance their conservation value.

Restoration of riparian and gully areas

A proportion of the remnants in the patch system targeted for study at Tumut are located within gullies. These areas, like the other remnants surveyed, support a wide range of mammals (see above) as well as many species of birds (Lindenmayer *et al.* 1999c). Indeed, riparian areas in many forest ecosystems have been found to be species-rich (e.g., Loyn *et al.* 1980; Recher *et al.* 1980; Naiman *et al.* 1993), although these areas do not provide suitable habitat for all forest-dependent taxa (Claridge and Lindenmayer 1994). Notably, in more recent decades, riparian native vegetation at Tumut has been exempt from clearing to establish Radiata Pine. Timber harvesting in areas established prior to this time is subject to special prescriptions in and near gully and riparian zones due to water quality and aquatic habitat considerations. Given these considerations, together with the potential importance of riparian areas for fauna, we recommend that native vegetation be restored in gully areas following the final clear-felling operation that completes a given logging rotation. Restoration of gully or riparian vegetation provides an opportunity to link retained areas within and across catchments and this may aid the dispersal of some species, particularly aquatic taxa (Wardell-Johnson and Roberts 1993) and those in which movement and/or dispersal is guide by particular topographic features such as streamlines (see Taylor *et al.* 1985; Rosenberg *et al.* 1997). Notably, other authors working in softwood production landscapes characterized by patches of remnant eucalypt forest have recommended that the original gully vegetation be restored to enhance the value of plantations for biodiversity (Recher *et al.* 1987).

Staggering the spatial configuration of clearfall harvesting

Although existing patches of remnant eucalypts within the boundaries of the softwood plantation at Tumut are exempt from cutting, the surrounding Radiata Pine is clearcut prior to the commencement of a new rotation. Radio-tracking of the Greater Glider at Tumut has indicated that animals use the Pine forest to move between eucalypt habitat patches (Pope and Lindenmayer 1997). Therefore, the clearfelling of stands of softwood trees surrounding eucalypt remnants may temporarily limit the movements of animals like the Greater Glider and isolate sub-populations in patches from those residing within other remnants in the patch system. One way to enhance functional connectivity (*sensu* Bennett 1998) between sub-populations that reside within spatially separated eucalypt patches would be to stagger the spatial configuration of clearfelled Radiata Pine stands adjacent to remnants. Thus, at any one time, some areas of advanced regrowth Radiata Pine in the matrix would link eucalypt remnants. Harris (1984) has made similar suggestions for guiding cutting regimes to buffer, and maintain connectivity between, areas of old growth Dougals Fir forest in the Pacific Northwest of the United States of America.

Conservation of components of understorey vegetation and ground cover

Careful treatment of understorey vegetation within stands of Radiata Pine could make an important contribution to the conservation of some species, including those considered to be at risk. For example, surveys of birds in our experiment have indicated that the Olive Whistler *Pachycephala olivacea* often occurs within stands of Radiata Pine where there are intact thickets of understorey *Leptospermum* spp. (Lindenmayer *et al.*, unpubl. data). Approaches that either minimize the disturbance of understorey vegetation or promote its regeneration following harvesting could help species like the Oliver Whistler to persist in softwood plantations. Notably, such understorey vegetation often occurs in gullies or wet seepage areas of swamps; places that could be readily captured within riparian zones.

Field observations completed at Tumut have indicated that a range of species of birds are closely associated with windrows of eucalypt logs that are the remains of the original stands cleared to plant softwoods. Several other studies have highlighted the importance for native fauna of windrows of cleared eucalypts within softwood plantations (e.g., Friend 1982; Curry 1991). Given their value, it is important that the decay of these windrows is not accelerated by damage from machinery during harvesting operations in stands of Radiata Pine.

CAVEATS

Work at Tumut has been underway for less than three years and detailed statistical exploration of datasets has been limited to those gathered for small mammals and arboreal marsupials. Extensive analyses of birds are likely to provide additional information and insights that will be important for softwood plantation design. This is because birds are the most species-rich assemblage at Tumut and they display a very wide range of different responses to fragmentation.

The plantation estate is presently being expanded at a rate of approximately 2 000 ha per year (D. Hobson, pers. comm.). This process will not be halted while new information is being generated. Moreover, the data gathered and analysed thus far indicate that the adoption of some of the design principles outlined in this chapter could substantially improve the value of softwood plantation for wildlife conservation. Thus, we believe that it is important to document some principles now although considerable further work needs to be completed. This is necessary to ensure that valuable existing conservation opportunities associated with the current establishment of new plantations are not foregone. However, there are few studies worldwide that have explored issues associated with plantation design to enhance the value of such areas for wildlife (Hanowski *et al.* 1997). On this basis, there is a clear need for more studies to study and monitor plantation forests (and the remnants of native vegetation they may contain) to help develop approaches that ensure they retain some nature conservation values. Indeed, baseline studies are required that examine the changes in species assemblages that occur when agricultural landscapes with areas of remnant native vegetation are transformed to landscapes dominated by softwoods encompassing fragments of the original native forest and woodland cover. These studies should be set within an adaptive management framework (Walters and Holling 1990; see Lindenmayer and Franklin, this volume) so that new data can be forwarded to softwood forest managers to better inform approaches for improved plantation design.

Our study has focused on native forest and woodland and their value for wildlife within softwood plantations. We have not considered remnant grassland habitats on grazed lands, but recognize that some areas targeted for planting may support these types of vegetation communities that have experienced substantial reduction and degradation since European settlement. Grassland communities warrant careful attention and the potential effects of plantation establishment on them as well as approaches for their conservation in such intensive production environments, should be a topic for further investigation.

ACKNOWLEDGEMENTS

This chapter is based on the results of a detailed field experiment at Tumut which has received financial and/or logistic support from a range of organizations including Land and Water Resources and Research and Development Corporation, Rural Industries Research and Development Corporation, NSW Department of Land and Water Conservation, Environment Australia, State Forests of New South Wales, New South Wales National Parks and Wildlife Service, VISY Industries, CSR Ltd, and the Canberra Ornithologists Group. Staff from the Tumut and Albury offices of State Forests of New South Wales have been most receptive to ideas on plantation design and have encouraged our thinking on these issues. Mr Ross Cunningham, Ms Christine Donnelly, Professor Henry Nix, Professor Hugh Possingham, Dr Robert Lacy, Dr Rod Peakall, Ms Heidi Hewittson, Mr Chris McGregor, Mr Craig Tribolet, Mr Bruce Lindenmayer, Mr Ryan Incoll, and Mr Andrew Wright have made major contributions to the work at Tumut. Their help, assistance and collaboration is most gratefully acknowledged. David B. Lindenmayer would like to thank Dr Rod Kavanagh, Dr David Norton and Dr Richard Hobbs for useful discussions on plantation design including some of the ideas outlined in this chapter. Mr D. Hobson, Mr M. Pettit and the staff of the Tumut office of State Forests of NSW have been most supportive of the fragmentation experiment taking place at Tumut and its applications to plantation design.

REFERENCES

Bennett, A. F., 1998. Linkages in the landscape. The role of corridors and connectivity in wildlife conservation. International Union for the Conservation of Nature: Gland, Switzerland.

Claridge, A. W. and Lindenmayer, D. B., 1994. The need for a more sophisticated approach toward wildlife corridor design in the multiple-use forests of southeastern Australia: the case for mammals. *Pac. Cons. Biol.* **1**: 301–07.

Clark, J., 1992. The future for native forest logging in Australia. Working Paper 1992/1. Centre for Resource and Environmental Studies, Canberra, Australian Capital Territory.

The Commonwealth of Australia and Department of Natural Resources and Environment, 1997. Comprehensive Regional Assessment — Biodiversity. Central Highlands of Victoria. The Commonwealth of Australia and Department of Natural Resources and Environment: Canberra, Australian Capital Territory.

Cunningham, R. B., Lindenmayer, D. B., Nix, H. A. and Lindenmayer, B. D., 1999. Counting birds in forests: a comparison of observers and observation methods. *Aust. J. Ecol.* **24**: 270–77.

Curry, G. N., 1991. The influence of proximity to plantation of diversity and abundance of bird species in an exotic pine plantation in north-eastern New South Wales. *Wildl. Res.* **18**: 299–314.

Disney, H. J. and Stokes, A., 1975. Birds in Pine and native forests. *Emu* **76**: 133–38.

Dunning, J. B., Borgella, R., Clements, K. and Meffe, G. K., 1995. Patch isolation, corridor effects, and colonisation by a resident sparrow in a managed pine woodland. *Cons. Biol.* **9**: 542–50.

Friend, G. R., 1982. Mammal populations in exotic pine plantations and indigenous eucalypt forests in Gippsland, Victoria. *Aust. For.* **45**: 3–18.

Harris, L. D., 1984. The Fragmented Forest. University of Chicago Press: Chicago, Illinois.

Hanowski, J. M., Niemi, G. J. and Christian, D. C., 1997. Influence of within-plantation heterogeneity and surrounding landscape composition on avain communities in hybrid popular plantations. *Cons. Biol.* **11**: 936–44.

Landsberg, J. and Wylie, F. R., 1983. Water stress, leaf nutrients and defoliation: a model of dieback of rural eucalypts. *Aust. J. Ecol.* **8**: 27–41.

Lindenmayer, D. B., 1994. Wildlife corridors and the mitigation of logging impacts on forest fauna in south-eastern Australia. *Wildl. Res.* **21**: 323–40.

Lindenmayer, D. B., Pope, M., Cunningham, R. B., Donnelly, C. F. and Nix, H. A., 1996. Roosting in the Sulphur-Crested Cockatoo *(Cacatua galerita)*. *Emu* **96**: 209–12.

Lindenmayer, D. B., Cunningham, R. B., Pope, M., Donnelly, C. F., Nix, H. A. and Incoll, R. D., 1997a. The Tumut fragmentation experiment in south-eastern Australia: the effects of landscape context and fragmentation on arboreal marsupials. CRES Working Paper, 1997/4.

Lindenmayer, D. B., Cunningham, R. B., Nix, H. A., Lindenmayer, B. D., McKenzie, S., McGregor, C., Pope, M. L. and Incoll, R. D., 1997b. Counting birds in forests: a comparison of observers and observation methods. CRES Working Paper. 25 Pp.

Lindenmayer, D. B., Cunningham, R. B., Pope, M. and Donnelly, C. F., 1999a. The Tumut fragmentation experiment in south-eastern Australia: the effects of landscape context and fragmentation of arboreal marsupials. *Ecol. Appl.*

Lindenmayer, D. B., Cunningham, R. B., Pope, M. L. and Donnelly, C. F., 1999b. A field-based experiment to examine the response of terrestrial mammals to landscape context and habitat fragmentation. *Biol. Cons.* **88**: 387–403.

Lindenmayer, D. B., Cunningham, R. B., Donnelly, C. F., Nix, H. A., Lindenmayer, B. D. and McGregor, C., 1999c. The distribution of birds in a fragmented landscape. CRES Working Paper 1999/1.

Loyn, R. H., Chestefield, E. A. and Macfarlane, M. A., 1980. Forest utilization and the flora and fauna in Boola Boola State Forest in south-eastern Victoria. Forest Commission of Victoria Bulletin **28**.

Naiman, R. G., Decamps, H. and Pollock, H., 1993. The role of riparian corridors in maintaining regional biodiversity. *Ecol. Appl.* **3**: 209–12.

Pawsey, C. K., 1966. Birds in relation to the Pine forests of the south east of South Australia. *Sth Aust. Ornithol.* **24**: 93–95.

Pope, M. L. and Lindenmayer, D. B., 1997. Use of a habitat patch network by the Greater Glider within an extensive exotic softwood plantation in south-eastern Australia. Poster Paper in Nature Conservation in Production Environments: Managing the Matrix. Taupo, New Zealand, December 1997.

Ray, S., Robinson, D. and Werren, G., 1983. Pines versus native forests: an annotated bibliography. Native Forests Action Council: Melbourne, Victoria.

Recher, H. F., Rohan-Jones, W. and Smith, P., 1980. Effects of the Eden woodchip industry on terrestrial vertebrates with recommendations for management. *Forest Commission of New South Wales Research Note* **42**: 1–85.

Recher, H. F., Shields, J., Kavanagh, R. P. and Webb, G., 1987. Retaining remnant mature forest for nature conservation at Eden, New South Wales: A review of theory and practice. Pp. 177–94 *in* Nature Conservation: the Role of Remnants of Vegetation ed by D. A. Saunders, G. W. Arnold, A. A. Burbidge and A. J. Hopkins. Surrey Beatty & Sons: Chipping Norton, New South Wales.

Resource Assessment Commission, 1992. Forest and Timber Inquiry. Final Report. Volume 1. Australian Government Printing Service: Canberra, Australian Capital Territory.

Rosenberg, D. K., Noon, B. R. and Meslow, E. C., 1997. Biological corridors: form, function and efficacy. *BioScience* **47**: 677–87.

Routley, R. and Routley, V., 1975. The fight for the forests: the takeover of Australian forests for pines, woodchips and intensive forestry. Research School of Social Sciences, The Australian National University: Canberra, Australian Capital Territory.

Smith, J. M., 1982. Ecological comparisons between pine plantations and native forests, Clouds Creek, New South Wales. Research Series in Applied Geography. University of New England: Armidale, New South Wales.

State of the Environment Australia, 1996. State of the Environment Advisory Council. CSIRO: Melbourne, Victoria.

Suckling, G. C. and Heislers, A., 1978. Populations of four small mammals in Radiata Pine plantations and eucalypt forests in north-eastern Victoria. *Aust. Wildl. Res.* **5**: 305–15.

Suckling, G. C., Backen, E., Heislers, A. and Neumann, F. G., 1976. The flora and fauna of Radiata Pine plantations in north-eastern Victoria. Forest Commission of Victoria Bulletin No. **24**. Forest Commission of Victoria: Melbourne, Victoria.

Taylor, R. J., Bryant, S. L., Pemberton, D. and Norton, T. W., 1985. Mammals of the Upper Henty River region, Western Tasmania. *Papers and Proceedings of the Royal Society of Tasmania* **119**: 7–15.

Tyndale-Biscoe, C. H. and Smith, R. F., 1969. Studies of the marsupial glider, Schoinobates volans (Kerr). III. Response to habitat destruction. *J. Anim. Ecol.* **38**: 651–59.

Walters, C. J. and Holling, C. S., 1990. Large scale management experiments and learning by doing. *Ecology* **71**: 2060–068.

Wardell-Johnson, G. and Roberts, J. D., 1993. Biogeographic barriers in a subdued landscape: the distribution of the Geocrinia rosea (Anura: Myobatrachidae) complex in south-western Australia. *J. Biogeog.* **20**: 95–108.

CHAPTER 6

Can Indonesian production forests play a nature conservation role?

P. Gunarso[1] and J. Davie[1]

As a result of institutional problems in forest resource management in Indonesia, wildlife refuges within production forests are not being effectively managed for the conservation of biodiversity. Regulation concerning conservation in wildlife refuges are incorporated in the contracts which govern the operations of forest concessionaires. During more than 20 years of production forestry under the current system of private industry concessions, the role of wildlife refuges for nature conservation has received little priority. Neither the Directorate-General of Forest Utilization, who have responsibility for the land under the concession, nor the Directorate-General of Forest Protection and Nature Conservation, has been prepared to assume responsibility. At the present time there is a shift in administrative functions within the Ministry of Forestry resulting in a strong decentralization of power to the local government level. We suggest that this shift in policy will create new opportunities for the nature conservation functions of wildlife refuges in forest concessions to be recognized and enforced. This chapter examines these new opportunities for a whole landscape approach to conservation using examples from production forest areas in Kalimantan and Sumatera.

Key words: Biodiversity conservation, Production forest, Landscape ecology, Wildlife reserve, Indonesia.

INTRODUCTION

INDONESIA is the global guardian for the conservation of an extraordinarily rich biodiversity with a high level of endemism. Indonesia covers only 1.3% of the Earth's surface, but it includes 10% of the world's flowering plant species, 12% of the world's mammalian species, 16% of all reptilian and amphibian species, 17% of the world's bird species, and 25% or more of the world's fish species (Ministry of National Development Planning 1993).

Nature conservation in Indonesia is primarily organized through a network of protected areas designated for this purpose (Ministry of National Development Planning 1993). This traditional reservation based approach has a long history extending back to colonial administrations.

The government, private agencies, and non-government organizations have embarked on an ambitious gazettal programme in the last two decades. To conserve the national wealth of biological diversity, the government of Indonesia has set-up a massive network of national parks and reserves. By the end of the *REPELITA VI* (sixth five-year plan) in 1998 it is intended that 19 million ha or 10% of the land surface will be in *Conservation Reserves*, 10 million ha will be in *Marine Reserves*, with a further 30 million ha zoned *Protection Forest* where limited use of secondary forest products by local people is permitted. There will be, for example, 35 national parks across the country covering a total area of more than 8.7 million ha.

Despite this impressive effort, the status of nature conservation in Indonesia remains tenuous. Land encroachment, illegal forest extraction and mining are non-compatible uses which have been inadequately regulated by the Directorate-General for Forest Protection and Nature Conservation, which remains inadequately funded and staffed.

These problems, as significant as they are, have grown even more important because of a disturbing pattern that has increasingly accompanied the management of timber production from forest concessions. Since the beginning of national forest management in 1967, Indonesian production forest

[1]School of Natural and Rural Systems Management, The University of Queensland, Gatton Campus, Queensland, Australia 4345.

Pages 50–59 *in* NATURE CONSERVATION 5: NATURE CONSERVATION IN PRODUCTION ENVIRONMENTS: MANAGING THE MATRIX ed by J. L. Craig, N. Mitchell and D. A. Saunders. Surrey Beatty & Sons, 2000.

had a dual statutory role, not only concerned with sustainable forest production but also providing habitat for wildlife. However, the conservation function has never received priority. As the concessions have been worked through, their value for conservation has been compromised. This situation has now been made even worse by the recent, widespread forest fires. Fragmentation and loss of production forest has thrown new pressures onto the conservation reserve system, which is suffering increased encroachment as a source of forest products for local people.

We believe that the problem consists of three main elements. The first is concerned with the way state forest land is allocated spatially and administered for various sectoral uses, whereas the second and the third relate to the way in which the specific functions of conservation and timber production are conceived and managed. Part of the solution may be found in legislation and policy changes currently being enacted in Indonesia, which transfer more power to local governments. However, the success of such legislation will also depend upon rethinking the current approach to conservation so that nature conservation values outside the hard boundaries of the formal reserves are acknowledged and integrated into the full range of productive land uses administered by other sectors of government.

This chapter elaborates on the problems facing nature conservation in the forest land of Indonesia by reviewing the system of administration of forest lands, and then examining the status of conservation in the protected forest network and in the forest concessions. It then discusses some of the opportunities that may arise from the new trends in government administration now in implementation.

ADMINISTRATIVE AND INSTITUTIONAL CONSTRAINTS TO CONSERVATION AND SUSTAINABLE FOREST USE

Environmental management falls within the mandate of several sectors of Indonesian central government. The State Ministry for Environment has direct responsibility for environmental policy and regulation. The Ministry of Forestry has responsibility as the management authority for protected area networks. Agency for Science and Research has responsibility for scientific documentation, and maintains museums and herbaria. Several other ministries also retain environmental responsibilities by virtue of the impact of their sectoral activities. These include the Ministry of Agriculture, Ministry of Mining and Energy, Ministry of Transmigration and Forest Squad Settlement, Ministry of Transportation, and Ministry of Tourism Post and Telecommunication. The legislation that defines the environmental mandate for each of the ministries is written in terms of its own sectoral responsibilities, and significant contradictions in philosophy and intent make it very difficult to identify co-ordinated policies (Haeruman 1997).

Despite this, the Ministry of Forestry is the main institution dealing practically with the use and conservation of forest resources. The mandate of the Ministry, under the *Forestry Act 1967*, involves a range of functions from assessment of forest resources and research through to commercial production, afforestation and plantations, and nature conservation. Each of these major functional areas is carried out by four Directorates-General. In practice, the Ministry of Forestry has not provided equal opportunities to its Directorates-General in terms of either budgets or human resources. The Directorates-General responsible for forest protection and nature conservation has consistently been under-resourced (Ministry of National Development Planning 1993).

State forest land is administered by three Directorates-General depending on its determined resource use. Resource use is defined under the *Basic Provisions for Forestry Act No. 5 of 1967* which recognizes four categories: *Production Forests, Protection Forests, Reserves*, and *Recreation Forests*. Each function occupies a distinct geographical space and has to be marked as a separate entity on maps produced for this purpose at a scale of 1:250 000. The system of spatial allocation of forest land uses is known as *Tata Guna Hutan Kesepakatan*. Under this system *Production Forests* are further divided into *Permanent Production Forests,* managed by state or private sector timber companies as concessions, and *Convertible Production Forest* which may be converted into Industrial Forest Plantation or any other land use. The network of forest concessions and Industrial Forest Plantation are administered by the Directorate-General for Forest Utilization. *Protection Forests*, *Reserves* and *Recreation Forests* are all administered by the Directorate-General for Forest Protection and Nature Conservation. On currently accepted figures *Permanent Production Forests* account for 64 million ha, *Convertible Forest* 30 million ha, *Protection Forest* 30 million ha and *Reserve and Recreation Forest* 19 million ha. Despite the enormous area of production forest, its potential role for biodiversity conservation has been afforded little importance by either of these two Directorates-General.

Permanent Production Forest is administered through the Ministry of Forestry which imposes a harvesting system known as the Selective Cutting and Replanting System which, when coupled to a requirement for the company to allocate a portion of the concession as wildlife refuge, is designed to ensure sustainable practice. The intent of *Act No. 5* is that *Permanent Production Forests* should retain their values for wildlife conservation while they are being logged. Concessionaires are required to identify some percentage of the forest concession for wildlife refuge, a term which is taken in this chapter to describe locations within production forests that provide shelter, and in the long term, habitat for wildlife to exist in this area.

In practice neither the Selective Cutting and Replanting System nor the system of wildlife refuges are often followed. Day to day management is the responsibility of the concessionaire, and limited resources and personnel, especially outside Java, mean that logging practice is not monitored in the field (Poffenberger 1996). Not surprisingly, the identification and retention of adequate wildlife refuges is considered to be a minor or non-essential activity. In the accounting guidelines for concessionaires, the activity of conservation is considered as a residual activity within the social roles of the concessionaires. It is termed *"other conservation activities . . ."*, a phrase which also encompasses analysis of local village development needs, roadside planting, bare-soil planting, environmental monitoring and environmental management (Kopkar Dephut 1996).

A direct solution to this problem is hampered by administrative blockages, which influence not only the availability of government staff but also their capacity to accept the problem within their job description. Staff of the Directorate-General of Forest Utilization, which is responsible for the administration of forest concessionaires, has no direct responsibility for biodiversity conservation, the remit of the Directorate-General for Forest Protection and Nature Conservation. In their work, *"other conservation activities"* is interpreted as management of environment and is mainly concerned with compliance to the requirements of environmental impact statements, which typically deal only in passing with nature conservation. The personnel of the Directorate-General for Forest Utilization have limited training and exposure to the issues of nature conservation.

In contrast, staff of the Conservation Directorate-General have no right and no direct access to forest concessionaires to monitor the presence of wild flora and fauna. Personnel from this agency are not allowed to enter the state forest under the operation of concessionaires without permission from the company's head office. Thus, while conservation personnel are prepared and well trained in the management of biodiversity conservation, they are prevented administratively from being involved in nature conservation activities within a production forest by the legal contract agreement.

CONSERVATION IN THE PROTECTED AREA SYSTEM IN INDONESIA

The more recent nature conservation legislation, *Act No. 5 of 1990 concerning the Conservation of Living Resources and their Ecosystems,* under which forest protection and conservation now operates, is also based on the principle of spatial land use zoning. In this act, the functions of *Reserves* and *Recreation Forests* are dealt with in a broader range of categories, aligned with accepted Conservation World Union recommendations. Where nature conservation in the absence of people is paramount, the categories of *Cagar Alam* (Strict Nature Reserve) and *Suaka Margasatwa* (Wildlife Sanctuaries) apply, while nature conservation areas that also meet compatible human uses are divided into *Taman Nasional* (National Parks), *Taman Hutan Raya* (Grand Forest Parks), and *Taman Wisata Alam* (Natural Recreation Parks).

Providing the goals of the sixth 5-year plan are met by the end of 1998, it is assumed that all types of habitat will be represented and securely conserved. An analysis of the existing reserves, however, show that lowland tropical forest, mangroves, and coastal areas remain under represented compared to montane forests as was the case when the first national conservation plan was devised (McKinnon and Artha 1982; Ministry of National Development Planning 1993). For example, in Sumatera lowland forest represents only 4% of the protected area, while montane forest represents 33%. Only 3% of the mangrove forests and other coastal vegetation is protected. In Kalimantan, 3% of the lowland rainforest is protected. While 39% of montane forest is protected, no mangrove and coastal areas are protected in Kalimantan. This deficiency is partly due to the fact that lowland tropical forests are largely zoned as economically important *Production Forest* and allocated to private sector forest logging concessions. Serious and wide-spread forest fire in the lowland of Sumatera and Kalimantan in 1994 and 1997 have further reduced the chance of obtaining greater ecosystem representation in the protected area system.

Notwithstanding the problem of skewed representation, the long-term success of the protected area network will also continue to be eroded by the demands for land and timber resources from an expanding population denied access to forest concessions, and the demands by local government for short-term financial returns on land use (Davie and Sumardja 1997). The pressures of encroachment on the 19 million ha of National Parks and other *Conservation Reserves* are severe (Blouch 1996; Carter *et al.* 1996; Wardhana 1997). However, the major threats to the protected area system are focused on the 30 million ha of *Protection Forest* because of the absence, in practice, of management in the field; and the low priority accorded to them in the protected area system. *Protection Forest* is reserved mainly for the purpose of watershed protection, by erosion control on steep slopes and fragile landscapes. Usually *Protection Forest* is selected according to geomorphological characteristics such as steep slope, high elevation, riverfronts or beachfronts. In Kalimantan and Sumatera, where the number of forest concessions are high, *Protection Forests* are under the most threat. In Sumatera destruction is most severe in Riau Province and in Jambi. *Protection Forest* has suffered severely in the current fires (Hendrizal and Noer 1997; Gunarso, unpubl. data 1997).

Like production forest, well-managed protection forest could form an important part of biodiversity conservation as buffer zones or wildlife corridors (Ministry of National Development Planning 1993).

Whereas management for *Protection Forests* remains the mandate of the Conservation Directorate-General, operationally their management has been devolved from the central government to the forest services of the provincial governments. The Provincial Forest Service, especially in provinces with a high number of forset concessions, have limited field staff to manage protection forest.

CONSERVATION IN THE PRODUCTION FOREST SYSTEM IN INDONESIA

In general, production forests are located in lowland rainforest, mangroves, and swamp forests, reflecting not only the high economic value of these forests but the importance of rivers in the transportation of the logs. The poor representation of these ecosystems within the formal conservation reserve network make it more important that all mechanisms by which the *Permanent Production Forests* might contribute to the conservation of biodiversity be explored.

Approximately 97% of the *Production Forest* is found outside Java (Ministry of Forestry 1998). Kalimantan and Sumatera are the two biggest islands in Indonesia's archipelago (Fig. 1) and contribute the most to the enormous commercial tropical forest wealth of the country. Approximately 60% of forest concessions are located in these islands, with species of dipterocarps the principal commercial focus (Ministry of Forestry 1998).

The essential problem for conservation in the Permanent Production Forests on Sumatera and Kalimantan is the same as that facing proponents of sustainable logging practice. During the past decade, it has become common practice on these two islands for many forest concessions to move from *Permanent Production Forest* to *Convertible Production Forest* following the initial cut-over by the logging company. With the transition of zoning the concessions can then be converted to various forest plantation uses, the most common being for fast growing species for pulp and paper, or for oil palm production. This trend clearly cuts across the conservation values of state forests that were considered to be an essential associated use when the original legislation allowing for wildlife refuges in production forest was drawn up. The mechanisms by which this disastrous situation has developed continue to evolve but are illustrated by the following case studies.

The general situation occurring in the forest concessions in Kalimantan and Sumatera is illustrated by recent developments in Sumatera north of the equator (Fig. 1). This is the area in Indonesia where natural forest conversion to plantation is occurring most rapidly. As in Peninsula Malaysia, this area is suitable for rubber and oil palm production, and forest plantation has been the major economic activity for this region since rubber was first planted by the Dutch towards the end of the 19th Century (Whitten *et al.* 1984).

On 1 August 1997 the national newspaper, *Kompas Daily,* reported that 761 820 ha from 15 forest concessionaires were to be released to the State Forestry Company, PT Inhutani IV. Of this total only 85 000 ha are still considered as potentially productive commercial forest. While the remaining area is still under evaluation by the Forestry Department, the initial result has revealed that at least 100 000 ha of these former concessions are highly degraded (*Kompas Daily* 1 Aug. 1997). Evaluation of a further 135

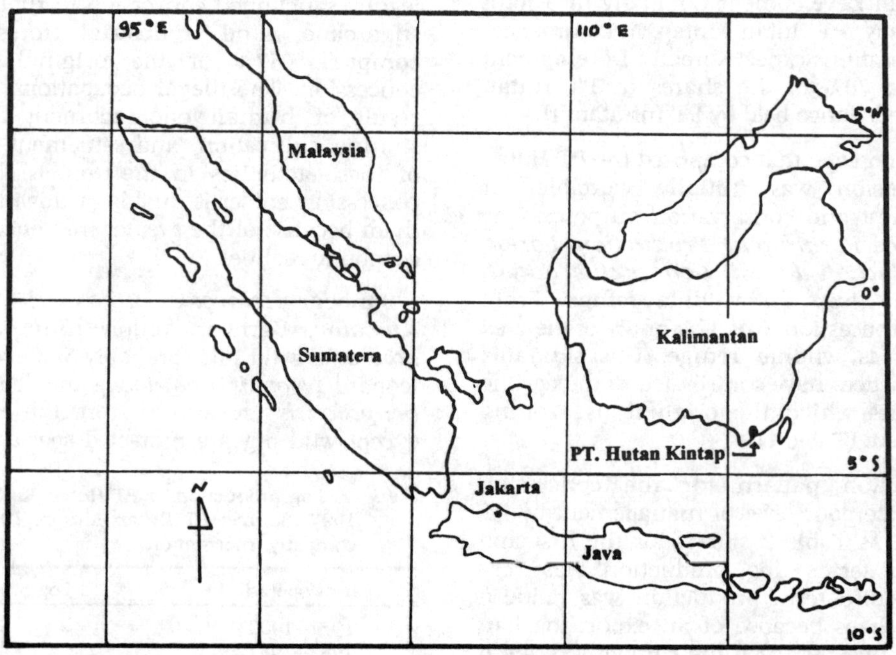

Fig. 1. Three major islands in the western part of Indonesia, showing location of PT Hutan Kintap.

concessions covering about 10 million hectares in Sumatera and Kalimantan reveal a similar situation (*Suara Pembaruan Daily* 7 Aug. 1997).

The decision to hand over this vast area of degraded forest to the State Forestry Company is momentous in its implications, but it is also consistent with Ministry of Forestry policy. Ministry of Forestry policy provides three alternative approaches for concession lease extension. If the forest is in a good condition at renewal time, a lease extension may be granted to the previous leaseholder. If the forest is moderately degraded but restorable given appropriate management, the government may provide technical and management assistance through the State Forestry Company, PT Inhutani, and the lease may be extended in the name of the original company. If the condition of the forest concession is highly degraded, the lease will be terminated and the area given to PT Inhutani to rehabilitate.

The decision for the State Forestry Company to take over a total of 150 concessions was an admission that all had been so poorly managed and degraded that the third policy option had to be applied. In consequence PT Inhutani IV (Northern Sumatera Province) has accepted responsibility for 1.49 million ha, while PT Inhutani II (mainly in Kalimantan) have received nearly 5 million ha (*Suara Pembaruan Daily*, 9 Sept. 1997). Management action over these areas is mainly rehabilitation.

The typical process by which concessions are degraded is well illustrated by the case of PT Hutan Kintap in Kalimantan (Fig. 1). This area is a state-owned forest concession, which was visited by the senior author in August 1997. The concession is managed by the state forest company PT Inhutani II, which has operated for 19 years, in co-operation with the Korean Development Company in a joint venture company, PT Hutan Kintap. At commencement of operations the Korean Development Company held 70% of the shares in PT Hutan Kintap with the balance held by PT Inhutani II.

The forest landscape that comprised the PT Hutan Kintap concession was initially valuable for production forest and conservation purposes. The area was zoned *Permanent Production Forest*, *Limited Production Forest*, *Convertible Forest*, and *Protected Area* for wildlife refuge. From 125 000 ha of concession, 6 400 ha or 5% of the area was assigned as wildlife refuge. Unsustainable production practices have contributed to the decline in forest values which the original use zoning sought to achieve (Table 1).

The production pattern for the concession demonstrates the poor level of management by PT Hutan Kintap. As Table 2 shows, for the first and second 5-year terms, log production was very high. In the third term production was reduced drastically, perhaps because of an export log ban introduced in 1985. In 1996 the company received its last annual cutting permit. The production during this period dropped further.

Table 1. Land use of the PT Hutan Kintap concession after 19 years of operation. Source: PT Hutan Kintap, 1996 (unpublished company information).

Land use	Initial Lease (ha)	Current Condition (ha)
Permanent Production Forest	79 500	16 618 *
Limited Production Forest	8 500	—
Convertible Production Forest	30 600	—
Wildlife Refuge	6 400	6 400 **
Sugarcane Plantation	—	4 505
Rubber Plantation	—	18 142
Industrial Forest Plantation	—	49 000
Mining	—	5 600
Transmigration	—	10 000
Animal Husbandry	—	2 000
Agriculture	—	1 400
Settlement	—	11 335
Total	125 000	

Notes: * No information on standing stock of this remaining area for future operation.
** Currently included as Martapura Pleihari nature reserve.

In 1997, just one year before the first term of the lease agreement expired, all Korean Development Company shares were inexplicably taken over by PT Inhutani II. The decision is clearly very poor in economic terms, as the area of production forest still available in the concession is no longer sufficient to support a commercial logging operation. After 19 years of operation PT Hutan Kintap has only 16 618 ha, 13% of the initial forest area, still capable of producing commercial timber.

The reduction in forest area under the management of PT Hutan Kintap has resulted from legal and illegal settlement, which has turned the land after logging into other non-forest land uses. The legally sanctioned conversion to rubber plantation, sugarcane, and industrial forest plantation comprises 57% of the original area of the concession. The illegal occupations have been the result of human encroachment for fishponds, shifting cultivation, and settlement. The presence of these activities in the area is illegal, yet the concession holder is unable or unwilling to exclude them because of the greater influence of social and economic realities.

Sunbear *Helarctos malayanus*, deer *Cervus unicolor*, Bornean yellow muntjac *Muntiacus atherodes* wild pig (probably *Sus scrofa*), clouded leopard *Neofelis nebulosa*, and leopard cat *Felis bengalensis* are known from the reserve and all except wild pig are protected species. Nevertheless,

Table 2. Log production of PT Hutan Kintap from 1977 to 1997. Source: PT Hutan Kintap, 1996 (unpublished company information).

Year of operation	Log production (m^3)
1977–1982	520 000
1982–1987	497 000
1987–1992	267 000
1992–1997	81 000

during the senior author's visit there was no evidence of any activities designed to conserve these species. The area of 6 400 ha zoned for wildlife refuge at the commencement of the concession has been excluded from the concession and is part of the Pleihari-Martapura Game Reserve. Within the concession, 300 ha have been apparently set aside to conserve germplasm. This is demarcated by a signboard but there is no information available about management. The company has also set aside about 200 ha for enrichment planting of seed trees and has created a vegetative propagation station over an area of 10 ha.

Sempadan Pantai (beachfront) and *Sempadan Sungai* (riverfront) are designated under the forestry act as critical lands, which need to be protected. However, these portions of the concession have also been encroached upon. Beachfront forest has been converted into fishponds, while most of the riverfront vegetation has been illegally cut. Stone and gravel extraction for building material and road construction also takes place in a number of locations, and it is only in these segments of the rivers that riverfront vegetation has been retained.

DISCUSSION: RESOLVING THE ISSUES

The management of production forests in Indonesia is seriously flawed. The private sector concessionaires and the Ministry of Forestry, whose task it is to monitor the activities of the logging companies, share the responsibility for this. *Permanent Production Forest* is being used unsustainably and nature conservation values are being squandered. Much greater levels of enforcement are clearly required; however, the problems of management go much further. In our opinion there are five issues that must be addressed.

Wildlife protection, buffer zones, and other co-functions of multipurpose *Permanent Production Forest* have been ignored. Economic aspects have overshadowed the social and environmental aspects of production forest. Private sector ignorance has not been addressed by the Ministry of Forestry partly because neither the Directorate-General for Forest Utilization nor the Directorate-General of Forest Protection and Nature Conservation has direct interest in the management of nature conservation on land designated as production forest.

The lack of recognition of co-functions in *Permanent Production Forest*, apart from commercial timber, results from a management approach that emphasizes spatial planning through zoning for single uses.

There are *de facto* local government powers that can be applied to the management of natural forest resources. However, in practice the implementation of these powers has not been acknowledged by central government.

Most of the forest concessions in Sumatera and Kalimantan have the potential to provide a valuable complement to the nature reserve network, as they strongly emphasize lowland forest ecosystems where nature conservation areas are poorly represented.

There are insufficient incentives and authorities available to the concessionaires to allow them to defend and protect their concession from both legal and illegal encroachment. Land use alteration by the state-owned company has reduced the concession area and threatened the sustainability of forest production.

These issues can be profitably considered in relation to four areas of discussion. The first involves a change in the approach to conservation, which will see it as an integrated part of the whole landscape and the people's use of it. The second is the need to resolve legislative priority. For example mining and forestry legislation both justify sectoral claims over the same piece of land without clearly giving priority to either bid. The third is the impact that will result from the shift in administrative emphasis and decision making in land and resource use from central government to local government. The fourth is the essential need to restructure forest concession contracts to clarify the responsibilities that must be borne by the concession holders.

Integrating forest production and conservation into the whole landscape

As the population grows and more land is opened up for commercial use there is a growing tension between conflicting land uses where each seek to maximize their single sectoral priority. The capacity of land, especially forest land, to support multiple uses is being compromised, with serious implications for local people, and also long-term ecological sustainability. The task of breaking down the hegemony of the dominant land resource sector in any location is made more difficult because current legislation seeks to resolve conflict through spatial separation. The administrative situation is made worse because each sector has its own Act, and each Act operates at the same level of authority in relation to land use decision making. Furthermore each Act is written in such a way as to legislate the basic principles only, with implementation dependent on Presidential and Ministerial Decrees, which have now become very numerous and frequently conflicting.

The effect of this approach is to skew spatial allocation to the resource sector, which obtains dominance through its perceived capacity to contribute the largest financial returns. Conservation reserves, which are not perceived to make the same economic contribution as other resource sectors, usually are given lower priority, and consequently potentially valuable timber producing forests are under-represented. For example, most production forests in Indonesia are located in lowland and tidal swamp areas. In Kalimantan and Sumatera only 3% and 4%, respectively, of lowland forest types are represented in conservation reserves.

The Kintap case study illustrates how a legal concession can be converted to several different land uses in conflict with its original purpose. Within 20 years the area has been converted from four land uses into ten different land uses. The area that was formerly ideal for sustainable production forest has been converted into secondary and degraded forest, unable to support any use sustainably. Co-functions, such as nature conservation, have been ignored in a rush to maximize short-term economic returns.

Several non-government organizations in Indonesia have sought to promote the nature conservation values of production forest by calling for a ban on further logging of natural forest. This proposal, although it may be appropriate in particular locations, might prove impractical because of political and economic considerations, since the country still needs foreign exchange from the timber industry. Nevertheless, it may spur further consideration of the advantages of forest management approaches that emphasize the importance of multiple-use.

Scientific forestry is based on the principle that production can be maintained in the long term by carefully regulated cutting cycles and extraction methods. The economic consequences of enhanced forest conservation might reduce short-term economic returns. Sustainable forestry assumes that an approach that preserves natural ecological processes (Davie 1997), including those that depend upon the biotic interactions of a rich native fauna and flora, will guarantee much greater long-term economic benefits from timber production. In addition such an approach will open up opportunities for new commercial endeavours from secondary forest products, which draw upon biodiversity resources.

The practical integration of conservation functions into production forest must be improved through modifications in concession management. Government controls must be tightened through better monitoring to ensure that the operation of forest concession is complying with the regulations acknowledged in the individual contracts. At present the regulatory functions that might be implemented by the Directorates-General of Forest Utilization and Forest Protection and Nature Conservation are only partially integrated. The direct role for Forest Protection and Nature Conservation in production forest zones is limited to evaluation of environmental impacts of developments and forest protection related to the use of secondary products. A nature conservation role for Forest Protection and Nature Conservation in production forest is not legally permitted in the job descriptions of these staff, as production forest is outside their mandate. Responsibility for nature conservation in production forest rests with the concession holders who deal only with staff from Directorate-General for Forest Utilization who have no expertise upon which to base advice.

Resolving the priority legislation

As forest land, and its biological and other natural resources, becomes increasingly diminished, the imperative for multiple use management is growing. The historical approach, in Indonesia, to the allocation of resources through their spatial separation by single use zoning, is under pressure through recent legislation, yet remains as the major obstacle to sustainable forest management.

During the 1960s, state land was separated functionally between its uses as forest (allocated to the Ministry of Forestry), and its uses for all other purposes: agriculture, human settlement, industry or special purposes such as for the military. As the major landholder in the country, the Ministry of Forestry has become the most powerful sectoral agency politically, and has been responsible for the biggest non-oil export product. The political and economic influence of the Ministry of Forestry has increasingly depended on its ability to defend its system of forest resource allocation, the *Tata Guna Hutan Kesepakatan,* against the land use needs of other sectors and other levels of government.

The wisdom of the *Tata Guna Hutan Kesepakatan* system has come under scrutiny in recent times for several reasons. One has been the cause of social justice and traditional resource rights of local people living within forest concessions. Another has been the inadequacy of the *Forestry Act* to deal with the mandate given to the Ministry for the protection of marine and coastal areas. New legislation is required for this purpose. *The Act No. 5 of 1990 Concerning Conservation of Living Organisms and their Ecosystems* is not sufficient, as issues dealt with under environmental legislation (*Act No. 4 of 1982*), administered by the State Ministry for Environment, are also of key importance.

The priority of the forest sector on state forest land has been strongly challenged by the *Spatial Use Management Act No. 24* introduced in 1992. The key roles in implementing this act are the responsibility of the Ministry of National Development Planning/National Development Planning Agency, the Ministry of Home Affairs and the State Ministry for Environment. This act sets out principles that have the intent of empowering local people and local government. Spatial planning process integrates forest resource use with other sectoral interests, to achieve a resource management system that acknowledges principles of ecological sustainability. This process requires ratification by the province governor. Significantly the implementation of this act is based on the premise that the forest sector should have the same status as all other resource sectors in relation to resource use decisions. In other words, the Ministry of Forestry, representing the forest sector, should be regarded as a *land user,* like agriculture or mining or industry, not, as previously, a forest land resource *decision maker*.

The *Spatial Planning Act* replaces the *Tata Guna Hutan Kesepakatan* system of agreed land use boundaries with a new system of boundaries identified in spatial plans, produced at a series of scales depending on the level of government.

Thus the planning cycle requires *Bupati* (district head) to have a *Kabupaten-DT II* (district level) plan produced at a scale of 1:50 000 to 1:100 000, which is then incorporated into a *Propinsi Dati I* (provincial level) plan requested and approved by the Governor at a scale of 1:250 000. These land use maps, while of course strongly influenced by the existing *Tata Guna Hutan Kesepakatan* system (mapped at 1:250 000 to 1:500 000), contrast with it by being at a much smaller scale and therefore much more accurate on the ground. Inconsistencies between the boundaries specified on the Spatial Plan and the *Tata Guna Hutan Kesepakatan* Plan are frequently so significant that there is conflict between the Ministry of Forestry and the Provincial authorities. In its struggle to protect its investment in boundary marking based on *Tata Guna Hutan Kesepakatan,* the Ministry of Forestry claims that the *Forestry Act* should have legal priority over the new act. The way in which this issue is resolved will be particularly interesting in view of the political imperative for decentralization, which is embodied in the modern *Spatial Planning Act.*

In practice it is the Provincial Governor (as head of province and concurrently as head of provincial government administration) and the *Bupati* (as head of district and concurrently as head of district government administration) who have the power to determine how land development proceeds, regardless of the Ministry of Forestry. The dual status of these positions, termed *penguasa tunggal di daerah* provides each with sole authority in their region and thereby greater power than the elected House of Representatives, at each respective level of government (*Act No. 5 of 1974*). The Governor and the *Bupati* are directly responsible to the President and to the Governor respectively, and not to their elected House. Consequently, recommendations by the Governor or Bupati are critical to the official process. As the scale of the land use maps is so small that land use and tenure conflicts have to be resolved on the ground, it is essential for this *de facto* decision making process within state forest lands to be acknowledged, if land and forest resource allocation is to be regularized, and responsive to professional as opposed to political recommendations. No matter what checks and balances are introduced at the national level to optimize multiple use approaches and sustainability, it will be local issues and requirements that continue to drive resource use decisions.

Shift in administration

The forestry sector is one of 19 resource use sectors, which have been required to devolve their tasks from central government to the *Kabupaten Dati II* (district level) under *Government Regulation No. 8 of 1995*. Under this instruction, specified functions are carried out by a civil service answerable through the province Governor to the Ministry of Home Affairs. The decentralization process must involve the reallocation of personnel, budget and infrastructure.

At a structural level the Ministry of Forestry has responded to the devolution order satisfactorily and has set up a pilot project in several *Kabupaten* (districts) in several provinces. In the selected administrations, autonomous status has been accorded and the forestry agency is called the *Dinas Kehutanan Dati II* (District Forest Service). Elsewhere, full devolution has not yet occurred and the agency is termed the *Dinas PKT Tingkat II* (Forestry and Land Conservation Service).

However, at a functional level the devolution is not yet working effectively, as Central Government staff obfuscate to maintain their own power bases. The principal outcome of this is that only five tasks have been devolved and these are not only marginal to the core business of forest production, for example bee-keeping, but provide no opportunity to increase local income. Not surprisingly, local governments are unwilling to assume full responsibility as there are insufficient new funds to support the greater level of responsibility.

There is no doubt that any opportunity to halt forest destruction, to conserve native biodiversity and to establish an internationally acknowledged sustainable forestry industry, will depend to a large degree on how the Ministry of Forestry faces the challenge of decentralization. The history and traditions of the civil service have long focused power and wealth in Jakarta. Nevertheless, unless trained staff and adequate infrastructure are located in the outer islands where the production forests exist, there can never be the level of regulatory control necessary to deal with the problems of illegal cutting, encroachment and conversion, and forest fires, which are now rapidly eating into the resource capital.

Restructure forest concession responsibilities

There are presently over 600 forest production management units in Indonesia, of which more than 500 are held as permanent production forest concessions. A further 100 are held by *Hutan Tanaman Industri* (industrial forest plantation) companies operating in the *Convertible Production Forest*. Until there are sufficient professional government foresters in the field, where they will be able to conduct the necessary level of supervision and monitoring, it will continue to be necessary to rely on the activities and records of the concessionaire. Incentives are required to encourage companies to take greater responsibility for the sustainability of their operations.

Several incentives have been suggested and these require serious consideration (Ministry of Forestry 1997). For example, forest concessions presently exist for only one cutting cycle and thus do not encourage any emphasis on implementing progressive, successful regeneration. It might be argued that the only way a concessionaire could guarantee a continuing income from his forestry activities is to engage in, or to encourage conversion to other uses particularly forest. Strip cutting rather than selective logging has also been suggested as an alternative approach.

Whatever policy is adopted the key to its effectiveness will depend on sufficient numbers of local staff. Longer contract agreements will require more intensive control, monitoring and supervision. Decentralization will put the forester in charge of monitoring and control near to the forest where time and cost will not limit his access.

CONCLUSION

If Indonesia is to meet its goals and international responsibilities for biodiversity conservation substantial reform is required in its approach to nature conservation. One of the most important areas for reform is in the management of its permanent production forest, where the inadequacy of current practices is resulting in the loss of forest habitat through conversion to a range of other non-sustainable uses.

Responsibility for the loss of forest habitat is shared between the private and public sectors. A greater level of responsibility by the private sector is constrained by organization and administration within the Ministry of Forestry, which remains ineffective through a centralization of expertise and power in Central Government.

Decentralization is a key process allowing for effective local level resource management. Despite the economic implications for small, centrally located interest groups, it is necessary for the Ministry of Forestry to show the national leadership which will hasten the shift in administrative power to the regions, where it is required. Continuing delay in power devolution will accelerate the loss of permanent production forest, removing any opportunity for a sustainable tropical forest industry to establish, prevent any future capacity to regulate or deal with the massive and growing problem of annual forest fire, and severely curtail the capacity of the country to reach its goal in the conservation of biodiversity.

Recent national trends to decentralization mean that district level government is the centre point for institutional and resource allocation reform. This level of governance is concerned with a landscape scale amenable to practical management because there is some degree of ecological, social and economic integration. At this level, land use planning that allocates priority land uses can be better understood in practical ways, and the manner in which land and its resources are used, can be regulated to achieve sustainable and multiple use functions. *Bappeda Dati II* (district development planning bureau) have the mandate, and in certain instances have demonstrated that they can think strategically to provide the sectoral co-ordination necessary to make such an approach work.

ACKNOWLEDGEMENTS

The research upon which this chapter is based has been supported by the International Tropical Timber Organization, which have provided a postgraduate studentship for the first author at the School for Natural and Rural Systems Management, the University of Queensland. The authors are grateful for the support offered by professional colleagues in Indonesia.

REFERENCES

Blouch, R. A., 1996. The Friends of Kutai National Park: Enlisting support of local industries in protected area management. Proc. of the Conference, *Conserving Biodiversity Amidst Growing Economic Prosperity*, Cisarua and Ujung Kulon National Park, 12–18 May 1996. Directorates General of Forest Protection and Nature Conservation, and Ministry of Environment, Republic of Indonesia, International Union for the Conservation of Nature and United Nations Education, Scientific and Cultural Organization: Paper No. 7, Technical Session II.

Carter, R. W., Kearny, S., Willis, D. and Lee, S., 1996 Ecotourism Strategy for Komodo National Park. University of Queensland (Gatton College) Department of Natural and Rural Systems Management, Lawes, Australia.

Davie, J., 1997. Is biodiversity really the link between conservation and ecologically sustainable management? A reflection on paradigm and practice. *Pac. Conserv. Biol.* **3**: 83–90.

Davie, J. and Sumardja, E., 1997. The protection of forested coastal wetlands in Southern Sumatera: a regional strategy for integrating conservation and development. *Pac. Conserv. Biol.* **3**(4): 366–78.

Gunarso, P., 1997. Fieldwork Report on Hutan Bakau Pantai Timur, Jambi Indonesia (non-published data).

Haeruman, H., 1997. Keterpaduan Konsep Pengelolaan Daerah Aliran Sungai (DAS) dan Hutan Lindung dengan Pengembangan Wilayah Dalam Hubungannya dengan Otonomi Daerah.

Hendrizal and Noer, MM, 1997. Mengungkap Jaringan Pencurian Kayu. *Sinar Weekly News* 2 August, 1997.

Kompas Daily, 1997. Rusak Total, 100,000 Hektar Hutan Eks HPH di Sumatera. *Kompas Daily* 1 Aug. 1997. Online at http://www.kompas.com/9708/01/Daerah/rusa.htm

Kopkar Dephut, 1996. Akuntansi Kehutanan dan Pedoman Pelaporan Keuangan Pengusahaan Hutan (Seri Cetakan). Koperasi Karyawan Departemen Kehutanan bekerja sama dengan Yayasan Adum Wanabangsa: Jakarta.

MacKinnon, J. and Artha, S., 1982. National Conservation Plan for Indonesia. Vol. 1 Introduction, Evaluation Methods and Overview of National Nature Richness. Report No. FO/INS/78/GOI(34). FAO: Bogor, Indonesia.

Ministry of Forestry, 1997. Contry Brief Indonesian Forestry Action Programme (IFAF). Ministry of Forestry Republic of Indonesia: Jakarta.

Ministry of Forestry, 1998. Statistik Kehutanan Indonesia 1996/1997.Biro Perencanaan, Departemen Kehutanan Republik Indonesia: Jakarta.

Ministry of State for Population and Environment Republic of Indonesia (MSPE), 1993. Act of The Republic of Indonesia Number 24 of 1992 Concerning Spatial Use Management.

Ministry of National Development Planning, 1993. Biodiversity Action Plan For Indonesia. Ministry of National Development Planning/National Development Planning Agency: Jakarta.

Poffenberger, M., 1997. Rethinking Indonesian Forest Policy. *Asian Survey* **XXXVII**(5): 453–69.

PT Hutan Kintap, 1996. Data dan Informasi Mengenai PT Hutan Kintap (unpublish company information). Disusun dalam rangka kunjungan Komisi IV Dewan Perwakilan Rakyat Republik Indonesia ke Propinsi Kalimantan Selatan, 23 Oktober 1996. PT Hutan Kintap, Sungai Baru, 1996.

Suara Pembaruan Daily, 1997. 135Eks HPH Swasta Diserahkan ke BUMN. Suara *Pembaruan Daily,* 9 September 1997.

Wardhana, Dadang, 1997. Nature Conservation and Neighbouring Communities. Unpublished M.Sc thesis, School of Natural and Rural Systems Management, The University of Queensland.

Whitten, A. J., Damanik, S. J., Anwar, J. and Hisyam, N., 1984. The Ecology of Sumatra. Yogyakarta: Gadjah Mada University Press.

CHAPTER 7

The short-term impacts of timber harvesting and associated activities on the abundance of medium-sized mammals in the Jarrah forest of Western Australia

Keith Morris,[1] Brent Johnson,[1] John Rooney[2] and Colin Ward[2]

The Jarrah forest and woodlands of the south-west of Western Australia have been subject to timber harvesting over the last 120 years and are now being managed by the Department of Conservation and Land Management for multiple use, including conservation, recreation, water harvesting and timber production. Several species of mammal that once had extensive distributions over much of Australia now persist only in the Jarrah forest. The Kingston study commenced in 1994 to determine the impact of timber harvesting and associated activities on the fauna and vegetation of the Jarrah forest. The impact on medium sized mammal abundance was assessed through a Before, After, Control, Impact study, with trapping grids established in three replicated silvicultural treatments (shelterwood, gaps with habitat trees, and gaps without habitat trees) with internal and external controls. Road trapping transects were necessary to adequately survey the wide ranging Chuditch and to assess mammal abundance at a broader, forest block level. Radiotracking of arboreal species was used to determine the refuge requirements and direct impact of timber harvesting on these species. Fox control was implemented across all sites.

This chapter reports on the short-term impacts of timber harvesting on medium-sized mammals in the Kingston forest. Fox control, using toxic baits, reduced fox activity and resulted in increased abundances of Woylie *Bettongia penicillata,* Quenda *Isoodon obesulus,* Brushtail Possum *Trichosurus vulpecula* and Chuditch *Dasyurus geoffroii.* After timber harvesting some declines in abundance occurred, however, none of the species disappeared from the study area. Shelterwood thinning reduced abundance of the arboreal Brushtail Possums in the short term, but had little impact on the ground dwelling Woylie and Quenda. Habitat tree retention was important in maintaining Brushtail Possum abundance in areas cut to gap. Although post harvest, regenerative burns reduced Brushtail Possum and Quenda abundance in unlogged buffers these are still considered important refuges in production forest and a source of recolonizing fauna, and their appropriate management is necessary for the maintenance of mammal diversity in the Jarrah forest. In the presence of high numbers of Woylies, trap saturation is an important consideration for monitoring less abundant species such as Chuditch. An apparent post harvesting reduction in Chuditch trap success was believed to be related to trapping methodology rather than logging or burning. Limitations in undertaking experimental research in production areas of the diverse Jarrah forest are discussed.

[1]CALMScience Division, Department of Conservation and Land Management, P.O. Box 51, Wanneroo, Western Australia, Australia 6065.
[2]CALMScience Division, Department of Conservation and Land Management, Manjimup, Western Australia, Australia 6258.
Pages 60–70 *in* NATURE CONSERVATION 5: NATURE CONSERVATION IN PRODUCTION ENVIRONMENTS: MANAGING THE MATRIX ed by J. L. Craig, N. Mitchell and D. A. Saunders. Surrey Beatty & Sons, 2000.

INTRODUCTION

JARRAH *Eucalyptus marginata* forests and woodlands once covered approximately 6 400 000 hectares in the south-west of Western Australia (Bradshaw *et al.* 1991). Although most have now been cleared for agriculture, the majority of the better quality Jarrah forest in the higher rainfall zones (Abbott and Loneragan 1983) has been kept in public ownership, either as multiple use State Forest (approximately 1 100 000 hectares) or conservation reserve (approximately 512 000 hectares), and is currently managed by the Department of Conservation and Land Management (Shea 1997). Logging of the Jarrah forests for its high quality hardwood has occurred since the establishment of the Swan River settlement in 1829, and the history and evolution of silvicultural systems employed is detailed in Bradshaw *et al.* (1991) and Heberle (1997). Over the last 10 years, 10 000–20 000 ha of Jarrah forest have been harvested each year (Department of Conservation and Land Management 1997). Current regimes of gap creation and thinning (shelterwood creation) have been operational since 1985 (Stoneman *et al.* 1989). Post harvesting burning is undertaken to remove logging debris and to create conditions favourable for regeneration of Jarrah and Marri. Since the 1950s, prescribed burning on a 7–10 year rotation has also been undertaken to reduce fuel loads. Approximately 10% of the Jarrah forest is prescribed burnt each year (Department of Conservation and Land Management 1997).

Twenty species of ground dwelling mammal are known from the Jarrah forest (Nichols and Muir 1989), and this area is important to mammal conservation in Western Australia as many of the medium-sized marsupials (35 g–5 500 g; Burbidge and McKenzie 1989) that were once widespread across Australia, are now restricted to it, or have core populations restricted to it. These include the Chuditch (Western Quoll) *Dasyurus geoffroii*, Woylie (Brush-tailed Bettong) *Bettongia penicillata*, Numbat *Myrmecobius fasciatus*, Tammar Wallaby *Macropus eugenii*, Quokka *Setonix brachyurus*, Quenda (Southern Brown Bandicoot) *Isoodon obesulus fusciventer*, Western Ringtail Possum *Pseudocheirus occidentalis* and Brush-tailed Phascogale *Phascogale tapoatafa tapoatafa*. No extinctions of mammal species are known from the Jarrah forest (Nichols and Muir 1989: Armstrong and Abbott 1995), unlike the adjacent Wheatbelt and Swan Coastal Plain regions where significant extinctions of mammals have occurred (Kitchener *et al.* 1978; Kitchener *et al.* 1980).

While there is an extensive literature on the impact of prescribed burning on Jarrah forest wildlife, including medium sized mammals (Christensen 1980; Christensen and Abbott 1989; Inions *et al.* 1989), little is known about the impact of logging and post logging regenerative burns on the mammalian fauna (Wardell-Johnson and Nichols 1991). The only published study to date is that of Rhind (1996) who examined the habitat tree requirements and the effects of their removal on the Brush-tailed Phascogale in the Kingston and adjacent forest blocks. She was able to quantify habitat tree requirements for this species and showed that phascogales ceased nesting in those parts of their territories affected by logging. Her criticism of Department of Conservation and Land Management for not leaving the correct number of habitat trees per hectare was challenged and found to be inaccurate (Armstong and Abbott 1996), and this was subsequently retracted.

The principles of ecologically sustainable forest use require the development and implementation of policies which maintain ecological processes, maintain biodiversity and optimize benefits to the community. Department of Conservation and Land Management's Forest Policy Statement (Department of Conservation and Land Management 1992) embodies these principles, viz *To manage the native forests of the south-west of Western Australia, in consultation with the community, so that they provide the values required by society while sustaining indefinitely, their biological and social diversity*. In order to implement this policy, there is a need for basic ongoing research to gain scientific knowledge of forest ecosystem processes and how these are impacted by forestry operations (Department of Conservation and Land Management 1992). There is also a need to monitor the implementation and impact of these operations to gauge the success of its current policies in regard to achieving sustainable forest use.

The Kingston study was established to investigate the ecological impacts of logging and associated burning operations on a range of fauna including invertebrates, small terrestrial vertebrates, birds and medium-sized mammals, as well as the flora and vegetation (Burrows *et al.* 1993). Through determining which factors contribute to observed impacts on this biota, prescriptions will be developed and/or modified to ensure the appropriate sustainable management of timber harvesting in the Jarrah forest. This chapter presents and discusses the results of the first three years of the study of the impacts of timber harvesting and associated activities on the abundance of medium-sized mammals in the Jarrah forest.

METHODS

Study site

This study was undertaken in the Kingston, Warrup, and Winnejup forest blocks (the Kingston Study area), approximately 25 km north-east of Manjimup, in the south-west of Western Australia (Fig. 1). These forest blocks cover approximately 15 000 ha and lie at the transition from high quality to low quality, Jarrah forest (Abbott and Loneragan 1983). Broadly they comprise open forest and woodland dominated by Jarrah *Eucalyptus marginata* and Marri *Corymbia calophylla*, however, Mattiske and Havel (1998) have identified seven different vegetation complexes within the study area. Annual rainfall is approximately 900 mm and soils are primarily lateritic with deeper loams in riparian zones. Previously, the area had been selectively cut over for good quality Jarrah timber in the 1920s and between 1940 and 1970 (Heberle 1997). Some areas remained unlogged. Prescribed fuel reduction burns on a 7–10 year rotation have been implemented since the mid 1960s.

This study site was selected as previous faunal surveys had shown that several species of medium-sized mammals occurred in the area at low abundances (G. Liddelow, pers. comm.), and an extensive fox control programme covering 30 000 ha had commenced in June 1993. Medium sized mammal species known to occur in the Kingston study area include Chuditch *Dasyurus geoffroii*, Quenda *Isoodon obesulus fusciventer*, Woylie *Bettongia penicillata*, Brushtail Possum *Trichosurus vulpecula*, Western Ringtail Possum *Pseudocheirus occidentalis*, Brushtail Phascogale *Phascogale tapoatafa tapoatafa* and Numbat *Myrmecobius fasciatus*. Currently the Chuditch, Western Ringtail Possum and Numbat are listed as threatened species under the Western Australian *Wildlife Conservation Act 1950*. At the start of this study (mid 1993) the Woylie and Quenda were also threatened species but have since been downgraded to a lower, more secure conservation status (Start *et al.* 1996; Morris *et al.* 1998).

Based on studies elsewhere in the Jarrah forest (Morris *et al.* 1995; de Tores 1998), fox control was expected to result in an increase in medium-sized mammal abundances before logging and burning treatments were imposed, thus enabling any impact due to these treatments to be detected. It was unlikely that any impact could have been detected if low mammal abundances had been retained through not implementing a fox control programme. Fox control now occurs throughout most of the Jarrah forest as part of the Western Shield fauna recovery programme (Bailey 1996). The site was also suitable because timber harvesting over the next decade or so would focus on similar landform types as found in the Kingston forest.

Timber harvesting coupes were selected by Department of Conservation and Land Management district and regional staff during 1993 using standard Departmental prescriptions (Department of Conservation and Land Management 1995). Trapping grid site selection and establishment followed this process and was completed by March 1994. Within the Kingston block, sampling sites were located in areas to be harvested to gaps and shelterwood. Attempts were made to sample the range of typical medium rainfall Jarrah forest habitat types found within that landform/soil unit. Control sites were established both within the Kingston forest block and in the adjoining Warrup and Winnejup forest blocks and these were withheld from harvesting for the duration of this study. Timber harvesting was programmed to commence at the end of 1994.

Fox control

Fox control using dried meat baits impregnated with 4.5 mg of sodium fluoroacetate (1080) commenced in the Kingston study area (and adjacent forest blocks) in June 1993. Until May 1997 baits were laid from the ground at intervals of 250 m along the network of forest tracks. After May 1997, fox baits were delivered by aircraft at a density of five baits per square kilometre. Baiting was undertaken every three months. Fox baiting intensity was equal across all impact and control sites.

The effectiveness of fox control was estimated using fox activity on sand pads (1 m × 1 m) as an index of abundance in unbaited forest (Alco and Thornton blocks; 25 km west of the study area) and baited forest (Kingston and Winnejup blocks). A network of sand pads was established: a) at sites on the forest/farmland boundary (edge), and b) at a minimum distance of three kilometres from forest/farmland boundary (core). Three transects,

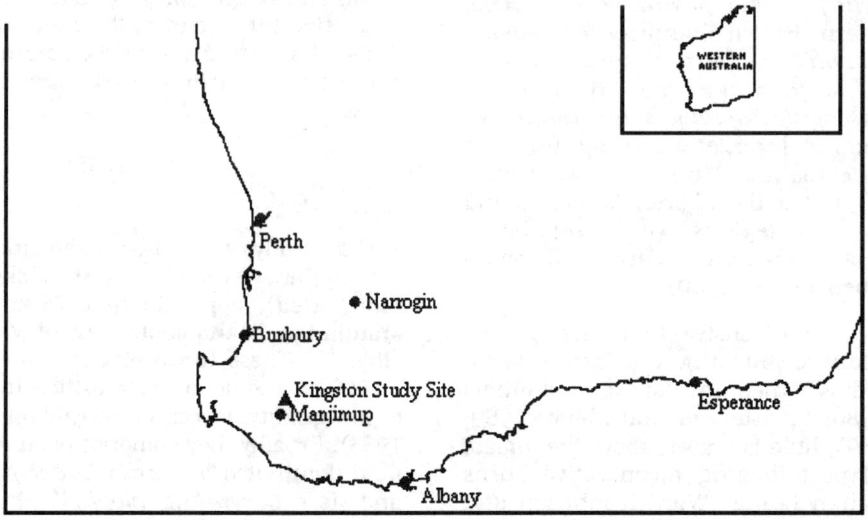

Fig. 1. Map of the south-west of Western Australia, showing the Kingston study site.

each of seven sand pads at 500 m spacings, were established on the edge and in core areas of baited and unbaited forest blocks. A non-toxic meat bait was buried in each sand pad and the number of sand pads disturbed by foxes recorded daily on all transects over a 10 day period in September 1996 and again in September 1997. The total number of sandpads disturbed in baited and unbaited forest blocks was compared using a Chi-square test.

Mammal sampling

The overall study design was of the Before, After, Control, Impact type, with replicates of each of four treatments. The four treatments imposed were: (i) Control, (ii) Shelterwood harvesting in 5–50 ha patches, (iii) Gaps of up to 10 ha created with three marked habitat trees retained per hectare, (iv) Gaps of up to 10 ha created without marked habitat trees retained. The controls were further divided into external controls, at least 1 km from the impact sites (no logging or burning), and buffer controls adjacent to impact sites (no logging).

Trapping grids to sample medium sized mammals and smaller vertebrates were established in impact and control sites and each given an alphanumeric code based on the logging block number (Table 1). Each treatment was replicated and all normal prescriptions pertaining to follow-up silvicultural work and prescribed burning were applied.

Each trapping grid comprised a core area of 80 × 80 m in which a 3 × 5 grid of pitfall traps (20 litre Rheem buckets; 25 cm wide, 40 cm deep) with a 7 m flywire fence, was established. Superimposed over this was a 3 × 5 grid of medium Elliott mammal traps (Elliott Scientfic Co., Upwey Victoria) at 40 m spacing and a further 3 × 3 grid of Sheffield wire cage traps (Sheffield Wire Co., Welshpool Western Australia) at 80 m intervals. Total grid area was 2.56 ha. All grids were sampled for three consecutive nights over a two week period. Because of time constraints and the need to clear traps before midday, the grids were split into northern and southern sectors, with the southern grids sampled in the first week and the northern grids sampled in the second week.

In addition to grid sampling, medium-sized mammals were sampled at a broader scale using road transects. Sheffield wire cage traps were set at 200 m intervals for 36 kilometres along tracks throughout the study area and operated for four consecutive nights during the same period as grid sampling. Nineteen per cent of trap points were adjacent to areas logged during this study. This technique is particularly suitable for sampling species such as Chuditch, whose movement areas greatly exceed those of the trapping grids. Road transects also provided an indication of the response of all medium-sized mammals to fox control.

This study examined only the Woylie (a mycophage), Quenda (an omnivore), Brushtail Possum (a foliovore/omnivore) and Chuditch (a carnivore). Because Western Ringtail Possums (a foliovore) rarely enter cage traps, the impact of timber harvesting on this threatened species is being studied by radiotelemetry in the Kingston 4 area (A. Wayne, pers. comm.). Abundance of Numbats (a termitophage) at Kingston is being assessed as part of the Numbat recovery plan (Friend 1994). Abundance of the Brushtail Phascogale (an insectivore) was examined during this study but will be reported on separately.

All captured animals were individually marked with ear tags, sexed, measured, weighed and released. Selected individuals were fitted with radio-transmitters (Biotrack two stage mortality transmitters) to determine home range, den usage, movement pattern and mortality data. Abundance was estimated using Known To Be Alive and percentage trap success rates. The Known To Be Alive method is highly suitable for studies such as this where capture and recapture rates are high, and frequent long-term sampling using the same trap effort, is undertaken (Krebs 1966).

All trapping and animal handling techniques were approved by the Department of Conservation and Land Management Animal Ethics Committee in October 1993 (CAEEC # 19/93).

The method used to assess and measure the impact on mammal abundance was based on the

Table 1. Treatments and sampling grids used in the Kingston study.

Treatment	Forest Block	Grid	AMG Location (Eastings/Northings)
External control	Warrup	C1	436585/6220888
(no logging or burning)	Warrup	C2	437344/6222685
	Winnejup	C3	438076/6229257
	Winnejup	C4	439398/6228809
With-in coupe control — buffers	Kingston	K1–3	438872/6228080
(no logging)		K5–3	441965/6221465
Shelterwood harvesting	Kingston	K3–1	441542/6227698
(logged and burnt)		K5–6	441943/6222783
Gap creation with habitat trees	Kingston	K1–4	437208/6227804
(logged and burnt)		K2–1	440742/6228145
		K5–4	441669/6223990
Gap creation without habitat trees	Kingston	K1–1	439287/6228095
(logged and burnt)		K5–1	441989/6221103

Before, After, Control, Impact method of Underwood (1981). Differences of Known To Be Alive were compared between treated and control sites, before, during and after impact (logging and regenerative burns), using single degrees of freedom contrasts. A 5% level of significance was used. The "before" period spanned six months between May 1994 and November 1994 (five sample points), the "during" period spanned 23 months between December 1994 and November 1996 (13 sample points), and the "after" period spanned six months between December 1997 and May 1997 (three sample points).

Timber harvesting operations

All harvesting prescriptions were set by operational staff according to standard Department of Conservation and Land Management silvicultural guidelines (Bradshaw 1986; Department of Conservation and Land Management 1995), except that at two sites (K1–1 and K5–1), all marked habitat trees were removed so that this study could examine the effects of habitat tree retention for arboreal species. Two major timber harvesting regimes are currently employed in the Jarrah forest. Gap creation is used to promote Jarrah lignotuber growth, and up to 10 ha of forest is clearfelled except for the retention of habitat trees of certain specifications (>70 cm dbh, senescence stage 4–6, Department of Conservation and Land Management 1995). Until 1996 habitat tree retention was at the rate of three per hectare. This was changed, as a result of findings of other aspects of the Kingston study, to four trees per hectare to cater better for the needs of arboreal species (Department of Conservation and Land Management 1995). Shelterwood cutting is employed where there is little Jarrah lignotuber development and the objective is to establish seedling regeneration. Habitat trees are also retained in shelterwood thinning.

Influence by researchers was limited to minor timetable adjustments to allow finalization of pre-impact data collection. Contact with logging contractors was limited to general observation and collection of anecdotal information. Contractors were advised to ignore any flagging or other signage that did not relate to harvesting so that no effort was made to avoid sites or refuge trees.

Timber harvesting commenced in December 1994. Most harvesting work ceased with the onset of rains during April/May 1995. The following summer saw some continuation of activity in coupes that had not been completed, with major new activity concentrated on grids within the K2 area. All timber harvesting relevant to grid sampling was completed during early 1996. Post harvesting burning (tops burn) was undertaken on all sites (except external controls) in December 1996.

RESULTS

Fox control

Fox activity was reduced in the baited forest compared with unbaited forest areas in both 1996 and 1997 (Table 2), however, this difference was only significant throughout (edge and core areas) the forest in 1996 (Chi-square = 58.25, $p < 0.005$). In 1997, the baited core areas did show a significant reduction in fox activity compared with unbaited forest (Chi-square = 4.33, $p < 0.05$) however, there was no difference in fox activity on the edges of baited and unbaited forest. The total number of sand pads disturbed by fox in the unbaited area declined from 84 in September 1996 to 38 in September 1997 and this apparent reduction in fox abundance in the unbaited area may have been associated with the commencement of fox baiting in adjacent forest blocks in May 1997 as part of the Western Shield programme (Bailey 1996).

Table 2. The total number of sand pads disturbed daily by fox in unbaited and baited forest blocks over a ten day period.

Treatment	September 1996	September 1997
Unbaited		
Edge	14	12
Core	70	26
Total	84	38
Baited		
Edge	6	11
Core	4	13
Total	10	24

Since fox control was implemented in the Kingston study area in 1993, there has been a steady increase in the abundance of all medium sized mammals (Table 3), with trap success rates increasing overall from 7% to approximately 60% over four years. Woylies were the most abundant species, accounting for approximately 66% of all captures and Chuditch the least abundant accounting for approximately 1.5% of all captures.

Refuge sites and mortality

BRUSHTAIL POSSUM *Trichosurus vulpecula*

Radiotracking of 49 Brushtail Possums in the forest prior to logging showed that 73% of refuge sites selected (91 sites) were hollows in standing trees and 27% (34 sites) were ground sites such as hollow logs, burrows and hollow stumps. An average of 3.2 trees were used by each possum in the period March 1994 to October 1995, and some sharing of trees with other Brushtail Possums and Ringtail Possums *Pseudocheirus occidentalis* occurred; however, the extent of this has not yet been quantified.

Mortality of Brushtail Possums before, during and after timber harvesting operations (logging component only) in control and logged areas is shown in Table 4. No mortality occurred in control areas during this period, while 67% of radiocollared possums died during or soon (within seven days) after shelterwood logging operations. All mortality appeared to be related to injury caused by logging operations i.e., crushed by equipment, or dead in fallen tree hollows. Lower mortality rates (33–35%)

Table 3. Trap success rates (%) for medium sized mammals along road transects during the Kingston timber harvesting study. Chuditch figures in brackets are trap success rates obtained using "Chuditch" bait, rather than universal bait.

Date	Treatment	Woylie	Quenda	Possum	Chuditch	Total
Aug. 1992		2.7	1.1	2.6	0.8	7.2
	Fox control commenced					
Apr. 1994		4.3	4.4	7.2	0.3	16.2
June 1994		9.0	8.6	5.4	1.8	24.8
Oct. 1994		15.9	7.7	7.7	1.1	32.4
Dec. 1994		18.1	9.4	12.8	0.6	40.9
	Timber harvesting commenced					
Feb. 1995		20.4	12.8	18.5	1.4	53.1
May 1995		18.5	11.7	16.7	0.8	47.7
Aug. 1995		26.7	13.9	13.8	0.7	55.0
	Timber harvesting completed					
Dec. 1995		32.0	10.9	18.6	1.9	63.4
Feb. 1996		31.0	12.5	24.2	1.0	68.7
May 1996		20.8	9.0	26.1	2.4	58.3
Aug. 1996		30.4	10.8	21.6	0.2	63.0
	Post logging burn					
Dec. 1996		36.9	8.7	16.1	0.7	62.4
Feb. 1997		43.2	10.5	5.6	0.2	59.5
May 1997		36.2	7.5	16.5	0.2	60.4
Aug. 1997		42.4	5.0	10.5	0.1 (1.4)	58.0
Dec. 1997		41.9	2.9	10.0	0.4 (0.8)	55.2

were recorded in areas of gap creation where habitat trees had been retained and where they had been removed.

WOYLIE *Bettongia penicillata*

Between March 1994 and October 1995 up to 10 Woylies at a time were radiotracked to 89 refuge locations. All were above ground nests constructed of bark, leaves and other vegetation. An average of 7.4 nests were used by Woylies during this period. Post logging refuge sites included the "tops" debris left after logging operations. No mortality was recorded during logging operations.

QUENDA *Isoodon obesulus fusciventer*

Because of problems associated with fitting radiocollars to Quenda for extended periods of time only up to four were radiotracked at any one time over the period April 1994 to October 1995. Fifty-four refuge sites were located; 67% were above ground nests, often under *Macrozamia* shrubs, 20% were in burrows, including disused rabbit warrens, and 13% were in hollow logs.

Two mortality events were recorded. One occurred prior to logging and was attributed to Chuditch predation on the basis of a Chuditch scat being associated with the carcass on a log, the other occurred after logging and was believed to have been the result of predation by a cat.

CHUDITCH *Dasyurus geoffroii*

A total of 18 Chuditch (up to 12 at any one time) were radiotracked during the period April 1994 to October 1995. Forty-three refuge sites were located; 67% were in hollow logs, 30% were in burrows and 3% were in above ground nests.

Five mortality events were recorded. One was predated (either as a live or dead animal) by a Wedge-tailed Eagle. The others were found intact and in unlogged habitat, but within 1 km of logged areas. One of these was found close to a road and may have been a road kill.

Abundance

BRUSHTAIL POSSUM *Trichosurus vulpecula*

A summary of the Brushtail Possum Known To Be Alive estimates before, during and after treatments are shown in Table 5. Numbers on the control grids increased throughout the experiment in response to

Table 4. Mortality of Brushtail Possums before and after logging.

Treatment	Number of possums tracked	Number dead before logging commenced on impact sites (% mortality)	Number dead following logging on impact sites (% mortality)
Control (no logging)	5	0	0
Gap with habitat trees	15	2 (13)	5 (33)
Gap without habitat trees	20	1 (5)	7 (35)
Shelterwood	9	0	6 (67)
TOTAL	49	3 (6)	18 (37)

Table 5. Summary of Known To Be Alive estimates for Brushtail Possums before, during, and after treatments (mean ± S.E). (*p* values * = significant at the 5% level).

Treatment	Known To Be Alive Before	Known To Be Alive During	Known To Be Alive After	% change Known To Be Alive (after-before)/ before × 100	ANOVA *p* value (after treatment *vs* control)
Control	3.8 ± 0.1	6.1 ± 0.1	6.8 ± 0.1	+79.8	
Buffer	10.4 ± 0.2	13.6 ± 0.2	8.9 ± 0.2	−14.4	0.027 *
Gap + HT	2.9 ± 0.2	7.9 ± 0.2	5.0 ± 0.2	+72.4	0.980 ns
Gap — HT	7.3 ± 0.2	10.0 ± 0.2	5.3 ± 0.2	−27.4	0.027 *
Shelterwood	7.4 ± 0.2	8.3 ± 0.2	6.0 ± 0.2	−18.9	0.011 *

fox control. During treatments (during and post logging, and pre-burn) possum numbers remained stable, compared with abundances in the control areas, in areas logged to gap without habitat trees retained, and in areas logged to shelterwood. Numbers increased significantly in areas logged to gap with habitat trees retained ($F = 5.54$, $p = 0.028$). After treatments, abundance declined significantly in the areas logged to gap with no habitat trees retained ($F = 5.66$, $p = 0.027$), in areas logged to shelterwood ($F = 7.75$, $p = 0.011$) and in the unlogged buffers ($F = 5.62$, $p = 0.027$). Compared with controls, there was no significant decline in abundance in areas logged to gap with habitat trees retained.

QUENDA *Isoodon obesulus fusciventer*

A summary of Quenda Known To Be Alive estimates before, during and after treatments is shown in Table 6. Quenda abundance on all trapping grids was generally low, usually between one and three individuals. Timber harvesting alone did not appear to impact Quenda abundance, however, there was a significant decline in abundance compared to controls in the unlogged buffers after regeneration burns had been implemented ($F = 7.11$, $p = 0.013$).

WOYLIE *Bettongia penicillata*

A summary of Woylie Known To Be Alive estimates before, during, and after treatments is shown in Table 7. Compared to controls there were no significant changes in Woylie abundance at any of the treatment sites. Woylie abundance increased at all sites irrespective of the timber harvesting treatment imposed.

CHUDITCH *Dasyurus geoffroii*

Chuditch trap success rates throughout this study were low, ranging between 0.2 and 1.5% (Table 3), while those for other species, particularly Woylies, were high (50–60%). Woylies, Brushtail Possums and Quenda are all readily attracted to universal peanut butter bait and usually occupy traps early in the evening. Chuditch, because of their lower numbers and wide ranging nature may not encounter a trap until later in the evening, by which time most traps were either occupied or disturbed, and not available for entry. A trial involving a comparison of Chuditch and other mammal capture rates on the standard universal bait of peanut butter/oats/sardines mix and a "Chuditch" bait of meat meal/fish oil/sardines/oats was undertaken in

Table 6. Summary of Known To Be Alive estimates for Quenda before, during, and after treatments (mean ± S.E). (*p* values * = significant at the 5% level).

Treatment	Known To Be Alive Before	Known To Be Alive During	Known To Be Alive After	% change Known To Be Alive (after-before)/ before × 100	ANOVA *p* value (after treatment *vs* control)
Control	1.0 ± 0.1	2.7 ± 0.2	2.9 ± 0.2	+190.0	
Buffer	2.7 ± 0.2	3.3 ± 0.2	2.0 ± 0.3	−25.9	0.013 *
Gap + HT	1.6 ± 0.3	2.6 ± 0.2	2.4 ± 0.2	+50.0	0.251 ns
Gap — HT	3.0 ± 0.2	3.0 ± 0.2	3.9 ± 0.3	+30.0	0.057 ns
Shelterwood	0.2 ± 0.2	0.9 ± 0.2	0.9 ± 0.2	+350.0	0.094 ns

Table 7. Summary of Known To Be Alive estimates for Woylies before, during, and after treatments (mean ± S.E).

Treatment	Known To Be Alive Before	Known To Be Alive During	Known To Be Alive After	% change Known To Be Alive (after-before)/ before × 100	ANOVA *p* value (after treatment *vs* control)
Control	1.8 ± 0.2	4.4 ± 0.1	6.1 ± 0.2	+238.8	
Buffer	0.6 ± 0.2	2.4 ± 0.2	3.6 ± 0.3	+500.0	0.720 ns
Gap + HT	3.2 ± 0.2	3.7 ± 0.2	4.7 ± 0.3	+46.8	0.281 ns
Gap — HT	0.8 ± 0.2	1.5 ± 0.2	2.9 ± 0.3	+262.5	0.440 ns
Shelterwood	4.0 ± 0.2	5.8 ± 0.2	4.8 ± 0.3	+20.0	0.139 ns

August and December 1997 to determine if Chuditch trap success rates were related to bait type.

In August 1997, one individual Chuditch was trapped using universal bait (0.1% trap success) and eight trapped using the Chuditch bait (1.4% trap success). In December 1997, 4 individuals were trapped using the universal bait (0.4% trap success) and nine trapped using the Chuditch bait (0.8% trap success). Trap success rates for other mammals were reduced from approximately 60% to 20–40%. These results (Table 8) suggest that the Chuditch specific bait does reduce capture rates of other, more trappable medium-sized mammals and enhances the trap success rate for Chuditch. This trial will be continued for at least 12 months to assess fully the value of the Chuditch specific bait.

Table 8. Preliminary results of a trial examining the value of different baits on trap success rates for all medium sized mammals and Chuditch.

Bait type	August 1997 Trap success rate (%) All mammals	August 1997 Trap success rate (%) Chuditch	December 1997 Trap success rate (%) All mammals	December 1997 Trap success rate (%) Chuditch
Standard universal bait	62.1	0.1	61.8	0.4
"Chuditch" bait	26.1	1.4	42.3	0.8

A summary of Chuditch Known To Be Alive estimates before, during, and after treatments is shown in Table 9. Because Chuditch ranged throughout the study area and were not restricted to treatment sites, Known To Be Alive estimates could not be derived for specific treatments. When only the results for trapping using universal bait were used there was a significant decline in Chuditch abundance following the regeneration burn ($F = 12.83$, $p = 0.002$). However, the increase in number of Chuditch trapped in August and December using Chuditch bait (Table 8) suggest that Known To Be Alive estimates derived using universal bait underestimates Chuditch abundance in the presence of high numbers of Woylies and other medium-sized mammals.

Table 9. Summary of Known To Be Alive estimates for Chuditch (mean ± S.E).

Known To Be Alive Before	Known To Be Alive During	Known To Be Alive After	Known To Be Alive % change (after-before)/ before × 100
9.4 ± 1.0	8.4 ± 0.9	3.4 ± 1.0 (universal bait)	−63.8
		6.0 ± 1.1 (Chuditch bait)	−36.2

DISCUSSION

The Kingston study is the first to examine the impact of timber harvesting on native mammals in the Jarrah forest. While some declines in abundance were observed, none of the species examined disappeared from the study area. Tree hollow-dependent species are considered to be among the most vulnerable fauna to the effects of timber harvesting in eucalypt forests (Gibbons and Lindenmayer 1997). Rhind (1996) reports a 17% mortality rate of Brushtail and Ringtail Possums after logging in an area adjacent to the Kingston forest. Of the species examined in this study, the arboreal Brushtail Possum was the most impacted by timber harvesting in the Jarrah forest. Shelterwood harvesting, in particular, resulted in the mortality of approximately two thirds of radiocollared possums, and following all timber harvesting activities, the population at these sites was reduced by approximately 20%. In areas of gap creation where habitat trees had not been retained, 35% of radiocollared possums died during or soon after logging, and the population declined by approximately 27%. However, in areas of gap creation where habitat trees were retained (the normal prescription) possum abundance did not decline despite a 33% mortality rate of radiocollared animals. This suggests that where high numbers of Brushtail Possums occur, such as in the Kingston forest, these levels of mortality are compensated for by immigration from the adjacent unlogged buffers, and road and stream reserves. The long-term management of these sites are obviously important in the maintenance of medium- sized mammal diversity in production forest areas. Appropriate fire management and length of logging rotation, in particular, are likely to be important.

Brushtail Possums are capable of utilizing a wide range of refuge sites including ground sites (Kerle 1984), particularly where foxes are absent or in low numbers. However, in the Jarrah forest, tree hollows are a significant proportion of the refuge sites and the maintenance of sufficient numbers of these are also important for the persistence of this and other arboreal species in the Jarrah forest. This study has demonstrated the importance of habitat tree retention for maintaining possum abundance and the retention of four habitat trees per hectare is standard operational practice for timber harvesting in the Jarrah forest (Department of Conservation and Land Management 1995). The rationale for increasing habitat tree retention from three per hectare to four per hectare was based on information gained using radiotelemetry on the density of possums, the number of habitat trees used by each possum, and the extent of overlap in habitat tree use by possums, and will be presented in a separate publication.

The ground dwelling Woylie and Quenda were not impacted by current operational logging practices of gap creation with retention of habitat trees and shelterwood thinning. However, post logging regenerative burns did reduce Quenda abundance in the unlogged buffer areas, but not in the adjacent logged areas. This species utilizes above ground refuge sites and could be regarded as being vulnerable to fire. In the logged areas refuge sites were protected beneath piles of logging debris while in the adjacent buffers, refuge sites were relatively flammable above ground nests. The low intensity regenerative burns do not consume all the logging debris piles, but would destroy above ground nests. This may have led to a high mortality of Quenda in the buffers. The maintenance of Quenda abundance in logged areas may also be related to an increase in food abundance in these areas compared to the unlogged, burnt buffers. Quenda have increased throughout the Jarrah forest, including production areas, in response to fox control and have now been removed from the state's threatened fauna list (Morris *et al.* 1998).

Woylies have the capacity to increase rapidly in abundance and this is one of the reasons they were removed from threatened fauna lists in 1996 (Start *et al.* 1996). This, and their liking for universal bait make them extremely trappable in many parts of the Jarrah forest, thus presenting problems monitoring other medium sized mammals with cage traps. Chuditch, which only occur at relatively low densities in the Jarrah forest even following fox control, are difficult to monitor in the presence of high numbers of Woylies. A preliminary trial undertaken as part of this study indicates that Chuditch trap success rates can be enhanced using traps baited with a meat meal based bait rather than a bait based on peanut butter. The capture of eight and nine individuals in August and September 1997 respectively following the completion of timber harvesting suggest that this disturbance has not had a detrimental impact on this threatened species. Persistence of Chuditch in other parts of the Jarrah forest which have been subject to timber harvesting in the past, also suggests that this species is not detrimentally impacted by this activity. As the largest native mammalian predator in the forest (Orell and Morris 1994), the presence of this species is a good indication of the productivity of the Jarrah forest ecosystem. Ground hollow logs, which are an important refuge site for Chuditch (Orell and Morris 1994) are retained in logged areas as part of the Jarrah forest silvicultural guidelines (Department of Conservation and Land Management 1995).

Fox predation has been shown to be a major factor in the decline of medium sized mammals in Western Australia (Kinnear *et al.* 1988; Friend 1995). This study and those of de Tores (1998) and Morris *et al.* (1995) have demonstrated that where fox control is implemented in the Jarrah forest, medium-sized mammal abundances increase significantly. Maintenance of fox control regimes is probably the most important management strategy for fauna conservation in the Jarrah forest. May and Norton (1996) suggested there may be a relationship between increased fox activity and disturbance in eucalypt forests. There was no evidence of increased fox predation of the species studied here after logging had occurred and abundances of Woylie, Brushtail Possum and Quenda along road transects increased during this period. Following the post logging burn in December 1996, fox activity did increase in the study area, but not significantly and trap success rates for all medium-sized mammals have been maintained at high levels (60 to 70%) throughout this period. However, other species, such as the Ringtail Possum, appear to be more vulnerable to fox predation following logging disturbance (A. Wayne, pers. comm.) and it may be appropriate to increase fox control in areas where such species occur, and where disturbances such as timber harvesting are planned.

Undertaking any study into the impact of operational activities, such as timber harvesting, on fauna in an ecologically diverse habitat such as the Jarrah forest is a difficult task. This study aimed to examine the impact of operational timber harvesting and as such, study sites were located where treatments were to be implemented — treatments were not implemented to suit study sites. Sampling periods, for example the length of the pre-treatment sampling period, are often a compromise between operational requirements and scientific protocol. The ability to set up a true experiment with impact and control sites identical in all respects except the treatment, is virtually impossible in the Jarrah forest because of its inherent variability. Seven different vegetation associations have been recognized in the Kingston study area by Mattiske and Havel (1998). Replicate sites can only be located in at broad vegetation types, usually with differing fauna abundances. Control sites that were not logged or burnt during this study, had been disturbed previously. Sizes of grids and number of replicates are a compromise between being adequate to sample the range of vertebrate fauna and the size and location of the area to be logged. When this study was established in 1993/1994, medium-sized mammal populations were still sparsely distributed in the Jarrah forest, and there were few sites available that were programmed to be logged that also contained a range of fauna in suitable numbers to detect change. All these factors need to be considered when planning and analysing the results of such a study.

While some short-term logging impacts have been observed during this study, medium-sized mammal diversity has not been reduced as a consequence of timber harvesting. Fauna monitoring will continue at Kingston indefinitely to fully evaluate longer term impacts of timber harvesting on medium-sized mammals in the Jarrah forest. It is anticipated that, providing fox control is continued, mammal abundances will return to pre-disturbance levels within a few years, as they have in other parts of the Jarrah forest that have previously been logged and burnt (for example Perup, Batalling, and Julimar forest blocks, and the sites of de Tores (1998) in the northern Jarrah forest). A retrospective study examining fauna abundances in forest of known disturbance age is also planned to commence in the next few years.

ACKNOWLEDGEMENTS

Gordon Friend and Peter Orell played an active role in the design, establishment and implementation of the first two years of this study. Adrian Wayne contributed to the latter parts of field work and analysis of results. Matt Williams provided advice on the experimental design and statistical analyses, and commented on drafts of this chapter. Department of Conservation and Land Management Manjimup District staff provided support with the design and implementation of this study. Brian Whitred, in particular has been a regular contributor. The study would not have been possible without the help of a vast number of volunteers who willingly gave their time, often in trying circumstances. The Forest and Wood Products Research and Development Corporation provided funding towards the Kingston Project in 1996 under Project PN 002.95. Figure 1 was prepared by Mr Peter Speldewinde, Department of Conservation and Land Management Science Division. Dr Ian Abbott, Mr Lachie McCaw and two anonymous referees provided useful comments on earlier drafts of this chapter.

REFERENCES

Abbott, I. and Loneragan, O., 1983. Growth rate of jarrah (*Eucalyptus marginata*) in relation to site quality in cut-over forest, Western Australia. *Aust. For.* **46:** 91–102.

Armstong, J. A. and Abbott, I., 1995. Sustainable conservation — a practical approach to conserving biodiversity in Western Australia. Pp. 21–28 *in* Conservation Through Sustainable Use of Wildlife ed by G. C. Grigg, P. T. Hale and D. Lunney. Centre for Conservation Biology: The University of Queensland.

Armstong, J. A. and Abbott, I., 1996. Habitat trees and logging — ecology of the Brush-tailed Phascogale in jarrah forest. Unpublished report, Science and Information Division, Department of Conservation and Land Management: Crawley, Western Australia.

Bailey, C., 1996. Western Shield — bringing wildlife back from the brink of extinction. *Landscope* **11:** 41–48.

Bradshaw, F. J., 1986. Silvicultural guidelines for virgin southern Jarrah forests. Technical Report No. 4. Department of Conservation and Land Management: Perth, Western Australia.

Bradshaw, F. J., Adams, R., Sneeuwjagt, R., Low, K., Havel, J., Bartle, J. R. and Stoneman, G. L., 1991. The Jarrah Forest: A case study in multiple use. *In* Forest Management in Australia ed by F. H. McKinnell, E. R. Hopkins and J. E. D. Fox. Surrey Beatty & Sons: Chipping Norton, New South Wales.

Burbidge, A. A. and McKenzie, N. L., 1989. Patterns in the modern decline of Western Australia's vertebrate fauna: causes and conservation implications. *Biol. Conserv.* **50:** 143–98.

Burrows, N., Friend, G., Morris, K., Stoneman, G., Craig, M. and Williams, M., 1993. An integrated study of the effects of timber harvesting on the Jarrah forest ecosystem. Unpublished report, Science and Information Division, Department of Conservation and Land Management: Perth, Western Australia.

Department of Conservation and Land Management, 1992. Management strategies for the south-west forest of Western Australia: A review. Department of Conservation and Land Management: Perth, Western Australia.

Department of Conservation and Land Management, 1995. Silvicultural practice in the Jarrah Forest. Silvicultural Guidelines 1/95. Department of Conservation and Land Management: Perth, Western Australia.

Department of Conservation and Land Management, 1997. Annual Report 1996–1997. Department of Conservation and Land Management: Perth Western Australia.

Christensen, P. E., 1980. The biology of *Bettongia penicillata* Gray 1837 and *Macropus eugenii* Desmarest 1817 in relation to fire. *Forest Department Bulletin* **91**, Western Australia.

Christensen, P. E. and Abbott, I., 1989. Impact of fire in the eucalypt forest ecosystem of southern Western Australia: a critical review. *Aust. For.* **52:** 103–21.

de Tores, P., 1998. Control and Ecology of the Red Fox in Western Australia: Prey Response to 1080 Baiting over Large Areas. Unpublished Report to Environment Australia — National Feral Animal Control Program, Canberra.

Friend, J. A., 1994. Recovery Plan for the Numbat. Report for the Endangered Species Program, Australian Nature Conservation Agency. Western Australian Department of Conservation and Land Management.

Friend, J. A. and Thomas, N. D., 1995. Reintroduction and the Numbat recovery programme. Pp. 189–98 *in* Reintroduction Biology of Australian and New Zealand Fauna ed by M. Serena. Surrey Beatty & Sons: Chipping Norton, New South Wales.

Gibbons, P. and Lindenmayer, D. B., 1997. Developing tree retention strategies for hollow-dependent arboreal marsupials in the wood production eucalypt forests of eastern Australia. *Aust. For.* **60:** 29–45.

Heberle, G., 1997. Timber harvesting of Crown Land in the south-west of Western Australia: an historical review with maps. *CALM Science* **2:** 203–24.

Inions, G. B., Tanton, M. T. and Davey, S. M., 1989. Effects of fire on the availability of hollows in trees used by the Common Brushtail Possum *Trichosursus vulpecula* Kerr 1792, and the Ringtail Possum *Pseudocheirus peregrinus* Boddaerts 1785. *Aust. Wildl. Res.* **16:** 449–58.

Kerle, J. A., 1984. Variation in the ecology of *Trichosurus:* its adaptive significance. *In* Possums and Gliders ed by A. P. Smith and I. D. Hume. Australian Mammal Society: Sydney, New South Wales.

Kinnear, J. E., Onus, M. L. and Bromilow, R. N., 1988. Fox control and rock-wallaby population dynamics. *Aust. Wildl. Res.* **15:** 435–50.

Kitchener, D. J., Chapman, A. and Baron, G., 1978. Mammals of the northern Swan Coastal Plain. Pp. 54–92 *in* Faunal Studies of the Northern Swan Coastal Plain: A Consideration of Past and Future Changes. Western Australian Museum and Department of Conservation and Environment.

Kitchener, D. J., Chapman, A. and Muir, B. G., 1980. The conservation value for mammals of reserves in the Western Australian Wheatbelt. *Biol. Conserv.* **18:** 170–207.

Krebs, C. J., 1966. Demographic changes in fluctuating populations of *Microtus californicus*. *Ecol. Monog.* **36:** 239–373.

Mattiske, E. M. and Havel, J. J., 1998. Vegetation mapping in the South West of Western Australia. Regional Forest Agreement — Vegetation Complexes, Pemberton, Western Australia (1:250,000). Department of Conservation and Land Management: Perth, Western Australia.

May, S. A. and Norton, T. W., 1996. Influence of fragmentation and disturbance on the potential impact of feral predators on native fauna in Australian forest ecosystems. *Wildl. Res.* **23:** 387–400.

Morris, K. D., Orell, P. and Brazell, R., 1995. The effect of fox control on native mammals in the Jarrah forest, Western Australia. Pp. 141–45 *in* Proceedings of the 10th Australian Vertebrate Pest Control Conference, Hobart, Tasmania. Department of Primary Industries and Fisheries: Tasmania.

Morris, K. D., Armstong, R., Orell, P. and Vance, M., 1998. Bouncing Back — Western Shield Update. *Landscope* **14**: 28–35.

Nichols, O. and Muir, B. G., 1989. Vertebrates of the Jarrah forest. *In* The Jarrah Forest: a Complex Mediterranean Ecosystem ed by B. Dell, J. J. Havel and N. Malajczuk. Kluwer Academic Publishers: Dordrecht, Boston, London. 408 Pp.

Orell, P. and Morris, K. D., 1994. Chuditch recovery plan. Wildlife Management Program No. 13. Department of Conservation and Land Management: Perth, Western Australia.

Rhind, S. G., 1996. Habitat tree requirements and the effects of removal during logging on the marsupial Brush-tailed Phascogale (*Phascogale tapoatafa tapoatafa*) in Western Australia. *West. Aust. Nat.* **21**: 1–22.

Shea, S. R., 1997. The integrated approach to conservation, public land and wildlife management and commercial forestry — case study Western Australia. Invited paper presented at the New Zealand Forest Research Institute's 50th Jubilee Forestry Celebration Day, 2 April 1997.

Start, A. N., Courtenay, J. and Morris, K. D., 1996. It's back — The return of the Woylie. *Landscope* **11**: 10–15.

Stoneman, G. L., Bradshaw, F. J. and Christensen, P., 1989. Silviculture. *In* The Jarrah Forest: a Complex Mediterranean Ecosystem ed by B. Dell, J. J. Havel and N. Malajczuk. Kluwer Academic Publishers: Dordrecht, Boston, London. 408 Pp.

Underwood, A. J., 1981. Beyond Before, After, Control, Impact: Experimental designs for detecting human environmental impacts on temporal variations in natural populations. *Aust. J. Mar. Freshw. Res.* **42**: 569–87.

Wardell-Johnson, G. and Nicholls, O., 1991. Forest wildlife and habitat management in southwestern Australia: knowledge, research and direction. *In* Conservation of Australia's Forest Fauna ed by D. Lunney. The Royal Zoological Society of New South Wales: Mosman, New South Wales.

CHAPTER 8

Road networks and forest spatial patterns: comparing cutting-sequence models for forestry and conservation

Richard T. T. Forman[1] and Andrew D. Mellinger[2]

Numerous ecological attributes of a forested landscape are affected by the spatial arrangement of not only logged areas, but also the road network. In searching for an ecologically optimum road network, we compared road density, routes for hauling logs, forest interior area, largest forest core, and total boundary length in models of five common or promising cutting regimes for a landscape. DISPERSED cutting (small cuts widely dispersed) had the greatest road density and boundary length, plus the least forest interior and smallest forest core, and therefore was ecologically the least desirable model. The CORRIDOR cutting approach (central cut strip followed progressively by parallel strips on opposite sides of it) was also poor. The EDGE model (parallel strips from one edge) represented the most ecologically desirable road network, whereas the JAWS model (L-shaped cuts spreading progressively from two adjacent edges) was best for forest spatial patterns. Overall the JAWS logging regime for a landscape appears to be best for maintaining desirable biodiversity, soil and water conditions. A combination of road network and forest pattern provides a better ecological evaluation of a cutting sequence or regime than does forest pattern alone.

Key words: Forest interior, Forest road, Forestry, Landscape ecology, Logging regime, Nature conservation, Road density, Road ecology, Road network, Spatial model, Spatial pattern.

INTRODUCTION

FORESTS are great reservoirs of species on land. They also hold the soil against raindrop energy and provide most of our clear streams and rivers. Yet, seemingly paralyzed, we watch the forests shrink before our eyes. In only one decade, the 1980s, world-wide forest land declined by 2% (World Resources Institute 1996), and the drumbeat of falling trees seems only to accelerate. That deforested 100 million hectares covered an area equal to Egypt, or to Prince Edward Island, Texas and Colorado combined. As global forest area shrinks the extinction of species, from plants to insects and amphibians to birds, continues unchecked (DeGraaf and Miller 1996). Furthermore, soil pours out from forest roads and clearcuts. Streams and rivers, the circulatory system of the land, are clogged with muddy sediment and contain impoverished natural fish populations.

Sustainability of forested landscapes requires a meshing of ecological and economic dimensions over the long term (Dale *et al.* 1994; Wiersum 1995; Chomitz and Gray 1996; Noble and Dirzo 1997). Increasingly, it appears that spatial arrangement will be at the heart of this meshing in any region (Forman 1995; Gallup *et al.* 1998). Short-term economics that keeps pace with an added billion people each decade now is the prime driving force for overall deforestation, as well as determining the spatial pattern of logging within forests.

We use the term ecological in the broad sense of "organisms interacting with their environment," which includes, e.g., trees and people affecting soil, water and vertebrates. Ecological evaluations of large forest areas undergoing rapid change increasingly capitalize on the landscape ecological focus on spatial patterns in the landscape. Thus

[1]Harvard University, Graduate School of Design, Cambridge, MA 02138, USA.
[2]Present address: Center for International Development at Harvard University, 79 JFK Street, Cambridge, MA 02138, USA.

Pages 71–80 *in* NATURE CONSERVATION 5: NATURE CONSERVATION IN PRODUCTION ENVIRONMENTS: MANAGING THE MATRIX ed by J. L. Craig, N. Mitchell and D. A. Saunders. Surrey Beatty & Sons, 2000.

the sizes, shapes and arrangements of cuts and forested areas are surrogates for the specific ecological attributes of biodiversity and natural process (Franklin and Forman 1987; Mladenoff *et al.* 1993; Wallin *et al.* 1994). Spatial processes in a changing landscape, such as habitat fragmentation, perforation, dissection, shrinkage and disappearance, are critical to the ecological effects (Forman 1995). Since logging patterns imprint the land for decades to centuries, understanding and guiding these patterns is critical (Wallin *et al.* 1994; Baker 1995).

A forest road network virtually always precedes and accompanies the logging process, yet curiously the road network is seldom included in the ecological evaluation of the logging process (Li *et al.* 1993; Reed *et al.* 1996; Tinker *et al.* 1998). Roads produce different and additional ecological effects than do the logging patterns (Forman and Alexander 1998). For example, increases in water runoff and length of surface water channels (Wemple *et al.* 1996), peak flows downstream (Jones and Grant 1996), intensive soil compaction (McNabb 1994), erosion and sedimentation (Rice and Lewis 1991; Gullison and Hardner 1993; Montgomery 1994), and sources and sinks for fire are typical (Forman and Hersperger 1996; Forman 1999).

Soil erosion bears special mention because roads commonly cause more erosion than the logged areas in a forest (Swanson and Dyrness 1975; McCashion and Rice 1983; Swanson *et al.* 1987; Rice and Lewis 1991; Grayson *et al.* 1993). In fact, it is not unusual for forest roads to produce 20 times more eroded sediment than cut areas. The arrangement of the road network is an important determinant of this ratio.

Road access into remote forested areas is of special significance because, while it facilitates the economic benefits of wood extraction, human access often leads to deforestation and disturbance of the remaining forest (Dale *et al.* 1994; Chomitz and Gray 1996; Forman *et al.* 1996). The arrangement of forest roads has frequently been evaluated for efficiency and economic benefits [Minamikata 1977 (cited by Cha *et al.* 1991); Peters 1978; Sessions and Yeap 1989; Cha *et al.* 1991]. Yet in a sustainability context we must also ecologically evaluate road networks. Indeed, sustaining both nature and wood extraction in a particular landscape over human generations requires careful conservation of its soil, water, microclimate, vegetation, food webs and ecological flows. Although techniques such as cable logging and helicopter logging that offer some valuable ecological advantages are used in spots, most of the world's disappearing forest land succumbs to road networks and large clearcuts.

Therefore in this chapter we will evaluate models of varied logging regimes from an ecological perspective, using both road network attributes and forest spatial patterns. The general approach is to develop simple spatial models of landscape change representing alternative logging regimes for a forest rotation. Such a model applied to dispersed-patch cutting has been used to evaluate diverse ecological attributes from fire and blowdowns to birds and biodiversity (Franklin and Forman 1987; Hansen *et al.* 1992; Li *et al.* 1993; Reed *et al.* 1996).

Additional models that mimic alternative spatial arrangements of cuts have been useful for comparisons, and have identified specific mosaic sequences (termed EDGE, JAWS, and JAWS-AND-CHUNKS) as especially suitable ecologically (Forman 1995; Forman and Collinge 1996). The only experimental test of such dynamic spatial models, using a changing micro-landscape and assaying grassland insects, emphasized the importance of maintaining at least one large natural-vegetation patch throughout the sequence (Collinge and Forman 1998).

Adding the road network to these forest pattern models should produce results significantly closer to those in real landscapes. A model of dispersed-patch cutting in the United States Pacific Northwest pointed out an increased level of aggregation of forest cuts when the road network was added (Li *et al.* 1993). In a forested landscape in Wyoming (United States of America) the road network created more forest edge area than did the dispersed-patch logging pattern, and significantly reduced the average size of forested patches (Reed *et al.* 1996). Roads appeared to have a greater impact on this landscape pattern than did clearcutting, an especially pronounced effect when roads were evenly distributed over the land rather than densely clustered in limited areas (Tinker *et al.* 1998).

Roads and highways cause a wide range of ecological effects (Reck and Kaule 1993; Forman 1995, 1999; Forman *et al.* 1997). These impacts extend different distances outward from a road, indeed from meters to kilometers. However, forestry roads are a special subset, characterized by mainly being narrow, unpaved, lightly travelled and remote (Forman and Alexander 1998).

The primary ecological impacts of forestry roads appear to be (1) habitat loss by road construction, (2) human access that increases disturbance in remote areas, (3) forest spatial patterns that alter biodiversity, and (4) sediment and other inputs to streams. The role of forestry roads in the spread of exotic species into nearby natural ecosystems and in disrupting animal movement patterns is poorly known (Saunders and Hobbs 1991; Forman 1995; Forman and Alexander 1998; Forman and Deblinger 1999). Additional factors such as road salt, heavy metals, roadkills, and so forth generally have minor or local ecological effects by forest roads (Reck and Kaule 1993; Forman 1995). Although minimizing all of these ecological effects is desirable, addressing the big four listed is most important for sustainably meshing forestry and conservation.

Thus the overall objective of this study is to gain insight into the ecological effects of combined road networks and forest patterns in landscapes with forestry. The specific goals are as follows:

1. Use measures of road density and haul routes in dynamic spatial models to ecologically evaluate alternative logging regimes.
2. Ecologically compare these models using forest interior, largest forest core, and boundary length.
3. Combine the results to identify the least damaging or most ecologically desirable logging regime, and determine whether evaluating road networks enhances or changes conclusions based on forest spatial patterns alone.

METHODS

Spatial models of logging regimes and road networks

We modelled forested landscapes using a 30 × 30 cell grid, with each cell representing 2.5 ha. The width of the landscape is 4.65 km and the area is 22.5 km^2 or 2 250 ha. The cell size, approximately 155 m wide, roughly accommodates the projected maximum distance that skidder machinery commonly drags a log from a stump to a logging road (Trimble and Weitzman 1953; Cha *et al.* 1991; Gullison and Hardner 1993). In the models a tiny strip along certain landscape boundaries requires skidding logs 220 m (the diagonal of a grid cell). The area surrounding the landscape is non-forested.

Five spatial models of landscape change were used to evaluate the ecological effects of logging road networks in forestry production environments. Each model has been described and diagrammed in the literature, and represents a different cutting sequence or type of logging regime. The DISPERSED model disperses small cuts evenly over the landscape by progressively locating the cuts to minimize the variance in inter-patch distance (Franklin and Forman 1987; Li *et al.* 1993; Forman 1995). The CORRIDOR model has a central cut strip followed progressively by parallel strips on opposite sides of it (Dale *et al.* 1994; Forman 1995). The EDGE model begins with a cut strip along one edge followed progressively by adjacent parallel strips (Franklin and Forman 1987; Forman 1995). The HALF-EDGE model has wider strips half-an-edge long alternating on opposite sides of a central line through the landscape (Li *et al.* 1993). The JAWS model has L-shaped cuts spreading progressively from two adjacent edges (Forman 1995; Forman and Collinge 1996).

The models can be used for either natural forests or plantations. Cutting strategies fall into two main categories: clearcutting (clear-felling) and selection (or group-selection) cutting (Food and Agriculture Organization of the United Nations 1978; Wallin *et al.* 1994; Noble and Dirzo 1997). The models mimic clearcutting though they can also be applied to selection and group-selection cutting. Selection cutting, especially group selection, is generally less ecologically damaging than clearcutting based on logging in equal-sized areas; however, when compared on the basis of wood products harvested, in some cases clearcutting is less damaging because a smaller area is logged (Kimmins 1997).

Regrowth is ignored in the models so the only options for a grid cell are forest or cut (no forest). Forest roads may be in any cell though the road bed is a tiny portion of a 2.5 ha cell [roadbed width averaging <12 m (Grayson *et al.* 1993; Gullison and Hardner 1993; McGurk and Fong 1995)]. For an evaluation of the amount of forest interior present, we classified certain grid cells near the perimeter of forest patches as forest edge, where environmental conditions are known to be modified (Chen *et al.* 1992; Forman 1995).

For each of the five landscape change models a road network is designed to provide efficient access to the entire landscape. All models have a single entrance/exit point located in the middle of one side of the landscape. Road networks were adapted from the basic types outlined by Food and Agriculture Organization of the United Nations (1978): a large loop; a large loop divided into smaller loops; and a dendritic network. The dendritic road system was omitted to avoid adding turn-around areas for logging vehicles at the ends of roads. Dead-end roads are only present at the initial stage of the JAWS model. The resulting road networks in all models are a series of large linear loops.

The simulation models describe a forest rotation of ten cutting stages. During each stage 10% of the landscape area is logged (90 of the 900 cells). In the DISPERSED patch-cutting model each cut is a 10 ha square (four contiguous cells). For each cutting stage the least amount of road needed to cut 10% of the forest is added (Fig. 1). Thus road networks were also chosen to minimize total road length. We assume that the logging regime and road network for the entire rotation over the landscape is planned. If roads were only planned for the next cutting stage, the resulting total road network might be of higher density and less efficiently laid out.

Assays for impacts of road and logging patterns

Landscape ecological impacts are measured at each time stage for all cutting models. The road network is evaluated by indices for (1) road density and (2) haul route. Road density is the average road length per square kilometer in the landscape (km/km^2). Haul route is calculated as the road-hauling distance for a log times the number of 90° turns in the route. Road-hauling distance is the average road length to haul out a log in a cutting stage, divided by the width of the landscape (a scaling factor that permits comparison of different sized or shaped landscapes). The number of 90° turns is the average number in the various routes to haul out logs during a cutting stage.

We assume that the lower the road density value, the less the forest is impacted and the better a road network is ecologically. For example, more roads mean more direct habitat loss by road construction and greater disruption of animal movement patterns. Also the lower the haul-route value the less the ecological impact caused by a road system. A

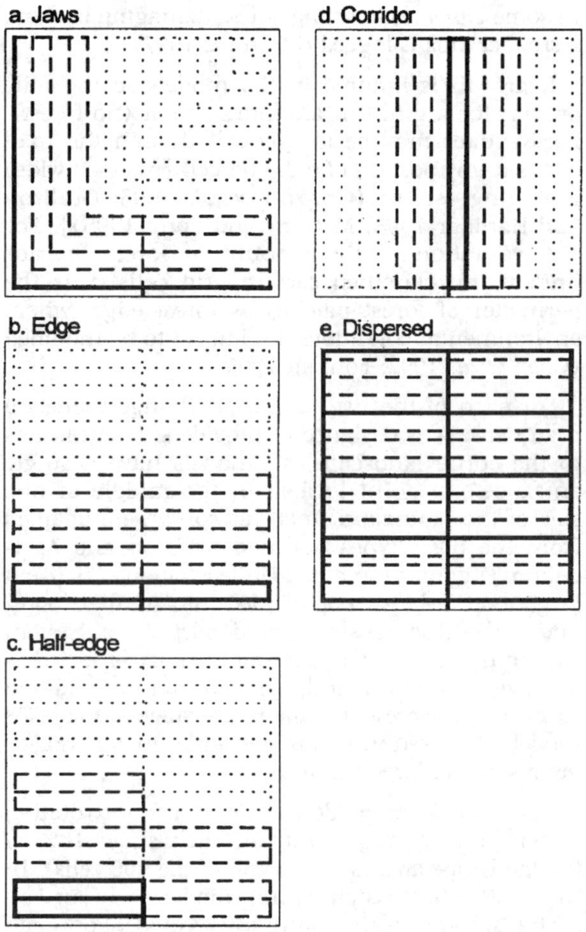

Fig. 1. Road networks in the five dynamic spatial models of forest cutting. Solid line = forest roads when landscape is 10% cut; dashed line = roads added to cut from 10 to 50%; dotted line = roads added to cut from 50 to 100% of the landscape. Each landscape is 4.65 km across and parallel roads are generally 310 m apart (see Methods). Models based on Franklin and Forman (1987), Li *et al.* (1993), Forman (1995), Collinge and Forman (1998) as described in Methods.

shorter haul route with fewer sharp turns means less disturbance due to vehicle noise, dust, pollutants and soil erosion. Of course, other ecological characteristics may be affected by the road network (Reck and Kaule 1993; Forman 1995; Forman and Alexander 1998; Forman and Deblinger 1999). However, these two attributes, road density and haul route, appear to be the major factors on forestry roads. In this case therefore, economics probably correlates with ecology for both road density and haul route: less is better.

Three forest spatial patterns are also evaluated at each cutting stage for each model: (1) total area of forest interior; (2) largest forest core within a patch; and (3) total boundary length. Total area of forest interior in the landscape is the area remaining after eliminating habitat lost to roads, logged areas, and forest edge (including around the perimeter of the landscape). We compared forest interior area using three forest edge widths, 40 m, 80 m and 200 m (Temple 1986; Forman 1995). The largest forest core is determined using the standard geography measure of the diameter of the largest circle that fits within a forested patch in the landscape. Boundary length or density is calculated as the average length of border between forest and logged areas per square kilometer for the landscape (km/km^2).

We assume that the more forest interior present the better it is ecologically, e.g., for protecting ground water, interior species, and so forth (Forman 1995). Similarly, a large forest core is considered ecologically better, for example, to protect rare vertebrates favoured by remoteness from human disturbance. Boundary length, on the other hand, has significant ecological attributes both negative and positive (Forman 1995). Longer boundaries (per unit area) tend to mean a higher density of edge and exotic species (Temple 1986; Hunter 1990; Chen *et al.* 1992). More edge length, however, also generally provides additional microhabitats for species richness and tends to support higher populations of many game species (Harris 1984; Hunter 1990). Movement of species between forest and open habitats may also be greater if edge length is increased by convoluting boundaries (Forman 1995).

RESULTS

Road network

Marked differences among the five models of alternative logging regimes emerge when their road density and haul routes are compared.

Road density. For road density the DISPERSED patch-cutting regime has a much steeper early increase, and reaches a higher level than the other four cutting regimes (Fig. 2). At the 50% cutting stage of the DISPERSED model the entire road network is in place. The CORRIDOR, HALF-EDGE, EDGE and JAWS models are essentially the same in road density throughout the rotation. Since road density indicates direct forest habitat loss due to road construction, for this variable the DISPERSED patch-cutting regime is considered least desirable ecologically.

Haul route. Overall the haul route, which combines average road length for hauling a log plus the number of sharp turns in the route, has the highest value in the JAWS model and is only slightly less in the CORRIDOR model (Fig. 3). Although the DISPERSED model looks good late in the rotation, it has a serious shortcoming in that logs are hauled from widely dispersed cut patches, so most of the road network must be maintained and used continuously throughout the rotation. Therefore based on the haul-route variable, the EDGE and HALF-EDGE cutting regimes are similar and appear to provide the greatest ecological benefits. The CORRIDOR and JAWS approaches are least desirable.

Combining these results for the overall road network, the EDGE and HALF-EDGE logging regimes appear to be best ecologically. The DISPERSED

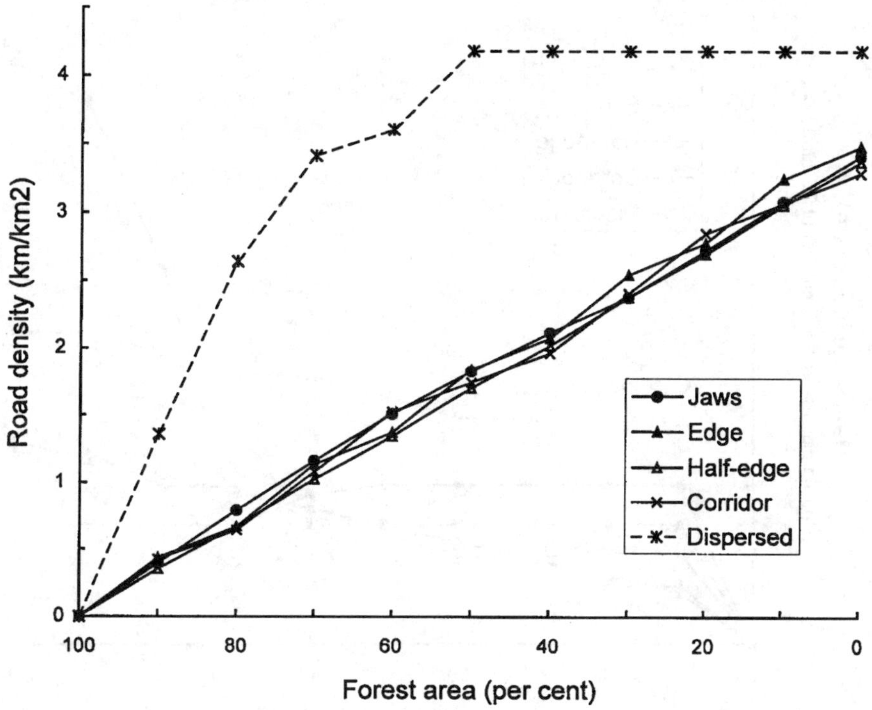

Fig. 2. Road density for the five cutting models. See Figure 1 for road networks and see Methods for descriptions of models.

approach is least desirable due to the large amount of forest habitat lost and the expected erosion and disturbance by vehicle usage.

Forest spatial pattern

In addition to the road network pattern, the spatial pattern of wooded areas strongly affects ecological conditions, such as diversity of interior species, groundwater, animals favored by remoteness from people, and abundance of forest edge populations. Three spatial measures of the forest pattern are used as indicators of such conditions: total forest interior; largest forest core; and boundary length. These forest assays also point to marked differences among the five logging regimes modelled.

Total area of forest interior. With a 40 m wide forest edge the DISPERSED model shows less forest interior than the other models throughout the rotation period (Fig. 4a). This same result is evident assuming an 80 m wide edge and a 200 m wide edge (Fig. 4b and c). Indeed, with grid cells 155 m wide essentially no forest interior remains by the time the forest is 50% cut.

The EDGE and HALF-EDGE models retain forest interior reasonably well through the rotation (Fig. 4). However, the JAWS model appears to be slightly better especially late in the rotation. In the JAWS logging regime L-shaped cuts spread progressively from two adjacent sides, which maintains a square forested patch with maximum forest interior throughout the forest rotation. Thus based on forest interior, the DISPERSED model is worst and the JAWS approach is considered ecologically best.

Largest forest core. The largest core within a forested patch in the landscape is consistently found in the JAWS model (Fig. 5). The smallest forest core remaining is in the DISPERSED and CORRIDOR models. The EDGE and HALF-EDGE models show intermediate and similar trends. The differences among the models are clearly established early in the cutting rotation. Based on maintaining a large forest core the JAWS cutting regime emerges as best ecologically, and the DISPERSED and CORRIDOR models poorest.

Boundary length. The length or density of boundaries (km/km^2) between forest and cut areas is much higher in the DISPERSED model than in the others (Fig. 6). The CORRIDOR and JAWS models both show a steep initial increase in boundary length, but through the rotation the CORRIDOR curve remains high while the JAWS curve drops progressively. The EDGE and HALF-EDGE models are generally similar and have the lowest amount of boundary present. The differences between these last two models might be ecologically important, especially early or late in the forest rotation.

Summarizing the results for forest spatial pattern we focus on forest interior and large core, because the ecological advantages are widely accepted and considered to be of greater significance than either the pros or the cons of boundary length (Harris 1984; Temple 1986; Hunter 1990; Forman 1995). Thus clearly the DISPERSED cutting regime is least desirable ecologically, and the CORRIDOR cutting approach is apparently second worst (Figs 4 and 5). The EDGE and HALF-EDGE approaches are

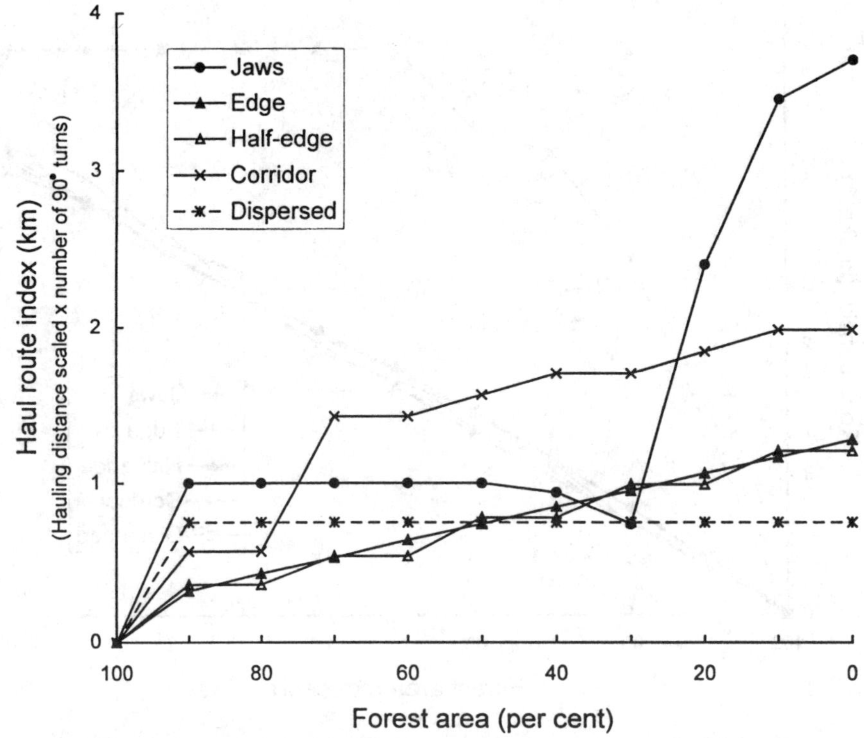

Fig. 3. Haul route index for the five cutting models. Haul route index = (average hauling distance for a log cut during one stage divided by the diameter of the landscape) × number of right-angle turns in the average route. See Figure 1 for road networks and see Methods for descriptions of models.

intermediate, and the JAWS model appears ecologically to be most beneficial.

As long as good forest interior and core conditions are present in the landscape, as in the JAWS model (Figs 4 and 5), the added ecological advantages of greater boundary length should outweigh any disadvantages. Thus the difference in boundary length among the JAWS, EDGE and HALF-EDGE models (Fig. 6) becomes quite useful in differentiating among them. The considerable boundary length in the JAWS model early in the rotation (Fig. 6) corresponds with good forest core conditions (Fig. 5), while later in the rotation boundary length decreases. Thus adding the three forest-pattern variables together (forest interior, core, and boundary) clearly suggests that, though the EDGE and HALF-EDGE approaches are promising, the JAWS cutting regime is ecologically the best.

Combining the results

Synthesizing the results of the road network and the forest spatial pattern measures provides a rather clear picture based on several lines of evidence. The DISPERSED model is least desirable

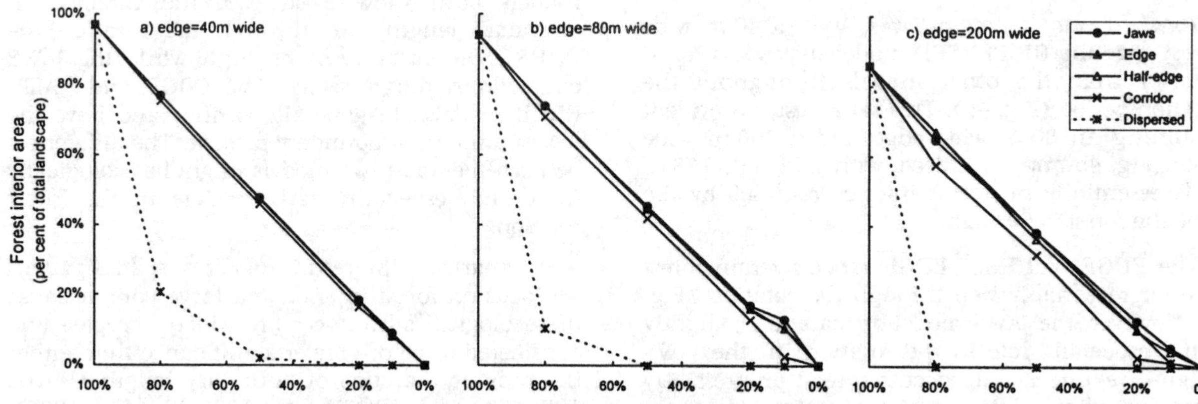

Fig. 4. Area of forest interior for the five cutting models. Forest interior is the total area after eliminating area covered by roads, logged areas, and forest edge area, assuming edges are 40, 80 or 200 m wide. Models based on Franklin and Forman (1987), Li *et al.* (1993), Forman (1995), Collinge and Forman (1998) as described in Methods.

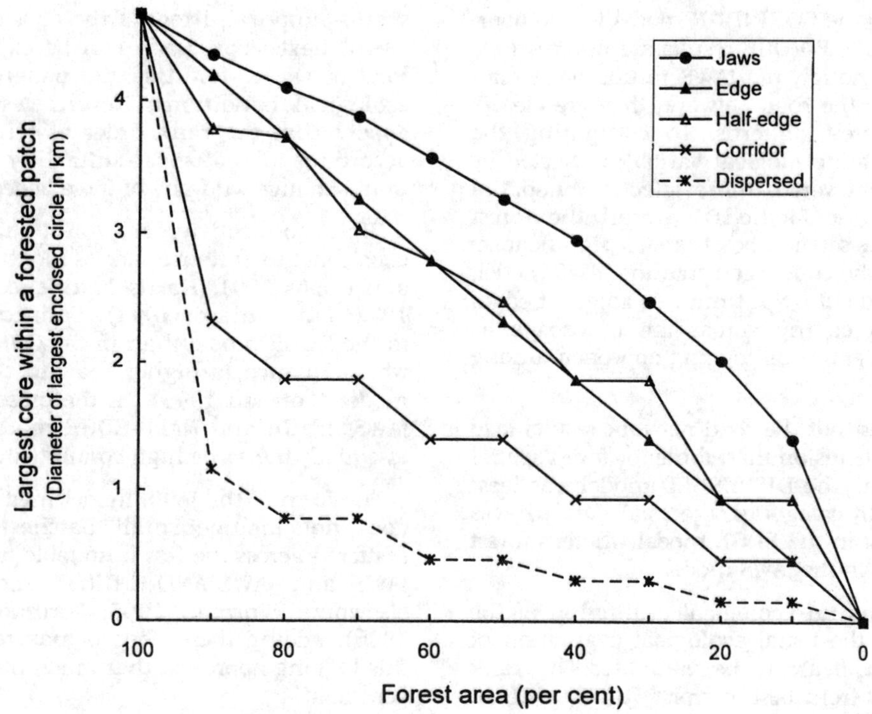

Fig. 5. Largest forested core area within the landscape for the five cutting models. See Methods for descriptions of models.

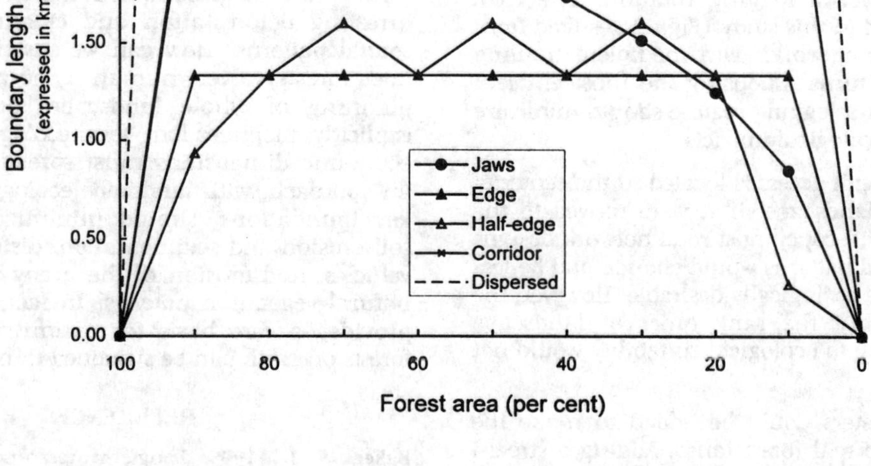

Fig. 6. Total forest boundary length for the five cutting models. See Methods for descriptions of models.

ecologically and the CORRIDOR model is also poor. The EDGE and HALF-EDGE results are intermediate and similar. Although the JAWS results are somewhat weaker for the road network, they are clearly the best for forest patterns. In comparing the significance of the ecological variables affected by the road network versus those affected by spatial forest pattern (see Methods), overall the forest pattern attributes seem to be of greater significance. This reinforces the conclusion that the JAWS model, that is, L-shaped cut strips from two adjacent edges progressively spreading across the landscape, is ecologically the optimum logging approach among these evaluated.

It is also clear that the road-network results lead to a different conclusion than those for forest spatial pattern. Although the DISPERSED model was least desirable in both categories, the road network was ecologically best in the EDGE model whereas forest pattern was best in the JAWS model.

Finally, adding an ecological evaluation of the road system to the usual ecological evaluation of forest pattern appears to be valuable. The rank order of models from best to worst (JAWS > EDGE = HALF-EDGE > CORRIDOR > DISPERSED) as determined by forest-pattern analysis did not change by adding the road system analysis. However, the combined analysis provided a fuller, more robust picture. The degree of difference between models changed a bit; for instance, the difference between JAWS and EDGE models narrowed by adding the road results. Finally, combining the results re-emphasized the importance and necessity of evaluating and ranking the relative significance of the many ecological attributes.

DISCUSSION

The model parameters in the present analysis are kept as simple as possible to facilitate the applicability of results to real landscapes. We assume that a road network is planned at the beginning of a forest rotation and is as efficient as possible to minimize road length. For each of the five models we initially designed possible alternative road networks for a logging rotation. The road systems analysed in this study (Fig. 1) resulted from eliminating the networks with inefficient hauling routes or extra turns. Choosing the most efficient road network for a logging regime should minimize both costs and ecological impacts.

The entrance/exit access is located in the center of one edge of a landscape. If it were moved to the corner of the landscape, most road network designs and forest spatial patterns would change and be less efficient and less ecologically desirable. However, the conclusions about the rank order of landscape models according to ecological suitability would not change.

Other parameters could be added to make the models closer to real forest lands. Adding a stream network, topographic relief, diagonal roads, regrowth areas and nature reserves (Li *et al.* 1993), and spatio-temporal irregularity or stochasticity are useful next steps. Each can be expected to affect road networks, forest spatial patterns, and overall ecological conditions. Nevertheless, it does not appear that the rank order of landscape models according to ecological suitability presented here would change with any of these added parameters.

High connectivity of forest habitat can be considered to enhance species movement (Saunders and Hobbs 1991; Harris and Scheck 1991; Hobbs 1992; Lindenmayer 1994). Connectivity is higher in the EDGE model than in the CORRIDOR model, which in turn is higher than in the DISPERSED model (Forman 1995). In the present analysis the JAWS, EDGE and HALF-EDGE models would have essentially the same high connectivity level.

The form of the JAWS model used in this analysis does not include small patches and corridors scattered across the less hospitable matrix, as in the JAWS and JAWS-AND-CHUNKS models presented elsewhere (Forman 1995; Forman and Collinge 1996). Adding these "bits of nature" would make this logging approach that much more ecologically beneficial.

SUMMARY AND CONCLUSION

Ecologically the JAWS model approach for logging a forested landscape (L-shaped cuts progressively spreading from two adjacent edges) is the best of the five approaches analysed. The DISPERSED and CORRIDOR approaches are least desirable and should be avoided. Ecologically the best road network and the best forest patterns were in different models. Combining the road network and forest pattern attributes provides a better ecological evaluation of forestry regimes than does one set alone.

Protection of soil, water and biodiversity depends on arresting habitat loss due to deforestation. These resources also depend heavily on the spatial pattern of logging the forest. Yet a crisis has arrived — wood products are demanded by an added billion people per decade — and it rapidly worsens.

How can we accomplish the two objectives of arresting deforestation and creating wise forest spatial patterns? How can we sustain both nature and forestry over human generations? Spatial planning of whole landscapes or regions that explicitly integrates long-term ecological with socio-economic dimensions must spread rapidly. Such an approach, with landscape ecology as one of the key foundations, should minimize habitat loss, soil erosion and sedimentation, disturbance due to vehicles, and erosion of the many values of large natural-vegetation patches. In fact, this approach provides a rare basis for optimism that natural forests on earth can be sustained through the future.

REFERENCES

Baker, W. L., 1995. Longterm response of disturbance landscapes to human intervention and global change. *Landsc. Ecol.* **10:** 143–59.

Bakowski, C. and Kozakiewicz, M., 1988. The effect of forest road on bank vole and yellow-necked mouse populations. *Acta Theriol.* **33:** 345–53.

Cha, D. S., Nakao, H. and Watahiki, K., 1991. A computerized arrangement of forest road using a digital terrain model. *J. Fac. Agricul., Kyushu Univ.* **36**(1–2): 131–42.

Chen, J., Franklin, J. F. and Spies, T. A., 1992. Vegetation responses to edge environments in old-growth Douglas-fir forests. *Ecol. Appl.* **2:** 387–96.

Chen, J., Franklin, J. F. and Spies, T. A., 1993. Contrasting microclimates among clearcut, edge, and interior of old-growth Douglas-fir forest. *Agricul. For. Meteorol.* **63:** 219–37.

Chomitz, K. M. and Gray, D. A., 1996. Roads, land use, and deforestation: a spatial model applied to Belize. *World Bank Econ. Rev.* **10:** 487–512.

Collinge, S. K. and Forman, R. T. T., 1998. A conceptual model of land conversion processes: predictions and evidence from a micro-landscape field experiment with grassland insects. *Oikos* **82:** 66–84.

Dale, V. H., O'Neill, R. V., Southworth, F. and Pedlowski, M., 1994. Modeling effects of land management in the Brazilian Amazonian settlement of Rondonia. *Conserv. Biol.* **8:** 196–206.

DeGraaf, R. M. and Miller, R. I. (eds), 1995. Conservation of Faunal Diversity in Forested Landscapes. Chapman and Hall: London.

Food and Agriculture Organization of the United Nations, 1978. Mountain Forest Roads and Harvesting: A Technical Report ed by R. Heinrich. Second FAO/Austria Training Course on Forest Roads and Harvesting in Mountainous Forests: Rome.

Forman, R. T. T., 1995. Land Mosaics: The Ecology of Landscapes and Regions. Cambridge University Press: Cambridge.

Forman, R. T. T., 1999. Horizontal processes, roads, suburbs, societal objectives, and landscape ecology. Pp. 35–53 *in* Landscape Ecological Analysis: Issues and Applications ed by J. M. Klopateck and R. H. Gardner. Springer-Verlag: New York.

Forman, R. T. T. and Alexander, L. E., 1998. Roads and their major ecological effects. *Ann. Rev. Ecol. System.* **29:** 207–31.

Forman, R. T. T. and Collinge, S. K., 1995. The "spatial solution" to conserving biodiversity in landscapes and regions. Pp. 537–68 *in* Conservation of Faunal Diversity in Forested Landscapes ed by R. M. DeGraaf and R. I. Miller. Chapman and Hall: London.

Forman, R. T. T. and Deblinger, R. D., 1999. The ecological road-effect zone of Massachusetts (USA) suburban highway. *Conserv. Biol.* In press.

Forman, R. T. T., Friedman, D. S., Fitzhenry, D., Martin, J. D., Chen, A. S. and Alexander, L. E., 1997. Ecological effects of roads: toward three summary indices and an overview for North America. Pp. 40–54 *in* Habitat Fragmentation and Infrastructure ed by K. Canters. Ministry of Transport, Public Works and Water Management: Delft, Netherlands.

Forman, R. T. T. and Hersperger, A. M., 1996. Road ecology and road density in different landscapes, with international planning and mitigation solutions. Pp. 1–22 *in* Trends in Addressing Transportation Related Wildlife Mortality ed by G. L. Evink, P. Garrett, D. Ziegler and J. Berry. Report FL-ER-58-96, Florida Department of Transportation: Tallahassee, Florida.

Franklin, J. F. and Forman, R. T. T., 1987. Creating landscape patterns by forest cutting: ecological consequences and principles. *Landsc. Ecol.* **1:** 5–18.

Gallup, J. L., Sachs, J. D. and Mellinger, A. D., 1998. Geography and economic development. Pp. 127–78 *in* Annual World Bank Conference on Development Economics. World Bank: Washington, DC.

Grayson, R. B., Haydon, S. R., Jayasuriysa, M. D. A. and Finlayson, B. L., 1993. Water quality in mountain ash forests: separating the impacts of roads from those of logging operations. *J. Hydrol.* **150:** 459–80.

Gullison, R. E. and Hardner, J. J., 1993. The effects of road design and harvest intensity on forest damage caused by selective logging: empirical results and a simulation model from the Bosque Chimanes, Bolivia. *For. Ecol. Manage.* **59:** 1–14.

Hansen, A. J., Urban, D. L. and Marks, B., 1992. Avian community dynamics: the interplay of landscape trajectories and species life histories. Pp. 170–95 *in* Landscape Boundaries: Consequences for Biotic Diversity and Ecological Flows ed by A. J. Hansen and F. di Castri. Springer-Verlag: New York.

Harris, L. D., 1984. The Fragmented Forest: Island Biogeography Theory and the Preservation of Biotic Diversity. University of Chicago Press: Chicago.

Harris, L. D. and Scheck, J., 1991. From implications to applications: the dispersal corridor principle applied to the conservation of biological diversity. Pp. 189–220 *in* Nature Conservation 2: The Role of Corridors ed by D. A. Saunders and R. J. Hobbs. Surrey Beatty & Sons: Chipping Norton, New South Wales.

Hobbs, R. J., 1992. The role of corridors in conservation: solution or bandwagon. *Trends Ecol. Evol.* **7:** 389–92.

Hunter, M. L. Jr, 1990. Wildlife, Forests, and Forestry: Principles of Managing Forests for Biological Diversity. Prentice-Hall: Englewood Cliffs, New Jersey.

Jones, J. A. and Grant, G. E., 1996. Peak flow responses to clear-cutting and roads in small and large basins, Western Cascades, Oregon. *Water Resour. Res.* **32:** 959–74.

Kimmins, H., 1997. Balancing Act: Environmental Issues in Forestry. University of British Columbia Press: Vancouver, Canada.

Li, H., Franklin, J. F., Swanson, F. J. and Spies, T. A., 1993. Developing alternative forest cutting patterns: a simulation approach. *Landsc. Ecol.* **8:** 63–75.

Lindenmayer, D. B., 1994. Wildlife corridors and the mitigation of logging impacts on fauna in wood-production forests in South-eastern Australia: a review. *Wildl. Resour.* **21:** 323–40.

McCashion, J. D. and Rice, R. M., 1983. Erosion on logging roads in northwestern California: how much is avoidable? *J. For.* **81:** 23–26.

McGurk, B. J. and Fong, D. R., 1995. Equivalent roaded area as a measure of cumulative effect of logging. *Environ. Manage.* **19:** 609–21.

McNabb, D. H., 1994. Tillage of compacted haul roads and landings in the boreal forests of Alberta, Canada. *For. Ecol. Manage.* **66:** 179–94.

Mech, L. D., 1989. Wolf population survival in an area of high road density. *Amer. Midl. Nat.* **121:** 387–89.

Minamikata, Y., 1977. On the essential road network to forestry and forest conservation. *J. Japan For. Soc.* **59:** 298–300.

Mladenoff, D. J., White, M. A., Pastor, J. and Crow, T. R., 1993. Comparing spatial pattern in unaltered old-growth and disturbed forest landscapes. *Ecol. Appl.* **3:** 294–306.

Montgomery, D. R., 1994. Road surface drainage, channel initiation, and slope instability. *Water Resour. Res.* **30:** 1925–932.

Noble, I. R. and Dirzo, R., 1997. Forests as human-dominated ecosystems. *Science* **277:** 522–25.

Peters, P. A., 1978. Spacing of roads and landings to minimize timber harvest cost. *For. Sci.* **24:** 209–17.

Reck, H. and Kaule, G., 1993. Strassen und Lebensraume: Ermittlung und Beurteilung strassenbedingter Auswirkungen auf Pflanzen, Tiere und ihre Lebensraume. Forschung Strassenbau und Strassenverkehrstechnik, Heft 654. Herausgegeben vom Bundesminister fur Verkehr: Bonn-Bad Godesberg, Germany.

Reed, R. A., Johnson-Barnard, J. and Baker, W. L., 1996. Contribution of roads to forest fragmentation in the Rocky Mountains. *Conserv. Biol.* **10:** 1098–106.

Rice, R. M. and Lewis, J., 1991. Estimating erosion risks associated with logging and forest roads in Northwestern California. *Water Resour. Bull.* **27:** 809–18.

Saunders, D. A. and Hobbs, R. J. (eds), 1991. Nature Conservation 2: The Role of Corridors. Surrey Beatty & Sons: Chipping Norton, New South Wales.

Sessions, J. and Yeap, Y. H., 1989. Optimizing road spacing and equipment allocation simultaneously. *For. Prod. J.* **39:** 43–46.

Swanson, F. J., Benda, L. E., Duncan, S. H., Grant, G. E., Megahan, W. F. and Ziermer, R. R., 1987. Mass failures and other processes of sediment production in Pacific Northwest forest landscapes. Pp. 40–78 *in* Streamside Management: Forestry and Fishery Interactions ed by E. O. Salo and T. W. Cundy. Contribution 57, Institute of Forest Resources, University of Washington: Seattle, Washington, USA.

Swanson, F. J. and Dyrness, C. T., 1975. Impact of clear-cutting and road construction on soil erosion by landslides in the western Cascade Range, Oregon. *Geology* **3:** 393–96.

Temple, S. A., 1986. Predicting impacts of habitat fragmentation on forest birds: a comparison of two models. Pp. 301–04 *in* Wildlife 2000: Modeling Habitat Relationships of Terrestrial Vertebrates ed by J. Verner, M. L. Morrison and C. J. Ralph. University of Wisconsin Press: Madison, Wisconsin.

Tinker, D. B., Resor, C. A., Beauvais, G. P., Kipfmueller, K. F., Fernandes, C. I. and Baker, W. L., 1998. Watershed analysis of forest fragmentation by clearcuts and roads in a Wyoming forest. *Lands. Ecol.* **13:** 149–65.

Trimble, G. R. Jr and Weitzman, S., 1953. Soil erosion on logging roads. *Soil Sci. Proc.* **17:** 152–54.

Van Dyke, F. B., Brocke, R. H., Shaw, H. G., Ackerman, B. B., Hemker, T. P. and Lindzey, F. G., 1986. Reactions of mountain lions to logging and human activity. *J. Wildl. Manage.* **50:** 95–102.

Wallin, D. O., Swanson, F. J. and Marks, B., 1994. Landscape pattern response to changes in pattern generation rules: land-use legacies in forestry. *Ecol. Appl.* **4:** 569–80.

Wemple, B. C., Jones, J. A. and Grant, G. E., 1996. Channel network extension by logging roads in two basins, Western Cascades, Oregon. *Water Resour. Bull.* **32:** 1195–207.

Wiersum, K. F., 1995. 200 years of sustainability in forestry: lessons from history. *Environ. Manage.* **19:** 321–29.

World Resources Institute, 1996. World Resources: A Guide to the Global Environment, 1996–97. Oxford University Press: New York.

CHAPTER 9

Conifers and cows: forest survival in a New Zealand dairy landscape

Bruce Burns,[1] Gary M. Barker,[1] Richard Harris[2] and John Innes[1]

Dairy farming occurs on fertile lowlands occupying 6% of New Zealand's land area but contributes 20% of its export income. Forest stands dominated by the indigenous conifer *Dacrycarpus dacrydioides* (kahikatea) are a conspicuous component of these landscapes. We examined the distribution, persistence, representativeness, history, and biodiversity of these stands in the middle Waikato Basin. In 1997, we located 116 stands occupying 126 ha (0.15% of the dairy landscape); five stands fewer than an equivalent survey in 1977. Approximately 70% of the 1997 stands were less than 1 ha in size. Basal areas of stands were high (mean = 102 m^2ha^{-1}) with kahikatea making up about 75%. The remainder was mostly contributed by a mix of angiosperms. Proxy evidence of former Waikato forests suggests that current forests represent only a part of the former diversity of forest types. Also, examination of kahikatea age structures suggests that most stands developed by recruitment of kahikatea about 125 years ago, probably around a few survivors after landscape modification for agriculture. Biodiversity of stands was sampled in four groups: plants, birds, beetles, and snails. There was a moderate indigenous biodiversity in all groups except birds, but few rare or regionally restricted species. Grazing of understories by cattle threatens persistence and indigenous diversity of stands. Stands possess a range of values, many inadequately recognized. These values will be slowly lost from future dairy landscapes if current trends continue. Here, we promote the concept of a dairy landscape integrating conservation and production as a desirable future requiring greater recognition.

Key words: Dairy farming, *Dacrycarpus dacrydioides*, Kahikatea, Biodiversity.

INTRODUCTION

DAIRY farming in New Zealand covers about 1.7 million ha (6% of New Zealand's land area), concentrated on alluvial flood plains and rolling hills on the North Island (Statistics New Zealand 1997). These lowlands have been in pastoral production since most of their original cover of indigenous forests and scrub was cleared and much of their wetlands drained in the latter half of the 19th and the early part of the 20th centuries (Ministry for the Environment 1997). Remnant forest patches from this clearance (Fig. 1) are almost all dominated by the conifer *Dacrycarpus dacrydioides* (A. Rich.) Laubenf. (kahikatea) and provide the main opportunity for conservation of indigenous biodiversity in these landscapes (conservation of remnant wetlands is perhaps the only other similar opportunity). In natural situations, kahikatea-dominant forest is a species of swamps, or of alluvial flats subject to stand destruction and regeneration by flooding (Wardle 1974; Duncan 1991, 1993).

Dairying in New Zealand is an intensive form of farming based on exotic grass and animal species, and imported phosphorus, potassium, and, increasingly, nitrogen fertilizers (Bolan and Podila 1996; O'Connor *et al.* 1996). Most dairy products are exported, contributing 20% of New Zealand's export income (Statistics New Zealand 1997). Currently, dairy farming provides relatively high economic returns per hectare compared with other land uses. For example, in the Waikato region, dairy returns are approximately NZ$1,500 per ha compared with NZ$307 per ha for nearby sheep and

[1]Landcare Research, Private Bag 3127, Hamilton, New Zealand.
[2]Landcare Research, Private Bag 6, Nelson, New Zealand.

Pages 81–89 *in* NATURE CONSERVATION 5: NATURE CONSERVATION IN PRODUCTION ENVIRONMENTS: MANAGING THE MATRIX ed by J. L. Craig, N. Mitchell and D. A. Saunders. Surrey Beatty & Sons, 2000.

Fig. 1. Kahikatea-dominated forest remnants near McGregor Road, Ngahinapouri, in the middle Waikato Basin dairy landscape.

beef farms (Ministry of Agriculture 1997). Therefore, where it occurs, the dairy industry is a potent economic force dominating societal infrastructures, local politics, and inevitably, landscapes. In such landscapes, recognition of the importance and values of indigenous biota has been largely negligible. In this chapter, we explore what conservation values remain in these landscapes (using kahikatea forest patches as a focus), and discuss possible future landscapes.

The specific objectives of this chapter are to:

1. describe the composition, extent and spatial distribution of kahikatea-dominant stands in dairy landscapes, and investigate changes in their extent over recent decades;
2. assess how well these forest patches represent the original forests of these landforms;
3. characterize their biodiversity (using data on vascular plants, beetles, snails, and birds);
4. develop scenarios of future alternative landscapes and their potential role in biodiversity conservation.

MIDDLE WAIKATO BASIN

We chose the dairy-dominated middle Waikato Basin as our study landscape. The middle Waikato Basin is an oval depression on North Island about 80 km north to south and about 40 km east to west almost entirely ringed by ranges (Fig. 2; Selby and Lowe 1992). It is a structural basin formed by faulting and differential uplift of bordering highlands, then infilled by fluvial, lacustrine, ignimbrite, and airfall tephra deposits, and dissected and shaped by the meanderings of the Waikato and Waipa Rivers (Selby and Lowe 1992). This has left an extensive, highly fertile alluvial floodplain. Low hills rise above the floodplain, and several restiad bogs (Komakorau, Rukuhia, and Moanatuatua), now almost totally drained, are present (Selby and Lowe 1992). Soils are gley or organic soils on the flood plains, derived from alluvium or peat, or, allophanic soils on the rolling hills, derived from airfall volcanic ash tephras (Orbell 1973; Hewitt 1993). The middle Waikato Basin experiences a warm temperate climate (Meurk 1984) with mean annual rainfall around 1 200 mm and mean annual air temperatures around 13°C (based on records from Ruakura Climatological Station; New Zealand Meteorological Service 1983). The region was occupied by substantial Maori populations up until the 1860s, but the present character of the landscape was largely developed during European settlement and establishment of farming in the twenty years immediately following the land-wars of 1863–1864 (Ministry of Works 1973; Barber 1978).

FOREST COMPOSITION, DISTRIBUTION, AND RECENT PATCH DYNAMICS

The former Waipa County (now mostly incorporated into the Waipa District during 1992 boundary changes) covers 113 000 ha, of which approximately 83 000 ha occurs in the middle Waikato Basin and, we considered, could be classified as dairy landscape. In 1977, the former Waipa County Council commissioned a survey of remnant forest patches within its administrative boundaries (Boubee 1978). This is an invaluable data set with which to quantify recent changes to the amount and distribution of forest remaining within a dairy landscape.

Examining only Boubee's (1978) data for Waipa County land within the middle Waikato Basin, 144 individual patches of indigenous forest were identified of which 121 (84%) were described as kahikatea-dominant. The few other patches identified were described as "gully" patches with abundant exotic *Salix* spp. but with kahikatea also present. From interpretation of 1995 aerial photos (New Zealand Aerial Mapping) combined with ground verification in 1997, we found that 116 (96%) of those kahikatea-dominant patches identified in 1977 were still present in 1997 (Table 1). Collectively, these stands covered an area of approximately 126 ha (Table 1), which represents only 0.15% of the 83 000 ha studied. Although the mean area of

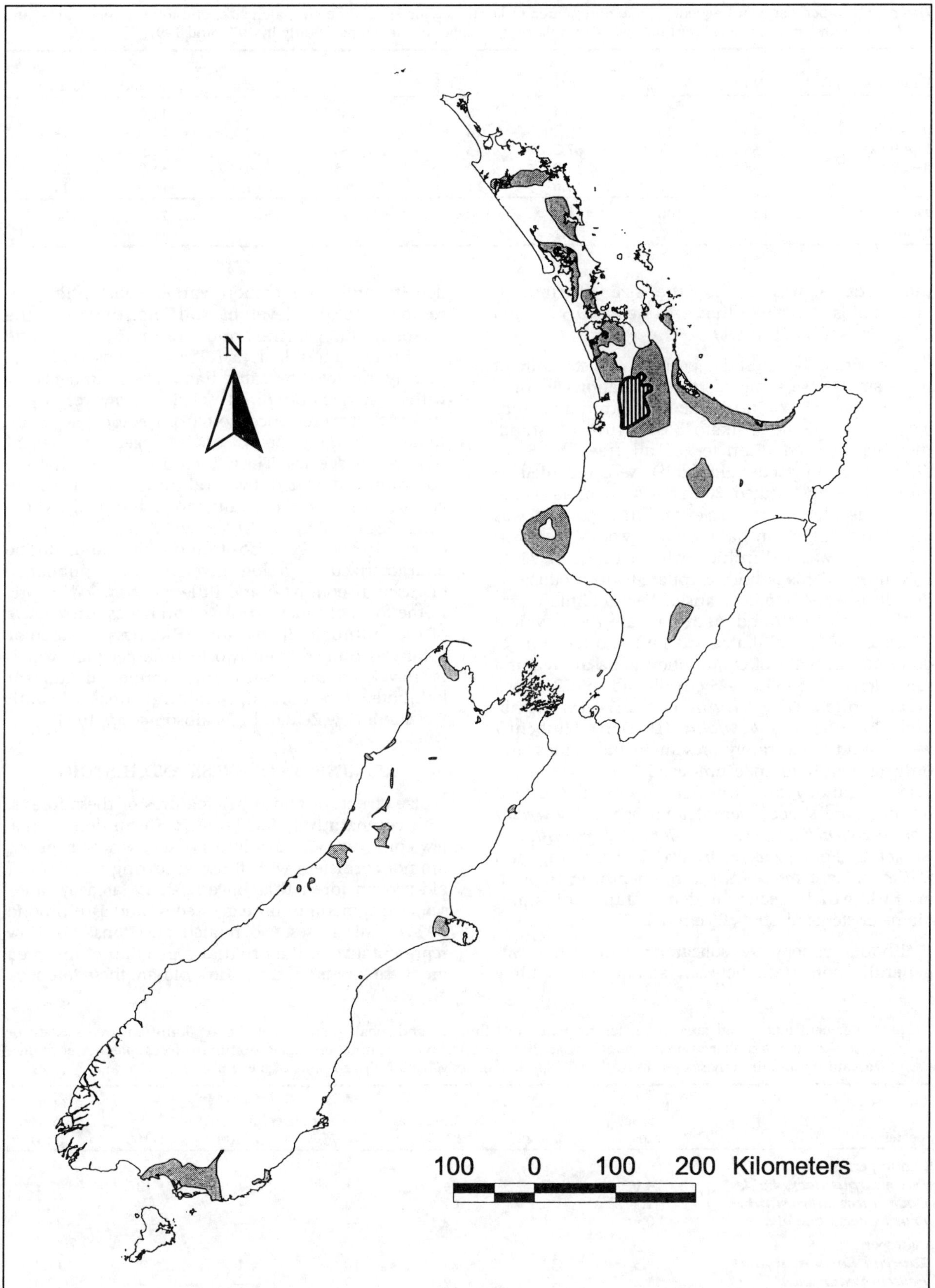

Fig. 2. Map of dairy-dominated farmland in New Zealand (shaded) with the middle Waikato Basin indicated (vertical lines) (from Fryer 1976).

Table 1. Number, per cent frequency, area, and % area of kahikatea forests of different patch size, and total number, area, and mean area of kahikatea forest patches within dairy landscapes of the Waipa County in 1977 and 1997.

	1977				1997			
Patch size (ha)	No.	% freq.	Area (ha)	% area	No.	% freq.	Area (ha)	% area
0–0.49	54	45	11.4	8	55	47	11.0	9
0.5–0.99	28	23	19.2	14	25	22	17.5	14
1–4.99	33	27	57.2	43	30	26	51.5	41
5–9.99	5	4	31.9	24	5	4	31.9	25
≥10	1	1	14.6	11	1	1	14.6	11
Total	121	100	134.3	100	116	100	126.5	100
Mean			1.1				1.1	

these stands was 1.1 ha, the size distribution of stands is skewed so that 47% were <0.5 ha and 69% were <1 ha (Table 1).

To describe the forest composition and structure of these stands, we sampled tree populations in nine of the larger (5–20 ha) kahikatea-dominant stands within the middle Waikato Basin. In each stand, the identity and diameter of all trees >2.5 cm diameter at breast height (dbh) were recorded in four randomly located 20 m × 20 m plots. A list of all vascular plant species within each plot was also compiled. Mean basal area of woody plants in these plots was 101 m^2ha^{-1}, with a range of 71 to 133 m^2ha^{-1}. This is high compared with podocarp/broadleaf forest in other parts of New Zealand where values between 40 and 70 m^2ha^{-1} are more typical (Wardle 1991). Kahikatea was found to be strongly dominant in terms of contribution to basal area and stem density of trees >25 cm dbh (Table 2). Two other conifers (*Dacrydium cupressinum* Lamb. and *Prumnopitys taxifolia* (D. Don) Laubenf.) were found to be rarely present in the stands and only as small to medium-sized trees (Table 2). The subcanopy and understorey are dominated by the more shade-tolerant angiosperms *Laurelia novae-zelandiae* A. Cunn., *Melicytus ramiflorus* J.R. et G. Forst., *Alectryon excelsus* Gaertn., and *Beilschmiedia tawa* (A. Cunn.) Benth. et Hook. f. ex Kirk which occur in dense stands of small diameter stems mostly <25 cm dbh (Table 2).

Although canopy and subcanopy composition was generally consistent between stands, understorey density and composition varied substantially in response to the level of soil moisture and the intensity of grazing and trampling by cattle (B. Burns, unpubl. data). Those stands that were heavily impacted by cattle had an open understorey with many exotic plant species. However, those stands that were fenced off from cattle had dense understoreys composed of a larger number of indigenous species. These included species that are commonly depleted by feral ungulates in more continuous tracts of indigenous forest elsewhere (e.g., *Asplenium bulbiferum* Forst. f., Mitchell *et al.* 1987). We also found the stands to be characterized by a low diversity and abundance of vascular epiphytes and little coarse woody debris in the form of fallen trunks or branches, irrespective of the anthropic disturbance. Richness in vascular epiphytes ranged from two to nine per plot, which is low compared with that found in ancient kahikatea forest in high rainfall areas of south Westland, New Zealand (Dickinson *et al.* 1993).

REPRESENTATIVENESS AND HISTORY

Size structure and other features of these forests, with comparatively few large (>50 cm dbh) stems, few epiphytes, and low levels of coarse woody debris are not consistent with those occurring in stands of old-growth forest that have been isolated by landscape fragmentation (e.g., Keddy and Drummond 1996). This raises two related questions: (1) How representative of the original vegetation of this area are these forests? and (2) How old are these forests?

Table 2. Per cent total basal area, diameter size class distribution, and mean diameter of the 10 dominant tree species for stems >2.5 cm diameter at breast height in 36 20 m × 20 m plots in nine kahikatea-dominant forest patches of middle Waikato Basin dairy landscapes. (A further 28 species collectively contributed 3.3% of basal area.)

Species	% basal area	No/ha in each diameter size class (cm) 2.5–25	25–50	50–75	75–100	>100	Mean diameter (cm)
Gymnosperms:							
Dacrycarpus dacrydioides	74.6	109	194	101	31	12	42.1
Dacrydium cupressinum	1.5	2.1	1.4	2.1	—	—	38.1
Prumnopitys taxifolia	0.5	1.4	2.1	—	—	—	25.1
Angiosperms:							
Laurelia novae-zelandiae	7.3	233	25	10	1	1	11.2
Beilschmiedia tawa	3.2	47	17	3	—	—	19.9
Alectryon excelsus	3.0	51	9	3	—	—	14.0
Elaeocarpus hookerianus	2.4	—	1.4	1.4	—	—	50.5
Melicytus ramiflorus	1.8	560	2	—	—	—	5.5
Knightia excelsa	1.2	1	1.4	1	—	—	38.0
Syzygium maire	1.2	16	1	—	—	—	13.4

Evidence for the past forest composition of the Waikato Basin is provided by palynological analysis of sediment cores taken from peat lakes (McGlone 1988; Newnham *et al.* 1989) and from identification of forest stumps and other macrofossils exposed as drainage lowers peat levels (Cranwell 1939). These studies suggest that the landforms of the middle Waikato Basin supported a range of forest types of which kahikatea-dominant forest similar to existing stands was only one component.

Palynological studies show that since the last glaciation (*c.* 11 000 years b.p.), the Waikato lowlands have been mostly forested, but with a few scattered swamp and bog communities in the wetter parts (McGlone 1988; Newnham *et al.* 1989). However, kahikatea was found to be poorly represented in the pollen deposits. Instead, the pollen deposits were dominated by the conifers *D. cupressinum*, *P. taxifolia*, *Agathis australis* (D. Don) Lindl., and *Phyllocladus trichomanoides* D. Don, by the angiosperms *Metrosideros* spp. and *Nestegis* spp., and by tree ferns of the genus *Cyathea*. Increases in charcoal in the peat cores about 800 years ago indicate clearance of forest for Maori agriculture (Newnham *et al.* 1989). Fire and cultivation at this time significantly transformed the original landscape (Jay 1997). When the region was first seen by European settlers in the 1860s, it was largely covered by the early successional species *Pteridium esculentum* (Forst. f.) Cockayne and *Leptospermum scoparium* J.R. et G. Forst, with large expanses of swamp and restiad bog (Jay 1997).

Many buried tree stumps have been and continue to be removed from former restiad bogs in the Waikato as these lands contract with drying and decomposition of the peat. Cranwell (1939) described stump belts around the major swamps up to 1.2 km wide. Species she identified from these stumps included *D. cupressinum*, *Metrosideros* spp., *P. taxifolia*, *Monoao colensoi* (Hook.) Molloy, *B. tawa*, *Knightia excelsa* R. Br., *A. australis*, kahikatea, and *L. novae-zelandiae*. Her interpretation was that kahikatea and *L. novae-zelandiae* with some *P. taxifolia*, *D. cupressinum*, and *M. ramiflorus* occurred in the wetter, lower lying, silt filled depressions and around the margins of restiad bogs. On the extensive drier surfaces, the forest was dominated by a different mix of species, e.g., *D. cupressinum*, *A. australis*, *P. taxifolia*, *B. tawa*, etc.

How then do we explain the predominance of kahikatea in extant forest patches of the Waikato landscape? Some indications may be gained by examining the age structure of kahikatea populations within these forests. We took samples of 10–20 tree cores at breast height from kahikatea populations within the nine stands previously mentioned, stratifying samples evenly across diameter size classes. Cores were taken with increment borers and, after mounting and surfacing cores in the laboratory following standard methods (e.g., Stokes and Smiley 1968; Duncan 1989), growth rings were counted to provide an estimate of individual tree age to coring height.

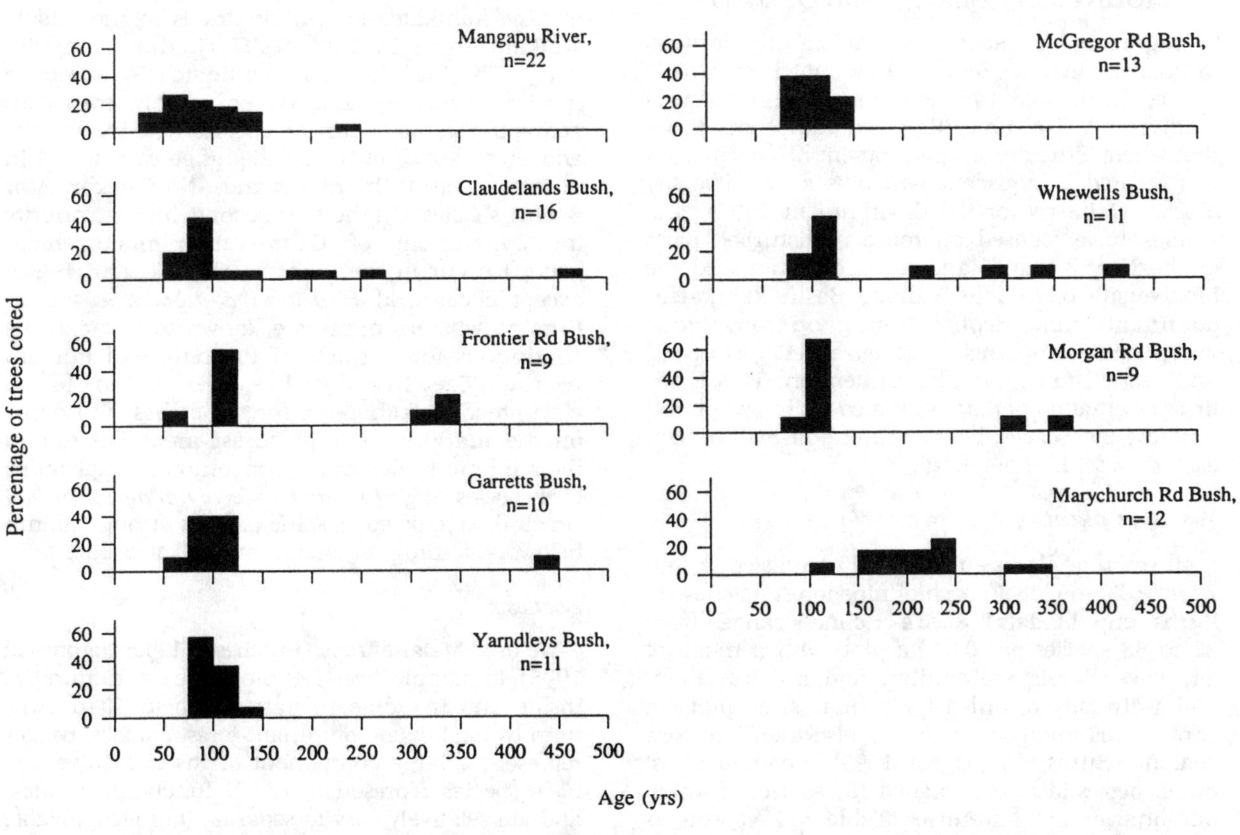

Fig. 3. Age distribution of tree core samples taken from nine kahikatea-dominant forest patches of the middle Waikato Basin. Samples were stratified across the range of sizes present.

Results suggest that kahikatea populations comprise two broad groups in almost all stands. A small group of large trees, 1–2 m dbh and 200–450 years old, form the core of most stands, and the majority of stems of 30–90 cm dbh and 75–125 years old surround this core. Only one of the nine stands returned a different pattern of tree ages. Most of the trees aged within a kahikatea-dominant stand on Marychurch Road were 150–250 years old.

It appears therefore, that the majority of the stands was formed by a relatively even-aged cohort of kahikatea establishing between 1870 and 1920 around individual or small groups of remnant trees (Cranwell 1939). Historically, this period coincides with the time of extensive land clearance and drainage in the Waikato (Jay 1997), disturbances which would have been advantageous for the establishment of kahikatea seedlings which are generally more shade-intolerant than most other trees including other conifers (Duncan 1993). Whaley *et al.* (1997) also report that at one site (Claudelands Bush), kahikatea dominance was even more enhanced by removal by logging of most large *P. taxifolia* and *D. cupressinum*. Further development of kahikatea-dominant stands will result in self-thinning of the even-aged kahikatea cohort and an increase in importance of other shade-tolerant tree species (e.g., *B. tawa*). Similar age-structures to those reported here have been reported for three of four kahikatea populations sampled by Arlidge (1992) in rural areas near Auckland.

BIODIVERSITY QUALITY AND QUANTITY

Fragments of natural systems in agricultural landscapes in New Zealand have not been widely studied for their contribution to regional indigenous biodiversity. The prevailing attitude seems to be that where agricultural development has occurred it has created a landscape which is a "biodiversity desert" (Ministry for the Environment 1997), and studies have focused on more undisturbed parts of the New Zealand landscape. We examined the biodiversity of middle Waikato Basin kahikatea-dominant stands, concentrating on four biotic groups, vascular plants, birds, terrestrial gastropods, and beetles, to explore this contention. This is the first presentation of data from a co-ordinated project on these landscapes; later contributions will present each data set in more detail.

Vascular plants

All vascular plants encountered were listed in four 20 m × 20 m plots in each of nine forest patches (B. Burns, unpubl. data). Species richness ranged from 20 to 58 species per 0.04 ha plot, with a mean of 34. These levels are similar, and not less than, alpha-diversity recorded for similar sized plots in continuous indigenous forest elsewhere in New Zealand (Burns 1995; Ogden 1995). A combined list of all species identified totalled 187 species of which one-quarter are adventive (Table 3). As well as many common weeds of roadsides and wastelands, some of the adventive species are clearly species with bird-dispersed fruits that have escaped from gardens or horticultural enterprises nearby (e.g., *Clerodendrum trichotomum* Thunb., *Actinidia deliciosa* (A. Chev.) Liang et A.R. Ferg., *Asparagus officinalis* L.). Of the indigenous species encountered, only one, *Myriophyllum robustum* Hook. f., an aquatic herb, is currently on the New Zealand threatened plant list (Cameron *et al.* 1995). Most are regionally common species. Whaley *et al.* (1997) found 79 indigenous and 60 adventive species in Claudelands Bush, a remnant within Hamilton City, with none nationally or regionally rare. Similarly, de Lange (1987, 1989) found 110 and 106 indigenous species respectively in two kahikatea-dominant forest patches in the Waikato Basin but did not record adventive species.

Table 3. Number of species and % adventive species of vascular plants, birds, beetles and terrestrial gastropods identified within kahikatea-dominant forest patches of Waikato Basin dairy landscapes.

Taxonomic group	Stands	Number of species	% adventive
Vascular plants	9	187	25
Birds	9	23	61
Malaise-trapped Coleoptera	2	91	26
Terrestrial gastropods	10	62	18

Birds

Birds were surveyed by conducting five-minute bird counts (Dawson and Bull 1975) and noting all birds heard or seen within and on the periphery of nine kahikatea-dominant stands in the middle Waikato Basin in 1996–1997 (J. Innes, unpubl. data). The bird fauna is dominated by adventive species (Table 3) and is remarkably consistent across the stands. Fourteen species (five indigenous and nine exotic) of the 23 identified were found in at least seven of the nine stands. No Category A or B bird species (highest or second highest priority for Department of Conservation management action) occur in the forest patches (Tisdall 1994), except occasional *Hemiphaga novaeseelandiae* Gmelin, whereas many are known to occur in the continuous forest tracts of Pirongia and Pureora nearby. The introduced species found to be widespread in kahikatea forest patches also occur on the margins of large forest areas but not in their interiors. Regionally uncommon frugivorous bird species (e.g., *Prosthemadera novaeseelandiae* Gmelin) were noted in some patches during autumn fieldwork, feeding on abundant kahikatea fruit.

Beetles

We used Malaise traps (Hutcheson 1990; Hammond 1995) to sample beetle (Coleoptera) communities inside and in adjacent pasture up to 50 m away from two kahikatea-dominant forest patches. Beetles represent a large component of insect biodiversity, have species represented in all functional groups, and are relatively easy to separate into recognizable taxonomic units. Malaise traps were sampled on a weekly basis for four weeks in December 1995

(R. Harris, unpubl. data). We identified 91 recognizable taxonomic units within the forest patches, with about one-quarter of these adventive (Table 3). Whereas the fauna of the forest patches was numerically dominated by indigenous species, the pasture beetle fauna was numerically dominated by a few introduced species. There was much greater richness of beetles within the forest patch than in the pasture and there was a greater richness in the forest patch with a more mature understorey (longer time since fenced off from cattle grazing). Most of the indigenous species sampled are known to be relatively widespread. For example, >75% of the species recorded had previously been found in a study of the beetle fauna of an Auckland suburb (Kuschel 1990). We found 27 unnamed species, but most are known from elsewhere and present in the New Zealand Arthropod Collection, Mt Albert.

Terrestrial gastropods

Ten stands were sampled for terrestrial gastropods by collection and sorting of litter samples. The total number of species recorded from the sampled stands was 62, with only 11 adventive species (Table 3). Richness ranged from 3 to 44 species per stand (G. M. Barker, unpubl. data), being low in degraded, grazed stands and high in stands with either a dense understorey (ungrazed) or with a higher importance of broadleaf trees contributing to litter. However, the richness of adventive species followed an inverse relationship both with grazing and litter quality. No gastropods were unique to the Waikato kahikatea stands, with communities being a subset of a Waikato regional terrestrial gastropod fauna that numbers around 150 indigenous and 25 adventive species. Only two indigenous gastropods have been recorded from Waikato dairy pastures (G. M. Barker, unpubl. data).

DISCUSSION

Indigenous forests have been almost entirely eliminated from the dairy landscapes of the middle Waikato Basin. However, over 100 small, discrete patches of kahikatea-dominated forest remain scattered across this landscape. These are mostly <1 ha in size and occur at a density of 0.15 patches per km^2. The number and size of these patches has reduced only slightly over the last 20 years. It is likely that this forest type represents only one of the former forest types of the pre-human landscape. Most of the kahikatea trees within these stands established within a few decades of 1870, corresponding to a period following major human-caused landscape modification. Therefore, there is a low variation in forest age-structures as well as in canopy species diversity.

The relatively even-aged structure of kahikatea in most stands suggests that populations of kahikatea will eventually decline unless disturbances at an appropriate scale occur to initiate renewed recruitment. The natural disturbances that once maintained a significant presence of kahikatea forest in the middle Waikato Basin no longer operate due to human control of waterways and flood frequency/severity. If kahikatea stands are to continue to be a feature of the landscape, there will have to be active management, either periodic planting of new stands or imposition of "disturbance" on "older" stands (e.g., clearing parts of stands for wood production).

Each stand sampled had a similar biota consisting of common species widespread in the North Island. Among the biotic groups sampled, we found no species unique to Waikato kahikatea stands and few are recognized as rare or threatened by conservation agencies. However, the biota encountered could not be described as depauperate and, except for birds, was dominated by indigenous species. The species richness of all taxonomic groups examined was lower and contained a higher exotic component in those stands degraded by understorey grazing by cattle.

Even though these stands represent most of the natural area with indigenous biodiversity conservation values in the middle Waikato Basin, few are formally protected (Wassilieff and Timmins 1984). Most stands depend for their persistence on the goodwill of landowners (generally farmers) and the inconsistent application of principles of sustainable management of natural resources embodied in the *Resource Management Act 1991* (particularly section 6c). Our observations suggest that, while most stands have persisted over the past 20 years, many are being progressively degraded by anthropic disturbance. The absence of regenerative capacity suggests an eventual collapse and loss of these disturbed remnants. There have been only minor attempts (e.g., approximately 1 ha at Lake Serpentine) to restore indigenous forests in this landscape. A survey of owner attitudes to protecting native forest in the former Waipa County (van Oostrom 1985) highlights the range of opinions present then and probably now. "Of the 153 properties visited in 1978, 35% of owners had favourable views towards conservation, a further 24% were conservation minded and active in this, 9% were indifferent, and 5% wished to see their bush removed." Without a co-ordinated plan to integrate forest areas into the dairy landscape, we believe a slow decline and elimination of stands is inevitable.

So what of the future? Two scenarios present themselves for future dairy landscapes. First, if the status quo continues, we foresee a dairy landscape which is even more monocultural than at present, largely a sea of grass and cows. In this scenario, society segregates production and conservation land, so that the environmental services of natural ecosystems to the dairy industry (e.g., potential surface and groundwater purification, biodiversity conservation, provision of soil benchmarks, sink for greenhouse gases, flood mitigation) occur on forested fringing hill country. Such a scenario would leave a landscape of higher "ecological risk" and the current market and non-market values of the current forest patches would be lost. We question whether such a scenario is sustainable at either individual farm or regional scales.

We suggest and advocate a second potential future landscape. In this, the dairy landscape is a mosaic of interlinked forests and grasslands so that production and conservation values are integrated in a single landscape. In this landscape, there would be increased on-site mitigation of any environmental impacts of dairy production, lower "ecological risk", and the current values of forests would be enhanced.

To achieve this vision, the conservation value of existing forest patches could be enhanced and these could then be linked to a network of newly-restored forest patches. Simply fencing off forest patches from grazing leads to an increase in indigenous biodiversity in understoreys. The suppression of adventive weeds is one reason given by farmers for continuing cattle grazing in remnants; our results suggest they are actually promoting weed invasion and abundance. New forest patches could be established by planting indigenous species on land not used or marginal for dairying. Taking evidence of past forests of the middle Waikato Basin as a model, restoration of both kahikatea and non-kahikatea dominant forests would be warranted to increase the diversity of forest types present and extend the range of patch ages represented. It is increasingly recognized that many components of indigenous biodiversity may be found in the understoreys of plantations of exotic tree species, e.g., *Pinus radiata* (Maclaren 1996) and such production woodlots could be used as nurseries for indigenous vegetation communities. Although this concept needs more investigation, a conservation strategy for enhancing dairy landscapes may include such woodlots as one component. We believe that achieving such a vision is in the interests of farmers and the dairy industry generally, as well as local authorities and conservation organizations (both public and private), and challenge these groups to develop and implement a conservation strategy for New Zealand dairy landscapes.

ACKNOWLEDGEMENTS

We thank a number of people for assistance in field and office including Malcolm Douglas, Grace Hall, Merrilyn Merrett, Jo Rees, Anne Rose, Mark Smale, Kathryn, Liam, and Patrick Whaley. We also thank the many landowners for permission to gain access and to sample their stands. Funds for this research were partly provided by the Foundation for Research, Science, and Technology under contract C09522.

REFERENCES

Arlidge, M., 1992. Phenology, dendrochronology, population structure: the ecology of sex in kahikatea (*Dacrycarpus dacrydioides*). Unpublished M.Sc. thesis, University of Auckland.

Barber, L. H., 1978. The view from Pirongia: the history of Waipa County. Richards: Auckland, New Zealand.

Bolan, N. S. and Podila, P. R., 1996. Environmental impact of increased fertiliser nitrogen use in pasture in New Zealand. *NZ soil news* **44:** 4–11.

Boubee, J. A. T., 1978. Native bush stands in the Waipa County. Unpublished report, Waipa County Council, Te Awamutu, New Zealand.

Burns, B. R., 1995. Environmental correlates of species richness at Waipoua Forest Sanctuary, New Zealand. *NZ J. Ecol.* **19:** 153–62.

Cameron, E. K., deLange, P. J., Given, D. R., Johnson, P. N. and Ogle, C. C., 1995. New Zealand Botanical Society Threatened and Local Plant Lists (1995 revision). *NZ Bot. Soc. Newsl.* **39:** 15–28.

Cranwell, L. M., 1939. Native vegetation. Pp. 23–30 *in* Soils and Agriculture of Part of Waipa County ed by L. I. Grange, N. H. Taylor, C. F. Sutherland, J. K. Dixon, L. Hodgson, F. T. Seelye, E. Kidson, L. M. Cranwell and P. W. Smallfield. Bulletin No. 76. Department of Scientific and Industrial Research. Government Printer: Wellington, New Zealand.

Dawson, D. G. and Bull, P. C., 1975. Counting birds in New Zealand forests. *Notornis* **22:** 101–09.

de Lange, P. J., 1987. The Orini kahikatea forest, Tauhei-Whitikahu. *Auck. Bot. Soc. Newsl.* **42:** 64–69.

de Lange, P. J., 1989. The indigenous vegetation of the McGregor Road kahikatea forests, Ngahinapouri, Hamilton Basin. *Auck. Bot. Soc. Newsl.* **44:** 50–54.

Dickinson, K. J. M., Mark, A. F. and Dawkins, B., 1993. Ecology of lianoid/epiphytic communities in coastal podocarp rain forest, Haast Ecological District, New Zealand. *J. Biogeogr.* **20:** 687–705.

Duncan, R. P., 1989. An evaluation of errors in tree age estimates based on increment cores in kahikatea (*Dacrycarpus dacrydioides*). *NZ Nat. Sci.* **16:** 31–37.

Duncan, R. P., 1991. Competition and the coexistence of species in a mixed podocarp stand. *J. Ecol.* **79:** 1073–084.

Duncan, R. P., 1993. Flood disturbance and the coexistence of species in a lowland podocarp forest, south Westland, New Zealand. *J. Ecol.* **81:** 403–16.

Fryer, W., 1976. Land Use. Pp. 142–43 *in* New Zealand Atlas ed by I. Wards. Government Printer: Wellington, New Zealand.

Hammond, P. M., 1995. Estimating diversity in speciose groups. Pp. 119–36 *in* Biodiversity Measurement and Estimation ed by D. L. Hawksworth. Chapman & Hall: London.

Hewitt, A. E., 1993. New Zealand soil classification. Landcare Research Science Series No. 1. Manaaki Whenua-Landcare Research New Zealand Ltd: Lincoln, New Zealand.

Hutcheson, J., 1990. Characterization of terrestrial insect communities using quantified malaise-trapped Coleoptera. *Ecol. Entomol.* **15:** 143–51.

Jay, M., 1997. A short history of the landscapes of central Waikato. Pp. 60–68 *in* People and Places in the Waikato Region ed by E. Stokes and M. Begg. Waikato Branch NZ Geographical Society, c/– Department of Geography, University of Waikato: Hamilton, New Zealand.

Keddy, P. A. and Drummond, C. G., 1996. Ecological properties for the evaluation, management, and restoration of temperate deciduous forest ecosystems. *Ecol. Appl.* **6:** 748–62.

Kuschel, G., 1990. Beetles in a suburban environment: a New Zealand case study. DSIR Plant Protection Division: Auckland, New Zealand.

Maclaren, J. P., 1996. Environmental effects of planted forests in New Zealand. New Zealand Forest Research Institute Ltd: Rotorua, New Zealand.

McGlone, M. S., 1988. New Zealand. Pp. 557–99 *in* Vegetation History ed by B. Huntley and T. Webb III. Handbook of vegetation science 7. Kluwer Academic Publishers: Dordrecht, The Netherlands.

Meurk, C. D., 1984. Bioclimatic zones for the Antipodes — and beyond? *NZ J. Ecol.* **7**: 175–82.

Ministry for the Environment, 1997. The State of New Zealand's Environment. Ministry for the Environment: Wellington, New Zealand.

Ministry of Agriculture, 1997. Farm Monitoring Report: North Region July 1997. Ministry of Agriculture: Hamilton, New Zealand.

Ministry of Works, 1973. Background. Pp. 17–21 *in* Waikato-Coromandel-King Country Region. National Resources Survey, part VIII ed by Town and Country Planning Division, Ministry of Works. Government Printer: Wellington, New Zealand.

Mitchell, R. J., Fordham, R. A. and John, A., 1987. The annual diet of feral goats (*Capra hircus* L.) in lowland rimu-rata-kamahi forest on eastern Mount Taranaki (Mt Egmont). *NZ J. Zool.* **14**: 179–92.

Newnham, R. M., Lowe, D. J. and Green, J. D., 1989. Palynology, vegetation and climate of the Waikato lowlands, North Island, New Zealand, since *c.* 18,000 years ago. *J. Roy. Soc. NZ* **19**: 127–50.

New Zealand Meteorological Service, 1983. Summaries of climatological observations to 1980. New Zealand Meteorological Service miscellaneous publications 177. Government Printer: Wellington, New Zealand.

O'Connor, M. B., Boddy, I. K., Morton, J. D. and Edmeades, D. C., 1996. Nutrient budgets — a key indicator of soil sustainability. *Proc. NZ Grassl. Assoc.* **58**: 219–22.

Ogden, J., 1995. The long-term conservation of forest diversity in New Zealand. *Pac. Conserv. Biol.* **2**: 77–90.

Orbell, G. E., 1973. Soils. Pp. 73–88 *in* Waikato-Coromandel-King Country Region. National Resources Survey, part VIII ed by Town and Country Planning Division, Ministry of Works. Government Printer: Wellington, New Zealand.

Selby, M. J. and Lowe, D. J., 1992. The middle Waikato Basin and Hills. Pp. 233–55 *in* Landforms of New Zealand, second edition ed by J. M. Soons and M. J. Selby. Longman Paul: Auckland, New Zealand.

Statistics New Zealand, 1997. New Zealand official yearbook 1997. Government Printer: Wellingon, New Zealand.

Stokes, M. A. and Smiley, T. L., 1968. An introduction to tree-ring dating. University of Chicago Press: Chicago, U.S.A.

Tisdall, C., 1994. Setting priorities for the conservation of New Zealand's threatened plants and animals. Department of Conservation: Wellington, New Zealand.

van Oostrom, A. J., 1985. A survey of owners' attitudes to protecting native bush in the Waipa County. Unpublished report, Waipa County Council, Te Awamutu, New Zealand.

Wardle, P., 1974. The kahikatea (*Dacrycarpus dacrydioides*) forest of South Westland. *Proc. NZ Ecol. Soc.* **21**: 62–71.

Wardle, P., 1991. Vegetation of New Zealand. Cambridge University Press: Cambridge, Great Britain.

Wassilieff, M. and Timmins, S., 1984. Register of protected natural areas in New Zealand. Department of Lands and Survey: Wellington, New Zealand.

Whaley, P. T., Clarkson, B. D. and Smale, M. C., 1997. Claudelands Bush: ecology of an urban kahikatea (*Dacrycarpus dacrydioides*) forest remnant in Hamilton, New Zealand. *Tane* **36**: 131–55.

CHAPTER 10

Integrated conservation and development in Papua New Guinea: some lessons from Wau Ecology Institute

H. Sakulas,[1] J. J. Bauer,[2] J. Birckhead[1] and T. DeLacy[3]

This chapter describes how action-oriented research, in biodiversity conservation and income generating sustainable development activities, can be integrated to achieve a productive environment in Papua New Guinea. The chapter also highlights Wau Ecology Institute's current and future plans for community-based conservation and sustainable development activities that are socially, and culturally compatible, ecologically friendly and economically viable. Shifting agriculture for subsistence food production and logging are major factors in tropical forest destruction. A modified approach to subsistence agriculture and tropical forest utilization can create a diverse and highly productive agricultural environment which satisfies most requirements for ecological sustainability and social and economic development needs. The destruction of tropical rainforests is also mirrored by an ongoing impoverishment of agricultural systems characterized by and tending towards single varieties and an increasing rate of clones. There is very little understanding of how these systems can be sustained in themselves and even less understanding of what role they might play as matrix for natural remnants. This trend corresponds with an increasing range of expectations regarding conservation. We attempt to provide some answers to these questions. Our model describes how local initiatives and a range of economic and subsistence sector activities have been integrated with tropical rainforest conservation at the Wau Ecology Institute.

Key words: Biodiversity conservation, Sustainable development, Subsistence agriculture, Wau Ecology Institute.

THE PAPUA NEW GUINEA ENVIRONMENT

PAPUA New Guinea has probably undergone more profound social change within one generation than any other country on Earth. Within less than 40 years a nation was formed out of a multitude of remote communities which were not only isolated by Papua New Guinea's mountains and rainforests but by more than 700 different languages, and deeply entrenched hostility and distrust between tribes. With four million inhabitants on a land area of 462 842 sq kilometres, 73% of it forested and rich in mineral deposits and biodiversity [5% of the worlds total species; (Saulei 1990)], Papua New Guinea is today considered one of the most resource-rich countries per capita in the world. Unlike most other nations in the world, Papua New Guinea's rural communities (over 80% of its populace) own 98% of the land. Despite this wealth, most communities have no road access, little or no infrastructure and an almost total reliance on subsistence agriculture. In its drive for development Papua New Guinea has attempted to utilize these natural resources in both traditional and modern ways (e.g., logging and mining). The recent rate of deforestation and forest conversion is estimated at about 350 000 ha per year (Saulei, in litt.). This rate of exploitative "development" would eliminate Papua New Guinea's useable timber resources within 15 years and involve enormous environmental impacts. Much of the converted land is being rapidly degraded due to high

[1]School of Environmental and Information Sciences, Charles Sturt University, New South Wales, Australia 2640.
[2]Environmental Studies Unit, Charles Sturt University, Panorama Avenue, Bathurst, New South Wales, Australia 2795.
[3]Griffith University Gold Coast, PMB 50, Gold Coast Mail Centre, Queensland, Australia 9726.

Pages 90–102 *in* NATURE CONSERVATION 5: NATURE CONSERVATION IN PRODUCTION ENVIRONMENTS: MANAGING THE MATRIX ed by J. L. Craig, N. Mitchell and D. A. Saunders. Surrey Beatty & Sons, 2000.

rainfall, extreme topography and subsequent high erosion rates. Given its current annual population growth of 2.7%, in 2020 if current trends prevail, approximately eight million people will have to live in a degraded and non-productive environment.

Papua New Guinea still has the opportunity to find its own approach to use of natural resources which will facilitate and ensure sustainable development. Access to natural resources by developers in Papua New Guinea requires the consent of the traditional landowners. This consent is also a prerequisite for conservation. Acknowledgement of traditional land rights is therefore the key issue for development and conservation in Papua New Guinea.

INTEGRATED CONSERVATION AND DEVELOPMENT

Approximately 8 000 protected areas occupy between 5 and 6% of the world's land surface. Barraclough (1995) recognized that communities surrounding these protected areas have in many cases lost traditional rights of access to resources and have therefore often opposed conservation. In response to this problem, projects linking conservation objectives with traditional utilization rights and development activities have been established, particularly in developing nations. These have been termed Integrated Conservation and Development Projects by Wells *et al.* (1992). These projects aim to incorporate economic and social development initiatives into the creation and management of protected areas, as incentives for the compliance and support of the local people. In their evaluation of integrated conservation and development projects however, Wells *et al.* (1992) concluded that the critical linkage between economic development and conservation is generally missing or unclear, and that the economic and social development incentive packages have not reduced pressure on the resources they were supposed to protect. Some authors including Uphoff (1985), Kottak (1985), Midgley *et al.* (1986) and Hobart (1993) have argued that many failures of these well-intended projects have resulted from the lack of genuine community participation as well as from the adoption of western conceived, and socially and culturally incompatible project designs based on inadequate analysis of community life and its complex characteristics and dynamics. An analysis of several Biosphere Reserves and United Nations Education, Scientific and Cultural Organisation's model for Integrated Conservation and Development further reveals that the concept of sustainability as applied during the development process does not necessarily include the maintenance of biodiversity, although this aim is stated over and over again.

Community "participation" can refer to many different actions. Uphoff (1985) described different levels of participation, and how they might be appropriate in different circumstances. He believed that the level of participation by the beneficiaries of a project in the project's development should be determined in consultation with them. He suggested that those people who are going to benefit from the project need to be acquainted with the project from the beginning, to elicit their ideas and suggestions, and to assist in the development of appropriate methods for institutionalizing participation at all stages of the project.

According to Uphoff (1985), real participation occurs when the beneficiaries are given the opportunity to comment on the actual project document at the planning or designing phase, and to ascertain whether the proposed activities are feasible and acceptable by the majority of the community as a whole. This will ensure appropriate modification is made at the beginning of the project. Such modification may include realistic time frames to implement the project in a society where flexibility of time is often required. In addition, it is crucial to make realistic financial allocation to what the community considers as vitally necessary to convince them and win their support from the beginning. Kottak (1985) also argued that real participation is when the communities are empowered and supported with the necessary resources to have a major say in how they want their natural resources to be managed. He emphasized that local people should be the ones to decide on what should be included in the project design and what they expect to gain from conservation-driven projects. That author emphasized that development intervention means eliciting the needs for change perceived by the local community; identifying culturally compatible goals and strategies for change, and developing socially appropriate, workable and efficient project designs. The local people must be able to make decisions and control the activities that affect their lives (Cernea 1991). This is why the community participatory process is so central to the success of integrated conservation and developments projects especially in societies where land and natural resources are communally owned and the expectation for incentives is high.

INTEGRATED CONSERVATION AND DEVELOPMENT IN PAPUA NEW GUINEA

In Papua New Guinea, projects which propose to integrate conservation with development are called Integrated Conservation and Development Projects. Ten of these Integrated Conservation and Development Projects have been developed by different stakeholders in Papua New Guinea since 1991 (J. James, in litt.). Most of these projects have been initiated by national and international non-government organizations. The first pilot project, led by the Department of Environment and Conservation and the United Nations Development Programme, was initiated in the Lak area of New Ireland Province in 1993. This project was funded by World Bank under the Global Environmental Facility Programme, a funding mechanism established by developed countries to assist developing countries to implement biodiversity conservation and sustainable development initiatives. Two years later the project was considered a failure. R. McCallum and N. Sehkran (in litt.) argued that lessons learnt from the Lak experience should be incorporated into future project designs. A second pilot project has now being initiated in the Ramu valley of Madang Province.

As stated earlier, in Papua New Guinea 80% of the country's four million people live in rural communities and own 98% of the land through customary clan ownership (Australian International Development Assistance Bureau 1994). In recent years many landowner groups in Papua New Guinea have permitted logging and mining activities in exchange for economic and social benefits offered by companies involved in these activities. Eaton (1992) believed that local people readily make the decision to allow companies access to their traditional lands in response to offers of easy cash benefits and the opportunity to acquire other services like roads, schools and medical aid posts. If Integrated Conservation and Development Projects are to succeed, the conservation initiatives offered must match the substantial benefits available from unsustainable short-term exploitative development. Initiators of Integrated Conservation and Development Projects must also appreciate that effective community participation, critical to the success of these initiatives (N. Grant, in litt.) rests on incentives offered, including the acknowledgement of traditional land utilization rights and the development of appropriate infrastructural facilities.

THE WAU ECOLOGY INSTITUTE MODEL

Our review suggests that the Integrated Conservation and Development model is a key process for the conservation of the world's ecological functions and the sustainable utilization of its resources. It also suggests that there is a wide gap between theory and practice in this model. It would seem that unless we can define the meaning of ecological sustainability and of biodiversity we will not be able to achieve long-term economic and social stability. The Integrated Conservation and Development model has defined the key feature of this process; the links between conservation, society and development. It has not however, provided the nature of these links nor has it sufficiently explored the process. Papua New Guinea appears to be a key country in the search for these unknowns, as it still has the social structures within its society for a community-based decision making process. In many other countries an all-embracing bureaucracy has became an almost insurmountable obstacle in the search for this process (see also Birckhead *et al.*, this volume). Remoteness in Papua New Guinea has become a buffer for communities antagonistic towards bureaucratic domination. In many other ways it is also its greatest weakness. The desire for development is overwhelming. Conservation efforts planned and driven from outside have neither the vision for greater indigenous people's involvement, nor the time and adequate resources to supply these functions. This could be equally said of the exploitative large scale development projects. There has also been too little effort and thought put into the strategic development of the capacity and involvement of indigenous peoples.

Many Integrated Conservation and Development model projects have failed because they are rarely comprehensive enough to fulfill all the functions needed and to link the integrated features of development, e.g., training, management, research and other community-based functions where the communities themselves have a major input. In most cases such projects have been used as an image promotion tool by developers and organizations to attract funding in the name of third world countries from bilateral and multilateral funding institutions. Many of these projects are unable to cite past practical achievements that are still sustainable today. Integrated Conservation and Development projects, however, can only be considered successful if they prove to be sustainable after the initial funding period is terminated, and are able to be maintained entirely by the national institutions or communities of those countries.

Papua New Guinea, which has been part of the global Integrated Conservation and Development process since 1991, has a long established and fairly successful Integrated Conservation and Development model of its own.

HISTORY OF WAU ECOLOGY INSTITUTE

Wau Ecology Institute was set up in the 1960s by the late Dr J. L. Gressitt as a field station for collecting biological specimens for the Bishop Museum of Hawaii and the Smithsonian Institution in Washington, DC. When Papua New Guinea's first House of Assembly was set up in 1964 members expressed a desire to establish a conservation and research institution. Dr Gressitt vigorously promoted the idea amongst his colleagues and a number of United States institutions. When these did not respond he took out a personal loan to buy the first properties from the European settlers who were beginning to leave Papua New Guinea due to political upheaval associated with self government. These properties were used as mortgage to purchase adjacent properties, until a total of 80 ha was acquired. Thirty ha of this land came from a degraded coffee plantation while 50 ha consisted of degraded mining leases.

Wau Ecology Institute was established at a time when there was little talk of conservation. It celebrated its 25th anniversary in 1995 and is now managed by Papua New Guinea nationals. In its operation, it has combined research, training, extension, management and even politics in a unique Papua New Guinean fashion. Today it is closely involved in many environmental activities with projects in several parts of Papua New Guinea. In so doing it has emerged as an important national and international player in conservation and sustainable development.

A major concern of this Institute is the destruction of Papau New Guinea's rainforests and the development of unsustainable farming practices and non-environmentally regulated mining operations. Much of what is presently known about New Guinea's biodiversity has originated from Wau Ecology Institute. Thus it has as its major mission the development of systems which do not require the full-scale destruction of rainforests and which are not simply concerned with conventional agriculture, but which also include the production of a range of other tropical rainforest products. These include insects for the global insect market, plants for medicinal use and the development of sensitive and sustainable nature and culture-based tourism.

Diverse activities such as reforestation, the establishment of an arboretum, the construction of extensive offices and laboratories, training and research complexes, residential areas, a zoo, a herbal medicinal farm, site stable subsistence agriculture demonstration sites, an environmental training centre, insect ranching and ecotourism facilities, and extension programme sites were set up to provide future options for self-funding.

Infrastructural facilities that support research have also been established. These include nationally significant zoological (invertebrates, vertebrates) and botanical collections (specialized in medicinal plants), an arboretum and a small specialized library. Laboratory space with basic and specialized (e.g., atomic absorption spectrometer to measure heavy metal residues due to mining) facilities aims at supporting field based ecological research for a range of habitats from 1 700 m upward to the less studied alpine grassland of New Guinea (Fig. 1).

Achievement of this extensive task convinced the Smithsonian Institution and the Bishop Museum to join the University of Papua New Guinea and the Papua New Guinea University of Technology in becoming share holders of the Institute, which was then incorporated in 1972 as Wau Ecology Institute. After the formation of the company under the country's *Company Ordinance Act* as a non-profit, non-taxable research and educational institution, it was dedicated to the study of fauna and flora of New Guinea and to provide appropriate training for the conservation of this biodiversity. Since 1982, and especially after the untimely death of the founding director in a plane crash, there has been a major shift to applied research in the areas of insect pests, herbal medicine, water quality testing, insect ranching and the diseases associated with subsistence agriculture. Therefore the undertakings of the Institute in the last 25 years have reflected its mission statement within a changing environment, which is to build a nationally managed research oriented institution that can function independently from government or external support.

During the past 25 years, Wau Ecology Institute's degraded environment has been transformed. By 1997 it contained all the major elements to fulfill the functions of research, training and production. The general appearance of the area is that of a tropical resort with a great variety of habitats, each of which sustain diverse populations of native vertebrates and invertebrates along with a diverse subsistence horticulture. The area almost recreates a traditional village in which Institute staff and a reservoir of casual workers reside, harvest and subsist on garden products including organically grown coffee which generate up to 80% of the Institute's running costs. A third generation of staff has now found employment and subsistence within a growing institutional capacity.

DESCRIPTION OF THE WAU ECOLOGY INSTITUTE RESEARCH PROGRAMME

Since its inception the Institute has promoted basic and applied research and conservation of the flora and fauna of New Guinea. During the 1970s research targeted the ecology and systematic biology of plants and animal life of New Guinea and the surrounding islands. The Institute became the centre for co-ordinating biological field surveys and field research for researchers in New Guinea and the islands of Melanesia. Researchers visited initially from the Smithsonian Institution and the Bishop Museum and later by scientists from Japan and Europe.

In recent years, research initiatives have increasingly addressed issues relating to multi-purpose land use systems in the subsistence sector, development impacts and mitigation studies, herbal medicine and community health care and community based integrated conservation and development initiatives. During this time more than 400 publications derived from research carried out at Wau Ecology Institute. Fourteen handbooks and a number of pamphlets on issues ranging from plants and animals to subsistence agriculture improvement programmes, medicinal plants and their use in rural communities, impacts of developments and soil conservation have been written and used in many training programmes for local communities and professionals in many of the community out-reach programmes. In describing several of the programmes in more detail we will adopt a conventional approach by listing them as separate entities (Table 1). However, in Wau Ecology Institute this is not really an appropriate approach. Research is closely linked with training, community extension, management of economic ventures, campaigning and policy making, so that all projects are interlinked as part of an integrated programme. While it is not possible to give a detailed description of these programmes here, we will attempt to give an account of some major activities of the Institute which can demonstrate the value and feasibility of its approaches.

FROM ECOLOGICAL AND BEHAVIOUR STUDIES IN ENTOMOLOGY TO INSECT RANCHING AS ECONOMIC INCENTIVES FOR CONSERVATION AND RESEARCH

The founding director's research interest in entomology has been the major factor contributing to the Institute housing the third largest entomological collection in the country. This collection includes ectoparasites, insect pests and predators found in subsistence food gardens, spiders,

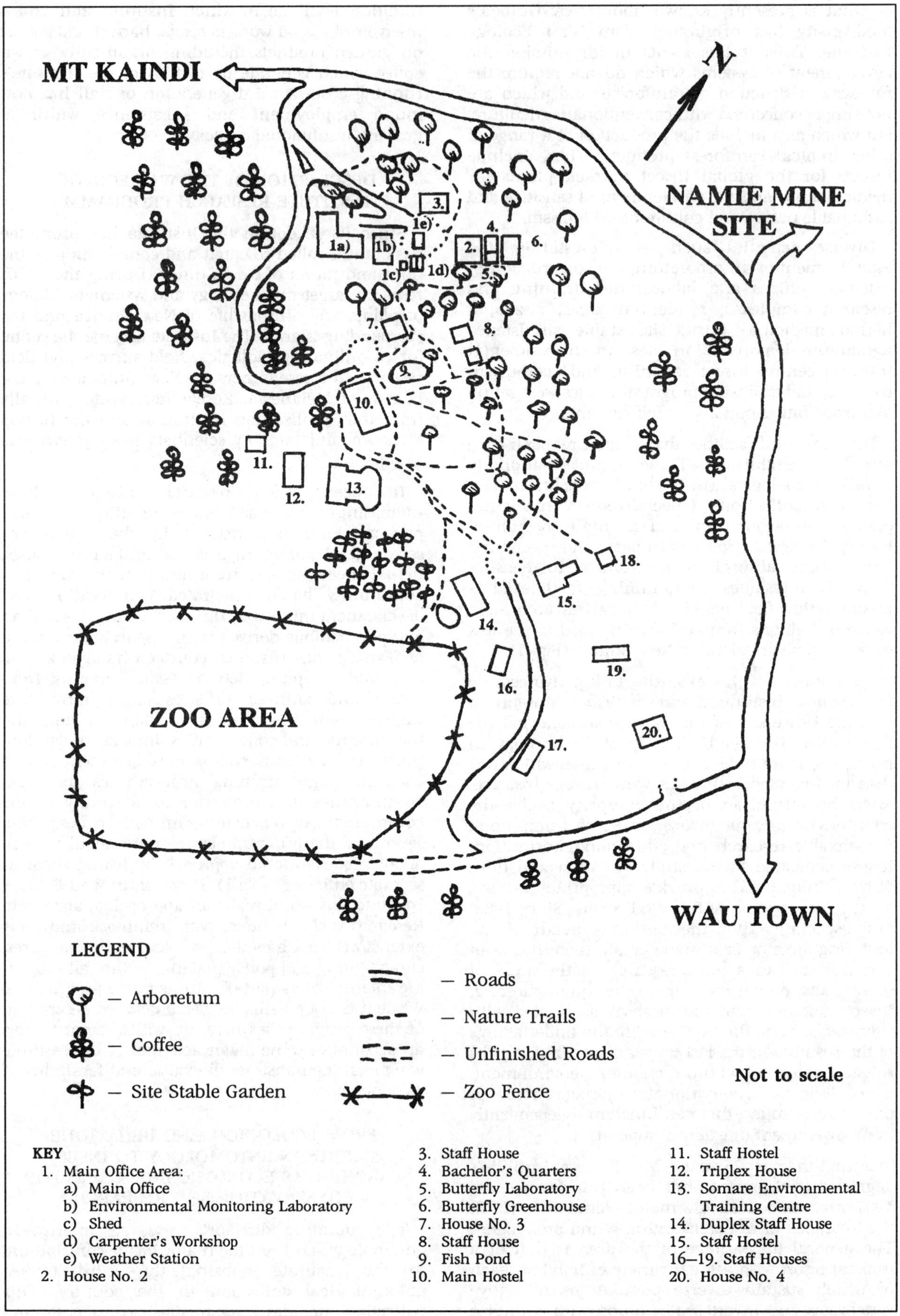

Fig. 1. Land use integration at Wau Ecology Institute.

Table 1. Wau Ecology Institute's research programme.

Managing land and water resources
1. Catchment management — from the mountain to the sea
2. Rehabilitation of degraded land
   2.1 Recovery and stabilization of eroded steep lands
   2.2 Rehabilitation of deserted coffee plantations
   2.3 Conversion of secondary grassland to productive land management systems
   2.4 Rehabilitation of mining sites
   2.5 Rehabilitation of creek and river systems

Agriculture
1. Stable site farming programme
2. Coffee planting in degraded rainforest

Forestry
1. Community forestry in the Prince Alexander conservation programme

Alternative land use
1. Permaculture programme
2. Insect ranching and breeding programme
3. Ecotourism
4. Inventory and screening of medicinal plants in Papua New Guinea and Solomon Islands

Conservation
1. Species inventory
   1.1 Inventory of medicinal plants in Papua New Guinea
   1.2 Inventory of insect species in Papua New Guinea
2. Captive breeding programmes
   2.1 Birds of paradise captive breeding programme
3. Conservation area programme
4. Environment impact assessment
   4.1 Mining impacts
   4.2 Logging impacts

lady bugs and moths found in cash crops such as coffee gardens, and economic insects and beetles which now stimulate much community interest in forest conservation. These specimens currently form part of the national reference collection for research in mid-montane conditions. Apart from the classified collections, Papua New Guinea's largest moth collection of some 12 000 specimens awaits identification and cataloging. More identification is also needed to complete the work on the other major collection of pests and predators of major subsistence food crops.

In addition to entomological surveys, classification and research, there is an increasing need to incorporate research on the commercial aspects of insect farming to promote it as a viable economic alternative to logging of tropical forest which destroys biological species and their habitats.

Observations from the Institute's insect ranching project have enabled the Institute to bring this information to the village communities. The Institute recognizes that the free ranching programme involves more people, providing them with opportunities for cash earnings, than a centralized highly technological captive breeding capitalist programme. This view has fostered the formation of partnerships with rural communities and has enabled the Institute to present to villagers the message of forest conservation and sustainable development.

Alternative income sources are demonstrated to the villagers by action rather than theory. This has had a profound impact on some villages, especially if the villagers can see an insect farmer from their village fetching 100 Kina for one pair of butterflies, and can confirm his monetary gains and achievement of wealth by selling insects to the Institute. Apart from the economic incentives in insect ranching the Institute has added new and rare specimens to the entomological collection with additional information on their geographic range as villagers have been encouraged to bring anything unusual and interesting. In a way this has in turn given an incentive to the village elders to pass on knowledge of particular species, their feeding plants and breeding cycles to their next of kin.

At the global level there is a long history of trade in entomological livestock in North America and Europe (e.g., Bauer 1993), estimated at approximately $US2 billion per annum in North America. Traders buy livestock for rearing for collections or for release in the wild, and deadstock for their collections. Deadstock from Taiwan alone, mostly "decorative", was estimated to account for $US20–30 million. While the decorative trade is not selective and deals mostly with "low quality" abundant species, the specialist trade supplies a market of hobbyists, serious collectors and even research workers. This type of collection could be directly related to the regional disappearance of a species. However, according to Morris *et al.* (1991) ". . . the trade in deadstock has been known to have a serious adverse effect on insect (mainly butterflies) populations only where the extent of the species habitat has already been severely reduced". While the authors emphasize the importance of legal protection of these species they do not advise indiscriminate bans as this would stifle the interest

Case Study 1: Insect Ranching as an Economic Incentive for Conservation and Research

In 1990 the Institute secured an AusAID grant to set up a captive breeding programme of bird wing butterflies. This programme was then further extended by the Institute to initiate the free ranching insect programme. This involved the propagation of insect food plants over the Institute's 80 ha property. From this it was observed that there was a high level of mortality of bird wing pupae found near food plants propagated near coffee trees and less forested areas than those within the more forested part of the arboretum, presumably due to parasitic attack associated with wasps and bees, common in coffee gardens and fruiting trees. The high technology needed to improve micro-climatic conditions and other factors in the greenhouse did not prove favourable for captive breeding. Instead these facilities are now used for studies of life cycles and food preferences of economically valuable species in a parasite free environment.

Villages collect and sell insects to the insect business section of Wau Ecology Institute, which is set up as a financially independent entity to prove its financial viability (Fig. 2). To many rural communities this often provides the only means of cash income. They are presented with an actual alternative as the Institute is able to buy their insects for more money than what they can get from large scale logging projects which are unlikely to exceed 20–50 Kina per ha.

As a comparison one villager can collect and sell forest insects in a single afternoon to the value of 100 Kina. For example, income from bird wing butterfly (Lepidoptera), hybrids of different species, or certain Buprestids (Coleoptera) can be several times the value of one ha forest destruction. In fact one of the farmers from Maprik, who earns 12,000 Kina per annum, cultivates butterfly food plants in his yam gardens. This is because the food plant *Aristolochia tagala* has cultural significance. In this area the plant is mixed with other plants to obtain a crude mixture used as traditional fertilizer for application on growing yam tubers, and is also said to be used for love magic.

To further strengthen and encourage the industry, training concerned with quality control for the international market is provided. Field officers who return from their buying trips recommend that keen farmers attend work attachment and training courses on collecting, drying and packaging, at the Institute. When the farmers return home they not only concentrate on looking for highly priced insects but know how to deliver good quality products.

The insect ranching and marketing project now employs one expatriate and six national staff with the hope of making a turnover of $AUD200,000 per annum by the year 2000.

of the trade which in its economical, scientific and educational aspects, "outweighs any small negative impacts", especially if seen within the context of full-scale forest destruction as an alternative.

More research is needed to document the results of the breeding and rearing activities carried out at the Institute's facilities. The economic value of insects is already convincing several village farmers to conserve their forest, which in turn suggests a future research component worthy of support.

FROM PLANT TAXONOMY TO ARBORETUM AND COMMUNITY FORESTRY AS ECOLOGICAL RESTORATION

Today much of the landscape seen at the Institute has been restored from its deforested state to a productive state. About 30 ha have been reforested with original native oak and pine forest and now provides suitable bird watching sites and habitat for the insect ranching programme. *Araucaria* pine planted as edges between blocks and along the roadside now provide much of the timber for the Institute's building and maintenance needs. Another 30 ha is planted with coffee which is a major income earner for the Institute. Fifteen ha consists of a medicinal plant garden, a zoo with a walk-through 30 m bird of paradise aviary, a site stable subsistence agriculture demonstration block, and arboretum containing plants brought in from other parts of the country. The remaining 5 ha houses the offices, laboratory blocks, Gressitt museum and Somare environmental training centre, staff quarters, ecotourism facilities and the insect ranching and trading centre. The Institute consists of a combination of ecological, economical and social components which make it a truely integrated environment.

Through having sufficient land of its own and a non-profit orientation the Institute has been able to initiate activities that cannot be done on a commercial basis. For example, Wau Ecology

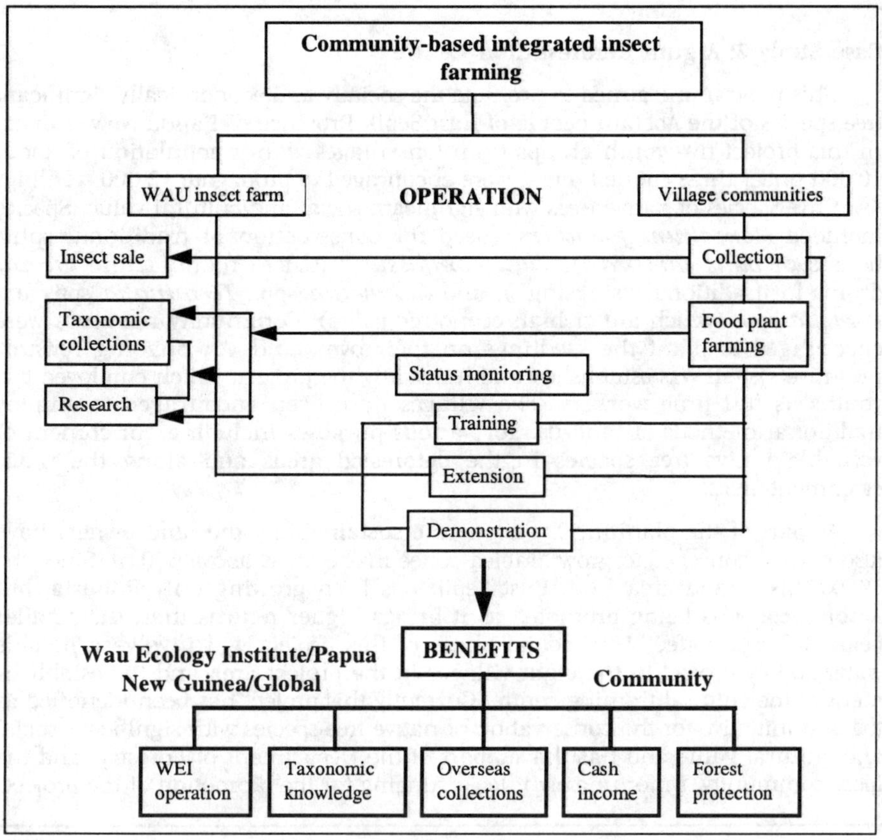

Fig. 2. The benefits resulting from the inter-relationship between research, education and development in the insect programme.

Institute has tried to integrate cash cropping with subsistence food production and fruit trees as shade instead of the usual introduced *Luceanna* species. This has been done by allowing the workers to grow bananas, taro and sweet potato in a controlled manner. Major fruiting tree species known to attract birds have been planted to incorporate bird watching aspect for visitors within the premises.

Wau Ecology Institute has been involved in many activities on more sustainable forestry. Its future role can make a significant contribution to development of a sustainable tropical forestry in Papua New Guinea. Case Study 2 describes the development of the Algum Reforestation centre as an example of the Institute's activities.

In Papua New Guinea vast tracts of tropical rainforest have survived the onslaught of the timber industry (e.g., Filer and Sekhran 1998). Due to a depressed timber market many logging activities have come to a standstill. This time is a respite for Papua New Guinea (F. Ahrends, pers. comm.). It can be used for training, for research and for redirection.

Tropical rainforests provide much more than timber. Similarly logged areas, which have become dominated by grasses, areas of regrowth and forest plantations (*Araucaria* sp. and *Pinus caribbean*) are all facets of Papua New Guinea's future landscape which need to be managed according to its potential and limitations in an integrated fashion. Wau Ecology Institute can be a model for this integration.

THE HEALTH, NUTRITION AND MEDICINAL PLANTS TRAINING PROGRAMME AS ALTERNATIVE FOR COMMUNITY HEALTH CARE

Plants already provide humans with a quarter of our medicines and they could produce many more. The United States National Cancer Institute, for example, has identified 3 000 species of plants which contain active anti-carcinogenic components. Seventy per cent of these come from rainforest. In 1980 the United States market for prescriptions containing active plant components already exceeded $US50 billion (Bird 1991).

Medicinal plants from rainforests are not just a vast treasure trove for the Western world and pharmaceutical firms, they still play an important role in community health in Papua New Guinea. In traditional and transitional Papua New Guinea society, plants may be used for food, medicine, building, clothing and many other purposes. Indigenous peoples have developed appropriate methods of utilization through millennia of trial and error. However, with the rapid conversion by Christianity of Papua New Guinea traditional communities to a reliance on imported Western medicine, their uses are being forgotten, despite the fact that modern medication is often hard to obtain, expensive to buy and problematic to administer through a lack of qualified health workers.

Case Study 2: Algum Reforestation Centre

This programme aimed to promote the socially and economically significant tree species of the Abelam people of East Sepik Province of Papua New Guinea. In this project five youth groups from ten villages with a population of about 10 000 under three council areas, were encouraged to propagate 12 000 seedlings from five species of native trees with significant social and cultural value. Species included *Homalium foetidum* (used for construction of traditional spirit houses or *haus tambran*), *Vitex comfassus* (used to manufacture *kundu* drums for traditional "singsing"), and *Canarium* spp., *Terminalia* spp. and *Pometia* spp. (which are of high economic value). Community members were encouraged to plant the seedlings on their own land. An ongoing reward-incentive system was established and funded by the project, which employed two youths as full time workers. The villages have been encouraged to employ traditional methods of land use for various purposes including enrichment of valuable native tree species in the deforested areas and along the main catchment areas.

As part of the planning for economic sustainability the land owners have also been encouraged to grow Arabica coffee in the areas above 600 m. Since the 1950s this inland area from East Sepik has been growing only Robusta, but Arabica coffee is being promoted as it brings higher returns than the smaller bean Robusta coffee. The second phase of this project will involve a potable water supply project to the eight villages in the project area and the establishment of the cultural training centre. Currently this project has been identified as the best initiative for the conservation of native tree species with significant social and cultural value and has the support of the Department of Forestry, and the local community. Unfortunately it lacks funding for the promotion of the project.

Case Study 3: Inventory and Screening of Medicinal Plants in Papua New Guinea and the Solomon Islands.

In this programme the original intention was to build up a specialized high altitude plant collection as a field reference for researchers working at that altitude. It now contains between five and ten thousand herbarium specimens with medicinal plants forming the core of the plant collection, and is the third largest collection in the country.

Plants that grow at an altitude between 750 and 1 800 m are brought to the Institute for propagation. A literature search is then carried out to see if some of these plants are also used elsewhere and could have potential for further studies. Besides surveying and recording, a total of 100–200 plants have been screened for crude alkaloid extraction.

The National Herbarium in Lae is the main depository centre of all the replicas. The programme has been diversified into plant screening for chemical compounds, the development of medicinal plant gardens in the Wau Ecology Institute compound and in other selected communities, and training programmes for medicinal orderlies in remote regions.

This research will be published in the first comprehensive catalogue of medicinal plants of Papua New Guinea and the Solomon islands, with information on the distribution, habitat requirements and medicinal use of approximately 400 species. Again, however, further training programmes and on-site screening of medicinal plants cannot be developed due to lack of funding.

This programme's aims are twofold: it attempts to record traditional knowledge of medicinal plants and thereby help to revive and maintain the use of this resource, and to raise the awareness of these cheap and accessible native remedies amongst the Western-trained nurses and medical officers. Additionally it seeks to evaluate the "modern" potential of promising species by chemical screening in a way which will secure optimal patenting conditions for local communities.

MONITORING AND ASSESSING MINING IMPACTS

As a consequence of the long history of mining in the Wau valley, mining impacts are visible everywhere. Many of these are ongoing. New Guinea

Goldfield near Wau Ecology Institute was the first mine that became operational (from 1926 to 1942) within the then Territory of New Guinea and was rebuilt after the Second World War. In 1988 the Institute initiated a heavy metal screening programme of the major mechanized open cut mining company as this mine was dumping most of its waste into the Bulolo River system. Mining operations ceased in 1991. Wau Ecology Institute is now the only institution in the country collecting data on other river systems that are going to be effected by the anticipated Hidden Valley Gold Mine. This involves collecting data on the flow rate of the river systems, water quality testing, collection of information on the freshwater fauna and climatic data such as the rainfall records, which are critical information for use by the government for water and waste disposal planning.

Another important part of this programme involved an assessment of the Environmental Impact System Studies and developmental agreements for the OK Tedi Mine. This study was presented on behalf of the land owners at the International Water Tribunal in The Hague, Netherlands (Tjamie and Sakulas 1992) leading to one of the largest compensation payments by a mining company to land owner groups. A range of educational material including a video production in pidgin on mining impacts, and training manuals in water quality monitoring and minesite rehabilitation have been produced and used in conjunction with a mobile training unit.

Again this programme has demonstrated the importance of linking research with training. In this way research activities have been linked to intensive community campaigns against destructive mining practices. This programme also cannot continue due to lack of funding.

INTEGRATING CASH CROPS AND EXPORT PRODUCTS WITH SUBSISTENCE AGRICULTURE

One of the core concerns of Wau Ecology Institute is economic sustainability through the establishment of export markets. The development of regional, high quality coffee production involving better land use management concepts is integral to this. The plan to service the local communities (especially the partners of Kuper Range conservation area) by processing their parchment coffee to green beans for added value for a minimal processing fee is designed to strengthen partnerships, and to assist the self-sufficiency effort of the Institute. In a second phase, the Institute could increase its coffee returns as coffee roasting, grinding and packaging facilities are developed.

Shifting subsistence agriculture is still a major contributor to the destruction of tropical rainforest in Papua New Guinea. It intensifies with increasing population and its impacts are very obvious in Wau Valley. This programme aims to encourage site-stable farming by linking subsistence, cash and export farming. It teaches improved farming practices in the region, much of it based on experiences gained at the Institute. Farmers are encouraged to apply cultivation methods which maintain and improve soil fertility, reduce soil erosion and reduce the impact of agricultural pests. A pilot programme carried out between 1980 and 1984 (supported by the German Government) focused on integrating subsistence food production with a cash crop (mostly coffee). The programme includes training courses conducted at the Institute and is linked with community projects in other parts of Papua New Guinea.

THE ECOTOURISM PROGRAMME

As with many other developing countries, Papua New Guinea has an unrealized potential for different types of tourism. As in many other countries, conservation, economic policy and local communities increasingly view ecotourism as the future saviour of biodiversity, culture and economy. However, there is no attempt being made so far to evaluate Papua New Guinea's potential as a tourist destination, to identify potential markets, to approach these markets and to deal with the vagaries of the Papua New Guinea environment (high cost economy, reliance on air transport, perceived high role of criminality) which are presently hindering this development. Most remote regions have insufficient infrastructural facilities to derive significant income from the few tourists who visit. Wau Ecology Institute's relatively good accessibility (several flights a day, two to three hours scenic drive from Lae), its good infrastructure and its central location to a range of Papua New Guinea's scenic, biological and cultural heritage, make it a suitable site for investing in ecotourism. Because its present facilities are not yet specifically aimed at the ecotourism market, it currently attracts only 200 tourists per annum. However, the Institute's programmes (insect programme, rainforest research station, community health care programme, conservation area programme) offer potential for commercially viable ecotourism which few other places in Papua New Guinea could match. There is every indication that ecotourism could become a substantial funding source for the Institute itself and its five Conservation Areas, if funds become available to develop this capacity fully. This type of tourism would also greatly benefit the local community, especially as the decline of mining in the area has resulted in high unemployment.

Ecotourism and research bases are also being set up by a Community Based Network in the four conservation areas. These facilities have been set up with the knowledge that ecotourism may not be a viable economic incentive for the local land owners, but provide well equipped facilities for youth ranger training programmes and an outreach centre for schools and religious camps.

INTEGRATED CONSERVATION AND DEVELOPMENT PROGRAMME

Wau Ecology Institute has increasingly taken on the role of motivating other communities and

Case Study 4: Coffee Planting in Degraded Rainforest

Coffee is the region's, indeed Papua New Guinea's, most significant cash crop. However, after the collapse of the local coffee industry (due to low profit margins in maintaining large coffee plantations) successful sales of coffee by the local growers have become difficult. Coffee planting has also contributed to a significant amount of land degradation. Traditionally, at lower altitudes, coffee plants require shading and are therefore an ideal understorey for open forests. This offers a potential for a multi-species system. Wau Ecology Institute has trialed natural rainforest species within its 80 ha compound plantation to find the optimum mixture for coffee production. Contrary to conventional practice two ha of coffee have been established under a canopy of native rainforest species lacking *Luceanna* species, which are the usual exotic shade trees for coffee growing. This system can support many hundreds of plant species and creates habitat for approximately one third (>200 sp.) of the regional species of birds and butterflies. There is also no need for control of insect pests and diseases. Although the success of this trial has never been assessed, it offers many promising avenues for future research. At present the sale of the Institute's coffee, which is grown without the use of chemical fertilizers, can almost cover 40% of the Institute's annual budget of approximately 200 000 Kina ($AUD160,000).

Integration of food crops, medicinal and culturally important plants, food plants for insect ranching, and cultivation of fuel wood and fruiting trees with the coffee plots are further attempts to assist farmers, especially women. Improvement of soil conditions by the cultivation of introduced and native leguminous trees and shrubs, and through soil conservation techniques incorporating terracing and organic composting enable more farms to be built on denuded grassland near the villages (Goeltenboth 1990). This would reduce the amount of time taken by women to travel between production sites, often carrying heavy loads.

About 100 courses were run by the Institute, mainly for women's groups, conservation partners and agriculture extension workers. This particular programme has been influential in the development of the first national policy on subsistence agriculture in Papua New Guinea. In 1987 the programme and its staff was transferred to the department of Morobe and is now an integral part of the division of agriculture. Plans are under way for the division to be upgraded and included as part of the Papua New Guinea National Agriculture Research Institute. Again, however, this programme has never been able to obtain a consistent long-term funding environment.

regions to develop similar integrated conservation and development approaches. In the last five years it has facilitated and promoted ecosystem conservation projects with land owner groups in six areas of Papua New Guinea. The programme was started in 1991, has gradually gained momentum and currently involves six partners (Table 2).

The Lake Kamu Project is jointly implemented with Conservation International and Foundation for the people of the South Pacific. The Institute's role in the partnership is to provide technical advice in conservation issues, assist the communities in the development of economic incentive programmes (e.g., insect farming, community forestry, site-stable farming), carry out biological inventories and protect their interests from exploitative timber and mining concessions. In 1995 the Institute assisted in acquiring 90 ha of degraded coffee farmland close to Wau township for the community who owns the Prince Alexander conservation area. Currently a village workforce is transforming this area into a mixed coffee plantation and stable-site farming system. Produce grown from the farm is being sold to a fast food outlet established by the Institute and leased to another land owner group (owners of the Kuper Range Conservation area in Wau). As an additional incentive to the villagers to engage in the programme, the Institute has provided a water supply for the community of Ilaru and has paid school fees for some of the land owners' children attending Wau High school. Economic development incentives for the Kuper Range land owners and the Prince Alexander land owners in the East Sepik Province are tangible results of a programme that aims to achieve future sustainability and viability.

For Schredder Range conservation area land owners the Institute has started working on their incentive programmes. More time is needed for the land owners of the least accessible areas of the Madang Province to decide on what economic activities to undertake. In the meantime many of them have been brought to Wau to gain first hand experience in what they could do. Many of the land owners have expressed interest in reviving the defunct coffee industry in the province. This seems a

Table 2. Partners with Wau Ecology Institute in Papua New Guinea conservation areas.

Land owner	Province	Status	Approximate size
Kuper Range	Morobe	Advanced	1 500 ha
Schrendder Range	Madang	Intermediate	2 000 ha
Prince Alexander Range	East-Sepik	Intermediate	1 000 ha
Bowuto Range	Morobe	Beginning	3 000 ha
Huon Peninsula	Morobe	Beginning	2 000 ha
Lake Kamu	Gulf	Advanced	3 000 ha

Beginning — Initial contact with land owners is established and consent is given.
Intermediate — Contact is well established, conservation area identified and staff deployed.
Advanced — The conservation area is about to be gazetted by the government.

viable proposition as the area is closer to Goroka where coffee production is quite high. Land owners of the Bowtu area and the Houn Peninsular of the Morobe Province are also taking their time deciding on what they intend to do for economic incentive.

Integrated projects for conservation and development in Papua New Guinea have been promoted since 1991. The official number of projects, which follow the Integrated Conservation and Development model presently promoted by the Papua New Guinea Goverment, World Bank and the United Nations Development Programme is ten. Wau Ecology Institute itself has not been identified as one of these sites, partly probably because it dates back to a time well before these areas and concepts had been established.

CONCLUSION

The process to identify, discuss, implement and continually improve integrated conservation and development is described in Figure 3. The model, which we call the Wau Ecology Institute model, describes an integrated process of incentives, participatory policy formulation and implementation. Its relevance rests in its historical evolution over 25 years and its documented success.

What makes the Institute special? More than any other project it has demonstrated that Papua New Guinea is capable of developing its own institutional models for sustainable development. It has been the only non-government organization engaged in conservation in Papua New Guinea for nearly 20 years.

Historically, the most significant aspect was its conversion from an externally driven institution to one which is now entirely driven by Indigenous people. It has also shifted from pure research to a spectrum of applied activities, from awareness campaigning to being able to offer alternative income generating activities from non-timber forest products. The Institute has managed to survive and grow within an environment which is being increasingly dominated by a vast global economy of conservation, sustainable development, community participation and centralized bureaucracy.

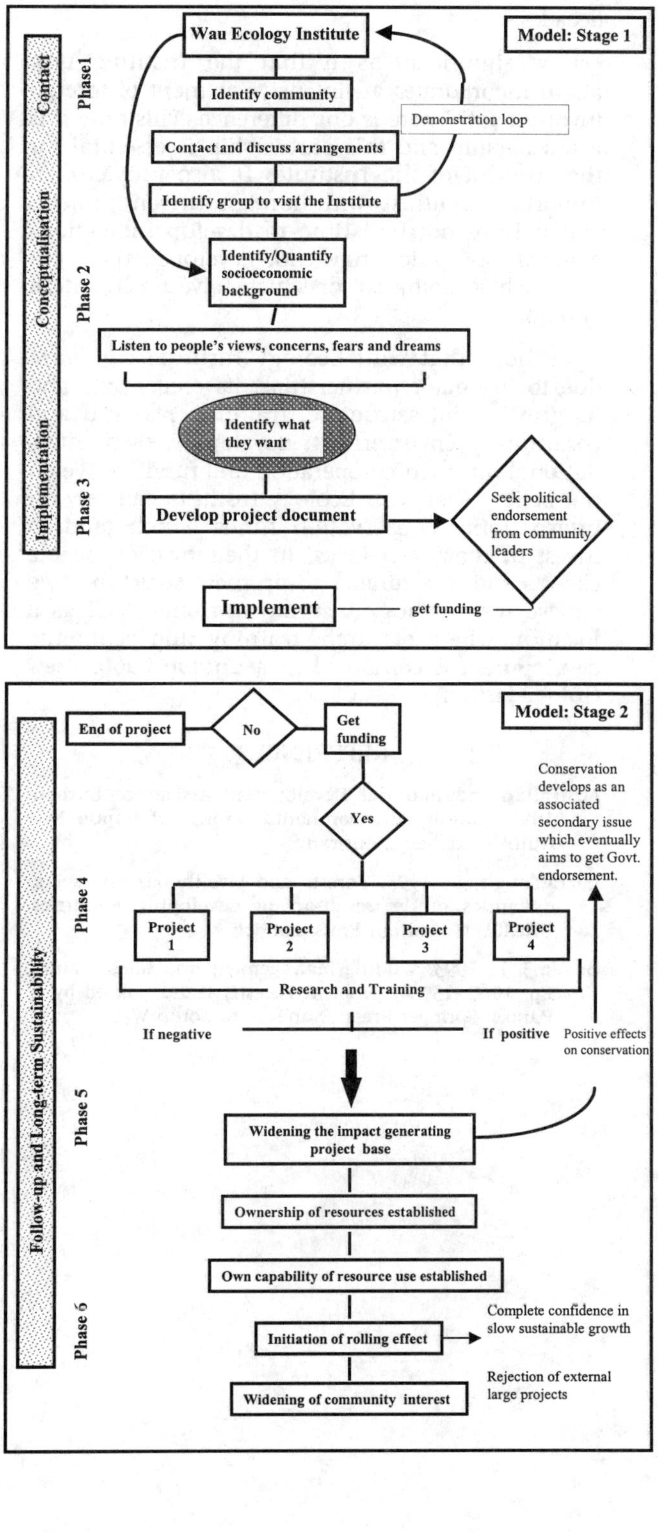

Fig. 3 (right). The Wau Ecology Institute model of Integrated Conservation and Development.

Integrated conservation, as well as the Biosphere model have emerged as the most promising ways to reconcile the growing needs of a developing world for natural reserves which at the same time need to be maintained. The reality for many of these models is not so promising. Hence, the survival of Wau Ecology Institute is of great importance. More than any other project it has indeed demonstrated that Papua New Guinea is capable of developing its own institutional models for sustainable development. And, most importantly, it has survived and grown for 25 years, during a period when the life-expectancy of most projects is measured in years rather than in decades.

How significant is it that the Institute now again incorporates an intensive element of foreign involvement? There is one difference. This time it is a partnership and this partnership is essential for the survival of the Institute. It also provides an important contrast with foreign driven projects which have not had time to develop indigenous concepts; or which have not developed the right approach in doing so, or which have no intention of doing so.

We hope that Wau Ecology Institute can now, due to its major partnerships, its credentials and its growing infrastructure, training, research and community environment, develop its own international outreach co-operation and funding. We do not believe that Wau Ecology Institute can rely on international conservation funds which tend to invest in their own ideas, in their own expatriate experts and in available government structures. We believe Wau Ecology Institute can offer itself as a location where research, training and economic development is combined in a unique Papua New Guinea fashion.

REFERENCES

Australian International Development Assistance Bureau, 1994. Country Environmental Profile of Papua New Guinea. Canberra, Australia.

Barraclough, L., 1995. Forests and Livelihoods: the social dynamics of deforestation in developing countries. UNRISD Macmillan Press: London.

Bauer, J. J., 1993. Wildlife Management and Conservation. Pp. 1060–139 *in* Tropical Forestry Handbook ed by L. Pancel. Springer Press: Sydney, New South Wales.

Bird, C., 1991. Medicines from the rainforest. *New Scientist* **131**(1782): 34–39.

Cernea, M. (ed), 1991. Putting People First: Sociological Variable in Rural Development. Oxford University Press: London.

Eaton, P., 1992. Indigenous People and Protected Areas New Guinea and Borneo: in Parks for Life: Report of the IVth World Congress on National Parks and Protected Areas ed by McNeely. International Union for the Conservation of Nature and World Wide Fund for Nature: Gland, Switzerland.

Filer, C. and Sekhran, N., 1998. Loggers, Donors and Resource Owners: Policy that Works for Forests and People. International Institute for Environment and Development: London.

Goeltenboth, F., 1990. Subsistence Agriculture Improvement, Manual for the Human Tropics. Wau Ecology Institute Handbook No. 10. Margraf Scientific Publications: Germany.

Hobart, M., 1993. An Anthropological Critique of Development: the Growth of Ignorance. Routledge: London.

Kottak, C., 1985. When people don't come first: some sociological lessons from completed projects. Pp. 325–56 *in* Putting People First ed by M. Cernea. Oxford University Press: New York.

Midgley, J., Hall, A., Hardiman, M. and Narine, D. P., 1986. Community Participation, Social Development and the State. Metheun: London.

Morris, M. G., Collins, N. M., Vane-Wright, R. I. and Waage, J., 1991. The utilisation and value of non-domesticated insects. Pp. 319–44 *in* The Conservation of Insects and Their Habitats ed by N. M. Collins and J. A. Thomas. Harcourt Brace Jovanovich: London.

Saulie, S. M., 1990. Forestry research and development in Papua New Guinea. *Ambio* **19**(8): 379–82.

Tjamie, L. and Sakulas, H., 1992. The case of OK Tedi Mine in Papua New Guinea. *Proceedings of the International Water Tribunal*, The Hague, 1992.

Uphoff, N., 1985. Fitting projects to people. Pp. 359–95 *in* Putting People First ed by M. Cernea. Oxford University Press: New York.

Wells, M., Brandon, K. and Hannah, L., 1992. People and Parks: Linking Protected area Management with Local Communities. The World Bank: WWF and USAID, Washington.

CHAPTER 11

Weed invasion in rural eucalypt forest remnants in eastern Australia

Jennifer Beer[1] and Marilyn Fox[1]

Native vegetation in production landscapes is threatened by vegetation clearing and invasion by exotic plant species. Forest remnants display gradients in microclimate and vegetation species composition from the forest edge to the interior. This study assessed vegetation in native forest remnants in rural New South Wales. Native and exotic plant species richness and cover were compared along edge-to-interior transects. The effect of edge aspect (north and south) and disturbance regime (reserved or grazed) were compared. Consistently higher numbers of exotic species were recorded at the forest edges, with polar edges relatively more weedy than equatorial edges. Fortunately, few exotic species associated with the pastures were found within the forest interior. In the absence of other sources of disturbance, these native forest remnants are resilient to weed invasion beyond the forest edge.

Key words: Edge effects, Eucalypt forest, Weed invasion, Aspect.

INTRODUCTION

VITOUSEK *et al.* (1996) identified the global exchange of biota and subsequent spread of invasive non-native species as an underestimated threat to global biodiversity. Australia, Canada and the United States each contain over 1 500 introduced plant species. Between 5 and 25% of the vascular flora in United States nature reserves are non-native introductions (Vitousek *et al.* 1996). Island biotas suffer greater impacts from species invasion than mainland biotas, and it has been found that more isolated islands are more seriously affected (D'Antonio and Dudley 1995). It is estimated that while 10% of the vascular flora in Australia is exotic (Michael 1994), the proportion of New Zealand's flora which is introduced is over 50% (Zwartz 1998).

Environmental change caused by species introductions is compounded by continued clearing and fragmentation of native vegetation communities (D'Antonio and Vitousek 1992). Plant and bird species have become locally extinct from isolated reserves of remnant vegetation (Drayton and Primack 1996; Turner *et al.* 1996; Turner 1996). These losses of native species are often accompanied by increases in the proportion of introduced species in the flora (Drayton and Primack 1996; Turner *et al.* 1996).

In rural areas extensive clearing leaves highly fragmented native vegetation "islands" in an agricultural mosaic. Australia's native forest cover has approximately halved since European settlement (Graetz *et al.* 1995), and the greatest clearance has been in some rural areas. The earliest inland colonies were in Central New South Wales where 75% of the forest and woodlands have now been cleared and many remaining forest patches are too small to sustain viable vegetation communities (Goldney and Bowie 1990; Goldney *et al.* 1995). Over 40% of the plant species listed as vulnerable in New South Wales are from the Central and Western Divisions (Benson 1991).

Whilst small native habitat remnants are still valuable for conservation planning, educational and scientific purposes (Shafer 1995), the ecological consequences of converting a landscape from continuous forest to fragmented vegetation are overwhelmingly negative. Habitat islands experience altered microclimate, wind and water fluxes, and the surviving species populations are small and genetically isolated (Saunders *et al.* 1991). Nutrients

[1]School of Geography, University of New South Wales, Sydney, New South Wales, Australia 2052.

Pages 103–10 *in* NATURE CONSERVATION 5: NATURE CONSERVATION IN PRODUCTION ENVIRONMENTS: MANAGING THE MATRIX ed by J. L. Craig, N. Mitchell and D. A. Saunders. Surrey Beatty & Sons, 2000.

from adjacent agricultural (Scougall *et al.* 1993) or urban (Rose 1997) developments can leak into the remnant. These ecological changes are modified by the time elapsed since a remnant has been isolated (Soulé *et al.* 1992) and the size of the remnant (Beer and Fox 1997).

An important consequence of fragmentation is the creation of edge zones between the remnant forest and the adjacent land use. The proportion of a remnant which is edge-affected will be determined by geometry, the size and shape of the remnant (Laurance and Yensen 1991) and response of various species to the edge. Table 1 shows that forest edges have been found to be warmer, lighter, drier and less humid environments than interior forest, and this seems to hold across a broad range of forest types. Microclimatic differences could account for differences in soil hardness, pH and soil nutrition, although these edge effects have not been as extensively studied. This suite of abiotic edge effects is accompanied by changes in the forest vegetation (Table 2). Many studies show a gradient of vegetation composition from edge-preferring to interior-preferring species (Wales 1972; Whitney and Runkle 1981; Matlack 1994; Jose *et al.* 1996). Forest edges are often characterized by high total species richness and high relative abundance of exotic species. The tree canopy is open at edges and the shrub layer is relatively denser (Table 2).

There have been no Australian studies to date comparing different aspects of forest edges, although edge aspect has been assessed in international studies. Microclimatic impacts of edge aspect were found in North America (Matlack 1993; Chen *et al.* 1995) and in New Zealand (Young and Mitchell 1994). Other North American studies have found

Table 1. Summary of abiotic edge effects in the atmosphere and soil of a range of vegetation types. Symbols indicate whether variables were higher (↑) or lower (↓) at the forest edge compared to the remnant interior.

Abiotic variable	Edge effect observed in different vegetation types
Air temperature	↑ Rainforest[1,2], Spruce[3], Douglas-fir[4], Oak-chestnut[5], Podocarp-broadleaf[6]
Soil temperature	↑ Rainforest[2,7], Douglas-fir[4], Acacia-eucalypt[8]
Light intensity	↑ Rainforest[2,9], Spruce[3], Douglas-fir[4], Oak-chestnut[5], Podocarp-broadleaf[6]
Wind speed	↑ Pine-forest[10], Douglas-fir[4]
Vapour pressure deficit	↑ Rainforest[1,11], Oak-chestnut[5], Podocarp-broadleaf[6]
Relative humidity	↓ Rainforest[2,9,12], Spruce[3], Douglas-fir[4]
Soil hardness	↑ Acacia-eucalypt[8]
Soil nutrients	↓ Rainforest[2] ↑ Acacia-eucalypt[8,13], Eucalypt[14]
Soil pH	↓ Rainforest[2]
Soil moisture	↓ Rainforest[2] ↓ and ↑ Rainforest[11], Acacia-eucalypt[8], Douglas-fir[4]
Litter moisture	↓ Oak-chestnut[5]

[1]Kapos 1989; [2]Jose *et al.* 1996; [3]Renhorn *et al.* 1997; [4]Chen *et al.* 1995; [5]Matlack 1993; [6]Young and Mitchell 1994; [7]Laurance and Yensen 1991; [8]Scougall *et al.* 1993; [9]Lovejoy *et al.* 1986; [10]Raynor 1971; [11]Camargo and Kapos 1995; [12]Williams-Linera 1990; [13]Hester and Hobbs 1992; [14]Cale and Hobbs 1991.

Table 2. Positive and negative responses of vegetation to remnant edge in a range of vegetation types. Symbols indicate whether variables were higher (↑) or lower (↓) at the forest edge compared to the remnant interior.

Vegetation variable	Variable response to remnant edge
Species richness	↑ Rainforest[1], Mixed hardwood[2], Chaco semi-arid[3]
Cover or relative importance of exotic species	↑ Rainforest[1], Acacia-eucalypt[4,5], Eucalypt[6,7] Mixed hardwood[2], Beech-maple[8]
Abundance of vines and lianas	↑ Rainforest[9,10]
Abundance of colonizer species	↑ Rainforest[1], Sugar maple-beech[11]
Vegetation density/basal area	↑ Podocarp-broadleaf[12], Chaco semi-arid[3]
Shrub cover/basal area	↑ Oak-chestnut[13], Chaco semi-arid[3]
Woody plant density/basal area	↑ Rainforest[14]
Tree mortality	↑ Douglas-fir[15]
Tree canopy cover/density	↓ Rainforest[14], Douglas-fir[15], Chaco semi-arid[3]
Groundlayer stem density	↓ Sugar maple-beech[16]
Abundance of rainforest/mesic species	↓ Rainforest[1], Sugar maple-beech[11]

[1]Fox *et al.* 1997; [2]Fraver 1994; [3]de Casenave *et al.* 1995; [4]Scougall *et al.* 1993; [5]Hester and Hobbs 1992; [6]Beer and Fox 1997; [7]Cale and Hobbs 1991; [8]Brothers and Springarn 1992; [9]Lovejoy *et al.* 1986; [10]Camargo and Kapos 1995; [11]Ranney *et al.* 1981; [12]Young and Mitchell 1994; [13]Matlack 1993; [14]Williams-Linera 1990; [15]Chen *et al.* 1992; [16]Palik and Murphy 1990.

that aspect influences vegetation composition (Wales 1972) and abundance of exotic plant species (Palik and Murphy 1990; Fraver 1994).

THE SUNNY CORNER STUDY

Previous study showed introduced plant species are more prevalent at the edges of eucalypt forest remnants, particularly invading small (<100 ha) remnants with edges adjacent to pastures (Beer and Fox 1997). This prompted further study into the depth of disturbed edge zones in small remnants of native vegetation.

The study area is the Sunny Corner/Yetholme district between Bathurst and Lithgow in the Central Tablelands of New South Wales, about 170 km west of Sydney. The climate is temperate with snowfalls in winter, annual precipitation is between 750 and 1 150 mm and prevailing winds are from the north-west (Anon. 1987), with summer storms often arising from the south-east (Jeans 1972). Variability in effective rainfall, available moisture and low winter temperatures are the climatic factors limiting plant growth in the area (Bowie 1987). Soils are red earths and red-brown earths of low to moderate fertility (Sommerville 1987). The forests in the area have the canopy dominated by eucalypts including *Eucalyptus dalrympleana* (mountain gum), *E. fastigata* (brown barrel), *E. pauciflora* (snow gum), *E. rubida* (candlebark) and *E. dives* (broad-leaf peppermint). The understorey is dominated by two species of Acacia, *Acacia melanoxylon* (blackwood) and *A. dealbata* (silver wattle). The groundlayer is grassy with most sites having abundant cover of *Poa sieberana* and a mixture of native dicotyledon species.

Six remnants were selected for the study, all less than 100 ha in area. Three of the remnants had livestock excluded, and were classified as having "minor disturbance". The other three study sites were subject to low intensity livestock grazing and were classified as having "moderate disturbance". Edge aspect was considered by sampling across edges facing both north (equatorial) and south (polar). To reduce confusion with Northern Hemisphere studies where edge aspect has been considered, edges will be referred to as either equatorial or polar rather than north-facing and south-facing. The edge orientation was difficult to replicate across the landscape so north-east facing edges were included in the equatorial category and south-west facing edges were included in the polar category. In two cases transects were replicated along the same remnant edge.

A total of 16 transects were surveyed across the six forest remnants, with four transects in each of four combinations of remnant disturbance and edge aspect categories (equatorial minor disturbance, equatorial moderate disturbance, polar minor disturbance and polar moderate disturbance). Vegetation surveys were conducted during the summer and autumn of 1997 (January to May). In the field the forest edge was taken as the treeline of the remnant. Remnant treelines usually followed existing or historical fencelines. Sampling transects were established perpendicular to the forest edge. Sampling began in the pasture 10 m from the forest edge and continued up to 45 m into the remnant. Distances from the forest edge are given as negative values for pasture samples, the forest edge is the 0 m distance, while positive values are used for sampling distances in the forest interior. Quadrats (1 m^2) were laid out on alternating sides of the measuring tape so that quadrats were not directly adjacent. Quadrats were sampled every metre up to 24 m into the forest when every second metre was sampled. In each 1 m^2 an estimate of percentage cover and the number of individual plants were recorded for each species present.

The vegetation transect data was grouped into sets of three contiguous 1 m^2 quadrats (as per Fox *et al.* 1997). Quadrat sets were located at –7, –4, –1, 2, 5, 8, 11, 14, 17, 20, 23, 26, 30 and 40 m from the forest edge. The distance of each quadrat set from the edge is given by the distance of the middle quadrat in the set of three. Mean relative cover per m^2 was calculated for each species in the quadrat set. Plant species were classified into six lifeforms; "exotics", "native grasses", "native groundlayer dicotyledons", "native monocotyledons", "native tree and shrub juveniles" and "native bracken fern".

Species richness of each lifeform group was determined for each set of three quadrats at each site. A relative cover score was calculated for each lifeform group by summing the individual cover values of all the species in the quadrat set belonging to each lifeform. Mean species richness and mean relative cover of each lifeform could then be calculated for quadrat sets with the same edge aspect and remnant disturbance to compare changes in lifeform composition across the edges of forests with contrasting aspect and disturbance regimes.

As most of the data did not approximate a normal distribution, non-parametric tests were applied to the richness and cover data. The Mann-Whitney U test was used to compare different aspects and the Kruskal-Wallis test was used to test for significant response to distance from the forest edge. All statistical tests were performed at the 0.05 significance level using Statview 4.5 for Macintosh (Abacus Concepts 1994).

Nine common exotic species were analysed in more detail. Percentage frequency of occurrence and mean relative percentage cover were calculated for each species using data from all 16 sites and grouping the quadrat data into four intervals (outside the forest, 0–9 m into the forest, 10–19 m into the forest and 20–29 m into the forest).

RESULTS OF VEGETATION SURVEY

Level of disturbance did not show significant effects in these analyses and so disturbance levels were pooled. There were then eight equatorial and eight polar transects. Figure 1 illustrates mean species richness and relative cover of different vegetation lifeforms with respect to distance from the forest edge. Data from equatorial and polar transects are easily compared. At equatorial edges, mean total species richness is highest at –7 to –4 m outside the forest and drops between 8 and 11 m into the forest (Fig. 1a). In contrast, species richness clearly peaks at the edge (–1 m) of polar sites and declines either side of the edge (Fig. 1b). The relative contribution of exotic species to species richness was highest in the pasture quadrats for both aspects and higher on average at the polar edges (Fig. 1a, 1b). The same trend was seen in mean relative cover (Fig. 2a, 2b). Polar edges had higher mean relative cover of introduced species than equatorial edges both outside the forest and just inside the forest edge (Fig. 2a, 2b). At polar edges relative cover of exotic species quickly declines once inside the forest, whereas the equatorial sites show a pulse of high relative cover of exotic species 11 m into the forest.

The contribution of native species lifeforms to the flora can also be compared from Figures 1 and 2. Native grasses contribute more to species richness and cover at equatorial edges than polar edges (Fig. 1a, 2a). Native groundlayer dicotyledon species contribute relatively more to plant species richness than to plant species cover. Conversely there were few species of grass contributing to the high percentage cover of this lifeform. Percentage cover of native grasses is low in the pastures and appears to be higher at pastures adjacent to equatorial edges (Fig. 2a), whilst cover for native grasses is low in the pasture near the polar edges and then increases into the forest (Fig. 2b).

Kruskal-Wallis tests (distance versus species richness) showed the edge-to-interior decline in exotic species was significant in both the equatorial ($p = 0.008$) and polar ($p < 0.001$) datasets, with the same significant trend seen in exotic species cover at both aspects. There was no significant edge-to-interior trend in native grass species richness, however, relative cover of native grasses was significantly higher in remnant interiors of polar sites compared to edges ($p = 0.008$). Tree and

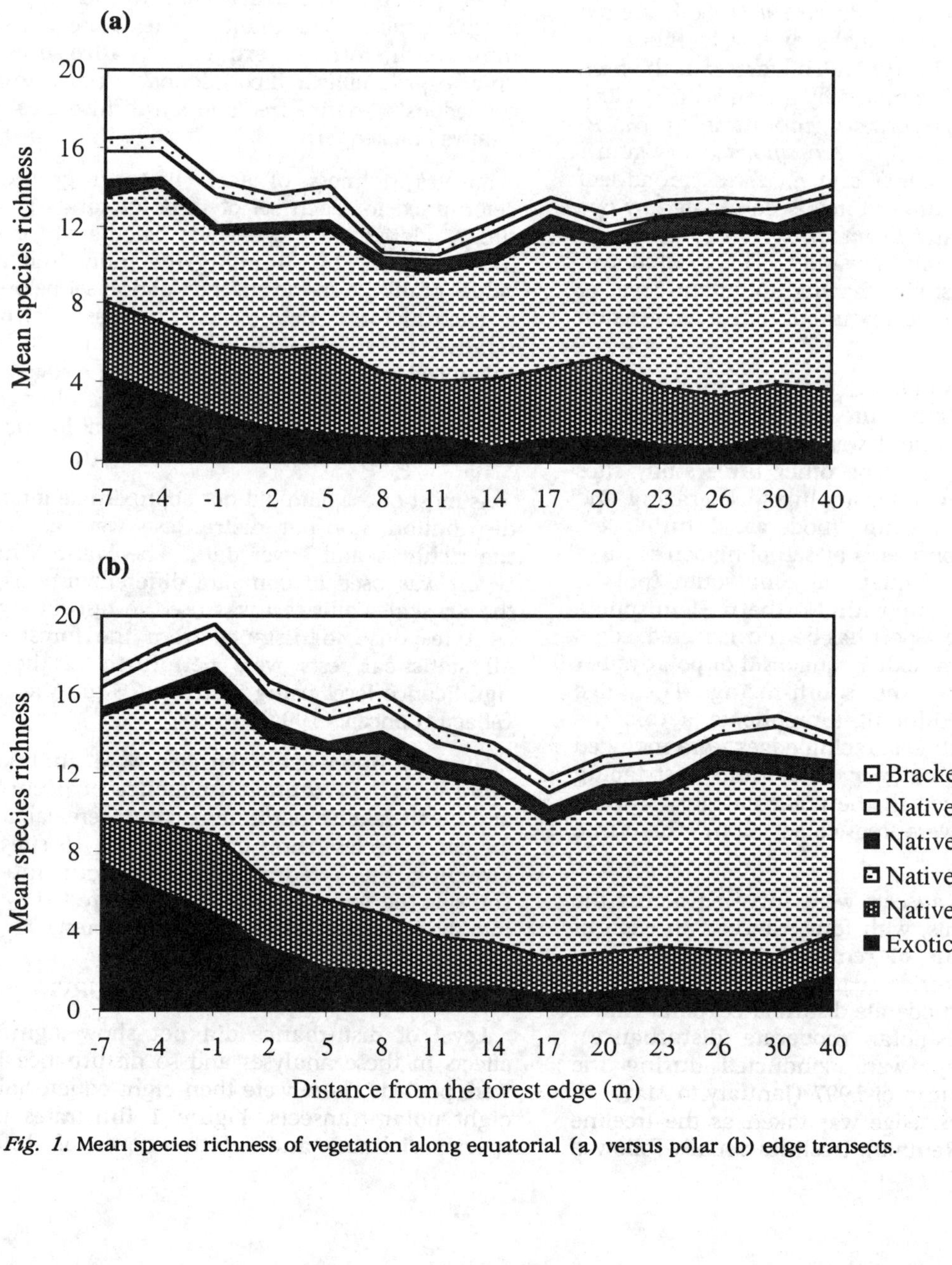

Fig. 1. Mean species richness of vegetation along equatorial (a) versus polar (b) edge transects.

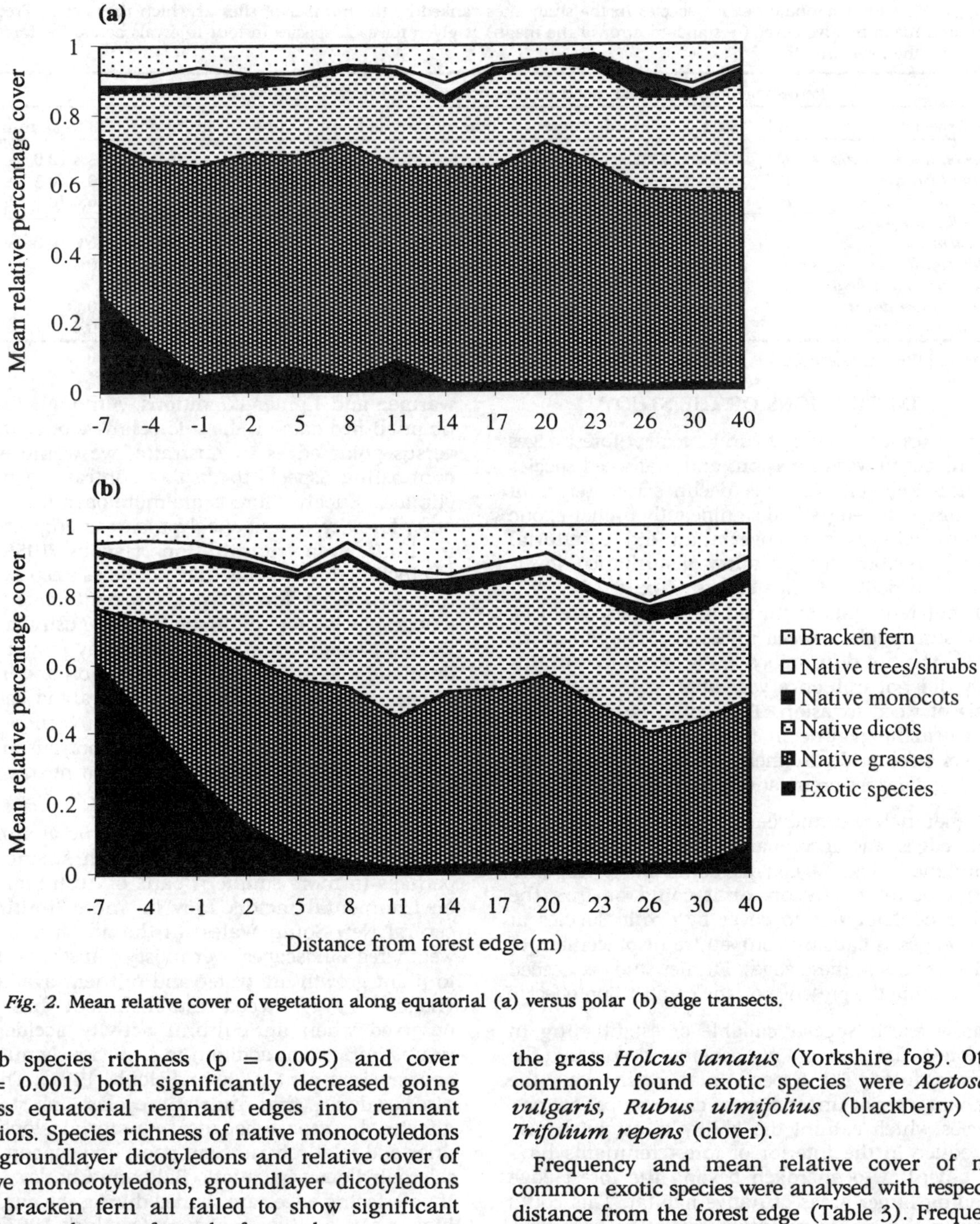

Fig. 2. Mean relative cover of vegetation along equatorial (a) versus polar (b) edge transects.

shrub species richness (p = 0.005) and cover (p = 0.001) both significantly decreased going across equatorial remnant edges into remnant interiors. Species richness of native monocotyledons and groundlayer dicotyledons and relative cover of native monocotyledons, groundlayer dicotyledons and bracken fern all failed to show significant response to distance from the forest edge.

Mann-Whitney U tests (aspect versus species richness) showed that exotics (p = 0.02), trees and shrubs (p = 0.001) and groundlayer dicotyledonyledons (p = 0.04) had significantly higher richness at polar edges overall, whilst native grasses had significantly higher richness at equatorial edges (p < 0.001). Analysis of relative cover gave the same significant trends.

A total of 24 exotic species were identified in the vegetation survey, including eight grass species. Half of the exotic species occurred at only one or two sites. Species occurring at most sites were the daisy species *Hypochaeris radicata* (flatweed) and the grass *Holcus lanatus* (Yorkshire fog). Other commonly found exotic species were *Acetosella vulgaris*, *Rubus ulmifolius* (blackberry) and *Trifolium repens* (clover).

Frequency and mean relative cover of nine common exotic species were analysed with respect to distance from the forest edge (Table 3). Frequency of occurrence of eight of these species was highest outside the forest (*Rubus ulmifolius* percentage frequency peaks at the forest edge). Only four species occurred up to 45 m into the forest, and of these *Hypochaeris radicata* was the only species to maintain percentage frequencies over 20%. The other three species were *Holcus lanatus*, *Trifolium repens* and *Rubus ulmifolius*. All of these four species had less than 5% mean relative cover beyond the forest edge and are not dominant components of the forest interior flora. Although *Hypochaeris radicata* had a high percentage frequency of occurrence across the distance divisions this species always had a relatively low mean percentage cover (less than 2%).

Table 3. Nine most abundant exotic species in the study sites ranked by the number of sites at which they occur. Frequency and mean relative cover (± standard error of the mean) is given for each species in four intervals across the forest edge into the forest interior.

			Percentage frequency and mean relative cover along intervals across the forest edge			
Exotic species	Family	Sites	Outside forest	0 to 9 m inside forest	10 to 19 m inside	20 to 29 m inside
Hypochaeris radicata	A	16	77% (4.0 ± 0.8)	65% (1.8 ± 0.3)	46% (1.1 ± 0.4)	63% (2.0 ± 0.3)
Holcus lanatus	P	15	55% (16.4 ± 3.0)	33% (3.8 ± 1.3)	19% (1.6 ± 0.1)	10% (0.3 ± 0.1)
Rubus ulmifolius	R	12	27% (1.0 ± 0.4)	29% (0.8 ± 0.4)	10% (0.6 ± 0.5)	6% (0.4 ± 0.2)
Acetosella vulgaris	Pg	12	38% (3.4 ± 1.6)	4% (0.05 ± 0.03)	2% (0.04 ± 0.04)	0%
Trifolium repens	F	11	39% (3.0 ± 0.9)	15% (0.3 ± 0.1)	4% (0.05 ± 0.03)	6% (0.06 ± 0.04)
Crepis capillaris	A	10	34% (1.6 ± 0.5)	8% (0.1 ± 0.1)	4% (0.1 ± 0.1)	0%
Centaurium erythrae	G	9	28% (0.5 ± 0.1)	6% (0.2 ± 0.1)	0%	0%
Dactylis glomerata	P	8	19% (3.8 ± 2.1)	4% (0.3 ± 0.3)	0%	0%
Agrostis capillaris	P	7	13% (7.6 ± 3.0)	0%	0%	0%

Key to families: A: Asteraceae, F: Fabaceae, G: Gentianaceae, P: Poaceae, Pg: Polygonaceae, R: Rosaceae.

IMPLICATIONS OF THE STUDY

These results show that rural eucalypt forest edges are subject to weed invasion, and that weed species richness and relative cover declines into remnant interiors. Polar edges had significantly higher exotic species richness and cover than did equatorial edges. Previous studies show disturbance levels impact on depth of edge effects (Fox *et al.* 1997) and that remnants disturbed by livestock could be more susceptible to weed invasion (Scougall *et al.* 1993). The two disturbance levels recognized in this study did not influence vegetation composition or levels of weed invasion. This could be due to the light grazing regime in these remnants, as land owners take pride in their "patch of bush" and stocking densities are relatively low.

Higher richness and density of exotic species at polar edges was accompanied by low native grass abundance (Fig. 2b). This could be a response to microclimatic factors or it could be that the higher contribution to cover by exotic species at polar edges is causing competitive displacement of native grasses at these edges. Further study is needed to determine the presence of any competition effect.

Many exotic species capable of establishing in the forest edge zone did not penetrate deeply into the remnant forest. These species may never invade the remnant interior under current disturbance regimes. Species which exhibit the ability to establish and reproduce in the interior of forest remnants have the potential to encroach beyond the forest edge invading larger areas of native habitat. This could threaten the ability of native vegetation to survive and reproduce.

The most striking difference between the results of this study and previous work on edge effects is the aspect effects upon vegetation composition, in particular the frequency and density of introduced species. Whilst other Australian studies have also found elevated weed invasion at native forest edges (Cale and Hobbs 1991; Hester and Hobbs 1992; Scougall *et al.* 1993; Beer and Fox 1997; Fox *et al.* 1997; Rose 1997), none of these studies found patterns related to aspect. In the North American studies where weed invasion was found to be enhanced at equatorial edges (Brothers and Springarn 1992; Fraver 1994), this was attributed to warmer and lighter conditions. Although there is no published data on the microclimate of equatorial versus polar edges in Australia, we would expect contrasting aspects to display contrasting microclimates. Eucalypt forests are more open than many other forest types and eucalypt leaves hang vertically to limit evapotranspiration (Jacobs 1955), so microclimatic gradients may not be as steep as those found in other forest types. On continents where topography is more dramatic than in Australia, such as New Zealand, remnant aspect may play a more dramatic role in vegetation composition and impact of vegetation fragmentation. New Zealand pastures on equatorial slopes receive up to 80% more solar radiation, have almost double evapotranspiration and the soils are relatively drier than nearby polar slopes (Radcliffe and Lefever 1981).

Whilst light and warmth are crucial limiting factors to vegetative growth in North America and perhaps in more southern parts of Australia, other environmental factors may be more limiting in central New South Wales. In the ancient, dry and weathered landscapes over most of Australia, limits to plant growth are water and nutrient availability (Beadle 1954). Weed establishment has been observed when agricultural activity accidentally results in nutrient addition to a previously nutrient-poor environment (Cale and Hobbs 1991). There is also evidence that nutrient cycles are strongly affected by aspect because equatorial slopes are drier and relatively nutrient-poor, and support more sclerophyllous vegetation (Bale and Charley 1994). If pole-facing edges are shaded during the day, have higher soil or litter moisture (Matlack 1993) and cycle nutrients more rapidly, then it is possible that these areas will be more vulnerable to weed invasion in dry Australian forests.

This research suggests that vegetation response to forest edge is not simply determined by one microclimatic variable (light intensity) as many papers have concluded. The availability of a suite of environmental resources play a role, which may be moderated by physiognomic characters of the landscape such as aspect. The availability of resources that limit plant growth in the landscape will be critical in determining the patterns and processes of exotic species invasion of natural systems. In some cases restricted resources will be light and warmth, but in other areas quite different resources are limiting and once these have been disturbed the opportunity for invasion is provided (Fox and Fox 1986).

MANAGEMENT QUESTIONS

Discussion with pasture researchers has revealed that some of the exotic species in this study are desirable for a productive pasture, and some of the native species are undesirable (David Kemp, Peter Dowling; Orange Agricultural Institute). This raises difficulties when deciding how to manage problem species across a mosaic landscape in which native vegetation is only a minor, fragmented element. Only three of the most common exotic species recorded in this study would be considered useful pasture species (*Trifolium repens, Dactylis glomerata* and *Agrostis capillaris*). Of these, only *Trifolium repens* was recorded invading further than 10 m into native forest in this study. Minimizing the populations of the other 21 exotic species is therefore desirable both for conservation of the forest remnant and for protecting the productivity of the adjoining pasture.

As forest-pasture boundaries appear to be favourable environments for weed establishment, these zones need to be targeted for integrated weed management across the landscape. Other zones providing resources otherwise limited in the landscape, such as drainage lines, will need to be monitored for weed infestation.

To facilitate weed management in production environments, the following questions need to be addressed.

- What are the long-term goals for ecologically sustainable management of the landscape? Can management be integrated to enhance both production and conservation goals?
- What are the invasive species in each land use element (native or exotic)?
- Which species are considered pests to both production and conservation areas? Can these species be targeted for control across the landscape? Is it appropriate to use the same methods of weed control in agricultural and protected areas?
- What areas across the landscape could support invasive species populations? What are potential sources of disturbance to the natural resource base? Is there a pattern of invasion which will allow targeted control of infestation and population spread?

These questions cannot be answered by looking at the goals of one property or only one stakeholder. Weed invasion influences biological diversity on a landscape and catchment scale and it is on these scales that the problem must be addressed.

ACKNOWLEDGEMENTS

This work was completed while Jenny Beer was in receipt of an Australian Postgraduate Award. Mary and Neville Pryke, Tom Beer and Pamela van Oploo assisted with field work. Jan de Nardi at the University of New South Wales John Waterhouse Herbarium identified plant specimens. Vern and Joyce Moffitt, Fred Taylor, Col and Phil Wilkinson, Julie Masters, Merv Berry, Mr and Mrs Hood, Mr and Mrs Bromfield and Mr Moss all kindly allowed us to work on their properties. National Parks and Wildlife Service, State Forests and the Crown Lands Department also gave permission to work on lands they administer. The authors appreciated discussions with David Goldney, Brian Stone, Andrew Fisher and Sylvia Cardale over the period of this work. Dr Scott Mooney made helpful comments on the manuscript.

REFERENCES

Abacus Concepts, 1994. Statview. Abacus Concepts Inc.: Berkeley.

Anon., 1987. Management plan for the Bathurst management area. Forestry Commission of New South Wales, Appendix 2.

Bale, C. L. and Charley, J. L., 1994. The impact of aspect on forest floor characteristics in some eastern Australian sites. *For. Ecol. and Mgt.* **67:** 305–17.

Beadle, N. C. W., 1954. Soil phosphate and the delimitation of plant communities in Eastern Australia I. *Ecol.* **35:** 370–75.

Beer, J. and Fox, M. D., 1997. Conserving vegetation in fragmented rural landscapes: edge effects and eucalypt forest remnants. Pp. 313–18 *in* Conservation Outside Nature Reserves ed by P. Hale and D. Lamb. Centre for Conservation Biology: University of Queensland, Queensland.

Benson, J., 1991. The effect of 200 years of European Settlement on the vegetation of New South Wales. *Cunninghamia* **2:** 343–501.

Bowie, I. J. S., 1987. Climate. Pp. 57–74 *in* The National Trust of Australia (NSW) Scenic and Scientific Survey of the Central Western Region. A Report to the Australian Heritage Commission. Volume I. Mitchell College of Advanced Education: Bathurst, New South Wales.

Brothers, T. S. and Springarn, A., 1992. Forest fragmentation and alien plant invasion of central Indiana old-growth forests. *Cons. Biol.* **6:** 91–100.

Cale, P. and Hobbs, R. J., 1991. Condition of roadside vegetation in relation to weed status. Pp. 353–62 *in* Nature Conservation 2: The Role of Corridors ed by D. A. Saunders and R. J. Hobbs. Surrey Beatty & Sons: Chipping Norton, New South Wales.

Camargo, J. L. C. and Kapos, V., 1995. Complex edge effects on soil moisture and microclimate in central Amazonian forest. *J. Trop. Ecol.* **11:** 205–11.

Chen, J., Franklin, J. F. and Spies, T. A., 1992. Vegetation responses to edge environments in old-growth Douglas-fir forests. *Ecol. App.* **2:** 387–96.

Chen, J., Franklin, J. F. and Spies, T. A., 1995. Growing-season microclimatic gradients from clearcut edges into old-growth Douglas-fir forests. *Ecol. App.* **5:** 74–86.

D'Antonio, C. D. and Dudley, T. L., 1995. Biological invasions as agents of change on islands versus mainlands. Pp. 103–21 *in* Islands: Biological Diversity and Ecosystem Functions ed by P. M. Vitousek, L. L. Loope and H. Adsersen. Springer-Verlag: New York.

D'Antonio, C. D. and Vitousek, P. M., 1992. Biological invasions by exotic grasses, the grass/fire cycle, and global change. *Ann. Rev. Ecol. Syst.* **23:** 63–87.

de Casenave, J. L., Pelotto, J. P. and Protomastro, J., 1995. Edge-interior differences in vegetation structure and composition in a Chaco semi-arid forest, Argentina. *For. Ecol. and Mgt.* **72:** 61–69.

Drayton, B. and Primack, R. B., 1996. Plant species lost in an isolated conservation area in metropolitan Boston from 1894 to 1993. *Cons. Biol.* **10:** 30–39.

Fox, B. J., Taylor, J. T., Fox, M. D. and Williams, C., 1997. Vegetation changes across edges of rainforest remnants. *Biol. Conserv.* **82:** 1–13.

Fox, M. D. and Fox, B. J., 1986. The susceptibility of natural communities to invasion. Pp. 57–66 *in* Ecology of Biological Invasions: An Australian Perspective ed by R. H. Groves and J. J. Burdon. Australian Academy of Science: Canberra, Australian Capital Territory.

Fraver, S., 1994. Vegetation responses along edge-to-interior gradients in the mixed hardwood forest of the Roanoke river basin, North Carolina. *Cons. Biol.* **8:** 822–32.

Goldney, D. C. and Bowie, I. J. S., 1990. Some management implications for the conservation of vegetation remnants and associated fauna in the Central Western Region of New South Wales. *Proc. Ecol. Soc. Aust.* **16:** 427–40.

Goldney, D. C., Bauer, J., Bryant, H., Hodgkins, D. and Watson, G., 1995. Winning battles but losing the war: the education marketing imperative. Pp. 574–88 *in* Nature Conservation 4: The Role of Networks ed by D. A. Saunders, J. I. Craig and E. M. Mattiske. Surrey Beatty & Sons: Chipping Norton, New South Wales.

Graetz, R. D., Wilson, A. M. and Campbell, S. K., 1995. Landcover disturbance over the Australian continent: a contemporary assessment. Biodiversity Series Paper No. 7, Biodiversity Unit, Commonwealth Dept. of Environment, Sport and Territories.

Hester, A. J. and Hobbs, R. J., 1992. Influence of fire and soil nutrients on native and non-native annuals at remnant vegetation edges in the Western Australian wheatbelt. *J. Veg. Sci.* **3:** 101–08.

Jacobs, M. R., 1955. Growth habits of the eucalypts. Forest and Timber Bureau, Dept. of Interior.

Jeans, D. N., 1972. An historical geography of New South Wales to 1901. Reed Education: Sydney, New South Wales.

Jose, S., Gillespie, A. R., George, S. J. and Kumar, B. M., 1996. Vegetation responses along edge-to-interior gradients in a high altitude tropical forest in peninsular India. *For. Ecol. and Mgt.* **87:** 51–62.

Kapos, V., 1989. Effects of isolation on the water status of forest patches in the Brazilian Amazon. *J. Trop. Ecol.* **5:** 173–85.

Laurance, W. F. and Yensen, E., 1991. Predicting the impacts of edge effects in fragmented habitats. *Biol. Conserv.* **55:** 77–92.

Lovejoy, T. E., Bierregaard Jr, R. O., Rylands, A. B., Malcolm, J. R., Quintela, C. E., Harper, L. H., Brown Jr, K. S., Powell, A. H., Powell, G. V. N., Schubart, H. O. R. and Hays, M. B., 1986. Edge and other effects of isolation on Amazonian forest fragments. Pp. 257–85 *in* Conservation Biology: The Science of Scarcity and Diversity ed by M. E. Soulé. Sinauer Associates: Sunderland, Massachusetts.

Matlack, G. R., 1993. Microenvironment variation within and among forest edge sites in the Eastern United States. *Biol. Conserv.* **66:** 185–94.

Matlack, G. R., 1994. Vegetation dynamics of the forest edge-trends in space and successional time. *J. Ecol.* **82:** 113–23.

Micheal, P. W., 1994. Alien plants. Pp. 57–84 *in* Australian Vegetation (2nd Ed) ed by R. H. Groves. Cambridge University Press: Cambridge.

Palik, B. J. and Murphy, P. G., 1990. Disturbance versus edge effects in sugar-maple/beech forest fragments. *For. Ecol. and Mgt.* **32:** 187–202.

Radcliffe, J. E. and Lefever, K. R., 1981. Aspect influences on pasture microclimate at Coopers Creek, North Canterbury. *NZ J. Agric. Res.* **24:** 55–66.

Ranney, J. W., Bruner, M. C. and Levenson, J. B., 1981. The importance of edge in the structure and dynamics of forest islands. Pp. 67–95 *in* Forest Island Dynamics in Man-Dominated Landscapes ed by R. L. Burgess and D. M. Sharpe. Springer-Verlag: New York.

Raynor, G. S., 1971. Wind and temperature structure in a coniferous forest and a contiguous field. *Forest Sci.* **17:** 351–63.

Renhorn, K. E., Esseen, P., Palmqvist, K. and Sundberg, B., 1997. Growth and vitality of epiphytic lichens. 1. Responses to microclimate along a forest edge-interior gradient. *Oecologia* **109:** 1–9.

Rose, S., 1997. Influence of suburban edges on invasion of *Pittosporum undulatum* into the bushland of northern Sydney, Australia. *Aust. J. Ecol.* **22:** 89–100.

Saunders, D. A., Hobbs, R. J. and Margules, C. R., 1991. Biological Consequences of Ecosystem Fragmentation: A Review. *Cons. Biol.* **5:** 18–32.

Scougall, S. A., Majer, J. D. and Hobbs, R. J., 1993. Edge effects in grazed and ungrazed Western Australian wheatbelt remnants in relation to ecosystem reconstruction. Pp. 163–78 *in* Nature Conservation 3: Reconstruction of Fragmented Ecosystems ed by D. A. Saunders, R. J. Hobbs and P. R. Ehrlich. Surrey Beatty & Sons: Chipping Norton, New South Wales.

Shafer, C., 1995. Values and shortcomings of small reserves. *BioSci.* **45:** 80–88.

Sommerville, W., 1987. Soils. Pp. 88–93 *in* The National Trust of Austalia (NSW) Scenic and Scientific Survey of the Central Western Region. A Report to the Australian Heritage Commission, Volume 1. Mitchell College of Advanced Education: Bathurst, New South Wales.

Soulé, M. E., Alberts, A. C. and Bolger, D. T., 1992. The effects of habitat fragmentation on chaparral plants and vertebrates. *Oikos* **63:** 39–47.

Turner, I. M., 1996. Species loss in fragments of tropical rain forest: a review of the evidence. *J. App. Ecol.* **33:** 200–09.

Turner, I. M., Chua, K. S., Ong, J. S. Y., Soong, B. C. and Tan, H. T. W., 1996. A century of plant species loss from an isolated fragment of lowland tropical rain forest. *Cons. Biol.* **10:** 1229–244.

Vitsousek, P. M., D'Antonio, C. M., Loope, L. L. and Westbrooks, R., 1996. Biological invasions as global environmental change. *Am. Sci.* **84:** 468–78.

Wales, B. A., 1972. Vegetation analysis of north and south edges in a mature oak-hickory forest. *Ecol. Monogr.* **42:** 451–71.

Whitney, G. G. and Runkle, J. R., 1981. Edge versus age effect in the development of a beech-maple forest. *Oikos* **37:** 377–81.

Williams-Linera, G., 1990. Vegetation structure and environmental conditions of forest edges in Panama. *J. Ecol.* **78:** 356–73.

Young, A. and Mitchell, N., 1994. Microclimate and vegetation edge effects in a fragmented podocarp-broadleaf forest. *Biol. Conserv.* **67:** 63–72.

Zwartz, D. (ed), 1998. New Zealand Official Yearbook 1998. 101st edition. Statistics New Zealand, GP Publications: Thorndon, Wellington, New Zealand.

CHAPTER 12

Managing the forest matrix: regrowth forests as bird habitat

Richard H. Loyn[1]

Eucalypt forests used for wood production in southeastern Australia are taken as a case study to examine the need to manage the matrix of regrowth forest for biodiversity conservation, and the value of the regrowth matrix as habitat for birds. Information is synthesized from published and unpublished Victorian studies.

It is argued that sympathetic management of the regrowth matrix is needed because of its extent, distribution and value as habitat. Evidence is presented that regrowth forests may help buffer reserves from adverse influences that operate when small patches of forest are surrounded by cleared land. Landscape dynamics add to the difficulty of conserving species solely in reserves, and an example is presented of species loss or decline from a well protected but isolated reserve.

Regrowth differs from mature forest at the coupe scale and managed forests differ from unmanaged forests at the landscape scale. Focus on these differences may have obscured the value of regrowth forests as habitat for many birds, which is much greater than for some alternative modified environments such as croplands or pasture. Some new approaches are suggested for improving the value of regrowth as habitat, along with the three strategies for conserving species that need elements of old forest (long rotations, retaining individual elements on coupes, or retaining selected stands).

An integrated approach is being taken to planning across the forest landscape through the Forest Management Planning and Regional Forest Agreement processes.

Key words: Forest management, Biodiversity conservation, Strategies, Eucalypt forest, Regrowth, Habitat, Birds, Matrix, Buffering, Dynamics.

INTRODUCTION

It has been common for conservation biologists to view the world as patches of high-quality habitat embedded in a matrix of hostile habitat, with theories of island biogeography applied to terrestrial reserve systems (MacArthur and Wilson 1963; Diamond 1975). However, there are many situations where habitats are not truly isolated and the matrix is not wholly hostile (Forman and Godron 1986). Increasing attention is being focused on these "variegated landscapes" and their management (e.g., McIntyre and Barrett 1992; Macleod and McIntyre, in litt.).

Forests used for wood production provide one example of a variegated landscape. They contain patches and strips of natural forest embedded in a matrix of regrowth from logging, but the regrowth provides some of the habitat values represented in the natural forest. A popular perception of logging is that it converts pristine old forest to devastated open clearings. However, most forest subject to current logging is not pristine, and each logging coupe regenerates with stands of regrowth which provide habitat for different species with values that may differ markedly both from those of the recent coupes and those of the original forest stands. These relative values and their changes over time need to be considered in any assessment of the effects of logging.

This chapter examines the eucalypt forests of Victoria (southeastern Australia) as a case study for assessing the need to manage the matrix for biodiversity conservation, and more specifically the value of the regrowth matrix as habitat for birds. The chapter takes an overview, based on experience during a wide range of forest wildlife studies. The Victorian Department of Natural Resources and Environment has an extensive programme of forest wildlife research, mostly conducted by the Arthur

[1]Arthur Rylah Institute, Department of Natural Resources and Environment, 123 Brown Street, Heidelberg, Victoria, Australia 3084.

Pages 111–19 *in* NATURE CONSERVATION 5: NATURE CONSERVATION IN PRODUCTION ENVIRONMENTS: MANAGING THE MATRIX ed by J. L. Craig, N. Mitchell and D. A. Saunders. Surrey Beatty & Sons, 2000.

Rylah Institute in association with the Centre for Forest Tree Technology, with funding from the Forests Service and Flora and Fauna Programme (both of Natural Resources and Environment) and Environment Australia (Federal Government). Information on bird habitat succession over many years has been obtained mainly from retrospective studies (e.g., Loyn 1980, 1985a, 1991) though some experimental sites have been monitored for up to 21 years after logging (unpubl. data). Recent work has focused on effects of scale and pattern of logging (e.g., Nelson *et al.* 1996).

The main aim of this chapter is to synthesize some published and unpublished information from these studies, to address three main sets of questions:

Do we need to manage the matrix and why?

What are the values of the regrowth matrix (as habitat for birds)? (How do they compare with values of land used for other purposes, e.g., agriculture?)

How can these values be improved by management? What are we doing to help manage the matrix?

It is hoped that some of the answers will stimulate further research and adaptive management, in Victoria and elsewhere.

VICTORIAN FORESTS AND PLANNING PROCESSES (CONTEXT FOR CASE STUDY)

Victoria has about a third of its land in public ownership, and 10% is formal conservation reserves (National Parks, State Parks, Flora and Fauna Reserves, etc.). About half of its public forests are in formal conservation reserves, with wood production permitted in the other half. Broadscale allocation of public land between these uses has been achieved in recent decades by a generally respected consultative process, conducted by the state Land Conservation Council. In State Forests where wood production is permitted, there is a more flexible system of planning land use at a finer scale, involving Forest Management Plans produced with public consultation, and Regional Forest Agreements between state and federal governments (Commonwealth of Australia 1997). Expert advice on biodiversity conservation is invited and used in all these processes. Forest management plans are expected to be revised on a five-year cycle (Natural Resources and Environment 1999) whereas formal land use allocations usually stand for longer periods.

The native forest industry is concentrated in the faster-growing, wetter eucalypt forests of eastern Victoria and the Otway Ranges, including forests of Mountain Ash *Eucalyptus regnans* in the Central Highlands, Alpine Ash *E. delegatensis* in the higher mountains and extensive mixed eucalypt forests in the foothills. Logging in these forests is usually by clear-felling coupes of up to 40 ha, with varying numbers of older trees retained according to local prescriptions. More selective logging systems are used in the drier, slower growing forests of western and northern Victoria. Salvage logging may be conducted on a broad scale after extensive wildfires. All logging is governed by a Code of Forest Practices (Natural Resources and Environment 1996). The state has managed extensive pine plantations on public land through the Victorian Plantations Corporation (recently privatized), but conversion of public native forests to pine plantations ceased in the 1980s. These planning processes operate within a legislative framework provided *inter alia* by the *Conservation, Forests and Lands Act 1987* and the *Flora and Fauna Guarantee Act 1988*. Victoria is a signatory to the National Forest Policy Statement (Commonwealth of Australia 1992), and has initiated a Timber Industry Strategy (1986) and a Biodiversity Strategy (1997), adopting global principles of biodiversity conservation and ecologically sustainable development.

The need to manage the matrix

Five main reasons can be identified for considering biodiversity conservation in managing the production forest mosaic, instead of concentrating efforts solely in formal reserves. They relate to the *extent* and *distribution* of the production forests, the *dynamics* of ecological communities, the need to *buffer* reserves from adverse impacts originating in adjacent land, and the ability of regrowth forests to provide *habitat*.

In terms of *extent*, timber production is permitted in about half of the public forest area (3.5 million ha, Fig. 1), and half of that is scheduled to be available for logging and hence managed as regrowth in the foreseeable future. This is too great an area to ignore in terms of current conservation values. These regrowth areas support high proportions of total populations of some species that are able to inhabit regrowth.

In terms of *distribution*, the formal reserves are concentrated in the least productive environments (e.g., mallee) and those with the greatest scenic values (e.g., the high alps) (Fig. 1). This is a consequence of historical and recent land allocation processes, with many productive environments cleared before effective conservation planning. This is a common situation world-wide, as discussed by Pressey (1994). In Victoria, the Land Conservation Council was always mindful of the need for a representative system of reserves (and this need has been reinforced by the National Forest Policy and the Regional Forest Agreement process, which demands a "Comprehensive and Representative" reserve system), but the Land Conservation Council also had to consider conflicting needs of industry and recreational land uses. The result is that wildlife species that are naturally concentrated in highly productive environments, are less well represented in reserves than those species which tend to avoid such environments.

Landscape *dynamics* add to the difficulty of attempting to conserve species solely in reserves. Even if a species seems adequately conserved in a single reserve at a particular point in time,

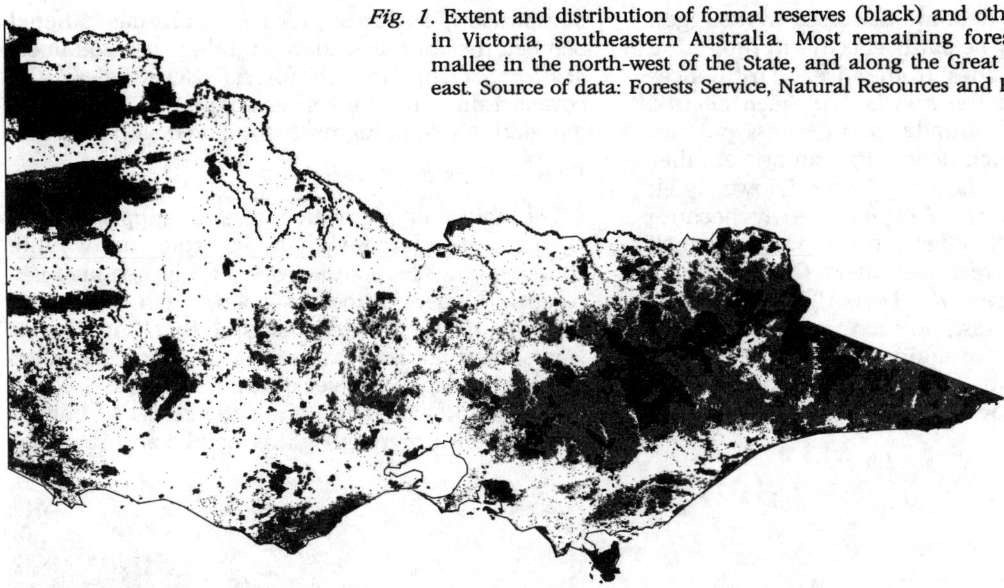

Fig. 1. Extent and distribution of formal reserves (black) and other forested land (grey) in Victoria, southeastern Australia. Most remaining forest is in the semi-arid mallee in the north-west of the State, and along the Great Dividing Range in the east. Source of data: Forests Service, Natural Resources and Environment.

unexpected stochastic events may come to threaten its security. This may be an obvious event such as a wildfire, or a more subtle process equally hard to manage. If this results in local loss of species, biodiversity will not be maintained unless those species are able to recolonize from nearby populations. It does not matter whether the nearby populations are in reserves or commercially productive forests, as long as the populations continue to flourish and there are no physical barriers to recolonization.

The need to consider dynamics is illustrated by some recent observations from northeastern Victoria. Chiltern Box-Ironbark National Park is a magnificent example of box-ironbark eucalypt forest on the dry inland slopes of the Great Diving Range. The Park has been regularly surveyed for birds (Traill *et al.* 1996) and has been well protected since extensive logging during gold rush days last century. One bird species has disappeared in the last decade (Crested Bellbird *Oreoica gutturalis*, Traill *et al.* 1996) and recent surveys by local naturalists, Birds Australia and Arthur Rylah Institute (unpubl. data) have shown that seven small insectivores are now rare or absent (Gilbert's Whistler *Pachycephala inornata*, Red-capped Robin *Petroica goodenovii*, Scarlet Robin *Petroica multicolor*, Buff-rumped Thornbill *Acanthiza reguloides*, Brown Thornbill *Acanthiza pusilla*, Speckled Warbler *Chthonicola sagittata* and Chestnut-rumped Heathwren *Hylacola pyrrhopygia*). In forests nearby, these species range in abundance from extremely rare (Gilbert's Whistler and Chestnut-rumped Heathwren) or scarce (Red-capped Robin, Speckled Warbler) to extremely common (Brown Thornbill, perhaps the most abundant bird in Victorian forests) (Emison *et al.* 1987). The Heathwren has always been scarce at Chiltern, but the other species seem to have declined markedly in recent years (E. Collins, pers. comm.). This group of birds includes no species that need tree hollows or other obvious elements of old forest. Gilbert's Whistlers and Brown Thornbills inhabit shrub thickets which tend to proliferate after fires before declining over time. The other species make varying use of shrub thickets and forage mainly from open ground with leaf litter among trees. Scarlet Robins and Buff-rumped Thornbill are known to favour early successional stages after fire (Reilly 1991; Loyn 1997). These species may be suffering from lack of recent extensive fires, though fire is thought to be rare in box-ironbark forest. Regardless of the cause, it is clear that species losses may occur even in a well protected but isolated reserve, especially if it is not practical to provide a diversity of management regimes. It is essential to manage areas in the matrix close to reserves, with adequate links, to provide sources for future recolonization.

This need is also evident from studies of turnover in bird species on islands (e.g., Saunders and de Rebeira 1985). In a study of fragmented forests in farmland, it has been estimated that bird species were being lost from forest patches at the rate of about one species per decade (Loyn 1987), in addition to effects of reduced area. It is expected that this rate could be influenced by managing the extent, connectivity and habitat values of nearby forest in the landscape, in accordance with biogeographical theory.

The need to *buffer* reserves is often quoted as a reason for sympathetic management of the adjacent matrix. This idea has been developed in relation to Victorian forest management by Shaw (1983) and Burgess *et al.* (1997), who advocate "feathering" retained forest with irregular boundaries or transition zones of selective logging, to mimic the structural variation of natural systems. Feathering increases the extent of edge in the landscape, and this can have both positive and negative effects on wildlife (Forman and Godron 1986; Hunter 1990). In ash forests, data suggest that such edge effects are related closely to habitat gradients (Nelson *et al.* 1996; Loyn 1998 and unpubl.).

Here some sets of data are discussed that suggest regrowth forests may be partly effective in protecting remnant forest patches from adverse influences that would apply if the matrix had been cleared for farmland. Bird abundance in forest patches in farmland is much lower in patches smaller than 10 ha than in larger patches (Howe 1984; Loyn 1987; Barrett *et al.* 1994), partly because of the adverse edge effects from grazing stock and aggressive native honeyeaters (Noisy Miners *Manorina melanocephala*: Loyn 1987; Grey *et al.* 1997, 1998). In contrast, no such trend was evident in patches of old Mountain Ash in a matrix of 1939 regrowth (from wildfire, often followed by salvage logging) (Loyn 1998). This is not a perfect comparison, as forest types and altitude differed between the various studies, and there are no Miners *Manorina* spp. in ash forests. Nevertheless, the comparison can be taken as suggestive evidence of a buffering role of regrowth compared with pasture.

Regrowth as habitat

The ability of regrowth forests to supply ***habitat*** is quite fundamental to its role in wildlife conservation. The popular view of a logged area is a recently logged coupe — open, often burnt and with few if any old trees left standing (Fig. 2a). It is less often realized that regrowth after a few years is usually much denser than the original forest (Fig. 2b), providing quite a different set of habitat values. Dense regrowth gains height and structure

(a)

Fig. 2. (a) Recently logged coupe in Mountain Ash *E. regnans* forest (1939 regrowth); (b) 5-year regrowth from logging in mixed eucalypt forest, with retained old trees; (c) 70-year regrowth from logging in mixed eucalypt forest, all in eastern Victoria, southeastern Australia.

(b)

(c)

and thins out as years progress (Fig. 2c) but still differs in many respects from old forest (e.g., Scotts 1991). Some middle-storey plants such as Silver Wattle *Acacia dealbata* have relatively short life-spans (Ashton 1976): they regenerate well after disturbance including fire, and become progressively less well represented in older stands. Some of the major differences relating to bird habitat in southeastern Australia are summarized below:

Mature forest typically has:

- many big old trees with hollows, fissured bark (used by treecreepers, especially Red-browed Tree-creepers *Climacteris erythrops*) and parasitic mistletoe (providing an important source of nectar and fruit for a wide range of birds);
- much sheltered open space below the canopy, providing foraging opportunities for aerial feeding insectivores, and facilitating movement for large birds such as owls;
- scattered shrubs and bare open ground (with bare open ground among trees needed by species such as Scarlet Robin *Petroica multicolor* and Spotted Quail-thrush *Cinclosoma punctatum*);
- diverse eucalypt species and uneven-aged structure (at least in foothill forests); and
- an ability to produce heavier crops of nectar than are typically observed in regrowth (though this needs to be tested further).

Regrowth forests typically have:

- much exposed bare ground in early stages (0–3 years), and very little later;
- dense shrub and eucalypt regeneration (used for foraging, roosting and nesting by many bird species);
- much damp ground below shrub layers (used for foraging by birds such as Superb Lyrebirds *Menura novaehollandiae*);
- abundant wattles including species favoured by many insectivorous birds (e.g., Rose Robin *Petroica rosea*) and some seed-eating birds (e.g., Brush Bronzewing *Phaps elegans*);
- ephemeral pulses of abundant food sources when many plants reach the same stage of development at the same time;
- dominance of particular eucalypt species, e.g., Silvertop *E. sieberi*; and
- even-aged stands with few old trees.

These differences apply at the coupe level, but there are also important structural differences at a landscape level between managed and unmanaged forests, as summarized below:

Natural landscapes tend to contain:

- more old forest;
- more uneven-aged forest stands; and
- extensive areas of same-age regrowth after severe wildfire, with varying numbers of older trees.

Managed landscapes used for wood production tend to contain:

- regrowth patches of many different ages;
- small numbers of retained older trees on coupes;
- strips and patches of retained older forest; and
- a pattern and scale of regrowth and older forest that can be managed, and is generally finer in scale but less varied than that produced by severe wildfire.

The successional changes are reflected in changes in bird abundance (Loyn 1980, 1985a, 1991, and unpubl. data from eastern Victoria). Several patterns have been observed (Fig. 3). Species that need open stands may prosper in the first two or three years after logging (3a), but be eliminated for many years from the dense regrowth which follows (e.g., Scarlet Robin *Petroica multicolor*, Flame Robin *P. phoenicea* and Buff-rumped Thornbill *Acanthiza reguloides*). The same pattern has also been observed after wildfire (Reilly 1991; Loyn 1997). Species that inhabit dense understorey may prosper in subsequent years, and species inhabiting middle layers become common at a later stage (3b or 3c). Many insectivorous birds feed extensively in wattles, with Silver Wattles being highly favoured by birds such as Rose Robin *Petroica rosea*, which follow pattern 3c and decline as the wattles thin out.

Honeyeaters often reach a temporary peak in abundance in regrowth aged 5–10 years after logging (3d), when they feed largely on manna, lerps and honeydew. The species that exploit this resource may differ from those that are most common in mature forest, where nectar from eucalypts and mistletoe is the main attraction. White-eared Honeyeaters *Lichenostomus leucotis* and Crescent Honeyeaters *Phylidonyris pyrrhoptera* are often common in young regrowth, whereas Yellow-faced Honeyeaters *Lichenostomus chrysops* and White-naped Honeyeaters *Melithreptus lunatus* tend to be more numerous in mature stands. If the dense regrowth includes thickets of Sunshine Wattle *Acacia terminalis*, a suite of species may be attracted to feed on the sweet glandular exudates produced while the plant is flowering in autumn. These birds include Striated Thornbills *Acanthiza lineata* and Silvereyes *Zosterops lateralis* as well as some honeyeaters. A similar flush of resources may be produced by older stands of eucalypts (10–20 years) as they shed bark from upper branches in autumn and winter. Birds seen feeding in this situation include Crimson Rosellas *Platycercus elegans* (probably taking scale insects), White-throated Treecreepers *Cormobates leucophaea*, Red-browed Treecreepers *Climacteris erythrops*, Crested Shrike-tits *Falcunculus frontatus* and Crescent Honeyeaters *Phylidonyris pyrrhoptera*. In general, dense stands of regrowth tend to produce abundant crops of particular resources for limited time-spans. Birds must be mobile to exploit these ephemeral flushes of resources.

The most common pattern (3b) involves a decline in abundance immediately after logging, followed

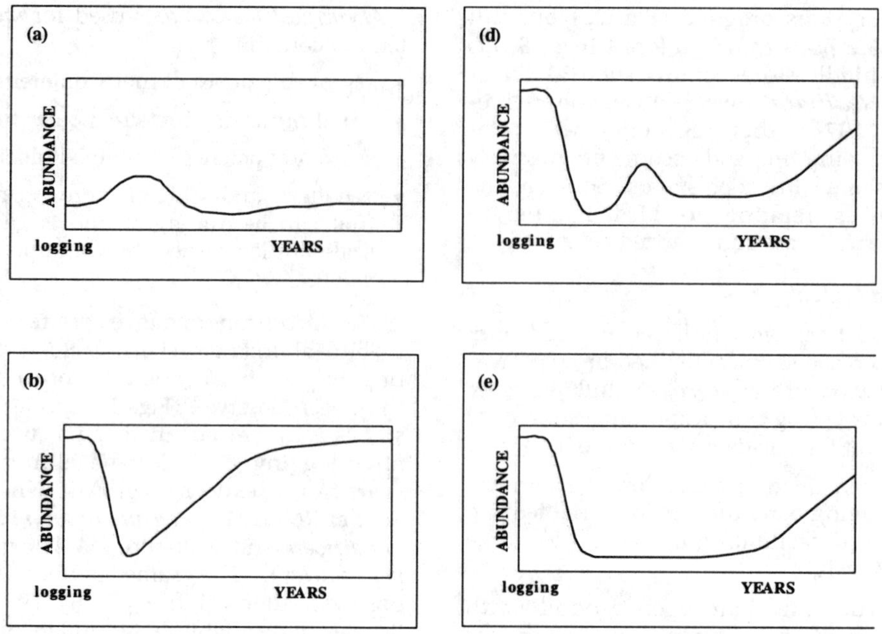

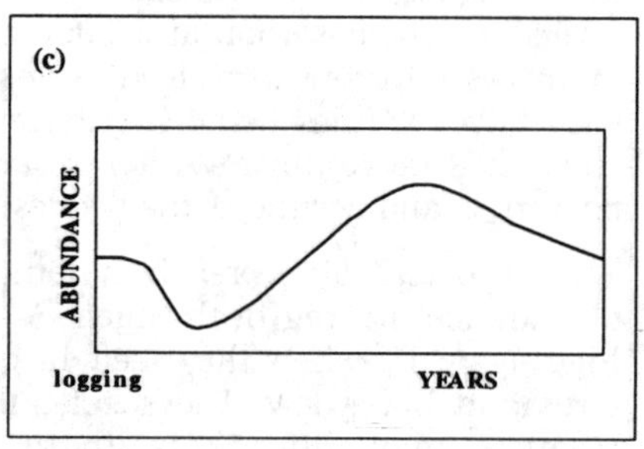

Fig. 3. Stylized curves for changes in bird abundance over time since logging for: (a) birds that feed from open ground; (b) birds that feed in understorey or canopy; (c) birds that feed in middle storey; (d) honeyeaters and (e) birds that need elements of old forest. Source of data: retrospective and experimental studies and general observations in eastern Victoria, southeastern Australia.

by progressive recovery over time as the forest regenerates. The main conservation issue in relation to logging concerns the species that fail to show significant recovery over a planned rotation (3e), or depend on some resource (e.g., hollow trees) that is not expected to regenerate in that time. As discussed by Loyn (1985b), there are precisely three strategies for conserving those species in productive forests. They are:

- to manage logged coupes on rotations long enough to regrow old trees or other elements of old forest;
- to retain selected trees (or other elements of old forest) on logged coupes, and regrow those elements over one or more rotations to replace them over time, and
- to retain selected stands of old forest (of varying shape, size, etc.) and younger patches to replace them over time.

The *long-rotation strategy* is unlikely to be of general use because by the time trees have grown old enough to supply resources such as large hollows, they will be unsuitable for wood production. It could be useful locally where the particular resources of concern are expected to develop before wood values deteriorate (as with open spaces below the canopy), or where individual trees can be expected or encouraged to develop hollows at the same time as other trees continue to produce wood. This may happen through damage from fire or lightning, or deliberate actions such as fungal inoculation (as considered in North America, Hunter 1990).

The *strategy of retaining trees or elements of old forest on coupes* has the advantage of reducing visual impact and adding structural diversity at the coupe scale, and more widely if retention rates are varied across the landscape. It has particular potential for species that use a mixture of old trees and young regrowth (e.g., Leadbeater's Possum *Gymnobelideus leadbeateri*: Macfarlane and Seebeck 1991; Smith and Lindenmayer 1992). The elements of old forest may include groups of retained trees, as well as other features such as understorey islands (Ough and Murphy 1996). However, some elements such as open ground below shrub layers are more practically conserved in retained stands than they are among regrowth on logged coupes. The same applies to combinations of elements, or minimum threshold levels of those elements that may be needed by particular wildlife species (e.g., Yellow-bellied Glider *Petaurus australis*, which needs large areas of suitable old forest, Milledge *et al.* 1991; Incoll *et al.*, unpubl.).

The *stand-retention strategy* has strong appeal and is a key element of current forest management planning. The focus of much current effort is to refine abilities to apportion land uses appropriately between retained and logged forest. This can bring dividends both for wood production and wildlife conservation by exploiting natural variations in values for each purpose over the forest landscape. Gullies are a classic example, as they are usually rich in wildlife (Loyn *et al.* 1980; Recher *et al.* 1980) and not available for logging, partly to protect water supplies. However, some species avoid gullies and need to be conserved elsewhere (e.g., Loyn 1980; Smith 1985; Recher 1991). Conflict inevitably arises when high values for each use coincide on the same parcel of land, and there are many examples of species which favour commercially productive environments (e.g., Braithwaite *et al.* 1984).

The Code of Practice for Timber Production (Natural Resources and Environment 1996) specifies minimum standards for tree retention on coupes, and these are often exceeded on a local basis. However, more work is needed to assess costs and benefits of different rates and configurations of tree retention and stand retention in a range of forest communities. All three strategies need to be considered together as options in developing prescriptions for logging coupes, and planning forest use at the landscape scale. Work continues to quantify the potential values of these strategies for wildlife, to help determine the costs and benefits of particular mixtures of them in different parts of the forest landscape.

Conservation biologists tend to emphasize the differences between regrowth and mature forest, perhaps leaving a public perception that regrowth has little value. However, regrowth forests may support bird populations which are more similar to those of mature forest than they are to cleared land. This is illustrated by sample results (Fig. 4), which show mean bird abundances in large numbers of sites in mature and regrowth forests, remnant forest in farmland, and cleared pasture or cropland. The values for bird abundance in pasture or cropland were at least an order of magnitude lower than for forest regrowth (Fig. 4). There were some small differences in species composition between mature forest and regrowth, and much greater differences in species composition between cleared farmland and remnant vegetation (unpubl. data). Many forest mammals also avoid pasture (Bennett 1990) and inhabit forest regrowth, though most arboreal species

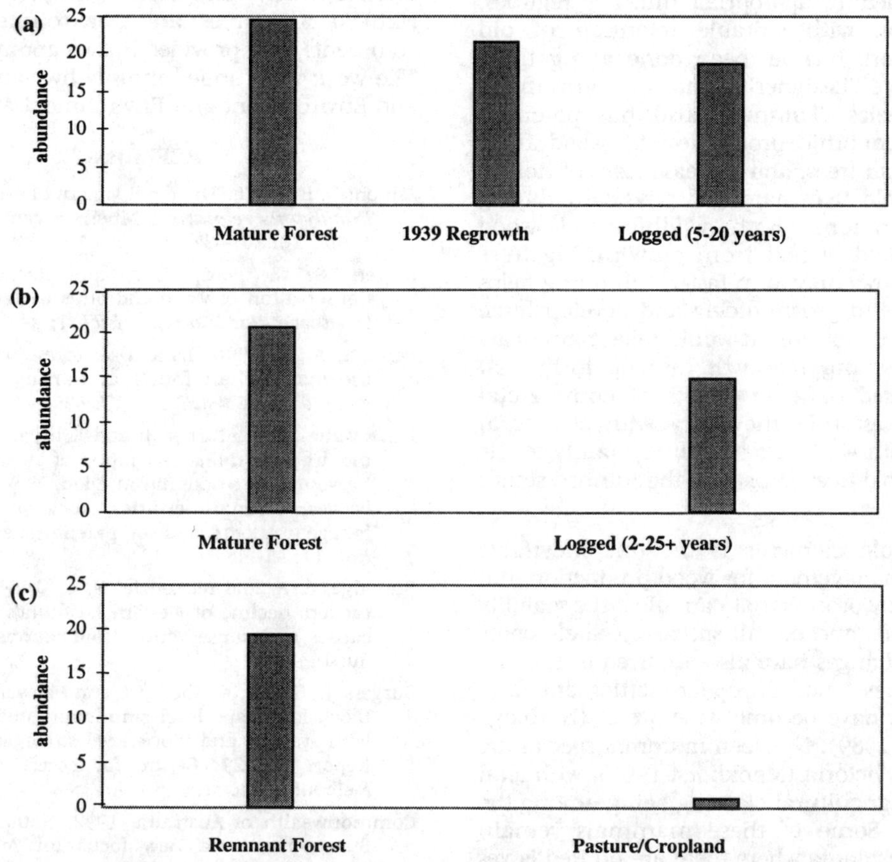

Fig. 4. Bird abundance (birds per count) in mature forest and modified environments, from three studies by area search (Loyn 1986). (a) in Mountain Ash *E. regnans* forests: old forest (n = 45), regrowth aged 55 years from 1939 fires (n = 73) and regrowth aged 5–10 years from logging (n = 16); (b) in mixed eucalypt forests: mature forest (n = 18), and regrowth aged 2–25+ years from logging (n = 45); (c) in remnant forests of mallee or box eucalypts (n = 67) and cleared pasture or cropland (n = 800). Source of data: unpublished Arthur Rylah Institute studies by R. Loyn, G. Cheers, G. Carpenter, J. Robinson and L. Clausen.

depend on old trees as den sites (e.g., Macfarlane 1988; Smith and Lindenmayer 1988; Lindenmayer *et al.* 1990; Nelson *et al.* 1996). Hence, if it is necessary to produce an economic crop from a parcel of land, there are much greater biodiversity benefits from regrowing forests than from converting them to pasture or cropland. It is also important to retain and regrow old forest because some species depend upon it, or are much more common there than elsewhere.

IMPROVING THE HABITAT VALUES OF MANAGED FORESTS

One way to improve the habitat value of managed forests is to address the main conflict identified above, and ensure a continuing supply of old forest elements in the landscape, using a mix of the three strategies discussed. Old trees are not the only element of old forest that needs attention in this respect. Wide but sheltered open spaces between trees are a characteristic element of old forest, that seems important for some birds (e.g., Sooty Owl *Tyto tenebricosa*) and mammals (e.g., Yellow-bellied Glider *Petaurus australis*). Those two species also need old trees *per se* (Milledge *et al.* 1991), but possibilities may exist to improve the value of regrowth for them by appropriate thinning regimes, in conjunction with suitable retention of old trees. Little work has yet been done along these lines to produce "designer" stands of regrowth for particular species. Thinning also has potential to increase economic production of wood from selected regrowth trees, and increase rates of hollow formation in old trees damaged physically during thinning operations. Both wildlife and wood production could benefit from growing big trees faster, and hollows may form faster if thinning helps particular trees to grow quickly and develop large lateral branches. However, it would take many years for thinning young regrowth to help form new hollows. Limited data on effects of commercial thinning suggest that they vary with structural complexity, with wildlife responding mainly to the availability of old trees in or near the thinned stands (Kutt 1996).

Decline of old elements is the one inevitable conflict with management for wood production, but it is not the only process that can affect the viability of forest wildlife, and not all species need elements of old forest. Changes have also occurred in fire and grazing regimes since European settlement, and feral predators have become widespread (Burbidge and McKenzie 1989). Nineteen mammal species are now extinct in Victoria (Menkhorst 1995), with feral animals and agricultural clearing being among the main causes. Some of these mammals remain common in Tasmania where there are no Red Foxes *Vulpes vulpes*. Better management of feral animals, and in particular local elimination of Red Foxes, could have major benefits for forest wildlife and no direct implications for wood production. There is some evidence that Dingoes *Canis familiaris dingo* may help reduce the prevalence of Red Foxes in some areas (Newsome *et al.* 1997).

Many of these points are being addressed to varying degrees in the Victorian forest planning process. Surveys of forest wildlife (initiated as part of the Regional Forest Agreement process) are conducted on a "tenure-blind" basis, in parks, reserves and state forests across the landscape. Data from these surveys are being used to model predicted distributions of key species (e.g., forest owls, now surveyed at 1 500 sites in eastern Victoria over two years). These models are then used along with other data to plan reserves or prescriptions for conserving owls in an integrated way across the forest landscape. Much remains to be done, but it is accepted that animals pay scant regard to artificial boundaries, and integrated management is the way forward. The need to manage the forest matrix is accepted, and our challenge is to make that management as effective as we can.

ACKNOWLEDGEMENTS

I would like to thank many colleagues who have assisted in this work at various times, and especially those from the Arthur Rylah Institute and Centre for Forest Tree Technology. Particular help in preparing this chapter was provided by Lucy Clausen (Arthur Rylah Institute, Natural Resources and Environment) and Rolf Willig (Forests Service, Natural Resources and Environment). Valuable comments were provided by two anonymous referees. The work was funded mainly by Natural Resources and Environment and Environment Australia.

REFERENCES

Ashton, D. H., 1976. The development of even-aged stands in *Eucalyptus regnans* F. Muell. in central Victoria. *Aust. J. Bot.* **23**: 867–87.

Barrett, G. W., Ford, H. A. and Recher, H. F., 1994. Conservation of woodland birds in a fragmented rural landscape. *Pac. Conserv. Biol.* **1**: 245–56.

Bennett, A. F., 1990. Land use, forest fragmentaton and the mammalian fauna at Narringal, South-western Victoria. *Aust. Wildl. Res* **17**: 325–47.

Braithwaite, L. W., Turner, J. and Kelly, J., 1984. Studies on the arboreal marsupial fauna of eucalypt forests being harvested for woodpulp at Eden, NSW. 3. Relationships between faunal densities, eucalypt occurrence and foliage nutrients, and soil parent materials. *Aust. Wildl. Res.* **11**: 41–48.

Burbidge, A. A. and McKenzie, N. L., 1989. Patterns in the modern decline of Western Australia's vertebrate fauna: causes and conservation implications. *Biol. Conserv.* **50**: 143–98.

Burgess, J., Cherry, K., Cleary, J. and Papworth, M., 1997. Ada Block landscape level simulation study. Part 1. Coupe level systems and block level strategies. VSP Technical Report No. 29. Centre for Forest Tree Technology: Melbourne, Victoria.

Commonwealth of Australia, 1992. National Forest Policy Statement — a new focus for Australia's forests. Advance Press: Perth. 52 Pp.

Commonwealth of Australia, 1997. Victorian statewide assessment of ecologically sustainable forest management. Joint Commonwealth and Victorian Regional Forest Agreements Steering Committee, Canberra.

Diamond, J. M., 1975. The island dilemma: lessons of modern biogeographic studies for the design of natural reserves. *Biol. Conserv.* **7**: 129–46.

Forman, R. T. T. and Godron, M., 1986. Landscape Ecology. John Wiley and Sons: New York.

Grey, M. J., Clarke, M. F. and Loyn, R. H., 1997. Initial changes in the avian communities of remnant eucalypt woodlands following a reduction in the abundance of Noisy Miners, *Manorina melanocephala*. *Wildl. Res.* **24**: 631–48.

Grey, M. J., Clarke, M. F. and Loyn, R. H., 1998. Influence of the Noisy Miner *Manorina melanocephala* on avian diversity and abundance in remnant Grey Box woodland. *Pac. Conserv. Biol.* **4**: 55–69.

Howe, R. W., 1984. Local dynamics of bird assemblages in small forest habitat islands in Australia and North America. *Ecology* **65**: 1585–601.

Hunter, M. L., 1990. Wildlife, forests and forestry: principles of managing forests for biological diversity. Prentice-Hall: 370 Pp.

Kutt, A. S., 1996. Bird population density in thinned, unthinned and old lowland regrowth forest, East Gippsland, Victoria. *Emu* **96**: 280–84.

Lindenmayer, D. B., Cunningham, R. B., Tanton, M. T. and Smith, A. P., 1990. The conservation of arboreal marsupials in the montane ash forests of the Central Highlands of Victoria, south-east Australia. 1. Factors affecting the occupancy of trees with hollows. *Biol. Conserv.* **54**: 111–31.

Loyn, R. H., 1980. Bird populations in a mixed eucalypt forest used for production of wood in Gippsland, Victoria. *Emu* **80**: 145–56.

Loyn, R. H., 1985a. Bird populations in successional forests of Mountain Ash *Eucalyptus regnans*. *Emu* **85**: 213–30.

Loyn, R. H., 1985b. Strategies for conserving wildlife in commercially productive eucalypt forest. *Aust. For.* **48**: 95–101.

Loyn, R. H., 1987. Effects of patch area and habitat on bird abundances, species numbers and tree health in fragmented Victorian forests. Pp. 65–77 *in* Nature Conservation: the Role of Remnants of Native Vegetation ed by D. A. Saunders, G. W. Arnold, A. A. Burbidge and A. J. M. Hopkins. Surrey Beatty & Sons: Chipping Norton, New South Wales.

Loyn, R. H., 1991. Effects of previous logging on bird populations in East Gippsland: VSP retrospective study. VSP Technical Report 18, Dept. Conservation and Natural Resources (now Natural Resources and Environment): Melbourne, Victoria. 62 Pp.

Loyn, R. H., 1997. Effects of an extensive wildfire on birds in far eastern Victoria. *Pac. Conserv. Biol.* **3**: 221–34.

Loyn, R. H., 1998. Birds in patches of old-growth ash forest, in a matrix of younger forest. *Pac. Conserv. Biol.* **4**: 111–21.

Loyn, R. H., Macfarlane, M. A., Chesterfield, E. A. and Harris, J. A., 1980. Forests utilisation and the flora and fauna in Boola Boola State Forest in south-eastern Victoria. Forests Commission Victoria. Bulletin 28, 80 Pp.

MacArthur, R. H. and Wilson, E. O., 1963. An equilibrium theory of insular zoogeography. *Evolution* **17**: 373–87.

Macfarlane, M. A., 1988. Mammal populations in Mountain Ash (*Eucalyptus regnans*) forest of various ages in the Central Highlands of Victoria. *Aust. For.* **51**: 14–27.

Macfarlane, M. A. and Seebeck, J. H., 1991. Draft management strategies for the conservation of Leadbeater's Possum, *Gymnobelideus leadbeateri*, in Victoria. Arthur Rylah Institute Technical Report No. 111.

McIntyre, S. and Barrett, G. W., 1992. Habitat variegation, an alternative to fragmentation. *Conserv. Biol.* **6**: 146–47.

Menkhorst, P. W. (ed), 1995. Mammals of Victoria: Distribution, Ecology and Conservation. Oxford University Press and Dept. Conservation and Natural Resources (now Natural Resources and Environment): Melbourne, Victoria.

Milledge, D., Palmer, C. L. and Nelson, J. L., 1991. "Barometers of change": the distribution of large owls and gliders in Mountain Ash forests of the Victorian Central Highlands and their potential as management indicators. Pp. 53–65 *in* Conservation of Australia's Forest Fauna ed by D. Lunney. Royal Zoological Society of New South Wales and Surrey Beatty & Sons: Chipping Norton, New South Wales.

Nelson, J. L., Cherry, K. A. and Porter, K. W., 1996. The effect of edges on the distribution of arboreal marsupials in the ash forests of the Victorian Central Highlands. *Aust. For.* **59**: 189–98.

Newsome, A., Pech, R., Smyth, R., Banks, P. and Dickman, C., 1997. Potential impacts on Australian native fauna of rabbit calicivirus disease. Australian Nature Conservation Agency, Canberra, P 47.

Natural Resources and Environment, 1996. Code of Forest Practices for Timber Production. Revision No. 2. Dept. Natural Resources and Environment (Victoria): Melbourne, Victoria.

Natural Resources and Environment, 1999. Proposed forest management plan for the North East. Dept. Natural Resources and Environment (Victoria): Melbourne, Victoria.

Ough, K. and Murphy, A., 1996. The effect of clearfell logging on tree-ferns in Victorian wet forest. *Aust. For.* **59**: 178–88.

Pressey, R. L., 1994. *Ad hoc* reservations: forward or backward steps in developing representative reserve systems? *Conserv. Biol.* **8**: 662–67.

Recher, H. F., 1991. The conservation and management of eucalypt forest birds: resource requirements for nesting and foraging. Pp. 25–34 *in* Conservation of Australia's Forest Fauna ed by D. Lunney. Royal Zoological Society of New South Wales and Surrey Beatty & Sons: Chipping Norton, New South Wales.

Recher, H. F., Rohan-Jones, W. and Smith, P., 1980. Effects of the Eden woodchip industry on terrestrial vertebrates with recommendations for management. Forestry Commission of NSW, Research Note 42, 83 Pp.

Reilly, P., 1991. The effect of wildfire on bird populations in a Victorian coastal habitat. *Emu* **91**: 100–06.

Saunders, D. A. and de Rebeira, C. P., 1985. Turnover in breeding bird populations on Rottnest Island: Western Australia. *Aust. Wildl. Res.* **12**: 467–77.

Scotts, D. J., 1991. Old-growth forests: their ecological characteristics and value to forest-dependent vertebrate fauna of south-east Australia. Pp. 147–59 *in* Conservation of Australia's Forest Fauna ed by D. Lunney. Royal Zoological Society of New South Wales and Surrey Beatty & Sons: Chipping Norton, New South Wales.

Shaw, W. W., 1983. Integrating wildlife conservation and timber production objectives in Australian forests. *Aust. For.* **46**: 132–35.

Smith, P., 1985. Effects of intensive logging on birds in eucalypt forest near Bega, New South Wales. *Emu* **85**: 15–21.

Smith, A. P. and Lindenmayer, D. B., 1988. Tree hollow requirements of Leadbeater's Possum and other possums and gliders in timber production ash forests of the Victorian Central Highlands. *Aust. Wildl. Res.* **15**: 347–62.

Smith, A. P. and Lindenmayer, D. B., 1992. Forest succession, timber production and conservation of Leadbeater's Possum (*Gymnobelideus leadbeateri* Marsupialia: Petauridae). *For. Ecol. Manage.* **49**: 311–32.

Traill, B. J., Collins, E., Peake, P. and Jessup, S., 1996. Current and past status of the birds of Chiltern — a box-ironbark forest in north-eastern Victoria. *Aust. Bird Watcher* **16**: 309–26.

CHAPTER 13

Occurrence of carabid beetles in retained unlogged strips within production forests of southern Tasmania

Robert Taylor,[1] Karyl Michaels[2] and Dick Bashford[1]

A previous study of the occurrence of carabid beetles in 2-25-year-old regrowth and mature stands of wet eucalypt forest in the southern forests of Tasmania identified four species (*Chylnus ater, Notonomus politulus, Pterocyrtus* sp.n., and *Stichonotus piceus*) that were either restricted to, or had a marked preference for, mature forest. Two of these species (*C. ater* and *N. politulus*) were frequently captured in mature forest, whereas the other two were rare or uncommon. In this study we resampled carabid populations in the regrowth and mature stands, but in addition sampled five mature forest sites that were present as retained strips along rivers or smaller streams. These strips ranged from 59 to 121 m in width. The number of species and individuals was greatest in the one-year-old regrowth, with survivors of the logging and new invading species both present. Three main patterns of occurrence were found: species present only in young regrowth; species present in both older regrowth and mature forest; and species with a marked preference for mature forest. Classification of the sites on the basis of their carabid faunas produced groups that consisted of a mixture of mature stands, strips and regrowth sites. Very young regrowth grouped separately from other stands. All four of the mature forest species were found in the retained strips, suggesting that strips may be an effective way of maintaining carabid species in the matrix between reserves and ensuring dispersal between them. However, monitoring of populations in strips is required to ensure such systems maintain viable populations in the long term, particularly of the rare species.

INTRODUCTION

GROUND dwelling carabid beetles have been found to be useful indicators of environmental conditions (Jarosik 1983; Tietze 1987; Holliday 1992; Luff *et al.* 1992), particularly in relation to the effects of forestry (Niemela *et al.* 1988; Niemela *et al.* 1993a,b; Spence *et al.* 1996; Atlegrim *et al.* 1997). Studies in different forest types in many parts of the world have shown that some species of carabids appear to be mature forest specialists which do not inhabit regenerating forest (Sustek 1981; Lenski 1982; Jennings *et al.* 1986; Niemela *et al.* 1988; Niemela *et al.* 1993a,b). A study of the effects of clearfelling in wet eucalypt forest in southern Tasmania also concluded that some species were mature forest specialists which were absent or at much lower densities in regenerating forest (Michaels and McQuillan 1995). These mature forest specialists are flightless and hence may be poor dispersers which suffer from fragmentation effects (Turin and den Boer 1988; Halme and Niemela 1993a; de Vries *et al.* 1996). In production forests in Tasmania, strips of uncut forest are retained to, among other things, maintain connectivity between larger areas of forest that are not subject to timber harvesting (Taylor 1991a). The present study involved repeating the sampling of carabid beetles in regrowth and mature forest undertaken by Michaels and McQuillan (1995) with additional sites sampled in retained strips of forest which linked larger areas of uncut forest. The aim of the study was to examine whether these strips provide suitable habitat for species of carabid beetles that are mature forest specialists and hence ensure that such species retained in blocks of unlogged forest would not suffer from isolation effects.

[1]Forestry Tasmania, G.P.O. Box 207B, Hobart, Tasmania, Australia 7001.
[2]School of Geography and Environmental Studies, University of Tasmania, G.P.O. Box 252–78, Hobart, Tasmania, Australia 7001.
Pages 120–27 *in* NATURE CONSERVATION 5: NATURE CONSERVATION IN PRODUCTION ENVIRONMENTS: MANAGING THE MATRIX ed by J. L. Craig, N. Mitchell and D. A. Saunders. Surrey Beatty & Sons, 2000.

STUDY AREA

The study was undertaken in a large area of production forest approximately 100 km south of Hobart in southern Tasmania. The forests are mainly tall eucalypt with a sclerophyllous or rainforest understorey. Most sites are dominated by pure stands of *Eucalyptus obliqua* with *E. regnans* dominant in smaller areas. Cool temperate rainforest dominated by *Nothofagus cunninghami* is present in some sites, mainly along steeper gullies that have survived past wildfires. Annual rainfall totals approximately 1 100 mm and is distributed relatively evenly throughout the year. Soils are either fertile clay-loams of variable depth developed on Jurassic dolerite or less fertile Triassic-Permian sandstone of the Parmeneer supergroup (Davies 1988).

Timber harvesting has been conducted in the area since 1920. In more recent times clearfelled coupes of around 50 to 80 ha have been dispersed through the landscape. Coupes are subject to a high intensity regeneration burn after logging. Typical timber yields are around 300 tonnes/ha of pulpwood and sawlogs (J. Traill, pers. comm.), with around 600 m^3/ha of coarse woody debris present in the regenerating forest (Meggs 1996). In the areas where the study took place, approximately 30% of scheduled coupes have been logged using clearfell and burn techniques. Streamside reserves are retained on all drainage lines with a catchment greater than 50 ha. The prescribed reserve width varies from 20 m either side of the creek on the smallest creeks to 40 m on the major rivers, although widths of 100 m are usually retained on the larger rivers on government-owned land. Strips of uncut forest (termed Wildlife Habitat Strips), that are often based on creeks but which also run up slopes and cross ridgetops, are retained every 3 to 5 kilometres (Taylor 1991b).

Sampling was undertaken in regrowth ranging from 1 to 26 years after clearfelling followed by regeneration burning and in unlogged forest with differing age structures, patch sizes and shapes. The age structure of unlogged forest is determined by the time since the last fire and the intensity of past fires. If a wildfire is very intense, all of the eucalypts are killed and an even-aged stand results. Often, however, fires are such that some trees survive and regeneration occurs resulting in a mixed age stand. The age of regrowth trees present in unlogged forest was estimated from the date of past wildfires and the presence of old trees is shown as OG in Table 1. These old trees would have been older than 200 years and some could have been up to 400 years old. The ground cover varied between and within sites. It was very open with little litter in very young regrowth. Dense weedy growth and *Gahnia grandis* was present in young regrowth and in mature forest stands the undergrowth was either open with dense litter or dense ferns or patches of *Gahnia* were present. The understorey was classified into three types depending on the dominance of rainforest species. "Wet" understoreys had no rainforest species, "mixed" had both sclerophyllous and rainforest species and "rainforest" had no or few sclerophyllous species (Table 1). Mature forest stands were present in two types of situations. They were either part of large blocks of forest that formed the major part of the forest matrix or they were present as strips of varying widths and lengths, all of which were based along rivers or streams. The landscape features of the strips are shown in Figure 1. The areas are coded as R, B or S, depending on whether they were regrowth (R), mature forest present as part of a large block (B) or as a strip (S). Each area within a category was then numbered sequentially (e.g., S1, S2 etc.) except for regrowth where the number assigned represents the age and an a or b differentiates two sites of the same age (e.g., R9b is the second of two regrowth sites which were nine years old).

Table 1. Details of the location (nine digit Australian Map Grid Reference), geology, vegetation and forest age for each site. R = regrowth, B = block of mature forest, S = strip of mature forest.

Site	Grid Reference	Elevation (m)	Geology	Dominant eucalypt	Age of eucalypts	Understorey	Ground cover
R1	4733 52165	200	Parmeena	*E. obliqua*	1 yr	wet	open
R2	4730 52188	200	Parmeena	*E. obliqua*	2 yr	wet	open
R5a	4726 52185	200	Parmeena	*E. obliqua*	5 yr	wet	open
R5b	4732 52287	200	Parmeena	*E. obliqua*	5 yr	wet	dense
R9a	4737 52146	300	Parmeena	*E. obliqua*	9 yr	mixed	patchy
R9b	4737 52146	300	Parmeena	*E. obliqua*	9 yr	mixed	patchy
R24	4794 52290	100	Parmeena	*E. obliqua*	24 yr	wet	patchy
R26	4794 52290	100	Parmeena	*E. obliqua*	26 yr	mixed	patchy
B1	4776 52274	100	Parmeena	*E. obliqua*	62, 90 and OG	wet	patchy
B2	4733 52157	300	Parmeena	*E. obliqua*	OG	rainforest	dense
B3	4777 52265	100	Parmeena	*E. obliqua*	62, OG	mixed	patchy
B4	4732 52287	200	Parmeena	*E. obliqua*	90, OG	mixed	dense
S1	4897 52167	210	Dolerite	*E. regnans*	OG	rainforest	open
S2	4822 52260	160	Parmeena	*E. obliqua*	OG	rainforest	open
S3	4742 52144	120	Parmeena	*E. obliqua*	OG	rainforest	dense
S4	4721 52367	325	Parmeena	*E. obliqua*	OG	rainforest	patchy
S5	4898 52311	360	Parmeena	*E. regnans*	90, OG	mixed	open

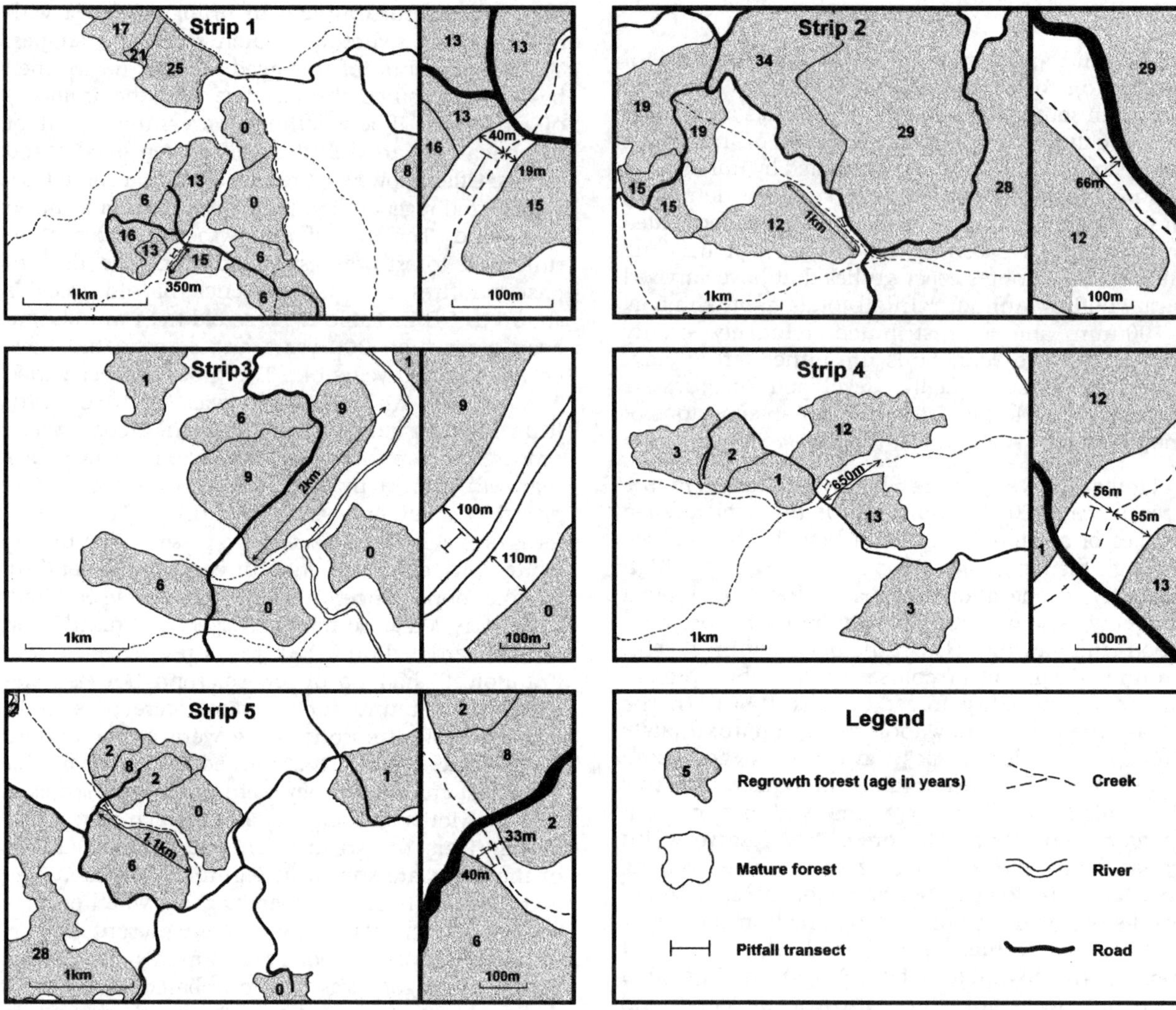

Fig. 1. Lengths and widths of the strips of mature forest sampled. The location around the pitfall transect is shown at a higher resolution on the right of each figure.

METHODS

Pitfall trapping of the regrowth and blocks of mature forest was carried out by Karyl Michaels and the retained strips were sampled by Richard Bashford. Trapping by Karyl Michaels was undertaken between 2 December 1994 and 2 March 1996 and by Richard Bashford between 19 December 1994 and 14 February 1996. Pitfalls were set out at 10 m intervals along a 50 m transect. Karyl Michaels had five areas (R1, R5a, R9a, R24, B1) with one pit at each interval along the transect (total of six pits), and five areas (R2, R5b, R9b, R26 B2) with one at four locations and three at each of two other locations (total of 10 pits). Sampling varied at these sites as the effects of trapping intensity on population depletion was being examined. As only two areas were sampled in large blocks of mature forest at the same time as the strips were sampled, the data for a further two areas in blocks of mature forest (B3 and B4) which had been sampled by Karyl Michaels in the previous twelve months (mid July 1993-mid July 1994) were included in the analysis. At these two areas three pits were placed at each site along the transect. At the strip sites two pitfalls were placed at each location along the transect. Michaels and McQuillan (1995) report the results of trapping in all of the areas monitored by Karyl Michaels, except R1, for a ten month period prior to the present study.

Each trap consisted of a plastic drinking cup, nine cm in diameter, inserted into a PVC plastic sleeve with the cup level with the soil surface. Ethylene glycol was used as a preservative. This chemical, like other preservatives, is probably differentially attractive to carabid species (Holopainen 1990). Pits were sheltered from the rain by a 12 cm diameter plastic lid supported a few centimetres above the litter with a tripod of sticks. Traps were emptied at approximately six week intervals.

The similarity of the carabid fauna at the different areas was assessed using the agglomerative clustering technique "flexible UPGMA" from the PATN software package (Belbin 1989). Since the numbers of traps used varied between areas, the abundance data were scaled at each area by dividing the number of each species collected by the number of individuals of the most frequently captured species at that area and multiplying by 100. The Bray-Curtis coefficient was used as a measure of dissimilarity between samples after standardizing data by subtracting minimum abundance of a taxon and dividing by its range to reduce the weighting of abundant taxa. Differences in the means of groups were examined using analysis of variance and least significant difference was used to test for differences between individual pairs of means if the analysis of variance was significant. Correlations between variables were examined using the Pearson correlation coefficient.

RESULTS

Twenty-nine species of carabids were captured. The similarity of areas in terms of their carabid faunas is shown in Figure 2. Four groups were indicated at a fusion level of 0.7. Areas with young regrowth (one, two and five-year-old) made up one group, five-year-old regrowth and two mature forest blocks and two strips made up the second group, the other older regrowth sites (>5 years) and one block and two strips made up the third, and the fourth group was composed of a block and a strip. Thus very young regrowth appears to have a distinctive fauna. However, the similarity of the fauna of other areas appears not to be primarily determined by the age or shape of the forest. None of the coarse habitat variables in Table 1 were associated with the pattern of grouping of areas. Geographical proximity explained some of the groupings with R9a, B2 and S3 being within two and a half kilometre of each other. However, B1 and B2 were also within two and a half kilometre of each other and they did not group together. The width of the strip was not associated since the two narrowest occurred in different groups, as did the two widest.

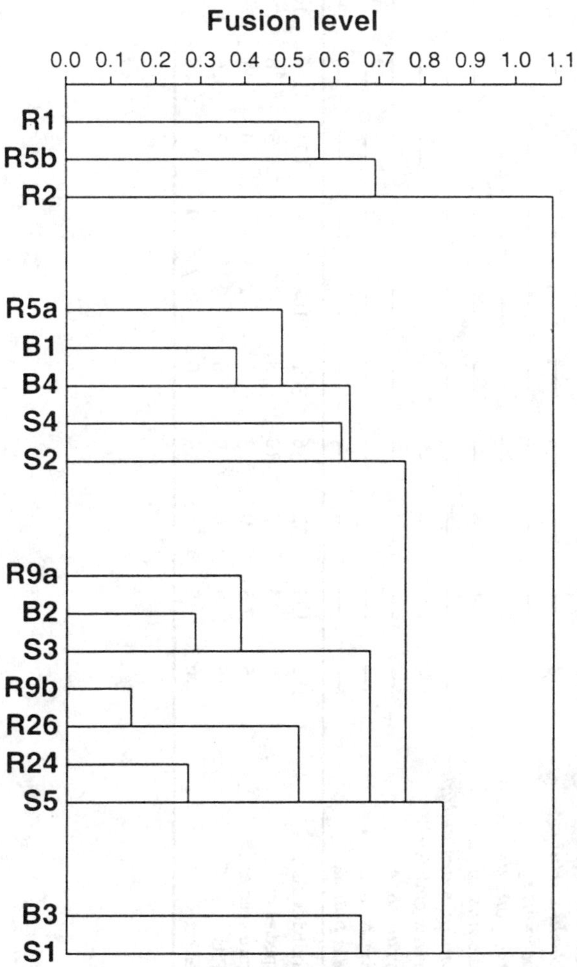

Fig. 2. Dendrogram for the areas trapped based on their species composition of carabid beetles.

The most abundant species, ***Rhabdotus reflexus,*** was common in all mature forest and regrowth areas except for five-year-old regrowth (Table 2). ***Mecyclothorax ambiguus*** was extremely abundant just after logging and was absent or uncommon in most other areas. ***Sloaneana tasmaniae*** was common at many of the older (>5 years) regrowth and mature forest areas. ***Chylnus ater*** and ***Notonomus politulus*** were generally abundant in mature forest and in low numbers in regrowth but at two of the mature forest areas (B3 and S5) numbers of both species were low. ***N. politulus*** was also absent from S1, the narrowest of the strips sampled. ***Promecoderus longus*** was only common in three of the mature forest sites and in one of the nine year regrowth sites. ***Scopodes sigillatus, Scopodes atterimus*** and ***Homothes elegans*** were generally more abundant in regrowth of nine years or younger and ***Percosoma carenoides*** was only present in areas 24 years or older but was absent from the area with 26-year-old regrowth.

Species with 11 or fewer individuals trapped were considered as rare for the purposes of analysis (Table 2). The number of rare species per site was significantly correlated with the total number of species per site ($r = 0.87$, $df = 16$, $p < 0.001$). When the areas were grouped into three classes (regrowth, blocks and strips) there were no significant differences between the groups in total numbers, numbers/pit, number of species or numbers of rare species. However, total numbers and numbers/pit were significantly greater ($df = 15$, both $p < 0.001$) in one-year-old forest than in other areas trapped. Total numbers captured were significantly correlated with numbers per pit ($r = 0.92$, $df = 16$, $p < 0.001$), despite the differences in numbers of pits at different sites. For analysis of trends in numbers of species, areas were grouped into three, i.e., very young sites (R1 and R2), old regrowth (R24 and R26) and all other sites. These three groups differed significantly in total numbers of species ($F_{(2,14)} = 24.3$, $p < 0.001$) and numbers of rare species ($F_{(2,14)} = 13.3$, $p < 0.01$). Least significant difference showed all three groups differed in total numbers of species ($df = 14$, all $p < 0.01$) with numbers being greatest in the youngest regrowth and lowest in the old regrowth group. For rare species inspection of the least significant difference analysis revealed that numbers were significantly higher in the very young regrowth than the two other groups ($df = 14$, $p < 0.01$ and 0.001).

Table 2. Numbers of each carabid species trapped at each area.

Species	R1	R2	R5a	R5b	R9a	R9b	R24	R26	B1	B2	B3	B4	S1	S2	S3	S4	S5	Total	No of Sites
Rhabdotus reflexus	36	12	9	1	174	110	31	122	26	58	22	14	62	47	193	117	49	1 083	17
Notonomus politulus	159	4	6	4	—	2	—	—	37	20	6	48	—	15	27	36	1	365	13
Mecyclothorax ambiguus	303	23	—	13	1	—	—	—	—	—	20	2	—	—	—	—	—	362	6
Sloaneana tasmaniae	11	—	1	3	90	39	5	31	17	0	28	14	53	19	4	40	5	360	15
Chylnus ater	3	14	5	—	—	1	—	—	20	16	2	42	21	22	54	14	1	215	13
Promecoderus longus	10	—	1	—	20	6	8	—	2	4	24	—	87	11	28	—	—	201	11
Scopodes sigillatus	8	32	8	7	8	8	—	1	—	3	4	2	—	—	—	2	—	83	11
Scopodes atterimus	30	21	2	1	—	—	—	—	—	—	—	—	—	—	—	—	—	54	4
Percosoma carenoides	—	—	—	—	—	—	10	—	4	4	2	—	2	2	—	7	17	48	8
Homothes elegans	7	19	—	1	4	—	—	—	—	—	—	—	—	—	—	—	—	31	4
Prosopogmus tasmanicus	—	1	—	—	—	—	—	—	4	—	18	—	—	—	—	—	—	23	3
Stichonotus piceus	2	—	—	—	—	—	—	—	6	2	—	—	1	—	—	—	—	11	4
Scopodes tasmanicus	5	2	—	—	—	—	—	—	—	—	2	—	—	—	—	—	—	9	3
Lestignathus cursor	1	—	—	—	3	—	—	—	—	1	—	—	—	—	3	—	—	8	4
Homothes guttifer	—	4	—	1	—	2	—	—	—	—	—	—	—	—	—	—	—	7	3
Lecanomerus verticalis	—	—	—	1	—	—	—	—	—	—	6	—	—	—	—	—	—	7	2
Trechodes sp.	—	1	—	4	—	—	—	—	—	—	—	—	—	—	—	—	—	5	2
Sloanella sp.	3	1	—	—	—	—	—	—	—	—	—	—	—	—	—	—	—	4	2
Lestignathus simsoni?	—	—	—	—	—	—	—	—	—	—	—	—	3	—	—	—	—	3	1
Trechistus humicola	—	—	—	—	—	—	—	—	—	—	—	—	—	3	—	—	—	3	1
Agonocheila curtula	2	—	—	—	—	—	—	—	—	—	—	—	—	—	—	—	—	2	1
Pseudoceneus sp.	—	—	—	—	—	—	—	—	—	—	—	—	1	1	—	—	—	2	2
Trechimorphus diemenensis	—	—	—	—	1	—	—	—	—	—	—	—	1	—	—	—	—	2	2
Amblystomus sp. *A.*	—	1	—	—	—	—	—	—	—	—	—	—	—	—	—	—	—	1	1
Notogonum marginellum	—	—	—	—	—	—	—	—	—	—	—	—	—	—	—	—	1	1	1
Pterocyrtus globosus	—	—	—	—	—	—	—	—	—	—	—	—	—	—	—	1	—	1	1
Pterocyrtus sp. *A.*	—	—	—	—	—	—	—	—	—	—	—	—	—	—	—	1	—	1	1
Pterocyrtus sp. *B.*	—	—	—	—	—	—	—	—	—	—	—	—	—	—	1	—	—	1	1
Sloanella pallida	—	—	—	—	—	—	—	—	—	—	—	—	1	—	—	—	—	1	1
Total individuals	580	135	32	36	301	168	54	154	116	108	134	122	232	120	310	218	74	2 894	
Total species	14	13	7	10	8	7	4	3	8	8	11	6	10	8	7	8	6		
Total rare species	5	5	0	3	2	1	0	0	1	2	2	0	5	2	2	2	1		
No of Pits	6	9	6	9	6	9	6	9	6	9	18	18	12	12	12	12	12		
Number/pit	96.7	15	5.3	4	50.2	18.7	9	17.1	19.3	12	7.4	6.8	19.3	10	25.8	18.2	6.2		

DISCUSSION

The numbers of carabid species captured was greater in this study (29) than in that of Michaels and McQuillan (1995) (18). However, our study was undertaken over a greater number of sites and over a longer time period. All common or abundant species were sampled in both studies, with the six most abundant species being the same. Michaels and McQuillan (1995) caught one species (two individuals of *Tasmanotrechus* sp.) which was not captured in the present study and we captured 13 species (all seven or fewer individuals) which were not captured by Michaels and McQuillan (1995).

Previous studies of the effects of logging on carabids have found a higher number of species in regenerating forest up to 27 years of age compared with mature forest (Niemela *et al.* 1988, 1993a,b). In this study, there was also an increase in the number of species after logging but the effect was very short-lived compared to the studies of Niemela *et. al.* (1988, 1993a,b), only occurring in the first two years after logging (Table 2). This increase in numbers of species was due to new species invading the disturbed areas and the species associated with older forest persisting for a short time after logging. Thus two different sets of species were coexisting in the early succession stages. One of the early succession stage species, *M. ambiguus,* occurred at the highest numbers for a species in an area (one-year-old regrowth). This is a typical response for an r-selected opportunistic species which can take advantage of short-lived conditions (Krebs 1978). The number of rare species was also greatest in the very young regrowth. This could have been related to a higher trappability in these areas due to reduced leaf litter and vegetation covering the ground after the regeneration burn (Greenslade 1964; Melbourne *et al.* 1997), or it may be a real effect in which greater numbers of vagrant species occur in the disturbed areas. The mature forest specialist species appear to survive the immediate impacts of logging but may not be able to reproduce and hence eventually disappear. A similar phenomena has been noted by Niemela *et al.* (1993a) for carabids in Canada. Numbers of species were lowest in the oldest regrowth (24 and 26-year-old). This may be due to the uniformity of the forest at this stage and to the high stem density with complete canopy cover producing heavy shade that leads to a reduced ground vegetation layer.

Species showed three major responses to logging. One group, consisting of small, winged species such as *M. ambiguus, S. sigillatus, S. atterimus* and *H. elegans*, showed a dramatic increase after logging and were absent or in low numbers in older regrowth and mature forest. Another group *(R. reflex, S. tasmaniae, P. longus* and *P. carenoides)* built up numbers in the older regrowth (9–26 years) and persisted in mature forest. A third group (*N. politulus, C. ater*) was most abundant in mature forest and, apart from high numbers very early on due to initial survival after the logging, had very low numbers in regrowth. These patterns are similar to those found in other studies (Niemela *et al.* 1993a,b). The large numbers of *N. politulus* and *C. ater* captured in very young regrowth could have resulted from a higher trappability associated with greater movements occurring in unfavourable habitats (Baars 1979).

Based on the classification of areas as depicted on the dendrogram (which utilized proportional representation of species) (Fig. 2), there appeared to be no consistent difference between strips and blocks of mature forest in terms of their carabid faunas. However, for the two most abundant species identified by Michaels and McQuillan (1995) as mature forest specialists, *N. politulus* and *C. ater*, the former was virtually absent from both of the two narrowest strips and the latter was virtually absent from one of these but common in the other. This tends to suggest that narrow strips may not be as suitable as large blocks for these species. However, one of the blocks of mature forest also had very low numbers of both these species. Without further sampling of narrow strips it is not possible to separate the possible effects of site differences and the effects of strip width. Both of the narrow strips had *E. regnans* as the dominant eucalypt whereas at all other sites *E. obliqua* was dominant. Sampling of narrow streamside reserves with *E. obliqua* dominant should be undertaken to see whether *N. politulus* is a dominant species in such sites.

Both of the two rare species which Michaels and McQuillan (1995) believed were likely to be mature forest specialists were caught in the strips. Only one individual of *Pterocyrtus* sp. *A.* was sampled (in S4) during the present study. *S. piceus* was captured in two of the four blocks (B1, six individuals; B2, two individuals) and one of the two narrowest strips (S1, one individual). Because of the rarity of these two species, it is difficult to assess whether the strips would support viable populations of these species. Local carabid populations frequently become extinct (den Boer 1981, 1985) and small forest fragments have been found to lack the specialized and rarer forest carabids (Niemela *et al.* 1988; Halme and Niemela 1993). The power of dispersal of a species is an important determinant of whether a species will occur in small isolated remnants (de Vries *et al.* 1996). The four mature forest specialists identified by Michaels and McQuillan (1995) are all wingless species who are thus likely to have poor powers of dispersal. Ensuring dispersal between retained mature forests stands in production forests is thus likely to be an important factor in maintaining complete faunas. Linear strips of habitat may provide for such dispersal (Vermeulen 1994; Plat *et al.* 1995; Sutcliffe and Thomas 1996). Presently in the Southern Forests, the two abundant carabid species which prefer mature forest are well represented in strips and the specialized rare species so far identified also occur here. Thus it is likely that some dispersal and genetic interchange is occurring between populations in mature forest blocks and strips.

It is presently unknown whether the mature forest specialists may in fact recolonize the regrowth forest before it is scheduled for logging at around 80 years of age. If these species were responding to the development of the litter layer populations could well recolonize by 50 years of age (Brown and Nelson 1992), as occurs for some other invertebrates (e.g., Bonham and Taylor 1997).

The strips sampled have only been created up to 13 years ago. Loss of species has been shown to increase as time since fragmentation increases (Butcher *et al.* 1981; Diamond 1984; Saunders 1989). Thus it will be important to continue to monitor populations in the strips to check whether species persist as a greater proportion of the mature forest is converted to regrowth and fragmentation increases. The continuity of the strips with larger blocks of forest should minimize such loses (Dunning *et al.* 1995; Petit and Burel 1998). Any consequences of reductions in area of habitat and fragmentation may be most severe for species that are rare and ecologically specialized (Diamond 1987; Halme and Niemela 1993a,b; Didham *et al.* 1996). Some of the mature forest carabids in other studies have been found to occur only in restricted habitat patches (Niemela *et al.* 1987, 1992, 1993b). Michaels and McQuillan (1995) thought that the rare species *S. piceus* was probably a habitat specialist. Further work on the habitat requirements and impacts of fragmentation on this species are justified given the possibility of its being particularly sensitive to future reductions in the occurrence of mature forest in production areas.

ACKNOWLEDGEMENTS

John Traill and Phil Rowe from the Huon District of Forestry Tasmania advised on the selection of sites. John Hickey assisted with collection of information on the vegetation. Figure 1 was prepared by Bill Brown.

REFERENCES

Atlegrim, O., Sjoberg, K. and Ball, J. P., 1997. Forestry effects on a ground beetle community in spring: selective logging and clear-cutting compared. *Entomol. Fennica.* **8:** 19–26.

Baars, M. A., 1979. Patterns of movement of radioactive carabidae beetles. *Oecologia* **44:** 125–40.

Belbin, L., 1989. PATN Users Guide. CSIRO Division of Wildlife and Ecology: Canberra, Australian Capital Territory.

Bonham, K. and Taylor, R. J., 1997. Distribution and habitat of the land snail *Tasmaphena lamproides* (Pulmonata: Rhytididae) in Tasmania. *Mollus. Res.* **18:** 1–10.

Brown, G. W. and Nelson, J. L., 1992. Habitat Utilisation by Heliothermic Reptiles of Different Successional Stages of *Eucalyptus regnans* (Mountain Ash) Forest in the Central Highlands, Victoria. VSP Technical Report, No. 17. Department of Conservation and Natural Resources: Melbourne, Victoria.

Butcher, G. S., Niering, W. A., Barry, W. J. and Goodwin, R. H., 1981. Equilibrium biogeography and the size of nature preserves: an avian case study. *Oecologia* **49:** 29–37.

Davies, J. B., 1988. Land Systems of Tasmania. Region 6: South, East and Midlands — a Resource Classification Survey. Department of Agriculture, Tasmanian Government Printer: Hobart, Tasmania.

de Vries, H. H., den Boer, P. J. and van Dijk, T. S., 1996. Ground beetle species in heathland fragments in relation to survival, dispersal, and habitat preference. *Oecologia* **107:** 332–42.

den Boer, P. J., 1981. On the survival of populations in a heterogeneous and variable environment. *Oecologia* **50:** 39–53.

den Boer, P. J., 1985. Fluctuations of density and survival of carabid populations. *Oecologia* **67:** 322–30.

Diamond, J. M., 1984. "Normal" extinctions of isolated populations. Pp. 191–246 *in* Extinctions ed by M. H. Nitecki. University of Chicago Press: Chicago.

Diamond, J. M., 1987. Bird survival in an isolated Javan woodland: island or mirror. *Conserv. Biol.* **1:** 132–42.

Didham, R. K., Ghazoul, J., Stork, N. E. and Davis, A. J., 1996. Insects in fragmented forests: a functional approach. *TREE* **11:** 255–60.

Dunning, J. B., Borgella, R., Clements, K. and Meffe, G. K., 1995. Patch isolation, corridor effects and colonization by a resident sparrow in a managed pine woodland. *Conserv. Biol.* **9:** 542–50.

Greenslade, P. J. M., 1964. Pitfall trapping as a method for studying populations of Carabidae (Coleoptera). *J. Ecol.* **33:** 301–10.

Halme, E. and Niemela, J., 1993. Carabid beetles in fragments of coniferous forest. *Ann. Zool. Fennici.* **30:** 17–30.

Holliday, N. J., 1992. The carabid fauna (Coleoptera: Carabidae) during postfire regeneration of boreal forests: properties and dynamics of species assemblages. *Can. J. Zool.* **70:** 440–52.

Holopainen, J. K., 1990. Influence of ethylene glycol on the numbers of carabids and other soil arthropods caught in pitfall traps. Pp. 339–41 *in* The Role of Ground Beetles in Ecological and Environmental Studies ed by Intercept Ltd: Andover, Hampshire.

Jarosik, V., 1983. A comparison of the diversity of carabid beetles (Col., Carabidae) of two floodplain forests differently affected by emmissions. *Ceskoslov. Spolec. Zool.* **47:** 215–20.

Jennings, D. T., Housewart, M. W. and Dunn, G. A., 1986. Carabid beetles (Coleoptera: Carabidae) associated with strip clearcut and dense spruce-fir forests of Maine. *Coleopter. Bull.* **40:** 251–63.

Krebs, C. J., 1978. Ecology. The Experimental Analysis of Distribution and Abundance. Harper and Row: New York.

Lenski, R. E., 1982. The impact of forest cutting on the diversity of ground beetles (Coleoptera: Carabidae) in the Southern Appalachians. *Ecol. Entomol.* **7:** 385–90.

Luff, M. L., Eyre, M. D. and Rushton, S. P., 1992. Classification and prediction of grassland habitats using ground beetles (Coleoptera, Carabidae). *J. Environ. Mgmt.* **35:** 301–15.

Meggs, J. M., 1996. Pilot Study of the Effects of Modern Logging Practices on the Decaying-log Habitat in Wet Eucalypt Forest in South-east Tasmania. Regional Forest Agreement — Tasmania, Forestry Tasmania: Hobart, Tasmania.

Melbourne, B. A., Gullan, P. J. and Su, Y. N., 1997. Interpreting data from pitfall-trap surveys: crickets and slugs in exotic and native grasslands of the Australian Capital Territory. *Mem. Mus. Vic.* **56:** 361–67.

Michaels, K. F. and McQuillan, P. B., 1995. Impact of commercial forest management on geophilous carabid beetles (Coleoptera: Carabidae) in tall, wet *Eucalyptus obliqua* forest in southern Tasmania. *Aust. J. Ecol.* **20:** 316–23.

Niemela, T., Haila, Y., Halme, E., Pajunen, T., Punttila, P. and Tukia, H., 1987. Habitat preferences and conservation status of *Agonum mannerheimii* Dej. in Hame, southern Finland. *Notulae Entomologicae* **67:** 175–79.

Niemela, J., Haila, Y., Halme, E., Lahti, T., Pajunen, T. and Punttila, P., 1988. The distribution of carabid beetles in fragments of old coniferous taiga and adjacent managed forest. *Ann. Zool. Fennici.* **25:** 107–19.

Niemela, J., Haila, Y., Halme, E., Pajunen, T. and Punttila, P., 1992. Small scale heterogeneity in the spatial distribution of carabid beetles in the southern Finnish taiga. *J. Biogeo.* **19:** 173–81.

Niemela, J., Langor, D. and Spence, J. R., 1993a. Effects of clear-cut harvesting on boreal ground-beetle assemblages (Coleoptera: Carabidae) in Western Canada. *Conserv. Biol.* **7:** 551–61.

Niemela, J., Spence, J. R., Langor, D., Haila, Y. and Tukia, H., 1993b. Logging and boreal ground-beetle assemblages on two continents: implications for conservation. Pp. 29–50 *in* Perspectives on Insect Conservation ed by K. J. Gaston, T. R. New and M. J. Samways. Intercept: Hampshire.

Petit, S. and Burel, F., 1998. Connectivity in fragmented populations: *Abax parallelepipedus* in a hedgerow network landscape. *Comptes Rendus De L Academie Des Sciences Serie Iii-Sciences De La Vie* **321:** 55–61.

Plat, S., Kuivenhoven, P. and van Dijk, T. S., 1995. Hedgerows: suitable corridors for ground dwelling forest carabid beetles? *Proc. Exper. and Appl. Entomol., N.E.V.* **6:** 73–75.

Saunders, D. A., 1989. Changes in the avifauna of a region, district and remnant as a result of fragmentation of native vegetation: the wheatbelt of Western Australia: A case study. *Biol. Conserv.* **50:** 99–135.

Spence, J. R., Langor, D. W., Niemela, J. K., Carcamo, H. and Currie, C. R., 1996. Northern forestry and carabids: the case for concern about old-growth species. *Ann. Zool. Fennici.* **33:** 173–84.

Sustek, Z., 1981. Influence of clear cutting on ground beetles (Coleoptera, Carabidae) in a pine forest. *Comm. Instit. For. Cechoslovaniae* **12:** 243–54.

Sutcliffe, O. L. and Thomas, C. D., 1996. Open corridors appear to facilitate dispersal by ringlet butterflies (*Aphanmtopus hyperantus*) between woodland clearings. *Conserv. Biol.* **10:** 1359–365.

Taylor, R. J., 1991a. The role of retained strips for fauna conservation in production forests in Tasmania. Pp. 265–70 *in* Conservation of Australia's Forest Fauna ed by D. Lunney. Royal Zoological Society of NSW: Mosman, New South Wales.

Taylor, R. J., 1991b. Fauna Conservation in Production forests in Tasmania. Forestry Commission: Hobart, Tasmania.

Tietze, F., 1987. Changes in structure of carabid beetle taxocenosis in grassland affected by intensified management and industrial air pollution. *Acta Phytopath. Entomol. Hung.* **22:** 305–19.

Turin, H. and den Boer, P. J., 1988. Changes in the distribution of carabid beetles in the Netherlands since 1880. II. Isolation of habitats and long-term time trends in the occurrence of carabid species with different powers of dispersal Coleoptera, Carabidae. *Biol. Conserv.* **44:** 179–200.

Vermeulen, H. J. W., 1994. Corridor function of a road verge for dispersal of stenotopic heathland ground beetles Carabidae. *Biol. Conserv.* **69:** 339–49.

CHAPTER 14

The status of threatened vertebrate fauna in New South Wales Cypress Woodlands: implications for State Forest Management

E. M. Date,[1] D. C. Goldney,[1] J. J. Bauer[1] and D. C. Paull[2]

A major survey of vertebrate fauna associated with White Cypress Pine in state forests was undertaken in western New South Wales over the period 1993–1995. Existing state forests include a broad range of disturbances from remnant patches reflecting many attributes of the pre-European community structure and composition to highly manipulated areas, all retained in a land system extensively cleared by European settlers since the early 19th century to facilitate agricultural land uses. This significant land system, occupying 33% of the land area of New South Wales, extends from the Queensland to Victorian borders and from the western escarpment of the Great Dividing Range to the eastern edge of the western plains. Seventy-five sites in 70 state forests were surveyed for vertebrate fauna and habitat features. Key functional groups of fauna were identified to facilitate the description of threatened fauna distribution and status. The multivariate factors contributing to habitat modification are described and fauna-environment interactions modelled as a basis for understanding the transitional nature of the region between coastal Bassian and continental Eyrean fauna, patterns of extinction and range restriction within the region, and cumulative and synergistic impacts on fauna groups. The management of key functional groups of fauna in this complex land system is addressed together with the potential for system restoration.

INTRODUCTION

MUCH has been documented of the decline of terrestrial vertebrate fauna in Australia (e.g., Recher and Lim 1990). Habitat alienation by clearing and fragmentation of native vegetation for agriculture and urbanization is a major cause of decline (Saunders *et al.* 1991; Saunders and Hobbs 1995), as is habitat simplification by pastoral and forestry impacts, fire regimes of low variability and the spread of feral carnivores. Grazing by introduced herbivores, mostly sheep, cattle, rabbits, goats and pigs, resulting in the extinction, decline and range restriction of many vertebrates, has had widespread impact (60% of Australia, Wilson 1990). Logging mostly occurs in high rainfall areas; an exception is the White Cypress Pine woodlands on the western slopes of New South Wales and southern Queensland (400–800 mm per annum). Logging has brought about declines of a number of forest species dependent on old growth habitats (Loyn 1985; Recher *et al.* 1987; Garnett 1992; Lindenmayer 1994). Simplification of fire regimes to infrequent wildfire and frequent fuel reduction burning has resulted in the decline of fauna dependent for the regeneration of food, shelter and breeding resources on fires of intermediate frequencies and intensities (Gill and Bradstock 1995). Cats and foxes, introduced to Australia last century, have benefited from habitat changes wrought by agriculture, urbanization and forestry and this has increased their impact on native vertebrates (Newsome 1994). Each of these factors impacts native terrestrial vertebrate fauna, but there is rarely a single factor affecting status, distribution and abundance. A few attempts have been made to examine their relative importance and interactive effects (e.g., Recher *et al.* 1987; Short and Turner 1994; Smith *et al.* 1992, 1994).

[1]Environmental Studies Unit, Charles Sturt University, Bathurst, New South Wales, Australia 2975.
[2]Department of Zoology, University of New England, Armidale, New South Wales, Australia 2350.
Pages 128–45 *in* NATURE CONSERVATION 5: NATURE CONSERVATION IN PRODUCTION ENVIRONMENTS: MANAGING THE MATRIX ed by J. L. Craig, N. Mitchell and D. A. Saunders. Surrey Beatty & Sons, 2000.

To date, assessment of agricultural and urban impacts has focused on largely contiguous and extensive alienated land, containing scattered, often degraded, unprotected remnants of native vegetation (How 1984; Loyn 1985, 1987; Goldney and Bowie 1990; Saunders *et al.* 1991; Barrett *et al.* 1994; Catterall *et al.* 1997). In contrast, impact assessment of forestry activities has focused on largely contiguous and extensive native forests in eastern and southwestern Australia (e.g., Loyn 1985; Recher *et al.* 1987; Lindemayer *et al.* 1990; Taylor 1991; Wardell-Johnson and Nicholls 1991; Garnett 1992; Smith *et al.* 1992, 1994). Large areas have been degraded (and some regenerated) to various degrees by logging and other forestry activities, and some are adjacent to and contain protected remnant old growth communities. The geographic scale of forestry and agricultural activities is similar. However, the type and intensity of impacts and the resulting landscape configuration of native vegetation are broadly different due to the difference in production technique. Agriculture requires replacement of native ecosystems by monocultures of introduced species. Within the production matrix of agricultural land the need for native ecosystem restoration is now critical (Goldney *et al.* 1995; Saunders and Hobbs 1995). In contrast, forestry in Australia operates within a functioning native ecosystem from which a resource is harvested, and many opportunities are available for ecosystem regeneration.

We examine the impacts of forestry activities in a region where approximately 85% of native ecosystems have been alienated and fragmented by agricultural impacts (Sivertsen 1993) and the "protection" of remnants for timber production has effectively preserved opportunities for nature conservation and ecosystem restoration that would otherwise have been lost: the White Cypress Pine woodlands of New South Wales, a region 268 000 km^2 in size or 33% of New South Wales. Here, fauna that might have been robust to forestry activities in a less fragmented ecosystem have declined and may be threatened (Bauer *et al.* 1999; Date and Paull 1999). We assess the combined impacts of habitat and landscape modification by forestry, agriculture and introduced mammals on threatened vertebrate fauna. Our studies were conducted as impact assessments of forestry activities affecting protected and threatened fauna in the north-west and south-west cypress and ironbark management areas of New South Wales under the *National Parks and Wildlife Act 1974,* as amended by the *Threatened Species Conservation Act 1995* (Bauer *et al.* 1999; Date and Paull 1999).

STUDY AREA

Geographic location

The study area roughly coincides with the Central Division of New South Wales with an easterly boundary on the western escarpment of the Great Dividing Range and a westerly boundary approximately where the Cobar peneplain and the floodplains of the Murray-Darling Basin begin. State forests and other Crown timber-lands in the study area occupy 7 783 km^2 or 2.9% and 49 843 km^2 or 18.5% of the region and represent a large and potentially rich resource for nature conservation in addition to the six large (Mt. Kaputar, Warrumbungles, Cocaparra, Goobang and Weddin Mountains National Parks, Pilliga Nature Reserve) and 36 small reserves, totalling approximately 3 000 km^2 or 1.1%. Our study area, the North-west and South-west Cypress and Ironbark Management Areas (Fig. 1), includes the majority of state forests in western New South Wales from which White Cypress Pine and ironbark timber products are extracted. Except for the Pilliga State forests (390 000 ha) and Goonoo State forest (62 500 ha), which make up two-thirds of the total state forest area, most state forests (237 of 265) are <5 000 ha in size (Table 1). In addition, there are timber reserves and large areas of leasehold land, over which State Forests of NSW controls the extraction of timber (Table 1). Almost all land in the Western Division, except state forests, National Parks and Nature Reserves, is held under long-term pastoral or cropping lease. Most land not in state forests, National Parks and Nature Reserves in the Central Division is private property (77.5%) with only 6% under lease. Much of the private property in the Central Division has been cleared for pasture and crops, particularly the wheatbelt, where cotton is now a competing industry (Dick 1992; Sivertsen 1993). The leasehold and private land are more difficult to access and assess for their nature conservation value.

We surveyed 75 sites in 70 state forests (Fig. 1). A site consisted of three separate transects, each 1 ha in size. State forests on the western plains in the vicinity of Nyngan and Ivanhoe were not included in the survey.

Biogeography

The study area forms a transitional zone between the Bassian and Eyrean regions of biogeographic origin (Spencer 1896; Schodde 1994; Bauer *et al.* 1999; Fig. 2), whereas it includes the Brigalow Belt South (New South Wales section) and New South Wales Southwestern Slopes, most of the Nandewar Range, plus the south-east corners of the Cobar Peneplain and Darling River Plain bioregions, identified by Thackway and Cresswell (1995, Fig. 2). The distribution of White Cypress Pine in New South Wales includes the study area and the bioregions to the west: Cobar Peneplain and Darling River Plain (Boland *et al.* 1984; Johnston and Jennings 1991; Fig. 2).

Vegetation communities

The study area includes dry sclerophyll forests of the western foothills of the Great Dividing Range and most of the semi-arid woodlands that do not occur on the Cobar peneplain and the floodplain of the Murray-Darling Basin (Dick 1992; cf. Sivertsen 1993). The large variation in vegetation structure and floristics across the study area has been classified by botanists and foresters (Beadle

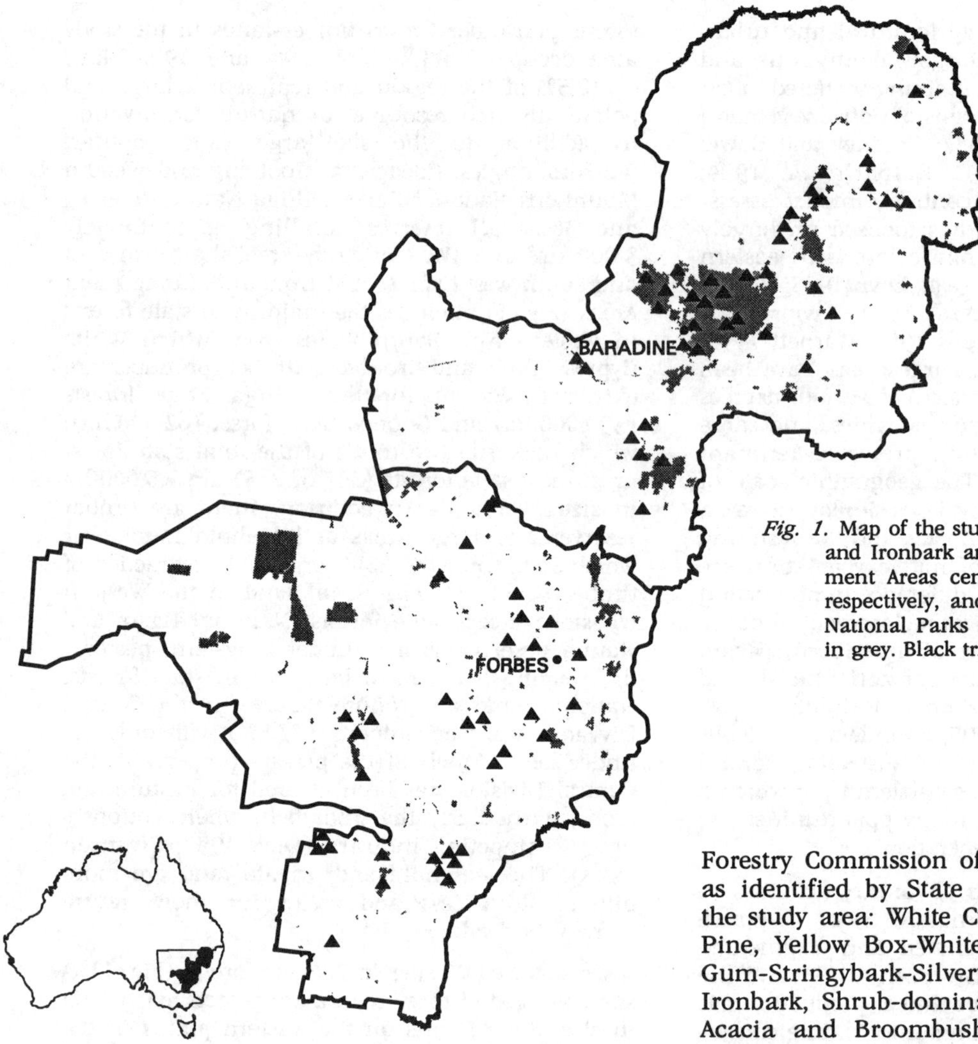

Fig. 1. Map of the study area, the North-west Cypress and Ironbark and South-west Cypress Management Areas centred on Baradine and Forbes respectively, and the study sites. State Forests, National Parks and Nature Reserves are shown in grey. Black triangles indicate study sites.

1948; Lindsay 1967; Moore 1970; Cardale 1987; Forestry Commission of NSW 1989; Sivertsen, pers. comm.). Vegetation in state forests is classified into five structural formations: grasslands, scrub or shrublands (1–4 m in height), grassy and/or shrubby woodlands; box woodlands and dry sclerophyll forests (Beadle 1948; Forestry Commission of NSW 1993a,b). Major vegetation associations or leagues have been defined by their dominant overstorey species (Beadle 1948; Moore 1953a,b; Forestry Commission of NSW 1989). Six leagues as identified by State Forests of NSW occur in the study area: White Cypress Pine, Black Cypress Pine, Yellow Box-White Box-Red Gum, Scribbly Gum-Stringybark-Silvertop Ash, Western Box-Ironbark, Shrub-dominant communities (Mallee, Acacia and Broombush). The leagues in turn include several vegetation types defined by their overstorey floristic composition.

Such a classification was useful for characterizing fauna habitats in the South-west Cypress Management Area where most state forests have been structurally and floristically simplified, and made more uniform by ringbarking of eucalypts and grazing to form cypress-dominated woodlands. Six structural and floristic types were defined within the White Cypress Pine League (Bauer *et al.* 1999).

Table 1. Area (ha) of state forest, timber reserve and leasehold land and number of state forests of three size classes in the Study Area according to the Contributing Management Plans in their years of publication. Numbers in parentheses represent areas of land held under lease in the Western (*vs* Central) Division of New South Wales.

Management plan	Year	State forest	Timber reserve	Leasehold land	>5 000 ha	1 000–5 000 ha	<1 000 ha
Narrandera	1981	33 640	146	2 089	1	5	32
Forbes	1986	83 504	1 102	116 698	3	18	47
Condobolin	1987	58 449	30 319	486 681 (+2 900 000)	4	12	19
Dubbo	1988	80 489	10 166	603 634 (+330 000)	1	6	8
Gilgandra	1984	20 160	96	24 952	0	6	18
Pilliga	1986	421 218	3 363	43 200	14	12	20
Gunnedah	1985	37 978	0	47 000	3	5	9
Inverell	1980	42 880	2 043	382 883	2	8	12
TOTAL		778 318	47 235	1 707 137	28	72	165

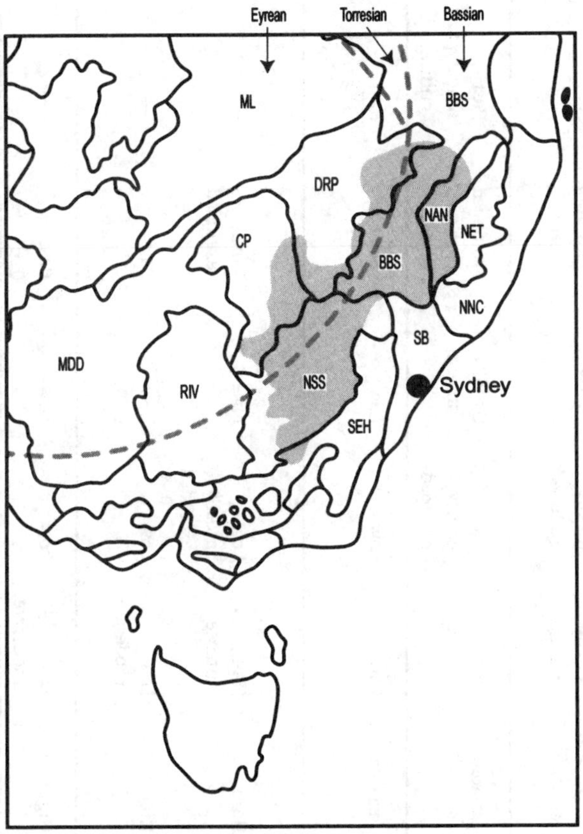

Fig. 2. Map of southeastern Australia showing the location of the study area in relation to the boundaries between Bassian (southern coastal), Eyrean (continental) and Torresian (northern coastal) biogeographic regions after Schodde (1994) and the boundaries of the relevant bioregions after Thackway and Cresswell (1995): BBS — Brigalow Belt South, CP – Cobar Peneplain, MDD – Murray-Darling Depression, ML – Mulga Lands, NAN – Nandewar, NET – New England Tableland, NNC – New South Wales North Coast, NSS – New South Wales Southwestern Slopes, RIV – Riverina, SB – Sydney Basin, SEH – Southeastern Highlands.

The classification system was not useful for characterizing fauna habitat in the North-west Cypress and Ironbark Management Area, which includes a broader array of vegetation communities and disturbances. We grouped sites by their floristic, structural and disturbance characteristics into nine types and found a strong association with the older state forests vegetation mapping types (Lindsay 1967; Date and Paull 1999), but a weak association with the more recent classification (Forestry Commission of NSW 1989; Date and Paull 1999). A compilation across the two management areas produced a classification of 15 community types (Table 2). For each community type, the dominant overstorey species are listed and a brief description of woodland understorey and disturbance history is given.

Threatened fauna — past and present

The White Cypress Pine woodlands include a diverse terrestrial vertebrate fauna, which is thought to have diverse biogeographic origins: Bassian or southern coastal, Eyrean or continental (west of the Great Dividing Range) and Torresian or northern coastal regions (Fig. 2). The study area is transitional between the Bassian and Eyrean regions and to a lesser extent the Torresian region and a large number of species reach the western, eastern and southern limits of their geographic ranges within the White Cypress Pine woodlands (Spencer 1896; Schodde 1994). However, at least two reptiles and seven mammals are thought to be endemic and a large number of birds and reptiles reach their highest abundances in the White Cypress Pine or Brigalow bioregion (Nix 1986; Paull and Date 1999; Robinson 1994; Thackway and Cresswell 1995). Twenty-five species of frog, 112 reptiles, 295 birds, 22 bats and 40 terrestrial mammals have been recorded in the study area in historical times. Seventy species, including two frogs, 10 reptiles, 32 birds and 26 mammals, are considered regionally threatened by human activities in the White Cypress Pine woodlands of New South Wales (*Threatened Species Conservation Act*). This does not include the 13 mammals considered extinct in the region. Nineteen frogs, 69 reptiles, 185 birds, 17 bats and 26 terrestrial mammals were observed during our surveys, including 4 reptiles, 14 birds and 11 mammals that are threatened in New South Wales.

To detect fauna we used standard fauna survey techniques: active search during the day and at night using a spotlight, playback, Elliot trapping on tree-brackets and on the ground, pit-trapping, scat and hair collection, sign records and opportunistic sightings (National Parks and Wildlife Service/State Forests NSW 1994). To document fauna habitat we used a structural and floristic proforma (National Parks and Wildlife Service/State Forests NSW 1994). We examined the patterns of occurrence among habitat types of all reptiles, birds and mammals recorded in our surveys, using pattern analysis (Belbin 1987; Date and Paull 1999). In addition, a habitat component analysis identified critical components associated with particular species assemblages and with broader characteristics of habitat structure and floristics (Date and Paull 1999; Paull and Date 1999). The habitat component analyses included frogs. We identified six functional groups of species by their frequent co-occurrence and frequent association with particular combinations of habitat features (Date and Paull 1999). They include, with vegetation communities in which they occur in parentheses: fauna dependent on mature eucalypts/hollow logs (Types 1, 2a,b, 3, 4, 6a,b, 7, 10), grassy woodland/grassland fauna (Types 1, 2a,b, 3, 4, 5, 8, 10, 14), fauna dependent on mixed understorey woodland (Types 2b, 3, 4, 5, 6b, 7, 8), fauna dependent on semi-arid scrub and shrubland (Types 13, 14, 15), terrestrial fauna dependent on old-growth groundcover and rocky terrain (Types 2b, 3, 5, 8, 9), frogs and waterbirds (Types 1, 2a, 10).

There were too few survey data for analysis of threatened species associations with habitat characteristics. Instead we aligned each species with the functional group whose habitat requirements were most similar, using natural history data

Table 2. Description of vegetation communities in the study area, documenting forest and Lindsay types (Lindsay 1967; Forestry Commission of New South Wales 1989), dominant overstorey species, understorey structure, logging, grazing and burning history and other disturbances.

Community type	Forest type	Lindsay type	Dominant canopy species	Under-storey type	Cypress logging	Cypress thinning	Ironbark logging	Wildfire	Fuel reduction burning	Grazing	Other
1. River Red Gum	199	R	*E. camaldulensis* *C. cunninghamiana* *A. floribunda* *C. glaucophylla*	Mostly grasses	Low	Low	Absent	Low	Low	Variable	Logging of River Red Gum
2. Blakely's Red Gum											
a) Dominant	178, 183, 192	BAP	*E. blakelyi* *A. floribunda* *C. glaucophylla*	Mostly grasses	Low	Low	Absent	Low-moderate	Low-moderate	Past	—
b) Mixed	205, 209	BCP	*E. blakelyi* *C. glaucophylla* *E. crebra* or *E. fibrosa* or *E. populnea* or *E. pilligaensis*	Shrub/Grass	Low	Low	Low	Low-moderate	Low-moderate	Past	—
3. Western Box-White Cypress Pine	193, 203	PfP, PgP, WP	*E. populnea* and/or *E. pilligaensis* and/or *E. microcarpa* *C. glaucophylla*	Shrub/Grass	High, Intense	Variable	Low, locally intense	Low, mostly excluded	Usually absent	Prolonged past and/or recent	—
4. White Box-White Cypress Pine	175	HP	*E. albens* *C. glaucophylla*	Grasses or Shrub/Grass	High	Variable	Low	Low, poorly recorded	Usually absent	Prolonged past and/or recent	Past eucalypt culling
5. Silver-leafed Ironbark-White Cypress Pine	207	MeP	*E. melanophloia* *C. glaucophylla*	Grasses or Shrub/Grass	High	Variable	Low	Low	Low	Variable	Past eucalypt culling
6. Narrow-leafed Ironbark											
a) Managed	208	COP	*E. crebra* *A. leuhmannii* *C. glaucophylla*	Variable	High, intense	High	High, intense	Low, mostly excluded	Absent	Mostly excluded	—
b) Mixed	204, 205	CPB	*E. crebra* *E. blakelyi* *C. glaucophylla*	Shrub/Grass	High	High	Low-moderate	Low	Low	Past	—
7. Mugga Ironbark-White Cypress Pine	206	SdP	*E. sideroxylon* *C. glaucophylla*	Shrub/Grass	High	High	Low-moderate	Low	Low	Past	—

Table 2 — *continued*

Community type	Forest type	Lindsay type	Dominant canopy species	Understorey type	Cypress logging	Cypress thinning	Ironbark logging	Wildfire	Fuel reduction burning	Grazing	Other
8. Caley's Ironbark-Hill Red Gum-White Cypress Pine	205	CaDP	*E. caleyi* *E. dealbata* *C. glaucophylla*	Grasses or Shrub/Grass	Low-high	Variable	Low	Low	Low	Past	—
9. Broad-leafed Ironbark-Black Cypress Pine	209	NTBp	*E. fibrosa* *E. trachyphloia* *C. endlicheri*	Dense shrub	Absent-low	Absent	Low	Moderate-high	Moderate-high	Absent	—
10. Yellow Box-White Cypress Pine	None	MP	*E. melliodora* *C. glaucophylla*	Grasses	High, intense	High	Absent	Low, mostly excluded	Variable	Prolonged past and/or recent	Past eucalypt culling
11. Casuarina-White Cypress Pine	212, 213	OP, Belah	*A. leuhmannii* or *C. cristata* *C. glaucophylla*	Not developed	High, intense	Low-high	Absent	Low, mostly excluded	Usually absent	Variable	Past eucalypt culling, casuarina thinning
12. White Cypress Pine	188	P	*C. glaucophylla*	De-pauperate	High, intense	High	Absent	Low, mostly excluded	Absent	Variable, usually frequent and intense	Past eucalypt culling
13. Broombush	223	Br	*M. uncinata*	Shrubland	Absent	Absent	Absent	Variable	Variable	Absent	Broom harvesting
14. Mallee	225	Mallee	*E. behriana* *E. dwyeri* *E. oleosa* *E. polybractea* *E. viridis*	Grasses or shrubs	Absent	Absent	Absent	Variable	Variable	Variable	Blue Mallee harvesting
15. Acacia	214, 224	Acacia	*A. harpophylla* *A. pendula* *A. cambagei* *A. aneura*	Grasses or shrubs	Absent	Absent	Absent	Low	Absent	Variable, usually absent	—

collated while compiling threatened species profiles (Bauer *et al.* 1999; Date and Paull 1999). The six groups are identified by their threatened members and the conservation status, biogeographic origin, historical presence, range reduction and critical habitat requirements of these threatened species in the study area are summarized in Table 3 (Blakers *et al.* 1984; Cogger 1992; Cogger *et al.* 1993; Garnett 1992, 1993; Smith *et al.* 1994; Strahan 1995; Australian Museum and NSW National Parks and Wildlife Service databases).

History of European land use

Prior to European settlement, the landscape of the western slopes of New South Wales was composed of continuous dry forests, woodlands, scrub and shrublands with some treeless areas of grassland. Structurally the woodlands varied among even-aged, mixed-age and mature overstorey with open grassy understorey, dense shrub and patchy mosaics of shrub and grass (Croft *et al.* 1997). Low intensity fires used by the Wiradjuri and Komilaroi Aborigines are thought to have shaped the vegetation of the creeks and rivers, Red Gum and Box communities, more than other vegetation communities, whereas wildfires ignited by lightning had more widespread effects. Within 30 years of European settlement, Aboriginal populations had declined and their way of life had disappeared, including their burning practices. From 1814 in the south-west and 1832 in the north-west, European pastoralists and sheep, cattle and horses, protected by the Europeans from

Table 3. Functional group, status as determined by the *Threatened Species Conservation Act 1995,* records of their occurrence in the study area in historical times (since European settlement) and during the survey period 1993–1995, biogeographic origin and impact factors for threatened species of the New South Wales White Cypress Pine woodlands, including a summary of status, biogeographic origin, historical and survey record and range change. Status: E — endangered, V — vulnerable, C — of concern, Ex — extinct. Biogeographic Origin: B — Bassian, E — Eyrean, En — Endemic to the inland slopes of southeastern Australia (i.e., Brigalow Belt South, Nandewar Range and New South Wales South-west Slopes), T — Torresian, U — Ubiquitous. Historical/Survey Record: Y — yes, N — no, ? — insufficient data. Range Change: N, NW, W, SW, S, SE, E, NE — contracted toward that compass point, C — contracted within survey area, F — fragmented, Ex — range extension due to our surveys, I — insufficient data, I(W) — insufficient data but thought to have contracted to the west, V — vagrant. Critical Habitat Component: A — aquatic grasses and sedges, D — large fallen debris, G — tall grass, Ht — tree hollow, Hl — hollow log, Ls — large shrub, Lt — large tree, M — abundant mistletoe in mature trees, N — abundant nectar in mature trees, R — complex rocky outcrop, Sg — abundant grass seed, Sh — structurally complex shrub layer, St — abundant seed crop in mature trees, W — temporarily inundated creeklines and wetlands. Impact Factor: A — agricultural practices (chemical use, irrigation, cultivation), B — broombush harvesting, C — clearing, fragmentation and decline of remnant health, D — ecological dysfunction, F — inappropriate fire regime, G — overgrazing by stock, H — overgrazing by feral herbivores (rabbits, goats and pigs), K — bee-keeping, L — logging of ironbark and/or White Cypress Pine, M — mallee harvesting, O — other factors, including road and dam construction, P — feral predators (cats, foxes and dogs), R — recreation, including shooting, poaching, etc., S — silvicultural practices (thinning of WCP and culling of eucalypts), W — collection of firewood.

Scientific name	English name	Status	Origin	Historical record	Survey record	Range change	Critical habitat	Impact factors
1. Fauna dependent on mature eucalypts/hollow logs								
Liasis maculosus	Eastern Children's Python	C	T	Y	Y	Ex(I)	Ht/Hl	CLSWGHFP
Morelia spilota	Carpet Python	V	U	Y	Y	I(F)	Ht/Hl	CLSWGHFP
Hoplocephalus bitorquatus	Pale-headed Snake	V	B	Y	Y	F	Ht/Hl	CLSWGHFP
Lophoictinia insura	Square-tailed Kite	V	U	Y	Y	F	Lt	CLGHFP
Cacatua leadbeateri	Pink Cockatoo	V	E	Y	Y	NW/F	Ht	CAGHLSFRDK
Calyptorhynchus banksii	Red-tailed Black Cockatoo	V	U	Y	Y	F	Ht/St	CAGHLSFRDK
Calyptorhynchus lathami	Glossy Black Cockatoo	V	B	Y	Y	F	Ht/St	CLSRK
Lathamus discolor	Swift Parrot	V	B	Y	Y	F	N	CLSRK
Polytelis swainsonii	Superb Parrot	V	E	Y	Y	C/F	Ht/Sg	CLSWRAFGHPK
Tyto novaehollandiae	Masked Owl	V	U	Y	Y	F	Ht	CLSWAPK
Ninox connivens	Barking Owl	C	U	Y	Y	F	Ht	CLSWAPK
Xanthomyza phrygia	Regent Honeyeater	E	En	Y	Y	S/F	N	CLSDK
Grantiella picta	Painted Honeyeater	V	E/T	Y	Y	NW	M	LSFG
Dasyurus maculatus	Tiger Quoll	V	B	Y	Y?	I(E/F)	Hl	CLSWGHFP
Phascogale tapoatafa	Brush-tailed Phascogale	V	B	Y	N	I(E/F)	Ht/Hl	CLSWGHFP
Phascolarctos cinereus	Koala	V	B	Y	Y	E/F	Lt	CLSFP
Petaurus australis	Yellow-bellied Glider	V	B	N	Y?	I	Ht/N	CLSFGHK
Petaurus norfolcensis	Squirrel Glider	V	En	Y	Y	I(F)	Ht/N	CLSFGHK
Acrobates pygmaeus	Feathertail Glider	C	B/T	Y	Y	I	Ht/N	CLSFGHPK
Cercartetus nanus	Eastern Pygmy Possum	C	B	Y	Y	I	Ht/N	CLSFGHPK
Chalinobolus dwyeri	Large Pied Bat	V	En	Y	Y	I	Ht	CLSPK
Chalinobolus picatus	Little Pied Bat	V	E	Y	Y	I	Ht	CLSPK
Nyctophilis timoriensis	Greater Long-eared Bat	V	E	Y	Y	I	Ht	CLSPK
Falsistrellus tasmaniensis	Great Pipistrelle	V	B	N	Y	I	Ht	CLSPK
Saccolaimus flaviventris	Yellow-bellied Sheathtail Bat	V	T	Y	Y	I	Ht	CLSPK
Conilurus albipes	White-footed Rabbit-Rat	Ex	B	N	N	I	Ht/Hl	CGHLPK

Table 3 — *continued*

Scientific name	English name	Status	Origin	Historical record	Survey record	Range change	Critical habitat	Impact factors
2. Grassy woodlands/grasslands fauna								
Aprasia parapulchella	Pink-tailed Worm-Lizard	V	B/En	Y	N	F	G/D	GHAFP
Delma impar	Striped Delma	V	B	Y	N	F	G/D	GHAFP
Botaurus poiciloptilus	Australasian Bittern	E	B	Y	Y	I	A	AGHP
Ardeotis kori	Australian Bustard	E	E/T	Y	N	N	G	AGHP
Burhinus grallarius	Bush Stone-Curlew	E	U	Y	Y	N	G/D	CAGHLSWFP
Pedionomus torquatus	Plains-wanderer	E	E	Y	N	W	G	AGHP
Geophaps scripta scripta	Squatter Pigeon	E	T	Y	Y?	N	G/Sg	GHAFPR
Neophema pulchella	Turquoise Parrot	V	B	Y	Y	F	G/Sg	CLSWPRGHAF
Tyto capensis	Eastern Grass Owl	V	T	Y	N	I	G	AGHFP
Poephila cincta cincta	Black-throated Finch	E	T	Y	N	N	G/Sg	CGHAPFR
Neochmia ruficauda	Star Finch	Ex?	T	Y	N	N	G/Sg	CGHAPFR
Dasyurus geoffroii	Western Quoll	Ex	U	Y	N	W	G/D	CGHP
Dasyurus viverrinus	Eastern Quoll	Ex	B	Y	N	S	G/D	CGHP
Macrotis lagotis	Bilby	Ex	E	Y	N	W	G	CGHP
Aepyprymnus rufescens	Rufous Bettong	V	B/T	Y	Y?	I(F)	G/D	CWSLGHFP
Bettongia lesueur	Burrowing Bettong	Ex	E	Y	N	W	G	CGHP
Lagochestes leporides	Eastern Hare-Wallaby	Ex	E	Y	N	I	G	CGHP
Vombatus ursinus	Common Wombat	C	B	Y	Y	I(E/F)	G	CGHFP
Lasiorhinus krefftii	Northern Hairy-nosed Wombat	Ex	En	Y (old bones)	N	N	G	CGHP
Rattus villosissimus	Long-haired Rat	V	E	Y	N	I(V)	G/D	CAGHLFP
Pseudomys australis	Plains Rat	Ex	E	Y	N	I	G/D	CGHP
Pseudomys gouldii	Gould's Mouse	Ex	E	Y	N	I	G/D	CGHP
3. Fauna dependent on mixed understorey woodland								
Pomatostomus temporalis	Grey-crowned Babbler	C	U	Y	Y	F	Sh/G	CLGHFP
Isoodon obesulus	Southern Brown Bandicoot	E	B	N	Y?	I(E)	Sh/G	CGHFP
Perameles bougainville	Western Barred Bandicoot	Ex	E	Y?	N	W	Sh/G	CGHP
Bettongia penicillata penicillata	Brush-tailed Bettong	Ex	E	Y	N	W	Sh/G	CGHP
Onychogalea fraenata	Bridled Nailtail-Wallaby	Ex	En	Y	N	N	Sh/G	CGHP
Macropus dorsalis	Black-striped Wallaby	E	B/T	Y	Y	N/F	Sh/G	CAGHLWFP
4. Fauna dependent on semi-arid scrub and shrublands								
Tiliqua occipitalis	Western Bluetongue	V	E	Y	Y	Ex(F)	Sh/D	CAMBWGHFP
Leipoa ocellata	Malleefowl	E	E	Y	Y	SW/F	Ls/Sh	CMBGHFPA
Hamirostra melanosternum	Black-breasted Buzzard	V	E/T	Y	N	NW	Ls	CAP
Falco hypoleucos	Grey Falcon	V	E/T	Y	N	I(V)	Ls	CAGP
Glossopsitta porphyrocephala	Purple-crowned Lorikeet	V	B	N	N	I(V)	Ht/N(Ls)	CMK
Sericornis cautus	Shy Heathwren	V	E	Y	N	W/F	Sh/G	CMBGHP
Drymodes brunneopygia	Southern Scrub-robin	V	E	Y	N	SW/F	Sh	CMBFP
Cinclosoma castanotum	Chestnut Quail-thrush	V	E	Y	N	I	Sh/G	CMBGHFP
Pachycephala inornata	Gilbert's Whistler	V	E	Y	Y	F	Ls/Sh	CMBLSGHF
Pachycephala rufogularis	Red-lored Whistler	E	E	N	N	F	Ls/Sh	CMBGHF
Certhionyx variegatus	Pied Honeyeater	V	E	Y	N	I(V)	N(Sh)	CMBGHFK
Antechinomys laniger	Kultarr	E	E	Y	N	SW	Sh/G	CAGHMFP
Sminthopsis crassicaudata	Fat-tailed Dunnart	C	E	Y	Y	I(W)	Sh/G	BWFGHP
Sminthopsis macroura	Stripe-faced Dunnart	V	E	Y	N	I(W)	Sh/G	BWFGHP
Ninguai yvonneae	Southern Ninguai	V	E	Y	N	I	Sh/G	CFGHMP
Chaeropus ecaudatus	Pig-footed Bandicoot	Ex	E	N	N	I	Sh/G	CGHP
Pseudomys pilligaensis	Pilliga Mouse	V	En	Y	Y	C	Sh/G	BFPGHLWCA
Notomys sp. nova	Unidentified Hopping-mouse	E?	En?	Y?	Y?	I	Sh	BFPGHLW
5. Terrestrial fauna dependent on old-growth groundcover and rocky terrain								
Underwoodisaurus sphyrurus	Border Thick-tailed Gecko	V	B/En	Y	Y	None	D/R	CLWGHFP
Anomalopus mackayi	Five-clawed Worm-Skink	E	En	Y	N	E	D	GAFLWP
Furina dunmalli	Dunmall's Snake	C	En	N	Y	Ex(F)	D	CAGHFLWP
Vermicella annulata	Bandy Bandy	C	U	Y	N	I(F)	D	CAGHFLWP
Petrogale penicillata	Brush-tailed Rock-Wallaby	V	B/En	Y	N	E/F	R	GHFP
Miniopterus schreibersii	Common Bentwing Bat	V	B/T	Y	Y	I	R	CP
Miniopterus australis	Little Bentwing Bat	V	T	N	Y	I	R	CP

Table 3 — *continued*

Scientific name	English name	Status	Origin	Historical record	Survey record	Range change	Critical habitat	Impact factors
6. Frogs and waterbirds								
Litoria alboguttata	Striped Burrowing Frog	C	T	Y	N	I	W	AGPO
Litoria raniformis	Southern Bell Frog	E	B	Y	N	W	W	CGPO
Oxyura australis	Blue-billed Duck	V	B/E	Y	N	I	W	AGHO
Rostratula benghalensis	Painted Snipe	V	U	Y	N	I	W	AGHO

Summary

	Status				Biogeographic origin#					Recorded		Range change			
	C	V	E	Ex	B	E	En	T	U	Past	Pres	F	C	Ex	I
Frogs	1	0	1	0	1	0	0	1	0	2	0	0	2	0	0
Reptiles	3	6	1	0	2	1	2	1	2	9	6	3	1	3	2
Birds	2	20	9	1	6	9	1	4	7	29	16*	9	14	1	8
Mammals	4	18	4	13	10	15	5	2	1	33	20**	3***	13	0	26

\# does not include the unidentified Hopping-mouse, four birds of mixed Eyrean/Torresian, four mammals of mixed Bassian/Torresian, one reptile and one mammal of mixed Bassian/Endemic and one bird of mixed Bassian/Eyrean origin.
* includes possible sighting of Squatter Pigeon.
** includes sightings of unidentified Bandicoot, Bettong, Quoll, Glider, Hopping-mouse.
*** these species have also undergone range contraction.

predation by Aborigines, underwent a population expansion during the 19th century. The new herbivore populations crashed at the end of the 19th century as a result of overgrazing (Flannery 1994), but during the 20th century have undergone partial recovery (Rolls 1981; Croft *et al.* 1997). Reduction of woodland area, simplification of vegetation structure and composition, the establishment and spread of rabbit, cat and fox populations and declines in fauna diversity accompanied the growth of the pastoral industry. The growth of the wheat and cotton industries and the recovery of the pastoral industry through pasture improvement increased the clearing of native vegetation and further declines of fauna diversity have occurred (Bauer *et al.* 1999; Date and Paull 1999).

Cessation of Aboriginal burning practices, exclusion of wildfires by pastoralists, overgrazing of grasses by stock and a series of mild wet summers, combined to produce favourable conditions for the "wheatfield" establishment of cypress seedlings. The broad-acre establishment of White Cypress Pine in marginal agricultural areas from 1860 to 1900 resulted in the protection of some White Cypress Pine woodlands for their timber values as state forests and timber reserves, though most woodlands were cleared and regeneration was prevented by grazing (Rolls 1981; Johnston and Jennings 1991; Norris *et al.* 1991). The expansion of rabbit populations and the recovery of the pastoral industry in the early 20th century then produced unfavourable conditions for the establishment of cypress seedlings until the introduction of myxomatosis reduced rabbit populations from the 1950s through the 1980s (Norris *et al.* 1991). Large areas of Crown land were leased to pastoralists under the conditions that culling of eucalypts and thinning of cypress were undertaken to promote cypress growth. Culling of eucalypts and thinning of cypress were also carried out in state forests reaching their peak intensity during the 1960s and 1970s (State Forests' district records).

Logging of cypress and ironbark has occurred extensively on state forests, Crown leases and private property of the study area since the late 19th century. Volumes logged from state forests have increased as resources on private property and leasehold land have become depleted (State Forests' district records). Over the same time period the average size of cypress logs has decreased from 60 cm diameter at breast height (dbh) to 25 cm with the minimum merchantable size at 18 cm dbh. Old-growth White Cypress Pine (>80 cm dbh), or "Old Greys", are virtually extinct in the study area. White Cypress Pine regeneration is dense when grazing and fire are excluded and overstorey density is low, conditions which have become prevalent in the remaining woodlands since European settlement of the study area. White Cypress Pine does not self-thin and management of dense regeneration includes thinning of saplings using machetes and small chainsaws and light, intermittent grazing of seedlings using stock. Burning of saplings using infrequent, low intensity fires, which is less costly and of higher conservation value, has been suggested but is not used (Johnston and Jennings 1991; State Forests of NSW personnel, pers. comm.). The size of ironbark logs cut was initially limited by that able to be handled manually. Very large (>100 cm dbh) ironbarks were not cut, though they were often ringbarked. By the 1980s, machinery became available for logging large ironbarks (for fence wood). Large (>60 cm dbh) ironbark stocks are now becoming depleted and logging of the extensive, regenerating, and dense (self-thinning) crop of small (<45 cm dbh) ironbarks has begun (State Forests' district records; State Forests' personnel, pers. comm.)

Some state forests of the South-west Cypress Management Area are plantation-like pre-1900 regrowth of White Cypress Pine with little other regeneration. This is due to the low frequency of regeneration events (about every 30 years) and the re-establishment of grazing under grazing permits following the reduction of rabbit populations. In the North-west Cypress and Ironbark Management Area where there are large areas of poor sandy soils unsuitable for cropping and only marginally useful for grazing (including by rabbits), large scale clearing began later and culling of eucalypts in state forests was less extensive. In addition, regeneration events occur every 5 to 10 years. As a result, few state forests in the north-west are plantation-like (Bauer *et al.* 1999; Date and Paull 1999).

CURRENT IMPACTS ON VERTEBRATE FAUNA AND THEIR HABITATS

Current impacts on vertebrate fauna and their habitats are due to clearing, fragmentation and degradation of woodlands for agriculture and other agricultural practices used in cropping and grazing (Goldney and Bowie 1990; Hobbs and Saunders 1992; Saunders *et al.* 1987, 1991, 1993; Saunders and Hobbs 1991). Impacts have been more severe in the south of the study area than the north with resultant differences in fauna diversity (Bauer *et al.* 1999; Date and Paull 1999). Feral herbivores, such as rabbits, goats and pigs and feral carnivores, that is, cats, foxes and dogs, have significant negative impacts on native fauna (Catling 1988; Newsome 1994; Barratt 1997). Forestry activities that impact fauna and their habitats in the study area include logging and other silvicultural practices, firewood collection, broombush and mallee harvesting, inappropriate fire regimes, grazing, bee-keeping, road-works, dam construction and recreation. Impacts are identified for threatened fauna in Table 3 and their habitats in Table 4 (for detailed discussion see Bauer *et al.* 1999; Date and Paull 1999). Ecological dysfunction, the "self-destructive" physiological and behavioural patterns engendered by the combined environmental stresses of impact factors and threatening processes, has also been demonstrated as a significant impact factor (Table 3; Emison and Caldow 1992, 1993; Ford *et al.* 1993 cf. Oliver *et al.* 1998; see also Grey *et al.* 1997).

A combined impacts assessment of threatened fauna

In this section we assess of the combined impacts of various agricultural and forest industry factors, feral animals and ecological dysfunctions on threatened native fauna, using a qualitative method, which evaluates the types of impact factors thought to be currently affecting them. The factors affecting each species are assessed by visual comparison for their similarity with factors affecting other species from the same and different functional groups (Table 3). The factors are:

A: the use of agricultural practices, such as cultivation, sowing, irrigating, fertilizing, biocide-spraying and harvesting of crops and pastures that have direct and indirect influences on nutrient cycles, hydrology and biodiversity of woodland remnants (Table 3; Strahan 1994);

C: agricultural impacts of clearing and decline of remnant woodland that cause loss and reduced continuity of habitat (Table 3; Saunders *et al.* 1991);

G: grazing of stock, which has significant effects on native fauna through competition for food and shelter and which reduces the structural and biotic diversity of understorey vegetation and

Table 4. Impacts of disturbances in the vegetation communities of the study area on a scale of "none", "low", "medium" ("med") and "high". (* – indicates thinning of Buloke).

Vegetation community	Cypress logging	Ironbark logging	Cypress thinning	Eucalypt culling	Broombush cutting	Mallee oil-harvesting	Firewood collection	Fuel reduction burning	Grazing	Bee-keeping	Road-works	Dam construction
1. River Red Gum	low	low	low	low	none	none	VARIABLE	low	m-h	high	high	high
2a. Blakely's Red Gum dominant	low	low	low	low	none	none		med	med	high	high	high
2b. Blakely's Red Gum mixed	med	med	med	low	low	none		l-m	l-m	high	l-h	l-h
3. Western Box-WCP	high	med	m-h	med	none	none		low	high	low	l-h	l-h
4. White Box-WCP	high	med	m-h	med	none	none		low	high	high	l-h	l-h
5. Silver-leafed Ironbark-WCP	high	med	m-h	high	none	none		low	med	low	l-h	l-h
6a. Narrow-leafed Ironbark-WCP-Buloke	high	high	l-h*	high	none	none		low	l-m	low	l-h	l-h
6b. Narrow-leafed Ironbark mixed	high	high	l-h*	high	low	none		low	l-m	low	l-h	l-h
7. Mugga Ironbark-WCP	high	high	l-h	high	none	none		low	l-m	high	l-h	l-h
8. Caley's Ironbark-Hill Red Gum-WCP	low	low	low	low	none	none		l-m	l-m	med	l-h	l-h
9. Broad-leafed Ironbark-Black CP	low	med	low	low	low	none		high	low	low	l-h	l-h
10. Yellow Box-White Cypress Pine	high	low	med	med	none	none		low	high	high	l-h	l-h
11. Casuarina-WCP	med	low	l-h*	high	none	none		low	med	low	l-h	l-h
12. White Cypress Pine	high	low	high	none	none	none		low	high	low	l-h	l-h
13. Broombush	none	none	none	none	high	none		l-h	low	med	l-h	l-h
14. Mallee	none	none	none	none	low	high		l-h	high	high	l-h	l-h
15. Acacia	none	none	none	none	none	none		low	med	med	l-h	l-h

soil in remnant woodlands (Table 3; Wilson 1990);

H: grazing by feral herbivores, which has negative effects on fauna similar to those of grazing by stock (Table 3);

K: keeping of bees for honey production, which results in competition for tree hollows of escaped bees with hollow-dependent fauna and competition for nectar of escaped and managed bees with native nectivores (Table 3; Paton 1996);

L: logging, which reduces the availability of present and future old-growth components of woodlands such as hollows and underbark refuges for hollow-dwelling species, seed, leaf, and nectar resources for seed-eaters, folivores and nectarivores and moist microclimates for burrowing amphibians and arboreal reptiles (Table 3; this study);

S: silvicultural practices, some of which have negative effects for conservation similar to logging, particularly the culling of eucalypts, and others of which may have positive effects for the restoration of grassy woodlands and fauna, such as thinning of dense cypress regeneration (Table 3);

B: broombush harvesting, which is a direct threat to young broombush stands and an indirect threat to the recruitment of old-growth broombush and dependent fauna (Table 3; Woinarski 1989, 1990);

M: eucalyptus oil-harvesting, which is a threat to all age classes of vegetation and fauna inhabiting Blue Mallee, particularly old-growth stands (Table 3);

W: removal of firewood, particularly large, naturally fallen logs, which constitutes the removal of critical habitat components for fauna dependent on mature (often hollow) logs and fauna dependent on mature ground-cover (Functional Groups 1 and 5, Table 3);

F: inappropriate fire regimes that threaten grassy and mixed understorey woodland, broombush and mallee and their fauna, which include infrequent, intense (wild) fires that can occur under an exclusion policy and very frequent, cool fires that are implemented under a fuel reduction burning policy (Table 3);

P: feral carnivores, which predate native prey populations and compete with native carnivores in a complex predator-prey relationship (Table 3; Newsome 1994);

R: recreation, in the form of shooting and poaching for the caged bird trade, which has negative effects on a broad range of fauna, including native parrots, large native carnivores and herbivores (except some macropods) and positive effects in the case of feral animal control (Table 3);

O: other factors related to road and dam construction, which increase human recreational activites, which are thought to increase the populations of feral animal populations in state forests (Newsome 1994) and which change water drainage patterns (Table 3: P is inclusive of O except where O indicates impacts on drainage patterns);

D: ecological dysfunction (Table 3).

Combined impacts: habitat perspective

Eighty per cent of threatened fauna in the study area are thought or known to be affected by clearing, fragmentation and subsequent decline of woodland vegetation. As this impact factor is so widespread, it could not be used to distinguish among functional groups. The adverse impacts of feral carnivores are also prevalent across all functional groups (83%), though less so among fauna dependent on mature trees/hollow logs (70%) and frogs and waterbirds (50%). Members of the former group are wholly or mostly arboreal and are thus less vulnerable to predation by foxes, which mostly hunt on the ground. They are vulnerable to predation by cats which affect birds, terrestrial and arboreal reptiles and mammals, including the Koala. Frogs and waterbirds are also less vulnerable to predation because feral predators are less likely to hunt in water. The other four functional groups are mostly composed of terrestrial and small fauna, which are vulnerable to predation because of their habitat and size.

Fauna Dependent on Mature Eucalypts/Hollow logs. All species in this group are adversely affected by logging and culling of eucalypts. Some 33% are also affected by the removal of large hollow logs for firewood, which function as shelter for them or for prey species. Two-thirds of the group are also affected by inappropriate fire and grazing regimes: fire exclusion or frequent fuel reduction and overgrazing by stock and feral herbivores. These factors may also have negative impacts on food supply, such as grass seed and acacia gum through the removal of food plants and prey species through the removal of shelter provided by understorey plants. Seventy per cent may be affected by competition with feral or managed bees. A small number of species are affected by agricultural practices that remove food plants (three parrots) or animals (two owls) through the use of pesticides. Five species of parrot are affected by poaching for the illegal bird trade (Table 3).

Grassy Woodlands/Grasslands Fauna. All species in this group are negatively affected by inappropriate grazing regimes, that is, prolonged intense grazing by high densities of stock and feral herbivores, which removes food and shelter requirements of these species (Wilson 1990). Some species may benefit from light and intermittent grazing or a combination of light, intermittent grazing and a regime of moderately frequent, low intensity fires (e.g., Bush Stone-curlew, Rufous Bettong, Filet *et al.* 1997), but there is no evidence from the study area. Fifty per cent, possibly more, are threatened by inappropriate fire regimes, principally fire exclusion and frequent (annual to triennial) fuel reduction burning. They are thought to require variable fire regimes, which include fires at various intervals (approximately 4–15 years) and various frequencies intermediate between fire exclusion and frequent fuel reduction (Date and Paull 1999). Over 50% are adversely affected by agricultural practices which alienate grasslands and the grassy understorey of woodlands by replacing them with introduced grasses and legumes that do not provide the food or shelter requirements of these species. Three species are also threatened by the removal of large (hollow) logs through firewood collection, culling of eucalypts and ironbark logging: the Turquoise Parrot requires tree hollows, stags or stumps to breed and the Bush Stone-Curlew and the Rufous Bettong require large logs among or within which to shelter (Table 3).

Fauna Dependent on Mixed Understorey Woodland. This group is also adversely affected by grazing of stock and feral herbivores, and is thought to be more sensitive than the previous group to all grazing regimes, including light intermittent grazing. Grazing simplifies or removes the small-scale mosaics of grasses and shrubs at various ages (since fire) that are required by these species to provide sources of food and shelter that are in close proximity. The group is also sensitive to regimes which include highly or moderately frequent low and high intensity fire. Low frequency (at intervals of >15 years), high and low intensity fires are thought to maintain the preferred mixed understorey woodlands of this group (Date and Paull 1999). Three species are also adversely impacted by the removal of shelter provided by large logs through logging, eucalypt culling and firewood collection (Table 3; Date and Paull 1999).

Fauna Dependent on Semi-arid Scrub and Shrublands. Most of the threatened fauna in this group (80%) are dependent on old-growth or intermediate seral stage broom or mallee with a well developed grassy or shrubby understorey that supports food (nectar, seed, insects) and shelter. The two birds of prey nest on the tallest of the old-growth trees in semi-arid scrub. Mallee and broombush harvesting, grazing and/or frequent low intensity fires, firewood collection and agricultural practices, which disturb soil structure and hydrology and introduce biocides, adversely affect these critical habitat components (Table 3; Woinarski 1989).

Terrestrial Fauna Dependent on Old-growth Ground-cover and Rocky Terrain. Clearing of woodland removes the foraging habitat of bats, which feed above the canopy and roost in caves, whereas other agricultural practices, logging and firewood collection (and garden rock collection) remove the shelter provided by large ground-cover (fallen logs and rocks) for the reptiles in this group. Grazing by stock and feral herbivores and frequent fire also reduce the old-growth ground-cover, litter debris and foraging habitat (Table 3).

Frogs and Waterbirds. Agricultural practices, grazing, road and dam construction remove or disturb the temporary wetlands on which this group depends.

Combined impacts: biogeographic perspective

The distribution pattern of 31% of threatened fauna in the study area is Eyrean, 22% Bassian, 9.5% Torresian and 12% are ubiquitous. Eleven per cent are endemic to the western slopes of southeastern Queensland, New South Wales and northern Victoria, that is, from the Brigalow Belt South, Nandewar Range and New South Wales South-west Slopes bioregions (Table 5; Fig. 2). A high proportion of the members of each functional group have similar distribution patterns (Table 5). Bassian species comprise 38% of the fauna dependent on mature trees and hollow logs, the group most vulnerable to logging (after the effects of clearing). Fauna dependent on rocky outcrops and old-growth groundcover, which can be regarded as specialists similar to fauna dependent on mature trees and hollow logs, are mostly species with strong Bassian influence (43%) and the primary threat (after clearing) is from logging, followed by inappropriate grazing and fire regimes. Eyrean species comprise 32% of grassy woodland, 33% of the mixed-understorey woodland and 72% of semi-arid scrub and shrubland fauna. The last three groups are most vulnerable to continued alienation of their habitat by agriculture and inappropriate grazing and burning regimes. Most (64.5%) of the extinct species are grassy woodlands species, 64.5% are Eyrean and 36% are both Eyrean and grassy woodlands species (Table 5).

Table 5. The percentage representation of biogeographic distribution patterns of threatened species within the functional groups. Biogeographic distributions include Bassian, Eyrean, Endemic to the inland slopes of southeastern Australia (i.e., Brigalow Belt South, Nandewar Range and New South Wales South-west Slopes), Torresian, Ubiquitous, B-mix (Bassian dominant) and E-mix (Eyrean dominant).

		Bassian	Eyrean	Endem.	Torres.	Ubiqu.	B-mix	E-mix
	N	18	26	9	8	10	8	4
Mature tree	26	38	15	12	8	19	4	4
Grassy woodland	22	23	32	4.5	18	9	9	4.5
Mixed understorey	6	16	33	16	0	16	16	0
Semi-arid shrubland	18	6	72	11	0	0	0	11
Rocky and old-growth	7	0	0	29	14	14	43	0
Frogs and waterbirds	4	25	0	0	25	25	25	0
Total	83	22	31	11	9.5	12	9.5	5

Of the species with sufficiently known distribution changes since European settlement (58%), 91% have undergone range contractions. Forty per cent of these have contracted to the north and west towards an Eyrean core and 27% have become fragmented in their distribution (as well as habitat) (Table 6). Since European settlement in the study area, Eyrean and grassy woodland faunas have declined more than other biogeographic and functional groups.

IMPLICATIONS FOR MANAGEMENT

Massive changes have occurred to vertebrate abundance, distribution, diversity and habitats since European settlement on the western slopes of New South Wales, as demonstrated by this study. Without major restoration efforts a large proportion of the remnant vertebrate fauna may disappear. The White Cypress Pine region of New South Wales is climatically and pedologically similar to other regions of the world that are affected by wheat and sheep farming, declining biodiversity and the first stages of desertification (e.g., Sahel zone of Africa, Inner Mongolia, pampas of Argentina). These regions, including the White Cypress Pine region, are currently unsuited for conventional agriculture with one year in five being a drought, broad-scale land degradation and poorly-supported agro-economies.

The landscape with which and within which State Forests of NSW and other agencies responsible for nature conservation must work cannot be returned to its pre-European state. But components of it may be maintained or restored as a viable framework for biodiversity conservation and human ecological sustainability (i.e., agriculture and forestry, *sensu* Saunders *et al.* 1993; Saunders and Hobbs 1995). Growing White Cypress Pine, in combination with other endemic products, such as broombush, Blue Mallee, ironbark, kangaroos and emus, is likely to be a long-term productive and sustainable alternative to agriculture. It is also likely to support a greater diversity of native plants and animals than conventional agriculture. At present, State Forests of NSW is too poorly funded to fulfil functions of advising farmers in agro-forestry, trialing new agro-forestry and forestry systems and nature conservation (Bauer *et al.* 1999; Date and Paull 1999).

In this section, we restrict our discussion to recommendations for the management (given adequate funding) of functional groups identified by our studies to be at risk of further declines on public and private land used for timber production (Bauer *et al.* 1999; Date and Paull 1999). We attempt to simplify such a large task by identifying four levels of management at which State Forests of NSW can participate:

- species, functional group and habitat management,
- operations management that incorporates fauna management prescriptions,
- extension and advisory services for ecosystem management within and outside state forests, and
- research into fundamental and goal-oriented ecological problems for species/habitat, operations and ecosystem management.

Each level should be undertaken by different personnel and require different expertise. Conservation actions for fauna groups, such as protection from and mitigation of forestry activities, need to be reviewed and updated regularly by trained ecologists, in consultation with foresters. Prescriptions for the various activities in state forests should be introduced or modified to incorporate conservation actions, and reviewed and updated regularly by trained foresters in consultation with ecologists. Extension and advisory services relating the relevance of nature conservation and forestry on farms to landscape and ecosystem management, such as expert contributions to Total Catchment Management, Greening Australia, Landcare and individual farm plans, require resource managers. Research needs and funding should be identified and sought by State Forests of NSW planning personnel in collaboration with in-house and consultant scientists and the research should be undertaken by scientists in consultation with State Forests' planners.

Species, functional group and habitat conservation

Conservation of threatened species, functional groups and their habitats requires reserve planning to ensure protection of sites that contain "diversity hotspots" for the functional groups or that contain habitats with potential to attract diverse groups of fauna, including threatened species. Enlargement of and connectivity among reserves can be easily achieved through the protection from logging and grazing of habitat "corridors" along drainage and ridge lines and between creeks and ridges. Locations where there are high diversities and abundances of

Table 6. The percentage representation of range contractions within the functional groups. Range change: N, NW, W, SW, S, SE, E, NE — contracted toward that compass point, C — contracted within survey area, F — fragmented, Ex — range extension due to our surveys, I — insufficient data, I(W) — insufficient data but thought to have contracted to the west, V — vagrant.

	0	N	NW	W	SW	S	SE	E	NE	C	F	Ex	I
Mature tree	—	—	7.5	—	—	4	—	4	—	4	23	7.5	50
Grassy woodl	—	27	—	18	—	5	—	—	—	—	14	—	36
Mixed u/storey	—	33	—	33	—	—	—	—	—	—	17	—	17
Semi-arid	—	—	5.5	5.5	17	—	—	—	—	5.5	11	5.5	50
Rocky and old growth	14.5	—	—	—	—	—	—	29	—	—	—	14.5	42
Frogs and w/birds	4	—	—	25	—	—	—	—	—	—	—	—	75

fauna from different functional groups, including threatened species, need to be reserved as core habitat from which regenerating and logged stands can be recolonized. Some have been identified (Bauer *et al.* 1999; Date and Paull 1999), but ongoing survey is required to identify and to monitor other biodiversity and population sources and sinks. Some sources could be reserved for a finite period until neighbouring sinks have rehabilitated, which may be achieved by lengthening logging cycles (see also Lindenmayer 1994).

Review of logging, fire and grazing prescriptions and the implementation of effective firewood and broombush and mallee harvesting prescriptions is required within and outside reserves to ensure adequate retention or maintenance of critical habitat components. Across the functional groups, retentions of dead (stags), senescent, mature and young trees are required during logging of ironbark and cypress, broombush and mallee cutting in commercial stands throughout the study area, especially until old-growth habitat components are re-established. Formulation of effective firewood collection regulations and log retention prescriptions is required to maintain and to restore this habitat component for terrestrial fauna dependent on large hollow logs and other large fallen timber. Management of understorey through fire and/or grazing management may be necessary for maintenance and recruitment of some threatened species. Sites managed for different functional groups may require different prescriptions. For example, light intermittent grazing and low intensity fires at variable intervals of 4 to 15 years are recommended for grassy woodlands and fauna, whereas management of the shrub/grass mosaic of mixed understorey woodlands and fauna may require exclusion of grazing and implementation of a fire regime of low and high intensity fires at intervals of 15 years or more, but further research is required (Bradstock *et al.* 1995; Date and Paull 1999).

Forestry operations

Inclusion of conservation management in forestry operations requires a change of focus from simplifying the ecosystem and obtaining a uniform timber "product" to retaining and restoring structural and biotic diversity while harvesting several different products. Retention of habitat components characteristic of old-growth woodlands is required as well as the ability to recruit old-growth components through the retention of a range of younger components. In addition, the maintenance of heterogeneous understorey structure is important. We identified operations which require review in order to incorporate conservation objectives.

Cypress and ironbark harvest management. Retention rates of cypress and ironbark that are mature and healthy and that are unsuitable for logging due to disease and malformation should be higher than the existing rates and logging cycles extended, allowing recovery of mixed-age structure and a broad range of habitat components, including old-growth components.

Silvicultural management of cypress, casuarina and ironbark. Discontinuing the culling of mature and large defective cypress, casuarina and ironbark and continuing the thinning of seedlings using grazing or burning should reduce the requirement for manual, non-commercial thinning, lengthen the cutting cycle and stimulate regeneration of diverse understoreys.

Implementation of mallee and broombush harvest management should ensure maintenance of stands at a range of ages, including old-growth.

Implementation of effective firewood management should ensure the adequate recruitment of large stags and logs.

Range management. To achieve priority of nature conservation and timber production goals over the requirements of the grazing lessee, grazing should be regulated so that regeneration can occur and structural and biotic diversity are maintained over space and time.

Fire management. Use of a variable fire regime over space and time should achieve similar results to regulated grazing.

Pest management. Continued and intensified control of feral animals, particularly rabbits, foxes and cats, and concurrent monitoring of native and feral fauna should aim to reduce feral and increase native populations.

Landscape and ecosystem management

Two processes with production and conservation objectives by which State Forests of NSW can contribute to landscape and ecosystem management are the restoration management of remnant vegetation and the establishment of agroforestry. State forests can contribute expertise, advice and on-ground assistance in the restoration management of the many small isolated state forests and other remnant woodlands in the study area through the implementation of highly regulated and monitored grazing, feral animal control, fire management and firewood collection (see also Goldney *et al.*, this volume; Windsor and Goldney, in litt.). In the current climate of Landcare, Greening Australia and sustainable land management on farms, there are opportunities for fostering agroforestry together with nature conservation through enlarging woodland habitat and the timber production estate on private property and leasehold land in consultation or partnership with State Forests of NSW (Nicholson 1997).

Research

We identified seven areas of research relevant to enhancing wildlife conservation in state forests of the New South Wales cypress woodlands, some of which also have long-term benefits for timber production.

Conservation value of White Cypress Pine. The accumulation of habitat components as individuals grow and become "Old Greys" within mixed-age stands and their contribution to fauna habitat can only be predicted. Long-term research is required into the changing conservation role and timber value of White Cypress Pine through its life history.

Habitat requirements of hollow and mature tree dependent fauna. Despite extensive research in coastal and mountain forests on the habitat requirements of hollow and mature tree-dependent fauna (e.g., Mackowski 1984; Lindenmayer *et al.* 1993; Gibbons and Lindenmayer 1997; Lindenmayer and Franklin 1997), little has been done to complement it on the western slopes (but see Traill 1991, 1993; Bennett *et al.* 1994). Similar research is required to determine more precisely mature tree-dependent fauna distribution and abundance patterns in the study area, the distribution and recruitment patterns of hollows and the development of nectar and other "non-wood" resources, the patterns of use of resources by the fauna and impacts of logging operations (Date and Paull 1999).

Targeted surveys for terrestrial fauna and their habitat requirements. Research into terrestrial fauna of grassy and mixed understorey woodlands is in a similar state to that for mature tree-dependent fauna (Lindemayer *et al.* 1994; cf. Date and Paull 1999; Paull and Date 1999). More extensive surveys of fauna populations of grassy and mixed understorey woodlands, the distribution and disturbance patterns of critical habitat components and the impacts of fire, grazing and logging regimes are required (Date and Paull 1999).

Experimental manipulation of grazing regimes for conservation. Experimental research into methods of enhancing critical habitat components for grassy woodland fauna is required. This should also lead to effective methods of controlling cypress regeneration and maintaining ecologically sustainable grazing loads for the maintenance of fauna populations and stock (see also Wilson *et al.* 1997). Reintroductions of threatened species should be considered. Co-operative arrangements between graziers and State Forests of NSW should be investigated to facilitate control of grazing for silvicultural and conservation objectives.

Experimental manipulation of fire regimes for conservation. Fire regimes appropriate for the maintenance and regeneration of grassy and mixed understorey woodland communities and for silvicultural management of White Cypress Pine should be determined through experimental manipulation of the intensity, seasonality and frequency of fire. Threatened species re-introduction programmes on experimental plots should also be considered.

Feasibility of mixed agro-forestry. The ecological and marketing feasibility of planting White Cypress Pine, Ironbark, Broombush and Blue Mallee on farms and degraded public land should be investigated.

Invertebrate Diversity and Abundance and Nutrient Cycling in Cypress Pine Woodlands. There is little information available on the diversity and function of invertebrates in White Cypress Pine woodlands, particularly as they relate to nutrient cycling. Litter accumulation is thought to asymptote at greater than 20 years post-fire (Hart 1995). Some eucalypt species may support more than 1 000 arthropod taxa (Recher *et al.* 1996). The reduction in abundance of eucalypts and mature cypress through logging and ringbarking along with the reduction and simplification of understorey vegetation and the premature removal of litter by fuel reduction burning and grazing may have severe negative impacts on invertebrate diversity and abundance. This has probably contributed to major changes in vertebrate diversity and abundance through vegetation growth and invertebrate prey abundance. Investigation of invertebrate diversity and abundance as it relates to vertebrate diversity and abundance and the function of invertebrates in nutrient cycling is required.

CONCLUSION

Given adequate funding, State Forests of NSW could play a leading role in the restoration of one of Australia's threatened ecosystems, the White Cypress Pine woodlands, while expanding its production base. If the four levels of management identified in our study: threatened species, forestry operations, landscape/ecosystem conservation and research, and specific recommendations outlined here and detailed by Date and Paull (1999) and Bauer *et al.* (1999) are implemented, we predict long-term positive impacts for nature conservation and timber production of the region. The four levels of management and suitable expertise to implement our recommendations already exist within State Forests of NSW. A change in philosophy towards integrating nature conservation with timber production in the White Cypress Pine woodlands has begun but requires continued commitment in order to be successful.

ACKNOWLEDGEMENTS

State Forests of NSW funded this study and provided much logistic and expert support. National Parks and Wildlife Service NSW contributed expert advice and review. Funding was also provided by the Department of Environment, Sports and Territories, Canberra. Unpublished data on bird distributions in the Pilliga were contributed by Liz Date, Harry Recher, Hugh Ford and Nick Reid from a study funded by the Australian Research Council. We would like to thank the biologists who worked with us in the field and lab: Clive Barker, Andrew Bryant, Roger Coles, David Costello, Neville Schrader, Brian Stone and Barbara Triggs. Janice Lowe at the GIS Branch of State Forests of NSW prepared the map of the study area.

REFERENCES

Barratt, D. G., 1997. Predation by house cats, *Felis catus* (L.), in Canberra, Australia. I. Prey composition and preference. *Wildl. Res.* **24**: 263–77.

Barrett, G. W., Ford, H. A. and Recher, H. F., 1994. Conservation of woodland birds in a fragmented rural landscape. *Pac. Conserv. Biol.* **1**: 245–56.

Bauer, J. J., Bryant, A., Goldney, D. and Costello, D., 1999. Fauna Report — State Forests within the South-west Cypress Zone. State Forest of NSW: Dubbo, New South Wales.

Beadle, N. C. W., 1948. The Vegetation and Pastures of Western New South Wales with Special Reference to Soil Erosion. NSW Dept. of Soil Conservation: Sydney, New South Wales.

Belbin, L., 1987. PATN Pattern Analysis Package, Technical Reference. Division of Wildlife and Ecology, CSIRO: Canberra, Australian Capital Territory.

Bennett, A. F., Lumsden, L. F. and Nicholls, A. O., 1994. Tree hollows as a resource for wildlife in remnant woodlands: spatial and temporal patterns across the northern plains of Victoria. *Pac. Conserv. Biol.* **1**: 222–35.

Blakers, M., Davies, S. J. J. F. and Reilly, P. N., 1984. The Atlas of Australian Birds. RAOU/Melbourne University Press: Melbourne, Victoria.

Boland, D. J., Brooker, M. I. H., Chippendale, G. M., Hall, N., Hyland, B. P. M., Johnston, R. D., Kleinig, D. A. and Turner, J. D., 1984. Forest Trees of Australia. Fourth edition. CSIRO: Canberra, Australian Capital Territory.

Bradstock, R. A., Keith, D. A. and Auld, T. D., 1995. Fire and conservation: imperatives and constraints on managing for diversity. Pp. 323–33 *in* Conserving Biodiversity: Threats and Solutions ed by R. A. Bradstock, T. D. Auld, D. A. Keith, R. T. Kingsford, D. Lunney and D. P. Sivertsen. Surrey Beatty & Sons: Chipping Norton, New South Wales.

Cardale, S., 1987. Vegetation. Pp. 111–30 *in* Scenic and Scientific Survey of the Central Western Region: a Report to the Australian Heritage Commission. Volume 1 ed by D. C. Goldney and I. J. S. Bowie. National Trust of Australia (NSW): Sydney, New South Wales.

Catling, P. C., 1988. Similarities and contrasts in the diets of foxes, *Vulpes vulpes*, and cats, *Felis catus*, relative to fluctuating prey populations and drought. *Aust. Wildl. Res.* **15**: 307–17.

Catterall, C. P., Storey, R. J. and Kingston, M. B., 1997. Reality versus rhetoric: a case study monitoring regional deforestation. Pp. 367–77 *in* Conservation Outside Nature Reserves ed by P. Hale and D. Lamb. Centre for Conservation Biology, The University of Queensland: Brisbane, Queensland.

Cogger, H. G., 1992. Reptiles and Amphibians of Australia. Reed Books: Chatswood, New South Wales.

Cogger, H. G., Cameron, E. E., Sadlier, R. A. and Eggler, P., 1993. Action Plan for Australian Reptiles. Australian Nature Conservation Agency — Endangered Species Program Project No. 24.

Croft, M., Goldney, D. and Cardale, S., 1997. Forest and woodland cover in the Central Western Region of New South Wales prior to European settlement. Pp. 394–406 *in* Conservation Outside Nature Reserves ed by P. Hale and D. Lamb. Centre for Conservation Biology, The University of Queensland: Brisbane, Queensland.

Date, E. M. and Paull, D. C., 1999. Fauna Survey of the Cypress/Ironbark Forests of North-west New South Wales. State Forests of NSW: Dubbo, New South Wales.

Dick, R., 1992. Arguing for adequate clearing controls in western NSW — the Culgoa River floodplain fauna study. *Nat. Parks J.* **36**: 13–18.

Emison, W. B. and Caldow, W. D., 1992. Studies on a threatened population of Red-tailed Black Cockatoos in southeastern Australia. Annual progress report 1991–92. Department of Conservation and Environment: Melbourne, Victoria.

Emison, W. B. and Caldow, W. D., 1992. Studies on a threatened population of Red-tailed Black Cockatoos in southeastern Australia. Annual progress report 1992–93. Department of Conservation and Environment: Melbourne, Victoria.

Filet, P. G., McCosker, J. C. and Osten, D., 1997. Management indicators for production and conservation in grazed woodlands of Queensland: a case study. Pp. 266–70 *in* Conservation Outside Nature Reserves ed by P. Hale and D. Lamb. Centre for Conservation Biology, The University of Queensland: Brisbane, Queensland.

Flannery, T., 1994. The Future Eaters. Reed: Sydney, New South Wales.

Ford, H., Davis, W. E., Debus, S., Ley, A., Recher, H. and Williams, B., 1993. Foraging and aggressive behaviour of the Regent Honeyeater *Xanthomyza phrygia* in northern New South Wales. *Emu* **93**: 277–81.

Forestry Commission of NSW, 1989. Forest Types in NSW Research Note No. 17. Forestry Commission of NSW: Sydney, New South Wales.

Forestry Commission of NSW, 1993a. Proposed Forestry Operations in the Cypress Pine/Ironbark Forests of North Western New South Wales. Fauna Impact Statement Draft. Forestry Commission of NSW: Sydney, New South Wales.

Forestry Commission of NSW, 1993b. Proposed Forestry Operations in the Cypress Pine/Ironbark Forests of South Western New South Wales. Fauna Impact Statement Draft. Forestry Commission of NSW: Sydney, New South Wales.

Garnett, S., 1992. The Action Plan for Australian Birds. Project Number 21. Endangered Species Program. Australian National Parks and Wildlife Service: Canberra, Australian Capital Territory.

Garnett, S. (ed), 1993. Threatened and Extinct Birds of Australia. RAOU Report No. 82. RAOU/Australian National Parks and Wildlife Service: Moonee Ponds, Victoria.

Gill, A. M. and Bradstock, R. A., 1995. Extinction of biota by fires. Pp. 309–22 *in* Conserving Biodiversity: Threats and Solutions ed by R. A. Bradstock, T. D. Auld, D. A. Keith, R. T. Kingsford, D. Lunney and D. P. Sivertsen. Surrey Beatty & Sons: Chipping Norton, New South Wales.

Gibbons, P. and Lindemayer, D., 1997. A review of prescriptions employed for conservation of hollow-dependent fauna in wood production forests of eastern Australia. Pp. 497–505 *in* Conservation Outside Nature Reserves ed by P. Hale and D. Lamb. Centre for Conservation Biology, The University of Queensland: Brisbane, Queensland.

Goldney, D. C. and Bowie, I. J. S., 1990. Some management implications for the conservation of vegetation remnants and associated fauna in the central western region of New South Wales. Pp. 427–40 *in* Australian Ecosystems: 200 Years of Utilization, Degradation and Reconstruction ed by D. A. Saunders, A. J. M. Hopkins and R. A. How. Ecological Society of Australia and Surrey Beatty & Sons: Chipping Norton, New South Wales.

Goldney, D., Bauer, J., Bryant, H., Hodgkins, D. and Watson, G., 1995. Winning battles but losing the war: the education marketing imperative. Pp. 574–88 *in* Nature Conservation 4: The Role of Networks ed by D. A. Saunders, J. L. Craig and E. M. Mattiske. Surrey Beatty & Sons: Chipping Norton, New South Wales.

Grey, M. J., Clarke, M. F. and Loyn, R. H., 1997. Influence of the Noisy Miner *Manorina melanocephala* on avian diversity and abundance in remnant Grey Box woodland. *Pac. Conserv. Biol.* **4:** 55–69.

Hart, D. M., 1995. Litterfall and decomposition in the Pilliga State forests, New South Wales, Australia. *Aust. J. Ecol.* **20:** 266–72.

Hobbs, R. J. and Saunders, D. A. (eds), 1992. Reintegrating Fragmented Landscapes: Towards Sustainable Production and Nature Conservation. Springer-Verlag: New York.

How, R. H., 1984. Local dynamics of bird assemblages in small forest habitat islands in Australia and North America. *Ecology* **65:** 1585–601.

Johnston, T. N. and Jennings, K. S., 1991. Management of Cypress Pine forests in Queensland. Pp. 182–87 *in* Forest Management in Australia ed by F. H. McKinnell, E. R. Hopkins and J. E. D. Fox. Surrey Beatty & Sons: Chipping Norton, New South Wales.

Lindenmayer, D. B., 1994. The impacts of timber harvesting on arboreal marsupials at different spatial scales and its implications for ecologically sustainable forest use and nature conservation. *Aust. J. Environ. Manage.* **1:** 56–68.

Lindenmayer, D. and Franklin, J., 1997. The importance of stand structure in off-reserve forest wildlife conservation: a case study from Victorian mountain ash forests. Pp. 506–15 *in* Conservation Outside Nature Reserves ed by P. Hale and D. Lamb. Centre for Conservation Biology, The University of Queensland: Brisbane, Queensland.

Lindenmayer, D. B., Cunningham, R. B., Donnelly, C. F., Tanton, M. T. and Nix, H. A., 1993. The abundance and development of cavities in montane ash-type trees in montane forests of the Central Highlands of Victoria, south-east Australia. *For. Ecol. Manage.* **60:** 77–104.

Lindemayer, D. B., Cunningham, R. B., Donnelly, C. F., Triggs, B. J. and Belvedere, M., 1994. The diversity, abundance and microhabitat requirements of terrestrial mammals in contiguous forests and linear strips in montane ash forests of the central highlands of Victoria. *For. Ecol. Manage.* **67:** 113–33.

Lindsay, A. D., 1967. Forest Types of the New South Wales Cypress Pine Zone. Technical Paper No. 8. Forestry Commission of NSW: Sydney, New South Wales.

Loyn, R. H., 1985. Birds in fragmented forests in Gippsland, Victoria. Pp. 323–31 *in* Birds of Eucalypt Forests and Woodlands: Ecology, Conservation, Management ed by A. Keast, H. F. Recher, H. Ford and D. Saunders. Royal Australasian Ornithologists' Union and Surrey Beatty & Sons: Chipping Norton, New South Wales.

Loyn, R. H., 1987. Effects of patch area and habitat on bird abundances, species numbers and tree health in fragmented Victorian forests. Pp. 65–77 *in* Nature Conservation: The Role of Remnants of Native Vegetation ed by D. A. Saunders, G. W. Arnold, A. A. Burbidge and A. J. M. Hopkins. Surrey Beatty & Sons: Chipping Norton, New South Wales.

Mackowski, C. M., 1984. The ontogeny of hollows in Blackbutt, *Eucalyptus pilularis* and its relevance to the management of forests for possums, gliders and timber. Pp. 517–25 *in* Possums and Gliders ed by A. P. Smith and I. D. Hume. Surrey Beatty & Sons: Chipping Norton, New South Wales.

Moore, C. W. E., 1953a. The vegetation of the south-eastern Riverina, New South Wales. I. The climax communities. *Aust. J. Bot.* **1:** 485–547.

Moore, C. W. E., 1953b. The vegetation of the south-eastern Riverina, New South Wales. II. The disclimax communities. *Aust. J. Bot.* **1:** 548–67.

Moore, R. M., 1970. South-eastern temperate woodlands and grasslands. Pp. 171–90 *in* Australian Grasslands ed by R. M. Moore. ANU Press: Canberra, Australian Capital Territory.

National Parks and Wildlife Service/State Forests of NSW, 1994. Guidelines For the Preparation of Fauna Impact Statements for Forestry Operations Within State Forests. 1: Fauna Survey. NSW National Parks and Wildlife Service/State Forests of NSW: Sydney, New South Wales.

Newsome, A. E., 1994. Vertebrate pests versus wildlife conservation in semi-arid New South Wales: a profound imbalance. Pp. 43–52 *in* Future of the Fauna of Western New South Wales ed by D. Lunney, S. Hand, P. Reed and D. Butcher. Royal Zoological Society of New South Wales: Mosman, New South Wales.

Nicholson, D., 1997. Managing Cypress Pine on Your Property. State Forests of NSW: Dubbo, New South Wales.

Nix, H., 1986. A biogeographic analysis of Australian elapid snakes. Pp. 4–15 *in* Snakes: Atlas of Elapid Snakes of Australia ed by R. Longmore. Australian Flora and Fauna Series Number 7.

Norris, E. H., Mitchell, P. B. and Hart, D. M., 1991. Vegetation changes in the Pilliga forest: a preliminary evaluation of the evidence. *Vegetatio* **91:** 209–18.

Oliver, D., Ley, A. J. and Williams, B., 1998. Breeding success and nest site selection of the Regent Honeyeater *Xanthomyza phrygia* near Armidale, New South Wales. *Emu* **98:** 97–103.

Paton, D., 1996. Overview of the Impacts of Feral and Managed Honeybees in Australia. Invasive Species Program, ANCA: Canberra, Australian Capital Territory.

Paull, D. C. and Date, E. M., 1999. Patterns of decline in the native mammal fauna of the north-west slopes of New South Wales. *Aust. Zool.* **31:** 210–24.

Recher, H. F. and Lim, L., 1990. A review of current ideas of the extinction, conservation and management of Australia's terrestrial vertebrate fauna. Pp. 287–301 *in* Australian Ecosystems: 200 Years of Utilization, Degradation and Reconstruction ed by D. A. Saunders, A. J. M. Hopkins and R. A. How. Ecological Society of Australia and Surrey Beatty & Sons: Chipping Norton, New South Wales.

Recher, H. F., Majer, J. D. and Ganesh, S., 1996. Eucalypts, arthropods and birds — on the relation between foliar nutrients and species richness. *For. Ecol. Manage.* **85:** 177–95.

Recher, H. F., Shields, J., Kavanagh, R. and Webb, G., 1987. Retaining remnant mature forest for nature conservation at Eden, New South Wales: a review of theory and practice. Pp. 177–94 *in* Nature Conservation: The Role of Remnants of Native Vegetation ed by D. A. Saunders, G. W. Arnold, A. A. Burbidge and A. J. M. Hopkins. Surrey Beatty & Sons: Chipping Norton, New South Wales.

Robinson, D., 1994. Research Plan for Threatened Woodland Birds of Southeastern Australia. Technical Report No. 133. Department of Conservation and Natural Resources: Melbourne, Victoria.

Rolls, E., 1981. A Million Wild Acres. Penguin: Sydney, New South Wales.

Saunders, D. A. and Hobbs, R. J. (eds), 1991. Nature Conservation 2: the Role of Corridors. Surrey Beatty & Sons: Chipping Norton, New South Wales.

Saunders, D. and Hobbs, R. J., 1995. Habitat reconstruction: the revegetation imperative. Pp. 104–12 *in* Conserving Biodiversity: Threats and Solutions ed by R. A. Bradstock, T. D. Auld, D. A. Keith, R. T. Kingsford, D. Lunney and D. P. Sivertsen. Surrey Beatty & Sons: Chipping Norton, New South Wales.

Saunders, D. A., Hobbs, R. J. and Ehrlich, P. R. (eds), 1993. Nature Conservation 3: The Reconstruction of Fragmented Ecosystems. Surrey Beatty & Sons: Chipping Norton, New South Wales.

Saunders, D. A., Hobbs, R. J. and Margules, C. R., 1991. Biological consequences of ecosystem fragmentation: a review. *Conserv. Biol.* **5:** 18–32.

Saunders, D. A., Arnold, G. W., Burbidge, A. A. and Hopkins, A. J. M. (eds), 1987. Nature Conservation: The Role of Remnants of Native Vegetation. Surrey Beatty & Sons: Chipping Norton, New South Wales.

Schodde, R., 1994. The bird fauna of western New South Wales: geography and status. Pp. 107–21 *in* Future of the Fauna of Western New South Wales ed by D. Lunney, S. Hand, P. Reed and D. Butcher. Royal Zoological Society of New South Wales: Mosman, New South Wales.

Short, J. and Turner, B., 1994. A test of the vegetation mosaic hypothesis: a hypothesis to explain the decline and extinction of Australian mammals. *Conserv. Biol.* **8:** 439–49.

Sivertsen, D., 1993. Conservation of remnant vegetation in the Box and Ironbark lands of New South Wales. *Vic. Nat.* **110:** 24–29.

Smith, A. P., Andrews, S. P. and Moore, D. M., 1994. Terrestrial Fauna of the Grafton and Casino State Forest Management Areas: Description and Assessment of Forestry Impacts: Grafton — Casino Management Areas EIS Supporting Document No. 1. Austeco Pty. Ltd. Armidale and State Forests of NSW: Sydney, New South Wales.

Smith, A. P., Moore, D. M. and Andrews, S. P., 1992. Proposed Forestry Operations in the Glen Innes Forest Management Area — Fauna Impact Statement for the Forestry Commission of NSW. Austeco Pty. Ltd: Armidale, New South Wales.

Smith, P., Pressey, R. and Smith, J., 1994. Birds of particular conservation concern in the western division of New South Wales. *Biol. Conserv.* **69:** 315–38.

Spencer, W. B., 1896. Summary of results. Pp. 139–99 *in* Report of the Work of the Horn Scientific Expedition to Central Australia ed by W. B. Spencer. Dulau: London.

Strahan, R., 1994. The dilemma of efficient agricultural practices. Pp. 27–30 *in* Future of the Fauna of Western New South Wales ed by D. Lunney, S. Hand, P. Reed and D. Butcher. Royal Zoological Society of New South Wales: Mosman, New South Wales.

Strahan, R. (ed), 1995. The Australian Museum Complete Book of Australian Mammals, second edition. Angus and Robertson: Sydney, New South Wales.

Taylor, R. J., 1991. Fauna management practices in state forests in Tasmania. Pp. 259–64 *in* Conservation of Australia's Forest Fauna ed by D. Lunney. Royal Zoological Society of New South Wales: Mosman, New South Wales.

Thackway, R. and Cresswell, I. D., 1995. An Interim Biogeographic Regionalisation for Australia: a Framework for Setting Priorities in the National Reserves System Cooperative Program. ANCA: Canberra, Australian Capital Territory.

Traill, B. J., 1991. Box-ironbark forests: tree hollows, wildlife and management. Pp. 119–24 *in* Conservation of Australia's Forest Fauna ed by D. Lunney. Royal Zoological Society of New South Wales: Mosman, New South Wales.

Traill, B. J., 1993. Forestry, birds, mammals and management in Box and Ironbark forests. *Vic. Nat.* **110:** 11–14.

Wardell-Johnson, G. and Nicholls, W., 1991. Forest wildlife and habitat management in southwestern Australia: knowledge, research and direction. Pp. 161–92 *in* Conservation of Australia's Forest Fauna ed by D. Lunney. Royal Zoological Society of New South Wales: Mosman, New South Wales.

Wilson, A. D., 1990. The effect of grazing on Australian ecosystems. Pp. 235–44 *in* Australian Ecosystems: 200 Years of Utilization, Degradation and Reconstruction ed by D. A. Saunders, A. J. M. Hopkins and R. A. How. Ecological Society of Australia and Surrey Beatty & Sons: Chipping Norton, New South Wales.

Wilson, A. D., Oxley, R. E. and Bratby, W. J., 1997. A Grazing Management Strategy for Buckingbong State Forest. Dept. of Land and Water Conservation and State Forests of NSW report.

Woinarski, J. C. Z., 1989. The vertebrate fauna of broombush *Melaleuca uncinata*. Vegetation in north-western Victoria, with reference to effects of broombush harvesting. *Aust. Wildl. Res.* **16:** 217–38.

Woinarski, J. C. Z., 1990. Broombush in the Victorian mallee: the vertebrate fauna of broombush vegetation in north-western Victoria and the environmental effects of the broombush harvesting industry. Land Conservation Council of Victoria: Melbourne, Victoria.

CHAPTER 15

The kangaroos of Yan Yean: issues for a forested water catchment in a semi-rural matrix

Graeme Coulson,[1] Phil Alviano,[1,2] Dan Ramp,[1,3] Sarah Way,[1] Natasha McLean[1] and Vural Yazgin[1]

The catchment of Yan Yean Reservoir is situated on the rural fringe of Melbourne, southeastern Australia, and supports a large population of Eastern Grey Kangaroos *Macropus giganteus*. The kangaroos of Yan Yean have been the source of problems within the catchment and in its agricultural matrix for five decades. We have begun an interdisciplinary study into the ecology of the kangaroo population, the impact of kangaroos on vegetation and water in the catchment, and the interaction between kangaroos and the surrounding rural matrix. The kangaroos are essentially sedentary, congregating at high densities on preferred feeding areas within the catchment. Kangaroos also cross the catchment boundary to feed on neighbouring farmland; the neighbours have mixed views on the presence of mobs of more than 100 on their properties. The catchment is closed to the public, and has two significant Reference Areas of remnant native vegetation. Kangaroos are the dominant grazer in the catchment, and grazed offtake is high. Grazing-induced disturbance is the major influence on floristic composition, overriding the effects of other environmental variables. Disturbance by kangaroos also accelerates run-off of surface water in the catchment, and yields more nutrients.

Key words: Yan Yean, Eastern Grey Kangaroo, *Macropus giganteus*, Population dynamics, Habitat use, Grazing disturbance, Water quality, Human dimensions.

INTRODUCTION

THE kangaroos of Yan Yean are Eastern Grey Kangaroos *Macropus giganteus*. This species is common through much of its range on the slopes of the Great Dividing Range and on the semi-arid plains of eastern Australia (Poole 1982). Eastern Grey Kangaroos occur in open forest, woodlands and grassland margins, favouring habitats with lateral cover offered by trees or shrubs, together with suitable forage provided by grasses in the ground layer (Caughley 1964; Bell 1973; Hill 1981; Taylor 1985; Southwell 1987).

High local densities of Eastern Grey Kangaroos can occur in rural settings where patches of woodland or forest are adjacent to crops (e.g., Hill *et al.* 1988) or improved pastures (e.g., Taylor 1984, 1985). In these situations, farmers often regard kangaroos as pests, reporting consumption of pasture and water provided for domestic stock, and damage to fences and crops (e.g., Dempster 1961; Lavery and Kirkpatrick 1985), but there has been little research into these issues. By contrast, the Western Grey Kangaroo *M. fuliginosus*, which is the sibling species of the Eastern Grey Kangaroo, has been studied extensively. Arnold and colleagues have investigated the ecology of Western Grey Kangaroos in remnants of native vegetation in the wheatbelt of Western Australia (e.g., Arnold and Maller 1987; Arnold *et al.* 1991, 1992), including the impacts of kangaroos on cropland (Arnold and Steven 1988; Arnold *et al.* 1989). Caughley and colleagues have conducted an equally comprehensive ecological study of Western Grey Kangaroos in the pastoral zone (Caughley *et al.* 1987), and Wilson (1991a,b) has quantified the impacts of this species in terms of pasture intake and sheep production.

[1]Department of Zoology, University of Melbourne, Parkville, Victoria, Australia 3052.
[2]Department of Geography and Environmental Studies, University of Melbourne, Parkville, Victoria, Australia 3052.
[3]School of Botany, University of Melbourne, Parkville, Victoria, Australia 3052.
Pages 146–56 *in* NATURE CONSERVATION 5: NATURE CONSERVATION IN PRODUCTION ENVIRONMENTS: MANAGING THE MATRIX ed by J. L. Craig, N. Mitchell and D. A. Saunders. Surrey Beatty & Sons, 2000.

High densities of Eastern Grey Kangaroos also occur in nature reserves where remnant stands of native forest, woodland or shrubland form a mosaic with patches of former farmland, as typified by Rotamah Island in the Lakes National Park, Victoria (Coulson and Raines 1985) and Tidbinbilla Nature Reserve in the Australian Capital Territory (Neave and Tanton 1989). Although the grazing pressure exerted on reserved areas by Eastern Grey Kangaroos is often claimed to have negative effects on other components of the biota, this issue has rarely been approached scientifically. Again, Western Grey Kangaroos have received more attention. Hattah-Kulkyne National Park, in northwestern Victoria, is the best-known example of the negative effects that Western Grey Kangaroos can have on ecosystem values in nature reserves (Cheal 1986; Walters 1986; Coulson and Norbury 1988; Coulson *et al.* 1990). Grazing impacts in this park are now being reduced by a large-scale culling programme (Sluiter *et al.* 1997). Similar problems are being addressed in the Flinders Ranges National Park, South Australia (Alexander 1997).

The kangaroos of Yan Yean encapsulate these issues. The Yan Yean catchment forms a matrix of significant remnants of native vegetation, and it is thought that the grazing pressure applied by kangaroos is degrading the botanical value of the area (Scarlett 1983). In addition, the forests and woodlands of the catchment lie within a matrix of farmland, and the kangaroos exploit this landscape, pushing under fences and feeding on pastures in mobs of more than 100 at times. The kangaroos of Yan Yean also introduce a novel issue for several reservoirs within the Melbourne water supply system. Eastern Grey Kangaroos are the dominant grazers and are a potential source of disturbance to soils and vegetation, with implications for the quantity and quality of the water yielded (Ecoplan Australia 1995).

The purpose of this chapter is to introduce a multi-disciplinary research programme on the ecology and management of the kangaroos of Yan Yean. The project began in 1992, taking advantage of this valuable research environment to explore the ecology and management of Eastern Grey Kangaroos in a multiple-use production matrix. This chapter covers the background to the study, and outlines the design and preliminary findings of the programme.

YAN YEAN RESERVOIR CATCHMENT

Yan Yean Reservoir Catchment (145°09'E, 37°32'S) is situated approximately 37 km north-east of Melbourne, Victoria, Australia (Fig. 1). The 2 250 ha catchment has the primary role of production and storage of water as part of the supply to the Greater Melbourne area, and is managed by Melbourne Water. The reservoir has a capacity of 30 000 ML, and has a surface area of 560 ha when filled to capacity at 185 m above sea level (Melbourne and Metropolitan Board of Works 1988, 1989). The climate is temperate, with a mean annual rainfall of 667 mm, and little difference in mean monthly precipitation throughout the year. February is the hottest month, having mean minimum and maximum temperatures of 16.3°C and 26.8°C respectively, and July is the coolest, having a mean minimum of 6.7°C and maximum of 12.9°C (Quin 1989).

The catchment is mostly surrounded by farmland, although it is being increasingly sub-divided for hobby farms and residential blocks. The north-east corner of the catchment abuts the township of Whittlesea (population 2 500). Much of the catchment is enclosed by a 1.8 m "Cyclone" chain-mesh security fence topped with three strands of barbed wire. Public roads run beside this fence along the northern, eastern and southern boundaries of the catchment, and along part of the western side (Fig. 1). The catchment is closed to the public, but there is an internal network of gravel roads maintained for management purposes.

Vegetation

The natural vegetation of Yan Yean was comprised of open eucalypt forest, but land use practices have modified the present distribution and composition of the flora. A number of prominent eucalypt associations occur in the catchment: Red Stringybark *Eucalyptus macrorhyncha*; Red Stringybark and Long-leaf Box *E. macrorhyncha-E. goniocalyx*; Red Stringybark and Candlebark *E. macrorhyncha-E. rubida*; Narrow-leaf Peppermint *E. radiata*; Swamp Gum *E. ovata*; Swamp Gum and White Sallee *E. ovata-E. pauciflora*; Yarra Gum *E. yarraensis*; and River Red Gum *E. camaldulensis* (Melbourne and Metropolitan Board of Works 1987). The majority of these associations form open forest on slopes and undulating lowlands, but pure stands of Swamp Gum and River Red Gum can be found on the lowest flats, forming open woodland.

Of a total of 310 plant species recorded within the catchment, 43 are natives trees and shrubs. The shrub layer, when present, is dominated by wattles (*Acacia implexa*, *A. mearnsii*, *A. melanoxylon* and *A. paradoxa*), Cassinias (*Cassinia aculeata*, *C. arcuata* and *C. longifolia*) and Cherry Ballart *Exocarpos cupressiformis*. The herbaceous flora is relatively rich (96 monocotyledon and 167 dicotyledon species). The ground flora is predominantly comprised of native grasses such as Weeping Grass *Microlaena stipoides*, Kangaroo Grass *Themeda triandra*, Wallaby Grass *Danthonia* spp. Spear Grass *Stipa* spp. and Tussock Grass *Poa* spp. The flora of the catchment also includes 66 exotic species, 19 of which are listed as serious invasive weeds in the Melbourne area (Carr *et al.* 1992). The invasive Sweet Vernal Grass *Anthoxanthum odoratum* is particularly widespread. A number of rare and restricted species are present within the catchment, such as Narrow-leaf New Holland Daisy *Vittadinia muelleri* and Clover Glycine *Glycine latrobeana*. Examples of restricted remnant communities are also present. A link between the abundance of invasive species and the decline of some rare species has been suggested (Scarlett 1983).

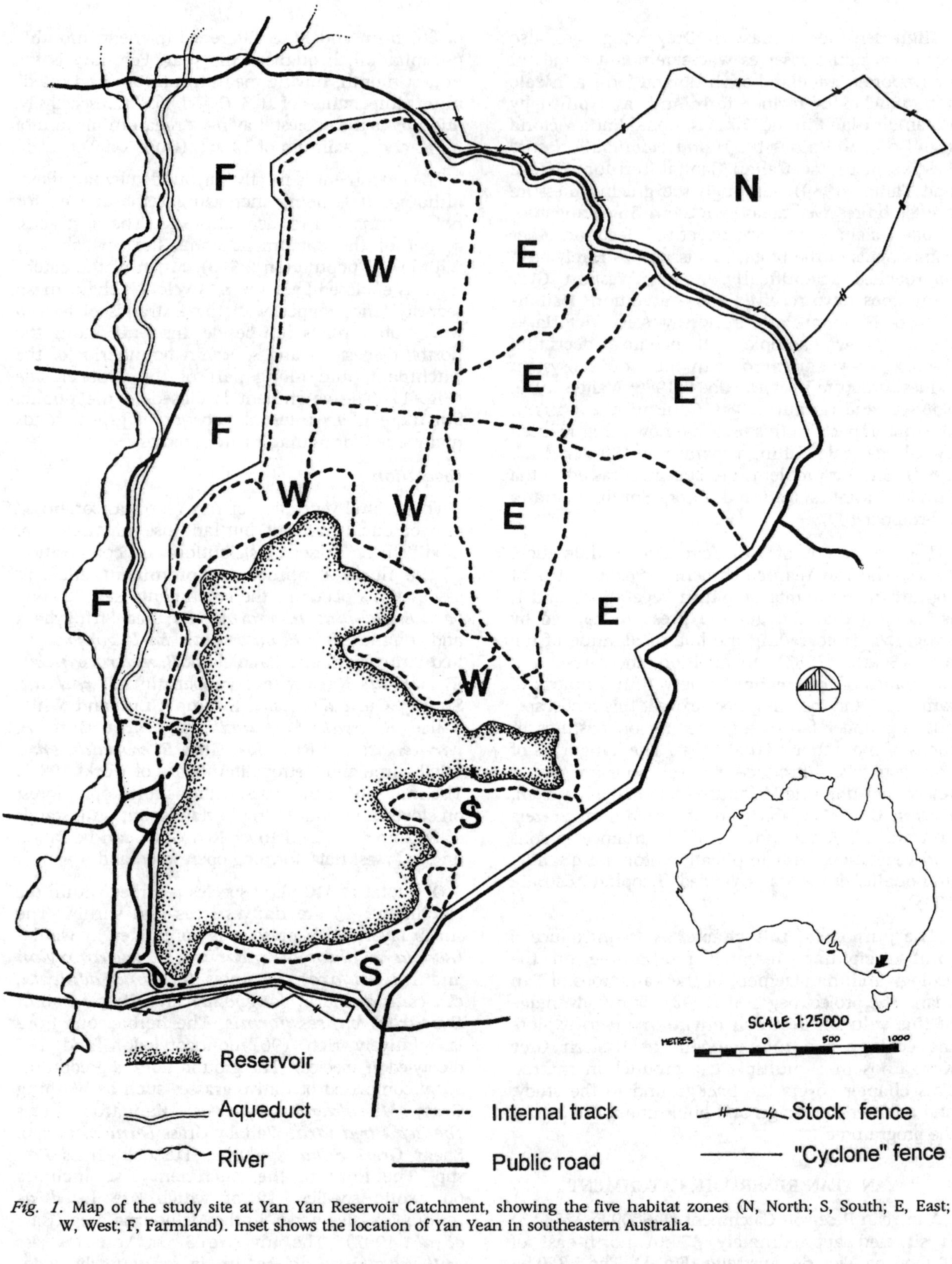

Fig. 1. Map of the study site at Yan Yan Reservoir Catchment, showing the five habitat zones (N, North; S, South; E, East; W, West; F, Farmland). Inset shows the location of Yan Yean in southeastern Australia.

Habitat zones

We have distinguished five zones of kangaroo habitat at Yan Yean, each reflecting the effects of differing land use. These zones are bounded by internal tracks, catchment fences, and other landmarks (Fig. 1). Three zones, East, West and South, form the major part of the catchment, which is surrounded by the "Cyclone" security fence. The North Zone has only a standard stock fence and is separated from East Zone by an unsealed public road. The Farmland Zone lies outside the catchment proper and provides additional habitat for kangaroos. It extends westwards from West Zone to the Plenty River and northwards to the outskirts of Whittlesea. The reservoir is also exploited at times by kangaroos, providing drinking water as well as a source of food around the margins when the water level falls. Water is also available in the other zones from permanent dams constructed for fire-fighting or domestic stock, and from several perennial creeks.

North Zone (345 ha). Most of the open forest in this steep, hilly zone is dominated by the Red Stringybark and Long-leaf Box association. The understorey varies between sparse *Cassinia* spp., *Acacia* spp. and a ground flora of *Poa* spp. on the ridges, to moderately dense undergrowth in the gullies dominated by Hop Goodenia *Goodenia ovata*. This zone is relatively undisturbed by human activity.

East Zone (710 ha). Two designated Botanical Reference Areas and their surrounding Management Zones (Melbourne and Metropolitan Board of Works 1987) comprise this zone. Yan Yean North Reference Area is representative of the remnant vegetation from low elevation open foothill forests which occur south of the Great Dividing Range, and is dominated by the Red Stringybark and Candlebark open forest association with an understorey of *Cassinia* spp. Yan Yean South Reference Area is comprised of Yarra Gum and Red Stringybark associations on the higher areas and Swamp Gum and White Sallee on the lower areas, with an understorey dominated by *Cassinia* spp. and *Acacia* spp. on the slopes, and dense patches of Austral Bracken *Pteridium esculentum* and Spiny-headed Mat-rush *Lomandra longifolia* along an intermittent stream. Anthropogenic disturbance of this zone has been relatively minor, as indicated by a low number of exotic species.

South Zone (145 ha). The vegetation of this zone has been heavily influenced by past land use, and contains a number of naturally regenerating plantations of Monterey Pine *Pinus radiata*, Blue Gum *E. globulus* ssp. *globulus* and Sugar Gum *E. cladocalyx* that have previously been harvested. It also contains a stand of a tree-form Silver Banksia *Banksia marginata*, which is unusual for this area.

West Zone (525 ha). A combination of past land use and continued kangaroo grazing has had a significant impact on the vegetation in this zone. The low-lying flats and surrounding rises are mainly covered by open woodlands of Swamp Gum and River Red Gum associations, and plantations of Monterey Pine and Sugar Gum provide the remaining tree cover. Cleared areas comprise approximately 20% of this zone and are most likely maintained by constant grazing pressure. Invasive exotic species dominate the understorey: Bulbil Watsonia *Watsonia meriana* cv. *bulbulifera*, Blackberry *Rubus* spp. and Sweet Briar *Rosa rubiginosa* are common, along with grasses such as Sweet Vernal Grass and Brown-top Bent *Agrostis capillaris*.

Farmland Zone (about 500 ha). This zone slopes down to the floodplain of the Plenty River. The managed pasture in this zone is dominated mainly by introduced grasses such as Lesser Quaking Grass *Briza minor*, Squirrel-tail Fescue *Vulpia bromoides* and Brown-top Bent (Moore 1995). Some native grasses persist, particularly Kangaroo Grass and Wallaby Grass, and some small stands of River Red Gum open woodland remain. Thickets of Hawthorn *Crataegus monogyna*, Blackberry and Sweet Briar are widespread.

Vertebrate fauna

A total of 37 mammal species have been recorded by the Atlas of Victorian Wildlife in a 5' × 5' grid cell centred on Yan Yean. Of these, 31 are native. Two native mammals are considered rare and are listed as threatened taxa (Schedule 2, *Flora and Fauna Guarantee Act 1988,* Victoria). These are a marsupial carnivore, the Brush-tailed Phascogale *Phascogale tapoatafa*, which has been recorded in North Zone, and a riparian bat, the Large-footed Myotis *Myotis adversus*, which has been recorded in the Farmland Zone. Herbivores other than Eastern Grey Kangaroos are uncommon: the native Common Wombat *Vombatus ursinus* and Swamp Wallaby *Wallabia bicolor*, and the introduced European Rabbit, *Oryctolagus cuniculus* occur at low densities in the catchment, and introduced Sambar Deer *Cervus unicolor* and Feral Pigs *Sus scrofa* are observed occasionally (Ecoplan Australia 1995).

A total of 155 bird species has been recorded by the Bird Observers Club of Australia since 1973 in annual surveys at Yan Yean, and only nine of these are introduced (G. Hosken, pers. comm.). The native species include several of conservation significance, such as the endangered Regent Honeyeater *Xanthomyza phrygia* and the rare White-bellied Sea-eagle *Haliaeetus leucogaster*. Sea-eagles have frequented Yan Yean since the 1960s, and a breeding pair nesting in a stand of Sugar Gum in the West Zone has been monitored closely in recent years (F. Muscolino, pers. comm).

Yan Yean also has a rich herpetofauna. Totals of 28 native species of reptiles and 18 native species of amphibians have been recorded in the Atlas of Victorian Wildlife for the 5' × 5' grid cell centred on Yan Yean. Of these, only the Glossy Grass Skink *Pseudemoia rawlinsoni* is of conservation significance as its status is insufficiently known.

HISTORY OF YAN YEAN

The Wurunjerri-baluk people inhabited the Plenty District when the reservoir site was covered mostly by swamp lands (Melbourne Parks and Waterways 1995). European pastoralists settled the area by the 1840s, grazing sheep and dairy cattle, growing crops of potoato and maize, and sowing pasture species (Edwards 1978). Timber was harvested in the catchment until the area was progressively reserved in the 1870s, cattle and horses were grazed on common land in part of the East Zone in the 1870s and 1980s, and orchards and a vineyard were cultivated in West Zone until 1890 (Payne 1975; Edwards 1978).

The last of the aboriginal people left the district when construction of the reservoir began in 1854 (Edwards 1978). This became the first reservoir supplying the rapidly-growing city of Melbourne. In 1872 the catchment was proclaimed a permanent reserve for water, managed by the Melbourne and Metropolitan Board of Works, and was proclaimed a native game protection area in 1917 (Melbourne and Metropolitan Board of Works 1987). Today, the catchment is managed by Melbourne Water, and Parks Victoria manages a small public park below the retaining wall in the southeastern corner of the study area (Melbourne Parks and Waterways 1995).

Kangaroo management

There has been a long history of concern about kangaroos from Yan Yean's neighbours (Melbourne and Metropolitan Board of Works 1982). From the early 1950s the Melbourne and Metropolitan Board of Works received complaints from neighbouring landholders adjacent to the catchment boundary, mainly from the Farmland Zone on the western side, but also from some properties along other boundaries. Most complaints were about crop damage, property damage and stock feed losses. One landholder was so affected by the kangaroos that he claimed damages from the Melbourne and Metropolitan Board of Works for financial losses incurred by kangaroos. The Melbourne and Metropolitan Board of Works denied responsibility for the kangaroos, arguing that they were wild animals living in the area long before the reservoir was built, and that the board was not responsible for wildlife on its land.

Neighbours claimed that the destruction permits issued by the Fisheries and Game Department to shoot kangaroos on their own properties were insufficient to deal with the problem, leading the Fisheries and Game Department to request the Melbourne and Metropolitan Board of Works to cull kangaroos within the catchment. Relocation was also suggested, but efforts by staff of Healesville Sanctuary and Melbourne Zoo in August 1952 resulted in the capture and removal of only six kangaroos from the population, which was estimated to be about 600 at the time. Although the Melbourne and Metropolitan Board of Works was reluctant to cull on the grounds of risks to water quality and public health, the caretaker of Yan Yean was permitted to shoot a total of nearly 600 kangaroos from 1954 to 1957. Despite more requests from neighbouring landholders, no further culling took place, although kangaroos were shot for scientific samples in the early 1960s as part of a research programme (see below). The Melbourne and Metropolitan Board of Works also considered a large-scale relocation programme to parts of the Plenty River Gorge to the south of Yan Yean, first in 1967, and again in 1981, but no action was taken.

Neighbours also made repeated requests to the Melbourne and Metropolitan Board of Works to erect kangaroo-proof fencing around the catchment (Melbourne and Metropolitan Board of Works 1982). Following complaints in September 1953 to the Prime Minister of the day, Mr Robert Menzies, an electrified 2-wire fence was erected a month later along the north-west boundary to protect a dairy property. This fence apparently kept the majority of the kangaroos within the catchment in that area at first, but its success was short-lived. By mid-1954 the Australian Primary Producers' Union was protesting to the Melbourne and Metropolitan Board of Works that the fence was being breached and there were more kangaroos than before. After years of such complaints, the Melbourne and Metropolitan Board of Works eventually offered to construct fencing if the cost was shared between the landholder and the board (Melbourne Water 1982). By 1981 the catchment fencing consisted of a variety of styles and heights: standard post and wire stock fence, 7-strand post and wire with barbed extension, 1.8 m sheep mesh and 1.8 m chain mesh. The Melbourne and Metropolitan Board of Works acknowledged that only 1.8 m chain mesh would contain the kangaroos, so by 1987 the entire catchment (excluding the North-East Zone) had been fenced with that configuration. However, this did not dispel all concerns. Some landholders claimed that kangaroos had been locked out of the catchment and were unable to leave their land after the fence was completed. Concerns were also expressed about the likelihood of increased mortality of the newly restricted population when forage became seasonally scarce, and the implications for water quality.

Previous research

In 1961, the Fisheries and Wildlife Department initiated a research programme into the population ecology of the kangaroos at Yan Yean, with particular emphasis on the pronounced die-off of young kangaroos (14–20 months) recorded during the winters of 1961, 1962 and 1963 (Dempster 1964). The size of the population in East and West Zones alone was estimated then to be 2 200–3 000 (Robertson 1975; Arundel *et al.* 1990). Quin (1989) subsequently analysed the demographic data collected by the Fisheries and Wildlife Department team in the 1960s. He found that fecundity was relatively low and concentrated in summer, and mortality was high, male-biased and concentrated in winter. Arundel *et al.* (1990) investigated the role of parasitic nematodes in the aetiology of the winter mortality pattern at Yan Yean in 1971 and 1972. They found that juveniles often carried heavy burdens of *Globocephaloides trifidospicularis*, *Rugopharynx australis*, *R. rosemariae*, *Pharyngostrongylus kappa*, *Macropostrongyloides baylisi*, *Paramacropostrongylus torlaiformis* and *Strongyloides* sp., particularly in winter, when pathological changes to the gut were evident and plasma protein, haemoglobin levels, and haematocrit values were low.

There has been less research into the vegetation of Yan Yean. The first floristic survey was conducted from 1981 to 1983 by Scarlett (1983). He carried out a reconnaissance of four sites within the catchment, identifying all vascular plant species present, and hypothesized that invasive weeds and kangaroo grazing were having deleterious effects on the native vegetation of the catchment. Melbourne and Metropolitan Board of Works staff followed up this work by traversing a 200 m grid over the East Zone to produce a distribution map of the major eucalypt associations within the two Yan Yean Reference Areas (Melbourne and Metropolitan Board of Works 1987).

CURRENT RESEARCH AT YAN YEAN

Population census

Previous surveys of the kangaroo population at Yan Yean have been infrequent and inconsistent in terms of the techniques used and the zones covered. In order to have a census regime that will allow seasonal and annual changes to be detected in the future, we have established 15 permanent, east-west transect lines, spaced 500 m apart, in the East, West, South and Farmland Zones. The first surveys were conducted by one of us (Natasha McLean) between October 1996 and February 1997, taking standard distance sampling measures with a hand-bearing compass and laser rangefinder during the day. We used the programme DISTANCE Version 2.0 (Laake *et al.* 1993), which has been shown to generate accurate and precise estimates of kangaroo density (Southwell 1994; Anderson and Southwell 1995). We estimated density for clusters of animals, using the Uniform, Half-normal and Hazard models (Buckland *et al.* 1993). The mean monthly estimate of total population size was 2.20 ha^{-1}. Importantly, the population was not distributed evenly across the study site. As an illustration, the estimated densities in November 1996 were: 1.90 ha^{-1} in East Zone, 1.67 ha^{-1} in West Zone, only 0.77 ha^{-1} in South Zone and as high as 2.49 ha^{-1} during the day in Farmland Zone.

Demography and mobility

We examined seasonality of breeding by observation of the reproductive status of the kangaroos. Surveys were conducted every second month from July 1993 to March 1994, beginning at first light. We made observations from a vehicle along roads in the East, West and Farmland Zones, using a window-mounted telescope, and clasified adult females according to Jaremovic and Croft's (1991) reproductive categories. The results conformed to the moderately seasonal pattern of breeding reported by Quin (1989) for samples collected in the 1960s at Yan Yean (see *Previous research*). In July, 80% adult females (n = 221) were carrying "medium" pouch young, and by November, 80% of adult females (n = 86) had either young-at-foot (42%) or had no young visible (38%). However, as Quin (1989) found, there were some females in each reproductive category throughout the year.

The structure of the adult population has been examined by routine collection of skulls of kangaroos that have died in the study area. Over 250 skulls have been collected so far. The skulls are aged by molar progression (Kirkpatrick 1967) and are sexed by reference to a set of skulls of known sex that was collected in the 1960s by Dempster and colleagues. Preliminary findings indicate that mean annual mortality after weaning is approximately 17%, mean age at death for adults is about 10 years, and longevity occasionally exceeds 20 years.

To examine patterns of mobility and habitat use, we have captured and tagged a sample of 79 kangaroos since 1992. Over half (39) of the sample were adult females at the time of capture, and the remainder were mainly sub-adults (10 females, 27 males). Eight adult females have been captured in East and West Zones using a dart gun; all other kangaroos have been captured in a draw-string trap (Coulson 1996a) as they were moving under the security fence on the boundary between West Zone and Farmland Zone. All kangaroos have been marked with paired colour combinations of reflective eartags. Many (36) have also been fitted with radio collars, and some with brightly-coloured collars for visual identification. We have accumulated over 2 500 sightings and radio fixes of these individuals since the project began.

Radio-tracking has shown that adult and sub-adult females are quite sedentary. Home ranges estimates for adult females, using 95% kernels (Worton 1989), are from 27 to 160 ha. Significantly, none of the adult females that were darted in East and West Zone ever moved out of the fenced catchment, whereas those caught at the fence continued regular movements between West Zone and Farmland Zone. Movements of young males were more variable: some were sedentary, others dispersed from the capture sites on the fence to East Zone and South Zone, and one has moved eastwards out of the catchment and established a home range in wooded farmland about 15 km east of the catchment.

Plant-animal interactions

We have begun an investigation of the floristic variation within the catchment, the effect of grazing intensity on the species composition and diversity of the flora, the threat of continued grazing to the conservation of the native flora, and the prediction of the effects of grazing intensity given changes to the management regime of the catchment. One of us (Dan Ramp) conducted a floristic survey of the North, South, East and West Zones to investigate the relative importance of environmental and exogenous disturbance variables in determining the spatial distribution and species composition of plants. A stratified-random procedure was used to locate 40 quadrats, at which seven site variables and all vascular plants were surveyed. Initial findings suggest that, while altitude and soil moisture significantly influence species composition on sites without a history of major anthropogenic disturbance, these variables are overridden by the impact of disturbance, as has been reported elsewhere in native Australian grasslands and woodlands (Leigh *et al.* 1989; McIntyre and Lavorel 1994; Tremont and McIntyre 1994; Pettit *et al.* 1995). The major disturbance variable at Yan Yean was the index of kangaroo grazing. At other sites, kangaroo grazing has been found to have significant impacts on ground flora by removal of annual and perennial grass biomass and by reducing species diversity through localized extinction (Robertson 1988; Leigh *et al.* 1989; Norbury *et al.* 1993; Tremont and McIntyre 1994).

We will also determine the extent of grazing pressure on the flora of Yan Yean through eight permanent exclusion plots, located in different zones. Grazing exclosures enable the biomass of vegetative matter being removed from each habitat to be quantified, as well as providing information on the recovery of plant species in the absence of grazing. The vegetation will be assessed (by Dan Ramp) on a seasonal basis and will be compared with adjacent control plots. Above-ground plant biomass will be estimated by a combination of the comparative yield and dry-weight-rank techniques (Jones and Hargreaves 1979; Friedel and Bastin 1988), which have been used to good effect in the arid rangelands of Australia (Norbury 1987; Catchpole and Wheeler 1992). Changes in species composition will be monitored by estimating the cover of all vascular plants. In addition, 39 small, portable exclosures have been established to determine the short-term offtake from different patches within the catchment. In conjunction with measurements of an index of kangaroo density at these points, predictions of the spatial distributions of kangaroos in relation to the distribution of available resources will be possible.

Water quality

Water quality is closely monitored by Melbourne Water because the catchment has a number of inherent problems. Clay soils found in much of the catchment are easily transported in the water column, resulting in increased turbidity, and the reservoir is shallow (no more than 9 m at maximum capacity), so wave action generated by windy conditions creates turbulence, remobilizing soil fines that had settled on the bottom. Consumer dissatisfaction with the colour and lack of clarity of the water led to the establishment of a water treatment plant on site in 1994. Light is also able to penetrate through most of the water column, providing conditions conducive to growth of micro-organisms such as *Dinobryon* spp. and *Synura* spp., which impart strange odours and tastes to the water, and *Aulacoseira* spp., *Synedra* spp. and *Rhizosloenia* spp., which are capable of causing blockages in the filtration systems (Bowles 1990). Despite the presence of these organisms, sometimes in high numbers, the reservoir has been taken off-line only in 1928, 1929 and more recently for three weeks in 1995. The closure each time was due to blooms of *Anabaena circinalis*, which can produce compounds that are toxic to humans (Water Ecoscience 1996).

The population of kangaroos in the catchment is thought to be another threat to the supply. Heavy grazing pressure may increase erosion and consequently have an adverse impact on water quality parameters such as turbidity and nutrient levels. Soil is compacted in a mechanical manner by animal movement, producing dust that blocks soil pores and is more susceptible to erosion, creating a harder, impermeable soil crust that allows less rainfall infiltration, resulting in greater surface flows and a greater chance for soil entrainment (Noble and Tongway 1986). Feeding activity also has indirect effects on the soil layer as vegetation is removed and the soil is exposed to the effects of rainsplash (Noble and Tongway 1986). Nutrients may also be concentrated and mobilized in faecal pellets deposited on the surface.

We are investigating some of these processes by measuring the effects of grazing by kangaroos at Yan Yean. One of us (Phil Alviano) has established 12 runoff plots, half of them fenced to exclude kangaroos, on moderate slopes at three heavily-grazed sites in West Zone. The plots have been in place for three years, the last of which was a dry El Niño year. After two years, above-ground biomass was six times higher in the ungrazed plots than the grazed plots. Structure and species composition also differs markedly between the two treatments: ungrazed plots are now dominated by Kangaroo Grass *Themeda triandra* tussocks, whereas grazed plots have a low sward dominated by Wallaby Grass *Danthonia* spp. Trends in the yield of water, solids and nutrients from open and closed plots have been in the expected direction, but have not reached statistical significance because of variation in the frequency and intensity of rainfall events. To eliminate these sources of error, we applied artificial rainfall to the plots (via sprinklers) at the end of the three-year period. There was a striking difference between treatments in water yield: the volume of water required to produced runoff was three times greater on ungrazed than on grazed plots. Runoff from ungrazed plots also had a higher concentration of nitrates and nitrites, but there was no difference in suspended sediment, dissolved solids or phosphates.

Impact on farmland

Despite the long history of complaints about kangaroos by landholders in the surrounding rural matrix, there has been no attempt to assess, in an objective way, what pressure is being exerted on the neighbouring properties. In 1998 we began a programme to measure kangaroo activity in the Farmland Zone. Initially, one of us (Vural Yazgin) surveyed the 19 km security fence around East, West and South Zones to determine the frequency and apparent level of use of holes that have been formed at the base of the fence by kangaroos and, to a lesser extent, by wombats and other species. We found 620 holes in total, equivalent to one every 30 m of fenceline. Of these, 60% were classified as actively used, and another 24% had been blocked recently by Melbourne Water staff or by neighbouring landholders. The highest intensity of use was along the fence between West Zone and Farmland Zone. As an illustration of the extent of this movement, we observed 109 kangaroos going through one well-used hole in 3 h one evening in March 1998.

We began seasonal monitoring of the number of kangaroos in the Farmland Zone in 1998. One of us (Vural Yazgin) surveys the paddocks from tracks and vantage points for at least 20 consecutive evenings each quarter. Preliminary results indicate that there are often high densities of kangaroos on these properties. In June 1998, for example, we recorded a mean count of 497 kangaroos (about 1 per ha), together with 209 head of cattle, 50 horses and six sheep. To quantify the extent of competition between kangaroos and domestic stock, we have designed and constructed short-term differential exclosures that will allow us to measure offtake by kangaroos and cattle separately. These have recently been set up in high biomass pasture that is now being grazed by both species, but results are not yet available.

Landholders' perceptions

In 1995 we conducted a survey of the landholders to investigate their perceptions of kangaroo issues. Each neighbour received a letter outlining the purpose of the survey, and was then contacted by telephone to arrange a time for an interview. All 16 neighbours who could be contacted agreed to be interviewed. Five undergraduate environmental science students conducted the structured interviews, using a standard questionnaire, in the homes of the respondents. The questionnaire comprised items of a descriptive nature about the properties, and a series of questions about numbers and origins of kangaroos on their properties, and the positive and negative aspects of the kangaroos as perceived by the landholder.

The properties range in size from 1 to 60 ha, and the five properties on the eastern boundary of the catchment are twice the size of the 11 on the western boundary (i.e., Farmland Zone). Neighbours on the eastern side have lived on their properties for a mean of 26 years, while those in the Farmland Zone had been there for only 10 years on average, reflecting the mainly residential nature of the western properties. Most properties carry beef cattle, some have sheep or horses, and one also has ostriches.

All but two of the neighbours in the Farmland Zone reported kangaroos on their properties; they estimated that they had 28 ± 33 (mean ± s.d.) kangaroos on their land, almost six times the number reported on the eastern side. All but one of the landholders with kangaroos believed that the kangaroos on their property came from the catchment, but were evenly divided on whether the numbers of kangaroos had increased or decreased in the time the neighbours had lived on the property.

Many (41%) of the neighbours had experienced no problems with kangaroos, except for creating a hazard while driving, especially at night. Six reported collisions with kangaroos. Those who experienced other problems cited mainly damage to their fences and crops, mostly in the Farmland Zone. One neighbour reported that a kangaroo had killed one of his dogs. The ability of large male kangaroos to drown dogs that pursue them into water is well known (Coulson 1996b), and dead dogs are removed from the reservoir by the caretakers occasionally (R. Dunn, pers. comm.). Surprisingly, four neighbours also claimed that they had been attacked by kangaroos.

The neighbours' attitudes to the kangaroos varied widely. Over half of the landholders said that they would not accept any kangaroos on their properties, but only one had considered shooting to reduce kangaroo numbers. Others saw aesthetic and ecocentric benefits in having kangaroos. None of the neighbours on the eastern side wanted to have the kangaroos confined to the catchment. On the western side, where kangaroos make extensive use of the Farmland Zone, half of the neighbours preferred to have kangaroos only in the catchment.

INTERIM CONCLUSIONS

Eastern Grey Kangaroos are widespread and common, and are protected throughout their range. At Yan Yean, the species poses a number of problems for the management of biodiversity within a closed water catchment that is, in turn, embedded in a matrix that is devoted mainly to agricultural production.

Kangaroos are rarely disturbed by humans or dogs within the fenced catchment of Yan Yean, and have probably enjoyed these conditions for decades. Population density is high, particularly in the zones that have been most altered by earlier forestry and agricultural activity so that they now provide open grassy patches. Fecundity is also high, but the rate of recruitment into the adult population is low as would be expected in a population that is apparently limited by intense intraspecific competition for forage. The consequent grazing pressure is evident throughout the catchment, and emerges as the dominant disturbance variable influencing the distribution and composition of the vegetation. The long-term impact of grazing disturbance on the high-value native vegetation reserved within the two reference areas in East Zone is yet to be determined. The many grassy slopes in West Zone, which are probably now maintained by grazing, yield runoff more rapidly after rainfall events than do sites where grazing is excluded, so kangaroo grazing is likely to be beneficial in terms of the volume of water delivered to the reservoir. Paradoxically, the quality of the water harvested from this catchment is likely to be poorer because kangaroo grazing mobilizes some nutrients, and probably also solids, leading to higher treatment costs and possible interruptions to supply.

We suspect that the population density of kangaroos within the catchment is elevated by the proximity to high quality pasture in the surrounding agricultural matrix. Although a proportion of the kangaroos remain within the catchment, many others move in and out frequently, often remaining in the Farmland Zone during the day unless disturbed. Construction of the security fence has not stopped these movements. The fence was poorly designed, being relatively easy to lift from the base, and has not been adequately maintained by Melbourne Water or neighbouring landholders. We have yet to determine the impact of these large numbers of kangaroos feeding in the Farmland Zone farmland, but some landholders believe that kangaroos cause significant production losses at times. However, attitudes towards kangaroos, and to population control, varies widely among neighbours. Two neighbours obtained permits to shoot kangaroos on their properties during an extended dry period in 1998, but many others do not favour, or do not bother with lethal control.

In the next phase of the research programme, we intend to develop a model, based on ideal free theory (see Tregenza 1995), to predict effects of specific management options on the dispersion of the kangaroos and their impact on vegetation at Yan Yean. Some options for control are localized culling (e.g., McNulty *et al.* 1997), electric fencing (e.g., Statham 1994) and fertility control (Tyndale-Biscoe 1991). The most promising options to emerge from the model will then be tested in field experiments, and will be evaluated in terms of ecological, economic and social outcomes in all components of the complex matrix occupied by the kangaroos of Yan Yean.

ACKNOWLEDGEMENTS

This work began as a collaborative research project with CSIRO Division of Wildlife and Ecology, funded by the Department of Conservation and Natural Resources, Victoria, to investigate benign methods for the control of free-ranging populations of grey kangaroos in southern Australia. Melbourne Water allowed us to conduct the study in the catchment and, together with Melbourne Parks and Waterways, provided funding for much of Phil Alviano's work. The Holsworth Foundation has supported Dan Ramp's component of the project. Animal capture and tagging has been conducted with the permission of the Department of Conservation and Natural Resources (permits No. RP-94-185, RP-95-150 and RP-97-136), and with the approval of the University of Melbourne Animal Experimentation Ethics Committee (projects No. 486 004 0 92 1157, 654 125 0 94 1617 and 654 158 0 95 1763).

Many individuals and organizations have assisted us. The caretakers at Yan Yean, Russell Dunn and Norm Pratt, have tolerated our comings and goings at all times of day and night, kept us informed, and extracted us from sticky situations. The research assistants, Jane McKenzie and Sally Troy, did much of the unrewarding preparatory work, particularly animal capture, radio-tracking and population surveys. David Middleton and Bryan Walters of Ecoplan Australia devoted many hours to development of capture methods and discussion of management issues. David Morgan gave us access to his early transect data and gave guidance with analysis. Juliet Bird and Peter Minchin have provided critical intellectual input throughout. Stuart Judd of Parks Victoria, and Carsten Berberich, Mike Chapman, Shane Hayden, Frank Lawless and John Riddiford of Melbourne Water, have assisted in a variety of ways. Graeme Hosken, of the Bird Observers Club of Australia generously provided the Challenge Count data.

Four Zoology Honours students, Rochelle Egan, Penny Fannin, Ben Moore and Miranda Wills, contributed ideas, data and hard labour, as did our visiting scholars from the Free University of Berlin, Petra Kretschmar and Andreas Gumbert. Five third-year B. Ed. (Environmental Science) students, Tania Douketis, Christo Ferguson, Simone Marshall, Dominico Scardamaglia and Ashley Wood, arranged and conducted interviews with owners of neighbouring properties. The neighbours, particularly the Barringtons, Boyds, Childs, Lawsons and Wilkinsons, have allowed us to wander all over their land and have tolerated decorated kangaroos in their paddocks. We finally thank the many friends, relatives, students, colleagues and other volunteers for help in the field and elsewhere, particularly Helena Bender, Catja Bier, Ruth Coulson, Phillip Fedele, Susannah Laird, Grainne Maguire, Jenny Martin and Teresa Mitakakis.

REFERENCES

Alexander, P., 1997. Kangaroo culling, harvesting and farming in South Australia —an ecological approach. *Aust. Biol.* **10:** 23–29.

Anderson, D. R. and Southwell, C., 1995. Estimates of macropod density from line transect surveys relative to analyst expertise. *J. Wildl. Manage.* **59:** 852–57.

Arnold, G. W. and Maller, R. A., 1987. Monitoring population densities of western grey kangaroos in remnants of native vegetation. Pp. 219–25 *in* Nature Conservation: the Role of Remnants of Native Vegetation ed by D. A. Saunders, G. W. Arnold, A. A. Burbidge and A. J. M. Hopkins. Surrey Beatty & Sons: Chipping Norton, New South Wales.

Arnold, G. W. and Steven, D. E., 1988. Variations in distribution of western grey kangaroos, *Macropus fuliginosus ocydromus*, in the Tutanning Nature Reserve and their impact on adjacent farmland. *Aust. Wildl. Res.* **15:** 119–28.

Arnold, G. W., Steven, D. E. and Weeldenburg, J. R., 1989. The use of surrounding farmland by western grey kangaroos living in a remnant of wandoo woodland and their impact on crop production. *Aust. Wildl. Res.* **16:** 85–93.

Arnold, G. W., Steven, D. E., Grassia, A. and Weeldenburg, J., 1992. Home-range size and fidelity of western grey kangaroos (*Macropus fuliginosus*) living in remnants of wandoo woodland and adjacent farmland. *Wildl. Res.* **19:** 137–43.

Arnold, G. W., Weeldenburg, J. R. and Steven, D. E., 1991. Distribution and abundance of two species of kangaroo in remnants of native vegetation in the central wheatbelt of Western Australia and the role of native vegetation along road verges and fencelines as linkages. Pp. 273–80 *in* Nature Conservation 2: the Role of Corridors ed by D. A. Saunders and R. J. Hobbs. Surrey Beatty & Sons: Chipping Norton, New South Wales.

Arundel, J. H., Dempster, K. J., Harrigan, K. E. and Black, R., 1990. Epidemiological observations on the helminth parasites of *Macropus giganteus* Shaw in Victoria. *Aust. Wildl. Res.* **17:** 39–51.

Bell, H. M., 1973. The ecology of three macropod marsupial species in an area of open forest and savannah woodland in North Queensland, Australia. *Mammalia* **37:** 527–44.

Bowles, B. A., 1990. Report on the biology of Yan Yean Reservoir. Unpublished report to Melbourne and Metropolitan Board of Works.

Buckland, S. T., Anderson, D. R., Burnham, K. P. and Laake, J. L., 1993. Distance sampling. Estimating abundance of biological populations. Chapman and Hall: London.

Carr, G. W., Yugovic, J. V. and Robinson, K. E., 1992. Environmental weed invasions in Victoria: conservation and management implications. Department of Conservation and Environment and Ecological Horticulture Pty. Ltd.: Melbourne, Victoria.

Catchpole, W. R. and Wheeler, C. J., 1992. Estimating plant biomass: a review of techniques. *Aust. J. Ecol.* **17:** 121–31.

Caughley, G., 1964. Density and dispersion of two species of kangaroo in relation to habitat. *Aust. J. Zool.* **12:** 238–49.

Caughley, G., Shepherd, N. and Short, J., (eds), 1987. Kangaroos: their Ecology and Management in the Sheep Rangelands of Australia. Cambridge University Press: Cambridge. 253 Pp.

Cheal, D., 1986. A park with a kangaroo problem. *Oryx* **20:** 95–99.

Coulson, G., 1996a. A safe and selective draw-string trap to capture kangaroos moving under fences. *Wildl. Res.* **23:** 621–27.

Coulson, G., 1996b. Anti-predator behaviour in marsupials. Pp. 158–86 *in* Comparison of Marsupial and Placental Behaviour ed by D. B. Croft and U. Ganslosser. Filander Verlag: Fürth.

Coulson, G. M. and Raines, J. A., 1985. Methods for small-scale surveys of grey kangaroo populations. *Aust. Wildl. Res.* **12:** 119–25.

Coulson, G. and Norbury, G., 1988. Ecology and management of Western Grey Kangaroos, *Macropus fuliginosus*, at Hattah-Kulkyne National Park. Arthur Rylah Institute for Environmental Research Technical Report No. 72, Department of Conservation, Forests and Lands, Victoria.

Coulson, G., Norbury, G. and Walters, B., 1990. Forage biomass and kangaroo populations in summer and autumn at Hattah-Kulkyne National Park. *Aust. Mammal.* **13:** 219–21.

Dempster, J. K., 1961. Kangaroo management in Victorian agricultural areas. *J. Agric.* **59:** 45–52.

Dempster, J. K., 1964. Control and conservation of kangaroos in Victoria. Pp. E 1–2 *in* Australian Vermin Control Conference 1964. CSIRO: Melbourne, Victoria.

Ecoplan Australia Pty. Ltd., 1995. Macropod and rabbit study report: Yan Yean Catchment and Reservoir Park, Plenty Gorge Park, Cardinia Catchment and Reservoir Park and Sugarloaf Catchment and Reservoir Park. Report for Melbourne Water and Melbourne Parks and Waterways, Melbourne, Victoria.

Edwards, D., 1978. Yan Yean: a history. Yan Yean School Council, Yan Yean, Victoria. 109 Pp.

Friedel, M. H. and Bastin, G. N., 1988. Photographic standards for estimating comparative yield in arid rangelands. *Aust. Rangel. J.* **10:** 34–38.

Geological Survey of Victoria, 1972. Yan Yean Mapsheet. Part 7922, Zone 55, Australia, 1: 63, 360.

Hill, G. J. E., 1981. A study of habitat preferences in the grey kangaroo. *Aust. Wildl. Res.* **8:** 245–54.

Hill, G. J. E., Barnes, A. and Wilson, G. R., 1988. The use of wheat crops by grey kangaroos, *Macropus giganteus*, in southern Queensland. *Aust. Wildl. Res.* **15:** 111–17.

Jaremovic, R. V. and Croft, D. B., 1991. Social organisation of the eastern grey kangaroo (Macropodidae, Marsupialia) in southeastern New South Wales. I. Groups and group home ranges. *Mammalia* **55:** 169–85.

Jeffery, P. J., 1981. A study of the land in the catchments to the north of Melbourne. Soil Conservation Authority, Melbourne. 70 Pp.

Jones, R. M. and Hargreaves, J. N. G., 1979. Improvements to the dry-weight-rank method for measuring botanical composition. *Grass and Forage Sci.* **34:** 181–89.

Kaufmann, J. H., 1975. Field observations of the social behaviour of the eastern grey kangaroo, *Macropus giganteus*. *Anim. Behav.* **23:** 214–21.

Kirkpatrick, T. H., 1967. Studies of the Macropodidae in Queensland. 6. Sex determination of adult skulls of the grey kangaroo and the red kangaroo. *Qld J. Agric. Anim. Sci.* **24:** 131–33.

Laake, J. L., Buckland, S. T., Anderson, D. R. and Burnham, K. P., 1993. DISTANCE user's guide v2.0. Colorado Cooperative Fish and Wildlife Research Unit, Colorado State University, Fort Collins, Colorado. 72 Pp.

Lavery, H. J. and Kirkpatrick, T. H., 1985. Kangaroos as pests. Pp. 103–32 *in* The Kangaroo Keepers ed by H. J. Lavery. University of Queensland Press: St Lucia, Queensland.

Leigh, J. H., Wood, D. H., Holgate, M. D., Slee, A. and Stanger, M. G., 1989. Effects of rabbit and kangaroo grazing on two semi-arid grassland communities in central-western New South Wales (Australia). *Aust. J. Bot.* **37:** 375–96.

McIntyre, S. and Lavorel, S., 1994. How environmental and disturbance factors influence species composition in temperature Australian grasslands. *J. Veget. Sci.* **5:** 373–84.

McNulty, S. A., Porter, W. F., Mathews, N. E. and Hill, J. A., 1997. Localized management for reducing white-tailed deer populations. *Wildl. Soc. Bull.* **25:** 265–71.

Melbourne Parks and Waterways, 1995. Yan Yean Reservoir Park Setting plan. Planning Services and Planning and Environment Division, Melbourne Parks and Waterways, Melbourne, Victoria.

Melbourne Water, 1982. Kangaroo problem: Yan Yean. Resume of File 57/9654. Operations Division, Watersheds Department, Melbourne Water, Melbourne, Victoria.

Melbourne and Metropolitan Board of Works, 1987. Yan Yean reference areas management plan. Forestry and Land Management Section, Melbourne and Metropolitan Board of Works, Melbourne, Victoria.

Melbourne and Metropolitan Board of Works, 1988. Water Supply System (map). Melbourne and Metropolitan Board of Works, Melbourne, Victoria.

Melbourne and Metropolitan Board of Works, 1989. Water for Melbourne (brochure). Melbourne and Metropolitan Board of Works, Melbourne, Victoria.

Neave, H. M. and Tanton, M. T., 1989. The effects of grazing by kangaroos and rabbits on the vegetation and the habitat of other fauna in the Tidbinbilla Nature Reserve, Australian Capital Territory. *Aust. Wildl. Res.* **16:** 337–51.

Noble, J. C. and Tongway, D. J., 1986. Herbivores in arid and semi-arid rangelands. Pp. 243–70 *in* Australian Soils — The Human Impact ed by J. S. Russell and R. F. Isbell. University of Queensland Press: St. Lucia, Queensland.

Norbury, G. L., 1987. Diet selection and demography of the western grey kangaroo *Macropus fuliginosus melanops* Desmarest in Hattah-Kulkyne National Park Victoria. Ph.D. thesis, Monash University, Australia.

Norbury, G. L., Norbury, D. C. and Hacker, R. B., 1993. Impact of Red Kangaroos on the pasture layer in the Western Australian arid zone. *Rangel. J.* **15:** 12–23.

Payne, J. W., 1975. The Plenty: A Centenary History of the Whittlesea Shire. Lowden Publishing: Kilmore, Victoria. 234 Pp.

Pettit, N. E., Froend, R. H. and Ladd, P. G., 1995. Grazing in remnant woodland vegetation: changes in species composition and life form groups. *J. Veget. Sci.* **6:** 121–30.

Poole, W. E., 1982. Macropus giganteus. *Mammalian Species* **No. 187:** 1–8.

Quin, D. G., 1989. Age structures, reproduction and mortality of the Eastern Grey Kangaroo (*Macropus giganteus* Shaw) from Yan Yean, Victoria. Pp. 787–94 *in* Kangaroos, Wallabies and Rat-kangaroos ed by G. Grigg, P. Jarman and I. Hume. Surrey Beatty & Sons: Chipping Norton, New South Wales.

Robertson, D., 1985. Interrelationships between kangaroos, fire and vegetation dynamics at Gellibrand Hill Park, Victoria. Ph.D. thesis, University of Melbourne, Melbourne, Victoria.

Robertson, G., 1988. Effect of rainfall on biomass, growth and dieback of pastures in an arid grazing system. *Aust. J. Ecol.* **13:** 519–28.

Scarlett, N., 1983. Floristic survey of grassy open forests in the Yan Yean catchment area. Report to Melbourne and Metropolitan Board of Works. La Trobe University Botany Department, Bundoora, Victoria.

Sluiter, I. R. K., Allen, G. G., Morgan, D. G. and Walker, I. S., 1997. Vegetation responses to stratified kangaroo grazing pressure at Hattah-Kulkyne National Park, 1992–1996. Flora andFauna Technical Report No. 72, Department of Natural Resources and Environment, East Melbourne.

Southwell, C., 1987. Macropod studies at Wallaby Creek II. Density and distribution of macropod species in relation to environmental variables. *Aust. Wildl. Res.* **14:** 15–33.

Southwell, C., 1994. Evaluation of walked line transect counts for estimating macropod density. *J. Wildl. Manage.* **58:** 348–56.

Statham, M., 1994. Electric fencing for the control of wallaby movement. *Wildl. Res.* **21:** 697–707.

Taylor, R. J., 1984. Foraging in the eastern grey kangaroo and the wallaroo. *J. Anim. Ecol.* **53:** 65–74.

Taylor, R. J., 1985a. Habitat use by the eastern grey kangaroo and wallaroo in an area of sympatry. *Mammalia* **49:** 173–86.

Taylor, R. J., 1985b. Effects of pasture improvement on the nutrition of eastern grey kangaroos and wallaroos. *J. Appl. Ecol.* **22:** 717–25.

Tregenza, T., 1995. Building on the ideal free distribution. *Adv. Ecol. Res.* **26:** 253–307.

Tremont, R. M. and McIntyre, S., 1994. Natural grassy vegetation and native forbs in temperate Australia: structure, dynamics and life histories. *Aust. J. Bot.* **42:** 641–58.

Tyndale-Biscoe, C. H., 1991. Fertility control in wildlife. *Reprod. Fertil. Dev.* **3:** 339–43.

Walters, B., 1986. Restoring Hattah-Kulkyne National Park. *Trees and Nat. Res.* **28:** 2–7.

Water Ecoscience, 1996. Yan Yean Reservoir Management Plan. Unpublished report to Melbourne Water.

Wilson, A. D., 1991a. The influence of kangaroos and forage supply on sheep productivity in the semi-arid woodlands. *Rangel. J.* **13:** 69–80.

Wilson, A. D., 1991b. Forage utilization by sheep and kangaroos in a semi-arid woodland. *Rangel. J.* **13:** 81–90.

Worton, B. J., 1989. Kernel methods for estimating the utilization distribution in home-range studies. *Ecology* **70:** 164–68.

CHAPTER 16

Sand plain forest fragmentation and residential development, Invercargill City, New Zealand

David A. Norton[1]

The Otatara area of Invercargill City contains the best remaining examples of coastal totara *Podocarpus totara* and totara-matai *Prumnopitys taxifolia* sand plain forests in New Zealand and as such their conservation is considered to be a matter of national importance in terms of Section 6(c) of the *Resource Management Act 1991*. These forests have been substantially affected by fragmentation, which has occurred in two phases. Between 1865 and 1947, and perhaps until 1967, forest clearance resulted in the reduction of total forest area, with remnants forming discrete patches of forest separated mainly by agricultural land. However, since 1947 and especially since 1967, the boundaries of these major forest areas have not changed substantially. Instead fragmentation has occurred primarily through residential development resulting in the opening up of the forest (perforation) leading to a reduction in area occupied by forest. The initial reduction in extent and the subsequent reduction in occupancy have both resulted in continued reductions in total forest cover, with forest in 1996 covering only 14% of the area it covered in 1865 and 52% of the area covered in 1946. This chapter discusses the process of fragmentation in these forests and considers options for their long-term conservation given their location on the edge of an urban area.

INTRODUCTION

THE traditional concept of fragmentation is of a spatial phenomenon at the landscape scale, best illustrated where a large intact natural area such as a forest has been divided up into several smaller units or remnants (Harris 1984; Saunders *et al.* 1987, 1991; Forman 1995). However, fragmentation is not just a spatial phenomenon, nor is it restricted to any particular scale. Fragmentation is simply the disruption of habitat continuity and occurs across a range of scales independent of the total amount of vegetation remaining (Lord and Norton 1990). For any given area, the vegetation will be at one stage along a continuum of reduction from intact natural vegetation to completely non-natural (alien) vegetation. However, this remaining vegetation can be spatially arranged, or dispersed, in different ways.

The classical concept of fragmentation, where a large intact area is divided up into several smaller intact units (Harris 1984; Saunders *et al.* 1987, 1991) is one extreme type of dispersion. Here, the spatial scale of remnants is large compared to the scale of the physiognomically dominant plants and has been termed "geographical fragmentation" (Lord and Norton 1990). At the other extreme of dispersion, intact vegetation remnants are small, at their extreme occurring at the scale of the physiognomically dominant plants. Here, remnants can simply be individual plants or small groups of plants embedded in an alien matrix. This situation has been termed "structural fragmentation" (Lord and Norton 1990). A related form of fragmentation occurs when an area of intact vegetation is opened up from within (e.g., through logging) and has been called "perforation" (Forman 1995). Although it is convenient to recognize extreme forms of dispersion, the more normal situation is for a range of different scales of fragmentation to occur in the same landscape (McIntyre and Barrett 1992).

While most forest fragmentation is geographical as more and more forest is cleared to make way for agriculture and forestry (Saunders *et al.* 1987), residential dwellings can affect forests in different ways. While residential development often has similar effects on forests to agriculture (i.e., geographical fragmentation; Forman 1995), the rise in

[1]Conservation Research Group, School of Forestry, University of Canterbury, Private Bag 4800, Christchurch, New Zealand.

Pages 157–65 *in* NATURE CONSERVATION 5: NATURE CONSERVATION IN PRODUCTION ENVIRONMENTS: MANAGING THE MATRIX ed by J. L. Craig, N. Mitchell and D. A. Saunders. Surrey Beatty & Sons, 2000.

interest in "life-style" living has led to an increase in the creation of small clearings within otherwise intact forests as people choose to live "close to nature". This results in structural or perforation fragmentation as the forest is opened up from within (Fig. 1), even though the extent of forest may remain unchanged.

This type of fragmentation is similar to that occurring when logging creates clearings in otherwise intact forests (Forman 1995), but appears to have been little described in urban situations. Given the growing trend amongst urban dwellers to move out of urban centres to life-style properties in adjacent non-urban areas, it would seem likely that there will be increasing pressure on forests close to urban centres from structural or perforation fragmentation. This study assesses the significance of the Otatara sand dune forests in terms of Section 6(c) of the *Resource Management Act 1991,* documents the nature of fragmentation at Otatara, presents an estimate of the minimum viable area of forest, and outlines options to reduce future impacts of residential development.

STUDY AREA

Otatara, on the western side of the New River Estuary between the Oreti River and Invercargill airport, is located in the Southland Plains Ecological District, Makarewa Ecological Region, and lies within the boundaries of Invercargill City (46°27'S, 168°18'E). The topography is generally flat, comprising sand dune systems less than 6 000 years old. Individual dunes rarely rise more than 5 m above the surrounding sand plains which are themselves quite extensive. Soils are mapped as Otatara yellow-brown sands (Anon. 1968), which have developed under forest. These soils have higher organic matter and cation-exchange capacity than yellow-brown sands in drier areas of New Zealand and soil pH is typically around six.

Climate information is available from Invercargill airport (Anon. 1982, 1986) adjacent to Otatara. The prevailing wind is from the west and wind speeds are high, with on average 28% of days experiencing gusts ≥63 km hr^{-1} each year. Mean annual rainfall is 1 042 mm with rain falling on *c.* 45% of days. Mean annual temperature is 9.6°C, with a January mean daily maximum of 18.4°C and a July mean daily minimum of 0.8°C. Screen frosts occur on average 46.3 days per year and ground frosts 111.1 days per year. Sunshine is experienced *c.* 42% of the possible time in summer and *c.* 35% in winter.

No systematic evaluations of ecological patterns in remnant indigenous vegetation at Otatara have been undertaken although published and unpublished botanical reports and species lists are available for several individual remnants (Johnson 1986a,b, 1987, 1988; Allen *et al.* 1989). The following summary is based on Norton (1996) and synthesizes these individual reports as part of a more general evaluation of the ecological values in this area.

The forest vegetation is dominated by conifer trees in the family Podocarpaceae, especially totara (a complex of forms grading between *Podocarpus totara* and *P. hallii*) and matai *Prumnopitys taxifolia*, although other podocarps including rimu *Dacrydium cupressinum* and kahikatea *Dacrycarpus dacrydioides*, and angiosperm trees including pokaka *Elaeocarpus hookerianus*, five finger *Pseudopanax colensoi* var. *ternatus*, makomako *Aristotelia serrata*, lemonwood *Pittosporum eugenioides*, kohuhu *Pittosporum colensoi* and matipo *Myrsine australis* are also present and can be locally important. Manuka *Leptospermum scoparium* is dominant in areas that have been recently disturbed. Three main forest types can be recognized (Norton 1996), totara, totara-matai and mixed podocarp, depending on the relative dominance of the different podocarp species.

Fig. 1. Oblique aerial photo of part of Otatara showing location of residential dwellings within forest remnant.

Three main factors influence the distribution of forest types; dune age, topographical position, and disturbance history (Norton 1996). There is a change in forest composition from totara dominance on the youngest dunes to mixed podocarp forest on older dunes. Matai is often codominant with totara, but does not appear to be as common on younger dunes, while miro *Prumnopitys ferruginea* was only seen as seedlings on older dunes. Rimu and kahikatea are also most abundant in forest on older dunes. These changes in podocarp distribution are paralleled by changes in angiosperm tree distribution, with pokaka rare or absent from young dunes, but common on older dunes, while kamahi *Weinmannia racemosa* was only observed at two sites on what are probably the oldest Otatara dunes.

This vegetation pattern is, however, modified by topography and disturbance history. Marked vegetation differences occur between dunes and slacks. On younger dunes, wetland vegetation occurs in slacks while on older dunes totara and matai are dominant on the dunes, while a more diverse mixed podocarp-angiosperm forest with rimu, kahikatea and pokaka, as well as totara and matai, is present in slacks and on sand plain areas. Human disturbances have also had a major impact on the forests, especially fire and logging and where the forest has not been completely removed, totara regeneration has occurred after disturbance.

METHODS

Fragmentation parameters

Quantification of forest fragmentation patterns was based on analysis of 1946, 1966, 1982 and 1996 aerial photos, although coverage was not complete in all years. For this reason, a study area of 850 ha was identified for analysis that was common to all photos. This area was fully forested in the 1860s (Heale 1865). The original aerial photos varied in scale from 1:7 960 to 1:16 165, and were adjusted to a common scale of 1:7 960. Data from the aerial photos were then recorded using a 16 × 16 m grid square into a data base that noted if each grid square contained forest or non-forest (residential or rural). 16 × 16 m was chosen as small enough to depict a clearing for a house in the forest. The data set for each aerial photo was then analysed using the Habitat Analysis and Modelling System, version 1.0 (Roseberry and Hao 1995) computer programme to quantify fragmentation patterns.

Fragmentation parameters derived included the number of remnants, total remnant area, mean remnant area, mean remnant edge:area ratio, distance to nearest remnant, contagion and fractal dimension of remnants. The contagion index is a measure of fragment clumping while fractal dimension measures the shape of the fragment based on Euclidean geometry. Fractal dimensions of around 1 represent simple shapes such as circles and squares, while fractal dimensions nearer 2 represent highly convoluted shapes.

Minimum viable forest area

The concept of minimum viable area provides an estimate of the minimum area that is required to sustain a particular species or ecosystem type. However, estimation of minimum viable area for a forest ecosystem such as that at Otatara is very problematic as it requires a detailed knowledge of edge effects, current and potential disturbance regimes and the many biotic (e.g., pollination and seed dispersal) and abiotic (e.g., nutrient cycling) interactions that characterize the forest ecosystem.

A simplistic approach to estimating minimum viable area is to only take into account edge effects (Murcia 1995) and disturbance regimes (Pickett and Thomson 1978). The minimum area of forest required to ensure that windthrow (the predominant disturbance in these forests) and hence subsequent forest regeneration is occurring continuously can be estimated based on the mean age of trees at windthrow (age), the proportion of trees that are windthrown (Pw) and tree density.

The proportion of trees windthrown each year Pt (based on Norton 1989) is:

$$Pt = (1/age)Pw$$

The number of trees (N) necessary to ensure that windthrow and subsequent regeneration are occurring continuously can then be calculated as:

$$N = 1/Pt$$

Based on tree density, it is then possible to estimate the area of forest required to support this number of trees. Edge effects are then added to this to provide a preliminary estimate of the minimum viable forest area.

RESULTS

Fragmentation parameters

The total area of forest declined from 230 ha (*c.* 27% of that present in 1865) in 1947 to 120 ha (*c.* 14%) in 1996 (Table 1, Fig. 2). The overall rate of forest loss was 2.2 ha^{-1} yr, which is substantially lower than the 7.6 ha^{-1} yr lost between 1865 and 1946. During the 1946–1996 period, forest loss was lowest in the 1982–1996 period (1.5 ha^{-1} yr). The number of forest remnants also decreased over the study period, from 61 in 1946 to 23 in 1996, with the rate of loss being constant through this period. Although the total forest area declined, the

Table 1. Fragmentation parameters for remnant forest patches at Otatara, Invercargill City, in four different years.

Year	1946	1966	1982	1996
Total area (ha)	230	183	142	120
Number patches	61	49	35	23
Mean patch area (ha)	3.7	3.7	4.2	5.2
Mean dist. to nearest neighbour (m)	27	37	52	62
Mean fractal dimen.	1.218	1.225	1.218	1.233
Contagion	0.40	0.45	0.52	0.57
Edge:area ratio	267	335	310	362

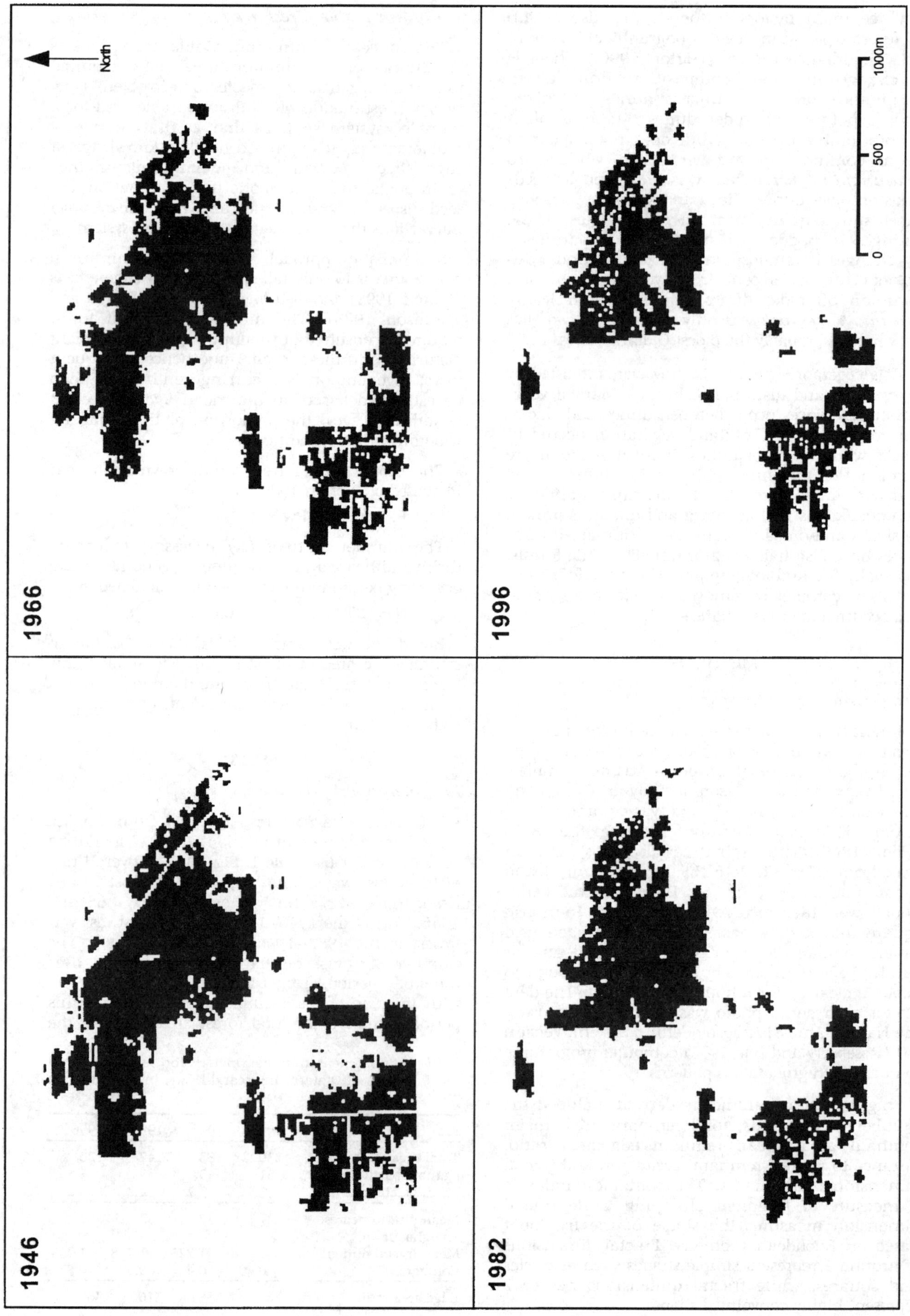

Fig. 2. Changing distribution of forest (black) and non-forest for part of the Otatara study area.

average size of remnants increased from 3.7 ha to 5.2 ha over the 1947–1996 period. As a result of the declining number of remnants, the mean distance between remnants increased from 27 m to 62 m over the 1947–1996 period (Table 1). Over the same time period contagion increased from 0.40 to 0.57, as remnants became more clumped. Overall remnant shape was fairly stable through the study period, with mean fractal dimension ranging from 1.22 to 1.23, suggesting that most remnants were fairly regular in shape (although this does not take into account forest loss within remnants). The edge:area ratio increased from 1947 to 1966, then declined slightly (1.5 m^{-1} ha^{-1} yr) to 1982, before increasing again to 1996 (Table 1). The increase from 1982 to 1996 was greater than from 1947 to 1966 (3.5 m^{-1} ha^{-1} yr cf. 3.2 m^{-1} ha^{-1} yr).

Minimum viable forest area

The proportion of trees windthrown was estimated as 0.0018 based on a mean age of a windthrown tree of 400 years (using totara; Rachel Ebbett, pers. comm. 1997) and the proportion of trees that are windthrown of 0.7 (based on data for terrace rimu forests in south Westland; Norton 1989). The number of trees necessary to ensure that windthrow and subsequent regeneration are occurring continuously was then estimated as 571. Assuming that mean tree densities are 625 trees per ha (based on a 4 m tree spacing), then an area of 0.91 ha of forest is required to allow sufficient trees for continuous forest turnover. If a 50 m edge zone (Young and Mitchell 1994) is allowed for around this, then the minimum viable forest area is 3.8 ha (195 × 195 m).

DISCUSSION

Significance of Otatara forests

Of all New Zealand's indigenous ecosystems, forested coastal dunes are among the least protected and most impacted upon by human settlement. Most of this impact has occurred during the last 150 years (Cockayne 1911; Park 1995; Geoff Park, pers. comm. 1996), but Maori society's focus on sandy coasts also led to substantial alteration prior to this. A circumnavigation of the New Zealand coastal zone focusing on its sand dune portions would reveal that the Otatara forests are, in this regard, nationally significant.

Sand dune systems are most expansive along the western coasts of Northland and the Manawatu, and in the north-west of South Island (Farewell Spit area). Elsewhere, the largest dunes are in the drier eastern areas of South Island, although small, local dune systems are present around all of New Zealand's coastlines including Stewart Island and offshore islands (Wardle 1991). While young, mobile and consolidating dunes would have been dominated by herbaceous and shrubland communities, extensive areas of forest would have occurred on stable and more sheltered dune systems and sand plains before human settlement (Cockayne 1928; Esler 1978). However, apart from the south-west of South Island (Dickinson and Mark 1994), the majority of these forests have now gone. For example, only 4% of the forest first seen by Europeans on the Manawatu dune systems remains (Esler 1978), while only one remnant remains of the extensive Horowhenua sand dune and sand plain forests (Park 1995). In the Hokitika Ecological District on South Island's West Coast, only 8% of the coastal landforms (including sand dunes) remain under indigenous vegetation (Awimbo *et al.* 1996). Given these dramatic reductions, it is not surprising that in a survey of New Zealand ecologists undertaken for the Department of Conservation, sand dune and sand plain ecosystems (including their forests) were identified as one of the most threatened ecosystem types in New Zealand (Geoff Park, pers. comm. 1996).

Given the extensive impact of human settlement on sand dune and sand plain forests, it is difficult to accurately reconstruct what former forest types were like. Northern sand dune and sand plain forests appear to have been dominated by diverse podocarp-angiosperm forests including species such as nikau *Rhopalostylis sapida*, titoki *Alectryon excelsus* and northern rata *Metrosideros robusta* (Park 1995), while in south Westland and on Stewart Island, sand dune forests are dominated by a less diverse podocarp-angiosperm forest dominated by rimu and kamahi (Wilson 1987; Wardle 1991; Dickinson and Mark 1994). In contrast, totara and matai dominated sand dune and sand plain forest remnants are present at some coastal sites along the south-east Otago coast (Johnson 1992), although podocarp-angiosperm forests more similar to those in south Westland are also present on dunes in this area (Smith *et al.* 1985). Esler (1978) suggests that totara may have been a component of the Manawatu sand dune forests, but the forest remnants are too fragmentary to accurately reconstruct the original forest pattern. Little is left of coastal forests in Canterbury, Marlborough or Nelson (Johnson 1992), or the eastern North Island (Partridge 1992). It may be that the totara and matai dominated forests of Otatara are more typical of the lower rainfall sand dune and sand plain systems of the eastern South Island and perhaps parts of the North Island than of higher rainfall western areas.

Several criteria have been proposed to determine "significance" of natural areas in relation to Section 6(c) of the *1991 Resource Management Act.* Whaley *et al.* (1995) suggest that "Significance, both in a conservation and biological/ecological sense, is a measure of an area's importance within the landscape; its representativeness, its value as a wildlife habitat, and its contribution to the biodiversity of the Ecological District, Ecological Region and New Zealand". The following points are important in considering the significance of the Otatara sand dune and sand plain forest remnants.

1. Nationally, sand dune and sand plain ecosystems have been amongst the most substantially impacted natural ecosystems by human activities

and most have now completely lost their indigenous vegetation cover, especially forest.

2. The Otatara area contains the best remaining example of coastal totara and totara-matai sand dune forests in New Zealand.
3. The Otatara area contains the only example of a sequence of totara, totara-matai and mixed podocarp forest remnants on different aged sand dune and sand plain surfaces.
4. Although much smaller in extent than they were historically, the Otatara forests are still of a sufficient size and still contain sufficient connectivity (including wetland-dryland connections) to ensure their future viability.
5. The Otatara forest remnants support populations of several endemic forest birds including tui *Prosthemadera novaeseelandiae novaeseelandiae*, bellbird *Anthornis melanura melanura*, fernbird *Bowdleria punctata* and kereru *Hemiphaga novaeseelandiae novaeseelandiae* that are strongly dependent on the forest remnants.
6. Although a large part of the Otatara forest is protected under the *Reserves Act* or *Conservation Act,* or through Queen Elizabeth II National Trust Covenants, substantial areas are also in private ownership. These areas are particularly important as they provide buffers to the protected areas, improve connectivity between different forest areas or provide important habitat in their own right. The loss of these areas could significantly compromise the overall viability of the Otatara-Sandy Point forest sequence.

Based on these considerations it would seem that the conservation of forest remnants in the Otatara area is a matter of national importance in terms of Section 6(c) of the *Resource Management Act.*

Fragmentation and its effects

Fragmentation at Otatara has occurred in two phases, an initial period that involved reduction in the extent of forest (especially up to 1967) and then a period of increasing opening up of the remaining remnants, without substantial changes in remnant extent. The earlier phase of fragmentation appears to have been driven primarily by land clearance for agriculture, while the more recent phase has been driven mainly by "life-style" residential development within the forest. The changing nature of forest fragmentation is shown clearly by the parameters that describe fragmentation (Table 1). While the total area of forest and the number of remnants have declined, mean remnant area has increased as the smallest remnants have been lost. Despite this increase, the edge:area ratio has also increased. This increase in the amount of edge relative to area reflects the increase in the number of openings within the remnants (Figs 1 and 2).

The earlier phase of fragmentation is similar to that documented widely elsewhere (Harris 1984; Saunders *et al.* 1987, 1991; Forman 1995) and is equivalent to what Lord and Norton (1990) called geographical fragmentation. However, the opening up of the remnants is a less well understood process. Lord and Norton (1990) used the term structural fragmentation to refer to an extreme form of dispersion of remaining habitat where remnants where at the same scale as the dominant plants (e.g., as might occur when every other tree is selectively removed from a forest); clearly fragmentation has not progressed this far at Otatara. However, the term perforation (Forman 1995) better describes the nature of fragmentation at Otatara where openings have perforated an otherwise intact area of forest.

Forman (1995) suggests that perforation is the most common way that land transformation begins and will eventually lead to a landscape dominated by small remnant patches of vegetation as cleared areas increase in size and eventually coalesce. Perforation also appears to be an increasingly common form of fragmentation in natural areas adjacent to urban centres (e.g., along the eastern coast of Australia). As urban dwellers increasingly seek opportunities for life-style living outside traditional urban residential areas, perforation of natural forest and woodland results as residential dwellings are located within natural areas to provide residents with a sense of closeness to nature. While the natural area appears relatively intact on the ground (e.g., as viewed from a residential dwelling), the view from the air shows the highly perforated nature of this form of development (Fig. 1).

Fragmentation is not a simple process and it is unusual to find an area that has only one form of fragmentation. Lord and Norton (1990) suggested that for any given amount of remaining original habitat, this habitat could be dispersed in different ways. It is quite conceivable that within a single area there will be examples of geographical fragmentation, structural fragmentation, and intermediate combinations (variegated landscape of McIntyre and Barrett 1992; and McIntyre and Hobbs, this volume). What is clear for Otatara is that there are several remnant forest patches which have themselves been perforated, giving a complex pattern of fragmentation with potentially complex effects on the indigenous biota.

The general effects of fragmentation have been widely documented (Harris 1984; Saunders *et al.* 1987, 1991; Forman 1995) and include effects associated with edges, decreased remnant size, increased remnant isolation, and altered disturbance regimes. In the Otatara situation, these effects are intensified by the areas' cool and windy climate and the nature of fragmentation itself. The effects of edges on forest microclimates and forest flora and fauna (Murcia 1995) are likely to be particularly important at Otatara because of the strong winds that characterize this area. Reduction of the total forest area will have led to loss of species either because the remaining remnants are too small (many obligate forest-interior species) or because of sampling effects (e.g., Norton *et al.* 1995) whereby the patchy distribution of species means that they are not necessarily represented (sampled) in the remaining remnants. On-going forest clearance will increase this effect.

Isolated remnants are also vulnerable to a variety of external influences (e.g., fire, altered water tables, increased nutrient levels, direct human impacts). The influence of these will increase as the forest becomes more fragmented with further clearing for housing, development of housing adjacent to the forest remnants and general forest clearance. Direct impacts may include predation by domestic cats on indigenous bird, reptile and invertebrate species (Graeme and Graeme 1997), and activities of land owners (e.g., weeding in forest remnants and rubbish dumping). Perforation also increases the chances of invasive plants impacting on the forests as it increases the proximity of houses and gardens to forest and increases the number and length of roads (Timmins and Williams 1991). The presence of a substantial seed source associated with housing within remnants means that "garden escapes" are likely to be an important source of new plants (Esler 1987). Human use also has an important influence on weed numbers, as humans disturb sites (e.g., through litter raking), facilitating weed growth and dispersing plants (Hobbs and Huenneke 1992).

Housing, both in clearings within forest remnants and in open land adjacent to remnants, can also have a number of indirect influences. These include increased nutrient status in the forest (e.g., due to fertilizer drift; Hobbs and Atkins 1988), altered hydrological conditions (potentially lowered water tables because of water extraction from bores and improved drainage around houses removing water from remnants; Brown and Weeber 1995), and herbicide impacts from control of unwanted plants on adjacent private land. However, adjacent land uses may also have some positive benefits (e.g., through increased tree planting in areas that are presently pasture) and where appropriate these activities should be encouraged (e.g., through the formation of a Bushcare group, provision of information on desirable trees to plant and rate rebates for restoration of private land).

Ongoing reductions in forest area and increasing isolation of forest remnants reduce the probability that the forest can recover after natural catastrophic disturbances such as windthrow (cf., Pickett and Thompson 1978). This occurs because seed sources become more distant from disturbed areas and as the probability of exotic plant species dominating after disturbance increases.

It is clear that the Otatara environment is a highly desirable place for residential dwellings. It is close to the central business district of Invercargill City but provides an attractive sheltered "bush" environment that is in sharp contrast to the normal urban residential suburbs. However, ongoing residential development is likely to have a major impact on the forest remnants as the forest becomes increasingly perforated. The concept of minimum viable area has been proposed in the conservation literature to provide an estimate of the minimum area that is required to sustain a particular species or ecosystem type. However, estimation of minimum viable area for a forest ecosystem such as that at Otatara is problematic as it requires a detailed knowledge of edge effects (Murcia 1995), current and potential disturbance regimes (Pickett and Thomson 1978) and the many biotic (e.g., pollination and seed dispersal) and abiotic (e.g., nutrient cycling) interactions that characterize the forest ecosystem.

This study has shown that a minimum viable forest area for this forest is about 3.8 ha, which would require a distance of 195 m between access roads and clearings. This estimate is, however, very simplistic and is very sensitive to variations in the different parameters used to derive it (e.g., proportion of trees windthrown and tree density) but it does provide a first approximation of the minimum viable forest area. However, it almost certainly underestimates the minimum area as it assumes continuous regeneration based on single tree windthrow events, rather than irregular larger-scale disturbance and subsequent regeneration that is more likely to occur in these forests. If irregular larger-scale disturbance is the main regeneration mechanism then a much larger area of forest would be required to ensure that all stages of tree regeneration are present. Nonetheless the model still provides a minimum estimate of the possible minimum viable area for totara forest at Otatara. Even using the 3.8 ha/195 m figure, it is already apparent that most distances between roads and clearings are less than this, except in the one large scenic reserve which is not available for residential development.

While it may seem that the best solution to the problems facing the Otatara forests is to prevent any more residential development, this will not solve the problems for two reasons. First, most of the land is in private ownership and there are serious issues of individual property rights that need to be addressed before a ban on future subdivision can be imposed. Secondly, perforation is already so extensive that stopping any further development may not alter an already severely modified ecosystem. It is therefore important to look for alternative solutions to ensure that indigenous biodiversity values are protected within a predominantly low-intensity residential area. Three elements appear important in achieving this.

1. Establishment of Bushcare groups to assist local residents to better manage forest on their properties through education, advice and support. This should include provision of information on how best to manage the forest itself (e.g., not raking up the dead wood and litter), appropriate species to plant in gardens to improve habitat for native birds, and advice on ways to buffer exposed forest edges.
2. Encouragement of restoration to in-fill unneeded gaps and to improve connectivity between remnants. Advice and assistance with restoration

could greatly improve the overall values of these remnants.

3. Introduction of financial incentives through the District Planning process to encourage land owners to limit the impact of future development. While it is tempting to introduce regulations that limit future residential development, it may be more profitable to work with developers through incentives and other mechanisms to look for solutions that result in a net conservation gain (e.g., through locating new residential dwellings in existing clearings and through restoration plantings) rather than through imposing blanket rules restricting any future developments.

It is clear that wherever forest occurs close to residential centres, urban dwellers seeking an alternative life-style are going to be drawn to these areas because of their natural values and life-style opportunities. While it may be possible and is probably desirable to control such developments based on a good knowledge of the ecology of these ecosystems prior to any development, in many situations such as at Otatara, development has been ongoing for many years. In these situations a regulatory approach to controlling development may result in fewer conservation gains than a combination of financial incentives to reduce impacts of future development and a proactive education based programme aimed at involving local residents in better managing the forest remnants.

ACKNOWLEDGEMENTS

I am very appreciative of the enthusiasm and funding provided by Ernest New for this work, discussions with Brian Rance about the Otatara remnants, assistance from Julia Sinnock in undertaking the analysis of fragmentation parameters and comments by Kath Dickinson, Alan Mark, Craig Miller and Brian Rance on a draft manuscript.

REFERENCES

Allen, R. B., Lee, W. G. and Johnson, P. N., 1989. Southland. Biological Survey of Reserves Series No. 19. Department of Conservation: Wellington, New Zealand.

Anon., 1968. General Survey of the Soils of South Island, New Zealand. Bulletin 27. Soil Bureau, DSIR: Wellington, New Zealand.

Anon., 1982. The Climatology of Invercargill Airport. Miscellaneous Publication 171. New Zealand Meteorological Service: Wellington, New Zealand.

Anon., 1986. Climate of New Zealand. Information Publication 15. New Zealand Meteorological Service: Wellington, New Zealand.

Awimbo, J. A., Norton, D. A. and Overmars, F. B., 1996. An evaluation of representativeness for nature conservation, Hokitika Ecological District, New Zealand. *Biol. Conserv.* **75:** 177–86.

Brown, L. and Weeber, J., 1995. Geology and groundwater. Pp. 37–58 *in* Riccarton Bush: Putaringamotu ed by B. P. J. Molloy. Riccarton Bush Trust: Christchurch, New Zealand.

Cockayne, L., 1911. Report on the Dune-Areas of New Zealand: Their Geology, Botany and Reclamation. Government Printer: Wellington, New Zealand.

Cockayne, L., 1928. The Vegetation of New Zealand, second edition. Engelmann: Leipzig.

Dickinson, K. J. M. and Mark, A. F., 1994. Forest-wetland vegetation patterns associated with a Holocene dune-slack sequence, Haast Ecological District, south western New Zealand. *J. Biogeog.* **21:** 259–81.

Esler, A. E., 1978. Botany of the Manawatu District, New Zealand. Government Printer: Wellington, New Zealand.

Esler, A. E., 1987. The naturalisation of plants in urban Auckland, New Zealand. 1. The introduction and spread of alien plants. *NZ J. Bot.* **25:** 511–22.

Forman, R. T. H., 1995. Land Mosaics: The Ecology of Landscapes and Regions. Cambridge University Press: Cambridge.

Graeme, B. and Graeme, A., 1997. Eco-subdivisions, living without cats and dogs. *Forest and Bird* **283:** 37–41.

Harris, L. D., 1984. The Fragmented Forest. University of Chicago Press: Chicago.

Heale, T., 1865. Map of the Southern Portion of the Province of Southland, New Zealand. Chief Surveyor: Wellington, New Zealand.

Hobbs, R. J. and Atkins, L., 1988. Effect of disturbance and nutrient addition on native and introduced annuals in plant communities in the Western Australian wheatbelt. *Aust. J. Ecol.* **13:** 171–79.

Hobbs, R. J. and Huenneke, L. F., 1992. Disturbance, diversity, and invasion: implications for conservation. *Conserv. Biol.* **6:** 324–37.

Johnson, P. N., 1986a. New River Estuary at Otatara: botanical report. Unpublished report, Botany Division, DSIR, Dunedin, New Zealand.

Johnson, P. N., 1986b. Forest on Bowman property, Otatara: botanical report. Unpublished report, Botany Division, DSIR, Dunedin, New Zealand.

Johnson, P. N., 1987. Bushy Point, Otatara: botanical report. Unpublished report, Botany Division, DSIR, Dunedin, New Zealand.

Johnson, P. N., 1988. Otatara Scenic Reserve. Unpublished report, Botany Division, DSIR, Dunedin, New Zealand.

Johnson, P. N., 1992. The Sand Dune and Beach Vegetation Inventory of New Zealand. II. South Island. Scientific Report No. 16. DSIR Land Resources, Christchurch, New Zealand.

Lord, J. M. and Norton, D. A., 1990. Scale and the spatial concept of fragmentation. *Conserv. Biol.* **4:** 197–202.

McIntyre, S. and Barrett, G. W., 1992. Habitat variegation, an alternative to fragmentation. *Conserv. Biol.* **6:** 146–47.

Murcia, C., 1995. Edge effects in fragmented forests: implications for conservation. *Trends Ecol. Evol.* **10:** 58–62.

Norton, D. A., 1989. Tree windthrow and forest soil turnover. *Can. J. For. Res.* **19:** 386–89.

Norton, D. A., 1996. Development and forest sustainability, Otatara, Invercargill. Unpublished report, Te Wai Pounamu Conservation, Christchurch, New Zealand.

Norton, D. A., Hobbs, R. J. and Atkins, L., 1995. Fragmentation, disturbance and plant distribution: mistletoes in woodland remnants in the Western Australian wheatbelt. *Conserv. Biol.* **9:** 426–38.

Park, G., 1995. *Nga Uruora: The Groves of Life*. Victoria University Press: Wellington, New Zealand.

Partridge, T. R., 1992. The Sand Dune and Beach Vegetation Inventory of New Zealand. I. North Island. Scientific Report No. 15. DSIR Land Resources, Christchurch, New Zealand.

Pickett, S. T. A. and Thompson, J. N., 1978. Patch dynamics and the design of nature reserves. *Biol. Conserv.* **13:** 27–37.

Roseberry, J. L. and Hao, Q., 1995. Habitat Analysius and Modeling System, Version 1.0. Cooperative Wildlife Research Laboratory, Southern Illinois University, Carbondale.

Saunders, D. A., Arnold, G. W., Burbidge, A. A. and Hopkins, A. J. M., 1987. Nature Conservation: the Role of Remnants of Native Vegetation. Surrey Beatty & Sons: Chipping Norton, New South Wales.

Saunders, D. A., Hobbs, R. J. and Margules, C. R., 1991. Biological consequences of ecosystem fragmentation: a review. *Conserv. Biol.* **5:** 18–29.

Smith, S. M., Allen, R. B. and Daly, B. K., 1985. Soil-vegetation relationships on a sequence of sand dunes, Tautuku Beach, south-east Otago, New Zealand. *J. Roy. Soc. NZ* **15:** 295–312.

Timmins, S. M. and Williams, P. A., 1991. Weed numbers in New Zealand's forest and scrub reserves. *NZ J. Ecol.* **15:** 153–62.

Wardle, P., 1991. Vegetation of New Zealand. Cambridge University Press: Cambridge.

Whaley, K. J., Clarkson, B. D. and Leathwick, J. R., 1995. Assessment of the criteria used to determine 'significance' of natural areas in relation to section 6(c) of the *Resource Management Act (1991)*. Unpublished report prepared for Environment Waikato by Landcare Research, Hamilton, New Zealand.

Wilson, H. D., 1987. Vegetation of Stewart Island. DSIR: Wellington, New Zealand.

Young, A. and Mitchell, N., 1994. Microclimate and vegetation edge effects in a fragmented podocarp-broadleaf forests in New Zealand. *Biol. Conserv.* **67:** 63–72.

CHAPTER 17

Nature conservation in agricultural landscapes: an integrated planning approach

Robert J. Lambeck,[1] Peter Cale,[1] John Ive[2] and Joe Leone[1]

Achievement of conservation outcomes in agricultural landscapes requires the integration of nature conservation objectives with other land use goals. This requires an understanding of what actions are needed to meet the conservation goal and a procedure for integrating these actions with other land uses. The habitat requirements for nature conservation in a catchment in the central wheatbelt of Western Australia were determined using a focal-species approach in which the needs of the most habitat-demanding species were used to specify the composition, area and placement of habitat patches in the landscape. It was assumed that habitat configurations that supported the species with the greatest demands for habitat would also support less demanding species. Birds were identified as being most sensitive to habitat area, although their needs for habitat were found to vary with isolation. For the study area, the minimum recommended patch sizes were 35 ha for both woodland and shrubland and 25 ha for heath although these sizes could be diminished as isolation decreased. The expansion of all patches less than these sizes up to the area specified would increase the proportion of the catchment under native vegetation by 6% from the current 7.4% of the catchment to 13.4%. This would create an additional 61 patches of habitat capable of supporting the most habitat-limited species. A decision support system was then used to develop landscape plans which integrated the recommendations for nature conservation with other land uses.

INTRODUCTION

AGRICULTURAL industries in Australia have generated significant wealth over many years. However, this wealth has not come without corresponding costs. Land degradation, loss of biodiversity, and in many cases, declining agricultural productivity are all symptoms of an industry that is not sustainable in its current form (Commonwealth of Australia 1966). If these trends are to be reversed and agriculture is to become ecologically sustainable, it will be necessary to develop new strategies which integrate multiple land use objectives (e.g., Fry 1991; Hansen *et al.* 1993; Hobbs and Saunders 1993; Hobbs 1997). To date, there has been limited progress towards achieving integrated approaches due to the absence of a conceptual framework, failure to articulate clearly the objectives of the exercise, and limited capacity to convert general management principles into specific management actions.

This chapter presents a case study in which the requirements for protecting nature conservation values in an agricultural landscape are integrated with other land uses. The chapter firstly outlines the approach taken to integrated planning and then presents a procedure for specifying the composition, amount and placement of habitat required to protect the native biota in the study area. A decision support system is then used to develop plans which combine nature conservation activities with other land uses.

THE CASE STUDY

The case study was based on 19 properties in the Wallatin Creek catchment. The 26 015 ha catchment is located to the north of the central wheatbelt town of Kellerberrin about 200 km east of Perth, Western Australia. Average annual rainfall is 339 mm and falls predominantly in winter. Unpredictable rainfall may also occur in some years during the hot summer months.

[1]CSIRO Wildlife and Ecology, Private Bag, P.O. Wembly, Western Australia, Australia 6014.
[2]CSIRO Wildlife and Ecology, Box 284, G.P.O. Canberra, Australian Capital Territory, Australia 2601.

Pages 166–78 *in* NATURE CONSERVATION 5: NATURE CONSERVATION IN PRODUCTION ENVIRONMENTS: MANAGING THE MATRIX ed by J. L. Craig, N. Mitchell and D. A. Saunders. Surrey Beatty & Sons, 2000.

The landscape is gently undulating with sand plains and ironstone gravel ridges in the higher parts of the catchment. The middle slopes are characterized by duplex soils where sand and gravel from higher in the landscape overlay loamy soils. The valley floors are broad and consist mostly of red loams. The catchment drains towards a saline lake system which has samphire communities around the margins, although these lie outside the project area. This landscape pattern is interrupted by occasional outcrops of granite which are surrounded by shallow sandy soils (Beard 1990).

The original vegetation consisted of tall, open woodland, dense shrubland, and low heathland which reflected the underlying mosaic of landforms and soil types (Beard 1983, 1990). Ninety-three per cent of the catchment has been cleared of its natural vegetation to make way for cereal cropping and sheep grazing. Most of the native vegetation that remains is confined to the higher, less arable portions of the landscape. The removal of native vegetation has resulted in significant losses and ongoing declines of plants and animals from the region (Saunders 1989; Saunders and Curry 1990; Hobbs *et al.* 1993) and a steadily rising, saline water table. The area affected by secondary salinity in the Kellerberrin shire has increased from 2 000 ha in 1962 to over 5 000 ha by 1990 (McFarlane *et al.* 1993).

THE PLANNING PROCESS

Land use planning at catchment scales may involve a range of people or groups with different goals. Therefore, it is essential that a formal planning framework is developed. The key elements of the planning process include: identification of the main land management issues; a clear statement of the management objectives; identification of the full range of potential land uses; assessment of land suitability/capability for those uses; and development of clear guidelines for land allocation. Each of these requirements are discussed in following sections.

We used the land allocation package Land Use Planning and Information System (Cocks *et al.* 1995) to help explore the consequences of alternative management strategies. The Land Use Planning and Information System package was used because: it is conceptually simple; it is not dependent on the quantification of all variables that are used in decision making; and it has the capacity to reflect social preferences through a transparent weighting system. Other approaches, such as mathematical programming, have been used to address multi-criteria planning problems (Field 1973; Cocklin 1989) but their computational requirements and conceptual complexity can alienate stakeholder groups.

To gain acceptance from the stakeholders, a decision-support tool must provide a transparent decision making process which reflects the interests of each group. Participants must be able to see how outcomes are arrived at and how their interests are performing. It is necessary to quantify the extent to which different objectives are being met under different scenarios and to determine the impact on all other objectives, of allocating a land unit to a particular objective. The Land Use Planning and Information System procedure provides this capacity. It generates land use plans from information that is presented by the various groups. The information is expressed in clearly stated guidelines and the effect of these guidelines is determined by a transparent weighting system. By altering the weights attached to each guideline, it is possible to reflect the preferences of the different stakeholder groups. If any guideline is not being adequately addressed, the weights are modified and the allocation rerun. Multiple iterations of this process result in convergence towards a more acceptable solution.

MANAGEMENT ISSUES

The primary issues relevant to the study area were: the need for a profitable and economically sustainable agricultural enterprise; the need to retain biological diversity; and the need to control land degradation which threatens both the production and conservation values of the catchment.

MANAGEMENT OBJECTIVES

Following from the above issues, the objectives for the catchment were identified as: (i) maximizing economic returns from agricultural activities; (ii) preventing the further loss of species from the area; and (iii) preventing further loss of agricultural production and biodiversity caused by a rising saline water table. The emphasis of this chapter is addressing the biodiversity objective. Consequently, productions and land conservation goals will be considered in less detail.

REQUIREMENTS FOR NATURE CONSERVATION

Determining what is needed to protect the biota in a production landscape is not a straightforward exercise. Ecological theories and principles available in the literature indicate that bigger remnants generally support more species than smaller ones (Freemark and Merriam 1986; Cowling and Bond 1991), that more connected landscapes are preferable to more fragmented ones and that landscapes with a greater diversity of habitats generally support more species than do more simple ones (Cody 1975; Roth 1976; Pickett and White 1985; Wiens 1989). Unfortunately, these principles are of limited help to landholders who want clear advice about the actions that they should take, the magnitude of the response required and the location of their actions in their landscape. These principles also fail to indicate exactly what conservation benefit we would expect to attain as a result of their implementation. In other words, general ecological principles can indicate directions in which to proceed but cannot specify the magnitude of the response required or the benefits that will accrue.

In order to specify the actions needed to protect the flora and fauna we need an understanding of the requirements of the biota that we are managing and of the threats that affect their ability to persist. In the fragmented agricultural landscapes of Western Australia the major threats to the native plants and animals are loss and fragmentation of habitat; the loss of critical resources; habitat degradation due to stock grazing and weed invasion; and inappropriate rates and intensities of ecosystem processes such as fire, nutrient cycling and predation (Hobbs *et al.* 1993). Because we could not assess the effects of these threats on all species, a focal-species approach (Lambeck 1997) was adopted whereby a suite of species were identified whose requirements, if met, should also meet the needs of all other species. Species which were considered susceptible to similar threatening processes were grouped and then, for each threat, the species that requires the most comprehensive response was identified (Fig. 1). The assumption was made that a landscape designed and managed to meet the needs of the most sensitive species will also meet the needs of those which are less sensitive.

The first dichotomy distinguishes between those threats which require the reconstruction of habitat and those that need better management of the existing habitat (Fig. 1). This chapter focuses on the threats that require habitat reconstruction as these are of greatest importance in a spatially explicit planning process. A reconstruction response must consider the amount of habitat needed to provide all of the resources required by the plants and animals in the catchment, the positioning of that habitat,

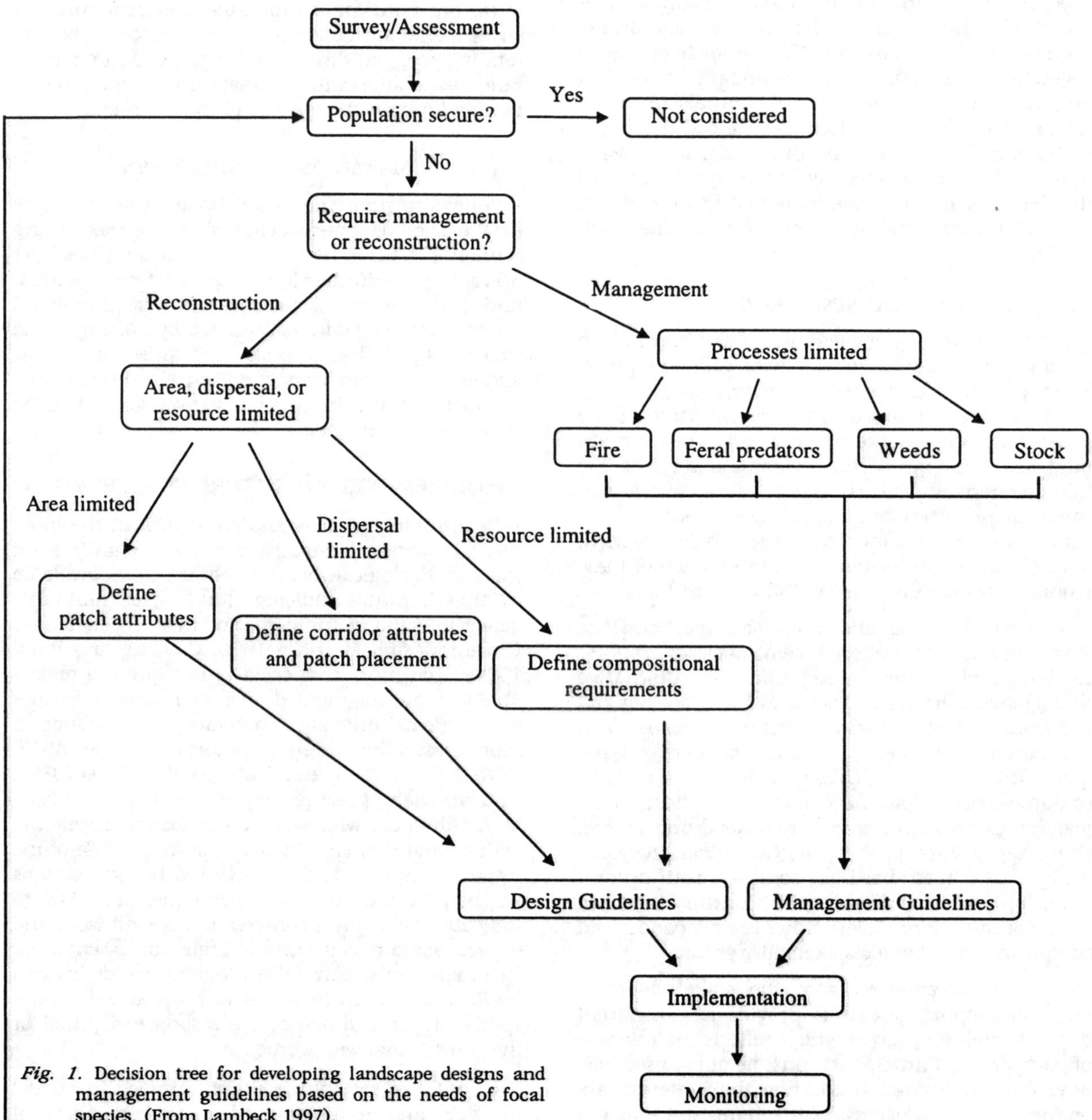

Fig. 1. Decision tree for developing landscape designs and management guidelines based on the needs of focal species. (From Lambeck 1997).

and the type of vegetation connecting the various habitat patches. The most area-limited species are used to define the minimum area of each patch type required, while the requirements of dispersal-limited species are used to specify the minimum inter-patch distances and the characteristics of connecting vegetation. This approach enables us to determine the minimum characteristics that patches of habitat must have in order to be occupied by the most sensitive species.

Species requiring habitat reconstruction were grouped according to whether they are limited by an inappropriate configuration of landscape units or the availability of particular critical resources.

Configuration-limited species

The distribution and abundance of species in a landscape are influenced by the amount of habitat and the spatial relationship between habitat patches. However, these two parameters are not independent. Where patches of habitat are larger the effect of isolation may be less or conversely, if patches are smaller isolation may play a more significant role.

Not all species respond to these parameters in the same way. Some species appear to be more sensitive to patch area while others appear more sensitive to habitat isolation. It is therefore necessary to develop models of species occurrence which consider these spatial components separately as well as in combination in order to determine which models best describe the distribution of each species.

All of the species that were considered sensitive to landscape configuration in this study were birds. Vulnerable mammals were primarily threatened by predation as were some reptiles. Other reptiles, amphibians and invertebrates were primarily threatened by loss of microhabitat caused by stock grazing and weed invasion. Vulnerable plants were also deemed to be most sensitive to stock and weeds or inappropriate disturbance regimes. All of these threats require altered management practices rather than modified landscape designs and hence are not considered in this chapter. On this basis, it was decided that it was possible to use the requirements of birds to specify the spatial attributes of the different patch types that have to be represented in the landscape. It is assumed that a landscape that is designed to meet the spatial needs of birds will also meet the spatial requirements of all other taxa. This is clearly a bold assumption that needs to be tested.

Previous work in the region suggested that ten species of birds were potentially vulnerable due to insufficient habitat (Table 1). All remnants in the catchment were thoroughly searched on two occasions during the breeding season (August–October 1997) to determine presence or absence of these species in the three dominant habitat types (woodland, shrubland/mallee, and heathland).

Logistic regression models (Tabachnick and Fidell 1996) were developed to describe the observed distribution of these species in the catchment. Variables considered in these models were vegetation condition, patch area and isolation.

Vegetation condition was classified on a score of 1 to 3 where category 1 indicates a highly degraded patch and 3 indicates a patch in good condition. Area was measured as the total area of each patch type in each remnant. Due to the low frequency of occurrence of some woodland bird species it was necessary to pool observations into area categories when developing regression models for woodland.

Ideally, isolation would be defined by the dispersal ability of each species. However, information on dispersal is available only for a limited number of bird species, most of which are common and highly mobile (Saunders and de Rebeira 1991). Very little is known about the movement capabilities of less common species which are the most vulnerable.

In the absence of direct measures of movement, estimates of potential dispersal ability were obtained by examining patterns of occupancy of suitable habitat patches with varying degrees of isolation. Isolation was measured as the mean distance to the five nearest suitable patches. This measure was considered more appropriate than a nearest neighbour measure as the latter does not distinguish between pairs and clusters of remnants.

The woodland species considered at risk in the current landscape were Jacky Winters, Sittellas and Brown-headed Honeyeaters. The logistic regression models indicated that, for these species, the most important characteristics of the landscape were the area and the condition of the woodland patches. Isolation variables were not good predictors of the patterns of occurrence of these three species suggesting that the current spatial configuration

Table 1. Bird species in the study area that were considered to be limited by spatial configuration of habitat patches.

Woodland	Shrubland/Mallee	Heathland
Jacky Winter (*Microeca leucophaea*)	Crested Bellbird (*Oreoica gutturalis*)	Field Wren (*Sericornis fuliginosus*)
Varied Sittella (*Daphoenositta chrysoptera*)	Western Yellow Robin (*Eopsaltria griseogularis*)	Blue-breasted Fairy Wren (*Malurus pulcherrimus*)
Brown-headed Honeyeater (*Melithreptis brevirostris*)	Southern Scrub-Robin (*Drymodes brunneopygia*)	Shy Hylacola (*Sericornis cautus*)
	Inland Thornbill (*Acanthiza apicalis*)	

of woodland patches in the catchment is adequate. For example, Figure 2 illustrates the distribution of Brown-headed Honeyeaters indicating that patches with isolation measures of up to 3 km remain occupied.

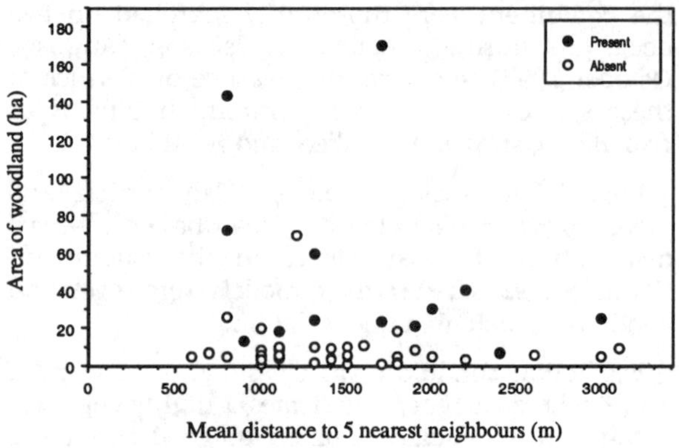

Fig. 2. Distribution of Brown-headed Honeyeaters in woodland patches of different sizes and configurations.

For each of these woodland species, the condition of the patch was correlated with the probability of occupancy, although the model for Jacky Winters was less reliable due to the small number of occupied patches. Sittellas had a probability greater than 0.70 of occurring in woodland patches in the 21–50 ha size range if the vegetation is in good condition (condition index = 3; Table 2). In order to have the same probability of occurring in more degraded patches (condition = 2) the patch area must exceed 50 ha. No patches in the study area with a condition of 1 would be expected to support Sittellas. Brown-headed Honeyeaters are capable of using smaller patches than are Sittellas, requiring 11–15 ha of woodland in good condition if they are to have a 70% probability of occurrence but require between 51 and 100 ha if the vegetation is degraded (Table 3).

Table 2. Estimated probabilities of Sittellas occupying remnants with different area and vegetation condition. Probabilities in bold indicate the combination of area and condition required to have a >70% probability of occupancy.

	Vegetation condition index		
Patch area	1	2	3
1–5 ha	0.01	0.06	0.22
6–10 ha	0.03	0.14	0.41
11–20 ha	0.08	0.29	0.65
21–50 ha	0.19	0.51	**0.82**
>50 ha	0.38	**0.73**	**0.92**

For Jacky Winters, it was only possible to develop a model based on woodland area alone. This model indicated that this species is only likely to be found in the largest patches. For example, 50 ha patches of woodland only have a 64% probability of being occupied.

Of the three vulnerable woodland species, the Jacky Winter appears to have the greatest demand for woodland area, requiring more than 50 ha of eucalypt woodland in good condition to have a reasonable probability of occupancy. However, the data to support this assertion are limited. This species has occasionally been recorded in remnants as small as 8 ha although the majority of patches of this size are not used. Of the remaining two species, the Sittella had the greatest demand for area, with a requirement for patches in the 21–50 ha size class if there is to be a 70% probability of occupancy.

Table 3. Estimated probabilities of Brown-headed Honeyeaters occupying remnants with different area and vegetation condition. Probabilities in bold indicate the combination of area and condition required to have a >70% probability of occupancy.

	Vegetation condition index		
Patch area	1	2	3
1–5 ha	0.01	0.05	0.21
6–10 ha	0.04	0.15	0.47
11–20 ha	0.11	0.38	**0.75**
21–50 ha	0.30	0.67	**0.91**
51–100 ha	0.59	**0.87**	**0.97**
>100 ha	**0.83**	**0.96**	**0.99**

In the absence of more reliable data for Jacky Winters, the value for Sittellas was used to specify the minimum acceptable area for patches of woodland. The mid-point of this size class (35 ha) was taken as the minimum area which had an acceptable probability (0.70) of being occupied. Because isolation was not identified as significant in the models it is considered that the configuration of woodland patches in the catchment is already adequate. It is therefore recommended that patches of woodland that are less than 35 ha be increased in area to at least that size.

Shrubland species considered sensitive to patch configuration were the Western Yellow Robin, Inland Thornbill and Crested Bellbird. Patch condition was not an important variable in the models for any of these species. The most important variables for the Inland Thornbill and Crested Bellbird were the area of shrubland patches and the mean distance to the five nearest shrubland patches, indicating that landscape configuration influences the distribution of these species. Both had similar distribution patterns, but Inland Thornbills were more sensitive to isolation as area increased. The effect of area and isolation on the pattern of occurrence of Inland Thornbills can be seen in Figure 3. Of these two shrubland species, the Inland Thornbill has the most demanding spatial requirements. For this species to have a probability of occurrence of >0.70 in a 50 ha remnant, it requires a mean distance to the five nearest neighbours of <1 800 m, and <750 m for a 20 ha patch (Fig. 4). For the Wallatin catchment the median value for this measure of isolation was 1 400 m. Consequently, patches of between 20 and 50 ha appear to be the most appropriate size for this species in the current landscape. Subsequent recommendations for habitat reconstruction are based on the midpoint of this size range (35 ha).

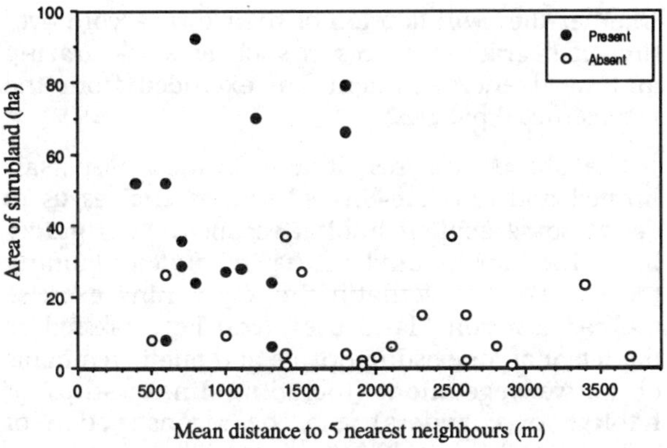

Fig. 3. Pattern of occupancy of shrubland patches by Inland Thornbills. All remnants greater than 40 ha were occupied regardless of spatial configuration, whereas smaller remnants remained unoccupied when the mean distance to the five nearest patches exceeded 1.3 km.

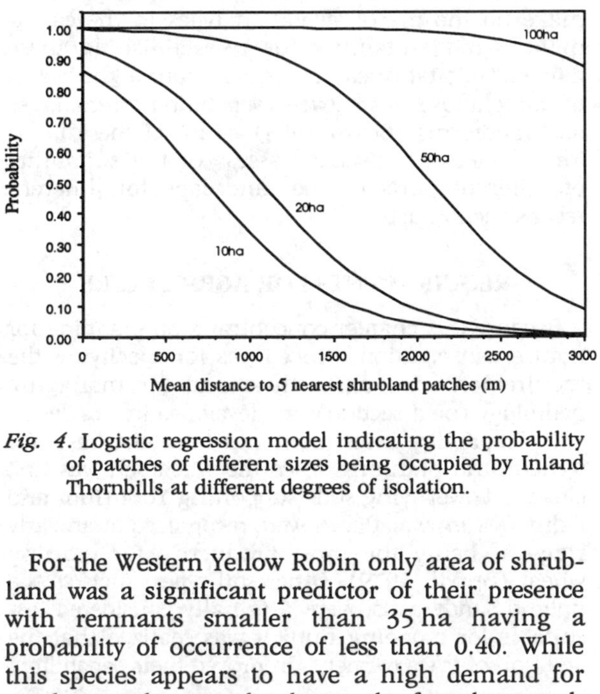

Fig. 4. Logistic regression model indicating the probability of patches of different sizes being occupied by Inland Thornbills at different degrees of isolation.

For the Western Yellow Robin only area of shrubland was a significant predictor of their presence with remnants smaller than 35 ha having a probability of occurrence of less than 0.40. While this species appears to have a high demand for patch area this may be due to the fact that much of the vegetation classified as shrubland may not be ideal habitat for this species. Yellow Robins do occasionally occur in much smaller patches suggesting that such patches may suffice if they better match the habitat needs of this species. A more detailed study would be required to clarify these requirements. Isolation did not appear to be important with several very remote remnants being occupied.

Heath vegetation is very poorly represented in the study area. Consequently it was not possible to determine the minimum suitable patch size for this habitat type. Shy Hylacolas were observed to use heathland but insufficient records were available to estimate their area requirements. This species was recorded in five patches, none of which were less than 25 ha. The Field Wren has only been recorded on the largest remnant in the catchment (1 100 ha) in a heathland patch that is approximately 25 ha in size. It is not possible to ascertain whether its presence on this remnant is attributable to the size of the remnant or to the size of the habitat patch as there are very few other remnants with equivalent amounts of heath. Blue-breasted Fairy Wrens have been occasionally recorded in the catchment in this vegetation type, although they occur more commonly to the north of the study area where heath is more abundant.

On the basis of the observed distribution of species considered limited by landscape configuration, the patch sizes considered necessary to support the most demanding species are greater than 35 ha for both woodland and shrubland/mallee given the current configuration of remnants. For shrubland species, this area can be reduced for remnants that are less isolated based on the regression models developed for each species (e.g., Fig. 4). While it was not possible to model the minimum area requirements for heathland species a minimum area of 25 ha has been recommended based on the observed occurrences of Shy Hylacolas.

All patches of woodland, shrubland and heath in the catchment that are less than these specified sizes were identified using standard Geographic Information System routines. The extent to which they need to be expanded in order to meet the size required by the most habitat limited species was then mapped (Fig. 5). If these patches are increased to the extent indicated they should support not only the most area-demanding species, but also all other species that use those patch types.

Because of our limited knowledge of the dispersal requirements of most species in the landscape, it is not possible to make unequivocal statements about the characteristics of connecting vegetation. Recommendations for corridors must therefore be based on a combination of data, where they are available, and general principles, where they are not. It is important to acknowledge that the following recommendations are subjective. Ideally further data should be acquired to provide more robust estimates. However, farmers are keen to undertake revegetation activities and require "best bet" estimates to guide these activities.

For the birds in the Wallatin catchment connecting vegetation should be sufficiently wide for interior habitat specialists to utilise them. In the case of Hylacolas and Field Wrens this may require corridor widths greater than 60 m. It is unlikely that the implementation of such a recommendation for all linkages between all remnants would be feasible in the short term. Hence it is suggested that linear vegetation of this quality should be preferentially established between sites occupied by these species and the nearest suitable habitat patches.

If linear vegetation is to act as habitat for small mammals, reptiles and invertebrates it must be ungrazed, relatively free of weeds and have a range of habitat types. Fencing is therefore a primary consideration. In order to reduce edge effects linear vegetation in this landscape should have a minimum width of 30 m if it consists of heath or shrubland species, with the width increasing as the plant density decreases. Where woodland species are used with little understorey, linear vegetation should ideally exceed 60 m. The linking vegetation should contain the structural and compositional characteristics of the local vegetation types. Where woodland species are used, substantial clumps of middle canopy and understorey species should be included. These recommendations should not be taken to imply that strips of vegetation less than 30 m have no value. Narrow strips can be used by a large number of species but they will simply have a lower probability of meeting the needs of the more dispersal-limited species or of species that use the linear vegetation as habitat.

Resource-limited species

Resource-limited species are those whose numbers are limited by the availability of critical resources. Resources that are commonly limiting include nest hollows for birds and mammals, nectar supplies for birds, insects and some small mammals, or appropriate microhabitats for invertebrates such as Mygalomorph spiders (Main 1987). These types of species play an important role in defining the compositional characteristics that have to be present in reconstructed habitat.

The only species in the region for which resources have been demonstrated to be limiting are honey-eaters (Lambeck 1995) and predatory invertebrates (Main 1987; Smith 1995). Populations of honey-eaters are restricted by the depletion of nectar resources in late summer and autumn. At this time, many of the nectar-dependent species leave the area while the more generalist honeyeaters become largely insectivorous. The capacity of the remnants to support honeyeaters at this time is lower than their carrying capacity when nectar is abundant. By increasing the availability of plants which produce nectar at the limiting times of the year it may be possible to gain substantial increases in populations for a limited reconstruction effort. For Wallatin Creek, the most appropriate species are those which flower in summer and autumn such as eucalypt mallees, the Acorn Banksia *Banksia prionotes* and *Astroloma serratifolium* (Lambeck and Saunders 1993; Lambeck 1995).

Some spiders and scorpions were considered limited by the availability of suitable micro-habitat. Of the vulnerable Trapdoor Spiders *Idiosoma nigrum* requires stable litter mats for burrow sites in York Gum/Jam Wattle (*Eucalyptus loxophleba*/*Acacia acuminata*) woodland. *Aganippe* sp. *D.* prefers flood-prone depressions and flats which support myrtaceous shrub heaths. *Teyl* sp. *B.* and two species of an undescribed diplurine genus prefer open patches within the litter matrix (Main 1987). An important issue that needs to be addressed is whether these microhabitat attributes need to be created as part of the habitat reconstruction or whether they will develop of their own accord over time if degrading processes such as stock grazing and weed encroachment are excluded from the reconstructed patches.

The above analyses of area-limited, dispersal-limited and resource-limited species enables us to derive some explicit habitat reconstruction guide-lines that can be used in the integrated planning process. For the Wallatin Creek planning exercise the conservation "land uses" can be expressed as the retention of existing private and public remnants of native vegetation (including linear strips of habitat, or corridors) and the reconstruction of habitat patches to a size and condition that meets the needs of the most sensitive species.

Because it has yet to be demonstrated whether it is possible to reconstruct all of the complexity of natural habitat, it is recommended that habitat reconstruction should initially be based on the re-establishment of the dominant species which make up the main vegetation types in the catch-ment. Land suitability for re-establishing these different habitat types was derived from a knowledge of the relationships between vegetation assemblages and landforms (Beard 1983). Maps of these land-form types were used to assess of the suitability of different parts of the landscape for different reconstruction actions.

REQUIREMENTS FOR AGRICULTURE

Because this chapter concentrates on planning for nature conservation, procedures for specifying the requirements for agriculture and for managing hydrology (next section) are developed in less detail. Land capability assessment has been the basis for agricultural planning since the wheatbelt was first cleared. Lower lying soils supporting York Gum and Salmon Gum woodlands were recognized at an early stage as being the most productive for growing wheat (Burvill 1979). Other soil types, such as the upland sandy soils, were originally considered less suitable for cropping until it was realized that the addition of trace elements improved their capability.

Soil types for the study area and their relationships with landform units were described by Hawkins (1990). The classification produced by Hawkins was used by farmers in the catchment to produce soil maps for each farm. These maps were digitised and incorporated into the regional Agriculture Western Australia geographic information system at Merredin. Farmers in the catchment then provided guidelines which indicated the suitability of each of these different soil types for various agricultural land uses (Appendix 1).

REQUIREMENTS FOR MANAGING HYDROLOGY

The increase in land and stream salinization throughout the south-west of Western Australia is caused by increasing groundwater discharge as a result of the clearing of native vegetation (Salama *et al.* 1994a). Possible responses to this increased

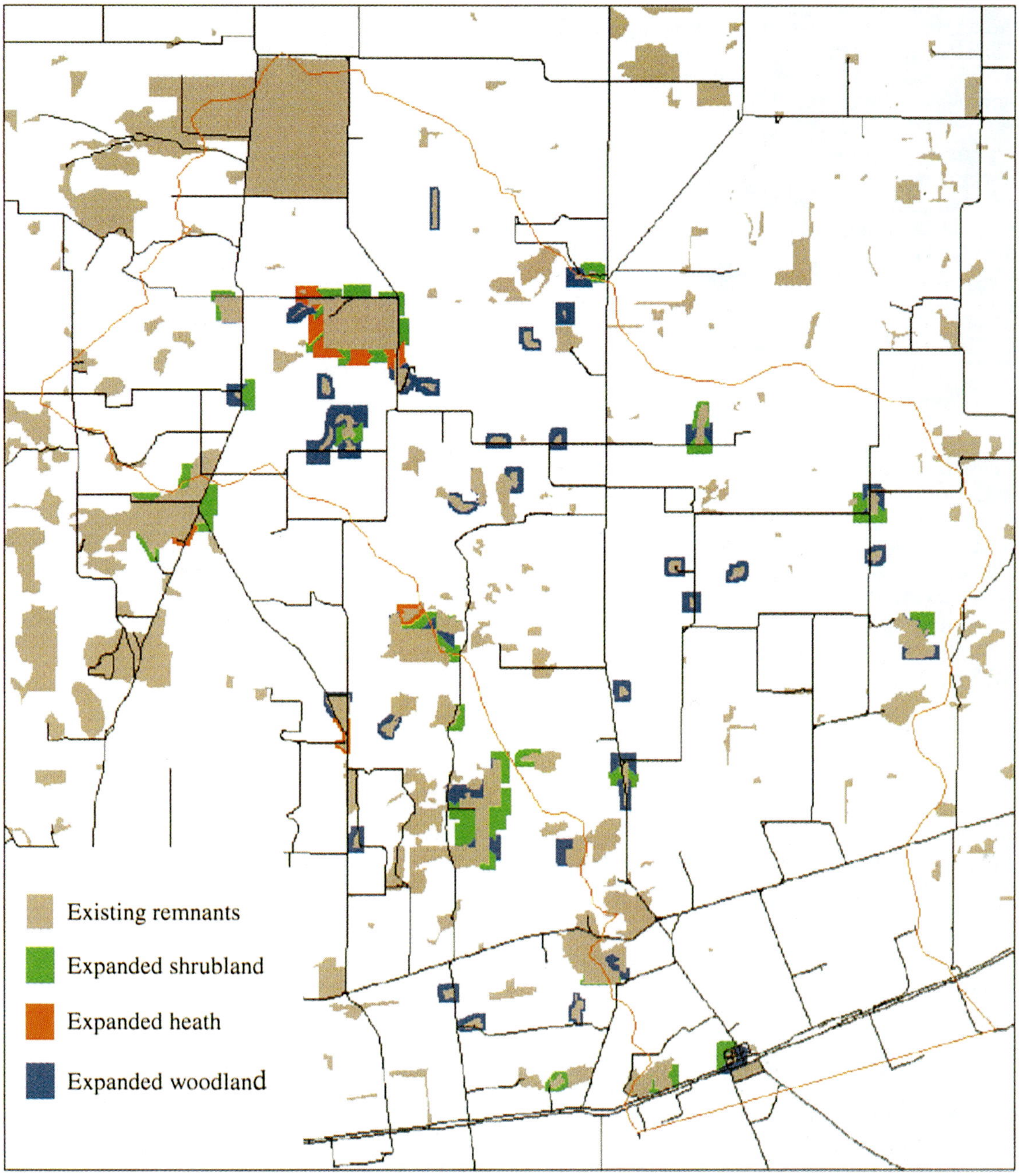

Fig. 5. Map of the Wallatin Creek catchment showing the areas that are most suitable for habitat expansion to meet the needs of the most habitat-limited species.

discharge include the re-establishment of woody perennial vegetation, development of high water-use crops, and implementation of drainage strategies. The type of response will depend on the level of understanding of the factors that influence catchment hydrology.

The Wallatin catchment represents a relatively complex hydrological system. It is bordered in the higher parts of the landscape by lateritic duricrust and granitic outcrops. The remainder of the catchment is dissected by dolerite dykes with associated basement highs, quartz veins and faults (Salama *et al.* 1994b). Thin sedimentary deposits occur throughout the catchment, particularly in the alluvial channels. Each of these features differs in its recharge and discharge characteristics and in its influence on salt storage.

There are two revegetation responses that can be adopted in the face of this complexity. The first of these is to acquire a detailed knowledge of recharge and discharge characteristics and strategically target revegetation to those areas where it will have the greatest impact. In situations where this knowledge is not available an alternative response is to distribute revegetation throughout the catchment in order to intercept water regardless of where it falls. The approach taken in this study represents a mix of these options. The detailed hydrological models that are required to produce precise revegetation recommendations were not available but there was sufficient expertise to produce specific guidelines for different parts of the catchment. On the basis of the hydrological characteristics of the Wallatin catchment a number of broad hydrological "land uses" were identified. These ranged from dense planting of perennial species, through alleys of perennial tree or shrub species interspersed with agricultural rotations and drains, to scattered plantings of perennial species around local recharge features. Guidelines for allocating these activities to different parts of the landscape are included in Appendix 1.

GUIDELINES FOR LAND ALLOCATION

When making decisions about what land uses are to be allocated to which portions of a landscape it is important to be clear about the basis for those decisions. By formalizing the "rules" whereby decisions are made and by making those rules transparent, all parties are aware of how decisions are being made and how outcomes are arrived at. The term "guidelines" is used in a Land Use Planning and Information System land-allocation exercise, rather than "rules" to reflect the fact that they are simply recommendations and that their relative importance can be changed.

Three types of guidelines are used in an allocation exercise. These are termed commitment, exclusion and preference guidelines. Commitment guidelines are used where there is a consensus among all stakeholders that a particular land unit can be used for one use only. For example, in the Wallatin catchment, there was general agreement that there should be no further clearing of remnant vegetation. This objective was captured in a commitment guideline which stated that all existing remnant vegetation be retained as natural habitat. Exclusion guidelines, as the name implies, exclude specified land uses from particular land units but allow other suitable uses. Preference guidelines allow any suitable land use to be applied to specified land units with the decision as to which use is allocated being determined by the relative weighting applied to each guideline.

The procedure adopted in a Land Use Planning and Information System exercise does not require the participants to agree about the content of those rules before the exercise commences. Each stakeholder group or individual can submit their own guidelines. The allocation process is then used to explore the implications of implementing the different guidelines or of altering their relative importance. A full list of the guidelines used in the current exercise is presented in Appendix 1. These guidelines are then used to allocate each of the land uses to different land units.

GENERATING LAND USE PLANS

The guidelines specified in Appendix 1 were converted to Quick Basic code (Halvorson and Rygmyr 1991) and incorporated into Land Use Planning and Information System. Weightings which reflect the social preferences of the stake-holders are then attached to each of the guidelines. On the basis of these weightings Land Use Planning and Information System calculates the relative suitability of each land unit for each land use and allocates each unit to the use with the highest suitability score (Cocks *et al.* 1995).

The distribution of different land uses in the Wallatin Creek catchment following an initial allocation based on equal weights for all guidelines is shown in Figure 6. The proportion of the landscape allocated to each land use under this scenario is displayed in Figure 7. For presentation purposes similar land uses were pooled to produce the six broad categories shown in Figures 6 and 7.

While Figure 7 indicates how much land has been allocated to each land use it does not show how well any particular guideline is performing relative to its potential. In other words it does not indicate how much of the land that is suitable for a particular use has actually been allocated to that use. Figure 8 is an extract from a Land Use Planning and Information System output which indicates the proportion of suitable land that has actually been allocated to the use specified in each guideline. It provides an assessment of the extent to which the various guidelines have been satisfied. Because different guidelines may represent the interests of different stakeholder groups it is possible for all stakeholders to assess how well their interests are being addressed.

A change in the performance of any one guideline as a result of changing its weighting will result in a change in the performance of one or more other guidelines. Consequently it is possible to assess the impact on all objectives of changing the weightings on guidelines pertaining to a particular objective. It is therefore possible to quantify the likely impact on production of addressing nature conservation and hydrology objectives.

The initial allocation produced a result in which the area designated for revegetation for nature conservation exceeded the requirements identified by the focal species analysis. This solution also resulted in much of the prime agricultural land being allocated to nature conservation. This solution is clearly inefficient and unacceptable to the landholders. Consequently the weightings on selected conservation guidelines were reduced and those for cropping on high value soils increased. Figure 9 shows the results of the new allocation which meets the requirements of the focal species analysis and also re-establishes patches of vegetation types that are under-represented in the catchment due to the preferential clearing of more fertile soils. This solution, if implemented would increase the proportion of the catchment allocated to nature conservation by 6%. When added to the existing remnant vegetation the area allocated to native vegetation would account for 13.4% of the catchment.

CONCLUSIONS

The procedures presented in this chapter represent a rational basis for developing landscape designs

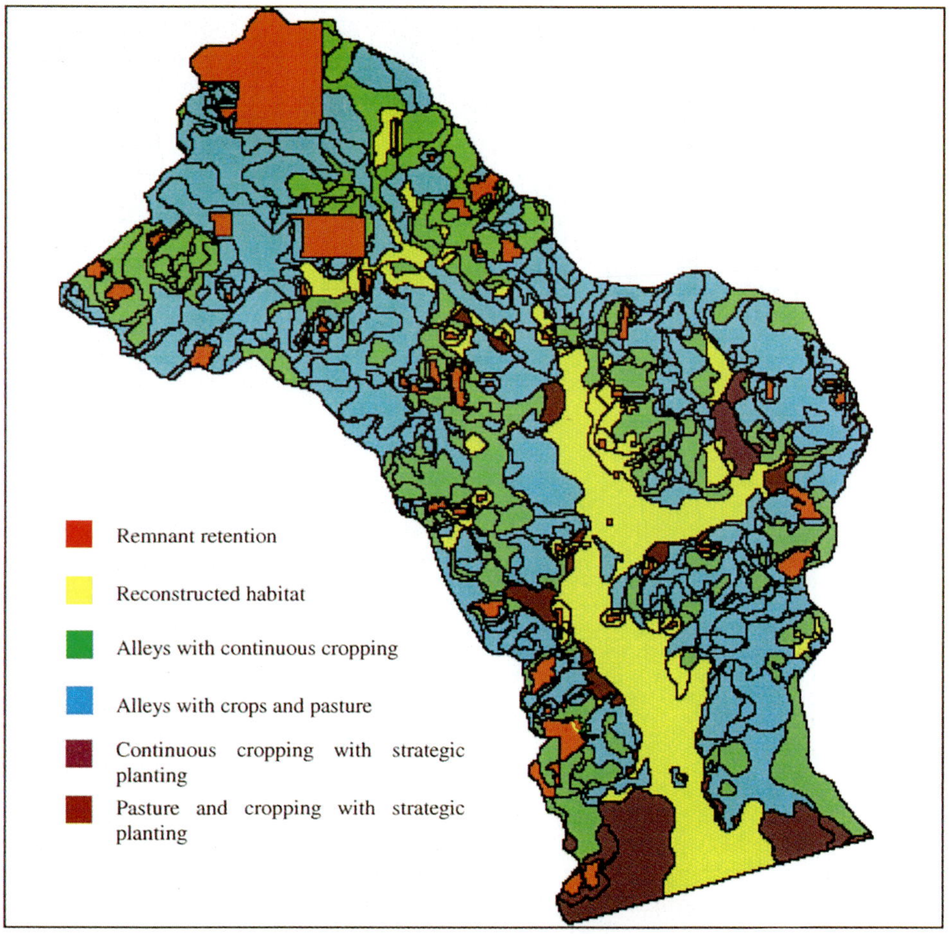

Fig. 6. Results of a land use allocation exercise for the Wallatin catchment based on an equal weighting for all guidelines.

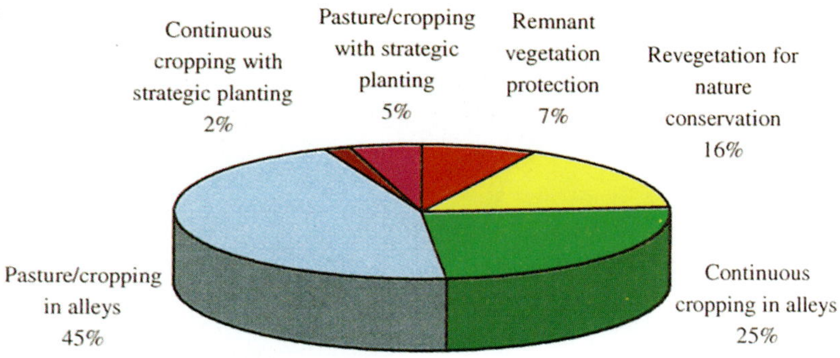

Fig. 7. Per cent of the Wallatin catchment allocated to each land use when all guidelines have equal weightings.

which aim to protect biological diversity in an agricultural landscape and for integrating these designs with other land use objectives. The procedure identifies the minimum amount and placement of habitat that is required by the most habitat-limited species. By focussing on the needs of a limited set of most sensitive species, the procedure assumes that other less sensitive species will also be protected. It is important to recognize however, that the solutions generated will not guarantee that the populations of all species will be viable. Wallatin catchment is not large enough to support viable populations of all species without jeopardizing the economic viability of the region. Rather than expecting a small group of farmers to solve their conservation problems on their own, the procedure identifies the contribution that land-holders can make towards protecting the flora and fauna that occur in their catchment.

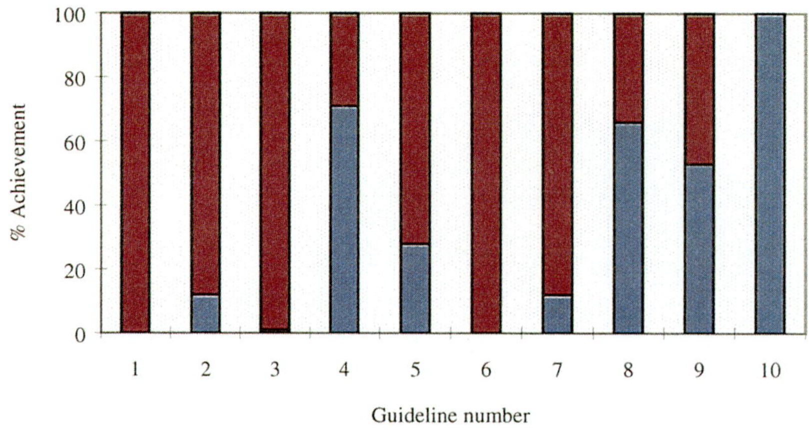

Fig. 8. Bar chart indicating the extent to which the different guidelines in a preliminary Land Use Planning and Information System exercise have been satisfied. Each column represents a different guideline. The height of the column represents 100% of the total area that could potentially have been allocated to the relevant land use as recommended by that guideline. The blue portion of the column indicates the proportion of suitable land that has been allocated, while the purple represents the area that is potentially suitable but which was allocated to an other land use competing for the same land unit. For clarity of presentation only the first 10 of the total of 32 guidelines have been shown. Guideline numbers correspond to those in Appendix 1.

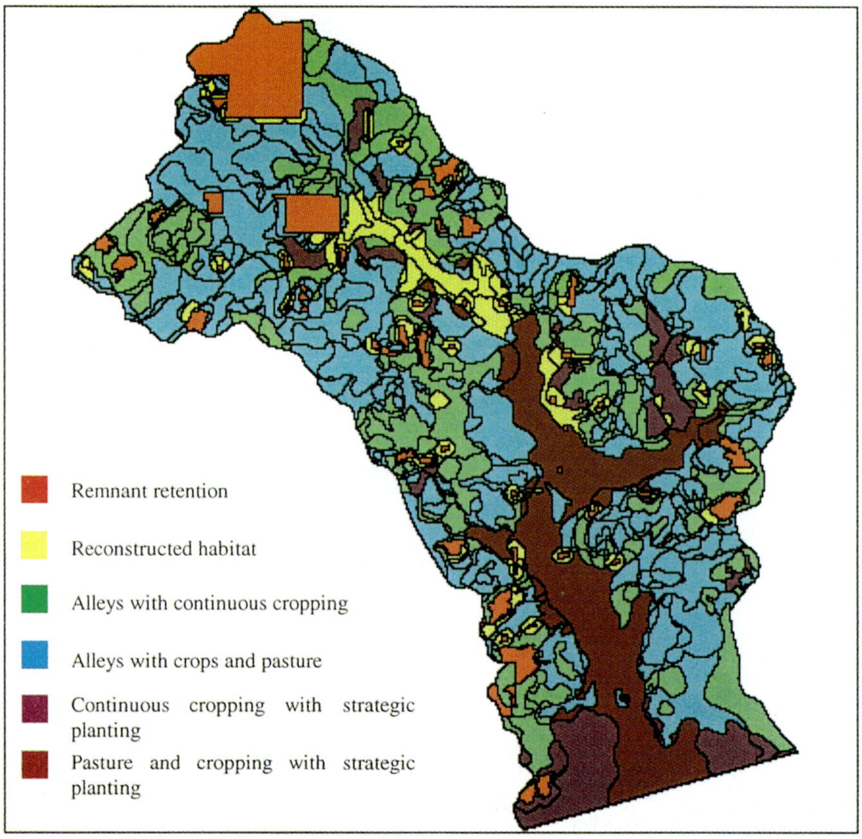

Fig. 9. Preferred land uses solution with additional conservation commitment in which all "inadequate" patches are enhanced to meet minimum size requirements and additional areas are allocated to Salmon Gum/gimlet reconstruction in recognition of the under-representation of this patch type in the catchment.

Failure of some species to persist will not be due to shortcomings in the Wallatin catchment but will be a consequence of the lack of an equivalent response in adjoining catchments. This highlights the need for nature conservation planning to be conducted at regional scales so that issues of long-term population viability can be addressed.

Another limitation of the study is that the recommendations were based on patterns of species distributions at a particular point in time. It is possible that these patterns may change over time, and patches that appear to be adequate under current circumstances may not be if conditions change. It is therefore necessary to view these recommendations as hypotheses which need to be tested. This will require the implementation of a monitoring strategy that not only tests the responses of the most sensitive species to any conservation activities, but also tests the assumptions that landscapes designed to meet the needs of a limited set of focal species will also meet the needs of less sensitive species.

If land managers are to implement nature conservation actions they will need to know how such actions will affect their other management objectives. The use of a decision support system such as Land Use Planning and Information System helps to define the issues to be addressed and clarify the actions that need to be taken. The structure of Land Use Planning and Information System encourages the participants, and more particularly their advisers, to convert their knowledge into statements of action. This is a critical factor in an environment where experts have invested significant amounts of resources into describing problems but failed to deliver solutions that landholders can act upon.

While Land Use Planning and Information System assists with the decision making process it does not take the decision making power out of the hands of the land-owners. It is a tool which organizes information and displays the results of the decision making process in a way which enables land managers to explore alternatives and examine the consequences of making different decisions. By linking the results of a land allocation exercise to economic models it will be possible in future to predict the economic implications of taking alternative courses of action. Such a capacity will be important for demonstrating the costs and benefits that accrue from integrating different land use objectives and can assist with estimating the magnitude of the barriers that may exist to adopting truly integrated and sustainable land use practices.

ACKNOWLEDGEMENTS

This project could not have been achieved without the support and cooperation of the farmers of Wallatin Ck. Spatial data was provided by the Spatial Resource Information Group, Agriculture Western Australia. Funding for the project was provided by Environment Australia's Biodiversity Section.

APPENDIX

Guidelines used in the Land Use Planning and Information System land-allocation exercise

Commitment guidelines

1. Ensure that all vegetation remnants (Public or private) remain as protected remnant vegetation.

Preference guidelines

1. Allocate landform type Booraan adjacent to patches of mixed woodland <35 ha to mixed woodland revegetation until an area of 35 ha is achieved.
2. Allocate landform type Ulva adjacent to patches of mixed shrubland <30 ha to mixed shrubland revegetation until an area of 30 ha is achieved.
3. Allocate landform type Merredin to revegetation with Salmon Gum/Gimlet woodland
4. Allocate landform type Belka to revegetation with Salmon Gum/Gimlet/mallee woodland
5. Allocate areas on gutless sands <20 ha to revegetation with Banksia woodland
6. Allocate Breakaways to revegetation with Wandoo woodland species
7. Allocate shallow clays to revegetation with Wandoo woodland species
8. Allocate Gravelly uplands to alleys of deep rooted perennials interspersed with lupin/cereal rotation
9. Allocate areas on gravelly uplands to alleys of deep rooted perennials interspersed with pasture/cereal rotation
10. Allocate Gravelly sands to alleys of deep rooted perennials interspersed with lupin/cereal rotation
11. Allocate Gravelly uplands to alleys of deep rooted perennials interspersed with pasture/cereal rotation
12. Allocate Booraan soils to alleys of deep rooted perennials interspersed with pasture/cereal rotation
13. Allocate Gutless sands to alleys of deep rooted perennials interspersed with pasture/cereal rotation
14. Allocate Gutless sands to alleys with pasture/pasture/ cereal rotation
15. Allocate Sandy loams to alleys of deep rooted perennials interspersed with lupin/cereal rotation
16. Allocate Sandy loams to alleys of deep rooted perennials interspersed with pasture/cereal rotation
17. Allocate areas of White Gum soils to alleys of deep-rooted perennials planted along the contour, interspersed with pasture/cereal rotation
18. Allocate areas on Mallee (duplex) shallow soils to alley farming with drains with pulses/cereals
19. Allocate areas on Mallee (duplex) deep soils to alley farming with drains interspersed with lupins/cereals
20. Allocate areas on White Gum duplex soils to alley farming with drains interspersed with pasture/cereals rotation
21. Allocate areas on Shallow grey duplex soils to alley farming with drains interspersed with pulses/cereals rotation
22. Allocate all areas of Salmon Gum/gimlet soils to rotation pasture/cereals with perennials around recharge features.
23. Allocate all areas of Salmon Gum/gimlet soils to rotation pulses/cereals/canola/cereals with perennials around recharge features
24. Allocate all areas of Salmon Gum/gimlet soils to rotation pulses/cereals/canola/cereals/pasture with perennials around recharge features
25. Allocate Salmon Gum/gimlet soils to pasture/cereals rotation with perennials around recharge features
26. Allocate Salmon Gum/gimlet soils to pulses/cereals/ cereals rotation with perennials around recharge features

27. Allocate Salmon Gum/gimlet soils to pulses/cereals/canola/cereals rotation with perennials around recharge features
28. Allocate areas of Danberrin to alleys of deep rooted perennials planted along the contour interspersed with pulses/cereals rotation
29. Allocate areas of Danberrin to alleys of deep rooted perennials planted along the contour interspersed with pasture/cereals rotation
30. Allocate areas of York Gum soils to alleys of deep rooted perennials planted along the contour interspersed with pulses/cereals rotation
31. Allocate areas of York Gum soils to alleys of deep rooted perennials planted along the contour interspersed with pasture/pasture/cereals rotation

REFERENCES

Beard, J. S., 1983. Ecological control of the vegetation of south western Australia. Pp. 66–73 *in* Mediterranean Type Ecosystems: the Role of Nutrients ed by F. J. Kruger, D. T. Mitchell and U. M. Jarvis. Springer-Verlag: Berlin.

Beard, J. S., 1990. Plant Life of Western Australia. Kangaroo Press: Kenthurst, New South Wales.

Burvill, G. H., 1979. Agriculture in Western Australia. University of Western Australia Press: Nedlands, Western Australia.

Cocklin, C., 1989. Mathematical programming and resources planning: new developments in methodology. *J. Environ. Manag.* **28:** 143–56.

Cocks, K. D., Ive, J. R. and Clark, J. L., 1995. Forest issues: processes and tools for inventory, evaluation, mediation and allocation: report on a case-study of the Batemans Bay area, New South Wales, Australia. CSIRO: Canberra.

Cody, M. L., 1975. Towards a theory of continental species diversities: bird distributions over Mediterranean habitat gradients. Pp. 214–57 *in* Ecology and Evolution of Communities ed by M. L. Cody and J. M. Diamond. Belknap: Cambridge, Massachusetts.

Commonwealth of Australia, 1996. Australia State of the Environment Report 1966. CSIRO: Collingwood, Victoria.

Cowling, R. M. and Bond, W. J., 1991. How small can reserves be? An empirical approach in Cape Fynbos, South Africa. *Biol. Conserv.* **58:** 243–56.

Field, D. B., 1973. Goal programming for forest management. *Forest Science* **19:** 125–35.

Freemark, K. E. and Merriam, H. G., 1986. Importance of area and habitat heterogeneity to bird assemblages in temperate forest fragments. *Biol. Conserv.* **36:** 115–42.

Fry, G. L. A., 1991. Conservation in agricultural ecosystems. Pp. 415–443 *in* The Scientific Management of Temperate Communities for Conservation. 31st symposium of the British Ecological Society ed by I. F. Spellerberg, F. B. Goldsmith and M. G. Morris: Blackwell, London.

Halvorson, M. and Rygmyr, D., 1991. Running MS-DOS Qbasic. Microsoft Press: Washington, DC.

Hansen, A. J., Garman, S. L., Marks, B. and Urban, D. L., 1993. An approach for managing vertebrate diversity across multiple-use landscapes. *Ecol. Appl.* **3:** 481–96.

Hawkins, A., 1990. Farm and catchment planning in the Kellerberrin Shire. Department of Agriculture, Western Australia and National Soil Conservation Program: Perth, Western Australia.

Hobbs, R. J., 1997. Future landscapes and the future of landscape ecology. *Lands. Urban Plan.* **37:** 1–9.

Hobbs, R. J. and Saunders, D. A. (eds), 1993. Reintegrating Fragmented Landscapes: Towards Sustainable Production and Nature Conservation. Springer-Verlag: New York.

Hobbs, R. J., Saunders, D. A., Lobry de Bruyn, L. and Main, A. R., 1993. Changes in biota. Pp. 65–106 *in* Reintegrating Fragmented Landscapes: Towards Sustainable Production and Nature Conservation ed by R. J. Hobbs and D. A. Saunders. Springer-Verlag: New York.

Lambeck, R. J., 1995. The Minimum Resource Requirements of a Honeyeater Community in a Fragmented Landscape. Ph.D. Dissertation. Curtin University of Technology: Perth, Western Australia.

Lambeck, R. J., 1997. Focal species: a multi-species umbrella for nature conservation. *Cons. Biol.* **11:** 849–56

Lambeck, R. J. and Saunders, D. A., 1993. The role of patchiness in reconstructed wheatbelt landscapes. Pp. 153–61 *in* Nature Conservation 3: The Reconstruction of Fragmented Ecosystems ed by D. A. Saunders, R. J. Hobbs and P. R. Ehrlich. Surrey Beatty & Sons: Chipping Norton, New South Wales.

Main, B. Y., 1987. Persistence of invertebrates in small areas: case studies of trapdoor spiders in Western Australia. Pp. 29–39 *in* Nature Conservation: the Role of Remnants of Native Vegetation ed by D. A. Saunders, G. W. Arnold, A. A. Burbidge and A. J. M. Hopkins. Surrey Beatty & Sons: Chipping Norton, New South Wales.

McFarlane, D. J., George, R. J. and Farrington, P., 1993. Changes in the hydrological cycle. Pp. 146–86 *in* Reintegrating Fragmented Landscapes: Towards Sustainable Production and Nature Conservation ed by R. J. Hobbs and D. A. Saunders. Springer-Verlag: New York.

Pickett, S. T. A. and White, P. S. (eds), 1985. The Ecology of Natural Disturbance and Patch Dynamics. Academic Press: New York.

Roth, R. R., 1976. Spatial heterogeneity and bird species diversity. *Ecology* **57:** 773–82.

Salama, R. B., Bartle, G. A., Farrington, P. and Wilson, V., 1994a. Basin geomorphological controls on the mechanism of recharge and discharge and its effect on salt storage and mobilisation — Comparative study using geophysical surveys. *J. Hydrology* **155:** 1–26.

Salama, R. B., Farrington, P., Bartle, G. A., Watson, G. D. and Hawkes, G. E., 1994b. Salinity investigations in selected catchments of Western Australia. 3. Base data for Wallatin Creek catchment. Technical Memorandum 94/6. CSIRO Water Resources: Perth, Western Australia.

Saunders, D. A., 1989. Changes in the avifauna of a region, district and remnant as a result of fragmentation of native vegetation: the wheatbelt of Western Australia. A case study. *Biol. Conserv.* **50:** 99–135.

Saunders, D. A. and Curry, P. J., 1990. The impact of agricultural and pastoral industries on birds in the southern half of Western Australia: past and future. *Proc. Ecol. Soc. Aust.* **16:** 303–21.

Saunders, D. A. and de Rebeira, C. P., 1991. Values of corridors to avian populations in a fragmented landscape. Pp. 221–40 *in* Nature Conservation 2: the Role of Corridors ed by D. A. Saunders and R. J. Hobbs. Surrey Beatty & Sons: Chipping Norton, New South Wales.

Smith, G. T., 1995. Diversity, habitat and conservation of scorpions in the Western Australian Wheatbelt. *Rec. West. Aust. Mus. Suppl. No.* **52:** 55–66.

Tabachnick, B. G. and Fidell, L. S., 1996. Using Multivariate Statistics. Third Edition. Harper Collins: New York.

Wiens, J. A., 1989. The Ecology of Bird Communities Vol. 1. Cambridge University Press: Cambridge.

CHAPTER 18

Functional mimics of natural ecosystems as a basis for sustainable agriculture

Ted Lefroy[1] and Richard Hobbs[2]

Many current agricultural systems are not sustainable in the long term due to the land degradation they cause and the high levels of external inputs they require. In Australia there is an increasing perception that achieving sustainable agriculture will require dramatic departures from current practices. We examine the concept of Natural Systems Agriculture which suggests that natural ecosystems might serve as models from which we can develop sustainable agricultural systems. By examining how components of natural ecosystems affect their function, it may be possible to develop functional mimics which retain the functions of the natural system while allowing sustainable production. We report on a workshop convened in Western Australia which brought agriculturalists and ecologists from around the world to explore this topic. Theoretical considerations were coupled with long-term case studies in an examination of the potential for developing functional mimics for use in the agricultural systems of southern Australia. Theoretical skepticism engendered by the traditional scientific approach was matched by pragmatic optimism from those that have used intuition to "have a go", and the combination of both may provide important signposts for the future of agriculture in Australia. A research agenda for Natural Systems Agriculture emerged that identified the steps required to replace as much as possible the growing external subsidies required of existing managed systems with services available from the co-operative functions of nature. These were 1) identify the ecosystem functions which are sub-optimal in current managed systems, 2) identify suites of species capable of carrying out these functions, 3) identify whether it is most appropriate to integrate or segregate these functions with production at field and landscape scales, and 4) assemble the suite of species required to achieve functional objectives within an adoption framework that has clear links to end users and includes socio-economic instruments to facilitate implementation.

INTRODUCTION

AN assessment of the state of the agricultural resource base in Australia put the value of losses due to land degradation at $AUD1.2 billion per annum (Industry Commission 1997). This is equivalent to 30% of total annual production costs. Of this, $AUD450 million is due to declining water quality, $AUD300 million to soil acidity, $AUD200 million to secondary salinity, $AUD200 million to soil structural decline, $AUD180 million to waterlogging and $AUD80 million to soil erosion. This is compounded by the fact that the $AUD900 million outlayed each year on fertilizers, accounting for 23% of total production costs, has a net efficiency of around 10% (Williams and Price 1997). The combined effect of this is not only diminished earnings but also a diminishing capacity to produce in future. On a global scale, nearly one-third of the world's arable land of 1 500 million hectares has been lost to erosion and continues to be lost at a rate of more than 10 million hectares per year (Pimental *et al.* 1995).

A common response to this is to take a compensatory approach. This involves accepting that erosion of the natural resource base is inevitable and determining what action will be necessary to compensate for this within the existing production systems. Projections using CSIRO's Stocks and Flows

[1]Centre for Legumes in Mediterranean Agriculture, University of Western Australia, Nedlands, Western Australia, Australia 6907.
[2]CSIRO Wildlife and Ecology, Private Bag, P.O. Wembley, Western Australia, Australia 6014.
Pages 179–87 *in* NATURE CONSERVATION 5: NATURE CONSERVATION IN PRODUCTION ENVIRONMENTS: MANAGING THE MATRIX ed by J. L. Craig, N. Mitchell and D. A. Saunders. Surrey Beatty & Sons, 2000.

Model suggests that it will take a 1% increase in crop yields per annum through germplasm improvement alone to compensate for Australia's lost productive capacity (Crome *et al.* 1998). While this level of yield increase has been achieved over the last 40 years through a combination of improvements in germplasm, fertilizer and machinery, the energy intensiveness of the latter two sources of yield improvement negate their contribution to sustainable production increase. It is also unrealistic to assume genetic improvement can continue to be expressed in an environment constrained by soil factors. From this perspective, conservation of natural heritage is not seen as an integral part of agriculture but as an additional cost.

Another response is to reject the inevitability of agriculture running down its natural resource base and damaging the off-site environment and to look for alternative approaches to producing food and fibre. This involves identifying the root causes of land degradation and designing agricultural systems that avoid them. In other words, designing agricultural systems where the conservation of land, water and to some degree biodiversity is a consequence of food and fibre production. There are some traditional agricultural systems that achieve this to a greater extent than industrial broadacre agriculture but they are under increasing pressure from alternative land uses and population increase to modify the aspects that impart sustainability, particularly rotation length (Altieri 1995). Permaculture (Mollison 1988) has been designed to conserve the quality of the resource base and the off-site environment within the context of a self-sufficient household. Food production is designed around a self-sufficient household, arranged in zones depending on management intensity and energy requirement. The challenge still remains to incorporate these same principles of land, water and energy conservation into broadacre food and fibre production systems.

NATURAL SYSTEMS AGRICULTURE

Natural Systems Agriculture takes the view that the best models for agriculture are natural ecosystems. The term was coined by scientists at Kansas State University to describe collaborative work being undertaken with The Land Institute in Salina, Kansas (Jackson 1996). Since 1978, research at The Land Institute, a private non-profit research and teaching institute, has been directed towards the development of high seed yielding perennial polycultures using the native mid-grass prairie as the model. In the words of their mission, they aim "... to develop ecological assemblies of edible seed-producing perennial polycultures featuring the four major guilds or functional units represented in native prairies and in roughly the same proportions" (Jackson 1996). While conventional agricultural monocultures grown in rotation are driven by the same fundamental natural processes as natural ecosystems, they do not mimic the structural and functional diversity sufficiently to ensure the integrity of some key processes. The shift from perennial to annual plants has altered the components of the water balance resulting in increased run off, soil erosion and nutrient loss for example. In the case of the Kansas prairie, the four functional groups being used as models by The Land Institute in the development of their seed producing perennial polycultures are the warm season grasses, cool season grasses, legumes and members of the sunflower family.

Table 1 shows land use systems classified in terms of their components (woody or herbaceous, annual or perennial), their output (fruits/seeds or vegetative) and their design and management (monoculture or polyculture). The eight types of polycultures could all be examples of Natural Systems Agriculture where there is functional similarity with natural ecosystems. Examples of the first five types can be found amongst traditional farming systems of the tropics. The only one that does not exist yet is type 7, The Land Institute's seed yielding perennial polyculture being developed as an alternative to high production annual crop monoculture (type 11).

The aim of developing ecological assemblies based on proportional representation of functional groups emphasises that in Natural Systems

Table 1. Land use types by life form, product and cultural system. Cultural types 1, 2, 5–14 from Jackson (1985).

CULTURAL TYPE	LIFE FORM		PRODUCT	LAND USE SYSTEM
Polyculture/ Monoculture	Woody/ Herbaceous	Annual/ Perennial	Fruit, Seed/ Vegetative	
1. Polyculture	Woody	Perennial	Fruit, Seed	Mixed Orchard
2. Polyculture	Woody	Perennial	Vegetative	Mixed Woodlot
3. Polyculture	Woody/Herb.	Pere./Ann.	Fruit, Seed	Alley Cropping
4. Polyculture	Woody/Herb.	Pere./Ann.	Fruit/Veg.	Forest Garden/Permaculture
5. Polyculture	Herbaceous	Annual	Fruit, Seed	Mixed Cropping
6. Polyculture	Herbaceous	Annual	Vegetative	Pasture/Companion Planting
7. Polyculture	Herbaceous	Perennial	Fruit, Seed	Multi-spp. Perennial Grain Crops
8. Polyculture	Herbaceous	Perennial	Vegetative	Pasture and Hay
9. Monoculture	Woody	Perennial	Fruit, Seed	Orchard
10. Monoculture	Woody	Perennial	Vegetative	Timber Plantation or Woodlot
11. Monoculture	Herbaceous	Annual	Fruit, Seed	High Production Cropping
12. Monoculture	Herbaceous	Annual	Vegetative	Silage
13. Monoculture	Herbaceous	Perennial	Fruit, Seed	Hay and Pasture Seed Crops
14. Monoculture	Herbaceous	Perennial	Vegetative	Hay Crops and Grazing

Agriculture a mimic is not constructed on taxonomic, biogeographical or aesthetic grounds (Lefroy *et al.* 1999). The aim is the pragmatic one of producing a functional mimic with commercially attractive harvestable products. This does not preclude the use of exotic species. The aim is to achieve functional mimicry regardless of the taxonomy or origin of the component plants. As an indication of this, the most promising high seed yielding perennials for inclusion in The Land Institute's first assemblies were a mixture of exotics, endemics and specially developed hybrids. Of the cool season grasses, *Leymus racemosus,* the giant wild rye of Central Asia, was the most promising. Amongst the warm season grasses, a hybrid between *Sorghum bicolor* and *S. halepense* and a native progenitor of corn, Eastern Gama Grass *Tripsacum dactyloides*, were the highest yielding. Legumes were represented by *Desmanthus illinoensis*, the Illinois Bundle Flower commonly found on roadsides in the Mid-West and sunflowers by the perennial *Helianthus maximilliani* (Jackson 1993). Of course, the free-trade approach to the use of plants from elsewhere in the world gives rise to the problem of these introductions becoming weeds (Hobbs and Humphries 1995). In finding solutions to one problem, we do not want to create greater unexpected problems in the future (Tenner 1996).

The major research questions that Jackson and his co-workers set out to address in 1978 are listed below (Jackson 1985).

1. Can perennialism and high seed yield go together?
2. Can a polyculture outyield a monoculture?
3. To what extent can such an ecosystem sponsor its own nitrogen fertility?
4. Does such an ecosystem have the ability to control insects, pathogens and weeds?

By 1995 they had affirmative answers to the first two questions (Piper and Kulakow 1993; Barker and Piper 1995). In the process of demonstrating high seed yield in perennials, their results challenge life-history theory, as selection for high seed yield did not compromise allocation of carbon to roots (Jackson and Dewald 1994; Pimm 1997).

By 1996, The Land Institute was able to answer the question whether a farming system was able to generate its nitrogen and energy from sunlight alone. A five-year whole farm experiment demonstrated that it would require 25% of the land area to produce the energy required for traction for conventional agriculture through oilseed production. A further 25% would need to be devoted to legume crops to sponsor nitrogen fertility (Jackson, pers. comm.). Table 2 shows estimates of the energy ratios for five stages in the evolution of agriculture. The introduction of tools increased energy returns from 6:1 to 15:1 while adding draft animals doubled the energy input for crop production, reducing the energy ratio to 6:1. Estimates for modern industrial wheat and corn production are typically in the order of 3:1 while Natural Systems Agriculture, assuming it requires only one third of the energy of conventional cropping, would be expected to have a ratio of 9:1.

Recent work has also given some support to the hypothesis that perennial polycultures can control pests and diseases. Studies of insect and viral populations in monocultures and polycultures provided evidence of the role of increased diversity in suppressing pests and diseases (Piper *et al.* 1997).

After 20 years, the research emphasis has now shifted to exploring community assembly rules. Is it possible to design diverse ecosystems that persist? In the first series of experiments, a four species mix comprised of one representative of each functional group failed to persist, being rapidly dominated by annual weeds. By adding a minimum of four other species from the next best selections in each functional group, a near weed free mix has persisted for three years after planting. Some encouragement for the development of community assembly rules comes from computer modelling of complex systems which suggests that although persistent complex mixes are statistically rare, they can be achieved through successive addition and loss of species (Pimm 1997).

CASE STUDIES

A critical test of the concept of Natural Systems Agriculture following The Land Institute's experience is whether the principle of perennial, multi-species agriculture based on functional mimicry of natural ecosystems has global application. To answer that question and explore the relevance of the concept to southern Australia, agriculturalists and ecologists were invited to an international workshop in Williams, a small town south-west of Perth in Western Australia, during September 1997. Those attending fell into three groups. The first group consisted of the keynote speakers who have spent a combined 30 years designing and assembling mimics of natural ecosystems. The second comprised the experimentalists whose work, while not deliberately setting out to develop mimics of natural ecosystems, has dealt with relevant aspects of the

Table 2. Energy ratios for five agricultural systems (from Jackson 1996).

Farming System	Energy Ratio (output:input)	Source
Semi-pastoral farming without tools	6:1	Cox and Atkins 1979
Semi-pastoral farming with tools	15:1	Cox and Atkins 1979
Modified draft horse farming	6:1	Cox and Atkins 1979
Modern industrial agriculture	3:1	Briggle 1980; Smil *et al.* 1983
Natural Systems Agriculture	9:1	Jackson and Bender 1980

concept such as ecosystem function, functional groups and interactions between species in mixtures. The third group were scientists from a range of disciplines who were asked to examine critically some of the underlying theoretical issues.

The two key note speakers described their experience assembling mimics of two very different natural ecosystems as a basis for sustainable agriculture. Wes Jackson described the work of The Land Institute in developing a productive mimic of a perennial grassland (Jackson 1985). Jack Ewel (Institute for Pacific Islands Forestry, Hawaii) described his work mimicking successional regrowth of tropical rain forest in an attempt to develop an alternative to shifting agriculture (Ewel *et al.* 1991).

Each had approached the challenge of assembling their mimics in slightly different ways. While Jackson and co-workers based their mimic on the functional groups present in the prairie with an emphasis on phenology, Ewel and co-workers based their mimic on structure. Following a detailed life-form analysis of a site at La Selva, Costa Rica prior to clearing, they set themselves the rule of substituting tree for tree, shrub for shrub and vine for vine with the added proviso that none of the species in the mimic could have got there by themselves. This exotic mimic was compared to three other treatments; (1) succession (i.e., the development of vegetation through the invasion and establishment of species from the surrounding forest), (2) a progressive monoculture (two years of corn, followed by cassava and a plantation of the tree species *Cordia alliodora*), and (3) enriched succession where the exotics in the mimic were added to the successional plots. The conclusion they came to was that where they mimicked the structure, they achieved the functional goal of retaining nutrients (Ewel *et al.* 1991). This work provided the first experimental evidence of functional similarity between a natural ecosystem and its mimic.

The work of Jackson and Ewel was the stimulus for a multi-disciplinary research project in Western Australia. Ted Lefroy, John Pate and Murray Unkovich (University of Western Australia), Richard Stirzaker (CSIRO Land and Water, Canberra) and Todd Dawson (Cornell University, USA) reported on a comparative study of water and nutrient cycling in a conventional cereal/legume crop rotation, a remnant of the pre-existing *Banksia prionotes* woodland, and an agroforestry system featuring crop rotations between wide spaced rows of the leguminous shrub *Chamaecytisus proliferus* (Lefroy *et al.* 1992). The hypothesis being tested is that the alley cropping system represents a satisfactory functional mimic. This team is making a direct functional comparison of the three ecosystems by tracking the fate of water and nitrogen through soil and vegetation using stable isotopes, direct measurement of water use with heat pulse sapflow sensors and predictive modelling (Lefroy and Stirzaker 1999). Rather than use vegetative structure or functional groups to design a mimic, this novel agroforestry system had been developed by farmers in an attempt to manage rising water tables, salinity and wind erosion (Lefroy and Melvin 1996).

In southern Australia, the development of agriculture has involved replacing woodland, heath and forest communities dominated by deep rooted summer active species with winter active, synthetic annual grasslands of crops and pastures. The difference in rooting depth and phenology between agricultural plant species and native vegetation has resulted in dramatic changes to the hydrologic cycle. While the deep rooted summer active component of the native vegetation draws on reserves of soil and groundwater that are recharged each winter, rainfall unused by annual crops and pastures during their six month winter/spring growing period accumulates through deep drainage and runoff to re-appear elsewhere in the landscape as waterlogging, salinity and problems associated with nutrient leaching such as acidity and eutrophication. Average drainage below the root zone of annual agricultural species in the drier areas (less than 750 mm yr $^{-1}$) has increased from <0.1 mm yr $^{-1}$ to >10 mm yr $^{-1}$ since clearing (George *et al.* 1997).

A primary reason for adopting the Natural Systems Agriculture approach in southern Australia is to develop crop and pasture production systems more closely tailored to the unique characteristics of Australia's climate and soils. The fact that very few new food plants have been domesticated in the last 200 years emphasizes the point that cultural habits do not change rapidly. Finding alternatives to the wheat that occupies 15 million hectares of Australia is not a realistic option. Rather than find a substitute for annual crops, the approach suggested by Natural Systems Agriculture would be to return deep rooted summer active species to the agricultural landscape in roughly the same proportions that existed prior to clearing. In some cases this will preclude broadacre farming, requiring forestry, horticulture or other land use systems based on woody perennials as shown in Table 1. Assuming the natural ecosystem is a good indicator of what is required in order to restore ecosystem processes, it is then up to the ingenuity of the designers to find commercially attractive products within those life form and functional constraints.

John Williams, Tom Hatton (CSIRO Land and Water) and Frank Dunnin (CSIRO Plant Industries) demonstrated the value of hydrologic modelling and ecosystem-scale measurements in providing comparisons of water use by agricultural and natural vegetation systems in southern Australia and how this could be used to re-design well adapted agricultural landscapes. It was clear from the data presented that southern Australian agricultural production systems are very leaky with respect to both water and nutrients. It is their leakiness in comparison with native ecosystems which lies at the heart of most land degradation problems confronting current agriculture. The challenge described was to redesign plant production systems so that sinks and sources of water and nutrients are more closely matched over both time and space (Williams and Price 1997). The daunting task facing those involved is that the modelling presented by Hatton is suggesting the need for a high proportion of woody species, from 30 to 70% of some landscapes.

The value of understanding below ground architecture of species in natural systems and potential candidates for mimic systems was emphasized by this team and echoed by the research of Pauline Grierson and Mark Adams (University of Western Australia) into nutrient cycling in Australian forest ecosystems. Deborah Neher (University of Toledo, USA) described research into the way components of below ground biodiversity are likely to be affected by different land use systems.

The Australian research strongly indicated that in the drier parts of Australia where open woodland was cleared for agriculture, and where recharge rates and the risk of salinity are currently highest, the re-introduction of woody perennials seems essential if any form of sustainable land use is to survive. In light of this, the African experience with agroforestry introduced a sobering note to the workshop. Chin Ong (ICRAF, Nairobi) cautioned against the uncritical adoption of agroforestry, citing the failure of alley cropping especially in the sub-humid tropics. He attributed this failure to simplistic assumptions about the complementarity of resource use between trees and crops. He pointed out that there is insufficient knowledge of the temporal and spacial patterns of resource capture for all but a handful of tree/crop combinations (Ong 1996). Brian Trenbath (Agriculture Western Australia) introduced a mathematical approach to describing the potential types of interactions between species in mixtures for those designing mimic systems (Trenbath 1983).

A long-term study of Europe's oldest continuous farming system led Richard Joffre (Centre Emberger, CNRS, France) to warn against un-realistic expectations of the productivity of mixed systems designed to persist under highly variable climatic conditions. The dehesa agroforestry system of southern Spain and Portugal features the Cork Oak *Quercus suber* at densities of 2 to 10 ha^{-1} in conjunction with pastures grazed by pigs, sheep and cattle. Joffre pointed out that this system had persisted for over 800 years because it was well adapted to local resource constraints and that this came at the cost of sub-optimal production (Joffre *et al.* 1988).

Meine van Noordwijk (ICRAF, Indonesia) also tempered any remaining enthusiasm for complex, multistrata agroforestry systems by reminding us of the naïve early view that all of the complex interactions in agroforestry systems would be positive. This view has been replaced by a more realistic appreciation of a mix of competition and complementarity. He also warned against the delusion that modest increases in biodiversity in agricultural systems would necessarily mean increased biodiversity conservation. He presented his modelling-based research into the question of whether it is better to integrate or segregate the production and conservation roles of land use, based on studies of rubber production in species diverse agroforests and intermediate and high-output monocultures (van Noordwijk *et al.* 1997). One question being asked is whether the objectives of biodiversity conservation and ecosystem functioning can be better met by segregating production. While agroforests may have high levels of diversity and be regarded as the ultimate in conservation farming, they are managed for production and cannot compete with a primary forest for biodiversity conservation.

These case studies showed that several different approaches are possible to the development of agriculture as a mimic of natural ecosystems, as summarized in Table 3. Of these five examples, the two that have survived the test of time both evolved through the selective exploitation and simplification of natural ecosystems with the

Table 3. Different pathways to achieving ecosystem mimics.

Natural Ecosystem	Agricultural System	Basis for Mimic Development	Mimic Status	Reference
Mid-grass prairie Kansas USA	Annual crop monoculture	Functional groups	Experimental seed-producing perennial polycultures	Jackson (1985); Barker and Piper (1995)
Humid tropical Forest Costa Rica	Shifting agriculture	Structure	Experimental successional mimic	Ewel *et al.* (1991)
Semi-arid Woodland SW Australia	Annual crop and pasture monocultures	Process (correcting hydrologic imbalance)	Alley farming: crops and pastures + *Chamaecytisus proliferus* treebelts	Lefroy and Stirzaker (1999); Dawson and Pate (1996); Pate *et al.* (1998)
Humid tropical Forest Sumatra, Indonesia	Shifting agriculture	Commercial exploitation of selected tree spp.	Damar and jungle-rubber agroforestry systems	van Noordwijk *et al.* (1997)
Mediterranean oak woodland, Southern Spain	Pastoralism	Commercial exploitation of *Quercus suber*	Dehesa agroforestry system	Joffre *et al.* (1988)

addition of some exotic plants and animals. The dehesa system essentially represents a weak mimic of the Mediterranean oak woodland with the woody understorey replaced by a pasture of annual and perennial herbaceous species and the tree component selectively reduced to one species of oak. Its persistence for over 800 years is an indication that within this reduced diversity there remains a representation of key functions. The persistence of the other three systems has not been tested. Their evolution however indicates that alternative pathways might be used to achieve the same ends. In the first case, the polyculture based on mid-grass prairie, life form and phenology were used as surrogates of function. In the second, the rainforest mimic, structure was used as the surrogate of function. In the third, structure was used by farmers as a likely clue to function (accessing water at depth over summer) while experimentally, function is being directly interrogated.

THEORETICAL CONSIDERATIONS

The third group invited were asked to address the conceptual issues implicit in the concept of mimicking natural ecosystems. Bert Main (University of Western Australia) was asked to address the question "How much biodiversity is enough?" He replied with "Enough for what?", and went on to explain that as ecosystem functions must be understood in evolutionary terms, the question of how much biodiversity is enough can only be answered from an appreciation of the geological, climatic, biological and social events that have shaped a particular region's history (Main 1992a). Wes Jackson elegantly made the same point, quoting Alexander Pope's Epistle to Burlington, "Consult the genius of the place in all", meaning to consult the *genius loci*, the spirit of place, or in Bert Main's terms, the history of place, in order to determine what is appropriate action for that location.

In quoting Pope, Jackson indicated that this idea of using nature as a standard or measure is not a new one. In practice it can be seen in many of the indigenous agricultural systems of the tropics. In the Western cultural tradition it can be found amongst the English Romantic poets and the writing of Virgil. Within the agricultural and scientific tradition, the concept has been advocated by writers such as Liberty Hyde Bailey *(The Holy Earth)*, Albert Howard *(An Agricultural Testament)* and J. Russell Smith *(Tree Crops)* from early this century (Jackson 1993).

Jackson also pointed out that the need to adopt this approach would be greatest on what he referred to as the least forgiving parts of the landscape, where inherent fertility and water holding capacity were least, and the erosion of ecological capital under monocultural agriculture greatest.

John Passioura (CSIRO Plant Industries) expressed this last point by suggesting that while the Natural Systems Agriculture concept would result in very different outcomes in different parts of the world, the common thread would be that sustainable agricultural landscapes would need to take on the characteristics of mosaics in space and time if plant production systems were to be well adapted to available resources.

Richard Hobbs and Steve Morton also pursued the question of how much biodiversity was enough by examining the recent debate in ecology on the ecosystem function of biodiversity. They reiterated Bert Main's point that one of the main functions of diversity was to provide resilience to change (Main 1992b), and to provide options for dealing with unusual and episodic events. The emphasis in ecology is shifting away from considering diversity of species to considering the diversity of kinds of species (Chapin *et al.* 1997), clearly indicating a convergence with the sorts of considerations being discussed in natural ecosystems agriculture.

Hobbs and Morton also discussed the changing paradigms in ecology, with greater emphasis on the dynamism and patchiness of natural ecosystems (Turner and Gardner 1991; Pickett *et al.* 1992; Fiedler *et al.* 1997; Ostfeld *et al.* 1997), and indicated the profound implications these changes had for the level of prediction possible in ecology, and for the idea of using natural ecosystems as a model. They summarized ideas presented by (Jordan 1995), who took "twenty great ideas in ecology" and interpreted these in a resource management framework. The important insight from this is that natural systems comprise many species with mutualistic or co-operative functions which are generally lost in intensively managed systems and replaced with external energy subsidies. Clearly, to reduce these subsidies again involves restoring key co-operative functions by increasing the species and/or landscape diversity.

While most presenters had concentrated on the biological and scientific constraints to developing agriculture based on natural systems, David Pannell (University of Western Australia) emphasized the social and economic constraints. He described the conditions necessary for farmers to adopt a farming systems innovation, and indicated that adoption rates were often very slow, even for clearly advantageous and profitable innovations. He concluded that with the types of complex innovations which would be necessary with Natural Systems Agriculture, the important challenges in developed countries were to find systems that are more profitable than the existing one, to find a way of assessing that they were in fact more profitable and to overcome the deep uncertainty about the technology. In developing countries there are the additional challenges of high interest rates and insecure or inequitable land tenure.

In summarizing the workshop Dawson and Fry (1998) identified three characteristics of a successful mimic system. They suggested that it should (a) be based on a scientific understanding of the functional characteristics of the main players in the model system. Ecologists, ecohydrologists, soil scientists, and others can help agriculturalists identify the key functions and functional groups; sustainability will arise from successfully mimicking the "processes" through which those functional groups interact. It is acknowledged that "natural" systems themselves are dynamic; the native vegetation is not an ideal but a model for adaptation to the constraints of that particular location. It should also (b) mimic variability at the landscape as well as the farm scales, i.e., in a mosaic across space and time. Diversity at a farm and landscape scale can help manage soil, nutrient and water cycles, stabilize and build the soil, and reduce "leakage" out of the system so that external inputs are minimized. A successful mimic should also (c) be designed within an adoption framework, with clear links to the farmers who will use it. There may be long time lags between implementation and payback, so part of the mosaic should supply short-term profits for the farmer. Here, a mismatch of scales needs explicit consideration. Conventional agriculture is focused on short-term gains, with key decisions being made at the farm scale and industry scale. Sustainability is a long-term goal that operates at different, generally larger, scales. Therefore we must identify the spatial and temporal boundaries within which the key functions of the agricultural system are to "match" the key functions of the natural system. Different kinds of decisions can be made by individuals within farm boundaries than can be made at, say, a catchment scale by some wider authority. A plan for adoption should provide a means of reconciling, in the words of Levins and Lewontin (1985), the "boundary of consideration" with the "boundary of causation".

To this should be added the use of simulation models that have powerful capability to ask the "what if?" questions that are so central to design. The on-farm checking of the predictive models against observation and experience is the basis of the adaptive management approach to wise use of natural resources and a way of addressing the tyranny of space (location) and time (season) variability.

CONCLUSIONS

Drawing together the experimental case studies and theoretical considerations, a broad research agenda for Natural Systems Agriculture emerged from this workshop. It attempted to identify the steps required to replace as much as possible the growing external subsidies required of existing managed systems with co-operative functions found in natural systems.

These steps are to 1) identify the ecosystem functions sub-optimal in current managed systems, 2) identify suites of species capable of carrying these out, 3) identify whether it is most appropriate to integrate or segregate these functions with production at field and landscape scales, and 4) assemble the suite of species required to achieve functional objectives within an adoption framework that has clear links to end users and provides socio-economic instruments to facilitate implementation.

A common theme that emerged with respect to the first point was the need for information on below ground plant architecture and activity in order to identify key functions and functional groups in natural and managed ecosystems. The second point extends the range of conventional economic botany or bio-prospecting to include consideration of the functional role of economic species in managed landscapes. The third point poses a question which may rule out ecosystem mimicry at the field scale due to the costs of competition for resources between species in polycultures. Where the cost of that competition is greater than the benefits of added ecosystem services or biodiversity conservation, the segregation of roles (production versus resource and/or biodiversity conservation) would be the better option. This question of scale emerged as a critical decision point for the application of Natural Systems Agriculture, particularly as maintenance of ecosystems functions such as water and nutrient cycling and biodiversity conservation operated at larger scales and were often in conflict with the economic decisions crucial to the survival of individual farmers and foresters.

There was, however, general agreement that the Natural Systems Agriculture concept represented a promising approach to the problem of matching agricultural systems to available resources, or in Jackson's words, "meeting the expectations of the land through the marriage of ecology and agriculture".

The workshop represented something of a watershed, in that it brought together scientists from a wide array of disciplines to explore the potential for progress in this area. The important questions to be asked do not fall neatly into the remit of any one of these disciplines; rather they inhabit the shady realms at the interface between many different disciplines. The workshop provided a useful opportunity to highlight some of these areas.

The workshop started with a field visit to the Melvin family farm near Dowerin in the wheatbelt of Western Australia. Dean Melvin provided the inspiration for the experimental work on ecosystem mimicry now underway in Western Australia. Struck by the mismatch between current agricultural practice and the way the native bush worked, he developed his alley farming system which then formed the prototype for the novel systems currently under close observation. The mix of lateral thinking coupled with the power to generalize from one system to the next derived from scientific modelling, testing and development, is very powerful and may well provide a signpost for the way ahead.

REFERENCES

Altieri, M., 1995. Agroecology: the science of sustainable agriculture. Westview Press: Boulder.

Barker, A. A. and Piper, J. K., 1995. Growth and seed yield if three grassland perennials in monocultures and mixtures. Pp. 193–97 *in* Proceedings, 14th North American Prairie Conference ed by D. C. Hartnett. Kansas State University Press: Manhattan.

Briggle, L. W., 1980. Introduction to energy use in wheat production. *In* Handbook of Energy Utilisation in Agriculture ed by D. Pimentel. CRC Press: Boca Raton, Florida.

Chapin, F. S., Walker, B. H., Hobbs, R. J., Hooper, D. U., Lawton, J. H., Sala, O. E. and Tilman, D., 1997. Biotic control over the functioning of ecosystems. *Science* **277:** 500–04.

Cox, G. W. and Atkins, M. D., 1979. Agricultural Ecology: An Analysis of World Food Production Systems. W. H. Freeman: San Francisco.

Crome, F., Foran, B. D. and Moore, L., 1998. Population growth and biodiversity loss in Australia. Pp. 77–85 *in* Proceedings of the Population Association Conference held in Canberra, 21–23 September 1997, Bureau of Population and Immigration Research.

Dawson, T. E. and Pate, J. S., 1996. Seasonal water uptake and movement in root systems of Australian phraeatophytic plants of dimorphic root morphology: A stable isotope investigation. *Oecologia* **107**(1): 13–20.

Dawson, T. E. and Fry, R., 1998. Agriculture in Nature's Image. *Trends Ecol. Evol.* **13**(2): 50–51.

Ewel, J. J., Mazzarino, M. J. and Berish, C. W., 1991. Tropical soil fertility changes under monocultures and successional communities of different structure. *Ecol. Applic.* **1**(3): 289–302.

Fiedler, P. L., White, P. S. and Leidy, R. A., 1997. The paradigm shift in ecology and its implications for conservation. Pp. 83–92 *in* The Ecological Basis of Conservation: Heterogeneity, Ecosystems and Biodiversity ed by S. T. A. Pickett, R. S. Ostfeld, M. Shachak and G. E. Likens. Chapman and Hall: New York.

George, R. J., McFarlane, D. J. and Nulsen, R. A., 1997. Salinity threatens the viability of agriculture and ecosystems in Western Australia. *Hydrogeol. J.* **5:** 6–21.

Hobbs, R. J. and Humphries, S. E., 1995. An integrated approach to the ecology and management of plant invasions. *Conser. Biol.* **9:** 761–70.

Industry Commission, 1997. A Full Repairing Lease; inquiry into ecologically sustainable land management. Draft report September 1997, Industry Commission, Canberra, Australian Capital Territory.

Jackson, W., 1985. New Roots for Agriculture. University of Nebraska Press: Nebraska.

Jackson, W., 1993. Meeting the Expectations of The Land. Pp. 1–16 *in* Visions for Agriculture, Proceedings of a workshop held at Tammin, Western Australian Department of Agriculture Miscellaneous Publication 6/93.

Jackson, W., 1996. Natural Systems Agriculture: The Solution to the Problem of Agriculture. The Land Institute: Salina, Kansas.

Jackson, W. and Bender, M., 1980. New Roots for American Agriculture. Office of Technology Assessment: US Congress, Washington.

Jackson, L. and Dewald, C., 1994. Predicting evolutionary consequences of greater reproductive effort in *Tripsacum dactyloides*. *Ecology* **75:** 627–41.

Joffre, R., Vacher, J., De Los Llanos, C. and Long, G., 1988. The dehesa: an agrosilvopastoral system of the Mediterranean region with special reference to the Sierra Morena area of Spain. *Agrofor. Syst.* **6:** 71–96.

Jordan, C. F., 1995. Conservation: Replacing Quantity with Quality as a Goal for Global Management. John Wiley and Sons: Chichester.

Lefroy, E. C., Salerian, J. and Hobbs, R. J., 1992. Integrating economic and ecological considerations; a theoretical framework. Pp. 209–44 *in* Reintegrating Fragmented Landscapes: Towards Sustainable Agriculture and Nature Conservation ed by R. J. Hobbs and D. A. Saunders. Springer-Verlag: New York.

Lefroy, E. C. and Melvin, D. P. C., 1996. The influence of alley farming and intercropping on the yield of lupins. Pp. 377–80 *in* Proceedings, 8th Australian Agronomy Conference. Australian Agronomy Society: Toowoomba, Queensland.

Lefroy, E. C. and Stirzaker, R. J., 1999. Agroforestry for water management in southern Australia. *Agroforestry Systems* **45:** 277–302.

Lefroy, E. C., Hobbs, R. J., O'Connor, M. H. and Pate, J. S., 1999. What can agriculture learn from natural ecosystems? Pp. 423–36 *in* Agriculture as a mimic of natural ecosystems ed by E. C. Lefroy, R. J. Hobbs, M. H. O'Connor and J. S. Pate. Kluwer: The Netherlands.

Levins, R. and Lewontin, R., 1985. The Dialectical Biologist. Harvard University Press: Cambridge, MA.

Main, A. R., 1992a. Landscape reintegration: problem definition. Pp. 189–208 *in* Reintegrating Fragmented Landscapes: Towards Sustainable Agriculture and Nature Conservation ed by R. J. Hobbs and D. A. Saunders. Springer-Verlag: New York.

Main, A. R., 1992b. The role of diversity in ecosystem function: an overview. Pp. 77–93 *in* Biodiversity of Mediterranean Ecosystems in Australia ed by R. J. Hobbs. Surrey Beatty & Sons: Chipping Norton, New South Wales.

Mollison, W., 1988. Permaculture: A Designers Manual. Tagari: Tyalgum, New South Wales.

Ong, C. K., 1996. A framework for quantifying the various effects of tree-crop interactions. Pp. 1–25 *in* Tree Crop Interactions: A Physiological Approach ed C. K. Ong and P. Huxely. CAB International: Oxford.

Ostfeld, R. S., Pickett, S. T. A., Shachak, M. and Likens, G. E., 1997. Defining the scientific issues. Pp. 3–10 *in* The Ecological Basis of Conservation: Heterogeneity, Ecosystems and Biodiversity ed by S. T. A. Pickett, R. S. Ostfeld, M. Shachak and G. E. Likens. Chapman and Hall: New York.

Pate, J. S., Jeschke, D. W., Dawson, T. E., Raphael, C., Hartung, W. and Bowen, B. J., 1998. Growth and seasonal utilisation of water and nutrients by *Banksia prionotes* Lindley. *Aust. J. Bot.* **46:** 511–32.

Pickett, S. T. A., Parker, V. T. and Fielder, P., 1992. The new paradigm in ecology: implications for conservation biology above the species level. Pp. 65–88 *in* Conservation Biology: The Theory and Practice of Nature Conservation ed by P. Fielder and S. Jain. Chapman and Hall: New York.

Pimentel, D., Harvey, C., Resosudarmo, P., Sinclair, K., Kurz, D., McNair, M., Crist, S., Shpritz, L., Fitton, L., Saffouri, R. and Blair, R., 1995. The environmental and economic cost of soil erosion. *Science* **267:** 1117–123.

Pimm, S., 1997. In search of perennial solutions. *Nature* **389:** 126–27.

Piper, J. K. and Kulakow, P., 1993. Seed yield and biomass allocation in *Sorghum bicolor* and F1 and backcross generations of *S. bicolor* x *S. halepense* hybrids. *Can. J. Bot.* **72:** 468–74.

Piper, J. K., Handley, M. K. and Kulakow, P. A., 1997. Incidence and severity of viral disease symptoms on eastern gamagrass within monoculture and polycultures. *Agricul., Ecosys. Environ.* **59**(3): 139–47.

Smil, V. P., 1983. Energy Analysis and Agriculture. Westview Press: Boulder. 191 Pp.

Tenner, E., 1996. Why Things Bite Back: Predicting the Problems of Progress. Fourth Estate: London.

Trenbath, B. R., 1983. The dynamic properties of mixed crops. Pp. 265–86 *in* Proceedings of the Indian Statistical Institutes Golden Jubilee International Conference "Frontiers of Research in Agriculture" ed by S. K. Roy. New Dehli.

Turner, M. G. and Gardner, R. H., 1991. Quantitative methods in landscape ecology. The analysis and interpretation of landscape heterogeneity. Ecological Studies. Springer-Verlag: New York.

Van Noordwijk, M., Tomich, T. P., de Foresta, H. and Michon, G., 1997. To segregate or to integrate: the question of balance between biodiversity conservation in complex agroforestry systems. *Agroforestry Today* **9**: 6–9.

Williams, J. and Price, P., 1997. Re-inventing Australian Plant Production Systems, CSIRO and Land and Water Research and Development Corporation, Operating Plan, 20 pp. Canberra, Australian Capital Territory.

CHAPTER 19

Wild geese and agriculture: achieving conservation priorities whilst minimizing agricultural damage

S. M. Percival[1] and M. Andrews[1]

Wild geese cause significant damage to forage, seed and root crops in many countries across the world. Almost all goose populations have increased greatly in line with agricultural intensification but although locally abundant, several of these species are vulnerable.

This chapter reviews the main management options for goose populations on agricultural land and assesses solutions to the conservation-agriculture conflict. Three main topics are addressed: the conservation status and population trends of the birds, the extent to which they cause agricultural damage and available management options. The best management option will be a combination of measures dependent on a number of factors, determined by the species' conservation status. For species with unfavourable conservation status there is a need to increase numbers. The provision of refuges could make a significant contribution to this. Populations with favourable conservation status need management to contain their growth. Scaring techniques and shooting may minimize adverse effects on the profitability of local agriculture.

To choose the best management options, data on the birds' population numbers and distribution and the factors that determine population dynamics are needed. These can be integrated into models that predict the consequences of the various management options. With continuous monitoring, a sustainable solution to the goose-agriculture problem is possible.

INTRODUCTION

TRUE geese (of the genera *Branta* and *Anser*) have a Holarctic distribution, breeding mainly in arctic and sub-arctic regions and migrating south to winter in temperate zones (Owen 1980). Though only indigenous to the Northern Hemisphere, there are numerous cases of introduction outside their usual range (Allan *et al.* 1995). All geese are primarily herbivorous and it is this characteristic that has brought many of them into conflict with agriculture. They compete directly with grazing stock for forage crops and cause loss of yield of forage, seed and root crops (Kuyken 1969; Kear 1970; Patton and Frame 1981; Ruger 1985; Owen and Pienkowski 1991; Patterson 1991). Almost all goose populations have increased greatly with agricultural intensification (Madsen 1991; Warren and Sutherland 1992; Greenwood 1993). Although locally abundant, several of these species are rare at a national or global level and are of conservation importance.

This chapter examines how the goose conservation-agriculture conflict might be managed. It focuses on species in their natural range but many of the management issues are equally relevant to introduced populations (although in general non-native species have much lower conservation importance) and indeed, other herbivores of conservation importance using agricultural land. Firstly, the conservation status and population trends of the different goose species are assessed. Secondly, the extent of agricultural damage caused by geese is evaluated. Thirdly, the management options available are reviewed and assessed to determine which are sustainable. Lastly, the ecological data and models required to evaluate the effectiveness of the management options are discussed.

[1]Ecology Centre, University of Sunderland, Sunderland SR1 3SD, United Kingdom.

Pages 188–97 *in* NATURE CONSERVATION 5: NATURE CONSERVATION IN PRODUCTION ENVIRONMENTS: MANAGING THE MATRIX ed by J. L. Craig, N. Mitchell and D. A. Saunders. Surrey Beatty & Sons, 2000.

CONSERVATION STATUS AND POPULATION TRENDS OF GOOSE SPECIES

Of a total of 15 goose species world-wide (Ogilvie 1978), the International Union for the Conservation of Nature red data book (Baillie and Groombridge 1996) lists four (Lesser White-fronted Goose *Anser erythropus,* Red-breasted Goose *Branta ruficollis,* Swan Goose *Anser cygnoides* and Nene or Hawaiian Goose *Branta sandvicensis*) as being of global conservation concern (Table 1). All four species have populations that are currently below previous recorded levels and one of them (Red-breasted Goose) also has a fragmented and decreasing range. The Nene has a very small population of less than 1 000 mature individuals.

Focusing on Europe, where populations have been particularly well-monitored, a recent review of bird populations (Tucker and Heath 1994) has classified two of the eight indigenous goose species as vulnerable (Lesser White-fronted Goose and Brent Goose *Branta bernicla*), two as localized (Red-breasted Goose and Barnacle Goose *Branta leucopsis*) and the remaining four as secure (Bean Goose *Anser fabalis*, Pink-footed Goose *Anser brachyrhynchus*, White-fronted Goose *Anser albifrons* and Greylag Goose *Anser anser*) at the European level (Table 2). Table 2 also indicates each species' recent population trend. A fourth category, "introduced", has been included to acknowledge that feral populations of Canada Geese, *Branta canadensis*, occur and have been recorded causing agricultural damage (Allan *et al.* 1995) but this species has no conservation value in Europe.

Thus four goose species have critical conservation importance on a global scale. At the European level, four species are of unfavourable conservation status, having either small populations or localized distributions. However, only one of these species, the Lesser White-fronted Goose is decreasing in numbers and most are increasing rapidly.

AGRICULTURAL DAMAGE BY GEESE

Agricultural land is now the main feeding habitat outside the breeding season for most wild goose populations. All those in Europe now use farmland during this period (Percival and Anderson 1998). Even some of the globally threatened species now feed on agricultural crops. Red-breasted Geese, for example, feed extensively on autumn-sown wheat fields in winter (Sutherland and Crockford 1993). In many cases, loss of traditional habitats means that current populations could not be supported on habitats other than farmland. The extent and scale of the damage caused to agriculture is related to both the population size and how it is distributed across its range. Small populations of conservation importance can have a locally important impact where they are concentrated at a small number of sites.

Table 1. Goose species of global conservation concern (Baillie and Groombridge 1996).

Species	IUCN category	Criteria	Range
Lesser White-fronted Goose	Vulnerable	A1	Albania, Bulgaria, China, Finland, Greece, Hungary, India, Iran, Iraq, Japan, Latvia, Norway, Pakistan, Romania, Russia, Sweden, Turkey, Turkmenistan, Yugoslavia
Red-breasted Goose	Vulnerable	A2, B1	Bulgaria, Greece, Hungary, Iran, Iraq, Kazakhstan, Romania, Russia, Turkey, Turkmenistan, Ukraine
Swan Goose	Vulnerable	A1+2	China, Japan, Kazakhstan, Mongolia, North Korea, Russia, South Korea, Taiwan
Nene	Vulnerable	D1	USA (Hawaii)

Criteria: A1 — reduction of at least 20% over the last three generations. A2 – a reduction of at least 20%, expected within the next three generations. B1 – Extent of occurrence estimated to be less than 20 000 km² and severely fragmented. D1 – Population estimated to number less than 1 000 mature individuals.

Table 2. Conservation status of the nine goose species that breed in Europe (source: Tucker and Heath 1994). Superscript indicates population trend during 1970–90; – = decline of 25–50%, + = increase of 25–50%, ++ = increase of more than 50%.

Species	European conservation status	Population size	Population trend
Lesser White-fronted Goose	Vulnerable	<2 500 pairs	–
Brent Goose	Vulnerable	<2 500 pairs; localized in winter	+
Red-breasted Goose	Localized†	76 000 individuals	++
Barnacle Goose	Localized†	175 000 individuals	++
Bean Goose	Secure‡	83 000 pairs	Stable
Pink-footed Goose	Secure‡	43 000 pairs	++
White-fronted Goose	Secure‡	50 000 pairs	Stable
Greylag Goose	Secure‡	56 000 pairs	Stable
Canada Goose	Introduced	110 000 individuals	++

† The category of localized includes species that have more than 90% of their population occurring at 10 or fewer sites, making them very susceptible to any changes that may occur at those few sites. All such sites have been defined as Important Bird Areas by Grimmet and Jones (1989).

‡ The category of secure includes species that exceed 40 000 individuals, are not declining and have a dispersed distribution.

Most studies that determined the extent of agricultural damage caused by geese compared yields of goose-grazed plots with those protected from grazing by exclosure cages (Patterson 1991). The main economic problems are associated with the effects of geese on grasslands and cereals. In the United Kingdom, silage yield losses of up to 38% have been reported as a result of winter goose grazing, with even greater losses of early spring growth (Percival and Houston 1992). In the Netherlands, Groot Bruinderink (1989) recorded silage yield reductions of 21–32% following winter goose grazing. Similar losses were recorded following spring goose grazing on pastures in Canada (Bedard *et al.* 1986).

In general, yield losses due to goose grazing are less for cereals than for grasslands. Summers (1990) recorded yield losses averaging 6%, 9% and 10% for cereal crops in the United Kingdom following winter grazing by Brent Geese over three years. Similar yield reductions have been found in other studies (Patterson 1991). Lower yield loss with cereals than with grasslands is usually because the cereals have longer to recover from the effects of the geese before harvesting. Nevertheless, the economic effect may be greater with cereals, as the yield is more valuable. Damage to other crops is less important in area terms but can still be locally important. For example, Pink-footed Geese cause damage to carrot crops in a small part of their range in north-west England (Owen and Pienkowski 1991). In this area they have a major impact on a small number of farms. Geese have also shown an adaptability to feed on Rape *Brassica napus*. The area sown of this crop has increased greatly in Europe in the last 20 years and has been used extensively (and damaged) by Brent Geese (McKay *et al.* 1993).

There is no doubt that goose grazing does cause considerable agricultural damage in many countries annually. However, the relationship between goose grazing density and the amount of damage caused is not straightforward, as many other factors can also affect yields. For example, in a field study carried out in western Scotland, a significant positive relationship was found between goose grazing intensity and silage yield loss. However, differences in levels of goose grazing could account for only 19% of the variance in yield loss (Percival and Houston 1992). Differences in the response of the grass to grazing were recorded during the two years of the study, and may have been related to differences in weather conditions. Groot Bruinderink (1989) also found considerable variability in the amount of damage caused by geese to grasslands between years, again apparently linked to weather conditions. The seasonal pattern of the goose grazing and the time between grazing and harvest appear to be important determinants of damage. Sometimes cereal crops substantially recovered from winter grazing. Removal of 75% of the standing crop in winter resulted in only a 6–10% yield loss at harvest (Summers 1990). The magnitude of the effect of the geese is highly variable between years and between sites, so producing quantitative predictions of any adverse effect is difficult (Patterson 1991).

In summary, geese undoubtedly cause substantial damage each year but the interaction between geese and agricultural crops is complex. Generalizations cannot be drawn with regard to the level of goose grazing and resultant yield loss. Ambient weather conditions, timing of goose grazing, crop being grazed and crop management regime are all important confounding factors that can determine yield. A multivariate approach based on long-term data will be necessary if reliable predictions of yield loss are to be developed.

MANAGEMENT OPTIONS AND SUSTAINABILITY

Agricultural land is important goose habitat but economic effects of yield losses need to be minimized. In this section we review the management options for goose populations and assess if a sustainable (i.e., long term) solution to the conservation-agriculture conflict is possible. The most commonly adopted definition of sustainability, the Bruntland one, has been used (Department of the Environment 1993). This states that "*for sustainable use a resource must be used at a level which does not exceed the capacity of that resource to renew or replace itself.*" Sustainable management seeks to conserve goose populations, without threatening either the viability of those populations or the profitability of the local agriculture. For small populations, management is required to increase numbers to ensure future stability. This includes the ability to withstand chance environmental perturbations. For large and increasing populations, numbers must be stabilized or even reduced to limit damage to agriculture. Any programme to reduce numbers must not threaten the viability of the population.

Five possible management options that may achieve the different objectives for different populations are examined below. These are discussed primarily from an ecological viewpoint. This would be an important component of the decision making process but not the exclusive issue. The socio-economics and politics of the problem would also contribute to the overall assessment of the best management options. In particular any solution must be acceptable to all stakeholders, including both conservation and agricultural interests.

1. Do nothing

The simplest approach to goose management would be to let the birds regulate their numbers without human interference. This is problematic for small and large populations. For small populations below their sustainable threshold, no conservation action may further threaten the population. With larger populations it may be equally inappropriate. If a large population did regulate itself then it would be clearly sustainable. Some populations have shown density-dependent regulatory behaviour (Cooch *et al.* 1989; Larsson and Forslund 1994) but most increase exponentially (Madsen 1991). With current resource levels and rapidly increasing numbers, one would expect equilibria to be reached at considerably larger populations than at present with resulting increased agricultural damage. Populations may also expand their range, releasing any density-dependent limitation. The Svalbard Barnacle Goose, for example, had been thought to have reached its equilibrium population (Owen and Black 1991) but in the last five years new breeding colonies have established and the population has returned to rapid growth (Pettifor *et al.* 1998).

There can also be conservation problems associated with excessive population size. Numbers of Snow Geese breeding in the Canadian arctic have increased at such a rate that they are now showing evidence of density-dependence (Cooch *et al.* 1989). High population density was already having a major deleterious effect on their breeding habitat, threatening many thousands of square kilometres of arctic tundra (Kotanen and Jefferies 1997).

2. Financial compensation

There are several schemes in the United Kingdom, the Netherlands and the United States under which farmers are paid financial compensation for allowing goose grazing on their crops. For rare species feeding on agricultural land, this effectively creates refuges and encourages population growth.

In the United Kingdom two main schemes, restricted to sites and species of exceptional conservation importance, are funded from the government conservation budget. One scheme has been established for Barnacle Geese on the Solway Firth in south-west Scotland, which includes the wintering grounds of the entire Svalbard-breeding population (Owen 1987). The other is for Barnacle and White-fronted Geese on the island of Islay, also in south-west Scotland. This site holds about 65% of the Greenland-breeding populations of both these species in winter (Percival 1996). The schemes are different. The Solway one has area-based payments, with a zonation into regions of different goose density and payment made according to the area of crops within each zone. On Islay the payment is made per goose, with counts carried out twice monthly through the winter and the final payment calculated from the average number of geese counted on each farm multiplied by the payment per goose (currently equivalent to US$15) (Percival 1996). With 40 000 geese over 4 000 hectares of farmland, this makes a very significant contribution to the farm economy on the island, and has resulted in a management strategy that is currently acceptable to both conservation and agricultural interests (though see comments below on its long-term viability).

In the Netherlands, the compensation scheme is different again, with a large part of the funding coming from goose hunters. Here the compensation scheme covers all species of geese. Part of the hunters' annual licence fee is paid into a game fund, which is used to contribute to payments to farmers (van Eerden 1990). An expert assessor evaluates the level of goose damage on each farm on an annual basis.

With forthcoming changes in European Union agricultural policy, particularly in the area of subsidies, at least some of the funding for goose compensation schemes may come from agricultural departments within government. This would be as payment for environmentally sensitive farming, which supported the goose conservation interest.

From the perspective of sustainability, payments linked to goose population size where numbers are increasing and with a limited budget, could not be sustained into the long term. This option needs to be linked with other management to achieve long-term sustainability.

3. Refuges

Provision of refuges has been used extensively in Europe to protect concentrations of geese (Owen 1990). Protected feeding areas for species with small populations are essential to increase numbers. As agricultural land is now such important habitat for most species, co-operation with farmers is fundamental to effective conservation of small populations. As numbers grow, refuges could also be a component of the strategy used to manage that increase. They could provide birds with alternative feeding areas and hence reduce grazing pressure on neighbouring agricultural land (Owen 1977, 1990). To support an increased number of geese, these areas would either need to be managed to increase food production per unit area or the area of refuge would need to be increased.

Much research has been carried out to determine how to manage refuges to make them attractive to geese and increase the density of birds that they can support (Owen 1971; Percival 1993; Vickery *et al.* 1994). Reseeding with productive grass species such as Ryegrass *Lolium perenne*, applying additional nitrogenous fertilizer and reducing disturbance effectively increase numbers of geese using refuges. Refuge provision usually increases local population density and often results in larger population size (Owen 1990) but not always. On a reserve on the island of Islay, Scotland, the area of reseeded pasture was increased 400% over a four-year period but there was no corresponding rise in goose numbers on the reserve (Percival 1988). The management attracted local birds to the reseeded fields but did not draw individuals from other parts of the island. This was explained by the birds' high site fidelity and small home range. There were effectively several sub-populations in different parts of the island. For a refuge scheme to be effective here, there would need to be a refuge within each sub-population range. Whilst refuge provision may provide a short-term solution for conserving geese and reducing their effect on agriculture, in the longer term it cannot be a sustainable solution on its own. Increased food provision without other management to control numbers would likely result in further population increase both inside and outside refuges. This could increase agricultural damage, potentially lead to over-grazing within the refuges and possibly adversely affect other conservation interests (Vickery *et al.* 1997).

4. Non-lethal scaring

Deliberate disturbance of geese to move them away from sensitive areas and/or disperse them over a wider area (and hence reduce local damage) may be a more politically acceptable solution than shooting. This management could be appropriate for both small and large populations, as long as there are other feeding sites that can accommodate the displaced birds. Scaring programmes effectively reduced local numbers of Snow Geese (Bedard and Belanger 1989), Brent Geese (Vickery and Summers 1992) and Barnacle Geese (Percival *et al.* 1997a).

Increasing disturbance can have undesirable side effects. For example, additional disturbance will result in the birds spending more time flying and increase their daily energy requirement (Belanger and Bedard 1990). This could reduce survival and productivity rates, potentially reducing the viability of rare populations. Scaring could also affect non-target species. A striking example of this occurred in the United States, where a disturbance programme for Ross' Geese reduced numbers of endangered cranes as well as geese using a refuge (Taylor and Kirby 1990). In addition the cost of carrying out a disturbance programme may not be economically viable (Percival *et al.* 1997a). Scaring may be a useful tool in an integrated goose management strategy, but would be unlikely to be effective on its own.

5. Shooting

Agricultural intensification was not the only reason behind the recent increases in many goose populations. Reduced hunting pressure coincided with the increase in many populations (Ebbinge 1985). Increased shooting may be appropriate to manage large populations that are increasing, and reduce damage to agriculture. It appears to be effective for the Icelandic-breeding population of Greylag Geese. Numbers of this population have levelled off and even started to decline in recent years (Kirby *et al.* 1995). This contrasts markedly with the Icelandic-breeding Pink-footed Goose population, which has been increasing at an annual rate of 6% (Kirby *et al.* 1995). Around 35 000 Greylag Geese are shot annually compared with about 5 000 Pink-footed Geese (A. Petersen, Icelandic Museum of Natural History, Reykjavik, Iceland and C. Mitchell, Wildfowl and Wetlands Trust, Welney, United Kingdom, pers. comm.). Greylag Geese spend a large proportion of their time in lowland agricultural areas where most hunting occurs. In contrast Pink-footed Geese spend most of their time in highland areas. Insufficient numbers of Pink-footed Geese are being shot to control populations.

Hunting of goose populations is also an important recreational pursuit and can have considerable economic value, apart from compensation to farmers. Positive benefits from hunting can include input to the local economy as well as reduced agricultural damage through lower goose numbers.

Goose hunting increases mortality and disturbance (Fox and Madsen 1997). This can keep geese away from susceptible crops and disperse populations but may increase damage by moving birds into new areas. The latter argument is one reason to keep Brent Geese off the list of hunted species, because they may be moved from traditional intertidal food resources on to agricultural land (J. Vickery, pers. comm.).

Shooting of scarcer species protected by conservation legislation would not be an appropriate option, so shooting to control populations would not be a universal solution. Over-exploitation by hunters is a major factor why the rare species are currently so scarce (Sutherland and Crockford 1993). Shooting could maintain populations at sustainable levels where they are increasing rapidly but it needs to be fully evaluated. The number of birds that need to be shot to restrict population growth needs to be known and populations of species important to conservation must not decline.

The response of a population to increased shooting mortality also needs to be considered. There are several examples where population control has been unsuccessful. In these cases density-dependent compensation may have been an important factor, for example through increased breeding success or increased survival rates outside the shooting season (Murton *et al.* 1974). The occurrence of such density-dependence in goose populations has been discussed above. Most populations are, however, increasing rapidly and would be unlikely to exhibit such behaviour.

ECOLOGICAL DATA FOR EFFECTIVE GOOSE MANAGEMENT

None of the management options alone will provide the solution to the goose conservation-agriculture conflict. Rather, a combination of approaches is required, tailored to the specific population being managed.

Whatever size of goose population is being managed, reliable data on its numbers, distribution, productivity and mortality are critically important, preferably with long time series, so that trends through time can be properly analysed. For small populations a complete census should be possible but for large ones, sampling may be necessary. In Europe, detailed goose monitoring has been carried out for several decades, with international co-ordination through the Wetlands International organization (Madsen 1991), linking into national monitoring schemes (Kirby *et al.* 1995). These provide population numbers and distribution with trends through time. The key aspect is to determine the factors affecting population numbers and distribution, so they may be manipulated to achieve the desired population size.

1. Factors affecting population numbers

Changes in bird numbers at the population scale result from the balance between gains to the population (productivity) and losses from it (mortality). Breeding productivity is regularly monitored in most goose populations (Madsen 1991). The usual method is to sample the proportion of juveniles in flocks on the wintering grounds. This is rapid but needs to account for variability between different parts of the population to obtain a reliable result (Lambeck 1990; Percival 1992). To determine breeding success, breeding histories of marked individuals are also important. These enable individual-level factors such as age, experience and social status to be incorporated into the analyses: this is particularly important when different parts of a population are managed in different ways (Black and Owen 1989).

Breeding success in geese is generally low, with most populations averaging an annual value of between 10 and 20% juveniles (Owen 1980). In some populations this value is consistent but in others it is highly variable among years (Fox and Gitay 1991). The Brent Goose population in northern Siberia suffers regular total breeding failures but will have as high as 50% young in other years (Summers 1986). These breeding failures occurred about every three years, and could reflect predator-prey interactions on the arctic breeding grounds. When arctic fox numbers are high but their main food supply (lemmings) scarce, foxes may switch to feeding on geese. However, there was a complex interaction of additional factors that also played an important part (Syroechkovsky *et al.* 1991). These factors include food supply in spring (Teunissen *et al.* 1985), individual feeding performance in winter and spring (Ebbinge 1989) and weather conditions in spring and summer (Fox and Gitay 1991).

From a management perspective, even though the geese breed in the arctic or sub-arctic regions, environmental conditions on the wintering and spring staging grounds can contribute to breeding success. Breeding success of small populations may be achieved by improving winter and spring feeding. Conversely, it may be possible to reduce the productivity of large populations by reducing the quality of their feeding habitats, and/or by increasing disturbance.

Annual rates of mortality have been estimated for most goose populations. The simplest method is to assume mortality to be the difference between the observed population change and breeding productivity. This incorporates count and breeding production error in the estimates but it is better than nothing and requires the least effort. A more reliable method is to use repeat observations of marked individuals (Clobert *et al.* 1987; Bell *et al.* 1993). As with breeding success, this enables individual status to be incorporated into analyses (Francis *et al.* 1992). An alternative approach is to mark birds with only metal rings and rely on returns from hunters and the public. This allows sampling over a wider area, which may be important for larger, more dispersed populations but can be biased as the sampling is less controlled (Aebischer and Coulson 1987). It is important that mortality rates are measured as accurately as possible, as these can have large consequences on rates of population change (Percival 1992). Confidence in the estimates enables precise predictions to be made of the effect of management on mortality rates and population change. Where populations are hunted, records of shooting bags should be kept to separate shooting and natural mortality.

Natural annual mortality rates are low, ranging from 5 to 15% (Owen 1980; Percival 1992; Bell *et al.* 1993) with some variation among years, attributable mainly to ecological conditions on the breeding grounds (Ebbinge 1985) and weather conditions during migration (Owen and Black 1989). In hunted populations, mortality rates can be up to 30% (Ebbinge 1991; Percival 1992; Madsen and Noer 1996).

The challenge is to utilize data on breeding success and mortality rates for increasing small populations and reducing large populations that cause damage. Population data are essential. If populations are left to achieve equilibria, data are needed to predict the point of stabilization. Provision of refuges resulted in the Svalbard Barnacle Goose population wintering on the Solway Firth in Scotland increasing from only a few hundred to over 20 000 individuals in 40 years (Owen 1987).

Understanding the birds' population dynamics is critical if population control is a management option. This requires knowledge of the numbers to be killed to restrict population growth but ensuring that there is a sufficient buffer in the population to withstand chance events such as a run of poor breeding seasons that could potentially threaten the sustainability of the population. Population viability analysis (Burgman *et al.* 1993) gives a quantitative assessment of the likelihood of various population outcomes, such as extinction or stability. Its accuracy is dependent on population data available (including numbers, mortality and productivity rates) and the reliability of those data.

2. Factors affecting population distribution

Movements between sites can have a major effect on the overall population dynamics (Hanski 1991; Harrison 1991). The only way to predict how dispersal patterns might change under different management scenarios is to obtain multiple resightings data from marked individuals. Though goose populations are generally site-faithful at most times of year, they do have a mobile component (Raveling 1979; Cooke 1987; Percival 1991; Wilson *et al.* 1991). This increases substantially in times of food shortage (Ebbinge *et al.* 1991). Also, although the birds that move between sites generally comprise a small proportion of most populations (15–25%), it is, on average, a greater proportion than the annual mortality or productivity rates. The number of geese at a particular site is likely to be affected more by the balance between immigration and emigration than by productivity and mortality.

The critical objective is to understand the factors that affect movement rates in relation to management. For refuges, a key aim is to increase local bird density by increasing immigration rates and decreasing emigration rates (Percival 1996). Scaring schemes aim to increase emigration whilst reducing immigration. One study, using marked individuals, showed that scaring did increase emigration but led to a slight increase in the immigration rate too, presumably as birds came in to occupy space left by emigrants. This reduced the effectiveness of the scaring scheme (Percival *et al.* 1997a).

One major factor affecting goose population distribution is the distribution of their food supply. The body of theory on how animals distribute themselves in relation to their food supply can be utilized to predict goose distribution. Two theories are relevant, optimal foraging (Krebs 1978) and the ideal free distribution (Fretwell 1972). The first predicts that the geese should select their feeding sites to maximize their food intake rate (taking into account quality as well as quantity). The second states that they should be distributed within their range in proportion to the availability of their food (again taking into account both quantity and quality). Both theories can set rules for feeding site selection in models to predict goose distribution (Sutherland and Allport 1994; Goss-Custard *et al.* 1995; Percival *et al.* 1996). A range of different factors, including food quantity and quality, have been shown to affect the birds' choice of feeding site within a population's wintering range. These include the area of preferred feeding habitats, such as improved grassland (Percival and Percival 1997), and proximity to suitable roost sites (Percival *et al.* 1997b). Field management, including reseeding and fertilizer regime can have a strong influence (Percival 1993), whilst proximity to traditional habitats is important in some species, including Brent Geese (Percival and Anderson 1998). Geese exhibit strong selection for particular plant species such as Ryegrass (Percival 1993) and Saltmarsh Herbs (Prins and Ydenberg 1985). Preferences are not fully understood but appear to be based on a range of factors, including the amount of food available, either at a specific moment in time (standing crop) or cumulative over time (Percival 1988; Vickery *et al.* 1995; Percival and Evans 1997), the nutritional quality of the food resource (Prins and Ydenberg 1985), requirements for specific amino acids (Sedinger 1984), avoidance of certain plant toxins (McKay *et al.* 1994) and plant digestibility (Boudewijn 1984; Buchsbaum *et al.* 1986; Anderson *et al.* 1997).

An understanding of the factors affecting field use is essential. This permits manipulation to fulfil the particular management objectives, including attracting birds to refuges and discouraging them from feeding on crops susceptible to damage. Simply locating the latter away from the birds' preferred feeding areas may reduce damage considerably. Physical features such as field size and boundary type affect goose distribution (Summers and Critchley 1990). Geese generally prefer large unenclosed fields with an open view. Disturbance levels are also important, affecting both the total numbers using a field and the seasonal pattern of use (Owens 1977; Percival 1993).

All of the factors that affect goose distribution are dynamic. For example, food preferences are not fixed in either space or time. They may vary among sites, dependent, for example, on the array of resources available and through time, dependent on, for example, changes in nutrient content of food plants and changes in nutrient requirements of the birds (Prins and Ydenberg 1985).

PREDICTING MANAGEMENT OUTCOMES: THE NEED FOR MODELLING

Ecological models are needed with data on factors affecting goose abundance and distribution to predict goose numbers and distributions under different management scenarios.

Population models, particularly in North America, have been used to determine sustainable harvests. These population models are of varying complexity, from simple through to complex metapopulation models (Hestbeck *et al.* 1991). The latter have the considerable advantage that they account for the spatial variation in populations, often essential for investigating management options. Figure 1 is a

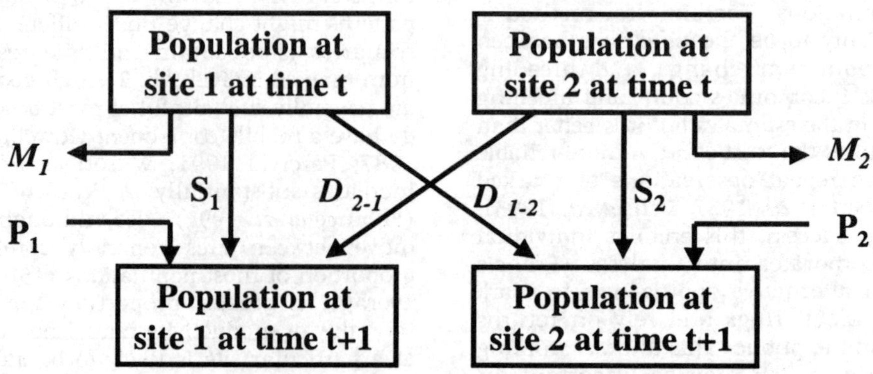

[M = mortality, P = productivity, S = survival, D = dispersal]

Fig. 1. Framework of a simple 2-site population model incorporating site-specific mortality and productivity and between-site dispersal.

simple representation of a population model that has a spatial and temporal component (i.e., population divided into two interacting sub-populations).

This has been used to predict Barnacle Goose population changes under a variety of scenarios on their wintering grounds. Prior to management on their main wintering area in 1992, modelling predicted that the change from mixed refuge and shooting to a total refuge would increase annual rate of change from +6% to +9% (Percival 1992; Table 3). Later analyses showed that the prediction was accurate (Percival 1996). The model did not account for habitat management but only shooting and disturbance. It produced reasonably accurate predictions only because there was no major change in habitat management during this time. For such models to work reliably under different scenarios, they need to incorporate all the factors affecting goose numbers and distribution.

Table 3. Predicted annual rates of Barnacle Goose population increase with different management scenarios (Percival 1992).

Possible scenarios	Annual population change
Mixed refuge/shooting	+ 6.0%
No winter shooting mortality	+ 7.0%
Reduced winter disturbance	+ 7.4%
All refuge (combined effect of reduced shooting on mortality and disturbance)	+ 8.7%

Some models of feeding site selection have also been constructed (Sutherland and Allport 1994; Percival *et al.* 1996, 1998). These can predict the effects of changes in food supply and bird numbers on distribution within a site and address local management but they do not predict effects on overall population size. A key challenge is to develop a framework to integrate all levels of information available. It is important to integrate management through a range of different spatial scales. This should bring population dynamics models together with those of feeding site selection to link habitat management to changes in numbers and population distribution. Models should have general applicability to the resolution of herbivore conservation-agriculture conflicts.

CONCLUSIONS

Different populations need different management. Small populations are likely to require positive conservation measures such as the establishment of refuges and/or payments to farmers to allow undisturbed feeding. Large populations may require control through shooting, the use of scaring schemes and/or compensation payments to farmers to reduce or ameliorate agricultural damage. Management for a particular population is likely to change if population size changes. A flexible approach is needed to be able to adapt to such changes. To determine the best approach or combination of approaches at a particular time needs continuous monitoring of population numbers and distribution, and of survival, dispersal and productivity rates. If this is done, then a sustainable solution to the goose-agriculture problem is possible. The management issues and possible solutions of the goose-agriculture problem could be equally applicable to other herbivore-agriculture conflicts where the herbivores feed together in large numbers on crops that are sensitive to grazing damage.

ACKNOWLEDGEMENTS

We are very grateful to Richard Kingsford and Grant Dumbell, who provided comprehensive reviews of our initial manuscript, and to Tracey Percival and Morag Andrews for their help throughout.

REFERENCES

Aebischer, N. J. and Coulson, J. C., 1987. Lower juvenile survival rates estimated from ringing returns: reality or artefact? *Ibis* **129:** 116–17.

Allan, J. R., Kirby, J. S. and Feare, C. J., 1995. The biology of Canada geese in relation to the management of feral populations. *Wildl. Biol.* **1:** 129–43.

Anderson, G. Q. A., Andrews, M., Percival, S. M. and Kirby, J. S., 1997. Nitrogen nutrition of brant, *Branta bernicla*, grazing on saltmarsh and pasture species. *Proc. XVIII Internat. Grassl. Cong.* **26:** 3–4.

Baillie, J. and Groombridge, B., 1996. The 1996 IUCN Red List of Threatened Animals. IUCN: Cambridge.

Bedard, J. and Belanger, L., 1989. Responses of staging Greater snow geese to human disturbance. *J. Wildl. Manage.* **53:** 713–19.

Bedard, J., Nadeau, A. and Gauthier, G., 1986. Effects of spring grazing by greater snow geese on hay production. *J. App. Ecol.* **23:** 65–75.

Belanger, L. and Bedard, J., 1990. Energetic cost of man-induced disturbance to staging snow geese. *J. Wildl. Manage.* **54:** 36–41.

Bell, M. C., Fox, A. D., Owen, M., Black, J. M. and Walsh, A. J., 1993. Approaches to estimation of survival in two arctic-nesting goose species. Pp. 141–56 *in* Marked Individuals in the Study of Bird Populations ed by J. D. Lebreton and P. M. North. Birkhauser: Basel.

Black, J. M. and Owen, M., 1989. Agonistic behaviour in barnacle goose flocks: assessment, investment and reproductive success. *Anim. Behav.* **37:** 199–209.

Boudewijn, T., 1984. The role of digestibility in the selection of spring feeding sites by brent geese. *Wildfowl* **36:** 97–105.

Buchsbaum, R., Wilson, J. and Valiela, I., 1986. Digestibility of plant constituents by Canada geese and Atlantic brant. *Ecology* **67:** 386–93.

Burgman, M. A., Ferson, S. and Akçakaya, H. R., 1993. Risk assessment in conservation biology. Chapman and Hall: London.

Clobert, J., Lebreton, J. D. and Allaine, D., 1987. A general approach to survival rate estimation by recaptures or resightings of marked birds. *Ardea* **75:** 133–42.

Cooch, E. G., Lank, D. B., Rockwell, R. F. and Cook, F., 1989. Long-term decline in fecundity in snow goose population: evidence for density dependence. *J. Anim. Ecol.* **58:** 711–26.

Cooke, F., 1987. Lesser snow goose: a long-term population study. Pp. 407–32 *in* Avian Genetics, a Population and Ecological Approach ed by F. Cooke and P. A. Buckley. Academic Press: London.

Department of the Environment, 1993. UK strategy for sustainable development. Her Majesty's Stationary Office: London.

Ebbinge, B. S., 1985. Factors determining the population size of Arctic-breeding geese, wintering in western Europe. *Ardea* **73:** 121–28.

Ebbinge, B. S., 1989. A multi-factoral explanation for variation in breeding performance of Brent Geese. *Ibis* **131:** 196–204.

Ebbinge, B. S., 1991. The impact of hunting on mortality rates and spatial distribution of geese wintering in the western Palearctic. *Ardea* **79:** 197–209.

Ebbinge, B. S., van Biezen, J. B. and van der Veot, H., 1991. Estimation of annual adult survival rates of barnacle geese using multiple resightings of marked individuals. *Ardea* **79:** 73–112.

Fox, A. D. and Gitay, H., 1991. Breeding success in Greenland barnacle geese wintering on Islay, Scotland. *Ardea* **79:** 359–64.

Fox, A. D. and Madsen, J., 1997. Behavioural and distributional effects of hunting disturbance on waterbirds in Europe: implications for refuge design. *J. App. Ecol.* **34:** 1–13.

Francis, C. M., Richards, M. H., Cooke, F. and Rockwell, R. F., 1992. Changes in survival rates of lesser snow geese with age and breeding status. *Auk* **109:** 731–47.

Fretwell, S. D., 1972. Populations in a seasonal environment. Princeton University Press: Princeton.

Goss-Custard, J. D., Caldow, R. W. G., Clarke, R. T. and West, A. D., 1995. Deriving population parameters from individual variations in foraging behaviour. II. Model tests and population parameters. *J. Anim. Ecol.* **64:** 277–89.

Greenwood, J. J. D., 1993. The ecology and conservation management of geese. *Trends Ecol. Evol.* **8:** 307–08.

Grimmet, R. F. A. and Jones, T. A., 1989. Important bird areas in Europe. International Council for Bird Preservation, Cambridge, UK. Technical Publication No. 9.

Groot Bruinderink, G. W. T. A., 1989. The impact of wild geese visiting improved grasslands in the Netherlands. *J. App. Ecol.* **26:** 131–46.

Hanski, I., 1991. Single-species metapopulation dynamics: concepts, models and observations. *Biol. J. Linn. Soc.* **42:** 17–38.

Harrison, S., 1991. Local extinction in a metapopulation context: an empirical examination. *Biol. J. Linn. Soc.* **42:** 73–88.

Hestbeck, J. B., Nichols, J. D. and Malecki, R. A., 1991. Estimates of movement and site fidelity using mark-resight data of wintering Canada geese. *Ecology* **72:** 523–33.

Kear, J., 1970. The experimental assessment of goose damage to agricultural crops. *Biol. Conserv.* **2:** 206–12.

Kirby, J. S., Salmon, D. G., Atkinson-Willes, G. L. and Cranswick, P. A., 1995. Index numbers for waterbird populations. III. Long-term trends in the abundance of wintering wildfowl in Great Britain, 1966/7–1991/2. *J. App. Ecol.* **32:** 536–51.

Kotanen, P. and Jefferies, R. L., 1997. Long-term destruction of wetland vegetation by lesser snow geese. *Ecoscience* **4:** 1895–898.

Krebs, J. R., 1978. Optimal foraging: decision rules for predators. Pp. 4–70 *in* Behavioural Ecology ed by J. R. Krebs and N. B. Davies. Blackwell Scientific Publications: Oxford.

Kuyken, E., 1969. Grazing of wild geese on grassland at Damme, Belgium. *Wildfowl* **20:** 47–54.

Lambeck, R. H. D., 1990. The applicability of age-ratio and brood size counts in population dynamic studies of brent geese. *Ardea* **78:** 414–25.

Larsson, K. and Forslund, P., 1994. Population dynamics of the barnacle goose in the Baltic area: density dependent effects on reproduction. *J. Anim. Ecol.* **63:** 954–62.

Madsen, J., 1991. Status and trends of goose populations in the Western Palearctic in the 1980s. *Ardea* **79:** 113–22.

Madsen, J. and Noer, H., 1996. Decreased survival of pink-footed geese *Anser brachyrhynchus* carrying shotgun pellets. *Wildl. Biol.* **2:** 75–82.

McKay, H. V., Bishop, J. D. and Ennis, D. C., 1994. The possible importance of nutritional requirements for dark-bellied brent geese in the seasonal shift from winter cereals to pasture. *Ardea* **82:** 123–32.

McKay, H. V., Bishop, J. D., Feare, C. J. and Stevens, M. C., 1993. Feeding by brent geese can reduce yield of oilseed rape. *Crop Protec.* **12:** 101–05.

Murton, R. K., Isaacson, A. J. and Westwood, N. J., 1974. A study of wood-pigeon shooting: the exploitation of a natural animal population. *J. App. Ecol.* **11:** 61–81.

Ogilvie, M. A., 1978. Wild Geese. T. and A. D. Poyser: Berkhamsted.

Owen, M., 1971. The selection of feeding site by white-fronted geese in winter. *J. App. Ecol.* **8:** 905–17.

Owen, M., 1977. The role of wildfowl refuges on agricultural land in lessening the conflict between farmers and geese in Britain. *Biol. Conserv.* **11:** 209–22.

Owen, M., 1980. Wild Geese of the World. Batsford Press: London.

Owen, M., 1987. The use of the Solway Firth by barnacle geese in relation to refuge establishment and increase in numbers. *Biol. Conserv.* **39:** 63–81.

Owen, M., 1990. The damage-conservation interface illustrated by geese. *Ibis* **132:** 238–52.

Owen, M. and Black, J. M., 1989. Factors affecting the survival of barnacle geese on migration from the breeding grounds. *J. Anim. Ecol.* **58:** 603–17.

Owen, M. and Black, J. M., 1991. Geese and their future fortune. *Ibis* **133:** 28–35.

Owen, M. and Pienkowski, M. W., 1991. Proceedings of the goose damage and management workshop, Martin Mere, UK. JNCC, Peterborough, UK. 86 Pp.

Owens, N. W., 1977. Responses of wintering brent geese to human disturbance. *Wildfowl* **28:** 5–14.

Patterson, I. J., 1991. Conflict between geese and agriculture: does goose grazing cause damage to crops? *Ardea* **79:** 179–86.

Patton, D. L. H. and Frame, J., 1981. The effect of grazing in winter by wild geese on improved grassland in west Scotland. *J. App. Ecol.* **18:** 311–26.

Percival, S. M., 1988. The grazing ecology of barnacle geese on Islay. Ph.D. thesis, University of Glasgow, Scotland.

Percival, S. M., 1991. The population structure of Greenland barnacle geese on the wintering grounds on Islay. *Ibis* **133:** 357–64.

Percival, S. M., 1992. Population Modelling of Barnacle Geese on Islay. Glasgow: Scottish Natural Heritage, Research, Survey and Monitoring reports.

Percival, S. M., 1993. The effects of reseeding, fertilizer application and disturbance on the use of grasslands by barnacle geese and the implications for refuge management. *J. App. Ecol.* **30:** 437–43.

Percival, S. M., 1996. An assessment of the impact of the Islay goose management scheme on Greenland geese populations on Islay. Report to Scottish Natural Heritage.

Percival, S. M. and Anderson, G. Q. A., 1998. Habitat use and site fidelity of Svalbard light-bellied brent geese At Lindisfarne: exploitation of a novel food resource. *Norsk Polarinstitut Skrifter* **200:** 295–304.

Percival, S. M. and Evans, P. R., 1997. Brent geese and *Zostera*: factors affecting the exploitation of a seasonally declining food resource. *Ibis* **139:** 121–28.

Percival, S. M., Halpin, Y. and Houston, D. C., 1997a. Managing the distribution of barnacle geese on Islay, Scotland, through deliberate human disturbance. *Biol. Conserv.* **82:** 273–77.

Percival, S. M. and Houston, D. C., 1992. The effect of winter grazing by barnacle geese on grassland yields on Islay. *J. App. Ecol.* **29:** 35–40.

Percival, S. M., Lilley, R. and Andrews, M., 1997b. Grassland management for geese: sustainable use for conservation and agriculture. *Proc. XVIII Internat. Grassl. Cong.* **15:** 1–2.

Percival, S. M. and Percival, T., 1997. Feeding ecology of barnacle geese on their spring staging grounds in northern Iceland. *Ecography* **20:** 461–65.

Percival, S. M., Sutherland, W. J. and Evans, P. R., 1996. A spatial depletion model of the responses of grazing wildfowl to the availability of intertidal vegetation. *J. App. Ecol.* **33:** 979–92.

Percival, S. M., Sutherland, W. J. and Evans, P. R., 1998. Intertidal habitat loss and wildfowl numbers: applications of a spatial depletion model. *J. App. Ecol.* **35:** 57–63.

Pettifor, R., Black, J. M., Owen, M., Rowcliffe, J. M. and Patton, D., 1998. Growth of the Svalbard Barnacle Goose winter population 1958–96. *Norsk Polarinstitut Skrifter* **200:** 145–64.

Prins, H. H. Th. and Ydenberg, R. C., 1985. Vegetation growth and a seasonal habitat shift of the barnacle goose. *Oecologia* **66:** 122–25.

Raveling, D. G., 1979. Traditional use of migration and winter roost sites by Canada geese. *J. Wildl. Manage.* **43:** 229–35.

Ruger, A., 1985. Extent and control of goose damage to agricultural crops. IWRB Special Publication No. 5. IWRB, Slimbridge.

Sedinger, J. S., 1984. Protein and amino-acid composition of tundra vegetation in relation to nutritional requirements of geese. *J. Wildl. Manage.* **48:** 1128–136.

Summers, R. W., 1986. Breeding production of dark-bellied brent geese in relation to lemming cycles. *Bird Study* **33:** 105–08.

Summers, R. W., 1990. The effect on winter wheat of grazing by brent geese. *J. App. Ecol.* **27:** 821–33.

Summers, R. W. and Critchley, C. N. R., 1990. Use of grassland and field selection by brent geese. *J. App. Ecol.* **27:** 834–46.

Sutherland, W. J. and Allport, G. A., 1994. A spatial depletion model of the interaction between Bean Geese and Wigeon with the consequences for habitat management. *J. Anim. Ecol.* **63:** 51–59.

Sutherland, W. J. and Crockford, N. J., 1993. Factors affecting the feeding distribution of red-breasted geese wintering in Romania. *Biol. Conserv.* **63:** 61–65.

Syroechkovsky, Y. V., Litvin, K. Y. and Ebbinge, B. S., 1991. Breeding success of geese and swans on Vaygach Island (USSR) during 1986–88; interplay of weather and Arctic fox predation. *Ardea* **79:** 373–82.

Taylor, J. P. and Kirby, R. E., 1990. Experimental dispersal of wintering snow and Ross' geese. *Wildl. Soc. Bull.* **18:** 312–19.

Teunissen, W., Spaans, B. and Drent, R., 1985. Breeding success in brent in relation to individual feeding opportunities during spring staging in the Wadden Sea. *Ardea* **73:** 109–19.

Tucker, G. M. and Heath, M. F., 1994. Birds in Europe: their conservation status. Birdlife Conservation Series No. 3. Birdlife International: Cambridge, United Kingdom.

van Eerden, M. R., 1990. The solution of goose damage problems in the Netherlands, with special reference to compensation schemes. *Ibis* **132:** 253–61.

Vickery, J. A. and Summers, R. W., 1992. The cost effectiveness of scaring brent geese from fields of arable crops by a human bird scarer. *Crop Protec.* **11:** 480–84.

Vickery, J. A., Sutherland, W. J. and Lane, S. J., 1994. The management of grass pastures for brent geese. *J. App. Ecol.* **31:** 282–90.

Vickery, J. A., Sutherland, W. J., Watkinson, A. R., Lane, S. J. and Rowcliffe, J. M., 1995. Habitat switching by dark-bellied brent geese in relation to food depletion. *Oecologia* **103:** 499–508.

Vickery, J. A., Sutherland, W. J., O'Brien, M., Watkinson, A. R. and Yallop, A., 1997. Managing coastal grazing marshes for breeding waders and over-wintering geese: is there a conflict? *Biol. Conserv.* **79:** 23–34.

Warren, S. M. and Sutherland, W. J., 1992. Goose populations — conservation, conflict and solutions. *Trends Ecol. Evol.* **7:** 71–72.

Wilson, H. J., Norriss, D. W., Walsh, A., Fox, A. D. and Stroud, D. A., 1991. Winter site fidelity in Greenland white-fronted geese, implications for conservation and management. *Ardea* **79:** 287–94.

CHAPTER 20

Towards a sustainable production environment in Xilingol Biosphere Reserve, China: adding wildlife to the equation

R. Thwaites,[1] J. Bauer[2] and T. DeLacy[3]

Xilingol Biosphere Reserve in Inner Mongolia, China, aims to protect the biodiversity of a typical steppe ecosystem, and to develop models of sustainable grassland resource use for the improved well-being of the local people. Within the reserve a substantial human population depends on animal husbandry for its economic development, and a variety of factors have contributed to serious degradation of the grassland. "Conservation" is largely discussed in terms of those components of the grassland ecosystem that provide a direct resource to the animal husbandry industry. Current proposals broadly aimed at increasing the productivity of the grassland include the increased use of fencing for rotational grazing, and for establishment of "artificial" (improved) grassland. The severely depleted populations of wild ungulate species may be further threatened by introduction of fencing on the open rangeland.

This chapter examines the inadequacies of attempts to integrate conservation and development which consider only those elements of biodiversity which provide a resource for development, but which ignore wildlife conservation. The attitudes and activities of local people within the reserve towards wildlife are described, and the potential to include planning for wildlife management and recovery in development plans for the reserve is discussed.

Key words: Sustainable development, Biosphere reserves, Steppe grassland, Land degradation, Sustainable grazing, Integrated conservation and development, Wildlife conservation.

INTRODUCTION

THE decline of productivity and biodiversity in pastoral regions is a widespread phenomenon with increasing pressure from expansion of livestock numbers and changed pastoral patterns. The development of more sustainable systems of pastoral land use has therefore become a fundamental issue for many countries. Arctic tundra, the steppes of Inner Asia, the sub-tropical savannah regions of Africa and Australia, the alpine highlands in the Himalayan region, the Andes, Europe and New Zealand are all similarly affected. Each of these regions must seek its own more diverse, productive and more sustainable use of the pastoral land, which will only be possible when the knowledge and planning of the agro-economic system is complemented by knowledge of the ecological, historical and social contexts.

The arid steppes of Inner Mongolia, China have suffered in recent decades from dramatic degradation including loss of soil, change of species composition and structure, and loss of productivity (Lin 1990; Zhang 1990; Longworth and Williamson 1993; Maxey and Lutz 1994). This degradation threatens the livelihood of local people living on the grassland.

This chapter is based on social and ecological research undertaken by the authors in Xilingol Biosphere Reserve, Inner Mongolia, China, between 1993 and 1997. The biosphere reserve has been

[1]Johnstone Centre of Parks, Recreation and Heritage, Charles Sturt University, P.O. Box 789, Albury, New South Wales, Australia 2640.
[2]Johnstone Centre of Parks, Recreation and Heritage, Charles Sturt University, Panorama Avenue, Bathurst, New South Wales, Australia 2795.
[3]Co-operative Research Centre for Sustainable Tourism, Griffith University, PMB 50, Gold Coast Mail Centre, Queensland, Australia 9726.
Pages 198–211 *in* NATURE CONSERVATION 5: NATURE CONSERVATION IN PRODUCTION ENVIRONMENTS: MANAGING THE MATRIX ed by J. L. Craig, N. Mitchell and D. A. Saunders. Surrey Beatty & Sons, 2000.

established over the top of an existing land management system. The grasslands are a production environment utilized by herders in state farms and what were once collective farms. In recent times, particularly since the introduction of economic reforms across China from 1978, the number of domestic livestock has increased dramatically, and grazing patterns have changed. Here, as elsewhere, there has been widespread degradation of the grasslands and loss of productivity, and some species of wildlife have all but disappeared from the grasslands. In this chapter, we reflect on grassland management and wildlife issues in Xilingol Biosphere Reserve, and consider the meaning of a "sustainable production environment".

"SUSTAINABILITY"

The title of this conference, "Conservation in a production environment" implies the need to achieve some kind of balance between conserving resources, and utilizing them for human development. This elusive balance has been discussed widely in recent years, and often the term "sustainable" is added to the description of the production activity to imply such a balance. Thus we hear of "sustainable production", "sustainable agriculture", "sustainable resource use" and so on.

The concept of sustainability is simple enough. That is the notion that an activity should be able to be sustained into the future, that it should not be self-defeating. The notion of "sustainability" has been adopted widely by different societies around the world, as a common sense objective for our survival and well being into the future. However, the notion can be very differently interpreted by different people or groups. Thus a resource user may see a problem of resource shortage in terms of "sustainable yield", isolating only the specific issues which impact upon the yield of the resource available. A broader view of the world is taken by those who promote the concept of "ecologically sustainable development". This takes account of a wide range of factors, which may impact on the manner in which the whole system functions — the social, cultural, economic, ecological factors, not just those directly related to production.

Sustainable development

Achieving a balance between conservation and utilization was raised as an international objective in the World Conservation Strategy in 1980, which described conservation as: "the management of human use of the biosphere so that it may yield the greatest sustainable benefit to present generations while maintaining its potential to meet the needs and aspirations of future generations" (International Union for the Conservation of Nature 1980). This strategy argued that both conservation and development are for people, and that while development aims to achieve human goals through use of the biosphere, conservation aims to achieve those same goals by ensuring that such use of the biosphere can continue indefinitely.

Since the publication of the World Conservation Strategy (International Union for the Conservation of Nature 1980), the notion of a reconciliation, or integration of conservation with development has received growing attention from scientists, planners and policy-makers. The concept of "sustainable development" was defined by the World Commission on Environment and Development as: "development that meets the needs of the present without compromising the ability of future generations to meet their own needs" (World Commission on Environment and Development 1987). This concept incorporates the notions of limits to environmental resources, conservation of biodiversity, meeting the needs of human populations in an equitable fashion, and respecting the rights of future inhabitants of the planet to resource use options available to us.

Following the United Nations Conference on Environment and Development, held in Rio de Janeiro in June 1992, the objective of sustainable development has achieved almost universal acceptance. Though there have been many definitions of sustainable development, as each nation grapples with the difficult problem of how to integrate conservation with development, similar themes arise. The Australian National Strategy for Ecologically Sustainable Development (Commonwealth of Australia 1992), describes three core objectives:

1. to enhance individual and community well-being and welfare by following a path of economic development that safeguards the welfare of future generations;
2. to provide for equity within and between generations;
3. to protect biological diversity and maintain essential ecological processes and life-support systems.

This strategy emphasizes two main features which distinguish this approach:

1. the need to consider in an integrated way, the wider economic, social and environmental implications of our decisions and actions for Australia, the international community, and the biosphere; and
2. the need to take a long-term rather than short-term view when taking those decisions and actions.

This description of Ecologically Sustainable Development, is then very different to simply adding the adjective "sustainable" to describe some production activity. It takes a holistic approach to the issues of resource use to meet human needs, and ensuring human survival and well-being into the future.

BIOSPHERE RESERVES

The model of biosphere reserves is well suited to the need for this holistic or integrated approach to conservation and development. The first biosphere reserves were designated under United Nations

Education, Scientific and Cultural Organisation's Man and Biosphere programme in 1975, with the intention of creating a world-wide protected area network of representative ecosystems. Biosphere reserves are designed to provide a framework for the establishment of protected areas which take account of the needs of local people; that is, providing for both conservation of biodiversity and for sustainable socio-economic development within the local community.

Biosphere reserves are described as having three complementary functions or roles:

1. Conservation role: biosphere reserves provide non-conventional protection of indigenous genetic resources, plant and animal species, ecosystems and landscapes of value for the conservation of the world's biological diversity;
2. Development role: biosphere reserves seek to combine conservation concerns with sustainable use of ecosystem resources through close co-operation with local communities, taking advantage of traditional knowledge, indigenous products and appropriate land management;
3. Logistic role: biosphere reserves are linked through a global network; they provide facilities for research, monitoring, education and training for local purposes as well as for international or regional comparative research and monitoring programmes (United Nations Education, Scientific and Cultural Organisation 1994).

Biosphere reserves are zoned to allow these different functions to be implemented within the landscape. The ideal model of a biosphere reserve described by Batisse (1986) includes core, buffer and transition zones, though some authors have used other terms for the various zones, such as "multiple use areas", "zones of co-operation" and "areas of influence" to more closely reflect the activities undertaken within these areas (e.g., Gregg and McGean 1985). A vision statement for the next century developed at the second International Conference on Biosphere Reserves held in Seville in March, 1995, directly related the biosphere reserve concept to the sustainable development objectives established by the United Nations Conference on Environment and Development in 1992. Biosphere reserves are described as:

> "examples that encapsulate the ideas of United Nations Conference on Environment and Development for promoting both conservation and sustainable development... Rather than forming islands in a world increasingly affected by severe human impacts, they can become theatres for reconciling people and nature, they can bring knowledge of the past to the needs of the future... Thus biosphere reserves are poised to take on a new role. Not only will they be a means for the people who live and work within and around them to attain a balanced relationship with the natural world, they will also contribute to the needs of society as a whole by showing a way to a more sustainable future" (United Nations Education, Scientific and Cultural Organisation 1996, p.5).

Clearly there is considerable overlap between the notion of sustainable development, and the biosphere reserve model.

RESEARCH IN XILINGOL BIOSPHERE RESERVE

The biosphere reserve concept has been embraced in China, with its focus on integrating conservation with development into the one programme. At the end of 1996, there were 61 nature reserves in the China Biosphere Reserve Network, 12 of which were recognized internationally under the United Nations Education, Scientific and Cultural Organisation Man and Biosphere programme (Fig. 1). In collaboration with the Inner Mongolia Grasslands Ecosystem Research Station of the Chinese Academy

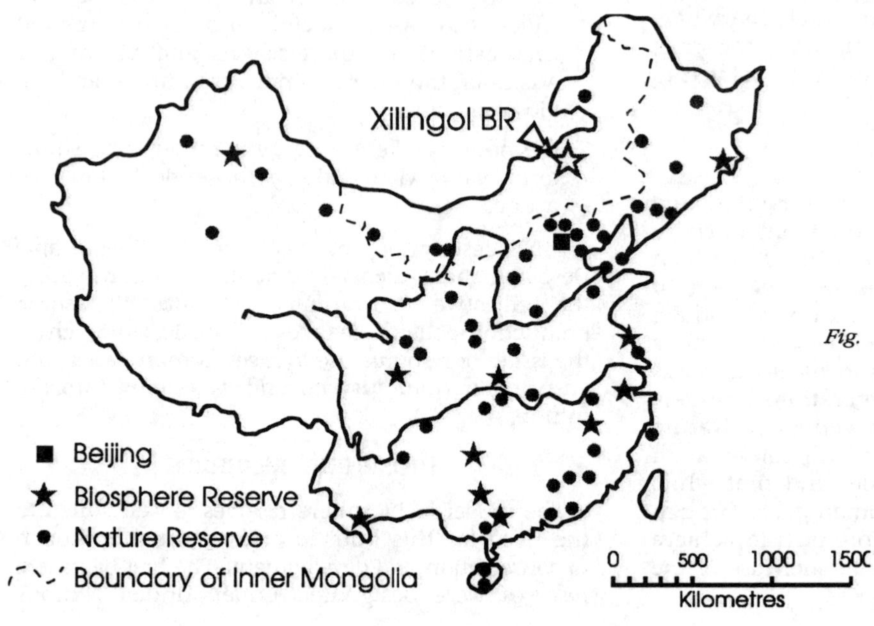

Fig. 1. Distribution of nature reserves of China Biosphere Reserve Network.

of Sciences, research has been undertaken to study the implementation of the biosphere reserve concept in the Xilingol Biosphere Reserve. Xilingol Biosphere Reserve is located about 600 km north of Beijing in the Inner Mongolia Autonomous Region, and was established in 1985 as China's first grassland nature reserve, joining the Man and Biosphere reserve network in 1987. Most of the previous research undertaken by Inner Mongolia Grasslands Ecosystem Research Station within the reserve has dealt with the fundamental structure and functioning of the grassland ecosystem and its biodiversity. This collaborative project was designed to cover a broader range of ecological and social research, to assist in the development of sustainable land management practices.

Visits to China were made by the authors each year from 1993 to 1996, to establish collaborative relationships with research partners, and to develop the research project and gather data. As very little was known by the authors about the existing situation in the Xilingol Biosphere Reserve, the social component of this research adopted a qualitative methodology to understand the complexity of the social, institutional, land management and grazing systems.

Data collection has included formal and informal workshops, semi-structured interviews with individuals and groups, site visits and observations, as well as gathering secondary data such as maps, reports, and policy and legal documents. Much of the research has been focused around sites within Xilingol Biosphere Reserve. However, interviews and observations have been undertaken further afield on the Xilingol grasslands, in Beijing, Hohhot, capital of Inner Mongolia, as well as in Mongolia itself to offer a comparative study of a different system of management. Most of the data has been collected during the short Inner Mongolian summers of 1995 and 1996, and particular emphasis has been placed on interviewing key informants from various government agencies, management authorities, researchers, as well as the grassland resource users, the herders and their families.

The ecological surveys were undertaken during the summers of 1995 and 1996. The survey methodology was based on a rapid grassland assessment method described by Bauer (1990), and modified for the specific of the Xilingol environment (Bauer *et al.*, unpubl. (a)). The survey attempted to obtain a measure of grassland composition, diversity and degradation across the landscape. Approximately 600 radial 12.64 m² plots were placed along walking transect lines sampling a range of environmental gradients (slope, aspect, altitude, soil, topography etc.). For each plot, information was gathered on vegetation (number and coverage of herb, grass and shrub species) and evidence of physical processes (erosion, trampling, defecation and so on). The sampling method took account of the accessibility of sampling points, as well as overlap with the sociological surveys described above. This rapid assessment method was complemented by specific habitat assessments (Bauer *et al.*, unpubl. (a)).

INTRODUCTION TO XILINGOL BIOSPHERE RESERVE

Xilingol Biosphere Reserve was established to protect the biodiversity of a typical steppe ecosystem and to develop models of sustainable grassland resource use for improved well-being of the local people (Li *et al.* 1995). The reserve covers an area of 10 786 km², corresponding to the catchment of the Xilin River, an inland flowing river in the south-east of the Mongolian plateau, (Fig. 2). The area experiences a cool temperate semi-arid climate, and covers a range of landscape types from 950 m to 1 500 m above sea level. The resultant variety of local climatic features, soil and vegetation types contains great ecosystem and species diversity, including over 650 species of flowering plants (Li *et al.* 1995).

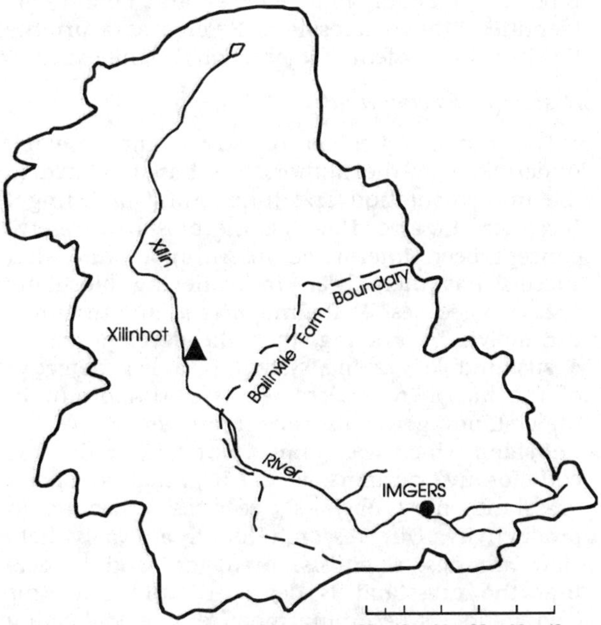

Fig. 2. Xilingol Biosphere Reserve.

Xilingol Biosphere Reserve also contains a substantial human population of around 200 000 people, living in the city of Xilinhot (about 150 000) and the surrounding livestock farms. The biosphere reserve has been established over an existing production environment and land management system. Though the Xilingol Biosphere Reserve Management Bureau has nominal responsibility for management of the reserve, it has no independent authority. The responsibility for management of the land rests with the resource users, the livestock farms, which focus mainly on production objectives independent of the reserve's objectives to integrate or balance conservation needs and values with production.

The main functional area of the reserve, including the areas nominated as core conservation areas, and buffer zones, is the Baiinxile Livestock Farm, which includes much of the headwaters of the Xilin River in the south-east of the reserve. This is a state-farm covering an area of 3 730 km², with the purpose of breeding fine varieties of domestic livestock, particularly fine-wool sheep. The human population of Baiinxile Farm in 1995 was 10 480.

The China Biodiversity Action Plan (Maxey and Lutz 1994) notes that the steppe grassland ecosystems in the semi-arid north of China are experiencing serious degradation caused by overgrazing, introduction of farming, and rodent plagues. While these steppes cover about one-third of the area of China, only 12 nature reserves to the end of 1991 had been declared covering 0.75% of the area of grassland ecosystems in China. Grasslands are clearly under-represented in designated nature reserves, and the importance of Xilingol Biosphere Reserve is reinforced by the fact that the area of this single reserve makes up 82% of the total area of grasslands protected. The China Biodiversity Action Plan (Maxey and Lutz 1994) identifies Xilingol Biosphere Reserve as a priority Category 1 ecosystem of international significance.

Research development

The central question on which the research undertaken by the authors was based related to the implementation and functioning of Xilingol Biosphere Reserve. How has the biosphere reserve concept been interpreted in Xilingol, and what success has there been in achieving biosphere reserve objectives? As the major land and resource-use activity is grazing, then the development of a sustainable grazing system is a key objective of the biosphere reserve. An issue important in understanding the use and management of the grassland which was pointed out by our Chinese collaborative partners, is the degradation of the grassland, most obviously seen as a decline in productivity. Our research found a widely held view amongst scientists, managers and herders that the grassland is degraded, and that this degradation is becoming more serious, indicating that the current grazing system is not sustainable. While the direct reason for this degradation has been given by a number of authors as overgrazing (Longworth and Williamson 1993; Maxey and Lutz 1994; Dolgipalam 1996), our research attempted to look at the underlying causes for overgrazing, at the factors which might cause certain decisions to be made. To understand the problems of implementing sustainable land and resource management practices, it was important to understand the existing grazing management system, and particularly to understand what factors influence the grazing decision making process.

Specific outcomes of our research, particularly related to factors influencing grazing management, have been discussed elsewhere in English and Chinese (Thwaites *et al.* 1996a; Thwaites *et al.* 1996b; Thwaites and DeLacy 1997; Thwaites *et al.* 1998). It is not the purpose of this chapter to discuss all the findings related to management of the landscape, or all the data from the ecological studies. Rather we intend to focus on one particular factor, the interpretation of "sustainable development", and one element of the land management system which results, wildlife management.

"SUSTAINABLE DEVELOPMENT" OR "SUSTAINABLE YIELD"?

In this chapter, we argue that in Xilingol Biosphere Reserve, and in particular, in Baiinxile Livestock Farm where most data was collected, planning is not considering the notion of sustainable development as an integration of diverse objectives. Despite the presence of the biosphere reserve with its international designation and objectives, and numerous policies and legislation from Beijing and Hohhot, management of the grasslands considers, at best, only "sustainable yield".

To support this hypothesis, we present the following evidence gathered during the field research programmes.

1. Economic decision making process in Baiinxile Farm. The Baiinxile Farm is a state farm, managed through the Communist Party hierarchy. As such it is required to respond to the "Five-year plans" from higher levels of the party with its own economic development strategy or five-year plan. The Ninth Five Year Plan for Baiinxile Farm, which runs until 2000, sets growth targets for economic and production outputs, as well as increasing livestock numbers. This economic planning process has taken no consideration of the ecological capacity of the grasslands but sets figures for "desired stocking rates" over the grasslands of the Baiinxile Farm. At the same time, these "desired stocking rates", or carrying capacities are based on total available grassland productivity, and disregard the grazing demands of wildlife and the need to maintain varied habitats to meet the needs of the different species.
2. Despite having some legal status and protection as a nature reserve under the National Environmental Protection Agency *Nature Reserve Act* (National Environmental Protection Agency 1994), and despite being identified by the China Biodiversity Action Plan (Maxey and Lutz 1994) as a high priority area for protection, the Xilingol Biosphere Reserve Management Bureau has no power at a local level to manage the land. The management of the land is vested with the livestock farms and collectives. It is these bodies which make most decisions relating to resource use, and production activities such as stocking rates. Since the introduction of economic reforms the individual herders have been given the right to own their own livestock, and in the collective farms much of the grassland has also been allocated on contracts to herder families. So the farms, and the herders make most decisions related to livestock and grassland management. The biosphere reserve management bureau is left with little authority to manage the major factors which directly influence the utilization and conservation of the grassland resource.

3. It can be seen from these two previous points that the production objectives over-ride the conservation objectives in the planning and management of grassland resource use. The primary objective of the Baiinxile Farm is to raise herder incomes, any objective of the biosphere reserve to conserve biodiversity is secondary to this primary development objective. This attitude is clearly indicated by statements made by local officials. The director of a sub-farm in Baiinxile Farm observed that the grass inside one of the small fenced core areas was very long, but "this is not useful to anyone, it cannot be utilized". The director of the biosphere reserve himself stated that "in China there are many people and nature protection is difficult. Nature protection is aimed at utilization. If we do not utilize the resource, but only protect it, it is not really nature protection." This utilitarian attitude to conservation also seems to imply that only production resources are worth conserving.
4. The degradation of the grassland is widely recognized, as is the threat this poses to the animal husbandry industry. However, this issue is not viewed in a holistic sense, in terms of the social, political and regulatory environment in which the degradation is occurring. An isolated view is taken of the problem as simply a lack of grass to feed the increasing numbers of livestock. As a result, the official solution is to increase the productivity of the grassland, thereby enabling the herders to increase their livestock numbers, and so raise their incomes. This one solution is expected to solve the problem of degradation of the grassland, and help the farm to meet its economic and production targets. This research has identified a range of factors outside the immediate grazing environment, which influence herder, and farm manager decisions, and therefore affect the use and condition of the grassland. These factors particularly relate to inconsistencies and uncertainties associated with government policies such as property rights. Changes in the productivity of the grassland alone will not deal with these causes of overgrazing, but will encourage herders to increase their livestock numbers, thus exacerbating the problem.
5. The proposed means of raising the grassland productivity include the establishment of "semi-artificial" and "artificial" grasslands, controlling livestock access to areas by fencing, and in some places the application of fertilizer and irrigation water. Artificial grasslands require the ploughing of the land, and sowing of a selected mix of species. There has been no attempt to consider the effect of these proposals on values of the grassland other than as a production resource, particularly the biodiversity of the grassland, and its associated fauna.
6. Discussions of "biodiversity" always seemed to focus on the variety of grassland species which provide a grazing resource. Though basic census of the species present has been completed, there has been little emphasis on turning this into detailed studies of the occurrence, status and distribution of species, and no work apparent on management plans for identified rare, threatened or endangered species. This is especially the case for animal species, which are typically viewed only in their role within the human ecology of the areas for damaging pasture, eating seeds, providing meat or fur, and predation (e.g., Zhou *et al.* 1988). Wildlife is seen officially as something outside or threatening to the production environment. Management related work has considered improving the productivity of the grasslands, but not at the same time looked at the biodiversity implications of current or proposed management practices.

XILINGOL AS A WILDLIFE HABITAT

Xilingol Biosphere Reserve covers a very diverse grassland system. It is not a typical steppe ecosystem, but is more diverse, with a range of landscape and vegetation types, and a combination of steppe and forest species.

Landscape geomorphology

The landscape of Xilingol Biosphere Reserve contains a great diversity of topography, landform, soil type, and climate, and thus offers a variety of habitat opportunities, significant for the evolution of the local flora and fauna. Landforms identified within the reserve include the following.

- Volcanic plateaus and hills ranging from about 1 000 m to 1 500 m, mostly in the upper catchment of the Xilin River to the south. These provide a diverse and climatically varied environment, including eroded drainage lines which offer protection for growth of tree and shrub species not present across the open steppe. These protected gullies also provide a preferred habitat for fauna species, including some birds, and the Roe Deer *Capreolus capreolus sibericus* to which they offer shelter, browse and escape cover from predators. Local herders also identified these gullies as likely sites to see predators, such as wild cats, wolf and fox. There is little water on the basalt plateau, so permanent grazing sites are isolated to the few places where water is found. The vegetation of the plateau is, however, cut extensively for hay to provide winter feed, and cultivated for growing crops.
- Sand dunes provide a special environment for both native biodiversity and domestic livestock, and occur in a strip about 10 km wide across the southern half of the reserve. The sand dune regions are very fragile, and susceptible to degradation. In Baiinxile Farm, the southern aspects of the sand dunes are typically eroding, with considerable bare soil, but supporting low shrub and herbaceous growth. The northern aspects support a relatively high amount of woody growth, in particular *Betula*, *Ulmus* and *Picea* sp. The narrow belt of dunes in Baiinxile Farm is very poor in water, and so does not offer a good

grazing site for summer. Open meadows are, however, cut for hay, and herds are brought into the sand dunes in winter for protected grazing.

- Hills of crystalline rock cover part of the reserve, particularly in the east and parts of the north. These generally experience a drier climate than the basalt plateau in the south, and though some gullies are found with shrubby vegetation and occasional trees, the landscape is generally more open with little woody vegetation, and particularly to the north of Xilinhot extends into wide plains.
- River meadows follow the wide valleys of the Xilin River and its tributaries. At the end of summer they are still densely grassed. However, they are heavily grazed by cattle and probably become an increasingly important food source in autumn and winter, as they retain biomass and are close to the villages and most of the herding sites which are focused around water supplies.
- Alluvial terraces are transitional habitats between the river meadows and wetlands, and the surrounding landscape of basalt, crystalline hills and sand dunes. These raised terraces are rather dry, and in a stage of light to moderate grazing contain a high biomass of diverse vegetation of herbs and grasses. As the herder population focuses around water supplies, these terraces are often heavily grazed and degraded.
- Aquatic habitats include shallow steppe lakes formed in depressions which harbour a diverse and significant avifauna, including migratory species such as five species of crane. These lakes, both within and outside Xilingol Biosphere Reserve have high conservation and tourism value, and provide a significant source of fish. The riverbeds and banks of the Xilin River and its tributaries are often badly eroded, but surrounding *Juncus* and *Carex* bogs with abundant flowering plants provide excellent waterbird habitat.

Vegetation

The grasslands of Xilingol are typical Eurasian steppes, and have been classified into 18 vegetation-soil-topography types, two of which are dominated by *Ulmus pumila* and *Prunus ansu* vegetation (Li *et al.* 1995). As well as a number of distinct grassland and vegetation types there are a great many transitional zones, which have been further modified by landform, and by grazing. The wide distribution of some tree species, and the presence of some forest remnants, suggest that, despite its low rainfall and semi-arid steppe climate, the area can support a healthy forest cover under particular microclimatic conditions, such as higher altitude, drainage lines and northern aspects. Spruce *Picea coreansis* forms a dense stand on the northern side of one large dune, considered by local botanical researchers to be a remnant of a once more extensive forest cover. Birch is widespread in the dune system, and also forms dense stands. These forests presumably play a role in providing a special habitat for species which are not typical steppe animals, and depend on shelter. The forests in Baiinxile Farm have been extensively harvested in the past, but are now protected from felling. Some small patches are fenced as core areas and protected from grazing, resulting in extensive regrowth. Similar birch forests to the east, outside the Biosphere Reserve are however being felled to provide timbers for coal mines close to Xilinhot.

Wild apricot trees occur in stands of varying density on the southern aspects of the sand dunes. The collection of kernels for making a sweet milk provide a locally important economic sideline to the herders. *Populus* sp. have been planted around some villages to provide shelter and emergency winter fodder for livestock, but also sheltering the villages and cultivated fields. Some of these plantations are many hectares in size, and have become quite diverse habitats offering shelter for wildlife.

GRAZING AND BIODIVERSITY

The grasslands of Xilingol have co-evolved over thousands of years with grazing by both domestic and wild animals. Research undertaken by Inner Mongolia Grasslands Ecosystem Research Station in controlled grazing plots suggests that maximum biodiversity of the grasslands is achieved with a relatively low grazing pressure, in the form of intense grazing for short periods of time followed by extended periods of no grazing (Li 1993). Transects of grazing sites undertaken as part of this research suggest that this system is relatively tolerant of grazing (Fig. 3). With an increase in grazing pressure above a certain level, we would expect an increase in exposure of bare soil, so, in effect, Figure 3 shows the relationship from 495 study plots between the number of herb species, and grazing pressure. At low grazing pressure, and no bare soil, we can see around 10 to 15 herb species identified per plot. As the exposure of bare soil increases, there is a significant decrease in species diversity, to about five to eight species. This species diversity however, is then maintained through increasing grazing pressure, and only falls again at very high grazing pressure. This data suggests that the grazing system is relatively stable from the point of view of species diversity under varying grazing pressures, but as with any system, has its limit. This data is, however, a compilation plot, the real situation being more complex, and varying considerably between landform types. The sand dune system for example does not show this same stability under increased grazing pressure (Bauer *et al.*, unpubl. (b)).

Figure 3 also shows nothing of the actual species recorded. With increasing grazing pressure the species composition and abundance changes, and some species are lost from the system. In any natural grazing system, there is considerable "patchiness" (e.g., Ludwig *et al.* 1997), with varying natural conditions of soil, topography, aspect, microclimate, and also varying grazing pressures experienced across the grassland, from very high to ungrazed. This patchiness provides for a wide range

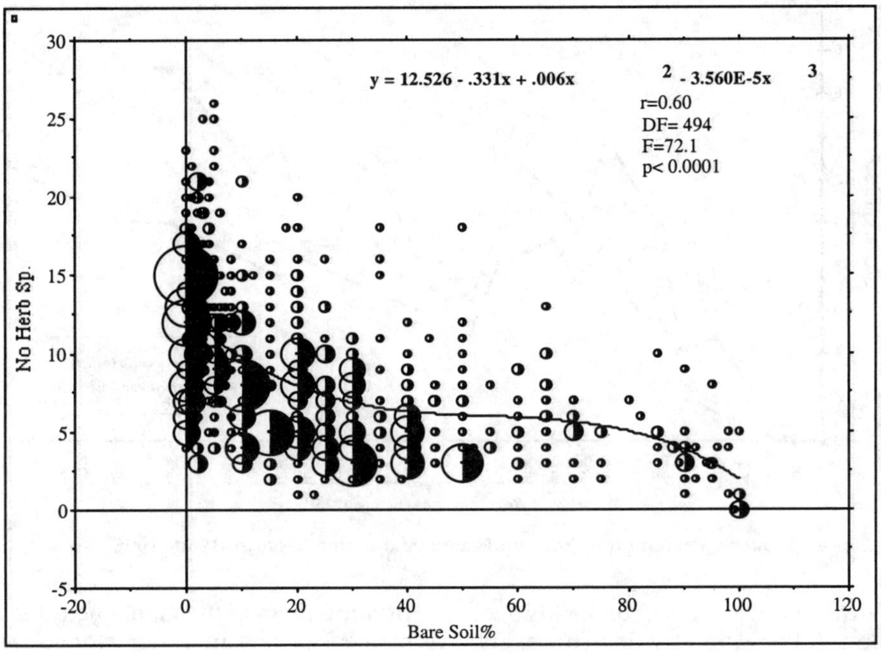

Fig. 3. Diversity of herbaceous plants and bare soil in Xilingol Biosphere Reserve.

of habitat conditions for different species to occur. However, as the grazing system moves towards a more uniform heavy grazing pressure, as would be experienced with changed grazing regimes from nomadic to sedentary and with increasing livestock populations and increasing hay-cutting, this patchiness, and therefore habitat and species diversity, is being reduced.

Fauna diversity

With its diversity of landform and vegetation, the grasslands of Xilingol Biosphere Reserve are able to support more than just a steppe fauna. Steppe lakes and water systems contain many species of waterfowl, waders and cranes, while the existence of some other species depends on the special habitat provided by the limited forest cover. Apart from an inventory of birds (Zhao and Feng 1991), and a review of the zoogeographical characteristics of mammals within Baiinxile Farm (Zhou *et al.* 1988), little is recorded of the vertebrate fauna of Xilingol. One exception is the long-term research being carried out by Inner Mongolia Grasslands Ecosystem Research Station on the pika (*Ochotona daurica*), a colonial rodent species which in response to vegetation changes caused by overgrazing, has become a significant economic pest species over large areas of Inner Mongolia.

Zhou *et al.* (1988) record 33 species of native mammal within their survey area, dominated by herbivorous animals, most of which are small burrowing rodents. The Pleistocene megafauna, which has been instrumental in the evolution of the steppe flora has virtually disappeared, and only very small numbers of two medium-sized native herbivores remain (*Capreolus c. sibiricus*, Siberian roe deer; and *Procapra gutturosa*, Mongolian gazelle). As would be expected, the predator fauna is specialized in the exploitation of small rodents, with four species (Mustela sp.) entirely depending on them and three species probably heavily exploiting them (*Vulpes vulpes*, red fox; *Vulpes corsac*, Corsac fox; *Meles meles*, badger). Rodents probably also play an important role in the diet of the two larger predators present, the wolf *Canis lupus* and the wild cat *Felis manul*.

PATTERNS OF FAUNA DECLINE

Nine species of domestic animals were also identified in field visits to Xilingol Biosphere Reserve including cat, dog, goat, sheep, cattle, horse, pig, donkey and camel. Livestock numbers have increased dramatically over the last forty years, as shown in Figure 4 for Baiinxile Farm and Xilingol League, the prefecture covering about 200 000 square kilometres in which the biosphere reserve is situated. The grazing patterns have also changed from nomadic or semi-nomadic, to sedentary. Grassland and steppe species are generally severely affected by increases in the population density of humans and livestock. Large species of herbivores are seen as competitors for feed, and offer a source of meat, or income from meat and skin. Predators are typically persecuted because of predation on livestock. In many pastoral regions, predators have been driven to extinction because of the need to protect livestock. Despite considerable hunting pressure, with organized hunting groups and modern weapons, the wolf is still present in reasonable numbers in parts of Xilingol Biosphere Reserve, although the interviews suggest that wolf and fox numbers have decreased significantly in recent decades.

The state of the ungulate fauna of the area has received little attention from scientific researchers, managers, or local herders. Our observations and

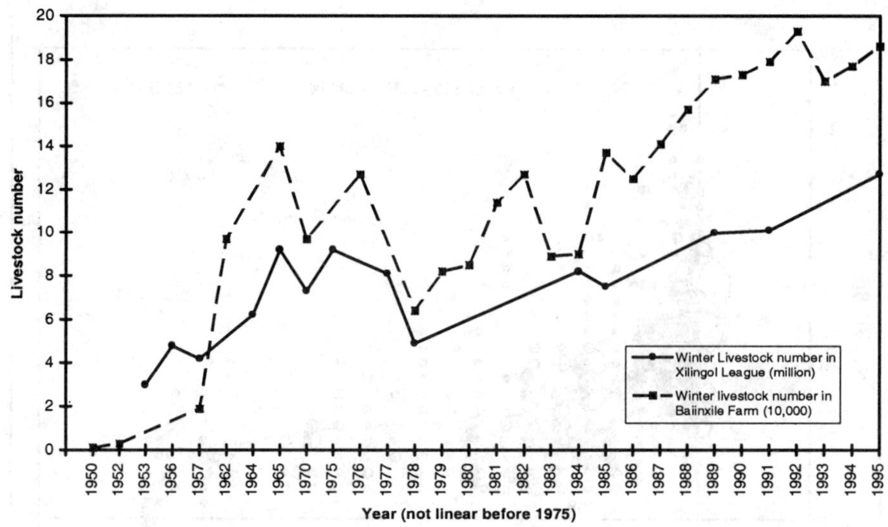

Fig. 4. Livestock population in Xilingol League and Baiinxile Farm, 1950–1995.

interviews verify that two species occur in Xilingol Biosphere Reserve in low to very low numbers, the Mongolian gazelle, locally called "huang yang" or "yellow sheep" *Procapra gutturosa*, and the Siberian roe deer, locally known as "paozi", *Capreolus capreolus sibericus*. The co-existence of these two species, one of which is a typical desert adapted, hardy steppe species and the other typical of the temperate boreal forests of the palae-arctic region, suggests that this region of Inner Mongolia contains transitional elements between the true steppes and temperate forests. These species may be remnants of a once more diverse ungulate fauna, and both could potentially hold economic value for local people. The present status and recent history of these species was reconstructed based on interviews with local herders, farm and reserve managers, and staff of the research station. We present here the example of the Mongolian gazelle.

Mongolian gazelle Procapra gutturosa — *a case study of faunal decline*

The numbers of Mongolian gazelle seen in the area of Baiinxile Farm have dropped dramatically in the last few decades (Fig. 5). Interviews with a number of people have suggested a dramatic decline in numbers of Mongolian gazelle on the grasslands, though the actual timing and pattern of decline may vary from place to place. In the past, Mongolian gazelle were migratory grazers, spending the summers in Mongolia, and coming in large numbers in winter to the Xilingol grasslands to graze on the longer grasses accessible through the snow cover. It seems likely from the interviews, that this migratory pattern has been broken. The large numbers of Mongolian gazelle still present in Mongolia are no longer coming south in the winter. Much of the steppe has become so degraded that it no longer provides good winter grazing, their habitat has been fragmented by increasing human and domestic livestock populations on the grassland, and in recent decades, the Mongolian gazelle have been under considerable hunting pressure to supplement the diet or incomes of local people in Inner Mongolia, and people in Chinese cities. Sightings described in the interviews also suggest that there has always been a smaller local or non-migratory population remaining in the area of Baiinxile Farm throughout the summer. With the migratory routes now cut off, the few remaining animals may be a remnant of this local sedentary population.

During the "Great Leap Forward", and the very bad seasons which accompanied this political event in 1960/61, there was widespread starvation, and organizations from Beijing and other big cities sent people on to the grasslands to shoot wild animals to feed their employees. One scientist who was sent in 1968 to the grasslands west of Xilinhot for ten years during the Cultural Revolution described a crash in the population of Mongolian gazelle during this time. He attributed this crash to the rapid degradation of the grassland resulting from a changed management regime, and to intensive hunting as a result of the policy of "self-sufficiency in food" which was promoted. From seeing groups of thousands of animals when he first arrived on the grassland, by the end of the Cultural Revolution, he said that there were only a few small groups, rarely seen.

The interviews indicated that the Mongolian gazelle was still relatively common on the Xilingol grasslands in the mid 1970s, with sightings of groups of 60 to 100 animals. Most herders in Baiinxile Farm who were asked, had never seen Mongolian gazelle, or had not seen them since the 1970s. The only herders who had seen them recently lived in isolated areas, and in diverse landscapes, such as the basalt plateau or sand dunes, not on the open steppe. Apart from one sighting in the southeastern corner of Xilingol Biosphere Reserve of about 20 animals in the winter of 1994/95, the group sizes of these sightings was small, up to seven animals, also suggesting a depleted population. One herder in the south of the biosphere reserve said that there were many

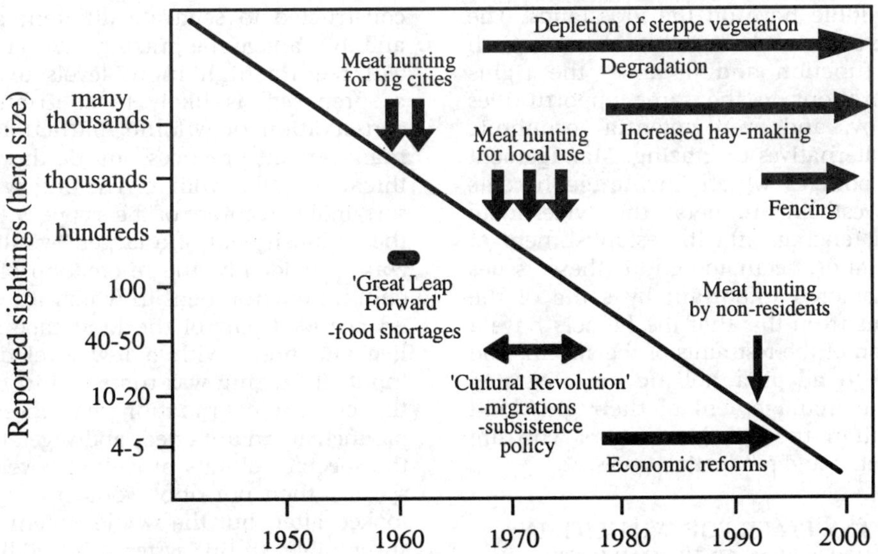

Fig. 5. The decline of the Mongolian gazelle in Xilingol (schematic indication from interviews with local people).

Mongolian gazelle only about five years ago. Hunting has been forbidden on the grasslands and local people from his village do not hunt, however, people coming from outside with cars and guns have killed many Mongolian gazelle. Now, he said, the numbers are very few. The hunting pressure this species has been exposed to is demonstrated by the report from one scientist who's home village is about 300 km from Xilinhot, near the Mongolian border. He described how large herds which come across the border in this area, are hunted from vehicles by local people with automatic weapons. One moonlit night two years before, four shooters killed about 400 Mongolian gazelle which were sold in Beijing. The migrating wildlife are being used by people in Inner Mongolia to quickly earn large amounts of cash.

ATTITUDES TO WILDLIFE

As has been described, official planning for landscape and grazing management within Xilingol Biosphere Reserve, and Baiinxile Farm takes no account of the presence of wildlife, or the impact of management activities on wildlife populations. Attitudes of herders to wildlife vary greatly, differing between species, and among different herders. The herders feel threatened by wolves, which take their livestock. In the past, each village in Baiinxile Farm employed a person to shoot wolves, but today wolf numbers are greatly reduced, and do not necessitate the employment of a person specifically to hunt them. Wolves are still allowed to be hunted, and groups of herders or people from outside sometimes work together to hunt the wolves. Other animals, including the Mongolian gazelle and roe deer are protected from hunting. Though most people interviewed said that herders do not hunt other species, it was clear from the interviews that roe deer, Mongolian gazelle, partridge and fox are hunted locally, particularly in winter.

Apart from the wolf which often aroused strong negative responses, the attitudes of herders towards wildlife were generally positive, placing the wildlife within a wider context of the grasslands and how they function, and the need to consider the future. One herder in Baiinxile Farm said that he likes to see the wildlife, and believed that it is important to both the herders and the country that wildlife is protected. "If we can protect the wildlife well, then tourists will come from outside to see it, and also, our own offspring will be able to see the wildlife". This notion of protecting the wildlife, or the grassland, for the sake of "our offspring" surfaced again and again in discussions with the herders. The Communist Party Secretary of Maodeng Farm, north of Baiinxile Farm, observed that the objectives of Maodeng Farm and the biosphere reserve were compatible. He said it is important to protect the grassland, the wildlife and precious plants "in order to leave plenty of resources for our offspring. If we do not protect the wildlife and natural resources, it will be a big threat to mankind". One young man in Baiinxile Farm said that, apart from the wolves, other wildlife does not do any harm, and that it is important to protect wildlife because they can "maintain the equilibrium", or ecological balance. Another herder said that it would be a very sad thing if Mongolian gazelle disappeared from the grassland. He said that protecting the grassland from degradation, and the wildlife from hunting is important for our offspring. Only one herder, an older Mongolian man, expressed a disregard for the loss of Mongolian gazelle from the grassland. He said he had not seen the "yellow sheep" since the 1970s, but he did not care if there were no yellow sheep.

It is interesting to compare these statements on wildlife, with the current and proposed management activities on the grasslands. Though wildlife was discussed with less than half of the 60 or so herders interviewed, there seemed to be some understanding of the place wildlife have on the grasslands. The herders showed some understanding of issues such as ecological function and balance, the rights of future generations to the same opportunities which we enjoy, and even potential economic benefits from alternatives to grazing. Management practices and policies which encourage herders to increase livestock numbers, the widespread construction of fencing, and the establishment of artificial grasslands, seem to ignore these issues which are considered important by some of the herders. It seems from this that the herders have a better recognition of the restraints of the system, and are more able to adopt a holistic or integrated approach to the management of their grassland environment, than the farm managers who are driven by economic and production goals.

IS THERE ANY PLACE FOR WILDLIFE IN THIS PRODUCTION ENVIRONMENT?

It seems that under the current management regime, there is no place for wildlife within this system. The impacts of proposed management practices on wildlife, and biodiversity in general, are not being considered. Many wildlife species are viewed as pests, either competing for feed, or preying on domestic livestock. Other wildlife provides easy money or food, even though hunting is banned. The pattern seen in other intensive grazing environments, with a loss of predator species and a decline of native grazers, is being repeated in Xilingol. The local herders themselves, however, recognize some potential values in maintaining the wildlife on the grasslands, in their role in the ecosystem, their existence value to future generations, and economic development potential.

The Xilingol Biosphere Reserve Management Bureau and Baiinxile Farm also recognize this potential economic value, as together they are constructing a "wildlife zoo". In an area of diverse habitat and good vegetative cover, where the Xilin River cuts through a rocky range and sand dune country, families have been removed and grazing excluded. An area of grassland two kilometres wide and seven kilometres long has been fenced. The intention is to re-introduce wildlife species into this area, such as roe deer, Mongolian gazelle and fox, providing for tourist viewing as well as educational and research opportunities. Another objective of this wildlife zoo is to be able to manage the grazing regime to maintain a high plant biodiversity.

There will, however, be difficulties faced balancing these objectives in a small area of grassland. Will the need to raise money by providing a good tourism experience outweigh the objective to control the grazing regime and maintain the biodiversity of the grassland? The managers are talking about capturing the wildlife elsewhere, perhaps even bringing the Mongolian gazelle in from Mongolia to put on show. With an area of 14 square kilometres, this zoo is unlikely to be large enough to enable the populations to be balanced, or self-sustaining. It is possible that fixed enclosures will need to be constructed to separate different animal species, and to manage the grazing rate. The "wildlife zoo" will require high input levels to maintain, and as proposed, is likely to contribute little to the conservation of wildlife, particularly while other management practices outside the zoo are directly threatening the wildlife. The only way to ensure the sustainable recovery of the steppe wildlife, is through the establishment of a larger "wildlife conservation zone", or ideally, the re-creation of a grazing and cultural environment in which the wildlife become an accepted part of the landscape again, and can live and breed with a low level of management input. If grazing was managed in such a way that the current degradation was reversed, grassland productivity maintained, and vegetation protected in the special habitats providing cover to a variety of species, then not only would the wildlife be better looked after, but the whole system would be more sustainable. In this system, the wildlife zoo could in fact play a role in helping to manage the wildlife species, providing a captive breeding base for the re-establishment of wildlife, and also for research.

The current planners and managers who are moving towards intensification of the grazing industry may ask how this would provide improved incomes and living standards to local herders. If the intensification road is followed, with widespread fencing, the replacement of native grasses, the development of fine breeds which are not suited to the local environment, and the application of fertilizers, then in this low rainfall and low fertility environment, it is most likely that without considerable political and structural changes, the livestock numbers will continue to increase, and degradation will continue. Not only will the local biodiversity be diminished, but the system will not be sustainable, and the ultimate result will be a failure to achieve the economic objectives of raised incomes and living standards.

A healthy system locally adapted to climatic, soil and other features, would not only provide the opportunity for the wildlife to co-exist with grazing, but would maintain its productivity, and provide for economic diversification. Wildlife could ultimately provide a resource to local people, through design of sustainable harvesting systems, as well as development of tourism potential. This vision of the future may appear idealistic, but it is clearly preferable to the alternative intensive production environment experiencing increasing degradation, species loss, declining productivity, and reduced economic returns. As degradation continues to get worse, it is unlikely that attempts to diversify into tourism will succeed. Will tourists travel to a degraded environment to view small numbers of wildlife inside a fenced enclosure? We believe that a more sophisticated experience is needed to attract enough visitors to provide real local economic benefit.

CONCLUSION

Imagine the Xilingol steppe how it must have once been. A productive and diverse grassland system, containing wildlife and nomadic herders and their livestock. It is a famous grassland in China for this romantic past. The local tourism agency still promotes it with a quote from an ancient poem: "Vast is the sky, boundless is the wilds, the bending grass reveals the animals as the wind blows". This grassland no longer exists. The wilds are becoming bounded by fences, and the grasses rarely grow to cover the animals' knees, let alone obscure them from view. The clock cannot be simply turned back to this time of long grass, diverse and numerous wildlife, low populations of herders and livestock. A sustainable production environment must be created, but not by considering only the productivity of the grassland as a resource. It is important that management planning considers the other elements of "sustainable development". The diversity of the landscape must be considered. It is important to maintain the special habitats, such as the forests in the sand dunes and in the gullies and gorges. Production based on the use of the grassland resource is a part of a wider ecological, social, economic and political system. Our understanding of the meaning of "sustainable development" suggests that if production planning does not take a more holistic view of the production process, then it is unlikely to remain sustainable in the long term. Even though the grazing theory might suggest that a production output, or yield may be sustainable in itself, if other elements of the landscape, ecological and socio-economic system are not sustainable, then the whole system will collapse.

The establishment of a wildlife zoo will not solve the problem of an empty steppe. The tourism value of the reserve today is not what is described in the ancient poem. The planners expect to develop the enormous tourism potential of the area to supplement the incomes from grazing, and yet the degradation of the grassland is continuing, habitats and species are being lost, the tourism value of the area is declining, and ultimately with degradation, the agricultural value will decline also. Leaving wildlife out of the equation not only results in the loss of that wildlife from the landscape, but reduces the future options, diminishing the potential of the tourism industry, and the diversity of local economic opportunities.

We would like to finish by presenting a simple model to help us visualize sustainability (Fig. 6). This presents the objectives of conservation and production. Two end points might be identified. The production end where economic development or production objectives predominate, and sustainability is interpreted only in terms of "sustainable yield", within the bounds only of immediate production considerations. Alternatively, sustainability could be ignored completely, creating a situation of "exploitation". The other end of the scale could be the preservation notion, where everything must be protected from development, where humans and human impacts are something separate from the environment. These two end points are often seen as mutually exclusive, or in conflict. Thus a line could be drawn connecting them, to show that in a land management situation, as production objectives become more dominant, the possibility of achieving conservation objectives is reduced, and similarly, as conservation objectives predominate, production potential is reduced (Fig. 6(a)). This is the trade-off description, as presented by Dyck (this volume). This is an over-simplification of the situation, and enshrines the attitude of a conflict between conservation and development,

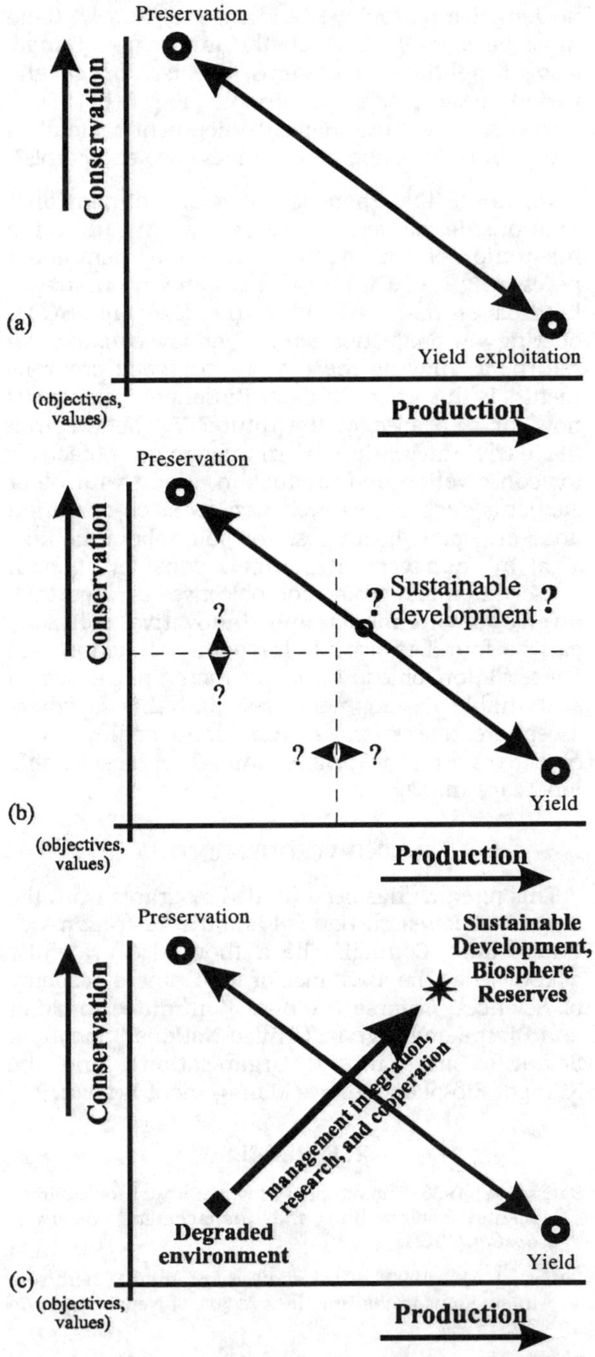

Fig. 6. Schematic model of relationships between conservation and production values or objectives, showing (a) the conflict or trade-off notion; (b) limitations imposed by the need to incorporate human development into conservation, and ecosystem processes into production; (c) opportunities for integration of conservation and production producing sustainable development.

which the notion of sustainable development sets out to reconcile.

So, where does this concept of sustainable development lie? One aspect of "sustainable development" is that people are a part of the system and must continue to survive. Any production must therefore be economically viable, so presumably there is a level which may vary, below which production is no longer viable or people are no longer a part of the landscape. Similarly, there must be a level at which the landscape change, loss of habitats, ecosystems, species, or genetic variation becomes unsustainable (Fig. 6(b)). Does this mean that "sustainable development" is really a compromise situation between these two end points?

We argue that there is not a one dimensional relationship between these two points, that the integration of conservation and development is not necessarily just a trade-off. Some poorly managed landscapes may fall below the line(Fig. 6(c)), offering low production return, and low conservation return, as may be the case in degraded environments. Is this where Xilingol Biosphere Reserve sits now, or is headed in the future? We believe that the early integration of management objectives for conservation and production, along with other elements such as research and local co-operation and participation, can push the point above the line. That by management planners considering both conservation and production objectives as important, diverse environments and innovative solutions can be found to provide for both production and conservation objectives to be met. This is where sustainable development lies, and this is where biosphere reserves must aim for to achieve their objectives of providing examples of sustainable landscape management.

ACKNOWLEDGEMENTS

This research has been funded by grants from the Australian Research Council and the Australian Vice Chancellors' Council. The authors also gratefully acknowledge the assistance of the Chinese Academy of Sciences, Chinese National Committee for Man and Biosphere Program (United Nations Education, Scientific and Cultural Organisation), and the Xilingol Biosphere Reserve Management Bureau.

REFERENCES

Batisse, M., 1986. The biosphere reserve: a tool for environmental conservation and management. *Environ. Conserv.* **9**(2).

Bauer, J. J., 1990. The analysis of plant herbivore interactions in the Himalaya region of Nepal. *Vegetatio* **90:** 15–34.

Bauer, J. J., Wiltshire, G. and Thwaites, R. (unpublished (a)). An assessment of steppe composition and steppe degradation in Xilingol Biosphere reserve, China, using a gradsec technique.

Bauer, J. J., Thwaites, R., Wiltshire, G. and DeLacy, T. (unpublished (b)). Understanding and managing steppe ecosystems: a case study from Xilingol Biosphere Reserve, China.

Commonwealth of Australia, 1992. National Strategy for Ecologically Sustainable Development. Australian Government Publishing Service: Canberra, Australian Capital Territory.

Dolgipalam, 1996. Intensive Management Model for Grassland Animal Husbandry. (In Chinese). China Agricultural Science and Technology Publisher: Beijing.

Gregg, W. P. Jr and McGean, B. A., 1985. Biosphere reserves: their history and promise. *Orion* 4(3), United States National Parks Service.

International Union for the Conservation of Nature, United Nations Environment Program, and World Wildlife Fund, 1980. World Conservation Strategy: Living resource conservation for sustainable development.

Li, Y., 1993. Grazing effects on the species diversity of *Stipa grandis* steppe and *Leymus chinensis* steppe. (In Chinese). *Acta Botanica Sinica* **35**(11): 877–84.

Li, Y., Yan, Y., Yong, S. and Chen, Z., 1995. Xilingol Grassland Reserve: role of scientific research in linking nature conservation with local development. *China's Biosphere Reserves — special English issue*. Chinese National Committee for Man and the Biosphere Program, Beijing.

Lin, X., 1990. Development of the pastoral areas of Chifeng City Prefecture. Pp. 74–92 *in* The Wool Industry in China: some Chinese Perspectives ed by J. Longworth. Inkata Press: Melbourne, Victoria.

Longworth, J. and Williamson, G., 1993. China's pastoral region: sheep and wool, minority nationalities, rangeland degradation and sustainable development. CAB International, UK.

Ludwig, J., Tongway, D., Freudenberger, D., Noble, J. and Hodgkinson, K., 1997. Landscape Ecology, Function and Management: Principles from Australia's Rangelands. CSIRO: Australia.

Maxey, C. and Lutz, J. (eds), 1994. China Biodiversity Action Plan. National Environmental Protection Agency: Beijing.

National Environmental Protection Agency, 1994. Regulations of the People's Republic of China on nature reserves. Order of the State Council of the People's Republic of China No. 167. National Enviornmental Protection Agency: Beijing.

Thwaites, R., DeLacy, T., Furze, B., Bauer, J., Li, Y. and Liu, X., 1996a. Xilingol Biosphere Reserve: management for people and the environment. Johnstone Centre Report No. 43. Johnstone Centre of Parks, Recreation and Heritage, Charles Sturt University: Albury, New South Wales.

Thwaites, R., Liu, X., DeLacy, T. and Li, Y., 1996b. The Sustainable Using Management of Grassland in Xilingol Biosphere Reserve. China's Biosphere Reserves, 3(1). (In Chinese) Chinese National Committee for Man and the Biosphere Program, Beijing.

Thwaites, R. and DeLacy, T., 1997. Linking development and conservation through biosphere reserves: promoting sustainable grazing in Xilingol Biosphere Reserve, Inner Mongolia, China. Pp. 183–89 *in* Conservation Outside Nature Reserves ed by P. Hale and D. Lamb. Centre for Conservation Biology: University of Queensland.

Thwaites, R., DeLacy, T. and Li, Y., 1998. Property rights, social change and grassland degradation in Xilingol Biosphere Reserve, Inner Mongolia, China. *Society and Natural Resources* **11:** 319–38.

United Nations Education, Scientific and Cultural Organisation, 1994. Biosphere Reserve Nomination Form. Man and Biosphere Programme, United Nations Education, Scientific and Cultural Organisation, Paris.

United Nations Education, Scientific and Cultural Organisation, 1996. Biosphere Reserves: The Seville Strategy and the Statutory Framework of the World Network. United Nations Education, Scientific and Cultural Organisation, Paris.

World Commission on Environment and Development, 1987. Our Common Future. World Commission on Environment and Development, Oxford Press.

Zhang, C., 1990. An overview of wool production and processing in the Inner Mongolia Autonomous Region. Pp. 57–73 *in* The Wool Industry in China: some Chinese Perspectives ed by J. Longworth. Inkata Press: Melbourne, Victoria.

Zhao, K. and Feng, L., 1991. Inventory of the birds in the west part of the Xilingol grasslands. *J. Inner Mongolia Normal Uni. (Nat. Sci.)* **2**: 47–57.

Zhou, Q., Zhong, W., Sun, C. and Wang, G., 1988. Zoogeographical characteristics of mammals in Baiyinxile area. (In Chinese with English abstract). Pp. 269–75 *in* Research on Grassland Ecosystem No. 3 ed by Inner Mongolia Grassland Ecosystem Research Station. Science Press: Beijing.

CHAPTER 21

Combining economic, conservation and cultural goals on Maori land in New Zealand: the Matakitaki-a-Kupe restoration project

G. F. Harris,[1] M. K. Burtenshaw,[1] R. Lucas[1] and Haami Te Whaiti[2]

The historic lands of Ngati Hinewaka, a South Wairarapa hapu (subtribe) of Ngati Kahungunu comprise 1 250 hectares surrounding Matakitaki-a-Kupe (Cape Palliser). The land contains some of the earliest evidence of human habitation in Aotearoa/New Zealand. It also includes significant natural landscape features as well as plant species that are classified as rare or endangered. The hapu, as well as being legal owners of the land, take their role of kaitiaki (caretakers) seriously. They are implementing a programme to revegetate the lands and restore the landscape while at the same time retaining productive grazing on the coastal terraces.

The programme involves establishing a nursery to propagate and grow locally sourced plants to enhance natural regeneration; training in techniques of plant propagation and production; revegetation planning and implementation; and restricting public access. The long-term aim is to provide sustainable management of this historically and botanically significant part of the coastline and horticultural skills for members of the hapu. The programme is being implemented in partnership with scientific staff of The Open Polytechnic of New Zealand.

INTRODUCTION

Whatu ngarongaro he tangata, toitu he whenua

— Man perishes, but the land remains

ON 14th December 1993 the New Zealand Government returned a 20 hectare block of land at Matakitaki-a-Kupe (Cape Palliser) to the Ngati Hinewaka hapu of the southern Wairarapa. This land which is in the centre of 1 250 hectares owned by the hapu, had been taken by the Crown in 1897 as a site for the Cape Palliser lighthouse. The return of the land provided the incentive for Ngati Hinewaka as the kaitiaki or guardians of the area, to consider an integrated plan for managing and developing their remaining ancestral lands. A strategic plan was developed after a series of hui (meetings). The Ngati Hinewaka me ona Karangaranga Charitable Trust was established to act as the legal and organizational umbrella for implementing the plan. The two main aims of the strategic plan are:

- Re-immersion and re-education of Ngati Hinewaka in their history and culture,
- Regeneration and rehabilitation of the ancestral lands centred on Matakitaki-a-Kupe.

Regeneration of the land includes re-establishment of the forest cover on the steep coastal hillsides, and preservation and conservation of the plants endemic to the area (some of which are classified as rare and endangered), and rehabilitation of the foreshore and immediate marine environment.

A fundamental component of the plan is a training programme for the hapu. To meet that objective, and to provide technical expertise, staff of The Open Polytechnic of New Zealand became partners in the project. Funds for the work were

[1]The Open Polytechnic of New Zealand, Private Bag 31914, Lower Hutt, New Zealand.
[2]Ngati Hinewaka me ona Karangaranga Trust. 60 Pownall Street, Masterton, New Zealand.

Pages 212–18 *in* NATURE CONSERVATION 5: NATURE CONSERVATION IN PRODUCTION ENVIRONMENTS: MANAGING THE MATRIX ed by J. L. Craig, N. Mitchell and D. A. Saunders. Surrey Beatty & Sons, 2000.

provided by the Environment and Heritage Fund of the New Zealand Lottery Grants Board.

MAORI OCCUPATION AT MATAKITAKI-A-KUPE

The early history of this area is associated with the explorer Kupe, who is said to have preceded the arrival of the ancestral canoes (*circa* 1000 AD) and circumnavigated both islands (Orbell 1985). Many of the place names and features attest to his exploits; one such place is Nga Ra o Kupe, two massive stone formations which represent the sails of Kupe which were hung up to dry (Te Whaiti 1994). Matakitaki-a-Kupe is a contraction of — *Te Matakitakinga a Kupe ki Kaikoura ki te waahi i haere ai te tamahine a Kupe* — "the gazing of Kupe towards Kaikoura the place where the daughter of Kupe had gone" (Te Whaiti ibid), and refers to the area of land in Palliser Bay between the Mangatoetoe Stream in the west and the Waitutuma stream in the east (Fig. 1).

When Europeans arrived in the Wairarapa in the 1840s, the Maori population was of the Ngati Kahungunu iwi (tribe) — descendants of forebears who came to Aotearoa on the Takitimu canoe. Before settling in the Wairarapa they had lived in Hawkes Bay, having moved there from Poverty Bay in the early fifteenth century. On arrival, they found the district occupied by the Rangitane people who were willing to cede ownership of the land in exchange for canoes and weapons. By 1650, Ngati Kahungunu were in control of southern Wairarapa (Aburn 1987).

Ngati Hinewaka, a hapu of Ngati Kahungunu, settled in the Palliser Bay area and co-existed peaceably with other hapu for some time. However, a dispute in which a woman of Ngati Hinewaka was killed led to the decimation of the hapu of the offender and eventual control of the Cape Palliser district by Ngati Hinewaka (Te Whaiti 1994). On 10 July 1871, the area, known as the Matakitaki block,

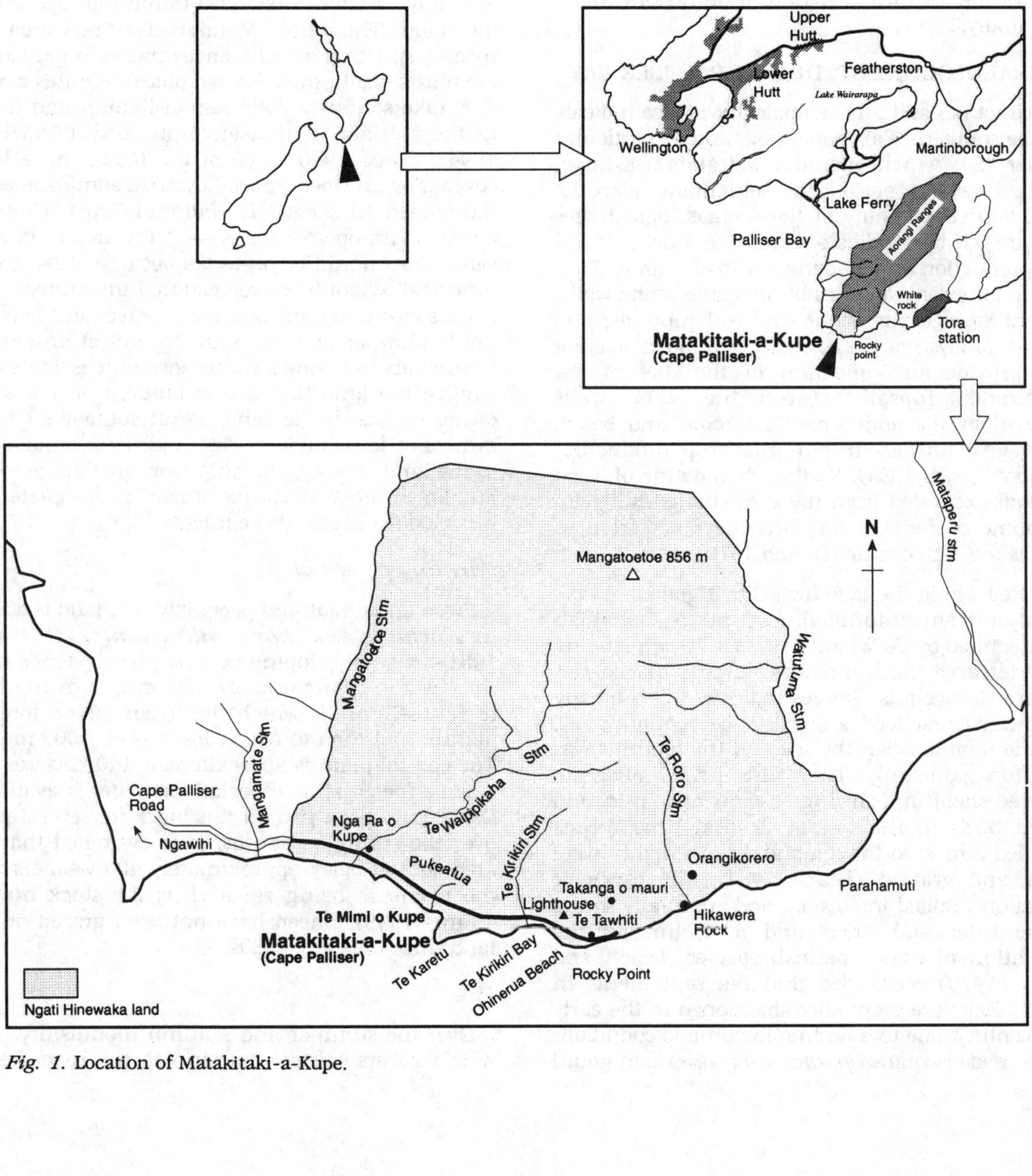

Fig. 1. Location of Matakitaki-a-Kupe.

was set aside by crown grant to ten owners of Ngati Hinewaka descent. This was a formal recognition by the government that these people were the local tangata whenua (people of the land) of this land (Leach 1991).

Most of the land, consisting of 1 250 hectares, is still owned by Ngati Hinewaka. That immediately surrounding the Cape Palliser lighthouse was recently returned to Ngati Hinewaka (14 December 1993) after being taken forcibly by the Crown in 1897 for the building of the lighthouse.

At a Maori Land Court sitting in Masterton (Ref. 48, Wairarapa minute book: Folios 21–6), the Crown stated that the land was being re-vested in the descendants of its former owners "at nil value in recognition of waahi tapu (sacred place) values that are present on this site". In addition the Historic Places Trust has designated the area, under the *1980 Historic Places Act* as an historic area, that is — "an area which contains an inter-related group of prehistoric or historic features which have historical value as a group even though some or all of the features may have little historic value individually."

ARCHAEOLOGICAL STUDIES AT PALLISER BAY

Place names and archaeological evidence indicate that the southern Wairarapa area, and in particular Palliser Bay which includes Matakitaki-a-Kupe, has had several periods of settlement over the past 1 000 years. Many of the archaeological sites that are visible indicate extensive horticultural crop production was undertaken in the area. They include an extensive series of protective stone walls, kumara storage pits, stone and soil mounds, and terraces. Analysis of the contents of middens (refuse pits) provide an indication of the diet of the inhabitants. Topsoil between the stone walls modified by the addition of charcoal and beach gravel give further insight into crop production practices (Leach 1984). Radiocarbon dating of wood charcoal excavated from the sites studied indicates that some of the sites had been occupied from as early as the 12th century (Leach 1979).

Located within the land owned by Ngati Hinewaka is a significant unfortified occupation site which was identified by Te Whaiti (1994) as Orangikorero. It is terraced and includes several raised-rim kumara storage pits. Studies indicate that the early inhabitants practiced a subsistence economy and that the land between the sea and the foothills was used for gardening. They also fished offshore, gathered shellfish, fished in streams and rivers and snared birds in the forests. At that time forests extended almost to the sea and the riverbanks were stable and grassed (Leach 1984). Clearance of vegetation resulted in erosion, and eventually silting of the inter-tidal areas and a decline in the population of some shellfish species. Leach and Leach (1979) concluded that the settlements in Palliser Bay were eventually abandoned in the early 16th century, due to a decline in climatic conditions which made kumara (*Ipomoea batatas*) and gourd (*Lagenaria siceraria*) culture marginal, in addition to the depletion of other important food reserves.

EUROPEAN SETTLEMENT OF THE AREA

From 1846 the Cape Palliser area was leased to European runholders. In 1853 major land purchases were made on behalf of the government, enabling runholders to freehold large areas. By then pastoralism was well established and flourishing. Since 1860 the Matakitaki block has been leased by a succession of farmers who cleared the vegetation on the coastal terraces and steep slopes and stocked first cattle and then converted to smaller holdings of sheep when refrigerated shipping in the 1860s opened up new overseas markets (Aburn 1987).

MATAKITAKI-A-KUPE TODAY

Site status

Much of the land at Matakitaki-a-Kupe is managed by a trust, whose members are Ngati Hinewaka. The land is divided into several blocks each of which is owned by whanau (extended family) groups within the hapu. The whole Matakitaki-a-Kupe area has special spiritual significance and is regarded by them as a waahi tapu (sacred place). For this reason they take seriously their responsibilities and duties as the kaitiaki (caretakers/protectors) (Te Whaiti 1994). Since 1986 three of the blocks have been covered by an Open Space Covenant administered by the Queen Elizabeth II National Trust. Under a covenant an open space is — "any area of land or water that should be protected because of its scenic, historical, scientific or recreational importance." An open space covenant is a legal agreement between the landowner and the Trust to protect an area of open space. An open space covenant is registered against the land title and is binding on the landowner or leaseholder and on all subsequent landowners or leaseholders. The landowner retains title to the land. The coastal strip from the Mangatoetoe Stream to the Waitutuma Stream is designated an historic area as detailed earlier.

Farming potential

From an agricultural perspective the land is classed as *North Island hard hill country* (Blaschke 1981). A gently sloping coastal plain extends from the sea for approximately 400 metres to the base of very steep hills which rise to an inland forested plateau and then to open tops at over 1 000 metres. The coastal plain is approximately 140 hectares and is ideal for grazing, especially as water is available. Cattle have been run on this block for several years over the summer months. It is estimated that the area should carry approximately 400 weaners with the numbers being reduced as the stock mature (Burns 1995). Sheep have not been grazed on the land since the early 1980s.

Fire

Over the summer and autumn months dry grass which covers the coastal strip to the road creates a fire hazard. On 5 February 1990, a camp fire swept through most of the Matakitaki blocks east of the lighthouse burning large areas of native vegetation. Details are recorded in an unpublished Department of Conservation debriefing report (Borger, n.d.). Access to the area is now controlled by the owners who restrict the number of campers. Grazing cattle on the coastal plain over summer keeps the grass relatively short and reduces the fire risk. An earlier fire in 1988 caused some minor damage to vegetation, but it was not nearly as extensive as the later 1990 fire. Controlling public access and camping has also reduced activities such as cutting of trees and shrubs for firewood, and driving recreational vehicles over the land. These were contributing to a general degradation of the landscape.

VEGETATION OF THE AREA

Druce (1971) writing of the flora of the Aorangi range commented about the southern part near Cape Palliser — *"If any area on the southern North Island coastline should be declared a National Coastline Park this is it!"*

Ngati Hinewaka land comprises the southern-most part of the Aorangi Ranges. The area includes the coastal strip between the Mangatoetoe stream in the west and the Waitutuma Stream in the east with a block of steep coastal ranges behind this area rising up to and including Mangatoetoe (856 m), the southern-most mountain in the Aorangi Ranges (see Fig. 1). The mountains rise steeply from the coastal strip and underlying greywacke, with faults running through NNE-SSW, are easily eroded. Natural erosion has been enhanced by deforestation from man-made fires starting in pre-European times and occurring at regular intervals up until the present day. Introduced animals including deer, goats, pigs, possums and rabbits, and the grazing of sheep and cattle have also contributed to erosion (Wardle 1967).

The original forest climax community of mahoe *Melicytis ramiflorus*, hinau *Elaeocarpus dentatus*, rewarewa *Knightia excelsa* hardwood forest with podocarps is presumed to have grown right down to the coastal strip, but this has long since been replaced by fire-induced seral plant communities. Kanuka *Kunzea ericoides* scrub regrowth is the main response to fire. Tall kanuka occurs in patches on the hillsides and on the older stable shingle fans at the base of the hills where regrowth has occurred. Prominent at former Maori habitation sites are groves of karaka *Corynocarpus laevigatus*. One was described by Leach 1991 as being "archaeologically significant". Recently located colonies of rengarenga lily *Arthropodium cirratum* are also considered indicative of early Maori occupation (Harris and Te Whaiti 1995).

The steep coastal hillsides most recently razed by fire are well colonized by European pasture grasses. Many native plants are starting to shade out the pasture grasses particularly in the more sheltered gullies and ravines. Ti kouka *Cordyline australis*, kowhai *Sophora microphylla*, mahoe, kanuka, tauhinu *Cassinia leptophylla*, ngaio *Myoporum laetum* and several species of *Coprosma* are the main shrub species regenerating on the steep hillsides.

Herbaceous species growing on the hillsides, and also in places amongst the coastal rocks, include wharariki *Phormium cookianum* and karamea *Aciphylla squarrosa* with scrambling lianas such as poananga *Clematis forsteri*, aka kiore *Parsonsia capsularis* and *Muehlenbeckia complexa* and *M. axillaris*. Shrubs occupying these sites include taupata *Coprosma repens*, Cape Palliser daisy *Brachyglottis greyii* and *Coprosma propinqua*.

This area of the coast contains a number of rare and endangered plants (de Lange 1993). Near Nga Ra o Kupe (the sails of Kupe) seven plants of *Muehlenbeckia astonii* can be found. Only 37 other plants of this species remain in the wild in the North Island. The survival of this species is complicated because the plants are monoecious, (having separate male and female plants); these survivors each grow some distance from a plant of the other sex (de Lange and Silberry 1992). Another rare species of Muehlenbeckia, shrubby tororaro *M. ephedroides,* is found growing in two colonies on the beach gravel. Other rare plants found in the area include a fern *Pleurosorus rutifolius*, and two grasses, *Rhytidosperma petrosum* and *Chionochloa beddiei* which is endemic to this coastal region.

Boxthorn *Lycium ferocissimum* is the only exotic plant that has become established to any significant extent in the area and Timmins (1996) identified this plant as being the only weed of particular threat to the revegetation/grazing programme. She noted that tall (2–4 m) wide bushes are common in the vicinity of the lighthouse and that localized patches have coalesced to form impenetrable thickets. All the boxthorn plants are mature specimens and it appears that occasional stocking with cattle has kept new seedlings in check. The bare shingle scree on the hill slopes appear to be free of boxthorn as are the hill slopes clothed in regenerating native cover.

THE RESTORATION PROJECT

The aims of the restoration project are:

- To restore the natural vegetation on the steep coastal hills that have been grazed and burned.
- To generate income by grazing cattle on the approximately 140 ha of coastal plain.
- To provide training for members of the Trust, in basic horticultural techniques including plant propagation and production using the nursery constructed for this purpose, and in native forest restoration.

In this plan, the steep coastal hills (approximately 200 hectares) are to be restored to natural vegetation cover. Prior to 1990, natural regeneration of vegetation on the coastal hills was taking place and some areas had a good cover of a mixture of trees including kanuka, manuka *Leptospermum scoparium*, mahoe, ti kouka, kowhai, ngaio and tauhinu. Karaka trees were well established on the lower slopes in addition to numerous herbaceous plants. The 1990 fire destroyed most of the trees on the steep hillsides although many of the karaka trees on the lower slopes survived. The fire did not reach as far as neighbouring Ngapotiki Station where vegetation, which includes the above species, is now well established. On some sections of the hillsides, natural regeneration has been rapid either from regrowth from plants that were damaged but not killed by the fire (for example ti kouka), or from seeds. Kowhai has re-established particularly well and many of the seedlings which have grown since the 1990 fire are now up to 2 m in height. Regeneration is most rapid in the more moist and shaded valleys and slower on the drier north facing slopes where exotic grasses are still dominant. However, even on these dry sites, kowhai seedlings are managing to emerge through the grass and make good growth.

Fencing

Porteous (1993) noted that the top priority in any programme to manage and restore forest remnants is to exclude stock. Cattle have been run on the Matakitaki-a-Kupe blocks over the summer months since 1991 to provide income for the owners. Although cattle avoid the steep slopes, they tend to move up the ravines in search of water and browse and trample the vegetation along the streambeds. Sheep which will graze steep slopes have not been farmed on these blocks for at least 20 years. In fact a condition of the Queen Elizabeth II National Trust covenant that covers parts of the land specifically excludes sheep from the covenanted areas. A priority of the revegetation programme will be to fence the base of the steep slopes to confine the cattle to the coastal plain. However, some very steep sections are unstable, and continuous erosion is preventing the establishment of vegetation. Here the technique of laying and pegging down capsule bearing branches of manuka on the steeper slopes as described by Porteous 1993 will be tried. Direct sowing of tauhinu seed will also be attempted.

It is interesting to compare Matakitaki-a-Kupe with parts of Tora Station, some 20 km to the north-east. Some of the steep coastal hills at Tora were fenced off in the mid 1970s as grazing by sheep and subsequent erosion had left them in an unstable condition. During Cyclone Bola in 1988, this area was subjected to very high rainfall over a short period. Within the fenced areas sufficient vegetation had established to prevent the severe erosion and slumping that badly affected parts of the coastal hills with no vegetation cover. The first plant to become established on these sites was tauhinu (Tyer 1996, pers. comm.).

Weeds

Timmins (1996) recommended that a priority for the revegetation programme should be to remove the boxthorn plants and monitor the area for regrowth and new seedlings. Other weed species identified in the area include; Cape Ivy *Senecio angulatus*, sweet briar *Rubus rubiginosa*, broom *Cytissus scoparius*, and thistles *Carduus nutans*, *Cirsium vulgare,* and *C. arvense*. Most of these species were not considered likely to present a significant problem but Timmins (ibid) recommended that they should be removed. No gorse *Ulex europeus* is present.

Pests

Rabbits are numerous on the coastal plain around and within the boxthorn thickets but not on the steep hillsides. Pindone baits will be laid to control them. Possum browsing in the coastal area does not appear to present a serious threat to the revegetation programme. Possums tend to take refuge during the day in clumps of wharariki. A trapping programme will be undertaken to ascertain their numbers. Possums are attracted to rengarenga and they will browse this plant in preference to others (Harris 1997, pers. obs.). Browsing on rengarenga is a good indicator of possums and regular inspections of the rengarenga colonies have not shown any significant damage.

Wardle (1966) reported that browsing by introduced herbivores including deer, goats and possums has modified the understorey in the mahoe-hinau-rewarewa forest that predominates further inland from the coast. The understorey now consists almost entirely of unpalatable species such as the fern *Blechnum discolor* and horopito *Pseudowintera colorata*.

The nursery

The nursery is adjacent to the one of the houses near the lighthouse. It occupies an area of 40 square metres and is surrounded by 2 m high windbreak cloth to give shelter from the frequent strong winds that blow from the south and the north-west. Two taupata trees were retained to provide shade. The floor has been terraced and covered with weed cloth and a 50 mm layer of 20 mm metal chip. The nursery includes a glass-clad cold frame for propagating plants from seed and cuttings, a potting shed and an outdoor potting bench.

The main purpose of the nursery is to provide training in plant propagation and production techniques for members of the hapu and to produce plants endemic to the area for landscaping around the houses. An interpretive centre will eventually be constructed here. To complement practical on site training, staff from The Open Polytechnic Natural Resources Centre will conduct a series of classes based on The Open Polytechnic *New Zealand Native Plants* course.

In keeping with established native forest restoration practices (Porteous 1993), seed is being only collected from plants in the area to be revegetated. Seed of the following species was collected on 1 March 1997: Taupata, ngaio, kowhai, poananga, harakeke *Phormium tenax*, ti kouka, kanuka, maukoro *Carmichaelia arborea*, kaikomako *Pennantia corymbosa*, karaka, rengarenga, and wharangi *Melicope ternata*. These were sown in trays on 4 June 1997. At the same time cuttings of shrubby tororaro, Cape Palliser daisy, puka *Griselinia lucida* and a koromiko *Hebe salicifolia* were taken. The cuttings were inserted into trays of a 50:50 sharp sand:peat mixture and placed in a cold frame.

Seedlings and rooted cuttings will be potted up and grown on until they are large enough to plant out. Commercial peat-based potting medium is being used at present; however, it is intended that a medium based on composted seaweed with added sand and nutrients will be prepared and trialed on site to attempt to reduce costs.

DISCUSSION

Maori as tangata whenua have a special relationship with the land. Just as the tribes are closely linked by a network of genealogical connections, so are they related spiritually to their ancestral lands (Patterson 1992). Firth (1972), stated that the entire Maori culture has been centred around issues relating to land. From their ancestral land, Maori find their place to stand (turangawaewae). Bennett (1979), recorded that — "the essence of turangawaewae is that the land is an outward and visible sign of something that is deeply spiritual and it is a source of nourishment to the inner man rather than to his physical needs. His identity belongs there, his sense of self awareness begins there and his sense of mana and importance belong there". Traditional Maori attitudes to the natural world reflect the relationships created through Papatuanuku (Earth mother) and Ranginui (Sky father). All living things are their descendants and are related. The interdependence between people and their environment creates a sense of belonging to nature. Humans are born from Mother earth and return to her on their death (Marsden 1977).

For Ngati Hinewaka, their turangawaewae is Matakitaki-a-Kupe — the place where many generations of their tipuna (ancestors) lie beneath its soil.

In this context, the restoration project is part of an overall kaupapa (plan) for Ngati Hinewaka to revive their traditional matauranga (knowledge and skills) so they can hold on to their past while making their way into the future. Ngati Hinewaka as kaitiaki or guardians of their land believe that they have a responsibility to care for and manage the land wisely, not only for future generations, but also for those who have gone before. Because the revegetation part of the project will allow the hapu to work towards re-creating the physical environment of their tipuna (ancestors), the project represents more than just restoring the forest cover on the land for future generations.

In addition to the spiritual considerations and the revival of traditional knowledge, the project provides other practical benefits. These include stabilizing the steep hillsides and preventing further erosion, improving the aesthetic appearance and attraction of the physical landscape, and protecting and proliferating rare and endangered plant species that grow in the area. The project will also provide social and educational benefits as it will provide training and skills for members of the hapu in horticulture, agriculture and conservation, and will allow Ngati Hinewaka to become better informed custodians of the land. Income necessary for costs associated with maintenance and restoration will be generated from grazing cattle. Grazing will provide the additional benefit of reducing fire risk over the summer months.

Long-term economic potential for the area could focus on tourism including elements of eco- and ethno-tourism. This might provide employment opportunities as well as generating income. Restoration of the physical landscape will make this possibility more likely. Ideally the whole area including the coastal plain should be restored to natural vegetation. However, in order to generate income from grazing, fencing off the steep hills for revegetation while allowing grazing on the flat land, is an acceptable compromise. It is an example of nature conservation in a production environment.

ACKNOWLEDGEMENTS

Ngati Hinewaka acknowledges the financial assistance provided by the Environment and Heritage Fund of the New Zealand Lottery Grants Board.

REFERENCES

Aburn, A., 1987. Pirinoa — People and Pasture. Roydhouse Publishing Ltd.: Carterton, New Zealand.

Bennett, Manuhuia, 1979. Te Kupu Whakamutunga: The last Word. Pp. 74–79 *in* He Matapuna — a source: Some Maori perspectives. The New Zealand Planning Council. Te Kaunihera Whakakaupapa mo Aotearoa: Wellington, New Zealand.

Blaschke, P. M., 1981. Analysis of New Zealand's vegetation cover using Land Resource Inventory data. *NZ J. Ecol.* **4:** 1–19.

Burns, R., 1995. Report on the visit to the Trust land at Palliser Bay on 16 January 1995. Unpublished report.

de Lange, P. J. and Silberry, T., 1992. Saving the shrubby tororaro (*Muehlenbeckia astonii* Petrie) — an urban approach to threatened plant conservation. *Proc. Roy. NZ Inst. Hort. Ann. Conf.* 1992 **1:** 99–105.

de Lange, P. J., 1993. Unpublished data on file, Wellington. Department of Conservation: P.O. Box 10–420, Wellington, New Zealand.

Department of Conservation Wellington Conservancy, 1990. *Cape Palliser Fire Debriefing: 19 February 1990.* Unpublished.

Druce, A. P., 1971. The Flora of the Aorangi Range, Southern Wairarapa. *Well. Botan. Soc. Bull.* **37:** 4–29.

Firth, R., 1973. Economics of the New Zealand Maori. (Reprint of first edition 1929). Government Printer: Wellington, New Zealand.

Harris, G. F. and Te Whaiti Haami, 1996. Rengarenga Lilies and Maori Occupation at Matakitaki-a-Kupe (Cape Palliser). *J. Polynes. Soc.* **105:** 271–84.

Leach, B. F., 1991. Matakitaki-a-Kupe Coastal Area. Submission to NZ Historical Places Trust. P.O. Box 2629, Wellington, New Zealand.

Leach, B. F. and Leach, H. M. (eds), 1979. Prehistoric Man in Palliser Bay. National Museum of New Zealand Bulletin 21.

Leach, H. M., 1979. Evidence of Prehistoric Gardens in Eastern Palliser Bay. Pp. 137–61 *in* Prehistoric Man in Palliser Bay ed by B. F. Leach and H. M. Leach. National Museum of New Zealand, Bulletin 21.

Leach, H. M., 1984. 1,000 Years of Gardening in New Zealand. A. H. and A. W. Read: Wellington, New Zealand.

Marsden, Maori, 1975. God, Man and Universe: a Maori View. Pp. 86–106 *in* Te Ao Hurihuri — The World Moves On ed by M. King. Hicks Smith and Sons/Methuen: New Zealand.

Orbell, M., 1985. The Natural World of the Maori. Collins: Auckland, New Zealand.

Patterson, J., 1992. Exploring Maori Values. Dunmore Press: Palmerston North, New Zealand.

Porteous, T., 1993. Native Forest Restoration — A practical guide for landowners. Queen Elizabeth II National Trust: Wellington, New Zealand.

Te Whaiti Haami, 1994. Matakitaki-a-Kupe. *Historic Places July 1994:* 28–30.

Timmins, S. M., 1996. Weed and Animal Pest Status of Ngati Hinewaka land at Matakitaki-a-Kupe: Ecologist's Report. Unpublished report.

Wardle, J., 1967. Vegetation of the Aorangi Range, Southern Wairarapa. *NZ J. Bot.* 5: 22–48.

CHAPTER 22

Winning battles and losing the war: a sequel

D. C. Goldney[1], S. Wakefield[1], G. Watson[2], D. Hodgkins[2] and B. Baldwin[2]

In 1995 we identified a failure by ecologists and land management agencies to engage with landholders in rural Australia in a culturally appropriate manner in the essential process of facilitating nature conservation on productive farmland. Our argument for a paradigm shift in educational strategies from "top-down" to "bottom up" approaches was illustrated by the development of a prototype kit.

The Commonwealth Government funded us to develop this process further as an "Australian model". In July 1997 we launched the "Save the Bush Toolkit" in the market place at the Mudgee Small Farm Field Days. The "Toolkit" consists of nine action learning kits which are designed to enable landholders to assess the health and sustainability of their bushland, scattered trees, farm watercourses, flora and fauna; to develop appropriate management strategies and to help facilitate the process of planning and implementing rehabilitation and management strategies on farms. The educational and marketing strategies employed are described as well as our interactions with landholders in formal focus groups and through our extensive networks, pre- and post-release. The success of the programme to date and its strengths and weaknesses are assessed as well as its relevance to other Australian land systems.

INTRODUCTION

IN 1995 we argued that in spite of the burgeoning Landcare Movement in Australia and the significant gains in conservation outcomes on farms that have occurred over the past decade or so, conservation losses in agricultural Australia as measured by land clearing and degradation, the non sustainability of woodlands and loss of biodiversity, were real causes for concern. We suggested that there was a need to move away from technocentric extension paradigms and to embrace "empowerment paradigms" as the basis for developing extension programmes for landholders (Goldney *et al.* 1995). Technocentric paradigms foster approaches which result in a one-way flow of information from the "information rich" scientific culture to the "information poor" landholders. More often than not extension outcomes based on such models have resulted in culturally inappropriate materials being produced. There have been and are, of course, many exceptions with valuable resource materials being produced by a range of agencies or groups (e.g., Breckwoldt 1983; Buchanan 1989; Davidson and Davidson 1992; Dorricut and Roberts 1993; Hussey and Wallace 1993; Brouwer 1997; Siepen 1997). In contrast the empowerment paradigm emphasizes processes and content and requires co-operative approaches between landholders and extension personnel in the development and implementation of relevant programmes. This approach values landholders as "owner researchers" whose behaviour and attitudes will change positively as the result of their observations, enabling them to initiate action plans for farm conservation. The empowerment process, we argued, should also be integrated with two strategic marketing strategies. Firstly, representatives of the client group should be involved as co-producers of extension material, providing feedback on aspects of the design and suitability of the developing educational material. Secondly "user testimonial" should ensure that clients have an extremely positive experience with the service we provide and that they themselves encourage other landholders to use the product.

[1]Environmental Studies Unit and the Johnstone Centre, Charles Sturt University, Bathurst, New South Wales, Australia 2795.
[2]University of Sydney — Orange Agricultural College, Orange, New South Wales, Australia 2800.
Pages 219–27 *in* NATURE CONSERVATION 5: NATURE CONSERVATION IN PRODUCTION ENVIRONMENTS: MANAGING THE MATRIX ed by J. L. Craig, N. Mitchell and D. A. Saunders. Surrey Beatty & Sons, 2000.

In this chapter we describe the "Save the Bush Toolkit" (Wakefield and Goldney 1997), we developed in 1996/97, the process used to develop the materials, including educational and marketing strategies, the involvement of landholders and other stakeholders in that process in focus groups or in other co-producing roles, the emerging feedback from the first purchasers of the Toolkit and an assessment of the programme to date, its immediate future and the longer term prospects, as well as the relevance of this programme for other regions in Australia. We conclude the chapter by offering some suggestions as to other critical educational material needed by landholders, based on our experiences in the Central Western Region of New South Wales.

THE "SAVE THE BUSH TOOLKIT"

The Toolkit is a professionally produced package with; a ring-back folder (A4 size) with the photograph of a local landholder with bushland in the background on the cover, the name of the product, "Save the Bush Toolkit", and a challenging message "At last we've got the tools . . . now let's get on with the job". Inside the Toolkit are nine individual integrated kits which may be used to: assess farm bushland; manage farm bushland; assess scattered farm trees; manage scattered farm trees; assess farm watercourses; manage farm watercourses; survey farm flora; survey farm wildlife; and plan for the integration of bushland and wildlife on the farm.

The kits are designed as stand alone units for landholders to assess their farm's natural resources. However, landholders will optimize outcomes for themselves if they begin with one of the assessment kits followed by the appropriate management kit. The last kit is the end point of the process and enables landholders to produce a farm plan which integrates nature conservation and agriculture outcomes. Landholders who live in the Central Western Region of New South Wales are also provided with a supplement to the wildlife kit that lists all vertebrate fauna found in the 1:100 000 map sheet where they live. Each Toolkit also has a brief introductory note describing how the Toolkit should be used and a kit evaluation proforma.

Twenty-three "Bush Info" notes covering a range of additional topics are included at the back of the Toolkit. Material ranges from detailed reference lists, information on bushland economics, fire as a management tool, farm nature conservation, farm ecology and sustainable agriculture.

KIT DESIGN

Each kit is designed using the same basic principles and layout and has the following sections.

Some background information

This section consists of an outline of the issues addressed by a particular kit, and ways that it will help address these issues on the user's property. The information provided across all kits is linked and builds "blocks" of information and conceptual ecological understandings through both observation and narrative as well as synthesizing these outcomes with the experiences of landholders. Layout and design features used include dot points, challenging questions, boxed messages and headings in bold, colour photographs and imaginative layout and design, familiar approaches in contemporary print media design.

Where to use this kit

This section guides the user to the appropriate situation for using the kit and explains the links between kits.

What you will do in this kit

This section describes the steps that the user will work through to complete the kit. The number of steps vary between five to nine and are designed to provide tasks that are achievable and reasonable from the user's point of view.

What and who you will need to complete this kit

This section describes the materials and resources that the user will need to complete the kit tasks. As far as possible, the use of specialized equipment or resources is avoided so as to minimize barriers to working through each kit. The only pre-requisite is that, in order to complete one of the "management" kits, the user must have worked through a corresponding "assessment" kit.

Detailed instructions for each step

This section enables the user to make a series of guided observations about bushland, scattered trees or riverine health and to be able to interpret their data in a way which makes sense. All of the assessment sheets are designed for use in the field and cannot be satisfactorily completed from the farm office. This encourages the user to look critically at his/her bushland, scattered trees and watercourses. To help optimize outcomes, users are provided with a key assessment sheet which lists a set of observations to be made, which require only a "True" or "False" answer. The greater the number of "True" scores, the better the condition of the bushland, scattered trees and farm watercourses. As far as possible, parameters included in the assessment sheets are factors which the user can rectify or improve upon if the response given during the assessment is "False". Parameters were selected in this way to avoid creating a sense of hopelessness if, for example, the user completes the assessment sheet and finds that his/her bushland/scattered trees or watercourses are highly degraded and feels that there is nothing that can be done about it. The following step then enables the user to determine the condition of the natural resource being assessed. A series of colour photographs with appropriate descriptions provide cues to help the user better understand the quality of their bushland, scattered trees and watercourses. Two subsequent steps enable the user to determine what his/her observations mean (Fig. 1) and what further actions might be considered (Fig. 2).

WHAT YOUR RESULTS MEAN

Bushland Health Rating	What it means
1. Quality bushland (colour code green)	Your bushland is in very good condition. Congratulations! You have a valuable asset. "Green" bushland patches should need little active management to keep them in good condition. Some suggested management actions for "green" patches are: • maintain fences to keep stock out; • keep rabbits under control; • control small weed outbreaks before they get out of hand; • keep foxes and feral cats under control.
2. Moderately disturbed bushland (colour code yellow)	The bushland is starting to become degraded, and you will need to take action to prevent it from degrading any further. If you act now, it will be easier and more cost effective to restore the bushland to good health than if you wait a few more years. Suggested management actions include: • fencing to exclude livestock; • controlling rabbits and weeds; • changing your burning off regime; • re-establishing native trees and shrubs.
3. Highly disturbed bushland (colour code red)	The bushland is highly degraded and you need to take action urgently to stabilize it and gradually improve its condition. Restoring degraded bushland is a slow process — you will not see results overnight — but if you do nothing, the bushland will eventually disappear altogether. Suggested management actions include: • fencing to exclude livestock; • controlling weeds and exotic grasses; • re-establishing native trees and shrubs.

Kit 2 in this series, *Managing Farm Bushland* will help you prepare management strategies to maintain and improve the condition of your bushland.

Fig. 1. What your results mean — Kit 1.

WHERE TO FROM HERE?

Having assessed the health and sustainability of your bushland patches you are now in a position to plan management strategies to maintain and improve their condition.

The options that we recommend are:

EITHER:

1. Go straight to Kit 2, *Managing Farm Bushland,* to develop management strategies to maintain or improve the health rating of your bushland patches.

OR:

2. Work through any or all of Kits 3, 5, 7 and 8 to obtain a better picture of the natural resources of your farm.

The titles of the Kits in this series are:

Kit 1: Assessing Farm Bushland	Kit 6: Managing Farm Watercourses
Kit 2: Managing Farm Bushland	Kit 7: Surveying Farm Flora
Kit 3: Assessing Scattered Farm Trees	Kit 8: Surveying Farm Wildlife
Kit 4: Managing Scattered Farm Trees	Kit 9: Farm Planning for Bushland and Wildlife
Kit 5: Assessing Farm Watercourses	

Other actions that you and your family might like to carry out are:

- Work out the percentage of your property covered by bushland patches.
- Calculate the area of bushland you could lose if your bushland continues to degrade.
- Discuss with your family, neighbours or Landcare Group the implications of further degradation of your bushland.
- Encourage your neighbours or Landcare Group to purchase and work through this Kit.
- Put your bushland map together with those of your neighbours to get a bigger picture about the health of bushland in your catchment.
- Work with your Landcare Group or neighbours to develop a plan of action for restoring the degraded bushland in your catchment.

Fig. 2. Determining the next step — Kit 1.

BUSHLAND MANAGEMENT GUIDE

Bushland Health Indicators	Management actions needed if the answer is "false"
1. Bushland is never or rarely grazed by stock.	Fence off or reduce grazing pressure. Control feral or pest species.
2. Seedlings or suckers less than 2m tall are present.	Encourage natural regeneration. Direct seed or plant tubestock in moderately or highly degraded patches.
3. Bush is burned infrequently, i.e., at intervals greater than 5 years.	Reduce the frequency of burning. Plough or slash firebreaks around the perimeter of the bushland.
4. Trees mainly healthy, with little or no dieback.	Inject selected trees with insecticide. Avoid herbicide drift and fertilizer application.
5. There are no rabbits in the bushland.	Control by fumigating, shooting or trapping.
6. Weeds are absent or found only around the edges.	Control by chipping or spot spraying.
7. Trees are not harvested for timber or firewood.	Avoid harvesting. Consider planting woodlots for future timber needs.
8. Free of land degradation e.g., soil erosion, salinity.	Stabilize erosion. Re-establish salt tolerant plant species.
9. Free of fertilizer application or herbicide drift.	Avoid herbicide drift. Don't overfly bushland during aerial spraying or fertilizing.
10. Less than 20% of trees have mistletoe.	Remove mistletoes; encourage predators of mistletoe.
11. Free of vehicle tracks (other than fire trail).	Rehabilitate unnecessary tracks.
12. Free of feral predators such as cats and foxes.	Control by shooting, baiting or trapping.
13. Native shrubs are present in the understorey.	Re-establish by natural regeneration, direct seeding or planting tubestock.
14. Old growth trees with hollows are present.	Put up nest boxes; avoid harvesting old trees.
15. Dead timber is left standing.	Avoid harvesting dead trees for firewood.
16. Fallen timber and logs are left on the ground.	Avoid harvesting fallen timber for firewood.
17. Ground flora is mostly native grasses and herbs.	Exclude stock; control rabbits, wild pigs and goats. Control numbers of native browsing animals (kangaroos and wallabies).
18. Ground is covered with litter of leaves, bark and twigs.	Reduce the frequency of burning off. Exclude livestock.
19. Bushland patch is greater than 0.6 ha in size.	Enlarge by allowing extra space inside the fenced off area. Encourage natural regeneration.
20. Joined to other bushland by bush corridor or scattered trees no more than 50 m apart.	Plant corridors to link bushland patches.

Fig. 3. Deciding management actions — Kit 2.

Whilst the structure of the three management kits is similar to that of the assessment kits some important differences can be identified. The management kits are divided into two sections. Firstly an overview of the management responses needed to convert "False" responses observed in the assessment kit into "True" outcomes (Fig. 3). Secondly a more detailed description of the remedial actions required are provided in subsequent pages and through linkages to other practical published materials. These kits do not provide in-depth information on management (e.g., how to plant trees, how to control pests) but suggest a variety of common sense management actions. It is up to the user to decide which actions he/she will adopt and then to seek out further information, if necessary. A list of reference material and names of people who may help is provided at the end of each kit to facilitate this. Additional steps in the management kits encourage the user to develop an action plan detailing actions required, estimated costs and target dates for completion and to develop monitoring strategies.

The two specialist kits, "Surveying Farm Flora" and "Surveying Farm Fauna" are included for the user who wishes to further his/her knowledge in these areas. While these two kits are not regarded as essential for all landholders to complete, they are included to encourage landholders to consider and assess these basic resources. Important linkages between all remaining kits are identified. The desired end point of the Toolkit is for landholders to work through and implement the planning strategies to develop a plan that integrates production agriculture with the conservation of bushland and wildlife.

Definitions of terms used in this kit

Scientific terms are avoided as far as possible in all of the kits. A definition of terms which are not in common usage, is included.

Where to from here

This section guides the user to other kits in the series, or to agencies or resource people for specialist advice.

Some extra information

This section points the user to additional specialist resources, such as aerial photographs of their property, and provides a list of relevant reference books.

TESTING OF KITS BY LANDHOLDERS

Pre-launch evaluations

Kit development took place over a period of 22 months, involving one or two primary authors who wrote a succession of drafts which were critiqued by members of the project team (comprising widely varying backgrounds and expertise), members of the Project Review Panel (a group of professionals from related agencies and rural groups), and by selected landholders who had expressed willingness to participate in assessing the kits.

Focus group based assessment

Draft versions of a selection of the kits were provided to seven Landcare Groups (35 landholders) across the Central Western Region of New South Wales, as well as individual landholders. Of these, five groups participated in the focus group discussions about the kits, while the remainder provided individual written feedback or were interviewed by telephone. The latter were asked the same questions regarding the kits as the focus group participants.

The focus group approach (Kreuger 1988) was selected for evaluating the prototype kits because it enabled us to gain in-depth insights into participants' responses on quite detailed issues of kit design and content. It also provided for a relatively informal context and a "client friendly" approach to landholder participation. It was hoped that landholders who were not members of the Landcare movement could be identified and persuaded to participate in the focus group programme, but all such landholders became members of Landcare groups during the process.

The focus groups were conducted by an independent facilitator. Groups were either provided with copies of the kit(s) through the district Landcare Co-ordinator or directly by mutual arrangement with the group secretary. A focus group session was then arranged within three to four weeks of receiving the kit, and participants were requested to attend the meeting having at least read and critically appraised the kit, and ideally having used the kit on their own property or in a group exercise on a selected district farm. Evaluation sheets with prepared questions were provided to guide feedback. Participants were given opportunities to submit written feedback and/or make verbal comments at the meeting. No direct contact or assistance was provided to participants after they received the kits for testing, although they were given a phone contact with a Project Team member to help answer any queries or concerns about the process.

The focus group sessions were arranged to follow the sequence of the kits, i.e., commencing with an "Assessment Kit", followed by its "Management" counterpart. The latter was given to the group at the completion of the "Assessment Kit" evaluation session. A further period of testing, followed by a second focus group meeting was held. Assessing and managing farm bushland, scattered farm trees and farm watercourses were evaluated in this manner. Surveying farm flora and wildlife and farm planning were not available in draft form in time for focus group analysis prior to the launch of the Toolkit. However, individual landholders provided feedback on draft versions of these kits.

The focus group sessions were facilitated using a semi-structured approach, emphasizing the use of open-ended questions (e.g., "Well, what were your experiences with the kit?") and occasional prompts to guide discussion towards the general categories of feedback sought. Discussions were audio-taped with the permission of attendees. Group participants (and all evaluation respondents) were asked to comment on kit design, readability, kit language and concepts, the suitability of the kit for use in the paddock, relevance of the kit information to the landholder's property and the district, and the perceived relevance and acceptability of the kit to other landholders generally.

Transcripts of the focus group sessions were then used to collate and summarize feedback to each of the evaluation elements (as noted above). The feedback was then presented to the Project Team and to kit authors. The draft versions of the kits were subsequently modified according to that feedback. The focus groups strongly endorsed the overall concept and content of the kits and supported the general design features of the kits. They found the kits to be basically "user friendly" for landholders, and endorsed the kits as highly relevant to rural landholder's requirements.

Within this framework of support there was a substantial amount of detailed, critical comment about all aspects of kit design, relevance, saleability and content. Extracts of some of the audio-taped group sessions were played back (anonymously) at Project Team meetings to give kit writers, and the Team generally, a direct "feel" for landholder's views and for the strength of these views. Many valuable suggestions for improvements to design and to content were made. Kit language and terminology was amended to better reflect landholder's reading levels and preferences.

Post-launch evaluations

Eighty-six phone interviews of landholders who purchased the Toolkit were carried out in August and September 1997, within two months of the launch of the Toolkit. Their views were sought about bush protection on farms and related issues, and about the kits, the impacts of the kits upon their own views and behaviours towards bushland and related issues. These landholders were different from the 35 that took part in focus group activities. At the time of the survey, 34 had read significant parts of the Toolkit and only one had used it in a paddock situation. Thirty-four had been positively influenced by the Toolkit and had altered perception of what is good "bush". Most found the design straightforward and easy to read. There was a desire by many for more specific information on their geographic location. All but one intended to use the Toolkit as a reference and in the field within 6–12 months and would recommend others to purchase the Toolkit.

MARKETING PHILOSOPHY FOR THE TOOLKIT

The principles and framework underpinning the communication and persuasion strategy for landholders have been previously identified and discussed by Goldney *et al.* 1995. A marketing frame of reference has been adopted for this project with the perception that: "marketing is the creation of appropriate relationships so that people are willing to make voluntary exchanges through time". This view distinguishes marketing from coercive selling and from top-down approaches that tend to ignore the longer term issues of client satisfaction and loyalty. The marketing of the Save the Bush Toolkit also recognizes that credible bush and natural resource use management is a complex product to package and sell. For instance, the value of conserved bushland is essentially long term in nature and not readily demonstrable in short-term farm profitability, productivity and sustainability. As well, the Toolkit is essentially offering knowledge and strategies that were previously unavailable in this written format for individual landholders to apply so there is no base of prior experiential involvement by these clients to appeal to in promoting the product.

The essence of the Toolkit is that it seeks to involve the client in co-producing the value that is ultimately delivered. Involvement means that the client is not a mere passenger, but he/she can take some ownership of the outcome and has the opportunity to be intrinsically rewarded by the learning process that takes place. A further extension of this notion is that user testimonials will be important in spreading the value of the Toolkit to other clients. The key to enabling this "word of mouth" process to happen effectively is to ensure that clients have a positive experience with the application of the Toolkit and become advocates in encouraging others to try it.

Under this marketing frame of reference, a contact strategy was devised that contained the elements of a marketing campaign. This involved a pre-launch period, the launch itself, an initial follow-up period after the launch, and a post-launch strategy.

THE MARKETING STRATEGY

The key elements of the marketing strategy are best described by relating them to an underpinning marketing model outlined by Watson (1996). This model is based on the notion that purchase behaviour can be essentially viewed as a progressive shift in client attitudes rather than a process driven primarily by logical client reasoning. To achieve this shift in attitude, the model contains the following primary activities, with the associated attitude shifts identified in brackets: crystallize the value proposition to the target landholder market segment; raise awareness of the product (change ignorance to awareness); create interest (convert indifference to interest); enable trial (convert scepticism to acceptance); facilitate changed behaviour (convert doubt to belief via facts, benefits, evidence) and support continued conservation action (reinforce conviction).

Raising awareness

The aim within this element of the marketing model is to move potential clients from ignorance of the product to awareness. A marketing campaign strategy built around the launch and pre-launch periods was developed which sought to promote this awareness via the following activities:

PRE-LAUNCH PERIOD

This included the development of a full colour brochure which utilized the cover photo and testimonial of a well respected, mid-30-year-old local landowner who had enthusiastically endorsed the Toolkit during its trial phase. The key message of this brochure was built on the tag line "At last we've got the tools now . . . let's get on with the job." This approach developed out of discussions with landowners during the trial phase who stressed that a broad "conserve your bush" approach was inappropriate as they considered that local landowner attitudes had already largely embraced the umbrella "need to conserve" message. Prior to the development of this brochure, a decision had been taken to develop a Save the Bush Central West logo and to build all communications to landowners and agencies around this overall Save the Bush theme.

Distribution of 26 000 copies of this brochure was carried out through *The Land,* the most widely read rural newspaper in the Central West of New South Wales, 10 days prior to the launch. The brochure enabled clients to purchase the Toolkit by post but additional "advertorial" in the paper encouraged them to visit the Save the Bush tent at the Mudgee Small Farm Field Days and participate in the formal launch of the Toolkit.

The creation of a professionally made video was undertaken during this pre-launch period which involved the same landowner describing the conservation imperatives he was already undertaking on his property and stressing the benefits to a landowner of having "tools that could fix his land". Scenes from this video were made available to a local television company to use in screening commercials for the Mudgee Small Farm Field Days, as the launch of the Save the Bush Toolkit was the central feature of this 1997 event.

The launch period

Awareness raising activities during this period centred on attracting clients from among the 30 000 visitors to the Mudgee Small Farm site. These field days have developed into a premier event for attracting a wide range of landowners within and beyond the Central Western Region and were selected for this reason for the launch activity. A moderately large 10 m × 10 m marquee was erected in a high traffic area with a suitable Save the Bush banner to identify it. The marquee contained bushland related displays, a video playback machine which ran the Save the Bush video and a sales desk where copies of the Toolkit could be inspected, discussed and purchased. Several Save the Bush members volunteered to circulate throughout the ground dressed in bush animal and bird costumes to encourage visitors to visit the Save the Bush marquee.

The official launch activity itself was designed to attract visitor attention. It was held in a separate part of the ground where seating was available, and involved a high profile personality, Mike Hayes, well known to landowners from his ABC radio series "The Prickle Farmer" and his contribution to profiling environmental issues in the Murray-Darling Basin.

Enabling trials

An important element in moving clients from an attitude of scepticism to an attitude of acceptance is to facilitate personal trial of the product. In this early phase of marketing the Toolkit, opportunities to enable personal trial have been limited to encouraging clients to inspect the individual kits and to stress that they have been developed in conjunction with landholders who provided strong support for their ease of use and practical approach.

Considerable debate among members of the Save the Bush team focused on whether to sell the kits individually or as a combined Toolkit package. The combined option was eventually decided upon for inventory control reasons but it was recognized that some clients might prefer to make a smaller investment in certain kits to enable them to "try and see" before making the larger investment in the remaining kits.

Seeking changed behaviour

This element of the marketing model addresses the issue of enabling clients to move from an attitude of doubt to an attitude of belief via the communication of facts, benefits and evidence about the product and its use in their situation. This framework provided the basis for the communication and promotion activities in the pre-launch and launch periods. Communication messages stressed the factual issues at stake, the benefits of purchasing and using the Toolkit and the testimonials of those landowners who had trialled the kits as evidence that they were applicable to landowner situations in the Central West.

Supporting continued action

This element of the marketing model recognizes the need to support clients during the post-purchase period in order to overcome any initial fears that they may have made a mistake in their purchase and to reinforce the conviction that the purchase will provide genuine value through time. A strategy of setting up an easily accessed telephone support line called "Bushline" was put in place to meet this need and to offer the option of placing telephone orders for the Toolkit.

Marketing outcomes and early lessons

Anecdotal feedback indicated that the marketing strategies described above appeared to generate substantial awareness of the existence of the product amongst visitors to the Field Days launch who purchased approximately 40 Toolkits. Some 6 months after the commencement of the marketing campaign, a further 210 mail orders had been processed. Clients were evenly split between landowners (both within and outside the New South Wales Central Western Region) and agencies/others. About 100 complimentary copies were distributed. By September 1998, approximately 430 kits had been sold.

One thousand Toolkits were printed. Since all material produced is in digital format it will be relatively easy to incorporate additional feedback and suggestions from regional landholders as this information comes to hand. We have strong anecdotal support for the concept, presentation and practicality of the Toolkit from a range of landowner and agency sources. The overall impression is that "here is a worthwhile concept and product but its acceptance in the landowner marketplace is still problematic."

STRENGTHS AND WEAKNESSES OF THE PROGRAMME

All purchasers of the Toolkit have had access to the "Bushline" or could provide feedback via a questionnaire included with the product. In addition we have the results of our focus groups, data from landholders that we have intentionally followed up by phone contact, scientists, post-graduate students, conservation groups, regional policy makers, Landcare groups, and agency personnel from many parts of Australia. The responses have been overwhelmingly positive, even effusive in their assessment. Only two landholders have recorded a very negative response through our formal processes. Some have indicated that the cost (AUD$60.00) is too much for landholders in tough economic circumstances. However, when we offered a "special" price at field days, there was no perceptible increase in purchases. We have had some criticism from a few agency personnel who thought that some Toolkit features should conform to their particular pre-conceived view. While we have been open to all suggestions, any major changes that were made during the development phase or that will be made in the future, have (or will be) been driven by landholder responses, since they are our target group.

A number of landholders in our region and from elsewhere have indicated the need for more specific information for their particular geographic locations. We did try to address such concerns through the use of Bush Info notes and by providing key references, but there is no realistic alternative for most local communities other than to gradually build up their own relevant natural resource inventories. Other landholders have identified "gaps" in the Toolkit coverage. Many times we have heard a plea for a kit to be provided for those landholders who have no bushland at all and who need to start from scratch. That issue will be the first addressed in 1999 when we add additional kits to the programme. Others will include catchment health, native grasslands, farm economics and nature conservation values, corridor and windbreak design, rare and endangered species and reintroduction of locally or regionally extinct species.

We readily recognize that the Toolkit is not an appropriate extension package for all landholders with bushland resources. It is designed for those with particular literacy skills. We do not understand the full range of landholder educational needs in our region, nor whether there are cost effective responses that can be made in all circumstances. Nor have we identified the major barriers that prevent positive conservation responses being made from various farming communities or individuals. In spite of having developed a sophisticated marketing programme, we are unsure whether we are eliciting an optimal response for our particular product. We were probably overambitious in our expectations of the rate of uptake and usage of the Toolkit. We have come to accept that our product requires a long-term commitment on our part to ensure optimal marketing penetration and more importantly changed behaviour on the ground.

RELEVANCE TO OTHER LAND SYSTEMS

We are currently negotiating with some groups the modification of the Toolkit for use in other catchments in New South Wales. The digital design should enable cost effective modifications to be readily made. A number of landholders and agency personnel are trialling the Toolkit in other land systems. We explicitly state that our product has been developed for the sheep-wheat belt and may not be appropriate for rainforest, mallee or range-lands. However, we see no reason why the approach we have developed could not in principle be used in any land system.

OVERALL IMPACT OF THE TOOLKIT

The post-launch phase calls for some refocussing of future directions. We believe that the key is to first explore how the target landowner market perceives the value of this product now that it is launched and available and its practical usefulness in strategically managing their farm operation. Then it will almost certainly be important to devise on-ground communication approaches that involve these landowners in physically using and trialling the Toolkit to build acceptance and to build conviction of its value and usefulness. We are currently planning such an approach which may include door to door selling in strategic areas and the use of targeted field days, making use of our extensive network, where the kits will be used by landholders. Another important strategy underway is the development of a schools Toolkit. This will contain identical materials to the landholder Toolkit but will include two additional kits, one explaining how the Toolkit can be used in the secondary curriculum and the other designed for use by primary and infant schools in New South Wales. This material is being prepared under contract by practicing teachers. Previous data we have gathered on the attitudes of landholders to bushland conservation (Hodgkins *et al.*, in press) have indicated the importance of the parent-child relationship in influencing parental behaviour. We are confident that the product we have produced is both attractive and reasonably priced (cost recovery only), and is innovative, fills a market niche, communicates relevant ecological theory and practice in a manner that is appropriate to many landholders. It is now up to us to convince landholders in our backyard that this product is essential to their long-term economic survival and for the future conservation needs of our regional flora and fauna.

ACKNOWLEDGEMENTS

We acknowledge the generous financial support of the World Wide Fund for Nature (Australia), the Land and Water Resources Research and Development Corporation and its dynamic director Dr Phil Price, Environment Australia and the National Resource Management Strategy. Post graduate students from the Environmental Studies Unit, Charles Sturt University, Bathurst Campus enthusiastically participated in our programme at every level and wrote many of the Bush Info Notes including Andrew Fisher, Steven Cox, Cilla Kinross, Donna Windsor, Marcus Croft and Chris Bloomfield. The following staff members from Charles Sturt University (Sylvia Cardale, Johannes Bauer, Brian Stone and Christine Riding also helped in numerous ways) and from the University of Sydney — Orange Agricultural College, David Carberry, Cilla Kinross and Hugh Haines. Most of the design and printing was carried out in house in the Charles Sturt University Printery at Bathurst. Particular thanks go to Joanna Barnes and printery staff. Many regional organizations and agencies also helped us, such as personnel from New South Wales Land and Water Conservation, media groups — particularly Prime TV and ABC Radio, the *Land* Newspaper in both allowing us access to national and regional papers, and many Landcare co-ordinators. Last but not least we would like to thank the numerous landholders who participated in focus groups and willingly gave of their time to ensure that the end product was genuinely co-produced. Special thanks go to Chris Blunt from the Emu Swamp Landcare group for his generosity in facilitating publicity for the project.

REFERENCES

Brechwoldt, R., 1983. Wildlife in the home paddock. Angus and Robertson: Melbourne, Victoria.

Brouwer, D.,1997. Managing waterways on farms. New South Wales Agriculture: Tocal, New South Wales.

Buchanan, R., 1989. Bush Regeneration. Technical and Further Education Student Learning Publication: Sydney, New South Wales.

Dorricut, K. and Roberts, B., 1993. Wildlife conservation on planned properties. Land Use Study Centre, University of Southern Queensland: Toowoomba, Queensland.

Davidson, R. and Davidson, S., 1992. Bushland on farms — do you have a choice? Australian Government Printing Service: Canberra, Australian Capital Territory.

Goldney, D., Bauer, J., Bryant, H., Hodgkins, D. and Watson, G., 1995. Winning battles but losing the war — the education marketing imperative. Pp. 574–88 *in* Nature Conservation 4: the Role of Networks ed by D. A. Saunders, J. L. Craig and E. M. Mattiske. Surrey Beatty & Sons: Chipping Norton, New South Wales.

Hodgkins, D., Goldney, D., Watson, G. and Tyson, G. (in press). The attitudes of landholders to a range of environmental issues, including the values of remnant bushland in the central western region of New South Wales. Pp. 336–50 *in* Temperate Woodlands in Australia ed by R. Hobbs and C. Yates. Surrey Beatty & Sons: Chipping Norton, New South Wales.

Hussey, P. and Wallace, K., 1993. Managing your bushland. Western Australia Department of Conservation and Land Management, Perth, Western Australia.

Kreuger, R. A., 1988. Focus groups — A practical guide for applied research. Sage, Newbury Park: California.

Siepen, G., 1997. Partnerships with nature. Managing your farm in an ecologically sustainable manner. Waratah and Landcare Australia.

Wakefield, S. and Goldney, D., 1997. Save the Bush Toolkit. Charles Sturt University: Bathurst, New South Wales.

Watson, G. K., 1996. Marketing issues for remnant vegetation conservation. Pp. 47–51 *in* Remnant Vegetation in the Central West — Winning Battles but Losing the War? ed by S. Wakefield. Charles Sturt University: Bathurst, New South Wales.

CHAPTER 23

Integrating conservation and production: lessons from Australian rangelands

David Freudenberger[1]

Conservation of biological diversity is at odds with ten thousand years of agricultural innovation and practice which has been based on simplification at a range of scales, from genes to landscapes. It is only recently that we have begun to understand how important biological diversity is to production systems, particularly in complex semi-natural systems such as rangelands. This chapter describes the ecological basis of livestock production in rangelands that relies on four critical components of diversity at contrasting scales: gut microbes, soil biota, multiple forage guilds, and landscape diversity as expressed by patchiness. Conservation of these critical elements of biological diversity is the basis of sustainable production in rangelands. However, there is a need to recognize, and acknowledge, that pastoralists cannot meet all conservation goals — spatial separation of uses is required. There will always be a need for areas free of agriculture at a range of scales and configurations within the production matrix.

INTRODUCTION

Is conservation of biological diversity essential to a farmer's enterprise? The title of this volume implies that it is, but a rigorous defence of the premise that biodiversity is essential to the farmer is neither simple nor readily apparent.

The ten thousand year history of agriculture is one of extreme simplification. Only a minute proportion of global biological diversity is used for agricultural purposes. Only a few species of plants have been cultivated as fruits, vegetables and grains. Only a dozen or so mammals and birds have been domesticated. Only one insect (the honey bee), a few fish (e.g., carp and tilapia), a few crustaceans (e.g., prawns), one mollusc (the terrestrial snail), and a single group of microbes (yeasts) have been widely exploited, while no reptiles have been developed for agricultural use. The recent history of agricultural evolution (past 150 years) has seen the dramatic reduction of the genetic diversity of these few plants and animals following rigorous selective breeding and distribution of relatively uniform germ plasma throughout the world. There has been an even greater simplification through the process of cultivation and husbandry of this small number of domesticated organisms. This has occurred through the clearing of vast forests, draining of wetlands and cultivation of grasslands.

This continuous strategy of simplification has been enormously successful in supporting increasingly large human populations. The advent of agriculture in various parts of the world has consistently heralded periods of rapid growth and concentrations of human populations (Ponting 1992). The world human population was probably in the order of 250 000 at this time (*circa* 10 000 Before Present) with annual growth rates between 0.0003–0.0005% (Cohen 1995). By A.D. 1, the human population was in the order of 170–330 million, representing a ten to several hundred fold increase in the population growth rate coinciding with the spread of agriculture. From A.D. 1–1650 the population rose to 500–600 million (0.02–0.05% annual growth rate). By 1850 the global human population had reached approximately 1.2 billion people (0.6% annual growth rate). Today, highly simplified food and fibre production systems still support a global population growth rate of 1.5% per annum with a cumulative population size nearing 6 billion (Cohen 1995).

[1]CSIRO Wildlife and Ecology, G.P.O. Box 284, Canberra, Australian Capital Territory, Australia 2601.

Pages 228–35 *in* NATURE CONSERVATION 5: NATURE CONSERVATION IN PRODUCTION ENVIRONMENTS: MANAGING THE MATRIX ed by J. L. Craig, N. Mitchell and D. A. Saunders. Surrey Beatty & Sons, 2000.

Some agricultural systems have become so simplified that a single food species (e.g., the tomato) is widely grown in a totally inorganic substrate in the complete absence of any other life form (i.e., hydroponics). Do farmers need to conserve biodiversity? Yes, but only when food and fibre is harvested from complex biological systems that are too costly or complex to simplify. Extensive grazing and cropping systems reliant on *in situ* incident rainfall are examples of agricultural ecosystems where the conservation of native biological diversity is fundamental to sustainable production. This essay defends this proposition by explaining the ecological basis of sheep production in arid and semi-arid rangelands of Australia.

WHY RANGELANDS?

The rangelands of Australia are the interior regions of the continent that are too dry for intensive agriculture. Nearly three-quarters of the continent are rangelands and 58% of this area is occupied by 6 000 pastoral enterprises running 5 million cattle (20% of the national herd) and 18 million sheep (13% of the national flock) (Anon. 1996). The rangelands support complex, semi-natural ecosystems comprising over 1 800 species of native flowering plants, 605 species of native vertebrate animals and an estimated 25 000 species of invertebrates (Anon. 1996). In contrast to "improved" pastures with perhaps 2–5 dominant plant species (Campbell *et al.* 1987), 150 species may interact in a single rangeland paddock (Stafford Smith 1994).

However, the rangelands have suffered the greatest number of extinctions in Australia (Flannery *et al.* 1990). Of the 72 species of mammals originally found by Europeans in the arid zone, 11 have become extinct, five have disappeared from the mainland and 15 more have declined dramatically in range (Morton 1990). Currently more than a third of Australia's threatened bird species and one-tenth of its threatened plants occur in the rangelands (Morton *et al.* 1995). The threats to native biological diversity in the rangelands are numerous and interacting, but stem primarily from grazing by feral, native and domestic herbivores which consume critical habitat and prevent native species from using their drought refuges. Introduced predators (foxes *Vulpes vulpes* and cats *Felis catus*) apply additional pressure to relic populations (Morton 1990). The provision of artificial watering points, an essential infrastructure in rangelands, is an ever increasing threat to native biological diversity. Today, there are few areas within the rangelands that are greater than 10 km from permanent water (James *et al.* 1995). However, a significant proportion of plant and animal diversity appears to survive only in areas distant from water. At six gradients from watering points located across the rangelands, 15–38% of a wide range of native plants and animals were found only at considerable distances (>6 km) from water and the grazing it supports (Landsberg *et al.* 1997).

Pastoral enterprises are clearly involved in the on-going loss of biological diversity in the rangelands, but does it matter to the productivity and profitability of these enterprises? This is a relevant and important question because if it does not matter to long-term productivity, then integration of conservation and production is a desire, rather than a necessity. I argue that conservation of some critical components of biological diversity are essential to sustained livestock production, however, conservation of all biological diversity may not be essential to pastoral enterprises. Holistic conservation of all components of biological diversity requires a partnership between private pastoral interests and the public through a diversity of mechanisms and institutions including publicly funded reserves free of artificial watering points.

LESSONS FROM THE OUTBACK

Sustainable rangeland enterprises are entirely dependent on four critical components of biological diversity: gut microbes, soil biota, multiple forage guilds, and landscape diversity as expressed by patchiness at a range of scales. Yet, these components do not encompass all aspects of biological diversity in the rangelands. If many of the species of birds or reptiles disappeared from the rangelands, it may not make much difference to sheep or cattle productivity. However, the loss of reptiles or birds may be indicators of profound changes to ecosystem processes that may well impact on livestock production. I explain the importance of some components of biological diversity in order to illustrate that conservation is critical to the farmer, at least in the semi-natural landscapes of the rangelands which are too erratically dry to succumb to gross simplification by grazed or cropped monocultures.

SHEEP LIVE ON GUT DIVERSITY

Livestock production is based on only two intensively bred livestock species, cattle and sheep, but these two species are walking menageries dependent on diverse and complex populations of bacteria, fungi and protozoa (Table 1). The mammalian herbivore lives off the waste products of anaerobic fermentation (short-chain fatty acids) generated by this diversity and vast number of microbes. The forestomach-fermenting herbivores

Table 1. The density and number of described species of micro-organisms that have been found in the rumen (Theodorou and France 1993; Orpin and Joblin 1988).

	Density (per ml rumen fluid)	Number of species
Bacteria	Billions (10^9–10^{10})	200+
Protozoa	Millions (10^5–10^6)	100+
Fungi	(Thousands) 10^3–10^5	10+ (8 genera)
Virus	Not known	10+

(ruminants, camels and kangaroos) also gain the majority of their protein from microbes through the digestion and absorption of microbial cells that flow from the forestomach to the small intestines (Van Soest 1982; Freudenberger *et al.* 1989).

Like many macro-ecosystems, the rumen micro-ecosystem (forestomach of sheep and cattle) is dominated by a few common species. *Bacteriodes succinogenes*, *Ruminococcus albus*, *R. flavefaciens*, and *Eubacterium cellulosolvens* are the dominant bacteria involved in the degradation of cellulose. This is the most common carbohydrate on the planet, but a compound whose bonds can only be broken by enzymes produced by certain classes of bacteria, fungi and protozoa (Hungate 1988).

The rumen may be dominated by a few microbial species at any one time, but the diversity of microbes enables livestock to digest their diverse diets. There are microbes that specialize in metabolizing the wide range of plant molecules found in any natural herbivore diet. There are bacterial species such as *Bacteroides amylophilus* which are obligate digesters of starch. *Megasphaera elsdenii* specializes on soluble sugars and *Bacteroides ruminicola* is one of the dominant metabolizers of plant proteins (Theodorou and France 1993). Some species persist at very low concentrations until a suitable dietary substrate is consumed by the host herbivore. For example, *Oxalobacter formigenes* uses oxalate, a toxin in some plants, as its sole energy source. Without large populations of this species, oxalate would be lethal to animals consuming plants with high concentrations of this compound (Dawson and Allison 1988).

Much of the microbial diversity of the herbivore forestomach may not be necessary under intensive production systems such as feedlots where livestock are subjected to a homogeneous and constant grain-based diet. The diversity of gut microbes provides the herbivore with the capacity to exploit highly heterogeneous environments with an ever-shifting composition of forage species that contain complex plant compounds including toxic ones.

Conservation of rumen microbial diversity is critical to livestock production, but it is the only component of agro-ecosystem biodiversity that is self-sustaining — microbes look after themselves. Over the past three decades, considerable research has gone into selectively enhancing or suppressing microbial populations in the rumen in order to improve the conversion of feed consumed to energy and protein absorbed by the animal (Van Nevel and Demeyer 1988). Little of this work has reached commercial application. An apparently simple manipulation such as a tightly targeted suppression of bacteria that generate methane (energy unavailable to the host ruminant) tends to cause all sorts of often adverse and unpredictable changes to protein and energy metabolism in the rumen and the whole animal (Van Nevel and Demeyer 1988).

The only success in manipulating rumen microbial populations has been enhancement of species richness by the introduction of a guild of bacteria to fill an otherwise empty nutritional niche. *Leucaena* spp. are leguminous browse species widely introduced to sub-tropical savannahs in Australia. Some species under certain conditions have high concentrations of the non-protein amino acid mimosine which is hydrolized in the rumen to a toxic pyridone, that can cause goitre in Australian cattle grazing *Leucaena* (Hegarty *et al.* 1976). Microbial populations in village goats in Hawaii were found to readily metabolize this pyridone toxin. Transfer of goat rumen fluid containing these microbes to cattle conveyed protection from high levels of mimosine and these microbes have now become widespread within Australian cattle populations (Jones and Megarrity 1986).

The complexity and resilience of the rumen has defied simplification that has become so dominant in macro agricultural ecosystems. However, 50 years of rumen ecosystem research has provided a better understanding of the nutritional requirements of the rumen and the whole animal.

PLANT DIVERSITY SUPPORTS SUSTAINED PRODUCTION

Rumen microbial diversity enables livestock to exploit a wide range of plant species which in turn allows livestock to exploit highly variable rangeland environments characterized by a diversity of soil types and erratic inputs of rainfall. Sustainable rangeland production is only possible by careful management of diverse guilds of forage plants. A reliance on just a few forage species, such as the reliance on ryegrass *Lolium perenne* and white clover *Trifolium repens* in some high rainfall regions of New Zealand and Australia, would be unsustainable in rangeland environments. I draw from my own experience in western New South Wales to illustrate this point.

Figure 1 is an example of the growth and persistence of four guilds of sheep forage (Table 2) measured in a semi-arid woodland described in detail by Tongway and Ludwig (1990) and Wilson (1991). Loss of any one of these species would greatly reduce the responsiveness of the system to rainfall and reduce the resilience of the system to grazing. For example, loss of *Digitaria brownii* and *D. coenicola* grasses would result in a failure of the most productive parts of the landscape (mulga *Acacia aneura* tree groves) to produce palatable forage following warm season (November-March) rainfall. These species are the only ones that grow in mulga groves at this time of the year (Freudenberger *et al.* 1999).

In contrast, other species of grasses such as *Thyridolepis mitchelliana* and *Monachather paradoxa* are capable of rapidly responding to rainfall at any time of the year, particularly during the cooler months (April-October). Ephemeral forbs (Table 2) contribute large quantities of biomass during abundant cool-season rainfall (Fig. 1). However, this production requires autumn

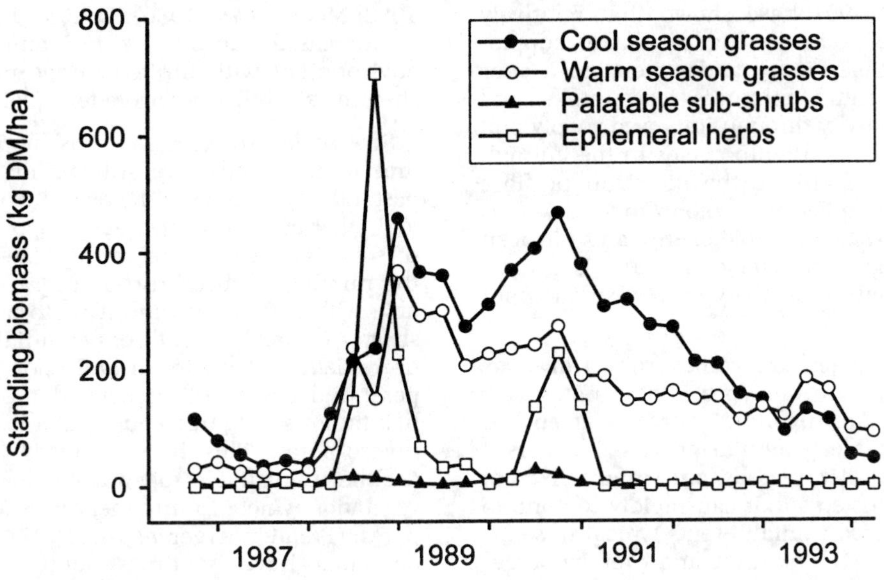

Fig. 1. Standing biomass of four groups of forages (composition listed in Table 2) in a semi-arid woodland in eastern Australia (adapted from Freudenberger *et al.* 1999).

Table 2. Composition of various forage guilds in a semi-arid woodland in western New South Wales (from Freudenberger *et al.* 1999).

Cool-Season grasses	Warm-Season grasses	Sub-shrubs	Ephemeral herbs
Monachather paradoxa	*Aristida contorta*	*Maireana villosa*	*Calotis cuneifolia*
Thyridolepis mitchelliana	*A. jerichoensis*	*Sida cunninghamii*	*C. inermis*
	A. obscura		*Erodium crinitum*
	Digitaria brownii		*Rhodanthe floribunda*
	D. coenicola		
	Eragrostis eriopoda		
	Themeda australis		

rains to initiate wide spread germination and subsequent rains through winter to sustain a large pulse of productivity in the spring.

The loss of *Erogrostis eriopoda* would result in reduction in the capacity of the landscape to recover from grazing. This long-lived grass adds little to livestock nutrition due to its low palatability and low nutritive value, but it persists in the landscape, particularly on the least fertile and dry run-off slopes (Freudenberger *et al.* 1999). This persistence maintains soil cover, reduces erosion, and enhances infiltration. *E. eriopoda* acts as a nursery plant allowing more palatable grasses and forbs to grow between tussocks (pers. obs.).

A single forage guild may dominate, but not for long. At CSIRO's "Lake Mere" study site in western New South Wales, forbs dominated during exceptionally wet periods of 1988, but this dominance never persisted. These forbs rapidly disintegrate during summer (Fig. 1). Single species may dominate in space, e.g., *Digitaria brownii* in mulga groves or *Erogrostis eriopoda* on run-off slopes, however no species in these complex and dynamic landscapes remained dominant through time (Table 3).

Table 3. Proportion of total forage biomass and green biomass attributed to each of the common species in a semi-arid wooded rangeland in western New South Wales over a cumulative seven year period in a lightly grazed paddock (from Freudenberger *et al.* 1999).

Dominant forage species	Proportion of total standing forage dry matter (%)	Proportion of total standing mass of green leaf (%)
Thyridolepis mitchelliana	39.8	19.6
Monachather paradoxa	7.7	9.1
Aristida jerichoensis	2.5	2.5
Digitaria brownii	12.0	10.3
Eragrostis eriopoda	17.9	8.6
Ephemeral herbs	13.3	38.0
Others forage species	6.7	11.9

GREEN LEAF FROM A DIVERSITY OF SOURCES

Livestock production in rangelands is closely related to the availability of digestible plant protein found in green leaves (Ash *et al.* 1995; Freudenberger and Noble 1997). Relatively small amounts of green leaf (e.g., 50 kg/ha) have a synergistic affect on sheep production by "fertilizing" the rumen. Numerous studies (see reviews by Leng

1990; Minson 1990) have shown that relatively small (<10% of total dry matter intake) supplements of digestible energy and protein, to a poor quality diet, greatly enhances feed intake and increases productivity through improved supply and balance of energy and amino acids to the animal. Green leaf may facilitate the digestion of fibre through facilitating bacterial colonization as well as providing readily fermentable energy and nitrogen for microbial growth. The end result is greatly enhanced animal productivity from a small input of green leaf.

The supply of green leaf comes from a diversity of forage species; no one species dominates over time (Table 3). An important source of green leaf comprises palatable perennial grasses that persist during frequent dry periods and maintain root systems and meristems that can rapidly respond to small and scattered rainfall events typical of semi-arid rangelands (Hodgkinson and Freudenberger 1997). However, these grasses can easily be killed by overgrazing and have been lost in many wooded rangeland landscapes (Grice and Barchia 1992; Hodgkinson and Cook 1995). Only annual forbs persist in many landscapes, but these cannot be relied upon for sustained production due to their ephemeral nature and reliance on infrequent periods of abundant cool season rainfall. A landscape with functionally different forage species maintains the diversity and abundance of rumen microbes that are so fundamental to livestock production. Conservation of forage diversity is not a luxury relegated to a few nature reserves, but essential to livestock production in the rangelands.

PLANT AND SOIL BIOLOGICAL DIVERSITY LINKED

The biological diversity supporting rangeland livestock production does not end with plants and gut microbes. There is a biological link to the soil which supports the plants. Inadequate soil water is the factor which limits plant survival and growth in the rangelands. Biological diversity plays a significant role in the transfer of rainfall to available soil water. The myriad of soil organisms, biotic structures such as roots, and biologically derived pores, found in any intact soil, play a particularly important role in creating and maintaining high rates of water infiltration into the soil profile. Many of the ancient rangeland soils of Australia have inherently low infiltration rates. At the "Lake Mere" site, Greene (1992) recorded infiltration rates as low as 25 mm/hr on relatively barren run-off slopes. Infiltration rates within the fertile and well vegetated mulga groves were 264 mm/hr. However, this high infiltration rate was not an inherent physical characteristic of the soil. When a negative pressure was applied to the disc permeameter used to measure infiltration rates, infiltration dropped to only 13 mm/hr in the groves and elsewhere. The applied negative pressure blocked all the macropores larger than 0.75 mm. These pores are formed by soil fauna, plants roots and cracks and fissures due to shrinkage of the soil (Beven and Germann 1982; Moore *et al.* 1986). Loss of these pores leads to increased run-off, erosion and a more arid environment with limited forage productivity for sheep or any other herbivore to exploit.

Fine-scale processes, such as biopore formation and maintenance, are still within the realm of pastoral management. The diversity of ants, termites and other macro-invertebrates, important generators of soil pores, cannot be manipulated directly, but the number of sheep run by a grazier is manageable with a few dogs and a motorbike. Too many sheep may lead to death of perennial vegetation by overgrazing (Hodgkinson and Cook 1995). Loss of perennial cover and associated roots is a loss of habitat for soil fauna leading to a loss of biological macro-pores. This in turn leads to a spiraling decline in available soil water which supports less vegetation which, in turn, generates fewer biological pores (Freudenberger *et al.* 1997). Management decisions have at times subtle, but profound influences in these rangeland environments.

PATCH DIVERSITY HOLDS LANDSCAPES TOGETHER

Conservation of biological diversity is more than maintenance of species assemblages. The spatial and temporal pattern of species distribution is just as important. Industrial agriculture homogenizes landscapes into neat rows or occasional curves if contours are followed. Even our science strives to seek statistical averages with minimal variance in order to detect and understand single phenomena. In the rangelands, heterogeneity and variation is the signal, rather than noise in the system. Patchiness is a striking indicator of rangeland landscapes functioning efficiently to capture scarce resources of surface-water and soil nutrients (Tongway and Ludwig 1994) (Fig. 2). Water and wind influence the availability of scarce resources through redistribution (run-off) and capture (as run-on) by landscape patches (reserves or sinks). Two attributes of landscapes, terrain (shape and slope) and patchiness (number, size and spacing), are indicators of the capacity of landscapes to function by capturing and concentrating water and soil nutrients (Tongway and Ludwig 1997a). In degraded, or dysfunctional landscapes, there is a loss in patchiness, that is, there are fewer traps to catch and retain water, sediments and litter (Fig. 2). When patches such as perennial grass tussocks, shrub hummocks, thickets, and groves are destroyed by over-grazing, so is conservative landscape function (Tongway and Ludwig 1997b).

Even in death, some plants can contribute to patch build-up and maintenance. It can take over 50 years for a dead mulga tree to topple over, and then it remains on the ground for decades supporting a diversity of organisms such as termites and small reptiles (Noble *et al.* 1989; Tongway *et al.* 1989). Patches build and survive by themselves if land management is sympathetic to maintaining a visually chaotic landscape of stumps, dead branches, old tussocks, patches of trees and

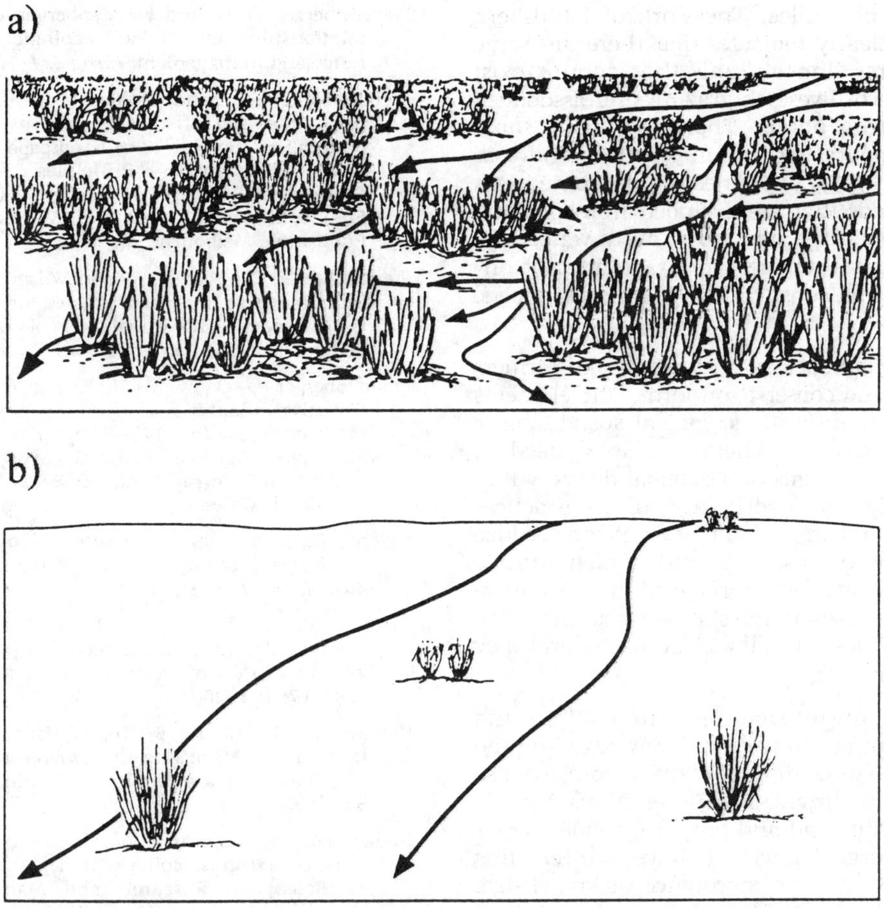

Fig. 2. (a) The tortuous flow of water and wind borne nutrients through a highly patchy (functional) landscape compared (b) to the flow of water and material through a dysfunctional landscape that has lost much of its patchiness (from Tongway and Hindley 1995).

shrubs (dead and alive), termite mounds and other "messy" features. For too long, a "clean" paddock has been considered a sign of honest hard work and diligence rather than a sign of gross simplification and landscape dysfunction.

CONCLUSIONS — LIMITS TO INTEGRATION

Conservation of biological diversity is at odds with agricultural innovation and practice that has evolved over the past ten thousand years. It is only in the past few decades that we have begun to understand how important biological diversity is to some production systems. It is easiest to perceive in complex semi-natural systems such as rangelands which have generally been too dry and infertile to simplify by traditional practices of fertilization and monocultures.

The rangelands still rely on monocultures of sheep or cattle, but there is a growing realization that these two intensively bred species rely on complex associations of species at a range of scales, from the micro-ecosystem of the rumen, to landscapes and the vital patchiness that maintains function.

This role of biological and landscape diversity is also being recognized in the simplified systems of agriculture in more mesic and fertile areas. It is now realized that treeless paddocks of rows of wheat *Triticum aestivum* or annual ryegrass *Lolium rigidum* are ecological time-bombs that leak large amounts of water into salt laden aquifers that are unable to cope with this increasing volumes of ground water. The rising water tables ultimately result in devastated landscapes pock-marked with surface salt (McFarlane *et al.* 1993).

However, conservation of all biological diversity is not necessarily essential to sustainable pastoral production. There is a need to acknowledge clearly the ecological and social limits of on-farm conservation. Conserving a diversity of plant species for their long-term forage value is in the best interests of rangeland pastoralists, but the conservation of every wildflower or species of moss may have little relevance to the farmer. Sheep production in the mallee rangelands will not collapse if the endangered mallee fowl *Leipoa ocellata* is lost through loss of old growth shrubland and predation by feral foxes (Priddel and Wheeler 1995). Rather, burning old growth temporarily improves the sheep grazing value of mallee. The work of Landsberg *et al.* (1997) clearly indicates that there are some species of plants and animals that cannot exist in the presence of livestock grazing and associated artificial watering points. There has to be some spatial and temporal separation of values and uses (Morton *et al.* 1995). There needs to be some areas of the rangelands that are excluded from pastoral influences. How large and how many water and livestock free areas are needed is a matter of on-going research and consultation (J. Landsberg, pers. comm.).

The challenge of society, researchers and farmers is to recognize and conserve, on-farm, core elements of biodiversity essential to ecological sustainability of production systems. There is also a need to recognize those elements of biological diversity that are unavoidably destroyed by agriculture practices and to design and implement reserves that reduce the probability of losing sensitive elements of biological diversity. The integration of nature conservation and agriculture should mean that both values are enhanced, but there are limits and they should be acknowledged.

The integration of nature conservation and production requires motivation, knowledge and the capacity to act. Motivation is derived from a sense of caring and willingness to develop sustainable relationships with land and people (Freudenberger and Freudenberger 1994). I have written this chapter to illustrate the importance of knowledge, and hopefully this kind of knowledge will contribute to more informed agricultural practices that maintain and enhance critical elements of biological diversity, while acknowledging that there are limits to agriculture's current capacity to conserve all elements of diversity in the landscape.

REFERENCES

Anonymous, 1996. Draft National Strategy for Rangeland Management. ANZECC/ARMCANZ: Canberra, Australian Capital Territory.

Ash, A. J., McIvor, J. G., Corfield, J. P. and Winter, W. H., 1995. How land condition alters plant-animal relationships in Australia's tropical rangelands. *Agri. Ecosys. Environ.* **56:** 77–92.

Beven, K. and Germann, P., 1982. Macropores and waterflow in soils. *Wat. Res. Res.* **18:** 1311–325.

Campbell, M. H., Hosking, W. J., Nicholas, D. A., Higgs, E. D. and Read, J. W., 1987. Establishment of perennial pastures. Pp. 59–74 *in* Temperate Pastures: Their Production, Use and Management ed by J. L. Wheeler, C. J. Pearson and G. E. Robards. CSIRO: Melbourne, Victoria.

Cohen, J. E., 1995. How Many People can the Earth Support? W. W. Norton: New York.

Dawson, K. A. and Allison, M. J., 1988. Digestive disorders and nutritional toxicity. Pp. 445–59 *in* The Rumen Microbial Ecosystem ed by P. N. Hobson. Elsevier Applied Science: London.

Flannery, T., Kendall, P., Wynn-Moylan, K., 1990. Australia's Vanishing Mammals: Endangered and Extinct Native Species. Readers Digest Press: Sydney, New South Wales.

Freudenberger, D. O. and Freudenberger, C. D., 1994. Good relationships: ethical and ecological perspectives on rangeland management. *Rangel. J.* **16:** 321–32.

Freudenberger, D. and Noble, J., 1997. Consumption, regulation and off-take: A landscape perspective on pastoralism. Pp. 35–48 *in* Landscape Ecology, Function and Management: Principles from Australia's Rangelands ed by J. Ludwig, D. Tongway, D. Freudenberger, J. Noble and K. Hodgkinson. CSIRO Publishing: Melbourne, Victoria.

Freudenberger, D., Palmer, R. and Wilson, A. D., 1999. The importance of perennial grasses for sheep production in the semi-arid woodlands of eastern Australia. *Rangel. J.* **21**(2): (in press).

Freudenberger, D. O., Wallis, I. R. and Hume, I. D., 1989. Digestive adaptations of kangaroos, wallabies and rat-kangaroos. Pp. 179–87 *in* Kangaroos, Wallabies and Rat-Kangaroos ed by G. Grigg, P. Jarman and I. D. Hume. Surrey Beatty & Sons: Chipping Norton, New South Wales.

Greene, R. S. B., 1992. Soil physical properties of three geomorphic zones in a semi-arid mulga woodland. *Aust. J. Soil Res.* **30:** 55–69.

Grice, A. C. and Barchia, I., 1992. Does grazing reduce survival of indigenous perennial grasses of the semi-arid woodlands of western New South Wales? *Aust. J. Ecol.* **17:** 195–206.

Hegarty, M. P., Court, R. D., Christie, G. S. and Lee, C. P., 1976. Mimosine in *Leucaena leucocephala* is metabolised to a goitrogen in ruminants. *Aust. Vet. J.* **52:** 490.

Hodgkinson, K. C. and Cook, J. D., 1995. The ecology of perennial grass collapse under grazing. Pp. 203–07 *in* Ecological Research and Management in the Mulgalands ed by M. J. Page and T. S. Beutel. University of Queensland Gatton College: Lawes, Queensland.

Hodgkinson, K. and Freudenberger, D., 1997. Production pulses and flow-ons in rangeland landscapes. Pp. 35–48 *in* Landscape Ecology, Function and Management: Principles from Australia's Rangelands ed by J. Ludwig, D. Tongway, D. Freudenberger, J. Noble and K. Hodgkinson. CSIRO Publishing: Melbourne, Victoria.

Hungate, R. E., 1988. Introduction: The ruminant and the rumen. Pp. 1–20 *in* The Rumen Microbial Ecosystem ed by P. N. Hobson. Elsevier Applied Science: London.

James, C. D., Landsberg, J. and Morton, S. R., 1995. Ecological functioning in arid Australia and research to assist conservation of biodiversity. *Pac. Conserv. Biol.* **2:** 126–42.

Jones, R. J. and Megarrity, R. G., 1986. Successful transfer of DHP-degrading bacteria from Hawaiian goats to Australian ruminants to overcome the toxicity of *Leucaena. Aust. Vet. J.* **63:** 259–62.

Landsberg, J., James, C. D., Morton, S. R., Hobbs, T. J., Stol, J., Drew, A. and Tongway, H., 1997. The effects of artificial sources of water on rangeland biodiversity. Final report to the Biodiversity Group, Environment Australia. Environment Australia and CSIRO: Canberra, Australian Capital Territory.

Leng, R. A., 1990. Factors affecting the utilisation of 'poor-quality' forages by ruminants particularly under tropical conditions. *Nutrit. Res. Rev.* **3:** 277–303.

McFarlane, D. J., George, R. J. and Farrington, P., 1993. Changes in the hydrologic cycle. Pp. 146–86 *in* Reintegrating Fragmented Landscapes: Towards Sustainable Production and Nature Conservation ed by R. J. Hobbs and D. A. Saunders. Springer-Verlag: New York.

Minson, D. J., 1990. Forage in Ruminant Nutrition. Academic: San Diego.

Moore, I. D., Burch, G. J. and Wallbrink, P. J., 1986. Preferential flow and hydraulic conductivity of forest soils. *Soil Sci. Soc. Ameri.* **50**: 876–81.

Morton, S. R., 1990. The impact of European settlement on the vertebrate animals of arid Australia: a conceptual model. *Proc. Ecol. Soc. Aust.* **16**: 201–13.

Morton, S. R., Stafford Smith, D. M., Friedel, M. H., Griffin, G. F. and Pickup, G., 1995. The stewardship of arid Australia: ecology and landscape management. *J. Environ. Manage.* **43**: 195–217.

Noble, J. C., Diggle, P. J. and Whitford, W. G., 1989. The spatial distributions of termite pavements and hummock feeding sites in a semi-arid woodland in eastern Australia. *Acta Oecol./Oecol. Gener.* **10**: 355–76.

Orpin, C. G. and Joblin, K. N., 1988. The rumen anaerobic fungi. Pp. 129–50 *in* The Rumen Microbial Ecosystem ed by P. N. Hobson. Elsevier Applied Science: London.

Ponting, C., 1992. A Green History of the World. Penguin Books: London.

Priddel, D. and Wheeler, R., 1995. The biology and management of the Malleefowl *Leipoa ocellata* in New South Wales. New South Wales National Parks and Wildlife Service: Sydney, New South Wales.

Stafford Smith, M., 1994. Sustainable production systems and natural resource management in the rangelands. Pp. 148–59 *in* Outlook 94. Vol. 2. Natural Resources. Australian Bureau of Agricultural and Resource Economics: Canberra, Australian Capital Territory.

Theodorou, M. K. and France, J., 1993. Rumen micro-organisms and their interactions. Pp. 145–64 *in* Quantitative Aspects of Ruminant Digestion and Metabolism ed by J. M. Forbes and J. France. CAB International: Wallingford, UK.

Tongway, D. J. and Hindley, N., 1995. Manual for Soil Condition Assessment of Tropical Grasslands. CSIRO Wildlife and Ecology: Canberra, Australian Capital Territory.

Tongway, D. J. and Ludwig, J. A., 1990. Vegetation and soil patterning in semi-arid mulga lands of Eastern Australia. *Aust. J. Ecol.* **15**: 23–34.

Tongway, D. J. and Ludwig, J. A., 1994. Small-scale resource heterogeneity in semi-arid landscapes. *Pac. Cons. Biol.* **1**: 201–08.

Tongway, D. J. and Ludwig, J. A., 1997a. The conservation of water and nutrients within landscapes. Pp. 13–22 *in* Landscape Ecology, Function and Management: Principles from Australia's Rangelands ed by J. Ludwig, D. Tongway, D. Freudenberger, J. Noble and K. Hodgkinson. CSIRO Publishing: Melbourne, Victoria.

Tongway, D. J. and Ludwig, J. A., 1997b. The nature of landscape dysfunction in rangelands. Pp. 49–61 *in* Landscape Ecology, Function and Management: Principles from Australia's Rangelands ed by J. Ludwig, D. Tongway, D. Freudenberger, J. Noble and K. Hodgkinson. CSIRO Publishing: Melbourne, Victoria.

Tongway, D. J., Ludwig, J. A. and Whitford, W. G., 1989. Mulga log mounds: fertile patches in the semi-arid woodlands of eastern Australia. *Aust. J. Ecol.* **14**: 263–68.

Van Nevel, C. J. and Demeyer, D. I., 1988. Manipulation of rumen fermentation. Pp. 387–443 *in* The Rumen Microbial Ecosystem ed by P. N. Hobson. Elsevier Applied Science: London.

Van Soest, P. J., 1982. Nutritional Ecology of the Ruminant. O&B Books: Corvalis, Oregon.

Wilson, A. D., 1991. The influence of kangaroos and forage supply on sheep productivity in the semi-arid woodlands. *Rangel. J.* **13**: 69–80.

CHAPTER 24

Impact of nitrogen flows from agricultural production environments on biodiversity of non-agricultural ecosystems in the Waikato region, New Zealand

I. K. G. Boothroyd,[1] J. R. Crush,[2] S. F. Ledgard,[2] B. Huser[3] and N. Selvarajah[3]

Intensive agriculture in the Waikato region of New Zealand is characterized by positive nitrogen nitrogen balances and leaching of nitrate nitrogen to shallow groundwater. On a per hectare basis, vegetable growing produces more surplus nitrogen than dairyfarming, but within any land use there is substantial variation in surplus nitrogen depending on a range of agronomic, edaphic and climatic variables. However, in general, agricultural land uses in areas with the presence of oxic shallow aquifers have resulted in elevated nitrate nitrogen in groundwater in the Waikato region. Since much of the region's surface water drains from shallow groundwater a significant proportion of the nitrogen in the regional surface waters originate from the groundwater. The balance of the surface water nitrogen is from direct discharges of effluent and run off from farm land. Collectively, the point and non-point sources of pollution from the agricultural ecosystems can have distant impacts on a variety of other aquatic ecosystems. However, in many cases, other factors (e.g., substrate, shade, gradient) are also important in determining the biological communities. The impacts of increased nitrogen inputs are most dramatic where nitrogen is the growth-limiting nutrient. In this situation any additional nitrogen input causes an immediate and often irreversible change in ecosystem function and structure. For example, adventive exotic aquatic weeds may smother the generally low-nutrient indigenous flora and vegetation of aquatic ecosystems. Excessive plant growths can reduce the diversity of the freshwater fauna but can increase the abundance of some species. In extreme circumstances, nuisance algal or plant growths can inhibit the use of water for other purposes.

Key words: Nitrogen, Nitrate leaching, Biodiversity, Aquatic ecosystems.

INTRODUCTION

THE Waikato region in the North Island of New Zealand (Fig. 1) is one of the nation's most productive agricultural regions. The area is characterized by intensive resource use from dairying, horticulture, market gardening and production forestry for both the domestic and export markets. In conjunction with secondary industry and human settlements this leads to significant point (direct discharge) and non-point (runoff) source wastes, and surpluses entering non-agricultural ecosystems (e.g., lakes, rivers, wetlands, estuaries). There is therefore considerable potential for impacts from agricultural production on these other ecosystems. In this chapter we briefly review the significance of excessive nitrogen in aquatic ecosystems, and combine data from production orientated research and aquatic conservation monitoring to highlight the interconnectedness of

[1]National Institute of Water and Atmospheric Research, P.O. Box 11–115, Hamilton, New Zealand.
[2]AgResearch, Ruakura Research Centre, Private Bag 3123, Hamilton, New Zealand.
[3]Environment Waikato, P.O. Box 4010, Hamilton, New Zealand.

Pages 236–45 *in* NATURE CONSERVATION 5: NATURE CONSERVATION IN PRODUCTION ENVIRONMENTS: MANAGING THE MATRIX ed by J. L. Craig, N. Mitchell and D. A. Saunders. Surrey Beatty & Sons, 2000.

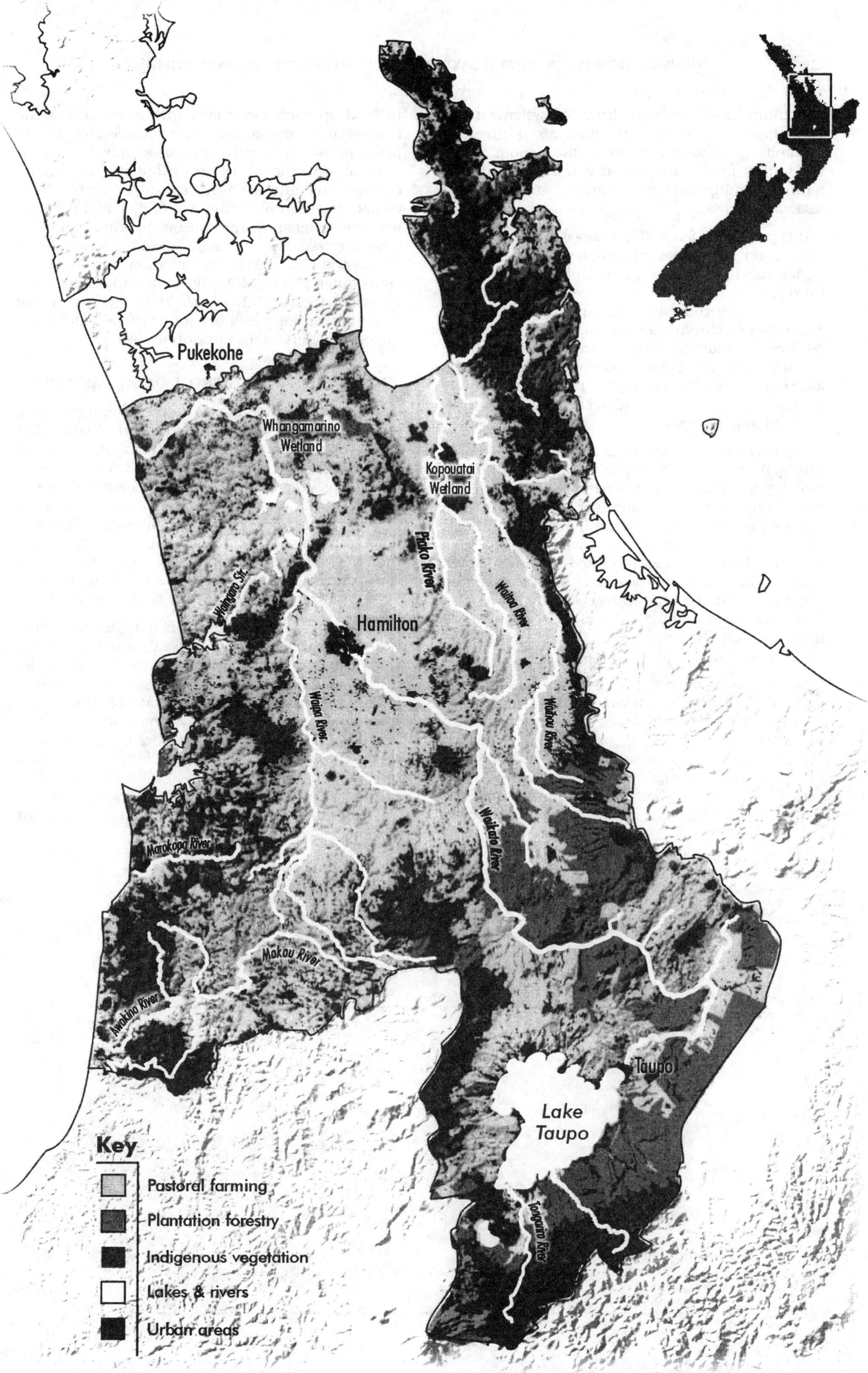

Pukekohe
Whangamarino Wetland
Kopouatai Wetland
Piako River
Waihou River
Waingaro Str.
Hamilton
Waipa River
Waikato River
Marokopa River
Mokau River
Awakino River
Taupo
Lake Taupo
Tongariro River
Key
Pastoral farming
Plantation forestry
Indigenous vegetation
Lakes & rivers
Urban areas

agricultural and non-agricultural ecosystems in the Waikato region. Particular emphasis is given to lowland agricultural systems in the region, which are highly productive but also retain significant remnant indigenous forest areas, wetlands, and restiad peat bogs.

It is possible to fence off conservation areas within production landscapes and avoid damage by some agricultural activities e.g., trampling by animals. However, conservation and productive areas are still linked by movement of surface and shallow groundwater carrying excess nutrients from agricultural production. The significance of these nutrient transfers between ecosystems need to be assessed when considering the long-term viability of conservation areas adjacent to, or within production landscapes.

Intensive agriculture is characterized by large nitrogen inputs, and large volumes of nitrogen moving rapidly between different components of the production system (O'Connor 1966). In contrast nitrogen cycling in the original forest vegetation of the region was probably characterized by small nitrogen inputs in nearly closed systems (Green 1982). Shrubland on peat soil in the Waikato presumably also had low nitrogen inputs and losses. The indigenous aquatic biota of the region would therefore have evolved under conditions of low nitrogen supply.

Movement of nitrogen across the boundaries of agricultural production systems occurs as dissolved nitrogen (nitrate in leachates and ammonium in effluent), gaseous nitrogen (N_2O, N_2 from denitrification; NH_3 by volatilization), and combined nitrogen in farm products and transported soil. Dissolved nitrogen will have the most immediate impact on aquatic biota, although organic nitrogen in food products eventually may enter an aquatic ecosystem as sewage and have considerable off-site impacts. Gaseous nitrogen losses may impact on aquatic ecosystems by contributing to climate change and ammonium deposited in rain. This chapter focuses on the impacts of dissolved nitrogen because intensive agriculture in the region is characterized by elevated groundwater nitrate concentrations, and the shallow groundwater contributes nitrate directly to streams, rivers and wetlands (Selvarajah *et al.* 1994). Discharge of treated farm dairy effluent into surface water is an important additional source of nitrogen.

NITRATE LEACHING FROM DAIRY PASTURES

Dairy farming represents the dominant intensive land use in the Waikato region on a land area basis. Inputs of nitrogen occur predominantly from the fixation of atmospheric N_2 by white clover in mixed pastures (Table 1). Most farms also use a low input of fertilizer nitrogen to increase grass growth during periods of feed shortage. However, among the farmers who use fertilizer nitrogen, more than 70% of dairy farmers apply less than 100 kg nitrogen/ha/year and only 2% apply more than 200 kg nitrogen/ha/year (Butler and Johnston 1997).

The "average" dairy farm in the Waikato region has an nitrogen surplus (total nitrogen inputs — nitrogen output in milk+meat) of 136 kg nitrogen/ha/year (Table 1, Ledgard *et al.* 1997). Detailed measurements of the fate of nitrogen in intensive dairy farmlet systems at the Dairy Research Corporation on free-draining soils with treatments varying in input of fertilizer nitrogen indicate that approximately 40–50% of the surplus nitrogen can be lost through nitrate leaching. This proportion is likely to be smaller on farms

Table 1. Estimates of nitrogen inputs and outputs (kg nitrogen/ha/year) for an average Waikato dairy farm (in 1995/1996) and for Dairy Research Corporation farmlets varying in nitrogen fertilizer application (nominally 0, 200 or 400 kg nitrogen/ha/year) and stocked at 3.3 cows/ha. Farmlet data are the mean of three years (Ledgard *et al.* 1997).

		Dairy Research Corporation farmlets		
	Average Waikato farm	0 nitrogen	200 nitrogen	400 nitrogen
Nitrogen inputs				
Fertilizer	53	0	215	413
N_2 fixation	140	174	117	40
Purchased feed	4	5	4	3
Atmospheric deposition	2	2	2	2
Total nitrogen inputs	199	181	338	458
Nitrogen outputs				
Milk	54	75	89	92
Meat	9	6	6	6
Surplus silage	0	0	12	26
Total nitrogen output	63	81	107	124
Nitrogen surplus	136	101	231	334
Nitrogen losses/removals				
Denitrification		5	17	25
Volatilization		15	41	65
Leaching		40	81	152
Transfer to lanes/sheds		57	78	84

with lower pasture utilization, such as the average Waikato dairy farm in Table 1, and on soils of poor drainage or derived from peat. The leaching losses in the farmlet treatments coincided with mean nitrate-nitrogen concentrations in the groundwater (at 1.5–3 m depth) of 6, 10 and 20 mg/L in the 0, 200 and 400 kg nitrogen/ha/year treatments, respectively (Ledgard *et al.* 1996).

NITRATE LEACHING FROM VEGETABLE GROWING

Vegetable cropping for both the domestic and export markets is a significant land use in the region. Nitrogen balance studies show large surpluses of nitrogen from vegetable growing in the Pukekohe district (Table 2, Crush *et al.* 1997). Winter crops produce larger nitrogen surpluses than summer crops, due to heavier fertilizer inputs and reduced off-take of nitrogen in products from the winter crop. Leaching of this surplus nitrogen has resulted in nitrate-nitrogen concentrations in the Upper Volcanic aquifer at Pukekohe increasing from 4 to 17 mg/l over the last 40 years (Cathcart 1996). This aquifer supplies streams which discharge high nitrate water into the Manukau Harbour.

EFFECTS OF NITROGEN ON LOWLAND AQUATIC ECOSYSTEMS

Water quality

The differing sources of nitrogen to rivers, streams, lakes, and wetlands can influence the impacts on water quality and aquatic communities. Nitrogen inputs from non-point source surface runoff are generally associated with soil organic matter, while groundwater seepage generally carries a much lower sediment yield. The bioavailability of sediment-associated nitrogen for aquatic plant growth is uncertain and depends on other factors such as sediment settling rates and seasonal rainfall and flooding. The concentration of nitrate in rivers generally follows closely that of river flow; nitrate levels are low during summer, even when fertilizer is being added, because the growing plants utilize nitrogen as soon as it becomes available (Mason 1981). Wilcock *et al.* (1995) found that nitrate levels in the polluted Whangamaire Stream, a lowland stream at Pukekohe, were generally 7–10 g m^{-3}, which were very high compared with baseflow values of other New Zealand rivers (median < 0.1 g m^{-3}; maximum 1.36 g m^{-3}, Close and Davies-Colley 1990). A large component of stream nitrogen in the Whangamaire Stream was organic-nitrogen, and likely to be derived from inputs of decomposing vegetation.

The Waikato River receives about 30 major permanent discharges from various industries; many are sewage outfalls (Environment Waikato 1997). Sewage outfalls can contribute large amounts of nitrogen to aquatic ecosystems, e.g., the Hamilton City sewage discharge contributes an average 1.3 t nitrogen/day. Estimates suggest that on average, 10.5 t dissolved inorganic nitrogen/day is discharged into the Tasman Sea at Port Waikato. As the volume of water flow is large in the Waikato River, the concentrations of dissolved inorganic nitrogen are about 0.3 mg L^{-1}. Small rivers and streams in the Waikato region carry much higher dissolved inorganic nitrogen levels; the Waitoa River dissolved inorganic nitrogen concentrations have been estimated at 4.0 mg L^{-1}, with an estimated 0.4 t dissolved inorganic nitrogen d^{-1} discharged by the Waitoa River into the Piako River in the Hauraki Plains (Selvarajah *et al.* 1994).

Excessive productivity

Water quality problems in New Zealand rivers and lakes are largely those of excessive biological productivity rather than of toxic materials (Rutherford *et al.* 1987). Excessive biological productivity largely results from the biologically available forms of nitrogen and phosphorus. This eutrophication has effects on algae and plant biomass and diversity, fish and other animal habitat requirements. It may also influence the success of invasions by exotic or undesirable species (and potential decline of desirable species), water appearance, water supply and water usage (Vant 1987).

Table 2. Potential contribution to groundwater nitrate of different land use activities in the catchment of the Upper Volcanic aquifer at Pukekohe (Crush *et al.* 1997).

Activity	Nominal area (ha) and % of total area		Surplus nitrogen (kg/ha)	Potentially leachable nitrogen (tonnes) and % of total nitrogen	
Dairying	377	8.2	157	59	9.6
Dry stock farming	1 884	40.9	30	57	9.3
Kiwifruit	133	2.8	166	22	3.5
Onions	892	19.4	118	105	17.1
Main crop potatoes	338	7.3	92	31	5.0
Early potatoes	506	11.0	429	217	35.4
Winter cabbage	94	2.0	242	20	3.2
Summer cabbage	94	2.0	168	16	2.6
Winter lettuce	35	0.7	286	10	1.6
Summer lettuce	35	0.7	186	7	1.1
Squash	105	2.2	238	25	4.0
Pumpkin	105	2.2	120	13	2.1
Sewage				30	4.9
Totals	4 598			612	

Phytoplankton (free drifting algae in standing or slow moving waters) yields in freshwaters is determined by the concentrations of total biologically available nitrogen and phosphorus; unlike periphyton (predominantly attached benthic algae) biomass in streams and rivers which is promoted by dissolved inorganic nitrogen and dissolved reactive phosphorus (Ministry for the Environment 1992). Growth limiting nitrate concentrations for periphyton appear to be between 40 and 100 mg NO_3-nitrogen m^{-3} (Ministry for the Environment 1992). Macrophytes (generally rooted higher plants) absorb nutrients from sediments but will absorb nutrients from the water column if high enough concentrations are available (Rattray *et al.* 1991). In standing waters, high nutrient levels causing increased phytoplankton growth may in fact decrease macrophyte biomass as a consequence of phytoplankton shading light (Phillips *et al.* 1978).

In the Whangamata Stream near Taupo, Vincent and Downes (1980) demonstrated the influence of seasonally variable macrophyte biomass on nutrient export. The winter decline of the adventive watercress *Nasturtium officinale* R. Br biomass resulted in winter peaks of dissolved nitrate concentration at the mouth of the stream. In contrast to the more stable baseflow conditions of the Whangamata Stream, the Purukohukohu Stream, also near Taupo, exhibits seasonal fluctuations in baseflow, but maintains a relatively constant macrophyte biomass (Cooper and Cooke 1984). Vincent and Downes (1980) also found that nitrate loss from the water varied markedly throughout the day and night, with maximum removal occurring at midday and minimum at midnight. This diurnal variation in nitrate removal was also closely correlated with changes in oxygen concentrations in the stream. Thus the variability in nitrate concentrations within rivers and streams may result more from in-stream processing and uptake, rather than input concentrations (Rutherford *et al.* 1987).

Excessive growth of phytoplankton (chlorophyll *a* concentrations 0.03–0.067 g m^{-3} (Ministry for the Environment guideline = 0.02 g m^{-3})) in the Waikato River was considered to result from the seasonally low summer values of inorganic nitrogen (dependence on flow and plant uptake) and the low nitrogen:phosphorus ratio, suggesting that phytoplankton growth was limited by nitrogen rather than phosphorus (Environment Waikato 1997). Nitrogen accounted for most of the variation in chlorophyll *a* in the Waikato River suggesting nitrogen as a useful indicator of chlorophyll *a* concentrations in the river. Three major point sources contributed more than 20% of the growing season mass flow of nitrogen to the Waikato River measured at Mercer in the lower Waikato, while seven monitored tributaries contributed a further 40%. Excessive macrophyte growth in the hydroelectric lakes causes problems for the hydrostation intake structures and also affects recreational use of lakes.

In all the above cases however, several other factors are important in determining productivity. Light, temperature, current velocity, water chemistry, sediment stability, grazing and applied controls (herbicide applications and weed cutting) are all significant and the degree of productivity will vary amongst locations as a result of variance in these factors.

Macrophyte biodiversity and eutrophication

New Zealand has a distinctive aquatic macrophyte flora but has experienced repeated invasions by adventive (exotic invading) species. Although aquatic macrophyte invasions have generally been poorly documented there has been a belief that the rapid growth and spread of adventive macrophytes has occurred as a result of increased nutrients (nitrogen and phosphorus) within water bodies. This belief arose from the observations of large biomass and standing crop of adventive species, and the assumption that these species require large supplies of nutrients (Howard-Williams *et al.* 1987).

While aquatic macrophyte standing crop can reach high proportions in New Zealand, site comparisons indicate that high adventive macrophyte biomass (per unit area) is not simply a function of nutrient status and is not necessarily the sole reason for loss of indigenous aquatic plants (Howard-Williams *et al.* 1987). High biomass values of invasive species have been reported from oligotrophic lakes in New Zealand (Clayton 1982), and these invasive species are not restricted to eutrophic lakes. Howard-Williams *et al.* (1987) concluded that the presence of adventives was not necessarily a function of eutrophication, and that large standing crops of macrophytes in New Zealand and elsewhere appeared to be primarily related to three factors: amount of shelter from wind and waves; lack of herbivory; and duration of growing season.

Nevertheless, the replacement of the native aquatic macrophyte assemblages in lakes, rivers, streams and wetlands, has been a feature of New Zealand for the past century. In lowland streams the rapid growth of adventives can have significant effects on the physico-chemical environment and on other aspects of the aquatic biota. In the presence of high nutrient concentrations, the adventive species are capable of high rates of nutrient uptake, and to spread over the surface of the stream and utilize nutrients directly from the water. The invasive species can outcompete and smother the generally low-nutrient indigenous vegetation of lowland aquatic ecosystems. The presence of submerged oxygen weeds (e.g., *Lagarosiphon major, Egeria densa*) often leads to exacerbated fluctuations in both pH and dissolved oxygen (Wilcock *et al.* 1995). Furthermore, many lowland streams in New Zealand often originally contained little or no aquatic macrophyte vegetation, so the introduction and excessive productivity of adventives has caused dramatic changes in the macrophyte and macroinvertebrate communities of these systems.

Macroinvertebrate diversity

Aquatic macroinvertebrates (benthic invertebrates) are commonly used as indicators of water quality in New Zealand, and in other parts of the world (Metcalfe 1989), and generally comprise aquatic insects (mayflies, stoneflies etc.), snails, worms and crustaceans. These animals are used because they are generally easy to sample, are relatively easy to identify, and have been well studied so that there is much information in the literature on their ecology and environmental requirements (Hellawell 1986; Rosenberg and Resh 1993). Macroinvertebrates are generally regarded as good integrators of environmental disturbance as many taxa have specific environmental and habitat requirements, occurring through many generations at the same site (Metcalfe-Smith 1996).

Macroinvertebrate diversity is less influenced directly by nitrogen additions (with the exception of ammonia), than by changes in other factors brought about by increased productivity from eutrophication or land use and habitat changes (Quinn and Hickey 1990). Increased algae and macrophyte production changes the substrate, refuge (shelter) and food availability for macroinvertebrates, and changes the structure and diversity of macroinvertebrate communities. Organic matter that enters streams and lakes can also have the same effects, although bacterial decomposition of organic material may reduce the amount of oxygen available with serious consequences for some species of aquatic insects in particular (Wiederholm 1984).

Numerous monitoring programmes and studies have assessed macroinvertebrate communities under a range of different nutrient and habitat conditions. Most lotic systems have a longitudinal gradient that can include the whole range from nutrient-poor to nutrient-rich waters, as well as changes in physical river size and habitat, all of which combine to influence species diversity (Vinson and Hawkins 1998). Differing degrees of nitrate and nutrient eutrophication will favour some organisms but will restrict others, and in extreme circumstances will favour one or two species only which can be extremely abundant. In recent reviews of freshwater quality in New Zealand, lowland streams in agriculturally developed catchments were identified as having reduced numbers of taxa, and abundance of macroinvertebrates, as well as lower numbers of sensitive macroinvertebrate species by comparison with streams with less than 30% of their catchments as improved pasture (Quinn and Hickey 1990).

Macroinvertebrates are a major source of food for native and introduced fish in New Zealand rivers and streams. Richardson *et al.* (1997) found that the diet of two New Zealand fish species (common smelt *Retropinna retropinna*, and inanga *Galaxias maculatus*) in a lowland stream reflected the composition of the macroinvertebrate community which lived on macrophytes. Lowland streams are important migratory pathways for many native fish species and the occurrences of stressful conditions (fluctuating temperatures, high ammonical-nitrogen, low dissolved oxygen), as well as the loss of suitable habitat and food sources has had an important impact on their survival (McDowall 1990).

NITROGEN AND MACROINVERTEBRATE DIVERSITY IN A LOWLAND RIVER ECOSYSTEM

The influence of nitrogen from agricultural production environments on aquatic macroinvertebrate diversity is illustrated in a study of the aquatic resources of the Piako River, a low-lying catchment (1 435 km^2, highest point 495 m) in the Hauraki Plains, in the north-east of the region studied, and which drains into the Firth of Thames (Fig. 1). The northwestern border of the catchment comprises sand and siltstone formations, while the remainder of the catchment comprises sedimentary and alluvial deposits, and the streams are generally deep and slow flowing through these stretches.

The lower catchment is characterized by peat deposits, the core of which forms the largely unmodified Kopouatai peat dome, covering an area of approximately 90 km^2. This is a raised restiad, ombriginous bog, recently classified and registered under the Convention on Wetlands of International Importance (Ramsar 1971). Additional wetland areas also occur within the lower catchment, and the mudflats of Firth of Thames are internationally significant as migratory bird feeding grounds. These areas are also classified under the Ramsar convention. Land use is predominantly pastoral (85%), but with some cultivation. Approximately 14% is covered by native vegetation, and 0.6% in exotic forest. The native vegetation is largely in the headwaters but remnant blocks of forest occur on the plains. The area supports an intensive, and highly productive dairy industry, with four major dairy factories, and a number of associated and rural-based industries.

Preliminary investigations of the water quality of the Piako River by Environment Waikato have indicated nitrate loads of 7–14 kgNO$_3$–Nha^{-1}yr^{-1}, and nitrogen loads to be 9–18 kgNha^{-1}yr^{-1}, considerably higher than the reported New Zealand average (Table 3). Winter averages are 8–10 times higher than occur during the summer; a result of increased rainfall and flows. Sediment losses from the Piako River were low compared to other New Zealand rivers, although suspended sediment concentrations were high. Nitrate levels increased along the length of the Piako River (Table 3) due, in part, to the geological characteristics of the catchment, intensive land use (approximately 85% of catchment land use is pastoral), numerous farm dairy effluent discharges and several industrial effluent discharges in the catchment.

Nitrate-nitrogen leaching from dairy pastures is considerable (see above). Nevertheless, despite the large and rapid soil nitrogen cycling little or no nitrate-nitrogen is found in the shallow aquifers in the Piako catchment. This is because most dairying in the Piako catchment is on peat soils and a large proportion of the nitrate-nitrogen produced in peat

Table 3. Export of nutrients (nitrogen, nitrate-nitrogen and phosphorus) loads from the Piako River catchment, 1989–1990. Year = yearly mean; S = summer mean; W = winter mean (n = 13). Piako River figures from Van Rossem (1990), New Zealand figures from Cooke (1980).

Subcatchment	Total nitrogen load ($kgNha^{-1}yr^{-1}$)			NO_3-nitrogen load ($kgNO_3ha^{-1}yr^{-1}$)			Total phosphorous load ($kgPha^{-1}yr^{-1}$)		
	Year	S	W	Year	S	W	Year	S	W
Mid-Piako River	9.9	1.6	19.1	7.2	0.8	15.1	0.5	0.2	0.6
Lower Piako River	15.8	3.6	25.1	11.2	2.3	18.7	1.3	0.4	1.7
Mid-Waitoa River	9.2	2.8	14.2	7.2	2.2	11.3	0.3	0.1	0.5
Lower Waitoa River	17.7	4.7	28.5	13.5	3.4	22.2	2.8	1.5	2.1
NZ streams draining pasture	5.0–14.0			0.7–5.0			0.3–1.6		

is denitrified in the soil or in anoxic shallow aquifers. Although the denitrification losses in groundwater were not measured directly the presence of reduced iron (ferrous form) in these aquifers indicates the conducive conditions for denitrification (Selvarajah *et al.* 1994). Therefore the nitrate-nitrogen increases along the length of the Piako River are probably due mainly to treated effluent discharges into streams and rivers in the catchment. Due to the slow flowing nature of the Piako River it is highly likely the effluent ammoniacal-nitrogen discharged is nitrified and resulting in nitrate level increase along the length of the Piako River.

A survey of the aquatic macroinvertebrate biota of the Piako River was undertaken by the senior author (I. K. G. Boothroyd) in 1988. Twenty-three sites were sampled for benthic and macrophyte associated macroinvertebrates using standard 250 μm mesh kick nets. The results indicated a decrease in taxa richness along its length from upland sites to the lower Piako (Table 4), while total numbers of macroinvertebrates increased. At the same time macrophyte cover increased from less than 20% in the upland sites to up to 70% in the mid-catchment and lowland sites (Table 4).

Macrophyte cover was a moderately significant environmental factor determining macroinvertebrate taxa richness in the Piako River, and it is likely that macrophytes form an important component of the substrate heterogeneity in lowland rivers. Rivers and streams with submerged and floating macrophytes often have a more taxa-rich macroinvertebrate fauna than non-macrophyte substrates in similar streams in the United Kingdom (Wright *et al.* 1992). Wright *et al.* (1992) also suggested that the emergent macrophytes are frequently the most diverse micro-habitats, not only in small streams, but also in the mid-lower reaches of larger rivers. In the Piako River, the macrophytes supported an abundant, if not diverse macroinvertebrate fauna, and substantially increased the macroinvertebrate taxa numbers of the mid-lower reaches (Table 4), but were less important in the upper reaches. Collier *et al.* 1998 found that macrophytes (covering an average of 26% of

Table 4. Macroinvertebrate taxa associated with macrophytes in the Piako River catchment, North Island, New Zealand. Site groups determined from TWINSPAN analysis of macroinvertebrate data: Group A = Upland sites below native bush; Group B = sites in foothills; Group C = Lowland stream and river sites; Group D = Lower Piako River sites.

Site Group	Site	Macrophyte Cover (%)	Total Taxa No.	% Taxa number on macrophytes	Total biota	% Total biota on macrophytes
A	WK1	0	29	0	2 226	0
A	PT1	0	22	0	1 651	0
B	PK1	0	41	0	4 220	0
B	PK2	15	35	0	3 126	0
B	TH1	5	30	0	3 067	0
B	TH2	0	21	0	3 665	0
B	NG1	5	38	0	9 634	0
B	NG2	10	29	20.7	3 349	10.9
C	WK2	0	21	0	3 982	0
C	WK3	5	20	55	2 371	29.9
C	WK4	10	15	86.7	984	43.6
C	WK5	10	15	0	1 977	0
C	PT2	20	15	0	8 004	0
C	PT3	15	16	37.5	8 505	2.3
C	W1	70	14	92.9	4 386	98.0
C	W3	40	19	63.2	8 862	61.9
C	W4	20	12	83.3	7 004	4.2
D	TH3	70	16	93.8	8 289	90.3
D	W2	40	13	61.5	7 082	66.2
D	P1	10	19	68.4	4 098	82.8
D	P2	10	10	0	885	0
D	P3	20	11	45.5	3 148	74
D	P4	0	8	0	686	0

streambed) contributed to the diversity of instream habitats for macroinvertebrates colonization in a range of lowland Waikato streams. In other studies, Jenkins *et al* (1984), and Ormerod (1988) found that river margins and vegetation support distinct macroinvertebrate assemblages.

MANAGEMENT AND POLICY

The *Resource Management Act (1991)* has placed an increased focus on managing the effects of resource use, and pastoral agriculture is no exception. Regional Policy Statements and Regional Plans developed by Regional Councils generally set out the policy and anticipated outcomes of resource management. The Regional Policy Statement by Environment Waikato emphasizes that there will be net improvement in water quality achieved across the region whilst maintaining the quality of outstanding waterbodies such as Lake Taupo. Currently regional plans are being developed through intensive consultations with stakeholders. The proposed draft regional plan has introduced a number of policies aimed at reducing the impacts of nitrogen loading in surface and groundwaters.

Most significant amongst these is that farm dairy effluent irrigation is now a permitted activity provided the nitrogen loading does not exceed 150 kg nitrogen ha^{-1} y^{-1}. This means that a resource consent (and associated application costs) are no longer required. In contrast, all effluent discharges to waterways are a discretionary activity (i.e., consents are required to discharge effluent into waterways and each consent will be assessed on a case by case basis). Continuous improvement on the quality of effluent discharge is expected from the resource users. Biosolids application to land will be a controlled activity with the emphasis on nitrogen and heavy metal loadings.

In addition to these regulations, environmental education is planned to promote water quality, and stream and riparian improvements, particularly through landcare groups. This includes supporting scientific research into improving water quality and ecosystem health, and supporting environmentally sustainable industrial guidelines and codes of practices (e.g., for fertilizer use) are intended. As a final resort, and where appropriate, enforcement will be used as a tool to minimize water pollution.

Despite the intentions of these policy directions many policies do not yet necessarily provide the focus and detail required to achieve the desired outcomes. In many cases this is the result of lack of knowledge (e.g., specific impacts of nitrogen on ecosystems, likely improvements in taxa numbers in lowland aquatic ecosystems, ecosystem structure, function and processes to set ecological sustainable environmental standards), and highlights the need for a close involvement of scientific research in policy setting and monitoring. Furthermore, implementation of these policies is not always easily achieved due to conflicting resource demands (human and technical resources), in addition to social, political and economic constraints.

FUTURE DIRECTIONS

High land values, labour costs and product prices interact, fuelling intensification of dairy farming and vegetable cropping in the Waikato Region. This is associated with a trend for larger nitrogen inputs and consequently larger nitrogen surpluses in the production sector. Market demand for "eco-labelled" food may counteract this trend, but the potential extent and impact of eco-labelling on production systems is uncertain. There is considerable scope to reduce the movement of nitrogen from farms into the wider environment e.g., on a typical Waikato dairy farm there is at least 0.5 kg nitrogen leaching down the soil profile for every kg nitrogen leaving the farm in products. The leached nitrogen has a fertilizer equivalent value of $NZ78/ha. Tightening up the nitrogen cycle of production systems so that they leak less nitrogen will improve the sustainability and profitability of farming as well as reducing adverse environmental impacts. Some approaches to reducing nitrogen movement across the production-conservation frontiers would involve the adoption of best management practices for fertilizer use that avoid inappropriate use of nitrogen fertilizers. The New Zealand Fertilizer Industry has developed a draft code of practice that will help reduce adverse environmental impacts. The establishment of local landcare groups of farmers and growers committed to improving land management techniques and farming sustainably will improve conservation practice e.g., in the Waikato region, the Franklin Sustainability Project is a collaborative effort by growers, consultants, researchers and Regional Councils to mitigate soil losses and nitrate leaching from intensive vegetable production (Basher *et al.* 1997). Additional measures include the use of farm dairy effluent as a nutrient resource for pasture rather than disposal of effluent into waterways. About 70% of Waikato dairy farmers now irrigate dairy effluent on to land instead of using older treatment systems that eventually released nutrients into water. Cattle urine patches, where 70% of dietary nitrogen can be returned to the soil in a very restricted area, are the principle source of nitrogen losses from pasture. Depending on the circumstances, nitrogen losses from urine patches could be reduced by manipulating the animals diet to increase the conversion of dietary nitrogen into product nitrogen. Inclusion of high energy supplements in a conventional grass/clover diet could reduce urinary nitrogen deposition by 10–20%. Destocking of pasture when leaching is likely to happen, and collecting effluent from the feed pad for even return across the pasture is currently being investigated as a management option for reducing nitrogen losses.

Instream management will reduce the impacts of nutrient enrichment. This includes riparian planting to reduce nitrate leaching (through uptake in riparian plants); riparian planting to increase shade and reduce macrophyte growth; improving instream habitat heterogeneity; and natural stream channel restoration. Selective aquatic weed removal to remove excessive macrophyte production, while retaining components suitable for biotic assmblages (refuge, food, reproduction) is rarely used in New Zealand, but is practised in other parts of the world. Buffer intercepts of surface water movements (even before riparian stream margins), including the protection of hillside seepage zones will also reduce nitrogen movements between ecosystems.

SUMMARY

Nitrogen losses from agricultural production activity (as well as secondary industry and human activity) in the Waikato region contributes to changes in species numbers in aquatic ecosystems. However, in most cases, it is not possible to target nitrogen as the sole factor responsible for these changes, and physical features such as substrate, slope, habitat, and other water quality characteristics such as dissolved oxygen and pH are also significant. The main influence of nitrogen in aquatic ecosystems is excessive plant and algal production, which in turn influences the species composition of biotic communities, particularly in slow flowing lowland streams and rivers, characteristic of the Waikato region. Careful development of policy and implementation strategies, based on sound scientific knowledge and understanding, are required for improvements to be achieved.

REFERENCES

Basher, L. R., Cathcart, S. N., Crush, J. R., Hart, B., Ross, C. W. and Williams, P. W., 1997. Soil and water management for sustainable vegetable production in a peri-urban area, Pukekohe, New Zealand. Pp. 321–26 *in* Wai Whenua/Water Land, Proceedings of the 24th Hydrology and Water Resources Symposium. New Zealand Hydrological Society and the National Committee on Water Engineering of the Institute of Engineers: Australia.

Butler, P. J. and Johnston, T. J., 1997. Fertliser use in the Bay of Plenty, Waikato and South Auckland regions. *Proc. NZ Grassl. Assoc.* **59:** 35–37.

Cathcart, S. N., 1996. An investigation of the nitrate contamination of an unconfined, shallow, fractured, basaltic aquifer at Pukekohe, South Auckland, New Zealand. M.Sc. thesis, University of Auckland.

Clayton, J. S., 1982. Effects of fluctuations in water level and growth *of Lagarosiphon major* on the aquatic vascular plants in Lake Rotoma from 1973 to 1980. *NZ J. Mar. Freshw. Res.* **16:** 89–94.

Close, M. E. and Davies-Colley, R. J., 1990. Baseflow water chemistry in New Zealand rivers. 1. Characterisation. *NZ J. Mar. Freshw. Res.* **24:** 319–41.

Collier, K. J., Wilcock, R. J., Meredith, A. S., 1998. Influence of substrate type and physicochemical conditions on macroinvertebrate faunas and biotic indices of some lowland Waikato streams. *NZ J. Mar. Freshw. Res.* **32:** 1–19.

Cooke, J. G., 1980. The effect of land use on water quality in New Zealand. *In* Landuse in Relation to Water Quantity and Quality. *Proceedings of conference, 1979.* Nelson Catchment Board.

Cooper, A. B. and Cooke, J. G., 1984. Nitrate loss and transformation in 2 vegetated headwater streams. *NZ J. Mar. Freshw. Res.* **18:** 441–50.

Crush, J. R., Cathcart, S. N., Singleton, P. and Longhurst, R. D., 1997. Potential for nitrate leaching from different uses of land in the Pukekohe area. *Proc. NZ Grassl. Assoc.* **59:** 55–58.

Environment Waikato, 1997. Waikato River Annual Report 1996: Waikato River Quality Monitoring Programme — Data Report 1996. Environment Waikato Technical Report No. 1997/5.

Green, T. G. A., 1982. Biological nitrogen inputs and turnover in native forests. Pp. 151–54 *in* Nitrogen Balances in New Zealand Ecosystems ed by P. W. Gandar. Department of Science and Industrial Research, Palmerston North, New Zealand.

Hellawell, J. M., 1986. Biological Indicators of Freshwater Pollution and Environmental Management. Elsevier: London. 546 Pp.

Howard-Williams, C., Clayton, J. S., Coffey, B. T. and Johnstone, I. M., 1987. Macrophyte Invasions. Pp. 307–31 *in* Inland Waters of New Zealand ed by A. B. Viner. Department of Science and Industrial Research Bulletin 241, Wellington, New Zealand.

Jenkins, R. A., Wade, K. R. and Pugh, E., 1984. Macroinvertebrate habitat relationships in the Teifi catchment and the significance to conservation. *Freshw. Biol.* **14:** 23–42.

Ledgard, S. F., Selvarajah, N., Jenkinson, D. and Sprosen, M. S., 1996. Groundwater nitrate levels under grazed dairy pastures receiving different rates of nitrogen fertiliser. Pp. 229–36 *in* Recent Developments in Understanding Chemical Movement in Soils: Significance in Relation to Water Quality and Efficiency of Fertiliser Use ed by L. D. Currie and P. Loganathan. Occasional report No. 9. Fertilizer and Lime Research Centre: Massey University, Palmerston North.

Ledgard, S. F., Penno, J. W., Sprosen, M. S., 1997. Nitrogen balances and losses on intensive dairy farms. *Proc. NZ Grassl. Assoc.* **59:** 49–53.

Mason, C. F., 1981. Biology of Freshwater Pollution. Longman: London.

McDowall, R. M., 1990. New Zealand freshwater Fishes. Heinemann Reed: Auckland, New Zealand.

Metcalfe, J. L., 1989. Biological water quality assessment of running waters based on macroinvertebrate communities: History and present status in Europe. *Environ. Poll.* **60:** 101–39.

Metcalfe-Smith, J. L., 1996. Biological Water-quality Assessments of Rivers: Use of Macroinvertebrate Communities. Pp. 17–43 *in* River Restoration ed by G. Petts and P. Calow. Blackwell: Oxford.

Ministry for the Environment, 1992. Water Quality Guidelines No. 1: Guidelines for the Control of Undesirable Biological Growths in Water. Ministry for the Environment: Wellington, New Zealand.

O'Connor, K. F., 1966. The improvement and utilisation of tussock grasslands: a scientist's viewpoint. Cycling nitrogen for production. *Proc. NZ Grassl. Assoc.* **28:** 59–78.

Ormerod, S. J., 1988. The micro-distribution of aquatic macroinvertebrates in the Wye river system: the result of abiotic or biotic factors? *Freshw. Biol.* **20:** 241–47

Phillips, G. L., Eminson, D. and Moss, B., 1978. A mechanism to account for macrophyte decline in progressively eutrophicated freshwaters. *Aquatic Botany* **4:** 103–26.

Quinn, J. M. and Hickey, C. W., 1990. Magnitude of effects of substrate particle size, recent flooding, and catchment development on benthic invertebrates in 88 New Zealand rivers. *NZ J. Mar. Freshw. Res.* **24:** 411–27.

Rattray, M. R., Howard-Williams, C. and Brown, J. M. A., 1991. Sediment and water as sources of nitrogen and phosphorus for submerged rooted aquatic macrophytes. *Aquatic Botany* **40:** 225–37.

Richardson, J., West, D. and Croker, G., 1997. Who ordered the *Austrosimulium* on *Egeria*? *Water and Atmosphere* **5**(3): 14–16.

Rosenberg, D. M. and Resh, V. H., 1993. Introduction to freshwater biomonitoring and benthic macroinvertebrates. Pp. 1–9 *in* Freshwater Biomonitoring and Benthic Macroinvertebrates ed by D. M. Rosenberg and V. H. Resh. Chapman and Hall: London.

Rutherford, J. C., Williamson, R. B. and Cooper, A. B., 1987. Nitrogen, phosphorus, and oxygen dynamics in rivers. Pp. 139–65 *in* Inland Waters of New Zealand ed by A. B. Viner. DSIR Bulletin 241, Wellington, New Zealand.

Selvarajah, N., Maggs, G. R., Crush, J. R. and Ledgard, S. F., 1994. Nitrate in ground water in the Waikato region. Pp. 160–85 *in* The Efficient Use of Fertilisers in a Changing Environment: Reconciling Productivity and Sustainability ed by L. D. Currie and P. Loganathan. Occasional report No. 7. Fertilizer and Lime Research Centre: Massey University, Palmerston North.

Van Rossem, J. E., 1990. Surface water quality issues in the Piako/Waitoa catchment —A literature review. Waikato Regional Council Technical Report 1990/27.

Vant, W. N., 1987. Major issues in Lake management. *In* Lake Managers Handbook ed by W. N. Vant. Water and Soil Miscellaneous Publication No. 103. Wellington, New Zealand.

Vincent, W. F. and Downes, M. T., 1980. Variation in nutrient removal from a stream by watercress (*Nasturtium officinale* R. Br.). *Aquatic Bot.* **9:** 221–35.

Vinson, M. R. and Hawkins, C. P., 1998. Biodiversity of stream insects: variation at local, basin and regional scales. *Ann. Rev. Entomol.* **43:** 271–93.

Wiederholm, T., 1984. Responses of Aquatic Insects to Environmental Pollution. Pp. 508 *in* The Ecology of Aquatic Insects ed by V. H. Resh and D. M. Rosenberg. Praeger: New York.

Wilcock, R. J., McBride, G. B., Nagels, J. W. and Northcott, G. L., 1995. Water quality in a polluted lowland stream with chronically depressed dissolved oxygen: causes and effects. *NZ J. Mar. Freshw. Res.* **29:** 277–88.

Wright, J. F., Blackburn, J. H., Westlake, D. F., Furse, M. T. and Armitage, P. D., 1992. Anticipating the consequences of river management for the conservation of macroinvertebrates. Pp. 137–49 *in* River Conservation and Management ed by P. J. Boon, P. Calow and P. E. Petts. John Wiley and Sons: London.

CHAPTER 25

Combining nature conservation with production through farm forestry in northern New South Wales, Australia

David J. Curtis[1], Mark Sandstrom[2] and David Carr[1]

Farm Forestry Projects that are being run by Greening Australia in northern New South Wales provide living examples of how conservation and production might be integrated at the farm level, and ultimately across the landscape. These projects have adopted a "conservation through utilization" approach which has been incorporated into a model for agro-forestry that is based on indigenous trees.

The Dorrigo Plateau, Northern Tablelands and North-west Slopes of New South Wales are rich in value for conservation, forestry and agriculture. About 130 tree species in the region have value for timber while providing important ecological services, and these projects are attempting to develop farm forestry models based on those indigenous species. Over 240 landholders have been involved and over 316 000 trees have been planted on about 130 properties. Plantings have included demonstrations of different types of woodlots and timberbelts as well as detailed species and provenance trials. Farm forestry plantings incorporate mixtures of timber species and habitat species, and use mostly indigenous trees throughout. Remnant forest areas are incorporated into plantations and are actively maintained. The plantings are incorporated into farm plans. As the projects develop natural forests are being linked, community networks are forming, and a social and ecological matrix is developing.

INTRODUCTION

RECENTLY we had the opportunity to play the game "Squatter". This game is a grazier's version of "Monopoly" and you win by being the first player to improve pastures, irrigate, and double your carrying capacity of sheep. The concepts forming the basis of this game, those of development of the land leading to increased production and increased income, are the concepts that our rural industry has been based upon. Barring occasional drought, flood and income tax, the inexorable drive to develop is rewarded with fatter wool cheques and better prices at sales. In the game "Squatter", except for the possibility of winning a trophy for soil conservation, conservation is not mentioned.

Scientists around the world are highlighting the problems associated with conservation of our biological estate in the modern world. E. O. Wilson (1992) describes the extraordinary richness of species that inhabit the world, considering that every species is precious and entitled to its place on the earth. Scientists have realized that the only way to conserve fauna is to maintain habitat across the landscape rather than relying on small reserves, that is whole landscape conservation (e.g., Barrett *et al.* 1994; Saunders 1996; Wiens 1997). Since much of the rural landscape is used for commercial farming whole landscape conservation must inevitably be incorporated with commercial farming. Wilson (1992) described ways of conserving species through their utilization — proposing that if a species is used, it will be valued economically and thus conserved.

Our chapter attempts to merge these two concepts — whole landscape conservation (which conserves biodiversity across landscapes) with conservation through utilization. By merging these concepts we

[1]Greening Australia, P.O. Box 1467 Armidale, New South Wales, Australia 2350.
[2]Greening Australia, P.O. Box 227 Dorrigo, New South Wales, Australia 2453.
Pages 246–57 *in* NATURE CONSERVATION 5: NATURE CONSERVATION IN PRODUCTION ENVIRONMENTS: MANAGING THE MATRIX ed by J. L. Craig, N. Mitchell and D. A. Saunders. Surrey Beatty & Sons, 2000.

hope to arrive at a way of successfully incorporating conservation with commercial farming and grazing enterprises. Perhaps in the future games like "Squatter" will reflect and absorb these ideas and become metaphors for a new way of obtaining produce from the land.

North-west New South Wales

The Northern region of New South Wales is dominated by the Northern Tablelands — a massive mountainous uplift formed 80 million years ago (White 1994). The Northern Tablelands range in altitude from 800 to over 1 500 metres with a variety of soils derived from ancient granites, sediments and more recent basalts. Annual rainfall varies from 1 000 to 1 500 mm in the east to 500 to 700 mm in the west. Volcanic activity 18 million years ago on the eastern edge of the region created the Dorrigo Plateau. The Plateau combines high rainfall (1 000–2 000 mm), mountainous terrain, varying altitudes (500–1 500 metres), with a variety of soils derived from basalts and the ancient underlying sediments.

In the west the Tablelands drop away into lower rainfall sloping country dominated by ancient sediments and granites. Further west the mountainous country gives way to the drier (300–500 mm) North-west Plains — enormous plains of alluvial clays with outcrops of sandy soils.

The region contains a wide diversity of environments and consequently a rich and diverse flora and fauna. Thirty-two forest types have been described on the Dorrigo Plateau, 48 on the Northern Tablelands, 46 on the North-west Slopes and 20 on the North-west Plains (Forestry Commission of New South Wales 1989). Tall rainforests and wet and dry eucalypt forests grow in the east. Before European settlement grassy eucalypt forests and woodlands dominated the Tablelands while scrubby eucalypt forests and scrublands dominated the slopes (Curtis 1989). The Plains mainly consisted of eucalypt woodlands interspersed with grasslands (Beadle 1948).

Conservation issues

The region has been developed to different degrees, depending on the particular land type. The flatter areas with fertile soils are highly developed and suffer from extreme tree decline, and in some places rural dieback (Williams and Nadolny 1981; Mackay *et al.* 1984; Nadolny 1984; Curtis 1991; Nadolny and Curtis 1994). Some parts of the region are still being cleared for crop production (Sivertsen 1994). Through these types of development, landholders in the region are now faced with the following environmental problems: soil salinity, soil erosion and soil structure decline, decreased water quality, chemical pollution, feral animals and

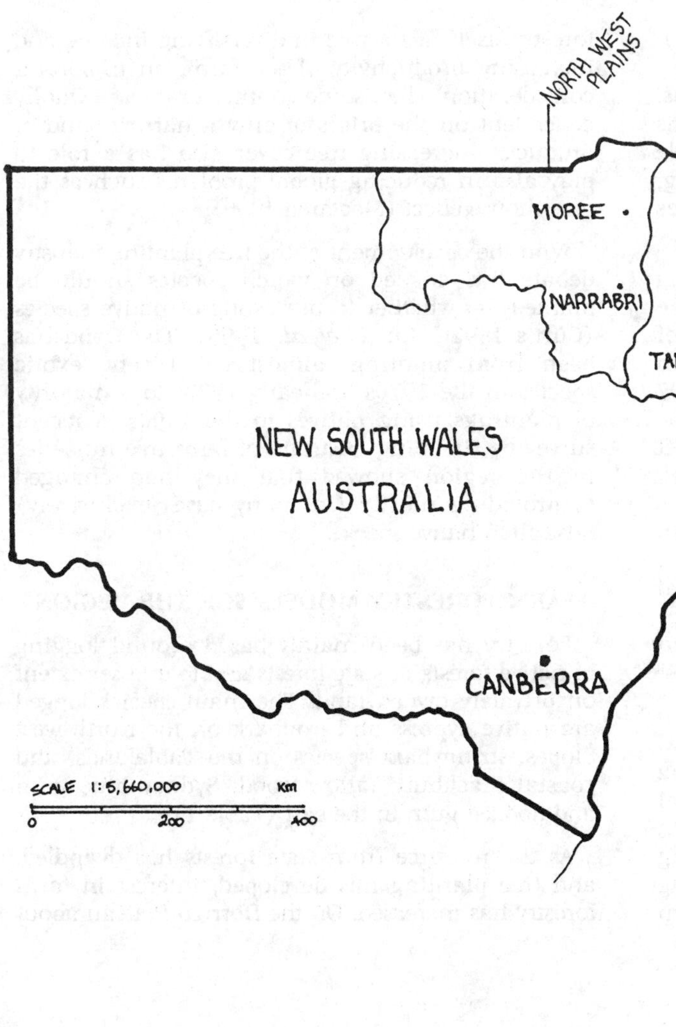

Fig. 1. The study region, encompassing the North-west Plains, North-west Slopes, Northern Tablelands and Dorrigo Plateau.

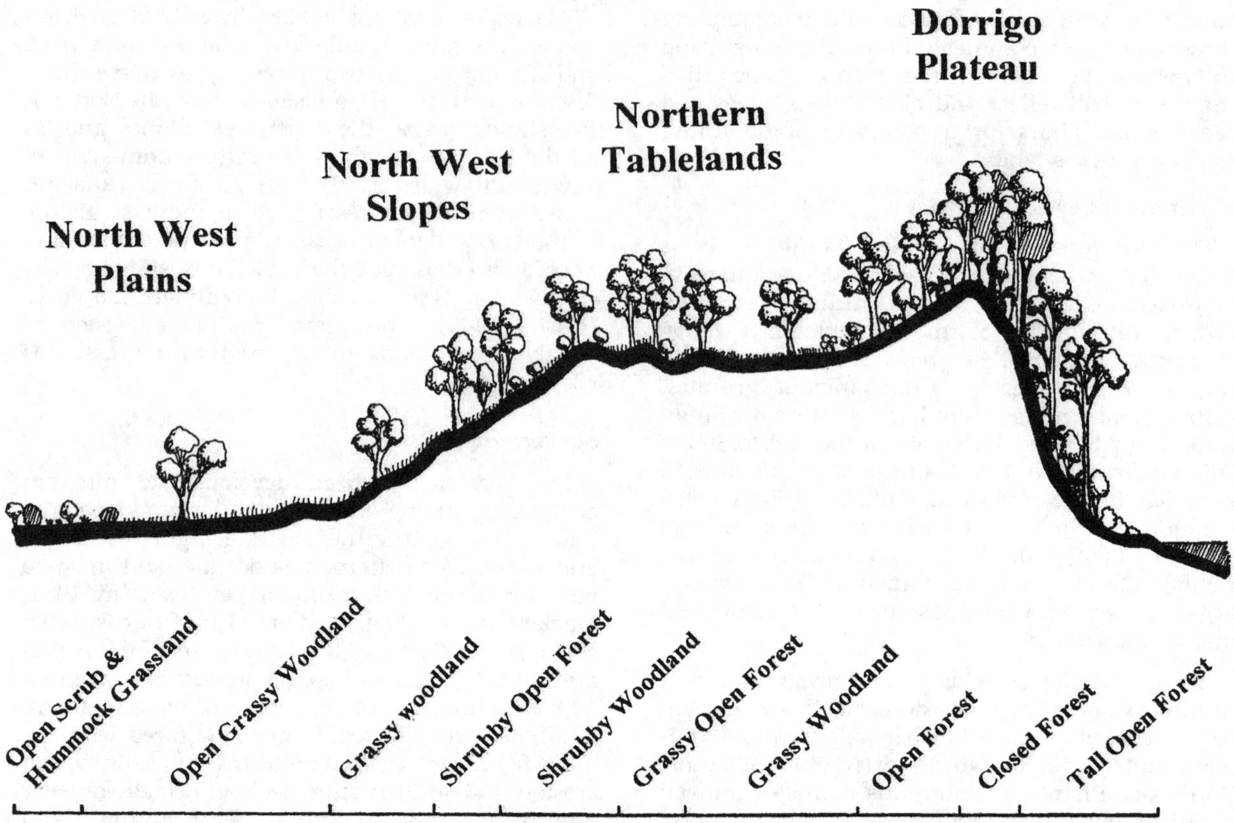

Fig. 2. The study region shown in relief, indicating the main types of natural vegetation.

plants, and loss of biodiversity (Donaldson 1996). Steep country remains forested and there are many National Parks and state forests in those regions, while hilly country with poorer soils often retains its tree cover and have been described as multiple use areas as they can support timber harvesting, grazing, catchment protection and other uses (Morgan and Terrey 1990).

In response to rural tree decline tree planting in the region has dramatically increased over the last decade (Curtis *et al.* 1995), with plantings of tubestock rising from around 300 000 per annum in 1989/90 to over 800 000 per annum in 1996/97 (Curtis and Nadolny, unpubl.). Coinciding with the increase in tree planting has been a trend to protect remnant bushland by fencing it from livestock grazing. Various government funded programmes have resulted in over 4 000 ha being fenced in the last few years (Curtis and Nadolny, unpubl.). Interest in biodiversity conservation has increased with the formation of various community groups and the development of rural conservation extension programmes (Trémont *et al.*, this volume).

If done correctly planting trees and shrubs can play a role in reversing many of the environmental problems besetting farmland as well as providing shade, shelter and reducing the environmental extremes on a farm (Bird *et al.* 1992). An unexpected benefit of maintaining or increasing tree cover is to improve the working and living conditions on the farm, and so reduce stress. Farm forestry itself has a role in diversifying income and increasing productivity of the farm, an important consideration when some country towns are wholly dependent on the prices of only a narrow band of products. Increasing tree cover also has a role to play also in reducing global problems such as the greenhouse effect (Maclaren 1996).

With the development of the tree planting industry debate has ensued on which species should be planted — whether to use exotic or native species (Curtis 1992a; Curtis *et al.* 1995). The trend has been from planting, almost exclusively, exotic species in the 1970s and early 1980s to a majority of plantings using natives in the 1990s. A recent survey by Greening Australia of farm tree nurseries in the region showed that they had changed to providing mostly (in many cases exclusively) Australian native species.

FARM FORESTRY MODELS FOR THE REGION

Forestry has been mainly based around logging of native forests in state forests and to a lesser extent on privately owned land. The main species logged are native cypress and ironbark on the North-west Slopes, stringybark species on the Tablelands, and coastal blackbutt, tallow wood, Sydney blue gum and flooded gum in the east (Table 1).

As the resource from state forests has dwindled, and tree planting has developed, interest in farm forestry has increased. On the Dorrigo Plateau about

Table 1. Main timber species logged in the region. *Source = Chiswell and Parker (1995), **Source = Thompson (1997).

Dorrigo Plateau*	Northern Tablelands**	North-west Slopes and Plains
Acacia melanoxylon	*Eucalyptus andrewsii*	*Callitris glaucophylla*
Araucaria cunninghamii	*Eucalyptus caliginosa*	*Eucalyptus crebra*
Corymbia maculata	*Eucalyptus laevopinea*	*Eucalyptus macrorhyncha*
Eucalyptus grandis	*Eucalyptus macrorhyncha*	
Eucalyptus laevopinea	*Eucalyptus microcorys*	
Eucalyptus microcorys	*Eucalyptus obliqua*	
Eucalyptus pilularis	*Eucalyptus saligna*	
Eucalyptus saligna	*Eucalyptus viminalis*	
Eucalyptus viminalis	*Eucalyptus youmannii*	
Pinus taeda		

80 000 m^3 was processed in 1995 (Roberts 1995), most of which went to one large processor (Curtis and Sandstrom 1995). Tablelands mills process 81 000 m^3/yr of sawlogs from state forests, private land and softwood plantations, while a further 43 500 m^3/yr is exported to mills at Dorrigo and on the coast (Thompson 1997). Since 1995 the quota from state forests servicing the mills on the Dorrigo Plateau has been reduced to about 16 000 m^3 (State Forests 1997).

Governments of both persuasions have indicated a willingness to establish many thousands of hectares of plantation forest (Department of the Prime Minister and Cabinet 1995) with the intention of tripling Australia plantation estate by 2020. The Federal government has initiated several programmes over several years to facilitate the increase in plantation establishment, among them the Vision 2020 and Department of Primary Industries and Energy Farm Forestry programmes (Ministerial Council *et al.* 1997). By the end of the 1998 planting season, State Forests of NSW established approximately 19 000 ha of hardwood plantations on the North Coast of New South Wales (R. Heathcote, Manager for hardwood plantations, pers. comm.).

Landholder surveys indicate an increasing interest in establishing trees for farm forestry although most landholders are more interested in growing trees for other purposes (Wilson *et al.* 1995; Nicoll and Dobbie 1996; New England Farm Forestry Programme 1997).

It is at present unclear which model of farm forestry should be developed in the region. In one model (e.g., New Zealand, central New South Wales and south-east South Australia) large industrial plantations, timberbelts and other plantings, consist primarily of exotic species. The remaining native vegetation mostly occurs within reserves and along watercourses. With an impending mill to process radiata pine sawlogs from the existing 10 000 ha in the region and a proposal for a further 36 000 ha of pine plantation (Rob Callaghan, Walcha Shire Council general manager, pers. comm.), pressure will inevitably be in that direction, at least on the Northern Tablelands (Fig. 3a).

If the farm forestry model depicted in Figure 3a were adopted widely in our region, the decline in native plant and animal species would continue. Small parts of the landscape might be kept as "natural" but native trees and attendant species would be lost from a large proportion of the landscape. Land considered not financially worth clearing may be cleared for forestry plantations. The monocultures which replace them will contribute less to biodiversity than the original forest. Whilst pine plantations have a larger number of native animals in them than totally cleared farm country and attract some native fauna (Anon. 1997) they still have a much lower number than natural forest because critical resources are missing (Driscoll 1977; Ford 1989). Superficially this model may offer the best returns to growers through uniform growth, silviculture and marketing strategies. However, in the long run, the full cost of biodiversity loss over the whole environmental, economic and cultural landscape may be negative.

Would it be possible to develop a farm forestry industry based around indigenous species and so integrate both forestry and conservation throughout rural landscapes? In previous papers we have described a farm forestry model which incorporates: plantations of selected high value indigenous species; timber belts and woodlots of selected indigenous timber trees; Landcare and other plantings (e.g., shelterbelts and gully plantings) using indigenous species (Curtis 1992b, 1997; Curtis and Sandstrom 1995). These plantings (mostly of native species) would link areas of native bushland — some of which might be managed for multiple use (i.e., mixed timber, grazing and conservation) or exclusively for conservation. Plantations of exotics, where they are used for specific economic outcomes, would be concentrated in specific sites (Fig. 3b).

There is a spectrum of options available to landholders ranging from purely economic plantations to plantings for biodiversity alone. With careful design, models and species can be developed to utilize production (e.g., timber, cut flowers, firewood, bushfoods, fodder) and habitat enhancement (e.g., riparian fencing, understorey planting) which suit individual properties.

There are many advantages in the farm forestry model described in Figure 3b from a conservation view, with native species conserved through their utilization. More varied habitats throughout the

AGROFORESTRY MODEL 1

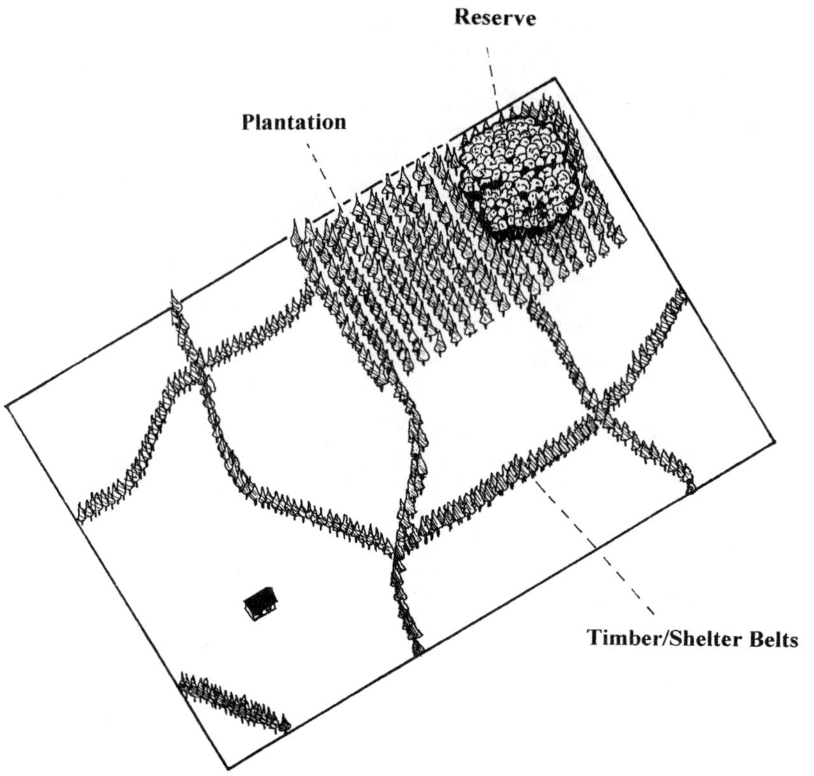

Fig. 3a. Farm Forestry Model 1, using mostly exotic species in plantations, timber belts and other farm plantings.

MODEL FARM PLAN

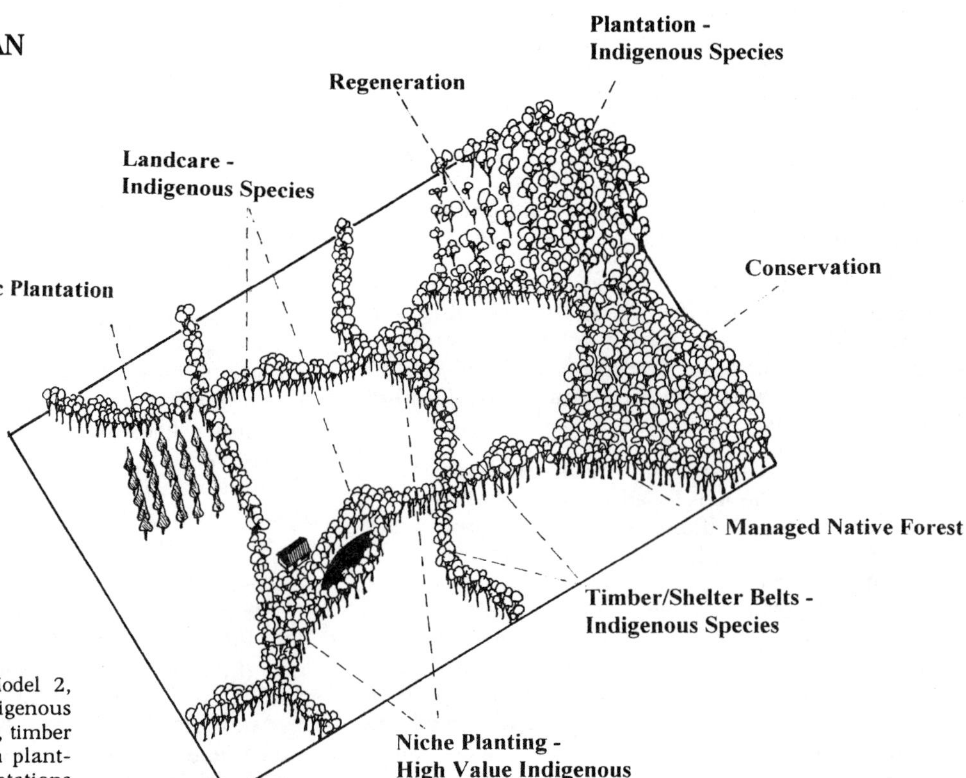

Fig. 3b. Farm Forestry Model 2, using mostly indigenous species in plantations, timber belts and other farm plantings with exotic plantations being confined to particular sites.

Incorporating Indigenous Trees & Timber into the Farm Plan.

farmland would better enable animal species to survive. If all the other components of habitat are incorporated (hollow trees, rotting logs, water, shrubs, corridors, seeding grasses, mistletoes, etc.) the outcomes for conservation could be high. Enthusiasm for the utilization of native species (and also less state forests being available for timber production) could, however, increase pressure on native forests on private land. The questions which flow from establishing such a model are therefore (1) could such an approach be economic?; and (2) what practical steps are needed to enact such a model with the best outcome for both production and conservation?

The Greening Australia farm forestry project

In 1995 Australia's Wood and Paper Strategy was released (Department of the Prime Minister and Cabinet 1995). The strategy aimed to reverse Australia's trade deficit in timber by increasing the plantation estate. As a central plank of the strategy Australian hardwoods were to be emphasized.

Greening Australia is a national non-government organization dedicated to helping Australians conserve and re-establish Australia's native vegetation. In 1994 we began a major farm forestry project on the Dorrigo Plateau, and in 1996 the project was extended to include the Northern Tablelands and North-west Slopes and Plains (Table 2). Chief elements of the project have been to establish species evaluation trials of local trees, plantings which demonstrate different ways of growing timber on farms (e.g., woodlots, timber belts, clumps and niche plantings), and extending methods of utilizing native forest on farms which maintain conservation values. As well as the trials and demonstrations the project is showing how trees can be managed and integrated in the farm plan to provide shelter, soil erosion control, salinity control, and habitat for wildlife. A feature of the plantations that have been established is that several species of timber trees, as well as understorey species are planted on each site. Remnant areas of native vegetation are often kept within the plantations. The work at Dorrigo is the most advanced and this paper will focus on results from that section of the project.

Species available for timber production

We have identified 127 tree species from 48 genera growing in the region which have a recognized timber value (Carr 1998; Sandstrom 1998). Only about 13% of the species are logged commercially. A further 85 native species on the North-west Slopes and Plains have a recognized food or medicine value (Carr and Hancock 1997).

Surveys were carried out of the majority of wood processors on the Dorrigo Plateau and the Tablelands to determine which species and which tree and log sizes were considered desirable by the industry (Chiswell and Parker 1995; Thompson 1997). The majority of the processors at Dorrigo and on the Tablelands produced sawn timber, while decking, furniture, veneer and poles were minor products. Table 1 lists the 18 species favoured by the Dorrigo processing industry. Most species came from native forests, only one came from plantations. An additional four species not currently utilized were identified as having potential based on their growth rates and utility.

In the Dorrigo survey the favoured logs had a centre diameter under the bark (cdub) of 60 to 80 cm and were pruned to 8 metres. Over half of the respondents favoured logs of 60 cm centre diameter or less and there was a trend to take smaller diameter logs. Sixty-four per cent of respondents would accept logs of 30 cm diameter and less as part of a load including larger logs. The logs that were preferred were single stemmed, straight and had no butt or stem damage. There was also an increasing interest in kiln drying and hence a move to high end-value sawn products for furniture, panelling and other decorative products. There appears to be an increasing market for smaller diameter trees and well pruned trees, both of which will favour farm forestry. Technology is also advancing to make more use of smaller, plantation-grown timber, such as radial sawing and portable kilns.

A preliminary study on the likely returns from the favoured species indicated that the gross value of products to the farmer for a realistic rotation length at current royalty values varied from $12,910 to $40,150 per hectare at Dorrigo (Chiswell and

Table 2. Works completed in Greening Australia Farm Forestry Project in north-west New South Wales, November 1997.

	Dorrigo Plateau	Northern Tablelands	North-west Slopes and Plains
No. species in evaluation trials	6	1	4
No. provenance trials	2	1	4
No. demonstration sites	102	16	10
No. landholders in the project	120	40	25
No. landholders where work has been carried out	52	14	6
No. of demonstration farms	17	14	6
No. trees planted	160 000	14 000	12 000
No. timber species evaluated	29	25	25
Area of farm forestry sites (ha)	121	13	25
Area of conservation works (ha)	39		

McDonald 1996; Table 3). The analysis did not suggest that all growers will achieve the returns projected. It is believed that the returns are realistically attainable and will easily be exceeded by some growers. Growth is directly related to the quality of inputs made by the grower and the quality of the site on which the trees are established. We are in the process of analysing both these returns and those projected from a case study of an eight hectare plantation that was established with a mixture of local timber species to determine the Net Present Value and the Internal rate of Return. Early indications are promising.

Preliminary analysis on the returns from several species have also been carried out for the Northern Tablelands and North-west Slopes and Plains. A comparison of the economic potential of silver wattle, New England stringybark, and Paulonia showed that Paulonia and silver wattle were economic as plantations while New England stringybark was not with the royalty rates then available (Harris 1995; Table 4). Since this study royalty rates for New England stringybark have risen considerably and it has been found that silver wattle is also suitable for veneers (McCombe *et al.* 1996; Ozarska *et al.* 1996) and has fodder potential, thus increasing its value. Preliminary data shows that silver wattle leaves are three times higher in protein than six common pasture grasses as well as being more digestible and having lower fibre, although tannins were not tested (C. Jones, University of New England, unpubl. data). Work is yet to be done on using regrowth stands of silver wattle and New England stringybark instead of plantations but it is likely that they would be more economic. Value adding the stringybark may also improve the outlook (e.g., the species has potential for internal panelling).

A study of the native timbers and the firewood market found the annual consumption of firewood of cities on the Northern Tablelands to be about 1 tonne per year per person (Wall and Reid 1993; Wall 1997). For Armidale City this amounted to over 20 000 tonnes pa. At present the market is supplied by trees killed earlier by ringbarking or dieback and the supply from farmland is unsustainable with current harvesting methods. Plantations of mixed fast and slow growing local native species (including silver wattle, fern leaf wattle, hickory wattle, black she-oak, river oak, white box, Blakely's red gum, New England stringybark, silvertop stringybark, yellow box, grey box, mugga ironbark), could become financially viable as prices inevitably rise with the firewood shortage (Wall 1997). Growth

Table 3. Financial returns on nine preferred timber species of the Dorrigo Plateau. Data from Chiswell and McDonald (1996). Royalty rates were based on State Forests royalties at the stump for the North Coast District in May 1996. Since that date the royalty rates increased by 20% in 1997 and were expected to rise again by another 10% in 1998. Assumed growth rates were based on data from clearfelled plantations from Bonney 1991, Neilsen 1990, Department of Land and Water Conservation 1997, Queensland Department of Natural Resources 1996, Queensland Forest Service 1992, West and Mattay 1993. Silvicultural regimes were designed by Chiswell with the aim of optimizing tree growth and producing target butt logs of 60 cm centre diameter and 6 m length within a reasonable time frame. The regimes also included pruning to enable the production of maximum clearwood, thus increasing the log values. The costs of establishment were based on real costs in forestry operations conducted by State Forests and the Greening Australia Farm Forestry project and amounted to $1,600 per hectare plus further ongoing management and maintenance costs of $100 ha^{-1} yr^{-1}.

Species preferred by wood processors (in order of preference).[1]	Rotation (years)	Gross yield $ ha^{-1}	Commercial Uses				
			Furniture	Poles	Veneer	Pulp	Saw logs
Eucalyptus pilularis	26	12 910	#	#	#	#	#
Eucalyptus pilularis	37	27 220	#	#	#		#
Eucalyptus microcorys	37	14 390		#	#		#
Eucalyptus saligna	37	28 645	#		#	#	#
Eucalyptus grandis	26	22 190	#		#	#	#
Acacia melanoxylon			#		#		#
Eucalyptus resinifera							#
Eucalyptus sp.							#
Araucaria cunninghamii	37	40 150	#		#		#
Eucalyptus sp.				#			#
Corymbia maculata			#	#	#	#	#
*Pinus radiata**			#	#	#	#	#
Eucalyptus andrewsii			#			#	#
Grevillea robusta	60	36 020	#				
Eucalyptus quadrangulata	30	14 425		#			#
Dysoxylum fraserianum			#				
Eucalyptus dunnii					#	#	#
Orites excelsa			#				
Syncarpia glomulifera				#			
Other species showing promise							
Eucalyptus nobilis					#	#	#
Eucalyptus nitens	26	22 115	#		#	#	#
Eucalyptus laevopinea			#				#
*Pinus taeda**			#	#	#		#

Notes: [1]All species from native forest except those grown in plantations (marked *).

Table 4. Financial returns on three timber species for the Northern Tablelands. Data from Harris (1995), assuming a discount rate of 7% and establishment costs of \$2,744 ha^{-1} for *Paulonia*, \$957 ha^{-1} for *A. dealbata* and \$1,134 ha^{-1} for *E. caliginosa*.

Species*	Rotation years	Yield \$ ha^{-1a}	Internal rate of return %	Current Market	F	P	V	Fi	Pu	S
					Commercial uses					
Acacia dealbata	30	8 851	12.8	No	#		#	#	#	
Eucalyptus caliginosa	40	−554	2.5	Yes		#		#		#
*Paulonia fortune**	15	16 777	17.6	No	#		#		#	

Notes: *Exotic species; [a]Net Present Value after tax; F = Furniture; P = Poles; V = Veneer; Fi = Firewood; Pu = Pulp; S = Saw logs.

rates of New England forest types were estimated as 1–2 m^3/ha/yr in an unmanaged condition and 3.5 m^3/ha/yr when managed for fuelwood production. Wall (1997) recommended that fuelwood production should be integrated with providing stock shelter, pasture enhancement, wildlife habitat, reserve pasture for drought and sawlogs.

On the Tablelands some species show promising growth rates, particularly silver wattle and manna gum (Curtis 1989; Russell 1995; Thorburn 1995; Jones 1996). However, these studies are yet to be compiled to assess possible financial returns. There is a potential production of some medicinal products from local native species such as eucalyptus oil (Alan Coates, pers. comm.; Hugh Harris, M.Res.Sci., in progress), and aesculin from blackthorn (Smith 1992). There are indications that there could be markets for specialist timbers from the region also. For example one landholder is getting high returns from hickory wattle, selling it to a manufacturer of door handles. On the slopes other local species may have potential for commercialization for fodder (e.g., from kurrajongs and wilgas), speciality timbers (e.g., certain species of wattle) and medicinals (e.g., aesculin from blackthorn). Further work is required to assess the commerciality of plantations of such products.

Management of thinned native forests

Thinned native forest can provide an alternative income to landholders, while retaining the benefits of tree cover. The Centre for Agricultural and Regional Economics is currently engaged in analysing the financial benefits of such an alternative and Greening Australia is researching best management practices. This income can then be used to establish plantings on other areas of the property where tree cover and native vegetation is less than optimum. Indications are that this will provide an important resource in maintaining a timber industry while plantations become established. For example, the Bellingen/Nymboida district has 360 000 ha of private forest. Assuming one eighth is timber producing there is 45 000 ha available for farm forestry. Native forests can produce 1 m^3/ha/year which means a production of 45 000 m^3 pa. This could be doubled with good silvicultural regimes. At present the small mills on the Dorrigo Plateau use about 20 000 m^3 each year so the resource could easily maintain those mills while a plantation industry became established. If 10% of the cleared land in Bellingen Shire is put down to native timber belts and plantations, 40 000 m^3 pa. could be supplied within 30 years on a sustainable basis, assuming a growth rate of 10 m^3/ha/yr. Managed plantations have a higher growth rate than native forests.

Native cypress pine is one of Australia's best structural timbers and grows extensively through the North-west Slopes. It is termite resistant and is used for interior wall panelling, flooring, furniture, weather boards, house frames, fence pickets, vineyard posts, shed poles and rafters, fencing and firewood. Native cypress saplings grow to a few metres high to a density of up to one million stems per hectare but do not self-thin. The dense regrowth suppresses growth of grasses and other species often exacerbating the erosion hazard on the steep sandy or the stony hillsides on which it usually grows (Nicholson 1997). Because of these growth characteristics (and despite its utility) many landholders consider the species to be a woody weed.

Thinning cypress regrowth has many benefits over leaving it alone or clearfelling it for pasture. Thinning to the recommended density of 6 × 6 m spacing allows grass and herb growth which eases the erosion hazard and allows some productive grazing, while suppressing secondary regrowth. A preliminary economic analysis by State Forests shows that the cost of thinning is compensated by a 3.37% Internal Rate of Return (Table 5). In this analysis the earnings for grazing were based on what State Forests earn from leasing their forests for grazing — (\$1.00/ha/yr) but is likely to be 5–10 times higher for the grazier. Thinning allows the income from both the timber and grazing. Leaving the stand unthinned prevents any income at all from the area from either grazing or timber production and has few conservation benefits. Clearfelling is likely to be almost as costly as thinning and grazing is the only likely income. A dense stand of cypress regeneration can result also. Understorey, eucalypts and other trees as well as other habitat components should be maintained in any managed cypress forest to maintain biodiversity.

On the Slopes and in hilly sections of the Tablelands and Dorrigo Plateau, many farms have sections of native forest. Frequently these areas of forests have been previously logged and have been left unmanaged since and often they have dense stands of regrowth. Common types on the

Table 5. Financial returns on different options of cypress management. Data from State Forests, NSW 1997. NCT = Non Commercial Thinning. Revenue based on a return of \$30 m^{-3}.

	Cost \$ ha^{-1}		Nett revenue \$ ha^{-1}						
Area thinned ha $person^{-1}$ day^{-1}	NCT	Year	1st thinning	Sawlog year	End cut	Sawlog year	Grazing \$ $cycle^{-1}$		Internal rate of return %
0.75	217	25	160	55	836	80	95		3.37
0.67	243	25	160	55	836	80	95		3.09

Tablelands and Slopes include regrowth forests of silver wattle, white cypress pine, New England stringybark, silvertop stringybark and silver-leaved ironbark. In many instances the density of the regrowth inhibits understorey development and prevents trees maturing and developing hollows and large branches. The issue for landholders in such situations is not whether they should plant trees, but whether they can manage their existing stands to get both the improved income and an improved conservation benefit that thinning might allow.

A farm forestry industry already exists around some of these forest areas, particularly on the North-west Slopes and the Dorrigo Plateau. The State Forests office at Inverell (on the border of the Tablelands and Slopes) has issued about 40 licences to portable sawmills (Richard Rienstra, pers. comm.) and there are 17 small mills on the Dorrigo Plateau and the adjacent area. Whether practices of these small operators are sympathetic to long-term sustainability of the timber resource and to conservation is unknown.

Some landholders have shown that it is possible to selectively log forests on a sustainable basis, graze livestock, provide drought fodder and maintain many conservation values (e.g., Curtis and Wright 1993). The Spark family's mixed grazing and cropping property near Tamworth, on the North-west Slopes and Plains contains areas of red stringybark forest. They have set aside some of the forest for conservation only and other parts of the forest for utilization through mixed grazing and sustainable timber harvesting. They have built conservation components into the utilization of this forest — thus hollow trees, rocks and logs are maintained for habitat for reptiles, mammals and birds, and grazing levels of cattle are kept low to ensure a good understorey of native shrubs, grasses and herbs is maintained. Preliminary costings of having the forest for both timber and grazing indicate a gross return of 3 to 30 times the level of just using the area for grazing, depending on the level of landholder input (Phil Spark, pers. comm.). If the area was just grazed with no income from the timber the pressure would be there to clear-fell it with a consequent reduction in conservation value.

CONCLUSION

With the predicted world shortage of timber, the need for wood producing trees will increase. This demand for wood provides an opportunity for the expansion of the farm forestry industry. Greening Australia is engaged in developing an approach to farm forestry for northern New South Wales which tries to incorporate whole landscape conservation with farm forestry, and features conservation of indigenous species through their utilization. Our work has arisen from the linked concerns of the decline in native biodiversity in the area combined with the need to provide a diversification of income and employment opportunities for rural families

The study region is diverse in land types, ecosystems and farming enterprises and different forms of farm forestry will be appropriate for different parts of the region. At this stage of the work we can only speculate as to the form of farm forestry that is most suitable for each part of the region and the economic viability of each.

On the high rainfall parts of the region as well as the sloping country to the south, north and west, there is already an existing timber industry based around indigenous species growing in mostly state owned, and often regrowth, forests. As state forests become less available for forestry it is likely that farm forestry using thinned native forests or woodlots and timberbelts will fill the vacuum. Some landholders have shown that it is possible to selectively log forests on a sustainable basis, graze livestock, provide drought fodder and maintain many conservation values. There is, however, no guarantee that farm forestry will be developed in a way that will foster conservation values. Joint Venture plantations for example tend to emphasize monocultures of a few species and remnant bush-land is usually cleared in advance of plantation establishment. Indigenous species in the high rainfall zone provide a range of excellent timbers, some of which appear to be economically viable to grow in farm woodlots, timberbelts or managing native forests, as well as food, spice and medicinal plants and fodder species with commercial potential.

The high sections of the Northern Tablelands is more complex than the sloping country fringing it. It has been more affected by tree decline and dieback, and tree growth tends to be slower. Like the slopes there are commercial products that are possible from local species, such as timbers, firewood, and medicinal products. At this stage, however, it is probably not economic to grow them solely for timber, and shelter and landcare benefits should be the paramount reason for growing them with any timber that is harvested considered as a bonus. Radiata pine grows well on the Northern Tablelands and has been extensively planted for timber. We have suggested a way woodlots of pine could be incorporated into productive grazing land where only small remnants of native vegetation remains. In this model a network of indigenous species is established in shelterbelts, gully protection and through the protection of remnants, and then woodlots of pines are established at a rate of 1–2 ha pa. Milling would commence after about 30 years (Curtis 1992b; Bird *et al.* 1996; Bird and Jowett 1996). The pines are essentially treated as a crop within a landscape where native species provide the framework.

To maintain conservation values through any farm forestry system the key thing is to maintain the building blocks of habitat (Barrett and Ford 1993; Ford and Barrett 1995), namely a diversity of native trees and understorey, corridors, big patches of forest or woodland, hollow trees, logs, creeks and waterways, nectar, mistletoe and seeding grasses. Such elements can be successfully incorporated into any productive farming landscape, and especially one that has farm forestry included.

We can foresee three criticisms to the approach we have described. Some will say that fostering native species for forestry is uneconomic because the trees grow too slowly, and that we would be better off developing an industry based around massive plantations of exotic species such as radiata pine. We believe if a full analysis is done of all the values that native vegetation provides, such an approach will ultimately be unsatisfactory and that any economic study should incorporate the ecological services provided by indigenous vegetation (Hobbs 1992).

From the point of view of conservation some people may argue that the promotion of the utilization of native species could lead to the over-exploitation of native species. With several of the species we are discussing we see little danger of this. Many exist in regrowth forests and have shown themselves quite capable to regenerate, having regenerated more than once since being originally cleared or logged. Indeed some species, such as blackwood and silver wattle, are fast growing pioneer species which are adapted to rapid growth and replacement. The major problem will be the encouragement of people to maintain the building blocks of habitat. Codes of practice and good extension programmes will be essential to ensure that conservation values are maintained. If native species are not given an economic value they run the risk of total intentional removal or being lost through neglect. Giving them a value should allow their retention.

A third criticism of the model could come from farmers concerned that if they plant native trees the may be prevented from harvesting later, particularly if they provide habitat. Harvesting guarantees will need to be provided for landholders. Any habitat damage at the time of harvesting should be countered by the benefits of having more native trees in the landscape.

We believe developing the industrial model for forestry based around a very small number of (mostly) exotic species will lead to negative conservation outcomes. A system that would be far more desirable would be one based around the diversity of our useful local species. Although the short-term returns may not be as high as with monocultures of high value species we believe the compensation will be in a more sustainable system. Certainly we believe the richness of the landscape and of the human experience will be much poorer if we do not work out ways of integrating the utilization of native species and their conservation into productive farms.

ACKNOWLEDGEMENTS

Sincere thanks to the following people who have assisted us in the preparation of this chapter and its presentation: Joy Kirby — typing; Jake Davies — figures; Chris Barnden and Jo Parker — photography; Ernie Chiswell, Chris Nadolny, Nick Reid, Rod Bird and Hugh Ford — reading drafts and providing helpful advice.

APPENDIX I — PLANT NAMES USED

Acacia dealbata	silver wattle
Acacia filicifolia	fern leaf wattle
Acacia implexa	hickory wattle
Acacia melanoxylon	blackwood
Allocasuarina littoralis	black she-oak
Araucaria cunninghamii	hoop pine
Brachychiton populneus	kurrajong
Bursaria spinosa	blackthorn
Callitris glaucophylla	white cypress pine
Casuarina cunninghamiana	river oak
Corymbia maculata	spotted gum
Dysoxylum fraserianum	rosewood
Eucalyptus albens	white box
Eucalyptus andrewsii	New England blackbutt
Eucalyptus blakelyi	Blakely's red gum
Eucalyptus caliginosa	New England stringybark
Eucalyptus crebra	narrow-leaved ironbark
Eucalyptus dunnii	Dunn's white gum
Eucalyptus grandis	flooded gum, rose gum
Eucalyptus laevopinea	silver top stringybark
Eucalyptus macrorhyncha	red stringybark
Eucalyptus melanophloia	silver-leaved ironbark
Eucalyptus melliodora	yellow box
Eucalyptus microcorys	tallowwood
Eucalyptus moluccana	grey box
Eucalyptus nobilis	white gum
Eucalyptus nitens	shining gum
Eucalyptus obliqua	messmate
Eucalyptus pilularis	coastal blackbutt
Eucalyptus quadrangulata	white top box
Eucalyptus resinifera	red mahogany
Eucalyptus saligna	Sydney blue gum
Eucalyptus sideroxylon	mugga ironbark
Eucalyptus viminalis	manna gum
Eucalyptus youmannii	Youman's stringybark
Geijera parviflora	wilga
Grevillea robusta	silky oak
Orites excelsa	prickly ash
Syncarpia glomulifera	turpentine
Paulownia fortunei	Paulonia
Pinus radiata	radiata pine
Pinus taeda	loblolly pine (plantations)

REFERENCES

Anon., 1997. Pine plantations can attract fauna diversity. *Aust. For. Grow.* **20:** 14–15.

Barrett, G. and Ford, H. A., 1993. Birds on Farms. *Greening Australia Field Note* 93/5, Greening Australia: Armidale, New South Wales.

Barrett, G. W., Ford, H. A. and Recher, H. F., 1994. Conservation of woodland birds in a fragmented rural landscape. *Pac. Cons. Biol.* **1**: 245–56.

Beadle, N. C. W., 1948. *The vegetation and pastures of western NSW, with special reference to soil erosion.* Department of Conservation of New South Wales: Sydney, New South Wales.

Bird, P. R., Bicknell, D., Bulman, P. A., Burke, S. J. A., Leys, J. F., Parker, J. N., Van der Sommen, F. J., Voller, P., 1992. The role of sheltering Australia for protecting soils, plants and livestock. *Agrofor. Sys.* **20**: 59–86.

Bird, P. R. and Jowett, D. W., 1996. Prospects for future farm forestry — the small scale option. Farm forestry and plantations — investing in future wood supply. 1996 Australian Forest Growers Conference, Mount Gambier, South Australia, 9–12 Sept. 1996.

Bird, P. R., Jowett, D. W., Kellas, J. D. and Kearney, G. A., 1996. Farm forestry clearwood production. *Technical report series*, January 1996, Agriculture Victoria.

Boland, D. J., Brooker, M. I. H., Chippendale, G. M., Hall, N., Hyland, B. P. M., Johnson, R. D., Kleinig, D. A. and Turner, J. D., 1984. Forest Trees of Australia. 4th ed. Nelson-CSIRO: Melbourne, Victoria.

Bonney, L., 1991. Growth of an *Eucalyptus grandis* plantation following intensive silvicultural treatments applied in the first six years. Forestry Commission New South Wales.

Bootle, K. R., 1985. Wood in Australia. McGraw-Hill: Roseville, New South Wales.

Carr, D. B., 1998. Species selection for farm forestry on the Northern Tablelands, Slopes and Plains of New South Wales. *Greening Australia Field Note* 98/3, Greening Australia: Armidale, New South Wales.

Carr, D. B. and Hancock, P., 1997. Bush food of the North-west Slopes and Plains of New South Wales. *Greening Australia Field Note* 97/4, Greening Australia: Armidale, New South Wales.

Chiswell, E. and McDonald, C., 1996. Financial aspects of farm forestry on the Dorrigo Plateau and Associated Areas. Unpubl. report, Greening Australia: Dorrigo, New South Wales.

Chiswell, E. and Parker, J., 1995. Survey of Wood Producers. Unpubl. report, Greening Australia: Dorrigo, New South Wales.

Cunningham, G. M., Mulham, W. E., Milthorpe, P. L. and Leigh, J. H., 1981. Plants of Western NSW. Soil Conservation Service: New South Wales.

Cremer, K. W. (ed), 1990. Trees for Rural Australia. Inkata Press: Melbourne and Sydney.

Curtis, D. J., 1989. Eucalypt Re-establishment on the Northern Tablelands of New South Wales. M.Sc. thesis, University of New England, Armidale, New South Wales.

Curtis, D., 1991. New England dieback — where to now? *Acres* **3**: 9–12.

Curtis, D., 1992a. Trees on Farms: the next ten years. *Aust. J. Soil Water Cons.* **4**: 9–15.

Curtis, D., 1992b. Webs and Honeypots — a design concept for rural areas. *Aust. J. Soil Water Cons.* **5**: 57.

Curtis, D., 1997. Best Management Practices for Native Vegetation on Farms. Pp. 77–93 *in* Proceedings SAWCAA Annual Conference Native Vegetation — "Managing the Change". Moree, New South Wales.

Curtis, D. and Sandstrom, M., 1995. Resolving Forest Conflicts and Providing for a Sustainable Forest Industry through Agroforestry. Pp. 63–71 *in* Ecological Economics Conference, Coffs Harbour, 19–23 November, 1995.

Curtis, D. J., Nadolny, C. and Ford, H. A., 1995. Lessons from a Network of Dieback Destroyers. Pp. 175–86 *in* Nature Conservation 4: The Role of Networks ed by D. A. Saunders, J. L. Craig and E. M. Mattiske. Surrey Beatty & Sons: Chipping Norton, New South Wales.

Curtis, D. J. and Wright, T., 1993. Natural regeneration and grazing management — a case study. *Aust. J. Soil Water Cons.* **6**: 30–34.

Curtis, D. J. and Nadolny, C. (unpubl.). Revegetation in Northwest New South Wales 1990–2000. Greening Australia: Armidale, New South Wales.

Department of Land and Water Conservation, 1997. Practical ways of integrating farm forestry, whole farm planning and nature conservation on North Coast farms. Workshop results. Dept. of Land and Water Conservation, New South Wales.

Department of the Prime Minister and Cabinet, 1995. Wood and Paper Industry Strategy. Forests Taskforce: Canberra, Australian Capital Territory.

Donaldson, S., 1996. Namoi Community Catchment Plan Stage 1: Situation Statement, March 1996. Prepared by Donaldson Planning and Management Services, Gunnedah for North West Catchment Management Committee, New South Wales.

Driscoll, P. V., 1977. Comparison of bird counts from pine forests and indigenous vegetation. *Aust. Wildl. Res.* **4**: 281–88.

Floyd, A. G., 1989. Rainforest Trees of South-eastern Australia. Inkata Press: Melbourne, Victoria.

Ford, H. A., 1989. Ecology of Birds: An Australian Perspective. Surrey Beatty & Sons: Chipping Norton, New South Wales.

Ford, H. A., Barrett, G. and Recher, H., 1995. Birds in a degraded landscape — safety nets for capturing regional biodiversity. Pp. 43–50 *in* Nature Conservation 4: The Role of Networks ed by D. A. Saunders, J. L. Craig and E. M. Mattiske. Surrey Beatty & Sons: Chipping Norton, New South Wales.

Ford, H. A. and Barrett, G., 1995. The role of birds and their conservation in agricultural systems. Pp. 128–34 *in* People and Nature Conservation: Perspectives on Private Land Use and Endangered Species Recovery ed by A. Bennett, G. Backhouse and T. Clark. Surrey Beatty & Sons: Chipping Norton, New South Wales.

Forestry Commission of NSW, 1989. *Research Notes* No. 17.

Harris, H., 1995. The potential of *Paulownia fortunei, Acacia dealbata* and *Eucalyptus caliginosa* timber plantations in the Armidale district. Thesis for Bachelor of Natural Resources, University of New England, Armidale, New South Wales.

Hobbs, R. J., 1992. Function of biodiversity in Mediterranean ecosystems in Australia: definitions and background. Pp. 1–25 *in* Biodiversity of Mediterranean Ecosystems in Australia ed by R. J. Hobbs. Surrey Beatty & Sons: Chipping Norton, New South Wales.

Jones, B. E., 1996. The contribution of a landcare group to local tree planting information — a study of information networks and planting success using an example from the New England Tablelands of New South Wales. M. Res. Science, University of New England, Armidale, New South Wales.

McCombe, B., Ozarska, B., Turville, G. and Manley, S., 1996. Decorative sliced veneer from East Gippsland Eucalypts — veneer manufacturing trials. CSIRO milestone report.

McIntyre, S. and Barrett, G. W., 1992. Habitat variegation, an alternative to fragmentation. *Cons. Biol.* **6:** 146–47.

Mackay, S. M., Humphries, F. R., Clark, R. V., Nicholson, D. W. and Lind, P. R., 1984. Native tree dieback and mortality on the New England Tablelands of New South Wales. *Research Paper* 3, Forestry Commission of NSW, Sydney, New South Wales.

Maclaren, J. P., 1996. Plantation forestry: its role as a carbon sink. Pp. 417–18 *in* Greenhouse: Coping with Climate Change ed by W. J. Bouma, G. I. Parman and M. R. Manning. CSIRO.

Ministerial Council on Forestry, Fisheries and Aquaculture Standing Committee on Forestry, Plantations Australia, Australian Forest Growers and National Association of Forest Industries, 1997. *Plantations for Australia: the 2020 Vision.* Plantation 2020 Vision Implementation Committee.

Morgan, M. and Terrey, J., 1990. Land Conservation for the New England Tableland: a regional plan. World Wide Fund For Nature Australia. P. 103 Final Report.

Nadolny, C., 1984. Eucalypt dieback on the Northern Tablelands of New South Wales. M.Sc. thesis, University of New England, Armidale, New South Wales.

Nadolny, C. and Curtis, D., 1994. Eucalyptus dieback: Moree area. Unpubl. report January 1994. Greening Australia: Armidale, New South Wales.

New England Farm Forestry Programme, 1997. Outcomes of stakeholder forums in Glen Innes and Walcha, 16/17 July 1997. New England North West Regional Development Board, Armidale, New South Wales.

Neilsen, W. A. (ed), 1990. Plantation Handbook. Forestry Commission: Tasmania.

Nicholson, D., 1997. Managing Cypress Pine on Your Property. State Forests, New South Wales.

Nicoll, C. and Dobbie, M., 1996. The CSIRO National Farm Tree Survey, Technical Memorandum 96.26. Division of Water Resources: CSIRO.

Ozarska, B., Turville, G. and Manley, S., 1996. The use of hardwoods for high value added wood products: Decorative sliced veneer from East Gippsland hardwoods. Production of veneered panels and profile wrapping, CSIRO Forestry and Forest Products, Divisional Report. No. DFFP522, 1996.

Queensland Department of Natural Resources, 1996. Overview of tree growing for timber production. Dept. of Natural Resources: Queensland.

Queensland Forest Service, 1992. Plantation volume growth summaries from growth plots in Queensland. Dept. of Natural Resources: Queensland.

Roberts, P., 1995. Dorrigo State Forests Annual Statistical Return, State Forests: Dorrigo, New South Wales.

Russell, A., 1995. An assessment of selected native trees and shrubs in shelterbelt plantings on the Northern Tablelands of New South Wales. Thesis Dip. Natural Resources, University of New England: Armidale, New South Wales.

Sandstrom, M., 1998. Trees for the Dorrigo Plateau. Greening Australia: Dorrigo, New South Wales.

Saunders, D. A., 1996. Does our lack of vision threaten the viability of the reconstruction of disturbed ecosystems? *Pac. Cons. Biol.* **2:** 321–26.

Sivertsen, D., 1994. The native vegetation crisis in the wheat belt of New South Wales. *Search* **25:** 5–8.

Smith, R., 1992. *Bursaria spinosa* and aesculin production. *Greening Australia Field Note 92/4.* Greening Australia: Armidale, New South Wales.

State Forests NSW, 1997. Proscribed Base Wood Supply Agreement — 1998. State Forests: New South Wales.

Thompson, D., 1997. Sawmill Survey of the New England Region. CARE: Armidale, New South Wales.

Thorburn, R. G., 1995. An assessment of native tree plantings conducted by the Stonehenge/Glencoe Landcare Group on the Northern Tablelands of New South Wales. Thesis, B. Nat. Res., University of New England, Armidale, New South Wales.

Wall, J. P., 1997. Sustainability of the Armidale fuelwood industry on the Northern Tablelands of New South Wales: resource yield, supply, demand and management options. Ph.D. thesis, University of New England, Armidale, New South Wales.

Wall, J. P. and Reid, N., 1993. Domestic fuelwood use in a rural township in Eastern Australia: evidence for resource depletion and implications for management. *Common. For. Rev.* **72:** 31–37.

West, P. W. and Mattay, J. P., 1993. Yield prediction models and comparative growth rates for six eucalypt species. *Aust. For.* **56:** 211–25.

White, M., 1994. After the Greening — the Browning of Australia. Kangaroo Press Pty. Ltd.: New South Wales.

Wiens, J. A., 1997. Metapopulation dynamics and landscape ecology. Pp. 43–62 *in* Metapopulation Biology: Ecology, Genetics and Evolution ed by I. Hanski and M. E. Gilpin. Academic Press: San Diego.

Williams, J. B. and Nadolny, C., 1981. A survey of eucalypt species involved in New England dieback. *In* Eucalypt Dieback in Forests and Woodlands ed by K. M. Old, G. A. Kile and C. P. Ohmart. Proceedings of a conference held at the CSIRO Division of Forest Research, Canberra, 4–6 August 1980, CSIRO, Canberra, Australian Capital Territory.

Wilson, E. O., 1992. The Diversity of Life. Penguin Books: Ringwood, Victoria.

Wilson, S. M., Whitham, J. A. H., Bhati, U. N., Horrath, D., Tran, Y. D., 1995. Trees on Farms, Survey of Trees on Australian farms: 1993–94. Australian Bureau of Agricultural and Resource Economics 95.7, Canberra, Australian Capital Territory.

CHAPTER 26

Wetland use and conservation in the agricultural environment: managing processes for the components

Margaret A. Brock[1] and Peter J. Jarman[2]

Agriculture uses natural, modified or created wetlands for free water and soil moisture and nutrients. It manipulates wetland systems to ensure harvestable production, sometimes by adding or subtracting components, but commonly by altering processes. For example, it often alters the pattern of flooding and drying to ensure a predictable water regime, which may be either wetter (damming) or drier (draining), and less variable, than "natural". These agricultural actions tend to reduce heterogeneity of wetlands across the landscape, and can reduce wetland values for both agriculture and conservation.

Changes to processes such as water regime alter the form and diversity of resources available to the biota. As a consequence, individual species (components) may be relatively favoured or disadvantaged. For example, many aquatic plants, frogs and waterbirds will be disadvantaged by maintaining persistent, deep water in dams; others will be disadvantaged by draining; some wetland-associated species also depend on the terrestrial matrix in some life-stages or for dispersal. The sometimes conflicting needs of biota make firm management guidelines difficult to formulate.

Targets for use of wetlands in an agricultural environment are usually very local and often privately determined, whereas species and ecosystem conservation requires planning and management at the landscape or even broader scale and is usually conducted in the national (i.e., public) interest.

These conflicts of scale and interest are best resolved by identifying needs and effects, and then management action, at the level of processes, not individual components. A defined system of appropriate wetlands might be conserved and managed by co-operation between agencies and willing landholders. This would depend upon interchange of facts, ideas and ideals between farmers, scientists and conservation managers.

Key words: Wetland components, Wetland processes, Flooding and drying pattern, Water regime, Agricultural land-use, Nature conservation.

INTRODUCTION

THE problems of conserving diversity of wetland types and wetland species in a production environment illustrate some general effects of production upon biodiversity. Consideration of these problems may lead to generalized concepts that can be applied to conservation in other parts of the agricultural environment.

In this chapter, we consider wetlands as areas of land that are occasionally to persistently covered with non-marine water that, when present, flows very slowly if at all. Our "wetlands" include lakes, lagoons, billabongs, ponds, dams, swamps, marshes, bogs and fens, and may be natural or artificial. We specifically consider such wetlands in a landscape used predominantly by agriculture.

[1]Botany, School of Rural Science and Natural Resources, University of New England, Armidale, New South Wales, Australia 2351.
[2]Ecosystem Management, School of Rural Science and Natural Resources, University of New England, Armidale, New South Wales, Australia 2351.
Pages 258–68 *in* NATURE CONSERVATION 5: NATURE CONSERVATION IN PRODUCTION ENVIRONMENTS: MANAGING THE MATRIX ed by J. L. Craig, N. Mitchell and D. A. Saunders. Surrey Beatty & Sons, 2000.

Table 1. A summary of what agriculture wants from, and consequently does to, natural wetlands.

What agriculture wants Values wanted:	What agriculture does Actions taken:	What agriculture loses Values lost:
• predictable water for stock and domestic use, or irrigation	• damming, deepening, storing	• soil to cultivate, wet meadows to be grazed; diverse biota
• predictably water-free soil	• draining	• periodic standing water; spatially heterogeneous plant growth
• irrigable soils	• levelling and draining	• spatially heterogeneous plant growth
• stock water and fodder	• giving stock access, hence grazing and trampling by stock	• plant community diversity

These wetlands are rich in species, and easily altered by or for agricultural land-uses; in the longer term, agriculture itself may lose other values that the original wetland offered (Table 1). We aim to show that those alterations affect wetland processes, and in turn affect their components. By processes we mean the dynamic interactions between the definable biotic or abiotic entities (organisms, substrates, water itself) that we call components. Conservation of wetlands usually requires more than mere replacement of physical or biotic components; it requires restoration of necessary processes. We focus upon flooding and drying processes and their consequences.

We also discuss the organizational level in human society at which decisions are made about managing wetlands for agricultural productivity, and contrast that with the levels at which management for conservation is desired and decided. We suggest that some effective compromises can be reached by bringing together those who are setting farm production strategies and conservation strategies.

THE BIOTA OF NATURAL WETLANDS

Wetlands in the agricultural landscape have an important place in conservation. They form distinct islands of wet habitat in a more extensive matrix of dry habitats. Consequently their biota often appear localized, isolated and different from those surrounding them. Because water, solutes and transportable organic and inorganic particles drain into them, wetlands are distinguished from their surroundings by higher nutrient status, more prolonged moisture availability, and often more alluvium-rich substrates. Higher and more prolonged plant productivity than in the terrestrial surrounds is encouraged by those characteristics (especially where rainfall seasonally limits dryland plant production), and may lead to the accumulation of humus in the wetland substrates.

Organisms differ in their ability to cross the hostile matrix between wetlands; the matrix is more permeable to some than to others. Those with highly mobile life stages may disperse across intervening matrix readily, while others cannot disperse without a connecting medium of water. For a highly dispersive species, separate but neighbouring wetlands contain local populations of their metapopulation. For water-bound species, separate wetlands contain truly isolated populations until joined by widespread flooding, and are more likely to show local endemism.

The biota we discuss represent several trophic types: cyanobacteria and phytoplankton; aquatic plants; invertebrates; frogs; turtles; and waterbirds. These have different degrees of dependence upon wetlands, and different mechanisms of dispersal between and persistence in them. Changes to the wetland system processes inevitably alter the form and diversity of the resources currently available to the biota.

HOW PROCESSES INFLUENCE WETLAND COMPONENTS: THE EXAMPLE OF FLOODING AND DRYING

Ecosystem processes cause wetland states to fluctuate in space and time. Most wetlands fluctuate in level; many Australian wetlands typically have wet and dry phases. The pattern of flooding and drying (inundation) is an important wetland process termed "water regime" (Bunn *et al.* 1997). Where, when and to what extent water is present in a wetland determines its biotic community. Diversity and complexity of water regimes reflect local and catchment climate as well as local and regional topographic factors. The inundation process influences whole wetlands and the components within them. The timing, frequency, duration, extent and depth, and variability of the presence of water are the attributes of the process that influence the presence, absence and abundance of the biotic components of a wetland, and hence of the communities they form. We consider how some of the biotic components respond to the wet and dry phases of wetlands created by the process of flooding and drying (Table 2).

Micro-organisms break down organic matter and are important in cycling nutrients within a wetland. Many bacteria are anaerobic and are active in sediments in deeper water with low oxygen concentrations. Aerobic bacteria function in shallow and well-stirred waters. As a wetland dries down, nutrients remain in the dried sediment. The flooding of a dry wetland releases a pulse of nutrients that together with light and water provide the resources

Table 2. How wetland components respond to the process of wetting and drying. The pattern of wetting and drying in each wetland has several attributes including timing, frequency, extent, duration and variability of flooding which will influence the components of the wetland.

Wetland component	Wetland processes Flooding and drying pattern		
	On flooding	While damp	On drying
Soil	Pulse of nutrients, siltation		Nutrients locked up, deflation
Micro-organisms	Process nutrients and organic matter		Change from anaerobic to aerobic bacteria, resistant stages
Cyanobacteria	Photosynthesize, reproduce, bloom	Fix nitrogen	Propagule bank of resistant spores
Phytoplankton	Photosynthesize in water column, light dependent, may bloom, reproduce	Stranded	Propagule bank of resistant sexual and asexual spores
Aquatic Plants — submerged	Germinate, grow and reproduce in water	Seed bank	Seed bank
Aquatic Plants — amphibious	Aquatic form or seed bank	Tolerate or respond to water fluctuation	Seed bank
Aquatic Plants — semi-terrestrial	Seed bank	May tolerate damp	Grow in damp areas
Invertebrates — zooplankton	Feed and breed		Propagule bank of resistant eggs
Invertebrates — benthic	Many larval stages of aerial adults; others, aquatic larvae and adults		Terrestrial adults, resistant eggs
Frogs	Eggs in or near water; tadpoles grow, mature, in water; many adults leave, but return to lay	Tolerate or use damp areas to move between wet and dry	Adults lay briefly resistant eggs in edge vegetation; some adults aestivate
Turtles	Lay and hatch out of wetland; grow, feed and mature in water; move between wetlands	Move between dry and wet wetlands	Eggs in terrestrial surrounds; some adults aestivate in cover; others leave to find water
Waterbirds	Feed, nest when habitat available		Disperse to other wetlands

for germination and growth of both micro and macro photosynthesizers and the hatching and development of invertebrates from the propagule bank. The open water becomes occupied by phytoplankton and zooplankton. A range of larger, rooted plants can grow submerged, floating or stranded at the wetland's edges. These components respond to the processes within the wetland, changing in space and time as they interact with the other parts of the systems, providing resources for some, using others as resources.

The species-richness of plant communities in wetlands is often maintained by flooding and drying. A species-rich community of amphibious plants occupies the edge of wetlands where their ability to tolerate or respond to fluctuations of wetting and drying maintains the community (Brock 1986; Brock and Casanova 1997).

Some aquatic plant species respond to fluctuations of water presence or absence by changing growth form, leaf type and modes of reproduction. For example, when stranded on mud the water milfoil *Myriophyllum variifolium* grows prostrate, has entire leaves and roots at each node, whereas in deeper water it has an upright growth form with divided underwater leaves. An individual plant can change form and behaviour as it is stranded or flooded. It can reproduce sexually and add numerous long-lived seeds to a persistent seed bank. Drying enhances seed germination. The plant may also reproduce prolifically asexually, from fragments. Dispersal over longer distances within a wetland occurs when wind moves floating clumps of plants from deeper water to become stranded on the shoreline. This species is adapted to maintain itself in space and time within a wetland by morphological plasticity and reproductive flexibility (Brock 1991). Other species have evolved similarly complex combinations of life cycle characteristics that allow flexible response to inundation and drying.

Species-rich seed banks in wetlands provide the potential for revegetation by most species after dry phases (Brock 1995, 1998). Longevity of seeds, and non-depletion of the seed bank in each germination event, both help populations maintain themselves (Brock 1998).

Many invertebrates associated with wetlands are obligate inhabitants of water, both as larvae and as adults. Some survive the dry phase of a wetland in a dormant state (e.g., as desiccation-resistant eggs, or as buried adults, Wiggins *et al.* 1980), usually in the sediments; but others have no such mechanism and will die out in that wetland if it dries completely. Yet others occupy water in the larval stage only, and as adults emerge into the air and potentially disperse. A flooding wetland can become occupied by invertebrates both that emerge from dormancy within its sediments, and that disperse to it from other wetlands.

Fishes of wetlands behave like non-dispersing, drought-intolerant invertebrates, in that almost all have no mechanism for surviving the drying out of a wetland. Drying out of a wetland usually means loss of its fish fauna, until it is reflooded from a water source containing fish. Thus temporary wetlands that fill from local rain rather than from a flooding river usually lack fish.

Freshwater turtles spend their juvenile and adult life in the water, emerging only to bask, or to lay their eggs in a nest chamber dug in dry ground away from water. Some southeastern Australian turtles (e.g., *Chelodina longicollis*) move up to 2 km overland from seasonally drying wetlands to one that is permanently wet. They disperse back to, and breed in, the temporary wetlands when floods or heavy spring rains fill them, stimulating a pulse of productivity from which the breeding turtles and their hatchlings benefit (Parmenter 1976). Their densities in the drought-refuge, persistent wetlands can be very high (Georges *et al.* 1986), and intra-specific competition can lower condition, retard growth and halt breeding (Kennett and Georges 1990). By dispersing overland to isolated, temporary wetlands, these turtles escape from competition from fish and other turtles in the permanent wetlands (Chessman 1984). These turtle species use the resources of more than one wetland and both persistent and temporary wetland types. The population's size and local density thus reflect the availability of diverse water regimes, habitats and resources at a landscape scale.

Some western and northern turtles (e.g., *Pseudemydura umbrina*, *Chelodina rugosa*, *C. steindachneri*, *C. novaeguineae*) aestivate when a wetland dries up, lying dormant in mud at the bottom or under vegetation and debris at the side, until the wetland refloods. *P. umbrina* and *C. oblonga* nest in spring or summer before their temporary wetlands dry out, but their eggs do not hatch until winter rains flood the wetlands. Freshwater turtle breeding is most successful when heavy rains or flooding coincide with the synchronized, seasonal hatching of their eggs.

Many adult frogs live well away from water, but mate and lay eggs in or next to water, in which the juveniles (tadpoles) swim and feed. Females may lay only once each year. The eggs and tadpoles take from one to seven months before leaving the wetland as small frogs (Roberts 1993), so a temporary wetland must stay wet that long for breeding to succeed. Many Hylidae respond to rain by emerging (often highly synchronously) from cover or burrows or descending from trees and moving to wetlands to lay eggs in or on the water in which their young develop (Tyler and Davies 1993). Some Myobatrachidae have evolved ways of breeding away from water. Others lay eggs on land in places (such as under leaf litter or in burrows) that flood during rain, either forming temporary pools or carrying the tadpoles to wetlands. The embryos of some of these species can delay development for many weeks until rains fall (Littlejohn *et al.* 1993).

Although breeding appears to be triggered directly by rainfall in many frog species, a closely studied community of five species was more influenced by changes in pond water-level than by rainfall itself. The timing of the reproductive peak thus varied greatly from year to year (Humphries 1979).

Frogs are vulnerable to desiccation and require rains at the right time of year to allow them to move overland through matrix habitat to wetland breeding sites. Some species may move a few hundred metres; others live as adults within less than 100 m of the wetland in which they breed, and are particularly sensitive to changes in its vicinity. Frogs are less likely than turtles to use a number of wetlands in their life, probably moving to and away from just one wetland. They may form locally discrete populations (as shown by genetic studies, e.g., Driscoll *et al.* 1995). Nevertheless, in agricultural landscapes at least some frogs commonly breed in a newly dug farm dam within its first wet season (pers. obs.), illustrating how widely some frogs disperse through the matrix.

Waterbirds associate with wetlands in many ways: some feed in the water or wetland, and breed away from it; others roost or nest securely on or over water, but feed away from it; yet others feed, roost and breed on, in or over water. All waterbirds leave a wetland that dries completely. Some return rapidly when it refloods, feasting on the seeds, eggs, and dormant organisms flushed from the sediments. Others follow soon after, to eat the pulse of phytoplankton, zooplankton and invertebrates that bloom after reflooding. Those that eat foliage of aquatic plants, or larger invertebrates, or fishes take longest of all to return, since their resource base takes longest to recover and develop the abundances that will sustain the birds.

Waterbirds also differ in their times to breed after reflooding. Those species whose nest sites merely require the presence of water can begin breeding straight away. Thus the ducks, herons, ibises, spoonbills and others that nest in trees standing in flood waters can nest as soon as flooding under the trees gives them protection from non-flying, non-swimming predators. However, those that nest in or on emergent or floating plants must delay until those plants have grown enough to provide nest materials, support or cover. More than turtles and frogs, waterbirds typically use many wetlands in a lifetime, depending upon their powers of flight to reach wetlands spread across ranges measuring hundreds or thousands of kilometres.

To generalize, wetlands offer these animals several resources: a physico-chemically buffered environment; security (against predators) provided by the water itself and supporting or hiding structures such as emergent or submerged plants; and food. Processes of flooding and drying directly alter some of these resources, and trigger secondary processes that alter others. Communities of biota in wetlands reflect both the recent history of flooding and drying of a wetland, and the longer-term regime of those processes that characterizes that wetland. The communities will also reflect availability of dispersing organisms in the environment and their ability to cross intervening matrix to reach that wetland. All of these may be altered directly or indirectly by land uses.

WHAT AGRICULTURE DOES TO WETLANDS

Agriculture can alter many wetland processes, of which we focus on water regime. Agriculture tends to use wetlands deliberately for either their water or soils. On the one hand, farmers may store water in wetlands to be extracted for irrigation or other water supplies, or they let stock into them to drink. To increase the predictability of stored water, the wetland may be dammed. On the other hand, if farmers want to grow crops or develop pasture predictably on the underlying wetland soil, rich in alluvium, nutrients and humus, they may first remove the "excess" water by draining. Once drained, alluvial wetlands are flat and easily cultivated and irrigated. Damming and draining are used to improve the predictability of the wetland's state that the farmers want for their planned production (Table 1).

Agriculture also accidentally affects wetlands because the landscape drains into them, carrying in soil, agricultural chemicals and crop and weed seeds. Sometimes farm waste (including chemicals) may be deliberately dumped in wetlands.

To obtain what they want from them, farmers often alter natural wetlands structurally, by deepening them to hold more water, or draining them to hold less. By altering them structurally, farmers also alter their temporal attributes, keeping them wet for a longer or shorter time than in their unaltered state, or altering the probability of their being wet or dry. Since water influx and efflux in many Australian wetlands are not constantly balanced, damming and draining inevitably alter fundamental processes of wetland flooding and drying, and the resultant dynamics of the plant and animal communities. However, the complete alteration of a wetland, to the detriment of its conservation value, may not always be necessary to achieve a farmer's aims.

By-products of agricultural use of wetlands include: compaction and pugging of soil and increases in soluble nutrients by stock grazing in or around the wetland, or drinking from it or wallowing and defecating in it; changes to vegetation through being eaten or trampled by stock (Payne 1992; Robertson 1997); and siltation of the wetland through run-off from agricultural land. The interaction of such by-products with altered water regime compounds the change and the consequences.

Increasing the predictability of agricultural benefits, by deliberately modifying wetland processes, often leaves the individual wetland more homogenous than it would naturally be and decreases wetland habitat diversity across the landscape. Homogenization may be as extreme as the physical homogenization of laser-levelling drained land for irrigation, or levee-building and damming to produce steep-sided, persistently deep water, or as subtle as the gradual biotic homogenization that occurs through community simplification.

Deliberate and accidental alteration of wetlands can both remove and add types of wetlands to the landscape. For example, the creation of a small, steep-sided dam for stock watering in almost every sheep paddock in Australia has added enormously to the availability of this type of wetland in the landscape. Unfortunately, most artificial wetlands, designed for agricultural purposes (especially irrigation storages and ditches), are structurally and biologically less complex and diverse than are natural wetlands in the same environment (Broome and Jarman 1983).

An example of changes to the diversity and distribution of wetlands across a landscape is provided by the historical impact of pastoralism on the wetlands of the high-altitude New England Tablelands of eastern Australia (Brock *et al.,* in press). Several (but not all) of the deeper, more persistently wet wetlands (locally called "lagoons") survive, although in the past most have been partly drained or dammed to provide water. However, the species-rich, temporary wetlands were seen by pastoralists merely as corners of paddocks that turned boggy in the rains, a haven for liver flukes and foot rot. Almost all have been drained and the remaining few are now wet for shorter periods (and hence unable to develop the full community that once they held) and also further apart (Brock *et al.* 1999). To the former array of natural wetlands have been added two new kinds: a few, deep, steep-shored water-supply reservoirs; and a plethora of small farm dams. The spatial attributes of the former system of wetlands have been altered, and the system's heterogeneity reduced.

Artificial wetlands do not compensate for the structural and biological diversity of natural wetlands lost through alteration for or by agriculture. Nor can they compensate for loss of locally endemic species from natural wetlands. A combination of homogenization of individual wetlands and artificial selection for or creation of particular wetland types has replaced a mosaic of heterogeneous wetland types in the landscape by a relatively less diverse system (Brock *et al.* 1999). The changes described here are not restricted to our New England Tablelands example. With local variants, these kinds of changes occur throughout agricultural Australia, and in New Zealand (Jay 1997).

IMPACTS OF WETLAND ALTERATIONS UPON THE BIOTA

Biota have inevitably responded to these alterations of structural, temporal, spatial and biotic attributes of individual wetlands and wetland systems across the landscape. Responses to structural change are obvious: biota adapted to shallow water may not flourish in artificially deepened water, for example. Responses to temporal change, however, may be at least as important for conservation. For example, human alteration to the pattern of flooding and drying in a wetland changes the frequency, timing, extent, duration and variability of flooding at both local and landscape scales (Tables 3 and 4).

Altering these hydrological processes eventually may diminish the farmer's ability to derive sustainable benefits from the land (Table 1). In many parts of Australasia, structural alteration has created direct or indirect problems for production, clean water and arable soils. Altering water regime to increase productivity in the short term may thus have long-term disbenefits for the farmer. By planned and moderate structural changes and wetland uses, nature conservation and agriculture may both be enhanced in the long term.

Increase in the spatial extent of flooding produces more extensive aquatic habitat, and this may encourage some plant species to become dominant (Table 3). Decrease in flooding extent creates more dry habitat in the matrix between neighbouring wetlands and decreases the amount of wet habitat in the landscape (Tables 3, 4).

Changes in the timing of attributes of flooding and drying have a marked influence on habitats and the biota occupying them. For example, artificial flooding at a time other than spring or early summer would have a severe impact on reproductive success in long-necked turtle *C. longicollis* populations in southeastern Australia. Some frog species whose breeding is triggered by rain would be disadvantaged if that rain failed to flood local wetlands for long enough for tadpoles to mature, because they had been drained. As described earlier, an artificial manipulation of water levels could profoundly affect reproductive success in local frog populations or whole communities (Humphries 1979).

Table 3. Predicted effects of human-induced change to a wetland process on the components of the wetland: the effects of human-induced increase or decrease in the frequency, extent, duration or variability of flooding on the wetland and its biota (adapted from Brock and Casanova 1997).

Alteration of process	Effects on components	
Attributes of flooding and drying pattern	Increase	Decrease
Frequency of flooding	Floods more often: • may encourage amphibious and fully aquatic species • has little effect on mobile species like turtles, some frogs and birds	Floods less often: • favours more terrestrial species and species with long distance mobility • selects species by their longevity in the propagule bank • flooding may be too short to complete life cycles to replace deaths in populations
Extent of flooding	More permanent-water and shallow habitat: • favours permanent-water species	More dry habitat: • reduces habitat diversity • favours deeper-water species
Duration of flooding	Longer flood events: • favour permanent-water species • favour domination of aquatic plant community by perennials, so reduce species and habitat diversity	Shorter flood events: • favour opportunistic species with short life cycles • cut short bird breeding and frog maturation, leading to massive juvenile mortality
Variability of flooding	More variable pattern: • decreases time scale of wetting and drying • favours flexible species which do not depend on particular length or timing of wetting • species richness may decrease • species with a persistent seed or egg bank will survive	Less variable pattern: • more predictable pattern • species richness and habitat diversity may decrease • favours species which prefer predictable flooding and drying • more permanently wet or dry habitat may favour most competitive species

Table 4. The predicted overall effect of human-induced change to flooding and drying patterns in wetlands at the local and landscape scale: increase or decrease of the frequency, extent and duration of water presence.

Alteration of process	Effects on components	
Attributes of flooding	Increase	Decrease
Local Scale	• each wetland more permanently wet • heterogeneity of habitats within a wetland is reduced • trophic complexity and species richness decrease • long-term survival of the full diversity of species depends on their resistant life cycle stages and dispersal abilities	• each wetland more permanently dry • more terrestrial (often weedy and colonizing) species encouraged • variety of aquatic habitats and species decreases • species survive that have long-lived resistant stages and disperse over longer distances
Landscape Scale	• more permanent and closer wetlands in the landscape • reduced heterogeneity of wetland types • long-term survival and maintenance of full biodiversity depends on species' abilities to disperse between and to recolonize wetlands, as well as to persist in individual wetlands	• fewer wetlands in the landscape; greater spacing between them • frog and turtle dispersal reduced • birds concentrated on fewer wetlands • change in availability of some wetland habitats • reduced heterogeneity • reduced species richness

Although plant species richness in wetlands is often maintained by the natural pattern of flooding and drying, plant communities may not be resilient to alteration to their flooding patterns. The species-rich suite of amphibious plants that occupies the edges of wetlands by tolerating or responding to water fluctuations (Brock 1986; Brock and Casanova 1997) will not compete well with submerged and terrestrial species under more prolonged flooding or drying (Table 3). This wet-meadow community, both while wet and when dry, importantly provides cover for adults and egg-laying sites for some frog species. If it is removed by heavy grazing or altered water patterns, some frogs will disappear along with the plants.

After a dry phase a wetland typically revegetates by germination from a species-rich and long-lived seed bank (Brock 1995, 1998). Across the landscape in New England (and elsewhere), wetland seed banks vary, with many species being specific to a single wetland and few species common to all sites (Brock, in press). The significance of this is that conservation of wetland species-richness in production environments cannot be limited to token wetlands. The landscape mosaic of wetlands with different wetting and drying regimes must be maintained to conserve the full array of species.

Draining and damming have reduced the number of types of wetlands in the landscape (Brock *et al.* 1999). The diversity of natural wetland types has been replaced in the pastoral landscape largely by one type of artificial wetland: the farm dam. Farm dams are often small, quite steep-sided, and often (not always) kept free of vegetation by stock. Many of these farm dams attract few waterbirds, and are not connected to other wetlands by surface flow; consequently they offer few avenues for dispersal for aquatic plants and non-flying invertebrates. At a landscape scale, the variety of habitats for water plants has been reduced, as has the plants' ability to disperse between wetlands. The influence of change in processes at the landscape scale is reflected in vertebrates as well.

Many waterbirds in Australia (ducks, herons, egrets, spoonbills, darters, cormorants) nest in vegetation in or over flooded wetlands, the standing water giving them security from terrestrial predators. They require standing water in the wetland for long enough to initiate courtship, mating, nesting and fledging of the young. While breeding attempts may be triggered by appropriate flooding, the success of the event will depend upon floods lasting long enough (Crome 1988; Briggs *et al.* 1997). Artificial shortening of flooding, by drainage of the wetland, will cut short the breeding event. Clearly there is a second level of temporal change. If floods that allow a successful breeding event happen too infrequently, because of drainage or river regulation, then each species may simply be unable to produce enough young to replace the losses since they last bred. Persistence of waterbird communities requires flooding of breeding sites for long enough and often enough to allow successful breeding frequently enough to maintain their populations.

Many frogs, even if the adults live out of the water, similarly require a wetland that stays wet long enough (one to 12 months) for their water-dependent tadpoles to mature. Draining both persistent and temporary wetlands may have accelerated the disappearance of frog species such as the rare Yellow-spotted Bell-frog *Litoria castanea* (or *flavipunctata*) from the New England Tablelands. Humphries (1979) reported rates of metamorphosis between zero and 2.7%, with mass mortality when wetlands dried out before tadpoles had metamorphosed. If draining decreases the length of time for which temporary wetlands hold water, it will inevitably lead to such desiccation in more years.

Changes to both temporal and spatial attributes of temporary wetlands may have affected Yellow-spotted

Bell-frogs. Not only are drained wetlands wet for shorter periods, those that hold any water at all are also further apart. Any wetland that loses its frog population through a local stochastic event is therefore less likely to gain another population through natural dispersal. For organisms with weak dispersal abilities and naturally small populations, an increase in the separation of appropriate wetlands dooms them to the classic problems of small and fragmented populations.

Even highly mobile wetland bird species may be affected by this combination of changes to the structural, temporal and spatial attributes of wetlands. Latham's Snipe *Gallinago hardwickii*, a key species protected by the Japan/Australia Migratory Bird Agreement, migrates between Japan and Australia. In the New England region it seasonally occupies temporary wetlands and tall, wet meadow communities on the edges of lagoons. Loss of persistent wetting of temporary wetlands through drainage, and loss of marginal wet meadow through increase in water depth of lagoons through damming, have much reduced the annual persistence, spatial extent, and dispersion of its habitat in historic times. Cattle grazing, and even worse sheep grazing in dry times, further reduce the vegetation cover needed by the birds.

Wetlands in any region of Australia are part of a continent-wide system of wetlands used by waterbirds, the highly mobile ducks, geese and swans in particular. The community of waterbirds seen on a wetland does not really "belong" to that wetland; they are at best temporary residents. Particular species may be present in a region persistently; others will appear at long intervals. This is illustrated by the occurrence of ducks and swans on six wetlands on the New England Tablelands, based on opportunistic records over the last 20 years (Table 5).

The lagoons in this sample range from ones that are more often dry than wet, to ones that dry for a few months perhaps one year in ten. Almost all waterbirds desert the lagoons when they are dry. However, one species, Australian Wood Duck *Chenonetta jubata*, is present persistently even when the lagoons are dry, because they associate with farm dams or stock water troughs. They have benefited (in terms of persistently available water) from pastoralism. Grey Teal *Anas gracilis* and Pacific Black Duck *A. superciliosa* return to lagoons within days, or even hours, of reflooding. Hardhead *Aythya australis* take a little longer to return. Black Swan *Cygnus atratus* take longer still and will not begin breeding until substantial aquatic plant biomass has developed. Musk Duck *Biziura lobata* and Blue-billed Duck *Oxyura australis*, species that dive for large invertebrate prey, may not return to the lagoons for many months or some years after they refill. The Australasian Shoveller *A. rhynchotis* and Pink-eared Duck *Malacorhynchus membranaceus* feed on dense plankton, which blooms some time after reflooding, depending upon water temperatures and nutrients. Their occurrence also reflects conditions elsewhere in Australia, and they visit the New England region usually when the inland is dry. The Whistle-ducks *Dendrocygna* species and Magpie Geese *Anseranas semipalmata* are tropical species, and the Chestnut Teal *A. castanea* a southern species, that are vagrants in the New England area. They illustrate waterbirds' powers of dispersal, and ability to colonize sites.

Such waterbird communities are obviously affected by any agricultural effects upon the persistence or frequency of flooding. They are also affected by availability of comparable wetlands in the landscape. The surviving wetlands are already only a fraction of those that existed before agricultural settlement; they are also only a selection of the range of types of wetlands previously available (Brock *et al.* 1999). Despite their high mobility waterbirds would reform communities less rapidly or readily if there were any further loss (and hence increase in separation) of these wetlands, or reduction in persistence of flooding. The diversity of the waterfowl community would also suffer if the wetlands became any more homogenous in structure and biota.

Persistent farm dams may be far less attractive than temporary wetlands for some seasonally

Table 5. Frequencies of presence (as number of years when birds were detected as present out of 20 years sampled opportunistically) of ducks, geese and swan on a set of six lagoons on the New England Tablelands, New South Wales, Australia.

Species	Frequency of presence	
Australian Wood Duck	20/20.	Dependent on dams, not lagoons
Pacific Black Duck	19/20.	Absent when lagoons dry
Grey Teal	19/20.	Absent when lagoons dry
Black Swan	19/20.	Absent when lagoons dry
Hardhead	18/20.	Absent when lagoons dry
Musk Duck	11/20.	Return 1–3 years after re-flooding
Blue-winged Shoveller	10/20.	Small groups, months after re-flooding
Blue-billed Duck	4/20.	Return unpredictably years after re-flooding
Chestnut Teal	2/20.	Small numbers, but stay for weeks
Pink-eared Duck	2/20.	During inland droughts
Plumed Whistle-duck	0.5/20	Large numbers, short periods; vagrants
Magpie Goose	0.1/20	Two birds, one month
Water Whistle-duck	0.02/20	One bird for one week

breeding frogs. For example, *Neobatrachus pictus* can occur at high density at ephemeral ponds, but only low density at farm dams (Roberts 1993) that presumably lack the seasonal pulse of productivity needed for tadpoles to thrive and grow fast.

In summary, agricultural alteration of wetland processes inevitably threatens components and, unless those diverse processes are conserved or restored, species of wetland-dependent plants and animals will be lost from individual wetlands and wetland systems across the landscape.

CONSERVATION CONTRIBUTIONS BY FARMERS

Creation of many small farm dams in the pastoral landscape has benefited a few species. Australian Wood Duck, which use the dams for loafing and security for young and adults, but feeds on improved pastures, have flourished. Some turtles may also benefit because they can disperse to such wetlands where they escape competition, especially from fishes. Long-necked turtles have been found at densities up to 160/ha in natural lagoons, but in farm dams at up to 240/ha in Gippsland (Chessman 1978) and 400/ha on the New England Tablelands (Parmenter 1976). These high densities probably reflect congregation of turtles in these permanent wetlands in the dry season (Georges *et al.* 1993). Nevertheless, permanently wet farm dams may maintain a well-dispersed metapopulation of long-necked turtles in a landscape that has otherwise lost its natural dispersion of a diversity of wetland types between which the species would have moved seasonally.

Agricultural effects on wetlands are not necessarily detrimental to conservation. For example, some types of wetlands would develop a floating plant community totally occupying the water surface if left undisturbed. This would greatly disadvantage birds of open water. Various disturbances (water-level fluctuations at the right frequency, for example) can maintain open water, as can moderate grazing and trampling in the flooded wetland by cattle. Low to moderate grazing and trampling can also help maintain species-rich wetland edge communities (Brock and Casanova 1997).

Many conservation values of wetlands do not require total cessation of agricultural activities. For example, if a wetland is being used to water stock, the detrimental effects of excessive stock trampling or feeding on aquatic plants can sometimes be avoided by fencing the stock out of all or most of the wetland, and piping water to troughs. Many farmers readily recognize the conservation value, not just of the water-filled wetland, but also of the wetland while dry (when dormant biota lie on the dry wetland bed) and of the habitats surrounding the wetland. Encouraging trees and other vegetation in, and fencing stock out of, a zone around the wetland may restore perching, roosting and nesting habitat for many wetland-associated birds, adult habitat, egg-laying sites and drought-refuge habitat (Osborne *et al.* 1996) for many frog species, and nesting sites for tortoises.

Many farmers are willing to use their earth-moving machinery and their skills to restore physical structure to wetlands, but may be less aware of the available techniques for restoring wetland vegetation. Many have tried to encourage waterbirds by building islands for roosting and nesting, and may have become disappointed when birds do not use them. This is often because the wetland as a whole is too steep-sided or deep to provide diverse habitats for aquatic plants to grow and encourage invertebrates, seeds and the birds that eat them.

Farmers can do simple things to aid conservation. For example, mosquito fish *Gambusia holbrooki* may locally damage frog populations by eating their tadpoles (Morgan and Buttemer 1996; Pyke and White 1996). *Gambusia* can be eradicated from a wetland merely by draining it and allowing it to dry completely, then allowing it to fill from a source free of *Gambusia*. Given such information, many farmers would willingly help restore frog populations and communities by actions on their own wetlands, especially if they know that theirs is part of a larger effort throughout a system of wetlands.

MANAGEMENT DECISIONS AND ACTIONS

The targets for wetland use for agricultural production are often local; the farmer is concerned with on-farm, or even within-paddock targets. Action to conserve wetland processes and components by one farmer is good and necessary, but will not be sufficient by itself to conserve wetland biotic diversity. Nature conservation relies on attributes at the landscape scale or greater. Some animals and plants may require the resources of more than one wetland to complete a life cycle or to reproduce without in-breeding. They need systems of inter-connectable wetlands to persist as populations. Thus the targets for conservation are at that same landscape scale. In addition, conservation responds to regional and national assessments of the status of species and communities. Consequently, its decision-making perspective has to be regional or national.

Local (and often private) interest in controlling wetland components or processes for agricultural production may conflict with regional or national (and often public) interest in conservation. Such conflicts are not well resolved by current laws controlling land-uses or conservation. Modern conservation laws now allow us to label particular ecosystems as requiring protection, but they do so most easily when the ecosystem can be described as occupying one or a few sites, and as being characterized by particular dominant plant species. An ecosystem that is as naturally fragmented and dispersed as wetlands, and each of whose fragments can exist in a number of different states, with ever-changing component species, is a regulator's nightmare. Selecting representative patches for conservation may be appropriate for organisms that live out their lives in just one wetland; but it will fail to conserve those organisms that in a lifetime use many wetlands, possibly for different purposes. Mobile organisms need a whole system of wetlands to be conserved in appropriate states, a point that drove the highly successful conservation programmes for seasonally migratory waterfowl in the United States and Canada.

Wetland types are not randomly represented in the different forms of tenure. Large and near-permanent natural wetlands are over-represented in the publicly owned conservation areas whereas small temporary wetlands are greatly over-represented on private lands (Corrick 1995; Brock *et al.* 1999).

When dispersed wetlands are owned by individual farmers or a variety of land-managing agencies, conservation has to be pursued by agreement rather than through purchase and ownership by the primary conservation agency. This is illustrated by the co-operation between private landholders and government agency to conserve the "easily drained shallow freshwater marshes" that are breeding habitat for rare Brolga *Grus rubicundus* in Victoria (McIntyre 1995). Freshwater wetlands formed the second greatest area (>4 000 ha) of any habitats conserved under voluntary agreements on 2 282 farms registered under Victoria's Land for Wildlife scheme in 1993 (Platt and Ahern 1995).

However, depending upon individual landholders to volunteer management of their wetlands for conservation will not guarantee a comprehensive and adequate system of conserved wetlands. National, regional and local needs for a heterogeneity of wetland types (e.g., representing a diversity of water regimes) must be researched and adequate systems for conservation identified. This process must aim to obtain a suitable dispersion of wetlands, and in some cases of connecting habitat between them; for example, wetlands that recharge by across-surface flooding must remain accessible to a source of floodwater. Ideal states for habitat on wetland margins and for the surrounding terrestrial matrix also need to be defined, especially for species such as turtles and frogs.

Wetlands representing all the types required in the system to be conserved should be identified on the ground. The conservation agency needs to identify alternative wetlands that could form such a system, and to select those that also happen to be owned by sympathetic landholders; this must be done by asking landholders to participate. Some will not want to; so the set of suitable wetlands may have to be reiteratively defined to ensure representativeness and connectedness in the selected system (Margules *et al.* 1988). Finally, the agency and the landholder should negotiate usage of the wetland that would satisfy both agricultural and conservation needs. Such wetland management might include restoring attributes (such as water regime or re-introduction of species). Whether agreement on use can be made legally binding and persistent, or reinforced by compensation will depend on circumstances and political will. We do not underestimate the skills and patience on all sides needed in this whole process.

We believe that agreement will follow recognition by farmers of the role of "their" wetland in the broader system, and by conservators of the legitimate needs of farmers. Above all, then, we see nature conservation in wetlands in the agricultural landscape as depending upon interchange of facts, ideas and ideals between farmers, scientists, and conservation managers.

ACKNOWLEDGEMENTS

We thank colleagues and students who have generated wetland data and insights, and those who discussed wetland management with us during UNE-supported study leave in 1997. We thank the referees who challenged us to extend our thinking. MAB thanks Land and Water Resources Research and Development Corporation for supporting her wetland research programmes.

REFERENCES

Briggs, S. V., Thornton, S. A. and Lawler, W. G., 1997. Relationships between hydrological control of river red gum wetlands and waterbird breeding. *Emu* **97**: 31–42.

Brock, M. A., 1986. Adaptation to fluctuations rather than to extremes of environmental parameters. Pp. 131–40 *in* Limnology in Australia ed by P. de Deckker and W. D. Williams. CSIRO. Dr W. Junk: Melbourne, Victoria.

Brock, M. A., 1991. Mechanisms for maintaining persistent populations of *Myriophyllum variifolium* in a fluctuating Australian shallow lake. *Aquat. Bot.* **39**: 211–19.

Brock, M. A., 1995. The role of seed banks in maintaining or rehabilitating vegetation in temporary wetlands. Final Report to Land and Water Resources Research and Development Corporation UNE7 Pp. 1–7. Australian Government: Department of Primary Industries and Energy Library, Canberra, Australian Capital Territory.

Brock, M. A., 1998. Are temporary wetlands resilient? Evidence from seed banks of Australian and South African wetlands. Pp. 193–206 *in* Wetlands for the Future: Proceedings of INTECOL's V International Wetlands Conference ed by A. J. McComb and J. A. Davis. Gleneagles Publishing: Adelaide, South Australia.

Brock, M. A. and Casanova, M. T., 1997. Plant life at the edges of wetlands; ecological responses to wetting and drying patterns. Pp. 181–92 *in* Frontiers in Ecology: Building the Links ed by N. Klomp and I. Lunt. Elsevier Science: Oxford.

Brock, M. A., Smith, R. G. B. and Jarman, P. J., 1999. Drain it, dam it: alteration of water regime in shallow wetlands on the New England Tableland of New South Wales. *Wetl. Ecol. Manage.* **7**: 37–46.

Broome, L. S. and Jarman, P. J., 1983. Waterbirds on natural and artificial waterbodies in the Namoi Valley, New South Wales. *Emu* **83**: 99–104.

Bunn, S. E., Boon, P. I., Brock, M. A. and Schofield, N. J., 1997. National Wetlands R & D Program Scoping Review. Land and Water Resources Research and Development Corporation Occasional Paper 01/97 Canberra, Australian Capital Territory.

Chessman, B. C., 1978. Ecological studies of freshwater turtles in southeastern Australia. PhD thesis, Monash University, Melbourne, Victoria.

Chessman, B. C., 1984. Food of the snake-necked turtle, *Chelodina longicollis* (Shaw) (Testudines: Chelidae) in the Murray Valley, Victoria and New South Wales. *Aust. Wildl. Res.* **11:** 573–78.

Corrick, A. H., 1995. Ownership of wetlands in Victoria and the occurrence of three species of waterbirds. Pp. 149–55 *in* People and Nature Conservation. Perspectives on Private Land Use and Endangered Species Recovery ed by A. Bennett, G. Backhouse and T. Clark. Transactions of the Royal Zoological Society of New South Wales: Mosman, New South Wales.

Crome, F. H. J., 1988. To drain or not to drain? Intermittent swamp drainage and waterbird breeding. *Emu* **88:** 243–48.

Driscoll, D., Wardell-Johnson, G. and Roberts, J. D., 1995. Genetic structuring and distribution patterns in rare southwestern Australian frogs: implications for translocation programmes. Pp. 85–90 *in* Reintroduction Biology of Australian and New Zealand Fauna ed by M. Serena. Surrey Beatty & Sons: Chipping Norton, New South Wales.

Georges, A., Limpus, C. J. and Parmenter, C. J., 1993. Natural History of the Chelonia. Pp. 120–28 *in* Fauna of Australia. Vol. 2A Amphibia & Reptilia ed by C. J. Glasby, G. J. B. Ross and P. L. Beesley. Australian Government Publishing Service: Canberra, Australian Capital Territory.

Georges, A., Norris, R. H. and Wensing, L., 1986. Diet of the freshwater turtle *Chelodina longicollis* (Testudines: Chelidae) from the coastal dune lakes of the Jervis Bay Territory. *Aust. Wildl. Res.* **13:** 301–08.

Humphries, R. B., 1979. Dynamics of a breeding frog community. PhD thesis, Australian National University, Canberra, Australian Capital Territory.

Jay, M., 1997. A short history of the landscape of central Waikato. Pp. 60–68 *in* Belonging to the Land, People and Places in the Waikato ed by E. Stokes and M. Begg. New Zealand Geographical Society, Waikato Branch, Department of Geography, University of Waikato: Hamilton, New Zealand.

Kennett, R. and Georges, A., 1990. Habitat utilization and its relationship to growth and reproduction of the eastern long-necked turtle, *Chelodina longicollis* (Testudines: Chelidae), from Australia. *Herpetologica* **46:** 22–33.

Littlejohn, M. J., Roberts, J. D., Watson, G. F. and Davies, M., 1993. Family Myobatrachidae. Pp. 41–57 *in* Fauna of Australia. Vol. 2A Amphibia & Reptilia ed by C. J. Glasby, G. J. B. Ross and P. L. Beesley. Australian Government Publishing Service: Canberra, Australian Capital Territory.

Margules, C. R., Nicholls, A. O. and Pressey, R. L., 1988. Selecting networks of reserves to maximise biological diversity. *Biol. Cons.* **43:** 63–76.

McIntyre, M. J., 1995. Conservation of the Brolga *Grus rubicundus* in Victoria: the role of private land. Pp. 166–69 *in* People and Nature Conservation. Perspectives on Private Land Use and Endangered Species Recovery ed by A. Bennett, G. Backhouse and T. Clark. Transactions of the Royal Zoological Society of New South Wales: Mosman, New South Wales.

Morgan, L. A. and Buttemer, W., 1996. Predation by the non-native fish *Gambusia holbrooki* on small *Litoria aurea* and *L. dentata* tadpoles. *Aust. Zool.* **30**(2): 143–49.

Osborne, W. S., Littlejohn, M. J. and Thomson, S. A., 1996. Former distribution and apparent disappearance of the *Litoria aurea* complex from the Southern Tablelands of New South Wales and the Australian Capital Territory. *Aust. Zool.* **30**(2): 190–98.

Parmenter, C. J., 1976. The natural history of the Australian freshwater turtle *Chelodina longicollis* Shaw (Testudinata, Chelidae). PhD thesis, University of New England, Armidale, New South Wales.

Payne, N. F., 1992. Techniques for wildlife habitat management of wetlands. McGraw Hill: New York.

Platt, S. and Ahern, L., 1995. Voluntary nature conservation on private land in Victoria — evaluating the Land for Wildlife programme. Pp. 211–16 *in* People and Nature Conservation. Perspectives on Private Land Use and Endangered Species Recovery ed by A. Bennett, G. Backhouse and T. Clark. Transactions of the Royal Zoological Society of New South Wales: Mosman, New South Wales.

Pyke, G. H. and White, A. W., 1996. Habitat requirements for the Green and Golden Bell Frog *Litoria aurea* (Anura: Hylidae). *Aust. Zool.* **30**(2): 224–32.

Roberts, J. D., 1993. Natural History of the Anura. Pp. 28–34 *in* Fauna of Australia. Vol. 2A Amphibia & Reptilia ed by C. J. Glasby, G. J. B. Ross and P. L. Beesley. Australian Government Publishing Service: Canberra, Australian Capital Territory.

Robertson, A. I., 1997. Land-water linkages in floodplain river systems: the influence of domestic stock. Pp. 207–18 *in* Frontiers in Ecology: Building the Links ed by N. Klomp and I. Lunt. Elsevier Science: Oxford.

Tyler, M. J. and Davies, M., 1993. Family Hylidae. Pp. 58–63 *in* Fauna of Australia. Vol. 2A Amphibia & Reptilia ed by C. J. Glasby, G. J. B. Ross and P. L. Beesley. Australian Government Publishing Service: Canberra, Australian Capital Territory.

Wiggins, G. B., Mackay, R. J. and Smith I. M., 1980. Evolutionary and ecological strategies of animals in annual temporary pools. *Archiv für Hydrobiologie Supplement* **58:** 97–206.

CHAPTER 27

Strategies for pond conservation in the agricultural landscape of North-west England

Andrew Hull[1]

This chapter explores ways in which landowners and land managers in the United Kingdom can be encouraged to maintain wildlife habitats in an intensive production environment and considers how the interests of farming and conservation might be reconciled. Specific reference is made to the work of the Farming and Wildlife Advisory Group which was set up in 1969 to promote the belief that modern agriculture and nature conservation need not be incompatible. With over 65 Farm Conservation Advisors working nationwide the organization is a prime believer in the notion of the voluntary principle; encouraging, supporting and advising farmers of the ways in which agricultural activity can be modified to meet the demands of nature conservation. An assessment is made of this approach. The chapter then considers a specific regional conservation initiative — the Pond *Life* Project — in which the Farming and Wildlife Advisory Group is one of 14 United Kingdom partners. The Pond *Life* Project, part-funded by the European Union, is seeking to develop a model to protect and manage small water bodies in the intensively farmed landscape of North-west England. The aims and objectives of the Project are examined, and the role of the Farming and Wildlife Advisory Group in the partnership is considered together with the benefits which accrue to both the Project and the organization.

Key words: North-west England, Farming and Wildlife Advisory Group, Ponds, Pond *Life* Project.

INTRODUCTION

Over many centuries, the landscape of lowland Britain has been transformed by the interaction of human activity and natural processes, that have been responsible for the evolution of an intricate and diverse patchwork that has come to characterize the British countryside. Throughout a long and ancient history, farming has been the essential ingredient of a rural landscape in which a multitude of agricultural practices have enhanced local distinctiveness: for example, field patterns, buildings, cropping systems and livestock. Physical changes to the shape and character of the landscape have also been instrumental in moulding regional landscapes. The drainage of the fenlands (waterlogged land) of East Anglia, for example, and the clearance of woodland for the grazing of sheep on the downlands (chalk uplands) of south-east England serve as a reminder of how change has come about. Inevitably, these changes have had an important impact on wildlife. In the case of the fens, drainage may be viewed as detrimental to the survival of wetland species; paradoxically, the creation of downland created an ideal opportunity for plants and animals suited to chalk habitats to gain an important foothold thereby adding to the nations' biota. Of course, when these changes were taking place, there was little or no appreciation of the impact that farming activities would have upon wildlife and habitat. In retrospect ancient agricultural practices, by and large, provided added-value — creating a range of new or previously under-represented habitats, rapidly exploited by adventitious plant and animal communities. By the beginning of the nineteenth century, however, the period of major landscape change had drawn to a close and small-scale, non-intensive mixed farming

[1]Liverpool John Moores University, 15–21 Webster Street, Liverpool, L3 2ET, United Kingdom.

Pages 269–77 *in* NATURE CONSERVATION 5: NATURE CONSERVATION IN PRODUCTION ENVIRONMENTS: MANAGING THE MATRIX ed by J. L. Craig, N. Mitchell and D. A. Saunders. Surrey Beatty & Sons, 2000.

predominated lowland Britain for the next 150 years. During this time, farming experienced a series of booms and slumps. From 1870 onwards the industry started to show a general decline, reaching its nadir during the Depression years of the 1930s, by which time many farms were abandoned and much of the countryside lay derelict.

Nineteen forty-seven marked a watershed in the history of British farming. The passing of *The Agriculture Act* meant that for the first time farmers could look forward to guaranteed prices and markets for almost everything they produced. Subsidies would be paid out of national taxation and would not be reflected in higher food prices. Whilst the benefits to both producers and consumers were immediately apparent, there were, however, significant environmental costs, including the transformation of a rural landscape which had evolved over the previous millennium. Today, the traditional British landscape has been significantly altered. The removal of hedgerows, increasing field size, the drainage of wetlands, water meadows and the destruction of woodland have simplified its complex structure. With the sudden loss of these traditional rural features, wildlife habitats were suddenly under threat and many have since disappeared, whilst those that remain have become increasingly fragmented. Alarmingly, species once considered common even 20 years ago are now themselves rare.

Nevertheless, farming based exclusively on increasing production is starting to be seriously challenged in the United Kingdom and there is an increasing awareness of the environmental impact of modern agricultural practices in general and upon wildlife in particular. This chapter will demonstrate the variety of ways in which nature conservation and production is being integrated in the United Kingdom and reviews the progress made through firstly, the introduction of a wide range of agri-environment schemes for which grant-aid is available and secondly, through the work of a specific farm conservation agency, working at the grassroots level, encouraging and supporting farmers to create space and place for nature on the farm.

TOWARDS MORE ENVIRONMENTALLY-FRIENDLY FARMING

As a result of increasing over-production and the continual need to maintain farm income there has emerged a range of agri-environment schemes in the United Kingdom which are supported by the Ministry of Agriculture, Fisheries and Food. Although accounting for only 5% of the total agricultural subsidy, they have been widely taken up.

The move towards environmentally friendly farming is being promoted via a disparate range of measures, both general and specific, directed towards: habitat; wildlife; countryside access and recreation; targeting specific environmental problem areas; and grant-aid.

In addition to the measures aimed at encouraging farmers to operate in a more environmentally friendly way, there has also been the publication of important legislation *(The Wildlife and Countryside Act, 1981)* and the formation of relevant agencies such as the Farming and Wildlife Advisory Group (established 1969) and, more recently, the Farming and Rural Conservation Agency (established 1997) which have been addressing some of the more important issues related to conservation and agriculture.

Today, there is a growing realization amongst the agricultural community that environmental management is becoming a necessary part of modern agricultural practice and that its adoption will be the only acceptable way of justifying continuing state subsidy of the industry. Increasingly, priorities, such as maximizing production, are being be reassessed by many members of the farming community and help and support from the organizations mentioned above represent a step in the right direction. Of these agencies, one of the most influential has been the Farming and Wildlife Advisory Group.

THE FARMING AND WILDLIFE ADVISORY GROUP

While nature reserves and other protected areas are clearly important for nature conservation, many species depend on areas outside designated conservation reserves, most of which is farmed and under private ownership. It follows, therefore, that farmers and landowners hold the key to effective conservation in the United Kingdom. If they can be encouraged to farm in an environmentally sympathetic manner, then the implementation of conservation initiatives will have a greater chance of success. For the farmed countryside, Davidson (1977), identified three short-term priorities: protection of the remaining natural or semi-natural habitats from agricultural intensification; the remedial management of agriculturally modified habitats to enhance their nature conservation value; and the creation of new elements to increase the ecological richness and scenic diversity of intensively farmed areas.

The task of informing farmers about the need to farm more sympathetically towards the environment has, for the past three decades, been championed by the Farming and Wildlife Advisory Group.

The Farming and Wildlife Advisory Group was set up by a range of government and non-government conservation and farming organizations to bring together those involved in farming and conservation to promote mutual understanding and co-operation. From the outset the Farming and Wildlife Advisory Group has been most closely associated with the third short-term priority identified by Davidson (1977) in the creation of new elements to increase the ecological richness and scenic diversity of intensively farmed areas. Arguably the least important from a conservation point of view it has been more than justified as an effective way of engendering the support and interest of the uncommitted farmer. More recently, the first two priorities of protection and management are being tackled with increasing enthusiasm.

The broad objectives of the Farming and Wildlife Advisory Group are, firstly, to raise farmers' awareness of the wildlife value of their land and, having achieved this, to suggest ways in which they can manage, sustain and improve upon the existing situation. In the first few years of operation in the early 1970s, the Farming and Wildlife Advisory Group was criticized by the environmental movement who viewed suspiciously, initiatives supported by the agricultural sector which changed the character of the British countryside. Furthermore, there was a commonly held view that much of what the organization achieved was little more than cosmetic; there being more to conservation than planting trees and dredging ponds. This view was compounded by the concern that re-creation of such secondary habitats led farmers to believe that they were doing the right thing as they continued to degrade important primary habitats.

Despite these reservations, the Farming and Wildlife Advisory Group was acutely aware of the criticism and as part of their mission emphasized the maintenance of existing landscape features is always more important than creating or restoring new ones (Farming and Wildlife Advisory Group 1982). Today the Farming and Wildlife Advisory Group has become an accepted and influential organization helping to change farmers' attitudes towards environmental management. Since the mid 1970s, a network of county Farming and Wildlife Advisory Groups has been set up to involve local farmers and landowners. With headquarters at the National Agricultural Centre in Stoneleigh, Warwickshire, there are now 65 Farming and Wildlife Advisory Groups across the United Kingdom (Fig. 1). The Farming and Wildlife Advisory Group's Technical Strategy Support Group, comprising leading academics in ecology, economics, arable and livestock production, meets bi-annually to provide the best strategic views on environmental priorities within the farming industry.

Contact with the farming community is impressive. The organization averages over 12 farm visits daily and has established over 30 000 contacts during the past 15 years. In 1996 the Farming and Wildlife Advisory Group advised on over 42 000 conservation issues from Scottish heather, moorland, grassland and wetland management to English hedges, hedgerow trees, scrub and ponds.

The underlying principle of the Farming and Wildlife Advisory Group has been achieved through promoting the belief that conservation and modern farming need not be incompatible and that with the provision of appropriate advice and encouragement, the loss of wildlife habitats through agricultural intensification can best be ameliorated by encouraging farmers to modify their practices. Through its national network of advisers, the Farming and Wildlife Advisory Group encourages landowners and land managers to incorporate business practices and working patterns that are sustainable both from a commercial and a conservation point of view.

Unlike state funded agri-environment schemes, the Farming and Wildlife Advisory Group is a prime expression of the voluntary principle in conservation (Cox *et al.* 1990). This has three components. These are: to stimulate and broadcast amongst farmers and landowners a social ethic concerning stewardship of the countryside; to promote an ideological defence of the autonomy of farmers and landowners from statutory controls, through an emphasis on the paramount need to retain their good-will and voluntary co-operation if workable remedies to conservation problems are to be found; and to provide an unusual but effective bottom-up approach to countryside management.

The Wildlife and Countryside Act 1981, stimulated a growing awareness of the environmental impact associated with modern farming practices and this, together with a growing concern over agricultural surpluses, generated conditions conducive to the emergence of a farming and conservation advisory agency.

The Farming and Wildlife Advisory Group filled the niche and immediately started to resist the encroachment of statutory controls over damaging agricultural activities. For the first time an organization existed which enabled farmers and conservationists to enter into effective dialogue, without conflict. Of course, the voluntary principle is not without its critics. Whilst some conservationists would say that it is little more than an ideological defence of the autonomy of the farming community, a more positive assessment would suggest that such an approach embraces a social ethic concerning the stewardship of the countryside.

THE FARMING AND WILDLIFE ADVISORY GROUP — TWENTY FIVE YEARS ON

The task of changing farmers' attitudes towards the environment is difficult, particularly if there is not the cash incentive for farming in a more sustainable way. Whilst there are encouraging signs from Brussels regarding the reform of the Common Agricultural Policy and the need to increase and extend agri-environmental subsidy significantly, the next two decades will be vital if we are to be assured of having any landscape worth conserving. The Farming and Wildlife Advisory Group, and similar organizations must avoid the criticism that they are doing little more than encouraging isolated opportunities or, more controversially, wildlife gardening.

According to Potter (1985), to have the greatest impact upon conservation, the Farming and Wildlife Advisory Group has to "lock into farmers' decision

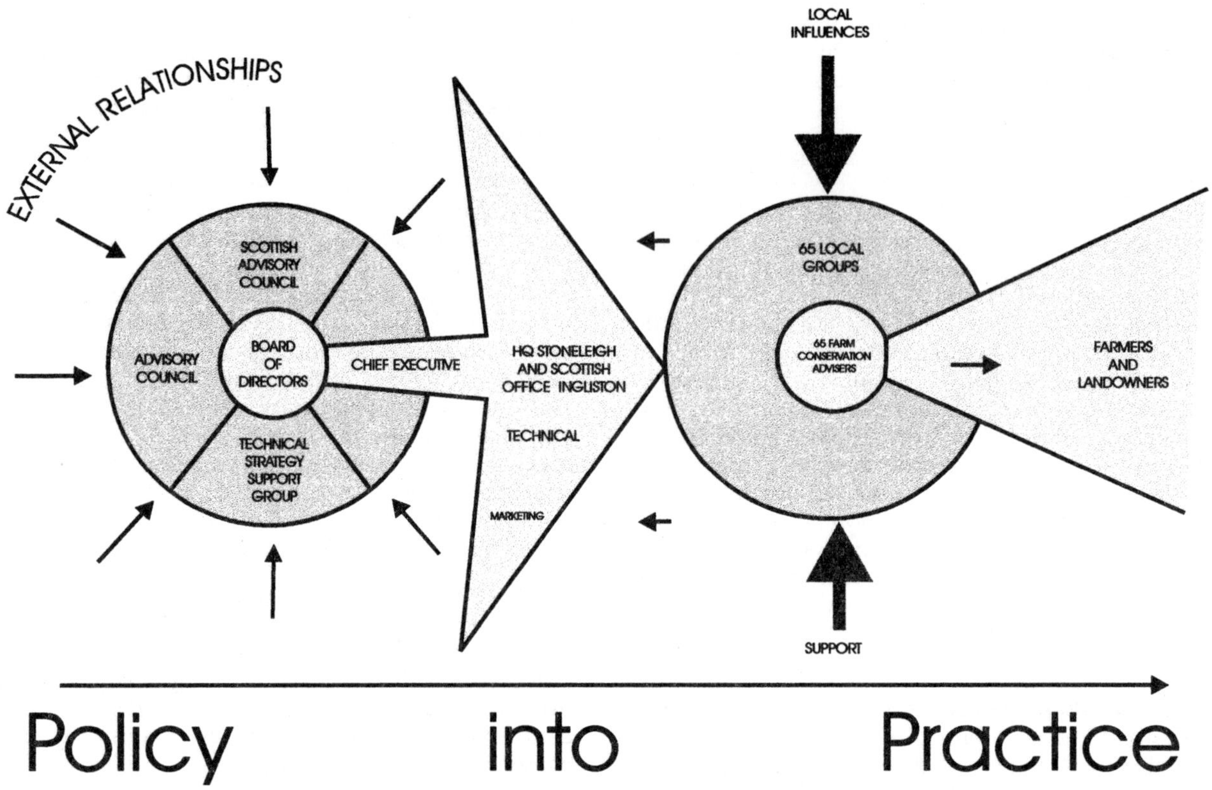

Fig. 1. Organizational structure of the Farming and Wildlife Advisory Group.

making at an early stage" for without this "their advice is limited to amelioration and enhancement once improvement has run its course." In an attempt to address this issue, the Farming and Wildlife Advisory Group has been instrumental in the development of Whole Farm Conservation Planning which considers all aspects of farm management and the farming system including: wildlife habitats; landscape features; environmental management (fertilizer and pesticide use, soil erosion, energy and water conservation, pollution control); and the economic management of the farm.

The Farming and Wildlife Advisory Group has devised a methodology for looking at conservation across the whole farm and developing it year by year. Landwise is a new initiative which gives farmers the opportunity to ensure that their farm is meeting this environmental challenge and making a positive contribution to national biodiversity targets.

Landwise offers: the opportunity for farmers to develop their own programme at their own pace; practical guidance on tackling priority issues; expert assistance from a network of qualified local advisers; advice on grant applications and incentives; and professional help at a low cost.

Farmers can choose any one of three Landwise options which is prepared and delivered by the local Farming and Wildlife Advisory Group contact. These include the Landwise Report, the Landwise Review and the Landwise Plan. The Report is the baseline study, the Review is for farmers wishing to update their plans, or develop a rolling programme of environmental improvement and the Plan is for the farmer who wants a highly detailed analysis and plan of action.

Agricultural land covers 80% of the United Kingdom. The stakeholders in this are the farmers and landowners who manage and farm the land, organizations whose policies relate to agriculture, the environment, and all consumers. The needs of each of these groups will be reflected in the advice given by the Farming and Wildlife Advisory Group and in future policy. Primarily, however, the Farming and Wildlife Advisory Group exists to help farmers, most of whom are aware of their unique and pivotal position, to run their businesses in a way which combines economic sense with environmental sensitivity. Nevertheless, the Farming and Wildlife Advisory Group remains committed to the local level and this will now be examined with reference to work being undertaken in North-west England to protect and manage the region's threatened pond landscape.

PONDS IN THE LANDSCAPE OF LOWLAND BRITAIN

Whilst the traditional village pond has served a multitude of purposes in the past, ponds in the wider agricultural landscape have had a more recent origin which can be traced back to the 18th and 19th centuries. Recently, the 19th century has been described as the "golden age of ponds" (Oldham and Swan 1997) when a whole series of industrial processes and agricultural practices required the excavation of raw materials to support the demands of a rapidly industrializing economy and a burgeoning population. Dependent largely upon draught animals and human labour, the extraction of minerals, such as brick and clay, left behind hundreds of thousands of small pits which, when exhausted, rapidly filled with water. By the turn of the century, many counties in England were characterized by a landscape containing tens of thousands of these small bodies of water, providing an opportunity for colonization by wetland plants and animals.

Ponds, unlike many other semi-natural habitats such as hedges and woodland, are temporary features in the landscape and their existence is relatively short-term unless vegetational succession is kept in check. Recent farming practice has largely ignored or found the pond an inconvenience on intensively-farmed land. Even on the mainly pastoral land of Cheshire and parts of Lancashire in North-west England, agricultural intensification has resulted in considerable pond loss despite the continuing demand for the watering of livestock (Hull and Boothby 1995).

The piping of clean water to the majority of fields has led to many field ponds being fenced-off. Without management, the transition from open water to dry land will occur, but the time-scale of this process varies with physical, biotic and human factors. On average, in lowland Britain, a typical pond, containing water for at least four months of the year, may exist for 200 years. From this, it follows that without intervention, the pond resource will soon be lost. Work in Cheshire, a county with the greatest pond density in England and Wales, has demonstrated that if current rates of loss continue, few ponds will remain in the agricultural landscape by the middle of the next century (Boothby and Hull 1997).

Furthermore, apart from the handful of ponds specifically assigned some form of conservation designation (such as Sites of Special Scientific Interest, Scheduled Ancient Monuments, Local or National Nature Reserves) there is no specific protection available, though certain pond-related species (such as the Great Crested Newt *Triturus cristatus*) do receive a measure of protection under *The Wildlife and Countryside Act 1981*.

The days have passed when farmland ponds were filled in as a matter of course. However, their continued existence is further threatened by current thinking on pond management which favours letting these semi-natural features proceed to their climax rather than maintaining a seral community (Biggs *et al.* 1994). Whilst acknowledging the validity of such an argument, if this formula is adopted without significant pond creation the resource will have largely disappeared by the middle of the next century. Herein lies a problem, not encountered to the same degree when considering other farm landscape features. Hedgerows, water meadows and ditch systems have traditionally been managed whilst ancient woodland has, by definition, time on its side. For field ponds, however, without management and without significant pond creation those that remain in the farmed landscape will be lost by the middle of the next century, together with the plants and animals that live in and around these small wetland habitats.

THE POND *LIFE* PROJECT

North-west England contains lowland Britain's densest pond landscape. However, little detailed information on the region's pond resource existed and comprehensive and precise locational and status information were unavailable except for a few biologically interesting sites. Importantly, existing conservation actions had tended to treat ponds on a site-by-site basis, a process exacerbated by rescue conservation initiatives.

In order to address this issue, a partnership of 14 organizations in North-west England are participating in the European Union funded Pond *Life* Project. This four year, £1 million project, aims to raise awareness about the region's nationally important pond landscape, the threats that it faces, and the immense wildlife value associated with these small bodies of water.

In distinct contrast to the work previously undertaken in the region, the Pond *Life* Project has adopted a strategic, landscape approach to pond protection, developing a project which has stressed the need to: develop a comprehensive database of sites to allow an overall review of the status of the resource; develop actions for the *pondscape* as well as for individual ponds; and create new ponds to regenerate the resource, but otherwise respect the ecological integrity of ponds at various stages of vegetational succession.

From its inception in 1995, the Pond *Life* Project, brought together a partnership of local authorities, nature conservation organizations, regulatory agencies, higher education and agri-environment organizations (Fig. 2). The distinctive ethos of the project is based upon six main considerations. These are: the need for a pro-active model for pond conservation; the scale of the problem — too vast to be left to experts and legal conservators; the need to develop grassroots skills and expertise; to develop a series of self-help initiatives; to grasp on-going changes in farming and farming policy resulting from the reform of the Common Agricultural Policy of the European Union; and to nest the project firmly within local communities.

Two Regional Pond Networks, comprising a broad membership of organizations whose remit impacted upon the pond landscape were set up. They brought

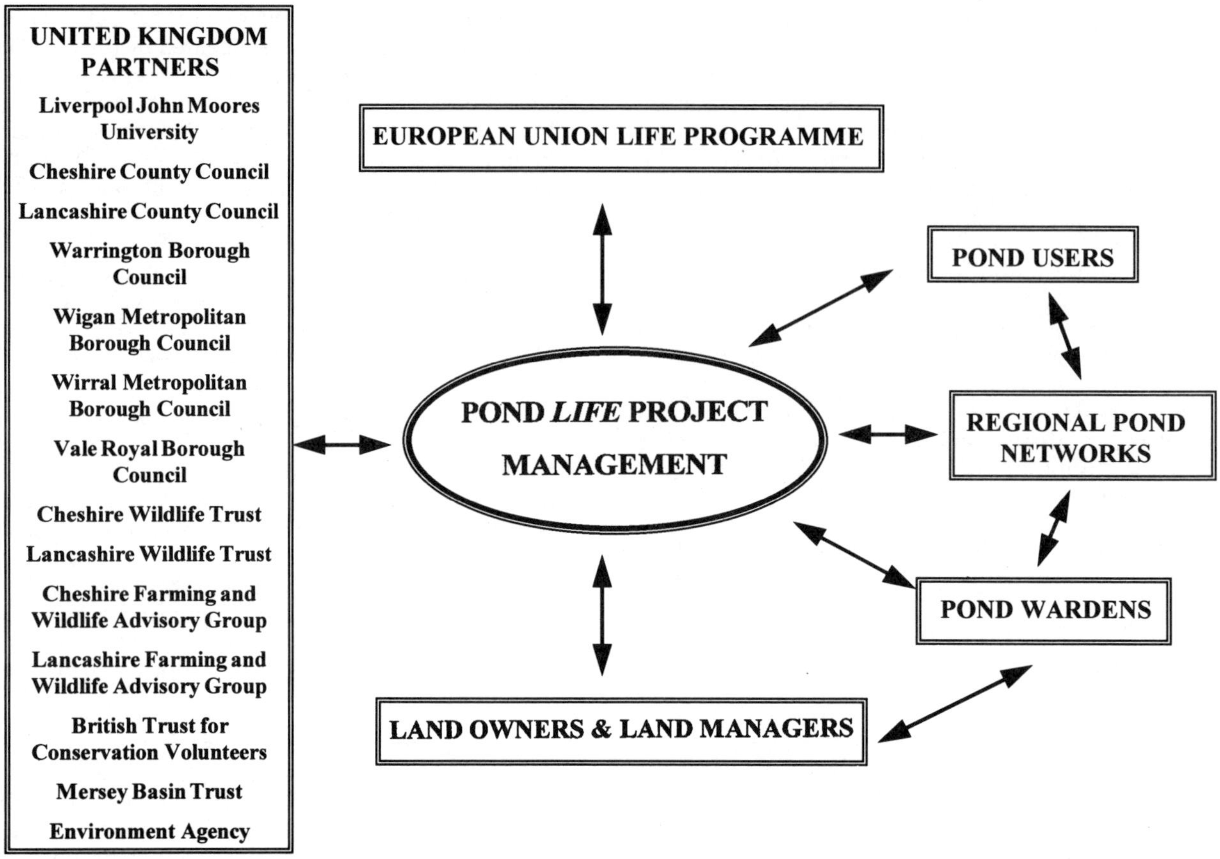

Fig. 2. Organizational structure of the Pond *Life* Project.

together agricultural interests and nature conservation organizations to discuss pond related issues and the ways that progress might be achieved in protecting these landscape features. At the same time, the Pond *Life* Project launched a community pond warden scheme throughout North-west England, encouraging people to get involved in pond protection in their locality. The pond warden occupies an intermediate role between the Project team and the farming community. To help them in their task, a series of six workshops on various aspects of pond ecology are held annually and they have been a major attraction for over 250 individuals and groups who are now participating in the scheme.

From the outset the message was not one of confrontation, but of reconciliation and respect. With this knowledge and, in many cases previous expertise, pond wardens were encouraged to get to know their local land owners and land managers, to gain their trust and friendship, to seek permission for access, to demonstrate the wildlife value of ponds and to start talking constructively about ways in which ponds on their land could be protected and managed and new ponds could be created.

This work has been supported by staff members of the Pond *Life* Project based at the project headquarters at Liverpool John Moores University. Two Pond Community Officers provide the link between the activists and the agricultural community. Workshops for both farmers and pond wardens, both in the classroom and in the field, have been extremely valuable for raising awareness about pond loss, and building friendship links.

To demonstrate the wildlife value of farmland ponds it was essential to undertake a biological survey of the resource. Although it was known from other studies that ponds were a rich wildlife habitat, local evidence was vital, to prove they were worth conserving. It was decided to carry out a detailed biological investigation of 250 ponds each year over the life of the Project. The results have proven conclusively that the farmland ponds of North-west England contain a great diversity of life forms (Guest 1997).

This biological record also provides data for the pond warden, to convince even the most uninterested of observers. Furthermore, it has demonstrated that a congregation of ponds is vital for many pond species and that their long-term persistence requires a comparatively high density of ponds. Increasing distances between ponds are leading to a functional fragmentation of the landscape, in which individual, isolated sites and their associated biota are particularly vulnerable (Ebenhart 1991). Moreover, this research has shown that although any individual pond may contain rare species, the diversity of the overall pond resource may perhaps best be protected by helping the pondscape survive. From these findings it is increasingly important to raise awareness about the pondscape rather than the individual pond. To do this effectively means raising the awareness of the principal stakeholders namely, the agricultural community.

THE FARMING AND WILDLIFE ADVISORY GROUP AND THE POND *LIFE* PROJECT

In practical terms, the development of actions to protect and manage the pond resource of North-west England (and elsewhere in the United Kingdom) is guided by several considerations. These are: the need to address pond conservation on private land; the need to identify and work with those organizations and agencies having relevant influence; the need to identify local action foci for a much wider regional problem; and to embrace and encourage the voluntary principle in conservation action.

The Farming and Wildlife Advisory Group was invited to join the Pond *Life* Project partnership. The presence of so many farmland ponds in North-west England, was already keeping both the Farming and Wildlife Advisory Group officers in the region extremely busy. Requests for pond advice often provided both Farming and Wildlife Advisory Group officers the opportunity to discuss pond conservation and the potential for developing other wildlife habitats on the farm.

THE BENEFIT OF PONDS TO THE FARMING COMMUNITY

It is well established that the British public has a strong affection for ponds. Over the centuries, ponds have played an important part in not only our economic, but also our social and cultural history (Probert 1989). It is not surprising, therefore, that ponds are still regarded by many members of the local farming community as special places even though they no longer serve any real economic value. Furthermore, ponds do have certain attributes which make them an ideal feature to demonstrate the benefits of wildlife conservation on the farm. These attributes are described below.

Immediacy — Any conservation activity which quickly produces dividends is highly desirable. New ponds can be created rapidly and will be colonized by plant and animal species from the beginning. In particular, high-profile taxa such as amphibians, dragonflies and wading birds favour new ponds.

Aesthetic — Ponds, in common with other water features, are attractive features of the landscape and have played an important part in rural culture and contribute to local distinctiveness.

Diversity — Biological survey of the pond resource of North-west England has clearly demonstrated the diversity of life forms associated with small water bodies in the region. Knowledge that a pond or particular pond type contains rare or endangered species can encourage appropriate management.

Visible — Ponds designed for wildlife are an attractive feature and their high visibility can be influential in persuading neighbours to adopt a more active approach to pond conservation and management.

Cost-effective — Creating new ponds and managing existing ponds for wildlife need not be expensive. Evidence has demonstrated that small, shallow bodies of water are as dynamic as ponds of a far greater size.

Self-regulating — Ponds, like many other landscape features have the ability to be self-regulating. Whilst hedgerows require annual maintenance, ponds can, by and large, be left to develop naturally. If management is to be undertaken, it can be carried out every fifth year or less frequently.

Local community support — Through the Pond *Life* Project there is an enthusiastic and growing workforce comprising local community volunteers. Many of these volunteers are now experienced in best management practices and are willing to help and advise wherever and whenever is necessary in their locality.

Potential economic value — There remains a largely unexploited economic value associated with small bodies of water. Ponds for angling, wildfowling or reservoirs can, through appropriate design, enhance overall wildlife value.

THE BENEFIT OF THE POND *LIFE* PROJECT TO THE FARMING AND WILDLIFE ADVISORY GROUP

Already the influence of the Pond *Life* Project on the pond landscape of North-west England has been far-reaching. Bringing together all stakeholders, through the Pond *Life* Project partnership and the Regional Pond Networks, has raised the issue of regional pond loss and many of the partners and network members, particularly the local authorities, have been instrumental in adopting a no-net-loss policy when giving planning permission for development (for example, Cheshire County Council). For other members, the biological survey results have provided vital information about the wildlife value of the resource; information which can be used to influence decision makers and, indeed, farmers themselves. For both Farming and Wildlife Advisory Group officers in the region there are benefits that include the following.

Best practice advice — Specialist advice is available through the Pond *Life* Project ranging from design, creation, restoration and management through to the most appropriate places to site new ponds. By the end of the Project in 1999 outputs will include a Best Practice Manual and accompanying video production which will illustrate pond management practice. Both outputs will be made freely available to farmers in the project area.

Limited grant-aid — Limited funding has been made available to community pond projects. In addition, there currently exists a wide range of community grant-aid for environmental projects, many of which have guidelines which do not exclude the farming community.

Raising awareness — Pond *Life* Project officers are involved in raising awareness about the issues confronting the region's pond landscape and handing out appropriate management advice. Farm visits account for a significant and growing number of contacts.

Biological survey results — Field survey is a vital part of the Pond *Life* Project. Without the co-operation of the farming community, this task would have been significantly more difficult to undertake. The survey work is undertaken with the results from each pond entered into the Project Geographic Information System and a copy of the survey results, together with a vegetation map, photographs and management advice about the pond, returned to each farmer. Additionally, pond management advice is outlined where appropriate.

Seminars and workshops — The Pond *Life* Project, in conjunction with the Farming and Wildlife Advisory Group, has organized a number of pond related events specifically geared specifically for the farming community both in the field and in the classroom.

Foot-in-the-door — Contact with the Pond *Life* Project often provides an initial "foot-in-the-door" and the Project officers will, where appropriate, comment on other areas of the farm where there may be a nature conservation gain.

NATURE CONSERVATION IN THE PRODUCTION ENVIRONMENT

As elsewhere in the developed world, the United Kingdom is capable of producing more than enough food now and in the foreseeable future. According to Green (1993), creative nature conservation can "... offer the opportunity to restore ecosystems and re-establish wild areas." providing an opportunity to, "... design, create and maintain new landscapes."

An opportunity not only to protect existing wild areas, but also to abandon land from farming and in so doing enhance its nature conservation value. He further suggests that there is the potential for restoring the 25% of semi-natural vegetation lost to post-war agricultural reclamation, and to create new wilderness areas from marginal farmland no longer needed for food production.

Such ideas are not new, however, as long ago as 1969 the Agricultural Commissioner of the European Community argued strongly that as much as 5 million hectares of farmland should be diverted to forestry, recreation and other uses. Over a quarter of a century later we have reached a situation whereby agricultural surpluses are no longer tenable and reform of the Common Agricultural Policy is inevitable. Quotas, set-aside and a series of measures designed to encourage the diversification of farm enterprises have been the main means of attempting to control production. In addition some subsidy has been switched from production to environmentally friendly farming. With an agricultural support budget of between £2 and £3 billion, about £100 million has been transferred to countryside management schemes every year this decade. Although a seemingly small amount of the Ministry of Agriculture, Fisheries and Food total budget it is, nevertheless, a significant injection of money into the environmental budget and compares to the government's financial support for both English Nature and the Countryside Commission (now the Countryside Agency).

Whilst such information may be reassuring to some, the stark reality remains that both species and habitats are still under threat. To pin hopes upon a "greener" Common Agricultural Policy may well be the solution in the long term; but what happens in the short term, the next few years or even decades? In order to safeguard our natural heritage, action must continue and, wherever possible, be extended at the grassroots level, talking, meeting, informing and persuading land owners and land managers of the benefits of environmentally-friendly farming. Evidence from the Farming and Wildlife Advisory Group over the past quarter of a century and from the more recent Pond *Life* Project has proven beyond doubt that the farming community is receptive, will listen and, more importantly, will act upon sound nature conservation advice. If this approach continues then the stage is set for a more sustainable agricultural policy in the United Kingdom where farming and conservation can be reconciled; in which wildlife habitats and amenity landscapes can be restored; where lower-input systems of agriculture can be introduced; and where habitat and countryside management can be an integral part of the intensive production environment.

ACKNOWLEDGEMENTS

Acknowledgement is given to the *Life* Programme of the European Union and to members of the Farming and Wildlife Advisory Group, in particular Richard Knight at Stoneleigh and Doug Hill, the Farming and Wildlife Advisory Group officer for Cheshire. Needless to say, the views expressed in this chapter are those of the author alone.

REFERENCES

Biggs, J., Corfield, A., Walker, D., Whitfield, M. and Williams, P., 1994. New approaches to the management of ponds. *Brit. Wildl.* **5**: 273–87.

Boothby, J. and Hull, A. P., 1997. A census of ponds in Cheshire, North-west England. *Aquat. Conserv.: Mar. Freshw. Ecosys.* **7**: 75–79.

Cox, G., Lowe, P. and Winter, M., 1990. The Voluntary Principle in Conservation: A Study of the Farming and Wildlife Advisory Group. Packard: Chichester.

Davidson, J., 1977. The prospects for action. Pp. 231–42 *in* Conservation and Agriculture ed by J. Davidson and R. Lloyd. Wiley: Chichester.

Ebenhart, T., 1991. Colonisation in metapopulations: a review of theory and observations. *Biol. J. Linn. Soc.* **42**: 105–21.

Farming and Wildlife Advisory Group, 1982. Farming and Wildlife Advisory Group Guide to Priorities: Wildlife on the Farm. Farming and Wildlife Advisory Group: Stoneleigh.

Green, B. H., 1993. The recognition and implementation of landscape management objectives for agriculture in the UK. Pp. 259–66 *in* Nature Conservation 3: Reconstruction of Fragmented Ecosystems ed by D. A. Saunders, R. J. Hobbs and P. R. Ehrlich. Surrey Beatty & Sons: Chipping Norton, New South Wales.

Guest, J. P., 1997. Biodiversity in the ponds of lowland North-west England. Pp. 49–58 *in* British Pond Landscapes: Action for Protection and Enhancement ed by J. Boothby. Pond *Life* Project: Liverpool.

Hull, A. P. and Boothby, J., 1995. Networking, partnership and community conservation in north-west England. Pp. 373–84 *in* Nature Conservation 4: The Role of Networks ed by D. A. Saunders, J. L. Craig and E. M. Mattiske. Surrey Beatty & Sons: Chipping Norton, New South Wales.

Oldham, R. and Swan, M., 1997. Pond loss and amphibians: a historical perspective. Pp. 3–16 *in* British Pond Landscape, Proceedings of the UK Conference of the Pond *Life* Project ed by J. Boothby. Pond *Life* Project: Liverpool.

Potter, C., 1985. From little acorns? The role of FWAG in the conservation of lowland landscapes. *Ecos* **6**(3): 6–10.

Probert, C., 1989. Pearls in the Landscape: The Conservation and Management of Ponds. Farming Press: Ipswich.

CHAPTER 28

Incorporating stream health into New Zealand hill-land farm management

John M. Quinn,[1] Kevin J. Collier[1] and Bruce Thorrold[2]

Conversion of hill-land use from native forest to pastoral farming alters a variety of fundamental stream habitat characteristics, including flow regime, channel morphology, light input, thermal regime, carbon sources, nutrient and sediment inputs. Fairly consistent patterns of change in stream fish and invertebrate biota have been associated with these land use related changes in different regions of New Zealand, some of which are perceived as undesirable. Avoiding, remedying or mitigating undesirable effects of farming on off-site ecosystems is an important aspect of sustainable land use as interpreted by the Resource Management Act. This chapter summarizes the effects on streams of farming at Whatawhata Research Centre, 20 km west of Hamilton, New Zealand. It reviews potential land management practices to enhance nature conservation values in pastoral streams and downstream aquatic ecosystems, and describes the approach being taken by the Whatawhata Catchment Management Group to (i) develop instream nature conservation goals for a well-managed, production, hill-land, catchment, and (ii) resolve conflicts between these goals and broader, economic, social and sustainability goals. There is potential for imaginative and practical options for land, vegetation and animal management to improve the ecological performance of hill-land catchments and aquatic ecosystems, in particular, as well as provide economic or farm performance benefits to farmers. Our experience to date indicates that active involvement with farmers is the key to realizing these benefits.

Key words: Sustainable agriculture, Riparian management, Streams, Benthic invertebrates, Fish, Water quality, Nutrients, Grazing, Multiple goals, Farmers, Economics, Conflict resolution.

INTRODUCTION

HEALTHY streams provide a range of goods and services that are valued by society. These include clean water for domestic and industrial use, detoxification and decomposition of catchment inputs, uncontaminated foods, conveyance of flood waters, sites for swimming, contemplation and spiritual renewal, and areas for conservation of indigenous biota (Meyer 1997). However, many of these attributes can be compromised by land use practices associated with production of food and fibre. Despite successful efforts to reduce point source pollution of waterways, there is clear evidence that water quality in New Zealand is often below the standards required for recreational and industrial use and the support of healthy aquatic ecosystems (Smith *et al.* 1993). Recently the New Zealand government produced a strategy for Sustainable Land Management which identified agricultural impacts on aquatic ecosystems as one of the three priorities for remedial action to achieve its desired outcomes (Ministry for the Environment 1996).

New Zealand's *Resource Management Act (1991)* promotes sustainable management of resources, to provide for the well-being of people now and in the foreseeable future, while "avoiding, remedying or mitigating any adverse effects on the environment" (subsection 5(2)(c)). With regard to adverse effects of land use on streams, New Zealand Government policy is to encourage land users to adopt practices

[1]National Institute for Water and Atmospheric Research Ltd, P.O. Box 11–115, Hamilton, New Zealand.
[2]AgResearch Ltd, Private Bag 3123, Hamilton, New Zealand.

Pages 278–89 *in* NATURE CONSERVATION 5: NATURE CONSERVATION IN PRODUCTION ENVIRONMENTS: MANAGING THE MATRIX ed by J. L. Craig, N. Mitchell and D. A. Saunders. Surrey Beatty & Sons, 2000.

voluntarily that improve stream health and water quality, rather than imposing regulations. This means land users need to balance stream protection measures with management actions to meet other multiple goals including farm financial performance, compliance with consumer and regulatory requirements, maintaining asset values and living a satisfying lifestyle. It is therefore important that technically-oriented management options, which will lead to improved stream health, are considered in the wider context of their impacts on the other goals of land users. Within the environmental goals there is also a need to consider wider effects, such as sedimentation or eutrophication of downstream water bodies, control of weeds and pests, and conservation of terrestrial species.

This chapter describes changes to New Zealand stream ecosystems resulting from agricultural use of hill-land catchments, and discusses approaches for restoring nature conservation values, and overall health, of hill-land pastoral streams while meeting the other goals of land users. We describe an approach currently being evaluated to resolve conflicts between the multiple goals that the community has for hill-land catchments, and the statutory requirement to avoid, remedy or mitigate adverse effects of pastoral land use on streams. This work is part of a research programme in the Mangaotama Catchment, at Whatawhata Research Centre (175°03'E, 37°47'S), 20 km west of Hamilton, in the Waikato region of the North Island, New Zealand that seeks to develop and test innovative management options to improve aquatic and terrestrial ecosystem health within the constraints of a production-oriented farming operation.

Land use change in New Zealand

National scale — Following the onset of European colonization in the 1840s, large tracts of lowland and hill-land forest were cleared for pastoral farming (Fig. 1). There are now over 13 million hectares, or around 50% of New Zealand's land surface, converted to grazing land compared with 1.54 million hectares in production forest and 91 thousand ha in horticulture (Department of Statistics 1992; Ministry of Forestry 1997). Agriculture is now the dominant land use in the middle and lower catchment areas of most of New Zealand's streams and rivers. Twenty-nine per cent of river channel length distinguished on 1:250 000 maps of the North and South Islands flows through pastoral land (Table 1), with the proportion greatest in the intensively farmed Hawke's Bay, Manawatu-Wanganui, Taranaki and Waikato regions where more than 40% of larger river channel length flows through pasture.

Development of land for agriculture has seen the introduction of around 60 million grazing animals (mostly cattle and sheep) which produce an estimated daily biochemical oxygen demand (BOD_5) loading of 10 000 tonnes (Taylor and Smith 1997). Recent changes in livestock populations, in particular, reductions in sheep numbers, and increases in pine forest planting have altered pressures on marginally productive pasture land. Even though the amount of agricultural land could decline by about 0.5% per year if pine forest conversion continues at the present rate of 70 000 ha y^{-1} (Ministry of Forestry 1997), agriculture will still remain the country's dominant land use into the foreseeable future.

Pastoral development has been accompanied by introduction of clovers, whose symbiotic nitrogen fixation adds about 1.1 million tonnes of nitrogen annually (Hedley *et al.* 1990), and by fertilizer application, which currently adds annually around 135 000 tonnes of nitrogen and 150 000 tonnes of phosphorus. Nitrogen fertilizer application, mostly in lowland dairying areas, increased 3-fold between 1990 and 1995 (Taylor and Smith 1997).

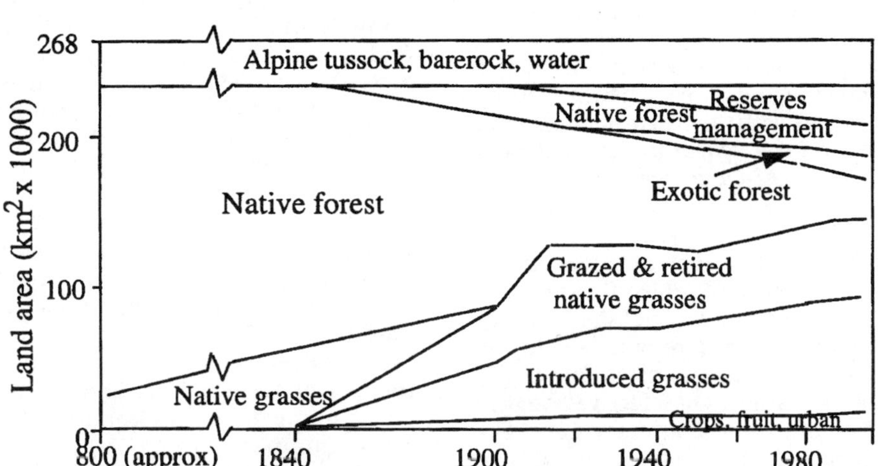

Fig. 1. Changes in vegetation cover of New Zealand with time (adapted from McColl and Ward (1987), based on Molloy (1980)).

Table 1. Length (km) of rivers and per cent of total river length flowing through improved and unimproved pasture in 14 geographic regions of New Zealand. Lengths were calculated from all rivers appearing on digitised 1:250 000 maps in NZMS 262.

	Improved pasture		Unimproved pasture		
	Length (km)	% of total	Length (km)	% of total	Total length (km)
North Island					
Auckland	1 153	38	12	<0.1	3 013
Bay of Plenty	1 049	13	146	2	8 244
Gisborne	1 821	28	166	3	6 448
Hawkes Bay	4 010	35	848	7	11 397
Manawatu-Wanganui	5 489	32	2 714	16	17 215
Northland	2 340	37	0	0	6 341
Taranaki	2 792	46	208	3	6 094
Waikato	5 752	37	601	4	15 374
Wellington	1 084	17	809	13	6 327
South Island					
Canterbury	6 585	24	239	1	27 723
Nelson-Marlborough	1 660	11	122	1	15 471
Otago	5 636	25	566	2	22 915
Southland	5 704	28	78	0	20 394
Westland	627	3	254	1	19 392
North and South Island total	45 702	25	6 763	4	186 348

Land development at Whatawhata Research Centre — Much of the information around which this chapter is based has been collected in the streams of the Hakarimata Ranges around the Whatawhata Research Centre. The geology and soils of this area are described in detail by Bruce (1978). Briefly, the soils of the rolling and steep land are derived from argillite and greywacke, while those of the flat to rolling land are pre-dominantly derived from volcanic ash. The podocarp-broadleaf forest present at the time of human settlement was cleared originally in the 1920s from much of the present research area and developed into pasture carrying sheep and both dairy and beef cattle. By the time that the Whatawhata site was purchased by the Department of Agriculture in 1949, considerable reversion of pasture to shrubby vegetation, including both native (manuka *Leptospermum scoparium*, kanuka *Kunzea ericodes*, bracken *Pteridium esculentum*) and introduced (gorse *Ulex europaeus*) species, had occurred (Farrelly 1986). The development of technology to apply phosphatic fertilizers from the air, allowing increased pasture fertility and vigour, combined with intensive fencing and weed control programmes, led to the current land use where sheep and beef cattle are carried at an average of 12 stock units per hectare. Current fertilizer use is 25 kg P ha^{-1}, as reactive phosphate rock, and 7 kg N ha^{-1}, as urea. The hormone sprays Grazon (triclopyr) and MCPA/MCPB are used for gorse and weed control at average rates of 0.3 litre ha^{-1}. In addition 0.06 litre of 2,4–D ha^{-1} is applied. In some areas native forest remains the dominant vegetation, although selective logging occurred there once in the last 100 years. Considerable areas of pasture in the Hakarimata Range have been planted in commercial pine *Pinus radiata* forest in the last 50 years. Annual rainfall at Whatawhata Research Centre is 1 608 mm and annual average temperature is 13.3°C, ranging from 18.1°C in February to 8.4°C in July.

Impacts of agricultural land use on stream ecosystems

National perspective — Recent nation-wide surveys of resource managers (Sinner 1992; Smith 1993) have highlighted widespread concern about the effects of agricultural land use on water quality, with sedimentation, nutrient contamination, alteration of physical characteristics, and faecal contamination of surface water identified as the most significant impacts. Comparative studies of stream ecosystems under pasture and native forest throughout New Zealand (Various 1990; Smith *et al.* 1993; Harding and Winterbourn 1995; Quinn and Cooper 1997; Townsend *et al.* 1997) have indicated fairly consistent patterns of stream habitat change, and biotic response, to conversion of native forest or grassland to pasture. Commonly observed changes in physico-chemical habitat include: (i) increased water yield (e.g., by 60% in a central North Island basin (Dons 1987)) and storm flows (Fahey and Rowe 1992); (ii) reduced shade in low order streams, with consequent increases in stream temperature and instream algae and macrophyte growth (Quinn *et al.* 1997a); (iii) narrower stream channels in small streams (<30 km^2 catchment area) (Davies-Colley 1997); (iv) lower instream leaf litter and woody debris, and higher instream plant production and algal and macrophyte biomass (particularly adventive species), due to less riparian vegetation (Howard-Williams and Pickmere 1994; Harding and Winterbourn 1995; Quinn *et al.* 1997a; Townsend *et al.* 1997); (v) often higher concentrations of suspended sediment, faecal contamination, nitrogen and phosphorus (e.g., Smith *et al.* 1993; Quinn *et al.* 1997a); (vi) loss of habitat diversity due to channelization to improve drainage (Quinn *et al.* 1992; Williamson *et al.* 1992).

The magnitude, extent and ecological effects of these changes vary, depending on the underlying catchment geology and climate (that both determine soils and native vegetation type), and the land management and stream engineering practices that have been undertaken. For example, changing tussock grassland to pasture causes less stream habitat change than native forest conversion to pasture, and the changes in invertebrate composition are also less (Townsend *et al.* 1997). Similarly, maintenance of riparian vegetation buffers along pasture streams can reduce the degree of change to stream characteristics and enhance diversity of sensitive aquatic groups (Quinn *et al.* 1992; Collier 1995).

Stream biota response to agricultural development at Whatawhata — The changes in stream habitat and the biotic responses of stream invertebrate and fish communities as a result of pastoral land use at Whatawhata are detailed in Boulton *et al.* (1997), Collier *et al.* (1997), Davies-Colley (1997), Hicks (1997), Hicks and McCaughan (1997) and Quinn *et al.* (1997a).

Pastoral land use had profound effects on the fish communities of Whatawhata streams (Hicks and McCaughan 1997). Pasture streams had 12-fold higher fish abundance than native forest streams, due largely to increased abundance of shortfinned eel *Anguilla australis* production. This is thought to be due to the greater instream primary production and water temperature in pasture than in forest streams. However, banded kokopu *Galaxias fasciatus*, an important component of the whitebait fishery in some rivers (Rowe *et al.* 1992), were present in forest but absent from pasture streams. This is consistent with patterns of land use effects in other regions (Rowe *et al.* 1999). Attempts to enhance populations of native fish and other organisms in pasture streams should consider fish access, because the majority of species are diadromous (i.e., migrate to and from the sea as part of their life cycle) and natural or artificial migration barriers can limit their ability to colonize hill-land streams (McDowall 1993).

Overall invertebrate species richness did not differ between the pasture, pine, and native forest streams, despite there being fewer species belonging to the sensitive mayfly (Ephemeroptera) and stonefly (Plecoptera) orders at pastoral sites (Quinn *et al.* 1997a). This change in composition reflects an increased number of more tolerant species, particularly midges (Chironomidae), oligochaetes and snails (mainly *Potamopyrgus antipodarum*) which are favoured by increases in streambed algal production in pasture streams (Quinn *et al.* 1997b). Indeed, the abundance of the latter groups resulted in 3-fold higher invertebrate abundance in the pasture than in the pine and native forest streams. Calculation of an Index of Biotic Integrity (Plafkin *et al.* 1989), that summarizes a variety of invertebrate community characteristics, indicated that the pasture streams are "moderately impaired" relative to native forest reference sites. This reflects the presence of fewer species sensitive to sedimentation and organic and nutrient enrichment effects, increased relative abundance of tolerant species, and lower proportions of the leaf-shredding invertebrates at pastoral sites.

The large number of stream characteristics that differ between the forest and pasture streams make it difficult to identify the causes of differences in invertebrate abundance and community composition. However, strong correlations between stream shade and a multivariate index of community similarity (Quinn *et al.* 1997a), and the results of shading manipulation experiments in streamside channels indicate that shade plays an important role in structuring benthic invertebrate communities at Whatawhata (Quinn *et al.* 1997b). The greater amount of fine sediment suspended in the water column and stored in the beds of pasture streams than forest streams appeared less important than shade effects, although landslides that deliver large quantities of sediment to the stream channel (Fig. 2) can cause severe reductions in downstream benthic invertebrate communities (JMQ, unpubl.

Fig. 2. Photograph of the landslide in the left branch of the Kiripaka Stream on 14 July 1995 that delivered approximately 11 000 tonnes of sediment to the stream in the year following the first slumping in April 1995 (photo Dr. Mike Scarsbrook).

data). Overall, the loss of shading by riparian shrubs and trees when forest is converted to pastoral land use tends to alter the conditions in headwater streams towards those naturally found much further downstream where stream channels are wide enough for light to reach the water surface directly, regardless of the presence of riparian vegetation (Quinn *et al.* 1992). The resultant loss of headwater stream habitat conditions and associated communities represent a reduction of regional-scale diversity, even though local species richness per unit of streambed may not be affected.

Goals and management options for nature conservation

Defining conservation goals — What are the management options for improving nature conservation values in hill-land pasture production systems? Efforts to enhance nature conservation of pasture streams may involve various goals, each requiring different management responses. For example, if we assume that nature conservation efforts should aim to maintain as wide a range of native species as possible within the stream, then our results at Whatawhata indicate that management actions should focus on restoration of shading riparian vegetation along headwater streams. However, if a narrower goal of enhancing production of eels was adopted, we might recommend maintenance of current practices because eel production is greater in pasture than forest streams (Hicks and McCaughan 1997). Alternatively, if the emphasis were on protection of a eutrophication-sensitive lake or estuary downstream, management would focus primarily on controlling sediment and nutrient yields from upstream catchments (Vant 1987; Williamson *et al.* 1996). In many situations, conservation goals will be multiple and there is a need to consider possible conflicts that could arise between the management responses to these different conservation goals.

Resolving conflicts between multiple goals in catchment management — Clearly, measures to enhance nature conservation in pasture catchments need to address agreed targets. In New Zealand, such measures will normally be implemented voluntarily by land users, and will therefore be balanced with other goals for the catchment/farm, including farm financial performance, compliance with consumer and regulatory requirements, maintaining asset values, and living a satisfying lifestyle. The Whatawhata Catchment Management Group has been established in an effort to maximize synergies, and minimize the conflicts, amongst the multiple goals facing landowners, and hence provide workable compromises. It has held workshops with a range of stakeholder groups and developed a pyramid describing the attributes of a well managed rural hill-land catchment (Fig. 3). Participants from Regional Councils, Department of Conservation, Federated Farmers and scientific organizations have agreed on the list of attributes and are currently developing a set of indicators to measure current status and monitor progress. For example, the effective farm financial surplus is likely to be accepted as an indicator of a viable business and mean annual suspended sediment concentration as an indicator of a healthy catchment ecosystem. The challenge in the next phase of the programme is to design a management plan that considers the weightings that the different groups give to the different attributes, and develop and test practical management options to meet the agreed goals.

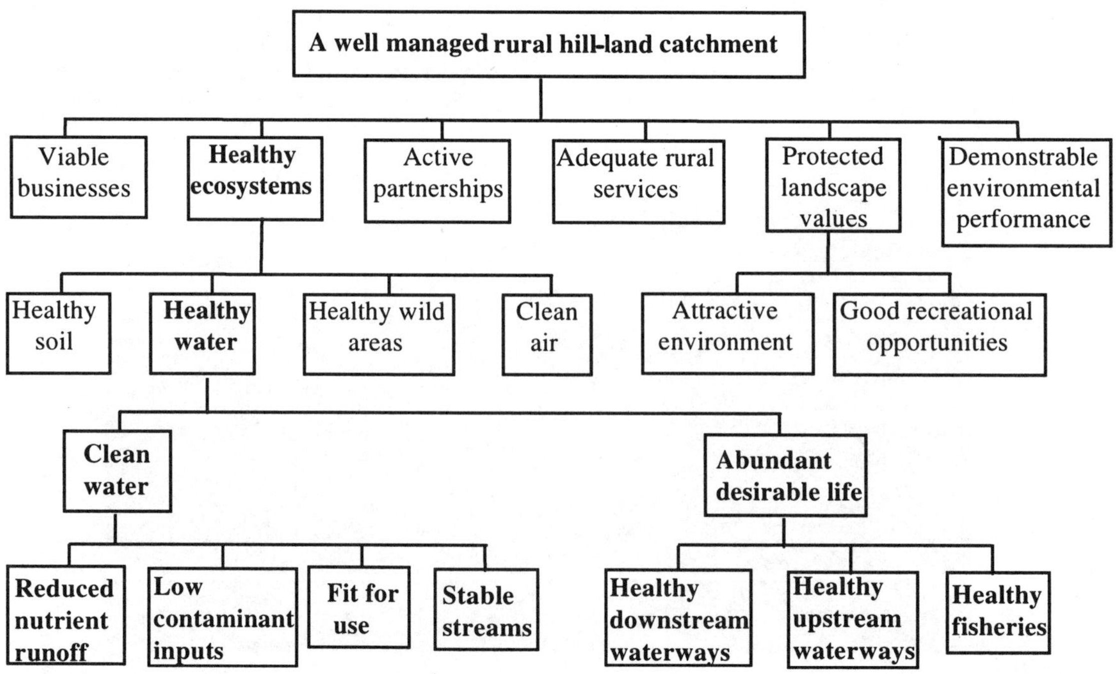

Fig. 3. Attributes identified by Whatawhata Catchment Management Group as key components of the pyramid describing a well managed hill-land catchment and third to fifth level attributes related to waterway management.

The following sections summarize options for managing stream water quality and habitat to enhance both local streams and downstream aquatic ecosystems. The relative value and practicality of these options will depend on the issues of concern and their importance relative to the other goals of the land users involved.

Managing sediment sources — Controlling hillslope and gully erosion has long-term benefits on pasture production and reduces damage to farm fences and roads caused by storms (Hicks 1995). These benefits relate to the Viable businesses, Healthy ecosystems, and Demonstrable environmental performance goals of the well managed rural hill-land catchment (Fig. 3). Potentially important sediment sources to hill-land pasture streams include (i) deep and shallow mass movement, (ii) surface erosion of pasture soils and roads, and (iii) streambank erosion (Hicks 1995).

The risk of deep erosion on hillslopes increases with slope, the degree of weathering of the underlying rock and the impermeability of the subsoil (Hicks 1995). Most of the sediment leaving one 2.7 km^2 Whatawhata catchment in 1995–1997 was derived from a massive hill side slip (Fig. 2) of approximately 0.41 ha eroding into the stream, which contributed about 11 000 tonnes of sediment between July 1995 and April 1996 (Dr Megan Balks, pers. comm.). Trees reduce erosion by increasing soil permeability and the strengthening effect of their interlocking root systems. Hillslope erosion rates can be reduced effectively by tree planting at densities over 100 stems per hectare for poplars (*Populus* spp.) (Hawley and Dymond 1988). No experimental evidence on the effect of tree density exists for other species, but pine plantations have been shown to be as effective as native bush in reducing hillslope erosion once canopy closure occurs (Hicks 1990; Marden and Rowan 1993). Timber production is an obvious commercial benefit from tree planting. While well developed and profitable systems exist for radiata pine, there are limitations of growth rate, rotation length, site specificity and timber quality that reduce the economic benefit from planting some other species (Thorrold *et al.* 1997). Similarly, Wall *et al.* (1997) found limited economic benefits to farm systems of planting poplars and willows other than erosion control and timber production. The possible roles of both native and exotic species for ecological restoration on farms are outlined by Simpson (1997). These roles include providing food sources and habitat for native fauna and providing corridors linking large areas of protected habitat.

Artificial debris dams across gully floors (e.g., structures constructed from half-round poles or supported wire mesh) are effective tools for control of gully erosion in combination with close tree planting (Hicks 1995).

Surface soil erosion control should aim to maintain a dense sward of groundcover by pasture improvement and management of grazing animal type, grazing regimes, and stocking rates. Pasture treading damage by cattle at Whatawhata during winter, when soil moisture is greatest, reduced water infiltration rate, leading to significant increases in sediment, nitrogen and phosphorus in runoff, particularly on steep (>25°) slopes and when soil surface damage exceeded 40% (Nguyen *et al.* 1998). Treading damage also reduced subsequent pasture growth at a time of year when feed supply is low (Sheath and Carlson 1998). Comparative studies suggest that cattle cause greater erosion and pasture yield reduction than sheep (Lambert *et al.* 1985). Good management of cattle grazing in winter can have beneficial impacts for both waterways and farm production. There are several options available to minimize the impact of treading damage (Sheath and Boom 1997). Cattle grazing should be avoided on wet soils near waterways and on steep slopes, these areas could be grazed with sheep or at times when the soils are dryer. Where farms contain a mix of soil types, cattle can be restricted to the more robust soils. At Whatawhata these are the soils derived from volcanic ash which tend to occur on ridges, well removed from the streams. Improved cattle growth rates can lead to earlier sale, reducing the number of heavy cattle held on farms over winter.

The importance of good road engineering to reduce sediment inputs into streams has been demonstrated in forestry operations (Wallis and McMahon 1994). Regular cutoffs, that divert road runoff to vegetated ground before flows become large enough to be highly erosive, are a key factor in controlling sediment input to streams from unsealed farm roads and tracks. All rural catchments have a combination of public and private roads and farm tracks of various standards. Little is known of the sediment generation from these sources relative to hillside erosion in pastoral catchments, and this is part of current investigations at Whatawhata. Well engineered and maintained roads and tracks are an important part of adequate rural services and internal access within farms, which are vital for intensively managed farm systems.

The key components of reducing streambank erosion include tree-planting, maintaining adequate ground cover and cattle exclusion. Cattle can destabilize streambanks, particularly in small streams (Williamson *et al.* 1992; Authors' unpubl. data), the impact of cattle can be reduced by better management of stock around stream edges, for example through fencing or providing alternative water sources and shade (Hicks 1992; Godwin and Miner 1996). Stream bank erosion can also be controlled by planting of trees to bind soils (Thorne 1990), although too high a density may shade out pasture grasses, whose roots have some stabilizing influence on streambank soils, and this may lead to a period of accelerated erosion as a new channel equilibrium is reached (Davies-Colley 1997). If these erosion rates are severe the plantings themselves may be placed at risk. Although native tree species have been advocated for streamside plantings to stabilize banks (Collier *et al.* 1995), and can be very effective (Hicks 1992), the ease of establishment and high soil-binding capacity of willows (*Salix* spp.) makes them particularly effective trees for stabilizing severely eroding banks. If tree rooting depth is less that the height of the eroding banks that they are intended to stabilize, then planting of trees on steeply sloping banks may exacerbate bank erosion due to the additional weight of the tree (Abernathy and Rutherfurd 1996). In contrast, on gently sloping banks the slope-normal contribution of tree weight is greater than the down-slope component, so it increases bank stability by increasing frictional resistance to deep erosion (Thorne 1990).

Nutrient management — Modest concentrations of nutrients, particularly nitrogen and phosphorus, enhance productivity in aquatic ecosystems, but in excess these can result in undesirable blooms of aquatic plants (phytoplankton in lakes, algae and macrophytes in stream and rivers) with a variety of nuisance effects on ecosystem health and human uses (Ministry for the Environment 1992) that are counter to the Whatawhata Catchment Management Group's goals for healthy water (Fig. 3). The evidence from Whatawhata is that nutrients are less important to local stream health than riparian shade and sedimentation (Quinn *et al.* 1997a,b). However, this result is tempered by the need, in other situations, to consider downstream cumulative effects of nutrients on rivers, lakes and estuaries.

Nitrogen and phosphorus export from pasture hill-land catchments average 3 to 10 times that of native forest (Wilcock 1986; Cooper and Thomsen 1988); similar differences between nutrient concentrations have been observed in many parts of New Zealand (Smith *et al.* 1993). The main causes of increased nutrient input to streams in pasture are: (i) increased soil fertility, from fertilization and increased nitrogen fixation by clover and legume crops; (ii) increased erosion carrying sediment-bound nutrients; (iii) direct inputs such as fertilizer drift, and dung and urine of grazing animals with free access to waterways; (iv) dung in surface runoff; (v) increased leaching (particularly of nitrate) due to greater fertility and local nutrient surplus due to dung and urine patches; and (vi) reduced denitrification following wetland and under-field drainage (Williamson and Hoare 1987; Cooke 1988; Cooper 1990; Hedley *et al.* 1990).

Measures to control hillslope and streambank erosion management, discussed above, also contribute substantially to phosphorus control (Williamson *et al.* 1996). Preventing stock access to streams (e.g., by fencing or altering their behaviour by providing alternative water supply and shade away from the stream and stock bridges) eliminates substantial direct inputs of phosphorus (Williamson and Hoare 1987). Grassy filter strips have been shown to be effective at removing particulates and associated nutrients from overland flow draining agricultural land (Smith 1989; Cooper *et al.* 1992; Muscutt *et al.* 1993; Hairsine 1996). Collier *et al.* (1995) provide information on the influence of slope, drainage and soil type on buffer widths requirements. Riparian wetlands can remove up to 90% of nitrate in groundwater by denitrification (Cooper 1990) and also retain sediment, nitrogen and phosphorus in surface runoff (Cooke 1988; Cooke and Cooper 1988). Controlling stock access to wetlands during periods when they are saturated should help maintain the nutrient retention capacity of these sites (Collier *et al.* 1995). Reductions in phosphorus loss will also be achieved where phosphorus fertilizer application to streams can be avoided (Cooke 1988). Developments of geographic information systems and global positioning systems technologies are reaching the stage where more precise aerial application of fertilizer will be possible. The cost of this will be significant and will need to be met by increased production from targeting fertilizer to responsive sites more effectively, rather than from the environmental benefits and cost savings from the phosphorus not lost to surface waters. Where fertilizers are spread from the ground then greater precision is already possible. Avoiding direct application of fertilizer to waterways and minimizing fertilizer runoff are part of the fertilizer industry's Code of Practice (FertResearch 1998).

Nutrient uptake by instream plants can also remove substantial quantities of nitrogen and phosphorus (Rutherford *et al.* 1987; Cooper 1990; Quinn *et al.* 1997b), resulting in downstream reductions in nutrient concentrations, at least temporarily under stable flows. The fate of nutrients in macrophyte tissue is poorly understood, but it is likely that substantial amounts are washed downstream and subsequently released during decay in settling areas (e.g., lakes). Increased riparian shade will reduce this instream uptake (Howard-Williams and Pickmere 1994; Quinn *et al.* 1997b) and might best be avoided if the primary concern about land use is related to nutrients causing summer periphyton blooms in a downstream river reach. However, if the focus is healthy water in the pasture streams, then there are clear benefits of riparian shading, as described below.

Riparian vegetation management — The focus of this section is on management of riparian vegetation to promote nature conservation in aquatic ecosystems and meet the Whatawhata Catchment Management Group's goals for healthy water (Fig. 3), acknowledging that there are also many potential benefits of well planned riparian plantings to plant diversity, habitat for terrestrial fauna, and connectivity of protected areas (e.g., Howard-Williams and Pickmere 1994; Collier *et al.* 1995; Simpson 1997).

Increasing stream shading and input of terrestrial carbon as leaf litter can be achieved easily in headwater streams through planting of native or exotic trees and shrubs, and/or the fencing of stream banks to allow grass growth and natural succession. Shade effects on stream temperature can be modelled (Rutherford *et al.* 1997) to provide shading targets related to information on the known thermal tolerances of fish and invertebrates (Quinn *et al.* 1994; Richardson *et al.* 1994), or to control nuisance periphyton blooms (Quinn *et al.* 1997b). Sufficient information exists for a range of tree types to estimate the planting density necessary to achieve a range of canopy densities, which varies considerably between tree species due to canopy structure (Power and Thorrold, unpubl. data). Not all parts of a stream are equally exposed to light due to the surrounding topography and stream direction relative to the angle of the sun. Selection of priority areas for shade management will help to increase the efficiency of any expenditure on tree planting.

Inputs of leaf litter from riparian trees and shrubs are trapped in streams and can increase the diversity of food resources available to stream biota. Small scale experiments indicate that, if retained long enough, leaf litter can play a role in enhancing invertebrate diversity and nutrient retention in Whatawhata streams (Quinn, unpubl. data). Both stream size and choice of tree type influence the importance of leaf litter to stream energy budgets. The litter trapping efficiency of streams decreases with increasing catchment area (and depth) (Webster *et al.* 1994), and thus leaves can be flushed from the system before being consumed by the biota. Trees with palatable, soft leaves (e.g., mahoe, *Fuschia*) decay more rapidly, and play a greater role in energy flow and nutrient retention than tough leaves (e.g., tawa *(Beilschmiedia tawa)*, rewarewa *(Knightia excelsa)* and Australian blackwood *(Acacia melanoxylon)* (Quinn *et al.* 2000). Invertebrates do not show preferences for native over exotic leaf litter of similar toughness (Parkyn and Winterbourn 1997), but the very pulsed input of deciduous tree leaves is likely to be less beneficial than the more continuous input of native or exotic evergreens.

Riparian tree planting will also influence microclimate conditions (Chen *et al.* 1994) which can be an important factor affecting the longevity of the adult stages of some aquatic insects (Jackson 1988; Smith and Collier 1997). This is likely to be important for the conservation of groups such as stoneflies (Plecoptera), that have relatively long adult phases (B. Smith, unpubl. data) and are much less abundant in pasture than forest streams at Whatawhata (Quinn *et al.* 1997a). Riparian vegetation overhanging the stream channel may also increase the delivery to the stream of terrestrial invertebrates which are important food sources for some native fish (e.g., Main and Lyon 1988; McDowall *et al.* 1996).

In relation to the multiple goals of the Whatawhata Catchment Management Group, simply increasing shading may have limited benefit because it may attract grazing animals, increasing the amount of time spent in stream and riparian areas, with consequent damage to streambanks, reduction of riparian groundcover by leaf litter and vegetation, and increased direct input of urine and dung to the stream and riparian zone. Considerably greater environmental benefits may be achievable through a combination of planting and retiring streamside areas from grazing. This includes the down-stream effects on water quality which have been identified by the group as an important indicator of Demonstrable Environmental Performance and Healthy Ecosystems (Fig. 3). Stream bank erosion was greatly reduced in the pumice soil Ngongotaha catchment, near Rotorua, by a combination of fencing and tree planting (Williamson *et al.* 1996). This result may be most significant in pumice soil areas though, as work at Whatawhata has shown that stream banks have narrowed under pasture and that, when these banks are heavily shaded by tree regrowth, stream bank erosion increases as the channel morphology readjusts to a forest light environment (Davies-Colley 1997). In the long term this will benefit stream biota by increasing the habitat area to that present under forest conditions, but it will give rise to a short-term increase in export of streambank sediments.

Simple systems for establishing trees in grazed or ungrazed riparian areas are available, so the solution is technically feasible. The cost of any option is a significant factor in farmers' decision making, and depends upon whether fencing covers all stream channels or just the larger streams (Fig. 4A). There are 70 m of stream length (including all the side tributary seeps) per hectare in the dendritic stream network of the Mangaotama catchment at Whatawhata (Fig. 4B), and the length of stream to be fenced increases exponentially as the stream, or catchment, size at which stream fencing begins decreases (Fig. 4A). Applying a simple system, involving poplar poles at 10 m spacings and a cattle proof electric fence along both sides of all streams, will cost around $NZ280 ha^{-1} ($NZ1.50 m^{-1} plus $NZ5 $tree^{-1}$ and 14 trees ha^{-1}). Alternatively, if the Whatawhata Catchment Management Group decided to focus on the health of larger streams (>50 ha upstream drainage of which there are 10 m stream length ha^{-1}), the costs for fencing the Mangaotama would be $NZ40 ha^{-1}. However, not fencing the small headwater streams will reduce greatly the control of stream temperature and contaminant inputs.

Discussions with farmers connected with the Whatawhata Catchment Management Group have identified a range of costs, and benefits associated with fencing-off stream banks and providing water in the paddocks in troughs. The benefits include: better feed utilization in big paddocks; reduced hill-slope erosion due to cattle tracking; easier mustering; reduced incidence of liver fluke *(Fasciola hepatica)*; improved ability to administer zinc treatment to prevent facial eczema; and reduced sedimentation of waterways. Costs identified include: fencing costs; lost grazing area; retired areas can become a source of weeds and pests; difficulties retrieving stock after getting into retired areas; and costs of installing water systems. Two points emerge from these observations. Firstly, even simple technology to improve water quality, and therefore

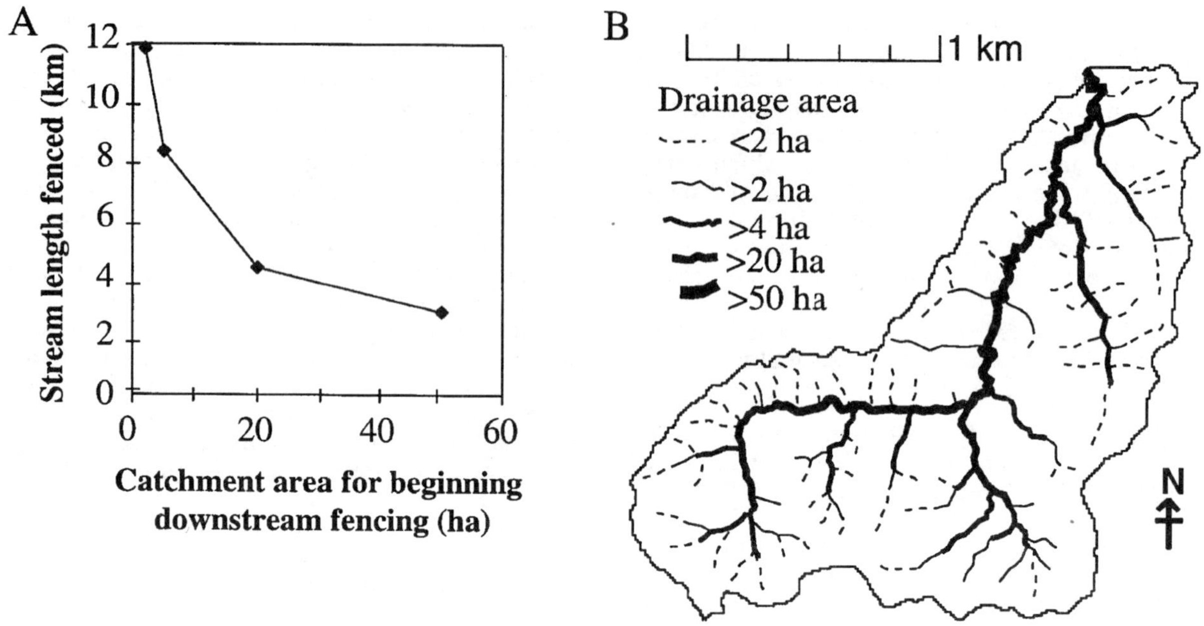

Fig. 4. A. Relationship between total stream length to be fenced and the drainage area at which fencing starts within the 2.7 km^2 Mangaotama Catchment at Whatawhata. B. Mangaotama drainage network showing distribution of stream lengths with progressively larger drainage area classes.

ecosystem health, can have a significant cost. Secondly, there are a number of benefits of a practice such as fencing riparian areas and managing water supplies which farmers can capture with a well-designed system. These benefits would often not be identified by researchers working in isolation, and may become obvious to farmers only as they experiment with systems on their own properties.

Resolving conflicts between multiple goals in catchment management — Results to date from the multi-stakeholder process at Whatawhata have shown that there is common ground between the users of hill-land catchments (farmers) and other organizations that have an interest in the economic and ecological consequences of land use. Surveys and discussions have revealed that there is potential for imaginative and practical options for land, vegetation and animal management to improve the ecological performance of hill-land catchments and aquatic ecosystems, as well as to provide economic benefits to farmers. Our experience to date indicates that active involvement with farmers is the key to realizing these benefits, and this is supported by the results from other participatory processes for sustainable land management carried out in the Waikato (Tarbotton *et al.* 1997). The benefits to farmers of improved environmental performance are likely to become more sharply focused as consumer concerns about farm products move past the current food safety issues into areas of animal welfare and sustainable management (Taylor 1997).

Measuring the long-term ecological and economic success of improved catchment management is an important part of the Whatawhata project. The results from our initial work have shown that there are important time lags in ecosystem response that cannot be well understood in short-term or small-scale experiments. This is true also of soils, vegetation and animal responses to changes in management where seasonal differences and the buffering capacity of animal condition, previous fertilizer input, and vegetation plasticity can all mask changes in the short term. At Whatawhata we plan to design and implement a catchment management scheme to test the long-term response to management change intended to promote healthy ecosystems, as well as meet other catchment management goals, particularly aspirations for maintaining viable businesses. This experiment by management should provide improved understanding of ecological function in hill-lands as well as a focus for debate over land use and land practise as stakeholders work towards implementation of the *Resource Management Act.*

ACKNOWLEDGEMENTS

We thank the members of the Whatawhata Catchment Management Group for their contributions to the ideas in this chapter, particularly Terry Parminter, the group facilitator. Morag Stroud produced the Geographic Information System map and calculated the data reported in Figure 4. The work was supported by funding from the New Zealand Foundation for Research Science and Technology for the National Institute for Water and Atmospheric Research programme on "River ecosystems and land use interactions" (C01516) and the AgResearch programme on "Managing hill-catchments to achieve multiple goals" (C10653). Thanks to Dr Bryce Cooper, Dr Denis Saunders, and an anonymous referee for helpful review comments.

REFERENCES

Abernathy, B. and Rutherfurd, I. D., 1996. Vegetation and bank stability in relation to changing channel scale. Pp. 213–19 *in* First National Conference on Stream Management in Australia, Merrijig, 19–23 February 1996. Cooperative Research Centre for Catchment Hydrology: CSIRO, Canberra, Australian Capital Territory.

Boulton, A. J., Scarsbrook, M. J., Quinn, J. M. and Burrell, G. P., 1997. Land use effects on the hyporheic ecology of hill country streams near Hamilton, New Zealand. *NZ J. Mar. Freshwat. Res.* **31:** 609–22.

Bruce, J. G., 1978. Soils of part Raglan County, South Auckland, New Zealand. New Zealand Soil Bureau Bulletin 41, Department of Scientific and Industrial Research: Wellington, New Zealand.

Chen, J., Naiman, R. J. and Franklin, J., 1994. Influences of riparian buffers on microclimate gradients across stream ecosystems. Pp. 59 *in* Proceedings of the International Workshop on the Ecology and Management of Aquatic-Terrestrial Ecotones ed by L. L. Link and R. J. Naiman. University of Washington: Seattle.

Collier, K. J., 1995. Environmental factors affecting the taxonomic composition of aquatic macroinvertebrate communities in lowland waterways of Northland, New Zealand. *NZ J. Mar. Freshwat. Res.* **29:** 453–65.

Collier, K. J., Cooper, B., Davies-Colley, R., Rutherford, J. C., Smith, C. M. and Williamson, R. B., 1995. Managing riparian zones: a contribution to protecting New Zealand's rivers and streams (Vol. 1 Concepts; Vol. 2 Guidelines). Department of Conservation: Wellington, New Zealand.

Collier, K. J., Smith, B. J. and Baillie, B. R., 1997. Summer light trap catches of adult Trichoptera in hill-country catchments of contrasting land use, Waikato, New Zealand. *NZ J. Mar. Freshwat. Res.* **31:** 623–34.

Cooke, J. G., 1988. Sources and sinks of nutrients in a New Zealand hill pasture catchment II. Phosphorus. *Hydrological processes* **2:** 123–33.

Cooke, J. G. and Cooper, A. B., 1988. Sources and sinks of nutrients in a New Zealand hill pasture catchment. III Nitrogen. *Hydrological processes* **2:** 135–49.

Cooper, A. B., 1990. Nitrate depletion in the riparian zone and stream channel of a small headwater catchment. *Hydrobiologia* **202:** 13–26.

Cooper, A. B., Smith, C. M. and Bottcher, A. B., 1992. Predicting runoff of water, sediment and nutrients from a New Zealand grazed pasture using CREAMS. *Trans. Amer. Soc. Agricul. Eng.* **35:** 105–12.

Cooper, A. B. and Thomsen, C. E., 1988. Nitrogen and phosphorus in streamwaters from adjacent pasture, pine, and native forest catchments. *NZ J. Mar. Freshwat. Res.* **22:** 279–91.

Davies-Colley, R. J., 1997. Stream channels are narrower in pasture than in forest. *NZ J. Mar. Freshwat. Res.* **31:** 599–608.

Department of Statistics, 1992. New Zealand in profile 1993. Department of Statistics: Wellington, New Zealand.

Dons, A., 1987. Hydrology and sediment regime of a pasture, native forest, and pine forest catchment in the Central North Island, New Zealand. *NZ J. For. Sci.* **17:** 161–78.

Fahey, B. D. and Rowe, L. K., 1992. Land-use impacts. Pp. 265–84 *in* Waters of New Zealand ed by M. P. Mosley. New Zealand Hydrological Society: Wellington, New Zealand.

Farrelly, E. J., 1986. Whatawhata Hill Country Research Station: An historical review. Eljay Enterprises: Hamilton, New Zealand.

FertResearch, 1998. Pastoral Farming User Guide, Code of Practice for Fertiliser Use. New Zealand Fertiliser Manufacturer's Research Association: Auckland, New Zealand.

Godwin, D. C. and Miner, J. R., 1996. The potential of off-stream livestock watering to reduce water quality impacts. *Biores. Tech.* **58:** 285–90.

Hairsine, P., 1996. Comparing grass filter strips and near-natural riparian forests for buffering intense hillslope sediment sources. Pp. 203–06 *in* First National Conference on Stream Management in Australia. Merrijig, 19–23 February Cooperative Research Centre for Catchment Hydrology: CSIRO, Canberra, Australian Capital Territory.

Harding, J. S. and Winterbourn, M. J., 1995. Effects of contrasting land use on physico-chemical conditions and benthic assemblages of streams in a Canterbury (South Island, New Zealand) river system. *NZ J. Mar. Freshwat. Res.* **29:** 479–92.

Hawley, J. G. and Dymond, J. R., 1988. How much do trees reduce landsliding? *J. Soil Water Cons.* **43:** 495–98.

Hedley, M. J., Tillman, R. W. and Ball, P. R., 1990. Nutrient losses from New Zealand agriculture. Pp. 62–91 *in* Proceedings of the New Zealand Fertiliser Manufacturers Research Association Conference-November 1990.

Hicks, B. J., 1997. Tracing food webs in forested and pastoral streams in the Waikato region, North Island. *NZ J. Mar. Freshwat. Res.* **31:** 651–84.

Hicks, B. J. and McCaughan, H. M. C., 1997. The distribution of fish and crayfish, and eel production, in streams in forest and pasture in the Waikato region, North Island. *NZ J. Mar. Freshwat. Res.* **31:** 635–50.

Hicks, D. L., 1990. Landslide damage to hill country under pasture, pine plantation, scrub and bush, in Taranaki. DSIR Land Resources Technical Report No. 31, 19 p. Wellington, New Zealand.

Hicks, D. L., 1992. Effect of soil conservation tree plantings on streambank stability. DSIR Land Resources Technical Record 118, 17 p. Wellington, New Zealand.

Hicks, D. L., 1995. Control of soil erosion on farmland. MAF Policy Technical Paper 95/4. Ministry of Agriculture and Fisheries: Wellington, New Zealand.

Howard-Williams, C. and Pickmere, S., 1994. Long-term vegetation and water quality changes associated with the restoration of a pasture stream. Pp. 93–109 *in* Restoration of Aquatic Habitats. Selected papers from the second day of the New Zealand Limnological Society 1993 Annual Conference ed by K. J. Collier. Department of Conservation: Wellington, New Zealand.

Jackson, J., 1988. Diel emergence and longevity of selected adult aquatic insects from a Sonoran desert stream. *Amer. Midl. Nat.* **119:** 344–52.

Lambert, M. G., Devantier, B. P., Nes, P. and Penny, P. E., 1985. Losses of nitrogen, phosphorus and sediment in runoff from hill country under different fertiliser and grazing management regimes. *NZ J. Agricul. Res.* **28:** 371–79.

Main, M. R. and Lyon, G. L., 1988. Contribution of terrestrial prey to the diet of banded kokopu (*Galaxias fasciatus* Gray) (Pisces: Galaxiidae) in South Westland, New Zealand. *Verh. Internat. Theoret. Andewandt. Ver. Limnol.* **23:** 1785–789.

Marden, M. and Rowan, D., 1993. Protective value of vegetation on Tertiary terrain before and during Cyclone Bola, East Coast, North Island, New Zealand. *NZ J. For. Sci.* **23:** 255–63.

McColl, R. H. S. and Ward, J. C., 1987. The use of water resources. Pp. 441–59 *in* Inland Waters of New Zealand, DSIR Bulletin 241 ed by A. B. Viner. DSIR Science Information Publishing Centre: Wellington, New Zealand.

McDowall, R. M., 1993. Implications of diadromy for the structuring and modelling of riverine fish communities in New Zealand. *NZ J. Mar. Freshwat. Res.* **27:** 453–62.

McDowall, R. M., Main, M. R., West, D. W. and Lyon, G. L., 1996. Terrestrial and benthic foods in the diet of the shortjawed kokopu, *Galaxias postvectis* Clarke (Teleostei: Galaxiidae). *NZ J. Mar. Freshwat. Res.* **30:** 257–69.

Meyer, J. L., 1997. Stream health: Incorporating the human dimension to advance stream ecology. *J. N. Am. Benthol. Soc.* **16:** 439–47.

Ministry for the Environment, 1992. Water quality guidelines No. 1: Guidelines for the control of undesirable biological growths in water. Ministry for the Environment: Wellington, New Zealand. 57 p.

Ministry for the Environment, 1996. Sustainable land management: a strategy for New Zealand. Ministry for the Environment: Wellington, New Zealand.

Ministry of Forestry, 1997. New Zealand country report: Montreal process criteria and indicators for the conservation and sustainable management of temperate and boreal forests. Ministry of Forestry: Wellington, New Zealand.

Molloy, L. F., 1980. Land alone endures. Department of Scientific and Industrial Research Discussion Paper No. 3 Wellington, New Zealand.

Muscutt, A. D., Harris, G. L., Bailey, S. W. and Davies, D. B., 1993. Buffer zones to improve water quality: a review of their potential use in UK agriculture. *Agricul., Ecosys. Environ.* **45:** 59–77.

Nguyen, M. L., Sheath, G. W., Smith, C. M. and Cooper, A. B., 1998. Impact of cattle treading on hill land. 2. Soil physical properties and contaminant runoff. *NZ J. Agr. Res.* **41:** 279–90.

Parkyn and Winterbourn, 1997. Leaf breakdown and colonisation by invertebrates in a headwater stream: comparisons of native and introduced species. *NZ J. Mar. Freshwat. Res.* **31:** 301–12.

Plafkin, J. L., Barbour, M. T., Porter, K. D., Gross, S. K. and Hughes, R. M., 1989. Rapid bioassessment protocols for use in streams and rivers: benthic macroinvertebrates and fish. U.S Environmental Protection Agency, Off. Water Regulations and Standards: Washington, DC.

Quinn, J. M. and Cooper, A. B., 1997. Land-water interactions at Whatawhata, New Zealand: Introduction and synthesis. *NZ J. Mar. Freshwat. Res.* **31:** 569–77.

Quinn, J. M., Cooper, A. B., Davies-Colley, R. J., Rutherford, J. C. and Williamson, R. B., 1997a. Land use effects on habitat, water quality, periphyton and benthic invertebrates in Waikato, New Zealand, hill-country streams. *NZ J. Mar. Freshwat. Res.* **31:** 579–97.

Quinn, J. M., Cooper, A. B., Stroud, M. J. and Burrell, G. P., 1997b. Shade effects on stream periphyton and invertebrates: an experiment in streamside channels. *NZ J. Mar. Freshwat. Res.* **31:** 665–83.

Quinn, J. M., Steele, G. L., Hickey, C. W. and Vickers, M. L., 1994. Upper thermal tolerances of twelve New Zealand stream invertebrate species. *NZ J. Mar. Freshwat. Res.* **28:** 391–97.

Quinn, J. M., Williamson, R. B., Smith, R. K. and Vickers, M. L., 1992. Effects of riparian grazing and channelization on streams in Southland, New Zealand. 2. Benthic invertebrates. *NZ J. Mar. Freshwat. Res.* **26:** 259–69.

Quinn, J. M., Burrell, G. P. and Parkyn, S. M., 2000. Influence of leaf toughness and nitrogen content on in-stream processing and nutrient uptake in a Waikato, New Zealand, pasture stream and streamside channels. *NZ J. Mar. Freshwat. Res.* **34:** in press.

Richardson, J., Boubee, J. A. T. and West, D. W., 1994. Thermal tolerance and preference of some native New Zealand freshwater fish. *NZ J. Mar. Freshwat. Res.* **28:** 399–408.

Rowe, D. K., Saxton, B. A. and Stancliff, A. G., 1992. Species composition of the whitebait (Galaxiidae) fisheries in 12 Bay of Plenty rivers, New Zealand: evidence for rivermouth selection by *Galaxias brevipennis* (Gunther). *NZ J. Mar. Freshwat. Res.* **26:** 219–28.

Rowe, D. K., Chisnall, B. L., Dean, T. L. and Richardson, J., 1999. Effects of land use on native fish communities in east coast streams of the North Island of New Zealand. *NZ J. Mar. Freshwat. Res.* **33:** 141–51.

Rutherford, J. C., Williamson, R. B. and Cooper, A. B., 1987. Nitrogen, phosphorus and oxygen dynamics in rivers. Pp. 139–66 *in* Inland Waters of New Zealand, DSIR Bulletin 241 ed by A. B. Viner. DSIR Science Information Publishing Centre: Wellington, New Zealand.

Rutherford, J. C., Blackett, S., Blackett, C., Saito, L., Davies-Colley, R. J. and Payne, G. P., 1997. Predicting the effects of shade on water temperature in small streams. *NZ J. Mar. Freshwat. Res.* **31:** 707–22.

Sheath, G. W. and Carlson, W. T., 1998. Impact of cattle treading on hill land 1. Soil damage patterns and pasture status. *NZ J. Agr. Res.* **41:** 271–78.

Sheath, G. W. and Boom, C. J., 1997. Impact of beef cattle grazing systems on treading damage and forage supply. *Proc. NZ Grasslands Assoc. Conf.* **59:** 87–92.

Simpson, P., 1997. Ecological restoration in the Wellington conservancy. Department of Conservation: Wellington, New Zealand.

Sinner, J., 1992. Agriculture and water quality in New Zealand. MAF Policy Technical Paper 92/17, Ministry of Agriculture and Fisheries: Wellington, New Zealand.

Smith, B. J. and Collier, K. J., 1997. Investigations into the diet and temperature tolerance of adult stoneflies. *NZ Limnol. Soc. Newsl.* **33:** 53–54.

Smith, C. M., 1989. Riparian pasture retirement effects on sediment, phosphorus, and nitrogen in channellised surface run-off from pastures. *NZ J. Mar. Freshwat. Res.* **23:** 139–46.

Smith, C. M., 1993. Perceived riverine problems in New Zealand, impediments to environmentally sound riparian zone management, and the information needs of managers. Water Quality Centre, National Institute for Water and Atmospheric Research-Ecosystems: Hamilton, New Zealand.

Smith, C. M., Wilcock, R. J., Vant, W. N., Smith, D. G. and Cooper, A. B., 1993. Towards sustainable agriculture: freshwater quality in New Zealand and the influence of agriculture. Wellington. MAF Policy Technical Paper 93/10, Ministry of Agriculture and Fisheries: Wellington, New Zealand.

Tarbotton, I. D., Wedderburn, M. E., O'Connor, M. B., Singleton, P., Paine, M. S. and Carylon, M., 1997. A method for improving farm practice; a case study of pugging. *Proc. NZ Grassl. Assoc. Conf.* **59:** 79–85.

Taylor, N. W., 1997. New Zealand and the world meat industry. *Proc. 43rd Internat. Conference Meat Sci. Technol.* Pp. 170–71.

Taylor, R. and Smith, I., 1997. The State of New Zealand's Environment. Ministry for the Environment/Government print Publications: Wellington, New Zealand.

Thorne, C. R., 1990. Effects of vegetation on riverbank erosion and stability. Pp. 125–44 *in* Vegetation and Erosion ed by J. B. Thornes. Wiley: Chichester.

Thorrold, B. S., Knowles, R. L., Nicholas, I. D., Power, I. L. and Carter, J. L., 1997. Evaluation of agroforestry options for three tree species. *Proc. NZ Grassl. Assoc. Conf.* **59:** 187–90.

Townsend, C. R., Arbuckle, C. J., Crowl, T. A. and Scarsbrook, M. R., 1997. The relationship between land use and physicochemistry, food resources and macroinvertebrate communities in tributaries of the Taieri River, New Zealand: A hierarchically scaled approach. *Freshw. Biol.* **37:** 177–91.

Vant, W. N., 1987. Eutrophication: An overview. Pp. 172–82 *in* Lake Managers Handbook, Water and Soil Miscellaneous Publication No 103 ed by W. N. Vant. Ministry of Works and Development: Wellington, New Zealand.

Various, 1990. Special issue on New Zealand River Classification Project. *NZ J. Mar. Freshwat. Res.* **34**(3): 277–440.

Wall, A. J., Mackay, A. D., Kemp, P. D., Gillingham, A. G. and Edwards, W. R. N., 1997. The impact of widely spaced soil conservation trees on hill pastoral systems. *Proc. NZ Grassl. Assoc. Conf.* **59:** 171–77.

Wallis, G. and McMahon, M., 1994. The impacts of forest management on erosion and sedimentation: A New Zealand review. *Log. Indus. Res. Organis. Rep.* **19:** 15 p.

Webster, J. R., Covich, A. P., Tank, J. L. and Crockett, T. V., 1994. Retention of coarse organic particles in streams in the Southern Appalachian mountains. *J. N. Amer. Benthol. Soc.* **13:** 140–50.

Wilcock, R. J., 1986. Agricultural runoff, a source of water pollution in New Zealand? *NZ Agr. Sci.* **20:** 98–103.

Williamson, R. B. and Hoare, R. A., 1987. Controlling nutrient loads and predicting resulting lake concentrations. Pp. 172–82 *in* Lake Managers Handbook, Water and Soil Miscellaneous Publication ed by W. N. Vant. Ministry of Works and Development: Wellington, New Zealand.

Williamson, R. B., Smith, C. M. and Cooper, A. B., 1996. Watershed riparian management and its benefits to a eutrophic lake. *J. Water Resour. Plng. and Mgmt.* **122:** 24–32.

Williamson, R. B., Smith, R. K. and Quinn, J. M., 1992. Effects of riparian grazing and channelisation on streams in Southland, New Zealand. 1. Channel form and stability. *NZ J. Mar. Freshwat. Res.* **26:** 241–58.

CHAPTER 29

Biograze — melding off-reserve conservation of native species with animal production in Australian rangelands

Craig D. James,[1] Mark Stafford Smith,[1] Jill Landsberg,[2] Alaric Fisher,[3] Rodger Tynan,[4] John Maconochie[4] and John Woinarski[3]

We summarize the results from recent studies that suggest that around a quarter of native plant and animal species may be declining in range and abundance as a result of grazing in rangelands of Australia. These results derive from studies in several vegetation types of differences in plant and animal assemblages along gradients of grazing intensity associated with distance from stock watering points. Species that were most prominent at sample points remote from watering points (>8 kms, hence were lightly grazed) were termed "decreaser species". Decreaser species may be at risk if the current trend of installing water points at 3–5 km intervals continues. The results suggest two strategies for improving conservation. Firstly, around three-quarters of our biodiversity may be able to coexist under current grazing management and their protection on-property is probably compatible with what are regarded as best-practice approaches to grazing management. Secondly, the remaining quarter or so of species need some sort of special action such as explicitly keeping some pockets of land protected from significant levels of grazing. We present an outline of a current project called "Biograze" in which we will develop some scenarios of landscape design that build-in a balance of land for production and conservation, and in which we will estimate the likely costs associated with this conservation planning. We present some preliminary and simplistic results of this work including some likely economic costs and look at these economic costs in the light of positive incentives and/or compensation where conservation goals meet public demand but compromise pastoral production.

INTRODUCTION

THE arid and semi-arid zones of Australia cover 70% of the land surface (5.3 million km^2). Within these zones rainfall is too low and unpredictable for regular agricultural use and the main land use by area (4.2 million km^2) is pastoralism. For this landuse, domestic animals are grazed on native pastures with little or no deliberate modification of the natural vegetation, for example, by clearing, cultivating, sowing exotic species or fertilizing.

Early settlers were quick to recognize the potential for grazing sheep on the native grasslands they encountered west of the Great Dividing Range in eastern Australia; they were the saltbush and blue bush shrublands across southern Australia and the woodlands and grasslands of central and northern Australia. The expectations of reliable production and resilience to grazing that presumably came from European experience were dashed in the unpredictable and sensitive environment of the Australian rangelands. Initially, large flocks of sheep and herds of cattle were built up only to collapse in drought a few decades on. This scenario of boom and bust was repeated across Australia in different regions at different times beginning as early as 1867

[1]Centre for Arid Zone Research, CSIRO Wildlife & Ecology, P.O. Box 2111, Alice Springs, Northern Territory, Australia 0871.
[2]CSIRO Wildlife & Ecology, P.O. Box 84, Lyneham, Australian Capital Territory, Australia 2602.
[3]Parks & Wildlife Commission of the Northern Territory [and Co-operative Research Centre for the Sustainable Use of Tropical Savannas], P.O. Box 496, Palmerston, Northern Territory, Australia 0831.
[4]South Australian Department of Environment, Heritage and Aboriginal Affairs, G.P.O. Box 1047, Adelaide, South Australia, Australia 5001.
Pages 290–300 *in* NATURE CONSERVATION 5: NATURE CONSERVATION IN PRODUCTION ENVIRONMENTS: MANAGING THE MATRIX ed by J. L. Craig, N. Mitchell and D. A. Saunders. Surrey Beatty & Sons, 2000.

(Anonymous 1867) in South Australia (Friedel *et al.* 1990). Associated with the fluctuations in numbers of domestic stock were plagues of rabbits, also responding to seasonal conditions (Friedel *et al.* 1990). Together, grazing by rabbits and stock in the first 50–80 years of pastoralism caused changes to the native vegetation, particularly the balance of perennial grass to shrub biomass, and the loss of productive value for grazing (Noble and Tongway 1983; Friedel *et al.* 1990). A complex set of changes caused by this period of heavy grazing is continuing today.

At the same time as the country was being grazed heavily by stock and rabbits, other forces of change were being introduced. The pattern of frequent burning of some vegetation biomes was reduced and predators new to the native animals of Australia were introduced (Thomson *et al.* 1987). These forces interacted to cause the extinction of 10 species of mammals that formerly occurred in the arid and semi-arid zones (Morton 1990; Short and Smith 1994; Smith and Quin 1996). A further 18 species of mammals are currently endangered or vulnerable (Flannery *et al.* 1990) and numerous species of birds and plants that occupy the arid and semi-arid zones are also endangered (Leigh and Briggs 1992; Reid and Fleming 1992). This dramatic record of recent extinctions and continuing decline of species has led to calls to modify the management of rangelands to include planning for the protection of the native biota (Morton *et al.* 1995). Conservation of critically endangered species is often a clear case of protection of small populations from sources of mortality (e.g., protecting mala, *Lagorchestes hirsutus,* from predation by foxes — Lundie-Jenkins 1993), or preserving favoured habitats. But for the majority of species, especially plants and invertebrates, little is known about the effects of past and current landuses on the likelihood of their persistence.

Morton *et al.* (1995) introduced a vision of incorporating nature conservation into rangeland management by having a hierarchy of reserves of different status. Their paper addresses the important issue of a network of representative reserves across rangelands landscapes, but observes that reserves alone will not be sufficient to ensure the persistence of all species in the arid and semi-arid zones because of the vastness of the landscape and the spatial and temporal variability in environmental conditions and therefore populations (see Stafford Smith and Morton 1990; Woinarski *et al.* 1992; and James *et al.* 1995 for further discussion). Conservation outside nature reserves is therefore necessary, and is likely to be effective and achievable in rangelands because of the relatively natural state of the ecosystems. James *et al.* (1995) discussed a possible research agenda to provide the principal information needed for the incorporation of nature conservation into the matrix of grazed rangelands. This agenda consisted of three themes: (1) understanding the spatial and temporal dynamics of populations of organisms; (2) understanding the effects of grazing on biota; and (3) developing technologies and tools to analyse layers of environmental, biological and cultural information and synthesize them to develop regional plans for conservation and production. While we have seen little empirical progress on the first theme, there has been substantial progress on the second and third. In this chapter we bring together our current results and then introduce plans for future work in a project entitled "Biograze: Waterpoints and Wildlife", which will provide information and tools for the development of regional plans that integrate nature conservation and pastoral production in arid and semi-arid zone rangelands. We present this information in four sections:

1. Effects of artificial sources of water, and the grazing they support, on native biota;
2. Implications for persistence of species in arid and semi-arid rangelands;
3. Field work planned to test hypotheses about distribution and abundance of species across regions rigorously; and
4. Preliminary and simplistic analyses of the potential economic costs of principles of nature conservation from the work presented above, into a regional plan that integrates production and conservation.

THE EFFECTS OF WATER POINTS AND GRAZING ON BIOTA

Approaches

Two recent studies have provided detailed data on how species assemblages change in response to different intensities of grazing along gradients away from permanent artificial sources of water (Landsberg *et al.* 1997; Fisher, unpubl. data). The design is predicated on the knowledge that cumulative grazing intensity diminishes with increasing distance from water because mammalian grazing species (stock, kangaroos, and feral species) need to drink regularly (see James *et al.* 1995 for a discussion of the rationale). The radial pattern of grazing impact that results around point sources of water has been termed the piosphere effect (Lange 1969) and it has been quantified for landscapes in central and southern Australia (Andrew and Lange 1986; Andrew 1988; Bastin *et al.* 1993; Pickup and Chewings 1994).

The recent studies of the effects of grazing on biodiversity were conducted at 11 sites in three of the more important vegetation biomes in the arid and semi-arid zone rangelands: the chenopod shrublands (4 sites), *Acacia*-dominated woodlands and shrublands of central and southern Australia (4 sites), and the Mitchell grasslands of northern Australia (3 sites) (Fig. 1). At each site, a gradient of at least six sampling points was surveyed, starting from close to a water point, where grazing intensities were highest, and extending to a reference sample point sufficiently far from water to be seldom grazed. Such reference sample points indicate the condition of the country prior to the provision of water for livestock, and so they may support biota that cannot persist closer to water. Such reference areas are now extremely rare in the productive rangelands so in some landscapes such as much of the Mitchell grasslands, the most water remote sites are within 10 km of watering points and are therefore subject to some grazing by cattle. Sample points along a gradient were chosen to occur in the same land-system and position in the landscape as the reference sample point. Some variables that may influence our results, but that could not be controlled for in the studies are grazing by rabbits (which are not reliant on drinking water), the history of grazing at a watering point, fire history, and the possibility that the most grazing-sensitive species may already have been eliminated from the landscape.

At each sample point along each gradient, the abundance of each species of a range of plants and animals was measured. Detailed descriptions of the taxa, sampling methods and sites are given in Landsberg *et al.* (1997). Two main analyses will be summarized here: change in species richness and change in assemblage structure along the gradients.

Changes in species richness

In the gradient studies of Landsberg *et al.* (1997) there was no single, simple response of all biota to the change in grazing with distance from water along the eight gradients studied. Species richness of taxa was correlated with distance from water in only 10 of 109 analyses (i.e., $p < 0.05$) across 13 or 14 taxonomic groups at each gradient (ground stratum perennial plants, upper stratum perennial

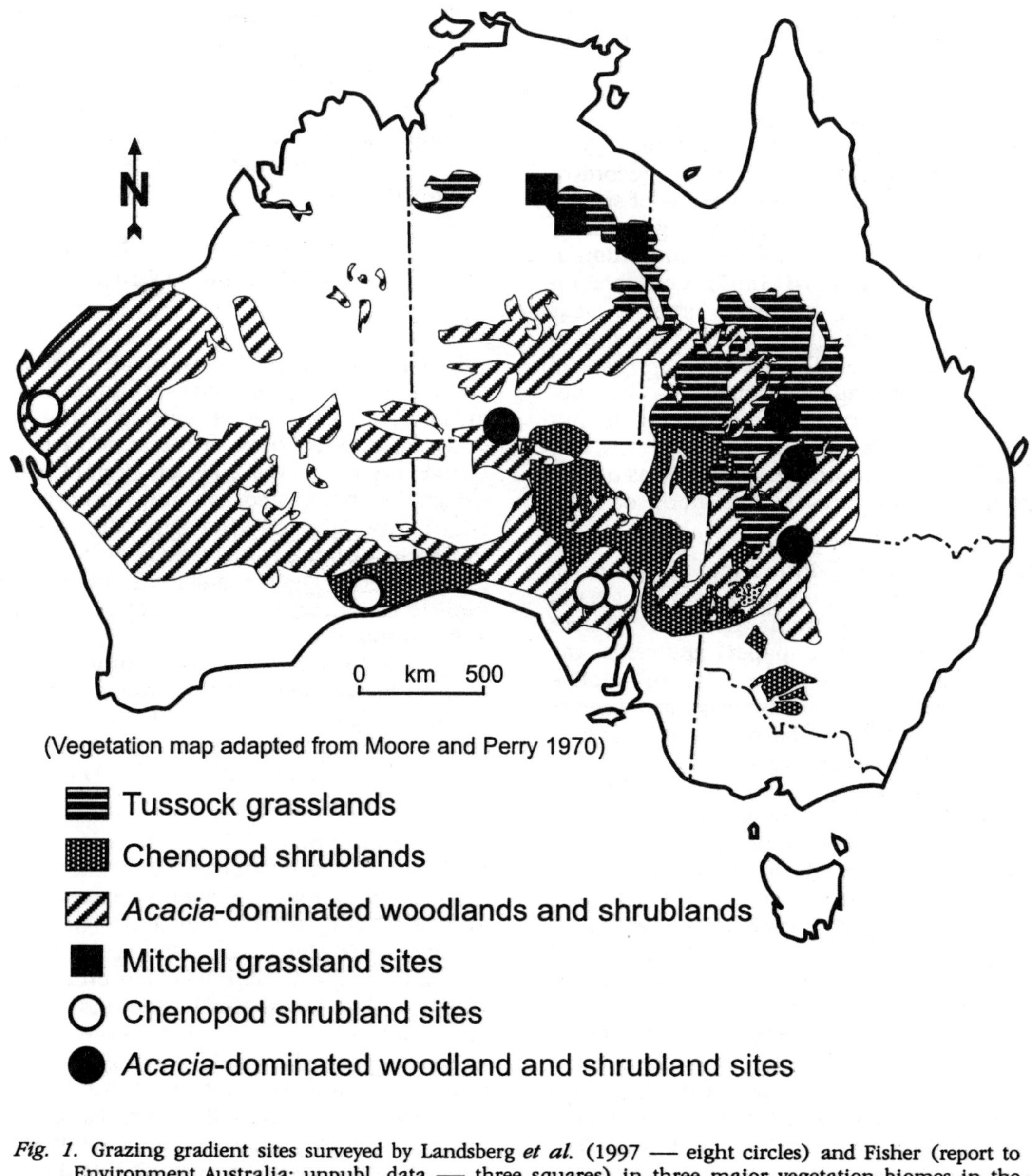

Fig. 1. Grazing gradient sites surveyed by Landsberg *et al.* (1997 — eight circles) and Fisher (report to Environment Australia; unpubl. data — three squares) in three major vegetation biomes in the Australian rangelands.

plants, seed bank plants on five gradients only, birds, reptiles, ants, beetles, springtails, grasshoppers and crickets, all plants, all vertebrates, all invertebrates, all animals and all biota). However, there were no consistent patterns of increasing or decreasing numbers of species, or bell-shaped changes in species richness with distance from water. We used the sequential Bonferroni correction for multiple comparisons (Rice 1989) to adjust the significance level of p to 0.0036 (i.e., 0.05/14) for each gradient and there was only one regression that was then judged to be significant: that for all biota on one gradient in the *Acacia*-dominate woodland biome ($r^2 = 0.97$; $p = 0.001$). Although this is a conservative test of biological significance, it highlights the lack of consistent pattern of change in species richness along gradients.

For the gradients in Mitchell grasslands there was a corresponding lack of consistent or statistically significant pattern of change in species richness with distance from water for mammals (report to Environment Australia; Fisher, unpubl. data). Species richness of reptiles increases slightly with increasing distance from water, and species richness of birds is higher at sites close to water points.

Changes in species composition

Species response groups were far more useful for interpreting the impact of grazing on plant and animal communities. Landsberg *et al.* (1997) used a two stage procedure involving correspondence analysis and log-linear regression of species abundance along a gradient, repeated for each taxon on each gradient. They identified three types of response that were common for most of the more abundant and diverse plant and animal taxa on most gradients: some species' abundance was higher at sample points closer to water ("increasers"); the abundance of a second group was highest at the reference sample point and decreased toward the water point ("decreasers"); and the abundance of the remainder was either neutral (abundance was similar at all sample points along the length of the gradient), or more variable (e.g., bell-shaped or parabolic curves). On average, between 10% and 33% of species in different groups were increasers, and between 15% and 38% were decreasers (Table 1). Nearly all the decreaser species were natives.

A number of species in the Mitchell grasslands also showed increaser and decreaser patterns of abundance in relation to water points, but the proportions of increaser or decreaser species were lower than they were for chenopod shrublands or *Acacia*-dominated woodlands. Six bird species, 10.7% of all bird species recorded, were only found within 5 km of water. No vertebrate species were *confined* to sites distant from water. Four vertebrate (one bird, one mammal and two reptile) species, 4.7% of the fauna, increased significiantly in abundance with greater distance from water. A single mammal species showed a significant increaser pattern. Patterns of response for invertebrates, especially ants, are still under study. This lower proportion of decreaser species may be because the gradients in the Mitchell grasslands did not extend as far from water as the gradients in the other two biomes, or that this ecosystem is more resilient to grazing (report to Environment Australia; Fisher, unpubl. data).

WATER DISTRIBUTION AND IMPLICATIONS FOR DECREASER SPECIES

The findings of the study of grazing gradients reported above, and the results of past work in semi-arid woodlands of eastern Australia (James, unpubl. data.; James and Hillery, unpubl. data), suggest that a sizeable proportion of the native biota is influenced by water points and the grazing associated with them. This response of biota to grazing may also have occurred around natural water points that supported populations of native herbivorous mammals (kangaroos and wallabies) prior to European settlement. But today, there has been a proliferation of artificial sources of water that allow much more sustained and widespread grazing than was possible in the past.

Landsberg and Gillieson (1996) examined the density of artificial sources of water in the arid and semi-arid zones of Australia and found that only the desert regions (Simpson, Tanami, Gibson, Great Sandy and Great Victoria deserts) had substantial areas that were more than 10 km from a named watering point. Furthermore, they analysed a sample of map sheets more closely to find that only 29% of watering points were named, suggesting that average distances between watering points were in fact substantially less than 10 km. We investigated this general result in more detail with geospatial data for water points in a 38 000 km^2 region of central Australia where cattle are the main livestock

Table 1. Proportions of increaser and decreaser species per gradient, calculated for all species in the more diverse and abundant taxa and averaged across all gradients surveyed by Landsberg *et al.* (1997). Values are shown as means ± standard errors, averaged across all eight gradients for most taxa, but only five gradients for seed-bank plants.

Group	No. species	% Increasers	% Decreasers
Understorey plants	90 ± 11	26 ± 4	38 ± 8
Overstorey plants	23 ± 5	10 ± 5	15 ± 7
Seed-bank plants	77 ± 8	33 ± 4	23 ± 7
Birds	30 ± 5	17 ± 7	23 ± 8
Reptiles	16 ± 1	18 ± 8	22 ± 7
Ants	76 ± 8	24 ± 8	26 ± 7

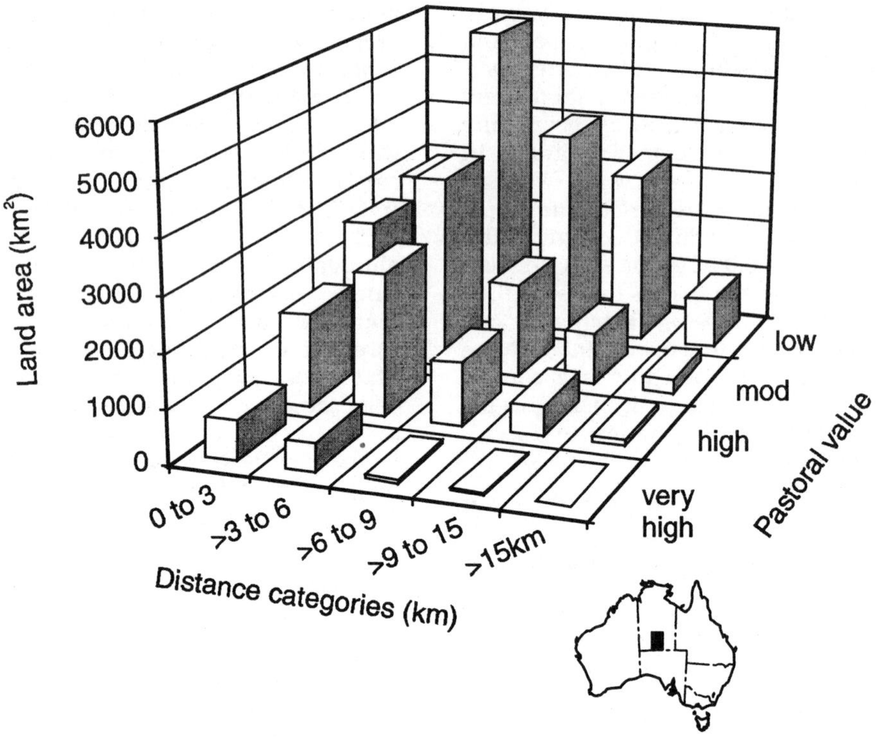

Fig. 2. Areas of land in different distance-from-water classes for land types grouped according to their value to the pastoral industry (see Bastin *et al.* 1993). In central Australia, where cattle are the main grazing animal, many decreaser species become prominent only at sample points >9 km from water. There is little land area remaining at distances >9 km from water in landsystems of moderate to high pastoral value.

animal (Bastin *et al.* 1993). Figure 2 shows the area of land at increasing distances from water, in land classed according to its value to pastoralism (Bastin *et al.* 1993). Clearly, for this region, there is a great imbalance in the area of land in distance categories favourable for decreaser species (>9 km) and the area favourable for increaser and neutral species (0–9 km); and this imbalance is most severe for landsystems that are of high pastoral value. There is now more "habitat" available for increaser species and less for decreaser species. Our concern is that decreaser species continue to be squeezed into smaller, increasingly fragmented parcels of land, as more artificial sources of water are established. A consequence may be that populations become so fragmented that they do not interact and localized extinction during severe environmental conditions may slowly reduce the geographic range of species. The trend of declining abundance and distribution of species of plants and birds outlined in the introduction may be related to this scenario.

We believe that future work needs to be undertaken to determine the regional consequences of the proliferation of artificial sources of water and the concomitant loss of water-remote habitat for decreaser species. It will also be important to involve producers in developing scenarios for maintaining a network of water-remote patches across regions, and to determine the potential economic costs of various options for achieving desired scenarios. This is in essence what the Biograze project will examine in the coming years.

TESTING HYPOTHESES AND DEVELOPING SCENARIOS: REGIONAL EXAMPLE

Building on the results of Landsberg *et al.* (1997) and Fisher (unpubl. data) described above, the Biograze collaborative project has been developed to explore ways of integrating conservation of biodiversity with pastoral activities in different regions. The project consists of three streams: (1) quantifying the regional distribution and abundance of species to determine the proportion of decreaser species; (2) identifying the shared characteristics of species that respond in similar ways to grazing, so that indicator species for monitoring can be identified more quickly; and (3) exploring management strategies and regional-scale designs, within the context of continued pastoral landuse, which are likely to ensure the persistence of all species including those most vulnerable to grazing and other disturbances associated with artificial water sources. Regional design strategies could include: selective adjustment of stocking rates at some watering points; adjustment of the total number of artificial watering points; the strategic spelling of paddocks; or shutting-off watering points at strategic times. Finally, the economic costs of different strategies will be estimated for two regions to illustrate the expenses.

Stream 1: Regional distribution of species

Results from the gradient studies by Landsberg *et al.* (1997) were consistent for a range of different biota even though the gradients occupied relatively small areas in widely separated localities. A more intense focus is needed, however, before the results can be applied for regional conservation planning. For example, it is possible that some species identified as decreaser species along carefully selected linear gradients may be more abundant or secure in other parts of the landscape or other paddocks with different management histories. To extrapolate from locality to region the following questions need to be addressed: (i) what proportion of a region's biodiversity can be identified as decreaser species, despite regional variation in landscapes and management history; and (ii) what proportion of a region's area provides suitable habitat for the persistence of these decreaser species?

These questions will be addressed with a regional survey of a few target taxa that showed strong responses in earlier studies. Using spatial databases, sites in four or more distance-classes within an extensive, relatively homogeneous ecosystem will be selected. We will restrict field surveys and subsequent analyses to a discrete area within a region (e.g., two 1:250 000 map sheets; approx 100 × 300 km) for logistic simplicity. We will sample around ten sites in each distance class to derive a statistically reliable estimate of the distribution and abundance of species. If sufficient data are available (e.g., on cover change, management history, feral animal distribution) selection procedures may be modified to take account of the likely differences in impact due to total (cumulative) stocking density at the sites. Taxa surveyed will be plants, birds and ants using methods developed during the gradient surveys (Landsberg *et al.* 1997). Analyses similar to those used for the gradient study (Landsberg *et al.* 1997) will be undertaken to identify increaser, neutral and decreaser response groups. Using a random stratified sampling design across the region will allow us to distinguish between changes in species composition due to geographic separation from that due to grazing gradients.

Inevitably, sampling at different distances from water to assess the effects of grazing involves spatially separated sites. It is possible that some of the variation in species composition that was detected along grazing gradients, and hence attributed to the effects of grazing and/or the water points *per se,* could be due to spatial variation in species composition. Therefore, another aspect of the Biograze field work will be to examine patterns of change in species assemblages on gradients that are perpendicular to and independent from grazing gradients. With the combination of a thorough sampling of biota in different land-systems that make-up a region and information on species turnover patterns independent of grazing, we will have a better understanding of the factors that affect the distribution of biodiversity at a regional scale.

Stream 2: Indicator response types

While it is theoretically possible to describe changes in biodiversity in terms of changes in the abundance of all the component species it is not logistically feasible to do so. This is a particular problem for species-rich taxa such as plants and invertebrates which may be represented by hundreds of species at a site and many thousands across a region. If generalizations are to be drawn from biodiversity surveys it is essential to focus on the most cost-effective and informative taxa, and to classify species into groups that relate more explicitly to the mechanisms underlying changes in their abundance and are less tied to species identities.

Indicator response types are groups of species that have similar responses to a disturbance (in our case grazing-related). We hypothesise that they share characteristics that, at least within a taxonomic group, lead them to have similar responses. We will identify the characteristics of groups of species within a taxon that respond in similar ways using data from previous surveys. This analysis will enable us to identify the likely responses of species in new regions without the need for comprehensive surveys. For plants, description and multivariate analysis of traits will be used following the approach outlined by Lavorel *et al.* (1997) to characterize the biological attributes of increaser and decreaser plant species. By starting with an explicit recognition of the differences among plant life forms and developing different lists of traits for different life forms, this approach aims to differentiate response types within life forms, to identify indicator types of decreasers for trees, shrubs, forbs or grasses, as appropriate. Functional group development for ants has tended to parallel plants although to date most analyses have focussed on the generic level (e.g., Andersen 1995). The proposed research will explore morphological sub-generic differences among Increaser and Decreaser ants at the level of species-groups. That for birds is already well developed from studies in temperate Australia (e.g., Robinson 1991).

Stream 3: Modelling economic costs of regional scenarios

Once the quantitative framework has been established by the field surveys proposed in Stream 1, it will be possible to model scenarios for changing stocking rates and/or the configuration of water sources in ways that enhance or increase water-remote refuge sites for biota undergoing decreaser trends. Stream 3 will: (1) analyse alternative ways in which theoretical paddocks might have management altered in order to protect Decreaser species; (2) develop generalized benefit/cost rules for different paddock sizes and spatial arrangements; and (3) apply these rules in exemplary studies of subregional-scale landscapes with realistic levels of biophysical and socio-economic complexity to give an estimate of the costs of changed management to industry and society.

Scenarios for regional conservation goals

The combination of conservation policies and the results of the field studies will imply that some minimum proportion of a district should not receive more than a given level of grazing pressure. This could be achieved by protecting areas from grazing totally or by ensuring areas remain remote from water even if still grazed; these areas could be of various sizes and spatial arrangements. Which approach is most appropriate for meeting this requirement will depend on: (1) the conceptual model of local extinction and re-invasion for organisms of most concern; (2) the cost of the changed management and other socioeconomic factors; and (3) its long-term reliability for conservation and viability for pastoralism. Real systems may benefit from a mixture of approaches.

Past work (Cridland and Stafford Smith 1993) shows that, from the point of view of grazing impact, distance to water and landscape heterogeneity are the over-riding factors in determining grazing impact. These operate at different scales, so that the scope for control of grazing patterns is greatly affected by paddock size: in large paddocks, manipulating access to water has a significant effect on access to pasture, whilst in small paddocks access to pasture can only be managed by fencing. A set of theoretical paddock arrangements will therefore be defined and the costs of different modes of protecting a given real proportion analysed. These will be net direct costs: that is, they will incorporate the cost of management and any potential productivity losses through reduced grazing areas or stocking pressures (i.e., the opportunity cost of not developing areas), as well as possible benefits in pest animal control or lower stocking rates. They will not assess the benefits to society of retaining the biodiversity. Rather, the purpose is to provide an assessment of how much society would *need* to value the maintenance of this biodiversity in order to implement any given management approach. The implications of different options will then be explored in realistic landscapes.

PRELIMINARY ANALYSES OF POTENTIAL ECONOMIC COSTS

Data such as those presented earlier in this chapter argue that at least two different strategies are needed to conserve biodiversity in the rangelands. First, landscapes which are grazed intensively will generally have few remaining sites remote from water. Conservation in these regions will generally require some acquisition of land for dedicated reserves, coupled where necessary with rehabilitation activities. In contrast, landscapes subjected to less intensive grazing are more likely to have maintained areas remote from water, and for these there will be a larger range of possible approaches. Options to reduce the risk of species loss in these may include fencing off distant corners of paddocks, or lowering general stocking rates. These represent approaches to what Morton *et al.* (1995) designated as "excised management units" (EMUs) when an area is fully fenced off, and "restricted use units" (RUUs) when use is controlled through some form of agreement.

The Biograze project will examine two regions (Mitchell grasslands of the Barkly Tableland, Northern Territory and chenopod shrublands in central South Australia) with their contrasting patterns of grazing, development and economics. Here we use data from a typical cattle property in the Sturt Plateau region of the Northern Territory for an initial exploration of these issues (Buxton *et al.* 1997). We consider a hypothetical property which comprises eight paddocks each of 64 km², such that each block of four is watered by a single bore (Fig. 3a). We assume that stocking rates are moderate for the region and that the pasture is homogeneous and reasonably palatable. Buxton *et al.* (1997) have explored the cost and returns of improving grazing access to all areas of the paddocks (Fig. 3b), which provides a theoretical optimum baseline against which to compare the cost of protecting these areas. The above strategies translate into three options (Fig. 3, Tables 2 and 3):

1. Remove a whole paddock permanently from production (Fig. 3c); this has obvious direct costs in reducing the total area of a property which is stocked. When the removal is for conservation purposes, there are additional costs associated with ensuring that waters are fully closed down and that fences with adjoining paddocks still in use are maintained. In terms of whole property economics, there may be some small savings in terms of reduced bore run distances and a long-term benefit if the closed paddock acts as a seed source to surrounding areas. The relative impact of this action depends greatly on the size of property. Here we assume that one of the eight paddocks on the example property is closed down with a consequent loss of 12.5% of the property carrying capacity, but we examine offsetting this by implementing the waterpoint developments recommended by Buxton *et al.* (1997) in the other paddocks. The cost of simply decommissioning a water point can be very small, but we allow $1,000 to ensure it is tidied up properly.
2. Fence off water-remote corners of paddocks (Fig. 3d); this means excising areas which are currently being used least for grazing production. Direct costs include the establishment and then maintenance of fencing, and the opportunity costs of not subsequently establishing a new water point in the area. Here we assume that a quarter of four of the eight paddocks on the example property is fenced off in such a way as to maximize adjoining areas and remove the same total area (64 km²) from production as in Option 1 (Fig. 3c). The spatial arrangement in Figure 3d requires 32 km of fencing (fencing parallel to the boundary requires the same), which means the property now contains 20 km more fencing than previously; we allow for maintenance on this.
3. Ensure that some corners of paddocks remote from water remain that way (Fig. 3e); this option

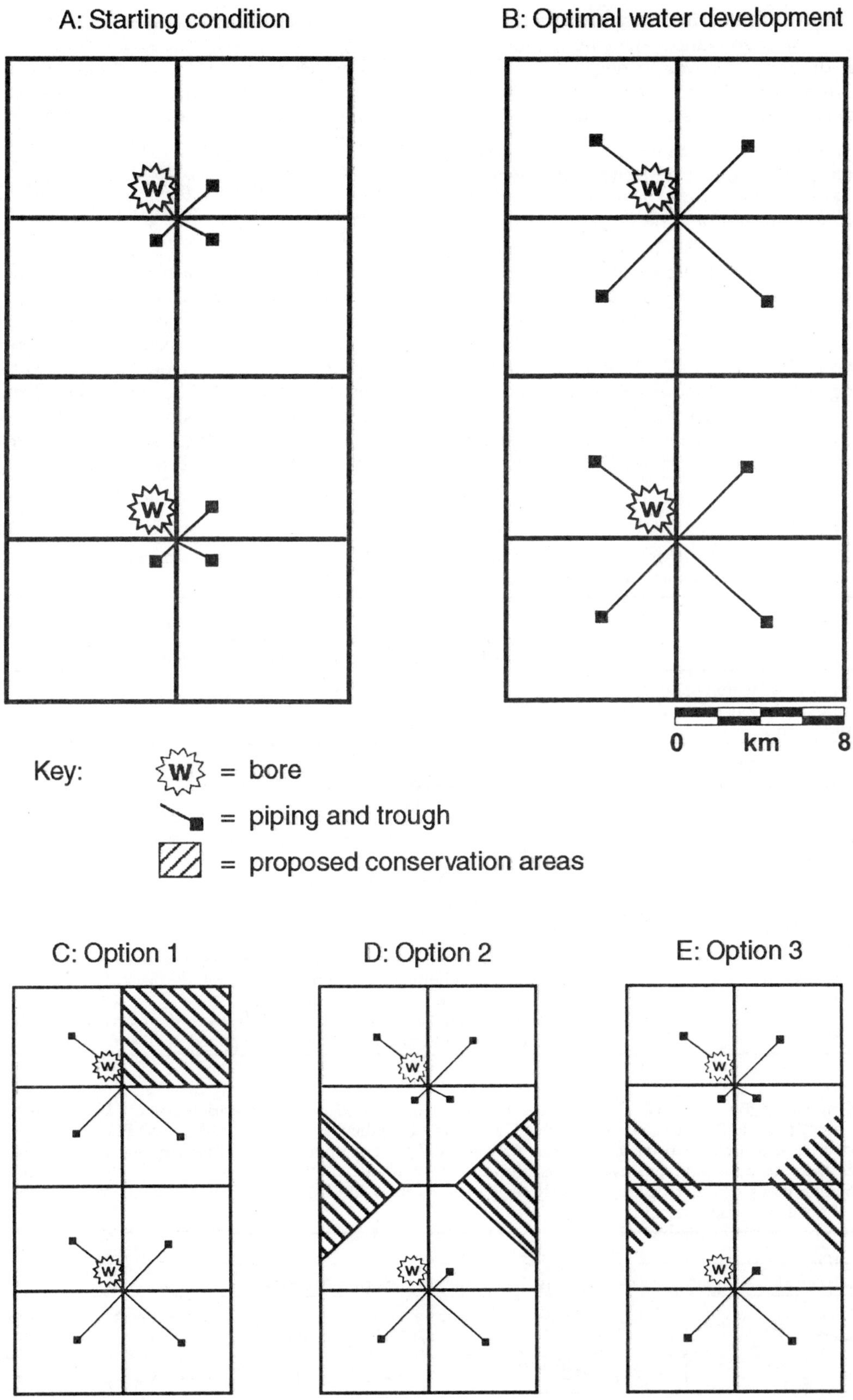

Fig. 3. Diagram illustrating the notional property layout explored in this study (see text and Table 2). A — current situation. B — improvement in water-point position for greater production. C — Option 1: Remove a whole paddock permanently from production (25% of property). D — Option 2: Fence off the most water-remote 25% of four paddocks. E — Option 3: Ensure that some water remote corners of paddocks remain that way; this option is equivalent to Option 2 but without the fencing.

is equivalent to Option 2 but without the fencing. Through some conservation agreement, producers commit themselves never to develop the areas remote from water in the same four paddocks, and also not to increase stocking rates.

Although cattle are able to walk out up to 20 km from a water point, the vast majority of their activity is within about 5 km of water (Foran *et al.* 1982). Their major consequent impact and source of production is therefore also in this zone. For the arrangement in Option 2, the shortest distance from the excised area to water is about 6 km. Data from central Australia (Fig. 2 in Pickup and Chewings 1988), as well as several model formulations (Pickup and Chewings 1988; Pickup and Bastin 1997), suggest that only about 2% of activity would occur outside 6 km from water in the paddocks described here, assuming modest stocking rates and reasonably palatable vegetation. Whilst this level of activity may be significant in terms of impacts on highly preferred pasture plants, it is of negligible importance for animal production. These results from central Australia are only indicative since animals may behave a little differently in northern Australia. However, for Option 2 we may neglect the loss of carrying capacity for grazing under the current paddock layout; it matters only as an opportunity cost when further water development is considered.

The opportunity cost represented by the comparison between Figure 3a and 3b is not insignificant; moving water from 1 km from the fenceline to the centre of an 8 km square paddock approximately triples the area of the paddock that is within 4 km of water. This can translate into a significant increase in the numbers of stock that could be carried without increasing the net impact on the resource, although there are absolute limits to how many

Table 2. Key cost factors allowed for in calculating the relative cost/benefit of the different options outlined in the text and Figure 3. AE stands for Animal Equivalents — a standard 450 kg castrated bull (steer).

Water development	
Cost	piping and equipping $AUD4,200/paddock
Benefit	estimated extra carrying capacity of 38 AE/paddock
Once-off costs of conservation options	
Water closure for Option 1	$AUD1,000
Fencing for Option 2	32 km @ $AUD1,500/km
On-going costs of conservation options	
Lost carrying capacity in the absence of water development	• 384 AE for Option 1 • <32 AE for Option 2 • none for Option 3
Paddocks in which water development cannot happen	• 1 for Option 1 • 4 for Options 2 and 3
Fence depreciation on extra fencing in Option 2 calculated on the basis of a 1-in-20 year replacement	• 20 km extra • replace 1 km/yr = $AUD1,500/yr
Other assumptions	• negligible changes in bore run costs • no change in employment needs • currently realistic market values outlined in Buxton *et al.* (1997) • the cost of all fence maintenance is applied against the pastoral concern

Table 3. Summary of the simulated once-off costs and on-going annual cash surpluses (constant 1997 $AUD) for different water development and conservation options (Fig. 3) for a hypothetical property on the Sturt Plateau, Northern Territory. The cost/benefits are shown for all options relative to the current undeveloped situation (upper panel), where negative values indicate costs and positive values surplus. More realistically the conservation options are combined with appropriate levels of water development (lower panel) to show the true cost/benefit after allowing for the opportunity costs of not developing waters in some paddocks.

	Actual values		Relative value	
Option for conservation without water development (i.e., ignoring opportunity costs)				
	Once-off cost	Annual surplus	Once-off cost	Annual surplus
Current situation	$0	$62,434	—	—
Option 1	$1,000	$41,560	–$1,000	–$20,874
Option 2	$48,000	$60,954	–$48,000	–$1,480
Option 3	$0	$62,434	$0	$0
Options for conservation assuming fullest possible water development				
	Once-off cost	Annual surplus	Once-off cost	Annual surplus
After water development	$33,600	$79,370	—	—
Option 1	$30,400	$56,396	+$3,200	–$22,974
Option 2	$64,600	$71,232	–$31,200	–$8,138
Option 3	$16,800	$72,732	+$16,800	–$6,638

stock should water from a single location. Here we adopt a conservative increase of 10% in stock numbers; this might also allow some improvement in growth, reproduction and mortality rates, which are not evaluated. We test this change for the baseline options with no conservation measures, and apply it to paddocks which are unaffected by the conservation measures, i.e., seven paddocks for Option 1, and four in Options 2 and 3.

The key calculations are shown in Tables 2 and 3. The costs relative to the baseline are effectively the cash flow implications for a producer who chooses to take the measures from their current state of development, whilst the costs relative to the developed baseline are the true costs and benefits. Developing the water points without any conservation measures results in once-off costs which are likely to be covered within about two years from increased cash flow (Table 3). Relative to this developed state, Option 1 results in a small once-off cost but a substantial on-going loss of production. Options 2 and 3 have a surprizingly modest on-going cost in terms of the lost opportunity to develop the water points; Option 3 saves the once-off establishment costs, but this benefit must be assessed against the greater reliability of long-term protection offered by Option 2. The effect of not considering the full opportunity cost makes a substantial difference to the results (Table 3); in particular Option 3 then appears to have negligible cost. Even though this may sometimes appear to paint a more favourable picture, failure to do the full analysis is likely to be contentious at a later date.

As mentioned earlier, there are some vegetation types where the impacts of grazing on biodiversity seem to be much less than others; if this indeed proves to be the case, then Option 3 may be a favourable approach with minimal establishment and opportunity costs to industry and society. Where it is impossible to be confident that even very low levels of grazing are compatible with the retention of decreaser species, then Options 1 and 2 are likely to be preferred on the grounds of greater certainty; for this particular scenario, the on-going costs associated with Option 2 seem reasonably modest.

The purpose of the study is to begin to assess what are the societal and private costs of preserving biodiversity in a grazed landscape, and consequently to what extent a lease-holder might reasonably be recompensed for such activities. The on-going maintenance costs (e.g., fences, controlling feral animals and weeds) for Option 2 have been left against the property in this analysis. In fact it may be reasonable to expect support for these from the public purse if the property were contributing to regional biodiversity conservation objectives, and it would also be reasonable for the public to meet the initial costs of fencing, which are not trivial. These and the opportunity costs, as assessed for today's costs and prices, might be covered through offering the lessee the first option to take on the role of conservation manager of the excised blocks, on a direct salaried basis.

This example is naïve in several respects, including: (a) paddock size and water point location (smaller or more evenly used paddocks will have a much greater proportion of their grazing occurring in the relatively remote corners); (b) management opportunities such as resting paddocks; (c) vegetation mixes — even quite large paddocks with a small proportion of highly preferred vegetation type may see a large part of that type being grazed at considerable distances from water, in ways which depend on the relative grain of landscape diversity and management infrastructure; and (d) the whole property-to-regional scale mix of landscapes (which will determine what pattern and connectedness of protected areas is likely to be most appropriate, and permit many possible local efficiencies where, for example, areas on adjoining properties could be concatenated). The example also ignores any economies of scale that might result from property re-structuring if some paddocks are taken out of production. All of these factors are being explored for some realistic landscapes in the Biograze project.

ACKNOWLEDGEMENTS

We thank Rosemary Buxton for providing economic data and Gary Bastin for providing data for Figure 2.

REFERENCES

Andersen, A. N., 1995. A classification of Australian ant communities, based on functional groups which parallel plant life-forms in relation to stress and disturbance. *J. Biogeog.* **22:** 15–29.

Andrew, M. H., 1988. Grazing impact in relation to livestock watering points. *Trends Ecol. Evol.* **3:** 336–39.

Andrew, M. H. and Lange, R. T., 1986. Development of a new piosphere in arid chenopod shrubland grazed by sheep. 2. Changes to the vegetation. *Aust. J. Ecol.* **11:** 411–24.

Anonymous, 1867. *Report of Commission Appointed by the Governor-In-Chief to Inquire into the State of the Runs Suffering from Drought; Together with Minutes of Evidence and Appendix.* Government Printer: Adelaide, South Australia.

Bastin, G., Pickup, G., Chewings, V. and Pearce, G., 1993. Land degradation assessment in central Australia using a grazing gradient method. *Rangel. J.* **15:** 190–216.

Buxton, R., Dyer, R. and Stafford Smith, M., Nov. 1997. Pastoralists from the region. *LUCNA Regional Report: Sturt Plateau and Victoria River District.* CSIRO: Alice Springs, Northern Territory.

Cridland, S. C. and Stafford Smith, D. M., 1993. Development and dissemination of design methods for rangeland paddocks which maximize animal production and minimize land degradation. *West. Aust. Depart. Agric. Miscell. Public.* **42/93.**

Flannery, T., Kendall. P. and Wynn-Moylan, K., 1990. *Australia's Vanishing Mammals: Endangered and Extinct Native Species.* Readers Digest Press: Sydney, New South Wales.

Foran, B., Bastin, G., Remenga, E. and Hyde, K. W., 1982. The response to season, exclosure and distance from water of three central Australian pasture types grazed by cattle. *Aust. Rangel. J.* **4:** 5–15.

Friedel, M. H., Foran, B. D. and Stafford Smith, D. M., 1990. Where the creeks run dry or ten feet high: pastoral management in arid Australia. *Proc. Ecol. Soc. Aust.* **16:** 185–94.

James, C. D., Landsberg, J. and Morton, S. R., 1995. Ecological functioning in arid Australia and research to assist conservation of biodiversity. *Pac. Cons. Biol.* **2:** 126–42.

Landsberg, J. and Gillieson, D., 1996. Looking beyond the piospheres to locate biodiversity reference areas in Australia's rangelands. Pp. 304–05 *in* Rangelands in a Sustainable Biosphere — Proceedings of the Fifth International Rangeland Congress, Volume 1, at Salt Lake City, Utah. Society for Range Management: Denver, Colorado.

Landsberg, J., James, C. D., Morton, S. R., Hobbs, T., Stol, J., Drew, A. and Tongway, H., 1997. The effects of artificial water sources of water on rangeland biodiversity. *Final report to the Biodiversity Convention and Strategy Section of the Biodiversity Group, Environment Australia, January 1997.* Environment Australia: Canberra, Australian Capital Territory.

Lange, R. T., 1969. The piosphere: sheep track and dung patterns. *J. Range Manage.* **22:** 396–400.

Lavorel, S., McIntyre, S., Landsberg, J. and Forbes, T. D. A., 1997. Plant functional classifications: from general groups to specific groups based on response to disturbance. *Trends Ecol. Evol.* **12:** 474–78.

Leigh, J. H. and Briggs, J. D., 1992. *Threatened Australian Plants: Overview and Case Studies.* Commonwealth of Australia: Canberra, Australian Capital Territory.

Lundie-Jenkins, G., 1993. Ecology of the rufous hare-wallaby, *Lagorchestes hirsutus* Gould (Marsupialia: Macropodidae), in the Tanami Desert, Northern Territory. I. Patterns of habitat use. *Wildl. Res.* **20:** 457–76.

Moore, R. M. and Perry, R. A., 1970. Vegetation. Pp. 59–73 *in* Australian Grasslands ed by R. M. Moore. Australian National University Press: Canberra, Australian Capital Territory.

Morton, S. R., 1990. The impact of European settlement on the vertebrate animals of arid Australia: a conceptual model. *Proc. Ecol. Soc. Aust.* **16:** 201–13.

Morton, S. R., Stafford Smith, D. M., Friedel, M. H., Griffin, G. F. and Pickup, G., 1995. The stewardship of arid Australia: ecology and landscape management. *J. Environ. Manage.* **43:** 195–217.

Noble, J. C. and Tongway, D. J., 1983. Pastoral settlement in arid and semi-arid rangelands. Pp. 217–42 *in* The Impact of Man on Australian Soils ed by J. S. Russell and R. F. Isbel. University of Queensland Press: St Lucia, Queensland.

Pickup, G. and Bastin, G. N., 1997. Spatial distribution of cattle in arid rangelands as detected by patterns of change in vegetation cover. *J. Appl. Ecol.* **34:** 657–67.

Pickup, G. and Chewings, V. H., 1988. Estimating the distribution of grazing and patterns of cattle movement in a large arid zone paddock: An approach using animal distribution models and Landsat imagery. *Intern. J. Rem. Sens.* **9:** 1469–490.

Pickup, G. and Chewings, V. H., 1994. A grazing gradient approach to land degradation assessment in arid areas from remotely-sensed data. *Intern. J. Rem. Sens.* **15:** 597–617.

Reid, J. R. W. and Fleming, M., 1992. The conservation status of birds in arid Australia. *Range. J.* **14:** 65–91.

Rice, W. R., 1989. Analyzing tables of statistical tests. *Evolution* **43:** 223–25.

Robinson, D., 1991. Threatened birds in Victoria: their distributions, ecology and future. *Vict. Natur.* **108:** 67–77.

Short, J. and Smith, A., 1994. Mammal decline and recovery in Australia. *J. Mammal.* **75:** 288–97.

Smith, A. P. and Quin, D. G., 1996. Patterns and causes of extinction and decline in Australian conilurine rodents. *Biol. Cons.* **77:** 243–67.

Stafford Smith, D. M. and Morton, S. R., 1990. A framework for the ecology of arid Australia. *J. Arid Environ.* **18:** 255–78.

Thomson, J. M., Long, J. L. and Horton, D. R., 1987. Human exploitation of and introductions to the fauna of Australia. Pp. 227–49 *in* Fauna of Australia Vol 1A: General Articles ed by G. R. Dyne and D. W. Walton. Australian Government Publishing Service: Canberra, Australian Capital Territory.

Woinarski, J. C. Z., Whitehead, P. J., Bowman, D. M. and Russell-Smith, J., 1992. Conservation of mobile species in a variable environment: the problem of reserve design in the Northern Territory, Australia. *Global Ecol. Biogeog. Letters* **2:** 108–18.

CHAPTER 30

Human impacts on landscapes: matrix condition and management priorities

S. McIntyre[1] and Richard Hobbs[2]

Human impacts on landscapes include both the complete removal or destruction of vegetation to create a mosaic of habitat and non-habitat, and various degrees of modification of the natural system. We argue that the classical fragmentation paradigm of patch-corridor-matrix does not capture the richness of possible landscape configurations, since it always assumes that the matrix is "non-habitat". We suggest instead that landscapes vary on the two axes of destruction and modification, and can be broadly partitioned into intact, variegated, fragmented and relictual types depending on the extent of habitat destruction and the extent to which the matrix is modified.

Management for conservation involves halting and sometimes reversing the trends of habitat destruction and modification. Primary management activities are the maintenance of existing good-quality habitat, the improvement of degraded habitat and the reconstruction of habitat where it has been destroyed. Priorities will differ according to the state of alteration of the landscape. The first priority will always be to maintain any existing good quality habitat. This includes the matrix in intact and variegated landscapes, but less intact habitat remains in fragmented systems, and it is likely that none remains in relictual landscapes. In these cases, improvement of degraded fragments is required. Relictual and fragmented landscapes also require "re-variegation", with priorities given to buffering fragments, restoring connectivity and rendering the matrix less hostile. In variegated landscapes, rehabilitation is less of a priority, but maintaining the condition of the matrix is important to halt trends towards fragmentation. Further development of this framework in the context of particular landscapes is now required.

Key words: Disturbance, Habitat fragmentation, Habitat variegation, Conservation, Biodiversity.

INTRODUCTION

HABITAT fragmentation is a term that has been used to describe the resultant effects of modern human society on ecosystems. Its enormous impact on the direction that conservation research has taken may, in part, be related to its simplicity i.e., a bifurcation of the landscape into habitat (fragment or patch) and non-habitat (matrix) categories. This is an extension of island biogeography concepts (MacArthur and Wilson 1967) into terrestrial landscapes, with habitat fragments representing islands and the destroyed matrix representing the ocean, a hostile barrier to terrestrial life. Developments in landscape ecology led to the identification of corridors, as well as patches and matrix (Forman 1995a), together with the concept that these elements varied in their suitability to the movement of biota (Forman 1995b).

Critics of the fragmentation model have been concerned that it is over-simplistic for some landscapes (McIntyre and Barrett 1992; Wiens 1994; Pearson *et al.* 1996). Since human impacts on landscapes can result in a continuum of variation, the delimitation of patches and identification of boundaries can be arbitrary (Kotliar and Weins 1990; Musick and Grover 1990). Even where habitat patches are clearly discernable, a continuum of levels of modification of these patches can be recognized.

[1]CSIRO Tropical Agriculture, 306 Carmody Road, St Lucia, Brisbane, Queensland, Australia 4067.
[2]CSIRO Wildlife and Ecology, Private Bag, P.O. Wembley, Western Australia, Australia 6014.
Pages 301–07 *in* NATURE CONSERVATION 5: NATURE CONSERVATION IN PRODUCTION ENVIRONMENTS: MANAGING THE MATRIX ed by J. L. Craig, N. Mitchell and D. A. Saunders. Surrey Beatty & Sons, 2000.

There is undoubtedly a continuing need for conceptual models that represent simplifications of complex landscapes. To refine these requires a balance of simplicity and realism to appropriately describe the range in variation of human impacts on landscapes. The boundaries these models impose on the landscape must represent significant differences in habitat condition, even if their exact location on the landscape is arbitarily delineated. The concept of landscape variegation (McIntyre and Barrett 1992) was such a refinement. It proposed to broaden the concept of fragmentation to highlight the fact that in some regions, the matrix is not composed of a hostile environment. Both models represent nodes on a continuum of ecosystem modification and/or destruction. This continuum has been described by Hobbs and Hopkins (1990) and McIntyre *et al.* (1996) who all recognized four levels of habitat alteration. Hobbs and Hopkins expressed these in terms of the prevalent land use — i.e., conservation of a more or less unmodified system, utilization of components of the system (e.g., forestry), replacement of the system with another type of system (e.g., agriculture, plantation forestry) and complete destruction (e.g., urban, mining). McIntyre *et al.* described the continuum of human impacts in terms of the patterns and severity of exogenous disturbances on natural ecosystems (Table 1).

The need to discriminate where a particular landscape lies on the continuum of human impacts has important implications as to how we go about assessing landscape condition (Aronson and Le Floc'h 1996). The purpose of this chapter is to consider how management priorities within a landscape might be related to its position in the alteration framework.

A FRAMEWORK TO DESCRIBE VARYING LEVELS OF LANDSCAPE ALTERATION

Our description of the range of human impacts on landscapes is essentially a depiction of two gradients of human impact on ecosystems: i) destruction and ii) modification. They can both be conceptualized as a continuum and each is associated with the effects of exogenous disturbance.

Following the concept of Fox and Fox (1986), we distinguish between disturbances to which the system has been exposed repeatedly through evolutionary time (termed endogenous disturbances), and disturbances which are novel to the system (termed exogenous disturbances). Exogenous disturbances are frequently the result of changing human activities, and tend to result in alteration of the ecosystem and irreversible loss of species. Exogenous disturbances can take the form of novel types of disturbance, changes to the endogenous disturbance regime or removal of endogenous disturbances. They can result in the destruction and modification of habitats as described below.

Habitat destruction

Habitat destruction results in loss of all structural features of the vegetation and loss of the majority of species. It results from the gross effects of exogenous disturbance and the gradient is represented as a continuum of spatial extent, four levels of which are depicted in Figure 1. Intact and relictual landscapes represent the extremes, while the two intermediate states, variegated and fragmented, represent landscapes where the habitat is progressively destroyed. In variegated landscapes, the habitat still forms the matrix, whereas in fragmented landscapes, the matrix comprises "destroyed habitat".

Each of the four levels described in Figure 1 is associated with a particular range of levels of habitat destruction. These thresholds are not entirely arbitary. The distinction between variegated and fragmented landscapes reflects suggestions by several authors that landscapes in which habitats persist on more than 60% of the area are operationally not fragmented, since they consist of a continuous cluster of habitat (Pearson *et al.* 1996; Wiens 1997). With (1997) has also recently indicated that a threshold in landscape connectivity may exist at between 55% and 60% habitat coverage.

Other than describe different extents of destruction, the model so far differs little from the habitat/non-habitat dichotomy of habitat fragmentation. By overlaying it with gradients of habitat modification, a more realistic scenario is built up.

Table 1. General associations between land use and levels of landscape alteration (synthesized from Hobbs and Hopkins 1990; McIntyre *et al.* 1996).

Four levels of landscape alteration along a continuum of levels of destruction	General form of land use associated with particular levels of destruction	Specific types of land use commonly associated with particular levels of destruction
Intact	Conservation	Lands maintained for nature conservation; where a land use remains undesignated; rangelands e.g., in arid regions.
Variegated	Utilization	Use of native ecosystems with some resulting modification and destruction e.g., native forestry, rangeland use, recreational use.
Fragmented	Replacement	Lands subjected to intensive uses e.g., plantation forestry, cropping or horticulture.
Relictual	Removal	Complete removal of vegetation e.g., urbanization, industry and mining.

Habitat modification

While the destruction gradient in Figure 1 represents a continuum of spatial extent of the severest effects of disturbance, habitat modification represents a continuum of intensity of disturbances. Habitat modification can occur in any of the situations depicted in Figure 1, and most of the land uses described in Table 1 are associated with modifications of the remaining habitat. Figure 2 illustrates this by depicting three levels of modification, although in reality, there is a continuum of modifications, and they may also be present as gradients. Modification acts to create a layer of variation in the landscape over and above the straightforward spatial patterning caused by vegetation destruction.

If a modifying disturbance is intense, or protracted enough, it will eventually lead to habitat destruction (Hobbs and Norton 1996; Yates and Hobbs 1997). Thus the patterns of destruction depicted in Figure 1 could be created quickly (e.g., through broad-scale vegetation clearance) or gradually over time as a result of the cumulative effects of intense modification.

The levels of modification depicted in Figure 2 reflect the tendency for habitats to become progressively more modified with increasing levels of destruction, owing to the progressively greater proportion of edge in remaining habitats. Theoretically, all combinations of habitat destruction and modification gradients could exist. In practice, it is highly unlikely to find a relictual, or even a fragmented landscape, containing unmodified remnants. This pattern is an important consideration for management planning.

HOW SHOULD WE PERCEIVE THE LANDSCAPE?

A human-oriented classification

The classification of spatial pattern and modification described above is a human-oriented one, in that it is based on the gross physical state of habitats (level of destruction) and an interpretation of endogenous and exogenous disturbances and their impact (level of modification). We suggest that this perspective represents some kind of average impact of humans on the biota, or at least the impact on the more visible parts of the ecosystem e.g., higher plants and animals. It is the perspective widely applied by researchers who describe landscapes as fragmented and, if we are considering the impacts of humans on habitats, it seems quite appropriate to "see" the landscape in terms of human perceptions. It is consistent with the scale and patterns of land use impacts.

The perspective of the biota

It is also necessary to go beyond the human perspective and consider how individual species

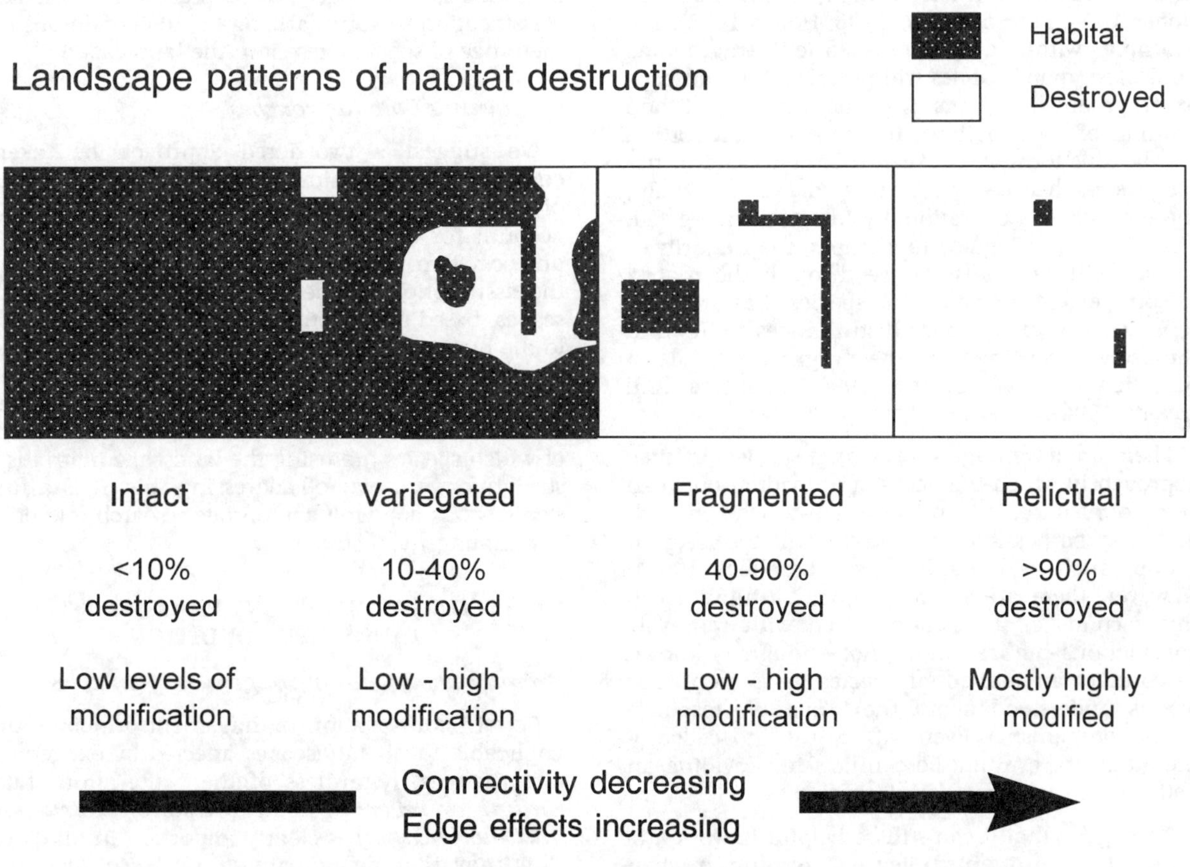

Fig. 1. Four levels of landscape alteration on a continuum of habitat destruction, each characterized by a range of proportions of habitat destroyed. Increasing destruction is generally associated with increasing levels of modification.

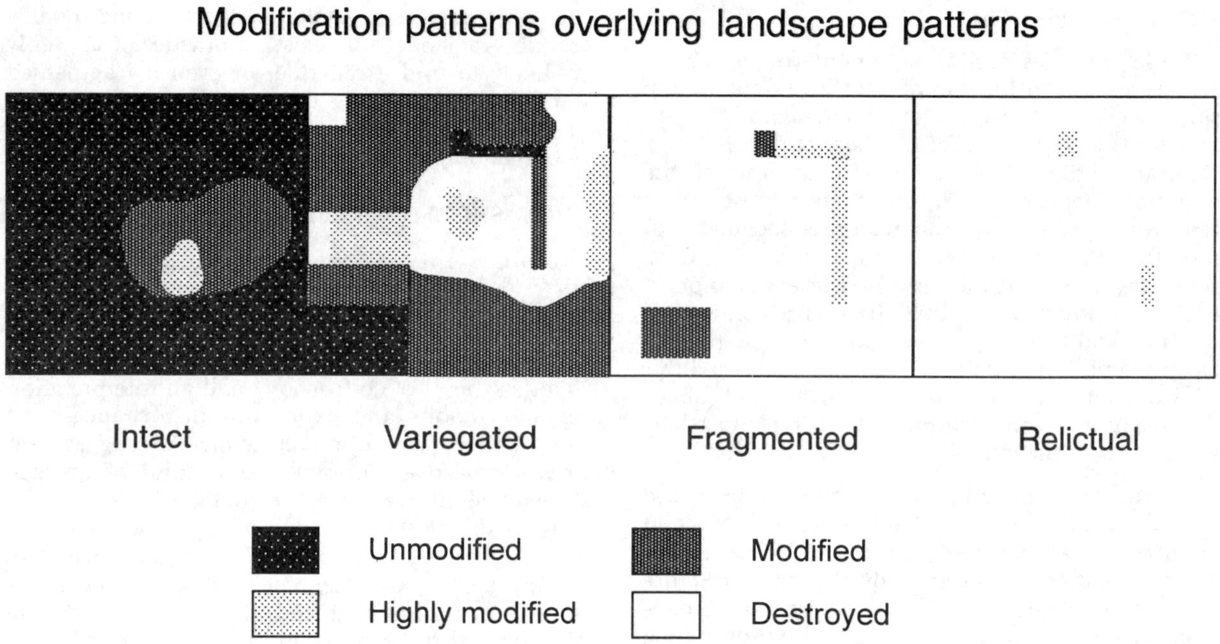

Fig. 2. Modifications resulting from exogenous disturbances overlying landscape patterns of habitat destruction. Although any combination of modification and destruction levels is theoretically possible, those considered to be typical of different destruction levels are illustrated.

respond to landscape alterations. An individual species' response will be determined by the spatial scale at which it perceives the environment and moves through it (Andrén 1994, 1996; Cale and Hobbs 1994; Pearson *et al.* 1996; Hobbs 1998). For example, within a (human-defined) fragmented landscape, many species will perceive the landscape as fragmented (plants with poor dispersal and animals of low mobility). In contrast, a proportion will be able to move between fragments and/or use the matrix between fragments so that for them, the landscape is operationally intact or variegated. Similarly, in variegated landscapes, the majority of species will be able to move through the matrix to some extent. However, the species that are most sensitive to exogenous disturbances will find themselves confined to the fragments of least modified habitat (Barrett 1994; McIntyre and Lavorel 1994).

There are advantages to taking a species-oriented approach in which the needs of individual taxa can be accounted for. It is the most fine-tuned approach and may be possible for well-known, species-poor groups such as birds (e.g., see Lambeck 1997). However, there are severe practical limitations to this second-tier assessment. Even with relatively conspicuous higher plants, not enough is known to conduct an individual species assessment. Far less is understood about the landscape response of microorganisms. Even with a full knowledge of the biota, there would be conflicts in providing an optimum environment for all species.

We suggest that it can still be helpful to work at a lower (human-oriented) level of resolution when considering biotic conservation at the scale of the landscape. This is particularly so when there is ignorance about the communities, or when a strategic planning/policy level is being considered. The important question that still remains to be fully answered is: how adequate is the human-oriented classification in summarizing habitat conditions for the range of species present in the landscape?

Combining both approaches

We suggest a two-tiered approach be taken, combining both human and biotic perspectives of landscape alteration. It is patently difficult to account for the individual needs of all species in an ecosystem. However, there is still a need to limit the assumptions made about how biota use landscapes, based on structural features that are obvious to the human eye only. Such assumptions can result in a mis-identification of landscape state e.g., when fragmentation models are inappropriately applied to variegated landscapes. An objective assessment of which species are using the landscape matrix is a straightforward way of addressing this. It is a first step towards applying appropriate research questions and management strategies.

FITTING MANAGEMENT PRIORITIES TO LANDSCAPE CONDITION

Broadly applicable management principles

Certain conservation management principles are applicable to all landscapes affected by exogenous disturbances, regardless of their alteration state. *Halting or reversing the processes of destruction and modification* is clearly important in all cases. This may include priority efforts to *protect and buffer areas in best condition* and if possible to *maintain or create connectivity* in the landscape.

What is the desired outcome in terms of the final landscape configuration?

Resources for management are generally limiting and consequently, not all landscapes are likely to be restored to an intact, unmodified condition, even if it were practically possible. However, it should be reasonable for intact landscapes to be maintained in an intact condition. Failing that, and because landscapes that have been severely altered by human activities will continue to be subject to some levels of human land use, it is useful to consider the minimal state of a landscape that will support viable habitat of the ecosystem subjected to disturbances. The following comments assume the previously noted downward trend in condition (i.e., increasing modification levels in landscapes with higher levels of destruction as illustrated in Figs 1 and 2).

Of the four alteration states described, variegated landscapes offer the greatest potential to support viable ecosystems in the face of human land use. Laurance and Gascon (1997) advocate "creative" fragmentation of forested landscapes that are subject to development, and incorporate design guidelines that would more likely result in a variegated rather than fragmented landscape. Viewing the matter from the opposite end of the destruction gradient, where landscapes are fragmented, there would be benefit from "re-variegation" — rendering the matrix less hostile, buffering fragments and increasing connectivity (e.g., Hobbs and Saunders 1993). In relictual landscapes, it should not be assumed that a variegated landscape is achievable, and options would range from abandonment of remnants considered non-viable, to major exercises in improvement and enlargement of remnants.

How should the framework classification affect management priorities in different landscapes?

We consider three actions that could be applied to habitats for conservation management.

1. *Maintain* the existing condition of habitats by removing and controlling the effects of exogenous disturbance. This is a baseline activity, and its importance lies in the fact that it is much easier to avoid the effects of degradation than it is to reverse them. Maintenance is therefore a priority in the less-altered landscapes (variegated and intact).
2. *Improve* the condition of habitats by reducing or removing exogenous disturbances. It may also be necessary to instigate more active management to initiate a reversal of condition (e.g., removal of exotic species, re-introduction of native species) in highly modified habitats. It may also be necessary to reinstate endogenous disturbances.
3. *Reconstruct* habitats where their total extent has been reduced below viable size using replanting and re-introduction techniques. As this is so difficult and expensive, it is a last resort action that is most relevant to fragmented and relictual landscapes. Restoration will not come close to restoring habitats to their unmodified state (Diamond 1987), reinforcing the wisdom of maintaining existing ecosystems as a priority.

Where on the landscape to allocate greater and lesser efforts towards different management actions?

Hobbs and Hopkins (1990) described the major impacts of different types of land use (as defined in Table 1) on ecosystem components and processes. They also provided pointers for the sorts of actions needed in order to reverse these impacts in order to restore functioning ecosystems. McIntyre *et al.* (1996) also provided a set of management priorities for systems with different degrees of alteration. In the former publication, the term reconstruction was used in the broadest sense; in the latter it was used narrowly, as defined above.

Table 2 summarizes this information in the context of the actions described above, but explicitly recognizing parts of the landscape (matrix, connecting areas, buffer areas etc.) in which these actions would be most effective. The table highlights the different priorities for management action in the different alteration levels of landscapes. The first consideration is the *maintenance* of elements which are currently in good condition. This will be predominantly the vegetated matrix in intact and variegated landscapes and the remnants which

Table 2. Management actions and location in the landscape where they may be most effective for conservation and ecosystem function. Actions and locations vary with the alteration state of particular landscapes. *Matrix* — the predominant part of the landscape and consists of habitat (in intact and variegated landscapes) or destroyed habitat (in fragmented and relictual landscapes); *Fragments* — restricted areas of habitat surrounded by areas of destroyed habitat (most relevant to fragmented and relictual landscapes); *Patches* — patches of least modified habitat against a background of more highly modified habitat (most relevant to intact and variegated landscapes); *Buffer areas* — occur around fragments or patches; *Connecting areas* — occur between fragments or patches.

	Landscape alteration level			
Action	Intact	Variegated	Fragmented	Relictual
1) Maintain	Matrix	Matrix	Least modified fragments	—
2) Improve	Patches, Connecting areas	Buffer areas, Connecting areas, Patches	Fragments, Matrix	Fragments
3) Reconstruct	—	—	Buffer areas, Connecting areas	Buffer areas

remain in good condition in fragmented landscapes. We assume that there will be no remnants left in good condition in relictual landscapes. Effective maintenance will entail the continuation of population, community and ecosystem processes which result in the persistence of the species and communities present in the landscape. However, this has to be considered in the context of the dynamic nature of ecosystems, which implies that community structure and composition will change over time and in response to disturbance and climatic events (Pickett *et al.* 1992; Hobbs and Norton 1996).

The second consideration is the *improvement* of critical elements in the landscape. In variegated landscapes, effort should be concentrated on buffer areas and/or corridors associated with patches of least modified habitat. In fragmented systems, improving the surrounding matrix to reduce degrading influences will be the priority. Note that in a fragmented system, it may be impossible to maintain fragments in good condition without also monitoring activities in the matrix to control landscape processes, such as hydrology. In relictual landscapes, improving the condition of fragments will be essential for their continued persistence.

Improvement may involve simply the removal of degrading influences such as stock grazing or feral predators, or active management to restore ecosystem processes, improve soil structure, encourage regeneration of plant species, or reintroduction of flora or fauna formerly present there (Hobbs and Yates 1997). An interesting question for fragmented and relictual landscapes is how to ensure the continuation of endogenous disturbance processes within fragments when the scale of previous disturbances may have been larger than the fragments.

Reconstruction is likely to be necessary only in fragmented and relictual areas. Primary goals of reconstruction will be to provide buffer areas around fragments and to increase connectivity with corridors, although the provision of additional habitat may also be a long-term goal (Hobbs 1993). While some basic principles of habitat reconstruction have been put forward (e.g., Porteous 1993; Recher 1993; Saunders and Hobbs 1995), the efficacy of such measures in creating effective buffers, corridors or habitat has rarely been tested. Questions remain about which characteristics of "natural" habitat are the most important to try to incorporate into reconstruction, and what landscape configurations are likely to be most effective.

Does all this really help?

The scheme we have outlined here attempts to provide a broad classification of landscapes as a basis on which strategic management decisions can be made. Identification of where a particular landscape sits in terms of its degree of destruction and modification is essential for deciding where the priorities for management should be. While this may appear obvious, a failure to recognize the different management needs in different landscapes can result in a significant mis-application of scarce resources into activities which are either inappropriate or unneccesary.

Having said that, we also recognize that the important next step is to develop more detailed management scenarios for particular landscapes. The extent to which we can generalize across landscapes within broad types is currently unknown. For instance, will the same management options be appropriate within a fragmented landscape with 50% native vegetation left and within one with 15% left? Clearly, more work needs to be done on assessing the importance of the extent and patterning of remnant vegetation, and changes in this, to overall biodiversity and selected elements thereof.

The only way we are likely to accumulate such information is to glean it from real landscapes and to utilize ongoing management activities as experimental opportunities. This involves a departure from traditional plot-based single-discipline science. An effective programme would ideally have the following components: on-ground works with stakeholder involvement; extensive inter-agency collaboration involving a large multi-disciplinary team, with a strategic plan, an operational plan, and three monthly milestones with well defined performance criteria; a communication plan (with properly identified stakeholders, appropriately targeted); a technical reference group, a management reference group and a research steering committee. Alternatively, we can struggle along as best we can, with limited time and resources, making alliances and collaborating where possible, and circumventing institutional turf battles. However we do it, we feel that the further exploration of the kinds of issues we have raised here will be essential for the effective conservation of biodiversity in production landscapes.

REFERENCES

Andrén, H., 1994. Effects of habitat fragmentation on birds and mammals in landscapes with different proportions of suitable habitat: a review. *Oikos* **71:** 355–66.

Andrén, H., 1996. Population responses to habitat fragmentation: statistical power and the random sample hypothesis. *Oikos* **76:** 235–42.

Aronson, J. and Le Floc'h, E., 1996. Vital landscape attributes: missing tools for restoration ecology. *Restoration Ecology* **4:** 377–87.

Barrett, G., 1994. Conservation of woodland birds in a fragmented rural landscape. *Pac. Conserv. Biol.* **1:** 245–56.

Cale, P. and Hobbs, R. J., 1994. Landscape heterogeneity indices: problems of scale and applicability, with particular reference to animal habitat description. *Pac. Conserv. Biol.* **1:** 183–93.

Diamond, J., 1987. Reflections on goals and on the relationship between threory and practice. Pp. 329–36 *in* Restoration Ecology: A Synthetic Approach to Ecological Research ed by W. R. Jordon III, M. E. Gilpin and J. D. Aber. Cambridge University Press: Cambridge, Great Britian.

Forman, R. T. T., 1995a. Some general principles of landscape and regional ecology. *Landsc. Ecol.* **10:** 133–42.

Forman, R. T. T., 1995b. Land mosiacs. The ecology of landscapes and regions. Cambridge University Press: Cambridge.

Fox, M. D. and Fox, B. J., 1986. The susceptibility of natural communities to invasion. Pp. 57–66 *in* Ecology of Biological Invasions: An Australian Perspective ed by R. H. Groves and J. J. Burdon. Australian Academy of Science: Canberra, Australian Capital Territory.

Hobbs, R. J., 1993. Can revegetation assist in the conservation of biodiversity in agricultural areas? *Pac. Conserv. Biol.* **1**: 29–38.

Hobbs, R. J., 1998. Managing ecological systems and processes. Pp. 459–84 *in* Ecological Scale: Theory and Applications ed by D. Peterson and V. T. Parker. Columbia University Press: New York.

Hobbs, R. J. and Hopkins, A. J. M., 1990. From frontier to fragments: European impact on Australia's vegetation. *Proc. Ecol. Soc. Aust.* **16**: 93–114.

Hobbs, R. J. and Norton, D. A., 1996. Towards a conceptual framework for restoration ecology. *Restor. Ecol.* **4**: 93–110.

Hobbs, R. J. and Saunders, D. A. (eds), 1993. Reintegrating Fragmented Landscapes. Towards Sustainable Production and Conservation. Springer-Verlag: New York.

Hobbs, R. J. and Yates, C. J., 1997. Moving from the general to the specific: remnant management in rural Australia. Pp. 131–42 *in* Frontiers in Ecology: Building the Links ed by N. Klomp and I. Lunt. Elsevier.

Kotliar, B. N. and Weins, J. A., 1990. Multiple scales of patchiness and patch structrure: a hierarchial framework for the study of heterogeneity. *Oikos* **59**: 253–60.

Lambeck, R., 1997. Focal species: a multi-species umbrella for nature conservation. *Conserv. Biol.* **11**: 849–56.

Laurance, W. F. and Gascon, C., 1997. How to creatively fragment a landscape. *Conserv. Biol.* **11**: 577–79.

MacArthur, R. H. and Wilson, E. O., 1967. The theory of island biogeography. Princeton University Press: Princeton, New Jersey.

McIntyre, S. and Barrett, G. W., 1992. Habitat variegation, an alternative to fragmentation. *Conserv. Biol.* **6**: 146–47.

McIntyre, S., Barrett, G. W. and Ford, H. A., 1996. Communities and ecosystems. Pp. 154–70 *in* Conservation Biology ed by I. F. Spellerberg. Longman: Harlow, England.

McIntyre, S. and Lavorel, S., 1994. Predicting richness of native, rare, and exotic plants in response to habitat and disturbance variables across a variegated landscape. *Conserv. Biol.* **8**: 521–31.

Musick, B. H. and Grover, H. D., 1990. Image textural measures as indices of landscape pattern. Pp. 77–104 *in* Quantative Methods in Landscape Ecology — The Analysis and Interpretation of Landscape Heterogeneity ed by M. G. Turner and R. H. Gardner. Springer-Verlag: New York.

Pearson, S. M., Turner, M. G., Gardner, R. H. and O'Neill, R. V., 1996. An organism-based perspective of habitat fragmentation. Pp. 77–95 *in* Biodiversity in Managed Landscapes: Theory and Practice ed by R. C. Szaro and D. W. Johnston. Oxford University Press: New York.

Pickett, S. T. A., Parker, V. T. and Fielder, P., 1992. The new paradigm in ecology: implications for conservation biology above the species level. Pp. 65–88 *in* Conservation Biology: The Theory and Practice of Nature Conservation ed by P. Fielder and S. Jain. Chapman and Hall: New York.

Porteous, T., 1993. Native Forest Restoration. A Practical Guide for Landowners. Queen Elizabeth the Second National Trust: Wellington, New Zealand.

Recher, H. F., 1993. The loss of biodiversity and landscape restoration: conservation, management, survival — an Australian perspective. Pp. 141–51 *in* Nature Conservation 3: Reconstruction of Fragmented Ecosystems ed by D. A. Saunders, R. J. Hobbs and P. Ehrlich. Surrey Beatty & Sons: Chipping Norton, New South Wales.

Saunders, D. A. and Hobbs, R. J., 1995. Habitat reconstruction: the revegetation imperative. Pp. 104–12 *in* Conserving Biodiversity: Threats and Solutions ed by R. A. Bradstock, T. D. Auld, D. A. Keith, R. T. Kingsford, D. Lunney and D. P. Siversten. Surrey Beatty & Sons: Chipping Norton, New South Wales.

Wiens, J. A., 1994. Habitat fragmentation: island *v* landscape perspectives on bird conservation. *Ibis* **137**: S97–S104.

Wiens, J. A., 1997. Metapopulation dynamics and landscape ecology. Pp. 43–62 *in* Metapopulation Biology: Ecology, Genetics, and Evolution ed by I. A. Hanski and M. E. Gilpin. Academic Press: New York.

With, K. A., 1997. The application of neutral landscape models in conservation biology. *Conserv. Biol.* **11**: 1069–080.

Yates, C. J. and Hobbs, R. J., 1997. Woodland restoration in the Western Australian wheatbelt: conceptual framework using a state and transition model. *Restor. Ecol.* **5**: 28–35.

CHAPTER 31

Perth's Bushplan — towards conserving the natural ecosystems of Western Australia's capital city

G. Whisson,[1] B. J. Keighery,[1] D. Nunn[2] and G. McGill[2]

Perth's Bushplan is an integrated programme to identify and conserve a comprehensive representation of the original biological diversity present in the Perth Metropolitan Area. The programme is established under a government policy — the Urban Bushland Strategy — that recognized the need for conservation of a target of at least 10% of each of the vegetation complexes present in the region. Perth's Bushplan represents a second generation review of biodiversity conservation requirements in the Perth area, building on the recommendations and experience arising from the System 6 report released in 1983.

Initially public submissions were called for areas to be considered for inclusion in this plan. The selection of areas was based on regional information on the remnant bushland in the Perth Metropolitan Area, vegetation complex mapping, identification of floristic community types, wetland mapping and classification, identification of threatened communities and species and details of each area's flora, and to a lesser extent fauna. Areas recognized include public and privately owned lands and range from small areas of several hectares (a threatened community) to comparatively large areas of several thousand hectares.

Significant lessons learnt from the experiences in implementing System 6 recommendations have influenced the approach adopted in Perth's Bushplan. This programme is an integrated exercise that combines: identification of bushland areas; development of a suite of mechanisms to conserve these areas; and the establishment of a co-ordinated process for implementing the recommendations.

Perth's Bushplan has evolved to be a whole of government and community initiative that is necessary if the still substantial unique natural areas of Perth are to be protected for now and the future. This chapter considers some of the lessons learnt in implementing System 6 recommendations and approaches developed from this experience for Perth's Bushplan.

Perth's Bushplan is to be released as a draft for three months public comment in late 1998.

Key words: Conservation, Bushplan, Integration, Vegetation complex, Floristic community types, Threatened Ecological Communities.

[1]Department of Environmental Protection, 141 St George's Terrace, Perth, Western Australia, Australia 6000.
[2]Ministry for Planning, 469 Wellington Street, Perth, Western Australia, Australia 6000.

Pages 308–17 *in* NATURE CONSERVATION 5: NATURE CONSERVATION IN PRODUCTION ENVIRONMENTS: MANAGING THE MATRIX ed by J. L. Craig, N. Mitchell and D. A. Saunders. Surrey Beatty & Sons, 2000.

INTRODUCTION AND REGIONAL CONTEXT

PERTH, Western Australia's capital, and much of its surrounding Metropolitan Area is superimposed on the diverse natural ecosystem of the Swan Coastal Plain Bioregion (Fig. 1). Though extensively cleared, substantial intact remnants of this ecosystem still exist within the boundaries of the Perth Metropolitan Area. For the past three years the State Government, through its planning and conservation agencies and the community of Perth, have been identifying regionally significant natural areas in a programme to review and update conservation recommendations for Perth and the surrounding region.

The Swan Coastal Plain Bioregion is a nationally recognized bioregion in the Interim Biogeographic Regionalization of Australia (Thackway and Cresswell 1995). It forms part of the South-West Botanical Province of Western Australia, a region influenced by winter rainfall that extends from Shark Bay on the west coast, through to the southern coastline at the western edge of the Nullarbor Plain. This Province has a rich and largely (70–80%) endemic flora of some 9 000 plant taxa, which is recognized nationally and internationally as one of the world's mega-diverse regions.

Almost 1 400 plant taxa have been identified on the Perth Metropolitan portion of the Swan Coastal Plain, including 21 Rare Flora and 74 Priority Flora (after Atkins 1998). Eight taxa appear to be endemic to the Metropolitan portion of the Plain. Many of the endemics on the Coastal Plain are restricted to the eastern side of the Plain, while another group are confined to soils on outcropping coastal limestone on the western side of the Plain.

Perth's Bushplan brings together regional vegetation patterning which has been determined in two ways. Both show that the pattern of plant communities on the Swan Coastal Plain is influenced by the broad landform units.

In the late 1970s vegetation complexes in the region were defined and mapped (Heddle *et al.* 1980) in relation to landform/soil units as determined by Churchward and McArthur (1980). The delineation of vegetation complexes is based on the concept of a series of vegetation communities forming regularly repeating complexes associated with a particular soil unit. Vegetation communities may occur in more than one vegetation complex, but the combination and relative proportions of the communities in a complex are unique. A total of 38 vegetation complexes are mapped on the Swan Coastal Plain, 26 of which occur in the Metropolitan Area.

More recently plot based floristic studies (Gibson *et al.* 1994) have identified that much of the floristic diversity on the Plain derives from often subtle interactions between soil types and variations in seasonal soil moisture from upland areas to seasonally saturated soils and areas subject to varying periods of inundation.

Four major floristic super-groupings have been found to be associated with: the heavy soils of the eastern side of the Plain; the old highly leached Bassendean Dunes; the younger, more coastal and often lime rich Spearwood and Quindalup Dunes; and a group associated with the seasonal wetlands and linked more to moisture conditions than soil type.

The highest plant species diversity is found in the woodlands and wetlands of the eastern side of the Plain (foothills and Pinjarra Plain), and the woodlands on Bassendean Dunes.

Vegetated wetlands and lakes are a prominent feature of the Swan Coastal Plain and account for a considerable proportion of its biological and landscape diversity. The distribution of fauna species is influenced more by the patterning of the major landform units, substrate types and wetlands than by the vegetation community types.

The Perth Metropolitan Area occupies approximately ⅓ of the Swan Coastal Plain Bioregion (Thackway and Cresswell 1995) (Fig. 1), covering an area some 110 km long, and up to 30 km wide. The Metropolitan Area also extends further inland onto the Darling Scarp and Plateau (Jarrah Bioregion) where extensive tracts of the bushland remain on State Forest, Water Catchment and Conservation Reserve lands, however, this study is confined to the Coastal Plain portion of the Perth Metropolitan Area.

PERTH'S URBAN BUSHLAND

Perth is very fortunate to still retain substantial areas of bushland in the Metropolitan Area. In all more than 55 000 ha (10% of the Metropolitan Area) is "reserved" in the Metropolitan Region Scheme as regional open space (this includes developed parkland areas as well as conservation areas). A further 15 000 ha is included in amendments to the Metropolitan Region Scheme currently in progress.

Overall some 30% of the Metropolitan portion of the Swan Coastal Plain retains its natural bushland. This clearing has been highly selective and many individual plant communities are highly cleared and their remnants are fragmented and poorly represented in protected areas.

The selective clearing patterns are a reflection of the extreme infertility of some of the sandy soils of the Swan Coastal Plain that are unsuitable for agriculture and still remain relatively intact beyond the expanding urban front. More recently, rapid growth in Perth's population has lead to significant clearing associated with urban expansion, particularly in near coastal development corridors and areas close to the city.

The more fertile damplands and the clay/loam soils of the eastern side of the Coastal Plain were extensively cleared for agriculture. Ironically, despite this selective clearing plant communities on the eastern side of the Plain are often somewhat

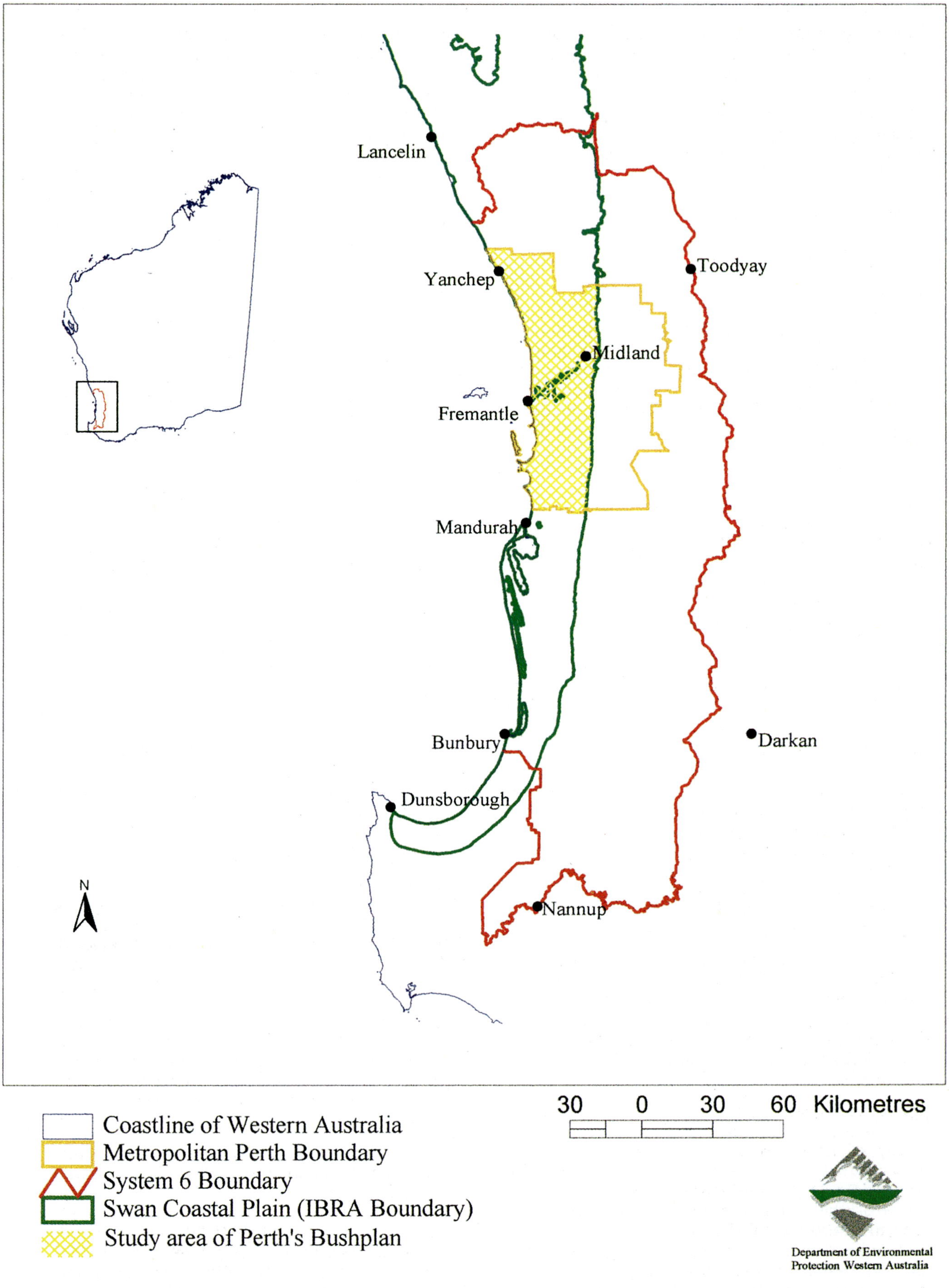

Fig. 1. Perth's Bushplan Study Area and associated boundaries.

better represented in the more intensively settled Metropolitan Area than on the remainder of the Coastal Plain. Outside the Perth Metropolitan Area more systematic clearing for agriculture has removed up to 97% of the original bushland (Burbidge and Rolfe 1987) from the heavy soil areas of the Swan Coastal Plain and is likely to have totally eliminated some communities. Over the 220 km length of the Swan Coastal Plain south of Perth, only two bushland corridors remain (both along road and rail reserves) as examples of the original vegetation changes across the eastern side of the Plain.

The Bushplan study area supports approximately 80% of the population of Western Australia and is subject to the most intensive land use pressures in the state. For this reason bushland in this region is particularly important for a variety of human needs, as well as conservation values. The bushland plays a crucial role in providing a "sense of place" and visual relief from the expanding built landscapes of the suburbs. In an urban environment and lifestyle that can be increasingly physically and culturally isolated from the natural world, bushland in urban areas also provides an important, enriching opportunity for contact with nature; and places in which to walk, play or find peace.

CONSERVATION IN THE PERTH METROPOLITAN AREA

System 6 Report

Between 1975 and 1983 the Environmental Protection Authority released a series of reports making recommendations for conservation reserves throughout Western Australia. For this purpose Western Australia was divided into 12 regions (referred to as Systems). System 6 covered most of the Coastal Plain and adjacent Darling Plateau (Jarrah Bioregion) representing the Perth Metropolitan Area and the surrounding region. The System 6 report (Department of Conservation and Environment 1983) was the first systematic study to identify areas of bushland, landscape and open space of regional significance in the Perth region.

Based on the information available at the time, System 6 made 108 recommendations for conservation areas in the Perth Metropolitan Area. A number of these recommendations identified areas of significance for landscape and public open space values more than biodiversity conservation significance.

Fourteen years on, the System 6 report can be seen as a landmark document for its time. With many of its recommendations implemented, it has resulted in the reservation of extensive areas and had a lasting influence on the retention of most of the remaining areas it recommended. System 6 also identified requirements for regional open space, many of which reflected the Metropolitan Region Planning Scheme, and introduced the concept of Regional Parks, which were intended to cover both private and public lands of conservation and public open space value. The Western Australian Government has recently committed to the establishment and funding of Regional Parks as a new category of reserve in the *Conservation and Land Management Act* and to vest Regional Parks in the National Parks and Nature Conservation Authority.

While System 6 has been a major success in facilitating the recognition and protection of its recommendation areas, this success has to be qualified by acknowledging that some important recommendation areas have been reduced in size and others lost.

Significant lessons have been learnt from the experience of implementing System 6. The System 6 area specific recommendations were accepted by government which approved the progressive implementation of the recommendations "as far as possible". The recommendation areas were not given statutory recognition, nor (at the time) was a systematic programme developed to co-ordinate their implementation. It was envisaged that they would be achieved progressively, primarily through ongoing planning processes.

These factors have complicated and protracted efforts to provide statutory protection for the recommendation areas. As a result the community was expressing increasing concerns about the time taken to implement recommendations, which they interpreted as a lack of success in protecting a number of the System 6 recommendations. The continuing proposals for development in some of these areas is also an issue of concern to the community. Moreover, addressing planning requirements and environmental impact assessment of these proposals requires a significant allocation of resources from the Department of Environmental Protection, the Environmental Protection Authority, the Ministry for Planning and other agencies.

In recent years a series of disputes arose over proposed developments involving the clearing of bushland areas that, while not identified for conservation in the System 6 report had become recognized by the community for their high conservation values. Substantial new biological information was available to support this recognition. These issues increased uncertainty for the community as to the adequacy of the conservation recommendations. The public disputes eroded confidence of developers over the status of their land and what development may or may not be permitted.

The environment, conservation, planning and water resource agencies in Western Australia responded independently to the need for improved information on regional biodiversity and conservation requirements by initiating a range of programmes that sought to gain a better understanding of various aspects of the biological diversity in the region and to update conservation recommendations to take account of new information.

Urban bushland conservation policy framework

Recent years have seen the development of more comprehensive policies for the conservation of biological diversity at state and national levels that reflect international treaty commitments under the Convention on Biological Diversity (United Nations Environment Programme 1992). This policy framework has been important, providing an approved basis for setting conservation objectives and targets.

Three policy documents are central to the objectives and recommendations in Perth's Bushplan:

National Strategy for Conservation of Australia's Biological Diversity (Commonwealth of Australia 1996) — This is the peak national strategy endorsed by the Commonwealth Government and the governments of each State and Territory of Australia. It establishes a goal, a number of principles, and a series of objectives with associated actions to conserve biological diversity throughout Australia.

In relation to Protected Areas, it sets the national objective:

> "To establish and manage a comprehensive, adequate and representative system of protected areas covering Australia's biological diversity." It also recognizes the need to "strengthen off-reserve conservation of biological diversity" and to provide for urban conservation.

Urban Bushland Strategy (Government of Western Australia 1995) — The Urban Bushland Strategy makes a number of policy statements framing conservation actions for regionally significant bushland in the Perth Metropolitan Area. The key statements for the purpose of establishing a regional conservation system is that of accepting that the State Government has responsibility for protecting regionally significant urban bushland, and establishing a target figure of retaining not less than 10% of each vegetation complex, in not less than five separate areas. This is the first time that specific biodiversity targets have been established in strategic planning for the Perth Metropolitan Area.

The Strategy also recognizes the importance of connectivity along corridors in the fragmented urban landscape, and the need for off-reserve conservation through alternative protective measures such as covenants.

Wetlands Conservation Policy for Western Australia (Government of Western Australia 1997) — Establishes several principle objectives including:

> "To prevent further loss or degradation of valuable wetlands and wetland types and promote wetland conservation, creation and restoration" and
>
> "To include viable representatives of all major wetland types and key wildlife habitats and associated flora and fauna within a statewide network of appropriately located and managed conservation reserves which ensures the continued survival of species, ecosystems and ecological functions."

This policy is significant as Perth's Bushplan protects a diverse range of wetlands, including two Ramsar listed wetlands of international significance (United Nations Education, Scientific and Cultural Organization 1971) in Thompsons Lake and Forrestdale Lake.

PERTH'S BUSHPLAN — AN INTEGRATED PROGRAMME TO UPDATE CONSERVATION RECOMMENDATIONS

The Urban Bushland Strategy (Government of Western Australia 1995) was the State Government's response to increasing community concerns over continuing losses of urban bushland. The strategy referred to the recently initiated System 6 update programme and also proposed to develop a regional bushland strategy.

While not required by the Urban Bushland Strategy, in 1996 the Department of Environmental Protection, Ministry for Planning, Department of Conservation and Land Management and the Water and Rivers Commission, took the step to integrate their various initiatives relating to urban bushland into one "whole of government" approach to conservation of a comprehensive, adequate and representative system of conservation areas for the Perth Metropolitan Swan Coastal Plain, which became known as Perth's Bushplan.

Perth's Bushplan has gained from the experience in implementing the System 6 recommendations which highlighted a need to:

- improve the regional context and area specific information on the conservation values of recommendation areas;
- improve co-ordination between key government agencies on biodiversity issues;
- revise approaches to implementation of conservation recommendations;
- provide statutory recognition to the conservation recommendation areas; and
- to broaden the toolbox of mechanisms available to conserve and manage regionally significant bushland.

Objectives of Perth's Bushplan

Perth's Bushplan is designed to become the primary mechanism for implementing the government's policy commitments for conservation of regionally significant bushland in the Perth region. When completed it will replace System 6 recommendations for this area.

The four objectives of Perth's Bushplan are:

- To develop a plan that meets the needs and aspirations of the community of Western Australia for the appropriate protection of bushland of regional significance in the Swan Coastal Plain portion of the Perth Metropolitan Area;

- To recommend a conservation system that is, as far as achievable, comprehensive, adequate and representative of the ecological communities and habitats of the region;
- To propose a range of measures to enable the implementation of Perth's Bushplan's recommendations for the protection of regionally significant bushland; and
- To bring greater certainty to the process of land use planning and environmental approvals by the early identification and protection of areas of regional conservation value.

Information sources

Perth's Bushplan is based on comprehensive information on the biodiversity of the Swan Coastal Plain. The main information sources used in the programme are as follows:

Floristic survey — Department of Conservation and Land Management, Department of Environmental Protection, Wildflower Society of Western Australia, Conservation Council and numerous other groups and individuals.

The principal regional information on patterning of plant communities came from an analysis of floristic data on over 1 100 quadrats located across the entire Plain. A major study (Gibson *et al.* 1994) provided the initial regional framework identifying 43 regional floristic community types. Subsequent studies by the Department of Environmental Protection, Griffin (1994) and G. Keighery (1996) which extended the area and supplemented the detail of this work identified a total of 66 regional floristic community types when compared with the groupings determined by Gibson *et al.* (1994). In addition, floristic information from a large number of published and unpublished studies on individual areas, ranging from detailed studies listing the total flora to description of the principal dominants were accessed wherever possible.

Mapping of Bushland in the Perth Metropolitan Area — Ministry for Planning with support from Department of Environmental Protection/Agriculture Western Australia.

The final bushland maps are derived from 1997, 1:20 000 digital orthophoto/cadastral maps, with field verification where vegetation condition or boundaries are uncertain. They provide fine scale resolution of bushland boundaries on a common cadastral base that ensures accurate registration with property boundaries. This mapping updates previous bushland mapping and condition assessment undertaken by the Ministry for Planning for the Perth Environment Project (Dixon *et al.* 1994 and an unpublished update in 1996).

Digitization of Vegetation Complex maps (Heddle *et al.* 1980) — Ministry for Planning.

Provides the relationship of recommendation areas with the regional Vegetation Complex maps that form the basis of the Western Australian Government's commitments in the Urban Bushland Strategy.

Wetland mapping and classification — Water and Rivers Commission (Hill *et al.* 1996a and b).

Detailed mapping of wetlands on the Swan Coastal Plain revealed over 15 000 wetlands. The wetlands in the Metropolitan Area were subsequently surveyed and assigned to categories of Conservation, Resource Enhancement, or Multiple Use (Hill *et al.* 1996). For the purposes of Perth's Bushplan, some Conservation Category wetlands were further field surveyed in 1997 to ensure that they had retained their values.

Threatened Ecological Community assessment — Department of Conservation and Land Management (English and Blyth 1997).

Four categories of Threatened Ecological Community are defined; Presumed Totally Destroyed, Critically Endangered, Endangered, or Vulnerable. On the Swan Coastal Plain 23 ecological communities have been classified as Threatened. Seventeen of these occur in the Perth Metropolitan Area, most of which were determined from the study by Gibson *et al.* (1994).

Threatened Species (Flora and Fauna) — Department of Conservation and Land Management and Western Australian Museum.

Location records of threatened flora were obtained from Department of Conservation and Land Management (Atkins 1998), and fauna from Department of Conservation and Land Management (1998) and the Western Australian Museum of Natural Science respectively and included when considering the regional significance of areas.

Selection of areas

Six criteria were developed as a basis for selection of regionally significant bushland (Table 1). The main objective is to protect, as far as is possible, a comprehensive, adequate and representative selection of the biological diversity of the metropolitan Coastal Plain.

In making these selections, regional information on floristic plant communities and mapping of vegetation complexes on the Coastal Plain enabled the selections in the Perth Metropolitan Area to be considered in the context of the biological diversity and conservation opportunities on the remainder of the Coastal Plain.

The initial selection of regionally significant bushland was achieved through an automated process based on selection of the largest, best condition, most diverse and best connected bushland areas in each vegetation complex required to protect at least 10% or a minimum of 400 ha of each vegetation complex (where possible).

These initial selections were subsequently refined by information from the sources described above and input from a panel of experts with specific knowledge of floristic plant communities, threatened

Table 1. Selection criteria (abridged).

Representation of ecological communities	A number of areas selected to represent the range of ecological communities and the places in which these communities merge. (Representation of not less than 10% of the original area of each vegetation complex)
Diversity	Areas with a high diversity of flora and/or fauna species or communities in close association.
Rarity	Areas containing rare or threatened communities or species, or species of restricted distribution.
Maintaining ecological systems or natural processes	Areas of importance in the maintenance of ecological processes or natural systems at a regional or national scale.
Scientific or evolutionary importance	Areas containing evidence of evolutionary processes either as fossilized material or as relic species and areas containing unusual or important geomorphological or geological sites. Areas of recognized scientific and educational interest as reference sites or as examples of the importance of environmental processes at work.
General criteria for the protection of wetland, streamline, and estuarine fringing vegetation and coastal vegetation	Conservation category wetlands, including fringing vegetation and associated upland vegetation. Coastal vegetation within the accepted coastal management zone.

ecological communities, rare flora/fauna, or other scientific, habitat or educational values related to specific criteria; and through consideration of constraining factors related to zoning or prior developmental approvals. Extensive consultation through a broadly based technical working group with representation from other government agencies, the local government association and umbrella non-government organizations conservation group provided a further review process.

The selection of regionally significant bushland for inclusion in Bushplan was influenced by land zoning. While areas were not generally excluded from consideration in Bushplan, areas on rural zoned land were selected in preference to areas zoned for more intensive uses such as urban or industrial zoned land, where these alternatives where possible. In practice this has not always been possible.

Similarly the target of achieving protection of a minimum of 10% of each vegetation complex is not achievable in all cases. Of the 26 vegetation complexes represented in the Perth Metropolitan Swan Coastal Plain, five are cleared to the extent that less than 10% of their original native vegetation remains. A further four vegetation complexes are extensively cleared and the remaining areas are subject to such intensive land use commitments that less than 10% of the bushland is available for conservation.

In addition to the target of achieving protection of 10% of each vegetation complex, a default target of protecting a minimum of 400 ha of each vegetation complex in the Metropolitan Area was also adopted for vegetation complexes with less than 4 000 ha in the Metropolitan Area. Again this is not universally achievable. One complex was only originally represented by 40 ha in the Metropolitan Area, four others have less than 400 ha remaining and a further four have just over 400 ha remaining. Where the complexes are mainly located outside the Metropolitan Area this minimum area is not a concern.

Bushplan has built on the foundation of existing bushland areas with some level of security for conservation, mostly areas Reserved for Parks and Recreation in the Metropolitan Region Scheme and areas identified in the System 6 recommendations.

In total Bushplan has identified approximately 52 000 ha for conservation, of which some 33 200 ha has some existing level of protection, with the remaining 19 000 ha of bushland and wetland without any current protection. The additional areas are required to provide comprehensive representation of the biological diversity of the Metropolitan Coastal Plain. The detailed map and area information has enabled an emphasis on selection of areas required to provide comprehensive representation of previously under represented ecological communities, consolidating existing areas and attempting to retain important linkages. As such effective representation of Threatened Ecological Communities has been treated as a priority, while the inclusion of areas supporting rare species has been sought where possible.

KEY FEATURES OF PERTH'S BUSHPLAN PROGRAMME

Perth's Bushplan is a second generation approach to the development of a regional conservation framework. Significant lessons learnt in the 14 years of implementing System 6 conservation recommendations have influenced the approach adopted for Perth's Bushplan in a number of ways.

Some key aspects of the programme are identified below.

Integration across key agencies and from science to successful implementation

Bushplan is designed to be a highly integrated programme, both across the key government agencies and "vertically", from the science of the biological survey right through to the statutory protection and management of a comprehensive regional framework of conservation areas.

This degree of integration of the key government agencies to jointly develop the conservation framework for a portion of the state is a first in Western Australia. It has facilitated the bringing together of a number of important information sets and diverse expertise that have been fundamental to the development of the comprehensive information on regionally significant areas compiled for this report.

The quality of information supporting recommendations has already been proven to be a major requirement for success in a region of intensive land use and competing interests such as the Perth Metropolitan Coastal Plain. Without detailed area based information, set in the context of a comprehensive regional framework, it is much more difficulty to argue the case for conservation, rather than development of expensive metropolitan land.

Implementation of the recommendations of Perth's Bushplan are proposed to be achieved through an integrated process co-ordinated and administered by the Ministry for Planning through a Bushland Co-ordinator and an implementation group comprising all key agencies and other major interest groups, with scope for other interests to be represented on a needs basis. The process will be regulated through a Memorandum of Understanding between the key agencies, identifying the role and resources that each will contribute to the process, and recognizing the statutory functions and mechanisms that each will use to achieve or regulate implementation decisions.

Statutory protection of recommendation areas will provide greater certainty of conservation outcomes and equally, greater certainty for landholders, developers and the conservation community. Importantly, it will also greatly simplify issues for decision makers. This will achieve one of the main objectives of Bushplan and overcome one of the main concerns associated with System 6.

Perhaps most importantly, vertical integration provides a clear framework and mechanism for setting and achieving conservation objectives designed to meet the policy commitments of State and Commonwealth Governments and community expectations.

A major difficulty associated with providing statutory protection for recommendation areas through reservation is the cost implications of purchasing all privately owned recommendation areas. In Perth, a Metropolitan Region Improvement Fund derived from Land Tax on metropolitan properties, provides a source of funds for acquisition of lands for public open space, including lands for conservation. The Metropolitan Region Improvement Fund is administered by the Ministry for Planning for the Western Australian Planning Commission, the statutory body responsible for planning decision making in the state.

This funding mechanism is vital in any attempt to achieve a comprehensive, adequate and representative system of conservation areas in the Metropolitan Area. It is equitable in that it returns a proportion of land taxes collected in the Metropolitan Area to the establishment of community open space, including conservation areas, and infrastructure needs in the same area.

Not unexpectedly, the cost of purchasing land in the Metropolitan Area for conservation is very expensive. If all areas identified in Bushplan were to be purchased (as is the current practice) the cost would be well beyond the capacity of the Metropolitan Region Improvement Fund to pay. Through selective application of off-reserve conservation mechanisms including the use of statutory covenants, negotiated solutions under planning legislation (applied to areas where conservation is constrained by urban or industrial zoning, or existing development approvals) and Environmental Protection Policies under the *Environmental Protection Act (1986)*, it is estimated that this cost could be reduced to less than $100 million, which is within the capacity of the Metropolitan Region Improvement Fund to fund.

Conservation on private lands — adding to the toolbox

In Western Australia as elsewhere biodiversity conservation has traditionally been achieved through a system of conservation reserves. In the Metropolitan Area the process has normally been initiated by "reservation" for parks and recreation in the Metropolitan Region Scheme, which enables the government to acquire the land funded through the Metropolitan Region Improvement Fund and formally reserve it for conservation (or other public open space purpose).

A complication arising from this "reservation" process is that the mechanism can require the government to purchase land that may be effectively protected through other mechanisms without owning it. This is costly and very restrictive if there is no need to reserve the land to protect it — particularly if the existing owner wishes to retain and conserve the bushland. The currently available mechanisms do not provide adequately for conservation on private lands.

It is now widely accepted that effective conservation requires conservation on private lands to complement the conservation reserve system. In the Metropolitan Area conservation on private lands is likely to be an important component of an effective protected area system.

The areas recommended for conservation in Bushplan necessarily involve a large number of separate bushland areas in order to achieve effective representation because of the intensity of land use and fragmentation of bushland in the Metropolitan Area. This creates a system of reserves or protected areas that would be difficult for a conservation agency to manage effectively. Conversely, the relatively small size of many of the recommendation areas makes them more suited to protection as individual areas of private land, managed for conservation by their owners and secured through a statutory covenant programme. This will be most applicable where public access is not an important consideration and the area does not have unique features or management characteristics that require specialist attention. The maintenance of a presence on the land provides further security in these intensively used lands.

The Western Australian Government, through the Department of Conservation and Land Management, is currently looking to establish an independent conservation trust with the ability to covenant. A similar covenanting programme not associated with a government agency is also being established by The National Trust of Australia (Western Australia).

To be effective, a covenanting system needs to include an active stewardship mechanism to provide advice, support, a feeling of involvement in part of a larger system, and where necessary the ability to act to secure the conservation values if the conditions of the covenant are not being complied with.

Incentives are important in encouraging private land owners to commit to covenanting land. They provide recognition of the community benefits to be derived from the actions of private landowners in setting aside and managing their land for conservation. Incentives being considered include rate relief (through reassessment of valuation) and relief from land tax.

Covenants can provide greater flexibility to achieve a form of conservation that suits land owners seeking a means of protecting the conservation values on their lands. However, in recommending that covenants have a wider role in providing for conservation of selected Bushplan areas, it is clear that covenants will not always be welcomed by the landowners, particularly where their intention may have been to subdivide. The question of how to equitably deal with land owners that are unwilling to manage their land for conservation through a covenant also needs to be addressed. It is considered that these equity issues can be addressed through the application of a "revolving fund". A revolving fund is simply a mechanism where a fund is established to purchase land and covenant it for conservation, then resell the covenanted land to recover costs to buy and covenant another area of land . . . so revolving the funds. In this case it would be applied where land owners do not wish to retain the bushland on their property, enabling the land to be purchased; covenanted and resold to an owner who is interested in conservation of the bushland.

In this way, not only can government be assured that the owners of the covenanted land are committed to managing the bushland for conservation, but much of the cost of securing the land can be recovered through the sale of the covenanted land and reused to covenant another block. As the revolving fund approach will be new in Western Australia, one significant area of current uncertainty is in estimating the degree of cost recovery that is likely to be achieved through the revolving fund.

Provision of an agreed process for updating information and proposals

One of the lessons from System 6 is that as information improves there is the potential for new areas of high conservation value to be identified. To maintain wide respect for the integrity of the conservation system there needs to be a process for periodically reviewing new information and assessing its significance against the benchmark established in Perth's Bushplan in making a case for additions to the protected area system. Bushplan is expected to have largely captured the biological diversity of the region but there is the potential for new work to identify the need for minor changes. It is proposed that Bushplan recommend an agreed process and criteria for assessing new information in order to maintain the currency and standing of Perth's Bushplan over time.

Replacement policy

It is recognized that situations may arise (particularly in the Metropolitan Area) in which there is an overriding public requirement to develop an area that has been identified for conservation. A statutory mechanism is provided for public review of such development proposal through environmental impact assessment in accordance with the *Environmental Protection Act (1986)*. In this situation the balance of representation of plant communities should be maintained by equivalent additions (or as near equivalent as possible) to Perth's Bushplan.

CONCLUSIONS

Perth's Bushplan is a second generation review of conservation requirements for the Perth Metropolitan portion of the Swan Coastal Plain. Overall Perth's Bushplan proposes the conservation of some 52 000 ha of bushland, on a combination of public and private lands, that are representative of the range of environments originally occurring in the region.

Several factors have been of critical importance in facilitating the development of Perth's Bushplan as a comprehensive conservation framework for Perth. These are:

- The existance of established biodiversity policy commitments, both specific State Government policy commitments to the conservation of biological diversity in the Perth Metropolitan Area, and a comprehensive national biodiversity policy signed by the State Governments;
- The ability to compile and present in Bushplan, consistent and detailed information on the characteristics and values of the Bushplan Sites, based on comprehensive region wide survey and map based data on the patterning of ecological communities;

- The existance of the Metropolitan Region Improvement Fund, to provide funds for the purchase of regional open space including conservation lands in Perth, sourced from taxes raised in the Perth Metropolitan Area; and
- Strong longstanding community support for conservation of urban bushland.

There are several features in the approach adopted in Perth's Bushplan which have been incorporated to address difficulties that have been experienced in implementing System 6 recommendations over the past 14 years. In particular, the introduction of a range of mechanisms to provide for conservation on private lands as part of a regional system of conservation areas, and the high degree of integration of the programme, both administratively across key agencies, and through all stages of the programme from biological survey to implementation of the recommendations. This represents a new level of integration for such programmes in Western Australia. Some of these approaches may have application singly or in combination elsewhere.

ACKNOWLEDGEMENTS

We would like to thank Gregory Keighery, Natalie Thorning, Lorraine Willock and Margaret Turpin for their comments and assistance in constructing this chapter.

REFERENCES

Atkins, K., 1998. Declared Rare and Priority Flora List for Western Australia. Department of Conservation and Land Management: Perth, Western Australia.

Commonwealth of Australia, 1996. The National Strategy for Conservation of Australia's Biological Diversity. Australian Government Publication Service: Canberra, Australian Capital Territory.

Churchward, H. M. and McArthur, W. M., 1980. Landforms and Soils of the Darling System. *In* Atlas of Natural Resources, Darling System, Western Australia. Department of Conservation and Environment: Perth, Western Australia.

Department of Conservation and Environment, 1983. Conservation reserves for Western Australia as recommended by the Environmental Protection Authority — 1983. The Darling System — System 6. Parts 1 and 2, Perth, (Pp. 229), Perth, Western Australia.

Department of Conservation and Land Management, 1998. Declared Rare and Priority Fauna List for Western Australia. Unpublished report to Department of Conservation and Land Management, Western Australia.

Dixon, J., Connell, S., Bailey, J. and Keenan, C., 1994. The Perth Environment Project and an Inventory of Perth's Remnant Native Vegetation. *In* A Vision for a Greener City. 1994 National Greening Australia Conference Proceedings, Perth, Western Australia.

English, V. and Blyth, J., 1997. Identifying and Conserving Threatened Ecological Communities in the South West Botanical Province. An unpublished report for the Department of Conservation and Land Management to Environment Australia, Perth, Western Australia.

Environmental Protection Act, 1986 (WA). Government of Western Australia, Perth, Western Australia.

Government of Western Australia, 1995. Urban Bushland Strategy. Government of Western Australia, Perth, Western Australia.

Government of Western Australia, 1997. Wetlands Conservation Policy for Western Australia. Government of Western Australia, Perth, Western Australia.

Government of Western Australia, 1998. (in preparation) Perth's Bushplan. Government of Western Australia, Perth, Western Australia.

Gibson, N., Keighery, B. J., Keighery, G. J., Burbidge, A. H. and Lyons, M. N., 1994. A Floristic Survey of the Southern Swan Coastal Plain. A report for the Australian Heritage Commission, prepared by the Department of Conservation and Land Management and the Conservation Council of Westerm Australia (Inc.), Perth, Western Australia.

Griffin, E. A., 1994. Floristic Survey of Northern Sandplains between Perth and Geraldton. Unpublished Report to the Heritage Council of Western Australia.

Heddle, E. M., Loneragan, O. W. and Havel, J. J., 1980. Vegetation of the Darling System. *In* Atlas of Natural Resources, Darling System, Western Australia. Department of Conservation and Environment: Perth, Western Australia.

Hill, A. L., Semeniuk, C. A., Semeniuk, V. and Del Marco, A., 1996a. Wetlands of the Swan Coastal Plain. Volume 1: Wetland Mapping Classification and Evaluation. A report for the Waters and Rivers Commission and the Department of Environmental Protection, Perth, Western Australia.

Hill, A. L., Semeniuk, C. A., Semeniuk, V. and Del Marco, A., 1996b. Wetlands of the Swan Coastal Plain. Volume 2: Wetland Mapping Classification and Evaluation — Wetland Atlas. A report for the Waters and Rivers Commission and the Department of Environmental Protection, Perth, Western Australia.

Keighery G. J., 1996. Plot records from Tuart (*Eucalyptus gomphocephala*) dominated communities. Unpublished plot information, Department of Conservation and Land Management: Perth, Western Australia.

Thackway, R. and Cresswell, I. D. (eds), 1995. An Interim Biogeographic Regionalisation for Australia: a framework for setting priorities in the national system of reserves, Version 4. Australian Nature Conservation Agency: Canberra, Australian Capital Territory.

United Nations Environment Programme, 1992. Convention on Biological Diversity, United Nations. Geneva, Switzerland.

United Nations Education, Scientific and Cultural Organization, 1971. Article 1, part 1, Convention on wetlands of international significance. Published in Australia, 1976, for Department of Foreign Affairs, Australian Government Publishing Service, Treaty Series 1975, No. 48.

CHAPTER 32

Native vertebrates as pests on subsistence farming systems in Jambi, Indonesia

Jean Balson,[1] Jim Davie[1] and Andy Gillison[2]

Agroforestry systems promote biodiversity and provide habitat for a number of native vertebrates. This chapter presents farmers' perceptions of the extent of damage, strategies used and population changes of vertebrate pests on agroforestry systems. Data were collected from interviews with smallholder farmers with rubber-based agroforestry systems in Jambi, Sumatra. Farmers perceive the damage caused by native vertebrate pests as a serious constraint to production. Wild pigs were reported to cause the highest level of depredation in all land uses. Monkey damage was also considered to be significant by farmers. The level of damage and population changes of vertebrates were found to relate to the distance from natural forest. Further research into native vertebrate pest management is required to find ways to improve productivity of agroforests, while maintaining the biodiversity values of these systems.

Key words: Agroforestry, Biodiversity, Vertebrate pests, Pigs, Monkeys, Habitat fragmentation, Rubber, Indonesia.

INTRODUCTION

THE increasing priority for local peoples to meet production needs of a rapidly growing human population, and the resultant need for increasing amounts of land to be put under agriculture, has led to an increase in conflicts between humans and native wildlife (Tchamba 1995). As well as the ecological changes from fragmentation of native vegetation, the development and introduction of modern agriculture into subsistence farming systems in the tropics has changed the perception, impact and control of vertebrate pests (Dove and Kammen 1997).

The impact of native vertebrate pests on subsistence farming systems is of particular relevance to the relationship between productivity and conservation. It is generally accepted that while conflicts between agriculturists and wildlife have always existed, the extent of the conflict has escalated (Hill 1997). An implication of this phenomenon is that the biodiversity of indigenous farming systems, of which vertebrates form a part, may be compromized. As a result, a number of ideological and practical difficulties exist for the co-management of conservation and production.

Many traditional farming systems promote the conservation of a high proportion of animal and plant diversity (Michon and de Foresta 1995). For example, agroforestry systems often have similar or higher densities of native vertebrates than primary forests. Primates, civets (Corlett 1996), squirrels and fruit bats (Thiollay 1995) maintain similar population densities to those observed in natural forests. Wild pig and deer populations increase with the fragmentation of forests, particularly in shifting cultivation systems, due to an increase in preferred food and secondary forest habitat (MacKinnon 1992). While this leads to a greater conflict between farmers and native vertebrates, agroforests can be seen as important for biodiversity conservation as deforestation continues to destroy natural habitat. Although, native vertebrates make up only a small part of the total biodiversity of the tropics, they form important links in the functioning of ecosystems through grazing, predation and the mechanical dispersal of seeds (McNeely 1995).

[1]School of Natural and Rural Systems Management, The University of Queensland, Lawes, Queensland, Australia 4343.
[2]Center for International Forest Research, Jl. CIFOR, Bogor, Indonesia.

Pages 318–25 *in* NATURE CONSERVATION 5: NATURE CONSERVATION IN PRODUCTION ENVIRONMENTS: MANAGING THE MATRIX ed by J. L. Craig, N. Mitchell and D. A. Saunders. Surrey Beatty & Sons, 2000.

Generally, there is a lack of adequate knowledge as to the scale of the impact of native vertebrate pests on subsistence farming systems. However, research on native vertebrate pests in the tropics has found that farmers consider crop raiding by vertebrates to be a major source of crop losses and a major constraint to the improvement of productivity (Hill 1997; Gauthier, unpubl.; Tchamba 1995). Twenty-three vertebrate pests were identified as frequently entering agricultural land in Uganda including baboons, pigs, vervet monkeys, birds and bushbuck (Hill 1997). In Sumatra six species have been identified as major pests to the crops and forest (Gauthier, unpubl.). In parts of Africa and Asia elephants are reported to cause considerable amounts of damage to crops of subsistence farmers (Tchamba 1995).

Primates and pigs are perceived by farmers to be the most common and destructive pests in the tropics. Both groups of animals attack a wide variety of plants and are of particular concern to subsistence farmers due to their ability to learn to raid crops and avoid control measures (Quick 1996). In Africa, baboons are considered to be very destructive due to the fact that they are opportunistic feeders and enter fields frequently and in large numbers (Hill 1997). Macaques and leaf monkeys have been reported to raid a variety of crop types in Indonesia (Gauthier, unpubl.), and in Thailand (Eudey 1996).

In Sumatra, vertebrate pest damage has been found to be a major constraint in improving the productivity of rubber agroforestry systems, and possibly the adoption of new planting materials (Tomich and van Noordwijk, unpubl.). Therefore it is important to understand farmers' perceptions towards pests and the strategies in place for their control. It is now recognized that traditional knowledge, attitudes, and strategies for control are essential in developing pest management strategies that are conducive to cultural settings, improvement of productivity and the conservation of species concerned (Gagriel 1989).

This chapter presents farmers' perceptions of native vertebrate pests in agroforestry systems in Jambi, Sumatra. Further it draws on local knowledge and experience of pest management strategies. Changes in vertebrate pest populations and reasons for these changes are also discussed. The purpose of this research is to provide some insight into the type and extent of vertebrate impact as a guide for further research.

METHODS

Study site

Bungo Tebo is an administrative District (Kabupaten) of Jambi Province, situated in the hinterland of central Sumatra (Fig. 1). The area is representative of both the piedmont and peneplain zones. The piedmont zone is characterized by hilly terrain and a soil composition of latisols and red yellow podzolics, whereas the peneplain zone is almost flat and the soil composition consists primarily of red-yellow podzolic soils (van Noordwijk and Tomich, unpubl.). Pockets of natural forest are scattered around the study area; however, most of the forest has been logged. The average temperature of the area is 26.6°C, however, temperatures can range from 22.7°C to 31.6°C. Monsoon begins in September or October and continues through to May.

Data collection

A total of six villages were included in the survey. These were selected by first stratifying the villages in the region according to their proximity to natural forest. The three strata included in the sample were (1) close to forest; (2) intermediate; (3) far from forest (Table 1). In villages classified as "close to the forest" agroforest plots are often adjacent to natural forest margins.

The people of the villages are representative of the Melayu ethnic group, the traditional farmers in the region. A total of 120 households were surveyed with 40 households included in each stratum. Households were selected from official records, or lists compiled by village heads where this information was not available.

Three different methods of data collection were applied. The primary method was a structured interview of farming households with the assistance of an interpreter. Topics covered by the questionnaire included land use, crop types, extent of native vertebrate damage, control methods and changes to native vertebrate populations. Informal interviews and discussions with key people (e.g., village heads and elders) complemented the questionnaire and were used as an independent check on the data.

The survey was tested in a village that was not selected in the sample. The aim of the test was to evaluate clarity, timing, appropriate sequencing of questions and validity of responses. Equal responses from male and female farmers were included in the

Table 1. Villages sampled according to strata.

Category	Distance from natural forest	Village	District
Close to forest	1–5 km	Lubuk Beringin	Rantau Pandan
	1–5 km	Rantel	Pelepat
Intermediate	10–15 km	Rantel Keloyang	Pelepat
	10–15 km	Tanjung Agung	Rantau Pandan
Far from forest	>20 km	Kotoyayo	Pelepat
	>20 km	Babeko	Muara Bungo

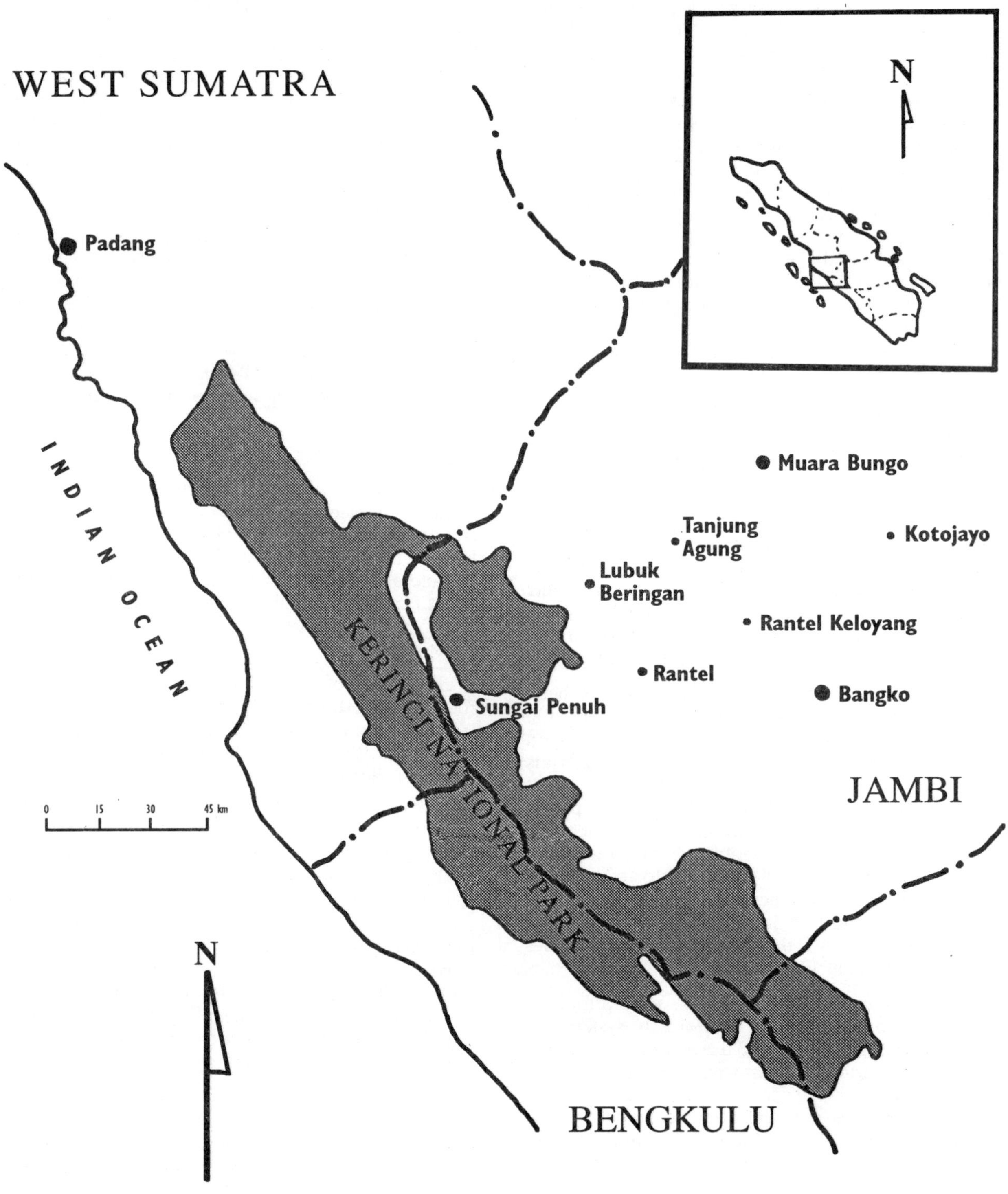

Fig. 1. Location of the study area.

sample to ensure women's knowledge and views were incorporated in the data.

RESULTS

Farming patterns

Land use patterns in the Muara Bungo region range from secondary forest displaying high biodiversity to systems practising more rigid monoculture. Indigenous Melayu farmers practice shifting cultivation, with rubber agroforestry (*kebun karet*) providing the principle source of income. After the clearing of a forest using slash and burn, upland rice is planted with rubber for the first year or, less commonly, two years. Other tree species, that provide timber and a variety of non-timber products, are often planted in adjacent rows and are left to mature amidst the natural regeneration of forest species. Wet rice fields (*padi sawah*), upland fields (*ladang*), bush fallow (*belukar*), and home gardens (*pekarangan*) make important contributions to subsistence needs. Collectively these land uses provide a variety of products including fruits, resins, rattan, honey, medicines and wood to meet or to subsidize income.

Damage by vertebrate pests

Farmers considered vertebrate pests as the cause of considerable damage in agroforestry and associated farming systems (Table 2). The extent of damage was influenced by land use, type of crop grown, and the characteristics of different vertebrate pests. Thirteen native vertebrate species were identified as pests in the rubber agroforestry system.

Wild pigs were perceived as a major pest damaging between 10 and 40% of crops planted in all land uses. After the wild pig, arboreal species were reported to be the next major pests. The banded leaf monkey, long-tailed macaque and squirrel caused high levels of damage in most land uses. Of the two macaque species found in the study area, the pig-tailed macaque was reported to be more destructive, with troops roaming in large territories, often not returning for months. Squirrels are considered a pest primarily due to a preference for jering and *petai (Parkia speciosa)*, which are important for the farming household.

Damage by sambar, tapir, sun bear and porcupine is restricted to the agroforest. In the upland fields porcupines are pests in the early seres, when pumpkin and other annuals are grown. Sambar, tapir and sunbears become pests as the agroforest matures. Tapir and sambar eat the sweet bark of the rubber tree, which often results in the death of the tree. Sun bear are reported to climb trees for durian (*Durio zibethinus*), jackfruit (*Artocarpus heterophyllus*) and champedak (*A. integer*)

Changes in population numbers

Wild pig, banded leaf monkey and long tailed macaque populations have been perceived by farmers to have increased dramatically in villages at all distances from forest. Sun bear and sambar populations have been perceived to have decreased at all distances from natural forest, but their populations are thought to be most stable close to forest (Fig. 2).

Respondents were asked to explain why they felt populations of different species had changed (Table 3). The most common explanation for increases in population numbers was the reproductive capacity of species, particularly pigs and monkeys. Deforestation was also a common explanation for the increase in species numbers, particularly in villages at intermediate and far distances from natural forest. The rationale given was that natural food sources had diminished, so animals were dependent on farmers' produce and the habitat provided by agroforests. Pest population numbers in some locations had reached such a level that farmers were resigned to high levels of predation.

Hunting was considered to be the cause of the declines in sun bear, sambar and porcupine populations. Sambar and porcupines are hunted to supplement the diet of local communities, while sun bears were reported to be hunted for their galls which are sold to the local Chinese community for medicinal purposes. In addition, sun bear population decreases were attributed to deforestation.

Land use categories

The distance of sites from the natural forest may affect the distribution of vertebrates, and therefore the amount of damage to farmers' crops. Damage was generally greatest in land uses intermediate from natural forest (Fig. 3), which is consistent with increases reported in primate populations at this distance. In the villages intermediate and far from the natural forest, respondents felt that it was the frequent human disturbance from schools, roads and neighbours, which reduced vertebrate predation pressure.

The upland (non-irrigated) fields sustain the highest level of damage from vertebrates. In this category, rubber is susceptible to attack from wild pigs and banded leaf monkeys. Upland rice and other annuals also attract pests. Immature agro-forests far from areas of natural forests receive less attention from predators because tapir and

Table 2. Percentage of damage by vertebrates reported by farming households in the agroforest system.

		Land use				
Common name	Species	Wet rice field	Home garden	Upland field	Immature agroforest	Mature agroforost
Wild pig	*Sus scrofa*	21–25%	36–40%	26–30%	21–25%	16–20%
Sun bear	*Helarctos malayanus*	0	0	0	11–15%	11–15%
Pig-tailed macaque	*Macaca nemestrina*	15–20%	0	11–15%	21–25%	26–30%
Long-tailed macaque	*Macaca fascicularis*	15–20%	16–20%	11–15%	16–20%	16–20%
Silvered leaf monkey	*Presbytis cristata*					
Banded leaf monkey	*Presbytis melalophos nobilis*	0	6–10%	11–15%	6–10%	16–20%
Black handed gibbon	*Hylobates agilis*	0	0	0–5%	0–5%	0–5%
Squirrel	Sciuridae	0	11–15%	6–10%	6–10%	6–10%
Porcupine	Hystricidae	0	0	6–10%	0–5%	0
Civet	Viverridae	0	0	0–5%	0–5%	46–50%
Malay flying fox	*Pteropus vampyrus*	0	16–20%	0	6–10%	11–15%
Tapir	*Tapirus indicus*	0	0	0–5%	6–10%	11–15%
Sambar	*Cervus unicolor*	0	0	0–5%	16–20%	0–5%

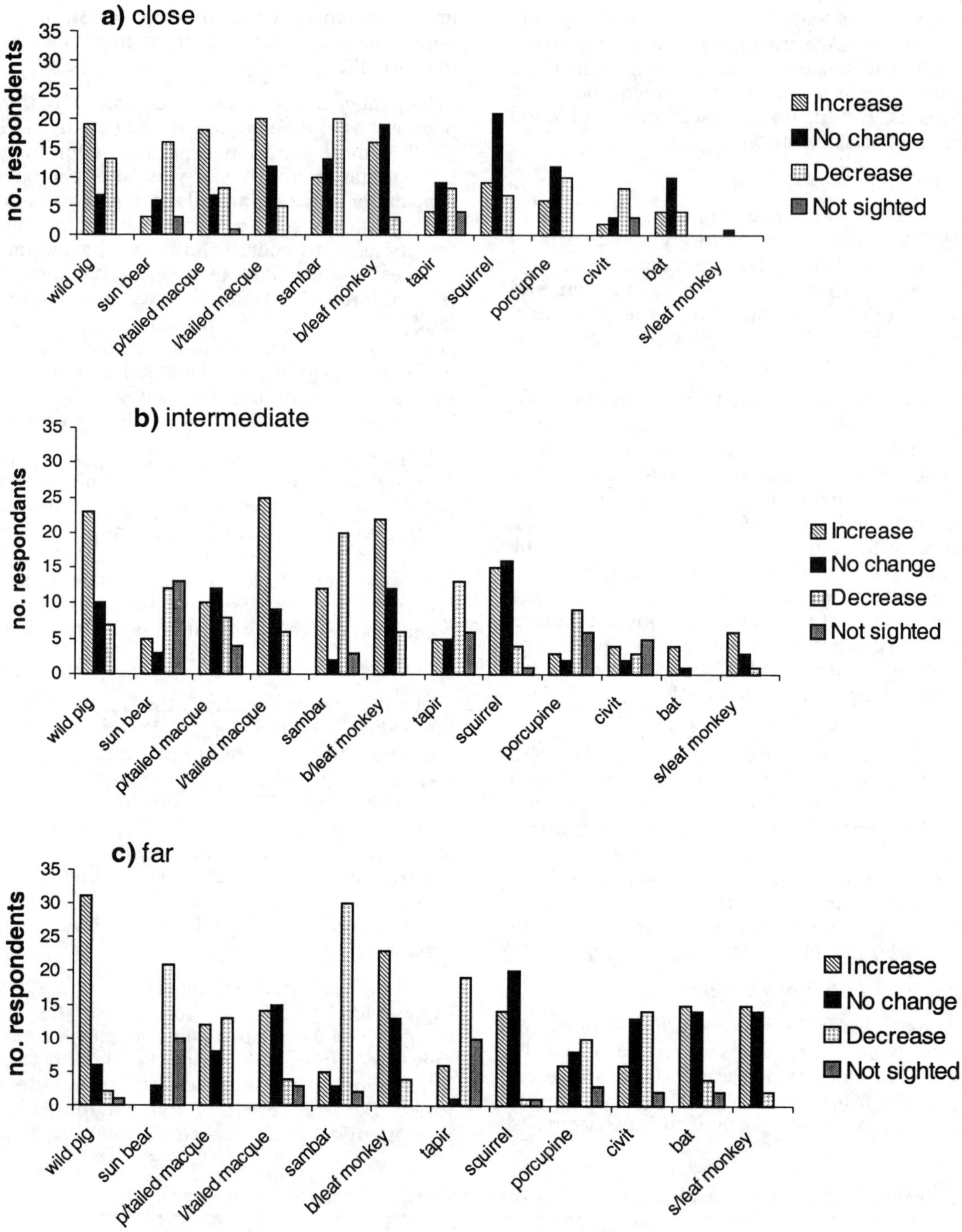

Fig. 2. Respondents perception of population changes of vertebrate pests in villages a) close b) intermediate and c) far from natural forest.

sambar populations are lower in more disturbed sites. The primary pests in mature agroforest are monkeys, which feed on fruit trees. Although reports of damage are high (between 11 and 15%), farmers are more tolerant of damage in these sites as fruits are not as important to families.

Local strategies for vertebrate pest control

The majority of farmers (74%) guarded their fields against vertebrate pests. At the time of the survey, thirty-three respondents were living in their fields, while nine respondents, with wet rice fields, reported sleeping in their fields to protect rice during harvesting. Respondents reported that they rely on human presence to deter most animals from entering a field.

Fencing is a common method used by farmers (70%) to protect crops. Fences are often innovative, showing the farmer's ingenuity in optimizing available resources. Wooden fences, made from logs

Table 3. Number of respondents reporting reasons for population changes in pest species.

	Increase				Decrease			
Animals	Breeding	Less natural forest	Not hunted	No effective control method	Hunting pressure	Less natural forest	Less attractive crops	Uncertain
Wild pig	28	11	13	9	14	1	1	7
Sun bear	3	1			19	7	1	15
Pig-tailed macaque	20	6	7	4	5	12	2	6
Long-tailed macaque	22	7	13	1	5	2	1	4
Sambar	9	10	3		52	2	2	7
Banded leaf monkey	28	8	14	5	10	2		9
Tapir	2	4	5	3	7	4	1	14
Squirrel	14	6	7		15			2
Porcupine	6		7	4	18	1	2	4
Civet	5		4	2	15			9
Malay flying fox	7	1	9			2		8
Silvered leaf monkey	9	2		2	3			2

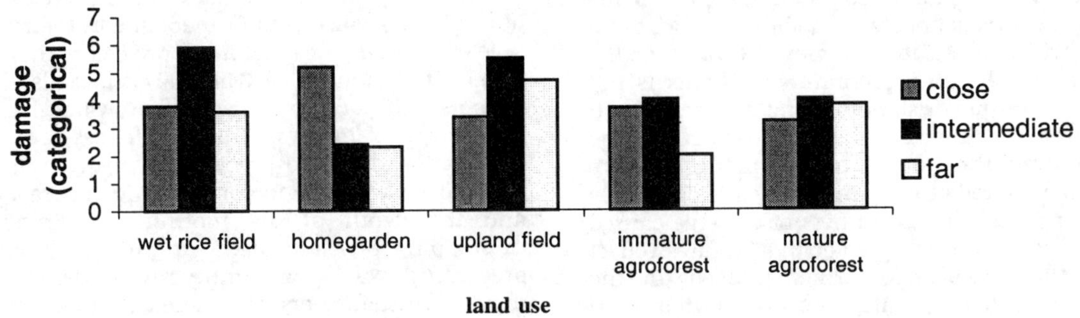

categories: 1 0-5% **2** 6-10% **3** 11-15% **4** 16-20% **5** 21-25% **6** 26-30%

Fig. 3. Percentage of damage by vertebrate pests on land uses at varying distances from the natural forest as reported by farmers.

left in the field after slash and burn are most effective in protecting fields from pigs. Often, plastic and aluminium cans are incorporated to frighten pigs and monkeys. In the wet rice field and home garden, string or plastic sheet fences are more commonly constructed, incorporating cans and plastic to scare animals.

Poison baits (51%) are mainly used for the control of pigs and monkeys. However, many respondents claim that recently pigs and monkeys are not taking bait because they learn to detect its presence. Poison is used sporadically, usually in the *sawah* and less commonly in *ladangs*. Traps, including snares and holes were used by 37% of respondents for a variety of vertebrate animals. Like fences, farmers create innovative traps and snares from bamboo, rattan, wood and grasses, to catch deer, monkeys, porcupines and pigs.

The use of guns (41%) is usually limited to small animals because regulations restrict the use of anything more powerful than an air rifle. However, according to farmers, if a banded leaf monkey is wounded, the troop often stays away from 5 to 14 days. Other than monkeys, squirrels are the only pests for which guns are considered effective. Regular community hunting with the use of spears and dogs (21%) was practised at one village situated close to the forest. Consequently, a majority of farmers in this village reported pig populations and impact levels to be decreasing. Other families kept dogs in their fields to warn of the presence of pigs and monkeys.

Scarecrows (18%) are primarily erected in wet rice fields to scare any intruding vertebrate pests. Effectiveness of scarecrows is limited as farmers stated that vertebrate pests learn they pose no threat. In addition to the methods described above, a few respondents planted cassava to attract pigs away from the young rubber. However, most farmers in the study area claimed that this method only encouraged the pig problem. For this method to be effective, most respondents felt continual re-planting was required, and this was said to be too time consuming.

DISCUSSION

Complex rubber agroforests provide easily attainable food sources and the protection of a forest environment for a number of native vertebrate pests. While this leads to a greater conflict between farmers and native vertebrates, agroforests can be seen as becoming more important for conservation of vertebrates as deforestation continues to destroy natural habitat. Farmers in the study perceived native vertebrates as major pests, causing significant damage in all land use types. Wild pigs were the most destructive vertebrate pest, impacting heavily on all farming systems. Monkeys were also identified as causing significant losses to farming systems. These findings are consistent with other studies of farmers' perceptions of vertebrates in the tropics (Gauthier, pers. comm.; Hill 1997).

Banded leaf monkeys and wild pigs are the major constraint to farmers establishing rubber tree plots. Research into the intensification of rubber agroforests in the study area has found that the main threat by pigs and leaf monkeys is to young rubber. Researchers are concerned that the severity of vertebrate damage will deter farmers from investing in higher yielding rubber planting stock (van Noordwijk *et al.*, unpubl. data; Gauthier, pers. comm.). The banded leaf monkey was considered a major pest because of the damage caused to foliage on young rubber plants. Leaf damage can greatly retard growth rates and cause additional horizontal growth, which needs to be removed to prevent the rubber tree branching prematurely. If this is not carried out, farmers risk not having a large enough trunk area for tapping. After the initial damage has occurred the young rubber plant is more susceptible to predation by the banded leaf monkey and other pests. The seriousness of damage caused by this species is much greater than figures reflect due to the importance farmers placed on the survival of rubber plants, and the difficulty in quantifying damage. Further research is needed to measure levels of damage by vertebrates directly to determine the significance of low yields in comparison to other contributing factors.

A result of agricultural expansion is the fragmentation of native habitats. Changes to population numbers and species diversity results from exploitation, transformation and modification of natural ecosystems. Studies by Hill (1997) and Tchamba (1995) found human conflicts with vertebrate pests are a result of increasing human populations and deforestation. While some species are totally displaced as a result of habitat destruction or alteration, many species are still abundant having adapted well to agricultural systems. Farmers indicated a species-specific trend in changes in vertebrate populations. Pigs and arboreal species have generally increased, while large, ground dwelling species have decreased. This result is consistent with research on vertebrate populations in agroforests which has found that many vertebrates have adapted well to the agroforest system, due to an increase in preferred food, and secondary forest habitat (MacKinnon 1992; Thiollay 1995; Corlett 1996).

The distance of the site from the natural forest appeared to influence the distribution and diversity of animals. General trends are that in villages close to the forest, numbers of pests have remained the most stable. The most significant changes in pest populations have occurred at distances furthermost from the forest where higher densities of wild pig, banded leaf monkeys, long tailed macaque and silvered leaf monkeys were reported. However, caution should be taken when the maintenance of diversity is derived from higher densities of one species and reduction of others (Thiollay 1995). Vertebrate diversity may therefore be best conserved in conjunction with areas of natural forest (including production forest).

Strategies used by farmers to protect crops against damage by vertebrates are restricted by limited technology and resources. Farmers rely on a number of different methods due to the diversity of vertebrate pests. Guarding is most common, and presently the most effective method for damage control of all species. The large amount of time and effort put into guarding is reported as necessary to protect against monkey damage during the day and pig damage during the night. Farmers reported that without guarding, their entire crop would be destroyed by animals. Primates in the study area are of particular concern to farmers due to their ability to learn to raid crops and avoid protection measures. Recent studies in Africa (Hill 1997) have also found farmers held similar perceptions towards baboons and vervet monkeys. Wooden fences are cost-effective, and in the process of construction also serve to clear the field for planting. Fence height and the depth of the footings are important in stopping pigs and deer jumping over, and in preventing wild pigs tunnelling under. Despite these efforts, farmers considered that no fence is totally pig proof.

Levels of damage are major constraints to farmers improving yields and are major factors that will influence farmers' decisions. Farmers displayed a general frustration towards efforts to mitigate extent of damage and many were resigned to huge losses. Further investigations are required to determine farmers' means and willingness to trial alternative control methods. Research needs to be targeted at methods most likely to be adopted by local farmers who are concerned that costs are surpassed by increases in yields.

Community hunting is the only control method reported to reduce pig numbers effectively. However, this was practised in only one village. In other villages, hunting sambar and porcupine for food effectively reduced population numbers of these species. Although wild pigs are a potential source of protein, the Muslim taboo on the consumption of pork may partly explain why they are not hunted. A potential, although limited, market exists for selling wild pig to the non-Muslim communities in nearby towns.

The rapid deforestation of remaining lowland forests means many wildlife conservation opportunities are being lost. By enhancing the productivity of smallholder agroforests and developing sustainable control strategies, these productive systems may offer an alternative approach to wildlife conservation. Unless the productivity of the systems described in this chapter are increased they may be replaced by large commercial monocultures of rubber and oil palm. In one village, private oil palm companies had already approached farmers to gain interest in the schemes. Although farmers said they prefer traditional agroforests because they provide product diversification and risk reduction, the importance of decreasing pest impact, and thereby improving productivity is becoming a higher priority.

CONCLUSION

This chapter illustrates the seriousness of vertebrate pest damage to rubber agroforests and associated farming systems. Increasing human populations and further fragmentation of the natural forest and the landscape have resulted in a decrease in the diversity of vertebrate pests, while at the same time creating niches for more serious pests. Farmers spend considerable amounts of time guarding fields from the incursions of vertebrate pests. A community approach may provide a more efficient and effective strategy for control of pests. Management decisions could be facilitated by provision of scientific knowledge on the impact and control of vertebrate pests, ecology of each identified pest species and the potential for their sustainable use.

ACKNOWLEDGEMENTS

The fieldwork, upon which this chapter was based, occurred during a field placement by the first author with the Center for International Forest Research. The authors are grateful to the Centre for this support and also to the School for Natural and Rural Systems Management at The University of Queensland, who provided logistic and travel support. Muhamad Zahri, a law graduate from the University of Jambi, interpreted for the village questionnaires and interviews, and provided cultural advice and friendship during the fieldwork.

REFERENCES

Corlett, R. T., 1996. Characteristics of vertebrate dispersed fruits in Hong Kong. *J. Trop. Ecol.* **12**: 819–33.

Dove, M. R. and Kammen, D. M., 1997. The epistomology of sustainable resource use: managing forest products, swiddens, and high yielding variety crops. *Human Organisa.* **56**(1): 91–101.

Eudey, A. A., 1996. Hill tribe peoples and conservation in Thailand: a preliminary assessment of the problem of reconciling shifting cultivation with conservation objectives. Pp. 237–48 *in* Primate Ecology and Conservation ed by J. G. Else and P. C. Lee. Cambridge University Press: Cambridge.

Gagriel, T., 1989. Pest control, pest management and the 'human factor'. *Trop. pest manage.* **35**(3): 254–56.

Hill, B. M., 1997. Crop raiding by wild vertebrates: the farmer's perspective in a agricultural community in western Uganda. *J. Integr. Pest Manage.* **43**(1): 77–84.

MacKinnon, K., 1992. The Wildlife of Indonesia. Penenbut PT. Gramedia Pustaka Utama: Indonesia.

Michon, G. and de Foresta, H., 1995. The Indonesian Agroforest Model: forest resource management and biodiversity conservation. Pp. 90–105 *in* Conserving Biodiversity Outside Protected Areas: The Role of Traditional Agrosystems ed by P. Halladay and D. A. Gilmour. IUCN: Switzerland.

McNeely, J. A., 1995. How Traditional agro-ecosystems can contribute to conserving biodiversity. Pp. 20–37 *in* Conserving Biodiversity Outside Protected Areas: The Role of Traditional Agrosystems ed by P. Halladay and D. A. Gilmour. IUCN: Gland, Switzerland.

Quick, D. L., 1996. Activity budgets and the consumption of human food in two troops of baboons, *Papio anubis*, at Gilgil, Kenya. Pp. 221–28 *in* Primate Ecology and Conservation ed by J. G. Else and P. C. Lee. Cambridge University Press: Cambridge.

Tchamba, M. N., 1995. History and present status of the human/elephant conflict in the Waza-Logone Region, Cameroon, West Africa. *Biol. Conserv.* **75**: 35–41.

Thiollay, J., 1995. The role of traditional agroforests in the conservation of rainforest bird diversity in Sumatra. *Conserv. Biol.* **9**(2): 335–53.

CHAPTER 33

Designing habitats at the landscape level to maintain viable populations of Barn Owls *Tyto alba* in pastoral farmland in southern Scotland

Iain R. Taylor[1,2]

Top predators are often highly sensitive to intensification processes in farmed landscapes and the populations of many of them have been severely reduced worldwide. The Barn Owl *Tyto alba* is typical of many of such species whose range and abundance have declined markedly in the Northern Hemisphere since the advent of modern farming practices. This chapter presents the results of a study from 1979 to 1993 in pastoral farmland in south Scotland whose aim was to specify a set of habitat characteristics at the landscape level that could maintain viable populations of the owls and be compatible with the objectives of intensive dairy and sheep production. The key habitat component was found to be strips of semi-natural grassland around small woodlands managed for on-farm timber use and as a cash crop. These grasslands have high densities of the owls' main prey species and are favoured hunting areas. From a simple population model it is estimated that an average annual production of 3.2 fledged young per female is required to maintain population stability. The owls' breeding performance is closely correlated with the amount of woodland edge around their nests and an average of 9–10 km length of edge is required within the owls' normal breeding season foraging range of approximately 1 km radius around the nest, to achieve required production levels. The provision of continuous suitable habitat at a landscape scale is impractical because of variations in topography and farmer interest, but it is suggested that groupings of at least five or six breeding season ranges into local patches in the landscape are needed to provide appropriate foraging conditions in winter when the birds range over larger areas in response to varying weather. The problems of providing adequate habitat at catchment levels to maintain viable populations of owls are discussed. The proposed landscape consisting of numerous small woodlands with appropriately managed edge habitats is compatible with government initiatives to diversify production by encouraging farm forestry.

INTRODUCTION

In most parts of the world two processes have occurred simultaneously in production landscapes: the loss and fragmentation of the original natural habitats and the progressive intensification of management of the cropped or harvested areas that have replaced these natural systems (Kitchener *et al.* 1980; Bennett 1987; Saunders *et al.* 1987). The dynamics of species change in relation to these processes are complex and far from fully understood (Simberloff 1995; Wiens 1995). Many species that are restricted to the original habitats have suffered population declines whereas others such as those whose natural habitats may have included ecotones or habitats similar to managed habitats have experienced population increases (Whitcomb *et al.* 1981; Opdam *et al.* 1985, 1995; Lynch 1987). However, as management has

[1]School of Biological Sciences, The King's Buildings, University of Edinburgh, Edinburgh, Scotland.
[2]Present address: The Johnstone Centre, Charles Sturt University, P.O. Box 789, Albury, New South Wales, Australia 2640.
Pages 326–35 *in* NATURE CONSERVATION 5: NATURE CONSERVATION IN PRODUCTION ENVIRONMENTS: MANAGING THE MATRIX ed by J. L. Craig, N. Mitchell and D. A. Saunders. Surrey Beatty & Sons, 2000.

intensified, many in this latter group have also declined (Green and Stowe 1993; Gates *et al.* 1994).

Predatory birds and mammals have often proved to be particularly sensitve to changes in land management for a variety of reasons. Usually they have large area requirements and some mammals in particular may be unable to traverse large gaps between habitat fragments so that populations become small, isolated, and in the long term, non-viable. Being at the top of food chains they are vulnerable to reductions in prey densities and also to the accumulation of agricultural chemicals in their body tissues with resultant decreases in survival and reproduction (Newton and Brogan 1978; Newton 1988; Nisbet 1988). Many are also perceived to be a threat to management objectives such as livestock production and game shooting and may suffer direct persecution (Elliot and Avery 1991; Hudson 1992). Consequently, the successful conservation of predatory species in productive landscapes can be highly challenging. However, it is worth attempting as it is likely that the habitat management for the target species may also conserve many other species.

The study reported here concerns Barn Owl *Tyto alba* populations in pastoral farmland in southern Scotland. It attempts to illustrate the reseach needed to specify a set of habitat characteristics that will support viable populations of the owls and at the same time be acceptable to the farming community. Barn Owls nest in cavities and prey on terrestrial small mammals which they hunt over relatively open areas of grassland. Their main natural habitats in Europe probably would have been the ecotones of forest/grassland or wetland edges (Taylor 1994). The original natural landscape of the study area before human settlement was mostly deciduous woodland interspersed with wetlands and river systems and it is likely that Barn Owls had a restricted distribution in this environment, as elsewhere in Europe. The advent and spread of agriculture created large areas of landscape with suitable combinations of foraging habitat and nesting places for the owls probably resulting in greatly increased densities. Following the rapid intensification of agriculture from the 1950s onwards Barn Owl populations have declined over most of Europe (Straeten and Asselberg 1973; Braaksma and de Bruijn 1976; de Jong 1983). Explanations for these declines are largely conjectural, as research into the species' ecology did not begin until the late 1970s. Nevertheless, major factors were probably reductions in the availablity of foraging habitat through the replacement of prey rich permanent rough grassland and hay meadows by short rotation, intensively managed grassland and silage which support few small mammals, the loss of edge habitats through the process of field enlargement and a reduction in on-farm grain storage and associated populations of mice and rats (Rowe *et al.* 1963; Taylor 1994). Nest sites have been lost through the felling of trees with suitable cavities and the removal of traditional farm buildings which provided many nesting places (Kaus 1977; Juillard and Beuret 1983; Taylor 1994). The increased use of insectides and rodenticides reduced survival and breeding success (Newton *et al.* 1990, 1991). The Scottish study area was one of the few remaining parts of Britain to retain relatively dense populations of the owls, for reasons discussed later.

Conservation of Barn Owls requires the management of habitat at a landscape scale but a major disadvantage of management at that scale is that it often not readily amenable to experimentation, mainly because of costs and time scales involved. Thus the approach adopted in this study was to develop a model based on investigations of the owls' responses to small scale variations in existing landscape characteristics rather than to manipulated landscapes. This has the disadvantage that only those elements of habitat that were already represented in the study area can appear in the model. The process of describing a suitable set of landscape characteristics to support viable populations of the owls involved a number of interrelated steps. As the birds nest in cavities whose availablity was in decline, the provision of artificial nest sites was an important component. Details of the other major habitat component, foraging habitat, had to be considered in relation to the locations of nest sites. On the question of foraging habitat, both qualitative and quantitative aspects were addressed; what kinds of habitat were needed and how much of it had to be provided? The latter involved questions of pattern; what density of habitat was needed and in what spatial arrangement? These were considered at the level of the requirements of individual birds or breeding pairs during both breeding and non-breeding seasons and at the level of whole populations. The meaning of the term "viable population" was considered in relation to environmental and demographic stochasticity and potential genetic effects as together these will determine required population sizes and hence the extent of the geographic area over which suitable habitat must be provided (Soulé 1987). Natural geographic variations in topography, soils and other factors would have prevented the establishment of continuous panmictic populations over very large geographic areas and the possibilty of conserving metapopulations with subpopulations interconnected by dispersing individuals was considered.

The study attempted to address all of these points by investigating the birds' foraging and ranging behaviour, habitat selection, population dynamics and relationships between habitat and population parameters.

STUDY AREA AND METHODS

From 1979 to 1993 a study of the population ecology and conservation of Barn Owls was undertaken in the entire catchment (1 600 km^2) of the Esk River system in south-west Scotland (Fig. 1). A section of the catchment (450 km^2), ranging from 50 to 150 metres above sea level, of enclosed

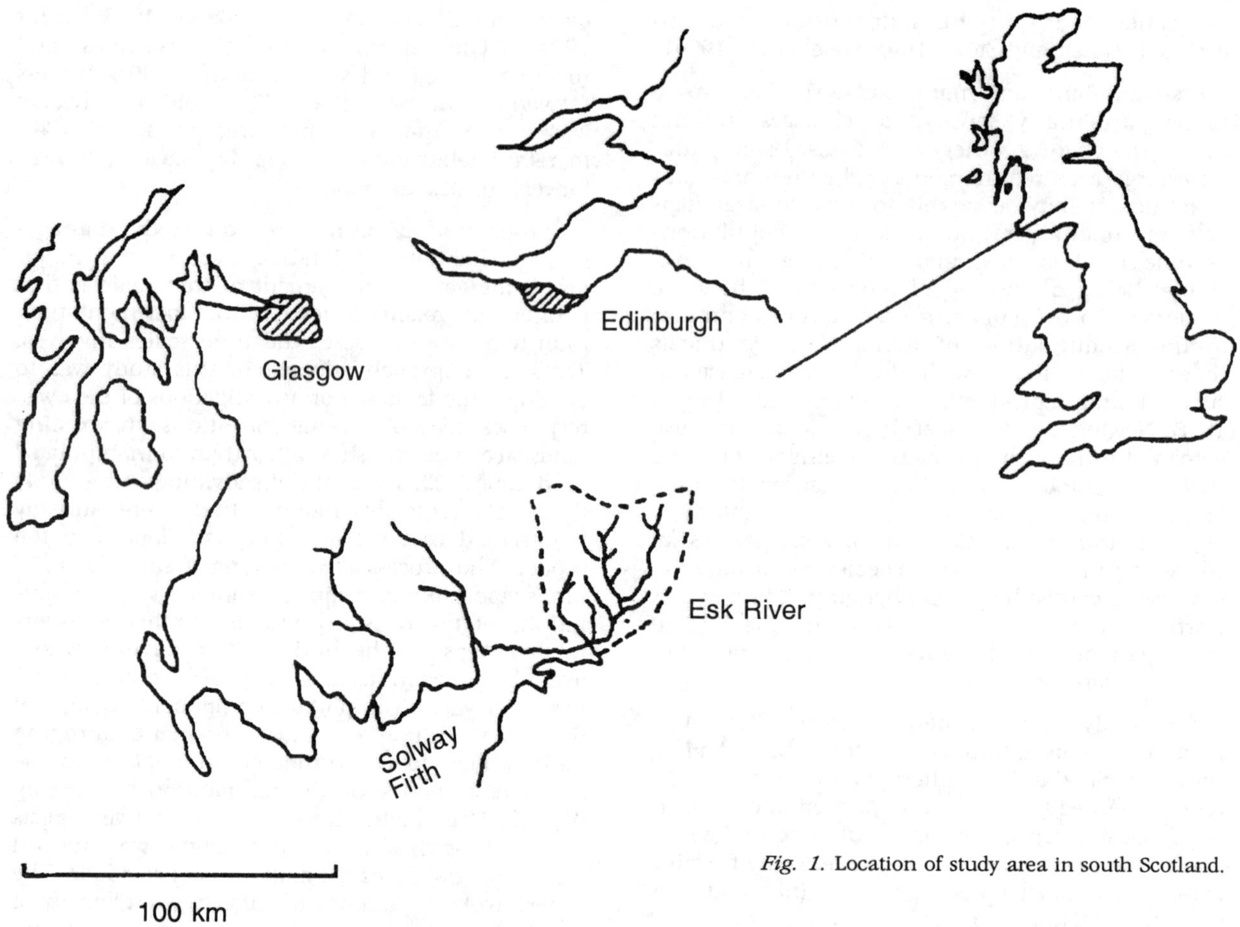

Fig. 1. Location of study area in south Scotland.

pastoral farmland was used for the work reported here. Rainfall in this section of the catchment occurred throughout the year, averaging approximately 120 mm per month from July to March and around 80–90 mm per month from April to June. Average daily air temperatures ranged from about 2°C in the coldest winter months of January and February to about 15°C in the warmest months of July and August. Snow, rarely exceeding a few centimeters depth, occurred between November and April, averaging 7–8 days with snow on the ground in January. Soils were mostly dense clays and topography was highly diverse as a result of numerous rivers and streams and glacial features.

Approximately 80% of the area was farmed, predominantly for dairy, sheep and beef production. Most fields were short rotation pastures, with silage and hay for winter feed. The climate was unsuitable for large scale cereal production although small areas were planted, mostly for oats, barley and turnips, as winter feed. The uneven topography prevented the development of large fields and about 20% of the area comprised small woodlands on rough or wet areas unsuitable for cropping or as woodland strips along water courses. Some of these were remnants of natural deciduous and pine forests but many were a mixture of exotic conifer and hardwood plantations. All woodlands were fenced from adjacent fields with a gap varying in width from about 1 m to 6 m and averaging 3–5 m, left between the woodland edges and the fencelines mainly to reduce any negative effects of shading on crop growth. These gaps developed dense, mainly grassland vegetation of native species. Field boundaries were either fencelines or hedgerows but because of the stock grazing in most of the fields there was little rank vegetation at the base of these structures.

Throughout the area most of the traditional stone farm buildings built during the 18th and 19th centuries were still intact although many were no longer in use other than for hay storage. As a result of farm amalgamations and reductions in the farm workforce during the 1960s and early 1970s, many traditional stone cottages and in some cases entire farms had been abandoned.

The owls' diet was determined each year by the analysis of pellet samples taken from each breeding pair. The pH of Barn Owl stomachs is higher than that of most other predatory birds so that most of the bones of ingested prey are untouched by digestion and pellet analysis has been shown to give an accurate representation of diet (Raczynski and Ruprecht 1974). Foraging ranges during the breeding season were determined over three years by the continuous monitoring of a sample of birds using radio transmitters attached to the central tail feathers. The birds were tracked using a three element "Yagi" antennae and AVM LA12 receivers. Range sizes were calculated using the convex polygon method. During tracking sessions the foraging locations of birds were also recorded and used to give an unbiased quantification of foraging habitat selection. Non-breeding season foraging ranges were determined from direct observations of a sample of 18 pairs of individually coded, plumage-dyed birds from 1980 to 1983. The birds hunted extensively during daylight hours and could also be located by spotlight at night. All breeding pairs within the study area were located each year and laying dates, clutch sizes and number of young fledged were recorded. As many of the fledged young as possible were banded each year and by the capture of these individuals in later years as breeding adults, dispersal distances from natal to breeding sites were estimated. About 85–95% of the breeding adults were trapped each year and used to give estimates of annual adult survival.

The entire study area was mapped by ground survey, recording locations of woodlands, fencelines, hedges, crop types and any other features and checks were made each year for any significant changes.

Comparisons were made of the abundance of small mammal prey species in all of the main habitats within the study area. Six sample sites, each with 20 trapping stations and two Longworth live traps at each, were selected randomly in each habitat. These were inspected twice daily over a period of five days and all captures of all species recorded.

RESULTS AND DISCUSSION

Prey species

Over 95% of the diet of all owl pairs during the breeding season consisted of only three species of terrestrial small mammals; the Field Vole *Microtus agrestis*, the Wood Mouse *Apodemus sylvaticus* and the Common Shrew *Sorex araneus*. Of these the Field Vole was the most numerous item, varying between 30% and 60% of the total, with upper values corresponding to peaks of vole density in the field during regular three-year cycles of abundance (Fig. 2). The average body mass of voles, mice and shrews was 26 g, 22 g and 10 g respectively, so that in terms of biomass, Field Voles were the most important prey at all times, contributing up to 70% of prey intake. Diets during the non-breeding season did not differ significantly from those in the breeding season. This preponderance of voles in the owls' diet has been recorded in almost all other studies of the species in northern Europe, so is a generally applicable requirement (Taylor 1994), and the provision of habitat suitable for voles is likely to be a critical aspect of habitat management for the owls. The voles' optimum habitat is known to be long grassland; they construct tunnels in the rank vegetation and their prefered diet includes the stems and shoots of the more nutritious and easily digestible grasses such as bents (*Agrostis* spp.), fescues (*Festuca* spp.) and rye grasses (*Lolium* spp.) in addition to a variety of dicotyledonous plants (Hansson 1971). In the study area the highest

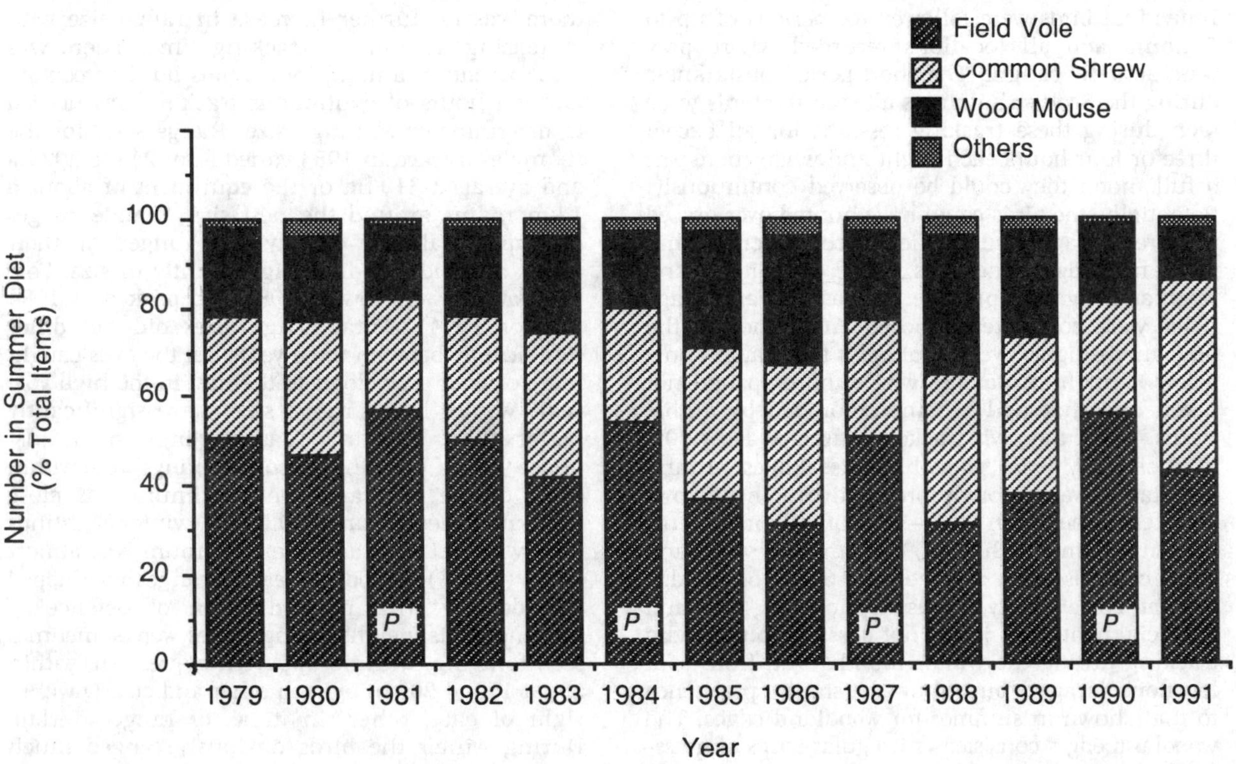

Fig. 2. Breeding season diet of Barn Owls in study area. The relative contribution of Field Voles to the diet varied cyclically in relation to cyclic variation in the abundance of the voles in the field. Years of peak vole abundance are indicated by the letter "P".

densities of Field Voles occurred in the rank grassland edges of the woodlands (Table 1). The type of woodland did not significantly influence these densities; similar values were obtained for the edges of exotic conifer plantations as for native deciduous woods. Uncultivated strips of grassland along hedges, where they had not been grazed by livestock, had slightly lower densities than the woodland edges. All of the cultivated grassland habitats and crops sampled as well as the interiors of both coniferous and deciduous woodlands had low Field Vole densities.

Table 1. Comparisons of the relative abundance of Field Voles in major habitat types within the study area, May to July 1981. Values are the mean numbers (± se) of voles trapped over five days from six sample sites in each habitat type, each with 20 trapping stations and two traps set at each station.

Habitat type	Mean No. voles trapped
Deciduous woodland grass edge	12.7 ± 2.4
Coniferous woodland grass edge	14.2 ± 2.6
Hedge "headland" grass edge	10.3 ± 2.2
Within coniferous woodland	0.5 ± 0.34
Within deciduous woodland	1.3 ± 0.7
Pasture	0
Silage	0
Hay	1.5 ± 0.6
Cereal	0.2 ± 0.2

Foraging habitat selection

Radio transmitters were attached to 23 owls during the breeding season between 1983 and 1985. Individual birds were followed for periods of up to 5 hours and all locations recorded where prey were actually caught. The short period of darkness during the Scottish summers allowed the birds to be seen during these tracking sessions for all except three or four hours each night and when there was a full moon they could be observed continuously. Potentially, the birds could have hunted over grazed pastures, silage and hayfields, cereal crops and along field edges, fencelines, hedgerows, ditches and woodland edges. In practice, the pastures and silage fields were completely ignored and although the birds made flights over cereal fields they caught only a few items from patches where the crop had not taken and that had become colonized by arable weeds. All of the owls tracked caught at least 90% of their prey from edge habitats. Edges around woodlands were strongly preferred over hedgerows and fencelines with 80%–90% of all prey items caught within this habitat, even though woodland edges comprised, on average, only 60.2% of all edge habitat. For a variety of reasons including time and financial contraints it was not possible to use radio tracking during the winter months but, from spot locations, foraging birds showed a similar preference to that shown in summer for woodland edges. The woodland edges consisted of irregular strips of grassland containing mixtures of native and cultivated species, up to about 40 cm deep at the height of the growing season, averaging 3–5 m, separated from the adjacent field by fencelines. These strips were not managed in any way but all of the woodlands had populations of Roe Deer *Capreolus capreolus* which browsed around the edges, inhibiting the growth of shrubs and maintaining the grassland.

The preference for hunting along woodland edges was clearly related in part to the high densities of preferred prey there but hedges that had only slightly lower prey densities were disproportionately avoided by the birds. Much of the owls' hunting was done from perched positions on the fenceposts between the woodland edges and the crop and this was especially important in winter (87% and 54% of observations in winter and summer respectively, $p < 0.05$, Taylor 1994) when the owls probably had to conserve energy more by adopting the lower energy demanding perched hunting compared with flight hunting (Village 1983). Suitable perches were not available to the owls along most of the hedgerows. Hunting along woodland edges also provided shelter against rain and wind both of which greatly increased the owls' rates of heat loss and energy consumption (McCafferty *et al.* 1997a) and probably also interfered with the effectiveness of the owls' prey location which is achieved mainly by use of auditory cues (Payne 1971).

Home range sizes

Data from the radio-tracking of foraging birds were used to determine the shapes and sizes of the home ranges of a sample of six adjacent pairs of owls during the breeding season in 1983 and four in the 1984 season using the minimum convex polygon method. Each individual was tracked until there was no further increase in range size with increasing amount of tracking time. There was some variation among individuals but between six and ten hours of continuous tracking was needed to determine final range size. Range sizes for the six males tracked in 1983 varied from 245 to 392 ha and averaged 319 ha or the equivalent of about a 1 km radius around the nest site. Female ranges overlapped almost exactly the ranges of their mates and did not differ significantly in size. Vole populations were low in 1983 and peaked in 1984 (Taylor 1994). Total range sizes did not differ significantly between these years but the owls caught more of their prey closer to the nest in the high vole year. Winter home range sizes were significantly larger than breeding season range sizes; just under 90% of breeding season sightings were within 1 km of the nest and the maximum was close to 2 km, whereas only 39.6% of winter sightings were within 1 km, and the maximum was almost 5 km (Fig. 3). Breeding season ranges overlapped considerably with no indication of defence of hunting areas; neighbouring males were sometimes seen foraging, with no apparent aggression, within about 150 to 200 m of each other and clearly within sight of each other, in areas of range overlap. During winter the birds obviously ranged much more widely, overlapping several of their neighbours' breeding season ranges. When not breeding they were not constrained by the need to make repeated feeding trips to the nest so, in the absence of clear territorial behaviour, they may simply have been free to exploit the best feeding areas around their roosts. Such areas may not have been constant; weather conditions such as wind speed and direction, rain and snowfall varied greatly often over short time periods and may have been as important as vegetation structure and prey density in determining the profitability of different sites. For example, a few hours of heavy snowfall accompanied by strong winds often resulted in deep drifts on the windward side of woods while the lee sides remained relatively free of snow.

Population dynamics and habitat structure

The conclusions from the above sections are that the owls fed almost exclusively on small mammals, especially the Field Vole which they caught mostly around woodland edge grassland strips, and that their range sizes were close to 1 km radius around the nest during the breeding season and up to 5 km during winter. On its own, this information does not adequately describe the habitat needed to maintain a stable population of the owls; further information is needed that links this to the owls' population dynamics. Details are needed on survival and natality rates and on dispersal behaviour. Relationships need to be quantified between these and the amount or density of the required habitat type within the defined distances around suitable nesting places.

A simple population model was constructed from survival, natality, recruitment and dispersal data collected from the population betwen 1979 and 1989, reported fully in Taylor and Masseder (1992). Average annual adult survival was close to 20% and average recruitment of fledged young to the breeding population was just over 15%. Neither of these rates were significantly correlated with the amount of woodland edge habitat within 1 km, 2 km or 5 km of the birds' nest sites. Mortality in all age classes occurred mainly during winter when weather conditions were poor (Taylor 1994). It was shown earlier that at this time the birds were highly mobile with large overlapping ranges and this probably explains the lack of any obvious correlation between these population parameters and habitat. However, there was a strong correlation between breeding performance and the amount of woodland edge habitat within the birds' normal breeding season ranges of 1 km around the nest site (Fig. 4). This relationship was based on 12 years of data at each site and within this time scale the occupants of the sites changed 4–6 times, on average, so that the relationship shown cannot be attributed to the qualities or ages of individual birds rather than to the qualities of the habitats around the nests.

The population model predicted that an average long-term productivity of 3.2 fledged young per breeding female per year would be needed to maintain a stable population of owls. This level would be achieved by having approximately 8–10 km length of suitable grassy woodland edge habitat (3–5 m wide) within the 1 km range of suitable nest sites. However, isolated patches of such habitat, even though suitable for breeding season productivity, may not be adequate for winter survival and consideration needs to be given to the spatial relationships among such patches. The ideal of continuous suitable habitat over large enough areas to support a viable population was not feasible because of variations in topography and interest among the farming community. The study

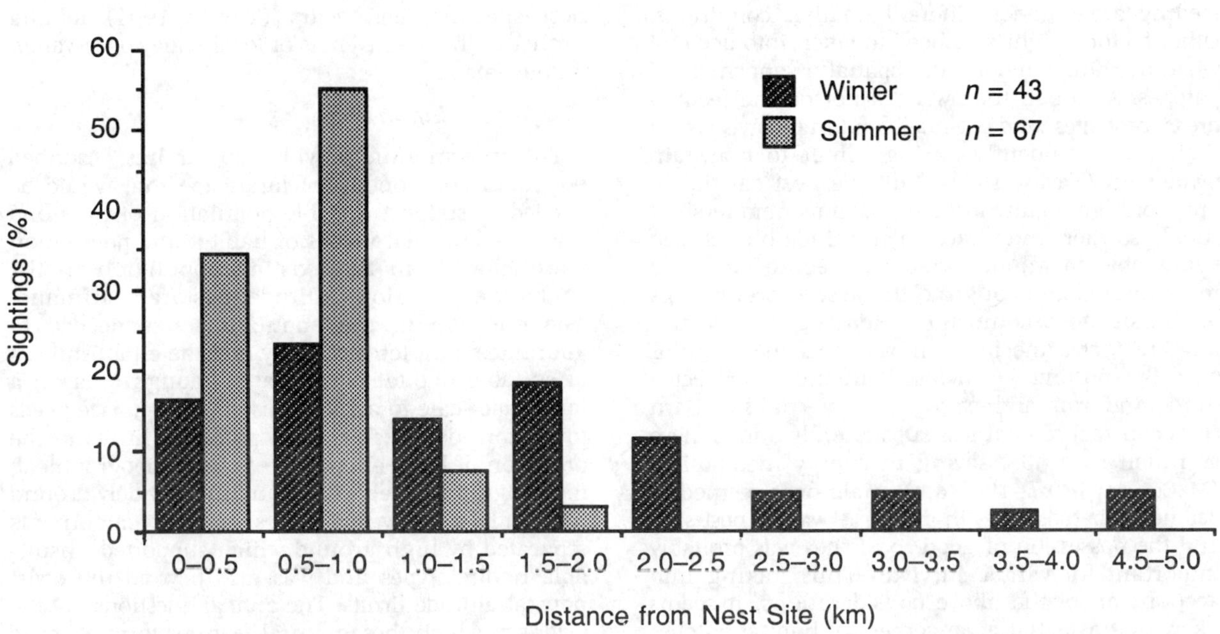

Fig. 3. Comparison of distances from the nest site at which colour-dyed Barn Owls were seen foraging in winter and in summer (Mann-Whitney *U*-test, $z = -4.45$, $p < 0.001$).

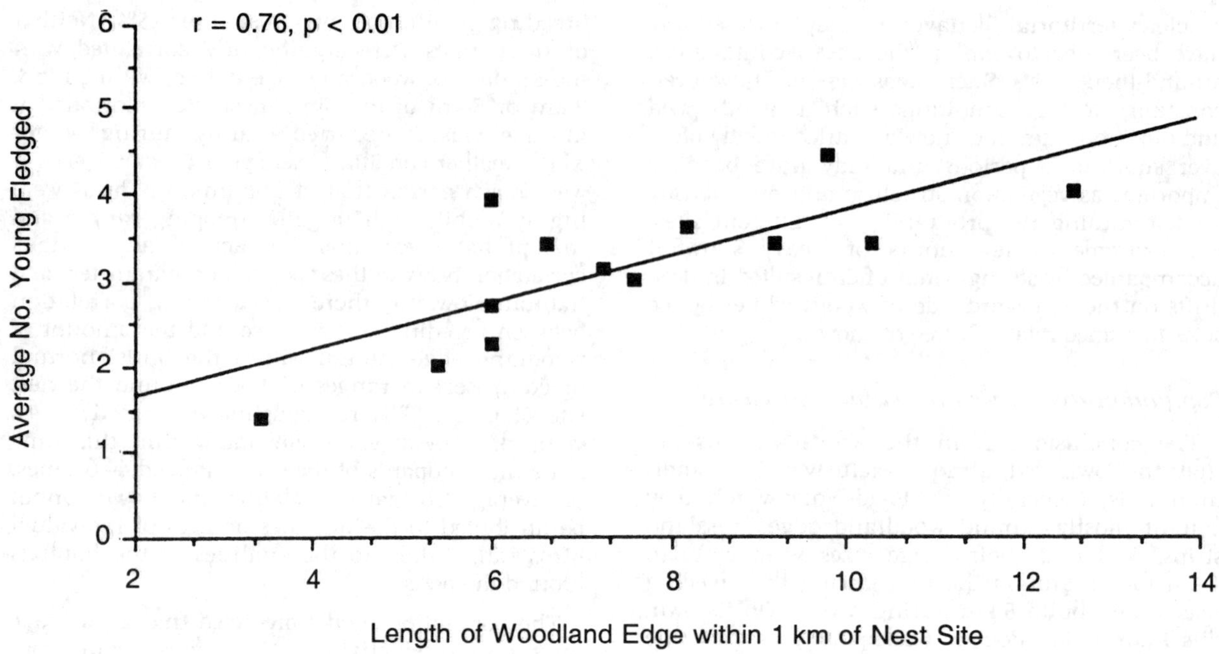

Fig. 4. The number of young fledged per year at 13 Barn Owl breeding sites, averaged over 12 years, in relation to the length of woodland edge habitat within a 1 km radius of the nest site.

showed that the birds moved up to about 5 km from the nest site in winter based on the existing configuration of habitat and nest site availablity in the study area, that is, in situations where several overlapping breeding ranges with suitable habitat were grouped together. As gaps between patches of foraging habitat of appropriate size for breeding are increased it must become progressively more difficult for the birds to move during winter to locate suitable feeding areas under different weather conditions. Other factors might also need to taken into account when deciding the optimum spatial arrangement of patches. Although the owls did not defend feeding areas, nest sites were defended, and male owls visited their nests frequently during winter to maintain ownership (Taylor 1994). Suitable nest cavities in areas of high quality foraging habitats were in short supply so there may have been a trade-off between being able to wander widely to secure food and remaining close enough to the nest to prevent loss of the site to a competitor. Roost sites may also have restricted the birds' movements in a similar way. By roosting in enclosed situations, protected from wind, rain and exposure to the cold sky, Barn Owls can reduce heat loss substantially and achieve as much as a 30% saving in energy metabolism (McCafferty *et al.* 1997 a,b). Male owls seemed to defend their roost sites in the same way as nest sites and the possession of good roost sites was probably important for winter survival. Thus, taking into account all of the above considerations, it seems likely that a spatial arrangement of habitat patches into units extending at least to the size of five or six breeding ranges would be better than isolated patches the size of individual breeding ranges.

Dispersal distances from natal to breeding sites were limited, with most individuals settling within 15 km of their hatching sites and the farthest dispersal distances close to 30 km (Fig. 5). From a conservation perspective this means that local habitat management would result in local population increases rather than having a high proportion of the offspring dispersing to other areas. Breeding adults were also almost totally faithful to their nest sites between years (Taylor 1994) adding further to the effectiveness of local scale conservation management.

Viable population sizes

The information provided so far has described the main components of landscape that would be needed to support a stable population of owls and was based on real values of habitat and population parameters from the existing population in the study area, in low altitude pastoral farmland. The question of the population size needed to guarantee long-term viability and the establishment of suitable habitat over a large enough area on a landscape scale to achieve this population size needs to be considered. The main potential limits to the provision of large enough areas were topographical; the geographic area within and more widely around the study area was a series of river catchments separated by high ground which supported unsuitable habitat types and was also beyond the owls' natural altitude limits. The critical questions are the extent to which these natural features form effective demographic and genetic barriers to the birds. The answers to these questions will determine whether viable populations need to be established within

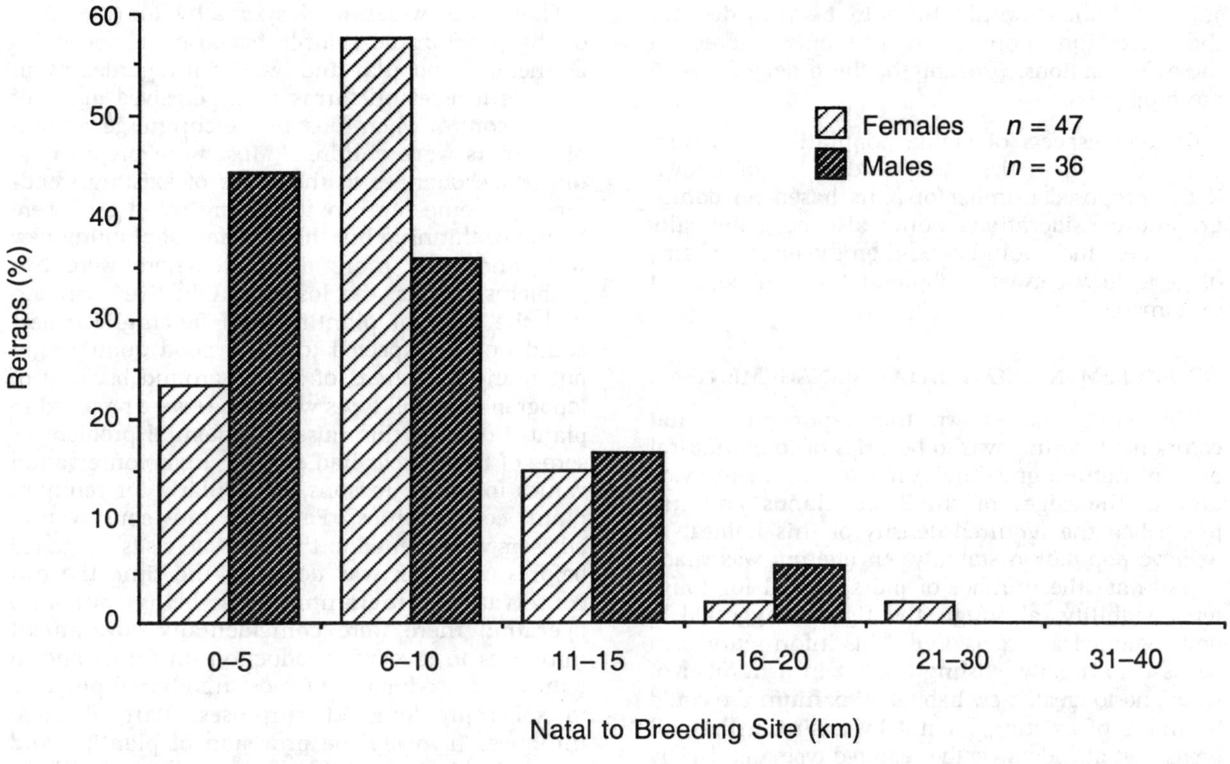

Fig. 5. Distances from natal sites to breeding sites of male and female Barn Owls within the study area. Females moved significantly farther than males (Mann-Whitney *U*-test, $z = -1.81$, $p = 0.035$).

individual catchments or whether a population can be considered to encompass several catchments. Distances between catchments generally exceeded the distances over which almost all individuals were recorded to disperse from their natal to breeding sites. Large numbers of young and breeding adults were banded over 12 years in three adjacent catchments but no instance of a movement from one catchment to another was recorded (Taylor 1994). Nevertheless, it could be argued that these dispersal distances were simply a function of the existing availablity of suitable foraging habitat and nesting places and that individuals would be capable of dispersing substantially farther, including significant movements among catchments, if suitable habitat occurred at lower density. This contention is difficult to refute in the absence of appropriate empirical evidence. Nevertheless, numerous instances were recorded of non-breeding individuals who failed to secure a suitable breeding site, yet remained close to their natal areas even when apparently suitable sites were available at greater distances within the same catchment (Taylor 1994). Dispersal distances estimated from banding recoveries throughout the whole of Britain were also small, even though almost all areas had declining owl densities which were significantly lower than those in the present study area (Percival 1990). The balance of evidence on dispersal suggests that from a demographic point of view, viable population sizes may have to be established within each catchment rather than over several. The nature of annual variations in population size are also relevant. Two variables were of greatest significance in the study area: food supply and winter weather conditions, both of which were shown to have significant effects on annual survival rates and variations in breeding population size. Field Vole populations followed a three-year cycle of abundance which was independent of weather conditions and annual survival of the owls varied around three-fold in response. Survival was further reduced by winters of exceptionally prolonged and deep snow cover with low temperatures. Fluctuations in owl abundance were correlated in adjacent catchments as both weather and Field Vole cycles were synchronized over wide areas (Taylor 1994). Thus, a conservation strategy that relied on dispersal from one catchment to another to replace losses seems unlikely to be successful. A more appropriate strategy would be to ensure that each catchment was numerically self-sufficient, supporting at least the minimum viable population size of the owls. Within the study catchment, over 15 years of study, the breeding population fluctuated between 15 and 55 pairs around a long-term mean of close to 30 pairs with no indication of long-term decline or increase in population size. The lowest levels occurred when low prey densities coincided with exceptionally severe winters. In the longer term it is conceivable that two or more such events might occur in successive years, possibly eliminating the population so that a long-term mean of around 30 pairs might well be below the viable population size. Possibly, 40 to 50 pairs might be closer to the required number. This mean level would be defined by cyclic fluctuations in population size so that suitable habitat would have to be provided for the maximum population size encountered in these fluctuations, probably of the order of 60–65 breeding pairs.

Genetic aspects of viable population size were not addressed in this study and it is not known if the proposed number of pairs based on demographic considerations would also be genetically adequate. Much would depend on the extent, if any, of gene flow between subpopulations in adjacent catchments.

IMPLEMENTING HABITAT MANAGEMENT

This study has shown the important habitat component for the owls to be strips of long, natural or semi-natural grassland, three to five metres wide around the edges of small woodlands and has prescribed the required density of this habitat to achieve population stability. An attempt was made to estimate the number of pairs needed for long-term viability, although the figure suggested is best regarded as provisional. This information can be used to modify existing habitat in a number of ways and to create new habitat. Maximum use could be made of existing habitat by ensuring that all areas that already have the required type and density of woodland edges have suitable nest sites for the owls. Areas that are close to the required habitat specifications and that would need the minimum of effort to bring them to the required levels could be identified and targeted for priority action. This could involve alterations to create edges of the desired structure or the planting of new woodlands to increase the density of edges.

The implementation of such a management strategy requiring widespread changes at the landscape level and involving many individual landowners must necessarily be a long-term process. Persuading farmers to implement these changes involves a number of steps from initial information sharing to the financial aspects of planting and fencing. Information sharing was done in a number of ways. As the study lasted 15 years in the area under investigation, it was possible to build up personal relationships and discuss all aspects of the work with individual farmers on their own properties. However, on a wider scale this approach was not feasible and information and advice was shared by arranging evening lecture/discussion sessions with groups of local farmers, foresters, game managers and natural historians. From these meetings, site visits were arranged to provide further advice. In conjunction with this a conservation booklet giving details of the owls' ecology, a conservation action plan and a practical conservation guide for landowners was produced and distributed freely (Taylor 1993). Articles were written for local newspapers and radio programmes presented to appeal for participation. Local representatives (Assistant Regional Officers) of the government conservation agency (Scottish Natural Heritage) were involved in the project and provided advice to landowners.

There was widespread sympathy for the aims of the programme mainly because the species is aesthetically popular and had been regarded as an important asset on farms as a perceived agent of rodent control when most of the current generation of farmers were children. Most were prepared to undertake changes to the edges of existing woodlands at some time in the future when they were felling or thinning but the question of planting new woodlands was more difficult. There were two problems in this: the loss of productive farmland and the costs of planting and fencing. Farmers could not be expected to plant good quality land but many had areas of rough ground because of topography or soil types which they were prepared to plant. However, this raised a potential problem as some of these areas had existing high conservation values for other reasons, particularly for remnant plant communities. Farmers and conservation advisors were alerted to this and site visits suggested before any work was done. At the time the owl conservation programme was being put into operation there were coincidentally government initiatives to diversify production on farms and to reduce the production of most traditional products to solve problems of surpluses. Part of these initiatives involved the provision of planting and fencing grants to establish more small woodlands on farmland. Thus, these programmes were entirely compatible with the aims of the owl conservation project and landowners who plan to take up these grants are being encouraged to incorporate suitable edge habitats around any new woodlands they establish. As yet it is too early to evaluate the success of this programme.

ACKNOWLEDGEMENTS

This study was funded by the Natural Environment Research Council, WWF, Scottish Natural Heritage, The Co-operative Society, Edinburgh University and the author. For general assistance in the field I thank Tom Irving, Mairi Osborne, Donna Kreft, Fiona Slack, Peter Bell, Ian Langford, Murray Grant Laurel Hanna and Sarah Taylor. Dr Mick Marquiss gave initial help with radio tracking and Dr Nigel Charles with small mammal trapping. I thank the many farmers and landowners who were involved in many ways and in particular the late Thomson McKnight. The study benefited from many discussions with Prof. Ian Newton and I also thank Dr Denis Saunders and two anonymous referees for their valuable comments on the draft.

REFERENCES

Bennett, A. F., 1987. Conservation of mammals within a fragmented forest environment: the contributions of insular biogeography and autecology. Pp. 41–52 *in* Nature Conservation: the Role of Remnants of Native Vegetation ed by D. A. Saunders, G. W. Arnold, A. A. Burbidge and A. J. M. Hopkins. Surrey Beatty & Sons: Chipping Norton, New South Wales.

Braaksma, S. and de Bruijn, O., 1976. De kerkuilstand in Nederland. *Limosa* **49:** 135–87.

de Jong, J., 1983. De Kerkuil. Kosmos, Utrecht.

Elliot, G. D. and Avery, M. I., 1991. A review of reports of Buzzard persecution. 1975–1989. *Bird Study* **38:** 52–56.

Gates, S., Gibbons, D. W., Lack, P. C. and Fuller, R. J., 1994. Declining farmland bird species: modelling goegraphic patterns of abundance. *In* Large Scale Ecology and Conservation Biology ed by P. J. Edwards, R. M. May and N. Webb. 35th Symposium of the British Ecological Society: Oxford, Blackwell Scientific Publications.

Green, R. E. and Stowe, T. J., 1993. The decline of the Corncrake *Crex Crex* in Britain and Ireland in relation to habitat change. *J. Appl. Ecol.* **30:** 689–95.

Hansson, L., 1971. Habitat, food and population dynamics of the field vole in south Sweden. *Viltrevy* **8:** 267–378.

Hudson, P. J., 1992. Grouse in space and time. The Game Conservancy Trust: Fordingbridge.

Juillard, M. and Beuret, J., 1983. L'emenagement des sites de nidification et son influence sur une population de Chouette Effraies *Tyto alba* dans le nord ouest de la suisse. *Nos Oiseaux* **37:** 1–70.

Kitchener, D. J., Chapman, A., Muir, B. G. and Palmer, M., 1980. The conservation value for mammals of reserves in the western Australian wheatbelt. *Biol. Conserv.* **18:** 179–207.

Kaus, D., 1977. Zur populationsdynamik, okologie, und brutbiologie der Schleiereule in Franken. *Anzeiger der Ornithologischen Gellsellschaft in Bayern* **16:** 18–44.

Lynch, J. F., 1987. Responses of breeding bird communities to forest fragmentation. Pp. 123–40 *in* Nature Conservation: the Role of Remnants of Native Vegetation ed by D. A. Saunders, G. W. Arnold, A. A. Burbidge and A. J. M. Hopkins. Surrey Beatty & Sons: Chipping Norton, New South Wales.

McCafferty, D. J., Moncrieff, J. B. and Taylor, I. R., 1997a. The effect of wind speed and wetting on thermal resistance of the Barn Owl (*Tyto alba*). I Total heat loss, boundary layer and total resistance. *J. therm. Biol.* **22:** 253–64.

McCafferty, D. J., Moncrieff, J. B. and Taylor, I. R., 1997b. The effect of wind speed and wetting on thermal resistance of the Barn Owl (*Tyto alba*). II Coat resistance. *J. therm. Biol.* **22:** 265–73.

Newton, I., 1988. Determination of critical pollutant levels in wild populations with examples from organochlorine insecticides in birds of prey. *Environ. Pollution* **55:** 29–40.

Newton, I. and Bogan, J., 1978. The role of different organochlorine compounds in the breeding of British Sparrowhawks. *J. Appl. Ecol.* **15:** 105–16.

Newton, I., Wyllie, I. and Freestone, P., 1990. Rodenticides in British Barn Owls. *Environ. Pollution* **68:** 101–17.

Newton, I., Wyllie, I. and Asher, A., 1991. Mortality causes in British barn Owls *Tyto alba*, with a discussion of aldrin-dieldrin poisoning. *Ibis* **133:** 162–69.

Nisbet, I. C. T., 1988. The relative importance of DDE and dieldrin in the decline of peregrine falcon populations. Pp. 351–75 *in* Peregrine Falcon Populations: Their Management and Recovery ed by T. J. Cade, J. H. Enderson, C. G. Thelander and C. M. White. The Peregrine Fund Inc.: Boise.

Opdam, P. G., Rijsdijk, G. and Hustings, F., 1985. Bird communities in small woods in an agricultural landscape: effects of area and isolation. *Biol. Conserv.* **34:** 334–52.

Opdam, P. G., Foppen, R., Reijnen, R. and Schotman, A., 1995. The landscape ecological approach in bird conservation: integrating the metapopulation concept into spatial planning. *Ibis* **137**: S139–S146.

Payne, R. S., 1971. Acoustic location of prey by barn owls *Tyto alba*. *J. Exp. Biol.* **54:** 535–73.

Percival, S. M., 1990. Population trends in British Barn Owls, *Tyto alba*, and Tawny Owls, *Strix aluco*, in relation to environmental change. *British Trust for Ornithology Research*, Report No. 57. British Trust for Ornithology, Tring.

Raczynski, J. and Ruprecht, A. L., 1974. The effect of digestion on the osteological composition of the owl pellets. *Acta Ornithologica* **15:** 1–12.

Rowe, F. P., Taylor, E. J. and Chudley, A. H. J., 1963. The numbers and movements of house mice *Mus musculus* living in corn ricks. *J. Anim. Ecol.* **32:** 87–97.

Saunders, D. A., Arnold, G. W., Burbidge, A. A. and Hopkins, A. J. M., (eds), 1987. Nature Conservation: the Role of Remnants of Native Vegetation. Surrey Beatty & Sons: Chipping Norton, New South Wales.

Simberloff, D., 1995. Habitat fragmentation and population extinction of birds. *Ibis* **137:** S105–S111.

Soulé, M. E., 1987. Viable populations for conservation. Cambridge University Press: Cambridge.

Straeten, E. van der and Asselberg, A., 1973. Het voedsel van de kefkuil in Belgie. *Giervalk* **63:** 149–59.

Taylor, I. R., 1993. Barn Owls: an action plan and practical guide for their conservation in Scotland. The University of Edinburgh: Edinburgh.

Taylor, I. R., 1994. Barn Owls: Predator-Prey Relationships and Conservation. Cambridge University Press: Cambridge.

Taylor, I. R. and Massheder, J., 1992. The dynamics of depleted and introduced barn owl populations: a modelling approach. Pp. 104–09 *in* The Ecology and Conservation of European Owls ed by C. A. Galbraith, I. R. Taylor and C. M. Percival. Joint Nature Conservancy, Committee: Peterborough, UK.

Village, A., 1983. Seasonal changes in the hunting behaviour of Kestrels. *Ardea* **71:** 117–24.

Whitcomb, R. F., Lynch, J. F., Klimkiewicz, M. K., Robbins, C. S., Whitcomb, B. L. and Bystrak, D., 1981. Effects of forest fragmentation on avifauna of the eastern deciduous forest Pp. 125–205 *in* Forest Island Dynamics in Man-dominated Landscapes ed by R. L. Burgess and D. M. Sharpe. Springer-Verlag: New York.

Wiens, J. A., 1995. Habitat fragmentation: island vs landscape perspectives on bird conservation. *Ibis* **137:** S97–S104.

CHAPTER 34

Biodiversity conservation in arable landscapes: greening arable support in the European Union

F. M. Burch,[1] B. H. Green,[1] J. Mitchley[1] and C. A. Potter[1]

Modern agriculture poses the greatest threat to Europe's biodiversity and yet it may also be seen as the industry most able to contribute to its conservation. Reforms of the European Union's Common Agricultural Policy now provide an opportunity to integrate environmental measures with support payments to farmers. This chapter presents a methodology for evaluating policy measures for integrating environmental concerns into arable production in the European Union. Policy options are identified and subjected to a hierarchy of conceptual filters: the scientific filter, where scientific evidence is examined to determine the validity of possible options; the advisory filter, in which discussion with advisory organizations is used to determine the likely environmental effectiveness of options and their administrative feasibility and the farmer filter, where the degree of farmer acceptability is gauged and practical problems of implementing the option identified.

We identified options which were specifically aimed at integrating biodiversity conservation into arable production. These were subjected to the three filters, using literature review (the scientific filter) and separate focus groups for advisors and farmers (the advisory and farmer filters respectively) in six European Union countries. The methodology resulted in modifications to the options tested and provided a means for identifying a number of key factors which are essential in translating theoretical options into practical policy measures.

Key words: Common Agricultural Policy, Cross-compliance, Equitability, Additionality, Winter fallows.

INTRODUCTION

Conservation within the arable landscape

THE major irony of modern agriculture in Europe is that while it may be perceived as the greatest threat to Europe's biodiversity it can also be seen as the industry most able to contribute to its conservation (BirdLife International 1997). Conservation in the wider countryside can be considered of equal importance as the protection of individual sites and species and there is a growing recognition of the value of the farmland matrix itself:

> Ecologist and conservationists should think not of "remnants of habitat being left amongst farmland" and more of a farmland biotope for which optimum management practices need to be developed. (Bignal and McCracken 1996).

Agri-environment initiatives so far have largely been targeted at habitats associated with livestock systems e.g., calcareous grasslands and upland moorlands maintained by grazing. However, there has been a growing awareness, in the United Kingdom at least, of the conservation importance of the arable biotope for three main reasons. Firstly, there is a mounting body of evidence concerning the rapid decline of many of Europe's formerly common farmland species as a result of changing arable practices (Fuller *et al.* 1995; Pain and Pienkowski 1997). Secondly, there has been a fundamental shift from a site-based conservation ethos aimed at the designation and protection of small areas of high conservation value to a recognition of the need to look to the conservation of the wider countryside and thus the wider farmland matrix (Pickett *et al.* 1992). Thirdly, the introduction of set-aside in

[1]Environment Department, Wye College (University of London), Wye, Ashford, Kent, TN25 5AH, UK.

Pages 336–44 *in* NATURE CONSERVATION 5: NATURE CONSERVATION IN PRODUCTION ENVIRONMENTS: MANAGING THE MATRIX ed by J. L. Craig, N. Mitchell and D. A. Saunders. Surrey Beatty & Sons, 2000.

Europe provided conservationists with a vision of how small and large-scale land diversion may be targeted to promote biodiversity within the arable landscape (Green 1993). Introduced initially as a voluntary measure in 1988, by which farmers were compensated for removing land from arable production, set-aside became an obligatory condition of the receipt of support payments for all but Europe's smallest farmers under the 1992 reforms of the Common Agricultural Policy.

Although introduced primarily as a production control measure, conservationists were quick to highlight the potential environmental value of the set-aside scheme (Green and Potter 1987; Newbold 1989; Andrews 1992). The large-scale removal of land from arable production offered opportunities for promoting diversity in the wider countryside and buffering, extending and linking remaining fragments of semi-natural habitat. Ecological research has documented the diversity of habitats which may develop on set-aside land. While many sites have predictably been dominated by weeds of arable production establishing from the soil seed bank, sites where fertility is low and colonization sources are nearby have developed species-rich swards colonized by many species typical of semi-natural habitats (Wilson 1992; Burch 1996, 1997). Set-aside fields have also proved beneficial to the farmland birds which have suffered as a result of agricultural intensification (Wilson *et. al.* 1995). The introduction of obligatory set-aside throughout the European Union under the 1992 Common Agricultural Policy reforms provided a powerful precedent both for the large-scale environmental enhancement of arable land in the European Union and for the concept of linking environmental conditions to the receipt of support payments.

The policy context

Since its inception, the European Union's Common Agricultural Policy has striven to improve agricultural production and efficiency, fuelling programmes of expansion and intensification. These changes have had severe consequences for Europe's biodiversity (Table 1). While farmer support remained directly linked to production through the provision of price subsidies, the signal to farmers was one of product maximization, through the upward adjustment of variable inputs and increases in the area of cropped land. Significantly, the 1992 reforms of the Common Agricultural Policy introduced a first step in removing this direct link between farmer support and production. The Arable Areas Payments Scheme, introduced under the 1992 reforms in combination with reduced price guarantees, offered payments to farmers based on the area in arable crops rather than as price support for the crops produced. This partial decoupling of farmer support meant that the link between subsidies and yields was much weaker than before, thus giving farmers a reduced incentive to intensify production. However, the subsequent high world prices failed to bring about the predicted reduction in input use.

The continued link between arable area and farmer support has provided only a weakly decoupled system, with farmers still retaining arable area in order to receive support and acting as a disincentive for many farmers to enter agri-environment schemes which were also introduced under Regulation 2078/92 of the 1992 reforms (House of Commons 1997). This Regulation enabled individual member states to introduce schemes, co-funded by the European Union, which offered incentives to farmers for management favourable to the environment.

For the next round of world trade talks in 1999, policymakers can expect to come under renewed pressure to re-examine these compensation arrangements in order to devise a more decoupled, and hence more politically defensible, system of support. Further, the switch to direct payments increased the transparency of farmer support thus rendering it open to the criticism that payments should be linked to direct tangible environmental benefits (Baldock and Mitchell 1995; Scottish Landowners Federation 1995). The need to meet future General Agreement on Tariffs and Trade requirements,

Table 1. The impact of arable cropping practices on biodiversity in the European Union.

Changes to arable cropping practices	Impacts on biodiversity
Simplification of rotations — monocultures	Reduced diversity of crop habitats. Increased use of pesticides leading to further loss of biodiversity.
Change from spring to winter sowing	Loss of winter stubbles — an important food source for farmland birds. Increased use of pesticides.
Land consolidation/increased field size	Loss of habitat, hedges and other important boundary features — decline and loss of biodiversity.
Increased use of inorganic fertilizers	Modification of semi-natural habitats due to nutrient run-off/deposition. Decline and loss of biodiversity in both terrestrial and aquatic ecosystems.
Increased use of pesticides	Loss of biodiversity both directly and through impacts of pesticides on the food chain.
Increased irrigation	Lowering of water table, leading to damage of wetland habitats. Conversion of low intensity arable systems to intensive arable production.
Increased mechanization	Loss of habitat as fields enlarged and marginal habitat brought into production.

together with the accession to the European Union of central and eastern European countries makes the Common Agricultural Policy in its current form unviable and a major reform is inevitable. Payments to farmers for environmental purposes already enjoy significant immunities under World Trade Organization rules. Generally regarded as trade non-distorting, it is widely expected that direct environmental remuneration will progressively take over from market support as the chief form of agricultural support, thus decoupling payments from the production of agricultural goods and effectively recoupling payments to the production of environmental goods.

However, we may still be a number of years from a fully integrated rural policy for Europe which takes full account of the environmental and social consequences of farming activities (Buckwell 1997). Meanwhile, interest has been expressed in attaching more environmental conditions to compensation payments which farmers are currently eligible to receive. The concept of cross-compliance: the attachment of environmental conditions to agricultural support policies (Baldock and Mitchell 1995) has been a regular feature in discussions on the greening of European agricultural policy (Jenkins 1990; Dixon 1993; BirdLife International 1997) and provides a significant opportunity to integrate environmental concerns directly into arable support systems under the Common Agricultural Policy.

Project rationale

The project described in this chapter was funded by the Environment Directorate of the European Commission in Brussels with the aim of evaluating options for the integration of environmental concerns into arable support policy. Working with partners in six European Union Member States, the first part of the project was to use literature review and consultation with acknowledged experts to review the current implementation of the Arable Area Payments Scheme in all 15 Member States of the European Union and the observed impacts on the environment, both negative and positive. Based on our findings, a range of potential policy measures was drawn up. In consultation with partners in Europe and the Commission in Brussels, the initial list was reduced to a manageable number of measures which would help solve the environmental problems identified in Europe's arable landscapes (Table 2).

These measures were then tested using focus groups in six study regions, ranging from core arable areas to those which were marginal for arable production, from a representative range of Member States (France, Germany, Italy, Netherlands, Spain and the United Kingdom). Focus groups were carried out separately with both farmers and advisors. The measures were presented in the same format to each group and the discussion took the form of a free flowing debate with the animator encouraging dialogue and steering the discussion along a path, while ensuring that the findings were truly the feelings of the group participants, expressed in their own words. Groups typically consisted of 8–10 delegates and the discussion lasted for approximately two hours. Written comments were also requested from some advisory organizations in the United Kingdom.

DEVELOPING A METHODOLOGY FOR POLICY FORMULATION

There are three fundamental principles which must be considered in policy formulation to facilitate the transition from initial concept to potential policy option: validity, feasibility and acceptability.

The potential environmental effectiveness of a policy measure depends on the first principle

Table 2. The full range of measures for environmental integration tested in focus groups.

Obligatory measures — linked to the receipt of support payments*
Maintenance and management of existing habitats
Control and maintenance of spraying equipment
Establishment of buffer strips adjacent to watercourses
Establishment of field margins
Autumn deadline for nutrient applications including organic fertilizers
Preparation of crop nutrition plans
Preparation of an Environmental Farm Plan
Voluntary measures — providing incentive payments for environmental management
Retention of winter fallows
Replanting of hedgerows/windbreaks
Establishment of field margins
Establishment of buffer strips
Targeted long-term (20 year) set-aside
Preparation of crop nutrition plans
Reduced pesticide applications e.g., by 50%
Unsprayed field margins
100% set-aside
Maintenance of extensive arable practices (tested in Spain only)
Preparation of an Environmental Farm Plan

*Note — some of measures were tested under both obligatory and voluntary options in order to gauge farmer and advisor response.

of scientific validity. This can be determined by drawing on evidence from the scientific literature and discussion with appropriate experts and organizations. In the absence of clear evidence of scientific validity, potential policy measures will gain little support or credibility.

The second principle relates to the feasibility of the measure in terms of implementation and administration. This includes the cost and feasibility of implementation, monitoring, policing, enforcing compliance and the necessity to provide advice and training linked to the measure. The logic here, is that while it may be possible to define prescriptions that are optimal from a scientific point of view, the most effective prescriptions will be those which balance scientific validity with effective implementation in the field.

The third principle concerns the degree of farmer acceptability in terms of the impact and applicability of the measure to existing farming systems and the costs of implementation. Equally, advisors who work with farmers on a day to day basis must both understand and be supportive of the objectives and prescriptions of a measure if it is going to succeed. Canvassing opinion from farmers and advisors at the early stages of policy formulation to assess likely acceptability and necessary adjustments to the proposed measure is crucial to success.

This approach is conceptualized as a hierarchy of filters through which each measure must pass before emerging, often modified or transformed, as a potential policy measure for implementation. Evidence from the scientific literature and from discussion with acknowledged experts is the scientific filter that identifies and validates suitable options for testing. Then, discussion with regional advisor groups who are involved in day-to-day discussions with farmers on agronomic and environmental issues, the advisory filter, provides insights into the likely farmer acceptability and administrative feasibility of the proposed options, as well as a judgement on the likely environmental effectiveness of the proposed measure in the region. Finally, it is the farmer who must ultimately implement the proposed measure and in the case of voluntary measures, make a decision whether to enrol or not. Discussion with farmer groups, the farmer filter, provides a gauge of farmer acceptability for the proposed options, identifying the practical problems of implementing the measure and its likely compatibility with existing farming practice.

CASE STUDIES

Two examples of the options tested on advisor and farmer groups during the project are presented in a common format to illustrate the methodology of policy formulation.

1. The maintenance and management of existing habitats on farmed land

SCIENTIFIC VALIDATION — THE SCIENTIFIC FILTER

In much of the European Union, remaining areas of semi-natural habitat, particularly in the intensively cultivated lowlands, have become increasingly fragmented and isolated as a result of agricultural activities. As these areas of semi-natural habitat have declined, biodiversity has become increasingly dependent on the peripheral features of the farmed landscape, for example hedgerows, ditches, small patches of woodland and grassland.

While larger areas of semi-natural habitat are afforded some protection, in the United Kingdom at least, as Sites of Special Scientific Interest for which restrictions and lengthy notification procedures prevent extensive loss, these small fragments of farmland habitat have limited or no protection. Boundary features and habitat fragments often serve little purpose in intensively farmed landscapes and are frequently removed or abandoned in favour of agricultural expansion. The increased mechanization of arable production and the size of the machinery has led to a substantial enlargement of arable fields and the concomitant loss of traditional boundary features such as hedgerows and ditches in the European Union (Baldock 1990; Jensen 1992; Beaufoy *et al.* 1994; Baldock *et al.* 1996).

Biodiversity is affected as a result of these losses, both through the direct loss of environmental capital, as well as the isolation of remaining fragments leading to greater vulnerability and risk of population extinction. The theory of metapopulation dynamics emphasizes the importance of retaining and extending existing habitat fragments in order to maintain the viability of local populations (Saunders *et al.* 1987; Fry 1994). Although controversy remains on the value of linear habitats such as hedgerows or grass strips as movement corridors (Simberloff *et al.* 1992), linear features such as hedges, banks and ditches may be important functional links between the remaining habitat fragments as well as valuable habitats in their own right (Saunders and Hobbs 1991). In addition, lack of appropriate management can lead to further decline in the wildlife and conservation value of remaining habitats and boundary features. A major consequence of the intensification of arable agriculture is the abandonment of marginal areas on the farm and the neglect of boundary features; habitat loss by default. Hedgerows, ditches, woodland fragments and other such farmland features are also often key determinants of the landscape character of the cultural landscapes of much of Europe. Thus, conserving these habitats is important from both a landscape and a biodiversity perspective.

The initial solution in this case is straightforward. The value of peripheral farmland features for both landscape and biodiversity highlights the need for the prevention of further loss or degradation of these habitats. Policy initiatives which aim at halting this trend must thus be seen as the first priority. Options to extend and enhance such habitats may also be seen as a desirable addition to the main policy. Linking this measure to arable support payments throughout the European

Union would maximize the conservation impact by targeting all farmers in receipt of support (Potter 1996).

THE POLICY

The following aim and proposed policy measure were defined as a result of the initial analysis of the scientific validation filter.

- *Aim:* To conserve the remaining fragments of semi-natural habitat and linear features on farms for both wildlife and landscape conservation.
- *Proposed policy measure:* Maintenance of all semi-natural habitats and features of landscape and conservation importance which remain on the farm, including hedgerows, banks, ditches, fragments of semi-natural grassland, as a compulsory cross-compliance requirement linked to the receipt of arable support payments.

ADMINISTRATIVE FEASIBILITY — THE ADVISORY FILTER

This measure achieved a positive response from most advisory groups, although French advisors saw it as a static approach to environmental enhancement which did little to address the problems of habitat loss in more intensively farmed areas. Administrative feasibility was not perceived to be a major problem as farmers were already filling in detailed information about their holdings in order to qualify for support payments. Extending this information to include those features which were targeted for conservation would be relatively easy to administer. The focus groups however, identified a number of key problems with the measure. These were: lack of equitability, problems of defining the features to retained and defining the management requirements.

The issue of equitability arose from the disparity between those farmers who had already removed the majority of features from the holding and had thus maximized the farmed area and those who had retained environmental features on the holding. Introducing this policy would effectively penalize the latter group, both by fixing the field size and farm structure in its current form and by imposing additional expense in terms of meeting management requirements for the remaining features. It could also be predicted that announcements of the proposed policy would stimulate a rapid removal of environmental features on farms to prevent incurring management costs. Thus, features which may have survived, partly as a result of neglect, might be rapidly lost unless the problems of inequity and management costs were tackled.

The definition of both the features to be retained and the management requirements for these features were clearly seen by many advisors as an issue to be tackled at the regional level. They argued that the variability in the character and extent of the farmland environmental resource between regions made regional determination of the target habitats essential. Equally, local traditions of management could be important in defining the individual quality of the region. To impose a uniform prescription for management would remove much of the landscape diversity of the European Union and thus management prescriptions should be determined and compliance monitored at the regional level.

FARMER ACCEPTABILITY — THE FARMER FILTER

This option was also well received by most of the farmers, given that many saw themselves as complying with the requirement already. Resentment, again, was greatest in France, where it was seen as a measure to reduce many farmers to little gardeners or park keepers with little chance of being competitive on world markets. United Kingdom farmers also drew attention to the cost of maintaining environmental features such as hedgerows or ditches. Despite these criticisms however, farmer response overall suggested that this measure would be acceptable, albeit with some adjustments made to take account of the issues raised in the farmer and advisor focus groups.

SUMMARY

The inequity identified by the advisory groups provided a major stumbling block to taking forward the policy option in its original form. Penalizing farmers who had retained environmental features, while doing nothing to reverse trends in habitat loss was clearly unacceptable. One solution to the problem might be to provide support payments (which we termed Environmental Resource Payments) for the retention of the target habitats in the same way that payments were received on a per hectare basis for arable crops. By registering environmental features on the same registration form as for farmed land, farmers could become eligible for an Environmental Resource Payment calculated in a similar way to arable compensation payments. This would reduce the financial disadvantage of retaining environmental features on the holding and would compensate to some extent for the increased management requirements highlighted by the farm groups. The eligibility of non-farmed land for resource payments could also provide a stimulus for enrolment into other proposed environmental schemes such as the creation of field margins or buffer strips adjacent to water courses. Area payments for the retention of environmental features would also go some way to redress the anomaly of the current area-based support payments which, by tying payments to areas of cropped land, effectively reward farmers who maximized their arable area at the expense of grasslands and other areas of conservation interest.

Where a strong case could be made for the removal of specific environmental features, a No Net Loss clause would allow for some flexibility in the retention of features and for the rationalization of field size or shape where necessary to accommodate new farming methods. Under this option, farmers would be able to remove areas of habitat or boundary features on the farm, provided the equivalent was replaced elsewhere on the holding, thus ensuring that within a farm, there would be no overall loss of wildlife and landscape features. This could be seen as a second-best option however, given the time taken for communities to develop and thus the loss in environmental value of replacement habitats.

In the light of discussions with farmers and advisors we proposed that the option should be considered as an obligatory cross compliance option linked to the receipt of arable support payments, but with eligibility for Environmental Resource Payments for the designated land to reduce problems of inequity. Derogations for the No Net Loss option would be considered for individual cases. While the basic principle would be applied across the whole European Union, definition of features to be retained and management requirements would be determined at the regional level, thus ensuring greater environmental effectiveness and acceptability of the measure amongst both farmers and advisors throughout the European Union.

2. *Retaining winter fallows*

SCIENTIFIC VALIDATION — THE SCIENTIFIC FILTER

There is a widespread and growing concern about the population declines of many characteristic birds of European farmland (Fuller *et al.* 1995; Pain and Pienkowski 1997). Changes in farming practices, in particular the switch from spring to winter sown cereals, the simplification of rotations, the intensification of grassland management and the increased use of pesticides during the 1970s and 1980s are thought to be the key reasons for this change (Fuller *et al.* 1995). Of these, the switch from spring sown to winter sown crops and the concomitant loss of winter stubbles has been cited as a key reason for the decline in farmland birds in the United Kingdom and northern Europe (O'Connor and Shrubb 1986; Fuller *et al.* 1995; Wilson *et al.* 1996).

While research has provided clear evidence for the benefits of winter stubbles for the retention of birds and for the reduction of soil erosion (Boardman 1990; Morgan and Rickson 1990), evidence for the potential to reduce nitrate leaching was less clear and provided a potential conflict with findings which suggested winter-sown crops to be an effective way of reducing nitrate run-off (House of Lords 1989). Following consultation with a number of acknowledged experts in the field, no research could be identified which made a direct comparison between winter sown crops and spring crops following a winter stubble in terms of nitrate leaching. However, pooling results from a number of studies (Hewitt *et al.* 1992; Burt and Haycock 1993; Haußmann 1993; Powlson and Davies 1993; Watson *et. al.* 1993; Webster and Goulding 1995) suggested that encouraging the retention of winter stubbles may reduce nitrate leaching in addition to the known benefits for biodiversity and the reduction of soil erosion.

THE POLICY

The aims and proposed measure were drawn up as a result of the analysis of the scientific validation filter.

- *Aims:* To provide valuable wildlife habitat, in particular for farmland birds. To reduce the incidence of soil erosion and of nitrate leaching associated with autumn cultivation for winter crops.
- *Proposed measure:* Incentive payment offered for the maintenance of a winter fallow by retaining the stubble of the previous crop until a given date in spring.

ADMINISTRATIVE FEASIBILITY — THE ADVISORY FILTER

The measure was favourably received in the United Kingdom, where much of the initial research on farmland birds had been conducted and where the need for winter stubbles was well known amongst most conservation organizations. However, comments from the French advisors highlighted the potential of this measure to promote further environmental damage rather than counteract it. Spring sown crops in the region of south-west France under discussion were typically those associated with the greatest intensity of production, such as maize and soya. Thus, through the encouragement to sow spring crops, this measure would effectively give grant aid to promote the planting of greater areas of environmentally damaging crops. Spanish advisors also indicated that the retention of winter stubbles was standard practice in the region and the measure would effectively pay farmers for something they were doing already, thus offering no additional environmental enhancement. It became clear from these discussions that this measure offered little or no benefit to some regions of the European Union.

FARMER ACCEPTABILITY — THE FARMER FILTER

Farmers also showed a mixed reaction to this measure. Those farmers who were already sowing spring crops or where soil conditions were suitable, received this option favourably, while those with heavy clay soils considered it unsuitable as it would be difficult to establish a seed bed in the spring. The suggestion of a fixed date before which no cultivation could occur was not well received. Farmers resented this restriction on their freedom to farm and the Italian and Spanish farmers considered it too late for a first cultivation.

SUMMARY

If introduced as a voluntary option, it would be important to ensure that the policy encouraged a change in cropping practices on suitable soils and did not merely pay farmers for a cropping system already in place. Payments should be restricted to stubbles which preceded a spring crop for which a winter sowing is also possible, such as cereals, thus preventing the encouragement of more environmentally damaging spring-sown crops. Equally, the option may only be suitable after crops such as cereals which are harvested sufficiently early to allow development of a good natural regeneration regrowth and which provide valuable winter bird food in the form of spilt grain. Regional definition of suitable preceding and following crops and of the permitted date for cultivation would thus be necessary to ensure the environmental potential of the option was maximized and that it gained acceptability amongst both advisors and farmers.

CONCLUSIONS — INTEGRATING ENVIRONMENTAL CONCERNS INTO ARABLE SUPPORT POLICY IN THE EUROPEAN UNION

Given the diversity of environmental issues and farming systems throughout the European Union, it is essential that a sound scientific basis to policy formulation is combined with effective communication and discussion of proposed measures. Our approach provides an appropriate method for formulating and testing policy measures and, through discussion with advisors and farmers, highlights a number of key issues fundamental to the successful integration of environmental concerns into arable support policy in the European Union. Issues of inequity and the need for regional definition reflect the problems of determining policy measures at a European Union wide level and highlight the importance of communication at all stages of the policy formulation process.

Tackling inequity

Equitability of policy impact across all European Union farm types must be seen as essential for all cross-compliance measures introduced as obligatory requirements for farmer support. Discussion of the measure to maintain and manage existing habitats highlighted the potential inequity for farmers who had so far refrained from removing peripheral habitats on their farms. Equally, other proposed measures, such as the creation of uncropped margins around fields or buffer strips adjacent to watercourses, would discriminate against those farmers who had retained smaller fields or watercourses on their land (Burch *et al.,* in litt.).

It is acknowledged that many of the agri-environment measures introduced under the 1992 reformed Common Agricultural Policy (Regulation 2078/92), have offered more opportunities to those farmers who have adopted intensive farming practices and removed features of conservation value in the past, than to those who have maintained high environmental standards and have thus retained environmental capital on the farm (House of Commons 1997). It is thus undesirable that measures introduced on a much wider basis though cross-compliance requirements should magnify this inequity. Indeed, many argue that farmers should be rewarded for management which has yielded environmental benefits (BirdLife International 1997; House of Commons 1997). Discussion of proposed measures with both advisors and farmers will identify potential problems of equitability prior to implementation. The proposed Environmental Resource Payment, which would compensate farmers for the retention of environmental features, provides one example of a simple measure to address inequity and reward those farmers who have retained environmental capital.

Ensuring additionality

While equitability is essential for obligatory cross-compliance, additionality must be ensured for all voluntary measures. In other words, incentive payments which are additional to farmer support must reward environmental management which is over and above normal farming practice. Here again, discussion, particularly with regional advisors, is necessary to identify potential anomalies and adjust policy proposals to maximize environmental gain.

The need for regional definition

Policies which are not well adapted to local conditions will inevitably meet with strong opposition from both farmers and advisors and thus a limited chance of environmental success. For the European situation, where centralized policies impact on countries of widely different geoclimatic conditions and diverse farming systems as well as cultural and administrative conditions, the principal of subsidiarity (determination by Member States on the implementation of broad policy goals) is essential to ensure both the effectiveness and acceptability of policy. At the other extreme however, complete regional autonomy in devising and administering schemes may lead to a patchy distribution of environmental measures and inequity for Europe's farmers in terms of the policy framework in which they operate and thus the balance of competitiveness between Member States. Weakened central administrative control may also increase the risk that environmental schemes are operated as a form of disguised income support.

A more effective position may be to consider the centralized definition of broad environmental measures for farmers, with the details of specific features and recommendations for the measures determined at the regional level. The establishment of a basic set of environmental rules would thus allow a degree of convergence in terms of targets and policy goals. More specific definition and implementation at the country or regional level would increase compatibility with the principles of subsidiarity and the political acceptability of the measure. Prescriptions adapted to local conditions would also be predicted to yield greater nature conservation benefit (Baldock and Mitchell 1995).

Regional definition of measures could take a number of forms. For example, environmental features to be conserved and issues to be addressed could be defined at a local level to ensure that resources, particularly discretionary voluntary payments, were deployed to ensure maximum environmental effectiveness. For arable options, the definition of suitable crops and rotations could be determined at the regional level, while local definition of management prescriptions would ensure that these were compatible with local cultural traditions and the landscape character of the individual region. The dissemination of information, advice and training for farmers to provide support in implementing new measures could also be most effectively organized at the regional or local level.

The importance of communication in policy formulation

The success of any measure which aims at integrating environmental concerns with support for production will depend on the commitment of both advisors and farmers to its effective implementation and management. Implicit in this commitment is the understanding from both parties that the concerns of farmers and advisors and regional variations in farming systems and environmental conditions have been fully considered. Without this understanding, criticism by farmers and advisors and resentment to new policy will significantly reduce its effectiveness and, in the case of voluntary schemes, jeopardize its likely uptake.

The tailoring of policies to meet the needs of both the environment and the farmer and to the variability of regional environments and farming systems, must be seen as fundamental. In order to give credibility to new policy and to maximize its chances of effective uptake and implementation it is essential that both advisors and farmers are consulted at the stage of policy formulation. Equally, it is essential that having determined new policy measures, adequate time and funding are given to communicating the aims and details of the measure to the wider farming audience and to monitoring policy implementation and its effectiveness in achieving environmental goals. Field visits and local meetings are essential to ensure that the implementation of conservation measures is informed by examples of best practice.

Postscript

The importance of integrating conservation into agricultural production is a subject under much discussion in Europe as the Common Agricultural Policy approaches further reform. Since completing this research, a number of new initiatives have continued to inform the debate. At the European scale, Agenda 2000 (Commission of the European Communities 1997) the proposal published by the European Commission for the reform of the Common Agricultural Policy, has done little to provide encouragement for conservationists. While environmental concerns are highlighted in the text, few concrete proposals are put forward to achieve progress and the emphasis remains firmly fixed on market support. An opportunity for Member States to attach environmental conditions to support payments is optional, while obligatory set-aside, despite its environmental potential, has been reduced to 0% for the year 2000.

In the United Kingdom, however, a pilot scheme, aimed at promoting environmental management on arable land, was introduced in early 1998. The Arable Stewardship Scheme offers incentive payments to farmers for options such as the retention of winter stubbles, establishment of field margins around cropped land or the environmental management of set-aside land. This marks the first significant step in the United Kingdom towards the greater integration of conservation concerns into the arable production matrix. If these first steps can be translated into broader policy measures integrated into farmer support across the European Union, the future for Europe's biodiversity in the arable landscape may look less bleak.

ACKNOWLEDGEMENTS

This research was funded by the Environment Directorate (DGXI) of the European Commission: Study Contract B4-3040/95/001042/MAR/D1. The ideas presented in this chapter do not necessarily represent the Commission's official position. This work could not have been completed without our partners: Erling Andersen, The Royal Veterinary and Agricultural University (KVL), Copenhagen, Denmark; Gerard Miclet, ENSAM/INRA, Montpellier, France; Brigitte Holz, University of Stuttgart, Germany; Dimitris Skuras, University of Patras, Greece; Luca Cesaro, University of Padua, Italy; Gerard Kolkman, Institute for Forestry and Nature Research IBN-DLO, The Netherlands. The focus group work was co-ordinated by Tony Houghton, Produce Studies Limited, Newbury, UK. FMB's attendance at the Taupo conference was made possible by support from the Royal Society, London and the C. Alma Baker Trust, Huntingdon.

REFERENCES

Andrews, J., 1992. Some practical problems in set-aside management for wildlife. *Brit. Wildl.* **3**: 329–36.

Baldock, D., 1990. Agriculture and Habitat Loss in Europe. World-wide Fund for Nature International: Gland, Switzerland.

Baldock, D. and Mitchell, K., 1995. Cross-compliance within the Common Agricultural Policy. A Review of Options for Landscape and Nature Conservation. Institute for European Environmental Policy: London.

Baldock, D., Bishop, K., Mitchell, K. and Phillips, A., 1996. Growing Greener: Sustainable Agriculture in the UK. World-wide Fund for Nature — United Kingdom: Council for the Protection of Rural England.

Beaufoy, G., Baldock, D. and Clark, J., 1994. The Nature of Farming. Low Intensity Farming Systems in Nine European Countries. Institute of European Environmental Policy: London.

Bignal, E. M. and McCracken, D. I., 1996. Low-intensity farming systems in the conservation of the countryside. *J. App. Ecol.* **33**: 413–24.

BirdLife International, 1997. A Future for Europe's Rural Environment. Reforming the Common Agricultural Policy. BirdLife International: Brussels.

Boardman, J., 1990. Soil erosion on the South Downs: a review. Pp. 87–105 *in* Soil Erosion on Agricultural Land ed by J. Boardman, I. D. L. Foster and J. A. Dearing. John Wiley and Sons: London.

Buckwell, A., 1997. Towards a common agricultural and rural policy for Europe. *European Economy Reports and Studies 5*. Office for official publications of the European Communities: Luxembourg.

Burch, F. M., 1996. The Establishment of Species-rich Grassland on Set-aside Land: Effects of Location, Establishment and Management on Species Colonisation and the Control of Weeds. Ph.D. thesis, University of London.

Burch, F. M., 1997. Colonisation of set-aside land by calcareous grassland species: a case study adjacent to Mount Caburn NNR, East Sussex. Pp. 297–300 *in* Species Dispersal and Land Use Processes. Proceedings of the 6th annual conference, International Association for Landscape Ecology (United Kingdom) ed by A. Cooper and J. Power. International Association for Landscape Ecology: United Kingdom.

Burt, T. P. and Haycock, N. E., 1993. Controlling losses of nitrate by changing land use. Pp. 341–67 *in* Nitrate: Patterns, Processes and Management ed by T. P. Burt, A. L. Heathwaite and S. T. Trudgill. John Wiley and Sons: London.

Commission of the European Communities, 1997. Agenda 2000. COM (97) 2000, Commission of the European Communities: Brussels.

Dixon, J. (ed), 1993. A Future for Europe's Farmed Countryside. Studies in European Agriculture and Environment Policy No. 1, Royal Society for the Protection of Birds: Sandy.

Fry, G. L. A., 1994. The role of field margins in the landscape. Pp. 31–40 *in* Field Margins: Integrating Agriculture and Conservation ed by N. Boatman. British Crop Protection Council Monograph No. 58, British Crop Protection Council Publications: Farnham.

Fuller, R. J., Gregory, R. D., Gibbons, D. W., Marchant, J. H., Wilson, J. D., Baillie, S. R. and Carter, N., 1995. Population declines and range contractions among lowland farmland birds in Britain. *Cons. Biol.* **9**: 1425–441.

Green, B. H., 1993. The recognition and implementation of landscape management objectives for agriculture in the United Kingdom. Pp. 259–66 *in* Nature Conservation 3: Reconstruction of Fragmented Ecosystems ed by D. A. Saunders, R. J. Hobbs and P. R. Erlich. Surrey Beatty & Sons: Chipping Norton, New South Wales.

Green, B. H. and Potter, C. A., 1987. Environmental opportunities offered by surplus production. Pp. 64–72 *in* Removing Land from Agriculture: The Implications for the Environment ed by D. Baldock and D. Conder. Council for the Protection of Rural England.

Haußmann, W., 1993. Abschlußbericht zu Projekt "Ökologische Begleituntersuchungen zur Flächenstillegung" — Untersuchungen zur Dynamik des mineralischen Stickstoffs. Institut für Bodenkunde und Bodenerhaltung der Universität Gießen.

Hewitt, M. V., Webster, C. P. and Bacon, E. T. G., 1992. The effect of cultivation and plant establishment on nitrogen conservation on set-aside land. Pp. 51–54 *in* Set-aside ed by J. Clarke. British Crop Protection Council Monograph No. 50. British Crop Protection Council Publications: Farnham.

House of Commons, 1997. Environmentally Sensitive Areas and other Schemes under the Agri-environment Regulation. Second Report, Vol. 1, The Stationary Office: London.

House of Lords, 1989. Nitrate in Water, Her Majesty's Stationary Office: London.

Jenkins, T. N., 1990. Future Harvests. The Economics of Farming and the Environment: Proposals for Action, Council for the Protection of Rural England and World-wide Fund for Nature: United Kingdom.

Jensen, J., 1992. Nature Management in Denmark, Ministry of the Environment, The National Forest and Nature Agency: Hørsholm.

Morgan, P. C. and Rickson, R. J., 1990. Issues on soil erosion in Europe : The need for a soil conservation policy. Pp. 591–603 *in* Soil Erosion on Agricultural Land ed by J. Boardman, I. D. L. Foster and J. A. Dearing. John Wiley and Sons: London.

Newbold, C., 1989. Semi-natural habitats or habitat re-creation: conflict or partnership? Pp. 9–17 *in* Biological Habitat Reconstruction ed by G. P. Buckley. Belhaven Press: London.

O'Connor, R. J. and Shrubb, M., 1986. Farming and Birds. Cambridge University Press: Cambridge.

Pain, D. J. and Peinkowski, M. J., 1997. Farming and Birds in Europe: The Common Agricultural Policy and its implications for bird conservation. Academic Press: London.

Pickett, S. T. A., Parker, V. T. and Fiedler, P. L., 1992. The new paradigm in ecology: implications for conservation biology above the species level. Pp. 66–88 *in* Conservation Biology. The Theory and Practice of Nature Conservation, Preservation and Management ed by P. L. Fiedler and S. K. Jain. Chapman and Hall: New York.

Potter, C. A., 1996. Decoupling by Degrees? Agricultural Liberalisation and its Implications for Nature Conservation in Britain. English Nature Research Reports No. 196: Peterborough.

Powlson, D. and Davies, B., 1993. Autumn and winter land management. Pp. 15–19 *in* Solving the Nitrate Problem: Progress in Research and Development. Ministry of Agriculture Fisheries and Food: London.

Saunders, D. A., Arnold, G. W., Burbidge, A. A. and Hopkins, A. J. M. (eds), 1987. Nature Conservation: the Role of Remnants of Native Vegetation. Surrey Beatty & Sons: Chipping Norton, New South Wales.

Saunders, D. A. and Hobbs, R. J. (eds), 1991. Nature Conservation 2: the Role of Corridors. Surrey Beatty & Sons: Chipping Norton, New South Wales.

Simberloff, D., Farr, J. A., Cox, J. and Mehlman, D. W., 1992. Movement corridors: conservation bargains or poor investments? *Cons. Biol.* **6**: 493–504.

Scottish Landowners' Federation, 1995. Supporting the Countryside — Future Opportunities. Scottish Landowners' Federation.

Watson, C. A., Fowler, S. M. and Wilman, D., 1993. Organic-N and nitrate leaching on organic farms. *J. Agricul. Sci., Camb.* **120**: 361–69.

Webster, C. P. and Goulding, K. W. T., 1995. Effect of one year rotational set-aside on immediate and ensuing nitrogen leaching loss. *Plant and Soil* **177**: 203–09.

Wilson, P. J., 1992. The natural regeneration of vegetation under set-aside in Southern England. Pp. 73–78 *in* Set-aside ed by J. Clarke. British Crop Protection Council Monograph No. 50. Farnham: British Crop Protection Council Publications.

Wilson, J. D., Evans, A., Grynderup Poulsen, J. and Evans, J., 1995. Wasteland or oasis ? The use of set-aside by breeding and wintering birds. *Brit. Wildl.* **6**: 214–23.

Wilson, J. D., Taylor, R. and Muirhead, L. B., 1996. Field use by farmland birds in winter: an analysis of field type preferences using re-sampling methods. *Bird Study* **43**: 320–32.

CHAPTER 35

The Smelter in the Park: using environmental care to improve business outcomes

John Hill,[1] Lynn Brake[2] and Ian Clark[2]

Conservation and production will not be successfully integrated into large industries until it is demonstrated that caring for the environment and dealing with the externalities of production result in positive outcomes which contribute to the core business. As long as conservation is driven solely by compliance to regulation, business leaders will view environmental care as an additional overhead rather than a necessary component in improving efficiency and effectiveness to boost profits. Increasing public concern for the environment may foster additional small conservation contributions from industry as part of the overall public relations expenditure, but these will remain marginal to the core business and are not likely to address the real conservation issues of production. Initiatives designed to substantially improve environmental practice in large industries must clearly contribute to the business objectives of the company. The "Smelter in the Park" programme was designed to make specific contributions to the core business at Portland Aluminium.

In the 1980s Portland Aluminium lifted the profile of environmental care in the operation of the Aluminium smelter at Portland in Victoria, Australia. The management of Portland Aluminium took this action in order to demonstrate their commitment to best standards in environmental operation and to make a positive contribution to work practices required to produce the best business outcomes from the leading-edge technology in the new smelter. Environmental care became one of the core strategies of the business plan and was regarded as an essential element in developing the commitment of employees to the continued success of the new industrial enterprise. Programmes were introduced to reduce waste and to integrate the operation of the smelter into a biologically diverse, aesthetically pleasing setting which made environmental care an integral part of day to day work practice. This chapter presents the outcomes of research undertaken to investigate the contribution made by the environmental care programme to the attitude, behaviour and work practice which resulted in important business outcomes between 1989 and 1996.

Key words: Environmental care, Best practice, Workplace reform, Business outcomes, Smelter in the Park.

A TECHNOLOGICALLY ADVANCED SMELTER

Construction of the largest, most technically advanced aluminium smelter in the Southern Hemisphere was completed in 1988 on a 600 ha site located on a peninsula which extends into the Southern Ocean a few kilometres from Portland, Victoria, Australia. Only 100 ha of the site is occupied by the buildings and general infrastructure of the smelter. The remaining company land and nearly 150 ha of adjacent public land was originally planned to be a buffer zone between the smelter and neighbouring grazing and residential land. The buffer distance was designed to minimize, and if possible eliminate, any possible impacts on the surrounding land users caused by emissions to air, noise, and the spoiling of the visual environment (Kinhill 1980).

Portland Aluminium commenced production during a time of rapid international corporate change in management structure, competitiveness

[1]Portland Aluminium, PMB 1, Portland, Victoria, Australia 3305.
[2]Centre for Environmental and Recreation Management, University of South Australia, The Levels Campus, Mawson Lakes, South Australia, Australia 5095.
Pages 345–59 *in* NATURE CONSERVATION 5: NATURE CONSERVATION IN PRODUCTION ENVIRONMENTS: MANAGING THE MATRIX ed by J. L. Craig, N. Mitchell and D. A. Saunders. Surrey Beatty & Sons, 2000.

and community relations (Schimidheiny 1992). Management theory and practice within respected business organizations was undergoing change in a quest for higher standards of efficiency and greater return on investment. Reduction in tariff protection exposed Australian industry to increased international competition. Poorly performing businesses disappeared and seemingly rock solid ones were subjected to down-sizing, take-over bids and asset stripping. This business environment stimulated the managers of the new green-field smelter at Portland to look for opportunities to strengthen their business in line with world's best practice.

The management of the new Portland Aluminium smelter was well aware of the competitiveness of the international Aluminium industry. The Aluminium industry demands a very high level of productivity, especially in Australia where labour and infrastructure costs are relatively high. The management knew that making the most of a smelter's technological potential requires the commitment and skill of the employees (McGreer 1989; Ramsey 1990). This commitment would depend on the employees' attitudes and the belief that pursuing company goals would lead to positive outcomes for themselves, their families and the Portland community.

RELATIONSHIPS — INTERNAL AND EXTERNAL

At the time the smelter was being planned and constructed people were becoming more aware of their dependence on the environment and were expressing concern about the environmental impact of modern industrial practices. This period was characterized by increasing community expectations that industry would use resources judiciously and control impacts to ensure the maintenance of natural ecosystems (The World Commission on Environment and Development 1987; Heaton 1989; Pagan 1989; Miller 1990).

The role of government and its relationship to industry was being questioned and redefined (House of Representatives Standing Committee on Environment 1994; Young *et al.* 1996). Environmental regulatory authorities were armed with new regulations legislated in response to the public concern for better industrial practice.

This concern for the environment was reflected in the local community. It was divided between those who welcomed the smelter and attendant commercial opportunities and those who questioned the environmental costs involved and resisted the cultural changes the development would bring (*Portland Observer*, December 28 1988, April 14 1989; *Warrnambool Standard*, July 8 1989). The environment became the focus for opposition to construction of the smelter and was used afterwards as an emotional rallying point for continued opposition and criticism.

By 1989 the work practices and business outcomes at Portland Aluminium was being negatively influenced by the internal relationships across the workforce and external relationships with the community.

Internally

The original management style was based on the traditional hierarchical model. Responsibilities were fixed and clearly delineated. Labour was recruited to perform clearly defined tasks and there was demarcation based on specialist skills. This resulted in a lack of trust and delegation. The workers depended on management, supervisors and unions for decision making (Portland Aluminium 1993).

Management decisions were made with little or no input from operators. Consequently people felt powerless and alienated from the company. There was sporadic industrial unrest; a legacy from the construction period. Staff turnover was high and there was concern among staff about health, particularly industrial asthma (Judd 1991, 1992).

Externally

There was suspicion and hostility in the community despite the economic advantages from the smelter. This was due to a number of factors including the following:

- The alienation of conservationists who opposed the loss of natural values and coastal vegetation that occurred during smelter construction.
- The social disruption caused by the construction of such a large industry in a small town and the resultant influx of people from outside the district.
- A property market distortion caused by a high demand for accommodation.
- The rough behaviour of construction people in a close-knit community where life had previously been based mainly on the farming and fishing industries.
- Industrial unrest during and after construction.
- Contract and smelter work going to outsiders.
- Concern about health risks among workers and the community; particularly asthma.
- Commissioning problems leading to visible emissions and odours (Judd 1991, 1992; Hughes 1993).

MANAGEMENT CHALLENGE

Negative perceptions and management systems led to a culture of confrontation, of barriers between different sections of the workforce and between the company management and the community. This resulted in low self esteem in the workforce and low individual commitment to productivity.

The management group understood that their primary objective was to operate the smelter efficiently to ensure a fair return to stockholders from the aluminium smelting process (Hedditch 1993). To meet this challenge they determined that they needed the following:

- Long-term, low cost operation. Most operating costs in a smelter are determined by the design

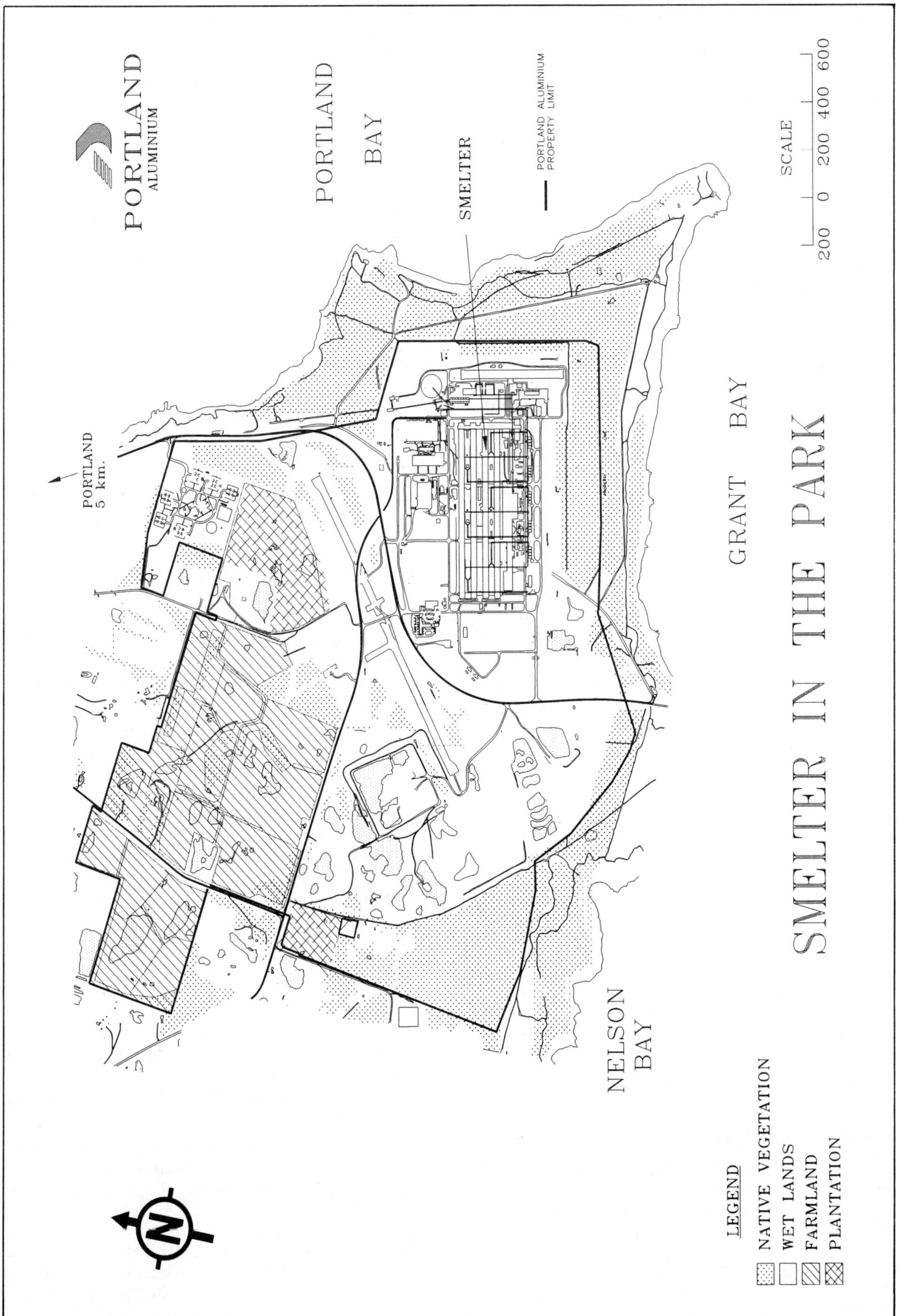

Fig. 1. Map of the Park.

of the smelter. These are relatively fixed. The greatest opportunity to influence costs is through the workforce. The ideal operation would have a minimum sized workforce with the individual members being responsible, flexible, responsive and safe.

- High quality results. Aluminium is a standard commodity. There is no difference between the product of competing smelters. To prosper a business must focus on quality control, process efficiency, customer satisfaction and raw material supply. This again involves the workforce. It infers a universal understanding of the smelter processes, joint ownership of policy and a high level of training.

A VISION FOR THE SMELTER

Management developed a vision of where they would like the company to be in order to meet this challenge. The vision was for an informed and supportive workforce drawn from an informed and supportive community operating efficiently and effectively to take full advantage of the advanced technology of the smelter. The mission and vision was as follows.

Mission

- We operate the Portland Aluminium smelter to consistently produce as much high quality aluminium as possible at the lowest sustainable cost.
- Our mission is carried out in accordance with our organization's values, and takes us towards our vision.

Vision

- We are a world class organization.
- We care for each other, our community, and the environment.
- We are healthy, active, and enthusiastic contributors to our business and society.
- We bring superior returns to the owners, our community and ourselves.

(Portland Aluminium 1996)

In this vision, concern for the environment would provide a focus for other organizational and work practice changes to motivate and unify company staff to work together—

- to solve operational problems,
- to implement new technology and
- to continually improve the smelting operation.

A concerted effort was put into making the organizational and work practice changes. The organizational structure was flattened and supervisory levels removed to ensure that individuals at the workface felt involved and accountable. Substantial investments were made in staff training, personal development, communication, industrial relations, community consultation, and capital infrastructure. Environmental care was used as a vehicle to build commitment and to encourage positive attitudes toward changes in work practice.

THE VISION AND ENVIRONMENTAL CARE

Management saw the community sensitivity to the impacts of the smelter as an opportunity to use the environment as a focus on which to help build a dedicated and efficient workforce. The company managers shared with the community and the workforce the desire to achieve efficiencies with positive environmental outcomes from the smelting process and to improved health and safety to create a better work environment. Care for the environment was used as a tool to help build the communications, co-operation and commitment required for the improvement of internal and external relationships. As such the investment in environmental care was to contribute better work practices and efficiencies which would ultimately improve the profitability of the company.

The managers expected substantive as well as work practice outcomes from environmental care through the reduction in waste and "end of pipe" externalities. Others have demonstrated similar expectations (Kelly and Pressman 1989; Prince and Denison 1991; Prokop 1992). At the substantive level, emissions and wastes were recognized as expensive raw materials lost. This loss of valuable resources affected both the profit and the environment. Conserving raw material would prolong industry life. Reducing waste and improving the environment would result in a clean and pleasant work place.

Surrounding the smelter with plants and animals thriving in a healthy environment would provide a tangible measure of success. It could be expected that this would develop a positive feeling of well-being and commitment in the workforce and the community in line with the experience of other workers (Kornbluh *et al.* 1985; Carlopio and Gardner 1992; Lukas 1993; McFadden and Demetriou 1993).

THE SMELTER IN THE PARK

The company needed a highly visible mechanism that would demonstrate to the community and company employees that they were serious about caring for the environment, improving work practice, addressing environmental health issues and breaking down barriers. Management decided that the concept of "The Smelter in the Park" could be developed to meet their needs.

In 1991 the managers commissioned academics from the Centre for Environmental and Recreation Management at the University of South Australia to work with them to help define how their objectives for work place reform and better business outcomes might be addressed through the development of a park around the smelter. The outcome of this co-operative effort was the development of the Master Plan for the Smelter in the Park (The Master Plan) (Brake *et al.* 1991).

The Master Plan proposed that the function of the Smelter in the Park should be:

> *"to integrate the operation of Portland Aluminium into a biologically diverse, aesthetically pleasing setting which makes environmental care an integral part of the work environment."*

The Smelter in the Park was designed to lift the profile of environmental care in the smelter and contribute to help bring business practice in line with their vision.

THE DEVELOPMENT OF THE PARK

An initial assessment as planning commenced for the Smelter in the Park indicated that the company had a number of assets that may have been of value in the original smelter design, but were of questionable worth in meeting management's goals for improved work practice outcomes from environmental care. Questionable assets included the large buffer zone, a landfill site for waste, large areas of sealed pavement, gravelled storage areas, and extensive water drainage systems. It was concluded that such assets could lead to undesirable attitudes and poor work practices.

1. A "buffer zone" encourages the attitude that the land surrounding the smelter is a place to pollute, where environmental values need not be considered too closely, and where the prying public eye should be discouraged. Ultimately this results in lax work practices, suspicious neighbours, and loss of pride and commitment.

2. A landfill site or dump leads to dumping. It is much easier to tip something out than to look

Fig. 2. View of the Smelter in the Park.

Fig. 3. Park wildlife.

for an opportunity to recycle it. Once things are in the dump the company discouraged people getting them out and reusing them for reasons of security and personal safety.

3. Hardstand storage areas result in the unnecessary storing of obsolete material and machinery and this leads to spillage, contamination, and general untidy clutter. Machinery and vehicles are driven everywhere, even off the pavements, with the result that nature is pushed further back and the site looks more and more untidy.
4. Extensive drains and drainage engenders the attitude that once-used processing and storm water are wastes which can be used to transport and dilute pollution; suggesting that "waste" water should be drained into the sea and away from the site as quickly as possible.

The Smelter in the Park project has been the focus of a concerted effort to avoid these attitudes and for people to adopt the vision of environmental care and world class performance. In the Master Plan it was suggested that negative assets be replaced.

1. The buffer zone to be replaced with a conservation/recreation park where wildlife habitats could be conserved and restored. The park would provide opportunities for research, education and enjoyment for the community. It would be open to the public at all times to break down barriers between the production environment of the smelter, the surrounding natural environment and the community.
2. The dump to be replaced with a transfer station where material would be sorted for re-use and recycling. The transfer station was to be developed in conjunction with a programme of reducing waste throughout the operation of the smelter. Employees were encouraged to suggest new ways for disused materials to be recycled within the smelting operations or to other businesses.
3. The hardstand storage areas to be replaced with gardens, wetlands, and wildlife areas. Access for machinery and storage areas were to be reduced to an operational minimum. If people have no place to "store" unused materials and unnecessary equipment work areas remain cleaner and people will find new ways to reduce waste, work efficiently and recycle.
4. Extensive common drains to be replaced with a system to harvest storm water and to recycle processing water. The use of water was to be considered carefully just like any other valuable resource. Water was to be harvested, recycled and stored on site. Wetlands were to be created for water storage, water quality improvement and wildlife habitat.

The Smelter in the Park strategy was designed to break down both physical and psychological barriers between the community, the workers and the smelter. The Master Plan suggested programmes be implemented which encouraged workers and community people to learn about environmental care and actively participate in the development of the park. All employees were to be encouraged to work efficiently, to avoid waste and to become committed to improving environmental performance. This improvement in the workplace environment and increased use of the park was designed to have very positive effects on work practice and community attitudes.

The development of the Smelter in the Park became the focus for environmental care at Portland Aluminium. As work place reforms were implemented the Park provided a range of opportunities for employee training, community involvement, improved communications, removal of barriers which contributed to improved work practices and efficiencies.

People from the production areas of the smelter, their families and other people from the community had been involved in the development of the Park. They had helped with monitoring bird and kangaroo numbers; in plant propagation and planting and in constructing amenities. Employees had participated in their own time and in company time. Garden groups had been formed to care for gardens near work areas. These groups took and continue to take a keen and at times a competitive interest in the gardens, helping with design, construction and maintenance.

During the early development of the Park, environmental staff regularly gave presentations to work crews on progress and future plans for the environmental care programme. These presentations were considered an essential element in raising environmental awareness and gaining the commitment of the workforce to resource conservation, environmental care and work practice efficiency.

OUTCOMES FROM ENVIRONMENTAL CARE TO 1996

By 1996 much of the physical assets of the Park had been created. The physical barriers between the productive environment of the smelter and the surrounding natural settings had been broken down.

The incorporation of environmental care as one of the core business strategies with the Smelter in the Park as a focus made key substantive and behavioural contributions that helped to improve the environmental performance and productivity in all aspects of the smelter operation. The substantive outcomes can be measured in terms of output and positive shifts in the externalities of production. Behavioural outcomes are reflected in better work practices and efficiencies.

ANALYSIS OF OUTCOMES: THE "CRITICAL ACTIVITIES" OF ENVIRONMENTAL CARE 1990–1996

Portland Aluminium assesses its performance by developing a list of "critical activities" for the operation of the smelter. These are published in an annual Operating Plan. Target values are set for the year for those critical activities that can be measured quantitatively and measurements of these values made on a monthly basis are referred to as Key Performance Indicators. Several Key Performance Indicators may be selected to measure performance in one critical activity. Both the Operating Plan and the Key Performance Indicators are widely distributed in the company.

The critical activities are used to measure performance and so it is possible to use them to demonstrate the way in which performance has been influenced by the programme of environmental care. The outcomes of the critical activities described here have been selected because they are the ones that are affected by work practice.

Fluoride control

The electrolytic cells (pots) in which aluminium is produced have the potential to emit quantities of undesirable off-gasses which contain fluorides. If this is not controlled these gases can cause problems to pot room workers and to vegetation in the vicinity of the smelter. Modern production facilities use precautions including cell hooding and scrubbing of gases to reduce the potential for fluoride emission (International Primary Aluminium Institute 1992). The efficiency of this gas removal is largely dependant on work practice of the operators.

When it was established the Portland smelter had to meet the strictest world standard fluoride emission level. The level of 0.73 kg/tonne of aluminium produced was imposed by the Environment Protection Authority and was over 30% lower than the emission levels of other Australian smelters at the time (Kinhill 1980). It was regarded by the company as close to the absolute minimum that efficient design and good operation could achieve. This was considered to be about 0.5 kg/tonne, (McGreer 1989).

The total annual output of fluoride since 1988 is about half of the Environment Protection Authority limit (Fig. 4). This has resulted in a new benchmark being set for the industry. The smelter today regularly runs at this level of fluoride emission which means that valuable production resources are being captured and recycled through the smelting process. This has been achieved not only because the equipment for scrubbing emissions is first class, but the operators conscientiously maintain the equipment and adhere to correct work practices.

Emissions are monitored monthly and any poor results are thoroughly investigated. As well as measuring total fluoride output, the company has an ongoing monitoring programme to test for fluoride contamination in the surrounding environment. Monitoring data is available on fluoride accumulation in: cattle owned by the company which graze on pasture adjacent to the smelter (Miller 1996, 1997); heathlands (Carr 1992; Doley 1994; Baldock 1996); and aquatic systems in the area both fresh and marine. This monitoring shows that fluoride emissions are having little effect on the surrounding environment

Safety statistics

Environmental awareness fitted into the Company's commitment to health and safety in the workplace. The initial concept of regulation and enforcement to achieve safe work practices was too narrow and did little to gain the individual operator's commitment.

Environmental care on the other hand suggests a holistic approach which includes health, lifeskills and the environment in which you live and work. A quality of life strategy was adopted as part of staff training to reduce the incidence of work related injury.

Environmental activity based on the Smelter in the Park fitted neatly into the Health and Safety programmes. Figure 5 shows that the frequency rate of disabling injuries has dropped by almost half and has stayed at that lower level since the implementation of the programme. The total number of injuries and the number of serious injuries have also dropped but still remain unacceptably high.

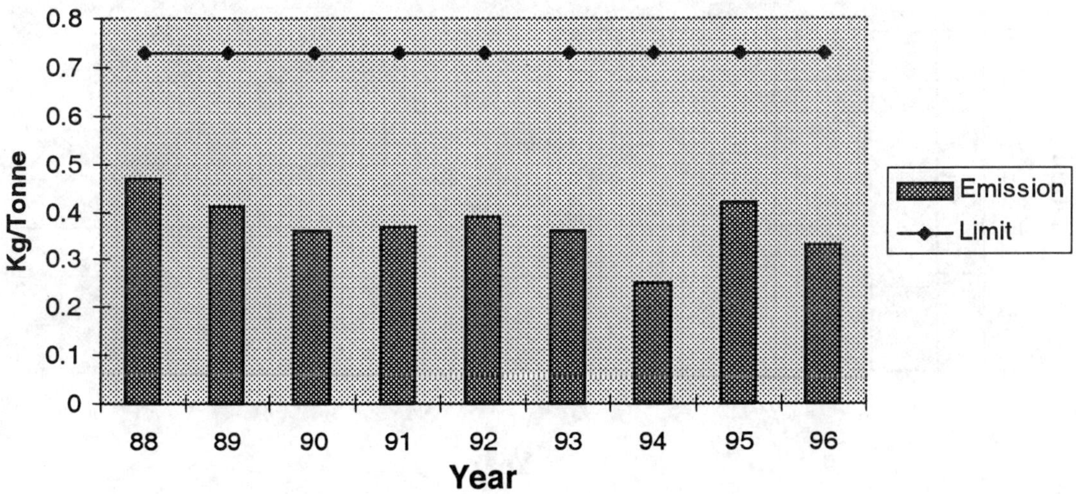

Fig. 4. Yearly total fluoride emission levels for the period 1988 to 1996.

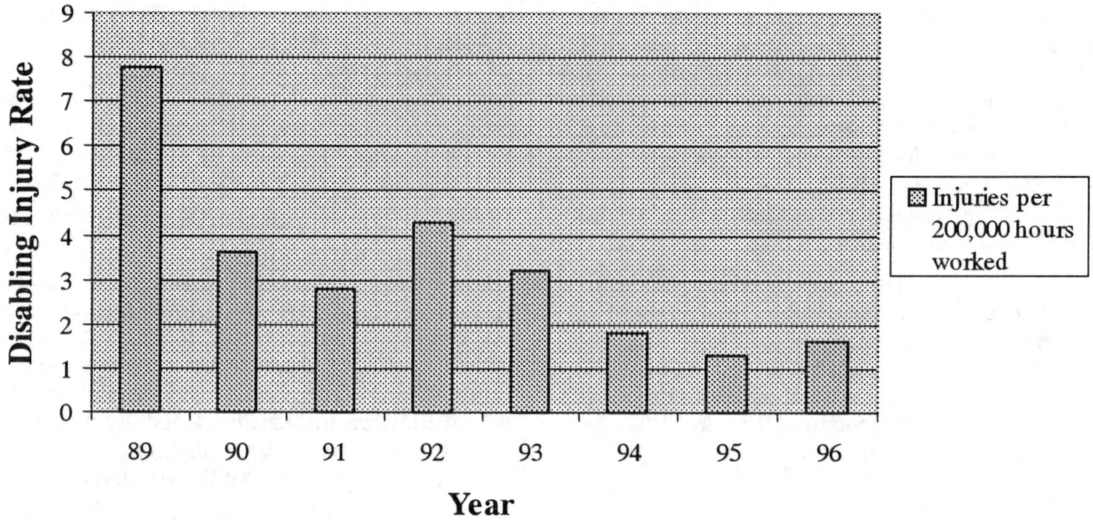

Fig. 5. Total disabling injury frequency rate since the inception of the environmental care programme.

Reportable environmental incidents

A Reportable Environmental Incident occurs when the smelter is not in compliance with the conditions of the Environment Protection Authority operating licence. Figure 6 shows that the total number of incidents has dropped since the implementation of the environmental care programme. Also the severity of the individual breaches has been steadily declining.

A good compliance record like this results in a cost saving. Less stringent monitoring conditions are imposed by the Environment Protection Authority with a consequent reduction in costs of monitoring air, water and vegetation. Good compliance also saves money by avoiding rectification costs which are expensive because of the need to clean-up and eliminate contaminated areas.

Alcoa environmental audit results

Alcoa maintains an environmental auditing system which although internal to the corporation is external to operational areas. All facilities are subject to a rigorous audit on at least a three-year cycle. The first audit under this system at Portland was held in October 1989. There were twelve issues which the audit team felt should be brought to the managers' notice. Three of these were positive and related to the spent potliner study, recycling and a general environmental review. Overall the team felt that there was a good environmental programme in place and that line management was the most environmentally aware and sensitive of any of the facilities they had audited to that time.

The second audit was held in October 1993. It found a high level of environmental awareness, responsibility and accountability and attributed this to the incorporation of environmental responsibilities into day-to-day operations and employee involvement at all levels. There were eighteen issues brought to the manager's notice, four of these were positive: two endorsed existing programmes and recommended that they be continued and two recognized the land management and waste

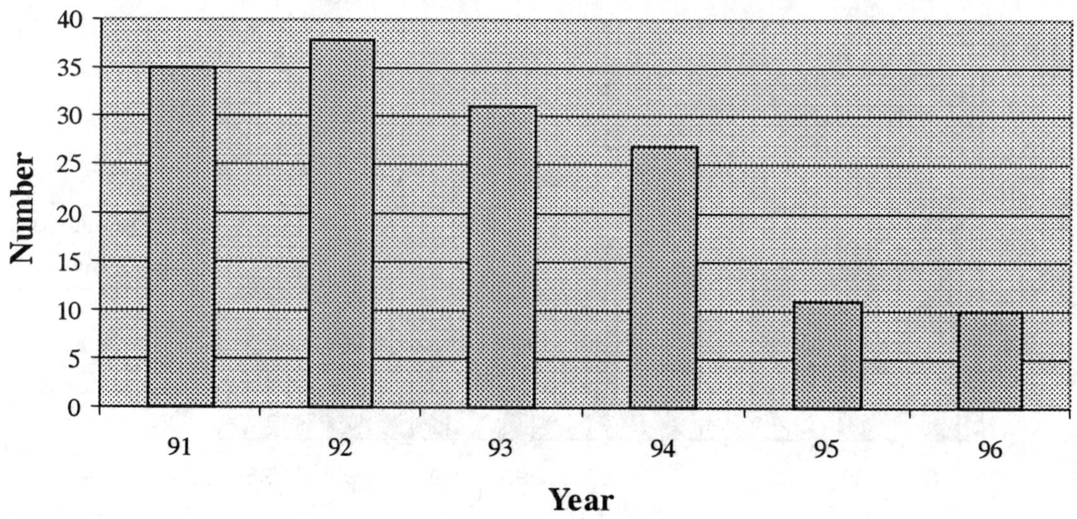

Fig. 6. Total number of reportable environment incidents.

minimization programmes as benchmark systems. There were no adverse findings. Overall it was a very satisfactory audit.

The third audit was held in October 1996 and found that Portland Aluminium had implemented strong environmental management systems to meet its regulatory and Alcoa requirements. There were nine issues brought to the manager's notice: four of these were positive, two endorsed existing programmes and recommended further development and two awarded benchmark status, one of which was for the implementation of the Smelter in the Park project. There were no adverse findings, again a very satisfactory audit.

Water use/waste water disposal

Purchased potable water is used in the smelter mainly for cooling with a smaller amount used for garden irrigation and domestic purposes. Total annual use peaked in 1994. A water-use evaluation which was part of the environmental care programme was completed in that year. Efficiencies and work practice changes were identified and implemented. Following the audit, water use started to decline for the rest of the study period (Fig. 7). This has been achieved by paying closer attention to the proper operation and maintenance of water using facilities.

The quality of waste water has also improved through improved smelter operation and good housekeeping. Other factors include the avoidance of water contamination by encapsulating external facilities during paint removal to catch dust and the avoidance of zinc based paints and corrosion inhibition treatments.

The water-use evaluation in 1994 was part of a larger study initiated with the view of developing a comprehensive water management plan. The improved water quality resulting from the Park development has made harvesting rain water from the smelter site and reuse of waste water in some smelter processes feasible. A cleaner site and processes means that excess waste water can be used to develop wetland habitat around the smelter and avoid disposal into the sea. Pilot trials and economic analyses are continuing. The ultimate goal of the water management plan is to eliminate "waste" water from leaving the site and to operate the smelter entirely from recycled processing and harvested storm water, rather than purchasing potable water for use in processing. The only potable water purchased would be for domestic use and some irrigation.

Waste minimization

An industry the size of the smelter has the potential to generate large quantities of waste, and this was the case during construction and the start-up phases of the 1980s. By 1989 a large area for landfill was being used and around 13 000 cubic metres was going into landfill annually. People were behaving wastefully, the easy option was often to simply throw things away; Alcoa was a big multinational company worth billions, why worry about the dollars and cents.

As a result of the minimization programme, the figure in 1996 had plummeted to less than 25 cubic metres, well on the way to the goal of zero landfill (Fig. 8). Internal surveys established that this change in behaviour was driven by environmental concern, not the financial savings for the company (see below for analysis of interviews). There is now a plant wide emphasis on housekeeping, recycling, reuse, avoidance of spillage and prevention of contamination of the soil. A turning point was the closing of part of the landfill and rehabilitating it for wildlife habitat. This forced the issue within the plant because there was no longer a place to hide unwanted material.

SUMMARY OF OTHER IMPORTANT OUTCOMES OF ENVIRONMENTAL CARE 1989–1996

A broad range of environmental programmes have been initiated in association with the Smelter in the Park project. These are separate from the operation

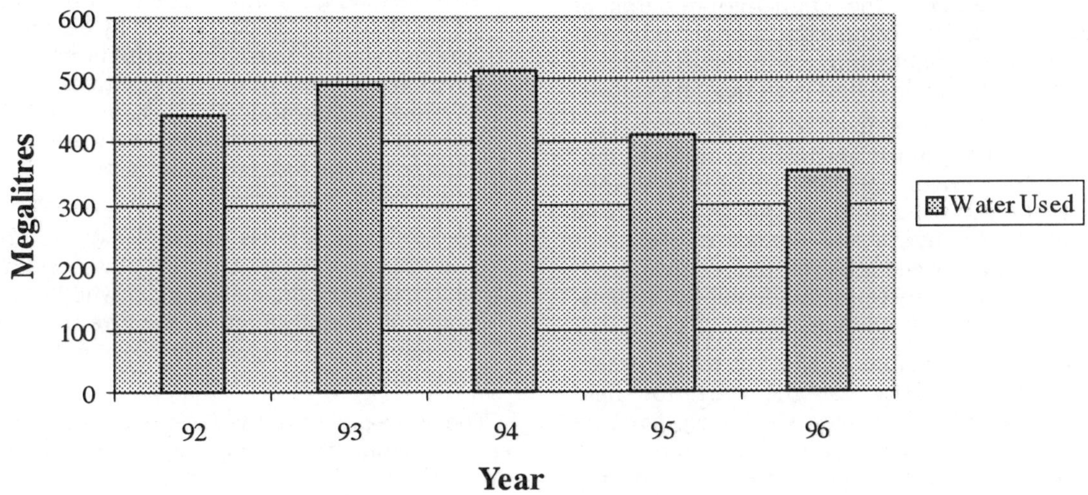

Fig. 7. Water use by the Portland Aluminium Smelter.

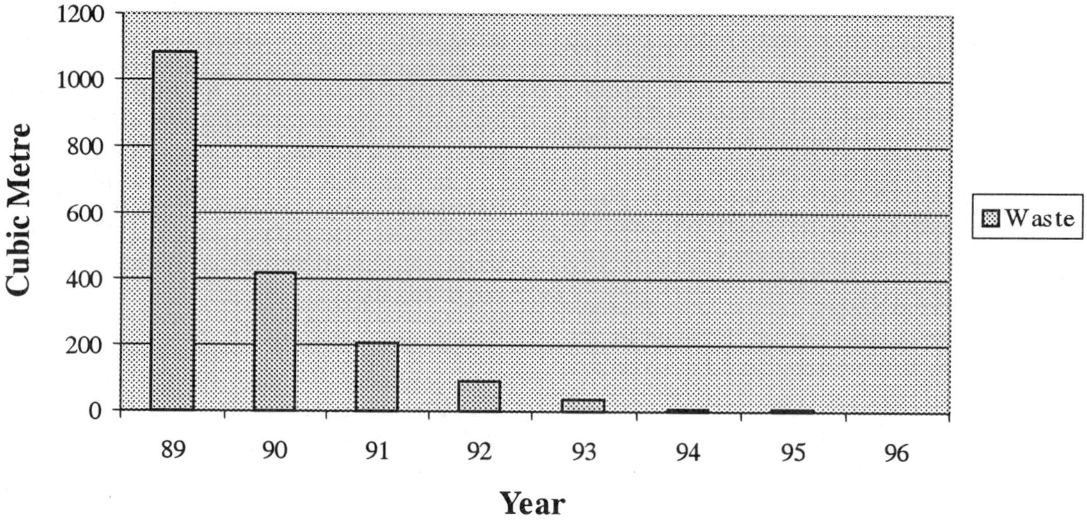

Fig. 8. Decline in the total amount of general waste going to landfill at the Portland Smelter for the period 1989–1996.

of the smelter and monitoring required under the operating licence. A summary is given to show the range of the programmes and what has been accomplished.

1. Vermin — A programme to control introduced predators and rabbits within the Park and surrounding lands has been operating for the past several years with the help of employees, surrounding land owners and members of the community. The density of rabbits, foxes and cats has been significantly reduced in the Park. This has resulted in increased numbers of ground dwelling native birds and animals. Sightings of introduced predators are now rare.
2. Revegetation — This programme had the ambitious aim of selecting areas in the Park to re-create some of the diverse plant communities that existed on Point Danger prior to clearing for agriculture and smelter construction. Associated with this has been the need to develop propagating facilities, knowledge and skills. Simplified communities have been created using locally collected seed but the challenge remains of introducing greater complexity to improve habitat values. The propagation of difficult species, especially terrestrial orchids is an ongoing study.
3. Weeds — Weed control is a very important corollary and often a precursor to revegetation. The Point Danger area had some very serious weeds such as cape tulip and gorse which have the ability to invade and replace native plant communities. Weed control plans have been completed and are being implemented systematically on a species and area basis.
4. Community Education — Students from primary school to university use the Park for field studies. Learning packages which help teachers to encourage students to study various facets of the Park have been developed to fit primary and secondary education curriculum. University students have helped with planning, implementation, monitoring, revegetation, and wildlife management in the Park as part of their studies.
5. Research — Scientists, university academics and post-graduate students have used the environmental care initiative in the Park and the smelter operations for a wide range of research projects. Many of these projects have been jointly funded by Universities and Portland Aluminium. Recently the government has also become a funding partner on several projects. Several research studies are on-going and opportunities exist to expand environmental research at Portland Aluminium.
6. Wildlife Management — Wildlife monitoring and research projects on population monitoring, ecology, reintroduction, and management have been an important part of the development of the Park. The wildlife projects on kangaroos, potoroos, reptiles and frogs are covered in associated chapters (e.g., see Coulson *et al.,* this volume).
7. Demonstration Farm — The farm is designed to promulgate the Landcare ethic by demonstrating productivity in harmony with nature. Selected areas have been planted with trees to provide livestock shelter and wildlife habitat, other areas are protected to maintain soil stability and wildlife values. Kangaroo numbers are rising and will need to be addressed.
8. Wetlands — Studies are underway to enhance and actively manage a variety of wetland habitats which will support a diversity of wildlife. The first stages of rehabilitation and active management has resulted in increased numbers and diversity of birds and aquatic insects.

The development of the Park, waste minimization, environmental compliance beyond that required by regulations and a wide range of other important environmental outcomes demonstrate some very positive substantive outcomes from environmental care. The ethos of environmental care exists at Portland Aluminium. The question remains, what effect has this had on work practice and community relations?

CHANGING COMMUNITY ATTITUDES — RESPONSE TO ENVIRONMENTAL CARE

A considerable body of information has been collected that supports the concept that Portland people are adopting a more positive attitude toward the company. Visitors also become aware very quickly of the efforts of the company to care for the environment.

Reports published in the local newspapers are monitored and assessed by the company librarian as positive or negative. Prior to 1989 many articles expressed concerns about health and environmental issues associated with the smelter. In later years the emphasis has shifted and there are often positive articles on the programmes at the smelter. In 1995 a regular phone survey was commissioned to monitor community awareness of Portland Aluminium and perception of performance. In this people responded positively to questions on the environmental performance of Portland Aluminium and trust was rated as high.

People from the community and visitors to Portland are now using the facilities in the Park. Local conservation and natural history groups are working collaboratively with smelter staff on such projects as feral animal control, coastal heath rehabilitation, native orchid propagation and bird counts. Local schools frequently make use of the education facilities and run curriculum-based activities within the Park.

The guided smelter tours operated for visitors have very high patronage. A questionnaire is administered to all tour participants before and after the tour. Analysis of this shows some initial environmental concern about the smelter operations before the tours begin, but the questionnaires indicate that this is being comprehensively overcome by the clean work environment, the park surrounding the smelter and the explanation of the environmental practice of the company.

CHANGING WORKPLACE RELATIONSHIPS — RESPONSE TO ENVIRONMENTAL CARE

Area supervisors, the Environmental Resource Group and operators throughout the plant were interviewed in help determine the impact of environmental care on work practice

Area supervisors

Area supervisors are in senior supervisory/operational positions in the company, but they are outside the management group and do not determine policy. In 1989 there were ten of them, by 1996 the number had been reduced to five. This reduction in supervisors was due in part to the improvement in work practice by plant operators.

Area supervisors are responsible for the day to day operation of an area of the plant (for example part of the Ingot Mill in 1989 or the full Ingot Mill in 1996) and are in an ideal position to observe work practices and people's attitudes to their work. Their major concerns are about relatively short-term issues such as operator performance and machine availability (up-time); they may judge the culture within the smelter differently from managers or operators. From this set of interviews a comprehensive description of work practice in the smelter was developed.

The area supervisors said that marked changes have taken place in work practice at the smelter. Significant workplace reforms included the adoption of twelve hour shifts and a new work place agreement which eliminated paid overtime.

According to the area supervisors the ethos of environmental care provided a focus for training which encouraged operators to be more responsible and aware of their work environment. People now are generally self starters who accept personal responsibility without supervision. Environmental care helped operators see the need to think for themselves, to be aware of the needs of other employees and to work carefully and efficiently to complete tasks within the time available.

From the area supervisors' point of view it is clear that the smelter is easier to run now, the people who work there are more productive and they are more co-operative. They felt that the programme of environmental care made a key contribution to these positive changes.

The Environmental Resource Group

The Environmental Resource Group was an important initiative for implementing environmental care into the operation of the smelter. It was formed with representation from all the business areas of the company specifically to achieve waste minimization.

Management felt that a co-ordinated drive for waste minimization in all facets of the smelter operation would encourage the workforce to adopt efficient and effective work practices. People try to avoid waste in their own lives so raising this issue at work would build on inherent belief about adopting the "right behaviour" in the workplace (Cohen-Rosenthal 1992). Early success with waste minimization would engender sense of achievement throughout the smelter that could be built upon and help build positive community relations (Vajda 1991).

From the results of these interviews it is clear that throughout the plant people have become aware of environmental issues and have modified their behaviour. According to the Environmental Resource Group, the initial motivation for this was to save resources, people were sickened by the waste: the view of the 1980s that "it belongs to the company it does not matter" led to such non-caring behaviour as material was dumped simply because it was the easiest option.

Other issues that were listed in the interviews as important in changing behaviour were a developing feeling of responsibility for the general environment and a concern for the landfill area itself, the sheer size of it, the pollution potential and the loss of land and wildlife values in that vicinity. People said that whole work areas took pride in the reducing material-to-landfill statistics and looked for ways to help.

Operators

The operators run the often complex machinery that supports the ongoing functioning of the smelter. They see that machinery is maintained and working effectively and they carry out the continuous fine tuning that ensures the smelter runs efficiently. It is essential that they are aware of the effects of their actions and are committed to high and improving standards.

The survey showed that these people have a very positive attitude towards achieving good environmental performance. For example the graph (Fig. 9) shows the pooled importance rating for a number of claimed outcomes for the Smelter in the Park project. Zero on this rating indicates the respondent thought the Smelter in the Park was not applicable to the claimed outcome, four indicates it was thought very important. The highest rating was given to clearly environmental questions such as those on conserving resources and preventing pollution or contributing to conservation by developing wildlife habitat. However, the more general propositions of encouraging safe work practices and building people's trust in the company were also rated highly.

The ethos of environmental care which developed among the operators provided a context for applying training and improving work practices to improve safety and increase efficiency.

Productivity of workforce

One of the main reasons for adopting environmental care as a key business strategy was to improve productivity through better work practices. There has been a steady reduction in the number of people employed at the smelter without a significant reduction in the total output of aluminium.

Some of the drop in staff numbers results from changed work practices and the fact that current staff are able to do tasks more quickly and with less help. This is due to their increased experience gained from working longer at the smelter, and their willingness to apply the training they have received to improve their work practices. This is an obvious cost saving since the same amount of product is produced for a considerably lower salary bill. Employees satisfied with their work environment and the value of their work provide a stable workforce which reduces the cost of recruiting and training new staff.

Our contention is that environmental care has been an important focus for the structural and work practice changes. It has provided a context for employees to develop personal responsibility and job satisfaction which has encouraged them to improve work practices. Employees are more comfortable in the improved surroundings of the smelter with tasks that contribute to a better environment and therefore are more likely to stay.

The need to quantify results

There were few precedents within industry for incorporating environmental care as a key business strategy at the time production began at Portland Aluminium (Washington 1991). Therefore, the inspiration did not come from other industries but rather from the emerging environmental sensitivity in society. Management looked outside the smelting industry for their inspiration and set about changing the accepted norm in their industry in a

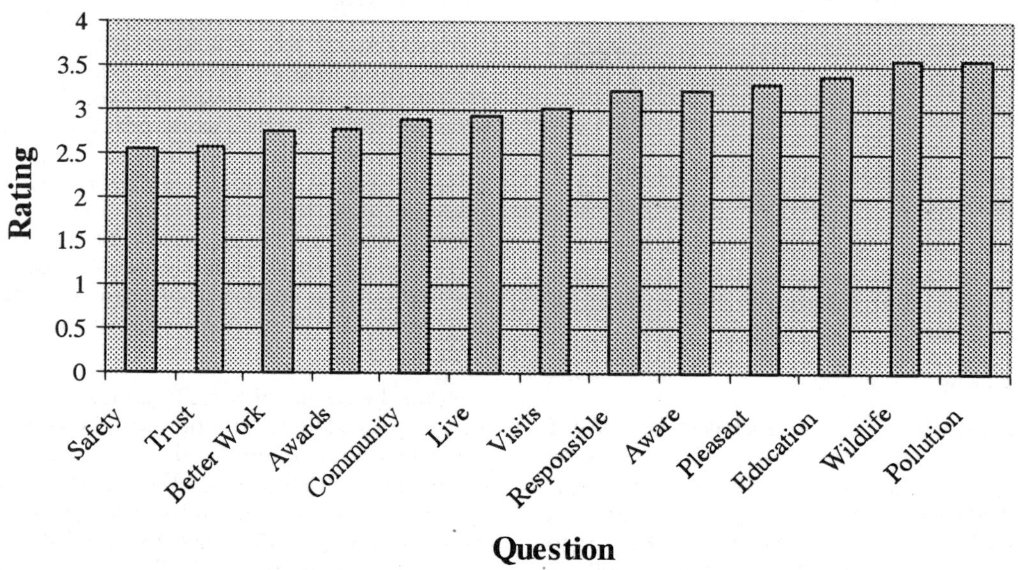

Fig. 9. Importance rating given by operators to claimed outcomes from the Smelter in the Park project.

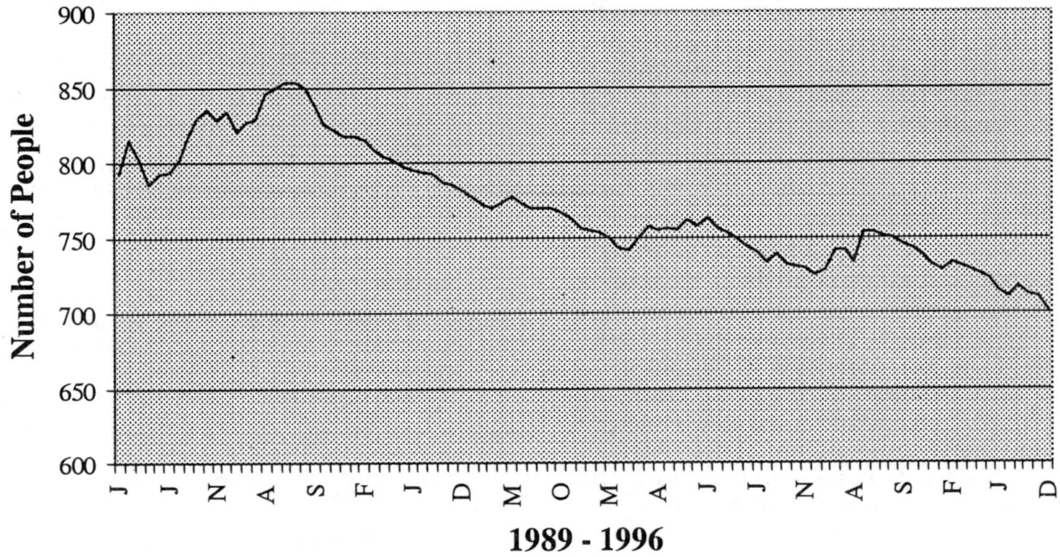

Fig. 10. Decline in the total workforce at the Portland Smelter for the period 1989–1996.

way that had long-term implications for Portland Aluminium.

National and international recognition for the design and development of the Smelter in the Park and for environmental best practice at the smelter could create the perception that the environmental aspects of the operation are well in-hand. This along with changing commodity prices and other pressures of the international business environment may lead to change in the emphasis for environmental care at Portland Aluminium.

The outcomes of environmental care do not always have direct tangible links to core products (Steward 1996). Such outcomes can easily be seen as peripheral to the main business. Difficulties often arise when trying to justify the expenditure of resources to care for the environment when the company budget is being driven by international competitiveness, efficiency and maximizing the return to share-holders (Paviour-Smith 1991; Newson 1996).

Expenditure beyond that strictly demanded by compliance to operating licences must show a return on the investment. Some environmental care programmes do have tangible outcomes. Waste reduction, parklands or improvements to the working environment can easily be seen and assessed, but these tangible outcomes are often hard to justify on a purely economic basis.

Many of the outcomes originally claimed as justification for the key role of environmental care at Portland Aluminium were and still are based on attitudinal and behavioural change. The contribution of these outcomes to better work practice and operational efficiency is often difficult to separate from the outcomes of other reform strategies.

Appropriate monitoring instruments were not designed and operated before and during the implementation of the environmental care programme, therefore little attempt was made to quantify the behavioural outcomes of environmental care or to show that they had in fact made a measurable, positive contribution to production and profits.

Any programme must demonstrate a return for the business if it is to survive rigorous scrutiny from successive generations of management; all of whom are charged with the responsibility of generating a fair return for the shareholders. However, this scrutiny needs to be based on the evaluation of monitoring legitimate indicators that can measure the return that the programme provides to the company.

CONCLUSION

Portland Aluminium made environmental care a core strategy to help strengthen other reforms implemented to improve the operation of the smelter. This contrasts with the approach taken in many industries where care for the environment is viewed as an impost related to government regulation; a cost to be minimized (Newman and Breeden 1992; Newson 1996).

Portland Aluminium management held the view that introducing a comprehensive programme of environmental care would make organizational and work practice changes more effective. To them environmental care was part of a comprehensive workplace strategy to introduce work practices which would operate the smelter at maximum efficiency with the least possible effect on the environment and also bring the best outcomes for the company.

There is little doubt that incorporating environmental care with the Smelter in the Park as a focus at Portland Aluminium has provided some important benefits to the operation of the smelter and to the profits of the company.

The exact nature and value of the returns to the company from the investment in environmental care have not been carefully monitored or sufficiently quantified. This makes it difficult to predict quantitatively what can be expected in the future if the investment in environmental care continues or the cost to the company if the commitment is wound back.

Environmental care is not a fixed point, which a company reaches, a range of products that can be measured, or a promotional concept for which a company receives awards. Environmental care needs to be one of the key principles in the on-going operation of a business. It requires continuous commitment from management and other employees to gradually improve the culture of the operation and the core business through actively caring for the environment. It also requires some clear indicators of the costs and benefits that accrue to the company as a result of improving environmental practices.

The pressures for international competitiveness and increased productivity can easily hide the less obvious and immediate work practice benefits that accrue from long-term investments in environmental care. Without a long-term commitment, the gains that have been made thus far at Portland Aluminium could easily be lost

Although the philosophy of using environmental care to help achieve company goals has not been widely implemented in industry, it has wide applicability. Many similar circumstances of environmental sensitivity exist around new and existing industries. Others may well benefit from the experience of Portland Aluminium and benefits will certainly continue to accrue to Portland Aluminium if the focus on environmental care can be shown to be truly a "best business practice".

REFERENCES

Baldock, R. N., 1996. A Portland Aluminium Initiative for Environmental Monitoring: A review of the Heathland Monitoring Project 1993–1996. Report for Portland Aluminium by Centre for Environmental and Recreational Management, University of South Australia.

Brake, L., Clark, I., Clay, R. and Savarton, F., 1991. The Master Plan for The Smelter in the Park. Report for Portland Aluminium by Centre for Conservation and Recreation Resources, University of South Australia.

Carlopio, J. R. and Gardner, D., 1992. Direct and Interactive effects of the Physical Work Environment on Attitudes. *Environ. Behav.* **24**(5): 579–601.

Carr, G. W., 1992. Review of Vegetation Data EPA Permanent Plots Portland, Victoria. 1982–92. Report for Portland Aluminium by Ecological Horticulture Pty Ltd. Clifton Hill, Victoria.

Cohen-Rosenthal, E., 1992. Labour-Management Partnership to Meet Workplace Ecological Challenges. *CIE Environ. Update* **Winter:** 9.

Doley, D., 1994. Site Inspection Portland Aluminium Limited Portland. Report for Portland Aluminium by Department of Botany, The University of Queensland.

Heaton, E. E., 1989. Americans Becoming Alarmed by Threats to the Environment and Their Personal Health. *Opinion Res. Corp. Issue Watch* **May**: 1–5.

Hedditch, D., 1993. The Rationale for the Organisation of Portland Aluminium. Portland Aluminium: Portland, Victoria.

House of Representatives Standing Committee on Environment, Recreation and the Arts, 1994. Working with the Environment. Opportunities for Job Growth. Commonwealth of Australia: Canberra, Australian Capital Territory.

Hughes, W., 1993. Productivity and Wildlife Programmes. Report for Wildlife Habitat Enhancement Council annual symposium by Portland Aluminium. Wildlife Habitat Enhancement Council, Richmond, Virginia.

International Primary Aluminium Institute, 1992. Primary Aluminium and Aspects of the Environment. Report by the International Primary Aluminium Institute, New York.

Judd, D., 1991. Portland Aluminium. Internal report for Portland Aluminium, Portland, Victoria.

Judd, D., 1992. Design Complementing Nature. Report for Wildlife Habitat Enhancement Council annual symposium by Portland Aluminium, Wildlife Habitat Enhancement Council, Washington, D.C.

Kelly, J. M. and Pressman, D. S., 1989. Wildlife Habitat Enhancement on Corporate Lands: Social, Economic and Corporate Benefits. Pp. 255–59 *in* Strategies for Meeting Natural Resource Needs ed by Richard E. McCabe. Wildlife Management Institute: Washington.

Kinhill, 1980. Alcoa Portland Aluminium Smelter Environmental Effects Statement and Draft Environmental Impact Statement. Kinhill Planners Pty Ltd: Melbourne, Victoria.

Kornbluh, H., Crowfoot, J. and Cohen-Rosenthal, E., 1985. Worker participation in energy and natural resources conservation. *Internat. Labour Rev.* **124**(6): 737–53.

Lukas, J. A., 1993. Improving the Project Controls Work Environment at Kodak. *Amer. Assoc. Cost Engin. Trans.* **37:** Q1.1–Q1.8.

McFadden, M. and Demetriou, E., 1993. The Role of Immediate Work Environment Factors in the Turnover Process: A Systematic Intervention. *Applied Psychol.* **42**(2): 97–115.

McGreer, J. P., 1989. Environmental Control in Smelting Trends in Environmental Control and Emissions Control. Third Australasian Aluminium Smelter Technology Course, Sydney. The University of New South Wales.

Miller, A., 1996. Livestock surveys for fluoride monitoring. Report for Portland Aluminium by Portland Veterinary Clinic, Portland, Victoria.

Miller, A., 1997. Livestock surveys for fluoride monitoring. Report for Portland Aluminium by Portland Veterinary Clinic, Portland, Victoria.

Miller, W. H., 1990. Earth Day: Twenty Years Later. *Industry Week* **March:** 86–97.

Newman, J. C. and Breeden, K. M., 1992. Lessons from Environmental Leaders. *Columbia J. World Bus.* **Fall and Winter:** 1–12.

Newson, M., 1996. Environmental Policy: Does it Satisfy your Stakeholder's Expectations? Pp. 72–75 *in* Australian Environmental Solutions ed by M. Mann. Environment Management Industry Association of Australia: Hampton, Victoria.

Pagan, R. D., 1989. A New Era of Activism. *The Futurist* **May–June:** 12–16.

Paviour-Smith, J., 1991. Guidelines for Industry in the Future. Prepared for Environmental Damage. Institute for International Research, Sydney, New South Wales.

Portland Aluminium, 1993. Model for Sustainable Organisational Change. Internal report for Portland Aluminium, Portland, Victoria.

Portland Aluminium, 1996. Mission, Vision, Values. Portland Aluminium Diary, Portland, Victoria.

Prince, J. and Denison, R. A., 1991. Launching a New Business Ethic: The Environment as Standard Operating Procedure at McDonald's and at other Companies. Prepared for Business Ethics and the Environment, Dartmouth College Hanover: New Hampshire.

Prokop, M. K., 1992. Managing Green: Defining an Organisation's Environmental Role. Pp. 275–84 *in* The 1992 Annual: Developing Human Resources. Pfeiffer and Company: San Diego, California.

Ramsey, P. J., 1990. Environmental Impact Assessment Portland Aluminium Smelter. Report for Portland Aluminium by Peter J. Ramsey and Associates Pty Ltd, Melbourne, Victoria.

Schimidheiny, S., 1992. Changing Course. Massachusetts Institute of Technology, Cambridge.

Steward, N., 1996. Towards best practice environmental management: the business perspective and opportunities. *The Quality Mag.* **April:** 64–68.

The World Commission on Environment and Development, 1987. Our Common Future. Oxford University Press.

Vajda, G. F., 1991. Integrating Pollution Prevention Into the Manufacturing Process. *Total Quality Environ. Manage.* **Autumn:** 61–70.

Washington, H., 1991. Ecosolutions Environmental Solutions for the World and Australia. Boobook Publications: Marrickville, New South Wales.

Young, M. D., Cunningham, N., Elix, J., Lambert, J., Howard, B., Grabosky, B. and McCrone, E., 1996. Reimbursing The Future. Commonwealth of Australia: Canberra, Australian Territory.

CHAPTER 36

The Smelter in the Park: managing wildlife for biodiversity

Graeme Coulson,[1] John Hill,[2] Jane McKenzie,[1,3] and Bryan Walters[4]

Portland Aluminium Smelter was constructed on a greenfields site adjacent to the City of Portland, on the southwestern coast of Victoria, Australia. A 200 ha buffer zone surrounding the smelter incorporated the city's former airport and surrounding farmland, together with remnants of natural coastal heathland and about 50 small wetlands. Portland Aluminium adopted a theme of "The Smelter in the Park", which aims to integrate the smelting operation into a biologically-diverse, aesthetically-pleasing setting by appropriate management of the buffer for conservation, rehabilitation, education, and recreation. A wildlife inventory has shown that species richness in this small park is high: 22 mammals, 8 reptiles, 6 amphibians and 98 birds, excluding seabirds. Seven species are listed as threatened in Victoria. Several others have become overabundant, conflicting with some management goals. Six mammal species and six bird species are introduced, most having pest status. A revegetation programme is underway to restore and enhance heathland habitats, and to control environmental weeds, and wetlands are being rehabilitated to improve waterbird habitat. Introduced mammalian predators and herbivores are controlled through an integrated programme with neighbouring private and public land. A series of reintroductions of species that have become locally extinct will further enhance biodiversity. Problems caused by overabundant native species are managed by training staff in co-habiting with wildlife, as well as by habitat modification and fertility control.

INTRODUCTION

IN 1979, Alcoa of Australia Limited put forward a proposal for the construction of a large aluminium smelter on the western coast of Victoria. The proposed location was a headland on the outskirts of the City of Portland, encompassing the city's airport and surrounding farmland, together with remnants of natural coastal heathland. As this was a major industrial development with the potential to cause significant environmental effects, Alcoa was required to prepare an Environment Effects Statement under the *Environment Effects Act 1978* (Victoria) and a Draft Environmental Impact Statement under the national *Environment Protection (Impact of Proposals) Act 1975*. These were presented as a single report (Alcoa and Kinhill 1980a), which was followed by a supplementary report after a period for public comment (Alcoa and Kinhill 1980b). These reports are referred to together in this chapter as the Environmental Impact Statement.

The Environmental Impact Statement included the results of a wildlife survey, conducted in July and August 1979 (Pescott and Kinhill 1979; Alcoa and Kinhill 1980a). Although the survey was brief, and limited to a season when many animals were inactive, two reptile species, four amphibians, 14 mammals and 51 birds (excluding seabirds) were recorded (Tables 1 and 2). Of these, a native rodent, the Heath Rat *Pseudomys shortridgei*, a marsupial carnivore, the Swamp Antechinus *Antechinus minimus*, and two heathland birds, the Southern Emu-wren *Stipiturus malachurus* and Rufous Bristlebird *Dasyornis broadbenti* were considered significant because of their limited distributions and rarity. The Environmental Impact Statement also identified a significant reptile, the Mourning Skink *Egernia coventryi* as a species that was potentially present at the site but may have been missed during the survey. An additional survey (Robertson and Kinhill 1980) was conducted in October 1979 specifically to search for this skink, and although

[1]Department of Zoology, University of Melbourne, Parkville, Victoria, Australia 3052.
[2]Portland Aluminium, Private Mail Bag 1, Portland, Victoria, Australia 3305.
[3]Present address: Department of Zoology, University of Tasmania, P.O. Box 252C, Hobart, Tasmania, Australia 7001.
[4]Ecoplan Australia Pty Ltd, P.O. Box 580, Hurstbridge, Victoria, Australia 3099.

Pages 360–71 *in* NATURE CONSERVATION 5: NATURE CONSERVATION IN PRODUCTION ENVIRONMENTS: MANAGING THE MATRIX ed by J. L. Craig, N. Mitchell and D. A. Saunders. Surrey Beatty & Sons, 2000.

Table 1. Records of mammal, reptile and amphibian species at Portland Aluminium, showing the species recorded as present by previous surveys and on the trapping grid at each site in the present study. The strongest evidence is presented for each species at each site, following the order (strongest to weakest): captured, seen, heard, indirect evidence and unconfirmed. Numerical values are the sum of captures in pitfall traps, "Elliott" traps and cage traps.

					Trap site									
Common name	Scientific name	EIS	R&K	ACF	1	2	3	4	5	6	7	8	Add.	All
Mammals														
Echidna	*Tachyglossys aculeatus*	•			i	s	s	s	i	s	i			•
Short-nosed Bandicoot	*Isoodon obesulus*	•		•			1			2	1			•
Eastern Grey Kangaroo	*Macropus giganteus*	•				s	s	s	s	i				•
Red-necked Wallaby	*Macropus rufogriseus*												•	•
Swamp Wallaby	*Wallabia bicolor*												•	•
Common Ringtail	*Pseudocheirus peregrinus*	•			s	s	s	s		s				•
Koala	*Phascolarctos cinereus*				s	i			s					•
Swamp Antechinus	*Antechinus minimus*	•	•										•	•
Heath Rat	*Pseudomys shortridgei*	•		•					18	4		3		•
Bush Rat	*Rattus fuscipes*	•	•	•					5	26	59	30		•
Swamp Rat	*Rattus lutreolus*	•		•			3	2	3	9	1	18		•
Black Rat *	*Rattus rattus*	•		•		10	1							•
House Mouse *	*Mus musculus*	•		•	1	2				22	5	2		•
Lesser Long-eared Bat	*Nyctophilus geoffroyi*	•			2	4								•
Southern Forest Bat	*Vespadelus regulus*					2								•
White-striped Freetail Bat	*Tadarida australis*					h	h							•
Common Bent-wing Bat	*Miniopterus schreibersi*												•	•
Little Red Flying-fox	*Pteropus scapulatus*												•	•
Red Fox *	*Vulpes vulpes*	•			s	s	s	s	s	s	s			•
Domestic Cat *	*Felis catus*	•								s				•
European Rabbit *	*Oryctolagus cuniculus*	•				s	s		s		s			•
Brown Hare *	*Lepus capensis*												•	•
Reptiles														
Striped Worm-lizard	*Aprasia striolata*									5				•
Wall Skink	*Lampropholis guichenoti*				25									•
White's Skink	*Egernia whitii*		•				s			3		1		•
Three-lined Skink	*Bassiana duperryi*		•											
Grass Skink	*Pseudemoia entrecasteauxi*				4		s	6	s	4	1	10		•
Blotched Blue-tongued Lizard	*Tiliqua nigrolutea*		•				1					1		•
Eastern Blue-tongued Lizard	*Tiliqua scincoides*				3	3		1	1					•
Copperhead	*Austrelaps superba*				s				s		s			•
Eastern Tiger Snake	*Notechis scutatus*				2	1	s	1	s	s				•
White-lipped Snake	*Drysdalia coronoides*		•										?	
Amphibians														
Southern Brown Tree Frog	*Litoria ewingi*	•	•		1	8	3	7	4	1	10			•
Growling Grass Frog	*Litoria raniformis*	•												
Pobblebonk (Banjo) Frog	*Limnodynastes dumerili*		•		2	16	16	3	1	10		1		•
Striped Marsh Frog	*Limnodynastes peroni*		•		6	6	4	5	1		4			•
Spotted Marsh Frog	*Limnodynastes tasmaniensis*		•											
Common Spadefoot Toad	*Neobatrachus sudelli*				3	3				1				•
Southern Toadlet	*Pseudophryne semimarmorata*	•	•											
Common Froglet	*Ranidella signifera*		•		1	7	9	12	5		14			•
Southern Smooth Froglet	*Geocrinea laevis*	•	•										•	•
Species richness		18	13	6	16	18	17	12	15	17	12	8	7	36

EIS — Pescott and Kinhill (1979) and Alcoa and Kinhill (1980a).
R&K — Robertson and Kinhill (1980).
ACF — Beauglehole, pers. comm.
Add. — Additional Incidental records collected since the beginning of this study.
All — All records obtained during this study, incorporating incidental records and trapping records.
• — present; s — seen; h — heard; i — indirect evidence; ? — unconfirmed; * — introduced species.

the target species was not detected, eight new species (four amphibians and four reptiles) were added to the species list (Table 1).

Once the project gained approval, construction began in 1980 and the first aluminium was poured in 1986. The smelter is operated by Alcoa in a joint venture trading as Portland Aluminium. The joint venture, which has varied in composition since start-up, now has Alcoa and the state-owned Aluvic Metal Sales Pty Ltd as major shareholders. Three other companies, CITIC Australia (Portland) Pty Ltd, Eastern Aluminium Ltd, and Marubeni Aluminium Australia Pty Ltd, each have a 10% holding.

During construction, approximately 70 ha of coastal heathland and shrubland was bulldozed, and some topsoil from this area was stockpiled. The former airport and three neighbouring farms were incorporated into a buffer zone surrounding the plant. An adjacent 39 ha area of partially-cleared heathland, known as Dean's Heath, was purchased by Portland Aluminium in 1983 and protected from further clearing by adding it to the buffer zone. Members of the Portland Chapter of the Australian Conservation Foundation and the Portland Field Naturalists Club surveyed the mammals of Dean's Heath in 1986 (C. Beauglehole, pers. comm.), recording one additional species for the area (Table 1).

In 1991, Portland Aluminium adopted a master plan for "The Smelter in the Park" (Brake *et al.* 1991). The plan aims to integrate the smelting operation of Portland Aluminium into a setting that is biologically diverse and aesthetically pleasing (Hill *et al.,* this volume). The plan designated five zones with differing, though overlapping, land use goals (Fig. 1). The largest zone (Zone D — Habitat for Wildlife) has the conservation of flora and fauna as its primary function, and a smaller zone (Zone C — Places for Learning) is primarily for education, research and monitoring of rehabilitation work. Two other zones (A and B) are mainly devoted to visitor information and recreation, and the fifth zone (E) is the core production area.

The aim of the present study was to assess the diversity of the wildlife in "The Smelter in the Park", and provide directions for the future management of wildlife in the park. We had five specific objectives —

1. Produce a comprehensive inventory of the wildlife of Portland Aluminium.

Table 2. Records of bird species at Portland Aluminium, showing the species recorded as present in the previous survey, the mean abundance of species in nine quarterly surveys, and the persistence of species, expressed as the percentage of quarterly surveys in which each species was recorded.

Common name	Scientific name	EIS	Add.	Persistence (%)	Abundance
Hoary-headed Grebe	*Poliocephalus poliocephalus*			100	9.3
Australasian Grebe	*Tachybaptus novaehollandiae*			89	10.1
Pied Cormorant	*Phalocrocorax varius*	•			
Little Pied Cormorant	*Phalacrocorax melanoleucos*			33	1.3
Great Cormorant	*Phalacrocorax carbo*	•		22	0.4
Little Black Cormorant	*Phalacrocorax sulcirostris*			22	0.7
Pacific Heron	*Ardea pacifica*	•		11	0.1
White-faced Heron	*Ardea novaehollandiae*	•		100	18.8
Cattle Egret	*Ardeola ibis*	•		56	36.6
Great Egret	*Egretta alba*			33	0.7
Rufous Night Heron	*Nycticorax caledonicus*			22	0.3
Sacred Ibis	*Threskiornis aethiopica*			67	12.4
Straw-necked Ibis	*Threskiornis spinicollis*			67	47.2
Royal Spoonbill	*Platalea regia*			11	0.3
Black Swan	*Cygnus atratus*			89	13.8
Austalian Shelduck	*Tadorna tadornoides*	•		11	0.2
Pacific Black Duck	*Anas superciliosa*			100	102.4
Grey Teal	*Anas gibberifrons*			78	21.6
Chestnut Teal	*Anas castanea*			100	7.9
Australasian Shoveler	*Anas rhynchotis*			56	6.8
Maned Duck	*Chenonetta jubata*			11	0.3
Musk Duck	*Biziura lobata*			44	1.6
Black-shouldered Kite	*Elanus notatus*			22	0.3
Brown Goshawk	*Accipiter fasciatus*	•		22	0.2
Marsh Harrier	*Circus aeruginosus*	•		44	1.2
Peregrine Falcon	*Falco peregrinus*	•		22	0.2
Australian Hobby	*Falco longipennis*			22	0.6
Brown Falcon	*Falco berigora*	•		33	0.3
Australian Kestrel	*Falco cenchroides*	•			
Buff-banded Rail	*Gallirallus philippensis*			11	0.2
Baillon's Crake	*Porzana pusilla*			11	0.2
Spotless Crake	*Porzana tabuensis*			11	0.1
Dusky Moorhen	*Gallinula tenebrosa*			56	3.2
Purple Swamphen	*Porphyrio porphyrio*			89	57.2
Eurasian Coot	*Fulica atra*			44	2.1
Brolga	*Grus rubicundus*		•		
Masked Lapwing	*Vanellus miles*	•		100	31.9

Table 2 — *continued*

Common name	Scientific name	EIS	Add.	Persistence (%)	Abundance
Banded Lapwing	*Vanellus tricolor*			11	0.4
Inland Dotterel	*Peltohyas australis*			11	0.1
Black-fronted Plover	*Charadrius melanops*			44	3.1
Latham's Snipe	*Gallinago hardwickii*			11	0.1
Common Bronzewing	*Phaps chalcoptera*			11	0.2
Brush Bronzewing	*Phaps elegans*	•		56	1.7
Yellow-tailed Black Cockatoo	*Calyptorhynchus funereus*			22	3.6
Galah	*Cacatua roseicapilla*			44	5.0
Long-billed Corella	*Cacatua tenuirostris*			11	2.8
Sulphur-crested Cockatoo	*Cacatua galerita*			11	0.1
Musk Lorikeet	*Glossopsitta concinna*			11	0.4
Crimson Rosella	*Platycercus elegans elegans*	•		89	2.7
Blue-winged Parrot	*Neophema chrysostoma*			33	1.7
Fan-tailed Cuckoo	*Cuculus variolosus*			11	0.1
Horsefield's Bronze-Cuckoo	*Chrysococcyx basalis*	•		22	0.2
Barn Owl	*Tyto alba*	?		11	0.2
Laughing Kookaburra	*Dacelo novaeguineae*			33	0.4
Skylark*	*Alauda arvensis*	•		33	0.7
Welcome Swallow	*Hirundo neoxena*	•		100	62.4
Fairy Martin	*Cecropis ariel*			11	0.4
Richard's Pipit	*Anthus novaeseelandiae*	•		78	3.6
Black-faced Cuckoo-shrike	*Coracina novaehollandiae*			22	0.3
White's Thrush	*Zoothera dauma*			44	2.9
Blackbird*	*Turdus merula*	•		100	11.0
Hooded Robin	*Petroica goodenovii*	•			
Eastern Yellow Robin	*Eopsaltria australis*	•		56	2.1
Olive Whistler	*Pachycephala olivacea*			11	0.1
Golden Whistler	*Pachycephala pectoralis*	•			
Rufous Whistler	*Pachycephala rufiventris*			11	0.3
Grey Shrike-thrush	*Colluricincla harmonica*	•		89	3.8
Grey Fantail	*Rhipidura fuliginosa*	•		100	11.2
Willie Wagtail	*Rhipidura leucophrys*			89	4.1
Clamorous Reed-Warbler	*Acroceophalus stentoreus*			22	0.2
Little Grassbird	*Megalurus gramineus*			56	1.1
Golden-headed Cisticola	*Cisticola exilis*			56	2.6
Superb Fairy-wren	*Malurus cyaneus*	•		100	56.2
Southern Emu-wren	*Stipiturus malachurus*	•		22	0.4
Rufous Bristlebird	*Dasyornis broadbenti*	•	•		
White-browed Scrubwren	*Sericornis frontalis*	•		89	7.0
Calamanthus	*Sericornis fuliginosus*	•			
Brown Thornbill	*Acanthiza pusilla*	•		100	12.9
Yellow-rumped Thornbill	*Acanthiza chrysorrhoa*	•		89	15.0
Striated Thornbill	*Acanthiza lineata*	•		89	8.9
Red Wattlebird	*Anthochaera carunculata*			100	9.4
Little Wattlebird	*Anthochaera chrysoptera*			100	18.9
Yellow-faced Honeyeater	*Lichenostomus chrysops*			78	8.7
Singing Honeyeater	*Lichenostomus virescens*	•		33	0.8
White-eared Honeyeater	*Lichenostomus leucotis*	•		67	2.4
New Holland Honeyeater	*Phylidonyris novaehollandiae*	•		100	40.0
Tawny-crowned Honeyeater	*Phylidonyris melanops*	•			
Eastern Spinebill	*Acanthorhynchus tenuirostris*	•		22	0.3
White-fronted Chat	*Ephthianura albifrons*	•		100	23.8
Spotted Pardalote	*Pardalotus punctatus*			11	0.1
Silvereye	*Zosterops lateralis*	•		78	33.6
European Goldfinch*	*Carduelis carduelis*	•		100	36.1
European Greenfinch*	*Carduelis chloris*			22	0.8
House Sparrow*	*Passer domesticus*	•		89	16.0
Red-browed Firetail	*Emblema temporalis*	•		78	14.6
Common Starling*	*Sturnus vulgaris*	•		100	70.7
White-winged Chough	*Corcorax melanorhamphos*		•		
Australian Magpie-lark	*Grallina cyanoleuca*	•		100	20.2
Dusky Woodswallow	*Artamus cyanopterus*	•		11	0.4
Masked Woodswallow	*Artamus personatus*			11	0.9
Australian Magpie	*Gymnorhina tibicen*	•		100	43.8
Grey Currawong	*Strepera versicolor*	•		67	2.7
Forest Raven	*Corvus tasmanicus*			67	18.6
Little Raven	*Corvus mellori*	•		100	27.1
Species richness		48	3		95

EIS — Pescott and Kinhill (1979) and Alcoa and Kinhill (1980a).
Add. — additional incidental records collected since the beginning of this study.

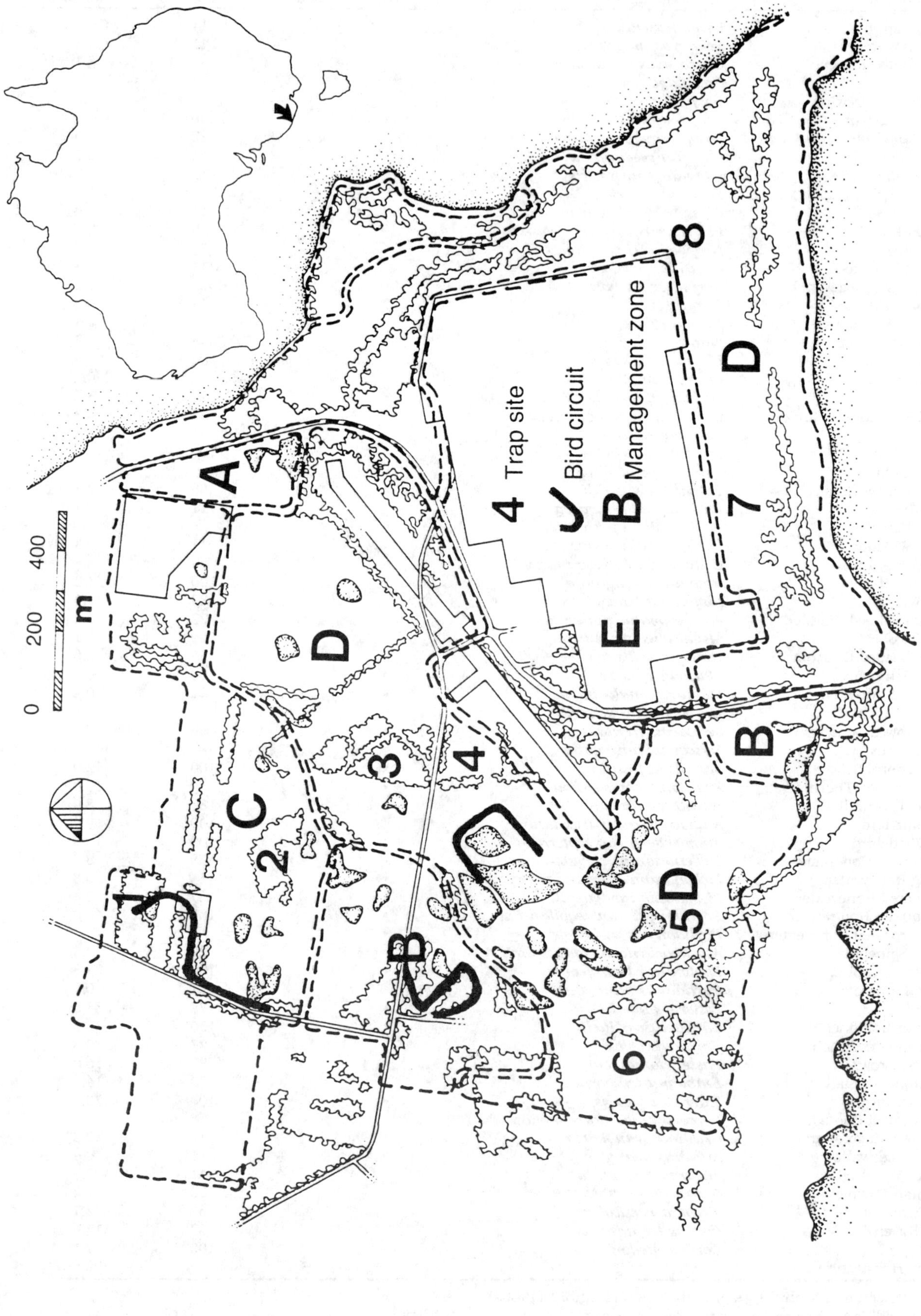

Fig. 1. Map of *The Smelter in the Park* at Portland Aluminium, showing management zones, trap sites and bird counting circuits. Inset shows the location of Portland in southeastern Australia.

2. Assess the effect of the construction and operation of the smelter on the wildlife of the site through a comparison, where possible, with the data available from the Environmental Impact Statement.
3. Establish a repeatable survey programme for the wildlife of Portland Aluminium, so that long-term trends can be detected and the effectiveness of management actions can be evaluated.
4. Identify species of conservation significance, including species that have become locally extinct, and provide guidance for the management of these species in order to maintain and enhance biodiversity.
5. Identify species that have negative impacts on biodiversity, human welfare or plant production, and provide guidance for the management of these species.

STUDY AREA

Portland Aluminium (38°23'S, 141°37'E) is located on the headland at the southern end of Portland Bay and the eastern end of Nelson Bay in south-western Victoria (Fig. 1). The site occupies an area of about 600 ha, of which about 100 ha are covered by the smelter buildings and a water retention pond (Zone E), and another 40 ha are given over to roads, hardstand storage, and waste disposal and transfer infrastructure. Twin 500 kV power lines cross the park from the north-east to a switchyard at the western end of Zone E. Most of the area comprises four of the zones for "The Smelter in the Park". These are bordered by the residential area of South Portland (including a golf course) to the north, and by farmland and coastal reserve to the west.

Zone C consists of farmland broken up by shelterbelts of non-local native trees and shrubs, as well as by small patches of remnant natural vegetation. A herd of beef cattle is maintained in this zone to monitor fluoride accumulation as part of the Environment Protection Authority's licensing requirements. Zone D encompasses areas of retired farmland, patches of native woodland, a partly revegetated heathland and, on the southern and western side of the smelter, there is a narrow band of remnant coastal heathland and shrubland contiguous with a more extensive coastal reserve around Nelson Bay to the west. Significantly, there are also approximately 50 small wetlands scattered throughout the park, mainly in Zones C and D. Most of these wetlands are ephemeral, but several have been artificially impounded or deepened so that they hold water permanently.

SURVEY METHODS

Mammals, reptiles and amphibians

We selected eight sites, sampling a range of habitats, for a systematic survey of the mammals, reptiles and amphibian within the park (Fig. 1).

1. *Predator-proof Enclosure* — This 4.5 ha enclosure, adjacent to the Salmon House Nursery in Zone C, was created by the erection of a 2 m high predator-proof fence in 1993. The fence was constructed of wire netting supported by plain wire fixed to treated pine posts, topped with a horizontal section of seven alternate live and earth wires. A stand-off live wire was also installed 200 mm above ground level on the outside, but was later disconnected from the energiser. Vegetation in the enclosure comprised a woodland of non-local eucalypts and acacias planted in 1980, together with areas of former pasture. Extensive patches of Gorse *Ulex europaeus* were removed from the understorey prior to trapping. The trapping grid was placed at the eastern end of the enclosure in an area of woodland with an open, grassy understorey.
2. *Farmland Shelterbelt* — This 4 ha woodland of non-local eucalypts and acacias was also planted in 1980. It is located roughly in the centre of Zone C and is surrounded by open pasture. The shelterbelt is fenced off from domestic stock, and has a dense understorey of Austral Bracken *Pteridium esculentum* and various pasture grasses. Part of the shelterbelt is used for storage of hay and fencing materials.
3. *Farmland-Laydown Area* — This triangular, 3 ha remnant of natural woodland and shrubland is in Zone D, on the north side of Madiera Packet Road. The dominant species is Shining Peppermint *Eucalyptus nitens*. There is a dense, shrubby understorey of Coast Wattle *Acacia sophorae* in some areas, and low heath in others. The remnant was last burnt about 20 years earlier but was occasionally disturbed by cattle. The trapping grid was in the eastern corner of the remnant, encompassing understoreys of both Coast Wattle and heath.
4. *Airport Heath* — This 3.5 ha area of dry heathland bordered by Coast Wattle is on the south side of Madiera Road, opposite Site 3. The area had been slashed every year prior to 1991, and part of it was burnt in the autumn of 1993. Invading Gorse has been sprayed regularly with herbicide. The trapping grid encompassed equal amounts of burnt and unburnt areas of heath.
5. *Heathland Revegetation Area* — In late 1980 and early 1981 the topsoil stockpiled from the smelter site was spread on to this 16 ha area of former pasture, roughly in the centre of Zone D, in an attempt to establish a natural heathland. The vegetation that has developed is a patchy, species-poor heathland with a dense herb cover in many places (G. Carr, pers. comm.). The trapping grid was positioned in the northwestern corner of the revegetation area.
6. *Dean's Heath* — Dean's Heath is a relatively extensive area of dry heath in Zone D, on the western side of the park. The heath is fairly uniform in maturity and has not been burnt for about 25 years. The trapping grid was located at the northern end of Dean's Heath, on a ridge above Thistle Road.

7. *Coastal Heathland I* — This wet heathland is situated on the southern boundary of the smelter complex in Zone D, at the western end of the narrow strip of remnant coastal vegetation. The heathland is quite tall (>1 m) and is dominated by tea-trees *Leptospermum* sp. and Rapier Saw Sedge *Gahnia trifida*, having numerous small depressions which hold water during the wetter months. The site has not been burnt for over 20 years. The trapping grid was located on the southwestern corner of the fence surrounding the smelter.
8. *Coastal Heathland II* — This dry heathland is situated at the western end of the same strip of coastal vegetation as was site 7. The heath is very low and dense. It was last disturbed about 15 years before construction of the smelter, when the heath was cleared and windrowed with a bulldozer, then burnt. The trapping grid was located on the southeastern corner of the fence surrounding the smelter.

Traps were located on a standardized grid (Fig. 2) permanently marked with steel posts. Each grid was divided into quarters by two lines of pitfall traps arranged in a cross configuration. Each arm of the cross consisted of nine pitfall traps (11 L metal drums) buried to ground level at approximately 5 m intervals, with a drift fence (insect mesh 20 cm high) running along the line of traps and extending 5 m further at each end. The grid in each quarter had 12 trap points marked by a wooden stake, 5 m apart in a 3 × 4 configuration. A total of 48 collapsible aluminium "Elliott" traps was set at each trapping grid, one trap placed at each trap point, allowing some flexibility (± 2 m) in trap placement to maximize trap success. Three wire cage traps were also placed within each quarter at favourable trap points.

The surveys were conducted every four weeks between November 1993 and March 1994. The sites were surveyed in pairs for four consecutive nights. All traps were checked and reset each morning and evening so that no animals remained in a trap for more than 12 h. Each evening a small amount of water was added to each pitfall to prevent dehydration of amphibians, and any excess rain water was sponged out in the morning to prevent drowning of animals captured during the day. The cage traps and "Elliott" traps were baited with a standard mixture of rolled oats, peanut-butter and honey. All animals were identified, sexed and weighed before release close to the point of capture.

Three spotlighting surveys were also conducted at each site, providing an opportunity to detect nocturnal species that were not trapped. Two people, each with a 25 W spotlight, walked at a steady pace searching all vegetation strata. The time spent spotlighting at each site varied due to differences in vegetation density. In addition, two harp traps were set up over a number of nights at a number of favourable positions within each site to capture bats in flight.

Birds

Quarterly bird surveys began in 1992. The data reported here are from the first nine surveys, from September 1992 to October 1994 (inclusive), to

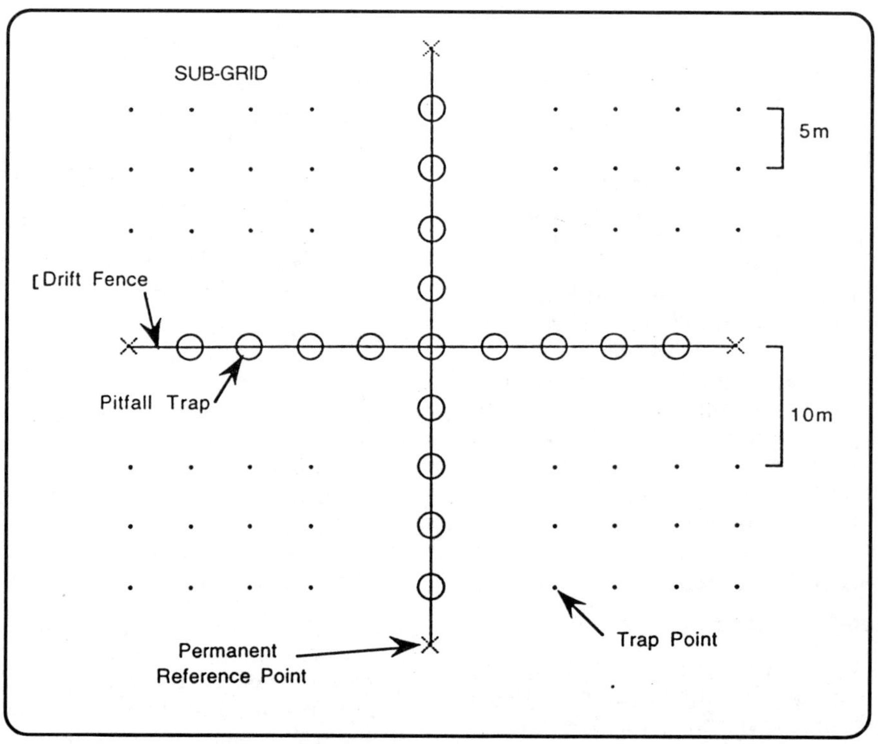

Fig. 2. Layout of the standard trapping grid used to survey amphibians, reptiles and terrestrial mammals at Portland Aluminium.

encompass the time of the survey for mammals, reptiles and amphibians. On each bird survey, a team of 4–6 experienced people walks along four standard circuits, recording the number of individuals of each bird species detected. The surveys are conducted in two parts, in the morning and evening of one day or on consecutive days. The circuits are placed specifically to survey the larger wetlands, but also sample some replanted areas and some patches of remnant native vegetation (Fig. 1).

Circuit 1 is in Zone C, in the north-east corner of the park. It includes a large, open wetland, belts of roadside plantings and the predator proof enclosure. Circuit 2, in Zone B, includes a diversity of wetlands. It also samples an area of remnant native woodland that is being developed mainly for barbecues and other recreational activities. Circuit 3 loops around an artificial wetland within the bund around a former waste disposal site in Zone D. This site was only partly filled and now operates as a waste transfer station (Hill *et al.*, this volume). Circuit 4 is adjacent to the smelter complex in the southwestern corner of Zone E. The circuit overlooks a deep settling pond which is used to treat process water before release into a small creek.

SURVEY RESULTS

A total of 134 vertebrate species were detected at Portland Aluminium during the study period: 22 mammals, eight reptiles, six amphibians and 98 birds, excluding seabirds (Tables 1 and 2). Our study was not designed to investigate any effects that electromagnetic radiation or fluoride emissions from the smelter might have on animal health (e.g., Boulton and Cooke 1994) or behaviour (e.g., Ganskopp *et al.* 1991), or on community composition (e.g., Letnic and Fox 1997). However, the high species richness, together with evidence that many species are breeding in the park, suggests that the smelting operation is having little negative effect on the wildlife in the "Smelter in the Park".

Prior to this study, the known vertebrate species richness was 80. We have added eight mammals, six reptiles, one amphibian and 49 birds to the species list (Tables 1 and 2). Three of these, the Great Egret *Egretta alba*, the Brolga *Grus rubicunda*, and the Common Bent-wing Bat *Miniopterus schreibersii*, are classified as threatened taxa in Victoria (Conservation and Natural Resources 1995). Eleven species that had been recorded previously were not detected in this study, but none is listed as threatened.

Most new bird records were obtained during the regular quarterly surveys, but we gained two species from incidental sightings (Table 2). A group of White-winged Choughs *Corcorax melanorhamphos* was seen occasionally on the northern side of Zone C, and a Brolga was observed descending to the creek in Zone D, to the south of the retention pond, in March 1994.

Most of the mammals, reptiles and amphibians were detected by trapping or sighting during the standardized surveys of the eight sites (Table 1). Three new bat species were recorded by other means (Table 1). We detected the White-striped Freetail Bat *Tadarida australis*, which is a high-flying species rarely caught in traps (Lumsden and Bennett 1995), by its distinctive, audible call at two sites. In January 1995 we made a more intensive effort to identify bats flying around floodlights at the nursery (adjacent to the predator-proof enclosure), using mist nets and additional harp traps, and captured some Common Bent-wing Bats. The Little Red Flying-Fox *Pteropus scapulatus* was identified from a recent photograph taken of an individual hanging from power lines.

Three other species might have been expected to be trapped, but were recorded in other ways (Table 1). A Swamp Antechinus was found as a roadkill on an unsealed road through the coastal heathland and shrubland to the south of the smelter. The Southern Smooth Froglet *Geocrinea laevis* was heard calling from a drain in Zone C in January 1993 (M. Littlejohn, pers. comm.). A small snake, which was probably a White-lipped Snake *Drysdalia coronoides*, was disturbed in the soil while the drift net was being installed at Dean's Heath, but escaped before a positive identification could be made.

We also obtained incidental records of two new macropod species (Table 1). Red-necked Wallabies *Macropus rufogriseus* are seen at times in clearings to the north of Dean's Heath. In 1991, an attempt was made to reinforce this population with surplus wallabies from a fauna park in Portland. Four (two of each sex) were released in the revegetation area with radio-collars, but two were found dead soon after and the others disappeared when their transmitters failed within a month. Two males were introduced to the predator-proof enclosure in March 1994 and were still present a year later. Swamp (or Black) Wallabies *Wallabia bicolor* have been sighted near the northern and southern boundaries of the park several times. This species is apparently expanding its range in southwestern Victoria (Bird 1992), and there have been recent sightings in coastal shrubland to west of Portland (C. Cooper, pers. comm.).

Spatial and temporal patterns

Amongst the mammals, reptiles and amphibians, some species were widespread and common, particularly the Echidna *Tachyglossus aculeatus*, Red Fox *Vulpes vulpes*, Grass Skink *Pseudemoia entrecasteauxi*, Southern Brown Tree Frog *Litoria ewingi* and Pobblebonk *Limnodynastes dumerilii*, which were recorded at seven of the eight trapping sites (Table 1). Another four species, the Swamp Rat *Rattus lutreolus*, Eastern Tiger Snake *Notechis scutatus*, Striped Marsh Frog *Limnodynastes peroni* and Common Froglet *Ranidella signifera* were recorded at six sites.

Species richness at the trapping sites ranged from 8 to 18 (Table 1). The faunal composition, particularly of reptiles and small mammals, also differed at these trapping sites, and was consistent with the different types of habitats at the sites. For example, Site 8 *(Coastal Heathland II)*, has not been disturbed for about 30 years and supported a typical heathland fauna in terms of small mammals and reptiles, but is apparently too dry to offer suitable habitat to any more than one species of frog. By contrast, the wetter heathland site (Site 7) had three species of amphibians. Of all the sites, Site 1 *(Predator-proof Enclosure)* has perhaps experienced the most disturbance since construction of the smelter began and lacks sufficient ground cover to support many small mammals, but has a good representation of skinks, snakes and frogs.

The bird surveys show that some species, such as the Pacific Black Duck *Anas superciliosus* and Common Starling *Sturnus vulgaris*, occur at high abundance year round. Others, such as the Hoary-headed Grebe *Poliocephalus poliocephalus* and Chestnut Teal *Anas castanea* are far less common but are also consistently present. Others occur less often, either because they are seasonal migrants, like the Blue-winged Parrot *Neophema chrysostoma*, or occasional visitors like the Pacific Heron *Ardea pacifica*.

Comparisons with the Environmental Impact Statement

The results of the Environmental Impact Statement are not directly comparable with the findings of the present study, since the surveys differed in spatial and temporal coverage. In spatial terms, the surveys conducted for the Environmental Impact Statement (Pescott and Kinhill 1979; Robertson and Kinhill 1980) were restricted to areas of coastal heathland and shrubland. Roughly half of this habitat was subsumed into the smelter complex (Zone E), while the remainder was retained, forming part of Zone D. By contrast, the surveys for mammals, reptiles and amphibians in the present study covered the entire park and sampled a greater range of habitats, including remnant native woodland, artificial shelterbelts and wetlands. Only Sites 6, 7 and 8 (Dean's Heath and Coastal Heathland I and II) sampled part of the area surveyed for the Environmental Impact Statement. Our records of common species such as the Koala *Phascolarctos cinereus*, Eastern Blue-tongued Lizard *Tiliqua scincoides* and Striped Marsh Frog *Limnodynastes peroni* undoubtedly reflect this wider coverage of habitats.

Our bird surveys also covered a wide range of habitats, concentrating on the wetlands, but did not sample coastal heathland and shrubland. Our new records of many waterbird species (e.g., grebes, ibis, ducks and crakes) correspond with the wider habitat coverage, as do the new records of common passerines that are associated with farmlands or wetlands, such as the Willie Wagtail *Rhipidura leucophrys* and Little Grassbird *Megalurus gramineus*. Our inability to detect the Hooded Robin *Petroica goodenovii*, Calamanthus *Sericornis fuliginosus* and Tawny-crowned Honeyeater *Philidonyris melanops* may be because the coastal habitats were not surveyed so these species were unlikely to be detected. The Calamanthus has subsequently been recorded once on a bird survey (February 1997). However, it is possible that some species have become locally extinct since construction of the smelter. The Tawny-crowned Honeyeater is a seasonal migrant to coastal areas in Victoria (Emison *et al.* 1987) and would probably continue to visit the park, but the Hooded Robin is a sedentary species and has undergone extinction in many parts of its range as a result of habitat loss and fragmentation (Robinson and Traill 1996).

The timing of the Environmental Impact Statement surveys also differed from ours. Environmental Impact Statement work is typically conducted under severe time constraints (Fairweather 1994), and the surveys for the proposed Portland smelter were no exception. The first survey (Pescott and Kinhill 1979) was conducted in winter, when the level of activity and detectability of some taxa, particularly bats and reptiles, would have been at a minimum. This methodological limitation was acknowledged in the Environmental Impact Statement report (Alcoa and Kinhill 1980a). The flaw was partly addressed by the supplementary survey for the Mourning Skink, conducted the following spring, which yielded a number of additional reptiles and amphibians. Our survey of mammals, reptiles and amphibians was conducted at the optimal time (summer). We trapped three new bat species in the park as a whole, and four new reptile species in the coastal heathland and shrubland sites (Sites 6, 7 and 8). Our bird surveys have been conducted quarterly so that mobile species are much more likely to be detected. For example, the Australasian Shoveler *Anas rhynchotis*, which is a mobile species in dry periods (Emison *et al.* 1987), and the Blue-winged Parrot, which is a seasonal migrant (Emison *et al.* 1987), were not recorded in the Environmental Impact Statement survey.

WILDLIFE MANAGEMENT

Threatened species

Two of the threatened species, the Great Egret and the Brolga, are apparently only occasional visitors to Portland Aluminium, using the wetlands in the park as a component of the broader landscape. Similarly, the Common Bent-wing Bats probably roost in caves at Bats Ridge State Faunal Reserve, 15 km to the west, and include the park in their foraging range. Conservation of this species depends critically on adequate protection of such roosting caves, and of large breeding caves such as the one near Narracoorte (South Australia) to the west of Portland and another near Warrnambool to the east (Menkhorst and Lumsden 1995).

The Heath Rat, Swamp Antechinus, Southern Emu-wren and Rufous Bristlebird are likely to be resident species, and therefore more influenced by the extent and quality of habitat within the park. All four species are habitat specialists, occurring almost exclusively in heathland and shrubland (Wainer and Gibson 1976; Cockburn *et al.* 1981; Emison *et al.* 1987). The most significant issue for the management of coastal heathland and shrubland in the Portland area is the interaction between reduced fire frequency and invasion of these habitats by Coast Wattle *Acacia sophorae* (McMahon *et al.* 1995). A series of small-scale burns have been conducted along the coastal strip to the south and east of the smelter, including Sites 7 and 8, and the effects on the vertebrate fauna will be assessed when these sites are surveyed again.

As expected, the Heath Rat was recorded in two heathland sites (Dean's Heath and Coastal Heathland II), but four individuals were also captured in the Heathland Revegetation Area (Site 7). Although this revegetated site is floristically depauperate, it clearly provides sufficient habitat for this species. The site is close to Dean's Heath (Site 6), which is the likely source of colonizers. The capture of five Striped Worm-lizards *Aprasia striolata* at Dean's Heath was also noteworthy. This is a burrowing species occurring in disjunct populations mainly along the southern coast of Australia (Cogger 1992). It is classified as "rare" in Victoria (Conservation and Natural Resources 1995), and Portland is apparently at the eastern limit of its range.

Missing species

One species that apparently no longer occurs in the park is the Growling Grass Frog. This species has undergone local extinctions in many parts of its range, and many of these extinction events apparently occurred during the 1982/83 drought (Osborne *et al.* 1996), although other factors such as an exotic viral disease (e.g., Laurance *et al.* 1996) may also have been involved. Growling Grass Frogs were recorded during the initial survey of the site in 1979, so it is possible that drought during the early stages of construction of the smelter was at least partly responsible for the loss of the species. The Growling Grass Frog is an ideal candidate for an experimental reintroduction to the park, since there is now ample wetland habitat and abundant prey in the form of other amphibian species.

A second species, which was not detected in the Environmental Impact Statement survey or in any subsequent survey, is the Long-nosed Potoroo *Potorous tridactylus*. This species occurs in six distinct population clusters in Victoria, including the south-west of the state near Portland (Seebeck 1981). Potoroos were last recorded in Portland South, about 1 km north of the park, in 1935 (J. Seebeck, pers. comm.). In an attempt to locate a possible founder colony for reintroduction to Portland Aluminium, we have conducted trapping surveys at a number of sites where potoroos have been reported and where there is still suitable habitat (e.g., Bennett 1993). Only two populations have been located within 15 km of Portland, and both appear to support only low densities of potoroos. While more distant founder populations are under investigation, the predator-proof enclosure (Site 1) is being re-vegetated to provide appropriate habitat for potoroos in preparation for an experimental reintroduction project.

Pest species

Six mammal species and six bird species are introduced, and all but one, the Brown Hare *Lepus capensis*, have been present in the park since the Environmental Impact Statement was conducted. Several introduced birds and mammals have pest status at Portland Aluminium. In particular, the Red Fox and Feral Cat *Felis catus* are likely to have a significant impact as predators on native birds and mammals. Sporadic control measures for foxes and rabbits were commenced in the buffer zone in 1984 with only intermittent success. In 1994 a Pest Animal Control Programme was prepared (Walters 1994) specifically for the control of foxes, rabbits and cats.

Control after 1994 relied in part on the written programme, which incorporated detailed prescriptions for systematic and staged actions, and prescriptions for recording and examining responses by the target species. Success was achieved through the training and dedication of nominated staff to aspects of the programme. Our demonstrable success within the park two years after commencement of the revitalized programme was the incentive for neighbouring landowners to join control efforts. Treatment now extends up to 10 km beyond the park.

Fox control was primarily through the buried-bait technique using 1080 poison in a commercially-available carrier (FOXOFF®), and occasionally with prepared meat baits to reduce the likelihood of bait wariness. Bait trails were selected strategically on the basis of known fox behaviour patterns. Continuous low-level deployment of non-lethal baits along the bait-trails was used as an index of fox activity. Cats were trapped initially with a high degree of success using a "bird-wing" lure at the trap site. Because of the solitary nature of cats, control is ongoing and action is on a sighting basis. Continuous vigilance is required.

This programme has been very successful, and sightings of foxes are now rare. Feral Cats also appear to be less common. There is strong evidence that some native species have responded rapidly to control of these exotic predators. Sightings of Short-nosed Bandicoots *Isoodon obesulus*, in particular, are now commonplace and widespread, whereas this species had previously been seen only occasionally and it was trapped at only three sites. The only colony of Australasian Gannetts *Morus serrator* on the Australian mainland occurs on Point Danger, less than 1 km south-east of the smelter, and the rapid expansion of this colony has also been attributed, in part, to the fox control programme (Norman *et al.* 1998).

European Rabbits are also controlled by poisoning in selected areas. Standardized monitoring lines were established, and an index of rabbit sightings was regularly recorded. Control action recommenced when the index indicated a rabbit density above a threshold of 5 rabbits per 20 ha. Sites with frequent high indices were examined for potential site modification or deployment of the anti-coagulant Pindone® on a localized basis. Poison is delivered at bait stations that offer security against non-target species. The successful control of rabbits has reduced grazing pressure, and rabbits are now seen much less often and are restricted to a few small patches. As with foxes, control did not rely on a single method, but integrated a number of control techniques designed to exploit key aspects in the target species' life strategies.

Overabundant species

A number of species have responded to habitat enhancement, either by increasing in population size or by utilizing microhabitats created in and around the Smelter. Several of these species have impinged in some way on the smelter activities or on the staff. As each of these overabundant species is identified, consultant expertise is sought and a written control strategy is formulated. So far, strategies for cohabiting with snakes, Masked Lapwings *Vanellus miles*, Australian Magpies *Gymnorhina tibicen* and Koalas have been written.

Australian Magpies have a tendency to construct nests amongst the 22 kV switchracks that regulate power to the smelter plant. Nesting material has caused flashovers, resulting in significant loss of power at times. This problem was addressed by modifying the surrounding habitat. Nearby lawns were converted to low shrubs to reduce the suitability of the area for foraging, and alternative feeding sites were provided further afield. Koalas have also entered the switchyard and threatened the power supply. Vegetation has been cleared around the existing security fence to reduce koala movement. Structural modifications to the fence are planned if another breach occurs.

Masked Lapwings *Vanellus miles* have caused concern through aerial attacks on Portland Aluminium staff. Although there have been no injuries reported, the conspicuous "bombing" behaviour of the birds causes anxiety as staff move to and from their workplaces. This problem has been addressed by instructing staff on the breeding behaviour and aggressive signals of lapwings so that the bird's responses can be better predicted. This informative approach is coupled with the careful relocation of specific nests to positions where less interaction with people will occur.

The population of Eastern Grey Kangaroos *Macropus giganteus* has increased from a single individual sighted during the Environmental Impact Statement survey in 1979 (Alcoa and Kinhill (1980a) to more than 100 in 1997. Kangaroos now occur in all management zones, but are particularly numerous in Zones C and D. Grey Kangaroos can attain high densities in nature reserves, such as "The Smelter in the Park", and can cause a range of management problems (Coulson *et al.,* this volume). There is concern that kangaroos are jeopardizing revegetation work and neutralizing the effectiveness of rabbit control in the park, and collisions with vehicles on the main access road through the park are increasing, although no serious injury has yet occurred to humans. Portland Aluminium is firmly opposed to lethal control of kangaroos, and instead has entered a collaborative arrangement with the University of Melbourne to conduct an experimental fertility control programme, using hormonal implants, beginning in 1998.

ACKNOWLEDGEMENTS

Many people assisted with trapping and spotlighting: Harry Behncke, Juliana Bongiorno, Dave Brown, Nick Gust, James Harris, Leona Knight and Andrew Pritchard. Many also helped with the bird surveys: Ivor Graney, Ruth Graney, Peter Manly, Lois Phillips, Max Phillips and Andrew Pritchard. A number of people contributed observations and ideas: Cliff Beauglehole, Andrew Bennett, Charlie Cooper, Penny Fannin, Murray Littlejohn, Lindy Lumsden, Ken O'Connor, Andrew Pritchard and John Seebeck. We are particularly grateful to Lindy Lumsden for extra help with the bats. Trapping was conducted with the approval of the University of Melbourne Animal Experimentation Ethics Committee (register number 654 108 0 93 1469), and under research permit (number RP-94-187) issued by the Department of Conservation and Natural Resources.

REFERENCES

Alcoa and Kinhill, 1980a. Alcoa Portland Aluminium Smelter. Environmental Effects Statement. Alcoa of Australia and Kinhill Planners: Melbourne, Victoria.

Alcoa and Kinhill, 1980b. Alcoa Portland Aluminium Smelter. Supplement to Environmental Impact Statement. Alcoa of Australia and Kinhill Planners: Melbourne, Victoria.

Bennett, A. F., 1993. Microhabitat use by the Long-nosed Potoroo, *Potorous tridactylus,* and other small mammals in remnant forest vegetation of south-western Victoria. *Aust. Wildl. Res.* **20:** 267–85.

Bird, P. R., 1992. Expansion of the range of the black wallaby in western Victoria. *Vic. Nat.* **109:** 89–91.

Boulton, I. C. and Cooke, J. A., 1994. Experimental fluoride accumulation and toxicity in the short-tailed field vole *(Microtus agrestis). J. Zool., Lond.* **234:** 409–21.

Brake, L., Clark, I., Clay, R. and Savarton, F., 1991. The Master Plan for the Smelter in the Park. Centre for Conservation and Recreation Resources, University of South Australia, and Franek Savarton and Associates.

Conservation and Natural Resources, 1995. Threatened fauna in Victoria — 1995. Department of Conservation and Natural Resources: Victoria.

Cockburn, A., Braithwaite, R. W. and Lee, A. K., 1981. The response of the heath rat, *Pseudomys shortridgei*, to pyric succession: a temporally dynamic life-history strategy. *J. Anim. Ecol.* **50:** 649–66.

Cogger, H. G., 1992. Reptiles and Amphibians of Australia. Reed Books: Frenchs Forest, New South Wales.

Emison, W. B., Beardsell, C. M., Norman, F. I., Loyn, R. H. and Bennett, S. C., 1987. Atlas of Victorian Birds. Department of Conservation, Forests and Lands, and the Royal Australasian Ornithologists Union: Melbourne, Victoria.

Fairweather, P. G., 1994. Improving the use of science in environmental assessments. *Aust. Zool.* **29:** 217–23.

Ganskopp, D., Raleigh, R., Schott, M. and Bracken, T. D., 1991. Behavior of cattle in pens exposed to ± 500 kV DC transmission lines. *Appl. Anim. Behav. Sci.* **30:** 1–16.

Laurance, W. F., McDonald, K. R. and Speer, R., 1996. Epidemic disease and the catastrophic decline of Australian rainforest frogs. *Cons. Biol.* **10:** 406–13.

Letnic, M. I. and Fox, B. J., 1977. The impact of industrial fluoride fallout on faunal succession following sand mining of dry sclerophyll forest at Tomago, N.S.W. — I. Lizard recolonisation. *Biol. Cons.* **80:** 63–81.

Lumsden, L. F. and Bennett, A. F., 1995. White-striped Freetail Bat. Pp. 168–69 *in* Mammal of Victoria. Distribution, Ecology and Conservation ed by P. W. Menkhorst. Oxford University Press: Melbourne, Victoria.

McMahon, A. R. G., Carr, G. W., Bedggod, S. E., Hill, R. J. and Pritchard, A. M., 1995. Prescribed fire and control of coast wattle (*Acacia sophorae* (Labill.) R. Br.) invasion in coastal heath south-west Victoria. *In* Fire and biodiversity. The effects and effectiveness of management. Biodiversity Series, Paper No. 8. Department of Environment Sport and Territories: Canberra, Australian Capital Territory.

Menkhorst, P. W. and Lumsden, L. F., 1995. Common Bent-wing Bat. Pp. 180–81 *in* Mammal of Victoria. Distribution, Ecology and Conservation ed by P. W. Menkhorst. Oxford University Press: Melbourne, Victoria.

Norman, F. I., Minton, C. D. T., Bunce, A. and Govanstone, A. P., 1998. Recent changes in the status of Australasian Gannetts *Morus serrator* in Victoria. *Emu* **98:** 147–50.

Osborne, W. S., Littlejohn, M. J. and Thompson, S. A., 1996. Former distribution and apparent disappearance of the *Litoria aurea* complex from the Southern Tablelands of New South Wales and the Australian Capital Territory. *Aust. Zool.* **30:** 190–98.

Pescott, T. and Kinhill Planners Pty Ltd, 1979. Terrestrial Vertebrate Fauna. Alcoa Portland Aluminium Smelter, Working Paper No. 13 (unpublished).

Robertson, P. and Kinhill Planners Pty Ltd, 1980. Mourning Skink Survey. Alcoa Portland Aluminium Smelter, Environmental Studies Report No. 1 (unpublished).

Robinson, D. and Traill, B. J., 1996. Conserving woodland birds in the wheat and sheep belts of Southern Australia. RAOU Conservation Statement No. 10, Hawthorn, Victoria.

Seebeck, J. H., 1981. *Potorous tridactylus* (Kerr) (Marsupialia: Macropodidae): distribution, status and habitat preferences in Victoria. *Aust. Wildl. Res.* **8:** 285–306.

Wainer, J. W. and Gibson, R. J., 1976. Habitat of the swamp antechinus in Victoria. *Vic. Nat.* **93:** 235–55.

Walters, B., 1994. A pest animal control programme for the Smelter in the Park. Unpublished report by Ecoplan Australia Pty Ltd, Hurstbridge, Victoria.

CHAPTER 37

Developing social and biological indicators of a sustainable viticultural landscape — case study the Limestone Coast Viticultural Zone in the south-east of South Australia. Part One: fundamental concepts and methodologies

Cecil Camilleri,[1] Jim Birckhead,[2] Johannes Bauer[3] and Brian Freeman[4]

The Australian wine industry has embarked upon a bold strategic business plan based on vineyard expansion and the marketing of Australia's unique cultural and natural heritage as a salient feature of wine quality. To cultivate this image the industry must work within the context of the Total Quality Environmental Management paradigm and the guiding principles of ISO 9000 and ISO 14000.

The wine industry and Charles Sturt University are collaborating on an initiative to identify social and biological indicators of sustainable production and the identification of synergisms between viticultural activities and biodiversity residing within vineyards. This adaptive management approach rests on the examination of baseline parameters of biodiversity, ecological sustainability and environmental attitudes.

INTRODUCTION — THE INTEGRATION OF SCIENCE AND ART

"At the broadest level the community is faced with decisions about whether land should be used for agriculture, grazing, irrigated agriculture or horticulture, forestry, urban and industrial purposes, preservation of flora and fauna for scientific purposes, recreation or open space, national parks and wildlife reserves. In modern society these forms of land use are competing. Some uses are incompatible and then decisions must be made as to which uses should receive priority. . . . Most economic uses of land preclude the preservation of native plant and animal communities for scientific purposes. Different types of land have different potentials for various uses". (Downes 1971)

THE south-east of South Australia is arguably the most "fertile" area in the state of South Australia. This area, which is arbitrarily defined by virtue of the position it occupies in the southeastern corner of the state, is dominated by a landscape typified by a system of dune ridges separated by flats with the

[1]"Yalumba", P.O. Box 10, Angaston, South Australia, Australia 5353.
[2]Johnstone Centre for Parks, Recreation and Heritage, Charles Sturt University, P.O. Box 789, Albury, New South Wales, Australia 2640.
[3]Johnstone Centre for Parks, Recreation and Heritage, Charles Sturt University, Panorama Avenue, Bathurst, New South Wales, Australia 2795.
[4]National Wine and Grape Industry Centre, Charles Sturt University, P.O. Box 588, Wagga Wagga, New South Wales, Australia 2678.
Pages 372–84 *in* NATURE CONSERVATION 5: NATURE CONSERVATION IN PRODUCTION ENVIRONMENTS: MANAGING THE MATRIX ed by J. L. Craig, N. Mitchell and D. A. Saunders. Surrey Beatty & Sons, 2000.

occasional ancient granite outcrop (Twidale 1983). Volcanic activity contributed significantly to the recent geology of the area. Evidence of Aboriginal activity on the sand dunes dates back to about 8 000 years ago (Cann *et al.* 1991). Whilst Aboriginals exploited the flora, fauna and mineralogy of the area, Caucasian occupation has been more intrusive: the flats have been drained, non-native plants and animals introduced and minerals added to the soils. As a result it is estimated that today only 7.3% of the original indigenous vegetation survives (South Eastern Drainage Board 1980).

In the south-east of South Australia, broad-acre cropping, pastoralism, forestry and tourism are major economic activities. This area, which includes the viticultural zone of the Limestone Coast (Australian Wine and Brandy Corporation 1993; Darling 1997), also boasts a vibrant viticultural industry centred on the townships of Coonawarra and Padthaway. In the wake of Vision 2025 (Australian Wine Foundation 1996), the wine industry's strategic plan for the next decade, this enterprising industry is opening up new viticultural areas around the locations of Robe, Cape Jaffa, Kingston and Wrattonbully.

Cultural landscapes may be defined as "*. . . areas where human activity has modified the original wild habitats and has created a new, and often very complex, mixture of natural features and human artefacts. . . .*" (Dower 1995). M. Dower, Director General of Britain's Countryside Commission explains that,

> "*. . . the flux of human activity over many centuries has caused changes in the landscape. The result of this great human saga is an immense variety of landscapes, ranging from the olive groves of the Mediterranean to high Alpine meadows, from vineyard terraces to the bocage of France and the green fields of Ireland. . . . Landscapes are important because they enrich the diversity of nature and the quality of human life . . . so we gain from the variety of human culture as expressed in landscape. Also, [landscapes] provide, for those who live within them, a sense of continuity, of human 'roots' in a place and in a tradition, and of a local distinctness. . . . Finally, they are a rich source of knowledge about past and present human interactions with land*" (Dower 1995).

The traditional wine industry of the Old World is characterized by a high level of historical mystique, a vision of "oneness" with nature! On the other hand, New World oenology and viticulture is often equated with large capital investments and intensive technology, the presumed prerequisites of landscape degradation. Therefore, modern viticultural landscapes, as cultural landscapes, may be thought of as being *depleted* of biodiversity. Landscape Ecology and Restoration Ecology, being two sides of the same coin, provide the necessary epistemology to explore this hypothesis (Naveh and Lieberman 1994).

However, the Australian viticultural landscape, or *terroir*, remains to a large extent undefined. Indeed the very concept of *terroir*, that much abused, misunderstood and mistranslated French word, is a tautological term that embodies the holistic reality captured by a glass of wine! Put another way,

Wine = f (vine genotypic and phenotypic factors, vineyard physiognomy, edaphic factors, meteorological factors, geographic factors, anthropological factors).

The anthropological factors include culture, heritage and technology, as well as organoleptic and olfactory preferences.

The high level of historical mystique, capital investment and technical know-how is a complex bio-, socio-economic mix with potential multiplier effects on the *Total Human Ecosystem*. The oenological and viticultural industry may potentially act as a local and national catalyst in the conservation of the ecosphere. A strategic and cumulative impact assessment may reveal a positive prognosis for biodiversity in those geographical areas with forward and backward linkages with the wine industry.

International competitiveness has motivated the Australian wine industry to embrace the concept of *Total Quality Management*. *Total Quality Management* is not a management tool but a holistic management philosophy based on a corporate culture motivated by the paradigm of *Quality*. If paradigms reflect underlying patterns, basic constructs, or fundamental ways of looking at the world, then *Total Quality Management* is based on management values that reflect *quality of life*. Profit is, in this instance, the product of this culture of quality. In terms of marketing management the corporate body identifies and targets the market niche it can supply with goods and services of a quality standard peculiar to that market segment. Perhaps more importantly, the firm identifies itself with the market segment that subscribes to the same quality values it adheres to (e.g., ISO 9000).

Furthermore, the age of consumerism also demands accountability — in particular with regards to resource use and other environmental ethical obligations (Thomas 1995). To be successful a culture of *Quality* requires a corporate environment based on probity, integrity, reverence and receptivity, especially when dealing with environmental issues. Humanity may therefore be thought of having reached the third stage in human development. The proceeding agricultural and industrial revolutions have been superseded by the *Environmental Revolution*, an ideological adaptation which will usher a new era of cultural sensitivity to the open and cultural landscapes — a concept integral to the study of *Total Ecodiversity* (defined as biological, ecological and cultural diversity). As a result, the *Total Quality Management* paradigm has now been replaced by Total Quality Environmental Management, a concept coined in 1991 by the business group Global Environmental Management Initiative to simultaneously address the concepts of environmental management and *total quality* (Global Environmental Management Initiative 1993).

Total Ecodiversity identifies human society as the missing link in the environmental equation. Mankind is not only identified with *Homo sapiens*, that is, a living entity belonging to a taxonomic grouping, but is also considered as a creature who has transcended the constraints of the biosphere and entered the noospheric realm of self-contemplation.

The challenge which faces humanity is thus to address the problem of neotechnological landscape degradation. In the words of Z. Naveh (1995),

> "... *larger areas of diverse natural and cultural landscape are converted within shorter time periods into vulnerable monotonous agricultural steppes and man-made technological deserts. Thereby, our open landscapes are losing not only their biological richness and ecological stability, but also their cultural wealth and scenic beauty*".

Studies undertaken in Central Europe and in the Mediterranean region have indicated that established natural and semi-natural, as well as rural production landscapes dependent on human perturbations show maximum *total landscape ecodiversity*. Therefore, non-equilibrium systems such as cultural landscapes can acquire long-term adaptive resilience and metastability. If the homeorhetic flow equilibrium is removed a cultural landscape will revert from a diverse and perturbation-dependent ecosystem to a succession-determined landscape with lower structural and biological diversity. Such a landscape is more typical of equilibrium systems. Therefore, the maintenance of the homeorhetic flow processes is not only responsible for the richness and evenness of biodiversity but also for ecological stability, cultural diversity and attractiveness.

INDUSTRY CONTEXT

> "*By any comparison the Australian wine industry has enjoyed a sparkling success*" (South Australian Government 1998).

The wine industry is currently being hailed as a shining example of successful agribusiness development in Australia. The industry's captains have breached the psychological bastion of parochialism and have embarked upon a proactive professional strategic plan (Australian Wine Foundation 1996) which has, at its epicentre, the promotion of Australia's unique cultural and natural heritage — that is, the perception of a cultural landscape set in a timeless scenario! The marketing plan is based upon the establishment of 40 000 hectares of new vineyards by the year 2025. This very "visible" expansion in growth (and profits) will undoubtedly attract the attention of risk averse consumers who will demand social and environmental accountability. Moreover, Strategy 2025 acknowledges the contribution of the Australian environment to the clean-and-green image conferred upon Australian wine products:

> "*Wine will meet the future consumer's demand for individualised, customised products which are natural, environmentally friendly and healthy*" (Australian Wine Foundation 1996).

Other food industry leaders have similarly acknowledged that "*. . . it is a reality that there will be a constant pursuit of a sustainable environment and culture*" (Clairs 1996). Unavoidably, the rural community will increasingly be looked upon as the *defacto* custodian of the environmental heritage.

A collaborative doctoral programme between Yalumba, the Johnstone Centre for Parks, Recreation and Heritage and the National Wine and Grape Industry Centre addresses the issues of sustainability and biodiversity as integral components of oenological and viticultural management (Camilleri 1998). This on-going programme, which commenced in 1995, anticipates to define the viticultural landscape in the study region in terms of past anthropological activities and to recommend and interpret the outcomes of future regional development policies for the company, particularly in the light of the National Strategy for Ecologically Sustainable Development (Commonwealth of Australia 1992), the National Strategy for the Conservation of Australia's Biological Diversity (Dept. Env. Sport and Territories) and the National Ecotourism Strategy (Commonwealth Department of Tourism 1994). The significance of these three national strategies to the wine/viticultural industry in the defined area will be analysed and their potential impact assessed. The contribution of the viticultural industry to ecological degradation and landscape engineering will also be assessed and quantified.

To participate in the national goal of ecologically sustainable development the wine industry will need to make fundamental changes in the way it conducts strategic business planning. It is suggested that *Strategic Environmental Assessment* and *Cumulative Impact Assessment* (Court *et al.* 1994) will be the "tools" to achieve sustainable industrial development. Whilst *Strategic Environmental Assessment* takes account of the environmental impacts of policies, plans and programmes the *Cumulative Impact Assessment* process accounts for the environmental impacts of actions which are individually acceptable, but may cumulatively have deleterious consequences because of synergisms, spatial and temporal crowding, "nibbling" and other indirect effects. The sustainability of the viticultural industry depends on the perception of vineyards as micro-cultural landscapes, or natural ecosystems dominated by human artefacts. As cultural *gestalt* systems, the individual elements that constitute "the vineyard" derive their nature and purpose from the whole and cannot be understood apart from it (Camilleri

1999). In order to manage a vineyard sustainably, it must be understood holistically. The forward and backward linkages between organizational culture and the "other" natural units of the vineyard ecosystem must be identified and recognized. To do this, the environmental beliefs, values and attitudes of an organization require definition in order to explain the rationale behind managerial decisions. It is common to observe the impact of the latter on the *Vitis vinifera* population; but traditionally the effect of human behaviour on other vineyard biota is not monitored. This broad concept implies that vineyards must be managed as ecosystems, thus integrating conservation and rehabilitation with grape production activities (Box 1).

Modern vineyard management practices must allow for the allocation of scarce resources amongst competing socio, economic and biological *values*. These values, which may include ecological, educational, recreational and aesthetic values as well as the generation of wealth, must be managed within the bounds determined by a unique set of ecosystem capabilities. The key criterion of modern vineyard management is not whether extractive use of resources occurs, but that a diversity of relatively compatible uses and values is promoted.

METHODOLOGY

The programme goals (listed in Box 2) and methodology were identified after an extensive review undertaken as part of the doctoral work and culminating in a transdisciplinary review on the dimensions of the concepts of sustainability and biodiversity (Camilleri 1997). This extensive study identified Farming Systems Research and Development as a holistic and integrated template to explore the "real-time" issues of biodiversity and sustainability in production matrices.

Farming Systems Research and Development is a systems approach to management studies, a holistic concept that organizes experimentation and modelling against the background of the complexity of systems and the variability of their behaviour. Farming Systems Research and Development is concerned with improving an individual's or community's ability and confidence in managing

BOX 1. THE VINEYARD AS AN ECOSYSTEM

- ❑ It is a microcosm or subset of a larger system — namely, the cultural landscape.
- ❑ It has flows of energy, species and materials moving out, into and within the vineyard ecosystem.
- ❑ It is dynamic — that is, it changes over time and space in response to inputs such as energy, introduced species, natural events, internal growth and development processes that may or may not be anthropogenically derived.
- ❑ It has a capacity to sustain *desired* conditions of diversity, ecological services, and resource uses and values as a result of a mix of variables, which include topography, soils, climate, biota, basic ecological processes, size and shape!
- ❑ The eco-capacity of the vineyard ecosystem determines what is sustainable in the short (7–15 yr) to medium term (15–25 yr). A longer time-frame (>25 yr) is probably not appropriate, as the basic capabilities of the system would have changed to an unpredictable status.

BOX 2. PROGRAMME GOALS

- ❑ Establish base-line data of vineyard biodiversity
- ❑ Identify a ground-based monitoring system
- ❑ Elicit environmental beliefs, values and attitudes
- ❑ Explore organizational culture
- ❑ Act as a catalyst for environmental awareness and training
- ❑ Develop an Environmental Management System
- ❑ Design a clean-and-green component to wine products

available limited resources efficiently (Simmonds 1985). It is an adaptive approach to resolving the inter-relationships that exist between the various rural enterprises and the institutional and physical environment. Farming Systems Research and Development is iterative in that it makes use of secondary data and other information generated by other research organizations. It encompasses quantitative and qualitative data. It is unique in that it is a comprehensive approach to research, and does not have a reductionist attitude towards problem resolution. This integrated modelling framework has a dialogic structure, that is, one that has several possible arguments and formulations. This type of knowledge system is both interpretative and integrative which therefore lends itself to accommodating socio-ecological data. This data tends to be both proximate and incomplete. A process of triangulation, using interpretative, qualitative and quantitative models, leads to a discursive environment in which dialogue between unrelated knowledge system, becomes the vehicle to build bridges between previously assumed mutually exclusive disciplines.

ADAPTIVE MANAGEMENT

Farming Systems Research and Development is process of *learning-by-doing*, a form of adaptive management technique. The adaptive approach to managing complex systems such as ecosystems proceeds on a basis of the best understanding of the system (McLain and Lee 1996; Dovers and Mobbs 1997). Given that mistakes can be expected, the strategies employed tend to be counter-active, thus making provision for management techniques to be adjusted as information and experience is increased.

Ground Based Monitoring is a methodology that assists adaptive management through the process of *learning-by-doing*. Monitoring is the process of repeated measurement to detect change over time, thus providing: 1. Feedback on the impacts of current and past land management techniques; and, 2. Baseline data to develop improved methods.

Ground-based monitoring programmes recognize the status of land and vegetation as a fundamental aspect of landcare. Ground-based monitoring is a system capable of providing information about the bio-physical environment. *Assessment*, which is the monitoring of biological diversity and pre-determined physical parameters, forms the basis of land management and allocation (that is, landcare!). *Indicator organisms* and physical parameters such as soil structure and stability provide an estimate, or trend, of sustainability levels under constant or varying management practices. Since perfect knowledge of ecological processes is unlikely, then the concept of sustainable land use will change as knowledge expands. Moreover, changing economic and social demands will also contribute to a dynamic definition of sustainability. Therefore, adaptive management can help vineyard managers with their decision making processes and in particular to achieve the following.

1. Revise management targets as the latter change in a dynamic cultural landscape.
2. Identify land management problems and to review the intuitive bases for their viticultural practices.
3. Evaluate their management skills.
4. Develop management strategies aimed at ecological sustainable development.
5. Learn about the intrinsic nature and needs of the vineyard ecology being managed.
6. Reduce risk and protect productivity.
7. Develop a knowledge base that facilitates dialogue with researchers and technologists.
8. Provide public accountability.

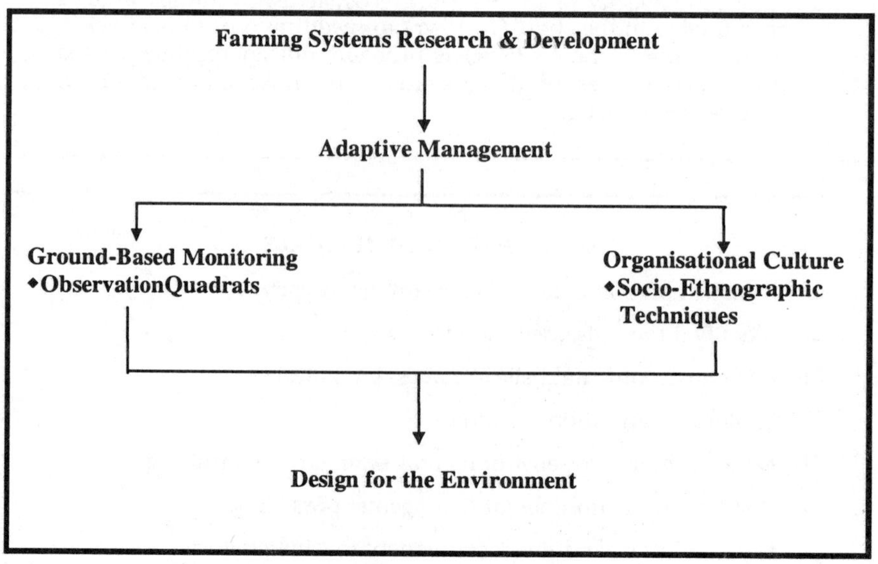

Fig. 1. Towards biological and social indicators of sustainability.

Irrespective of the clear advantages of adaptive learning, the successful implementation of ground-based monitoring will depend on the following specific problems being preferably addressed at the pre-implementation stage.

1. The relationship between cause and observed effect should be established as closely as possible. For example, vineyard managers may find it difficult to separate the influence of managerial, site and climatic factors in the interpretation of recorded change.
2. Goals must be clearly established.
3. Goals need to have clearly defined indicators of success or failure. Moreover, these indicators must have recognizable values.
4. The strengths and weaknesses of the methodology must be considered explicitly.
5. Vineyard managers must relate the frequency and sophistication of their observations to variation in the rate, frequency and magnitude of landscape degradation processes. For example, irregular events of great magnitude, such as disease outbreaks and heavy frosts, require mangers to make more observations than do regular events of small magnitude. In practice, the reverse tends to be the norm!
6. The need to properly relate the information provided to that required in order to detect change at varying temporal and spatial scales. Particular attention should be given to the following pitfalls:
   (a) The low representativeness of site data due to landscape variability.
   (b) The extrapolation of data from a few points to a whole landscape.
7. The effects of changes in biodiversity on productivity are generally unquantified and the linkages undefined.
8. The stakeholders must perceive the data collated from monitoring systems as relevant.
9. A monitoring system must incorporate the needs and decision making styles of the stakeholders.
10. Formal data must be matched with human (subjective) judgement.
11. A monitoring system must have flexibility built into the process.
12. A monitoring system must recognize the complementary value of "hard" and "soft" technologies.
13. The stakeholders must be suitably satisfied that the location of the site is representative of the general landscape: this is important in the perception of "usefulness".

The above criteria clearly emphasize the need for the end-users to be active participants in the design process.

OBSERVATION QUADRATS

Given the above-mentioned properties of adaptive management techniques, the methodology used to sample the environment within and outside the vineyard ecology was carefully designed in order to avoid the pitfall of non-representation. A permanent quadrat for monitoring purposes was designed using the one described by Hopkins *et al.* (1987) as a template. The quadrat, illustrated by Figure 2, has the following properties.

1. The quadrat, 15 rows wide (33 m–49.5 m, depending on vineyard row width) and *circa* 100 m long, is permanently marked.
2. A minimum of five years of observations may be necessary to distinguish long-term trends in biota from year-to-year variability and changes resulting from short-term cycles.
3. The design of the quadrat is suitable for the monitoring of a wide range of organisms, irrespective of abundance and size.
4. A photo-point is located 10 m off a predetermined fixed point.
5. The quadrat may be marked off in a series of nested quadrats. This may facilitate the recording of presence/absence data.

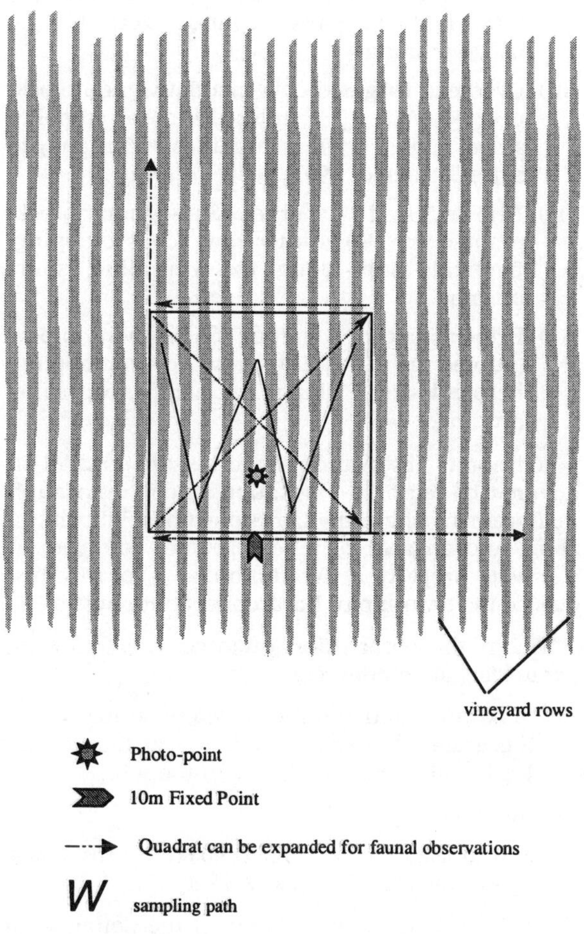

Fig. 2. Quadrat design.

6. Canopy measurements may be made as intercepts along two sides and two diagonals of the quadrat, or randomly along a *W*-shaped path.
7. For systematic observations of avi-fauna, the quadrat may be extended to an area of one hectare. This extended area would include the original quadrat and would share a common corner.
8. The quadrat is located *within* a community and is inappropriate for the measurement of changes at an ecotone.
9. Ten pit-fall traps are located along a *W*-shaped path for invertebrate sampling.

The observation quadrats are located within production blocks of a commercial vineyard known as *The Menzies* and owned by Yalumba, a company of S. Smith & Son Pty Ltd. The property is located in the Limestone Coast Zone, near the township of Coonawarra, South Australia. The production blocks reflect a diversity of viticultural practices and each has one observation quadrat.

(i) Dryland Cabernet Sauvignon block; planted *circa* 1974; inter-row and undervine total herbicide weed control; zero tillage; hedge-pruning.

(ii) Overhead-irrigated Cabernet Sauvignon block; planted *circa* 1974; inter-row and undervine total herbicide weed control; zero tillage; hedge-pruning.

(iii) Overhead-irrigated Cabernet Sauvignon block; planted *circa* 1974; inter-row cover cropping since August 1997; "organically" managed since September 1997; hedge pruning.

(iv) Drip-irrigated Cabernet Sauvignon block; planted in 1994; inter-row cover cropping; herbicide-strip along the vine line; spur-pruning.

(v) Drip-irrigated Cabernet Sauvignon block; planted in 1996; inter-row cover cropping; herbicide-strip along the vine row; spur-pruning.

Additional observation quadrats are located in an area of remnant vegetation under restoration management within the Limestone Coast Viticultural Zone, and an area of relic vegetation situated on a private property at Wrattonbully, a locality just outside the township of Naracoorte, South Australia.

Within the observation quadrats the following parameters are monitored.

1. Maximum and minimum temperatures (i) at the grapevine canopy level; (ii) at the ground level; and (iii) *circa* 15 cms in the soil.
2. Rainfall.
3. Soil moisture at 30, 60 and 90 cm depths, using gypsum blocks (Skinner *et al.* 1997).
4. Soil bacterial activity using (i) the Cotton Strip (Shirley Soil Burial Test Fabric) Assay, adapted from Kirkby (1997) and Harrison *et al.* (1988) and (ii) the Most Probable Number technique (Alexander 1982).
5. Soil macro and micro nutrient status.
6. Soil invertebrate activity, namely surface dwelling, or epigeic invertebrates sampled by using pitfall traps (Pankhurst 1997), and epigeic and endogeic earthworm populations using a mustard suspension to flush them to the soil surface (Gunn 1992; Buckerfield 1997).
7. Soil structure stability, using the slaking and dispersion technique of Cass *et al.* (1996).
8. Soil texture analysis and classification using the finger method technique.
9. Qualitative assessment of humus condition using the method adapted by Brinton (1983).
10. Survey of avifauna within the remnant vegetation under restoration management (method adapted from Recher *et al.* 1983).

In addition to the above parameters, the following is also monitored within the vineyard observation quadrats.

1. Grapevine phenological stages as modified from Lorenz *et al.* (1995) and Coombe (1995).
2. Grapevine nutrient status, using the petiole analysis technique.
3. Grapevine disease incidence.
4. Selected invertebrate populations within the grapevine canopy with special reference to Light Brown Apple Moth *Epiphyas postvittana* and lacewing species (*Micromus* sp. and *Plesioohrysa* sp.).
5. Vineyard management records, including spray schedules.
6. Meteorological data from the vineyard weather station.
7. A register of avifauna observed on the Menzies property is maintained.

The details of the methodologies employed are described in Camilleri (1997).

SOCIAL ENVIRONMENTAL INDICATORS OF A SUSTAINABLE ORGANIZATIONAL CULTURE

"Successful corporations have . . . values and beliefs to pass along — not just products" (Deal and Kennedy 1988).

In the early sections of this chapter, it was argued that successful organizations must cultivate a culture of reverence, probity, receptivity and integrity when dealing with environmental issues. This is not only a social responsibility but also a marketing imperative.

In this research, the environmental attitudes of the wine and grape industry are explored by using a case study approach. The organization Yalumba is therefore the primary focus of study. Jary and Jary (1991) describe an organization as an example of collectivity established for the pursuit of specific aims or goals, and characterized by a formal structure of rules, division of labour, a structured line of authority, and controlled membership or admission.

An organization has culture, that is, it has a particular configuration of norms, values, beliefs and ways of behaving that characterize the manner in which groups and individuals collaborate within the organization. In this view, culture is the sum of a social group's observable pattern of behaviour, customs, and way of life.

Culture is the broadest socio-ethnographic concept. *Ethnography* is the art and science of describing a group or culture (Fetterman 1989). In writing about the routine, daily lives of people the researcher must keep an open mind about the group or culture being studied (Furze *et al.* 1996). The foci of inquiry tend to be the more predictable patterns of human thought and behaviour. Socio-ethnographic work is therefore

1. *A holistic perspective* — necessary to gain a comprehensive and complete picture of a social group.
2. *Contextual* — that is, it places observations into a larger perspective.
3. *An emic view of reality* — that is, the research considers the insider's or employee's perspective of reality. An emic perspective leads to the recognition and acceptance of multiple realities, and therefore crucial to an understanding of why individuals or organizations think and act in the different ways.
4. *An etic view of reality* — or the external (researcher's), social-scientific perspective on reality. The researcher must attempt to view "reality" objectively, that is, without making value judgements.
5. *A Non-judgemental view of reality* — The process of socialization makes the problem of ethnocentricity difficult to eliminate completely; however, the researcher must acknowledge or guard against the most obvious personal tastes, beliefs and biases.

Fieldwork is the heart of the socio-ethnographic research design (Furze *et al.* 1996). In the field, basic sociological concepts, data collection methods, techniques, and analysis make up the fundamental elements of "doing ethnography". The case-study approach aims at a holistic understanding of the phenomena being studied. For this reason it makes use of a variety of techniques. The following are the principal techniques that constitute the methodology for this study.

Observation: Artefacts, or the cultural phenomena that one sees, hears and feels, are the most visible products of a group or individual, but the most difficult to decipher. Such artefacts may include: behaviour, organizational processes into which behaviour is made routine, art and architecture, language, technology, manufactured goods or services, clothing, manners of address, emotional displays, myths and stories about the organization or the individual, and observable rituals and ceremonies.

Interviews, or conversations with a purpose: The interviewer decides on the purpose of an interview. The interview is therefore asymmetric. By definition, the interview belongs to a class of methods that yields subjective data; in other words, direct descriptions of the world of experience. Interviews explain and put into a larger context what the researcher sees and experiences. They require verbal interaction, and language is the commodity of discourse. Words and expressions have different values in various cultures. People exchange these verbal commodities to communicate. The interview may be more or less structured according to the purposes of the researcher. In the *structured* interview, the interviewer will often make use of an interview schedule containing the precise questions he or she wishes to put to the respondent. In the *unstructured* interview, however, the interview may only have a list of subjects to cover. In the structured interview, the interviewer is seen as a source of error that must be controlled: standardized questions are used in order to minimize the researcher's involvement with the respondent and hence, it is argued, to obtain valid data.

In the unstructured interview, the rationale is reversed. The source of error is seen precisely as any formalization or standardization of the situation, and the interviewer is free to adjust his approach according to the respondent's reactions.

Formally *structured and semi-structured* interviews are verbal approximations of a questionnaire with explicit research goals. These interviews generally serve comparative and representative purposes, i.e., comparing responses and putting them in the context of common group beliefs and themes. *Informal interviews* are the most common in ethnographic work. They seem to be causal conversations, but where structured interviews have an explicit agenda, informal interviews have a specific but implicit research agenda. An informal interview is different from a conversation, but it typically merges with one, forming a mixture of conversation, and embedded questions. *Retrospective interviews* can be structured, semi-structured, or informal. The ethnographer uses retrospective interviews to reconstruct the past, asking informants to recall personal historical information.

Content analysis, or the study of written material: Documents and written materials contain data relating to the variables that are of interest to the socio-anthropologist. The technique of content analysis requires that these variables be translated by operational definition into categories of content, whose occurrence in the document can be counted.

Questionnaires: Studies of environmental beliefs, values, and attitudes are complicated by the ambiguous meanings of both the objects of study and the instruments used (Gooch 1995). The environment is, technically speaking, anything that exists outside the self (Herberlein 1981a) and conceptually, can range from the very local to the global.

Whether a separate biospheric value orientation or a "new ecological paradigm" is in the process of emerging in the general population cannot be determined without time series analysis of value structures. In the case of attitude theory and measurement, it appears that many of the conceptual, definitional and measurement difficulties that have occurred in other fields of application have also arisen in investigating environmental attitudes: there is little agreement about the definition of *attitude* and hence what aspects of attitudes are worth measuring. There is also a need for research to focus on specific environmental issues rather than on generalized environmental concern, and for a more precise conceptualization of such concern that recognizes cognitive as well as demographic determinants.

A common approach in obtaining measures of environmental attitudes has been to construct *Likert* scales (Likert 1932) composed of a number of attitude statements or items to which respondents indicate their level of agreement or disagreement. By assigning a score to each level and summing the scores across items in the scale a measure can be obtained of generalized environmental concern, or of attitudes to specific environmental issues. Respondents are required to read and respond to the items *instinctively* without reflecting at length upon the issues raised in the items. *Cronbach's alpha* for a particular scale is a measure across all respondents of the consistency in their responses to the items comprising the scale. Bivariate correlations are used to identify relationships between the summed scores on the scales and other variables. Considerable attention is devoted to ensuring that items are measuring a single attitudinal dimension so that responses by any one respondent will tend to be consistent across the items comprising a scale — *unidimensional attitudinal response*.

Respondents who vary markedly in their responses to the items of the scales and whose responses are cognitively influenced or inconsistent due to comprehension problems weaken the assumption of unidimensionality so fundamental to the Likert Scale construction. It is these "inconsistent" respondents who lower the reliability and alpha value of the scale, add random "noise" to the summed scores for the scale and blur any regression relationships that might be sought between the scale score and other variables of interest. Cluster analysis and discriminate analysis indicate that even internally inconsistent respondents can be divided into definable categories, and differences in response patterns may relate significantly to other variables such as business structure, commodities being produced, attitudes to external influences in decision making and to conservation generally (Reeve and Black 1994).

Cross-checking information against other "bits" of information obtained is fundamental to the process of deciphering a group's culture. This process continues until a pattern begins to emerge. During the process dialogue between the insider and the outsider, or the emic and etic view of reality, is crucial. This is the process of ***triangulation***.

Triangulation is the method that tests one source of information against another in order to strip away alternative explanations and prove a hypothesis. Typically, the researcher compares information sources to test the quality of the information (and the person sharing it), to understand more completely the part an actor plays in the social drama, and ultimately to put the whole situation into perspective. Generally the following apply.

- Triangulation works with any topic, in any setting and on any level. The trick is to compare comparable items and levels during analysis.
- Triangulation improves the quality of data and the accuracy of socio-ethnographic findings.
- Triangulation can occur naturally in conversation as easily as in intensive investigatory work.

The role of the researcher in this research programme will vary from the classical definition of the ethnographer, particularly since the principal researcher (Cecil Camilleri) is an employee of Yalumba, the subject of the case study. In this study, the researcher is not only attempting to study organizational culture from the "descriptive" view point, but is also involved in the process of culture change. Indeed, an important contribution to cross-checking of data is to test the etic insight with the resultant emic behaviour and intervention. Lewin (1947), who stated that if a one wants to understand a system then one must try to change it, espoused such a method. This principle is fundamental to action research and the study of diagnostic activities in all human systems. It is also particularly relevant to the diagnosis of cultural elements. This type of research may be referred to as clinical research (Schein 1987a; Schein 1991). Its distinguishing features are as follows.

(i) The organization initiates the process of culture analysis and change.

(ii) Data is voluntarily contributed from the members of the organization.

(iii) The researcher is psychologically licensed by the client to undertake the research and to develop the research focus.

(iv) There is close collaboration between the researcher and *Key Actors* or *Gatekeepers* within the corporate organization. The latter tend to be more articulate and culturally sensitive than other members of the organization. They also tend to be the repository of detailed historical data, knowledge about contemporary interpersonal relationships (including conflicts), and a wealth of information about the nuances of everyday corporate life.

(v) The researcher takes on the role of the participant observer, in other words the researcher is directly involved in the social behaviour of the organization. In the clinical research model, the observer's emotional reactions and biases are primary data to be analysed and understood. The researcher's contributions to organizational culture development become artefacts that must be acknowledged and taken into account.

The fundamental difference between the ethnographical method and the clinical model is that the aim of the ethnographer is generally to leave the system being studied intact! This assumes that interaction is minimal and that the culture can be deciphered if sufficient time is allocated to observation. On the other hand, the clinical model assumes that culture will not reveal itself easily and that one must be proactive to determine where stable rituals, espoused values, and basic shared assumptions are located. Culture in a mature group comes to cover all aspects of life. Because it becomes so extensive its analysis becomes difficult. Culture change, in the sense of changing basic assumptions is, therefore, difficult, time consuming, and highly anxiety provoking. This point is particularly relevant for the leader who sets out to change the culture of the organization. The researcher has the professional obligation to fully understand what the potential consequences of the research are.

TOWARDS A WINE PRODUCT DESIGNED FOR THE ENVIRONMENT

"*No business that strives to remain competitive, open to new markets and new opportunities, can afford to ignore the global demands of environmental quality . . . Australian products intended for export have to meet . . . new global-market standards for environmental quality . . . Local products have to compete against imports with an 'environmental edge'* " (Gertsakis *et al.* 1997).

The socio-ecological programme described in this chapter is a longitudinal, transdisciplinary contribution to environmental credibility in the market place. The market place of the future will increasingly expect and demand of companies demonstrable environmentally sound practices. The vineyard of the future will be managed as an ecosystem. A sustainable viticultural landscape will incorporate a socio-ecologically-integrated wine and grape industry that produces and markets environmentally sustainable products with minimal impact on the natural and cultural landscape. This concept falls within the realm of Industrial Ecology (Allenby 1998), a holistic discipline that determines how resources and energy flow in a commercial enterprise, how resources and energy are transformed into consumer products and services, how by-products of production and marketing effect the environment, how institutional and geo-biological factors interact with entrepreneurial activity, how local production processes influence regional and global issues, and how to make ecologically sound decisions in risky commercial environments?

In the late 1960s and early 1970s, the need to provide evidence of adequate environmental stewardship culminated in the development of a methodology that led to the following.

- *Life cycle thinking*, a holistic environmental view of a product or service that takes into consideration all the raw materials used during production and marketing.
- *A Life cycle perspective*, an outlook that encourages organizations to consider all environmental aspects of their operations and thus integrate environmental issues into the overall decision making process.

The *cradle-to-grave* or Life Cycle concept of product assessment is summarized by Figure 3. This framework, used by the ISO/TC 207's Subcommittee 5 of the ISO 14000 Environmental Management System series, has been developed by the Society of Environmental Toxicology and Chemistry and includes the following phases.

(i) Bounding the Assessment — that is, defining the scope and goals.

(ii) The Life Cycle Inventory Analysis, which involves measuring the inputs, or resources and energy used, and the environmental outputs, or releases, that arise along the entire continuum of the product.

(iii) The Life Cycle Impact Assessment, which examines the actual and potential environmental and human health effects associated with the use of resources and with the environmental releases that occur.

(iv) The Life Cycle Improvement Assessment, which focuses on those parts of the system that can be changed in order to reduce the impact of the service or product on the environment. This process is also referred to as *Design for the Environment*.

Design for the Environment — attempts to design or redesign processes, products or services that achieve the following (i) Minimize or eliminate pollution, (ii) Lower the costs associated with hazardous waste disposal, (iii) Lower the costs associated with compliance with regulations, and (iv) Encourage businesses to use *full cost* or *total cost accounting* as a tool to assist decision makers to measure and quantify the benefits of pollution prevention. This accounting method allocates direct and indirect environmental costs to a service, product or activity.

This attitude towards industrial processes will require a revolutionary change in organizational culture. Cultural change can, however, facilitate the development of an *Environmental Declaration*.

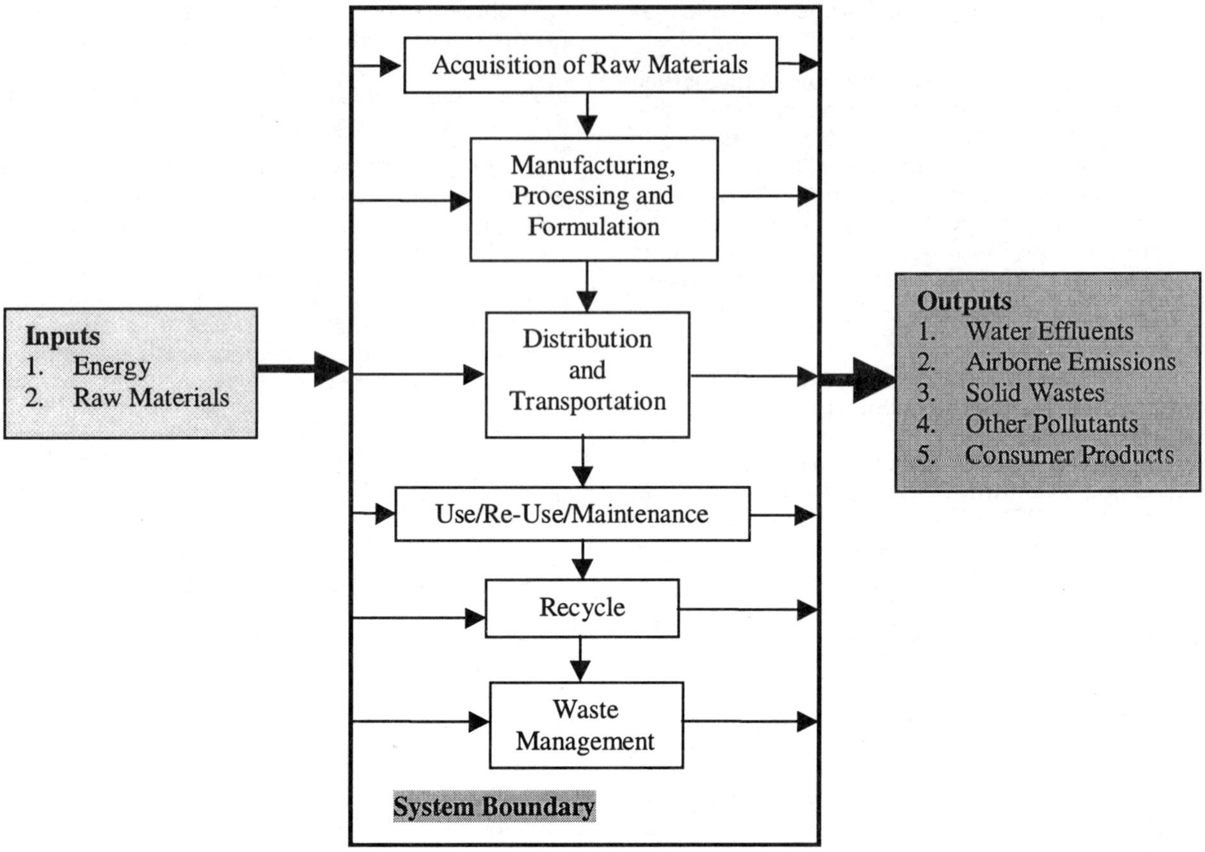

Fig. 3. Life cycle assessment process.

This declaration may also take the form of an *Eco-Label*, one that defines the "environmental attributes of a product and service [and] may take the form of statements, symbols, or graphics on product or package labels, product literature, technical bulletins, advertizing, publicity, etc." International Organization for Standardization (1998). Failure to realize its environmental responsibilities to the international community will jeopardize the grape and wine industry's future export growth scenario. At retail prices, the latter is currently valued as *circa* A$120 m, and is projected to increase to A$2.5 billion by the year 2025.

This research programme will contribute to the projected success of the industry. At the corporate level, the commercial spin-offs are more specific. In addition to contributing to the design of an ecologically sustainable wine product, the following additional contributions are anticipated.

- Extend the concept of Total Quality Management to include Total Quality Environmental Management.
- Build and strengthen "green" commitment from senior management and employees.
- Formalize corporate commitment to ecological sustainability in an environmental ethic and policy.
- Develop and maintain strategic alliances with suppliers of goods and services.
- Build community and consumer trust at the local and international level.
- Capitalize on a competitive "green" edge.
- Explore training needs to inculcate the community of management and employees with an environmental understanding and commensurate work ethic.
- Explore employee participation through consultation and Environmental Quality Circles.

Superficially, the programme appears ambitious. However, in reality, the full scope of the Farming Systems Research and Development programme cannot be achieved without the approval and participation of the community of employees. From this point of view, the Farming Systems Research and Development programme is contextual and participative. It may be described as an ***in-house landcare movement***, a model for an effective corporate culture to assist the move towards more sustainable resource use and transformation.

ACKNOWLEDGEMENTS

A special recognition of contribution made by the technical and management staff of Yalumba Coonawarra Vineyards Pty Ltd; namely, Ms Anna Hurn, Mr Michael Dear, Mr Peter Freckleton, and Mr Mick Muller. Without their help and continued contribution this project would not be feasible.

REFERENCES

Alexander, M., 1982. Most Probable Number Method for Microbial Populations. Chapter 39 *in* Methods of Soil Analysis, Part 2. Chemical and Microbiological Properties. Agronomy Monograph No. 9. (2nd Edition). Soil Society of Agronomy, Inc./Soil Science Society of America, Inc.: Madison, USA.

Allenby, B. R., 1998. Industrial Ecology: A Policy Framework and Implementation. Prentice Hall: United Kingdom.

Australian Wine and Brandy Corporation, 1993. Viticultural Boundaries to be Established in Law. *Aust. NZ Wine Indust. J.* **8**(3): 192.

Australian Wine Foundation, 1996. Strategy 2025 — The Australian Wine Industry. Winemakers' Federation of Australia: Adelaide, South Australia.

Australian Wine Foundation, 1996. Chapter 2: The Success Story of Australian Wine. *In* Strategy 2025 — The Australian Wine Industry. Pp. 2 Winemakers' Federation of Australia: Adelaide, South Australia.

Brinton, W. F., 1983. Chapter 15: A Qualitative Method for Assessing Humus Condition. Pp. 382–93 *in* Sustainable Food Systems ed by D. Knorr. AVI Publishing: Westport, Conn.

Buckerfield, J., 1997. CSIRO Division of Soils, Glen Osmond, South Australia, pers. comm.

Camilleri, C. S., 1997. Dimensions of Biodiversity and Sustainability — Landscape Ecology and the Viticultural Industry. Books 1 and 2. The Johnstone Centre for Parks Recreation and Heritage: unpublished.

Camilleri, C. S., 1997. Integrated Sustainable Vineyard Management — Principal Methods of Sustainability. Johnstone Centre for Parks Recreation and Heritage: unpublished.

Camilleri, C. S., 1998. The Viticultural Landscape in the South East of South Australia — An Integrated Approach to the Sustainable management of its Ecosystem. Pp. 37 *in* The Johnstone Centre 1998 Workshop Abstracts ed by A. Curtis and L. Wilson. Charles Sturt University: Albury, New South Wales.

Camilleri, C. S., 1999. Managing the Vineyard as a Micro-Cultural Landscape — A Gestalt Experiment in Landscape Ecology. Pp. 237–38 poster paper *in* Proceedings of the 10th Australian Wine Industry Technical Conference, 2025 — Meeting the Technical Challenge. Australian Wine Research Institute: in press.

Cann, J. H., De Deckker, P. and Murray-Wallace, C. V., 1991. Coastal Aboriginal Shell Middens and their Palaeoenvironmental significance, Robe Range, South Australia. *Trans. Roy. Soc. SA* **115**(4): 161–75.

Cass, A., Mckenzie, N. and Cresswell, H., 1996. Physical Indicators of Soil Health. Chapter 7 *in* Indicators of Catchment Health: A technical Perspective ed by J. Walker and D. J. Reuter. CSIRO: Melbourne, Victoria.

Clairs, R., 1996. Market-Led. *In* Professional Responsibility for the Environment, Proceedings of 1996 AIAS Annual Conference. Australian Institute of Agricultural Science: Victoria.

Commonwealth of Australia, 1992. National Strategy for Ecologically Sustainable Development. Australian Government Publishing Service: Canberra, Australian Capital Territory.

Commonwealth Department of Tourism, 1994. National Ecotourism Strategy. Australian Goverment Publishing Service: Canberra, Australian Capital Territory.

Coombe, B. G., 1995. Adoption of a System for Identifying Grapevine Growth Stages. *Aust. J. Grape Wine Res.* **1**: 104–10.

Court, J. D., Wright, C. J. and Guthrie, A. C., 1994. Assessment of Cumulative Impacts and Strategic Assessment in Environmental Impact Assessment. Commonwealth Environment Protection Agency: Canberra, Australian Capital Territory.

Darling, G., 1997. Defining Our Grape Regions. Australian Viticulture, October/November, 6–7.

Deal, T. and Kennedy, A., 1988. Corporate Cultures: the Rites and Rituals of Corporate Life. Penguin: Harmondsworth.

Dower, M., 1995. Europe's Landscape and the Bern Convention. *Naturopa* **77**: 24–25.

Downes, R. G., 1971. Land, Land Use and Soil Conservation. Pp. 25 *in* Conservation ed by A. B. Costin and H. J. Firth. Penguin: United Kingdom.

Department of Environment, Sport and Territories (undated). National Strategy for the Conservation of Australia's Biological Diversity. Canberra, Australian Capital Territory.

Dovers, S. R. and Mobbs, C. D., 1997. An Alluring Prospect? Ecology, and the requirements of Adaptive Management. Chapter 4 *in* Frontiers in Ecology: Building the Links ed by N. Klomp and I. Lunt. Ecological Society of Australia and Charles Sturt University. Elsevier: Australia.

Fetterman, D. M., 1989. The First Step: An Overview. Pp. 11–25 *in* Ethnography Step by Step. Applied Social Research Methods Series Volume 17. Sage Publications: London.

Furze, B., De Lacy, T. and Birckhead, J., 1996. Using Knowledge from the Social Sciences. Pp. 31–47 *in* Culture, Conservation and Biodiversity. John Wiley and Sons Ltd: England.

Furze, B., De Lacy, T. and Birckhead, J., 1996. Using Methods from the Social Sciences. Pp. 48–92 *in* Culture, Conservation and Biodiversity. John Wiley and Sons Ltd: England.

Gertsakis, J., Lewis, H. and Ryan, C., 1997. The Importance of EcoReDesign. Pp. 5–6 *in* A Guide to EcoReDesign — Improving the Environmental Performance of Manufactured Products. Centre for Design at RMIT: Melbourne, Victoria.

Global Environmental Management Initiative, 1993. Total Quality Environmental Management: The Primer. Global Environmental Management Initiative: Washington, DC.

Gooch, G. D., 1995. Environmental Beliefs and Attitudes in Sweden and the Baltic States. *Environ. Behav.* **27**(4): 513–39.

Gunn, A., 1992. The use of Mustard to estimate Earthworm Populations. *Pedobiologia* **36**: 65–67.

Harrison, A. F., Latter, P. M. and Walton, D. W. (eds), 1988. Cotton Strip Assay: An Index of Decomposition in Soils. ITE Symposium No. 1. Institute of Terrestrial Ecology/Natural Environment Research Council: Grange-Over-Sands, United Kingdom.

Herberlein, T. A., 1981a. Environmental Attitudes. *Zeitschrift Umweltpolitik* **4**: 241–70.

Hopkins, A. J. M., Brown, J. M. and Goodsell, J. T., 1987. Chapter 33. A Monitoring System for Use in the Management of Natural Areas in Western Australia. *In* Nature Conservation: The Role of Remnants of Native Vegetation ed by D. A. Saunders, G. W. Arnold, A. A. Burbidge and A. J. M. Hopkins. Surrey Beatty & Sons: Chipping Norton, New South Wales.

International Organization for Standardization, 1998. Environmental Labels and Declarations — General Principles. ISO14020:1998. Page 2. Published by the International Organization for Standardization: Geneva.

Jary, D. and Jary, J., 1991. Collins Dictionary of Sociology. Harper-Collins: United Kingdom.

Lewin, K., 1947. Group Decision and Social Change. *In* Readings in Social Psychology ed by T. N. Newcomb and E. L. Hartley. Holt, Rinehart and Winston: Troy, Mo.

Likert, R., 1932. The Method of Constructing an Attitude Scale. *Archives of Psychology* **140:** 44–53.

Lorenz, D. H., Eichhorn, K. W., Bleiholder, H., Klose, R., Meier, U. and Weber, E., 1995. Growth Stages of the Grapevine. *Aust. J. Grape Wine Res.* **1:** 100–03.

McLain, R. J. and Lee, R. G., 1996. Adaptive Management: Promises and Pitfalls. *Environ. Manage.* **20:** 437–48.

Naveh, Z., 1995. Interactions of Landscapes and Cultures. *Lands. Urban Plan.* **32:** 43–54.

Naveh, A. and Lieberman, A., 1994. Supplement 3, Remote Sensing, Geographic Information Systems, and Landscape Ecology. *In* Landscape Ecology, Theory and Application (Second Edition) ed by Z. Naveh and A. Lieberman. Springer-Verlag: New York.

Reeve, I. J. and Black, A. W., 1994. Understanding Farmers' Attitudes to Land Degradation: Some Methodological Considerations. *Land Degr. Rehab.* **5:** 179–89.

Recher, H. F., Milledge, D. R., Smith, P. and Rohan-Jones, W. G., 1983. A Transect Method to Count Birds in Eucalypt Forest. *Corella* **7**(3): 49–54.

Schein, E. H., 1987a. The Clinical Perspective in Fieldwork. Sage: Newbury Park, California.

Schein, E. H., 1991. Legitimating Clinical Research in the Study of Organizational Culture. Massachusetts Institute of Technology. Sloan School of Manaement Working Paper, No. 3288–91.

Simmonds, N., 1985. Farming Systems Research: A Review. World Bank Technical paper 43. The World Bank: Washington, DC.

Skinner, A., Hignett, C. and Dearden, J., 1997. Resurrecting the Gypsum Block for Soil Moisture Measurement. Australian Viticulture, October/November, 13–19.

South Australian Government, 1998. The Way for Wine. Adelaide, South Australia.

South Eastern Drainage Board, 1980. Environmental Impact Study on The Effect of Drainage in the South East of South Australia. Ref. SEDB 28/76, 13/79. South Eastern Drainage Board: Adelaide, South Australia.

Thomas, R. I., 1996. The EPA and the Australian Wine Industry. Pp. 178–81 *in* Maintaining the Competitive Edge, Proceedings of the Australian Wine Industry Technical Conference ed by C. S. Stockley, A. N. Sas, R. S. Johnstone and T. H. Lee. Winetitles: Adelaide, South Australia.

Twidale, C. R., Campbell, E. M. and Bourne, J. A., 1983. Granite Forms, Karst and Lunettes, Chapter 3 *in* Natural History of the South East ed by M. J. Tyler, C. R. Twidale, J. K. Ling and J. W. Holmes. Royal Society of South Australia Inc.: South Australia.

CHAPTER 38

Tourism: a glue in the production/conservation matrix

Paul M. Blaschke,[1,2] Camilla Cox[1,3] and Carol Lough[1]

Tourism has the potential to facilitate nature conservation in production environments. It is a generally non-extractive land use which can enhance nature conservation by: preserving wildlife or habitat as a tourist attraction; providing diversified employment and income for preservation or conservation; or changing attitudes to conservation. To achieve this potential, especially at a regional level, there is need for a vision of more sustainable tourism that is wider than just a concept of "nature tourism" or "eco-tourism". The vision should include the whole pattern of tourism in a district or region, including both its urban and rural parts.

Two main groups of factors can adversely affect the ability of tourism to achieve this potential. The first includes adverse effects from tourism developments, e.g., on conservation resources, wilderness and amenity values, through infrastructure demands. The second includes factors in the system of tourism management, e.g., insufficient attention to the maintenance of environmental quality or discrepancies between the marketing of tourism product and the ability of the system to manage the resulting level of tourism.

These concepts are discussed and illustrated with reference to several examples in New Zealand showing different relationships between tourism activities, the regional economy, and nature conservation.

Key words: Tourism, Nature conservation, Environmental effects, Regional planning, New Zealand.

INTRODUCTION

THE concept of nature conservation in production environments presents an immediate dichotomy, perhaps a contradiction. Nature conservation seeks to preserve natural environments, whereas primary production activities, by their very nature, change environments and can cause adverse effects on them. In the background notes to this conference, the term "production environments" is not defined, and it is easy to think of "production" only as primary production from land and sea. However, in most western societies, and increasingly in Third World countries as well, the majority of people are not directly engaged in such primary production. In New Zealand for example, only 10% of the adult population earn their livelihood from the primary industries. Most people's means of production are indirect, e.g., in the service industries, and are predominantly in urban areas (Statistics New Zealand 1998). Therefore, a wider definition of "production environments", which is the one adopted here, is "environments from which people gain their livelihoods". In this context, both the concept of nature conservation in production environments, and the potential contribution of tourism to nature conservation in production environments, can be more clearly appreciated.

Tourism is now one of the most important service industries both world-wide and in New Zealand. In 1996, there were an estimated 16 400 private businesses involved in tourism in New Zealand, employing about 90 000 full-time job equivalents and about the same number indirectly (Statistics New Zealand 1998). Notwithstanding a significant downturn in 1998, the sector is the country's biggest earner of foreign exchange, contributing $4.78 billion in foreign exchange in 1996, excluding international airfares (New Zealand Tourism Board 1996). The total contribution by tourism to national Gross Domestic Product in 1988/89 was $3.2 billion, of which domestic tourism contributed about 60%.

[1]Office of the Parliamentary Commissioner for the Environment, P.O. Box 10–241, Wellington, New Zealand.
[2]Present address: Boffa Miskell Ltd, P.O. Box 11442, Wellington, New Zealand.
[3]Present address: Ministry of Fisheries, P.O. Box 1020, Wellington, New Zealand.

Pages 385–94 *in* NATURE CONSERVATION 5: NATURE CONSERVATION IN PRODUCTION ENVIRONMENTS: MANAGING THE MATRIX ed by J. L. Craig, N. Mitchell and D. A. Saunders. Surrey Beatty & Sons, 2000.

Tourism is here defined as "the provision and use of goods and services for and by tourists", tourists being "all persons travelling to a place within New Zealand other than their usual place of residence (whether this be in New Zealand or another country) for a period of not less than 24 hours or one night, for purposes of either leisure or business" (Parliamentary Commissioner for the Environment 1997a, based on standard World Tourism Organization definitions). Tourism is a productive activity in the sense of producing income, and it is in essence non-extractive, i.e., it does not directly change the environment in which it takes place. The business of catering for both domestic and international visitors is a well-established part of the New Zealand economy and way of life.

The linkage that tourism provides between nature conservation and production can be best understood if tourism is seen as a land use, either directly through the land required for tourism facilities or infrastructure, or more importantly indirectly, as the setting for tourism activities. Almost all outdoor tourism activities rely on the natural environment, especially in New Zealand (New Zealand Tourism Board and Department of Conservation 1993; New Zealand Tourism Board 1996). This means that for a sustainable tourism industry the environmental qualities on which the industry depends must be maintained and in some cases enhanced, which in turn implies that tourism and nature conservation not only can co-exist, but can also be mutually beneficial (Brandon 1996; Ribout 1997).

This chapter arises out of a recent report by the Parliamentary Commissioner for the Environment into the management of environmental effects associated with tourism (Parliamentary Commissioner for the Environment 1997a). The investigation was undertaken because of widespread concerns about the effects of tourism on natural environments, and it identified loss of quality of some parts of New Zealand's natural environment as a significant potential adverse environmental effect associated with tourism. It was noted that visitor pressure in some "icon" attractions (e.g., Waitomo Caves, Milford Sound) cannot be sustained even in the medium term without major attention being given to reducing adverse visitor effects (see also Kearsley 1997).

The principal recommendation of the report was for the Minister of Tourism to facilitate and resource a widely consultative strategic development process for the tourism sector in New Zealand, with objectives of, among other things:

- reducing strategic risks to the industry by maintaining high environmental quality;
- focusing on environmental awareness in every part of the sector;
- recognizing the appropriate complementary roles of the private and public components of the tourism sector; and
- promoting an industry-government partnership to incorporate quality in all aspects of environmental management in the New Zealand tourism sector.

The implication of the Parliamentary Commissioner for the Environment's conclusions and recommendations was that we have some way to go before we achieve the potential for tourism to enhance nature conservation that is claimed above. However, the potential benefits of such enhancement are considerable. This chapter outlines some ways in which this can happen and explores some factors which influence the relationship between nature conservation and tourism. Its emphasis is on terrestrial environments in New Zealand.

HOW CAN TOURISM FACILITATE NATURE CONSERVATION?

In our interpretation we identify three main ways in which tourism can facilitate nature conservation, but in reality these three ways closely interact (see Brandon 1996 for another perspective of these interactions).

1. Providing diversified income and employment opportunities

The reason that international tourism has been so strongly supported and promoted by all recent New Zealand governments is because of its national economic benefits of foreign exchange earnings and employment potential. This is directly reflected in New Zealand Tourism Board goals of $9 billion in foreign exchange earnings and 180 000 full-time jobs (or part-time equivalents) by the year 2000. Nationally these foreign exchange earnings maintain the trade balance and various tax revenues (private and corporate income tax, Goods and Services Tax, etc.) which provide the Crown revenue for its investment in maintaining environmental quality, including expenditure on nature conservation. It has been estimated that the industry is responsible for more than $1 billion of revenue annually to the government in taxes. The numerous calls for increased or special taxes for international tourism such as an international visitors' levy (e.g., Sage 1995), recognize this connection.

At the local government level, rating income from tourism businesses provides revenue that can contribute to local government expenditure on nature conservation, such as planning provisions, incentives for conservation, pest control, local and regional parks, etc.

All private conservation initiatives need to be supported by available private income. Tourism increasingly provides such income especially in rural areas, either directly through the provision of tourism infrastructure e.g., "farmstay" accommodation, or rural tourism activities; or indirectly through off-farm employment in the tourism sector. Generally, tourism in a number of rural regions has added value and new business opportunities to an often declining agricultural economic base.

Some regions which in the past have been heavily dependent on extractive industries such as forestry and mining, e.g., the South Island West Coast and the Coromandel Peninsula, are now generally more protective of their natural environment and seek to promote income opportunities which are seen as less extractive and ultimately more sustainable. The growth of nature-based tourism in both of these regions has been a strong element in that change, and tourism has been vigorously promoted by councils and individuals on that basis. At least for the West Coast, it is hard to envisage the extent of protected areas that now exists being agreed to by local interests without at least the potential of an alternative tourism industry.

Tourism is also increasingly seen as an important vehicle for sustainable economic development for Maori. Increasingly, Maori are seeking to develop their own tourism opportunities while at the same time to more actively express their culture and their role as kaitiaki (guardians) of natural resources. For example Whale Watch Kaikoura Ltd, the award-winning whale-watching business based in Kaikoura, which brings substantial economic benefits to the whole local community, is Maori-owned and operated and is also seen by many as an expression of Maori culture (Butcher *et al.* 1998; Henley *et al.* 1998). In Northland, the process of developing a sustainable tourism industry for Maori is less advanced (Cloher 1998) but Maori surveyed as part of this process have shown clear preferences for tourism development provided tourist numbers are limited and local control over development is maintained. Various types of nature-based tourism were viewed with approval and there was also strong support for Maori-owned tourism to avoid negative environmental effects and damage to the natural environment (Johnston 1998).

2. Direct conservation of wildlife or habitat as tourism attraction

Cause-and-effect relationships between tourism and nature conservation in New Zealand are complicated, and not well documented. They have been better documented for historic heritage in urban areas (Bohlin and Hanefors 1994; Hall and McArthur 1996). Often National Parks or other reserved areas have become popular tourism destinations only after their gazettal and the establishment of some infrastructure. However, the potential for tourism has often been a factor in achieving reserve status: this has occurred throughout the history of National Parks in New Zealand from their inception in 1887 to the current discussions about creation or extension of new National Parks or equivalent reserves in Northland, the Hauraki Gulf and Whanganui (Thom 1987; Collier 1994; Kearsley *et al.* 1997).

Thus the environmental benefits from tourism are mainly indirect. For example, on the South Island West Coast, tourism revenues did not directly pay for the main network of national parks and conservation reserves, but the potential for tourism revenue has been a major factor in its development in the last decade, e.g., the establishment of the Paparoa and Kahurangi National Parks and the gazettal of the South West New Zealand World Heritage Area. Revenues from concessions in public conservation areas are available for conservation spending in the West Coast conservancy of the Department of Conservation.

Wildlife viewing is a well-established tourism activity overseas and has undoubtedly been a factor in nature conservation in southern Africa and Central America, but in New Zealand is relatively recent as a mainstream tourism activity. In the most well-known recent wildlife viewing attraction, whale-watching at Kaikoura, the activity has not been shown to have a detrimental effect on the whales or the marine environment although debate and research on the thresholds of use continue. Marine mammal based tourism has stimulated some further research on marine mammal behaviour (Gordon *et al.* 1992; Slooten and Dawson 1995; Constantine, in litt.). In some circumstances, tour guides are significant providers of information for marine conservation scientists.

At the level of individual farms or properties, the provision of farmstay accommodation, which has grown rapidly in the last decade, may in some cases have resulted in increasing investment by the property owners in nature conservation. The incentive to provide an attractive environment for visitors, plus the income from these visitors, encourages such measures as provision of habitat for birds, and retention and better management of pockets of native forest. There is an important potential for local government to encourage such measures through district and regional planning incentives and provision of information and expertise, at relatively little public cost.

Penguin Place on the Otago Peninsula is an example of a significant tourism venture which has grown out of a private conservation project. Over a decade ago, the landowner began planting native trees, placing nesting boxes and trapping predators to protect yellow-eyed penguins on their farm. The resulting increase in penguin numbers has attracted visitors and a tourism venture based around strictly controlled access to viewing the penguins and other wildlife on the property has grown rapidly in the past few years. The development of Penguin Place has created employment opportunities and has contributed to the attractiveness of the peninsula as a whole to tourists. It has also created research opportunities and encouraged people away from uncontrolled penguin nesting sites where they can interfere with the birds. Farming remains the primary land use over most of the property, although, economically, the tourism operation is a significant enterprise in its own right.

In a further South Island tourism venture run by a former director of New Zealand's largest environmental non-government organization, two upmarket "wilderness lodges" are set in natural surroundings and offer a variety of eco-tourism experiences (Hartley 1997). One is situated amid native forest at Lake Moeraki, Westland. The other is situated on a 2 400 ha sheep station adjacent to Arthur's Pass National Park. More than a quarter of this property has been retired from grazing and is managed as a conservation reserve, and the whole property is managed for conservation values as well as being maintained as a working sheep farm and providing tourist experiences.

The above examples certainly do not yet represent the norm. However, in a number of regions, declining farm incomes and farming activities on more marginal land (over several decades) have led to an increase in natural vegetation communities regenerating on previously cleared farmland. In some of these regions (e.g., Banks Peninsula, south-east Otago) there has also been a more recent increase in farm and nature-based tourism, predominantly driven by the need to augment and diversify farm incomes, as discussed in the previous section. It is possible that these two trends have become mutually reinforcing as the nature conservation values of the previously farmed land increase. On Banks Peninsula, several permanent covenants have recently been established on natural areas traversed by a private land walkway which has become a significant tourist destination.

3. Changing attitudes to conservation

Perhaps because of the above two more economically-founded factors, tourism may be a catalyst in changing attitudes to nature conservation from host communities and landowners. Natural resources may be seen not just as something to be mined or exploited but something that has values which some people are prepared to pay to visit in an undeveloped state.

Changing attitudes to conservation are not just associated with host communities. Tourists who participate in tourism in natural environments may experience changes in attitudes towards conservation, and be more inclined to support conservation after their visit. This factor is often cited as a benefit of wildlife viewing activities, even when the direct benefits to wildlife of tourism are uncertain. (It is also cited as a justification for tourism in near-pristine environments, e.g., Antarctica and the sub-Antarctic islands.)

Combination of these mechanisms

A current project in Wellington City shows how all the above mechanisms may apply in some cases. A proposal has been developed over the last six years to a develop a wildlife sanctuary in the Karori Reservoir catchment, only three kilometres from the city centre (Karori Sanctuary Wildlife Trust 1997). The 250 ha reserve, formerly a water catchment, comprises various, mainly indigenous, regenerating vegetation communities and animal species subject to various past and present degradation processes. The proposal is essentially to erect a state-of-the-art pest-proof fence around the catchment, in order to allow the permanent eradication of known pests, restoration of the habitat and reintroduction of a variety of native species which would not otherwise be able to survive in the wild.

Although conceived as a conservation project, it was realized at the outset that the project had considerable economic potential for tourism, as well as recreation and education. As the implementation of the project requires a multi-million dollar investment, the importance of tourism and recreation to the project's financial viability became more widely appreciated, as well as its potential to be one of the most important tourism attractions in the Wellington region. The current draft management plan and resource consent are based on an anticipated annual visitation of between 100 000 and 150 000 people. The plan proposes active management of visitor access, zonation of types of use and development, and a monitoring programme, in order to prevent the risks of overuse of the restored catchment.

THE RELATIONSHIP BETWEEN TOURISM AND NATURE CONSERVATION AT THE REGIONAL LEVEL

For tourism to facilitate nature conservation, on more than an isolated local basis, probably all the above factors need to be operating to some degree, as illustrated in the last example. To see how they interact, one needs to examine the regional pattern of tourism (Dowling 1993). Not all tourism activities will directly enhance nature conservation, but in order to provide a potential for enhanced nature conservation outcomes, all should be at least neutral in their direct environmental effects (see below) and make a positive economic effect at the regional level, so that nature conservation can be enabled somewhere else in the region. This general point is illustrated by the three contrasting regional sketches of tourism and the natural environment which accompany this section (Boxes 1–3). The word "regional" is used in this chapter to denote spatial scales broader than the purely local. It is used roughly in the sense of New Zealand ecological regions (McEwen 1986). In terms of local government administration units, the ecological region may not equate to the regional government unit (i.e., regional councils). For example, the spatial scale of the examples discussed in Boxes 1–3 is that of the territorial local authority.

The relationship between tourism and landscape values is not straightforward. In some instances, tourism is a factor in protecting landscapes from development as indicated by the West Coast example noted in the previous section. However, tourism can result in undesirable landscape change, often incremental, where developments are placed in pristine and scenic areas. Also, tourism can be used as an argument to prevent other land uses (e.g., forestry and farming) from changing landscapes. If tourism potential is the reason for "preserving" a landscape, that may not allow the farm/forestry sector to generate the income for private nature conservation initiatives. A recent report commissioned by the Tourism Policy Group of the Ministry of Commerce examined rural landscape from a tourism perspective, noting the range of values attached to landscape (Boffa

Miskell 1997). A workshop following the report's publication concluded that landscape is dynamic, reflecting past and present land uses and culture. It also noted that tourists come to New Zealand to see New Zealand culture, so a changing landscape will not necessarily lose its tourism potential. Another important point raised at that workshop is that landscape should be managed for New Zealanders, reflecting our aspirations and lifestyles.

FACTORS REDUCING THE POTENTIAL FOR TOURISM TO FACILITATE NATURE CONSERVATION

Adverse tourism effects

It is recognized that tourism has wide-ranging physical, social, economic and cultural effects some of which have the long-term potential to seriously damage both the environment and the industry itself. These factors have been the subject of considerable, if *ad hoc*, study in New Zealand and elsewhere, including a number of syntheses and inventories of effects (Devlin *et al.* 1995; Tourism Policy Group 1995, 1996; Ward and Beanland 1996; Department of Conservation 1997; Parliamentary Commissioner for the Environment 1997a,b). Many of these effects were identified in interviews with and submissions from interested parties during the Parliamentary Commissioner for the Environment's investigation.

The Parliamentary Commissioner for the Environment identified three broad negative environmental effects which were felt to be of particular importance. They were:

- loss of quality of some relatively unspoilt parts of New Zealand's natural environment (both physical and intrinsic values);

BOX 1. ROTORUA DISTRICT

Rotorua is one of New Zealand's main tourism destinations for both domestic and international visitors. In 1996/97, an estimated 1.3 million people visited Rotorua, of which 55% were New Zealanders (Tourism Rotorua, unpubl. 1997). In 1992/93, visitors are estimated to have contributed $235 million to the local economy (Rotorua District Council 1994).

In the Rotorua District, visitors are concentrated in a few areas, mainly around the lakes, geothermal areas and Rotorua itself. The district is largely farming and forestry based, being essentially rural, although tourism is now the largest earner of revenue. Many parts of the district are not directly affected by tourism, and economic benefits also tend to be concentrated.

Tourism has had both positive and negative effects on nature conservation in Rotorua District. In some cases, natural areas have been protected on account of the tourism potential. The geothermal area at Whakarewarewa is a good example. Here, users of geothermal power and water in the area surrounding Whakarewarewa had their use drastically curtailed when it was realized that commercial and residential utilization of the resource was damaging the geothermal area and threatening its attractiveness to visitors. Department of Conservation-administered land and some places run under concession leases, such as the Waimangu volcanic valley, are also protected. Waimangu is carefully managed to minimize the impacts of visitors on the area; protecting both its natural and tourism values.

There are also cases where tourism has resulted in pressures on natural areas. Most widely noted is the incremental pressure that tourism-related developments place on the unique Rotorua Lakes environment. Lack of conservation monitoring information was widely noted as a significant constraint to the management of environmental effects. A recent proposal (now not proceeding) for a big hotel and golf course development on undeveloped land between Lakes Okareka and Tarawera was seen by many in the community to epitomize the threat that tourism poses for natural areas and amenity values. A proposal to build a rack railway to the top of Mount Tarawera, is also seen by some to be another blow to the integrity of the natural environment, although others see the proposal as a sensible mechanism for controlling access and thereby the effects of visitors on the mountain environment.

In Rotorua, the effects of tourism on conservation and natural areas are also related to the aspirations of Maori. Local Maori have a strong interest in the management of natural resources and a long involvement in tourism, and have both development and conservation objectives.

- loss of amenity values from incremental development (in both rural and urban environments);
- pressure on infrastructure, such as sewerage and roads, often seasonal, resulting in significant costs to small communities.

Both of the first two effects, especially the first, are very significant for nature conservation. The Commissioner discussed several qualifications on the environmental effects identified, including that many of the effects identified are perceptual, much of the evidence for them is anecdotal, and the actual extent, scale, severity and nature of the effects are much harder to determine than the perceptions. He noted that social and cultural effects are widely perceived as being of more concern than biophysical effects at the present time.

Some environmental effects of tourism are of particular significance to Maori, including some which affect natural resources that also have very high natural conservation values generally, e.g., mountains and waterways (Parliamentary Commissioner for the Environment 1997a). These effects must be recognized and respected, and appropriate provisions for dealing with them integrated into assessment, planning and management for the tourism sector.

This section has concentrated only on negative effects associated with tourism. A number of positive effects were also identified in the Commissioner's report, as discussed in previous sections of this chapter. It is of course the positive effects which allow the potential for nature conservation in production environments, as long as the potential for negative effects is also recognized and that these effects, where present, are avoided or mitigated.

Inadequate system of tourism management

This was a principal focus of the Parliamentary Commissioner for the Environment's investigation (Parliamentary Commissioner for the Environment 1997a). The central Term of Reference of that investigation was to review and assess the allocation of functions to public authorities involved in the tourism sector, and the implementation of those functions, with particular reference to the effectiveness of public authority management of environmental effects associated with the tourism sector.

A principal finding in this regard was that the government system for management of tourism is fragmented, especially between agencies which promote tourism and those which manage its effects. There are few mechanisms for communication and co-ordination between different parts of the tourism sector, which is itself very fragmented.

Furthermore, the management of environmental effects in the tourism sector is more difficult because New Zealand's strategic approach to tourism has been primarily focused on maximizing the benefits from increased international visitor numbers. Government agencies have very little ability to influence the direction of the tourism industry and thus its environmental effects. Since the formation

BOX 2. WAITOMO CAVES

The village of Waitomo Caves is a settlement of about 200 people in the Waitomo District, northern King Country, North Island. It is principally known for its famous limestone caves, especially the Glow-worm Cave. In 1995/6, 450 000 people visited the Glow-worm Cave, mostly on day package tours, but including an increasing number of independent travellers, many of whom stay in the area longer and participate in other activities such as walks, garden and farm visits, cave rafting, visits to other caves, and horse trekking. Such activities are strengthening links between tourism and nature conservation, which have not been strong in the past. Also, because of the preponderance of day visitors, the district has probably experienced fewer adverse social and cultural effects than other districts with comparable visitor numbers.

The relationship between the Waitomo Caves and their surroundings is now increasingly recognized after a number of problems in the last 20 years with aspects of the cave environment. Some of these problems were linked to flooding and siltation and this has illustrated the links between the caves and their surroundings. It is now generally recognized that to safeguard tourism in the area, attention needs to be given to land management. Landowners in the caves, catchment and the regional council have now acknowledged this by forming a landcare group which takes an active role in managing properties to protect the catchment and the caves. This should have benefits for nature conservation as the landcare group (and other individuals and groups) promote and implement catchment protection practices such as retiring land, planting riparian strips, and using buffer zones for plantation logging. There is also heightened concern about farm practices that might harm the caves, such as chemicals used on the soil and plants and the risks of chemical spillage.

BOX 3. QUEENSTOWN-LAKES DISTRICT

Queenstown is one of the main tourism destinations in New Zealand. Although the district is predominantly rural, tourism is by far the largest source of income. It is estimated that 95% of people in Queenstown are dependent either directly or indirectly on tourism. In 1996, there were 630 000 visitors to Queenstown itself. International visitors alone are estimated to have spent $260 million in the Southern Lakes area in 1996 (New Zealand Tourism Board 1997).

The entire Queenstown-Lakes district is heavily dependent on tourism. A wide range of activities, including skiing, sightseeing, adventure activities, homestays and general relaxation, are undertaken throughout the district.

As with Rotorua, tourism has had mixed impacts on nature conservation in the district. The heavy use of national parks and conservation areas by tourism operators in the district confirm their reliance on nature and amenity conservation. However, there are concerns that tourism itself is at risk of eroding the very natural values upon which it relies. The use of helicopters to take visitors to the top of the Remarkables for hiking trips, the huge number of buses plying the Milford Road, the noise of small aircraft, and the visual impact of skifields and their access roads are all examples of the negative effects that tourism can have on the natural environment.

Where the conservation estate is being used by concessionaires, revenue generated is used within the conservancy. However, the amount raised is relatively small, and there is no obligation for the monies to be used directly for nature conservation. Some systematic monitoring of environmental effects is undertaken. In general, there is little indication that money generated by private tourism operations has been used for nature conservation on private land.

In Queenstown-Lakes District, productive land is being subdivided for urban and peri-urban residential development, largely due to the rapid growth of tourism experienced recently. There are concerns that such pressures may also adversely affect the open landscape qualities that for many are a strong element of the district's scenic appeal. Because of the open nature of the landscape — largely pasture, tussock and mountains — there is not the same pressure to conserve small portions of it as it is the whole nature of the district that gives it unique character and so much of the environment has already been extensively modified by farming and goldmining practices. Notwithstanding this, the recreation and tourism potential and use of the Kawerau River were strong arguments used in the recent application for (and subsequent granting of) a water conservation order for the river. The revenue generated by a concession for bungy jumping off the historic Kawerau Bridge was used by the Department of Conservation to restore and maintain the bridge.

Tourism potential is a strong factor in the management of land use in the district. Limitations placed on the planting of forest in the Wakatipu Basin in the current Queenstown-Lakes District Plan are largely due to the impact that this would have on the landscape.

of the New Zealand Tourism Board in 1991, its effective marketing activities have largely driven the New Zealand tourism sector, to an extent that may have outstripped the capacity of the sector to manage the environmental consequences of tourism growth. This may be related to the setting of New Zealand Tourism Board objectives in isolation from more wide-ranging objectives for the sector as a whole.

The result of this situation is a fragmented system where some important areas of tourism management "fall through the cracks". In the context of this chapter, the most notable example is where environmental effects are not managed in a way that enables tourism management and nature conservation objectives to be met simultaneously. Overall the system is not good at doing this, although the Commissioner concluded that Department of Conservation, which has the primary responsibility for managing the environmental effects associated with tourism on the conservation estate, should retain responsibility for visitor services, including concessions management, as part of its management responsibilities. Of all national agencies, the Department of Conservation is the only one with the statutory and functional ability to co-manage tourism and nature conservation functions.

An example of conflicting agency functions may be seen in the drive throughout the early to mid-1990s throughout the tourism sector to provide more hotel accommodation for the rapidly growing international tourism market. At times this may have been without due regard for the effects of new hotel developments on environmental values including the natural environment (see Box 1). In such situations territorial local authorities are often faced with conflicting internal roles for both economic development promotion and for environmental management and regulation. They have sought to resolve this conflict, with varying success, through transparent separation of functions, and through statutory processes under the *Resource Management Act 1991,* which have recently been supported by a useful guide to the *Resource Management Act* for the tourism sector (New Zealand Tourism Board *et al.* 1996).

In some areas funding issues need attention. This is poorly documented in spite of the vociferous debate in recent years about the adequacy of resources for conservation management. In the tourism management area it appeared that the management of tourism effects on the conservation estate was only one of the key problem areas; the others identified were provision of infrastructure at the local level, and the gathering of data for tourism management (Parliamentary Commissioner for the Environment 1997a). Research and analysis are needed to accurately determine the current levels of investment by government agencies in the tourism sector, and appropriate means of targeting resources to ensure effective management of the environmental effects associated with tourism.

Given that tourism in New Zealand is overwhelmingly driven by the private sector, and that many of the positive initiatives for linking tourism and nature conservation are private ones, it may be that strengthening these linkages is not simply a matter of directing more resources at government agencies, but rather in ensuring that those agencies are able to support and at times facilitate other organizations and individuals. Local government (including local and regional tourism organizations) obviously have a vital role in this respect. The Commissioner concluded that outside the conservation estate, local authorities have the primary responsibility for managing the environmental effects associated with tourism. Their performance in this role varies widely.

ECO-TOURISM AND SUSTAINABLE TOURISM

Eco-tourism is defined as "nature-based tourism that involves education and interpretation of the natural environment and is managed to be ecologically sustainable" (Commonwealth Department of Tourism 1994). This definition recognizes that "natural environment" includes cultural components and that "ecologically sustainable" involves an appropriate return to the local community and long-term conservation of the resource.

There is obvious potential for eco-tourism in linking tourism and nature conservation (e.g., Penguin Place and many other new ventures). Eco-tourism's focus on the environment, education and positive contributions to communities are recognized as creating opportunities for nature conservation and economic production to coincide. The potential has been recognized by the New Zealand Tourism Industry Association special award for environmental tourism which has usually been awarded to eco-tourism ventures. However, eco-tourism is not inevitably favourable for nature conservation. Eco-tourism has its own management implications. By taking visitors into natural, and sometimes fragile or pristine areas, the potential for damage to those areas is increased. Also, on a regional or national scale, there is also little point having exemplary eco-tourism, no matter how positive its nature conservation contribution or outcomes, if this is dwarfed in scale by "mass tourism" that has significant adverse environmental effects.

It is therefore more important to focus on the sustainability of tourism generally to ensure that all tourism is a sustainable land use and activity. Such tourism would at least not detract from or interfere with nature conservation, and eco-tourism would be a subset of sustainable tourism (Blangy 1997). Some suggested components of sustainable tourism are listed in Table 1. These principles go beyond a set of nature conservation principles and in fact we suggest that sustainable nature conservation projects would need to display a similar set of components.

Table 1. Suggested components of sustainable tourism.

- The resources of the environment must be protected.
- The environment has an intrinsic value beyond its value as a tourism asset. The process of tourism development should respect this intrinsic value.
- Visitors must be given a quality experience.
- Local communities must benefit both economically and in quality of life.
- Sustainable tourism management involves protecting and enhancing a natural attraction for future generations, which means its long-term survival must not be prejudiced by short-term considerations.
- Tourism should encourage a positive relationship between the attraction, the visitors and the host community.
- Tourism activities and developments should respect the scale, nature and character of the place in which they are located.
- Tourism sustainability also involves meeting the needs of present visitors and of host communities — which may be identified Maori interests, a rural community, a town, or New Zealand in general.

Source: after Sowman 1992; Blangy 1997.

CONCLUSION

In this chapter we have introduced some ideas about tourism as a land use and an activity, and its potential for nature conservation. We have been positive about this potential but mindful of the factors that can negatively affect it. The linkages between tourism and nature conservation appear at present to be mainly concerned with economic values.

The examples used in this chapter indicate that in important respects the potential of tourism for nature conservation is not yet realized in New Zealand, except in very localized situations, and perhaps in a very general "macro-economic" sense of generating income to allow for the maintenance of certain levels of public and private expenditure on nature conservation. To better realize the potential will require: closer focus on the desired environmental outcomes for both tourism and the natural environment; more attention to the system for the management of environmental effects associated with the tourism sector; more research on these effects (including development of environmental indicators); and, above all, the development and realization of a vision and strategy for sustainable tourism for the whole of the sector in New Zealand (Parliamentary Commissioner for the Environment 1997a).

Perhaps, then, it is going too far to interpret tourism in New Zealand at present as a strong nature conservation activity ***in itself***. Maybe its potential lies most as an activity that through its economic benefits integrates and binds traditional production and conservation, hence the "glue" analogy alluded to in the title of this chapter.

To conclude by expanding on this analogy, the glue (tourism):

- must not contain harmful ingredients;
- must be correctly applied (not too thickly);
- must be applied in the right places (not the most fragile ones);
- should be applied to a large (regional) matrix containing a number of different elements;

but if these conditions are met,

- then the matrix can be strong, resilient, and good for nature conservation!

ACKNOWLEDGEMENTS

We thank Stephen Fuller, Harvey James, Howard McGrouther, David Norton, Julie Warren and Bruce Watson for information and insights provided to us, and Dick Brathwaite, Rochelle Constantine, Ronda Cooper and Robert Sowman for constructive comments on earlier drafts.

REFERENCES

Blangy, S., 1997. A few concepts and definitions. Pp. 7 *in* Sustainable Tourism Development ed by J-P. Ribout. *Naturopa* No. 84.

Boffa Miskell, 1997. Landscape from a Tourism Perspective: Background Paper. Tourism Policy Group, Ministry of Commerce: Wellington, New Zealand.

Bohlin, M. and Hanefors, M., 1994. Heritage management in New Zealand and Australia. *Ann. Tour. Res.* **21**: 2.

Brandon, K., 1996. Ecotourism and Conservation: a Review of Key Issues. World Bank Environment Department Papers No. 033.

Butcher, G., Fairweather, J. R. and Simmons, D. R., 1998. The Economic Impact of Tourism on Kaikoura. Tourism Research and Education Centre (TREC), Lincoln University, Report No. 7.

Cloher, D. U., 1998. A sustainable Maori tourism in Northland, New Zealand. *In* Proceedings New Zealand Tourism and Hospitality Third Biennial Research Conference, December 1998, Akaroa, New Zealand.

Collier, A., 1994. Principles of Tourism: a New Zealand Perspective (third edition). Longman Paul: Auckland, New Zealand.

Commonwealth Department of Tourism, 1994. National Ecotourism Strategy. Commonwealth of Australia: Canberra, Australian Capital Territory.

Department of Conservation, 1997. Impacts of Visitors on Natural and Historic Resources of Conservation Significance. Part 1 Workshop Proceedings. Part 2 Research and Information Needs. *Science and Research Internal Report* Nos 156 and 157. Department of Conservation: Wellington, New Zealand.

Devlin, P. J., Corbett, R. A. and Peebles, C. J., 1995. Outdoor Recreation in New Zealand. Vol. 1: A Review and Synthesis of the Research Literature. Department of Conservation and Lincoln University: Lincoln, New Zealand.

Dowling, R. K., 1993. An environmentally based planning model for regional tourism development. *J. Sustain. Tour.* **1**: 17–37.

Gordon, J., Leaper, R., Hartley, F. G. and Chappell, O., 1992. Effects of whale-watching vessels on the surface and underwater acoustic behaviour of sperm whales off Kaikoura, New Zealand. *Science and Research Series Report* No. 52. Department of Conservation: Wellington, New Zealand.

Hall, C. M. and McArthur, S., 1996. Heritage Management in New Zealand and Australia: The Human Dimension (second edition). Oxford University Press: Auckland, New Zealand.

Hartley, P., 1997. Conservation Strategies for New Zealand. New Zealand Business Roundtable: Wellington, New Zealand.

Henley, M., Poharama, A., Smith, A., Simmons, D. R. and Fairweather, J. R., 1998. The Impact of Tourism on the Maori Community in Kaikoura. Tourism Research and Education Centre (TREC), Licoln University, Report No. 8.

Johnston, C. S., 1998. Resident perceptions of tourism among Maori in North Hokianga: a report of survey results. *In* Proceedings New Zealand Tourism and Hospitality Third Biennial Research Conference, December 1998, Akaroa, New Zealand.

Karori Wildlife Sanctuary Trust, 1997. Karori Wildlife Sanctuary Draft Management Plan. Karori Wildlife Sanctuary Trust: Wellington, New Zealand.

Kearsley, G. W., 1997. Managing the consequences of over-use by tourists of New Zealand's conservation estate. Pp. 87–98 *in* Tourism Planning and Policy in Australia and New Zealand: cases, issues and practices ed by C. M. Hall, J. Jenkins and G. Kearsley. McGraw Hill Australia: Sydney, New South Wales.

Kearsley, G. W., Hall, C. M. and Jenkins, J., 1997. Tourism planning in natural areas: introductory comments. Pp. 66–74 *in* Tourism Planning and Policy in Australia and New Zealand: cases, issues and practices ed by C. M. Hall, J. Jenkins and G. Kearsley. McGraw Hill Australia: Sydney, New South Wales.

McEwen, W. M., 1986. Ecological Regions and Districts in New Zealand. Department of Conservation: Wellington, New Zealand.

New Zealand Tourism Board, 1996. Tourism in New Zealand: Strategy and Progress. New Zealand Tourism Board: Wellington, New Zealand.

New Zealand Tourism Board, 1997. Draft Southern Lakes Tourism Strategy. New Zealand Tourism Board: Wellington, New Zealand.

New Zealand Tourism Board and Department of Conservation, 1993. New Zealand Conservation Estate and International Visitors. New Zealand Tourism Board and Department of Conservation: Wellington, New Zealand.

New Zealand Tourism Board, Ministry for the Environment and New Zealand Tourism Industry Association, 1996. Tourism's Guide to the Resource Management Act. New Zealand Tourism Board, New Zealand Tourism Industry Association and Ministry for the Environment: Wellington, New Zealand.

Parliamentary Commissioner for the Environment, 1997a. The Management of the Environmental Effects Associated with Tourism in New Zealand. Parliamentary Commissioner for the Environment: Wellington, New Zealand.

Parliamentary Commissioner for the Environment, 1997b. The Management of the Environmental Effects Associated with Tourism in New Zealand: Review of Literature on Environmental Effects. Parliamentary Commissioner for the Environment: Wellington, New Zealand.

Ribout, J-P. (ed), 1997. Sustainable Tourism Development. *Naturopa* No. 84.

Rotorua District Council, 1994. Tourism and its Impact on the Rotorua District Economy. Updated from December 1991. Rotorua District Council: Rotorua, New Zealand.

Sage, E., 1995. Beyond the picture postcard. *Forest and Bird* No. **275:** 20–31.

Slooten, E. and Dawson, S. M., 1995. Conservation of marine mammals in New Zealand. *Pac. Conserv. Biol.* **2:** 64–76.

Sowman, R., 1992. Tourism sustainability — a discussion paper. Ministry of Tourism: Wellington, New Zealand.

Statistics New Zealand, 1998. New Zealand Official Yearbook 1998. GP Publications: Wellington, New Zealand.

Thom, D. A., 1987. Heritage: the Parks of the People. Lansdown Press: Auckland, New Zealand.

Tourism Policy Group, Ministry of Commerce, 1995. Tourism Research Bibliography 1985 — mid 1994. Ministry of Commerce: Wellington, New Zealand.

Tourism Policy Group, Ministry of Commerce, 1996. Tourism Research Bibliography 1994–1996. Ministry of Commerce: Wellington, New Zealand.

Ward, J. C. and Beanland, R. A., 1996. Biophysical Impacts of Tourism. Information Paper No. 56. Centre for Resource Management, Lincoln Environmental, Lincoln University: Lincoln, New Zealand.

CHAPTER 39

Nature conservation in New Zealand marine production environments

R. G. Cole,[1] K. R. Grange,[1] D. J. Morrisey,[2] M. Gibbs,[2] M. R. James[3] and A. Ross[3]

Marine production via harvest fisheries and aquaculture is an important activity in the coastal zone of New Zealand, with the potential to conflict with conservation. We contrast terrestrial and marine environments and organisms, particularly with respect to the use of "islands" as conservation entities. Focusing on aquaculture, we consider how nature conservation may be affected by human activities, considering the legislation involved, reviewing the effects of different types of aquaculture, the possibility of food web effects, conflicts between aquaculture and native fauna, the "green-ness" of aquaculture, and the role of aquaculture in spreading invasive species. We review opportunities for enhancement of nature conservation in production areas, consider linkages with the terrestrial environment, note current technologies which assist production and finally consider the costs and benefits of aquaculture relative to alternative sources of protein for human consumption.

INTRODUCTION

CONCERNS about loss of habitat and biological diversity are all-pervasive for terrestrial habitats. Understanding such matters is proceeding much more slowly below the high tide mark, however, which is of concern since global fish and shellfish production and consumption continue to increase. The aquaculture industry has grown markedly over the last three decades, producing ever-increasing proportions of world fish production relative to capture fisheries. In 1995, aquaculture produced almost 21 million MT of fish and shellfish, worth ~US$36 billion, 18.5% of total seafood supply, and 27% of seafood consumed by humans (Goldburg and Triplett 1997). However, there have been social and environmental costs of this, leading to moratoria on the development of salmon farm sites in Ireland and Norway, and legal proceedings to remove shrimp farms in India (Goldburg and Triplett 1997). Salmon and shrimp mariculture are notoriously resource-intensive (Beveridge *et al.* 1994). Beveridge *et al.* (1997) point out that environmental problems cost Asian aquaculture at least US$ 3 billion per annum. Here we examine the possibilities for nature conservation in marine production environments in New Zealand, focusing particularly on aquaculture. Marine conservation in New Zealand was recently reviewed by Creese and Cole (1995). Slooten and Dawson (1995) reviewed the conservation of marine mammals, which potentially conflicts with some usages of the marine environment, though those authors make no mention of aquaculture *per se.*

Definitions

Here we consider conservation to be "sustainable use" *(Resource Management Act 1991)*. Production environments are defined to be those in which organisms are harvested commercially from environments which are managed. This explicitly excludes fishing production, which occurs in "natural" or at least unmanaged environments (see Andrew and Pepperell 1992; Dayton *et al.* 1995 for reviews of fishing effects). Aquaculture is the cultivation of organisms for human usage in aquatic environments (we will only consider marine and estuarine environments).

[1]National Institute of Water and Atmospheric Research Ltd, P.O. Box 893 Nelson, New Zealand.
[2]National Institute of Water and Atmospheric Research Ltd, P.O. Box 11–115, Hamilton, New Zealand.
[3]National Institute of Water and Atmospheric Research Ltd, P.O. Box 8602, Riccarton, Christchurch, New Zealand.

Pages 395–404 *in* NATURE CONSERVATION 5: NATURE CONSERVATION IN PRODUCTION ENVIRONMENTS: MANAGING THE MATRIX ed by J. L. Craig, N. Mitchell and D. A. Saunders. Surrey Beatty & Sons, 2000.

DIFFERENCES BETWEEN TERRESTRIAL AND MARINE ENVIRONMENTS

The marine environment differs from terrestrial systems in a number of important ways that have meant that nature conservation has lagged behind that on land. The marine environment in general is less well understood than terrestrial systems, and the lack of visual information means that dramatic subtidal changes may occur undetected by humans. Endemism is thought to be high in both marine and terrestrial environments in New Zealand (Towns and Ballantine 1993).

There is high connectivity among sites via currents, and few barriers to dispersal. Sessile marine organisms must cope with movement of a very dense medium to remain attached to the substratum; the nearshore environment is often a harsh one and size may be limited by purely physical constraints (e.g., Denny *et al.* 1985). The density of seawater also permits it to move other objects which may pose further hazards for sessile organisms (e.g., logs — Dayton 1971, stones — Shanks and Wright 1986). However, the density of seawater also means that there is much less requirement for structural support tissue. The obvious terrestrial contrast is the extensive human use of cultivated forests for timber — plant structural support tissue. On shallow temperate rocky subtidal shores, the habitat-forming organisms (large brown seaweeds) have a much shorter lifespan than terrestrial habitat-formers such as trees — 3–10 years for seaweeds, 20–500+ years for trees — making marine systems more resilient (*sensu* Connell and Sousa 1983) to disturbances. However, other systems may be more vulnerable (e.g., seagrasses — de Casabianca *et al.* 1997, sponges and corals — Dayton *et al.* 1995). Large sessile marine organisms (e.g., some seaweeds) escape the mechanical limits to size by becoming flexible (Denny *et al.* 1985). The characteristics of marine and terrestrial environments are thus very different on the basis of the physical nature of the media alone.

Many marine areas are subject to intermittent harvesting by methods that disrupt the habitat. For example, it has been estimated that the Dutch sector of the North Sea is trawled once or twice per year (Lindeboom 1995). Because of the system of public ownership it will always be most profitable to harvest a species until returns drop, and then move on to another species (Botsford *et al.* 1997).

DIFFERENCES BETWEEN TERRESTRIAL AND MARINE ORGANISMS

Much marine primary production stems from phytoplankton, which differs from terrestrial primary production in that it is mobile, and may undergo massive intra- and inter-annual variations in its abundance and quality. A few sedentary marine primary producers are cultured (see Schiel and Nelson 1990 for New Zealand review), but these require a substratum to grow on and such artificial substrata are expensive to maintain. The *Macrocystis* kelps are cropped from natural substrata in the United States of America (North 1994), and other macroalgae are harvested in South America (Castilla and Bustamante 1989, Edding *et al.* 1994). Because *Macrocystis*' surface canopy only is cropped, and most of the organisms that depend on it live in the water column, there is little evidence that such cropping has any effects on other organisms. However, other macroalgal species are harvested in their entirety, rendering space on the substratum available for colonization by other species. There may be colonization differences due to the timing of harvesting (e.g., Schiel 1988), suggesting that management may be required to maintain macroalgal species composition. Some of the greatest difficulties for macroalgal culture lie with our lack of knowledge of their taxonomy and ecology. For example, the red alga *Curdiea coriacea* produces an agar product with very desirable thermal properties. However, an extensive research project over three years failed to develop a successful culture technique for this species due, in part, to the extreme scarcity of male plants in natural populations (fewer than five found over five years) (R. G. Creese, pers. comm.). In contrast to terrestrial farming, where very few secondary consumers are cultured (poultry being a notable exception) many cultured aquatic species are secondary consumers (e.g., shrimp, salmon, snapper).

Most cultured marine organisms have some type of dispersive life history stage, which may be in the water column for periods from seconds to months. Many cultured primary consumers are immobile and depend on water motion to supply food particles. This has practical advantages for aquaculturists (e.g., no fencing required) but also means that structures must be provided for organisms to dwell on. The reproductive characteristics of marine invertebrates are also distinct from most cultivated terrestrial organisms; many are capable of reproducing both sexually and asexually. This obviously has advantages for husbandry, allowing faster selection for individuals with desirable characteristics. Generally asexual reproduction involves short dispersal distances and sexual reproduction long dispersal distances, but there may be exceptions (e.g., antipatharians (black corals) — Miller and Grange 1997; Miller 1997, 1998).

Because of this dispersive capability, Carlton *et al.* (1991) suggested that most marine invertebrates (at least) have large effective population sizes, and hence are relatively invulnerable to extinction. Only one marine species (a limpet with a specialized habitat) is known to have become extinct in recent times (Carlton *et al.* 1991, but see Tegner *et al.* 1996). However, 80% of the 148 marine fishes evaluated by a 1996 workshop were considered "threatened" and were proposed for inclusion in the 1996 International Union for the Conservation of Nature Red List (Vincent and Hall 1996). There are conflicting views concerning the likelihood of marine extinctions (e.g., Malakoff 1997). Vermeij (1993) suggested that regions of high productivity should be a priority for protection from over-exploitation and habitat destruction. However, if exotic species are to be cultured, the dispersive life history stage will potentially allow them to colonize new areas rapidly.

In marine systems there have probably been few extirpations (but see Malakoff 1997); changes in the relative abundances of species, and perhaps population size structure are more likely. Generally it is difficult to establish the effects of human intervention, because we know little of the marine environment prior to the advent of SCUBA in the 1960s (though there are some historical records for species visible from the surface, e.g., *Macrocystis* forests) (Dayton *et al.* 1998). In the terrestrial environment deforestation is obvious, and there are large tracts of land where deforestation has not occurred with which we can make comparisons. There are no unexploited marine areas (apart from recently established marine reserves) with which to compare our coastal environments, all of which have been fished intensively, most since at latest the 1960s. This means that expectations regarding our coastal ecosystems are reduced (Dayton *et al.* 1998).

Further, we are unsure how to "enhance" the environment. This involves a value judgement regarding desirable qualities of habitats. For example, the early literature regarding sea urchin grazing of kelp forests in California demonstrated such an attitude. Here, kelp forests were judged to have more value than unvegetated areas (e.g., Leighton 1962; Leighton *et al.* 1966), partly because of the kelp's commercial value. It is not clear what is the "natural" condition of rocky reef environments, even for the relatively well-studied reefs of northeastern New Zealand. The productivities of different subtidal rocky habitats in northeastern New Zealand were compared by Taylor (1998), who suggested that finely structured seaweeds provided habitat for small mobile macroinvertebrates, which contributed about 80% of total annual secondary productivity on reefs. The establishment of kelp forests will bring about some predictable changes in the species composition of fishes on rocky reefs (Jones 1984; Choat and Ayling 1987). However, the feasibility and effects of enhancement of other natural habitats are largely unknown, with the possible exception of seagrasses (e.g., Fonseca *et al.* 1988, 1996a,b) and kelps (Schiel and Foster 1992).

ISLANDS AS CONSERVATION ENTITIES

Islands (areas permanently emersed but separated from other areas by water) cannot be used as conservation entities for marine organisms in the same way as they can for terrestrial systems. New Zealand has an illustrious career of restoring islands as terrestrial reserves (e.g., Clout and Saunders 1995; Mansfield 1996). Islands are not islands underwater however — they are constantly being invaded by larvae. There are advantages to marine reserves at island sites; they may themselves have unique flora and fauna (e.g., Choat and Ayling 1987; Kingsford *et al.* 1989; Cole *et al.* 1992). On the other hand, re-establishing a population of a species that has widely-dispersive planktonic larvae may see reproductive output going "offshore". This may in itself be advantageous; it might be possible to establish islands as propagule sources for some species. For example, the Three Kings Butterfish *Odax cyanollix* probably occurs in northern New Zealand (F. J. Brook, unpubl.) because of larvae from source populations at the Three Kings Islands. Island reserves are also easy to enforce and control.

Conservation in marine systems has emphasized preservation of habitat rather than preservation of species. Ecosystem conservation is easier because of the constant rain of larvae; single species restorations are unlikely (see below for exceptions). However, marine conservation to date has essentially involved merely preventing further harvesting, with little or no attempt at restoration. Restoration of populations is probably an option for a limited range of taxa and techniques. Laminarian macroalgae, ascidians, sponges, and abalone are the most likely candidates for this, because they generally have limited larval dispersal, and populations are likely to become self-sustaining due to localized recruitment. Only sedentary organisms might be expected to remain on islands in moderate numbers if they were re-introduced, although homing behaviour may be (unexpectedly) demonstrated by organisms. Thompson (1983), for example, found that transplanted small (~10 cm total length) reef fish homed over 1 km or more. Pilot restoration projects are under way at several New Zealand sites, with hatcheries being used to supply juvenile bivalves to restore shellfish beds, and test replanting of *Zostera* being undertaken. For seagrasses, recent studies suggest that reduced genetic diversity may result from transplants (Williams and Davis 1996), but compared to some of the genetic bottlenecks of New Zealand birds (review in Lambert and Millar 1995), for example, these may be trivial.

Production environments may themselves act as island equivalents. Marine farming often involves the provision of hard substrata which plants and animals can grow on, usually in an environment which is not suitable for their growth otherwise e.g., pelagic, or sandy sediments. Marine farms could therefore act as larval sources for areas downstream, thus contributing to diversity. Diseases of marine organisms may have large effects (e.g., Lafferty and Kuris 1993) and reports of mass mortalities of organisms are increasingly common (e.g., Butler *et al.* 1995). Consequently, having farmed populations of valuable species might be a double-edged sword. Farmed populations might act as refuge populations if wild populations approach extinction (Malakoff 1997), but considerable difficulties might arise via transporting diseases with stocks. In New Zealand the Mussel Industry Council has funded research investigating whether toxic phytoplankton might be transported with mussel spat from Kaitaia to the Marlborough Sounds.

MARINE PRODUCTION ENVIRONMENTS — AQUACULTURE

Annual growth rate of global cultured finfish and shellfish production increased from 5 to 6%

in 1990/91 to 14% in 1994/95, whereas capture fisheries have had near zero or negative growth (Fisheries and Agriculture Organization 1996). With fisheries production approaching full exploitation, aquaculture offers an attractive solution (e.g., Kearney *et al.* 1996). The importance of managing aquaculture to achieve biological diversity as well as food production has been noted (Mathias and Moyle 1992; Beveridge *et al.* 1994). Most aquaculture environments are nearshore, due to operational and economic considerations, and aquaculture structures overlie, but do not occupy, areas of seafloor. This potentially places them in conflict with several other sectors; recreational use of the coastal marine area may be affected by aquaculture structures, capture fisheries may lose potential fishing area to aquaculture structures, and there may be a wide variety of other impacts (visual, environmental, food chain, benthic), though intertidal farming of cockles and oysters does physically occupy the benthic environment itself.

New Zealand's regulatory environment for marine conservation and aquaculture

Marine areas are regulated under several statutes, most notably the *Resource Management Act 1991* (reviewed by Rosier and Hastie 1996). Waters to 12 miles offshore, including areas used for aquaculture, are the responsibility of unitary and Regional Councils, though Ministry of Fisheries retains responsibility for issuing licenses to harvest cultured stock under the *Fisheries Act 1996*. A small percentage of the coastline falls within marine protected areas (under the *Marine Reserves Act 1971*), the majority of which are not available for production (though organisms that leave the marine protected areas are able to be harvested). There are particular areas which have their own regulations (e.g., Tasman Bay bryozoan beds protected under the *Fisheries Act,* marine parks established by local authorities). Marine mammals are protected by the *Marine Mammal Protection Act 1978.* Seabirds are generally protected under the *Wildlife Act 1953.* A resource consent for the structure and a MFish permit to harvest stock are required for aquaculture. Appeals against decisions by councils may be made to the Environment Court. Important features of this new framework for coastal planning include its emphasis on the effects of activities, and the greater responsibility it places on industry and (human) communities to improve coastal management (Rosier and Hastie 1996).

Mechanics of aquaculture

A range of structures are used in aquaculture in New Zealand. Racks are used to grow oysters intertidally, while various longline systems are used to grow mussels, sponges and oysters. Longline systems involve rows of buoys on the surface, with rope droppers on which the mussels grow, extending about 12 m down into the water. Farming of paua is usually carried out in tank structures onshore, with seawater being either recirculated, or (more commonly) pumped fresh from the sea nearby. Cage farming of finfish is very localized, and typically involves large floating structures from which nets are suspended, and often supporting facilities to store the large volumes of food required. Since feeding of stock is a continuous process, and considerable maintenance must be undertaken, there is also often accommodation for staff at such farms.

Most published studies of the effects of farming of finfish have been done at sites shallower than the 25+ m used for salmon farming in New Zealand (Ye *et al.* 1991: 12 m; Lumb 1989 — most <20 m; Johnsen *et al.* 1993 — 11–14 m; Holby and Hall 1991 — 18–21 m; Hargrave *et al.* 1993 — 12–14 m). Similarly mussel culture is mainly carried out in areas deeper than 20 m in New Zealand, but in much shallower sites elsewhere (e.g., Hatcher *et al.* 1994 — 7 m). The greater water depth may lead to a diminution of the severity of the impact but an increase in the area affected.

IMPACTS OF AQUACULTURE ON THE ENVIRONMENT

Visual impacts

There may be visual impacts of structures for all types of aquaculture, though the extent of these will depend on the crop, and the level of husbandry required. For species which eat phytoplankton, buoy (subtidal species) or rack (intertidal species) systems are used to suspend the stock in the food-bearing current. The number and size of these structures will depend on the crop, with much lower crop weights — and hence lower buoy sizes and numbers — being possible for sponge culture than for mussels. Visual impacts will depend greatly on locality, and there are complex considerations.

Occupation of a site by a marine farm prevents fishing and other usages, which may damage habitats more than farming itself. The impacts of farms on benthic fauna must be weighed against these other effects. In the Marlborough Sounds marine farms may be an effective barrier to scallop dredging of the nearshore sandy slopes. Such dredging may be more physically damaging (e.g., Thrush *et al.* 1995) and affect a greater area than the small area actually impacted by shell drop under mussel farms. Aquaculture may also reduce harvest of wild stocks, and thus protect both the environment and the stock itself. Remedying loss of sandy or muddy habitat would be difficult; reefs may be offered as mitigation for such losses, doubly reducing the area of muddy habitat. In general very few harvestable species are associated with muddy habitats (flatfish being one of the few exceptions), and addition of shell to the mud below is widely believed to be "advantageous" because it increases the area of habitat for species which are desirable angling targets. In the United States of America shell is used to provide settlement habitat for clams; the only parallel here we are aware of is a similar usage to encourage settlement of dredge oysters *Tiostrea chilensis* in Foveaux Strait.

Food web effects

We are unaware of evidence of important effects of aquaculture on food webs in the New Zealand marine environment, though there could be problems with phytoplankton supply in years of low primary productivity. It has been suggested that intensive aquaculture may have altered "community" structure in one intensively-cultivated Spanish embayment, but there is little evidence that such effects occur elsewhere (Beveridge *et al.* 1994). Natural fluctuations in availability of phytoplanktonic food are probably of a magnitude similar to that which farming might induce. Current mussel farming practices are such that only the upper 12 m of the water column is affected, and phytoplankton densities usually return to background levels within 100 m of mussel farms (NIWA, unpubl. data). It has been suggested that aquaculture systems themselves may act as "mine canaries", and that environmental effects are most likely to emerge as an effect on the crop (Stewart 1997). It does not, however, necessarily follow that because a crop is at high density the effects on that species will emerge first, because other species may be more sensitive to food limitation than the crop. Stewart's (1997) discussion suggests that few initially successful aquaculture operations have been terminated due to environmental damage, but this may reflect ineffective environmental monitoring rather than the non-importance of the effects.

In some tropical areas, shrimp larvae and gravid females are in decline in the wild, because of harvesting pressure to supply culture ponds. Such pressure also impacts on other parts of the food chain; 10 kg of larvae and fry may be discarded to obtain 1 kg of shrimp post-larvae (Silas 1987, cited in Beveridge *et al.* 1994). Scallop and mussel spat are wild-caught in New Zealand at present, but there is little evidence that this is having any detectable effect on wild stocks. Loss of habitat because of aquaculture has not yet been widespread in New Zealand, but the widely-publicised loss of mangal associated with growth of shrimp aquaculture in south-east Asia provides a salutary note. This is important because mangal habitat is thought to act as a nursery for small prawns and shrimps. Robertson and Phillips (1995) suggested that 2–22 ha of *Rhizophora*-dominated mangal were required to filter the effluent of 1 ha of shrimp farm. Larsson *et al.* (1994) estimated that a semi-intensive tropical shrimp farm required an area 35–190 times larger to support it.

A combination of over-exploitation of mussel beds and poor recruitment years for mussels in the Dutch Wadden Sea led to effects on populations of oystercatchers *Haematopus ostralegus* and eider *Somateria mollissima* (Beukema 1993; Dankers and Zuidema 1995). Again there is little evidence that this might be occurring in New Zealand.

Impacts on the seabed

On an areal basis, benthic productivity may be diminished by the presence of farms if they smother infauna. Loss of primary productivity may occur via effects on primary producers (e.g., seagrass). Fish cultures were suggested to have impacted seagrass habitats in a Mediterranean study (Delgado *et al.* 1997). Detritivorous species have been trialed in polyculture systems to consume debris and reduce the nutrient load of effluent (e.g., Krom *et al.* 1995; Porter *et al.* 1996). Everett *et al.* (1995) documented negative effects of commercial stake and rack culture of Pacific oysters *Crassostrea gigas* on *Zostera marina*. The mechanisms of impact varied for stake and rack culture methods; stake culture increased sedimentation and direct physical disturbance of husbandry, whereas racks were thought to increase erosion and perhaps act to increase shading.

Shell drop may provide large areas of "hard" substratum below mussel farms, and this may act as habitat for other, potentially desirable, species. Impacts on the seabed may occur via deposition of shell and uneaten food and pseudofaeces. These may also provide opportunities for enhanced fish feeding, and provision of hard benthic habitat (Kaspar *et al.* 1985). The artificial reef literature has explored the link between abundance and the area of desirable habitat (e.g., Pickering and Whitmarsh 1997; Grossman *et al.* 1997). Those reviews have focused on what has become known as the "attraction *vs* production" debate. It has been suggested that some artificial reef programmes are little more than excuses to dump debris (Cognetti 1987), and there is little evidence regarding whether mobile species are merely concentrated by structures, or whether total abundance (or production) is actually increased (Polovina 1989; Pickering and Whitmarsh 1997). There may be effects of artificial reefs on nearby benthic communities; Carbines (1993) found that spotties aggregated near mooring blocks for marine farms, while Davis *et al.* (1982) noted alterations to the benthic fauna around artificial reefs, which may have been partly due to fish foraging.

Effects of provision of habitat

Increasing the area of habitat may enhance population abundance where demographic bottlenecks occur in the life cycle. For example, juvenile lobsters appear to be habitat limited at some times, and provision of habitat ("casitas") may allow greater survivorship (Eggleston *et al.* 1990, 1992, 1997; Wahle and Steneck 1991). This is particularly relevant for the Florida Keys, where plankton blooms largely destroyed the biotic habitat on which the juvenile lobsters depended, whereupon they became more dependent on habitat constructed by humans (Butler *et al.* 1995). We are unaware of any quantitative information regarding valuable native species being secondarily enhanced by aquaculture in New Zealand. Recreational fishers certainly target mussel farms for desirable fish species (e.g., snapper) in Pelorus and Kenepuru Sounds. These are probably short-term responses to cleaning mussel lines, rather than a response to provision of habitat, and may be subject to the same controversy of attraction *vs* production as discussed above. There are anecdotal accounts of the protected spotted black groper *Epinephalus damelii* recruiting to mussel farms in northern New Zealand, but the effect of this on local populations is unknown. It is quite likely that pelagic/epifaunal secondary production (mainly ascidians, bryozoans, hydroids) on a per area basis, is considerably enhanced by the vast surface area of habitat that mussel ropes and mussels provide. Most of the species are not of interest to humans, but we note that generalizations regarding the negative impacts of aquaculture on biodiversity (e.g., Beveridge *et al.* 1994) may not apply to those situations.

A further result of harvesting of wild populations may be alterations to dispersion patterns. Spawning biomass may have to be of a certain size for populations to persist (but see comments on relative invulnerability of marine invertebrates to extinction discussed above (Carlton *et al.* 1991)). This has been hinted at for *Haliotis cracherodii* whose populations crashed as a result of a disease (e.g., Richards and Davis 1993). Fertilization ecology in the marine environment (e.g., Denny and Shibata 1989; Levitan 1991; Levitan *et al.* 1991) is also becoming better understood and it may be that the high densities of stock in the farming environment enhance larval output and hence recruitment to other areas.

Conflicts between aquaculture and native fauna

Aquaculture structures may provide safe feeding and roosting sites for seabirds (shags, terns, and in some New Zealand areas, feeding areas for the endangered yellow-eyed penguin; pers. obs.), shelter in bad weather, and a source of food, as structures likely aggregate small fish (authors, pers. obs.). We are unaware of data indicating that marine farming increases the carrying capacity of the environment for birds, but Glahn and Stickley (1995) suggested that the channel catfish aquaculture industry in the Mississippi region lost in the order of US$2M per year to double-crested cormorants *Phalacrocorax auritus*. Pemberton and Shaughnessy (1993) reviewed conflicts between seals and marine fish-farms in Tasmania, such conflicts also occur in New Zealand. The most effective way of controlling the impacts (net damage and fish consumption) appears to be to cage seals out. Shooting was done on a permit system in Tasmania, and was therefore under some legal control, but proved to be largely ineffectual (Pemberton and Shaugnessy 1993).

Other impacts

The other impacts of aquaculture on wildlife are less well-described. The starfish *Coscinasterias calamaria* (a predator of bivalves) is removed from areas under management for scallop enhancement. The effects of this are unknown, but are probably trivial compared to the effects of scallop dredging.

Aquaculture — clean and green?

Impacts from chemical contamination via aquaculture seldom arise in New Zealand; because most species are grown for food, environmental quality is paramount, and the aquaculture industry here currently uses virtually no therapeutic or other chemicals. Beardmore *et al.* (1997) noted that no reliable data concerning pollution resulting from aquaculture were known to them. Natural contaminants such as toxic phytoplankton are currently the greatest marine source of concern in production areas, and it is possible that farmed stocks could act as warning systems for such problems. For this to be of general use to the remainder of the ecosystem, the farmed species would have to be more sensitive than other species, however.

The exception to this lack of effects is nutrient release from benthic organic enrichment under marine finfish farms (Lumb 1989), which has been implicated as a nutrient source for toxic phytoplankton blooms (Pridmore and Rutherford 1992). There is considerable information regarding the impacts of fish farming (reviews of Gowen and Bradbury 1987; Wu 1995). More recent research suggests that there may be physiological effects whereby increased algal enrichment may enhance uptake of pollutants such as polychlorinated biphenyls (Gilek *et al.* 1996). Polyculture has been suggested and used in some overseas systems as a means of slowing impacts or of rehabilitating areas (e.g., Folke and Kautsky 1989; Krom *et al.* 1995; Neori *et al.* 1996; Porter *et al.* 1996). For example, Stirling and Okumus (1995) found slightly faster mussel growth on salmon farms than on mussel farms without finfish nearby, and suggested that there were advantages of integrating bivalve culture with salmon cage farming.

Terrestrial contamination on the other hand is of considerable concern — NIWA's NIFTY system (see below) informs aquaculturists whether harvesting is permitted by Department of Health regulations. It relies on the relationship between rainfall in a catchment and levels of faecal coliforms at shellfish harvesting sites to predict whether poor sanitation prevents harvesting (Brock *et al.* 1985).

Invasive species

There are numerous overseas examples of the consequences of invasive species; the catastrophic invasion of the Great Lake system by the mussel *Dreissenia* (Roberts 1990; Johnson and Carlton 1996), the formation of dense stands of the toxic alga *Caulerpa taxifolia* in the Mediterranean (Meinesz *et al.* 1995; Chisholm *et al.* 1997), as well as toxic phytoplankton blooms. Some invasive species (e.g., Pacific oysters *Crassostrea gigas*, possibly *Undaria pinnatifida*) are themselves important aquaculture species. Expansion of the distribution of *Undaria* is probably linked to the aquaculture industry, though not necessarily through aquaculture practices themselves. In the Marlborough Sounds at present there is little evidence that it is spreading to adjacent shores in Pelorus Sound, an area of intensive aquaculture (C. Hay, pers. comm.), and it is certainly spreading on shores where there is little aquaculture (e.g., Queen Charlotte Sound — R. Cole, pers. obs.).

Disease risks are mainly associated with importation of foreign animals, but regional spread of pests/parasites could be of concern. There is considerable movement of vessels and materials round the New Zealand coast, though the natural movements of organisms and water bodies may also be considerable. Antibiotics are generally not used in New Zealand salmon farms, but overseas they have been found to be a potential source of mortality for benthic organisms at farm sites (Black *et al.* 1997).

OPPORTUNITIES FOR ENHANCEMENT OF NATURE CONSERVATION IN PRODUCTION AREAS

Davis (1985) discussed the use of artificial structures to mitigate marina construction impacts; there is the possibility that such options could be built into the planning process to mitigate for habitat occupation or destruction (e.g., Ambrose 1994). Freshwater crayfish aquaculture has been suggested to have increased population numbers of colonial wading birds in the southern United States (Fleury and Sherry 1995) and there are other examples of seabirds perhaps either becoming dependent on fisheries or suffering when prey species are harvested (e.g., Monaghan 1992). Aquaculture as practised in New Zealand clearly has an effect on benthic food webs, but few rare organisms occur there. One example of nature conservation and production being enhanced simultaneously is the trial culture of a rare sponge for pharmaceuticals in the Marlborough Sounds. The distribution of the species in the wild is limited to a small area which is inaccessible to trawling because of bottom topography. Thus the persistence of this rare species might be enhanced because it is cultured for human use; ultimately restoration of wild populations could occur.

Marine-terrestrial linkages

"Marine pollution from land-based activities poses one of the most serious threats to the quality and productivity of the coastal and marine environment" accounting for over three-quarters of marine pollution (Williams 1996). Marine production environments are almost invariably downstream of terrestrial ones and there is generally a unidirectional flow of contaminants. Terrestrial agriculture is a major source of nutrients for primary production, and hence input to marine farming. Nutrient impacts on the marine environment are greater for pastures than for forestry. Fahey and Coker (1992) documented runoff from forestry roads in the Marlborough Sounds region. Brock *et al.* (1985) found faecal coliforms to be strong related to rainfall, river flow and salinity in Pelorus Sound. These relationships are used as the basis of a shellfish sanitation programme in which rainfall is used to determine harvesting (see below). As outlined above, sanitation is so critical to aquaculture that corner-cutting would be commercial suicide.

The converse process, terrestrial pollution from marine activities, has localized impacts, such as terrestrial disposal of mussel shells, and perhaps lost ropes and floats washing ashore. The Mussel Industry Council has established environmental guidelines for disposal of litter, and generally this litter appears insignificant relative to other sources, such as harvest fisheries and public recreation (R. Cole, pers. obs.).

Terrestrial reserves and sanctuaries are generally considered safe in marine production environments. For example, the bird sanctuary Maud Island is located in Pelorus Sound, an area of intensive aquaculture. Domestic animals or rats from boats could potentially be a hazard to the island's wildlife, but domestic animals are more likely to originate from pleasure craft, and rats would be anathema to the hygiene-conscious seafood industry.

TECHNOLOGICAL ADVANCES

There are several examples of technological advances contributing to the maintenance of the quality of production environments. Contamination of mussels by terrestrial faecal coliforms is a problem for the aquaculture industry in the Marlborough area. NIWA's NIFTY system relies on predictive relationships between rainfall in catchments and faecal coliforms in shellfish offshore from those catchments to inform producers whether or not harvesting is permitted under Ministry of Health guidelines. There is further scope for using computer models to predict paths of falling particles in order to predict the "footprint" of various debris (e.g., pseudofaeces, uneaten food particles) from a marine farm. Such information could be used by councils to manage the coastal environment, by determining carrying capacities etc.

BALANCING THE COSTS AND BENEFITS OF AQUACULTURE

While negative aspects of aquaculture make it an easy target for environmentalists, the alternative way of obtaining marine protein for human use is via capture fisheries. Capture fisheries are generally damaging to the environment and other species, almost universally in decline, and there is evidence that many are not sustainable (Botsford *et al.* 1997). Aquaculture is generally a more benign method of obtaining marine protein, more readily controlled, subject to lower levels of bycatch, and frequently produces a higher quality product. If properly managed, it is potentially sustainable. The development of the New Zealand aquaculture industry over the last 25 years has seen numerous gains in efficiency, and alterations to the species cultured. Generally we believe that controlled future developments in aquaculture will enhance rather than harm marine conservation. The areas in which conflicts may arise are the culture of introduced species and interactions between protected species (e.g., seals) and farmed stocks. Technologies which will further both conservation and aquaculture are available, and increases in their usage will contribute to a balance between these two.

ACKNOWLEDGEMENTS

Thanks to Big Glory Seafoods and the Mussel Industry Council for supporting our research, R. G. Creese, D. Keuskamp, R. B. Taylor and the referees for comments on the manuscript, and the long-suffering NIWA Greta Point library staff.

REFERENCES

Ambrose, R. F., 1994. Mitigating the effects of a coastal power plant on a kelp forest community: rationale and requirements for an artificial reef. *Bull. Mar. Sci.* **55:** 694–708.

Andrew, N. L. and Pepperell, J. G., 1992. The by-catch of shrimp trawl fisheries. *Oceanogr. Mar. Biol. Annu. Rev.* **30:** 527–65.

Beardmore, J. A., Mair, G. C. and Lewis, R. I., 1997. Biodiversity in aquatic systems in relation to aquaculture. *Aquacul. Res.* **28:** 829–39.

Beukema, J. J., 1993. Increased mortality in alternative bivalve prey during a period when the tidal flats of the Dutch Wadden Sea were devoid of mussels. *Neth. J. Sea Res.* **31:** 395–406.

Beveridge, M. C. M., Phillips, M. J. and Macintosh, D. J., 1997. Aquaculture and the environment: the supply of and demand for environmental goods and services by Asian aquaculture and the implications for sustainability. *Aquacul. Res.* **28:** 797–807.

Beveridge, M. C. M., Ross, L. G. and Kelly, L. A., 1994. Aquaculture and biodiversity. *Ambio* **23:** 497–502.

Black, K. D., Fleming, S., Nickell, T. D. and Pereira, P. M. F., 1997. The effects of ivermectin, used to control sea lice on caged farmed salmonids, on infaunal polychaetes. *ICES J. Mar. Sci.* **54:** 276–79.

Botsford, L. W., Castilla, J. C. and Peterson, C. H., 1997. The management of fisheries and marine ecosystems. *Science* **277:** 509–15.

Brock, R. L., Galbraith, G. R. and Benseman, B. A., 1985. Relationships between rainfall, river flow, and salinity to faecal coliform levels in a mussel fishery. *NZ J. Mar. Freshwat. Res.* **19:** 485–94.

Butler, M. J. IV, Hunt, J. H., Herrnkind, W. F., Childress, M. J., Bertelsen, R., Sharp, W., Matthews, T., Field, J. M. and Marshall, H. G., 1995. Cascading disturbances in Florida Bay, USA: cyanobacteria blooms, sponge mortality, and implications for juvenile spiny lobsters *Panulirus argus*. *Mar. Ecol. Prog. Ser.* **129:** 119–25.

Carbines, G. D., 1993. The ecology and early life history of *Notolabrus celidotus* (Pisces: Labridae) around mussel farms in the Marlborough Sounds. MSc thesis, University of Canterbury.

Carlton, J. T., Vermeij, G. J., Lindberg, D. R., Carlton, D. A. and Dudley, E. C., 1991. The first historical extinction of a marine invertebrate in an ocean basin: the demise of the eelgrass limpet *Lottia alveus*. *Biol. Bull.* **180:** 72–80.

Castilla, J. C. and Bustamante, R. H., 1989. Human exclusion from rocky intertidal of Las Cruces, central Chile: effects on *Durvillaea antarctica* (Phaeophyta, Durvilleales). *Mar. Ecol. Prog. Ser.* **50:** 203–14.

Chisholm, J. R. M., Fernex, F. E., Mathieu, D. and Jaubert, J. M., 1997. Wastewater discharge, seagrass decline and algal proliferation on the Cote d'Azur. *Mar. Pollut. Bull.* **34:** 78–84.

Choat, J. H. and Ayling, A. M., 1987. The relationship between habitat structure and fish faunas on New Zealand reefs. *J. Exp. Mar. Biol. Ecol.* **110:** 257–84.

Clout, M. N. and Saunders, A. J., 1995. Conservation and ecological restoration in new Zealand. *Pac. Conserv. Biol.* **2:** 91–98.

Cognetti, G., 1987. Artificial reefs: myth or reality? *Mar. Pollut. Bull.* **18:** 367–68.

Cole, R. G., Creese, R. G., Grace, R. V., Irving, P. and Jackson, W. R., 1992. Abundance patterns of subtidal benthic invertebrates and fishes at the Kermadec Islands. *Mar. Ecol. Prog. Ser.* **82:** 207–18.

Connell, J. H. and Sousa, W. P., 1983. On the evidence needed to judge ecological stability or persistence. *Amer. Natur.* **121:** 789–824.

Creese, R. G. and Cole, R. G., 1995. Marine conservation in New Zealand. *Pac. Conserv. Biol.* **2:** 55–63.

Dankers, N. and Zuidema, D. R., 1995. The role of the mussel (*Mytilus edulis* L.) and mussel culture in the Dutch Wadden Sea. *Estuaries* **18:** 71–80.

Davis, G. E., 1985. Artificial structures to mitigate marina construction impacts on spiny lobster *Panulirus argus*. *Mar. Ecol. Prog. Ser.* **37:** 151–56.

Davis, N., VanBlaricom, G. and Dayton, P. K., 1982. Man-made structures on marine sediments: effects on adjacent benthic communities. *Mar. Biol.* **70:** 295–303.

Dayton, P. K., 1971. Competition, disturbance, and community organization: the provision and subsequent utilization of space in a rocky intertidal community. *Ecol. Monogr.* **41:** 351–89.

Dayton, P. K., Tegner, M. J., Edwards, P. B. and Riser, K. L., 1998. Sliding baselines, ghosts, and reduced expectations in kelp forest communities. *Ecol. Appl.* **8:** 309–22.

Dayton, P. K., Thrush, S. F., Agardy, M. T. and Hofman, R. J., 1995. Environmental effects of marine fishing. *Aquat. Conserv. Mar. Freshwat. Ecosystems* **5:** 205–32.

de Casabianca, M.-L., Laugier, T. and Collart, D., 1997. Impact of shellfish farming eutrophication on benthic macrophyte communities in the Thau lagoon, France. *Aquacul. Internat.* **5:** 301–14.

Delgado, O., Grau, A., Pou, S., Riera, F., Massuti, C., Zabala, M. and Ballesteros, E., 1997. Seagrass regression caused by fish cultures in Fornells Bay (Menorca, Western Mediterranean). *Oceanol. Acta* **20:** 557–63.

Denny, M. W. and Shibata, M. F., 1989. Consequences of surf-zone turbulence for settlement and external fertilization. *Amer. Nat.* **134:** 859–89.

Edding, M. E., Fonck, E. A., Macchiavello, J. E., 1994. *Lessonia*. Pp. 407–46 *in* Biology of Economic Algae ed by I. Akatsuka. SPB Academic Publishing: The Hague, The Netherlands.

Eggleston, D. B., Lipcius, R. M., Miller, D. L. and Coba-Cetina, L., 1990. Shelter scaling regulates survival of juvenile Caribbean spiny lobster *Panulirus argus*. *Mar. Ecol. Prog. Ser.* **62:** 79–88.

Eggleston, D. B., Lipcius, R. M. and Miller, D. L., 1992. Artificial shelters and survival of Caribbean spiny lobster *Panulirus argus*: spatial, habitat and lobster size effects. *Fish. Bull. US* **90:** 691–702.

Eggleston, D. B., Lipcius, R. N. and Grover, J. J., 1997. Predator and shelter-size effects on coral reef fish and spiny lobster prey. *Mar. Ecol. Prog. Ser.* **149:** 43–59.

Everett, R. A., Ruiz, G. M. and Carlton, J. T., 1995. Effect of oyster mariculture on submerged aquatic vegetation: an experimental test in a Pacific Northwest estuary. *Mar. Ecol. Prog. Ser.* **125:** 205–17.

Fahey, B. D. and Coker, R. J., 1992. Sediment production from forest roads in Queen Charlotte Forest and potential impact on marine water quality, Marlborough Sounds, New Zealand. *NZ J. Mar. Freshwat. Res.* **26:** 187–95.

Fisheries and Agriculture Organization, 1996. Recent trends in global aquaculture production: 1984–1995. http://www.fao.org/waicent/faoinfo/fishery/trends/aqtrends/contrf.htm.

Fleury, B. E. and Sherry, T. W., 1995. Long-term population trends of colonial wading birds in the southern United States: the impact of crayfish aquaculture on Louisiana populations. *Auk* **112:** 613–32.

Folke, C. and Kautsky, N., 1989. The role of ecosystems for a sustainable development of aquaculture. *Ambio* **18:** 234–43.

Fonseca, M. S., Kenworthy, W. J. and Thayer, G. W., 1988. Restoration and management of seagrass systems: a review. Pp. 353–68 *in* The Ecology and Management of Wetlands. Vol. 2: Management, Use and Value of Wetlands ed by D. D. Hook *et al.* Timber Press: Portland, Oregon, USA.

Fonseca, M. S., Kenworthy, W. J. and Courtney, F. X., 1996a. Development of planted seagrass beds in Tampa Bay, Florida, USA. 1. Plant components. *Mar. Ecol. Prog. Ser.* **132:** 127–39.

Fonseca, M. S., Meyer, D. L. and Hall, M. O., 1996b. Development of planted seagrass beds in Tampa Bay, Florida, USA. 2. Faunal components. *Mar. Ecol. Prog. Ser.* **132:** 141–56.

Gilek, M., Bjoerk, M., Broman, D., Kautsky, N. and Naef, C., 1996. Enhanced accumulation of PCB congeners by Baltic Sea blue mussels, *Mytilus edulis*, with increased algae enrichment. *Environ. Toxicol. Chem.* **15:** 1597–605.

Glahn, J. F. and Stickley, A. R. Jr, 1995. Wintering double-nested cormorants in the delta region of Mississipi: population levels and their impact on the catfish industry. *Colon. Waterbirds* **18:** 137–42.

Goldburg, R. and Triplett, T., 1997. Murky waters: environmental effects of aquaculture in the United States. Preface and Executive Summary. http://www.edf.org/pubs/Reports/Aquaculture.

Gowen, R. J. and Bradbury, N. B., 1987. The ecological impact of salmonid farming in coastal waters: a review. *Oceanogr. Mar. Biol. Annu. Rev.* **25:** 563–75.

Grossman, G. D., Jones, G. P. and Seaman, W. J., 1997. Do artificial reefs increase regional fish production? A review of existing data. *Fisheries* **22**(4): 17–23.

Hargrave, B. T., Duplisea, D. E., Pfeiffer, E., Wildish, D. J., 1993. Seasonal changes in benthic fluxes of dissolved oxygen and ammonium associated with marine cultured Atlantic salmon. *Mar. Ecol. Prog. Ser.* **96:** 249–57.

Hatcher, A., Grant, J. and Schofield, B., 1994. Effects of suspended mussel culture (*Mytilus* spp.) on sedimentation, benthic respiration and sediment nutrient dynamics in a coastal bay. *Mar. Ecol. Prog. Ser.* **115:** 219–35.

Holby, O. and Hall, P. O. J., 1991. Chemical fluxes and mass balances in a marine fish cage farm. II. Phosphorus. *Mar. Ecol. Prog. Ser.* **70:** 263–72.

Johnsen, R. I., Grahl-Nielsen, O. and Lunestad, B. T., 1993. Environmental distribution of organic waste from a marine fish farm. *Aquaculture* **118:** 229–44.

Johnson, L. E. and Carlton, J. T., 1996. Post-establishment spread in large-scale invasions: dispersal mechanisms of the zebra mussel *Dreissena polymorpha*. *Ecology* **77:** 1686–690.

Jones, G. P., 1984. Population ecology of the temperate reef fish *Pseudolabrus celidotus* Bloch and Schneider (Pisces: Labridae). I. Factors influencing recruitment. *J. Exp. Mar. Biol. Ecol.* **75:** 257–76.

Kaspar, H. F., Gillespie, P. A., Boyer, I. C. and McKenzie, A. L., 1985. Effects of mussel aquaculture on the nitrogen cycle and benthic communities of Kenepuru Sound, Marlborough Sounds, New Zealand. *Mar. Biol.* **85:** 127–36.

Kearney, R. E., Andrew, N. L. and West, R. J., 1996. Some issues in the management of Australia's marine and coastal fisheries resources. *Ocean Coastal Manage.* **33:** 133–46.

Kingsford, M. J., Schiel, D. R. and Battershill, C. N., 1989. Distribution and abundance of fish in a rocky reef environment at the subantarctic Auckland Islands, New Zealand. *Polar Biol.* **9:** 179–86.

Krom, M. D., Ellner, S., van Rijn, J. and Neori, A., 1995. Nitrogen and phosphorus cycling and transformations in a prototype "non-polluting" integrated mariculture system, Eilat, Israel. *Mar. Ecol. Prog. Ser.* **118:** 25–36.

Lafferty, K. D. and Kuris, A. M., 1993. Mass mortality of abalone *Haliotis cracherodii* on the California Channel Islands: tests of epidemiological hypotheses. *Mar. Ecol. Prog. Ser.* **96:** 239–48.

Lambert, D. M. and Millar, C. D., 1995. DNA science and conservation. *Pac. Conserv. Biol.* **2:** 21–38.

Larsson, J., Folke, C. and Kautsky, N., 1994. Ecological limitations and appropriation of ecosystem support by shrimp farming in Colombia. *Environ. Manage.* **18:** 663–76.

Leighton, D. L., 1962. Locusts of the sea. *Skin Diver* **August 1962:** 38–39.

Leighton, D. L., Jones, D. G. and North, W. J., 1966. Ecological relationships between giant kelp and sea urchins in southern California. Pp. 141–53 *in* Proceedings of the 5th International Seaweed Symposium ed by E. G. Young and J. L. McLachlan. Pergamon Press: New York.

Levitan, D. R., 1991. Influence of body size and population density on fertilization success and reproductive output in a free-spawning invertebrate. *Biol. Bull.* **181:** 261–68.

Levitan, D. R., Sewell, M. A. and Chia, F.-S., 1991. Kinetics of fertilization in the sea urchin *Strongylocentrotus franciscanus*: interaction of gamete dilution, age, and contact time. *Biol. Bull.* **181:** 371–78

Lindeboom, H. J., 1995. Protected areas in the North Sea: an absolute need for further research. *Helgol. Meeresunters.* **49:** 591–602.

Lumb, C. M., 1989. Self-pollution by Scottish salmon farms? *Mar. Pollut. Bull.* **20:** 375–79.

Malakoff, D., 1997. Extinction on the high seas. *Science* **277:** 486–88.

Mansfield, W., 1996. Moving from successful restoration of islands to ecosystem restoration on mainland New Zealand. http://www.doc.govt.nz/issues/iucn.htm

Mathias, M. E. and Moyle, P., 1992. Wetland and aquatic habitats. *Agric. Ecosyst. Environ.* **42:** 165–76.

Meinesz, A., Benichou, L., Blahier, J., Komatsu, T., Lemee, R., Molenaar, H. and Mari, X., 1995. Variations in the structure, morphology and biomass of *Caulerpa taxifolia* in the Mediterranean Sea. *Bot. Mar.* **38:** 499–508.

Miller, K. J., 1997. Genetic structure of black coral populations in New Zealand's fiords. *Mar. Ecol. Prog. Ser.* **161:** 123–32.

Miller, K. J., 1998. Short distance dispersal of black coral larvae: inference from spatial analysis of genotypes. *Mar. Ecol. Prog. Ser.* **163:** 225–33.

Miller, K. J. and Grange, K. R., 1997. Population genetic studies of the antipatharian black coral *Antipathes fiordensis* from Doubtful and Nancy Sounds, Fiordland, New Zealand. Pp. 353–63 *in* Proceedings of the 6th International Conference on Coelenterate Biology, 16–21 July 1995 ed by J. C. den Hartog. Nationaal Natuurhistorisch Museum: Leiden, The Netherlands.

Monaghan, P., 1992. Seabirds and sandeels: the conflict between exploitation and conservation in the northern North Sea. *Biodiver. Conserv.* **1:** 98–111.

Neori, A., Krom, M. D., Ellner, S. P., Boyd, C. E., Popper, D., Rabinovitch, R., Davison, P. J., Dvir, O., Zuber, D., Ucko, M., Angel, D. and Gordin, H., 1996. Seaweed biofilters as regulators of water quality in integrated fish-seaweed culture units. *Aquaculture* **141:** 183–99.

North, W. J., 1994. Review of *Macrocystis* biology. Pp. 447–527 *in* Biology of Economic Algae ed by I. Akatsuka. SPB Academic Publishing: The Hague.

Pemberton, D. and Shaughnessy, P. D., 1993. Interaction between seals and marine fish-farms in Tasmania, and management of the problem. *Aquat. Conserv.: Mar. Freshwat. Ecosyst.* **3:** 149–58.

Pickering, H. and Whitmarsh, D., 1997. Artificial reefs and fisheries exploitation: a review of the "attraction versus production" debate, the influence of design and its significance for policy. *Fish. Res.* **31:** 39–59.

Polovina, J. J., 1989. Artificial reefs: nothing more than benthic fish aggregators. *Rep. Calif. Coop. Fish. Invest.* **30:** 37–39.

Porter, C. B., Krost, P., Gordin, H. and Angel, D. L., 1996. Preliminary assessment of grey mullet (*Mugil cephalus*) as a forager of organically enriched sediments below marine fish farms. *Bamidgeh* **48:** 47–55.

Pridmore, R. D. and Rutherford, J. C., 1992. Modelling phytoplankton abundance in a small enclosed bay used for salmon farming. *Aquacult. Fish. Manage.* **23:** 525–42.

Richards, D. V. and Davis, G. E., 1993. Early warnings of modern population collapse in black abalone *Haliotis cracherodii*, Leach, 1814 at the California Channel Islands. *J. Shellfish Res.* **12:** 189–94.

Roberts, L., 1990. Zebra mussel invasion threatens U.S. waters. *Science* **249:** 1370–372.

Robertson, A. I. and Phillips, M. J., 1995. Mangroves as filters of shrimp pond effluent: predictions and biogeochemical research needs. *Hydrobiologia* **295:** 311–21.

Rosier, J. and Hastie, W., 1996. New Zealand coastal planning: an issue-based approach. *Ocean Coastal Manage.* **33:** 147–65.

Schiel, D. R., 1988. Algal interactions on shallow subtidal reefs in northern New Zealand: a review. *NZ J. Mar. Freshw. Res.* **22:** 481–89.

Schiel, D. R. and Foster, M. S., 1992. Restoring kelp forests. Pp. 279–342 *in* Restoring the Nation's Marine Environment ed by G. W. Thayer. NOAA: College Park, Maryland.

Schiel, D. R. and Nelson, W. A., 1990. The harvesting of macroalgae in New Zealand. *Hydrobiologia* **204/205:** 25–33.

Shanks, A. L. and Wright, W. G., 1986. Adding teeth to wave action: the destructive effects of wave-borne rocks on intertidal organisms. *Oecologia* **69:** 420–28.

Slooten, E. and Dawson, S. M., 1995. Conservation of marine mammals in New Zealand. *Pac. Conserv. Biol.* **2:** 64–76.

Stewart, J. E., 1997. Environmental impacts of aquaculture. *World Aquaculture* **March 1997:** 47–52.

Stirling, H. P. and Okumus, I., 1995. Growth and production of mussels (*Mytilus edulis* L.) suspended at salmon cages and shellfish farms in two Scottish sea lochs. *Aquaculture* **134:** 193–210.

Taylor, J. J., Southgate, P. C. and Rose, R. A., 1997. Fouling animals and their effect on the growth of silver-lip oysters, *Pinctada maxima* (Jameson) in suspended culture. *Aquaculture* **153:** 31–40.

Taylor, R. B., 1998. Density, biomass and productivity of animals in four subtidal rocky reef habitats: the importance of small mobile invertebrates. *Mar. Ecol. Prog. Ser.* **172:** 37–51.

Tegner, M. J., Basch, L. V. and Dayton, P. K., 1996. Near extinction of an exploited marine invertebrate. *TREE* **11:** 278–80.

Thompson, S. M., 1983. Homing in a territorial reef fish. *Copeia* **1983:** 832–34.

Thrush, S. F., Hewitt, J. E., Cummings, V. J. and Dayton, P. K., 1995. The impact of habitat disturbance by scallop dredging on marine benthic communities: what can be predicted from the results of experiments? *Mar. Ecol. Prog. Ser.* **129:** 141–50.

Towns, D. R. and Ballantine, W. J., 1993. Conservation and restoration of New Zealand island ecosystems. *TREE* **8:** 452–57.

Vermeij, G. J., 1993. Biogeography of recently extinct marine species: implications for conservation. *Conserv. Biol.* **7:** 391–97.

Vincent, A. C. J. and Hall, H. J., 1996. The threatened status of marine fishes. *TREE* **11:** 360–61.

Wahle, R. A. and Steneck, R. S., 1991. Recruitment habitats and nursery grounds of the American lobster *Homarus americanus*: a demographic bottleneck? *Mar. Ecol. Prog. Ser.* **69:** 231–43.

Williams, C., 1996. Combatting marine pollution from land-based activities: Australian initiatives. *Ocean Coastal Manage.* **33:** 87–112.

Williams, S. L. and Davis, C. A., 1996. Population genetic analyses of transplanted eelgrass (*Zostera marina*) beds reveal reduced genetic diversity in Southern California. *Restor. Ecol.* **4:** 163–80.

Wu, R. S. S., 1995. The environmental impact of marine fish culture: towards a sustainable future. *Mar. Pollut. Bull.* **31:** 159–66.

Ye, L.-X., Ritz, D. A., Fenton, G. E. and Lewis, M. E., 1991. Tracing the influence on sediments of organic waste from a salmonid farm using stable isotope analysis. *J. Exp. Mar. Biol. Ecol.* **145:** 161–74.

CHAPTER 40

Naturalizing the city: a case study of Waitakere City, New Zealand

Dorothy Wilson[1]

In 1993, Waitakere set the objective for becoming an eco city. This arose in response to the perception that prevailing forms of urban growth degraded the natural and social environment. Agenda 21, the 1992 Earth Summit document, provided the framework for the Council's strategic approach to reversing unsustainable practices. Promoting a more "compact city" is seen as critical to enabling the conservation, repair and restoration of natural ecological systems. This is achieved through an integrated planning process where strategic, environmental (District Plan) and annual plans are linked to advance the Green Network. The Green Network is one of Council's eight strategic pathways to progress the City's sustainability agenda over a 20 year period. The Green Network represents the programmes to restore, protect and increase natural ecological areas. By a combination of regulation, systematic proactive Council investment and community based programmes aligned to Green Network objectives, stream corridors and areas of significant and outstanding vegetation are being identified, protected and restored. Key initiatives include staff upskilling, partnerships with Maori, the community and business to deliver the Green Network. It integrates traditional investment in parks, recreation and stormwater management to achieve broader ecological objectives. Priority in the future will be given to increasing community involvement — so as to increase public participation in the prevention, damage and the mainstreaming of good practice.

PERSPECTIVE

THE Council of Waitakere City in 1993 set the objective of becoming an eco city. The Council's 20 year strategic plan identifies how progress towards that goal will be made in collaboration and partnership with nature and with the City's diverse communities. My comments reflect then an explicit position — of commitment to achieving eco city goals through political processes and community agency.

Waitakere City, the fifth largest city in New Zealand (170 000) is the western part of the Auckland regional conurbation. It is socially (14% Maori, 11% Pacific Island Polynesian, 7% Asian, rest mainly European) and environmentally diverse (one-third suburban, one-third rural, one-third regenerating wilderness). Its location is spectacular, bounded by the Tasman Sea, and the Manukau and Waitemata Harbours. Like many cities in Australasia and North America the style of development has placed great pressure on natural systems — and on communities.

"Naturalizing the City" in our experience serves social and economic justice and wellbeing as well as environmental restoration.

CITIES AND ENVIRONMENTAL IMPERATIVES

In New Zealand for some decades — but particularly since the mid 1980s we are counting the costs of our style of car-based, low density urban development. As a small nation in a relatively isolated location globally we have tended to be complacent. While there are articulate lobbyists, as a whole we "forget" about the ozone layer gap over Antarctica, and about global climate change and its potential consequences world-wide and for our own region, here in the South Pacific. There is little evidence that individuals (or communities) see (and act on) the links between their day to day activities and global issues.

[1]Waitakere City Council, Private Bag 93109, Henderson, Waitakere City, New Zealand.

Pages 405–10 *in* NATURE CONSERVATION 5: NATURE CONSERVATION IN PRODUCTION ENVIRONMENTS: MANAGING THE MATRIX ed by J. L. Craig, N. Mitchell and D. A. Saunders. Surrey Beatty & Sons, 2000.

The symptoms of the impacts of human activity on the landscape are visible in the landscape, e.g.:

- air pollution in Auckland and Christchurch
- degraded streams and harbours
- degraded soil — and the loss of capacity for production or pleasure
- loss of bush, species of wildlife, animals and birds, of whole ecosystems.

And the causes? The current styles of ordinary everyday living, farming and "development".

Large areas of our cities are "developed", tamed, "denaturalized". In cities like Auckland or Waitakere about one-third of the urban area is in roads and carparks, about one-third taken up by buildings and much of the rest in predominantly grass monoculture or some exotic species.

The preference for the "quarter acre paradise", the desire for each family to have their own garden, playground and individual space, has created significant if unforeseen environmental impacts — an experience shared in other frontier societies such as the USA. Similarly subsidized housing (some 70% are homeowners) and car-based transport systems, enabled this "desire" to be realized by a high proportion of the community.

"The State of New Zealand's Environment 1997" report, characterizes this development style as a "quarrier economy" (*Ministry of the Environment 1997:9*). It describes the assault on the land, as reflecting a "slash and burn" mentality, and crystallizes the need for action at all levels within our society. Our clean green image may be overstated and certainly will be at risk if we maintain our current habits and patterns.

World-wide there are moves to create and foster sustainable cities. In Waitakere, and over 1 000 cities and towns world-wide from Toronto to Beijing and Berlin, where Agenda 21 (the document of the Earth Summit in Rio de Janeiro in 1992) has been adopted, these aspirations have become practical, visible initiatives. This involves getting to grips with "unsustainable practices". For example reducing (and where practicable) eliminating waste — of water, soil and other key resources.

Water, for example, is seen as plentiful, as a free good — and to have an apparently endless capacity to "self clean". A key requirement then is enabling people to gain a sense of the value, the preciousness of the resource. This can, and in cities like Waitakere, does involve a complete rethink of the wise use of water, how the resource is provided, used, and paid for. It requires an holistic, ecosystem type appraisal and a willingness to shift from an "end of pipe", engineering, "supply at all costs approach". It is easy to talk of the need for significant shifts in mindsets. However, when such mindsets are reinforced by established infrastructure, by specialist knowledge (engineering), by ideological fashions (consumer sovereignty), hostility to or ignorance of Maori ecological beliefs, by existing economic imperatives, then the enormity of the challenge becomes clear.

There is no lack of information concerning the harmful effects of contemporary urban development on social and "natural" environments. Newman and Kenworthy (1989) have catalogued not only the problems but also a range of initiatives that would result in a more benign set of human/ environmental relationships in cities. Their research, conducted over nearly 20 years, demonstrates beyond doubt that low density, car-dependent urban development (compared with the opposite) means:

- higher levels of CO_2 and greenhouse gas emissions per capita
- higher costs for infrastructure provision
- poor public transport, with ongoing negative social and environmental consequences
- fragmented, weaker communities. (Newman and Kenworthy 1989).

The prospects for moving towards sustainable cities rests on more than techniques. It rests on grasping a set of philosophies and values about how humans should relate to or cohabit with nature. It draws on the vision of people like Ian McHarg (1969) who argued for ecologically based design processes, involving an understanding of nature, of the culturally diverse ways humans can and do live, and of different ways of setting economic priorities.

In Waitakere we take the view that the post-war patterns of urban growth cannot be sustained. Central government politicians often argue that current state-supported economic and social regimes are no longer viable (housing, welfare, health) but only partly accept the need for significant change in terms of human impacts on the environment. The "denaturalizing" of the city, is a consequence of conventional ways of living and doing business both directly (housing, economic activity, car use/roading) and indirectly (excessive demand for energy, food, and services, land use changes, chemical control use etc both within and beyond the City limits). This, in a nutshell, is the New Zealand cities' experience — in contrast to the "traditional" cities of Europe or Asia. (NB: those cities confront huge problems also but under conditions of greater population and development pressures and often more limited financial and technical resources).

ACTION AREAS FOR SUSTAINABLE CITY DEVELOPMENT

The development and embedding of a comprehensive long-term approach or strategy is critical to maintaining a drive towards creating more sustainable cities. That strategy will depend on identifying environmental risks and their causes, requisite historical and socio-economic conditions, ideological assumptions and a range of other factors often of a local nature.

In Waitakere the core strategies are:

- to develop a compact city of urban villages (higher densities, greater variety of housing types, mixed land use) focused on transport nodes (and premised on the active promotion of alternatives to car use for social and economic as well as environmental reasons). This focuses development, conserves energy and resources and limits pressures on the urban edges.
- to foster and extend the network of green spaces, for landscape protection, as water cleansers, and air purifiers, as wilderness areas and wildlife habitats, recreation areas, bank vaults of biodiversity; to link the ranges to the sea and in the long term to "enclose" the urban areas.

In New Zealand, Australia and the USA this approach is called "new urbanism". But for those with an interest in the history of cities and urban planning these are not new ideas. They reflect the way European cities grew in the 19th and 20th centuries, and the philosophies of the garden city movement.

For Waitakere the *Resource Management Act* provided a supportive rationale and a process for developing an understanding of environmental objectives, the scale and impact of environmental problems and the means to discuss and develop proactive strategies with the community. (Waitakere City Council 1995).

The provision of water, treatment of wastewater (grey and brown) and disposal of stormwater forms a central component of Waitakere City's 20 year sustainable development strategy. The actions derive from an analysis which looks at water holistically from a conservation ethic point of view. This is reflected in all aspects of the management of the resource in particular in matching demand to supply. Areas of activity such as this (and transport) involve huge capital expenditure to maintain, extend or modify assets by public authorities as well as by individual households and companies. Influencing these expenditures systematically has the potential to embed substantial change long term.

As part of the long-term sustainability strategy initiatives which are often treated in public policy making as separate, can be drawn together to create a strong base for change.

In Waitakere — as in other Agenda 21 cities — there is a wide range of complementary programmes. These include:

- cleaner production approaches to waste generation
- public transport promotion together with other clean transit technologies (bikes, walking)
- promotion of local economic development that is environmentally benign
- community strengthening, including local access to health, educational and social services
- support for cultural diversity and in particular tangata whenua rights.

Fundamental to Waitakere's approach are the principles and processes of community involvement identified in Agenda 21. (Waitakere City Council 1996).

BRINGING NATURE BACK — THE KEY METHODS

Waitakere has attempted an holistic approach to environmental restoration. In part this reflected recognition that all too often positive initiatives can be negated or cancelled out by actions elsewhere. For example, in District Plans, Councils can seek to eliminate weeds by regulation while garden centres sell many such species as decorative plants. Similarly targets can be set by regulatory authorities to decrease air pollution while elsewhere in the society, other measures such as road and petrol pricing and land use planning encourage car use.

The core actions taken by Council to protect existing green areas and to promote restoration include:

- the District Plan — which provides the knowledge base to set targets to:
  - identify and control development around stream systems;
  - define environmental bottom lines;
  - identify and protect areas of outstanding and significant vegetation and fauna habitats;
  - provide a mechanism, (through consent processes) to encourage householders and developers to protect, replant valued species and require this for developments in specific restoration areas throughout the city and in defined circumstances, provide rate incentives to promote ecological objectives.
- Increase organizational and community capacity:
  - by upskilling staff to persuade developers and businesses and encourage good practice;
  - by providing information and guidelines to improve mainstream knowledge and practices.
- Council's investment priorities and operational conduct, e.g.:
  - parks and reserves strategies, e.g., comprehensive planting, restoration and priority setting systems;
  - water course and catchment management to specific ecological, safety and water management objectives;
  - priority given to natural, relative to engineered infrastructure e.g., impermeable surface areas increased and maximized, even in built up areas (to clean streams and filter stormwater);
  - to prioritise planting locally appropriate plants and undertake pest eradication to enable native wildlife to re-establish and flourish.

In these initiatives particular attention is given to:

Maori

- to increase the management options and add to the knowledge base on the local ecosystems.

Businesses

- to require business to minimize environmental damage and to encourage the development of environmentally benign or positive construction and operational practices (e.g., West City a shopping mall with ecological features or ecomall and a stream restoration project). We recognize in this respect we are a long way behind the requirements in parts of Europe and Canada when if negative environmental consequences cannot be avoided now (e.g., generating significant car traffic) then significant compensation from the retail chain is required.

Community

- through a diverse range of mechanisms ranging from information provision, to school, arts and volunteer projects, involve people directly in locally significant restoration projects. Of these a Trees for Babies (families of all babies born each year participate in tree planting projects in their area) programme is most widely valued by the community.

Council

- in addition to projects described earlier:
  - long-term street greening programme within ecological parameters
  - influence over all town centre/major subdivision design through its property company (good practice as at Harbour View Estate), through major town centre, community design processes (as at New Lynn, Glen Eden, Te Atatu Peninsula etc.), and staff time allocated to influencing developer proposals to achieve better results than the *Resource Management Act* requires.

GREENING THE CITY — CREATING GREEN NETWORKS

The New Zealand State of the Environment report suggested that attaining sound environmental health would require 20% of lowland and wetland areas to be restored to their "natural" state. At this stage Waitakere has not set a specific area target although the provisions of the District Plan, the Parks Strategy and the Greenprint (the Strategic Plan) would suggest a figure approaching that level. The Green Network strategy is one of the eight sustainability pathways adopted by the City.[1] *(Waitakere City Council Greenprint 1996).* It involves:

1. An ecological knowledge base.
2. Clear objectives of linking the ranges to the ocean by restoring streams, wetlands, bush and coastal areas, on a scale sufficient to recreate natural ecosystems.
3. Core elements include:
   - specific short and long-term objectives including 21 specific targets;
   - careful long-term planning (including funding) to integrate the eight pathways towards the sustainability objectives;
   - co-ordination of Council actions and efforts to persuade and integrate actions of other public and private agencies;
   - promote understanding and good practice;
   - publicise (and involve community in) successes;
   - maximize community involvement;
   - demonstrate what works.

The value of the programme has been alluded to earlier in this chapter. In essence it:

- creates larger and more diverse natural habitats, bush and wildlife restoration
- brings nature closer to people, integrates the City's natural and production habitats
- offsets CO_2 generation (carbon sinks)
- improves water management (ie reduced resource consumption, soft engineering measures implemented)
- contributes to more efficient financial management in part by fuller scrutiny of costs over product life cycles
- protects stream margins and habitats.

It is our view that green accounting systems would demonstrate the financial as well as the environmental superiority of a more ecologically based approach to urban development.

These programmes have been in place for little more than four or five years. There are, however, some signs of success:

- the presence of the kereru (wood pigeons) in the urban fringes
- wetlands extended (Harbour View) and birdlife increased
- historic landscapes restored along stream banks (but a long way to go)
- effective marriage of human creativity and stream cleanups as a prelude to restoration (e.g., Rewarewa Bridge).

The regeneration of the Waitakere Ranges since the early part of the 20th century remains both an icon of what can be achieved and a reminder of the threats to full restoration that remain.

[1]The eight pathways are in effect work programmes designed to move the city from the status quo to a more sustainable position. This involves the co-ordination of budgets, work programmes both within and external to Council. The key pathways are Urban Villages, Public Transport, Green Network, Sustainable Business, Social Wellbeing, Eco Water, Solid Waste, Democracy.

The City has benefited from creative leadership from its own environmental groups such as the Waitakere Ranges Protection Society and the New Zealand Forest and Bird Society — and many others. People like Di Lucas from Christchurch (who produced the revegetation guidelines for the City), Virginia King who designed the Rewarewa Bridge and Professor John Argue of Adelaide who assisted our advisors and decision makers to rethink the approach to water and stormwater management, have provided stimulus and practical alternatives to status quo methods. Such people provide the drive, the inspiration as well as the practical know-how that enables change to occur.

But it needs also to be said that this strategy has been adopted and progressed because of bold political leadership which embraces the community's enthusiasm and passion for the surf, the outdoors and the bush (locally called westie culture). This occurs in a national environment where politics is seen often as venal self interest and where citizens are characterized as apathetic and atavistic.

THE NEXT CHALLENGES

The focus in the next 3–5 years will be on extending current programmes and more fully embedding key strategies. At one end are large scale plans such as the 50 year strategy for the West Coast wilderness area (and the pressures derived from growing visitor numbers). This involves all key stakeholders and is founded on partnerships with Maori, the communities of the area and other major agencies such as the Auckland Regional Council. At the other end of the scale is the endeavour to increase the level of participation in the ethos, practices and planting regimes by householders. To be successful ultimately all 50 000 households of the city need to be practising sustainable land use management.

Widespread community involvement is a key to the long term target of conserving and enhancing both the quality and quantity of land and resources available to future generations.

The maintenance of the compact city/urban villages strategy through a combination of regulation, persuasion and direct investment by Council remains pivotal to protecting and enhancing the natural foundation of urban life. This requires skilled staff and partnerships with specialist design professionals and applied scientists amongst others. The shift needs to occur quickly enough to support public transport initiatives, without on the other hand degrading or over-stressing amenity or social networks and services in the locality.

This suggests key challenges will lie in achieving the co-ordination of activities over which the Council has limited influence, e.g., provision of education, health and public transport services or the generation of local employment. Equally important will be bringing home to individuals, households and businesses the real costs and consequences of some current and accepted practices (e.g., chemical overuse, waste generation, car overuse, water supply and wastewater disposal).

Enabling people to adopt often small scale, low cost solutions will be a necessary part of the behaviour change mix (e.g., household size worm farms to process organic waste). There are significant issues of equity and access which current user pays rationing regimes do not address and to which current political agendas do not appear to provide answers. It could be counter productive if larger and sometimes poorer than average households disown resource conservation imperatives due to higher costs arising from user charge regimes.

At Waitakere further emphasis will be placed on:

- promoting "green thinking" — particularly demonstrating the economic sense of green strategies.
- demonstrating the economic development potential of sustainable systems — from design through to "final" products. This approach is being picked up in Europe and more recently in Japan. This in part reflects the view that cleaning up environmental problems will be a 21st century growth industry and in part recognition that new products will need to meet higher environmental standards. The quest of the Japanese car industry for cars that will not pollute is just one example.

However, attaining the objective of greater sustainability, depends on increasing community involvement. Central to this are:

- better communication programmes which avoid the printed word, involve face to face discussion and debate, working models, demonstration projects, visual methods (photos, cartoons etc.), radio and television
- joint decision making and programme development by more extensive community involvement in devising solutions to problems they can identify and understand.

It will involve campaigns, not only in an advertising sense but in sustained, community based programmes which involve informed debate and decision making at a local level.

CONCLUSION

The agenda described in this chapter can and does work. Nature is an affordable answer. It makes sense. It brings jobs — as well as nature back into our cities and our lives.

The Waitakere experience demonstrates the importance of supporting household and local community actions as well as those of central and local government.

Waitakere's programmes to "naturalize the city" arose from analysis of the degradation of the natural environment and of the impact of the prevailing forms of urban development. The consequential core strategies to address these issues can appear paradoxical. One focus is on transforming urban form, to achieve a different mix and density of urban activities, not only to decrease the net environmental impact and strengthen communities but also to decrease urban pressure on vulnerable natural environments. The other focus is directly on the natural environment and on strategies to restore species and ecosystems. They in fact reinforce each other.

At Waitakere, strategic environmental and annual plans are aligned to this end. While in the mid 1990s efforts focused on regulation and direct Council investment in catchment, stream restoration, park systems and related programmes, more recently efforts have focused on enabling communities and households to engage in environmental restoration and enrichment. The provision of planting advice, guidelines, the alignment of reserve management plans and the introduction of rating incentives are illustrations of this approach.

Thus at the core of the "naturalizing" programme lies urban consolidation, a coherent, targeted plan to reshape city form.

ACKNOWLEDGEMENT

This chapter reflects the vision and commitment of my colleagues the Mayor, Councillors, Community Board members and staff at Council, and the wider community of Waitakere City. The assistance of Ann Magee, Senior Management Group, Waitakere City Council, is appreciated in preparing this chapter.

REFERENCES

Ministry for the Environment, 1997. The State of New Zealand's Environment. Ministry for the Environment: Wellington, New Zealand.

McHarg, I., 1969. Design with Nature. John Wiley and Sons: New York.

Newman, P. W. G. and Kenworthy, J., 1989. Cities and Automobile Dependence. Gower: New York.

United Nations Conference on Environment and Development, 1992. Outcomes of the Conference (Agenda 21) New York.

Waitakere City Council, 1993 and 1996. Greenprint (Draft) Waitakere City, New Zealand.

Waitakere City Council, 1995. Our City Our Future. Proposed District Plan, Policy Section: Waitakere City, New Zealand.

CHAPTER 41

The planning and ecology of green corridor networks in tropical urban settlements: a case study

Clive Briffett,[1] Navjot S. Sodhi,[2] Lily Kong[3] and Belinda Yuen[1]

In Singapore there is a plan to connect all the existing national, regional and neighbourhood parks using linear strips primarily of native vegetation over the next 20 years. These strips will constitute a nation-wide park connector (corridor) network. The primary objective of the park connectors is to provide humans with recreational and an aesthetically pleasant environment in highly urbanized Singapore. Ancillary to this the park connectors may enhance and facilitate wildlife movements.

We carried out a pilot study to evaluate human and wildlife (bird) uses of corridors. In one completed corridor we found that the majority of human users live locally and use it to walk or jog. About one-third who visit, do so 2–3 times a week in the early morning or evening period. There was a high appreciation for natural surroundings, fresh air and the freedom of open space and a love for nature. Focus group discussions revealed that people's relationships with nature is highly textured and nuanced and had to be understood in the context of their lives.

Birds were used as indicators of biodiversity and we recorded 95 species of birds across five corridors. It is possible that habitat diversity affected bird density more than human activity or presence. Birds indulged in various behaviours including feeding and courtship activities indicating that the corridor met certain critical needs. Human and wildlife interactions were difficult to measure and this is the subject of further ongoing research for which we propose a more refined research methodology.

Key words: Green corridors, Park connector, Greenways, Recreation, Urban birds, Landscape ecology.

INTRODUCTION

Background

URBAN green corridor systems or greenways have been researched and implemented in many parts of the world. As early as 1963, the potential for using redundant railway lines in the United Kingdom as greenways was suggested (Dower 1963). In the United States of America in the 1960s, green corridors usually termed greenways and related landscape ecology techniques were of increasing scientific interest (McHarg 1969; Schwarz 1994). Similar corridors are also being extensively identified, planned, designed and created in Europe (Andrews 1993), Africa (Roberts 1994), and Australia (Saunders and Hobbs 1989).

Such corridors have been considered for the more extensive open spaces of rural and wilderness areas (Saunders and Hobbs 1989), the small and limited habitats of islands (Terborgh 1976), as well as heavily urbanized built up areas (West Midlands 1984; Tyne and Wear 1988; London Borough of Lewisham 1989). Because of varying historical and natural succession factors and differing land use constraints and planning policies, the designs of corridors are extremely diverse and often unique.

[1]School of Building and Real Estate, National University of Singapore, 4 Architecture Drive, Singapore 117566, Republic of Singapore.
[2]School of Biological Sciences, National University of Singapore, Kent Ridge, Singapore 119260, Republic of Singapore.
[3]Department of Geography, National University of Singapore, 1 Arts Link, Singapore 117570, Republic of Singapore.

Pages 411–26 *in* NATURE CONSERVATION 5: NATURE CONSERVATION IN PRODUCTION ENVIRONMENTS: MANAGING THE MATRIX ed by J. L. Craig, N. Mitchell and D. A. Saunders. Surrey Beatty & Sons, 2000.

Corridors are subject to many variable development pressures and are utilized for numerous recreational purposes. Their designs have, however, embraced alternative and improved land uses which, it is claimed, can meet many ecological, environmental and recreational benefits never previously considered in urban locations. In some countries, the main aim of creating or adapting areas to form corridors is to facilitate human recreational use where habitats are viewed only for aesthetic enhancement as greening strategies (Briffett *et al.* 1996). In others, more emphasis is placed on accommodating wildlife needs and improving biological diversity by maintaining and creating habitats and ecosystems (Forman 1983).

The view that corridors may act as wildlife conduits was first mooted by Preston (1962) and was later seen as a special application of the island biogeographic theory developed by MacArthur and Wilson (1967). This theory states that a patch of habitat or island of a larger size will hold more species especially if it is near possible colonists such as other patches, than if it is small and distant. The significance of this theory for corridors is that such linear connections can possibly assist in improving accessibility for species between smaller fragmented patches to increase re-colonization.

More recent work, however, relates to meta-population theory (Hanski 1989). This theory states that the numbers of any given species may fluctuate greatly in small patches of habitat. This is because the local extinction of species in one patch may be replaced by re-invasion by other similar populations from other adjoining collections of patches. The rate of establishment of new populations has to exceed the rate of local extinctions for the meta-population to survive. In corridor research, it is therefore useful to ascertain the numbers of viable patches present and the nature of their connections to assess the likelihood of re-colonization up to the natural carrying capacity of the habitat.

Another research area relevant to the suitability of corridors as wildlife habitats is landscape ecology. This comprises the study of landscape structure, function and the change of scale of entire landscapes as well as the application of these results to the design and management of both natural and human dominated areas (Forman and Godron 1986). Important impacts can therefore be determined by considering this area of interest (Forman and Godron 1981; Turner 1989).

A substantial body of research suggests that corridors are effective as conduit systems for wildlife and that such links should be retained and created (Diamond 1974; Wilson and Willis 1975; Noss 1983; Bennett 1991; Moore 1991). Much of this has not met formal, rigorous hypothesis testing requirements and has thus been heavily criticized by others (Dawson 1994). This is because many of the findings are consistent with alternative explanations, and the few conclusive studies only confirm a conduit function in a limited range of circumstances. Several studies actually conclude that corridors are not particularly effective in providing routes for local migration of animals (Helliwell 1975; Simberloff and Cox 1987) and that disturbance by humans in such habitats is significant (Groome 1987).

Existing research either assesses the suitability of corridors for human recreational uses (Mostyn 1984; Harrison 1987) or for wildlife habitation and survival (Andrews 1993), but not both. The environmental impact of humans on wildlife survival and use has been researched for parks and reserves (Hooper 1971; Diamond 1974; Driver 1987; Kuss 1990; Vaske 1994) and human responses to vegetation and landscapes have also been ascertained (Ulrich 1986). However, most studies on corridor systems have been done in temperate locations and it is unknown if these can be applied to tropical urban areas.

Local context

Singapore is an highly urbanized tropical city with one of the world's highest population density. Due to urbanization, many native habitats within Singapore are now greatly reduced. In 1991 the Urban Redevelopment Authority of Singapore published the *Revised Concept Plan for Singapore* that included a "green and blue plan". This plan identified areas of open space throughout the island that will be designed to meet the recreational needs of human users, to protect the natural environment and to improve biodiversity (Liu 1991). The Plan also incorporated earlier proposals of areas suitable for nature conservation as identified in the *Master Plan for the Conservation of Nature in Singapore* published by the Nature Society, Singapore (Nature Society, Singapore 1990).

Following the adoption of the Revised Concept Plan, the National Parks Board (NParks) published a proposal for an island-wide park connector network (*Straits Times* 1992; Fig. 1). This was essentially designed to meet the perceived growing needs of the human population for various alternative recreational facilities. The park connectors were mainly planned and located along existing water drainage culverts and river systems. These were supposed to connect existing national and regional parks, local neighbourhood parks and coastal areas. They comprise an open space facility for off-road cycling, jogging, exercise and walking. On completion, it was envisaged that users would be able to traverse the entire island without having to experience the dangers, inconveniences and pollution of the main roads. To achieve this vision, there is considerable work required in connector construction, landscaping and particularly in the construction of road crossovers. These may be in the form of "at grade" (same as road level) road cycle crossings, tunnels or bridges (*Straits Times* 1996).

Originally the total park connector programme was envisaged to take up to 30 years to complete and to eventually extend for a total distance of approximately 300 km (*Straits Times* 1995). More

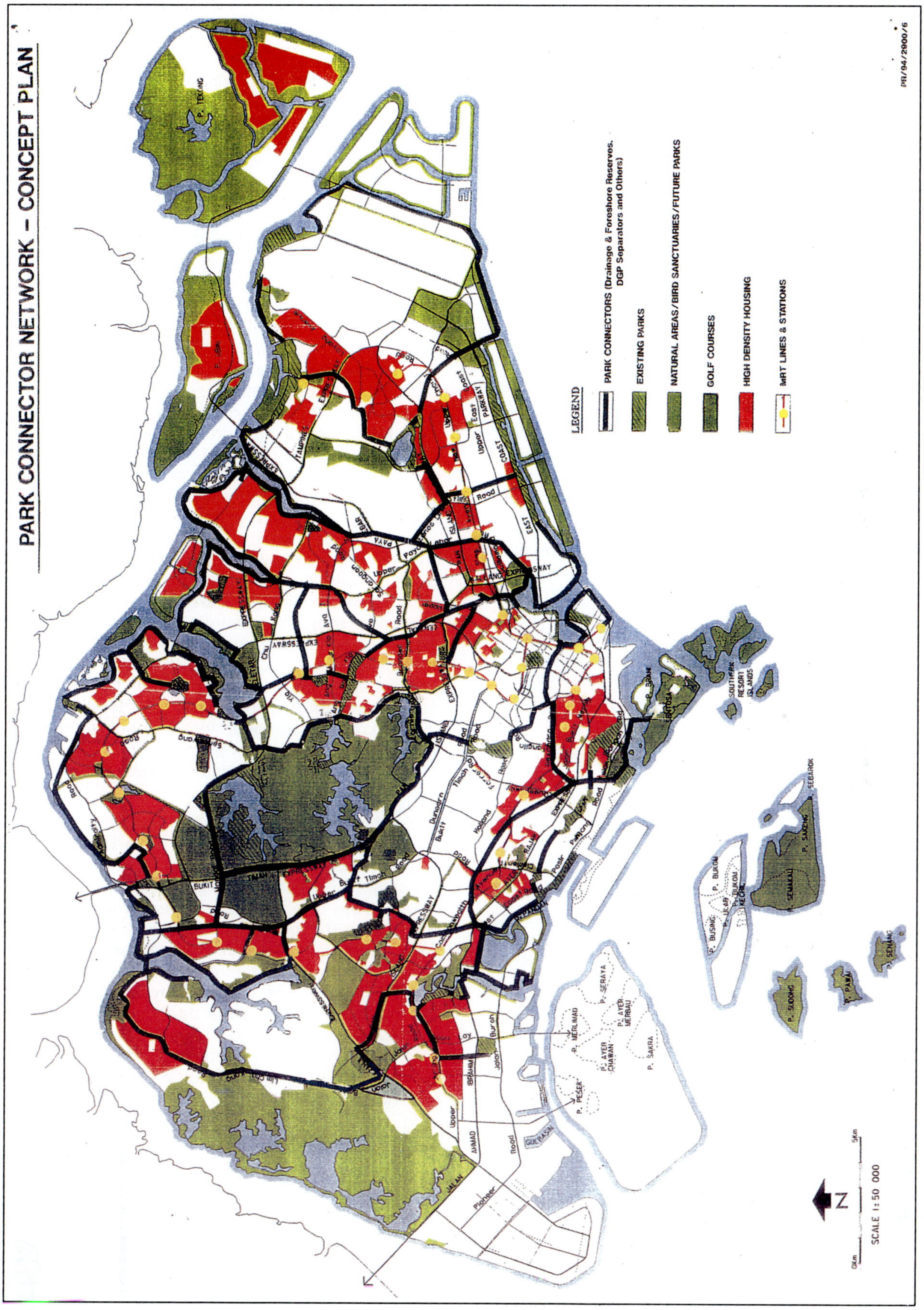

Fig. 1. Proposed park connector network of Singapore. Source: National Parks Board Singapore.

recently it has been revised to a twenty-year programme. To date, four routes are almost complete and 14 more are planned for completion by the year 2000. There are longer-term plans for at least another 30 connectors. Most of the existing linear parks included within the network are already complete but there are further plans to upgrade these with additional recreational facilities.

Apart from the need to cater to human recreational needs, the connectors' potential to cater to wildlife was first mooted in the *Singapore Green Plan — Action Programmes* in 1993. The original Green Plan was prepared for the Rio Summit by the Ministry of the Environment as a contribution to a blueprint for a global action programme on sustainable development. The green network in this plan for Singapore was described as "a tapestry of green to make parks and nature sites more accessible to the public and to provide corridors for the movement of bird life" (Ministry of Environment 1993:51).

AIMS AND OBJECTIVES OF STUDY

As the island-wide park connector scheme is still in its early implementation stages, it was considered an opportune time to evaluate planning and design criteria. The main aim of this study was to make a preliminary assessment of park connectors (corridors) in Singapore in meeting the mutual needs of wildlife and humans. To meet this requirement the following broad objectives were set:-

- to determine human recreational use patterns and behaviours;
- to identify appropriate design techniques that may maximize recreational uses;
- to evaluate methodologies that can facilitate an understanding of human investment of meanings and values in nature;
- to assess the diversity and composition of wildlife (birds); and
- to make preliminary recommendations for improving existing and proposed habitats for wildlife occupation.

Resulting from this pilot and other ongoing research we hope to produce planning and design guidelines that will be useful for developers, town councils, community groups, non-government organizations and government departments.

To determine human recreation use patterns in corridors, we conducted recreational and household surveys in one fully completed connector, Ulu Pandan Canal. To further determine what humans would like in the corridors in particular and other open areas in general, focus discussions were carried out. For determining wildlife use of corridors, bird surveys were made in five linear areas. Ongoing studies elsewhere have examined the migration of birds in corridors. In addition, studies using birds as indicators of habitat and ecosystem health, quality and diversity have been conducted for tropical forests (Round 1988), small temperate woods (Opdam *et al.* 1985), hedges (Osborne 1984), estuaries (Peck 1993) and city parks (Gilbert 1989; Lock 1996), farmland (Lack 1992), and urban areas (DeGraaf *et al.* 1991). Birds are therefore considered a suitable subject for monitoring corridors due to their relatively high level in the food chain and the comparative ease of their identification in survey work.

STUDY SITES AND METHODS

Study sites

Ulu Pandan Canal: The total length of this fully completed jogging and cycling track is 5 km and it aligns with an existing water culvert. The connector is generally narrow ranging from 30 to 100 m wide and has a combined walking/cycling trail only 6 m wide. Several road and railway bridges including a Mass Rapid Transit line cross the trail (see Fig. 2). The main arterial, Clementi Road, divides the trail into two distinct lengths. Areas adjoining the eastern length of this trail comprise undeveloped secondary forest to the south with private housing to the north. To the west of Clementi Road there is an area of high-tech industrial property on the north side that faces directly to a public housing estate on the south. Some previously vacant land areas to the far-western stretch are currently being developed for more residential condominiums. To the north side of the far western section there are more extensive, wilder areas of secondary vegetation. The concreted water culvert in this corridor is usually partly filled with incoming seawater and extends to the sea beyond the end of the cycling trail under a road expressway. There is no trail connection across this expressway at present.

Kent Ridge Park: This highly elevated park is located to the western end of a long ridge of hills and covers an area of 47 hectares. The upper section rises from private housing areas on the south side where there are numerous recreational fitness facilities. In the lower section on the northern slopes there are many alternative contour paths extending to a large and a small pond. Two carparks are situated on the ridge accessed by a steep road on the south slope and pedestrian access is possible from all sides of the park. The park contains small patches of secondary rain forest locally termed *belukar* mainly comprising species tolerant of nutrient poor, acidic soils.

East Coast Park: Located on the south-east coast of Singapore this popular park comprises a longitudinal area of land reclaimed from the sea approximately 20 years ago. It is 13 km long with a maximum width of 300 m. Fringed by imported sandy beaches to the east, this park is heavily used by large numbers of recreationalists and pleasure seeking public, particularly at weekends. There are a large number of water sports facilities, tennis, squash and ten pin bowling clubs, a golf driving range and a learner car driving centre. There is also a crocodile farm and numerous food outlets.

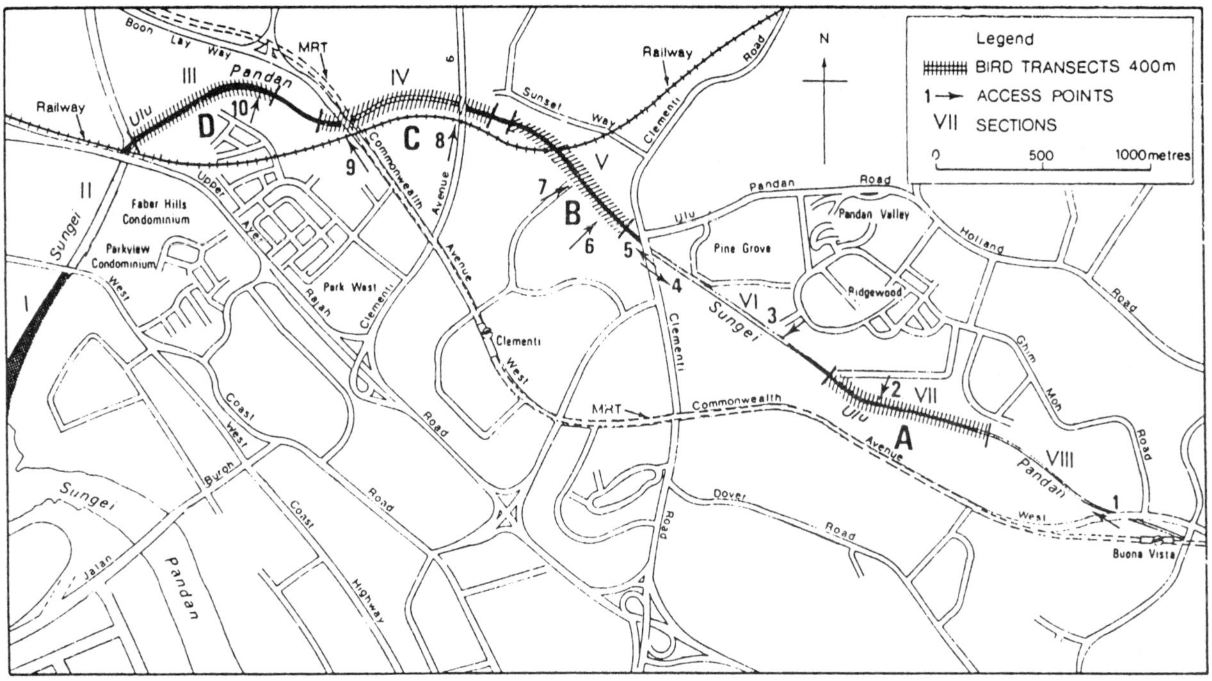

Fig. 2. Plan of Ulu Pandan Canal corridor showing transects, access points and sections.

Most of the park vegetation comprises short, mown grass areas interspersed with isolated planted trees of both introduced and indigenous species. The trees closest to the seashore are predominately Casuarina *Casuarina equisetifolia* and Coconut Palm *Cocos nucifera*. Most of the park areas alongside the expressway edge are planted out with thicker clumps of rain trees, flowering shrubs, and flower beds. A selected number of locations were established as "bird sanctuary" islands about five years ago. These comprise wilder long grass vegetation and secondary growth with a thicker foliage of medium height trees and low shrubs. There are numerous car parks, access roads, cycleways throughout the park and pedestrian tunnels connect to housing areas on the opposite side of the expressway. Extensive barbecue facilities are also located near the seashore.

Bishan Park: This eight-year old, linear park is formed between public housing and private condominiums and is approximately centrally located on the main island. It is encircled by major highways but beyond these to the west is the Central Water Catchment National Park. The corridor covers a total area of 52 hectares and is divided into three sections separated by cross roads. There are a number of ponds used for intensive fishing activity by the public and a concrete water culvert extends through the full length of the park. Other recreational facilities include bike hire, children's playgrounds, fitness centres, open spaces for kites and jogging and cycling trails. Several small bird sanctuary areas similar to those in the East Coast Park are established alongside the central water culvert.

Dairy Farm Trail: This is located along the edge of primary and secondary rainforest to the Bukit Timah Nature Reserve in the central area of the island. The trail was recently adapted for mountain bike use, joggers, and walkers. It is aligned through the denser forest areas commencing at Dairy Farm Quarry and crosses an internal access road to an orchid farm (Dairy Farm). It continues through the forest to an open grassed pipeline reserve and extends alongside an expressway. There are few built developments in close proximity to this trail but it is frequently used at weekends by cyclists, mountain climbers, and other picnic groups especially in the quarry vicinity.

Methods

Recreational Surveys: Three on-site surveys were conducted in the Ulu Pandan corridor in April/May 1996 with the help of student assistants. They were: observations at access points to the connector, observations along the connector and user questionnaire surveys. In order to obtain a comprehensive picture of the use of the park connector, the surveys were conducted over both weekdays (one-third) and weekends (two-thirds). Observations were carried out during the day, between 0600 and 1900 hours. No observation was conducted at night because certain sections of the park connector were not lit. Student helpers were briefed on the conduct of the surveys in order to maintain consistency in interviewing and observation techniques and to minimize error. Pre-designed observation forms and questionnaires were used in the surveys.

- Observations at access points to the connector — Observers were stationed at nine designated access points along the connector to observe and record the number of people entering the connector at those points.

- Observations along the connector — Observers were asked to systematically walk through the park connector recording the various activities and behaviours of people as well as their socio-demographic profile. The connector was divided into eight sections and an observer was assigned to walk each section.

User questionnaire surveys — Interviewers were employed to find out what people on the connector do, feel and think about various aspects of the connector. A questionnaire parallel to that used in the household survey (but shorter in order to minimize the interruption to people's activities) was drawn up for the on-site user surveys. A total sample of 303 questionnaires was completed and the basis of selection was random.

The collected data were coded and then entered into a personal computer for processing using SPSS-PC. Descriptive statistics were generated in the first instance to determine the general patterns in usage and experiences.

Household Surveys: A door-to-door survey involving 300 individuals was conducted in May 1996. The respondents were drawn from the housing estates around Ulu Pandan Canal, namely, the private housing estates in the West Coast, Faber Heights and Holland vicinities, and the Housing and Development Board estate of Clementi. The intention was to obtain information from those who used, as well as those who did not use, the Ulu Pandan Canal. Attempts were made to ensure that the sample's ethnic and socio-economic composition (using housetype as a surrogate index) approximated the profile of the national population. Each questionnaire was administered in about 15 minutes, and the results obtained were analysed using SPSS-PC. Descriptive statistics were generated to provide insights into general patterns of behaviour.

Focus group discussions: In-depth focus groups were convened to explore more deeply and fully the personal meanings and values that people invest in nature, particularly within the full context of their everyday lives. A total of five groups: a users group, an all-student group, a community all-woman group, a young adults group, and an expatriate group were arranged. The discussions extended beyond these immediate concerns to include an expression of views regarding the experience of nature and other landscapes (for example, urbanscapes and countryside) in general.

The users group comprised members identified through both the household and on-site user questionnaire surveys, the student and young adults groups were drawn from educational institutions nearby, the community all-woman group was convened through a neighbourhood church, while the expatriate group was drawn from personal contacts. The size of the group ranged from four to eight persons per group, and all groups were of mixed gender, except for the community all-woman group.

While both the household and on-site user questionnaire surveys were well-suited to establishing broad-based general data about trends and patterns of behaviour, and in providing some descriptive material about people's usage of the corridor, they were rather less suited to probing the often intangible and inchoate, and sometimes subliminal experiences of nature. While there were some open-ended questions which revealed interesting observations, it was difficult to include too many of such questions in the questionnaires without lengthening the time needed to administer each questionnaire inordinately, thus risking a high rate of incomplete returns. Thus, to collect more qualitative data, the more qualitative method of in-depth focus group discussions was used.

The merits of such a method in selected research situations have been dealt with at length elsewhere (Walker 1985; Eyles and Smith 1988). As Walker (1985) pointed out, qualitative research is appropriate when the topic is sensitive and complicated, and when the phenomenon is not measurable. It is also appropriate when one is concerned with the "understanding of meanings in specific contexts" (Eyles 1988), and with capturing the "wholeness" of an individual (Donovan 1988) in the words of the people themselves. For these reasons, a qualitative method seemed appropriate for collecting information on the symbolic meanings and values invested in nature.

Bird Surveys: The study of bird life and behaviour was conducted using bird census and activity surveys. These were carried out between April 1996 and March 1997. At the Ulu Pandan Canal, 16 transects of 100 m length each were identified and marked. These 16 transects were arranged into four 400 m blocks, each consisting of four transects of 100 m arranged consecutively (see Fig. 2). Each 400 m block of transects was separated from the nearest block by 100 m. In Kent Ridge, twelve 100 m long transects (three 400 m transect blocks) were established. In other corridors (Bishan Park, East Coast Park, and Dairy Farm Trail) eight 100 m transects (two 400 m transect blocks) were set out. The transects were chosen to represent different types of vegetation/habitat within each site. All transects were surveyed once during each of nine months (April, May, June, September, October, November, December, January and March). Birds within 50 m of both sides of the transect were recorded. The aim was to determine the extent to which all corridors were successful in attracting birds. All birds seen or heard between 0700 hours and 0900 hours and 1630 hours and 1900 hours were counted on days without rain or heavy winds.

In addition, a walking survey was conducted by one observer along the transects recording bird activities into four broad categories: flying, feeding, perching, and singing. The activity of each sighted bird was recorded. For an individual bird, only the first noticed activity was recorded. Birds displaying any activities other than those stated above were ignored in the analyses presented. The aim was to determine the bird's habitat requirements that the corridors can fill. Because the interest is in the overall value of the corridor to the bird community, data from different species were pooled in the analysis. Bird activity surveys were conducted between 0700 hours and 0900 hours and between 1630 hours and 1900 hours on fair weather days.

The bird species diversity was calculated by using the Shannon-Wiener formula

$$H' = \Sigma (pi \times \log pi)$$

where pi is the proportion of species i in the corridor. This index gives a more appropriate index of bird diversity at a site because it takes into account abundance of each species recorded at a site.

To determine bird community similarity among the five connectors surveyed, a cluster analysis was performed using the programme PC-ORD. The cluster analysis used the nearest neighbour pair-group clustering and Euclidean distance coefficients (McCune and Mefford 1995). Data matrix for the cluster analysis consisting of five corridors × 95 bird species were used (see Table 1). The maximum number of individuals recorded for each species were used in this matrix. The data matrix was log-transformed to approximate normal distribution. The cluster analysis gives a graphical representation that can be used, in this case, to determine how similar bird community composition is among different corridors.

RESULTS AND DISCUSSION

Recreational surveys

During the access point survey 4 087 people were recorded as entering the connector. Of these, 70% were male and 79% were Chinese. The majority (70%) were walking or jogging and 25% cycling. Most of the users were under 50 years old (80%). The single largest age group of users (25%) were between 21–30 years while the smallest age group of users (7%) were over 60. Fifty-five per cent came to the connector by themselves.

In the on-site trail observation surveys 43 different behaviours were noted and these included passive activities such as looking (17%), conversing (10%) and listening (10%) as well as the active recreational activities noted above and other affectionate and problem behaviours.

In the on-site user questionnaire surveys, 303 people were interviewed. Eighty-three per cent lived in public housing near the connector. Their reason for visiting the canal was predominantly for getting exercise (68%) while others cited the place as suitable for relaxation (7%), and as convenient in terms of access from home (9%). Users liked the connector for a variety of reasons including quietness (14%), good air (11%), and presence of vegetation (10%). Dislikes were that the canal was smelly (16%), that too much construction was going on (5%), that there were too few rain shelters

Table 1. Mean number of bird species recorded in five corridors of Singapore. Mean numbers are followed by standard error. UP = Ulu Pandan Canal, KR = Kent Ridge Park, EC = East Coast Park, DF = Dairy Farm Trail and BP = Bishan Park.

No.	Species		UP	KR	EC	DF	BP
1	Purple Heron	*Ardea purpurea*	0.17±0.07	0	0	0	0
2	Striated Heron	*Butorides striatus*	0.33±0.12	0.07±0.05	0.11±0.07	0	0.33±0.17
3	Chinese Pond-heron	*Ardeola bacchus*	0.19±0.11	0	0	0	0
4	Cattle Egret	*Bubulcus ibis*	1.86±0.70	0	0	0	0.28±0.28
5	Little Egret	*Egretta garzetta*	0	0	0	0	2.33±1.04
6	Yellow Bittern	*Ixobrychus sinensis*	0.03±0.03	0	0	0	0.11±0.07
7	Cinnamon Bittern	*I. cinnamomeus*	0	0	0	0	0.06±0.06
8	Black Baza	*Aviceda leuphotes*	0	0.11±0.11	0	0	0
9	Oriental Honey-buzzard	*Pernis ptilorynchus*	0.03±0.03	0.07±0.05	0	0	0
10	Brahminy Kite	*Haliastur indus*	0.14±0.06	0.07±0.07	0.06±0.06	0.11±0.07	0
11	White-bellied Fish-eagle	*Haliaeetus leucogaster*	0	0.04±0.04	0	0	0
12	Grey-headed Fish-eagle	*Ichthyophaga ichthyaetus*	0	0	0	0.06±0.06	0
13	Japanese Sparrowhawk	*Accipiter gularis*	0	0	0	0	0.06±0.06
14	Slaty-breasted Rail	*Gallirallus striatus*	0.06±0.06	0	0	0	0
15	Red-legged Crake	*Rallina fasciata*	0	0	0	0.06±0.06	0
16	White-breasted Waterhen	*Amaurornis phoenicurus*	0.28±0.08	0.44±0.11	0	0.22±0.15	0.56±0.18
17	Wood Sandpiper	*Tringa glareola*	0.06±0.04	0	0	0	0
18	Common Sandpiper	*T. hypoleucos*	0.44±0.16	0	0.11±0.11	0	1.00±0.34
19	Little Tern	*Strena albifrons*	0.08±0.08	0	0	0	0
20	Pink-necked Green-pigeon	*Trenon vernans*	1.33±0.35	3.81±0.83	1.50±0.78	1.00±0.51	0.94±0.35
21	Rock Pigeon	*Columba livia*	0.61±0.19	0	0	0	0.56±0.24
22	Spotted Dove	*Streptopelia chinensis*	2.31±0.35	0.59±0.17	0.72±0.24	2.67±0.70	2.11±0.81
23	Zebra Dove	*Geopelia striata*	0.97±0.21	0	0	0.06±0.06	0.33±0.56
24	Emeraldd Dove	*Chalcophaps indica*	0	0	0	0.06±0.06	0
25	Long-tailed Parakeet	*Psittacula longicauda*	0.08±0.06	0.11±0.11	0	0.17±0.12	0.22±0.15
26	Lesser Sulphur-crested Cockatoo	*Cacatua sulphurea*	0.11+0.06	0	0	0	0
27	Eclectus Parrot	*Eclectus roratus*	0	0.04±0.04	0	0	0
28	Indian Cuckoo	*Cuculus micropterus*	0.03±0.03	0.04±0.04	0	0	0
29	Banded Bay-cuckoo	*Cacomantis sonneratii*	0.03±0.03	0	0	0.11±0.07	0
30	Tanimbar Cockatoo	*Cacatua gottini*	0	0.04±0.04	0	0	0
31	Drongo Cuckoo	*Surniculus lugubris*	0	0	0	0.11±0.11	0

Table 1 — *continued.*

No.	Species		UP	KR	EC	DF	BP
32	Asian Koel	*Eudynamis scolopacea*	0.39±0.08	0.63±0.21	0.61+0.22	0	0.67±0.28
33	Chestnut-bellied Malhoka	*Phaenicophaeus sunatranus*	0	0	0	0.28±0.22	0
34	Greater Coucal	*Centropus sinensis*	0.03±0.03	0.18±0.11	0	0.11±0.07	0
35	Lesser Coucal	*C. benalensis*	0.03±0.03	0	0	0	0
36	Swiftlet spp.	*Collocalia* sp.	0.97±0.56	0.37±0.20	0	0.28±0.19	0.06±0.06
37	House Swift	*Apus nipalensis*	0.06±0.04	0.26±0.18	0	0.11±0.11	0.22±0.22
38	Common Kingfisher	Alcedo atthis	0.11±0.08	0.07±0.07	0.06±0.06	0	0.61±0.27
39	White-throated Kingfisher	*Halcyon smyrnensis*	1.14±0.15	0.52±0.13	0.11±0.07	0.89±0.18	1.39±0.29
40	Black-capped Kingfisher	*H. pileata*	0.03±0.03	0	0	0	0
41	Collared Kingfisher	*Todirhamphis chloris*	0.11±0.06	0.18±0.08	0.39±0.07	0.22±0.17	0.28±0.12
42	Blue-tailed Bee-eater	*Merops philippinus*	0.08±0.06	0	0	0	0
43	Blue-throated Bee-eater	*M. viridis*	0.11±0.07	0.11±0.08	0	0.44±0.24	0
44	Dollarbird	*Eurystomus oreintalis*	0.19±0.08	0.22±0.12	0	0.06±0.06	0
45	Coppersmith Barbet	*Megalaima haemacephala*	0.06±0.06	0.04±0.04	0	0	0
46	Laced Woodpecker	*Picus vittatus*	0	0	0	0.11±0.07	0
47	Banded Woodpecker	*P. mineaeceus*	0.06±0.04	0.15±0.06	0	1.22±0.34	0
48	Common Flameback	*Dinopium javanense*	0.03±0.03	0	0	0.11±0.11	0.06±0.06
49	Sunda Woodpecker	*Dendrocopus moluccensis*	0.08±0.04	0	0.06±0.06	0	0.11±0.07
50	Barn Swallow	*Hirundo rustica*	0.03±0.03	0.37±0.37	0	0.50±0.44	0
51	Pacific Swallow	*H. tahitica*	1.64±0.34	0.15±0.10	1.44±0.39	0.44±0.21	1.50±0.37
52	Pied Triller	*Lalage nigra*	0.06±0.04	0.04±0.04	0	0.06±0.06	0.06±0.06
53	Common Iora	*Aegithina tiphia*	0.83±0.29	0.78±0.30	0.06±0.06	0.50±0.33	1.39±0.44
54	Straw-headed Bulbul	*Pycnonotus zeylanicus*	0	0	0	0.89±0.48	0
55	Yellow-vented Bulbul	*P. goiavier*	4.53±0.80	3.81±0.53	0.72±0.32	1.22±0.25	5.50±1.35
56	Olive-winged Bulbul	*P. plumosus*	0	0.22±0.11	0	1.06±0.40	0
57	Greater racket-tailed Drongo	*Dicrurua paradiseus*	0.08±0.06	0	0	4.00±0.75	0
58	Black-naped Oriole	*Oriolus chinensis*	1.69±0.25	3.74±0.59	2.17±0.55	3.56±0.60	2.17+0.73
59	Asian Fairy-bluebird	*Irena puella*	0	0	0	0.11±0.07	0
60	House Crow	*Corvus splendens*	2.61±0.84	1.44±0.75	6.67±1.96	0	4.28±1.37
61	Large-billed Crow	*C. macrorhynchos*	0.36±0.30	0.33±0.12	0.11±0.11	0.06±0.06	0.17±0.17
62	Short-tailed Babbler	*Malacocincla malaccensis*	0	0	0	0.22±0.12	0
63	Abbott's Babbler	*M. abbotti*	0.06±0.06	0.26±0.15	0	0.06±0.06	0
64	Chestnut-winged Babbler	*Stachyris erythoroptera*	0	0	0	0.06±0.06	0
65	Striped Tit-babbler	*Macronous gularis*	0.03±0.03	0.30±0.13	0.06±0.06	1.89±0.45	0
66	Oriental Magpie-robin	*Copsychus saularis*	0.03±0.03	0.22±0.08	0	0.17±0.08	0
67	White-crested Laughing-thrush	*Garrulax leucolophus*	0	2.56±0.96	0	0.22±0.22	0
68	Hwamei	*G. canorus*	0.03±0.03	0.59±0.55	0	0	0
69	Golden-bellied Gerygone	*Gerygone sulphurea*	0.17±0.11	0.07±0.07	0	0	0.17±0.08
70	Arctic Warbler	*Phylloscopus borealis*	0.06±0.06	0.11±0.11	0.06±0.06	0.28±0.19	0.11±0.11
71	Oriental Reed-warbler	*Acrocephalus orientalis*	0	0	0	0	0.44±0.24
72	Common Tailorbird	*Orthotomus sutorius*	0.28±0.16	0.15±0.08	1.00±0.51	0.44±0.27	0.22±0.12
73	Dark-necked Tailorbird	*O. atrogularis*	0.06±0.06	0.18±0.08	0.06±0.06	1.28±0.49	0.44±0.26
74	Ashy Tailorbird	*O. ruficeps*	0	0	0.06±0.06	0	0
75	Rufous-tailed Tailorbird	*O. sericeus*	0.03±0.03	0	0	0.44±0.21	0
76	Yellow-bellied Prinia	*Prinia flaviventris*	0.08±0.04	0.07±0.05	0	0	0.06±0.06
77	Asian Brown Flycatcher	*Muscicapa dauurica*	0	0	0	0.06±0.06	0.06±0.06
78	Pied Fantail	*Rhipidura javanica*	0.03±0.03	0	0	0	0
79	Brown Shrike	*Lanius cristatus*	0.03±0.03	0	0	0	0.61±0.29
80	Tiger Shrike	*L. tigrinus*	0	0	0	0.06±0.06	0
81	Long-tailed Shrike	*L. schach*	0.22±0.10	0	0	0	0.44±0.15
82	Asian Glossy Starling	*Aplonis panayensis*	3.97±1.41	3.85±2.06	1.28±0.81	8.89±5.24	0.94±0.50
83	Purple-backed Starling	*Sturnus sturninus*	0.08±0.08	0.56±0.39	0	0.06±0.06	0
84	Common Myna	*Acridotheres tristis*	0.56±0.28	0.11±0.06	2.33±0.90	0.44±0.44	3.00±0.95
85	White-vented Myna	*A. javanicus*	13.78±1.97	6.18±2.10	11.28±2.68	11.61±2.25	11.33±2.11
86	Hill Myna	*Gracula religiosa*	0	0	0	0.06±0.06	0
87	Plain Sunbird	*Anthreptes simplex*	0.06±0.04	0.30+0.13	0.06+0.06	0.61±0.30	0.89±0.89
88	Purple-throated Sunbird	*Nectarinia sperata*	0	0	0	0.17±0.12	0
89	Olive-backed Sunbird	*N. jugularis*	0.72±0.25	1.59±0.28	0.50±0.22	0.17±0.12	1.06±0.39
90	Crimson Sunbird	*Aethopyga siparaja*	0	0.11±0.08	0	0.44±0.19	0
91	Little Spiderhunter	*Arachnothera longirostra*	0	0	0	0.94±0.40	0
92	Orange-bellied Flowerpecker	*Dicaeum trigonostigma*	0	0	0	0.28±0.17	0
93	Scarlet-backed Flowerpecker	*D. cruentatum*	0.42±0.13	0.37±0.090	0	0.17±0.08	0.06±0.06
94	Eurasian Tree-sparrow	*Passer montanus*	0.22±0.17	0	0.83±0.53	0	0.61±0.55
95	Scaly-breasted Munia	*Lonchura punctulata*	0.75±0.35	0	0	0.22±0.22	0

(5%) and not enough vegetation to provide shade (5%). Concern was expressed for inconsiderate cyclists (8%) and insufficient lighting (4%) but the latter has since been rectified. Suggestions for improvements included more shelters and lights, separate trails for cyclists, installing more exercise facilities, telephones and seats and planting more vegetation. Cleaning up the canal was also a popular suggestion.

Household surveys

Of the three hundred respondents interviewed, 59% confirmed they had never used the connector. Of these, 52% were unaware it existed and 16% stated it was too far to walk although most lived within a few hundred metres in the Clementi area. The household survey users' results were similar to the on-site user surveys for types of activities. Frequency of visits varied but 30% visited once every 2–3 days and weekday and weekend visits were fairly evenly spread. Times of visit were highly concentrated in the early morning (27%) and early evening (61%). The majority of people spent up to two hours in the connector during each visit.

A large percentage of users (83%) felt that nature contributed to their sense of well-being and 90% liked to know there were green areas nearby whether they visited them or not. Other preferences for nature including contact with greenery and wildlife and walking through vegetation were expressed by well over 50% of people surveyed. On the scenic beauty of the canal, an overwhelming proportion of users (97%) evaluated it positively. In a more detailed question about nature awareness, over 80% confirmed that trees, shrubs, grass, plants, water and birds were all noticed from afar. Few people (less than 10%) however, were keen to have voluntary contact with any of the natural features listed. Most respondents (77%) rated the canal as a safe place to visit.

The survey results relating to the design of paths produced some useful observations. Whilst interviewees were generally ambivalent about whether a path was straight or curved, a circular path was certainly preferred to a one-way path. Seventy-five per cent of people surveyed expressed a need for a wider path and 71% desired separation for cyclists. Respondents were generally keen on having a path with plenty of access points but were unconcerned whether some sections were flat or sloped. The need for having paths well lined with trees and shrubs was firmly supported and a preference for ponds or streams to be located alongside the path was also expressed.

Focus group discussions

The preliminary analysis completed so far and included in this chapter is confined to women's views. Further analysis is ongoing. Women generally confirmed that they enjoyed contact with nature although emotions aroused by such experiences could range from joy to fear. Three themes emerged in the discussions — nature as a living classroom, nature as a recreational space and nurturing a love for nature. This is because many of the women saw their roles as teachers and mothers. The sense of freedom and openness was a recurrent theme in the discussions and interest in nature was often stated to be the derivation of a bond established in early childhood experiences (Kong *et al.* 1997).

As mothers, the women saw their role as awakening their children to the wonders and workings of nature. Melanie, for example, takes her grandchildren for walks along Ulu Pandan Canal and points out plants to them, naming them and explaining the processes of flowering and fruiting, and other natural processes. She also spoke about how nurturing children's knowledge of and interest in nature does not come only from information imparted, but from curiosity and love developed through doing. She therefore gets her grandchildren interested in her garden and involved in her gardening.

In as much as the women saw their roles as teachers to their children of knowledge about the natural world, they also spoke of the opportunities for recreation that nature afforded to their children. Such opportunities were thought to be healthy and necessary for children to grow up into robust, well-adjusted adults. As Grace and Agnes pointed out, children needed plenty of space to run about, to expend their energy, and so they would bring their children to the public parks and gardens for those precise reasons. However, beyond the openness of such spaces, nature also offered other recreational opportunities to children: to feed the fishes, climb trees, pick shells, build sand castles, or even to go hunting for eggs that hens had laid!

A third theme that recurred in the women's discussion group focused on two related emotions: the joy of seeing something grow as one nurtures and tends it, and the love for nature that develops out of this bond with it. Alicia's delight as she recalled her joy in seeing leaves sprouting from her seeds was palpable. Grace goes one step further and suggests that because children today have little contact with and are not involved in tending and nurturing trees and plants, there is little love for nature. She was sure that if a survey was done with children about what television programmes interested them, they would not mention nature documentaries at all. She felt strongly that without the opportunities for children to *"grow with nature"*, they would never *"have this love for it"*.

Apart from the association with the nurturing process, women's relationships with nature are also finely balanced between ambivalent experiences of enjoyment and fear. While communing with nature is exciting and provides an almost transcendental link to creation, the experience can also be deemed threatening and frightening. Enjoyment stems from the sense that nature represents unpolluted environments, the *"green and clean air"*, which human beings can enjoy. Such environments are thought to be untainted by the contagion of human actions. However, in the context of Singapore where there are few such untouched areas, views expressed were sometimes qualified, consciously or otherwise. Agnes, for example, was careful to add a caveat about the wonders of the "countryside", couched in terms of her lack of scientific knowledge.

The sense of freedom and openness associated with nature was another theme that emerged frequently. This was generally thought to be lacking in highly urbanized Singapore and many interviewees spoke about *"escaping"* to neighbouring Malaysia. The converse of an unfettered relationship with nature that is anchored in freedom and openness is a relationship grounded in fear and a cognizance of danger. What is the nature of fear amongst women in Singapore? In Kong *et al.* (1997), we draw contrasts with the British context, but in this report, we share briefly the observation that the fears that Singapore women confront are varied and encompass both socially and naturally based fears. Socially based fears are evident in two ways: first, fear of personal crimes, particularly sexual violence (flashing, rape, molest) but also property crime (snatch thieves), and second, fear of an absence of human company. In turn, naturally based fears are evident in the fear of wildlife. Such threats featured significantly for some women, and snakes, in particular, posed a threat to various group members. Indeed, Alicia expressed the view that she was not afraid of people but of snakes. Others expressed the view that they did not know what animals they might meet and how dangerous these might be.

The need to feel in control, to know what to anticipate and how to handle the circumstances was extremely important to many women. As Agnes went on to elaborate, she could handle *"well-planned parks"* that were *"properly arranged"* because she would know the *"layout"* and directions to take for escape. However, if nature areas were not of the *"well-planned"* variety, *"familiarity would help"*, so having experienced and knowledgeable *"others"* to bring her and show her around at least once would assuage her fears.

As the above discussion indicates, people's richly textured experiences of nature, including contradictions and ironies, as well as deeply-held beliefs and values, can be accessed through focus group discussions. As more and more groups participate in the discussions and the analysis gathers momentum, the project promises to produce a better understanding of human-nature interactions and relationships.

Bird surveys

A total of 95 bird species were recorded in all sites combined (Table 1). The number of bird species recorded in various sites ranged from 27 to 67. However, this number is not directly comparable among sites because of uneven sampling. The sites with water culverts (Ulu Pandan and Bishan Park) contained numerous waterbird species (e.g., the Purple Heron; Table 1). Similarly, the site fringing the forest (Dairy Farm Trail) contained some true forest bird species such as the Short-tailed Babbler and Chestnut-bellied Malkoha (Table 1). The Pied Fantail which occurs frequently in mangroves was also found in the Ulu Pandan Canal. Numerous migratory bird species such as the Japanese Sparrowhawk, Arctic Warbler, and Oriental Reed-Warbler were also recorded at the sites (Table 1).

The highest bird diversity index was found in Ulu Pandan Canal connector (5.7), followed by the Dairy Farm trail (3.5), Kent Ridge Park (2.9), Bishan Park (2.7), and East Coast Park (2.4). However, the number of transects surveyed in Ulu Pandan Canal corridor was double that at most other sites, and this may have inflated the bird diversity index here. If a correction is made for the extra sampling effort in Ulu Pandan by dividing the diversity index by two, the value of 2.8 is obtained.

The cluster analysis revealed that the bird species composition of the Ulu Pandan Canal is closer to that at Bishan Park and East Coast Park, than at the Dairy Farm and Kent Ridge Park (Fig. 3). This suggests that the type of bird species attracted to three linear corridors were similar. However, bird communities at the forest fringe site (Dairy Farm) and an established park containing a patch of secondary forest (Kent Ridge) were distinct.

The birds carried out a variety of activities in all corridors, with feeding as the most frequently performed activity in Ulu Pandan Canal, East Coast Park, and Dairy Farm (Fig. 4). Perching and flying was the most frequently performed activity in Bishan Park and Kent Ridge Park, respectively. Because singing behaviour was observed, it is likely that some birds may use corridors as a part of their territory and may nest there. The frequency of activities performed by birds differed among sites ($\chi2 = 180.59$, df = 12, P = 0.0001). Activities were performed by birds at various structures of the habitats within the corridors (Fig. 5). The ground was the most frequently used substrate by birds to carry out activities in Ulu Pandan and East Coast Park. Whereas the tree was the most frequently used substrate in Dairy Farm and Kent Ridge (Fig. 5). In Bishan Park, birds were more frequently observed on fences/railings (Fig. 5). The frequency of use of substrates (shrub, tree, and overhead) to carry out activities by birds also differed significantly among sites ($\chi2 = 107.59$, df = 8, P = 0.0001).

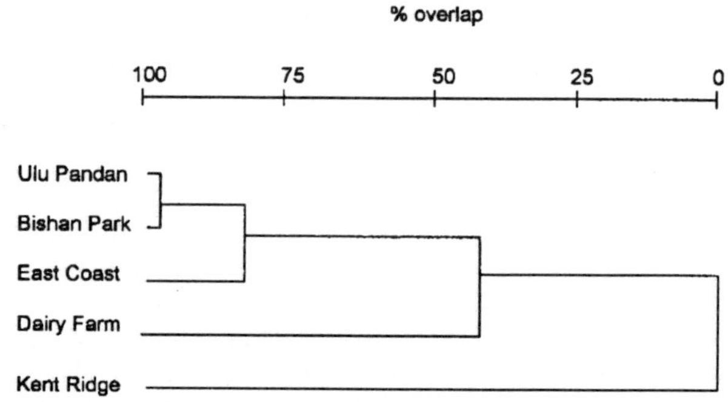

Fig. 3. A cluster dendrogram showing bird community similarity among five different corridors of Singapore.

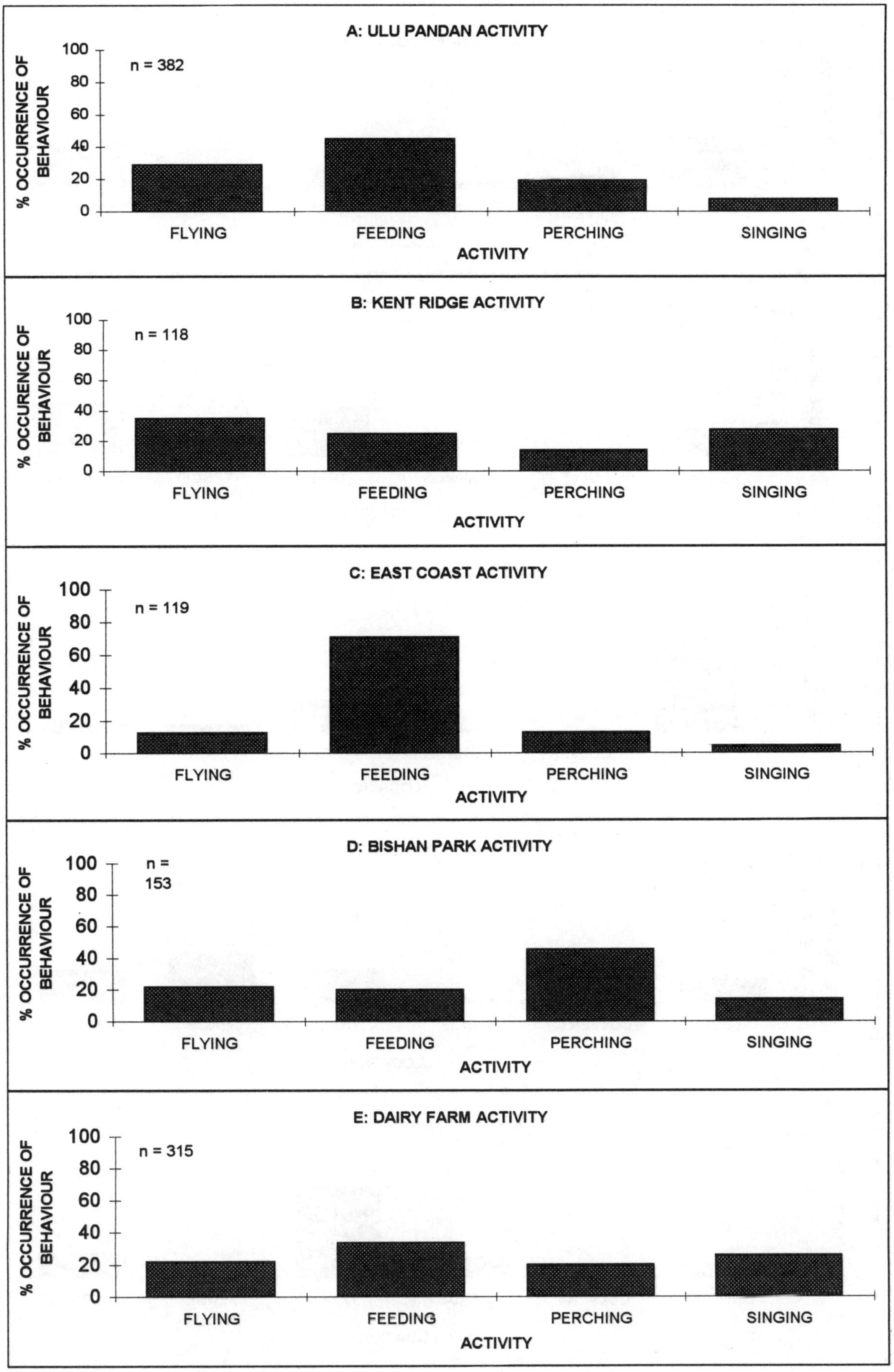

Fig. 4. Activity of birds observed at five corridors in Singapore. N indicates the number of all bird individuals observed.

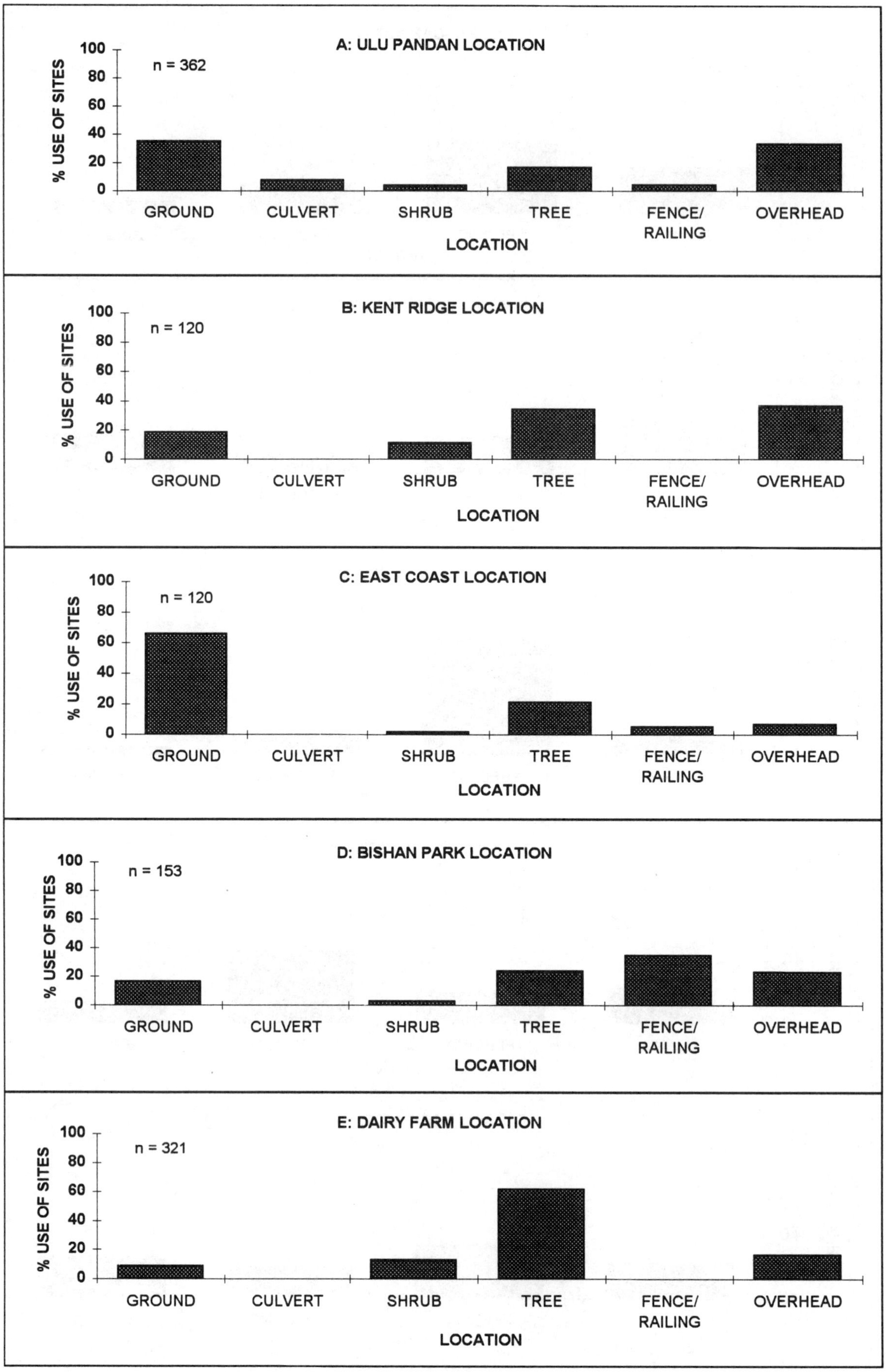

Fig. 5. The sites where bird activities were observed at five corridors in Singapore. N indicates the number of all bird individuals observed.

The number of bird species recorded in different transect blocks of the Ulu Pandan corridor ranged from 29 to 35 (Table 2). The transect blocks closer to each other had similar number of bird species (A and B versus C and D) (see Fig. 2). Transects A and B appeared to have similar habitat (e.g., closer proximity of a secondary forest), which may result in a similarly diverse bird community there. The same may be true for transects C and D.

Table 2. The number of bird species and maximum number of bird individuals recorded in 400 m blocks of transects (each transect 100 m in length) in Ulu Pandan Canal park connector. The number of humans recorded in these transect blocks is also given.

	Transect block			
	A	B	C	D
Number of bird species	35	34	29	29
Maximum number of bird individuals recorded	174	122	103	79
Number of humans recorded	99	1 053	663	250

The number of bird individuals recorded in all transects combined ranged from 79 to 174 (Table 2). However, no clear relationship was apparent between the number of humans recorded in the transect blocks and the number of bird species and individuals recorded there (Table 2). Our above comparison is based on the assumption that the number of humans within transect blocks remained similar over the months (human numbers were recorded only during April/May 1996).

Anthropogenic habitat changes (e.g., forest loss and fragmentation) have been experienced in all urban settlements. Such human-caused habitat destruction might be more devastating for wildlife in the tropics because of high levels of biodiversity (Brooks *et al.* 1997). In Singapore, because of extensive urbanization, native habitats have been heavily reduced. For example, only 5% and 6% of original native tropical rainforest cover and mangrove forest cover remains in Singapore, respectively (Turner *et al.* 1994; Hilton and Manning 1995). Although there have been numerous local extinctions of flora and fauna within Singapore (Ng and Wee 1994), it is poorly understood how well some of the native wildlife has adapted to human changed or made habitats. Such an understanding would be critical in determining or predicting the success of habitat enhancement programmes such as the island-wide park connector scheme currently being implemented.

Our study shows that human-made corridors or linear habitats are successful in attracting a large number of both migratory and resident bird species. However, the habitat within and immediately adjoining these corridors may affect the bird species composition of a particular corridor. We found that bird species carried out a variety of activities within the corridors suggesting that these corridors serve as functional habitats for bird species. The extent of this functional use of corridors for birds is uncertain at this point. However, we can safely say that the corridors at least do serve the birds' critical needs of foraging and singing activities indicative of engaging in mating behaviours and/or territoriality. But the degree of functionality likely vary for each corridor, again depending on factors such as habitat within the corridors.

Based on our preliminary data, there appeared to be no correlation between bird diversity and number and human presence in one of the corridors (Ulu Pandan Canal). It is unclear at this time if the abundance of a particular bird species is affected by human use. It is possible that human activity rather than human number affects bird diversity in transect blocks. Similarly, it is possible that bird numbers may be more strongly affected by habitat structure than human use. These aspects are planned to be addressed in subsequent research and analyses.

CONCLUSIONS AND RECOMMENDATIONS

Overview

The results of this pilot study show that the Ulu Pandan Canal park connector in Singapore is meeting a variety of recreational needs for a wide range of local users. In a comparison of five connectors including Ulu Pandan Canal wildlife (birds) requirements were assessed using bird census and activity surveys. A good range of both migratory and resident bird species were present. There was little correlation between numbers of users and bird species in the transects at Ulu Pandan. Comparisons made between sites however suggest that where there were more users present and increased area of managed vegetation as found in East Coast Park bird diversity was low. Alternatively where user numbers were low and habitats were more diverse bird species richness increased as at the Dairy Farm trail. In connectors with adjoining areas of mature, unmanaged vegetation as at Ulu Pandan increases in bird diversity were also recorded. The degree of influence habitat diversity has relative to user population and activity on birdlife occurrence is an issue currently being studied in ongoing research.

Household surveys revealed that there are still many local residents unaware of the facilities offered in the connector but there was good support for natural surroundings and a high degree of appreciation for nature. The focus groups reinforced an enthusiastic interest in and involvement with nature. For the older members these feelings and attitudes mainly derived from childhood experiences.

The interaction of humans and wildlife is a complex issue as it embodies many varied dimensions involving attitudes, values, beliefs and behaviours. Peoples' relationships with nature are highly textured and nuanced and must be understood within the contexts of their lives. In general, people enjoy contact with nature and the recreational potential of park connectors. This raises the question of whether human use may create disturbance for wildlife. From the evidence so far, it would seem that the interaction of humans and birds requires much more detailed observation work before any meaningful recommendations on design and planning of connectors can be made. In future work, we will need to consider the likelihood that human impacts on wildlife and habitats may relate as much to the type of recreational use as to the intensity of use.

Whist there was limited data on the preferences for planning and design issues some useful suggestions for improvements to infrastructure and facilities were received. These included requests for wider trails, with separation for cyclists, more shade trees and circular routes. Night lighting was popular as dusk is always early in Singapore and outside evening temperatures are pleasant. A large proportion of users enjoyed contact with greenery and wildlife. They were also impressed with the scenic beauty of the Ulu Pandan Canal connector.

Efforts have been made in this research to discover the nature and effect of the interacting behaviours between humans and wildlife. Whilst these have been interesting and informative, it is our belief that many more intangible concerns are waiting to be observed and understood. It is hoped that what we have learned in this pilot study will assist in further efforts to narrow down the human and wildlife behaviours to be evaluated.

Future research work

As this is a pilot study, the surveys conducted at the Ulu Pandan Canal revealed several useful insights and lessons for the future refinement of the research methodology. In retrospect, some modifications would help to refine what were essentially adequate methods. Some entail modifying practical arrangements; others involve refining the questionnaires and the treatment of data sets; while others entail introducing new methods.

Practical arrangements: For the on-site observations along the connector, the pilot study revealed the need to have two observers walking each section of the connector rather than one. This enables the extra person to help in the recording of observations during peak hours and also to act as a relief for the observers at the access points. The pilot study also revealed the importance of proper pre-survey briefing and training of student helpers. In particular, training of student helpers was identified as an important and beneficial pre-requisite to the smooth execution of data collection to ensure minimal bias. In surveying bird species for assessing human impacts there is also considered value in surveying areas off trail as well as on trail to compare the difference.

Refining observation forms and questionnaires: The pilot study provided an opportunity to test out the pre-designed observation forms and questionnaires. In particular, it has provided feedback in the revision of the forms and questionnaire, in terms of both structure and content. For example, certain questions in the questionnaire were found to be redundant; others could be improved with better phrasing; and yet others which had not been asked deserved to be included to provide further information on people's knowledge and experience of nature. A more detailed set of questions have thus been included in the revised questionnaires.

Refining the treatment of data sets: While our analysis in this chapter has dealt with the household survey results and on-site user survey results separately, we feel that in our subsequent surveys, the information on users from both questionnaires needs to be combined and analysed together as one data set. This has the advantage of enlarging the sample size and ensuring that a spread of users' views are captured. The data from each of the questionnaires will then be matched to complement each other. Such combined analysis is only possible if both the questionnaires cover the same ground for the section on users, which was not done in this pilot study. The on-site questionnaire did not include all the questions in the household questionnaire in the interest of time. We were concerned that too long a questionnaire on-site may turn away potential respondents. Based on the pilot study, this does not appear to be a major problem, and so, for the main study, we propose to include the same questions for users for both questionnaires.

Introducing new methods: The focus group interviews revealed extremely valuable and rich information. They confirmed our belief that such situations provided better opportunities than structured questionnaire surveys to explore in-depth, people's rich and detailed experiences with nature, which are intangible and highly nuanced. However, we also realized that the interviewees were removed from the immediate natural surroundings they talked about. The research team therefore decided that, in addition to these specially convened groups, it would also be fruitful to interview users on-site, walking through the park connector with them and hearing, in an immediate first-hand manner, their views and experiences. This would complement the on-site observations, which, while offering all the merits discussed, do not always reveal whether visitors are listening to birds, looking at greenery in the distance, or engaging in other such activities. Such on-site in-depth user interviews will be used in the next stage of research.

The need for modification of the bird survey methodology became apparent during discussions among research team members and other scientists and after preliminary data analysis. It was found that the transect survey would be the most appropriate method to sample birds in linear strips of habitats such as park connectors. Bird activity surveys give important information about bird use patterns of park connectors. However, to provide more detailed information the team will modify bird activity surveys by making prolonged observations of individual species within the Ulu Pandan Canal and other connectors.

It was also found that obtaining accurate quantitative habitat information from park connectors was difficult because of their narrow linear arrangement. It would be better to obtain and assess qualitative habitat information from the connector and then attempt to correlate this with bird diversity, abundance and activity.

Setting new lines of investigation: Potential questions that still need to be answered and which will be addressed in the forthcoming research on birds include the following:

- Are birds in the connectors better off (e.g., in terms of foraging and reproductive success) than in other non-linear habitats (e.g., mangrove and lowland forests)?
- Are some species of birds using this (and possibly other connectors) as corridors to reach other larger habitats?
- Are birds in higher danger (e.g., of predation) within this (and other such connectors) than in larger habitats?
- Can birds be used as overall indicators of biodiversity enhancement of the park connector system?

Work has already commenced by us on further studies of more corridors. All these survey studies were to be completed by June 1998. With the enlarged data sets collected using more refined methodologies, more detailed results will be available and greater understanding achieved. It may then be possible to develop more precise planning, design and management recommendations.

ACKNOWLEDGEMENTS

We wish to thank the Hong Kong Bank and the National University of Singapore for funding the above research and the many volunteers of the Nature Society, Singapore and the National University of Singapore students for assisting in the survey work. We are also indebted to the National Parks Board for their assistance and interest in the research.

REFERENCES

Andrews, J., 1993. The reality and management of wildlife corridors. *Brit. Wildl.* 5(1): 1–7.

Bennett, A. F., 1991. What types of organisms will use corridors? Pp. 407–08 *in* Nature Conservation 2: the Role of Corridors ed by D. A. Saunders and R. J. Hobbs. Surrey Beatty & Sons: Chipping Norton, New South Wales.

Briffett, C., Kong L., Sodhi, N., Yuen, B., 1996. Of corridor connectors and conduits, *Landscape East.* Institute of Landscape Architects Singapore **5**: 20–23.

Brookes, T. M., Pimm, S. L. and Collar, N. J., 1997. Deforestation predicts the number of threatened birds in insular southeast Asia. *Conserv. Biol.* **11**: 382–94.

Dawson, D., 1994. Are habitat corridors conduits for animals and plants in a fragmented landscape? A review of scientific evidence. *London Ecology Unit.* English Nature.

DeGraaf, R. M., Geis, A. D., Healy, P. A., 1991. Bird population and habitat surveys in urban areas. *Landscape and Urban Plan.* **21**: 181–88.

Diamond, J. M., 1974. Relaxation and differential extinction on land bridge islands: applications to natural preserves. *Proc. 16th Int. Orni. Cong.* 616–28.

Donovan, J., 1988. "When you are ill, you've gotta carry it" Health and illness in the lives of black people. Pp. 180–96 *in* Qualitative Methods in Human Geography ed by J. Eyles and D. M. Smith. Polity Press: Cambridge

Dower, M., 1963. 'Greenways. A positive future for Britain's cast-off railways.' *Architec. Rev.* Sept. 387–93.

Eyles, J. and Smith, D. M. (eds), 1988. Quality Methods in Human Geography. Polity Press: Cambridge.

Eyles, J., 1988. Interpreting the geographical world: Qualitative approaches in geographical research. Pp. 1–16 *in* Qualitative Methods in Human Geography ed by J. Eyles and D. M. Smith. Polity Press: Cambridge.

Forman, R. T. T., 1983. Corridors in a landscape: their ecological structure and function. *Ecology* (CSSR) **2**: 375–87.

Forman, R. T. T. and Godron, M., 1981. Patches and structural components for a landscape ecology. *Bioscience* **31**: 733–40.

Forman, R. T. T. and Godron, M., 1986. Landscape Ecology. John Wiley and Sons: New York.

Gilbert, O. L., 1989. City Parks. The ecology of urban habitats. Chapman and Hall: London. Pp. 197–205.

Groome, D., 1987. Recreational corridors. A Study of the use of the Middlewood Way and Macclesfield Canal. Dept of Town and Country Planning. University of Manchester.

Hanski, I., 1989. Metapopulation dynamics: does it help to have more of the same? *TREE* **4**: 113–14.

Harrison, C., Limb, M. and Burgess, J., 1987. Nature in the city — popular values for a living world. *J. Environ. Manage.* **25**: 347–62.

Helliwell, D. R., 1975. The distribution of woodland plant species in some Shropshire hedgerows. *Biol. Conserv.* **7**: 61–72.

Hilton, M. J. and Manning, S. S., 1995. Conservation of coastal habitats in Singapore: indications of unsustainable development. *Environ. Conserv.* **22**: 307–22.

Hooper, M. D., 1971. The size and surroundings of nature reserves. Pp. 555–61 *in* The Scientific Management of Animal and Plant Communities for Conservation ed by E. Duffey and A. S. Watt. Oxford: Blackwell.

Kong, L. L. L., Yuen, B., Briffett, C., Sodhi, N. S., 1997. Nature and nurture, danger and delight: urban womens experiences of the natural world. *Lands. Res.* **22**(3).

Lack, P., 1992. Birds on Lowland Farms. HMSO: London.

Liu, T. K., 1991. Concept Plan Singapore in the 21st Century. City Trans Asia Singapore.

Lock, N. Y., 1996. Breeding Biology of Birds in Urban Parks of Hong Kong. Research Paper. University of Hong Kong.

London Borough of Lewisham, 1989. Green Spaces. Directorate of Environmental Services.

MacArthur, R. A. and Wilson, E., 1967. The Theory of Island Biogeography. Princeton University Press: Princeton, New Jersey.

McCune, B. and Mefford, M. J., 1995. PC-ORD. Multivariate analysis of ecological data, version 2.0. MjM Software Design, Gleneden Beach, Oregon, USA.

McHarg, I., 1969. Design with Nature. Natural History Press: Garden City, NY. Reprinted by John Wiley. 1992.

Ministry of the Environment, 1993. Nature Conservation. The Singapore Green Plan — Action Programmes. Singapore 51.

Moore, N. W., 1991. Observe extinction or conserve diversity. Pp. 1–8 *in* The Conservation of Insects and their Habitats ed by N. M. Collins and J. A. Thomas. 15th Symposium of the Royal Entomological Society of London.

Mostyn, B. J., 1984. Personal benefits and satisfactions derived from participation in urban wildlife projects. A qualitative evaluation undertaken by social and community planning research on behalf of the Nature Conservancy Council.

Ng, P. K. L. and Wee, Y. C., 1994. The Singapore Red Data Book. The Nature Society: Singapore, Singapore.

Noss, R. F., 1983. A regional landscape approach to maintain diversity. *Bioscience* **33:** 700–06.

Nature Society, Singapore, 1990. Master plan for the conservation of nature in Singapore. Nature Society. Singapore formerly Malayan Nature Society, Singapore Branch. C. Briffett (ed).

Opdam, P., Rijskijk, G., Hustings, R., 1985. Bird communities in small woods in an agricultural landscape: effects of area and isolation. *Biol. Conserv.* **34:** 333–52.

Osbourne, P., 1984. Bird Numbers and habitat characteristics in farmland hedgerows. *J. Appl. Ecol.* **21:** 63–82.

Peck, K., 1993. Estuaries inventory — research towards a better understanding of the interactions between birds and human activities in UK estuaries. *RSPB Conserv. Rev.* **(7):** 42–46.

Preston, F. W., 1962. 'The canonical distribution of commonness and rarity: Part 1. *Ecology* **43:** 185–215.

Roberts, D. C., 1994. The design of an urban open-space network for the City of Durban (South Africa). *Environ. Conserv.* **21**(1) **Spring:** 11–17

Round, P., 1988. Resident forest birds in Thailand. Their status and conservation. ICBP Monograph No. 2. International Council for Bird Preservation: Cambridge.

Saunders, D. A. and Hobbs, R. J., 1989. Corridors for conservation. *New Scientist* **28th Jan.:** 63–68.

Schwarz, L., 1994. Greenways: A guide to planning, design and development. The Conservation Fund. Island Press: USA. 122.

Simberloff, D. S. and Cox, J., 1987. Consequences and costs of conservation corridors. *Conserv. Biol.* **1:** 63–71.

Straits Times, 1992. First stage of nature corridor project to link parks opens. The *Straits Times*. Singapore. 15th August.

Straits Times, 1995. Island wide green network to link parks. The *Straits Times*. Singapore. Sat. 18th November.

Straits Times, 1996. New bicycle crossing for park connectors. The *Straits Times*. Singapore. Friday 12th April.

Terborgh, J., 1976. Island biogeography and conservation: strategy and limitations. *Science* **193:** 1029–030.

Turner, I. M., Tan, H. T., Wee, Y. C., Ibrahim, A. B., Chew, P. T. and Corlett, R. T., 1994. A study of plant species extinction in Singapore: lessons for the conservation of tropical biodiversity. *Conserv. Biol.* **8:** 705–12.

Turner, M. G.,1989. Landscape ecology: the effect of pattern on process. *Ann. Rev. Ecol. System.* **20:** 171–97.

Tyne and Wear, 1988. Nature Conservation Strategy. Nature Conservancy Council.

Ulrich, R. S., 1986. Human responses to vegetation and landscapes. *Lands. Urban Planning* **13:** 29–44.

Walker, R. (ed), 1985. Applied Qualitative Research. Aldershot, Hampshire: Gower.

West Midlands County Council, 1984. The Nature Conservation Strategy for the County of West Midlands. West Midlands County Council: Birmingham.

Wilson, E. O and Willis, E. O., 1975. Applied biogeography. Pp. 522–34 *in* Ecological Structures of Species Communities ed by M. L. Cody and J. M. Diamond. Harvard University Press: Cambridge, Mass.

CHAPTER 42

How does urbanization affect bird communities in riparian habitats? an approach and preliminary assessment

James R. Miller,[1,2] John A. Wiens[2] and N. Thompson Hobbs[3]

Riparian areas in the western United States are important sources of natural diversity and contain some of the most diverse avifaunas in North America. Watercourses are also centres of human settlement, and this may negatively impact the riparian bird communities. Human settlement often involves the replacement of the native flora with exotic vegetation and leads to increases in the density of human commensal species. Such species often prey upon or compete with native species.

We investigated the effects of development on bird communities in lowland riparian areas along the Front Range of Colorado. At 16 sites arrayed along gradients of urbanization on four drainages, we conducted bird censuses from 1995 to 1997. We also investigated changes in predation rates along these gradients using experimental nests and by monitoring natural nests. Preliminary results indicate that overall species richness and the number of migrant birds increase from developed to rural sites. At highly developed sites, species richness increased with riparian width, but was still lower compared to rural sites with relatively narrow riparian zones. Species that are characteristic of surrounding habitats were more evident in urban riparian areas than in rural areas. Daily mortality rates on experimental nests at rural sites were significantly higher than in developed areas on all four drainages. Daily mortality rates on experimental nests also increased with distance from recreational trails in the riparian zone. Daily mortality rates on active nests were not higher in developed versus rural areas, but were actually lower on one drainage. Overall, predation rates were substantially lower for active nests compared to experimental nests, suggesting that parental nest defense may be an important factor in determining nest success.

These preliminary results suggest that both human activities in the riparian zone and land use in the surrounding landscape matrix influence ecological processes affecting bird communities in riparian habitats. Surrounding land use may partially negate the effect of local habitat characteristics, such as riparian width. Because land use decisions are often made at the lowest level of government and at very small scales relative to riparian systems, successful conservation of these habitats and the biota they harbour will require not only explicit statements of management objectives, but also careful integration with conservation priorities at larger scales.

INTRODUCTION

In arid and semi-arid regions of the world, water is life. The natural landscape surrounding a watercourse is often a mosaic of grasslands and shrublands intersected by tendrils of riparian woodlands. Such riparian areas, which include the alluvial floodplain on either side of a stream or river and the vegetation growing there (*sensu* Ohmart

[1]Department of Zoology, University of Wisconsin, Madison, Wisconsin 53706, U.S.A.
[2]Department of Biology/Graduate Degree Program in Ecology, Colorado State University, Fort Collins, Colorado 80523, U.S.A.
[3]Colorado Division of Wildlife, 317 West Prospect, Fort Collins, Colorado 80526 and Natural Resource Ecology Laboratory, Colorado State University, Fort Collins, CO 80523, U.S.A.
Pages 427–39 *in* NATURE CONSERVATION 5: NATURE CONSERVATION IN PRODUCTION ENVIRONMENTS: MANAGING THE MATRIX ed by J. L. Craig, N. Mitchell and D. A. Saunders. Surrey Beatty & Sons, 2000.

1994), provide habitat for a large number of species not found in the sparser, dryer vegetation and may act as corridors that enhance population and landscape connectivity (Naiman *et al.* 1993). Riparian zones are also centres of human settlement; throughout history, humans have been attracted to watercourses as places to live. This juxtaposition of human settlement and riparian habitats may result in conflict between the need to maintain the ecological integrity of riparian systems and the need to provide for human welfare.

This potential for conflict is particularly apparent in the western United States, where riparian zones are critical sources of natural diversity (Thomas *et al.* 1979) with unusually high value for vertebrate faunas (Johnson 1989; Terborgh 1989; Finch and Ruggiero 1993). Although riparian zones at lower elevations comprise less than 1% of the total land area in the western United States (Bottorff 1974; Knopf *et al.* 1988), up to 80% of the terrestrial vertebrate species in this region depend on them for at least part of their life cycles (Chaney *et al.* 1990). These riparian habitats contain some of the most diverse avifaunas in North America (Johnson *et al.* 1977; Stevens *et al.* 1977; Ohmart 1994), serve as critical habitat for migrating birds (Stevens *et al.* 1977; Knopf *et al.* 1988) and provide habitat for more species of breeding birds than any other western plant community (Stamp 1978; Knopf and Samson 1994). In northeastern Colorado, 82% of breeding bird species occur in riparian vegetation (Knopf 1985).

As human populations have grown and become increasingly concentrated in urban areas, however, the potential impacts on natural environments have increased dramatically. Urbanization involves profound alteration of natural ecosystems, producing massive changes in the flows of energy, water, and materials (Douglas 1994). In arid and semiarid regions, human demands on water are especially great, and the pressures on riparian areas and the biota that they harbour may be severe. Recognizing these pressures (and in some cases adding to them), land-use managers, residential developers, and landscape architects in many urban areas of the western United States have made riparian systems an integral part of their planning efforts (Adams and Dove 1989; Binford and Buchenau 1993). In many instances these efforts have been part of broader attempts to control or direct growth.

Various planning tools may be used to regulate development (Duerkson *et al.* 1997). At the county or municipal level in the United States of America, for example, zoning may restrict the types of growth that can occur or set limits on residential density in a given area. Floodplains are by definition subject to flooding, especially in semiarid areas where rainfall events may be sporadic but intense, and for this reason zoning regulations often limit development in riparian habitats. In Colorado, many counties and cities are also implementing open space or greenway programmes (Smith and Hellmund 1993) or are actively seeking to expand their land-holdings (e.g., City of Fort Collins 1992; Larimer County 1993; Boulder County Planning Commission 1997; City of Boulder Planning Department 1997). Non-governmental organizations have also become involved in acquiring land in growth areas for purposes other than development. Land acquisition and zoning regulations are often intended to benefit people living in the immediate vicinity — the protection of mountain views or provision of recreational opportunities, for example. Increasingly, however, a primary consideration in such endeavours is the preservation of biological diversity. Similar deliberations are carried out elsewhere in the world where urbanization is rampant, such as Australia (Adam 1995), New Zealand (Wilson, this volume), and Singapore (Briffett *et al.,* this volume).

Conservation is problematic in areas undergoing rapid growth, and decisions are often made in the absence of baseline information. Such information is generally lacking because ecologists have traditionally avoided areas of human habitation when selecting research sites (Pickett *et al.* 1992). Indeed, although some ecologists have called for a greater focus on populated areas (McDonnell and Pickett 1993; Platt *et al.* 1994), there is still some debate in conservation circles as to whether this type of research should be a priority (Wuerthner 1994; Siegel 1996; but see Keeley 1993). Whatever the relative merits of these opinions, the lack of data has resulted in a situation in which land use planning is often founded more on land availability or its perceived conservation value (based on data collected in unpopulated areas) than on an understanding of the relationship between urban activities and the natural functioning of ecosystems. This gap in understanding is especially great in the case of riparian areas.

Here, our first objective is to generate predictions about the effects of urban and suburban development on bird communities in riparian areas. We do this by examining the salient features of developed lands with regard to birds in the context of previous studies of urban avifaunas. Next, we describe our approach to testing these predictions in lowland riparian areas in Colorado and present preliminary results of our studies. Finally, we comment on research and conservation priorities in such urban-riparian interfaces.

URBANIZATION

Habitat alteration

To develop predictions about the effects of urbanization on riparian bird communities, we can first examine these effects in a broader context. By definition, urban and suburban development involves the replacement of relatively natural habitat with a human-dominated environment. Post-development native habitats often exist as small remnants that are contiguous with a wide variety of anthropogenic landscape elements (Keeley 1993; Nassauer 1997). Buildings, paved surfaces, and traffic may serve as urban barriers for some birds that continue to occupy such remnants, and impose a degree of isolation on native populations that is disproportionately great in relation to the actual distances involved (Soulé *et al.* 1992; Foppen and Reijnen 1994). This isolation may prevent recolonization of habitat fragments in the event of local extinction, to which small populations are especially prone (Shaffer 1981). Our understanding is quite limited, however, with regard to the ways in which the multitude of fragmenting features in urban environments affect various native species (Engels and Sexton 1994). For this reason, the application of the habitat fragmentation paradigm, which has been largely derived from landscapes where extractive activities predominate, is not straight-forward in settled areas.

It is important to recognize that the impacts of settlement on natural habitats are in some respects qualitatively different from impacts brought on by the types of land use changes that have served as the traditional focus of applied ecology. The latter (logging, agriculture, water development) often drive fundamental conversions of land cover, from forest to clear cut, from grassland to cultivated field. These conversions produce correspondingly dramatic changes in the operation of ecological processes, and this has been the traditional domain of applied ecological research. High-density urban areas represent a substantial conversion of cover types to a patchwork of alternative land uses, but this may not be the case with all forms of development (Duerkson *et al.* 1997). Instead, existing land cover may be "perforated" with urban structures, as is often the case in lower-density areas and in wooded habitats, rather than isolated by them (e.g., Norton, this volume). The effects of such perforation are far more subtle than those which accrue from the large-scale conversion of cover types.

In arid and semi-arid areas, urban and (especially) suburban development are often accompanied by the planting of trees and shrubs. Along the Front Range in Colorado, xeric shortgrass prairie is converted to the rough equivalent of a wooded savannah, and open grasslands and shrublands are perforated by mesic deciduous woodlands. In towns and suburbs, the native flora is further reduced by extensive cultivated lawns.

In addition to the physical alteration of habitat, post-development environments are characterized by new suites of species. The most obvious and prolific, of course, is humans. Humans and their activities may adversely affect some species, particularly during critical phases of an animal's life cycle such as the post-natal or breeding period (Hockin *et al.* 1992). Furthermore, animals that are commensal with humans tend to reach high densities in urban areas and often compete with or predate the native fauna (see below).

Urban avifaunas

How do these habitat changes affect birds? Clearly, some species respond to the transformation from native to non-native vegetation. In the United States, a strong correlation between reductions in native vegetation and native birds has been reported in the north-east (DeGraaf 1987), east (Geis 1974), midwest (Beissinger and Osborne 1982), and south-west (Mills *et al.* 1989). Such species tend to persist, however, in areas where a large proportion of native trees and shrubs are left undisturbed, such as cemeteries (Lussenhop 1977) and golf courses (Terman 1994).

While some avian species decline in response to habitat alteration, others thrive and may reach higher densities than exist in the predevelopment landscape. Although overall bird density is often reported to be higher in urban versus less intensively-used areas, fewer species account for most of the density (Batten 1972; Emlen 1974; Geis 1974; Beissinger and Osborne 1982; DeGraaf 1987). Some of these species, such as the introduced European Starling *Sturnus vulgaris* and House Sparrow *Passer domesticus* in the United States, are able to displace other species by virtue of sheer numbers and aggressive behaviour (Johnston and Garrett 1994). Several workers have reported shifts from native to exotic species and from migrant species to year-round residents with increasing urbanization, both temporally and spatially (Walcott 1974; Aldrich and Coffin 1980; DeGraaf and Wentworth 1986; DeGraaf 1987; Friesen *et al.* 1995; Blair 1996). Many of the "increaser" species have had a long history of cohabitation with humans and can use artificial structures for nesting (Geis 1974; Weber 1975; DeGraaf and Wentworth 1986). Birds that nest on or near the ground may be particularly vulnerable in urban environments (Emlen 1974; Guthrie 1974; Weber 1975; Vale and Vale 1976), perhaps as a result of predation, human activity, or a lack of low vegetation. Ground-gleaners, seed-eaters, and omnivorous species that can exploit expansive lawns and bird-feeders often tend to dominate the avifauna, particularly in winter (Emlen 1974; Beissinger and Osborne 1982; Bezzel 1985; DeGraaf 1987; Mills *et al.* 1989; Brittingham 1991).

Nest predation

Nest predation is thought to be the primary cause of reproductive failure in birds (Ricklefs 1969; Best and Stauffer 1980; Martin 1988), although evidence to support this claim in developed areas is largely circumstantial. In one of the few studies investigating this phenomenon, Wilcove (1985) found that predation rates were higher on artificial nests in suburban woodlots than in more remote areas.

One might expect nest predation rates to be greater in areas that are more developed for several reasons. Fragmented landscapes tend to be associated with elevated levels of nest predation (Andrén *et al.* 1985; Herkert 1994; Robinson *et al.* 1995), especially near habitat edges (Gates and Gysel 1978; Angelstam 1986; Donovan *et al.* 1997).

The fragmented nature of most urban areas, coupled with the high degree of edge associated with the native areas that remain, suggests one reason why nest predation rates may be high there.

This scenario is exacerbated by the tendency for a number of nest predators to reach higher densities in human-dominated environments. Examples include Pied Currawongs *Strepera graculina* in some Australian cities (Major *et al.* 1996), raccoons *Procyon lotor* and striped skunks *Mephitis mephitis* in the United States (Hoffman and Gottschang 1977; Haspel and Calhoon 1989; Rosatte *et al.* 1991), or dogs *Canis familiaris* and domestic cats *Felis catus* world-wide (Churcher and Lawton 1987). Increases in predator densities may be attributable to a combination of the availability of supplemental food and shelter. Moreover, Soulé *et al.* (1988) noted that populations of small to medium-sized predators may reach higher densities in the absence of larger carnivores, and large carnivores tend to be absent from urban areas. The implication is clear: higher densities of nest predators will result in greater levels of nest predation (Soulé *et al.* 1988; Harris and Silva-Lopez 1992; Noss 1993).

At a finer scale, one might also expect to find higher predation rates at particular locations within human-dominated environments, again attributable to locally high concentrations of potential predators. For example, people often allow their dogs to run loose in parks or along hiking trails, and omnivorous animals such as corvids or raccoons may be attracted to garbage or food scraps at these sites. In developed areas, the mere presence of certain predators may be sufficient to exclude sensitive species from otherwise suitable habitat (Engels and Sexton 1994).

Levels of nest predation in urban environments probably change over time. As the composition of the avifauna changes, species with nesting strategies that are better suited to developed areas may be selected for. Martin and Clobert (1996) documented lower rates of nest predation for human-dominated areas in Europe in comparison with those in the United States; they suggested that this difference may be a function of the former's longer history of large-scale human disturbance.

Predictions

How might these features of urban/suburban habitat changes and avifaunas apply to bird communities in riparian zones that are embedded in developed areas? Riparian habitats in arid or semi-arid regions are usually more or less linear and are relatively narrow, so they may be particularly susceptible to detrimental influences from beyond their boundaries (Janzen 1986). These effects might be especially great in urban areas. Likewise, Backhouse (1987) has observed that long, narrow reserves may be particularly susceptible to internal disturbance resulting from human activities, such as might be associated with recreational trails. On the other hand, the wooded overstorey of these riparian habitats may blend into the vegetation of the surrounding landscape more smoothly in developed (especially residential) areas than in more natural settings, facilitating the spread of some species into the riparian zone. These observations lead us to predict that:

- Avifaunas in riparian habitats associated with development are characterized by a decrease in species richness compared with less-developed areas.
- Riparian areas associated with development will contain bird species that are also characteristic of the surrounding habitats, while this will be less evident in less developed settings. Migrants will be less abundant in riparian habitats near developed areas.
- Nest predation rates will increase with increased development in the surrounding matrix.
- Nest predation rates will increase with proximity to recreational trails in the riparian area.

APPROACH

How can one test these predictions? One could consider urban and suburban areas as "treatments" to be compared with less-developed "controls," but there is so much variation among either developed or less-developed areas that true replication would be difficult. We chose instead to consider riparian habitats along gradients of urbanization and to follow a gradient-based design (Austin and Heyligers 1991; Wiens and Parker 1995). Because streams flow along topographic contours, they provide a well-defined gradient along which sampling may be conducted. We arrayed our samples along several riparian watersheds that intersected a spectrum of development, from city centres to suburbs to rural countryside.

Study area

Our study sites are comprised of 16 parcels situated along Coal Creek, South Boulder Creek, and Boulder Creek in Boulder County, and the Cache la Poudre River in Larimer County, Colorado, (Fig. 1). These sites are located on the shortgrass steppe at the base of the Front Range of the Rocky Mountains and were chosen to represent a gradient from relatively high levels to low levels of development on each drainage. Elevations range from approximately 1 640 m at the foothills/plains interface to approximately 1 550 m in eastern Boulder and Larimer Counties. The overall study area is essentially that of the Plains Grassland life zone as described by Marr (1961, 1964). The riparian woodland overstories are dominated by Plains Cottonwood *Populus deltoides*, Crack Willow *Salix fragilis*, and Hybrids *Populus* x *acuminata* of Plains Cottonwood and Narrowleaf Cottonwood *Populus angustifolia*, while Snowberry *Symphoricarpus occidentalis,* Chokecherry *Prunus virginiana,* and Sandbar Willow *Salix exigua* are typical of the understorey.

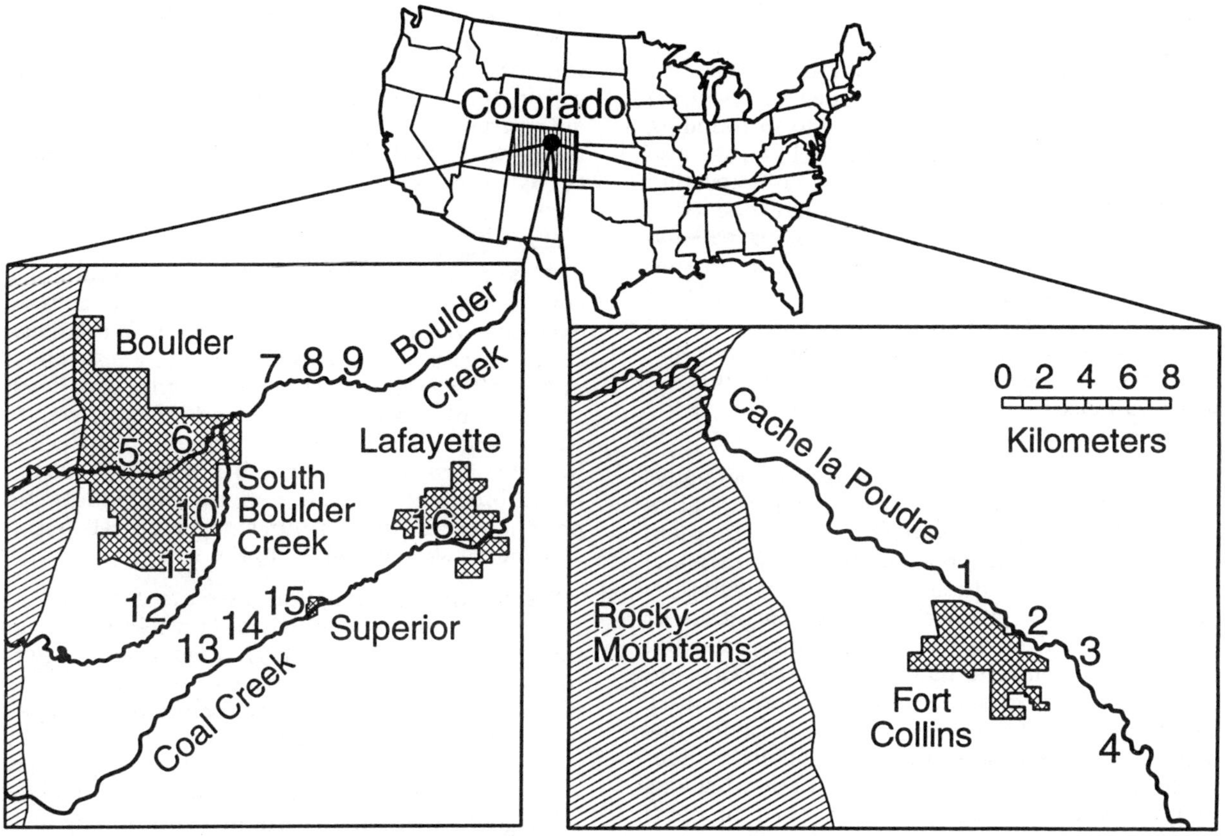

Fig. 1. Map of the study area. Numbers indicate study sites, cross-hatched areas indicate the Rocky Mountains, and grey areas indicate cities and towns.

Methods

Avian Communities. We sampled the avifauna at each site three times during the breeding season in 1995, 1996, and 1997, using point transects. Each transect had a randomized starting point and census points 110 m apart (Bibby *et al.* 1992). Individual trained observers rotated visits to a given site in order to minimize observer bias, and sampling effort was distributed as evenly as possible among sites that were associated with different levels of development (Verner 1985). Bird surveys were conducted between sunrise and approximately 0830 on days with no rain and windspeeds <32 km/h. At each point, a motionless observer recorded the species and number of all birds seen and heard during an 8-minute sampling period, including the best approximation of exact distance, to be used for estimating densities (Buckland *et al.* 1993), and an indication of whether the bird was seen or heard, or both. Birds in flight were recorded and annotated as such. At the beginning of each transect, the observer recorded pertinent site information on temperature, estimated wind speed, and estimated cloud cover.

Because avian species richness and composition have been reported to be strongly correlated with vegetation structure in riparian areas (Stauffer and Best 1980; Rice *et al.* 1984) and in urban areas (Beissinger and Osborne 1982; Mills *et al.* 1989), we estimated the per cent coverage for the canopy, subcanopy, shrub, and ground layers at each census point. We also quantified tree diameters and heights as well as species composition for trees, saplings, and shrubs, and the number of dead stumps. The width of the riparian zone has been shown to have a strong influence on the composition and abundance of bird species found there (Stauffer and Best 1980); it was assessed at each census point on the basis of the extent of woody riparian vegetation using aerial photos and a Geographic Information System.

Nest Predation. To study nest predation, ideally one would locate and monitor a large number of active nests. This is difficult, however, when the objective is to compare a number of sites for a number of species. Alternatively, experimental nests provide a means of investigating patterns and causes of nest loss while standardizing factors such as nest appearance and egg size (Martin 1987; Haskell 1995), and this method has been used to describe patterns of nest predation across sites in different landscape contexts (Andrén *et al.* 1985; Wilcove 1985; Ratti and Reese 1988; Donovan *et al.* 1997). Not all variables that potentially influence predation rates on natural nests are controlled, however (e.g., parental activity at the nest). Predation rates on experimental nests should therefore be viewed as an index of predator activity and relative predation risk among sites and interpreted with caution.

We established transects of experimental nests (20 per transect in 1995 and 30 per transect in 1996 and 1997) at 12 of our study sites between late May and early July in order to examine differences in predation rates related to surrounding levels of urbanization. In 1995 and 1996, we also investigated more localized effects on nest predation related to human activity in the riparian zone. Here, we chose two sites (treatments) that had recreational trails adjacent to the stream and two control sites on the same streams that did not have trails. At each of the treatment sites we placed one transect of nests (n = 20) on the trail side of the stream and one on the opposite (non-trail) side. A nest transect was also placed on a randomly chosen side of the stream at each of the control sites.

For both experiments we used commercially-available wicker canary nests modified with grass and mud in order to resemble natural nests more closely. Such modifications are thought to facilitate the recognition of these nests as prey items by predators (Martin 1987; Sieving 1992). Each transect of experimental nests had a randomized starting point. Nests were alternately placed 10–20 m on either side of an imaginary line at 30 m intervals, resulting in a between-nest distance of approximately 40 m. This arrangement was chosen in order to simulate natural nest densities in these habitats and to avoid repeated predation by a predator following a linear course, or "traplining." Nests were placed 0.5–2.5 m above ground in trees and shrubs and baited with two Japanese Quail eggs *Coturnix coturnix* plus one egg made of modeling clay. The latter served two purposes; first, it was closer in size to songbird eggs than were the quail eggs, thus avoiding biases that may be associated with the use of quail eggs (Roper 1992; Haskell 1995; DeGraaf and Maier 1996), and second, it allowed us to obtain tooth/beak impressions of the nest predators (Möller 1988; Major 1991; Nour *et al.* 1993). The clay egg was tethered to the nest by a thin piece of wire so that it could be recovered after a predation event (Major *et al.* 1994). Every effort was made to place nests in locations similar to those in which real bird nests are found. We checked nests every third day for 15 days, which simulates the combined egg-laying and incubation periods for most passerines (Martin 1987). At each check we noted the number of eggs remaining as well as egg and nest condition. Experimental nests were considered to be predated when one or more eggs were disturbed.

We also monitored natural bird nests at four sites in 1996 (n = 149) and again in 1997. These data provide a basis for comparison with predation rates on experimental nests. We concentrated on finding nests of the American Robin *Turdus migratorius* because this species was present on all areas and was thought to exist in numbers great enough to constitute adequate sample sizes. Nests of other passerine species, however, were also monitored when found. Natural nests were checked approximately twice per week from the time we found them until they either fledged young or failed.

For examinations of nest predation where the level of development surrounding a site was of interest, we summed the number of buildings within 100 m of the watercourse. We did this using a Geographic Information System and land use maps made available by the offices of the county tax assessors.

Data analysis

When quantifying species richness and community composition, birds flying over an area were excluded. We also excluded nocturnal and crepuscular species, aerial insectivores (e.g., swallows and swifts), raptors, shorebirds, and waterfowl, because point counts are not an effective way to census such species (Bibby *et al.* 1992). A species was classified as a migrant if it travels >300 km to its breeding grounds and no individuals are present year-round on our study areas (Peterson 1990; Andrews and Righter 1992).

We quantified predation pressure for nests at each site using the methods of Mayfield (1961) as modified by Johnson (1979). Predation is determined using the total number of nests depredated and the total days that observed nests were exposed to predation, and is expressed as a daily mortality rate, or the probability of predation per day (Mayfield 1961). We conducted an overall test for homogeneity in daily mortality rates among sites on a given drainage using Chi-Square analyses and multiple comparisons (Sauer and Williams 1989). When the null hypothesis was rejected, we conducted tests on a pairwise basis using a sequential Bonferroni correction for multiple comparisons (Rice 1989).

RESULTS AND DISCUSSION

This study is continuing, and therefore both the analyses and results presented here should be considered as preliminary. We report and discuss the results of our initial analyses, which were conducted on data collected in 1996 unless otherwise noted.

Avian communities

For purposes of describing avian community composition, we focus on three sites per drainage that we have designated as high, medium, or low, reflecting the level of development in the surrounding landscape (Table 1). Overall, the number of bird species tends to increase as development decreases along a given drainage (Fig. 2), lending support to our first prediction. This trend is most striking along the Boulder Creek drainage. The high-urbanization end of the Boulder Creek gradient is our most urbanized site overall (Fig. 1, site 5), with many buildings and a heavily-used recreational trail in close proximity to the creek. This site also has the longest history of urbanization, and several other workers have reported a gradual decrease in species richness with time since development (Batten 1972; Walcott 1974; Aldrich and Coffin 1980).

Table 1. Average width of the woody riparian zone for three sites on each drainage. High, medium, and low refer to the relative level of development surrounding each site.

Drainage	Site	Average Width (m)
South Boulder Creek	High	56
	Medium	96
	Low	48
Coal Creek	High	73
	Medium	53
	Low	51
Boulder Creek	High	76
	Medium	66
	Low	127
Poudre River	High	115
	Medium	130
	Low	120

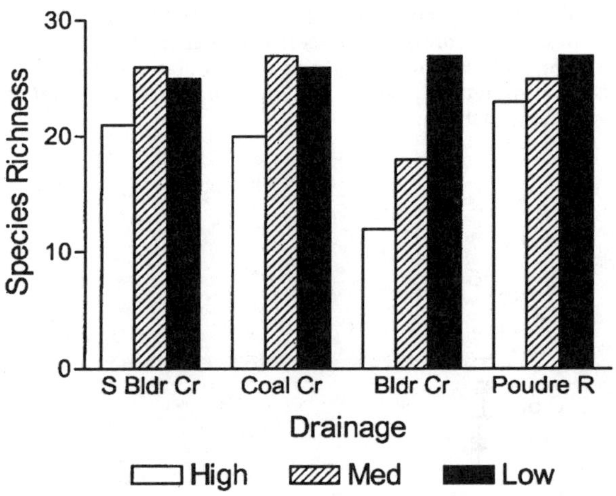

Fig. 2. Avian species richness in 1996 for three sites each on South Boulder Creek, Coal Creek, Boulder Creek, and the Poudre River. Only species known to breed in riparian areas have been included, while nocturnal and crepuscular species, aerial insectivores (e.g., swallows and swifts), raptors, shorebirds, and waterfowl have been excluded. High, medium, and low refer are relative designations for the level of urbanization in the landscape surrounding each site.

South Boulder Creek and Coal Creek also showed increases in richness with decreased urbanization but there were slightly more species at the sites with intermediate levels of development than at the low-development sites, and overall differences along these gradients were not as pronounced as on Boulder Creek. It is important to remember that the designations of high-medium-low were made relative to other sites on the same drainage. On Coal Creek for example, one end of the transect at the medium site is within several hundred meters of a residential area, whereas there is no development within 6 km of the low site. The relative designations are therefore correct, but it may be that the differences are not sufficient to affect breeding birds in the riparian zone. Furthermore, development at the most urban sites on these drainages is both more recent and less extensive than on Boulder Creek. These observations underscore some of the limitations of a categorical approach to describing urbanization. We have begun to quantify the type and extent of land use in the matrix surrounding each site using digital maps made available by county planners. Such an approach will ultimately allow us to treat development as a continuous rather than a categorical variable.

Species richness in the study areas along the Poudre River increased with decreasing urbanization, but here the differences were less pronounced than on the other drainages. This may be attributable to the relatively greater width of the woody riparian zone along the Poudre (Table 1), and thus greater distances to nearby development. Several other studies have shown a strong correlation between the width of the riparian zone and bird species richness (Best and Stouffer 1980; Darveau *et al.* 1995; Hodges and Krementz 1996). This relationship appears to be influenced by landscape context, however, as evidenced by the medium and low sites on Coal Creek. These areas were among the narrowest in terms of riparian width, but compared favourably with wider, more-developed sites in terms of the number of avian species that they support during the breeding season. This result lends further support to the observation that narrow riparian habitats have conservation value for some species (Darveau *et al.* 1995; Fisher and Goldney 1997).

Species richness is one measure of avian response to urbanization, but it is not the only one. It may actually mask important variation if the number of species at two sites is similar, but the composition differs considerably. In our initial examination of the composition of the bird assemblages at our study sites, we asked whether species that breed in adjacent areas also occurred in riparian habitats. On each gradient, we examined the riparian avifauna at the highly developed site for birds that are characteristic of towns and cities in general (Andrews and Righter 1992). Likewise, we examined the avifauna at rural riparian sites for species that generally breed in the surrounding grasslands. The results support our prediction that riparian areas associated with development will contain bird species that are also characteristic of the surrounding habitats, while this will be less evident in less developed settings (Fig 3a). We suggest that this is a function of the planting of trees in urban areas that once existed as shortgrass prairie. Riparian vegetation is therefore less distinct in urban settings and provides habitat for human commensals in much the same way that a park or tree-lined street might.

Migration strategy was the second lens through which we viewed compositional changes in riparian avifaunas. The percentage of migrant species tended to increase with decreasing levels of urbanization (Fig. 3b), and again, this trend is supported by our 1995 data. Thus, not only do urban areas generally have fewer species, but a smaller percentage of those

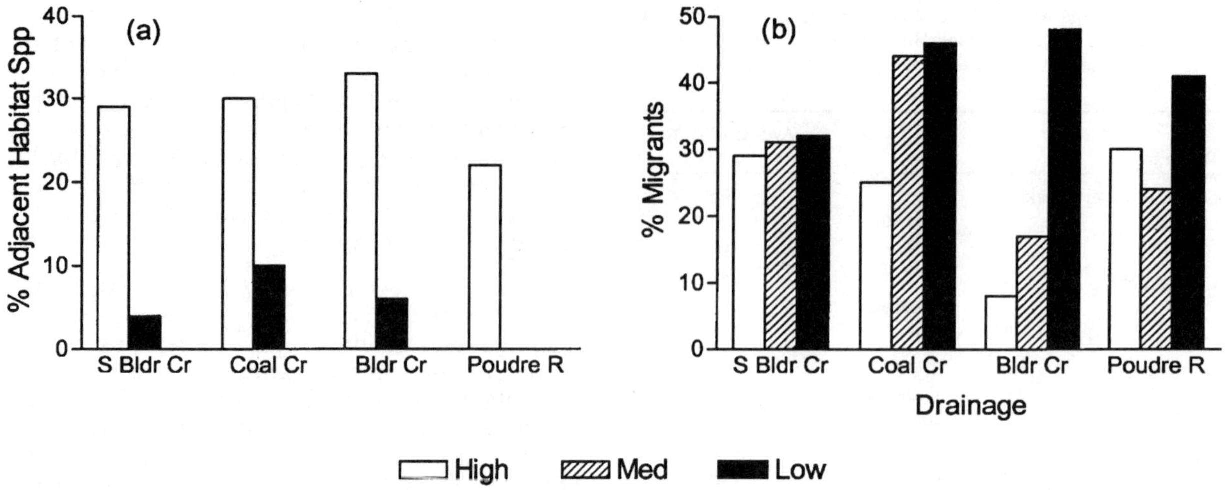

Fig. 3. For South Boulder Creek, Coal Creek, Boulder Creek, and the Poudre River during the 1996 avian breeding season (a) the percentage of the bird community consisting of species characteristic of urban habitats (for the high urbanization site) and grassland habitats (for the low urbanization site) as described by Andrews and Righter (1992), and (b) the percentage of the bird community at each of three sites per drainage that consists of birds that winter in the neotropics but breed in Colorado's lowland riparian areas. For (b), we have excluded nocturnal and crepuscular species, aerial insectivores (e.g., swallows and swifts), raptors, shorebirds, and waterfowl. High, medium, and low are relative designations for the level of urbanization in the landscape surrounding each site.

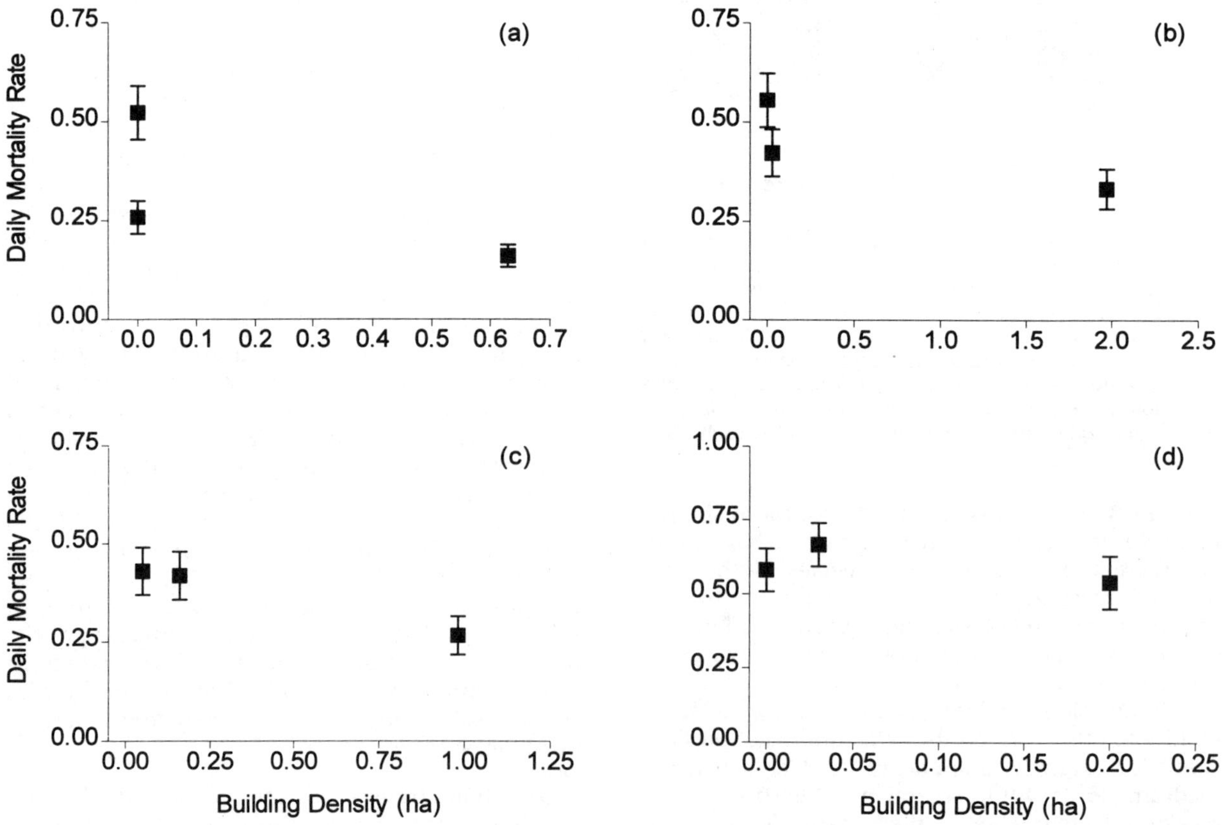

Fig. 4. Daily mortality rates ± 1 SE *vs.* building density per ha for experimental nests at each of three sites on (a) South Boulder Creek, (b) Coal Creek, (c) Boulder Creek, and (d) the Poudre River in 1996. Building density is based on the number of buildings within a 100 m radius buffer surrounding the study reach of each watercourse. We tested for overall homogeneity in rates on each drainage (a: $p < 0.001$; b and c: $p < 0.05$; d: $p < 0.01$), and conducted pairwise tests for rates on each drainage for most- *vs.* least-developed sites (a and b: $p < 0.001$; c: $p < 0.03$; d: $p < 0.02$).

species tend to be migrants. What factors might account for this observation?

Other researchers have also noted decreasing numbers of migrant species in urban areas when compared with rural sites (Walcott 1974; DeGraaf 1987; Friesen *et al.* 1995). One explanation is that resident species in urban areas are simply more tolerant of a substantial human presence (Cooke 1980; Burger and Gochfeld 1991). A second, though not mutually exclusive, possibility is a function of the flora often found in urban areas. Non-native and ornamental trees and shrubs may support fewer insects (Southwood 1961). Migrant species, which are often insectivorous, would thus be selected against when compared with omnivores and seed-eaters. A more thorough examination of the differences in ecological and life-history traits of these avian species, plus consideration of the structural and floristic attributes of riparian vegetation at these sites, is necessary to understand the patterns of habitat occupancy that we have observed.

Nest predation

We predicted that we would find greater daily mortality rates at the urban end of our gradients, in accordance with commonly accepted dogma regarding nest predation (e.g., Soulé *et al.* 1988; Harris and Silva-Lopez 1992; Noss 1993). We extended this thinking to areas near recreational trails, and reasoned that people often allow their dogs to wander near trails and that picnic areas and garbage cans associated with such sites might attract potential nest predators. Wilcove (1985) found predation rates on experimental nests to be nearly 150% higher in woodlots near suburban development as compared to more rural sites, and attributed this difference to increased densities of human commensals such as Blue Jays *Cyanocitta cristata*, Common Grackles *Quiscalus quiscula*, raccoons *Procyon lotor*, grey squirrels *Sciurus carolinensis*, dogs *Canis familiaris*, cats *Felis catus*, and rats *Rattus norvegicus* near suburban areas.

Contrary to our predictions, we found no evidence that predation rates on experimental nests were higher at human-dominated sites. Rather, when we compared predation pressure as a function of building density, rates were significantly lower for high-development sites compared with rural sites on all four drainages (Fig. 4). The case was much the same regarding sites with recreational trails. Daily mortality rates increased with distance from trails on both drainages and in both years (Fig. 5a and b). Furthermore, there was no evidence that daily mortality rates for natural nests were greater at developed sites (Fig. 6). Rather, there was no difference in predation rates on Coal Creek, and the low-development site on South Boulder Creek had a higher rate than did the high-development site.

What factors might account for the discrepancy between our study and the prevailing dogma? As stated earlier, ecologists have often avoided areas of high human activity when selecting research sites, and thus biotic interactions in such areas are relatively unexplored (Pickett *et al.* 1992; McDonnell and Pickett 1993). Moreover, little is known about the ecological integrity of riparian systems that occur in an urban or suburban matrix, or about the impacts of recreational activities that often occur in these corridors on wildlife species (Adams and Dove 1989). Consequently, the prevailing dogma regarding nest predation in human-dominated areas is largely untested.

First and foremost, we stress that one must exercise caution when interpreting results based on experimental nests. Although experimental nests provide a means for gathering data on nest predation in a controlled manner, there are several important differences between these nests and active bird nests, such as the lack of nestlings, differences in scent, etc. Rather than dismissing this method as a result of its limitations, we suggest that careful comparisons of experimental nest results with data

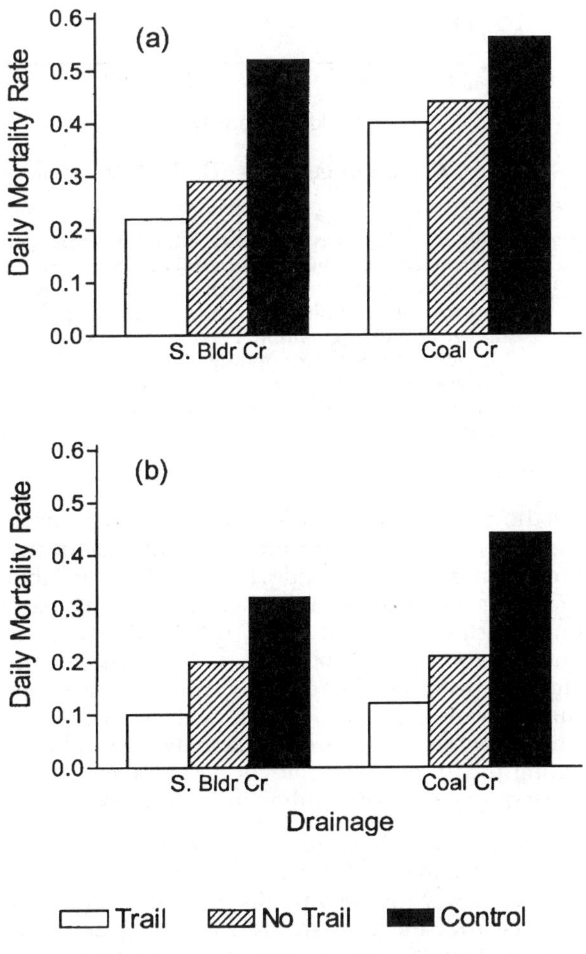

Fig. 5. Daily mortality rates on experimental nests at each of two sites on South Boulder Creek and on Coal Creek in 1996 (a) and 1995 (b) as a function of distance from recreational trails. "Trail" refers to nests on the trail side of a stream, "No Trail" refers to nests on the opposite side of the stream, and "Control" refers to nests at a site on the same stream that is not associated with a trail in the riparian zone.

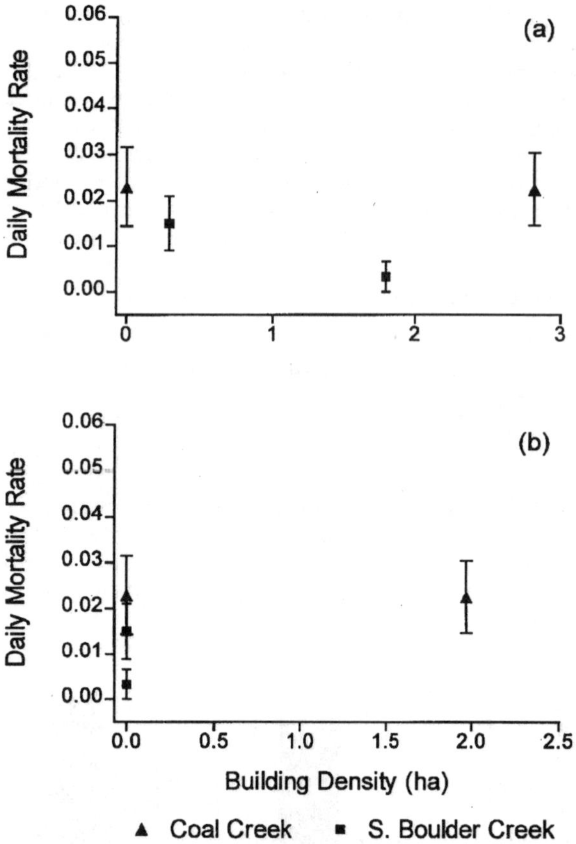

Fig. 6. Daily mortality rates ± 1 SE vs. building density per ha for American Robin *Turdus migratorius* nests at each of two sites on Coal Creek and South Boulder Creek in 1996. Building density is based on the number of buildings within 1 000 m (a) and 100 m (b) radius buffers surrounding the study reach of each stream. We tested for DMR homogeneity on each drainage (Coal Creek, $p > 0.5$; South Boulder Creek, $p < 0.1$).

on the nest success of select species may provide insights into the mechanisms that determine predation rates and guide future efforts in this area. For example, we found daily mortality rates on experimental nests to be much higher than on active American Robin nests. This species is a vigorous defender of its eggs and young, as we were made aware each time we checked nest contents. This suggests that parental activity may play a strong role in reducing rates of nest predation, and suggests potentially fruitful areas for future study.

It is possible that human activities that occur near recreational trails influence predation rates, but not in the way that we initially predicted. Rather, it may be that the human presence in these areas actually serves to dampen predation rates for species able to nest there. Some support for this notion is derived from studies of avian breeding success in environments that are associated with various levels of human activity. Osborne and Osborne (1980) found that Blackbirds *Turdus merula* nesting near sidewalks or buildings at the University of Exeter not only enjoyed significantly higher breeding success, but were much less discriminating in nest site characteristics than were birds in more remote locations. Similarly, the amount of nest predation for birds breeding in two urban parks in Poland was negatively correlated with the frequency of human visits (Tomialojć and Profus 1977).

Finally, it is possible that our most rural sites are not remote enough, as even the most rural area is within 6 km of suburban development. It would be interesting to measure predation rates in riparian areas even farther from human activity. This is not only logistically difficult, but as one moves farther from developed areas in Colorado, other factors increasingly confound comparisons, such as differences in vegetational, geomorphological, and topographical features. Ultimately, there are no pristine riparian sites at low elevations to use as baselines in the western United States (Ohmart 1994), one reason being that conservation interest in these areas has only developed over the last two decades.

Ecology, urban planning, and conservation

At this point, it is premature to offer detailed prescriptions for urban planners and natural resource managers on the basis of this study. There are some suggestive patterns in our data, however, that bear closer scrutiny. It does seem that, other things being equal, wider riparian areas support more species than narrow ones. It also appears that the nature of the surrounding landscape matrix influences species richness, so that a narrow riparian habitat in favourable surroundings may harbour more species than a wider area in a less suitable setting. Our data suggest that human activities in the riparian zone may also play a role in determining which species persist there. It may be that certain advantages are conferred on birds that are able to tolerate a human presence — not all species may be able to do so. More research in addition to further analyses of our data is needed in order to gain a better understanding of both local and extrinsic factors that affect avian conservation in riparian habitats.

By their very nature, riparian systems transcend political boundaries. A single stream or river may bisect numerous cities and towns, and flow through a number of counties or, in some cases, states. When one also considers a network of associated tributaries, it soon becomes apparent that such a dendritic system poses a host of conservation issues. For this reason, the implementation of policy based on sound ecology is challenging because authority for decisions on land use is often relegated to the lowest possible level of government. Land use change over very large areas can develop as the result of many fine-scale decisions. Often these decisions are made "one at a time" with little or no consideration for their overall impact at larger scales. Because these choices are numerous and diffuse in time and space, it is difficult for ecologists to support them in a meaningful fashion. It is therefore critical that agencies or groups whose mission is conservation not only cooperate across political boundaries, but also be quite explicit about what it is that they are trying to conserve.

Knopf and Samson (1994) note that generalist species tend to dominate bird communities at most riparian sites in the west. On a regional scale, this would seem to be true at our study sites. It is also true, however, that these may be the only habitats where some of these species occur over smaller scales, such as the county. As human populations shift toward urban areas, connections to nature become increasingly tenuous. Because conservation action, even in remote or rural areas, may depend on the support of people in urban areas, the preservation of native bird species as well as other components of biological diversity in areas where people live takes on added importance. It may be our best hope for fostering a greater appreciation of our natural heritage.

ACKNOWLEDGEMENTS

We thank the City of Boulder Open Space Department, the Colorado Division of Wildlife, Boulder County Parks and Open Space, and the Graduate Degree Program in Ecology at Colorado State University for supporting this project. For assistance with field work and data entry, we thank Stephanie Bestelmeyer, Christian Hagen, Janet Jorgensen, Joe LaFleur, Jolene Newth, Sara Shields, Soley Snyder, Jill Winn, and especially Scott Severs. For logistical and technical assistance, we appreciate the help of Jon Osborne, Dave Theobald, and especially Clint Miller and Michael Sanders. Peter Cale, Richard Hobbs, and Denis Saunders provided helpful comments on the manuscript. JRM would like to thank the Australian-American Fulbright Commission, CSIRO Division of Wildlife and Ecology, and the Graduate Degree Program in Ecology at Colorado State University for enabling him to attend the meeting in Taupo, at which this chapter was presented.

REFERENCES

Adam, P., 1995. Urbanization and transport. Pp. 55–75 *in* Conserving Biodiversity: Threats and Solutions ed by R. A. Bradstock, T. D. Auld, D. A. Keith, R. T. Kingsford, D. Lunney, and D. P. Sivertsen. Surrey Beatty & Sons: Chipping Norton, New South Wales.

Adams, L. and Dove, L., 1989. Wildlife reserves and corridors in the urban environment. National Institute for Urban Wildlife: Columbia, Maryland.

Aldrich, J. W. and Coffin, R. W., 1980. Breeding bird populations from forest to suburbia after thirty-seven years. *American Birds* **34:** 3–7.

Andrén, H., Angelstam, P., Lindstrom, E. and Widen, P., 1985. Differences in predation pressure in relation to habitat fragmentation: an experiment. *Oikos* **45:** 273–77.

Andrews, R. and Righter, R., 1992. Colorado Birds. Denver Museum of Natural History.

Angelstam, P., 1986. Predation on ground-nesting birds' nests in relation to predator densities and habitat edge. *Oikos* **47:** 365–73.

Austin, M. P. and Heyligers, P. C., 1991. New approach to vegetation survey design: gradsect sampling. Pp. 31–36 *in* Nature Conservation: Cost Effective Biological Surveys and Data Analysis ed by C. R. Margules and M. P. Austin. CSIRO: Australia.

Backhouse, G. N., 1987. Management of remnant habitat for conservation of the Helmeted Honeyeater *Lichenostomus melanops cassidix*. Pp. 287–94 *in* Nature Conservation: The Role of Remnants of Native Vegetation ed by D. A. Saunders, G. W. Arnold, A. A. Burbidge and A. J. M. Hopkins. Surrey Beatty & Sons: Chipping Norton, New South Wales.

Batten, L. R., 1972. Breeding bird species diversity in relation to increasing urbanisation. *Bird Study* **19:** 157–66.

Beissinger, S. R. and Osborne, D. R., 1982. Effects of urbanization on avian community organization. *Condor* **84:** 75–83.

Best, L. B. and Stauffer, D. F., 1980. Factors affecting nesting success in riparian bird communities. *Condor* **82:** 149–58.

Bezzel, E., 1985. Birdlife in intensively used rural and urban environments. *Ornis Fennica* **62:** 90–95.

Bibby, C. J., Burgess, N. D. and Hill, D. A., 1992. Bird Census Techniques. Academic Press: London.

Binford, M. W. and Buchenau, M. J., 1993. Riparian greenways and water resources. Pp. 69–104 *in* Ecology of Greenways ed by D. S. Smith and P. C. Hellmund. University of Minnesota Press: Minneapolis.

Blair, R. B., 1996. Land use and avian species diversity along an urban gradient. Ecol. App. **6:** 506–19.

Bottorff, R. L., 1974. Cottonwood habitat for birds in Colorado. *American Birds* **28:** 975–79.

Boulder County Planning Commission, 1997. Boulder County Comprehensive Plan. Boulder, Colorado.

Brittingham, M. C., 1991. Effects of winter bird feeding on wild birds. Pp. 185–90 *in* Wildlife Conservation in Metropolitan Environments ed by L. W. Adams and D. L. Leedy. National Institute for Urban Wildlife: Columbia, MD.

Buckland, S. T. anderson, D. R., Burnham, K.P. and Laake, J. L., 1993. Distance Sampling: Estimating Abundance of Biological Populations. Chapman & Hall: London.

Burger, J. and Gochfeld, M., 1991. Human disturbance and birds: tolerance and response of resident and migrant species in India. *Environ. Cons.* **18:** 158–65.

Chaney, E., Elmore, W. and Platts, W. S., 1990. Livestock Grazing on Western Riparian Areas. U.S. Environmental Protection Agency: Washington, DC.

Churcher, P. B. and Lawton, J. H., 1987. Predation by domestic cats in an English village. *J. Zool.* **212:** 439–55.

City of Boulder Planning Department, 1997. Boulder Valley Comprehensive Plan. Boulder, Colorado.

City of Fort Collins, 1992. City of Fort Collins Natural Areas Policy Plan. City of Fort Collins Natural Resources Department, Fort Collins, CO.

Cooke, A. S., 1980. Observations on how close certain passerine species will tolerate an approaching human in rural and suburban areas. *Biol. Cons.* **18:** 85–88.

Darveau, M., Beauchesne, P., Belanger, L., Huot, J. and Larue, P., 1995. Riparian forest strips as habitat for breeding birds in boreal habitat. *J. Wildl. Man.* **59:** 67–78.

DeGraaf, R. M., 1987. Urban wildlife habitat research — application to landscape design. Pp. 107–11 *in* Integrating Man and Nature in the Metropolitan Environment ed by L. W. Adams and D. L. Leedy. National Institute for Urban Wildlife: Columbia, MD.

DeGraaf, R. M. and Wentworth, J. M., 1986. Avian guild structure and habitat associations in suburban bird communities. *Urban Ecol.* **9:** 399–412.

DeGraaf, R. M. and Maier, T. J., 1996. Effect of egg size on predation by white-footed mice. *Wilson Bull.* **108:** 535–39.

Donovan, T. M., Jones, P. W., Annand, E. M. and Thompson, F. R. III, 1997. Variation in local-scale edge effects: mechanisms and landscape context. *Ecology* **78:** 2064–075.

Douglas, I., 1994. Human settlements. Pp. 149–69 *in* Changes in Land Use and Land Cover: A Global Perspective ed by W. B. Meyer and B .L. Turner II. Cambridge University Press: Cambridge.

Duerksen, C. J., Elliott, D. L., Hobbs, N. T., Johnson, E. and Miller, J. R., 1997. Habitat Protection Planning: Where the Wild Things Are. American Planning Association Planning Advisory Service Report 470/471, Chicago, IL.

Emlen, J. T., 1974. An urban bird community in Tucson, Arizona: derivation, structure, regulation. *Condor* **76:** 184–97.

Engels, T. M. and Sexton, C. W., 1994. Negative correlation of blue jays and golden-cheeked warblers near an urbanizing area. *Cons. Biol.* **8:** 286–90.

Finch, D. M. and Ruggiero, L. F., 1993. Wildlife habitats and biological diversity in the Rocky Mountains and northern Great Plains. *Nat. Areas J.* **13:** 191–203.

Fisher, A. M. and Goldney, D. C., 1997. Use by birds of riparian vegetation in an extensively fragmented landscape. *Pac. Cons. Biol.* **3:** 275–88.

Foppen, R. and Reijnen, R., 1994. The effects of car traffic on breeding bird populations in woodland. II. Breeding dispersal of male willow warblers (*Phylloscopus trochilus*) in relation to the proximity of a highway. *J. App. Ecol.* **31:** 95–101.

Friesen, L. E., Eagles, P. F. J. and MacKay, R. J., 1995. Effects of residential development on forest-dwelling neotropical migrant songbirds. *Cons. Biol.* **9:** 1408–414.

Gates, J. E. and Gysel, L. W., 1978. Avian nest dispersion and fledging success in field-forest ecotones. *Ecology* **59:** 871–83.

Geis, A. D., 1974. Effects of urbanization and type of urban development on bird populations. Pp. 97–105 *in* Wildlife in an Urbanizing Environment ed by J. H. Noyes and D. R. Progulski. University of Massachusetts: Amherst.

Guthrie, D. A., 1974. Suburban bird populations in southern California. American Midland *Naturalist* **92:** 461–66.

Harris, L. and Silva-Lopez, G., 1992. Forest fragmentation and the conservation of biological diversity. Pp. 197–237 *in* Conservation Biology ed by P. K. Fielder and K. J. Subodh. Chapman and Hall: New York.

Haskell, D. G., 1995. Forest fragmentation and nest predation: Are experiments with Japanese quail eggs misleading? *Auk* **112:** 767–70.

Haspel, C. and Calhoon, R. E., 1989. Home ranges of free-ranging cats (*Felis catus*) in Brooklyn, New York. *Can. J. Zool.* **67:** 178–81.

Herkert, J. R., 1994. The effects of habitat fragmentation on midwestern grassland bird communities. *Ecol. App.* **4:** 461–71.

Hockin, D., Ounsted, M., Gorman, M., Hill, D., Keller, V. and Barker, M. A., 1992. Examination of the effects of disturbance on birds with reference to its importance in ecological assessments. *J. Environ. Man.* **36:** 253–86.

Hodges, M. F. J. and Krementz, D. G., 1996. Neotropical migratory breeding bird communities in riparian forests of different widths along the Altamaha River, Georgia. *Wilson Bull.* **108:** 496–506.

Hoffman, C. D. and Gottschang, J. L., 1977. Numbers, distribution, and movements of a raccoon population in a suburban residential community. *J. Mammal.* **58:** 623–36.

Janzen, D. H., 1986. The external threat. Pp. 286–303 *in* Conservation Biology: The Science of Scarcity and Diversity ed by M. E. Soulé. Sinauer Associates Inc.: Sunderland, Massachusetts.

Johnson, A. S., 1989. The thin green line: riparian corridors and endangered species in Arizona and New Mexico. Pp. 35–46 *in* Preserving Communities and Corridors ed by G. Mackintosh. G.W. Press: Washington, DC.

Johnson, D. H., 1979. Estimating nest success: the Mayfield method and an alternative. *Auk* **96:** 651–61.

Johnston, R. F. and Garrett, K. L., 1994. Population trends of introduced birds in western North America. *Stud. Avian Biol.* **15:** 221–31.

Johnson, R. R., Haight, L. T. and Simpson, J. M., 1977. Endangered species vs. endangered habitats: a concept. Pp. 68–79 *in* Importance, Preservation and Management of Riparian Habitat: a Symposium. General Technical Report RM-43 ed by R. R. Johnson and D. A. Jones, Jr U.S. Forest Service: Rocky Mountain Forest and Range Experiment Station, Fort Collins, CO.

Keeley, J. E (ed), 1993. Interface Between Ecology and Land Development in California. Southern California Academy of Sciences: Los Angeles, CA.

Knopf, F. L., 1985. Significance of riparian vegetation to breeding birds across an altitudinal cline. Pp. 105–11 *in* Riparian Ecosystems and Their Management: Reconciling Conflicting Uses. First North American Riparian Conference, April 16–18, 1985, Tucson, Arizona. General Technical Report RM-120 ed by R. R. Johnson, C. D. Ziebell, D. R. Patton, P. F. Ffolliott and R. H. Hamre. U.S. Department of Agriculture: USFS, Rocky Mountain Forest and Range Experiment Station, Fort Collins, CO.

Knopf, F. L., Johnson, R. R., Rich, T., Samson, F. B. and Szaro, R., 1988. Conservation of riparian ecosystems in the United States. *Wilson Bull.* **100:** 272–84.

Knopf, F. L. and Samson, F. B., 1994. Scale perspectives on avian diversity in western riparian ecosystems. *Cons. Biol.* **8:** 669–76.

Larimer County, 1993. Larimer County Parks Comprehensive Parks Master Plan. Larimer County Parks Department: Fort Collins, CO.

Lussenhop, J., 1977. Urban cemeteries as bird refuges. *Condor* **79:** 456–61.

Major, R., 1991. Identification of nest predators by photography, dummy eggs, and adhesive tape. *Auk* **108:** 190–95.

Major, R. E., Pyke, G. H., Christy, M. T., Gowing, G. and Hill, R. S., 1994. Can nest predation explain the timing of the breeding season and the pattern of nest dispersion of New Holland honeyeaters? *Oikos* **69:** 363–72.

Major, R. E., Gowing, G. and Kendal, C. E., 1996. Nest predation in Australian urban environments and the role of the pied currawong, *Strepera graculina*. *Aust. J. Ecol.* **21:** 399–409.

Marr, J. W., 1961. Ecosystems of the East Slope of the Front Range in Colorado. University of Colorado Studies, Biology Series 8. University of Colorado Press: Boulder.

Marr, J. W., 1964. The vegetation of the Boulder area. Pp. 34–42 *in* Natural History of the Boulder Area ed by H. G. Rodeck.. University of Colorado Museum: Boulder.

Martin, T. E., 1987. Artificial nest experiments: effects of nests appearance and type of predator. *Condor* **89:** 925–28.

Martin, T. E., 1988. Habitat and area effects on forest bird assemblages: is nest predation and influence? *Ecology* **69:** 74–84.

Martin, T. E. and Clobert, J., 1996. Nest predation and avian life-history evolution in Europe versus North America: a possible role of humans. *Amer. Nat.* **147:** 1028–046.

Mayfield, H., 1961. Nesting success calculated from exposure. *Wilson Bull.* **73:** 255–61.

McDonnell, M. J. and Pickett, S. T. A., 1993. Scope and need for an ecology of subtle human effects and populated areas. Pp. 1–5 *in* Humans as Components of Ecosystems ed by M. J. McDonnell and S. T. A. Pickett. Springer-Verlag: New York.

Mills, G. S., Dunning, Jr, J. B. and Bates, J. M., 1989. Effects of urbanization of breeding bird community structure in southwestern desert habitats. *Condor* **91:** 416–28.

Möller, A. P., 1988. Nest predation and nest site choice in passerine birds in habitat patches of different size: a study of magpies and blackbirds. *Oikos* **53:** 215–21.

Naiman, R. J., Decamps, H. and Pollock, M., 1993. The role of riparian corridors in maintaining regional biodiversity. Ecol. App. **3:** 209–12.

Nassauer, J. I. (ed), 1997. Culture and landscape ecology: insights for action. Pp. 1–14 *in* Placing Nature: Culture and Landscape Ecology ed by J. I. Nassauer. Island Press: Washington, DC.

Noss, R. F., 1993. Wildlife corridors. Pp. 43–68 *in* Ecology of Greenways ed by D. S. Smith and P. C. Hellemund. University of Minnesota Press: Minneapolis.

Nour, N., Matthysen, E. and Dhondt, A. A., 1993. Artificial nest predation and habitat fragmentation: different trends in bird and mammal predators. *Ecography* **16:** 111–16.

Ohmart, R. D., 1994. The effects of human-induced changes on the avifauna of western riparian habitats. *Stud. Avian Biol.* **15:** 273–85.

Osborne, P. and L. Osborne., 1980. The contribution of nest site characteristics to breeding — success among blackbirds *Turdus mercula*. *Ibis* **122:** 512–17.

Peterson, R. T., 1990. Field Guide to Western Birds. Houghton Mifflin Company: Boston.

Pickett, S. T. A., Parker, V. T. and Fiedler, P. L., 1992. The new paradigm in ecology: implications for conservation biology above the species level. Pp. 65–88 *in* Conservation Biology ed by P. L. Fiedler and J. K. Jain. Chapman & Hall: New York.

Platt, R. H., Roundtree, R. A. and Muick, P. C. (eds), 1994. The Ecological City: Preserving and Restoring Urban Biodiversity. University of Massachusetts Press: Amherst.

Ratti, J. and Reese, K., 1988. Preliminary test of the ecological trap hypothesis. *J. Wildl. Man.* **52:** 484–91.

Rice, J., Anderson, B. W. and Ohmart, R. D., 1984. Comparison of the importance of different habitat attributes to avian community organization. *J. Wildl. Man.* **48:** 895–911.

Rice, W. R., 1989. Analyzing tables of statistical tests. *Evolution* **43:** 223–25.

Ricklefs, R. E., 1969. An analysis of nesting mortality in birds. *Smith. Contrib. Zool.* **9:** 1–48.

Robinson, S. K., Thompson, F. R. III, Donovan, T. M., Whitehead, D. R. and Faaborg, J., 1995. Regional forest fragmentation and the nesting success of migratory birds. *Science* **267:** 1987–990.

Roper, J. J., 1992. Nest predation experiments with quail eggs: too much to swallow? *Oikos* **65:** 528–30.

Rosatte, R. C., Power, M. J. and MacInnes, C. D., 1991. Ecology of urban skunks, raccoons, and foxes in metropolitan Toronto. Pg. 31–38 *in* Wildlife Conservation in Metropolitan Environments ed by L. W. Adams and D. L. Leedy. National Institute for Urban Wildlife: Columbia, MD.

Sauer, J. R. and Williams, B. K., 1989. Generalized procedures for testing hypotheses about survival or recovery rates. *J. Wildl. Man.* **53:** 127–42.

Shaffer, M. L., 1981. Minimum population sizes for species conservation. *Bioscience* **31:** 131–34.

Siegel, J. J., 1996. "Subdivisions versus agriculture": From false assumptions come false alternatives. *Cons. Biol.* **10:** 1473–474.

Sieving, K. E., 1992. Nest predation and differential insular extinction among selected forest birds of central Panama. *Ecology* **73:** 2310–328.

Smith, D. S. and Hellmund, P. C. (eds), 1993. Ecological Greenways. University of Minnesota Press: Minneapolis.

Southwood, T. R. E., 1961. The number of species of insect associated with various trees. *J. Anim. Ecol.* **30:** 1–8.

Soulé, M. E., Bolger, D. T., Alberts, A. C., Wright, J., Sorice, M. and Hill, S., 1988. Reconstructed dynamics of rapid extinctions of chaparral-requiring birds in urban habitat islands. *Cons. Biol.* **2:** 75–92.

Soulé, M. E., Alberts, A. C. and Bolger, D. T., 1992. The effects of habitat fragmentation on chaparral plants and vertebrates. *Oikos* A. C. 39–47.

Stamp, N. E., 1978. Breeding birds of riparian woodlands in southcentral Arizona. *Condor* **80:** 64–71.

Stevens, L., Brown, B. T., Simpson, J. M. and Johnson, R. R., 1977. The importance of riparian habitat to migrating birds. Pp. 156–64 *in* Importance, Preservation and Management of Riparian Habitat: a Symposium. General Technical Report RM-43 ed by R. R. Johnson and D. A. Jones, Jr U.S. Forest Service: Rocky Mountain Forest and Range Experiment Station, Fort Collins, CO.

Stauffer, D. F. and Best, L. B., 1980. Habitat selection by birds of riparian communities: evaluating effects of habitat alterations. *J. Wildl. Man.* **44:** 1–15.

Terborgh, J., 1989. Where have all the birds gone? Princeton University Press: Princeton, New Jersey.

Terman, M. R., 1994. The promise of natural links. *Golf Course Management.* **December:** 52–59.

Thomas, J. W., Maser, C. and Rodiek, J. E., 1979. Riparian zones. Pp. 40–47 *in* Wildlife Habitats in Managed Forests: the Blue Mountains of Oregon and Washington ed by J. W. Thomas. U.S. Department of Agriculture: Forest Service, Pacific Northwest U.S. Forest and Range Experiment Station, Portland, OR.

Tomialoj_, L. and Profus, P., 1977. Comparative analysis of breeding bird communities in two parks of Wroclaw and in an adjacent Querco-Carpinetum forest. *Acta Ornithologica* **16:** 117–77.

Vale, T. R. and Vale, G. R., 1976. Suburban bird populations in west-central California. *J. Biogeog.* **3:** 157–65.

Verner, J., 1985. Assessment of counting techniques. *Current Ornithol.* **2:** 247–302.

Walcott, C. F., 1974. Changes in bird life in Cambridge, Massachusetts from 1860 to 1964. *Auk* **91:** 151–60.

Weber, W. C., 1975. Nest-sites of birds in residential areas of Vancouver, British Columbia. *The Can. Field–Nat.* **89:** 457–60.

Wiens, J. A. and Parker, K. R., 1995. Analyzing the effects of accidental environmental impacts: approaches and assumptions. *Ecol. App.* **5:** 1069–1083.

Wilcove, D. S., 1985. Nest predation in forest tracts and the decline of migratory songbirds. *Ecology* **66:** 1211–214.

Wuerthner, G., 1994. Subdivisions versus agriculture. *Cons. Biol.* **8:** 905–08.

CHAPTER 43

Uruamo Headland (North Shore City, Auckland): an urban remnant conserving ecological goods and services

M. P. Galbraith[1]

Ecosystem goods and services are the patterns and processes of natural ecosystems that provide life-support systems, "goods" for harvest, and aesthetic and cultural components that sustain human societies. A common global trend is the extensive modification and fragmentation of natural environments associated with settlement, with the subsequent loss or degradation of ecological goods and services. The Tamaki ecological district, centred in Auckland, New Zealand, is largely urban and reflects this global trend of urbanization, with less than 30% of the region's original indigenous terrestrial habitats remaining intact.

Within this ecological district, the Uruamo headland complex of North Shore City consists of approximately 240 hectares of vegetated land in an urban production environment, with tenure including local authority, government, private industrial and private residential. Although the headland is recognized as having regional ecological significance, only 30% of the vegetated land has legal protection. There is a risk that the remaining unprotected vegetation may be lost or fragmented through commercial or residential development, following the pattern of past urban growth.

There is potential for an integrated management scheme for the headland that recognizes and conserves the ecological goods and services it provides. This may represent a management model for similar urban situations. A description of the area, its regional significance and recommendations for restoration and management are discussed.

Key words: Ecological goods and services, Urban remnants, Urban restoration, Uruamo Headland.

INTRODUCTION

ECOLOGICAL services are the natural processes that sustain human societies (Costanza *et al.* 1997; Daily *et al.* 1997). They maintain biodiversity and the production of ecosystem "goods", the products that are harvested and traded as an integral part of the human economy such as food for humans and livestock, timber, biomass fuels, natural fibre, and many raw materials for pharmaceuticals and industry. Ecosystem services are also the actual life-support functions such as recycling and renewal, plus intangible aesthetic and cultural components of human societies. Environmental modification degrades and disrupts the provision of these goods and services.

In urban production environments, human society creates an ever-increasing demand for the very life-sustaining processes that are reduced through extreme environmental modification. In New Zealand, urban expansion is at the expense of highly productive wetlands, estuaries and fertile soils of floodplains (Taylor *et al.* 1997).

The rich and varied natural resources of the Auckland region have attracted human settlement for approximately 1 000 years (the Maori name for the area, *Tamaki-makau-rau*, land of a thousand lovers, reflects the high demand for land from which to access these rich resources). Urban encroachment of natural ecosystems has accelerated over the past five decades, and currently Auckland is both

[1]School of Environmental and Marine Sciences, University of Auckland, Private Bag 92019, Auckland, New Zealand.
Current address: Landscape and Plant Science, Unitec Institute of Technology, Private Bag 92025, Auckland, New Zealand.

Pages 440–47 *in* NATURE CONSERVATION 5: NATURE CONSERVATION IN PRODUCTION ENVIRONMENTS: MANAGING THE MATRIX ed by J. L. Craig, N. Mitchell and D. A. Saunders. Surrey Beatty & Sons, 2000.

the largest and fastest growing urban area in New Zealand (Taylor *et al.* 1997). The cummulative effects of human settlement has been the extensive modification and depletion of the natural resources. While urban development is spread over a number of ecological districts of the region, the Tamaki ecological district is centred on the city (see Fig. 1), and has thus been subjected to the extremes of environmental modification. The Auckland Regional Policy Statement (Auckland Regional Council 1995a) cites 92 local plant extinctions from the Tamaki ecological district as an illustration of continuing loss and degradation of the region's natural areas.

The depletion of ecological goods and services, is recognized in the Auckland Regional Policy Statement: "The heritage of the Auckland region has been depleted and continues to be under threat" (Auckland Regional Council 1995a).

The modification of the natural environment of the region includes loss of freshwater and saline wetlands, local extinctions and increasing rarity of many plant and animal species (that remain common elsewhere in the country), extensive fragmentation and isolation of the remaining remnants of natural environment, extensive modification of the region's coastal and freshwater environments, and deterioration of habitat quality in the region due to invasive introduced plant and animal species (Auckland Regional Council 1995a).

The extent of modification of the terrestrial ecosystems for the Tamaki ecological district is extensive, with wetlands suffering the most (Table 1). While figures on the loss of coastal habitats are not available, modification to the coastal environment is thought to be extensive (Auckland Regional Council 1995a). What remains is a matrix of ecosystem fragments of varying sizes embedded in an urban mix of industrial, commercial and suburban areas.

Table 1. Current terrestrial indigenous vegetation cover in the Tamaki ecological district, Auckland Region, expressed as a percentage of total land area of the ecological district (59 900 ha). Figures in brackets are estimated percentages of possible vegetation cover in pre-European times based on historical commentaries (after Auckland Regional Council 1995a.)

Forest	Scrub	Freshwater wetland	Total	Total protected
1.1 (<5)	5.2 (85)	0.03 (10)	6.3 (98–100)	1.8

With few natural ecosystems intact within this urban production environment, the demands for ecosystem goods and services are likely to outweigh their availability in most of the matrix. There is a need to protect and manage the on-going functioning of the ecosystem goods and services that remain. This is identified in the Regional Policy Statement (Auckland Regional Council 1995a), which recognizes the need for:

- preservation of the natural character of the coastal environment, wetlands, and lakes and rivers and their margins;
- protection of outstanding features and landscapes;
- protection of areas of significant indigenous vegetation and significant habitats of indigenous fauna;
- achieving integrated management of all heritage resources in the region.

A stated objective to "preserve or protect a diverse and representative range of the Auckland region's

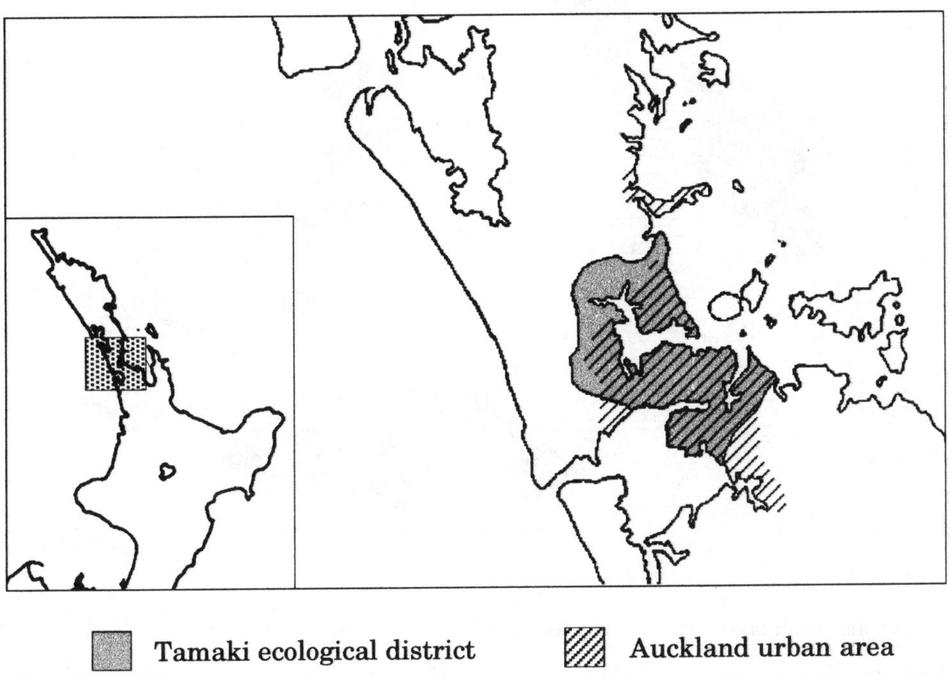

Fig. 1. Auckland region showing the Tamaki ecological district (McEwen 1987) and the extent of urbanization.

heritage resources" (ibid.) has difficulties, however, when there are limited sites representing the original habitats and resources.

Policies of urban authorities only partially recognize the importance of the natural environment, but, in reality, management action is failing to maintain even the status quo, including public reserves, and accords minimal responsibility on private owners to maintain the quality of their vegetated land. For example, the Regional Plant Pest Management Strategy (Auckland Regional Council 1997) classifies many of the weeds in reserves as "regional surveillance plant pests". This classification has the objective of arresting further spread of the plants, but confers "no requirement of the land occupiers to control established infestations" (ibid.).

Within the Tamaki Ecological district, the Birkenhead/Northcote Ward of North Shore City (Fig. 2) contains extensive areas of remnant vegetation, uncharacteristic of an urban situation. Only half of this vegetation, however, is protected under statutory legislation as small fragmented public recreational reserves. There are 39 bushed reserves with an average size of 5.4 ha, with the predominant ecosystem type being forested stream margins in valley floors where residential development is most difficult. The Uruamo Headland is an exception to this pattern. The headland consists of a section of the Waitemata Harbour coastline and hinterland that consists of approximately 240 ha of contiguous vegetation. This area includes protected reserves, special-purpose crown land and privately owned industrial and residential land (Fig. 2).

REGIONAL SIGNIFICANCE OF THE URUAMO HEADLAND

The regional significance of the Uruamo Headland was recognized by Campbell *et al.* (1982), and identified more recently to contain "terrestrial sites of natural significance" by the Auckland Regional Growth Forum (1997). The components encompass a variety of habitats and vegetation types that are representative of the ecological district, with the most significant features being the collective area (*c.* 240 ha) and the non-fragmented state of the vegetation.

The North Shore City District Scheme (North Shore City Council 1995) recognizes the following characteristics of the headland: "this area contains the richest diversity of habitats, and consequently biota, in North Shore City". This scheme classifies the headland as a "Site of Special Wildlife Interest" with a moderate conservation ranking. This ranking is given where "all habitats supporting good numbers of species which are typical of that

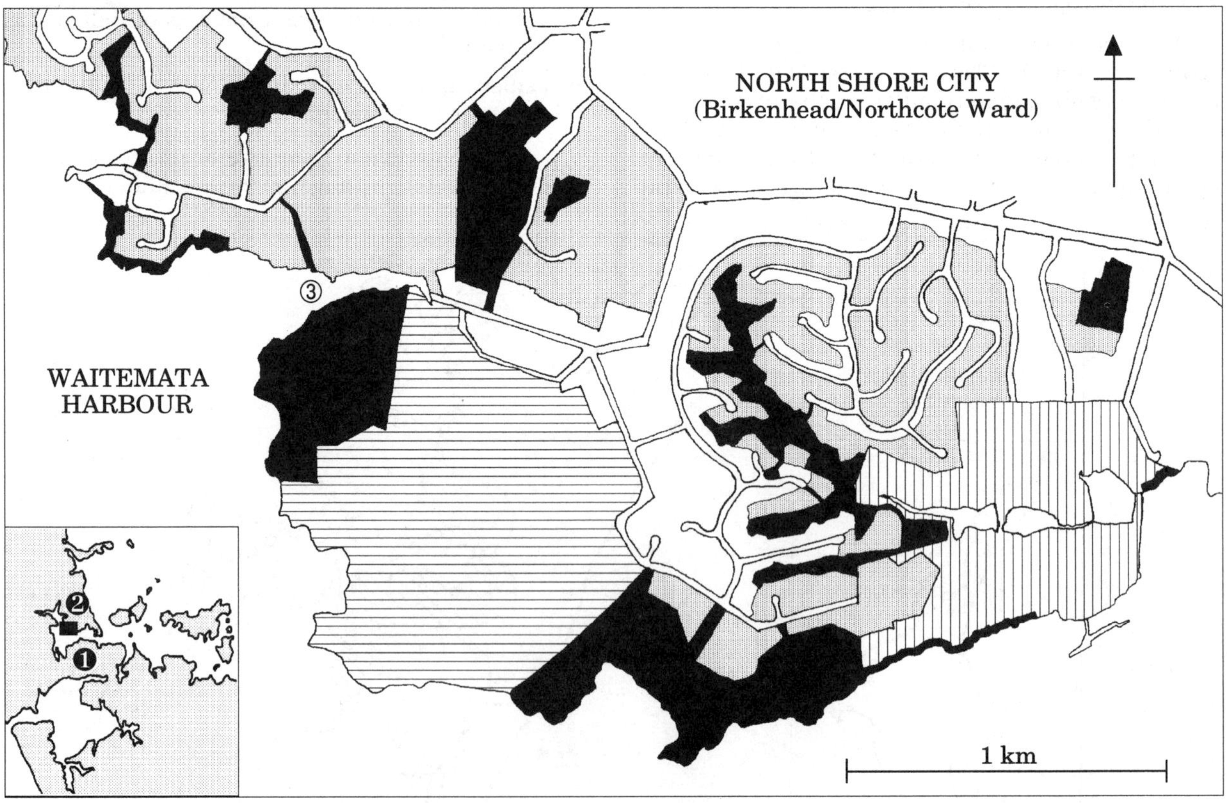

Fig. 2. Uruamo headland, North Shore City, showing vegetated land and the mosaic of tenure.

Key: ① Auckland City ② North Shore City ③ Soldiers' Bay ■ recreational reserve (protected) ⊟ defence, Royal New Zealand Navy Armament Depot ◫ Chelsea Estate ░ private — residential (natural heritage zoning) □ private — other zoning.

particular habitat within an ecological region and which have not been heavily modified by man's influence" (North Shore City 1995).

This substantial area of continuous vegetation provides a wide range of contiguous habitats, including harbour to ridge continuum and the complete successional seres from sea to land in a natural state, not normally found intact in an urban setting. The range of habitats is reflected by the 50 species of avifauna recorded within the headland or associated with its coastline, 31 of which are indigenous. This rich diversity of biota justifies the Site of Special Wildlife Interest classification of the District Scheme.

As an area of intact vegetation, the headland provides a range of ecosystem services lost or degraded in the remainder of this urbanized ecological district. These services (after Costanza *et al.* 1997) include the following.

Erosion control and sediment retention — the coastal margins of the headland are cliffs of Waitemata sandstone and classified as "high hazard: slopes unstable" (Auckland Regional Growth Forum 1997). The existing vegetation cover is providing essential protection to this soft sediment. This protection is also linked to the quality of harbour water and hence the quality of the associated fisheries. The presence of native fish (*Galaxias* spp.) in the streams, species that migrate between fresh and salt water habitats at some stage of their life-cycle, is an indicator of both high water quality and intact interfaces between natural water courses and the harbour (McDowall 1990; Hume 1993). The juvenile form of galaxids are taken in commercial and recreational fishing as whitebait.

Gas regulation — the presence of extensive vegetation, with potential to provide valuable carbon-sink and oxygen-source services, located centrally within urban sources of carbon emissions, is rare. A carbon-sink service of the vegetation of the headland can be inferred from carbon monoxide emission data published by the Auckland Regional Growth Forum (1997). This shows the headland within an emission zone of 10–40 tonnes per annum, but surrounded by emission zones of greater than 400 tonnes per annum.

Nutrient cycling — soil plays a key role in the regulation of major element cycles (Daily *et al.* 1997). The headland is providing an extensive range of interrelated soil services not available urban areas dominated by impervious ground cover and buildings.

Pollination — approximately 97% of flowering plants rely on faunal pollination services for successful fruit formation and/or seed production (Nablhan and Buchmann 1997). Most New Zealand forest plants depend on small flies, beetles and moths for pollination (Brockie 1992), with a native species of short-tongued bee (*Lassioglossum* sp.) that is present in the headland known to be an important pollinator of fruit trees (Miller 1971). Many of these species that provide a pollination service to urban gardens, including the introduced honey and bumble bees, require undisturbed nesting and foraging habitats adjacent.

Refugia — the headland provides a nursery or habitat for species for which the urban environment is unsuitable for viable populations. This is a source of biodiversity uncommon in an urban setting, and offers an opportunity to return native heritage to local residents. This includes stable breeding populations of many indigenous species which act as a reservoir for species that range into urban gardens but do not breed there. Species of specific interest that have been recorded within the headland are:

- forest gecko *Hoplodactylus granulatus* and peripatus *Peripatoides novaezealandiae* (Bergquist 1992);
- cave weta *Deincrenida* sp. (E. Hobbs, pers. comm.) and glow-worms *Arachnocampa luminosa*, usually associated by the public with limestone caves;
- North Island kaka *Nestor meridionalis septentrionalis* and the long-tailed cuckoo *Eudynamys taitensis* which have seasonal migration patterns through the Auckland isthmus. Both of these are dependent on extensive forested areas for food along the migration pathways;
- the fern *Loxsoma cunninghamii*, an endemic genus ("living fossil") associated with kauri forest and is a nationally restricted species (Given 1981), and found within at least one of the protected reserves (Norton 1990);
- two vegetation types that are rare locally. They are hard beech *Nothofagus truncata* and kauri *Agathis australis* association, a relict of a past cooler climate, and a succession stage in which the grass-tree *Dracophyllum sinclairii* is abundant;
- the sundew, *Drosera* sp. (Bergquist 1992), and a freshwater rush, *Eleocharis sphacelata* (Ogden 1981).

Recreation and cultural — through a network of tracks, 50% of the headland is accessible to the public (includes both public and private land). Observations indicate that there is wide use of these tracks (B. Norton, pers. comm.). The headland also has aesthetic value as a natural vista as viewed from the harbour and Auckland City. Campbell *et al.* (1982) refer to the area as "the most prominent tree clad headland in the Waitemata harbour". The scientific value of a tract of continuous vegetation of this magnitude close to the city also falls within this category.

STATUS AND DESCRIPTION OF THE HEADLAND COMPONENTS

The headland complex can be considered as having five components. The land tenure and area is shown in Table 2.

Table 2. Components of the Uruamo Headland and their land area.

Component	Approx. area
Recreational reserve *(protected)*	70 ha
Coastal reserve *(protected)*	3 ha
Crown, special purpose *(defence)*	80 ha
Chelsea Estate, private *(various zonings)*	52 ha
Residential, private *(Natural heritage zoning)*	35 ha
Total	240 ha

Recreational reserves

Ten recreational reserves are components of the headland. These reserves, administered by the North Shore City Council, include a wide variety of habitats and vegetation state ranging from mature Kauri forest to regenerating forest dominated by pines (*Pinus radiata* and *P. pinaster*) and other invasive exotics.

Management of the reserves also varies from an intense level administered by a community-based voluntary committee to a minimal level where the Council has sole responsibility. The minimal level appears to consist principally of grass mowing and rubbish removal, with little input into weed control. This contrasts significantly with the management by the community where the committee acts as in an advisory role to the Council, and volunteer input includes revegetation, weed control and track maintenance. None include control of herbivorous and predatory mammals.

At least 25 plant species designated as environmental weeds are established within the reserves. Despite a regional plant pest management strategy (Auckland Regional Council 1997), systematic weed control is carried out in two reserves only, and then solely by volunteers. A number of weeds pose a significant threat to indigenous vegetation because of their difficulty to control and their ease of dispersal by wind or birds. These are wandering jew *Tradescantia fluminensis*, climbing asparagus *Asparagus scandens*, pampas *Cortaderia* sp., kahili ginger *Hedychium gardnerianum* and Japanese honeysuckle *Lonicera japonica*. There is no obligation for these plants to be controlled under the Regional Plant Pest Management Strategy.

Where residential properties adjoin reserves, issues arise with respect to the reserve management. Problems include dumping of household rubbish and garden clippings into the reserve, gradual encroachment of clearing and cultivation into reserves, and opposition to vegetation management (B. Norton, pers. comm.).

There is potential for an education campaign to increase the awareness of residents neighbouring reserves of the importance of the management of their own properties as an integral part of the conservation matrix of an urban area.

Coastal reserve (Soldiers Bay)

Although Soldiers Bay is a relatively small estuarine inlet, the existence of a succession sequence from mangrove to terrestrial forest is of regional significance (Campbell *et al.* 1982). Bergquist (1992) describes the bay as being "of great significance as it demonstrates the full succession from mangrove sandflats, shell bank, salt marsh, through bulrush wetland to coastal forest". The estuary supports a one of the few remaining populations of banded rail *Rallus philippensis assimilis* in urban Auckland, a species that is under increasing pressure in urban situations due to habitat loss and predation from cats and dogs.

The estuary is listed as a regional Coastal Protection Area (Auckland Regional Council 1995b). This classification recognizes the importance of maintaining ecosystem services and biodiversity, preventing fragmentation, and the vulnerability to inappropriate development. Despite this recognition of significance, however, no management is carried out within the estuary and progressive damage from human use is evident. Sewage pollution in the bay is common, the result of a pumping station unable to carry fluid capacities at peak loading. Multiple tracks, makeshift boardwalks and an unsafe bridge are clear indicators of the impact of people, and invasive exotic plants (especially pampas) are encroaching into the estuarine vegetation.

Crown land, special purpose

This land serves as an armament depot for the Royal New Zealand Navy. The bulk of the land is vegetated and forms a buffer zone between the armament storage area and neighbouring public land and residential properties. A central portion of this land is fenced with very limited public access through a single entry point that has a constant security presence.

The Navy has initiated a revegetation programme following the removal of pines deemed to be a fire risk (G. Blue, pers. comm.). The remainder of the vegetation consists of indigenous podocarp forest or regeneration stages dominated by indigenous tea-tree (*Leptospermum scoparium* and *Kunzia ericoides*), and introduced *Hakea* and *Acacia* species. Other invasive exotics are present, but tend to be localized. Outside the security fence, the vegetation is continuous with two of the headland's receational reserves.

As a fenced vegetated area with restricted access, this land has potential for management as a mainland "island" (Saunders 1990), a concept recognized by some Navy personnel (B. Fotheringham, pers. comm.). This ecological management need not compromise the security requirements of an armament depot, and could introduce significant conservation value to the buffer zone surrounding the depot. A successful model for this management is Garden Island (Western Australia) administered by the Royal Australian Navy where the management vision is to conserve and protect the environmental values while "providing for defence use and public appreciation" (Garden Island Environmental Advisory Committee 1996).

Chelsea Estate (Chelsea Sugar Refinery)

This property contains New Zealand's sole sugar refinery established in 1884. The refinery covers *c.* 5 ha adjacent to a deep water wharf. Of the remaining 47 ha, most is vegetated with regenerating native forest that encircles four freshwater dams. There are pockets of old pines present, and extensive invasions of exotic weeds, of which minimal control is being undertaken (P. Simpson, pers. comm.). Some invasive exotic species now present were not recorded in a 1981 botanical survey of the property (Ogden 1981), and have altered community composition considerably. This spread of invasive plant pests and lack of their control questions claims by the company to have maintained a "high standard of stewardship" for the estate (Rooney 1997).

As a direct result of the extensive freshwater habitats, the refinery property contributes a significant level of biodiversity to the headland — 41 of the 50 bird species recorded within the headland matrix have been recorded from the Chelsea Estate. The bushed land provides a protection buffer around the dams, and also links two of the adjacent reserves. It is a critical factor in maintaining the continuity of vegetation across the headland. The largest of the freshwater dams has nesting colonies of pied and little shag (*Phalocrocorax varius* and *P. melanoleucos*). These colonies are expanding, in contrast to other colonies within the Tamaki ecological district that are contracting or collapsing (M. Taylor, pers. comm.).

Current zoning of this as yet undeveloped land includes reserve, industrial, commercial and residential options. The refinery management has indicated a desire to expand both commercial land use and residential land use within the vegetated and open spaces portion of the property (Rooney 1997). The Company has already initiated these changes, and has sought to have Site of Special Wildlife Interest status lifted from the land, enlarged commercial and residential zones, and industrial zoning placed over the freshwater dams. These management initiatives do not recognize the ecological value of the intact vegetated area. Any development would impact on the biota by reducing the total area of vegetated land and perpetuate the historical pattern of fragmentation of the natural environment.

A project in Portland (Victoria, Australia) may be a model for management of the the Chelsea Estate — this is the development of an aluminium smelter where the company, supported by employees and the local community, has successfully implemented the concept of "Smelter-in-the-Park" (Hill *et al.*, this volume). While the scale is different, many of the "park" components already exist at Chelsea. This approach could be marketed by the company to enhance their public image, and perhaps justify their claims of a "high standard of stewardship" of the land.

Extensive parts of this property would be better managed as reserve to ensure protection of the ecological services, and should be considered for purchase by the local or regional authority.

Private residential

Most of the residential land bordering the larger tracts of bush have a "Natural Heritage Protection" zoning. This zoning recognizes "good quality native bush in relatively undisturbed tracts", and minimizes vegetation disturbance by restricting the site area per residential unit (North Shore City 1995). This zoning has the effect of gradual grading of the vegetation between the undeveloped bush tracts and high density residential land, and significantly increases the area of vegetation providing ecological services.

There are areas of bush in private ownership on steep slopes abutting terrestrial and coastal reserves that should be considered for more definite protection (as reserves or through covenants) given their importance to the conservation matrix of the headland, and their role as a buffer between residential zones and sensitive ecological areas.

RESTORATION AND MANAGEMENT POTENTIAL

The current moderate Site of Special Wildlife Interest ranking of the headland is inappropriate as it reflects neither the regional importance of the headland nor the potential for ecological restoration. The headland has the potential to meet all of the objectives for management of Heritage Resources listed in the Regional Policy Statement (Auckland Regional Council 1995a) — protecting and enhancing biodiversity, public access and landscapes of the Auckland region. The potential for joint management of land is recognized as a means of gaining protection for land that is not public (Auckland Regional Council 1998), although this option has not yet been initiated for any area of regional parkland.

Management of invasive exotic plants is a critical restoration action for the area. The reserves in which weed control is already occuring will be under constant threat of reinvasion if neighbouring properties remain infested with weeds. Although weed control will be addressed ultimately when management plans are written for all reserves, the effectiveness of such control will be limited by the potential for invasion of weeds from private land. Hence it is essential that to consider all components of the matrix for effective management of this problem. Initiatives by local authorities are already raising public awareness of the threats and need for control of weeds on private property.

There is potential for the refugia service to be increased through reintroduction of indigenous birds. The Regional Policy Statement comments that at least eight bird species have been lost from the region over the past 150 years (Auckland Regional Council 1995a). Whitehead *Mohoua albicilla* and North Island tomtit *Petroica macrocephala toitoi* could establish viable populations without specific restoration requirements as they already co-exist with mammalian predators elsewhere on the mainland.

In addition to the large number of New Zealand islands where effective habitat restoration allows reintroduction to used extensively as a management tool for threatened species, restored mainland sites are sought increasingly for recovery programmes for appropriate species (Saunders 1994; A. Saunders, pers. comm.). Even limited control of mammalian predators may create an environment within the headland suitable for reintroduction of other indigenous species (T. Lovegrove, pers. comm.) — e.g., North Island weka *Gallirallus australis greyi*, North Island fernbird *Bowdleria punctata vealeae*, brown teal *Anas aucklandica chlorotis*, bellbird *Anthornis melanura melanura*, and yellow-crowned parakeet *Cyanoramphus auriceps auriceps*.

Mammalian predators present are rats, mice, brushtail possum *Trichosurus vulpecula* and feral cats. There are unconfirmed mustelid observations (B. Norton, pers. comm.), with descriptions suggesting the presence of ferret *Mustela furo*. Control of the brushtail possum is erratic and not systematic. There are no control measures undertaken for other predators. Dogs are also permitted on reserve land provided they are under the control of their owners, although most observations are of dogs not on a leash. The Council promotes a "responsible owners" policy for dogs, which could be extended to include cats to minimize their impact on indigenous fauna.

Since the headland is surrounded in its inland borders by residential land, there is limited opportunity for reinvasion of predators following control. Current evidence suggests that even limited predator control can be effective in improving an environment for sustainability of populations of indigenous birds (T. Lovegrove, pers. comm.). The management of the defence land as a mainland island located centrally within the headland would provide a secure "reservoir" for these species to disperse out into the surrounding reserves.

Community involvement in restoration projects is now widespread. With a residential pool of potential volunteers within easy access of all components of the headland, initiating a community-based restoration programme should not be difficult. Advocacy generated through species reintroductions may stimulate increased community interest and involvement in management and restoration of the area as has occurred elsewhere (Craig *et al.* 1995).

CONCLUSIONS

Ecological goods and services are the patterns and processes we see in natural ecosystems, but with the dependence of human societies on them emphasized. In the extensively modified environments of cities, provision of ecological goods and services is degraded and fragmented. Remnants of natural environments within an urban matrix warrant recognition for the critical role they have in sustaining city communities.

The Uruamo Headland in Auckland is an example of an extensive urban remnant, and its regional environmental significance is well documented. The size of the vegetated tracts and the ecosystem services provided (including biodiversity) are unsurpassed in an urban situation in New Zealand.

Two reserves have comprehensive management and there is a statutory obligation for the other reserves to have management plans. Community involvement is already in place, and there is considerable potential for this to be extended. The Royal New Zealand Navy acknowledges the ecological value of their land, and will consider proposals for further restoration. The New Zealand Sugar Company has indicated that it may consider a voluntary "code of management" of land for ecological goals.

While an ideal management strategy may be the protection of more land as reserve, this requires both the allocation of funds by urban authorities to purchase land, and the willingness of private owners to sell land. Without such protection, however, there is still potential for integrated management (through covenants or joint management) of the components of the headland under common goals to meet the natural heritage objectives of the Regional Policy Statement, and to prevent further fragmentation and loss of bushed areas and their associated ecological services. Such management need not compromise the needs of individual owners, and will be essential if ecological services are to be maintained within the urban matrix as proposals to develop extensive portions of the private land are considered.

There is considerable potential for the ecological restoration of the headland to be intitiated as a community-based project, with resulting advocacy not only for the regional environment, but for conservation of urban remnant ecosystems in general.

ACKNOWLEDGEMENTS

I wish to acknowledge the Science and Technology Teacher Fellowship Scheme administered by the Royal Society of New Zealand, and my hosts under the scheme, Prof. John Craig and Dr Neil Mitchell (School of Environmental and Marine Sciences, University of Auckland) for their assistance in researching and preparing this chapter.

In addition, I wish to thank the following for their assistance in collating information about the Uruamo Headland: Sir Hugh Kawharu (Ngati Whatua Trust Board); Dr John Ogden (School of Environmental and Marine Sciences, University of Auckland); Dr Denis Saunders (CSIRO, Canberra); Shona Myers and Dr Tim Lovegrove (Auckland Regional Council); Alan Saunders (Department of Conservation); Richard Hollier, Jane Aitken and Dr Carol Bergquist (North Shore City Council); Lt Cdr Brett Fotheringham and Graham Blue (Royal New Zealand Navy); Peter Simpson (NZ Sugar Refinery); Eric Hobbs (Northcote/Birkenhead Community

Board) and Brian Norton (Kauri Point Centennial Park and Chatswood Reserve Management Committee).

REFERENCES

Auckland Regional Council, 1995a. Proposed Auckland Regional Policy Statement. Auckland Regional Council: Auckland, New Zealand.

Auckland Regional Council, 1995b. Proposed Auckland Regional Plan: Coastal. Auckland Regional Council: Auckland, New Zealand.

Auckland Regional Council, 1997. Auckland Regional Plant Pest Management Strategy. Auckland Regional Council: Auckland, New Zealand.

Auckland Regional Council, 1998. Auckland Regional Parks Land and Recreation Access Strategy. Auckland Regional Council: Auckland, New Zealand.

Auckland Regional Growth Forum, 1997. Natural and physical resource constraints. Auckland Regional Growth Forum: Auckland, New Zealand.

Bergquist, C., 1992. Areas of significant indigenous vegetation and significant habitats of indigenous fauna in North Shore City. Unpublished report, North Shore City Council.

Brockie, R., 1992. A living New Zealand forest. Bateman: Auckland, New Zealand.

Campbell, G., Mann, B. and Muir, M., 1982. The Natural Environment of the Auckland Region: a bibliography and inventory. Auckland Regional Authority: Auckland, New Zealand.

Costanza, R., d'Arge, R., de Groot, R., Farber, S., Grasso, M., Hannon, B., Limburg, K., Naeem, S., O'Neill, R. V., Paruelo, J., Raskin, R. G., Sutton, P. and van den Belt, M., 1997. The value of the world's ecosystem services and natural capital. *Nature* **387:** 253–60.

Craig, J., Mitchell, N., Walter, B., Walter, R., Galbraith, M. and Chalmers, G., 1995. Involving people in the restoration of a degraded island: Tiritiri Matangi Island. Pp. 534–41 *in* Nature Conservation 4: the Role of Networks ed by D. A. Saunders, J. L. Craig and E. M. Mattiske. Surrey Beatty & Sons: Chipping Norton, New South Wales.

Daily, G. C., Matson, P. A. and Vitousek, P. M., 1997. Ecosystem services supplied by soil. Pp. 113–32 *in* Nature's Services: Societal Dependence on Natural Ecosystems ed by G. C. Daily. Island Press: Washington.

Garden Island Environmental Advisory Committee, 1996. Report of activities 1995–1996. HMAS Stirling: Rockingham, Western Australia.

Given, D. R., 1981. Rare and endangered plants of New Zealand. Reed: Wellington, New Zealand.

Hume, K., 1993. 'bait!. *New Zealand Geographic* **17:** 50–68.

McDowall, R. M., 1990. New Zealand Freshwater Fishes: a natural history and guide. Heinemann: Auckland, New Zealand.

McEwen, W. M., 1987. Ecological Regions and Districts of New Zealand. NZ Biological Reseources Centre Publication No. 5. Department of Conservation: Wellington, New Zealand.

Miller, D., 1971. Common insects in New Zealand. Reed: Wellington, New Zealand.

Nabhan, G. P. and Buchmann, S. L., 1997. Services provided by pollinators. Pp. 133–50 *in* Nature's Services: Societal Dependence on Natural Ecosystems ed by G. C. Daily. Island Press: Washington, New Zealand.

North Shore City Council, 1995. North Shore City Proposed District Plan.

Norton, D. A., 1990. Preliminary report on the vegetation and flora of Chatswood Reserve, North Shore City. Unpublished report: School of Forestry, University of Canterbury, New Zealand.

Ogden, J., 1981. Report on a botanical survey of the vegetation of the Chelsea Estate. Unpublished report: University of Auckland, New Zealand.

Rooney, E., 1997. Rezoning plan for park land. *North Shore Times Advertiser* 4 December 1997.

Saunders, A., 1990. Mapara: island magament "mainland" style. Pp. 147–49 *in* Ecological Restoration of New Zealand islands ed by D. R. Towns, C. H. Daunherty and I. A. E. Atkinson. Department of Conservation: Wellington, New Zealand.

Saunders, A., 1994. Translocations in New Zealand: an overview. Pp. 43–46 *in* Reintroduction Biology of Australian and New Zealand Fauna ed by M. Serena. Surrey Beatty & Sons: Chipping Norton, New South Wales.

Taylor, R., Smith, I., Cochrane, P., Stephenson, B. and Gibbs, N., 1997. The State of New Zealand's Environment. Ministry for the Environment: Wellington, New Zealand.

CHAPTER 44

Minimizing impacts of mosquito management in ecologically sensitive urban areas: a multi-scale approach

A. Muhar[1,2] and P. E. R. Dale[1]

Mosquito-borne diseases are increasing in coastal regions of Australia. In south-east Queensland, more than 5 000 cases of Ross River virus infections were reported for the years 1990 to 1996. Although traditional mosquito management methods such as spraying of pesticides or habitat alteration have been further developed recently to be more target-specific, they may still cause a broad range of side-effects on ecosystems. This chapter presents a multi-scale approach to mosquito management: based on a risk analysis for individual areas, a strategy has to be developed, which integrates measures from the household level up to the town planning level. Particular emphasis is given to the integration of mosquito risk in planning decisions (e.g., zoning of residential areas) in order to avoid the need for intrusive management measures in ecologically sensitive areas.

INTRODUCTION

FROM the perception of many conservationists, mosquito management is often associated with spraying of pesticides or draining of wetlands and therefore considered as an adverse human intervention in the landscape. On the other hand, there is a need to manage mosquito populations which are vectors of disease and which are close to human populations. This chapter deals with the environmental planning aspects of mosquito management in Australia, where more than 90% of the human population live in coastal areas. European settlements were usually founded around sheltered harbours at river estuaries or calm water bays protected from the open seas by islands.

The choice of these sites for settlement by the early colonists created the need for mosquito management: extensive coastal intertidal areas are usually associated with mangrove forests, salt marshes, and tidal mudflats, which all favour mosquito breeding, and the freshwater rivers, whose estuaries form the natural harbours are surrounded by swamps and Melaleuca forests, which provide favourable habitats for freshwater mosquitoes. As more and more people moved closer to the natural mosquito habitats, mosquitoes have been regarded not only as a nuisance but also as a threat to public health.

The geographic scope of this chapter is south-east Queensland, the planning issues relate to Queensland legislation; nevertheless it can be assumed that similar problems can be found in other regions of coastal Australia.

RELEVANCE OF MOSQUITO-BORNE DISEASES IN SOUTH-EAST QUEENSLAND

In south-east Queensland, a number of arthropod borne viruses (arboviruses) can cause diseases. Each virus is usually associated with a limited number of mosquito species, which themselves also have restricted ranges. If mosquito management aims at disease control rather than the treatment of a nuisance, a detailed knowledge of the ecology of the vector species is essential.

Ross River virus

The most prevalent mosquito borne disease in south-east Queensland is epidemic polyarthritis

[1]School of Environmental Planning, Faculty of Environmental Sciences, Griffith University Brisbane, Nathan, Queensland, Australia 4111.
[2]Institut für Freiraumgestaltung und Landschaftspflege, Universität für Bodenkultur Wien, Peter Jordan-Straße 82, A-1190 Vienna, Austria.
Pages 448–55 *in* NATURE CONSERVATION 5: NATURE CONSERVATION IN PRODUCTION ENVIRONMENTS: MANAGING THE MATRIX ed by J. L. Craig, N. Mitchell and D. A. Saunders. Surrey Beatty & Sons, 2000.

caused by the Ross River virus, commonly called Ross River Fever (Curran *et al.* 1997). In the years 1991 to 1996, more than 5 000 cases were recorded. Ross River fever is not a life-threatening disease, although it can cause sickness lasting for many months, thus generating enormous social costs (Boughton 1994).

Vector species and their breeding habitats: In south-east Queensland, Ross River virus is primarily transmitted by two mosquito species with distinctive habitats, the saltwater species *Aedes vigilax*, inhabiting saltmarshes, tidal flats, and mangroves, and the freshwater species *Culex annulirostris*, which can be found in freshwater wetlands like Melaleuca forests. Other mosquito species such as *Aedes funereus* and *Aedes procax* have been found to carry the virus, although their significance as vectors is still unclear. Also, local and seasonal variations have to be considered (Russell 1993; Ryan and Kay 1997; Ryan *et al.* 1997). Mammals play an important role in the transmission process: as opposed to other "classic" mosquito borne diseases such as Malaria, it is believed that there is little or no direct transmission of Ross River virus from human to mosquito to human, i.e., there seems to be a need for intermediate hosts. Marsupials are thought to be important intermediate hosts, and the role of other mammal species (e.g., bats, see Ryan *et al.* 1997) as well as birds is still to be investigated in detail. Regarding the close relationship between the virus, the vectors and the intermediate hosts, Ross River fever can be seen as a "landscape disease".

Spatial distribution of infections: Ross River fever is a notifiable disease, therefore address data of cases have to be reported to the Queensland Health Department. This detailed database has been used to determine differences in the infection density in individual suburbs of Brisbane (Dale *et al.* 1998). Figure 1. shows the great variability of infection densities and their relation to major wetland areas. Although flight ranges of up to 40 km have been reported for the saltmarsh mosquito *Aedes vigilax* (Russell 1993), the data illustrated in Figure 1 suggest that there is a very close correlation between the location of cases and the proximity to wetlands.

Other mosquito borne infections

Barmah Forest virus: In the early 1970s another arbovirus from the same group as RRv was detected. This Barmah Forest virus causes symptoms similar to Ross River fever, though usually less severe. As with Ross River virus, the main vector species are *Aedes vigilax* and *Culex annulirostris*, with many other species seen as potential vectors (Hawkes *et al.* 1987).

Barmah Forest virus infections are probably less frequent than Ross River virus (Mackenzie and Smith 1996), yet the number of infections is difficult to determine, as before 1995 it was not listed as a

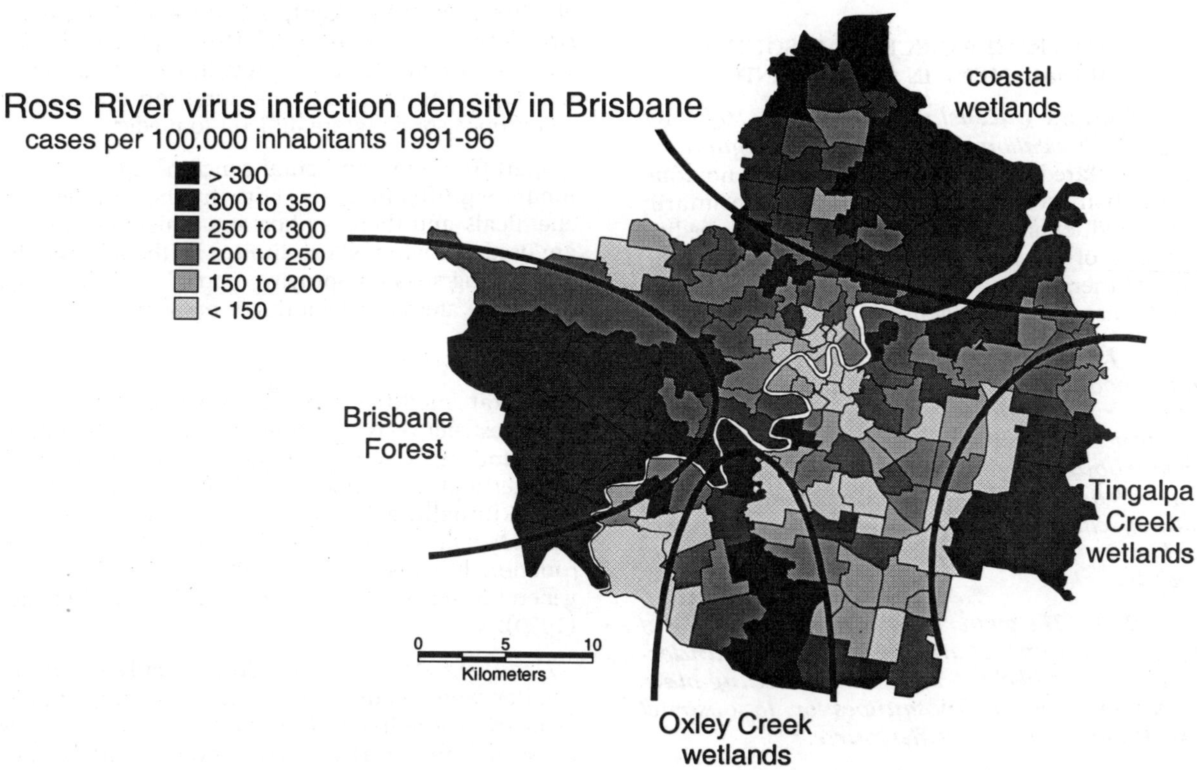

Fig. 1. Spatial distribution of RRv in Brisbane (Dale *et al.* 1998). (2 851 geocodable cases, taken from Queensland Health records. The total number of cases was approx. 50% higher, as many cases could not be assigned to geographic locations.)

separate arbovirus category in the records of the National Notifiable Diseases Surveillance system (Curran *et al.* 1997).

Dengue virus: Dengue fever is a potentially serious disease, which has caused many epidemics in tropical and subtropical Australia. Dengue fever is currently restricted to northern Queensland, although there were outbreaks in south-east Queensland earlier in this century (Mackenzie *et al.* 1994). The vector ecology of Dengue is completely different from Ross River virus: It is transmitted by *Aedes aegypti*, a relatively immobile species that usually disperses less than 400 m. *Aedes aegypti* can breed in artificial containers e.g., water tanks, tyres, undrained flower pots. The Dengue virus can be transmitted directly from human to mosquito to human, which can result in major epidemic outbreaks.

Arbovirus encephalitis: There are many different viruses which can cause severe symptoms of encephalitis, e.g., Murray Valley Encephalitis virus or Kunjin virus, quite often with fatal consequences (Boughton 1994; Mackenzie *et al.* 1994). Infections do not occur very frequently, although the vector species (e.g., *Culex annulirostris*, see Russell 1993) are present in south-east Queensland.

Malaria: Malaria is caused by *Plasmodium* species and transmitted by various mosquito species from the *Anopheles* genus. Although Australia is now regarded as Malaria-free (Russell 1993), there is still a risk of re-introduction from other tropical countries.

LEGAL FRAMEWORK FOR MOSQUITO MANAGEMENT IN QUEENSLAND

Based on the *Queensland Health Act 1937*, the *Mosquito Prevention and Destruction Regulations 1982* declared all mosquito species as noxious. Although most of the regulations aim primarily at the reduction of domestic mosquito breeding (covering of water tanks, emptying of buckets etc.), some of them could have a severe impact on urban ecosystems, if applied literally:

10 (1): "Where there is on any premises any pond, pool, swamp, or other accumulation of water or other liquid, whether permanent or not, which is likely to serve as a breeding-place or harbourage for mosquitoes if it is not drained or filled in, the owner of the premises shall effectively drain or fully fill in such pond, pool, swamp or other accumulation of water of other liquid."

12 (1b): "The owner of any premises shall cut down and remove any vegetation or undergrowth that is likely to serve as a breeding-place or harbourage for mosquitoes or to obstruct any gutter, drain, or watercourse."

It is the responsibility of local governments to ensure that the regulations are enforced. In practice this usually means that the local authority itself undertakes mosquito control, at least insofar as killing mosquitoes or modifying their habitat is concerned.

There are further considerations in applying these regulations. In marine parks *(Marine Park Act, 1982)* and areas under the jurisdiction of *the Fisheries Act (1994)*, which latter includes all intertidal lands, mosquito control activities require a permit and may require an Environmental Impact Statement. In areas protected under the *Nature Conservation Act (1992)* mosquito control is only allowed if a serious threat to human health is shown and the environmental effects are not significant. Further, and this has not been tested, complying with the requirements of the regulations could lead to severe environmental harm and result in prosecution under *the Environmental Protection Act (1994)*.

MOSQUITO MANAGEMENT METHODS

Current mosquito management practices aim primarily at the reduction of mosquito breeding activities, either by killing the larvae (larviciding) or by modification of breeding habitats. Spraying of adult populations (adulticiding) usually occurs only as an emergency measure after severe outbreaks.

Larviciding

Direct larviciding involves spraying of pesticides either from the ground or from helicopters. One method is to apply insect growth regulators which are synthetic hormonal agents preventing larvae from developing into functioning adults. A biochemical control agent, *Bacillus thuringiensis var. israelensis,* is also used. Organophosphates are seen to have greater non-target impacts than these, although in any case the long-term effects of continuous spraying are difficult to predict.

Apart from environmental concerns, spraying also incurs significant costs, both in the purchase of chemicals and their application. Unless helicopters are used, spraying is also limited by the accessibility of breeding sites for spraying from the ground, e.g., mangrove areas can be a major problem.

Habitat modification

Habitat modification can take several forms. Most destructive is ditching to drain a wetland, or impounding to create a permanent water body. In intertidal areas Open Marsh Water Management and Runnelling, which is a minimal form of Open Marsh Water Management, maintain marsh function but also reduce mosquito breeding. The general impacts were reviewed in Dale and Hulsman (1990).

Draining: Mosquito breeding usually occurs in smaller water bodies rather than in larger ponds or lakes, where there are more predators as well as physical limitations to larval development (waves, wind). Over centuries, draining of wetlands has been applied in the Western world to reduce mosquito populations and to eliminate mosquito-borne diseases such as Malaria. On the other hand, the environmental impact of draining on wetland ecosystems is devastating (Dale and Hulsman 1990). Apart from the generally known effects on biodiversity, in the coastal areas of Queensland this type of habitat modification can cause other severe side-effects: acid sulphate soils, once drained, release large amounts of acid into the drainage system, thus causing extensive fish killings and also damaging concrete foundations of buildings and infrastructure (Soukup and Portnoy 1986; Saffigna *et al.* 1997).

Impounding: This method is mainly used in intertidal wetlands to reduce populations of mosquitoes from the *Aedes* genus, which lay eggs on damp earth. It is destructive of wetland function and causes great changes to the ecosystem. A modified form of impoundment called Rotational Impoundment Management maintains some tidal exchange and is less destructive than permanent impoundment (Dale and Hulsman 1990; Carlson and Carroll 1985).

Open Marsh Water Management: Developed in the USA in the 1960s Open Marsh Water Management aims mainly at restoring the tidal exchange to previously ditched marshes (Ferrigno and Jobbins 1968; Meredith *et al.* 1985). Its impacts have recently been reviewed and summarized by Wolfe (1996). One pilot project is under way in Queensland (Dale, unpubl. data). In general runnelling, described below, is preferred to Open Marsh Water Management whenever practicable, as its environmental impacts appear to be less severe.

Runnelling: For mosquito breeding sites in tidally influenced areas (mangroves, saltmarshes) a more environmentally sensitive habitat modification method has been developed. Small, shallow ditches are created to connect tidal ponds to the open water. As opposed to draining, runnelling does not aim to permanently lower the water table. The main purpose of runnels is to flush larvae into open water after high tide and to provide access for predators (Hulsman *et al.* 1989).

Runnelling has been applied in many different coastal regions of Australia, and monitoring of runnelled sites has shown that side effects on other species are much smaller than those caused by draining (Dale *et al.* 1993). It is supported as an alternative to the use of chemicals by the Queensland Department of Primary Industries, which encourages the phasing out of pesticides in intertidal wetlands (Department of Primary Industries 1996). Nevertheless it is a modification of an ecosystem and the impacts during the construction process itself cannot be neglected.

Other control methods: Many other methods have been suggested, e.g., stocking of ponds with predatory fish or release of large numbers of sterilized males. In practice, with the exception of some specific applications, these methods have not proved to be successful. The introduction of the "mosquitofish" *Gambusia affinis* to Australia has had severe consequences for the native fish fauna (Dale and Hulsman 1990).

PROBLEMS OF MOSQUITO MANAGEMENT IN EXPANDING URBAN AREAS

As human populations grow it is important that decisions about settlement design and location are also based on public health related issues such as mosquito breeding. To minimize disease and discomfort it is necessary to direct actions towards not only control of the pest, but also control of human activities (Dale 1993). The following discussion focuses on the major issues.

Organization of vector control

Mosquito management is organized at the local level, many local governments in coastal Queensland having their own vector control units. In the administrative hierarchy of councils, these units are usually linked to public health or works departments, and not to town planning or conservation. As mosquito control officers have to handle large volumes of potentially hazardous pesticides, vector control units are usually physically separated from other council administration departments, reducing chances for informal contact.

As a consequence of this organizational structure, there is little awareness of mosquito control issues in other departments, and vector control staff are usually not parties to any town planning or zoning decision that affects mosquito management issues. Thus mosquito management in most local governments of south-east Queensland is mainly reactive rather than proactive. The following paragraphs illustrate this deficiency using practical examples from different local governments:

Residential developments near breeding sites in natural open spaces

South-east Queensland, in particular the coastal strip between Brisbane and the Gold Coast, is one of the fastest growing regions in Australia. Residential development takes place primarily in former agricultural land close to the coast. In these areas there are many remnants of natural bushland, swamps and mangrove forests. Even if these open spaces are preserved, they can be heavily affected indirectly by urbanization. New residential areas are quite often up-market estates, where buyers expect a high-quality lifestyle without mosquito nuisance. The traditional rural population is much more used to living with mosquitoes (which does not prevent it from being infected), and therefore does not complain at the council administrations as often as the newer residents. As land developers and real estate agents do not want to lose their clientele, they impose political pressure on local governments to increase mosquito control in adjacent natural areas.

Accessibility of natural areas for conventional treatment

Conventional mosquito control by spraying from the ground relies on accessibility. During the peak breeding season in summer, potential sites have to be checked and treated, if necessary, at intervals of no more than five days. This is only feasible if these sites can be accessed by some motorized vehicle, e.g., all-terrain motorbikes. Residential development close to natural mosquito breeding sites can therefore result in an increased need to build access roads.

Many mosquito breeding sites in coastal areas are linked to mangrove forests, where vehicular access is practically impossible. In these cases major modifications of the landscape may be necessary to ensure acceptable levels of mosquito control in adjacent residential areas.

Agriculture and mosquito management

Grazing land in tidally influenced areas of south-east Queensland can be significant for mosquito breeding, in particular when there has been some drainage work and these drains are not maintained regularly. Even hoofprints serve as temporary breeding sites. As they are ephemeral structures, monitoring is very time-consuming.

In other parts of Australia, flood irrigation practices cause intensive mosquito breeding, in particular of *Aedes* spp., with all the associated arbovirus risks (Dhileepan 1997).

Land uses sensitive to mosquito management

Prawn farming is a growing industry in south-east Queensland. Prawn farms are usually located along the coastline or near tidally influenced rivers. These areas also require mosquito management, in particular with regard to saltmarsh mosquitoes. As prawns are extremely sensitive to some of the pesticides used in mosquito control (e.g., organo-phosphates), the establishment of new prawn farms can have heavy impacts on mosquito management, requiring a change to more expensive products or a move from spraying to habitat modification. Again, mosquito management units are rarely involved in decisions about such developments.

FRAMEWORK FOR INTEGRATED MOSQUITO MANAGEMENT

If mosquito management is to be seen as more than just a treatment of problem areas, with all the potential of serious impacts on the environment, it is necessary to develop a strategic framework for the integration of mosquito management into the decision making processes which steer the development of a landscape. The objectives of such a strategy at the local government level would be:

- to secure a long-term perspective for public health and quality of life
- to prioritize prevention measures (e.g., avoidance of artificial breeding sites) over treatment measures (e.g., spraying)
- to promote public awareness of mosquito management issues
- to improve the formal and informal communication between vector control units and other local authority units
- to minimize impacts on natural and anthropogenic ecosystems.

The implementation of these objectives will vary between local government areas, nevertheless some general aspects should be regarded in the development of such a strategy:

Risk assessment for individual areas

Before management measures can be specified, the infection risk has to be assessed. Although many components of the transmission process are still unknown (e.g., the role of intermediate hosts), the probability of the interaction between vectors and recipients can generally be defined by the probability of their coincidence (see Fig. 2). The probability of an infection in humans is highest when there are many people at the same time in the same location as infected adult female mosquitoes.

In many cases, the necessary base data are not available. This refers in particular to information about ephemeral mosquito breeding sites and intermediate hosts. In this case, inferences can be made from available data with some indicator function for mosquito management, e.g., vegetation maps (Barnes *et al.* 1997).

New remote sensing methods, e.g., the use of Synthetic Aperture Radar sensors have opened new perspectives for the detection of mosquito breeding sites under forest canopy (Dale *et al.*

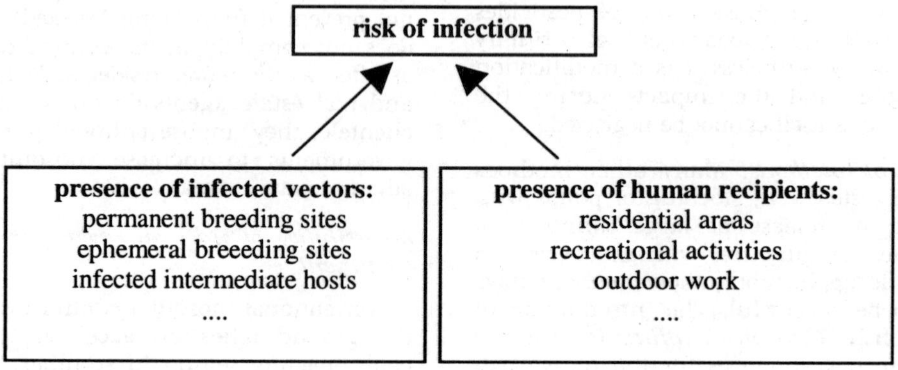

Fig. 2. Determinants of vector risk (adapted from Barnes *et al.* 1997).

1998). Using Geographic Information Systems technology, the various layers of information can be aggregated into a risk model.

Integration of mosquito management into town planning

The most important aspect of integrated mosquito management is the integration of vector control aspects into urban planning at the local government level. This is where most of the decisions for future development of an area are made and where mosquito management issues have to be considered equally with other public interests such as the regional economy, tourism and conservation.

This integration has to be carried out both on a formal and on an informal level. Formally, mosquito management experts should be involved in all relevant town planning projects (zoning plans, local area plans, development control plans, strategic plans etc.) as well as in individual development approval processes. This would also require more training for mosquito control officers to understand their new role. Depending on the situation, this formal integration could be achieved by reassignment of the mosquito control unit within local government administration. On the other hand, there is also a need for better informal contact between vector control unit staff and staff from other units to overcome the usual physical separation described above.

Integration of mosquito management into urban open space planning

Some goals of urban open space planning can conflict with integrated mosquito management. In order to avoid these conflicts, co-operation at an early stage is essential, e.g., in the preparation of wetland management strategies or bushland conservation strategies.

Traditionally, open space planning aims at linking natural areas to urban ecosystems via corridors. These corridors are quite often formed by rivers and creeks or by dense vegetation. Although very little is known about the function of these corridors for the migration of mosquitoes from breeding sites into urban areas, some authors recommend the breaking up of continuous vegetation corridors in order to inhibit this migration (Tweed Shire Council 1993). This is an area where more research work is required to find better understanding of mitigation measures and possible compromises.

Education

Educational programmes play a key role in the promotion of public health issues. In the past decade, awareness of arbovirus risks has increased in the general public, yet there is still a need for the introduction of educational programmes both at schools and at workplaces with special arbovirus infection risk, e.g., construction sites, national parks etc.

Regional co-operation

Mosquitoes do not respect local government boundaries, therefore local governments need to co-ordinate their management activities. In some parts of Australia, Contiguous Local Authorities Groups have been formed to co-operate in mosquito control at a regional scale, which can include joint operation of helicopters, funding of research projects and development of information material (Bell 1989). While this co-operation between various vector control units is usually very successful, little integration exists between Contiguous Local Authorities Groups and other regional planning activities, e.g., strategic plans for tourism or residential development.

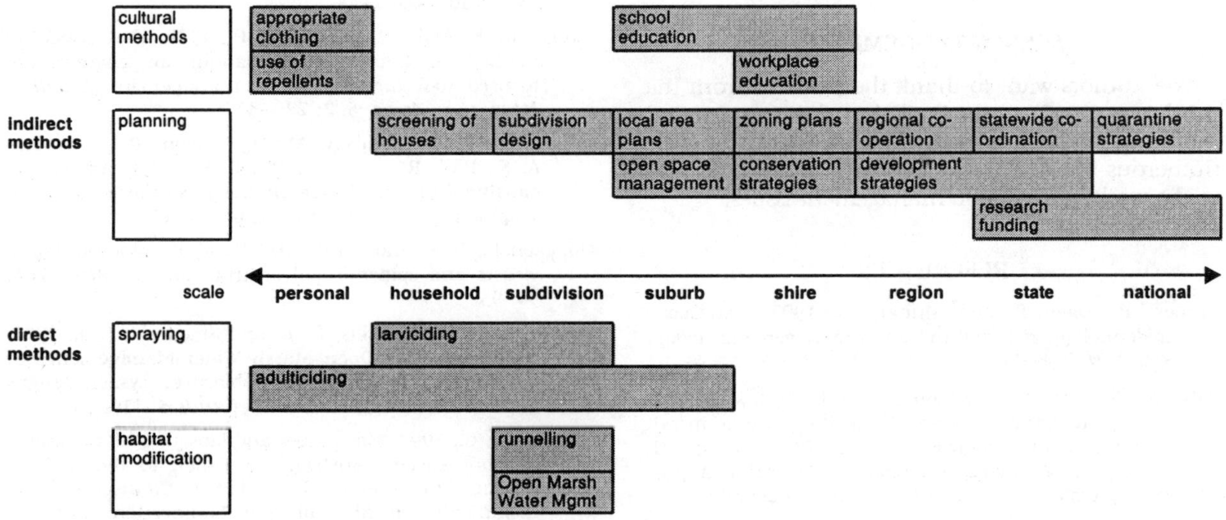

Fig. 3. Framework for the integration of mosquito management at different scale levels.

Integration of mosquito management at different scale levels

As demonstrated in the previous paragraphs, issues of mosquito management cannot be solved without co-ordination with other disciplines. There is also a need for integration and co-ordination between different scale levels. Development decisions at regional level can have severe impacts on mosquito management at the local level. On the other hand, the risk of infection can be minimized by measures taken at the household level, e.g., provision of mosquito screens. Practical activities of vector control units are usually carried out at the level of subdivisions, however, measures at other scale levels are necessary to complement these methods and, ideally, to reduce the need for direct mosquito control as much as possible. The responsibility for these measures can be found at any scale level from the individual person up to the national government (Fig. 3).

CONCLUSIONS

This chapter has illustrated, with reference to subtropical Queensland, how an integrated multi-scale approach to mosquito management could potentially reduce the risk of disease to humans whilst also minimizing adverse impacts on the environment. This is achieved in two ways. One is to minimize human-mosquito contact at local to regional scales by careful design and location of human settlements and activities, as well as, at the individual level, encouraging personal protection. The second is to manage the mosquito populations. This is usually carried out at a local level, with the preferred method being minor modification of the breeding habitat for long term effectiveness, supplemented by spraying with highly specific agents when necessary.

If authorities can implement integrated practices such as described, conflicts between mosquito management and conservation would be minimized, as they would be persuing overlapping goals.

ACKNOWLEDGEMENTS

The authors want to thank the students from the third year (1997) in the Bachelor of Environmental Planning Programme at Griffith University for their numerous inputs during classroom discussions and studio work on mosquito management issues.

REFERENCES

Barnes, P., Dale, P. and Muhar, A., 1997. Landscape epidemiology, risk and mosquito management. *Arbov. Res. Aust.* **7**: 8–11.

Bell, K. M., 1989. Development and Review of the Contiguous Local Authority Group Programme on salt marsh mosquito control. Pp. 168–70 *in*. Proc. 5th Symposium Arbovirus Research in Australia, August 28–September 1 1989. CSIRO: Brisbane, Australia.

Boughton, C. R., 1994. Arboviruses and disease in Australia. *Med. J. Aust.* **160**: 27–28.

Carlson, D. B. and Carroll, J. D., 1985. Developing and implementing impoundment management methods benefiting mosquito control, fish and wildlife: a two year progress report about the Technical Subcommittee on Mosquito Impoundments. *J. Flor. Anti-Mosq. Assoc.* **56**: 24–32.

Curran, M., Harvey, B., Crerar, S., Oliver, G., D'Souza, R., Myint, H., Rann, C. and Andrews, R., 1997. Australia's notifiable diseases status 1996. Annual report of the National Notifiable Diseases Surveillance system. *Communic. Dis. Intell. (Aust.)* **21**: 281–307.

Dale, P. E. R., 1993. Australian Wetlands and Mosquito Control — Contain the Pest and Sustain the Environment. *Wetl. Aust.* **12**: 1–12.

Dale, P. E. R. and Hulsman, K., 1990. A critical review of salt marsh management methods for mosquito control. *Aqua. Sci.* **3**: 281–311.

Dale, P. E. R., Dale, P. T., Hulsman, K. and Kay, B. H., 1993. Runnelling to control saltmarsh mosquitoes: long-term efficacy and environmental impacts. *J. Amer. Mosq. Con. Assoc.* 9, **2**: 174–81.

Dale, P. E. R., Muhar, A., Thalib, L. and Arito, E., 1998. The distribution of Ross River virus disease in Brisbane and its relationship with wetlands and vegetation. Paper presented at the Mosquito Control Association of Australia 3rd Conference, September 23–25, 1998, Gold Coast, Queensland.

Dale, P. E. R., Ritchie, S. A., Territo, B. M., Hulsman, K. and Kay, B. H., 1998. Use of remote sensing to map intertidal wetland water distribution. Pp. 581–95 *in* Wetlands for the Future ed by A. J. McComb. Proceedings INTECOL International Wetlands Conference, September 1996, Perth, Western Australia. Gleneagle Press: Australia.

Department of Primary Industries Queensland, 1996. Departmental Procedures For Permit Applications Assessment And Approvals For Insect Pest Control In Coastal Wetlands. Fish Habitat Management Operational Policy FHMOP 003, Queensland Government, September 1996.

Dhileepan, K., 1997. Mosquito seasonality and arboviral disease incidence in Murray Valley, southeast Australia. *Med. Vet. Entomol.* **10**: 375–84.

Ferrigno, F. and Jobbins, D. M., 1968. Open Marsh Water Management. *Proc. New Jersey Mosq. Exterm. Assoc.* **55**: 104–15.

Hawkes, R. A., Boughton, C. R., Naim, H. M., Myrick, B. A. and Ramsay, L. G., 1987. Barmah Forest virus infections in humans in New South Wales. *Med. J. Aust.* **146**: 569–73.

Hulsman, K., Dale, P. and Kay, B. H., 1989. The runnelling method of habitat modification: an environment focused tool for salt marsh management. *J. Amer. Mosq. Con. Assoc.* 5, **2**: 226–34.

Mackenzie, J. S., Lindsay, M. D., Coelen, R. J., Broom, A. K., Hall, R. A. and Smith, D. W., 1994. Arboviruses causing human disease in the Australasian zoogeographic region. *Arch. Virol.* **136**: 447–63.

Mackenzie, J. S. and Smith, D. W., 1996. Moquito-borne viruses and epidemic polyarthritis. *Med. J. Aust.* **164**: 90–93.

Meredith, W. H., Saveikis, D. E. and Stachecki, C. J., 1985. Guidelines for "Open Marsh Water Management" in Delaware's salt marshes — objectives, system designs and installation procedures. *Wetlands* **5**: 119–33.

Russell, R. C., 1993. Mosquitoes and Mosquito-borne disease in Southeastern Australia: a Guide to the Biology, Relation to Disease, Surveillance, Control and the Identification of Mosquitoes in Southeastern Australia. Department of Medical Entomology: Westmead Hospital, Westmead, New South Wales.

Ryan, P. A. and Kay, B. H., 1997. Ross River virus ecology is complex. *Arbov. Res. Aust.* **7**: 247–51.

Ryan, P. A., Martin, L., Mackenzie, J. S. and Kay, B. H., 1997. Investigation of gray-headed flying foxes (Pteropus poliocephalus) (Megachiroptera: Pteropodidae) and mosquitoes in the ecology of Ross River virus in Australia. *Amer. J. Trop. Med. Hyg.* **57**: 476–82.

Saffigna, P. G., Hey, K., Holland, G., Dale, P. E. R. and Mulder, G., 1997. Impact of acid sulfate soils on habitat modification for mosquito control in coastal Australia. *Arbov. Res. Aust.* **7**: 252–57.

Soukup, M. A. and Portnoy, J. W., 1986. Impacts from mosquito control-induced sulfur mobilisation in a Cape Cod estuary. *Environ. Cons.* **13**: 47–50.

Tweed Shire Council, 1993. Development Control Plan — Biting Midge and Mosquito Control in Tweed, New South Wales.

Wolfe, R. J., 1996. Effects of Open Marsh Water Management on selected tidal marshes: a review. *J. Amer. Mosq. Con. Assoc.* **12**: 701–12.

CHAPTER 45

Community involvement in urban bushland conservation

Margo O'Byrne[1]

Conservation of terrestrial and marine reserves throughout Western Australia is based on the development of a reserves system. The 1983 "System-6 Report" encompassing the Perth metropolitan area and 80% of the state's population presented recommendations for a system of conservation through primarily terrestrial reserves within an area approximately 1% of Western Australia. Difficulties in implementation and lack of awareness led to community concern about the viability of a conservation estate based upon the recommendations contained in the System-6 Report for urban conservation.

To renew community awareness and assist implementation of "System-6" recommendations, the Environmental Protection Authority and conservation groups developed the "Ecoplan" community involvement programme. Objectives were to develop public ownership of reserve areas, raise community awareness of System-6, create or motivate existing action groups to become involved in urban conservation and provide a co-ordinated framework for the implementation of System-6 recommendations.

Ecoplan presents a model for supporting individuals and existing community groups involved in conservation activities while enabling new volunteers to understand the issues and become involved in urban conservation. Community activities include bush regeneration, fencing, flora and fauna surveys, nursery operations, bushland management and sourcing funding opportunities. In meeting Ecoplan objectives System-6 reserves have greatly benefited from community involvement developed through volunteer training, workshops, assistance to volunteer groups and access to the conservation network.

INTRODUCTION

THE State of Western Australia covers more than 2.5 million square kilometres, which is a third of the Australian continent's landmass and coastline. It is one of the oldest landscapes on the planet and is characterized by great diversity, in landform, climate, flora and fauna. The South West land division of Western Australia stretching from Shark Bay in the central west coast through to Esperance in the Great Australian Bight is one of the most significant botanical areas in the world.

Perth is the capital city of Western Australia situated at 31°56'N and 115°50'E. The city is fortunate to have substantial areas of bushland remaining (approximately 50% in the metropolitan area). This is due in part to the fact that the highly infertile soils of the Swan Coastal Plain remained uncleared because of their unsuitability for agriculture and in part to the vision of early planners. The consistent and collective efforts of farsighted scientists, naturalists, governments, and community members have enabled the retention of bushland in the Perth area. Bushland has long been appreciated and valued by the community. Laws have been invoked for many years that prohibit the picking of wildflowers on public lands and these are generally well observed. Tracts of land have been set aside for the protection of flora and fauna in conservation areas such as Kings Park Bushland, Local Government Reserves, National Parks and Nature Reserves (Keighery and Gray 1993).

Conservation through reserves and System 6

In the late 1970s a system of conservation through reserves was set up in Western Australia. The state was divided into 12 regions or systems on the basis of geography and human activities. Each

[1]Community Involvement Co-ordinator, Department of Environmental Protection, Westralia Square, 141 St Georges Terrace, Perth, Western Australia, Australia 6000.

Pages 456–61 *in* NATURE CONSERVATION 5: NATURE CONSERVATION IN PRODUCTION ENVIRONMENTS: MANAGING THE MATRIX ed by J. L. Craig, N. Mitchell and D. A. Saunders. Surrey Beatty & Sons, 2000.

region was examined to define areas, which should be set aside for their particular flora, fauna, landscape, geology or recreation value. While the majority of the reserves are terrestrial, there are a smaller number of marine and estuarine reserves as well as many wetland and riverine systems.

Perth is in System 6. Development pressure in System 6 is significant as it represents 1% of the area of the state yet it supports 80% of the state's population. The System 6 study proposed 108 recommended reserves in the metropolitan area and 101 in the country part of System 6. The size of reserves range from large non-metropolitan (country) areas, such as the Julimer Management Priority Area of 47 672, to very small metropolitan reserves, such as Point Heathcote, on the banks of the Swan River which is less than one hectare. Recommendations applied to areas on both public and private lands.

In the 1980s there was widespread community consultation and promotion of the values of these recommended areas and the Cabinet of the State Parliament endorsed the recommendations of the System 6 report in 1984. However, there was no legislation to support the protection of these reserves and recommendations were not always specific and often referred to "recreation and conservation" without much understanding of the potential for conflict in years to come.

The primary means of protecting these areas is to "reserve" them for parks and recreation under the Metropolitan Region Scheme. Parks and recreation reservation enables the state planning agency to purchase the area for public open space. Many areas were not secured and the cost to government of the purchase and management of so many areas is still beyond budget allocation. In addition some areas (local government and private land) are not suitable, or available, for purchase and management support is more appropriate. However, there is no system of rates or taxing specifically designed for the protection and management of bushland in Western Australia.

This has led to much confusion. Conservationists believe that the land should be used for the conservation of flora and fauna while others seek to use the reserves for active recreation such as horse riding. The definition of public open space has not given these reserves a conservation status. Many reserves have been subject to the pressures of urban expansion, either from development proposals within the reserve, such as the construction of an adventure park in the Beelier Wetlands, or the ongoing demands of utilities for service corridors, easements and roads.

By the early 1990s, there was concern about the lack of security for many of the System 6 areas close to Perth. Some of these areas were losing their special values through lack of management, and others were coming under pressure to be developed. Active recreational use by a growing population had affected their condition. This led to several of Perth's conservation groups approaching the Department of Environmental Protection to see how the matter could be redressed.

A strategic approach was called for. The System 6 recommendations needed to be implemented, updated and stakeholders included in ongoing discussions. At the same time, there was a need to develop a programme to extend community awareness of the issues affecting urban bushland.

In 1991, the Department of Environmental Protection set up a process involving stakeholders throughout state and local government together with a community representative to review the recommendations of the System 6 study. This update programme evolved to become Perth's Bushplan (Whisson *et al.,* this volume).

ECOPLAN

Although the original System 6 study had received extensive publicity, resources were limited and the apparent low profile given to the issue led to a community belief that it was no longer relevant.

At the same time, many groups were forming to respond to issues affecting their local bushland area, only some of which were included in System 6. As well as a review and update of the recommendations for System 6, there was a need to raise community awareness about the special values of these bushland reserves.

A working group of government officers and representatives of the community conservation movement was established. Through fortnightly meetings for six months, a strategy evolved which became known as Ecoplan. The Department of Environmental Protection funded a full time Community Involvement Co-ordinator in April 1991, and the Minister for the Environment and the President of the Conservation Council of state environmental groups jointly launched Ecoplan (Keating 1992).

Objectives of Ecoplan

The specific objectives of Ecoplan are:

- To develop public ownership (support and backing for the recognition) of System 6 areas;
- To raise community awareness of System 6;
- To create or motivate existing action groups to seek and/or get involved in the implementation of System 6 recommendations; and
- To have a co-ordinated framework for the implementation of System 6 recommendations.

Ecoplan has sought to implement these objectives by:

- Conducting community training programmes and workshops;
- Increasing the community skills base;
- Assisting focus groups;
- Seeking partnerships in funding opportunities;
- Publishing a regular newsletter; and
- Assisting with appropriate community publications.

Establishing a relationship with the conservation community

Valuable work in caring for the bushland areas of Perth was already being carried out by various conservation groups. It was important for Ecoplan to establish contact with and gain the support of these groups. The purpose was to raise awareness about, and gain support for Ecoplan.

Response to Ecoplan at most of these encounters was initially cynical. Many people felt they had been working for years to save these areas. It appeared that government approved development within System 6 reserves without due consideration for ecological process or fauna or flora values. There was a need to overcome the resistance of the community to assistance from public servants (Masters 1995). Despite this criticism, many members of established groups became staunch allies in the promotion of Ecoplan.

Involving the broader community

Early in the programme the need to recruit new volunteers became apparent. Members of existing conservation groups were occupied with issues relating to their particular focus and recruitment for these groups was limited to those with an existing knowledge of environmental issues. In order to achieve community ownership of System 6 areas, new volunteers were needed especially those in the neighbourhood of the reserves. A recruitment drive was conducted to involve community members in looking after their local environment.

Involving schools in the conservation of bushland

Initial contacts were made with individual teachers through environmental education organizations and also with the Ministry of Education. An information kit, known as an Ecopack was produced. These kits contained a video about the region, a teaching guide and details of the special values of each System 6 reserve. These were delivered to schools throughout the metropolitan area. Response from schools varied according to the interest of particular teachers and bushland proximity.

Regional co-ordinators

Early in the programme regional co-ordinators were identified to play an important leadership role in Ecoplan. They collected and co-ordinated information and material for use by other volunteers in the local area, contributed information about the region to the quarterly newsletter and acted as a mentors for those wanting to form new groups. Many of these leaders are now contracted to conduct information sessions for newly establishing groups.

Community training

The response to advertisements placed in local community newspapers was high and a training programme was set up to inform and educate new volunteers.

Training is divided into the following three aspects.

- Introductory training or "getting started" targeted at new volunteers and members of existing community conservation groups.
- Focused training or "keeping going" is directed towards members of community conservation groups and planned in consultation with community members on a needs basis.
- Leadership training or "extending the vision" takes the form of sponsoring group leaders at appropriate conferences and workshops.

Introductory training — Subjects covered in the training sessions include issues affecting urban conservation, introduction to the flora and fauna of an area, developing management plans for reserves, working successfully with the local authority, principles and practices of bush regeneration and the roles and responsibilities of various government departments. In all, there are 12 hours of information delivered over four sessions, with at least one field trip to observe work on a local conservation project.

Ecoplan strives to attract interesting and informed speakers who will present up to date scientific information in a way, which is stimulating for new volunteers. Local government and state government agencies with an interest in the nearby reserves are asked to speak. Where possible, speakers who are themselves members of community conservation groups are contracted for the presentation. Such speakers have an immediate rapport with the audience and this ensures that valuable problem solving information is shared. Paying a speaker's fee to community volunteers, values the information and experience gained from working in the area of community based conservation. Existing conservation groups in the local area are involved in the training sessions and in many cases, act as a host group for new volunteers. Interest in these sessions is high, and the retention rate at training sessions is high.

Focused training — Topics covered in focused training include information for effective conservation such as plant identification, wetland ecology, fauna trapping and identification, weed identification and weed control. Other topics relating to issues faced by community groups have included applying for and administering funding, planning for groups, conflict resolution within community organizations, setting up displays and presenting ideas to the media. These sessions are conducted either by the Ecoplan staff or contracted presenters.

Bush regeneration training — Ecoplan encourages groups to carry out bushland restoration and regeneration according to the best principles and practices. The Bradley method of bush regeneration (Bradley 1988) has been carried out in New South Wales since 1971. The principles and practices of this method, together with knowledge and understanding of local conditions are taught in a course conducted in the Perth area. Ecoplan has sponsored places for volunteers in this course and, in addition, has sponsored opportunities for the course to be conducted at particular reserves. This has served to focus on the different aspects of regeneration needed for wetland revegetation or regeneration of bushland on specific soil types.

Western Australia now has an established network of bush regenerators and relevant information is circulated through a newsletter to all members. Ecoplan assists with the distribution of this newsletter.

Plant survey training — In 1996 Ecoplan sponsored aspects of an existing community based bushland survey programme (Keighery 1994) run by the Wildflower Society of Western Australia. This initiative provides both botanical survey and plant identification experience for volunteers and information on the flora and vegetation of the surveyed bushland areas. It also assists volunteers to understand the broader context of scientific research and to develop the skills to undertake a bushland plant survey in local reserves (Keighery 1994).

Leadership training — Conferences and seminars provide valuable opportunities to increase knowledge and understanding of issues affecting nature conservation and to maintain an up to date resource base. Ecoplan provides sponsorship for community group leaders at such events. Many volunteers are not in a position to pay to attend conferences and seminars, and by providing this sponsorship Ecoplan reinforces the value of these people and provides the many opportunities available for information gathering and networking.

Key ingredients for a successful training programme

- Engage informed and enthusiastic speakers who can present the topic in an interesting way.
- Ensure there is a friendly and informal atmosphere, which allows participants to ask questions, seek information and to share their own stories.
- Provide an opportunity for networking during breaks.
- Good food is an important ingredient for successful training.
- Evaluate each workshop and learn from this. Flexibility is important as is responding to requests for specific information.

Training outcomes

- There is a consistently high interest in Ecoplan training. Attendance is from 20 to 60 participants, the majority of whom have not previously been involved in conservation.
- A review of Ecoplan conducted after 5 years of the programme showed that 75% of those who had become involved with Ecoplan had volunteered their time to an urban bushland project.
- Ecoplan provides social opportunities as well as information sessions.

ACTIVITIES OF ECOPLAN TRAINEES

Once participants have undertaken the initial introductory training and joined with an existing or new conservation group they become involved in many conservation related activities in their local reserves. This involves such things as bush regeneration, weeding, erosion control and land stabilization, rubbish removal, planting, growing small trees and native shrubs, collecting seed, building paths, bird hides and nesting boxes and erecting signs to bring attention to the conservation value of the area.

Groups are encouraged to work within the guidelines of a management plan, if one exists and to develop a regeneration plan for the reserve. All work needs to be undertaken with the support of the State or Local authority in which their area of interest is vested. In many cases there is positive support for the newly forming group, but sometimes the emergence of a group interested in the conservation and management of the reserve can be interrupted as a threat to the vesting authority. All efforts are made to minimize conflict and ensure that the management agency and the community are working together.

Bushland care workdays

As in most community organizations, the physical work of bush regeneration is often undertaken by the same small number of dedicated people, with the subsequent risk of burnout and depletion of energy. Ecoplan seeks ways to assist the committed members of small conservation groups and negotiated with the Australian Trust for Conservation Volunteers to assist with bushland care in the urban area.

The Australian Trust for Conservation Volunteers is national, non-profit organizations who recruit a pool of volunteers from the community for specific conservation projects. The majority of volunteers involved with Ecoplan tend to be in the 35 years and over age group with an interest in ongoing involvement in a particular area. Australian Trust for Conservation Volunteers attracts mainly younger volunteers, many of whom are students or travellers who choose to work on specific projects on a once only or casual basis. The cost for an Australian Trust for Conservation Volunteers work team is approximately $300 per day and includes a supervisor.

Ecoplan trialled a series of 10 partnership days involving Ecoplan, Australian Trust for Conservation Volunteers and community groups. They became known as Bushland Care Workdays. Friends groups were encouraged to nominate a particular task that they may not be able to complete alone, and to advertise these workdays to the local community through newspapers and radio. The aim of a partnership day was to complete the nominated task by involving more volunteers and the broader community thus attracting more local support and providing encouragement to the dedicated members of the community conservation group. It also meant that the Australian Trust for Conservation Volunteers became involved in urban bushland projects. An evaluation was undertaken to look at the benefits of these workdays (Elder 1995).

Following the success of the trial, the programme was expanded to 12 workdays the following year sponsored by Ecoplan and another four sponsored by local government. An opportunity was identified to gain corporate sponsorship for this programme thereby extending the Ecoplan budget and fostering the relationship between community groups and a corporate sponsor. AlintaGas sponsored the 1997 Bushland care workdays. The success of these days is dependent on the ability of the community group to identify an appropriate project, attract project management skills for the day and market the activity to the local community. A stipulated condition of the workdays is that the group is working towards best practices in bush regeneration and within the guidelines of a regeneration plan for the reserve.

Publications

Ecoplan News — Communication of ideas and information is vital for community members and the publication of the quarterly newsletter "Ecoplan News" has filled this role. The newsletter is circulated to a range of people including volunteers, local members of parliament, local councils, and interested members of the public. It has a circulation of 1 000 readers. The newsletter highlights the positive contribution to the environment made by community groups, gives a list of upcoming events relating to conservation in the region, reviews recently published resources and writes seasonal ecological notes. Each edition profiles a particular group, their work and objectives (Ecoplan News). Ecoplan assists community groups and non-government organizations publish information gained from their experience. Examples include Tizard 1993, Scheltema and Harris 1995 and Wake and Reeves 1995.

Weed education

The presence of environmental weeds in urban bushland is a continual threat to the viability of reserves. Many community conservation groups spend much of their time removing weeds.

Ecoplan has sought to highlight the issue of environmental weeds by:

- Working with the Australian Association for Environmental Education and the Western Australian Municipal Association to co-ordinate the *Weeding Western Australia* forum for land managers;
- Commissioning consultants to prepare a data base of information on the 40 most common and damaging environmental weeds and control methods for these weeds. This became the basis of a detailed weed identification section in the publication Managing Perth's Bushland (Scheltema and Harris 1995).
- Assisting the community based Environmental Weeds Action Network and the Australian Association of Bush Regenerators (WA) with the printing and distribution of newsletters and information.

OUTCOMES

In June of 1996 a review of the Ecoplan programme was conducted by the Department of Environmental Protection. The review took the form of a questionnaire sent to 300 people who had been involved in the Ecoplan introductory training programme. Of these, 124 responses were received.

Recommendations of the review suggested that Ecoplan should continue the work it is already doing, encourage new groups, generate more involvement in existing groups and raise the public profile of Ecoplan and urban bushland issues. Some 75% of respondents have volunteered their time to an urban bushland project since becoming aware of Ecoplan, which suggests that the training sessions are successful in moving people from passive receivers of information to active participants in bushland care. Eighty-six per cent of respondents had a sense of ownership of the area, thus one of the primary objectives of Ecoplan has been achieved. Those who become involved and associated have a long-term commitment with 62% of people remaining committed to ongoing involvement.

Lessons

There are a number of lessons learned from Ecoplan over the past 6 years. Staff consistency has ensured that the programme knowledge has been retained, and the networks developed within the conservation community, government and industry are fostered. The following are some lessons for government workers and community members.

SUGGESTIONS FOR GOVERNMENT WORKERS

- Listen.
- Link new community groups or individuals with existing conservation groups and leaders.
- Respond to the issues raised by those who are working actively in the community and feed back information.
- Assist groups to present their story of the reserve. This may mean help with information, maps, copies of background reports, etc.
- Recruit paid and volunteer staff who can relate to people. Workers in community support programmes must have good people skills. Technical information and other skills can be acquired or purchased.
- Do not burn out your paid staff or your volunteers.
- Seek opportunities for formal and informal information sessions delivered in a common language. Have teachers address teachers, local government speak to other local government workers. This promotes an atmosphere of safe sharing of knowledge.
- Look for ways to publicly acknowledge volunteer contribution, such as nominating volunteers for awards.

- Work out of normal office hours, and develop a system within your agency, which supports community extension workers.
- Spend time feeding the information back to the relevant officers within your own departments. Seek opportunities to incorporate this information into the policy and strategies of relevant departments.
- Make sure that other staff members in your agency have a sense of ownership of the community programme by circulating newsletters and good news stories of which they can feel proud.
- Gain political support for your project. Ensure there are good publicity opportunities for your Minister or CEO.

LESSONS FOR MEMBERS OF COMMUNITY CONSERVATION GROUPS

- Community groups need to be persistent.
- Keep your long-term goal in view at all times and resist the temptation to score points by exposing the shortcomings of the system.
- Be positive even in the face of difficulty.
- Be creative; you have a lot more flexibility than government departments.
- Do not only look for simple solutions. Issues are complex, often involving many players and the delicate balance of interactions between players.
- Look towards sustainability. Sustainability of yourself, your group, and the environment.
- Seek allies.
- Use networking opportunities to present your group in a good light. Develop public relations skills, or find someone in the group who is good at this.
- Retain a sense of humour.
- Avoid burnout.
- Present a united front, and seek assistance to resolve conflicts within the group before they become major issues.

CONCLUSION

Apart from the intrinsic values of bushland for its assemblage of unique flora and fauna, bushland in an urban setting has many values which include the creation of a sense of locality, aesthetic, conservation, scientific, educational, recreational and tourism values.

Managing bushland in an urban environment is not easy. Assuming the best possible co-ordination between the vesting agency, managing body, individuals and community groups there are often difficulty. Programmes such as Ecoplan create an opportunity for ordinary members of the community to become actively involved in the protection and care of their own environment. Such experiences empower people, increase knowledge and link individuals and communities.

Managing this complex matrix with limited resources, while safeguarding the extraordinary biodiversity of the area is an enormous challenge and must involve community participation.

ACKNOWLEDGEMENTS

I would like to acknowledge the valuable contribution made to the Ecoplan programme by the many wonderful volunteers who give tirelessly in their effort to preserve and protect the environment of Western Australia. I thank my co-workers, Jon Elder, Jillian Harris and Roxanne Shadbolt. I would like to thank Bronwen Keighery and Jon Elder for reviewing this chapter. I wish to thank the Department of Environmental Protection and the Conservation Council of Western Australia in particular Joan Payne, for their ongoing support. Finally my family, for their support and understanding.

REFERENCES

Bradley, J., 1988. Bringing Back the Bush ed by J. Larking, A. Leening and J. Walker. Lansdowne Press: Sydney, New South Wales.

Elder, J., 1995. Ecoplan/ATCV Bushland Care Days June–October 1995. Department of Environmental Protection: Perth, Western Australia.

Ecoplan News, 1991–2000. Department of Environmental Protection: Perth, Western Australia.

Keating, F., 1992. Urban Conservation: Working together to look after Bushland. Australian Institute of Urban Studies (WA): Perth, Western Australia.

Keighery, B. J., 1994. Bushland plant survey. A guide to plant community survey for the community. Wildflower Society of WA: Perth, Western Australia.

Keighery, B. J. and Gray, C. M., 1993. Towards an Urban Bushland Policy. National Trust of Australia (WA): Perth, Western Australia.

Masters, B. K., 1995. Why community networks fail: the role of public servants and the community. Pp. 199–204 *in* Nature Conservation 4: The Role of Networks ed by D. A. Saunders, J. L. Craig and E. M. Mattiske. Surrey Beatty & Sons: Chipping Norton, New South Wales.

Scheltema, M. and Harris, J. (eds.), 1995. Managing Perth's Bushlands. Greening Western Australia: Perth, Western Australia.

Tizard, B., 1993. Reality, Dreams and Pathways: A Manual for Preparing and Implementing Management Plans. Armadale Settlers' Common Field Study Center and System 6 Ecoplan: Perth, Western Australia.

Wake, D. and Reeves, N., 1995. Discover Your Local Bushland. Quinns Rocks Environmental Group Inc: Perth, Western Australia.

CHAPTER 46

Using product certification to promote the sustainable use of natural resources

R. C. Nias,[1] M. Rae,[1] L. Jedreski[2] and S. Towle[2]

The emerging concepts of product certification and eco-labelling are important new tools for promoting sustainability in the use of natural resources. By harnessing the growing awareness of consumers, the trend toward greater corporate reporting on environmental matters, and the precedent established by independent financial auditing, a new area of opportunity is established to develop patterns of sustainable resource use.

We discuss some of the reasons why product certification is gaining interest as a market-based mechanism to promote sustainable resource use, and some of the assumptions inherent in the approach. Two natural resources in particular seem well suited to the product certification approach — timber and fish. Recent developments in two major certification schemes — the Forest Stewardship Council and the Marine Stewardship Council — are presented and their potential application in Australia and New Zealand is highlighted. The extension of the product certification approach to other resources, including minerals, is also discussed.

INTRODUCTION

SUSTAINABLE resource use and the concept of sustainable development have at their foundation, the limits to consumption of natural capital set by the Earth's biological processes. The principle of ecological sustainability therefore transcends a purely economic approach and attempts to integrate economics within the boundaries set by ecological constraints. Sustainability also emphasizes the qualitative aspect of development, as opposed to the more traditional emphasis on quantitative growth captured in measures such as Gross National Product. Growth implies a simple increase, whereas many would argue that real human development must also encompass qualitative improvement in material and non-material well being (e.g., Brown *et al.* 1992).

The incorporation of ecologically sustainable development as a fundamental goal of economic and social decision making is probably the greatest challenge that faces humanity. The scale of human numbers, resource use and waste production is now sufficiently great to be considered a global problem, and not one confined to a particular location, region or nation. Damage to the protective atmospheric ozone layer, the measurable depletion of the world's supply of freshwater, and anthropogenic climate change are environmental impacts that would have been unimaginable to previous generations. Much of this impact is driven by the natural resource consumption of the developed world, where the role of the free market has been heavily expanded in recent decades.

Three propositions about markets and sustainability

One of the most significant aspects of the late twentieth century has been the influence of capitalist economics and the subsequent growth of the private corporation as a global influence in human affairs, not the least as a factor in environmental degradation. The development of huge transnational corporations, globalization of market places, and liberalization of world trade will ensure that this influence continues to grow into the future, surpassing the power of many nation-states to limit its effects. The challenge for environmentalists, and the international community, is to determine how this influence can be used to promote sustainability as a desirable outcome. The aim of this chapter is to briefly

[1]WWF Australia, 5/725 George Street, Sydney, New South Wales, Australia 2001.
[2]WWF New Zealand, Education and Environment Centre, Botanic Gardens, Gelnmore Street, Wellington, New Zealand.
Pages 462–67 *in* NATURE CONSERVATION 5: NATURE CONSERVATION IN PRODUCTION ENVIRONMENTS: MANAGING THE MATRIX ed by J. L. Craig, N. Mitchell and D. A. Saunders. Surrey Beatty & Sons, 2000.

consider some simple assumptions and models for how this might be achieved. In particular we consider how the power of the consumer can be used to create new markets for sustainably produced products.

One of the most significant factors that has lead to the over-exploitation and wastage of resources is the inability to value and incorporate the costs of such actions into the marketplace. Given that biological diversity underpins much of our agriculture, industry and medicine, then every individual would appear to have a stake in its conservation. However, the true value of natural systems has not been recognized and accounted for in economic systems. The costs of degradation of the biological resource, the process of extinction and the destruction of natural and other productive ecosystems, have been externalized. The costs of not conserving natural capital (such as the expensive rehabilitation programmes needed to address land degradation, or the costs of pollution remediation) are generally passed on to others. In many cases these costs are borne by the poor and future generations. The costs of pollution remediation and land degradation, for example, are frequently borne by subsequent generations, long after the economic benefits have been absorbed elsewhere or dissipated. In some cases, "dirty" production methods are transferred to developing countries where environmental regulations are often lower than in developed countries, and where the need for economic development over-rides environmental considerations.

This chapter therefore poses three propositions about the role of market forces in promoting ecological sustainability:

1. That market forces are a potentially powerful but over-looked factor in efforts to promote sustainable resource use.
2. That mechanisms exist, or can be developed, to harness this power in favour of sustainability.
3. That major new gains in sustainability will be made most readily with support from the market.

THE MARKET PLACE AND SUSTAINABLE DEVELOPMENT

In most western societies, environmentalists, and indeed governments, have pursued the goal of sustainable natural resource use primarily through legislative command and control mechanisms. Natural resource managers concentrate on the "supply-side" — the quality of the product, in order to meet legislated standards for pollution levels, stock levels, area protected, or other similar "quality" measures. Relatively less importance is placed on modifying demand for these resources. While the management of the supply-side through regulation remains of major importance, the potential for the demand-side to promote sustainable resource use is increasingly being recognized. There are many reasons for this.

1. Some governments seem to have reached "reform fatigue" in terms of major new legislation on the environment.
2. Decision making on environmental issues is being devolved from centralized government agencies to local agencies and the community.
3. Governments are becoming less the manager of the environment, and more the standard setter and compliance monitor.
4. The market place may often provide novel, effective and cheap solutions to environmental problems.
5. The power of business and commerce is growing rapidly, particularly in developing economies, both in terms of its ability to capture and mobilize vast amounts of natural resources, and its influence on decisions on the allocation of these resources.
6. Consumers are becoming more aware of the environment and increasingly more discerning in their purchasing.
7. Competitive pressures are forcing business to differentiate their products in the market place.
8. Adverse environmental effects arise from the sum of all of the individual resource consumption decisions made by consumers, and might be reversed in the same manner.

There is a growing reluctance by some western governments to develop major new legislative programmes for the environment. Indeed, most emphasis in Australia in the late 1990s appears to be on "stream-lining" current environmental legislation, ostensibly to make it simpler to apply, rather than creating new legislative powers. Similarly, the trend to decentralization of environmental decision making and management may act to "empower" communities to manage their resources but may also have the side-effect of promoting localized issues at the exclusion of national or international goals.

It is clear that many of the formal mechanisms designed to ensure sustainable resource use have partially or wholly failed, or are incomplete. In part, these failures may stem from an inability of governments to control the actions of major corporations (in developing economies), small and medium sized operations in developed countries (such as private forestry and agriculture, small-scale fisheries, small mining companies etc.) or individual consumers. Additional approaches are therefore required which recognize the new reality of a global economy, hundreds of large transnational corporations, millions of small-scale enterprises, and the combined impact of billions of product consumers.

How can market forces be harnessed in favour of sustainable resource use?

Progress toward sustainability, even though limited, has been encouraged by public pressure through purchasing decisions, through adverse publicity, or from government requirements. Increasing urbanization has alienated many people from the natural environment, and a direct role in conserving nature, and large numbers are looking for practical ways in which they can assist in environmental protection. By making use of their power as consumers, an opportunity exists to reconnect people with the environment, and their impact upon it. It is encouraging to note that some corporations recognize that their power in the market place might also bring responsibilities. A number of companies are starting to make their decision making more transparent and open to influence. There is widespread adoption of environmental codes of conduct within companies and within particular sectors of the economy. Several large corporations now publish their own environmental reports. It is important to note however that such steps, while welcome, are merely beginnings.

One promising approach to these opportunities is the promotion of products and services that are more ecologically sustainable than their competitors in the market place. The objective of the environmentalist is to create a market for better environmental products, through environmental education and awareness. The eventual aim is to use market forces to eliminate demand for the worst products, create additional demand for the best products, and shift the average for the industry towards sustainability.

Some of the assumptions inherent in this approach are that:

- environmentally better alternatives exist, or can be created,
- that, with appropriate market incentives, research and development will be directed towards more sustainable products,
- that, with independent and cost-effective certification and labelling, sustainable products can be readily identified in the market place,
- a significant proportion of consumers will recognize environmentally better products and choose these preferentially,
- market competition will act to drive the whole product category toward sustainability.

Two cases in particular are discussed here — fisheries and forest products. The approach is also discussed in relation to other consumer products, such as packaged foods and electricity, as well as non-consumer products such as minerals.

FORESTS AND FISH

Forest and fisheries products are obvious candidates for this approach for a number of reasons. The production of timber and other wood products such as paper is probably the largest consumer of natural wild resources, and indirectly through deforestation, is a major cause of biodiversity loss (Dudley *et al.* 1995). Similarly, fisheries remain the largest source of wild caught protein and through direct and indirect means (e.g., by-catch and trawling damage) represent a major environmental impact on marine species and ecosystems. Independent certification has the potential to promote a variety of positive outcomes in both the timber and fisheries sectors, including:

- an objective and verifiable means to test environmental claims,
- the creation of markets for sustainably produced products,
- the widespread adoption of sustainable development indicators.

Both timber and fish are well represented (though not totally) at the end of the process chain by consumer goods (e.g., furniture, paper, canned and frozen fish). Consumers are generally aware that these products have a "natural" origin, and this is often reinforced through advertising. As a result of this direct linkage between the product and its natural origin, it is easy for consumers to make the link between the product and its potential environmental impact. For example, it is easier to visualize the link between a teak or mahogany table and logging in tropical forests, than it is between a computer and coastal dune mining for mineral sands.

Given the relatively high level of environmental impact within many forestry operations and fisheries, major environmental gains should be achievable with even a limited improvement in performance. Furthermore these gains would seem to be possible with existing technologies. In both areas the role of the private corporation, and hence the market, is well established. Many products are harvested from the wild state and marketed with limited processing or product mixing (e.g., whole frozen fish or sawn timber). The marketing of these products may be dominated by a few key corporations, often transnationals that may be sensitive to the demands of the environmentally sophisticated markets such as those in western Europe or North America. This sensitivity may be as a result of many factors, ranging from a genuine desire to promote sustainability, a desire to avoid negative publicity, or seeking to exploit a "green" niche market.

The Forest Stewardship Council (FSC)

In an effort to validate the claims of timber certifiers and to avoid market confusion, a diverse group of representatives from environmental organizations, foresters, timber traders, indigenous peoples organizations, community forest associations, and forest product certification institutions established the Forest Stewardship Council (FSC) — an independent body with a mandate to accredit forest certification programmes.

The goal of the FSC is to promote environmentally responsible, socially beneficial and economically viable management of the world's forests, by establishing a world-wide standard of recognized and respected Principles of Forest Management. The basic approach involves third-party auditing of the quality of forest management against performance standards; a "chain of custody" tracing the forest product from the audited forest to the market; and a label allowing the consumer to identify the certified product (Jedreski and Rae 1996).

The FSC has developed ten Principles and Criteria For Natural Forest Management which are broad based guidelines relating to forest and plantation management issues (FSC 1995). By themselves, the FSC Principles and Criteria are not designed to be used as the basis for certification in the field, but to provide a consistent framework for the development of locally determined forest management standards. To date however, all FSC field certification has been done directly against the ten Principles and Criteria. In order to ensure the consistency and integrity of standards in different regions around the world the FSC formally endorses those standards which clearly meet all FSC requirements, including the process leading to their development.

The FSC encourages the development of national and/or regional FSC Working Groups which provide support to the development of forest management standards, and are responsible for recommending standards to the FSC for approval. At the time of writing (1997), several hundred thousand hectares of commercial native and plantation forests had been certified in a dozen countries. A number of major retail chains, and wood users, particularly in western Europe had agreed to source their timber products only from FSC certified forests.

There is a growing interest in certification and labelling as measures to promote sustainable forest management (see DPIE 1997 for a substantial review of current progress and views on timber certification). Timber seems to satisfy the assumptions made for a commodity that could be made more sustainable through the use of independent certification and labelling, viz:

- the level of sustainability within forest management is variable across operations and better managed forests can be identified,
- the World Bank and many national governments are investing in research and development toward achieving sustainable forest management,
- FSC labelled products are being successfully marketed at low cost through major retail chains (Knight 1997),
- FSC labelled timber products are being preferentially sought by retailers and larger consumers in many instances.

It is too early to assess the final assumption, namely that the whole product category will be moved toward sustainability, as FSC timber, despite a rapidly growing market share, still represents only a tiny fraction of all timber products being sold.

The Marine Stewardship Council

The MSC approach came into existence as a result of initial work carried out by WWF with Unilever — the owner of some of the world's largest selling frozen-fish brands and one of the world's largest consumers of fish products. The basis for the co-operation was that Unilever and WWF shared a common view, the need to ensure long-term sustainable supplies of fish. Full details on the MSC, including the Principles and Criteria for Sustainable Fishing, have now been published (MSC 1998).

In basic terms, both the FSC and MSC have a similar approach.

1. An independent body is created with ownership of an appropriate eco-label or logo.
2. Stakeholder discussions are held to ascertain the basic concerns from an industry, government, consumer, social and environmental perspective.
3. A system of basic global environmental performance standards is developed that will underpin the application of the logo.
4. The standards are refined and tested on real examples and in particular situations for both general applicability and ability to deal with specific circumstances.
5. Local or sector specific standards may then be developed in consultation with major stakeholders.
6. Independent certifiers are trained and accredited to apply the standards to an operation or product and authorize the use of the eco-label following an audit process.
7. The eco-label is promoted in the market place.
8. Periodic re-testing ensures that standards are being maintained.

The MSC was launched in 1997 and developed substantially as a separate organization during 1998. At the time of writing, trials were being conducted to test and refine the basic global criteria.

Application of the FSC in Australia and New Zealand

In order to stimulate demand and increase awareness of the FSC initiative an Australian WWF Buyers Group has been successfully established in 1997. The following is a summary of the operation of the FSC and its potential application in Australia and New Zealand (a more complete assessment is given by Jedreski and Rae 1996).

The WWF Buyers Group currently has four member companies (as of December 1997) with a total turnover of $66m and it is anticipated that more companies will join in the new year. The purpose of the Buyers Groups is to develop a partnership between WWF and companies determined to play their part in improving the quality of forest management around the world. WWF Buyers Groups have been formed in over 12 countries, including United Kingdom, Belgium, Germany, United States of America and Canada. Member companies are committed to purchasing forest products from sources independently certified as being well managed and to promoting these products in the market place. The underlying philosophy of the WWF Buyers Group is one of continual improvement, with members phasing out the trade in uncertified timber, replacing such sources with certified product. WWF Buyers Groups are not about offering customers a choice between timbers of unknown origin and the certified product. The aim is to only offer certified timber — a product the trader knows has come from a well managed source.

Proposals for two field trials of the FSC model of timber certification in Australia have been identified and developed. The proposals have been produced in close consultation with industry and NGO partners and are designed to prove the applicability of the FSC system to Australian native forest and plantation management.

The highly politicised forest debate in Australia, however, has hampered the implementation of the field trials of the FSC system thus far but WWF believes that the FSC system will prove to be successful in Australia and is continuing to advocate its adoption. In New Zealand a group known as the ITTG, the Independent Tropical Timber Importers Group, comprising major forestry companies, importers, wholesalers, retailers and environmental NGOs has expressed interest in becoming *de facto*, the buyers group in New Zealand. So far, these approaches have been successful in raising awareness in Australia and New Zealand of the issues of timber certification and labelling, and the Forest Stewardship Council. Several New Zealand plantation companies have been investigating FSC certification for their operations.

In relation to fisheries, consultations with relevant stakeholders have taken place around the world, including in Australia and New Zealand. The draft criteria for sustainable fisheries have been developed from these consultations, and from existing codes such as those developed by the UN Food and Agriculture Organization. A series of field trials has followed to test and refine the model in practical applications. One of Australia's major seafood companies, Kailis and France Pty Ltd, has already endorsed the MSC approach and an implementation trial is being planned with the Western Rock Lobster fishery. It is planned to launch the MSC logo in 1998, and have MSC labelled products available by late 1998.

A WIDER APPLICATION FOR INDEPENDENT CERTIFICATION?

Similar approaches, based on independent standards and certification, might also be considered in other natural resource categories where similar opportunities are present. For example, labelling schemes are already in existence for some food products, designating cruelty-free production, organic growing, or similar attributes likely to be desirable by consumers.

Household electricity can be generated from a variety of sources with differing degrees of environmental impact. Allowing consumers to choose the mix of their electricity source can shift energy production toward more sustainable and less polluting sources. The promotion of "green electricity" is developing rapidly within the Australian electricity market. Interestingly, this trend seems to have been spurred on by the corporatization of state-owned electricity authorities and their subsequent need to be more competitive and market oriented.

Minerals and mining

Although somewhat removed from the final consumer, the minerals and petroleum sectors are not isolated from competitive pressures with an environmental focus. Competition for access to resources is very keen and companies that can readily prove their environmental credentials to communities and regulators are more likely to win community sanction and licences to operate than competitors who cannot, or are renowned poor performers.

While not marketed directly to household consumers, manufacturers may be persuaded that sourcing their raw materials from environmentally sensitive operations might be a useful device for differentiating their products. This is already a feature of some manufacturing processes operating under environmental management systems certified to the international standard, ISO 14000. Some car manufacturers for example, identify recycled metals content in their cars.

A credible system capable of independently certifying the on-ground environmental performance of mining and petroleum companies would be very valuable in improving environmental performance and communication between stakeholders. For companies it would provide a mechanism to prove environmental probity, which could result in competitive advantage through faster and cheaper exploration and exploitation approvals. For local communities and NGOs, such a system would allow for more faith to be placed in company environmental reports.

Financial services

A number of "green investment" services are now being provided which give consumers options over where their money is invested. In New Zealand, some banks are exploring preferential lending rates to demonstrably green clients, for example those that have cleaned up their contaminated sites. This is an important recognition that the risks associated with non-compliance of environmental laws become a risk management factor for investors. A similar idea has already been advanced in relation to major insurance companies concerned about future claims resulting from climate change induced disasters. Indeed, independently verified environmental performance might eventually become part of standard accounting practice as an important consideration in investment and lending by risk sensitive financial institutions.

Product eco-labels — national versus local?

The costs of developing and meeting national standards is relatively high and returns on the small domestic market may not always be compatible with the expense involved. Large importers may source product from dozens of countries and do not want the marketing costs or confusion that would arise from consumers faced with a multitude of different schemes. Global labels therefore seem to be the most sensible approach. Global voluntary labels should also help counter problems relating to non-tariff barriers and disputes within the GATT framework.

Several countries, including New Zealand, are attempting to develop national eco-labels. Such schemes in small, exporting nations are difficult and some have failed. The Australian scheme Environmental Choice, developed in the 1980s, was an obvious failure. There may be many reasons for this, not the least being a major credibility gap. The Environmental Choice scheme, for example, did not set specific product guidelines, but merely sought to verify any environmental claims made on the product packaging or in advertising. Thus a product could achieve status for being phosphate free, or using only recycled paper, when these were irrelevant to the products main environmental impact. A major review of the benefits and costs of eco-labelling schemes has been published by the OECD (OECD 1997).

How does the community perceive these initiatives?

The use of independent certification and eco-labelling is not universally supported by major vested interests, regulatory authorities, or even other environmental organizations. The reasons for this are varied. Some common criticisms from various sources include:

- the use of market-based mechanisms may be seen as *de facto* endorsement of capitalism and consumerism (this may be a strength or a weakness depending on the prevailing political ideology),
- it assumes that sustainable processes of production are achievable (for example, many environmentalists do not believe that sustainable forestry in native forests is achievable),
- for corporations, it introduces and element of openness, and potential loss of control over the production and marketing of their product,
- it threatens vested interests that believe they have more to gain from confrontation than compromise.

SUMMARY

The promotion of market-based incentives and mechanisms for sustainable development is a rapidly expanding area of interest in the late 1990s. In particular, those resources that can be traced through the production chain in a relatively unaltered form are ideally suited to consumer awareness campaigns. It may be possible to extend the approach to more elaborate manufactured goods, energy, and even services such as tourism and finance. However, as with any new approach, there will likely be mistrust of the proponents, and abuses by the supporters. Distrust of the motives behind "green consumerism" is often well founded, and there is always the possibility that corporations will simply exploit and manipulate consumer desire for environmentally friendly alternatives (see Beder 1997 for an in-depth discussion of this topic). Clearly the proponents of the product certification approach must demonstrate the potential of these approaches before they will be widely accepted.

The market is a powerful force in the allocation of resources. In terms of sustainablitity goals, it seems possible to make use of consumer demand to ameliorate negative impacts, and perhaps even to create new markets for sustainable products that replace existing ones. The use of independent product certification is at an early stage in the timber and fisheries industries. Nonetheless, successful precedents in comparable situations, such as independent auditing of financial accounts; the success of other consumer labelling approaches, and the desire for many corporations to be able to verify their environmental claims in the market place, suggest that product certification could be a powerful tool in promoting the sustainable use of natural resources.

REFERENCES

Beder, S., 1997. Global Spin. The Corporate Assault on Environmentalism. Scribe Publications: Melbourne, Victoria.

Brown, L. R., Flavin, C. and Postel, S., 1992. Saving the Planet. How to shape an environmentally sustainable global economy. Worldwatch Institute. Earthscan: UK.

DPIE, 1997. International Conference on Certification and Labelling of Products from Sustainably Managed Forests. May 1996, Brisbane. Published at http://www.dpie.gov.au/dpie/conference/icocal/papers.

Dudley, N., Jeanrenaud, J-P. and Sullivan, F., 1995. Bad Harvest? The Timber Trade and the Degradation of the World's Forests. Earthscan Publications Limited: UK.

FSC, 1995. FSC Process Guidelines for Developing Regional Certification Standards. FSC: Oaxaca, Mexico.

Jedreski, L. and Rae, M., 1996. Forests for Life. Timber Certification in Australia and New Zealand. WWF Australia Discussion Paper (2nd Edition), Melbourne, Victoria.

Knight, A., 1997. Retail Perspective. In DPIE (1997) International Conference on Certification and Labelling of Products from Sustainably Managed Forests. May 1996, Brisbane. Published at http://www.dpie.gov.au/dpie/conference/icocal/papers.

MSC, 1998. Principles and Criteria for Sustainable Fishing. Marine Stewardship Council, 119 Altenburg Gardens, London SW11 IJQ, United Kingdom.

OECD, 1997. Eco-labelling: Actual Effects of Selected Programmes. OECD: Paris, 1997.

CHAPTER 47

Who pays? Habitat protection on private land

Peter Davis[1] and Chris Cocklin[2]

The extensive clearance of habitat for agriculture and forestry in New Zealand means that much of what remains on private land needs to be retained and protected. A range of mechanisms have been introduced to reduce habitat losses. The benefits of such protection are mostly public, the costs are not. This chapter considers distributional issues related to the approaches taken to retain and protect habitat in New Zealand. The uneven distribution of habitats means there are spatial and social implications arising from the mechanisms adopted. Those with the highest proportion of their land providing habitats, landowners on economically marginal holdings and Maori, are least financially able to carry these additional responsibilities. A similar concern is also evident at the local and regional levels of government; the areas of New Zealand that have high conservation priorities have smaller economic bases from which to support protection measures. The chapter argues that reducing habitat loss by placing additional responsibility on habitat owners penalizes the very landowners who are providing the desired public good. Instead of construing the problem as being habitat loss, it needs to be reframed as being one of habitat scarcity. It then follows that the responsibility for protecting remaining habitats should be related to cleared land not the land that is still in habitat. If the government's goal of "no net habitat loss" is to be achieved, these factors need to be addressed by recognizing the public good of habitat provision and rebalancing the costs and benefits of habitat protection at both the individual and local government level. The concerns are illustrated with findings from a case study of the Northland region.

Key words: Habitat protection, Distributive issues, Mechanisms, Regulation, Northland.

INTRODUCTION

THE question of how the costs of retaining and protecting habitats should be distributed both within and between communities is critical if biodiversity conservation is to be achieved in an effective and equitable manner. The public good nature of habitat protection and the costs involved (opportunity costs, fencing, pest control) mean that it is rarely in the financial interest of habitat owners to conserve their habitats. In order to protect as much of the remaining habitat as possible moral and regulatory pressures are being applied, often in ways which place responsibility for such protection on the habitat owners.

Concerns that such approaches are both inequitous (penalizing those who provide the benefit) and self-defeating (as those with habitats are likely to be less financially able to bear the costs) are relevant at various scales. Environmental advocates in the intensively developed habitat-poor countries have pressured habitat-rich developing countries to forego opportunities such as developing timber or agricultural resources. These have been labelled as hypocritical and have been countered with calls for the habitat-poor North to fully compensate for the costs of such protection (Shiva *et al.* 1991; Panayotou 1994; Gunn 1995). Within developed countries, urban dominated populations are demanding that public environmental goods associated with rural land, such as wildlife and amenity factors, be maintained. Various mechanisms, such as planning restrictions, production subsidies or grants tied to environmental criteria, fees-for-service and compensation for foregone opportunities are being employed to achieve the socially desired level of biodiversity benefits (Bishop and Phillips 1993; Young *et al.* 1996). Those which place restrictions on landowners without compensation are often perceived as infringing on property rights while

[1]Department of Geography, University of Auckland, Auckland, New Zealand.
[2]Department of Geography, Monash University, Melbourne, Australia 3168.
Pages 468–79 *in* NATURE CONSERVATION 5: NATURE CONSERVATION IN PRODUCTION ENVIRONMENTS: MANAGING THE MATRIX ed by J. L. Craig, N. Mitchell and D. A. Saunders. Surrey Beatty & Sons, 2000.

burdening landowners with the costs of providing public environmental goods (Lowe *et al.* 1993; Marsden *et al.* 1993).

The way that the costs (and benefits) of habitat protection programmes are distributed within and between communities depends on the mechanisms used and the geographic scale at which they are applied. A fundamental difference between the various approaches is the way that landowners' rights and duties with respect to habitats are construed. Retaining habitats can be seen as providing a public good that deserves recompense; alternatively habitat clearance can be judged to generate a public bad which should be penalized (Hodge 1991a, 1991b). How the problem is construed in turn affects the acceptability and effectiveness of habitat protection programmes.

This chapter focuses on these issues in the context of New Zealand, a country which proudly promotes a clean green image, but which has lost around 70% of its forest cover, 90% of its wetlands and over 50 endemic species during its relatively short history of human colonization (Department of Conservation 1994; Ministry for the Environment/Ministry of Foreign Affairs and Trade 1995; Ogle 1994). We consider the question of how the costs of habitat protection in New Zealand are distributed under a framework that devolves regulatory control to the regional and local councils and addresses environmental problems by managing externalities.

The first two sections provide historical context and an overview of the regulatory regime relevant to habitat protection. The following two sections address the question of how appropriate the current approach to habitat protection is in terms of the distribution of the costs and benefits at both the individual landowner and the local government level. The analysis is illustrated with a case study of the Northland region, an area that is habitat-rich, income-poor and which has a large number of threatened species. Plans to protect areas of significant habitats have drawn strong opposition from landowners who feel they are being burdened with the cost of providing environmental benefits. The final section suggests alternative approaches that address some of the inequities identified.

HISTORICAL CONTEXT

The clearance of indigenous ecosystems during New Zealand's relatively short human history has been rapid and extensive (Fig. 1). In pre-settlement times around 80% of land was forested. By the time of European colonization this had been reduced to about 53%, 6.7 million hectares having been cleared by Maori (Molloy 1980). Following European settlement the rate of depletion increased as state policies aimed at developing a pastoral-based rural economy and imported technologies accelerated the conversion of large swaths of native forest into pasture. As a result of habitat clearance, over-hunting and the introduction of pest and weed species, more than half the endemic bird species have become extinct and over 400 species of plants and animals are listed as threatened with extinction, including the kiwi (*Apteryx* spp.), one of New Zealand's national symbols (Ministry for the Environment 1995).

Times have changed since the pioneering days when bush clearance was virtually an obligation on landowners. In particular, over the last decade significant steps have been taken to reduce the loss of indigenous habitats; most publicly owned indigenous forests are now protected, public funding for protecting private habitats has increased, and most felling of indigenous forests by private corporations has ceased. However, although the public conservation estate covers over a quarter of the land area (over eight million hectares), proportionally it is unrepresentative of certain indigenous ecosystems, especially the lowland forests and wetlands (on which many threatened species are largely or solely reliant) that are typical of the more accessible and fertile lands (Molloy 1980; Ministry for the Environment 1995).

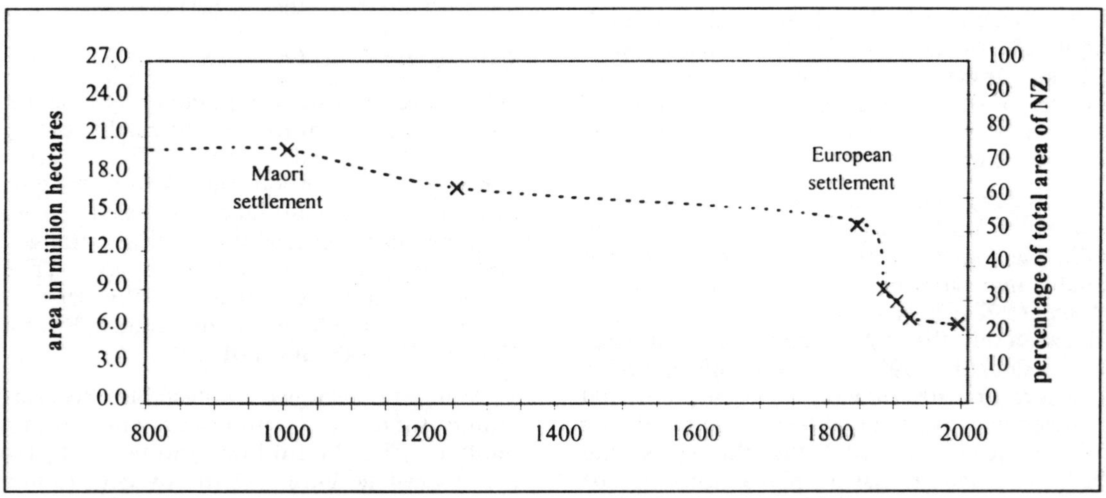

Fig. 1. Changing extent of indigenous forest cover in New Zealand. Source: Data from Molloy 1980; Memon and Wilson 1993.

About 20% of the remaining habitats are on private land, and many of the habitat types and species that are under-represented in the public estate depend upon these for their survival. Few of these habitats have the formal protection afforded by covenants or management agreements, and while some have been intentionally retained and cared for by the owners, their ongoing preservation may depend on economic factors or the attitudes of future landowners. Few if any habitats are unaffected by the insidious damage caused by stock, weeds and pests, and some remnants are effectively on "death row", retaining their current form only while the standing canopy survives (MacGibbon 1996:20). There is an urgent need to reduce both active and passive forms of destruction, and increase the level of legal, permanent protection so that the biodiversity and ecosystem values of these habitats can be retained (Cutting and Cocklin 1992).

REGULATORY CONTEXT

Historically, vegetation clearance in New Zealand has been largely unregulated. Constraints on felling native forests were seen as politically unacceptable and undesirable during the expansionary phase of land development. Measures that were introduced to control erosion and flooding problems were rarely enforced (Memon and Wilson 1993). However, as the area of habitat decreased, demand for its retention for amenity, recreational and biodiversity conservation purposes increased. In the late 1980s, with the introduction of sustainable management principles into legislation, landowners' rights to destroy habitats were significantly reduced (Memon 1993).

Sustainable management and habitat protection in a regulatory context

The *Resource Management Act 1991* transformed the regulation of resource use from an approach based on prescriptive zoning, to one that regulates the environmental effects of human activities. It inscribed "sustainable management" as the guiding principle for decisions over the allocation and use of natural and physical resources, creating an externalities-based regime that upholds the rights of individuals to do as they choose with their land, provided their choices do not reduce the welfare of others or undermine the ability of the environment and resources to be managed on a sustainable basis.

Power is delegated to the lowest administrative level at which matters can be properly addressed, so while the *Resource Management Act* defines sustainable management (s.5(2) *Resource Management Act 1991*), it is interpreted at the regional and local level, through negotiation and contestation (Cocklin 1996). To overcome conflicts between levels of government, regional plans and policy statements must not be inconsistent with any national policy statements, and the rules and policies contained in district plans must not be inconsistent with those at either the regional or national level.

The *Resource Management Act* requires that significant areas of indigenous vegetation and habitats of fauna be protected (s.6(c)), though in keeping with the devolutionary principle of the Act, interpretation of "significant" is determined by the regional and district councils. In addition land may not be "used" in a manner that contravenes district or regional plans (or proposed plans) unless a resource consent is granted by the appropriate authority, with the term "use" covering the destruction of, damage to, or disturbance of, the habitats of plants or animals in, on, or under the land (s.9(4c)). Both levels of local government can establish rules to control the clearance of habitats, though with different foci; regional councils for the purposes of soil conservation and management of water quality and quantity (s.30(1)), district councils for land use management.

Educational and motivational approaches can also be used to protect habitats, as well as development incentives that are conditional on habitats gaining formal protection, such as non-transferable subdivision permits and development bonuses which allow for additional site coverage (Cocklin and Doorman 1994; Far North District Council 1997). Rates relief can be also used as an incentive. Councils must remit at least 50% of rates for most land owned for conservation purposes (e.g., under the *Conservation Act* or *Reserves Act*) and may remit up to 100% of these rates. Councils may also remit or postpone rates on habitats with other forms of protection such as a Queen Elizabeth II Covenant or informal protection (Bellingham 1992; Department of Conservation 1995).

Logging of native forests is controlled by the *Forest Amendment Act (1993)*. Apart from a few exceptions, timber can only be harvested under a sustainable management system approved by the Ministry of Forestry which specifies extraction rates and methods and ecosystem maintenance requirements. However, the Act does not prevent forests being cleared for other land uses (though this may be covered under the *Resource Management Act*), nor does it oblige owners to maintain their forests or protect them from stock or pest damage.

Agencies and mechanisms

In addition to these regulatory approaches there are several mechanisms by which habitats on private land can be formally protected. The Department of Conservation has various mechanisms for protecting habitats on private land, including covenants, management agreements, leasing arrangements, and land exchanges and purchases. The other primary mechanism used is the Open Space Covenant (1) offered by the Queen Elizabeth II National Trust (Saunders 1996).

There are also two contestable, criteria-based public funds for purchasing and covenanting habitats. The $4 million annual budget of the Forest Heritage Fund (2) mostly goes to proposals from the Department of Conservation (generally for land purchases) and the Queen Elizabeth II

Trust (mostly for covenants on native forests). The second fund, Nga Whenua Rahui (3), is used to protect habitats on Maori land through kawenata (covenants) which enable tino rangatiratanga (sovereignty) to be upheld (Piripi 1991). These can incorporate review options that allow future generations to make their own decisions regarding their resource (Young 1991).

In addition, there are a number of non-governmental initiatives, both national and local, to conserve and protect indigenous habitats. These range from those with a specific focus of habitat protection and maintenance to industry-related initiatives promoting the sustainable management of native forests.

While this raft of mechanisms is indicative of the growing commitment to habitat protection, their effectiveness is often restricted by budgetary and political factors. In addition, inadequate monitoring of habitat loss means that the ability of the current mix of mechanisms to reduce habitat loss is still largely unknown.

HABITAT PROTECTION IN NORTHLAND

The Northland region (Fig. 2) covers some 1 260 000 hectares, of which 59% is pasture, 14% scrub, shrubland and dune vegetation, 13% indigenous forest, 10% exotic plantations and 1.8% is wetlands, lakes and rivers (Northland Regional Council 1997). A very small portion is in horticulture, unvegetated dunes and urban land use.

The regional economy is predominantly land-based, with farming contributing some $1 billion to the region's economy (dairy $630 million, sheep and beef $370 million) (Northland Regional Council 1997a) and employs around 20% of the workforce (Statistics New Zealand 1993). Dairying and forestry are the main growth industries currently; dairying increased from 123 000 hectares to just under 131 000 hectares between 1992 and 1995, and over 20 000 hectares of plantation forest, mainly radiata pine, were established in the three years to 1995, bringing the total area to around 150 000 hectares (Macmillan 1995). As a result of this land based economy the region has one of the highest proportion of workers dependent on farming (20%), however, it is not a "rich" farming area. The proportion of those occupied in agriculture and fisheries who earn under $20,000 per annum is the third highest in the country (1991 Census data). Low incomes are not confined to the agricultural sector, as the region also has the lowest per capita income.

In terms of biodiversity, Northland has a high diversity of endemic species and is one of the most habitat-rich regions of New Zealand with around 30% of the land area still in indigenous habitats. However, largely as a result of habitat loss, it also has the largest number (24) of Category A (4) threatened species and the third highest total (101) of all three categories (Department of Conservation 1995; Northland Regional Council 1997a). The Department of Conservation's network of protected areas covers some 150 000 hectares. Though extensive, it does not adequately represent the natural values of Northland, nor does it provide sufficient protection for the endangered species. There is an urgent need to reduce the loss of the habitat remaining on private property (Department of Conservation 1995; Northland Regional Council 1997a). Of the 200 000 hectares of habitats (5) on private land, only a tiny proportion has formal protection under the Queen Elizabeth II Trust or other mechanisms. For the great majority of habitats the only formal protection is that provided by the rules laid down by the regional and district councils.

The habitats that remain on private land are not evenly distributed among different types of landowners. A survey of Northland Federated Farmers found that beef and sheep farmers have more habitat on average than dairy farmers, both in terms of extent and proportion of land area. Beef/sheep farms had 32 hectares of bush or 11% of the farm on average compared to dairy with 12 hectares or 8% of farm area (Davis and Cocklin 1997). It is also significant that beef and sheep farmers generally have a lower income than dairy farmers and are less financially able to undertake habitat protection measures such as fencing (Blunden *et al.* 1997).

The other factor that needs to be considered here is the large Maori population in the region. At the district level the figures are 41% in the Far North, 24% in Whangarei and 22% in Kaipara (compared with a national figure of 14%). National figures show Maori landowners have a high proportion of habitat remaining on their land. Maori land includes customary and freehold Maori land, but does not include Maori owned General land (i.e., alienated customary land that has subsequently been purchased by Maori). There are around half a million hectares of native forest on Maori land out of a total Maori land area of 1.5 million hectares (Ministry for the Environment 1997; Te Puni Kokiri 1997). The proportion of habitat remaining on Maori land in Northland has not been ascertained, but it is likely to reflect this trend by being higher than that retained on land in general.

Table 1. Summary of land use in Northland.

	Far North	Whangarei	Kaipara	Northland
Total area (ha)	684 100	272 000	303 000	1 260 000
Pasture, exotic forestry, horticulture (% total area)	56%	71%	80%	65%
Forest, shrubland, wetlands (% total area)	38%	20%*	12%*	28%
Department of Conservation estate (% total area)	17%	5%	7%	12%

Sources: Statistics New Zealand various years; Far North District Council 1996; Kaipara District Council 1997a; Whangarei District Council 1997. * estimate based on various sources.

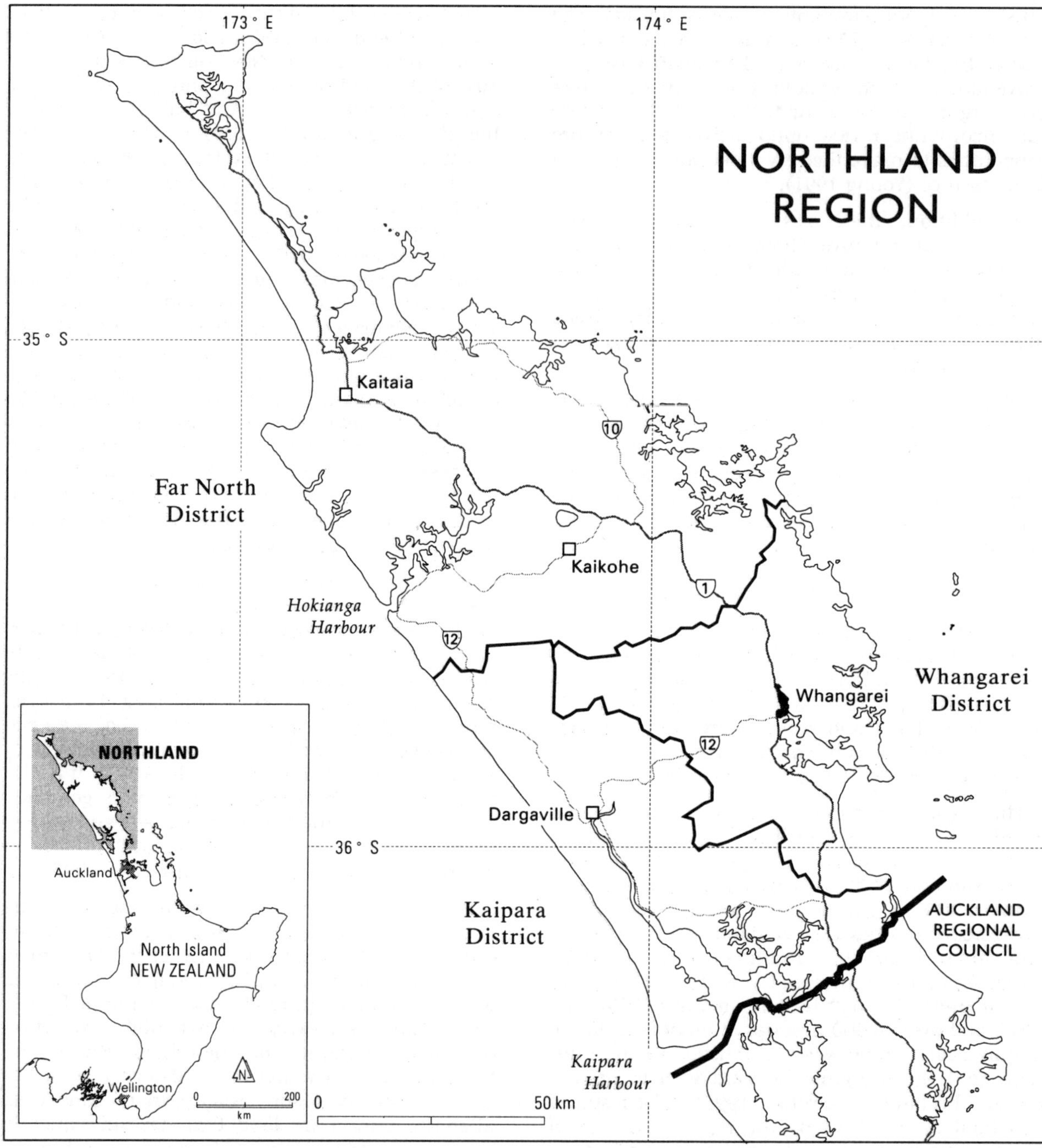

Fig. 2. Northland region with regional and district government boundaries. Map by Jan Kelly, Department of Geography, University of Auckland.

Policies, rules and protecting habitats

The Northland region is covered by one regional council and by three district councils (refer Fig. 2). As explained in an earlier section, the two levels of local government have different responsibilities for protecting habitats. In fulfilling its obligations, the Northland Regional Council has established policies for the protection of significant habitats (Northland Regional Council 1997a) and for protection of habitats for water and soil quality control (Northland Regional Council 1997b). It has determined that significant habitats shall be those ranked as having Moderate (local), Moderate-High (regional), High (national) and Outstanding (international) significance according to Sites of Special Biological Interest and Protected Natural Areas criteria (Northland Regional Council 1997a).

The approaches that the three district councils have taken to discharging their duties under the *Resource Management Act* in terms of habitat protection have all involved developing schedules of significant sites, determining which categories of significance deserve protection and then establishing rules to achieve the desired goals (Table 2). The survey undertaken for the Far North District

Council identified 884 Significant Natural Areas covering a total area of 242 000 hectares (35% of the district). Nearly two-thirds (63%) of this area is in private ownership (Far North District Council 1997a). In Whangarei District, 319 Sites of Ecological Significance were identified, covering just under 60 000 hectares, or 22% of the district (Boffa Miskell 1995). The vast majority of these (94% by area) are of High or Outstanding ecological significance. Approximately a third of these are under the Department of Conservation or Council protection (L. Chester, pers. comm. 1997). Kaipara District Council relied upon a Wildlife Service survey undertaken in the late seventies that focused on wildlife rather than fauna or habitats. One hundred and eighty of these sites still exist, covering just under 40 000 hectares (13% of the district). Around 40% of this area is on private land. Over 95% of the sites have High or Outstanding ecological significance. The Significant Natural Area survey that the Department of Conservation are currently undertaking is expected to significantly increase the number of sites (P. Anderson, pers. comm. 1997).

The regional council's definition of significant suggests that over 95% of the habitats identified in the various surveys, over a quarter of the total land area, require some protection. Around 60% of this is on private land.

Regulatory controls at the district level

The three district councils have each reached different stages (6) establishing their rules in law and have proposed a range of controls on vegetation clearance (Table 3). The Far North District Council did not differentiate between types of Significant Natural Areas and proposed that consents be required for activities affecting all Significant Natural Areas. However, in the face of massive public protest the proposed plan has been withdrawn and will be rewritten (refer next subsection). The Whangarei District Council have proposed that private landowners will only require consents for activities in Outstanding Sites of Ecological Significance (4% of significant habitat area). Owners of High Sites of Ecological Significance will have the option of having their habitats protected under the plan if they so choose, while all lower ranked habitats would have no protection under the District Plan (Whangarei District Council 1997). This approach appears to be inconsistent with the Regional Policy Statement and may therefore change. In Kaipara District, the Operative Plan requires consents for clearances of indigenous vegetation of over five hectares in any one year. A proposed change would require consents for activities likely to adversely affect sites of Moderate significance or higher. More stringent conditions would apply to sites of High and Outstanding significance.

Controlling passive damage to habitats

Deliberate clearances of habitats is not the only problem facing indigenous species in Northland. They are also under threat from stock grazing and pests. The regional council has adopted a Pest Management Strategy which recognizes that the conservational benefits of pest control are shared by landowners and the public (Northland Regional Council 1995). Regional rates therefore fund an initial knock down of possum numbers while responsibility for maintaining low numbers rests with the landowners. Protecting habitats against stock is more difficult. Existing use rights mean

Table 2. Summary of significant habitats in Northland.

	Far North	Whangarei	Kaipara [a]	Northland
No. of significant habitats (exc. estuarine)	870	302	176	1 348
Total area of Significant Forest, Shrublands or Wetlands [b] (ha)	242 000	48 000	40 000	330 000
Proportion of district with significant habitats (%)	35%	22%	13%	27%
Area of Significant Forest, Shrublands or Wetlands [b] on private land (ha)	147 000	32 000	15 000	194 000
% of district that is private Significant Forest, Shrublands or Wetlands	21%	12% [c]	5%	15%

[a] Kaipara figures are expected to increase with the new Significant Natural Area survey. [b] Significant Forest, Shrublands or Wetlands of Moderate significance or higher as per Northland Regional Council (1997). [c] Estimate provided by the Whangarei District Council.

Table 3. Vegetation clearance controls imposed by District Councils.

District Council — *Status of decision* — Rules
Far North — *Variation to Proposed District Plan* (Plan now withdrawn) — In Significant Natural Area habitat clearance is *discretionary* if over 500 m² and for any activity that might affect a scheduled wetland. Outside Significant Natural Areas habitat clearance is *permitted.*
Whangarei District — *Draft District Plan* — All Outstanding Sites of Ecological Significance to be considered significant. Landowners with High Sites of Ecological Significance will have the option to request their habitat be covered by the rules.
Kaipara District — *Proposed Change No. 1* — The proposed change would mean activities likely to affect sites of Moderate-High and Moderate significance will be *discretionary*, those of Outstanding and High significance will be *non-complying*. In the interim the Operative Plan treats clearances of over five hectares of indigenous vegetation in any one year as a discretionary activity.

Source: Far North District Council 1997a, Kaipara District Council 1997b; Whangarei District Council 1997.

councils cannot require landowners to exclude stock from habitats to which they have previously had access. While constraints can be placed on new grazing (e.g., Far North District Council 1997) such an approach is likely to be applicable in very few cases.

Variations within a region

The different approaches taken by the councils mean that the proportion of landowners with significant habitats who will be affected by the rules varies; the more stringent rules proposed by the Far North District Council meant a higher proportion of habitat owners were affected than is the case in Whangarei District under their draft proposals. However, even if the same criteria (those of the Northland Regional Council) are applied across the region the proportion of ratepayers affected by the rules varies considerably; from around 14% in the Far North to around 5% in the urban dominated Whangarei District (Table 4).

None of the district councils have completed the process in respect of rules governing habitats. The eventual level of protection that will be in their Operative Plans could well change in response to submissions or public pressure. The Far North District provides a good example of this. The proposed controls on clearance of significant habitats and landscapes that were in the Far North District Council *Proposed Plan* (October 1996) were published prior to any public consultation and without the public being informed of the existing vegetation clearance restrictions, the need for protection or the methods being proposed. These were subsequently loosened in the *Variation* (March 1997) largely as a result of public pressure. Since then over 4 000 people have made submissions to the Far North District Council, most of which challenge the designation of areas as significant and the rules covering these sites (Far North District Council 1997b). As a result of the vociferous public campaign the Council has withdrawn the *Proposed Plan* and will redraft it (*New Zealand Herald* 1997). This will most likely result in the form of controls and their coverage being reduced.

THE COSTS OF PROTECTION

The actual costs incurred by owners of significant habitats as a result of the rules restricting development options will vary widely depending on various factors; location, topography, adjacent land use, etc. The foregone opportunities will depend on the potential use of land. These opportunities are factored in to the price of land. Most undeveloped land in Northland ranges between $500 and $2,000 per hectare, though a more typical range would be between $800 and $1,200 (G. Eady, pers. comm. 1996). Land with high amenity value, such as coastal areas and those close to towns, can be much higher. Ministry of Forestry figures show forestry returns in Northland to be around 7 to 13% per annum (Ministry of Forestry 1994). Lease arrangements can return an annual payment of 7% of the land value plus a share of the stumpage returns (K. Prime, pers. comm. 1996). Restrictions on habitat clearance are likely to reduce the relative value of uncleared land except in more highly sought after residential areas where retained habitats can improve at least the saleability, if not the value of the properties (V. Winiata, pers. comm. 1996; Young 1996).

Rates are paid on all privately owned land, though councils can provide rate relief for habitats with formal and informal protection. Not surprisingly, many landowners object to paying rates on habitats that have been identified as significant and which therefore have constraints placed on them. Maori have been particularly concerned about this because of the disproportionate amount of their land that is undeveloped (Spencer 1993; Mutu 1994). Their concern has been partially addressed by the recent Appeal Court ruling that multiple-owned Maori land should be valued differently to other land for the purposes of rating (Waikato Times 1997). The payment of rates on significant habitats also concerns the Department of Conservation as landowners use the fact that they are having to pay rates on such areas to support their applications for consents to clear them (P. Anderson, pers. comm. 1997).

While rate relief is promoted as an incentive for habitat protection by many councils, they usually require that some formal protection mechanism be used, such as an Open Space Covenant. However, the cost of establishing these far exceeds that saved by avoiding the rates bill. Rate relief is less of an incentive than a token reward for those who wish to protect habitats for other reasons, such as personal commitment or to secure subdivision options (Davis and Cocklin 1997).

Estimates for the costs that can be associated with a typical 30 hectare area of bush on a Northland farm are summarized in the table below. The cost of owning the land would average around $30,000 and the opportunity costs for land that had the potential to be "developed" for forestry would be around $1,500 to $2,000 per year. Rates would cost another $200 per annum. Landowners are under no

Table 4. Landowners affected by habitat protection policies in Northland.

	Far North	Whangarei	Kaipara[a]	Northland
Approx. No. of landowners with Significant Forest, Shrublands or Wetlands[b]	2 200	1 000	450	3 650
% of ratepayers[c] potentially affected by Regional Policy Statement	14%	5%	8%	8%

[a] Kaipara figures are expected to increase with the new Significant Natural Area survey. [b] Significant Forest, Shrublands or Wetlands of Moderate significance or higher requiring protection as per Regional Policy Statement (Northland Regional Council 1997). [c] Estimated using number of dwellings from 1991 Census.

Table 5. Costs involved in retaining and maintaining habitats.

Approximate costs for 30 hectares of bush in Northland		
Land value	$30,000	Ranges from $15,000 to much higher in desirable residential areas.
Fencing	$9,000	Lasts 20–30 years. Based on average fencing costs involved in creating a Queen Elizabeth II covenant. (Contract fencing costs are $10–11 per metre, 60% labour 40% materials.)
Pest control	$300 p.a.	Based on contractor costs for possum control only.
Rates	$200 p.a.	Based on land value using rate of .007 (i.e., 0.7 cents in the dollar).
Opportunity costs	$1,500–2,000 p.a.	Based on figures for forestry development.

Sources: Ministry of Forestry 1994; Queen Elizabeth II 1994, 1996; K. Prime, pers. comm 1996; Department of Conservation 1997.

obligation to undertake fencing, though doing so would often contribute to the health of the habitat. They are, however, under some obligations to undertake pest control. Possum control alone could add another $300 on the annual costs.

Landowners, property rights and responsibilities for biodiversity conservation

Landowners' traditional rights to develop land by clearing bush and draining wetlands evolved whilst indigenous habitats and species were relatively abundant. As habitats have become more scarce and the demand for biodiversity protection has increased, the rights of landowners to destroy habitats have come under increasing scrutiny. The *Forest Amendment Act* and the *Resource Management Act* substantially redefined landowners' (presumed) rights to externalize their environmental costs, including their rights to clear habitat. In Northland, the regional and district councils are relying primarily on regulatory mechanisms by placing restrictions on significant habitats. The owners of these habitats are being told that the (public) benefits that their habitats provide are important and should not to be terminated.

Many farmers and other rural landowners claim that this approach takes away their property rights and gives them an unfair share of the costs associated with biodiversity conservation (e.g., Bailey 1997; Upton 1997a). These objections have some justification. The reason owners of significant habitats find themselves facing constraints is largely because of the cumulative effect of all previous land development in that area. Their habitats would be less likely to be judged as "significant" and in need of protection if an abundance of such habitats remained. The regulatory changes have, in effect, retrospectively allocated rights to clear habitat on a "first-in-time first-in-right" basis. The owners of the remaining habitats face additional obligations to conserve biodiversity for the public good compared to their habitat-free neighbour or urban cousin and are in effect being penalized for previously unrestrained clearance.

The social and economic implications of this are important. Some sectors of society, particularly Maori and less financially viable landowners, are being disproportionately affected by policies aimed at solving a problem for which they are not especially responsible.

Who pays at the collective level?

As has been mentioned earlier, Northland is a habitat-rich area. Around 15% of private land in the region has habitats that have been identified as being of national significance. The importance of conserving these therefore extends beyond the regional level. It is also worth noting that Northland is relatively economically disadvantaged. The income levels of both the agriculture sector and the region as a whole are among the lowest in New Zealand (Statistics New Zealand 1997). By contrast, many urban and intensively farmed parts of New Zealand have very little habitat left, while the economic bases upon which habitat protection initiatives can be based are much stronger. It should be noted here that much of the support for conservation issues has come from urban voters (Pawson 1992).

There are also substantial differences within the region in terms of the extent of habitat requiring protection and the proportion of ratepayers affected by the rules. The economic differences within the region are quite distinct. The Far North has the third lowest per capita income for the whole country ($17,000). The figures for Kaipara ($18,700) and Whangarei ($19,400) are both significantly higher, though still well below the national average ($21,600) (Statistics New Zealand 1997). Based on the number and extent of significant habitats identified, the Far North District Council has the highest conservation need of the three districts. It is no coincidence that it has drawn such strong challenges to its proposed plan. While it has been argued that the processes adopted by the Far North District Council has contributed to the protests (Upton 1997b), the fact that the district has the highest proportion of ratepayers affected by habitat protection policies should not be ignored.

The Minister for the Environment has urged councils to consider non-regulatory approaches to protecting habitats such as the payment of incentives or compensation (Upton 1997a). The ability of councils to offer these would clearly vary depending on the need for habitat protection and the funding base of the councils. One simple initiative that would be supported by landowners would be a more liberal rate relief policy on habitats that have been identified as significant (Blunden *et al.* 1997). The district councils in Northland collect around one million dollars of rates from

Table 6. Estimated land values and rates for significant habitats in Northland.

	Far North	Whangarei	Kaipara	Northland
Significant Forest, Shrublands or Wetlands [a] on private land	147 000 ha	32 000 ha	15 000 ha [b]	194 000 ha
Approximate land value ($/ha)	$500–800	$1,000–1,200	$800–1,000	—
Land value of Significant Forest, Shrublands or Wetlands ($ m)	$74–118 m	$32–38 m	$12–15 m	$118–171 m
Rate [c] ($ per $ of land value)	.0079	.0062	.0060	—
Rates on private Significant Forest, Shrublands or Wetlands [d] ($000)	$580–930 k	$200–240 k	$72–90 k	$852–1,260 k
Rate strike (1995/96)[e] ($ m)	$22 m	$25 m	$6 m	$53 m
Rates on private Significant Forest, Shrublands or Wetlands as % of total rate strike	2.5–4.1%	0.8–1.0%	1.2–1.5%	1.6–2.4%

[a] Significant Forest, Shrublands or Wetlands of Moderate significance or higher as per Northland Regional Council 1997a. [b] Based on old Site of Special Wildlife Interest database; will increase with new Significant Natural Area survey. Kaipara District Council and Whangarei District Council have a single general rate for rural land. Far North District Council has several rates, the Kaikohe rate has been used as it is mid-range (in areas with lower land values a higher rate is used, so the median rate is likely to understate the amount of rates paid) (A. Melville, pers. comm. 1996). [d] Some rate relief is already provided for protected areas but is insignificant e.g., less than $40,000 in the Far North District Council. [e] Data provided by Local Government New Zealand.

the 200 000 hectares of significant habitats on private land. Table 6 provides some estimates for the amounts that may be involved, using a land value that would be toward the lower end of the price range. If rate relief was provided for significant habitats, this amount would need to be regained by increasing the rates across the district, a move that is unlikely to gain voter support. The differences between the districts are again noteworthy, varying by a factor of three.

The Department of Conservation has identified 150 key sites (70 000 hectares) in Northland that it deems are in greatest need of protection and which have international or national significance (Department of Conservation 1997). If protection were achieved using covenants or acquisition by the Department of Conservation the rates currently collected from that land would be foregone (Department of Conservation currently do not pay rates). This alone would necessitate the redistribution of around $300,000 across the ratepayers.

With obligations for habitat protection resting at the local level, there is cause for concern that those districts (and regions) with the greatest need for habitat protection may be too economically or politically vulnerable to achieve the desired level of protection. Councils with small economic bases will be less able to provide rate relief and other incentives, and gaining political support for policies to protect significant habitats will be harder if those policies affect a large section of the community. There is some reason, therefore, to ask whether it is appropriate that the financial consequences of protecting these habitats should rest primarily at the local/regional tier of government.

ALTERNATIVE APPROACHES

If the goal of reducing habitat loss is to be achieved equitably and effectively other approaches need to be tried which overcome the disparities and inequities identified in this chapter. Such approaches need to reward those who provide the desired environmental goods rather than penalizing them; they need to rebalance the allocation of benefits and costs between districts, and they need to encourage habitat owners to maintain their habitats as an asset rather seeing them as a burden. The current approach invites landowners to ignore their habitats or, worse, actively degrade them, such as by encouraging grazing, so that the habitat's ecological significance is lost and the constraints on developing the site are removed.

While there is an urgent need to reduce habitat loss, it is necessary to reframe the problem from the current conception that it is one of habitat loss to seeing it as one arising from habitat scarcity (Table 7). Having done this, it then makes sense to link responsibility for biodiversity conservation to land that provides no habitat benefits rather than the other way round. This would, we suggest, overcome some of the inequities in the existing system and result in a more acceptable and effective approach. However, there would clearly still be problems of operating this at the local level if no redistribution occurred at higher levels of government.

One option would be a price-based approach that involves charging a levy on all cleared land to provide funding to pay habitat owners for the biodiversity services provided. Such a fee would provide an incentive to retain and maintain the habitat (presuming some form of criteria that included maintaining the health of the habitat were used). It would also encourage some landowners to retire marginal land and actively restore indigenous habitats on that land. A similar notion of a greenkeeper fee has been proposed in New Zealand as a way of making the urban population pay for the rural landscapes that they wish to have retained (Christian 1996).

Another approach would involve tradeable habitat obligations. These would require a reference level being set for the proportion of land that should be retained in indigenous ecosystems for biodiversity conservation purposes. Landowners with less than the reference level would compensate those who provide an excess of biodiversity benefits (Bromley 1996). A reference level approach has been used in

Table 7. Reframing the problem of allocating responsibility for habitat protection.

	Current approach	Alternative approach
Problem	Loss of additional habitat.	Scarcity of habitat.
Approach	The clearance of habitat incurs environmental costs.	Habitat retention has environmental benefits.
	The consequences of additional clearances are "managed" by restricting the rights of habitat owners to undertake further clearances.	The costs associated with protecting sufficient habitat for biodiversity conservation are internalized as a cost of owning cleared land.
	Rights to clear habitat are allocated retrospectively on a "first in time — first in right" basis.	Habitat owners receive a fee for the public benefit provided by retaining and managing their habitats.
Outcome	Responsibility is linked to the amount of habitat one owns.	Responsibility is inversely related to the amount of habitat provided.
	Habitat-rich individuals and districts/regions incur most of the costs.	Habitat-poor individuals and districts/regions incur most of the costs.
	An incentive is provided for owners of habitats to degrade them to the point at which their biodiversity benefits are lost.	An incentive is provided for owners of marginal lands to retire them and undertake habitat restoration.

Western Australia, however, their system does not overcome some of the inequities discussed earlier. Their approach involves a reference level of 20% indigenous vegetation cover at the property and shire level (Young *et al.* 1996). Where habitat cover is below this at the shire or property level, a would-be developer has to prove that further habitat clearance will not threaten the biodiversity. Where the shire level is above 20%, habitat-rich landowners can clear down to the reference level. Landowners who are already at or below the 20% are under some obligation to retain their habitat. However, where the shire level is below 20%, those with an abundance of habitat may not be able to undertake any clearance, and are thus disadvantaged because of the high level of clearance that has already occurred on other properties.

CONCLUSION

The current approach to reducing the loss of habitats on private land is based on the premise that those with habitats on their properties have obligations to provide public environmental benefits, obligations which owners of cleared land avoid. By penalizing those who, intentionally or otherwise, provide conservation benefits, the current approach creates injustices, provokes unnecessary conflict and undermines the potential for achieving the desired goal. The fact that income levels and habitat levels tend to show opposite trends, at both the individual and collective level, exacerbates these inequities and undermines the potential to achieve the goal of no net habitat loss.

The devolution of responsibility to the local level embodied in the *Resource Management Act*, and championed by its architects, also creates disparities. At a regional level, the integration of urban and rural populations that occurred as part of the legislative overhaul may ensure that the environmental demands of the urban populations hold sway to the cost of the more rural districts within the region. Alternatively, habitat-rich districts, such as the Far North, could find themselves in no-win situations, with the large numbers of ratepayers affected by policies designed to protect nationally significant habitats creating substantial opposition to the proposed plan, leading to a lessening of the restrictions at an ecological cost to the nation.

Alternative approaches that recognize the value that habitats provide and which rebalance the costs and benefits of habitat protection between landowners and across jurisdictional boundaries are necessary to overcome these problems. At the same time, such approaches could address both the need for ongoing maintenance of habitats and the need to encourage owners of marginal lands to consider retiring the land from production and providing, instead, conservation benefits.

ACKNOWLEDGEMENTS

The research on which this chapter is based benefited substantially from the financial support of the Department of Conservation (Grant No. 2152). It also benefited from the access provided to work carried out under the project funded by the Foundation for Research, Science and Technology: "Definition and analysis of sustainable land-based production", based in the Department of Geography, University of Auckland (Foundation for Research, Science and Technology Contract No. UOA 509). The authors are particularly thankful for the assistance and input of Dr Greg Blunden. The support of the Northland Regional Council's Sustainable Management Scholarship for research undertaken as part of an M.A. thesis is also gratefully acknowledged.

ENDNOTES

1. Just over a thousand covenants have been established, protecting some 41 000 hectares of native habitats (Royal Forest and Bird Protection Society of New Zealand 1997).
2. By mid-1996 the Trust had committed a total of $26 million to protect just under 100 000 hectares, around half of which involved purchasing (A. McKenzie, pers. comm. 1996).
3. By August 1996, Nga Whenua Rahui had funded the protection of around 75 000 hectares at a cost of just over $12 million (A. McKenzie, pers. comm. 1996).

4. Threatened species are categorized using seventeen criteria to assess five factors; taxonomic distinctiveness, status of the species, threats facing the species, vulnerability of the species, and human values. Species are allocated to three categories according to their score. Those in Category A are the Highest Priority Threatened Species (Department of Conservation 1994).
5. In this discussion estuarine habitats have been excluded as they do not fall within landowners or district council jurisdictions.
6. Plans and policy statements go through several stages; an optional *draft* stage, they are then publicly notified as a *proposed* document (at which point the rules have some legal standing) and finally, following the resolution of further submissions, they become *operative*.

REFERENCES

Bailey, M., 1997. New fund will defend eroded property rights. *Nat. Bus. Rev.*, July **4:** 54.

Bellingham, M., 1992. Protection of Land, Plants and Animals. Pp. 205–34 *in* Handbook of Environmental Law ed by C. D. A. Milne. Royal Forest and Bird Protection Society of New Zealand: Wellington, New Zealand.

Bishop, K. D. and Phillips, A. A. C., 1993. Seven steps to market — the development of the market-led approach to countryside conservation and recreation. *J. Rural Stud.* **9**(4): 315–38.

Blunden, G. and Cocklin, C., 1995. Farmers' perspectives on sustainable land use and rural communities in Northland. Working Paper No. 1, Department of Geography, University of Auckland: Auckland, New Zealand.

Blunden, G., Davis, P. R and Cocklin, C., 1997. The Resource Management Act and pastoral farmers in Northland. Working Paper No.7, Department of Geography, University of Auckland: Auckland, New Zealand.

Boffa Miskell Ltd, 1995. Whangarei District — Sites of Ecological Significance. Report prepared for Whangarei District Council.

Bromley, D. W., 1996. The environmental implications of agriculture. Paper presented to OECD seminar on environmental benefits from a sustainable agriculture: issues and policies, Helsinki, 10–13 September 1996, (Bromley, University of Wisconsin-Madison).

Christian, G., 1996. Greenkeeper fee due farmers. *New Zealand Herald* **November 11:** A11.

Cocklin, C. and Doorman, P., 1994. Ecosystem protection and management in New Zealand: a private land perspective. *Appl. Geog.* **14:** 264–81.

Cocklin, C., 1996. The Resource Management Act: Issues of interpretation and meaning. Working Paper No. 2, Department of Geography, University of Auckland: Auckland, New Zealand.

Cutting, M. and Cocklin, C., 1992. Planning for forest conservation in the Auckland region, New Zealand. *Lands. Urban Plan.* **23:** 55–69.

Davis, P. R. and Cocklin, C., forthcoming, Protecting Habitats on Private Land: Perspectives from Northland. Department of Conservation: Wellington, New Zealand.

Davis, P. R., 1997. Cows, Kiwis and Chainsaws — Protecting Habitats on Private Land. Unpublished thesis for Master of Arts (Geography), University of Auckland, Auckland, New Zealand.

Department of Conservation, 1994. Setting Priorities for the conservation of New Zealand's threatened plants and animals (2nd ed.). Department of Conservation: Wellington, New Zealand.

Department of Conservation, 1995. Draft Conservation Management Strategy for Taitokerau Northland. Department of Conservation: Whangarei, New Zealand.

Department of Conservation, 1997. Identification of key ecological places off the conservation estate — Northland Conservancy. Unpublished paper (May 1997), Department of Conservation: Whangarei, New Zealand.

Far North District Council, 1996. Proposed Far North District Plan. Far North District Council: Kaikohe, New Zealand.

Far North District Council, 1997. Proposed Far North District Plan — Variation No. 1. Far North District Council: Kaikohe, New Zealand.

Far North District Council, 1997. Summary of submissions. Far North District Council: Kaikohe, New Zealand.

Gunn, A. S., 1995. Environmental ethics and tropical rain forests: should greens have standing? *Environ. Ethics* **16:** 21–40.

Hodge, I., 1991a. Incentive policies and the rural environment. *J. Rural Stud.* **8**(3): 207–18.

Hodge, I., 1991b. The provision of public goods in the countryside: How should it be arranged. Pp. 179–96 *in* Farming and the Countryside: an Economic Analysis of External Costs and Benefits ed by N. D. Hanley. CAB International: Oxford.

Kaipara District Council, 1997a. Operative District Plan. Kaipara District Council: Dargaville, New Zealand.

Kaipara District Council, 1997b. Operative District Plan 1997 — proposed change No. 1. Kaipara District Council: Dargaville, New Zealand.

Lowe, P., Murdoch, J., Marsden, T., Munton, R. and Flynn, A., 1993. Regulating the new rural space: the uneven development of land. *J. Rural Stud.* **9**(3): 205–22.

MacGibbon, R., 1996. Native forest restoration: Active management or fence it and leave it. *Tree Grower* **May:** 19–21.

Macmillan, S., 1995. North's changing face of farming. *The New Zealand Farmer,* **December 14:** 22.

Marsden, T., Murdoch, J., Lowe, P., Munton, R. and Flynn, A., 1993. Constructing the Countryside. UCL Press: London.

Memon, P. A., 1993. Keeping New Zealand Green — Recent Environmental Reforms. GP Print: Wellington, New Zealand.

Memon, P. A. and Wilson, G. A., 1993. Recent indigenous forest policy issues in New Zealand: an annotated bibliography. Environmental Policy and Management Research Centre, Publication No. 2, University of Otago: Dunedin, New Zealand.

Ministry for the Environment, 1995. Environment 2010 Strategy: a statement of the government's strategy on the environment. Ministry for the Environment: Wellington, New Zealand.

Ministry for the Environment, 1997. The state of New Zealand's environment. Ministry for the Environment: Wellington, New Zealand.

Ministry for the Environment and Ministry of Foreign Affairs and Trade, 1995. New Zealand Report to the Commission on Sustainable Development. Ministry for the Environment and Ministry of Foreign Affairs and Trade: Wellington, New Zealand.

Ministry of Forestry, 1994. Regional Studies — Northland. Ministry of Forestry: Wellington, New Zealand.

Molloy, L., 1980. Land alone endures: land use and the role of research. Department of Scientific and Industrial Research: Wellington, New Zealand.

Mutu, M., 1993. Current Valuation and rating practices. Pp. 97–120 *in* Conference Proceedings of the New Zealand Institute of Valuers Conference, 26–28 November 1993, Auckland, New Zealand.

New Zealand Government, 1991. Resource Management Act. New Zealand Government: Wellington, New Zealand.

New Zealand Government, 1993. Forest Amendment Act. New Zealand Government: Wellington, New Zealand.

New Zealand Herald, 1997. Far North replace its unpopular plan. *New Zealand Herald*, **December 18:** A12.

Northland Regional Council, 1995a. Pest Management Strategies. Northland Regional Council: Whangarei, New Zealand.

Northland Regional Council, 1997a. Regional policy statement for Northland. Northland Regional Council: Whangarei, New Zealand.

Northland Regional Council, 1997b. Proposed Regional Water and Soil Plan for Northland Regional. Northland Regional Council: Whangarei, New Zealand.

Ogle, C. C., 1994. Recognition of ephemeral wetlands from their plant species assemblages. Department of Conservation: Wellington, New Zealand.

Panayotou, T., 1994. Conservation of biodiversity and economic development: The concept of transferable development rights. *Environ. Res. Econ.* **4:** 91–110.

Pawson, E., 1992. Protected Landscapes. Pp. 207 *in* Changing Places in New Zealand: A Geography of Restructuring ed by S. Britton, R. Le Heron and E. Pawson. New Zealand Geographic Society: Christchurch, New Zealand.

Piripi, H., 1991. Nga Whenua Rahui. *Conserv. Quorum* **5:** 4–6.

Queen Elizabeth II National Trust, 1996. Funding submission to Minister of Conservation for 1996/7 financial year. Queen Elizabeth II National Trust: Wellington, New Zealand.

Royal Forest and Bird Protection Society of New Zealand, 1997. Queen Elizabeth II notches 1,000. *Forest and Bird* **May:** 6.

Saunders, C., 1996. Conservation covenants in New Zealand. *Land Use Policy* **13**(4): 325–29.

Shiva, V., *et al.*, 1991. Biodiversity — Social and Ecological Perspectives. Zed Books: London, and World Rainforest Movement: Penang.

Spencer, A. D., 1993. Rating. Pp. 85–96 *in* Conference Proceedings of the New Zealand Institute of Valuers Conference, Auckland, 26–28 November 1993.

Statistics New Zealand, 1997. Census 1996 on SuperMap3. Statistics New Zealand: Wellington, New Zealand.

Statistics New Zealand, various years, Agriculture Statistics. Statistics New Zealand: Wellington, New Zealand.

Te Puni Kokiri/Ministry of Maori Development, 1997. www.maoriland.govt.nz.

Upton, S., 1997a. Monitoring the Resource Management Act. Speech to the 10th Annual Conference of Local Government New Zealand, 9th July 1997.

Upton, S., 1997b. The legislation simply provides a framework. *New Zealand Herald*, **August 5:** A11.

Waikato Times, 1997. Group to study rating ruling. *Waikato Times,* **October 17:** 2.

Whangarei District Council, 1997. Draft Whangarei District Plan. Whangarei District Council: Whangarei, New Zealand.

Young, D., 1991. A Standing Order. *Terra Nova* **6:** 16–18.

Young, D., 1996. Trees add to saleability — valuers. In *Rural News*, **August 19:** 16.

Young, M. D., Gunningham, N., Elix, J., Lambert, J., Howard, B., Grabosky, P. and McCrone, E., 1996. Reimbursing the Future, An evaluation of motivational, voluntary, price-based, property-right, and regulatory incentives for the conservation of biodiversity. Part 1 and Part 2 — Appendices. Biodiversity Series, Paper No. 9, Biodiversity Unit, Department of the Environment, Sport and Territories: Canberra, Australian Capital Territory.

CHAPTER 48

An environmental and economic framework for evaluating development proposals to clear native vegetation in Western Australia

R. D. Safstrom[1]

Clearing of native vegetation for development is controversial. Land clearing is reducing environmental values but short-term economic pressures are driving increases in clearing applications. Frameworks and principles by which land clearing applications can be assessed, which incorporate environmental, social and economic issues, have not been developed in Australia.

A framework for evaluating land clearing applications should be based on sustainability principles which consider environmental issues, social issues and economic issues. Clearing should only proceed if these principles are not contravened.

This chapter discusses two land degradation principles, 13 biodiversity principles, four cultural issues and six principles covering private financial and external economic considerations. Social issues are not considered in this chapter. A net present value analysis is used for evaluating private financial considerations and a methodology based on cost of offset vegetation is developed to cost external economic considerations (externalities). Options for achieving management of native vegetation for maintenance of nature conservation values instead of development for agriculture are also discussed.

The methodologies developed in this chapter could be adapted for use where decisions regarding development conflict with protection of native vegetation systems.

INTRODUCTION

LAND clearing for development is a controversial issue in Australia in both urban and agricultural environments due to conflicts between development, increasing land degradation and loss of biodiversity. For these reasons state wide clearing regulations have been introduced in South Australia, Victoria, New South Wales and Western Australia but not in the other states and territories (Australia and New Zealand Environment and Conservation Council 1996). Land clearing remains a national issue, despite controls in some states, as approval can be gained for clearing in specific circumstances. Issues surrounding land clearing include urban development, business viability/profitability, biodiversity losses, increased salinisation and property rights.

The issue of the right of private landowners to clear native vegetation on their land is important. Some landowners feel they have an unfettered right to decide how to manage their land, including clearing native vegetation, yet private property has always been subject to public interest (McLeod 1996). As concern for the maintenance of natural systems increases we can expect greater restrictions on management of land by private landowners (Bates 1992).

The purpose of this chapter is to develop clear principles for evaluating land clearing decisions so that public interest can be accommodated. The chapter is based on the development of clearing regulations in agricultural areas of Western

[1]Box 616, Mundaring, Western Australia, Australia 6073.

Pages 480–93 *in* NATURE CONSERVATION 5: NATURE CONSERVATION IN PRODUCTION ENVIRONMENTS: MANAGING THE MATRIX ed by J. L. Craig, N. Mitchell and D. A. Saunders. Surrey Beatty & Sons, 2000.

Australia. In Western Australia land clearing controls were introduced in 1986 under the *Soil and Land Conservation Act (1945–1991)*. These controls were based on land degradation criteria alone but by 1996 were considered to be inadequate due to rising water tables, increasing salinity, soil erosion and acidification of areas already cleared. The importance of protecting nature conservation values was recognized due to the significant decline in nature conservation values since settlement. Nature conservation principles have now been introduced (Government of Western Australia 1997) and incorporate land degradation and nature conservation principles (Safstrom 1996). There is no consideration of private financial benefits or of external economic issues when land clearing proposals are assessed in Western Australia.

This chapter develops a framework for considering land clearing issues, discusses environmental principles (nature conservation and land degradation) and economic principles. Social issues, such as effects on demographics, are beyond the scope of this chapter. The chapter demonstrates the practical application of the principles in a hypothetical case study on the south coast of Western Australia. The chapter concludes with a discussion of mechanisms for the management of native vegetation retained as a result of non-approval of land clearing applications.

ENVIRONMENTAL ISSUES AND PRINCIPLES

State and national governments have recently adopted a national strategy for the conservation of Australia's biodiversity, with its goal to protect biological diversity and maintain ecosystem processes and systems (Government of Australia 1996). The strategy considers biological diversity at three levels:

- genetic diversity — the variety of genetic information contained in all of the individual plants, animals and micro-organisms that inhabit the earth. Genetic diversity occurs within and between the populations of organisms that comprise individual species as well as amongst species;
- species diversity — the variety of species on the earth; and
- ecosystem diversity — the variety of habitats, biotic communities and ecological processes.

The maintenance of all levels of biodiversity must consider the surrounding production landscape. Biodiversity will not persist if the landscape is degrading and the agricultural component will be less stable without the ameliorating influences of remnant native vegetation (Hobbs and Saunders 1992). The various aspects of sustainability (ecological, social and economic) must also be determined at their appropriate level. For land degradation we have the paddock, the whole farm and the landscape scale (Lefroy and Hobbs 1992). For nature conservation the appropriate levels are the remnant itself, its relationship with other remnants and appropriate landscape scales for each species in the remnant.

Developing principles for nature conservation is difficult as there is limited research in the area. The principles discussed in this chapter therefore draw on information from studies on a few species and ecosystems in Western Australia. The principles were originally developed as criteria for small conservation reserve selection (Safstrom 1995). They have been reworked to be applicable for developing principles and criteria for assessing land clearing. The principles are illustrated in Safstrom (1996) and justified in greater detail in Safstrom (1997). The precautionary principle should be applied where knowledge is limited.

Ecological sustainability, inclusive of all macro and micro biota, on a landscape scale is an overarching principle. The specific principles are considered under the following headings:

- Regional processes — importance of native vegetation in maintaining ecological processes (Table 1);
- Uniqueness — role in conserving the genetic diversity (Table 2);
- Viability — survival of natural values (Table 3); and
- Cultural values — natural features of importance for cultural reasons (Table 4).

Viability — survival of natural values

Viability considers factors which can be identified as having a high likelihood of resulting in serious degradation of a remnant over the next 50 years. Viability is related to robustness which is related to threatening processes and soil type. In Western Australia simpler communities such as woodlands are less robust as they are more accessible to threatening (human) processes and have more fertile soils favourable for weed establishment. Species rich communities have higher robustness, as they are often on poor soils unfavourable to weeds (Craig and Safstrom 1997). Areas which are degraded and considered not viable may be valuable if restorable or provide a seed source and source of micro-biota for habitat reconstruction.

It is recognized that remnant vegetation may have several levels of viability. It may be viable for its current complement of species for 50 years, then viable for half of them for a further 50 years. The factors considered when determining if serious degradation is likely to occur are discussed in Table 3.

ECONOMIC ISSUES AND PRINCIPLES

The decision by a landowner to clear and develop a property depends on an analysis of the net benefits, including the net returns from agriculture or subdivision, the opportunity costs from losing the native vegetation, and taking into account the relevant taxes and subsidies. A landowner may not

Table 1. Regional processes — importance of native vegetation in maintaining ecological processes.

No.	Issue	Principle Native vegetation should be retained if . . .	Justification
1.	Water	clearance of that vegetation is likely to cause deterioration in surface and groundwater catchments which result in increases in salinity and eutrophication.	Clearance of vegetation should not be considered if the result is an incremental rise in the water table, which is likely to be harmful to the environment, for example loss of valley woodlands, or increasing eutrophication (Fig. 1). Any incremental increase in recharge in catchments known to have rising ground water may be undesirable as extra amelioratory works will have to undertaken elsewhere in the catchment to make up for the increase.
2.	Soil	clearance of that vegetation is likely to contribute to an increase in soil erosion, water logging or flooding.	Remnant vegetation plays a role in preventing both waterlogging and soil erosion by wind and water. Native vegetation needs to be retained where there is a high likelihood of increased soil degradation if the land is cleared.
3.	Corridor	the land provides a corridor or stepping stone between areas of conservation land. The width of vegetation required should be based on good principles for corridor design.	Native vegetation close to other remnants and with good connecting corridors has greater viability for many species (Safstrom 1995). Corridors should preferably maximize species richness (include soil boundaries and catenary sequences), wider is better for viability and suitability for use by a greater number of species (Fig. 2). Minimum corridor widths will depend on the plant community and threatening factors such as edge effects. Refer Craig and Safstrom (1997) for suggestions, but recognize that narrow corridors can be useful and corridors can have disadvantages such as sinks, predator traps and in reducing movement.
4.	Buffers	it provides a buffer or is an inlier to areas reserved for conservation.	Native vegetation adjacent to conservation reserves improves the viability and conservation values of the reserve by providing larger core areas, buffers the reserve from edge effects, sometimes consolidates boundaries and sometimes add plant communities not represented or under-represented in the reserve (Fig. 2). The width of buffers required will depend on the robustness of the vegetation communities, suggestions for Western Australia vegetation communities can be found in Craig and Safstrom (1997).

give consideration to the external effects of clearing his property. Such external effects are known as externalities and some externalities can be costed. In many of the agricultural areas of Western Australia it is now accepted that traditional farming land uses following clearing of native vegetation results in both land and stream salinization, impacts on water resources infrastructure as well as environmental and social costs (Government of Western Australia 1996(a)). The problems of land degradation in Australia illustrate the divergence between the landowner's goals of profit maximization, adverse long-term environmental impacts on the economy and the much broader social goals of nature conservation (Thompson 1986).

Total social benefits of clearing and development are the sum of the private benefits minus private costs plus community benefits minus community costs. If total benefits are negative then there is a cost to society from development. If the native vegetation is retained for the public good and private benefits are reduced then there is a question of who should pay for the public good (Gross 1993). Before the question of who should pay is addressed, estimates of the external benefits of vegetation retention have to be made. For example, the salinity control benefits attributable to the withholding from clearing of the next 1% of the Loddon catchment in Victoria (4 179 ha) were estimated to be $365,904 which is equivalent to $88 per hectare and is the cost of restoring the river to its original salinity level by pumping and removal of the additional saline groundwater (Greig and Devonshire 1981).

Clearing land and development of agriculture in Western Australia has resulted in serious economic implications due to increasing salinity, soil erosion, soil degradation and water quality reductions. The capital value of land lost to production due to salinity, which includes an opportunity cost of production lost, is $1,445 million. If the current rate of salinity expansion continues, the resulting annual agricultural loss will be $64 million each year. In the Kent River catchment alone, the off-site net present value of costs of salinity to infrastructure such as road and rail systems, towns, farm buildings, pipelines, culverts and bridges is $210 million (over 20 years). These costs do not take into account other environmental and social costs (Government of Western Australia 1996(a)).

This section explores firstly a principle and method of costing one external cost of land clearing as an example and then explores private financial principles for evaluating land clearing.

Table 2. Uniqueness — role in conserving genetic diversity.

No.	Issue	Principle Native vegetation should be retained if . . .	Justification
1.	Threatened flora, fauna, plant communities and flora, fauna and plant communities of special interest.	it contains or is likely to contain threatened flora, fauna and plant communities or flora, fauna and plant communities of special interest at state and regional significance levels.	Native vegetation which contains threatened flora and fauna species, threatened communities, species of special interest, such as species at the edge of their range and disjunct populations, are a high priority for protection (Fig. 3). Protection should not be restricted to species and communities of state significance. Regionally threatened species are also important to maintain regional representation of flora and fauna. Relictual Gondwanan genera/habitats may also be important in Western Australia. Refer to the Department of Conservation and Land Management in Western Australia for lists of Declared Rare and Priority flora and fauna.
2.	Wetlands	it contains wetlands of international, national or regional significance. All other wetlands should be thoroughly assessed for their conservation significance.	Wetlands (and their surface and groundwater catchments) of international, national and regional significance have a high priority for protection (Fig. 3). Other wetlands are also likely to be a priority for protection as significant areas of wetlands have been lost, or substantially altered, in the agricultural landscape in Western Australia.
3.	Ecological representation	there is less than 20% of the original occurrence of a plant community represented in conservation reserves of secure tenure such as national parks, nature reserves, or under covenant.	A minimum of 20% of the original cover of each plant community should be retained in line with the 20% rule for retention of remnant vegetation within farms, catchments and Shires in Western Australia although there is no scientific evidence for this figure. In the Western Australia wheatbelt, stands of species rich plant communities on ancient land surfaces need to be represented at least at intervals less than 15 kilometres due to the rate of speciation and changes in plant community composition over that distance (Burgman 1988) (Fig. 4). In other areas the distances over which ecological representation is required will depend on rates of change in plant community composition.
4.	Significant habitats for wildlife.	it has significance as habitat for wildlife or if a loss of diversity by clearing partof the land will adversely impact on fauna dependent on a mosaic of vegetation types.	Some areas are particularly valuable as habitats for wildlife, for example nest hollows in woodlands and autumn food sources and if removed or significantly reduced then a regional population decline of a species could occur (Fig. 5).

External economic issues

The scope of this chapter does not allow the analysis of a full range of externalities from land clearing. Some externalities such as loss of a corridor for wildlife are very difficult to cost in real financial terms and alternative methods for costing non measurable costs such as contingent valuation are not explored. Similarly, costs to the community which may result from land clearing such as increasing eutrophication, soil erosion and flooding are difficult to measure although remedial actions such as treatments for eutrophication, soil erosion control mechanisms and larger culverts to cope with flooding can be costed, but they are not within the scope of this chapter.

Rather, a method is explored of calculating a readily measurable, external, community cost of land clearing where additional water table rises and salinity occur as a result of land clearing. The purpose is to demonstrate the methodology as an example and it is recommended that the more difficult to measure external economic outcomes be explored in depth in other studies.

There is no current requirement for external costs of land clearing to be considered by individuals or by government when decisions to clear land are made. All such external costs should be taken into account as part of the decision making process to clear native vegetation for agriculture (Table 5).

One method of costing the external impact of land clearing

In order to take into account the externalities that may result from clearing native vegetation a method of costing needs to be derived. The following discussion derives a method for costing one externality which may result from conversion of native vegetation to traditional agriculture.

The method assumes that actions elsewhere in the catchment will be taken to reduce the externality of additional water entering the water table due to clearing and conversion to traditional agriculture. The method depends on estimating the cost of the action to utilize the additional water. It is argued that the cost of the actions required elsewhere in the catchment to utilize the excess water is a direct

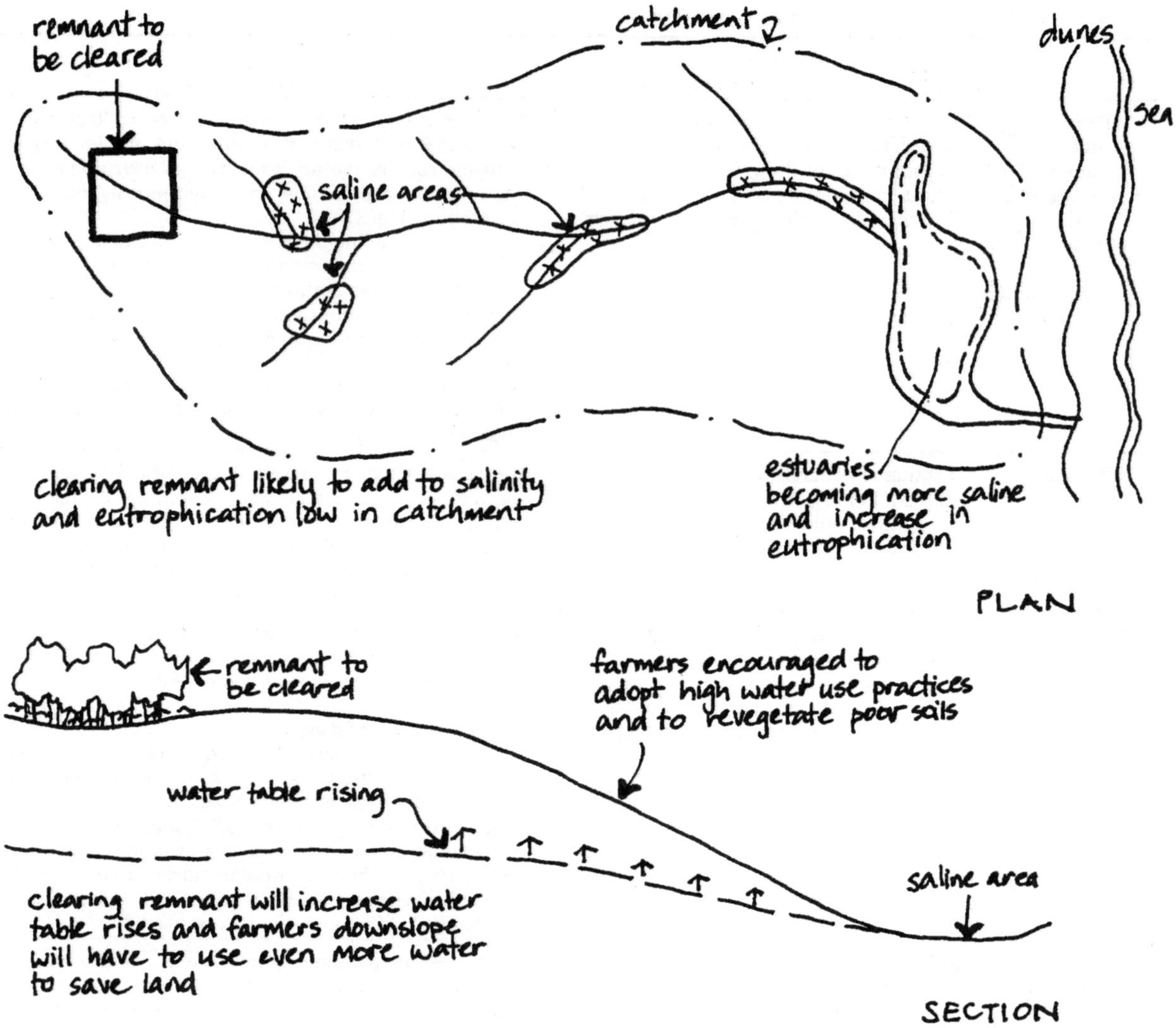

Fig. 1. Native vegetation should be retained if clearance of that vegetation is likely to result in increases in salinity and eutrophication.

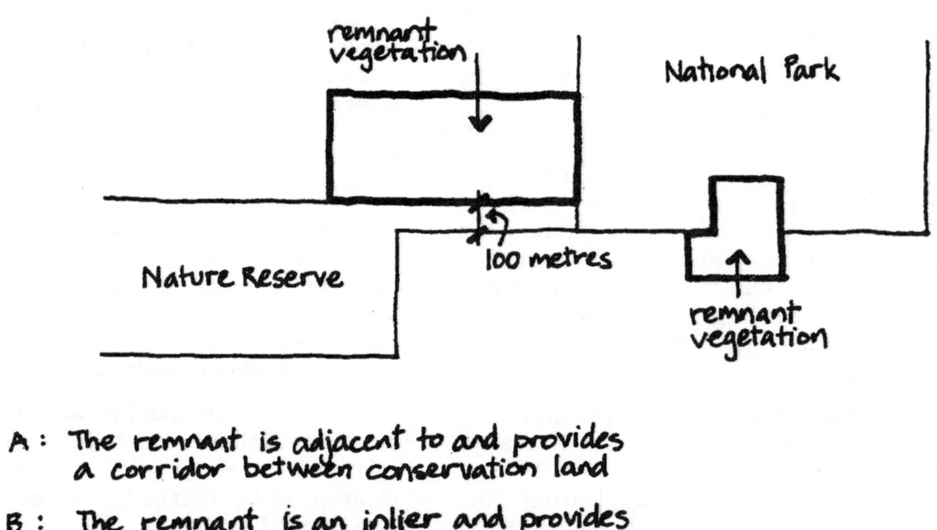

Fig. 2. Native vegetation should be retained if the land provides a corridor, stepping stone between areas of conservation land or a buffer to areas reserved for conservation.

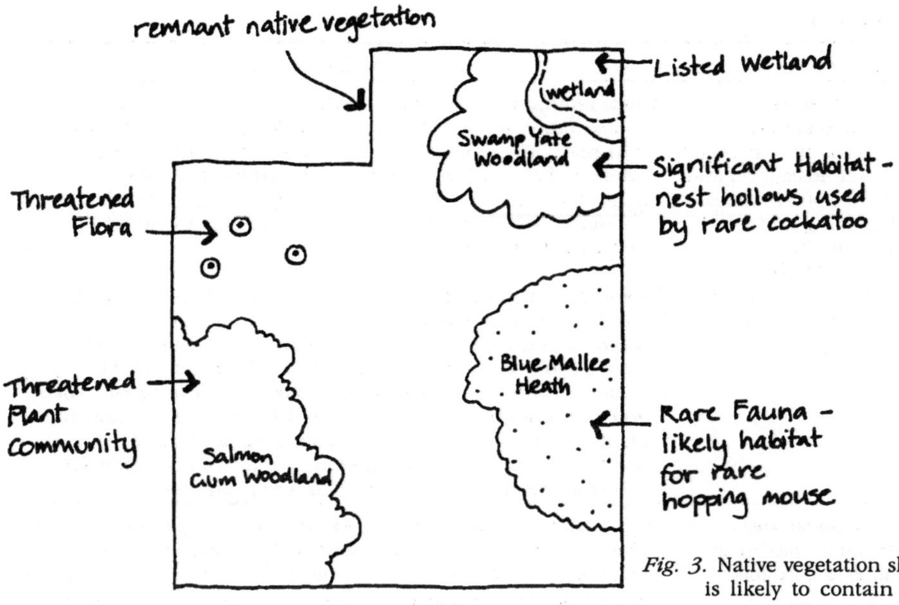

Fig. 3. Native vegetation should be retained if it contains or is likely to contain threatened flora, fauna or plant communities at state and regional significance levels.

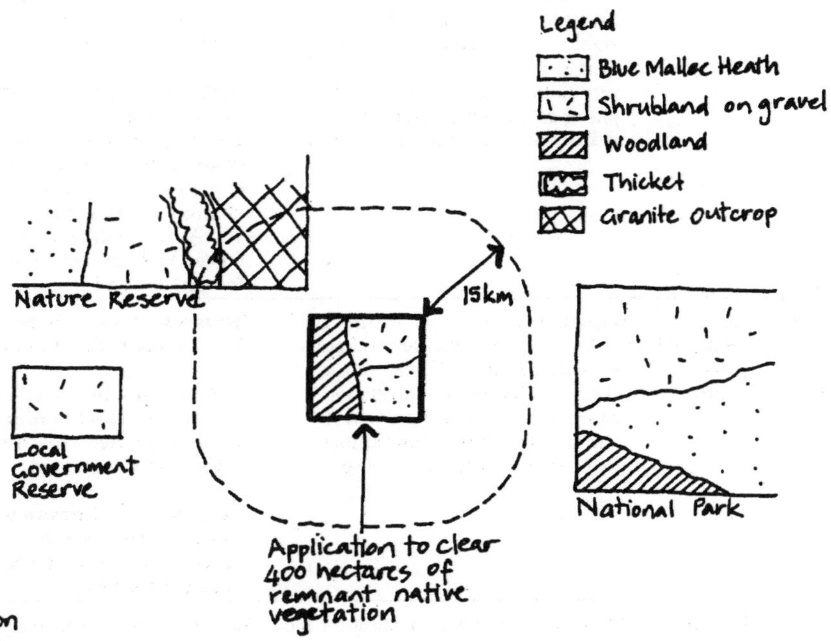

Fig. 4. In agricultural areas of Western Australia native vegetation should be retained if there is less than 20% of the original occurrence of a plant community represented in secure conservation reserves within 15 kilometres of the native vegetation.

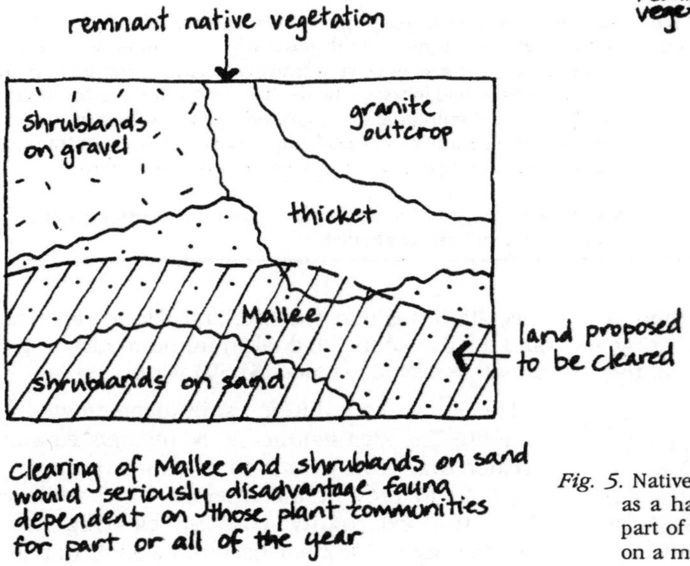

Fig. 5. Native vegetation should be retained if it has significance as a habitat for wildlife or if loss of diversity by clearing part of the land will adversely impact on fauna dependent on a mosaic of vegetation communities.

Table 3. Viability — survival of natural values.

No.	Issue	Principle	Justification
1.	Area of native vegetation	Maximize the area of native vegetation retained to enhance viability. Retain all remnants greater than 1 500 hectares. Recognize that areas smaller than 1 500 hectares are valuable depending on the conservation goal and that groups of small remnants should also be retained where they support fauna able to move between remnants.	Viable remnant size is dependent on species requirements, the configuration of vegetation in the landscape, the threats and the time scale. The protection of the totality of biota may require a number of reserve sizes. In this chapter it has been assumed that larger remnants (>1 500 ha), have higher conservation values and are more likely to be viable for a range of fauna than small remnants (Kitchener *et al.* 1980). There appears to be little agreement on the minimum size of remnants for conservation purposes. Craig and Safstrom (1997) noted size ranges for specific Western Australia fauna ranging from 2 000 hectares for 50 dunnarts; 20–50 hectares for honey possums; 30 hectares for lizards; 25 hectares for trapdoor spiders; and 5 hectares in good condition for invertebrates. Threatened plants can sometimes persist in quite small areas. Any remnant can be a source of local germplasm and micro-organisms for landscape reconstruction and act as a stepping stone.
2.	Remnant shape	Maximize widths of linear areas of native vegetation to maximize viability and value as corridors.	Remnants with small edge to area ratios are likely to be better for nature conservation than remnants with large edge to area ratios, and the shape of a remnant is likely to be more important in small and linear remnants as more edge habitat and edge disturbances are created (Safstrom 1995).
3.	Intact area	Remnant vegetation with large intact areas is more likely to be viable than remnant vegetation with small intact areas or with very difficult to control invasive plants capable of permanently destroying or disrupting ecosystem.	Remnants with large intact areas (free of major disturbances) are likely to have better viability and be of greater value for wildlife than remnants with small intact areas (Fig. 6). Weeds, grazing by stock and vermin and other disturbances, reduce the area of intact vegetation by changing species compositions and creating soil degradation. Soil degradation, such as soil compaction, processes in woodland remnants may make seed bed conditions unsuitable for species recruitment leading to gradual changes in vegetation structure with intact woodlands (Yates *et al.* 1994). Scougall *et al.* (1993) found major differences between fenced and unfenced remnants of York Gum *Eucalyptus loxophleba* and Jam *Acacia acuminata* woodland.
4.	Disease, pests and invasive plants	Vegetation is worthy of retention if it is free of major diseases and pests such as Dieback (*Phytophthora* species). Disease free vegetation is more important for retention if similar vegetation communities in nearby reserves are diseased.	Diseases such as Dieback (*Phytophthora* species) can have a big impact on a vegetation community and will reduce viability. In some cases disease will be present or likely to spread further in reserves but is yet to impact on private remnants. In these cases the value of the remnant to retain disease free examples of the original vegetation is increased. The presence of invasive plants and pests with potential to cause modification to species richness, species abundance or ecosystem function or to totally and permanently destroy an ecosystem will reduce viability.
5.	Adjacent land uses	Adjacent land uses impacting on the viability of the remnant vegetation must be considered.	Adjacent land uses impacting on remnant vegetation, particularly edges, include: saline water disposal into a remnant, wind and water borne soil and weed seed transported into remnant vegetation from adjacent paddocks, pesticide drift, irrigation causing water table rises and eutrophication, ground water lowering by water extraction, in some places acid rain, human impacts of trampling and rubbish dumping and nutrients from waste systems. Where such impacts are severe controls can be considered and impacts on viability evaluated.

external cost of the land clearing. The method is applicable to areas where water tables are still rising, have not stabilized and are detrimental to the community.

Safstrom (1997) found that where 10% of the total area cleared was revegetated (offset revegetation) in the most optimal position in the landscape to utilize the excess water there was an externality cost of $155 for every hectare cleared. The costing assumed that native vegetation was direct seeded, and allowed for fencing and the cost of the land taken from production giving a total revegetation cost of $1 550 per hectare.

The above calculations should be treated with caution. The area required to be revegetated may be greater than 10% and if three times the area was required to prevent incremental water table rises then the externality cost of clearing to the community is $462 per hectare cleared. Caution is

Table 4. Cultural values.

No.	Issue	Principle	Justification
1.	Species richness	Areas of native vegetation with very high species richness should be retained.	Species richness (the number of species present in a particular area) is sometimes used as a measure of nature conservation but it is considered to be a cultural rather than an environmental issue. An example is to compare the species richness of the Kwongan (a species rich Western Australia plant community) with tropical rainforest. Species richness is not necessarily equated with high nature conservation value. Species richness is an artefact of where the work has been done and does not predict faunal richness. Simple communities can be just as valuable for nature conservation and are often very productive systems, compared with species rich communities (Craig and Safstrom 1997).
2.	Landscape values	Native vegetation should be retained if it has high landscape (aesthetic) values.	The familiar rural landscape of farmland fringed and dotted with trees and patches of bush can only be maintained with positive action. Retain vegetation with high scenic quality, strongly defined vegetation patterns, unique specimen stands, areas of high plant diversity which display distinctive textural and colour patterns and dramatic displays of seasonal colour. In Western Australia refer to Revell (1994).
3.	Special physio-graphic features	Native vegetation on areas with special physiographic features should be retained.	Special features on the land that may be of cultural interest are outcropping dolerite dykes, granite outcrops and breakaways. These features also often contain unique vegetation communities and habitats such as rock outcrops and seasonal water supply.
4.	Significant Aboriginal and European cultural sites	Native vegetation on sites of Aboriginal and European significance should be retained and the sites not disturbed.	Presence of Aboriginal and European sites on the land. The significance of such sites needs to be determined by an appropriate authority.

Table 5: External economic issues.

No.	Issue	Principle	Justification
1.	Externalities	Community benefits and costs (externalities) of clearing native vegetation should be taken into account.	Total social benefits of clearing and development are the sum of the private benefits minus private costs plus community benefits minus community costs. If total benefits are negative then there is a cost to society from development. If the native vegetation is retained for the public good and private benefits are reduced then there is a question of who should pay for the public good. Without an analysis of the economic benefits and disbenefits of land clearing, society is not in a position to make sensible analysis of clearing proposals.

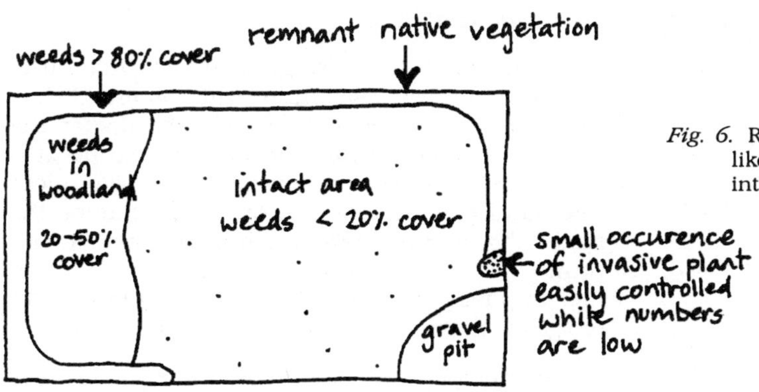

Fig. 6. Remnant vegetation with large intact areas is more likely to be viable than remnant vegetation with small intact areas or with difficult to control invasive weeds.

also required as tree water use is affected by soil type, rainfall, species, provenance, tree health, planting density, adaptation to the site and depth to ground water. Trees on lower slopes and discharge sites have given encouraging results but long-term viability on these sites is a cause for concern (Scott 1993). Trees are also not an instant panacea. It takes several years for a revegetated area to reach its water using potential (Farrington and Salama 1995). The water table is now at a critical depth in many areas. Further clearing now may tip the delicate balance. In this circumstance no further clearing should take place in the catchment until the offset revegetation has established its water use potential and calculations show that an equivalent or greater quantity of water is being used.

It is recognized that there are different options for using the additional water generated as a result of clearing and conversion to agriculture. Deep rooted perennials can be established and crops and cultural practices can be improved to utilize more water (Government of Western Australia 1996(b)). In some areas the offset revegetation could provide direct economic benefits while addressing water use objectives through the production of timber, fodder crops, seed production, wildflowers and minor forest products and there would be no external cost (Force 1995). In many of these cases the revegetation would be of minimal value for nature conservation.

It is clearly not in the community interest to allow clearing of native vegetation when there is a private gain but a net community cost. If community costs can be estimated then there is a strong argument against clearing. To date there has been no attempt to calculate these costs and it is recommended that more work be done in this area.

Where there is a community cost from land clearing the community has the following options:

- bear this cost;
- purchase the land proposed for clearing;
- compensate the landowner for not being able to develop his land;
- require the landowner to make a contribution to the community equivalent to the calculated externality; or
- prohibit clearing and do not provide compensation.

Private financial issues

There is no government requirement for a landowner to demonstrate that clearing native vegetation will provide positive personal financial outcomes prior to clearing being approved. This section looks at all the options a landowner could consider instead of clearing and investigates a method for a landowner to calculate the financial returns as a result of clearing.

It is suggested that landowners should go through a thorough process to compare options and to calculate likely returns. In particular, landowners should consider all the costs of land clearing as many of these costs are easily hidden in other farm costs or become costs at a later time.

The following principles (Table 6) cover economic alternatives to clearing the land and the financial viability of developing the land.

EVALUATION OF LAND CLEARING MALLEE ROAD SUMP CATCHMENT

A hypothetical study based on land clearing for cropping on the south coast of Western Australia has been developed to demonstrate the application of the principles in this chapter. In Western Australia there are no incentives to clear land such as subsidies but it is recognized that some landowners may claim some clearing costs as a tax deduction. Where prices and costs are used they are based on Agriculture Western Australia (1997).

Environmental evaluation

The case considers an application to clear 400 hectares of native woodland, mallee and shrubland vegetation for development of cropping enterprises (Table 7). The hypothetical example relates reasonably well to local vegetation communities and to the extent of clearing in the region. The native vegetation on the land and in the nearby reserves is free of pests and diseases such as Dieback, is mostly intact being generally free of weeds and disturbances, the woodlands are a high priority for protection and Declared Rare Flora have been found in Blue Mallee Heath communities in the district.

The hypothetical catchment is a good example of a catchment, with a broad flat valley floor, with continuing water table rises. The water table is rising at 0.2 metres per year, with higher rises on the floor of the valley. It is thought that in the virgin state, the rainfall was balanced by transpiration from the native vegetation, evaporation from a number of swamps and a small amount of recharge. Since clearing, the transpiration has decreased, the swamp evaporation has remained static and the recharge has increased. Under most of the valley floor the water table is now approximately two metres from the surface, a critical depth at which salt accumulation in the root zone can seriously effect crop and pasture growth. In such a hydrological setting it is essential to lower the water table in order to maintain productivity (Siewert and Abbott 1991).

A desk study using the information provided above will determine if any of the environmental principles developed in this chapter would be contravened if the land were to be cleared.

Economic evaluation

In this case study it is assumed that offset perennial native vegetation will be established elsewhere in the catchment to make up for the increase in water infiltration due to clearing at a cost of $1,550 per hectare as developed under externalities above and the cost per hectare on the land to be cleared is $155 per hectare. There are no commercial revegetation options in the area. Calculations of externalities other than offset revegetation are beyond the scope of this chapter.

The principles developed above for economic uses for uncleared bushland, sustainability of economic uses and sale of bushland have not been investigated in this case study. Principles for purchase of developed land, costs of clearing and developing land and a long-term financial analysis are discussed below.

An economic evaluation of land clearing in the hypothetical catchment has been developed in (Safstrom 1997) utilizing a Net Present Value

Table 6. Private financial issues.

No.	Issue	Principle	Justification
1.	Economic use options for bushland	(a) Consider alternative economic use options for uncleared bushland before seeking to clear a potential resource. (b) Consider the sustainability of nature conservation values whilst utilising bushland for economic uses.	The potential value of bushland as an economic resource in its own right needs to be explored. There are uses of bushland being developed, many of current economic value, which could be considered (Bradby 1984). Options range from seed, wildflower and timber production to nature based tourism. While the report lists the above uses as sustainable (that is a regular or periodic supply of products over a long period) there are few studies to indicate at what level of utilization of these products the nature conservation values will also remain viable. The profitability, suitability and sustainability in each particular circumstance needs to be evaluated.
2.	Sale of bushland	Consider the financial return from the sale of bushland to conservation interests or to others who may combine utilization with conservation provided such sales do not disadvantage existing farm management.	An option for a landowner considering clearing native vegetation could be sale of the uncleared land. Sale would provide a cash return which would assist in purchasing developed land or could be used for an alternative investment. It could be argued by some that their bushland does not have financial value due to lack of buyer demand. There is evidence of increasing buyer interest in bushland because it is seen as a disappearing resource (Safstrom 1997). In some cases sale may not be feasible due to subdivision restrictions, access constraints, costs of subdivision and the fragmented nature of remaining native vegetation.
3.	Purchase of developed land	Consider the purchase of developed land as an alternative to clearing land for agriculture where fixed costs are not substantially increased.	A landowner will need to compare the costs of developing his own bushland with purchase of already developed land. If the bushland can be sold then this sale offsets the purchase of developed land. A landowner may perceive a benefit in clearing his own land even if development costs are greater than purchasing developed land because: land is not available nearby, land is only available in larger less affordable holdings, economies of scale may be achieved (Herbert 1993) or travel and other costs are reduced. The decision to develop land or purchase developed land will depend on the particular circumstances of the landowner.
4.	Costs of clearing and developing land	Take into account the full costs of clearing and developing land for agriculture including clearing costs, fencing, stock water, fertilizer and grade banks if required.	A landowner will need to consider the full costs of clearing bushland and developing the land for agriculture to determine whether the financial returns warrant the investment. Costs include: clearing, raking, cultivation, fertilizer and trace element application, water supply, access, landcare works such as grade banks, fencing uncleared areas such as streams and pasture or crop establishment.
5.	Net present value of land clearing and land development for agriculture	Conduct a financial analysis by determining the net present value of the costs and returns of land development, to determine the likely long-term profitability of the development and a comparison with alternative investments.	Calculating the net present value of an enterprize option is an accepted method for determining which of a number of competing land use options is likely to yield the greatest return. Net present value is defined as the present value of future income less the present value of the expenditures giving rise to the income. The investment having the highest net present value is the most profitable (Eckersley 1993). Difficulties in applying the net present value method include: specifying an appropriate rate of interest, specifying the time horizon, future returns are uncertain. Net present value is only as good as the assumptions made but it does provide a comparison between competing options.

Table 7: Evaluation of nature conservation principles in a hypothetical catchment.

Principle	Evaluation	Determination
1. Water	Saline groundwater is rising at 0.2 metres a year and land clearing in the shrubland and Blue Mallee Heath communities, both thought to be intake zones, is expected to result in an incremental water table rise.	The shrublands and Blue Mallee Heath vegetation should not be cleared.
2. Soil	It is not known if clearance of native vegetation is likely to contribute to soil degradation.	Field survey and further data required.
3. Corridors	The land does not provide a corridor but as it lies between a Nature Reserve and a National Park it could provide a stepping stone for some species moving between the Nature Reserve and the National Park.	Part or all of the land should not becleared. Specialist advice on species movements in this area should be sought.
4. Buffers	The land is not adjacent to a reserve so is no value as a buffer.	This principle does not preclude land clearing.
5. Threatened flora fauna and plant communities and flora, fauna and plant communities of special interest	Blue Mallee Heath could contain Declared Rare Flora. The presence of threatened fauna is not known.	Field survey required to look for threatened flora and fauna.
6. Wetlands	There are no wetlands on the land.	This principle does not preclude clearing.
7. Ecological represen-tation	The plant communities on the land, shrublands on gravel, Blue Mallee Heath and woodland are not represented in National Parks, Nature Reserves or on Crown land within 15 kilometres of the remnant.	The land should not be cleared as the plant communities have poor reservation status within 15 kilometres.
8. Significant habitats for wildlife	Significance as habitat not known.	Field survey required.
9. Area of native vegetation	400 hectares.	400 hectares is considered to be reasonably viable as the vegetation is largely intact.
10. Remnant shape	Square shape of remnant has a small edge to area ratio and the core area is maximized.	The shape does not detract from long-term viability.
11. Intact area	The remnant is generally weed free with few disturbances, it is largely intact.	There are few weeds or disturbances on the land to detract from long-term viability.
12. Disease, pests and invasive plants	The land is free of diseases and invasive plants; diseases do not occur on the adjacent reserves. Pests such as foxes are present.	Confirm with field survey.
13. Adjacent land uses	Farming uses in this area have little impact on adjacent remnant vegetation except where water flows from farmland into remnant bushland.	Check with field survey.

Table 8: Evaluation of economic principles in a hypothetical catchment.

Principle	Evaluation	Determination
1. Externalities	Offset revegetation is required as there will be an incremental water table rise if the land is cleared.	The landowner or the community will need to pay the cost of the offset revegetation if the rate of water table rises are not to increase.

Table 9. Net present value analysis of land clearing in a hypothetical catchment.

Principle	Evaluation	Determination
1. Net present value of land clearing and development for agriculture	It is uneconomic to develop land in the hypothetical catchment if only average yields and commodity prices are achieved. If the cost of offset revegetation is added to development costs then profitability of the enterprise is further reduced.	Clearing is only warranted in the hypothetical catchment and districts where similar returns and costs apply if yields are substantially higher than the district ten year average, commodity prices increase substantially or if the farmer can develop the land very cheaply.

methodology. Land development costs including clearing, fencing, grade banks, water and fertilizer have been calculated for both low development costs ($349 per hectare) and medium development cost ($463 per hectare) options. The value of developed land to purchase varies from $200 to $425 per hectare indicating that in many cases it could be cheaper to purchase developed land than to clear land.

A net present value analysis assumes a cereal grain legume rotation beginning with two rotations of wheat which is the best current option in the district. Gross margins assume ten year average yields for the district (1.02 tonnes per hectare for lupins (*Lupinus* sp.) and 1.36 tonnes per hectare for wheat), 1.25 times and double average yields and current commodity prices (lupins $151 per tonne, wheat $126.64 per tonne). Gross margins nett of machinery costs were -$6.84 per hectare for lupin and $24.46 for wheat at average yields. Two discount rates are used, 8% based on the long-term government bond rate and 12% to determine the sensitivity to possibly increasing interest rates.

The analysis indicates the following (Table 9):

- It is uneconomic to develop land in the hypothetical catchment for a range of development costs if only average yields are achieved. It is, however, economic if consistent yields of double the district average are achieved.
- Profitability is moderately sensitive to a 4% increase in interest rates with net present value being reduced between 9% and 30% depending on the yield and commodity price options.
- Profitability is poor over a five year time period unless higher than average yields are produced.
- A 1.25% increase in average yield converts a loss situation to a modest profit and a 20% commodity price increase still leaves a loss for medium development costs but a profit is realized if development costs are low.

The above analysis has not taken tax into consideration.

Land clearing in the hypothetical catchment has been analysed from the principles developed in this chapter. There appear to be sound reasons that the land should not be cleared.

- Land degradation principles of incremental rises in water tables would be contravened.
- Nature conservation principles of stepping stones, priority plant communities and local representation would be contravened.
- There is an external cost to the community of clearing the land because there is likely to be a net economic loss to the community.
- Private financial principles are contravened because it may be cheaper to purchase already developed land and the outlook for long-term profitability is poor.

OPTIONS FOR ACHIEVING MANAGEMENT OF NATIVE VEGETATION

Use of the principles developed in this chapter is likely to recommend against the majority of proposals proceeding in the agricultural areas of Western Australia due to the highly fragmented landscape, the low percentage of remnant vegetation remaining in most areas, the high risk of increasing salinity and the low productivity of soils remaining uncleared. This is the experience in Western Australia where the environmental principles have been applied since 1997.

The land for which clearing has not been approved requires management but a landowner for whom an application to clear has been refused is unlikely to be motivated to manage the land and could even passively clear by allowing grazing and fertilizing. Therefore a strategy is required to develop options for the future management of the land for its environmental values.

Agriculture Western Australia has developed an adjustment process which goes part way to resolving the problem for landowners who have been refused permission to clear within a given time frame. The adjustment process is administered by the Rural Adjustment and Finance Organization. The adjustment process requires that a Conservation Covenant be registered on the property title and Rural Adjustment and Finance Organization may undertake any of the following.

- Make a payment to the landowner to retain the vegetation. This payment may be based on the difference between the value of the land as if it could be cleared and its value without clearing approval.
- Assist in achieving a separate title for the land and in negotiating the sale of the land to a third party.
- Purchase and then on-sell the land.

Rural Adjustment and Finance Organization have recently demonstrated this process by assisting a farmer to subdivide his bushland and in negotiating a sale to an adjacent owner with bushland.

While the adjustment process will assist some landowners there will be significant areas for which sustainable conservation management will not be achieved. If remnant vegetation could achieve a high land value for its conservation values and if it became a sought after resource, then market mechanisms would achieve good management with sales to sympathetic purchasers. This is beginning to be achieved and there appears to be increasing interest by the general public in purchasing bushland for its conservation value. This process is being facilitated by raising awareness of the potential conservation market with real estate firms. Planning impediments to the subdivision of bushland from farming land need to be overcome for bushland to become available to private purchasers in Western Australia.

There are already significant examples where bushland has been purchased for its conservation values by individuals, corporations, and by community groups and the Australian Bush Heritage Fund. These activities need to be encouraged. If significant non farmer management of conservation land in rural areas can be achieved then the process can bring people and expertise into rural communities and partly redress the downward slide in population and services.

SUMMARY AND RECOMMENDATIONS

There is increasing concern that continuing land clearing for agriculture in Western Australia is adversely affecting nature conservation and exacerbating land degradation. This community concern is reflected in new government administrative arrangements to protect biodiversity values and growing community awareness that ownership of land does not infer the right to clear if it adversely affects community interests. But it is clear that private economic pressures such as increasing commodity prices or new industries are driving an increase in clearing applications which may not be in the best long-term interests of the local community or the state.

This chapter has provided principles to assess land clearing applications. Land clearing should only proceed if there are no serious environmental, social or financial and economic concerns.

Demonstration of the principles developed in this chapter is achieved through a hypothetical case study on the south coast of Western Australia, demonstrating practical use of the environmental, financial and external economic principles. The case study found that land clearing for cropping contravened environmental principles relating to nature conservation and land degradation. The case study demonstrated that, for the given assumptions, the land should not be cleared for private financial reasons and if clearing proceeded there is likely to be a negative economic impact on the community.

Land not approved for clearing requires long-term stewardship if it is to remain valuable for the conservation estate. Owners who have not had approval to clear may not be motivated to provide that stewardship. An adjustment process such as the provision of incentives, facilitating the sale of the land or acquisition by the state.

Recommendations:

1. The principles developed in this chapter be tailored for local conditions and be thoroughly assessed by personnel with the appropriate expertise prior to clearing approvals.
2. Social issues in relation to land clearing need to be developed.
3. Where the environmental and external economic analysis does not demonstrate serious concerns, that landowners provide a detailed financial analysis of their land clearing proposal, prior to acceptance of the proposal by the responsible authority. Each of the financial principles developed in this research should be considered. Only proposals that are likely to be profitable given realistic analysis and for which there are no economically attractive alternatives, such as purchasing developed land, should be further considered for approval.
4. Every effort should be made to provide the community with advice regarding the likely profitability of land clearing and to actively promote alternate uses of the land which retain the native vegetation.
5. That a study be undertaken to provide a thorough analysis of externalities in relation to land clearing with a focus on developing methodologies for costing these externalities based on real costs to the community. Costing of externalities as a result of land clearing applications should be carried out.
6. The question of who should pay, the landowner or the community, for degradation as a result of land clearing, be investigated.
7. That the cost of evaluating land clearing proposals be borne by the proponent and the cost added to the financial analysis. The profitability of many proposals would be weakened.
8. Mechanisms are required to encourage individual and community group management of conservation land and remove impediments such as restrictions on subdivision of bushland from farmland.

REFERENCES

Agriculture Western Australia, 1997. 1997 Farm Budget Guide. *Farm Weekly*, Perth.

Australia and New Zealand Environment and Conservation Council, 1996. Working Group on nature conservation on private land. Nature conservation on private land. Commonwealth, State and Territory legislation and programmes. Australian Nature Conservation Agency: Canberra, Australian Capital Territory.

Bates, G. M., 1992. Environmental Law in Australia. Butterworths: Brisbane, Australia.

Bradby, K., 1984. A rationale for respect. Diversity or Dust. A review of the impact of agricultural land clearance programmemes in south west Australia. Australian Conservation Foundation: Hawthorne, Aust. 33–37.

Burgman, M. A., 1988. Spatial analysis of vegetation patterns in southern Western Australia: implications for reserve design. *Aust. J. Ecol.* **13:** 415–29.

Craig, G. F. and Safstrom, R. D., 1997. Improving the criteria for environmental evaluation of native vegetation in south west Australia. Summary of interviews with scientists, literature review and project proposals. Department of Environmental Protection: Perth, Western Australia.

Eckersley, P., 1993. Interest rates in farm decisions. Agriculture Western Australia: Perth, Western Australia.

Farrington, P. and Salama, R. B., 1995. Controlling dryland salinity by planting trees in the best hydrological setting. *Land Degrad. Develop.* **7:** 183–204.

Force, F. F. T., 1995. Farm Forestry in Western Australia. Farm Forestry Task Force: Perth, Western Australia.

Government of Australia, 1996. The national strategy for conservation of Australia's biodiversity. Commonwealth Department of Environment, Sport and Territories: Canberra, Australian Capital Territory.

Government of Western Australia, 1996a. Salinity a situation statement for Western Australia. Government of Western Australia: Perth, Western Australia.

Government of Western Australia, 1996b. Western Australia salinity action plan. Government of Western Australia: Perth, Western Australia.

Government of Western Australia, 1997. Memorandum of Understanding between the Commissioner for Soil and Land Conservation, Environment Protection Authority, Department of Environment Protection, Agriculture Western Australia, Department of Conservation and Land Management and Water and Rivers Commission for the protection of remnant vegetation on private land in the agricultural region of Western Australia. Government of Western Australia: Perth, Western Australia.

Greig, P. J. and Devonshire, P. G., 1981. Tree removals and saline seepage in Victorian catchments: some hydrologic and economic results. *Aust. J. Agric. Econ.* **25**(2): 134–48.

Gross, R. E., 1993. A green evolution. Retooling agriculture policy for greater sustainability. *J. Soil and Water Conserv.* **July August 1993:** 285–88.

Herbert, A., 1993. Rural land value. Agriculture Western Australia: Perth, Western Australia.

Hobbs, R. J. and Saunders, D. A., 1992. Introduction. Pp. 3–9 *in* Reintegrating Fragmented Landscapes, Towards Sustainable Production and Nature Conservation ed by R. J. Hobbs and D. A. Saunders. Springer-Verlag: New York.

Kitchener, D. J., Chapman, A., Muir, B. G. and Palmer, M., 1980. The conservation value for mammals of reserves in the Western Australian Wheatbelt. *Biol. Conserv.* **18:** 179–207.

Lefroy, T. and Hobbs, R., 1992. Ecological indicators for sustainable agriculture. *Aust. J. Soil and Water Conserv.* **5**(4): 22–28.

McLeod, P., 1996. Review of natural resource management and viability of agriculture in Western Australia. A discussion document. Agriculture Western Australia: Perth, Western Australia.

Revell, G., 1994. Reading the Remote: Landscape Characters of Western Australia. Department of Conservation and Land Management: Perth, Western Australia.

Safstrom, R. D., 1995. Conservation values of small reserves in the central wheatbelt of Western Australia. Department of Conservation and Land Management: Perth, Western Australia.

Safstrom, R. D., 1996. Environmental evaluation of native vegetation in the wheatbelt of Western Australia. Principles and criteria used to appraise land clearing proposals. Department of Environment Protection: Perth, Western Australia.

Safstrom, R. D., 1997. A framework for evaluating proposals to clear native vegetation for agriculture in the transitional rainfall zone of Western Australia. Faculty of Agriculture. University of Western Australia: Nedlands, Western Australia.

Scott, P. R., 1993. Trees to control groundwater — some examples from Western Australia. *In* The Role of Trees in Sustainable Agriculture. A national conference. Agriculture Western Australia: Perth, Western Australia.

Scougall, S. A., Majer, J. D. and Hobbs, R. J., 1993. Edge effects in grazed and ungrazed Western Australian wheatbelt remnants in relation to ecosystems reconstruction. Pp. 163–68 *in* Nature Conservation 3: Reconstruction of Fragmented Ecosystems ed by D. A. Saunders, R. J. Hobbs and P. E. Ehrlich. Surrey Beatty & Sons: Chipping Norton, New South Wales.

Siewert, R. and Abbott, S., 1991. The Mallee Road Catchment Project. Agriculture Western Australia: Perth, Western Australia.

Thompson, N. J., 1986. Fiscal incentives for Australian bushland. *Environ. Manage.* **10**(5): 591–97.

Yates, C. J., Hobbs, R. J. and Bell, B., 1994. Landscape scale disturbances and regeneration in semi-arid woodlands of southwestern Australia. *Pac. Conserv. Biol.* **1:** 214–21.

CHAPTER 49

Voluntary protection of natural areas: an effective conservation tool in the productive landscape

Tim Porteous[1]

One of the major conservation challenges, if not *the* conservation challenge, in New Zealand is how best to ensure that those areas of value for nature conservation on private land can be protected for the benefit and enjoyment of future generations. In New Zealand there have been a number of institutional responses to this challenge, including regulation, incentives and provision for voluntary legal protection. This chapter examines the importance of private land conservation in New Zealand and the protective mechanisms available to private landholders. The benefits of voluntary conservation are examined in greater detail.

INTRODUCTION

NEW Zealand has a natural history that extends far back into antiquity. The islands making up New Zealand are a remarkable biological storehouse with many plants and animals having an ancestry that can be traced back 200–150 million years ago when New Zealand was part of the super continent Gondwanaland.

Eighty million years ago, as a result of the movement of the earth's tectonic plates, New Zealand gained its physical independence from the super continent. It is this isolation that makes New Zealand's biota so unique. Jared Diamond (1990), an American scientist, considered that New Zealand was as close as it is still possible to get an opportunity to study life on another planet.

Our history of human occupation is, by contrast, relatively short; the first Polynesian explorers arrived about 1 200 years ago and the first Europeans a little over 150 years ago. The New Zealand that confronted the first Polynesian immigrants was mostly covered in forest. The remaining 20% of land included the alpine zone, and the grasslands of the central North Island volcanic plateau and the Southland plains. Fires, both deliberate and accidental, had reduced forest cover to 50% by the time the first European settlers arrived.

From 1850 onwards, New Zealand underwent a period of landscape change unequalled in speed and impact. Vast areas of forest were cleared either through milling and burning, swamps were drained to produce flat fertile pastures, and our tussock grasslands were repeatedly burnt. Many unique plant communities were lost, and habitats for native fauna destroyed and fragmented. The destruction of habitats contributed to the extinction of many native bird species (Enting and Molloy 1982). Soil erosion was widespread. A mixture of ignorance and sentiment led to the introduction of a long list of undesirable plants and animals from around the world — rabbits, hares, possums, deer, goats, wallabies, cats, rats, ferrets, weasels, stoats, gorse, broom, heather, briar and others as well as the unintentional introduction of diseases and parasites.

The Ministry for the Environment's (1997) recent all-encompassing publication "The State of New Zealand's Environment 1997" provides a succinct summary of the effects of this colonization on our biodiversity. Examples include:

1. **Fauna**

   32% of the endemic land and freshwater birds are now extinct.

   75% of the remaining birds are now classified as threatened species.

2. **Indigenous forest**

   As noted, this originally covered 80% of New Zealand's land area. This area was reduced to 23% or 6.2 million hectares.

---

[1]Queen Elizabeth II National Trust, P.O. Box 3341, Wellington, New Zealand.

Pages 494–96 *in* NATURE CONSERVATION 5: NATURE CONSERVATION IN PRODUCTION ENVIRONMENTS: MANAGING THE MATRIX ed by J. L. Craig, N. Mitchell and D. A. Saunders. Surrey Beatty & Sons, 2000.

3. **Wetlands**

   Only 10% of our wetlands remain, reflecting the extensive conversion of these areas by drainage for fertile farming land.

4. **Dunelands**

   Most of the 50 000 hectares of duneland ecosystems have been heavily modified by grazing, fires, drainage, coastal development and introduced plants.

A second and important legacy of our agricultural forefathers is the distribution of our remaining ecosystems.

New Zealand has, by international standards, an enviable public conservation estate, protecting some 30% of the total land area. However, the public protected lands are, in general, concentrated on lands of higher altitude unsuited for agriculture and not representative of the wide range of New Zealand's landscapes, natural areas and ecosystems. For example, estuaries, freshwater wetlands, shrublands, tussock grasslands and particularly lowland forest are under-represented in the public protected estate. Over 90% of the lowland areas have been converted to farmland with a corresponding detrimental impact on biodiversity. But despite this degree of modification, significant areas containing natural values remain scattered throughout the lowland areas. Studies and surveys such as those conducted as part of the Protected Natural Areas programme continue to identify a range of such areas. Frequently these areas are key ones for the maintenance of biodiversity. Lowland forests, for instance, support the greatest diversity of plants and animals of all New Zealand forest ecosystems. Many apparently degraded and tattered remnants continue to support a surprising range of plant and animal species, frequently those considered to be threatened and endangered.

It is therefore now widely accepted that the most significant conservation gains will involve the protection and enhancement of privately owned conservation resources which, in general, are located in the lowland areas and form part of our working and productive landscapes.

RESPONSES TO THE CHALLENGES

How can society best bring about protection for these areas?

Let us briefly examine three generalized responses to the challenges:

1. *Rules and regulations* — Through legislation such as the *Forests Act 1949* and the *Resource Management Act 1991,* it can be more difficult to destroy or deliberately modify natural areas. But rules and regulations cannot offer protection to a natural area in the widest sense. Because of the fragmented nature of many of our lowland ecosystems and the resulting effect of plant and animal pest invasions, stock grazing, wind damage and absence of the full range of indigenous plant and animal species, some form of active management is required to prevent total collapse. Rules and regulations do nothing to halt this more subtle decline and at best they can be regarded as temporary holding mechanisms.

2. *Purchase of the landowners' interest in the conservation area* — There will always be a legitimate role for the Crown or local authorities to purchase and manage natural areas to ensure their protection. However, changing political philosophies and resultant reducing levels of government spending have resulted in this option only "scratching the surface". Concerns have also been expressed recently about the Crown's ability to adequately manage its existing estate and have raised questions about the wisdom of further adding to the estate (for instance, animal pests are estimated to pose a threat to biodiversity in some 1.8 million hectares of Department of Conservation land) (Ministry for the Environment 1997).

3. *Voluntary permanent protection* — A third and highly successful approach to protecting native conservation on private land is that of voluntary protection by way of a legal agreement, most commonly a covenant. The Queen Elizabeth II National Trust is one very popular option in New Zealand for landowners wishing to protect areas on their land. To date, over 1 070 covenant agreements have been registered protecting some 43 000 hectares of private land.

   A further 336 covenants are progressing towards registration protecting 59 500 hectares. Land subject to a covenant does not become the property of the Trust, rather the landowner retains title to the land. An open space covenant is usually in perpetuity. Covenants are registered against the land title and are binding, not only on the present landowner or leaseholder, but all subsequent owners or leaseholders.

What are the advantages of promoting the use of covenants?

1. *Cost effectiveness* — Covenants secure a perpetual interest in the land and ensure protection **without the need to purchase an owner's interest outright.** The savings to the taxpayer by covenanting rather than purchase are considerable.

   Covenants are **voluntary** landholder-initiated agreements, therefore a minimum of negotiation time is needed. "Compensation" expectations are also minimized.

   The work of the Trust is based on a partnership between the private landholder and the Trust. The manifestation of this partnership is the sharing of the cost of establishing a covenant. Typically, the Trust and the landholder share the cost of fencing the area to be covenanted on a 50/50 basis. The cost of surveying the area, necessary to facilitate registration on the land title, is borne by the Trust. No compensation is paid for production potential foregone.

2. *On-site caretaker* — Day-to-day management to legally enforceable established standards remains the responsibility of the landholder with oversight by the Trust. The savings to the taxpayer in on-going management costs for natural areas protected by Trust covenants are considerable when these costs essentially remain the responsibility of the landholders.

   The covenant agreement stipulates those actions that are and are not permissible within the covenanted area. Most covenants protect areas of native forest and are, therefore, fenced to exclude grazing animals.

   Each covenant is inspected by a representative of the Trust every 24 months to ensure compliance, including maintenance of the protective fence in working order. Should there be a breach of covenant conditions, legal action would be taken against the landholder.

   As landholders are the on-site caretakers, the level of management applied to covenants is high. For instance, weed and pest control actions are frequently integrated into normal farm management operations.

   Advice and assistance provided by the Trust ensures that management problems can be identified and remedied at an early stage before problems become insurmountable. As a result a strong partnership develops between the Trust and landowner. This explains why, in twenty years of operation, there has never been a breach of covenant conditions.

3. *Community support* — Establishing a covenant often results in the engendering of support from the community to assist in this costs of establishment and on-going management.

   A number of Trust covenants are used on a regular basis by local schools as outdoor classrooms. Subjects as diverse as drama, science and outdoor skills are taught.

   A further increasing use of some covenants is that of eco-tourism. A number of landowners have constructed tracks to allow public usage of their covenant. Others have established accommodation units adjacent to covenanted areas to allow guest enjoyment of the protected area.

4. *Permanence and security* — The National Trust is an independent statutory body with perpetual succession.

   The National Trust is funded from a government grant, earnings of investments, membership subscriptions, income from properties, donations and bequests.

   Open space covenants are registered against the title of the land and are in nearly all cases *in perpetuity*.

   A covenant cannot be revoked by the landowner. It requires the unanimous agreement of the full Trust Board of Directors to agree to the uplifting of a covenant.

CONCLUSION

Voluntary conservation has developed a considerable momentum in New Zealand. It is a powerful and successful answer to the problem of securing permanent and robust protection for our remaining precious natural areas, habitats and biodiversity where they occur on private land.

Voluntary protection neatly encompasses the dual aspects of securing permanent protection cost effectively and ensuring adequate on-going management in one package.

Voluntary conservation is the way of the future.

REFERENCES

Diamond, J., 1990. New Zealand as an archipelago: an international perspective. Pp. 3–8 *in* Ecological Restoration of New Zealand Islands ed by D. R. Towns, C. H. Daugherty and I. A. E. Atkinson. Department of Conservation: Wellington, New Zealand.

Enting, B. and Molloy, L. F., 1982. The Ancient Islands: New Zealand's Natural Environments. Port Nicholson Press: Wellington, New Zealand.

Ministry for the Environment, 1997. The State of New Zealand's Environment 1997. GP Publications: Wellington, New Zealand.

CHAPTER 50

Matching landholders learning styles to the formulation of nature conservation information

Greg Siepen[1] and Kay Dorricott[2]

Rural landholders usually learn aspects of farm management through experiential learning situations, which are reinforced by demonstrations, field days, audio-visual presentations and written information. Landholders in Queensland have requested more local information about nature conservation, which they can apply to the management of their properties (Siepen and Stone 1997). To meet these needs a series of learning tools have been trialed.

This chapter reports on the formulation of these learning techniques and evaluates their success in the light of theories of communication, adult learning and extension.

Key words: Communication, Adult learning, Action learning, Landholder attitudes.

INTRODUCTION

MANY government agencies involved in nature conservation recognize that extension and education programmes for rural landholders (mostly primary producers) are an important process of their goals for protecting the survival of natural biodiversity in their state or region. At the same time these programmes are often not well funded. As a consequence extension and education programmes to rural landholders are often not very effective.

Additionally, research in Queensland by Siepen and Stone (1997) indicates that needs analyses are often not undertaken in rural areas. This means that there is a general failure to identify the most effective communication methods for these landholders based on their learning styles (see Honey and Mumford 1986).

In this chapter we outline two independent projects we undertook during 1995/96 to shed light on the need to identify learning styles of rural landholders. Dorricott conducted field days and produced a manual as part of a kinesthetic learning approach. Siepen researched and compiled a video and handbook (visual-auditory learning style). Our aim was to demonstrate that extension and education programmes might be more appropriately targeted by government natural resource agencies using appropriate communication techniques and information.

LEARNING STYLES AND ACTIVITIES

Each of us has a variety of learning styles, which we use to acquire knowledge and skills. In most of us one or more styles predominate. For example, some people learn by immersing themselves in new experiences. Honey and Mumford (1986) call this type of learning style "activist". Others learn primarily by sitting back and pondering experiences. They analyse situations in detail and consider all implications. They are "reflectors" *(ibid)*. The other learning styles are "theorists" *(ibid)* who make decisions objectively, rationally and logically, and "pragmatists" *(ibid)* that are very practically oriented and "learn by doing".

Associated with learning styles are the various activities by which we learn. These are by visual (learning through seeing), auditory (learning through hearing) and kinesthetic (learning by doing) means. Studies in Mackay and Rockhampton, Queensland (Martin, pers. comm. 1996) have shown that farmers often score very high on the kinesthetic approach as their primary learning activity. Visual and auditory means are used to reinforce and corroborate information received and decisions made.

---

[1]School of Natural and Rural Systems Management, University of Queensland, Gatton, Queensland, Australia 4343.
[2]Queensland Parks and Wildlife Service, P.O. Box 834, Atherton, Queensland, Australia 4883.
Pages 497–501 *in* NATURE CONSERVATION 5: NATURE CONSERVATION IN PRODUCTION ENVIRONMENTS: MANAGING THE MATRIX ed by J. L. Craig, N. Mitchell and D. A. Saunders. Surrey Beatty & Sons, 2000.

It is clear that different learning styles may be present in any community. It is important, therefore, that the primary learning style of a group is identified before designing or implementing extension and education activities.

Siepen and Stone (1997) have documented that rural landholders rank learning from field days and local nature notes more highly than learning from watching videos. In our project we trialed these two different learning techniques: the field days exercises run by Dorricott emphasize kinesthetic learning, while the video and booklet produced by Siepen emphasize visual and auditory learning. We now outline the two different methods we used, and in the discussion, which follows, we discuss the benefits and pitfalls of each method.

COMMUNICATION METHODS

Kinesthetic learning

Dorricott selected properties for field days where the relationships between ecology, property management and economic management could be shown easily. Properties were surveyed in four shires in southern Queensland (Fig. 1) to ascertain the quality of native vegetation and animal communities occurring on them; the interests, attitudes and beliefs of the landholders in relation to nature conservation; the practical applications that were being undertaken in relation to conservation of the native plant and animal communities; and the standing of the landholders within the community.

1. A field day was held in each shire to introduce the concept of integrating productive agriculture and conservation, determine landholders' desired outcomes and needs for nature conservation information and call for property nominations as case study sites.
2. Nominated properties were inspected and case study sites selected using a stratified sampling design. Properties were selected on the basis of typical enterprise representation for the shire and sampling sites were selected to include a range of land types and land uses (Dorricott *et al.* 1997).
3. Small-scaled fauna and habitat surveys were conducted. Landholders were encouraged to participate. Anecdotal records of historical changes in faunal abundance and diversity were also compiled. Past, present and future property management strategies were documented and landholders were asked to identify tangible and intangible benefits and disadvantages of retaining native timber. Farm forestry potential was also assessed.
4. Property plans for selected properties were drawn up in consultation with landholders, with provision for nature conservation within the context of maintaining a sustainable production enterprise being emphasized.
5. A second field day on balancing productivity with nature was held in each shire to demonstrate the values of native vegetation on farms for conservation, farm forestry and other uses such as bush food, pharmaceutical products and nature-based tourism. Landholders' initiatives on integrating nature conservation with production were highlighted. The landholder at each property on which the field day was held took a leading role in the day's presentation and discussion.
6. A nature conservation manual for landholders in the study area has been produced. It features profiles of case study properties and contains a series of questions and answers on nature conservation based on the information needs identified by landholders at the first field day.

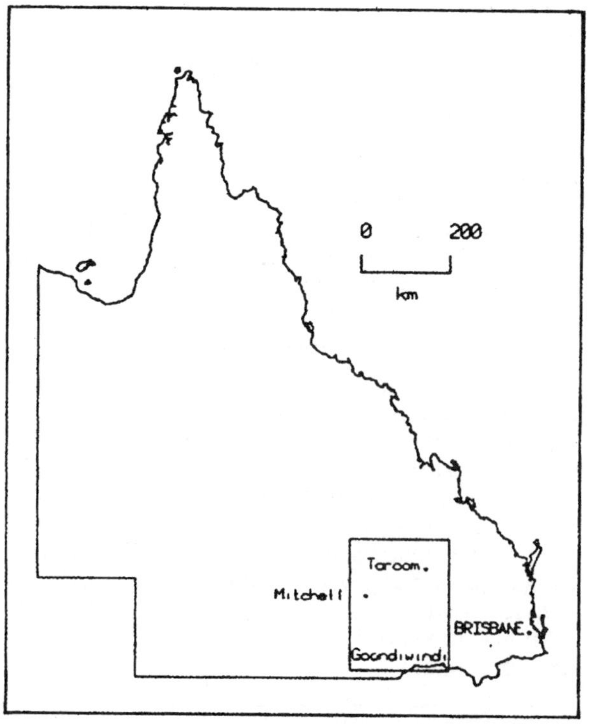

Fig. 1. Area in which field days were held to show relationships between ecology, property management and economic management.

What has been learned

The Case Study approach was a kinesthetic learning experience for both landholders and researchers. Key aspects are summarized below. These approaches could be applied in further work with landholders:

1. It strengthened the respect between both the landholder and the researcher.
2. It involved the landholder very early in the analysis.
3. Landholder knowledge was given "credit" and equal to scientific knowledge.
4. Landholder observational skills were highly regarded in providing accurate information about nature.
5. It showed that "nature conservation" was not a fundamental motivator for the majority of landholder producers.

6. Multiple use was identified as the main motivator for retaining native bushland.
7. Local work and local information is regarded more highly than information from outside the area/region.

The introductory field day and subsequent surveys created a great deal of positive interest in the project with ten properties being nominated at the first field day in one shire. Additionally, much information was gained on the distribution and abundance of native fauna and flora, which was unknown to the landholders. Species were identified which can be used to demonstrate how management strategies affect species abundance and diversity (Dorricott *et al.* 1997 P. 237).

The benefits of communicating using kinesthetic learning approaches are that

- individuals can be involved at an early stage (if the individual is willing to be involved)
- individuals can see first hand actions undertaken and results achieved
- the landholder on whose property the demonstration/field day is occurring is telling a great deal of "the story"
- individuals can take anything they have learned, reflect on it and trial it.

Disadvantages include

- individuals not becoming involved or participating
- this method does not "reach" the individuals who have other learning styles as their primary style
- experts can dominate and use the exercise as technology transfer
- timing of the demonstration/field day may preclude some individuals.

Visual and auditory learning

Taking into consideration landholder learning styles and methods of communication preferred by landholders (Siepen and Stone 1997), Siepen designed a booklet and accompanying video. The aim of the video tape and booklet was to motivate rural landholders to undertake nature conservation activities in association with primary production. The video and booklet provided information on topics such as

- what is ecology
- basics of retention and replanting native bushland
- judging the condition of bushland
- habitat corridors for wildlife
- planning for conservation and
- fencing programmes for conservation.

Rural landholders from various states, who had successfully integrated nature conservation and primary production, were chosen as examples to show others their strategies. The handbook was designed to present both ecological information and examples of successful strategies in a small format for storage in a farm vehicle glovebox. It was also designed to demystify scientific terms such as "biodiversity", "ecology" and to recognize the knowledge held by landholders.

The video documented several case studies by way of personal interviews. Landholders told their "story" on video relating the conservation projects undertaken on their properties to costs involved and outcomes. This was then translated into the booklet by way of formal quotes. Additional information and examples were presented in the booklet to present a full picture of integrating primary production and nature conservation. The handbook and video were designed as a package, each complementing the other and both complementing field days, practical case study work and other conservation activities on rural properties and in landcare groups.

What has been learned

The handbook and accompanying video have not been formally evaluated since their release in March, 1997. Many positive comments have been forthcoming from landholders, landcare co-ordinators and primary industry extension officers. For example, one landcare co-ordinator has found the package very useful and informative and she has made it an essential part of the property planning workshops dealing with identifying and understanding the natural resources occurring on properties. (C. Rhinehart, pers. comm. Oct. 1997).

The video and booklet was distributed to all 3 500 landcare groups throughout Australia (at time of release) for use at workshops, meetings seminars and field days.

This more reflective style of communication and learning has many benefits. It is easily accessible at a time that is suitable to the individual. Like the field days and demonstrations people learn from their colleagues not experts. It is like a field day in your living room. It allows for plenty of reflection before implementing any actions. It allows several of the steps in kinesthetic learning to occur, but it leaves out the practical hands-on perspective.

DISCUSSION

Kinesthetic learning

The case study approach (kinesthetic learning style) is a mixture of adoption-diffusion and farmer-first extension models. Initially, the researcher sets the framework but at the introductory field day the landholders identify their concerns and issues in regard to nature conservation. That is, the landholders can set the agenda. The flora and fauna surveys follow in which the landholders are encouraged to participate and share their knowledge with the researcher's knowledge. The final field day is run by the scientist to share the information to all and a manual is compiled for use in decision making by those landholders.

According to Vanclay and Lawrence (1995:101) when farmers are given responsibility for setting the agenda and determining priorities for extension activities, "barriers to adoption do not logically exist, since there no longer is a normative reason (within the perspective) why adoption ought to occur."

The value of living on a property while doing ecological research and involving the landholder in that research was very beneficial in building closer relationships, understanding and respect between researchers and rural producers (Woodhill 1992). This partnership is created whereby both parties are interested in the research outcomes instead of a "them" and "us" situation which is evident in many parts of rural Queensland (Dorricott, pers. comm.) This setting also allows for input by the landholder in determining the research directions and reduces the potential for creating barriers to adoption (Vanclay and Lawrence 1995). It also facilitates the researcher's credibility, which allows information to be accepted by the landholder (Garnett and Crowley 1997).

In the case of nature conservation it has been shown that a desire to conserve native plant and animal communities is not the primary motivator for landholders to retain native bushland on their properties. The concept behind such retention and management is one of multiple use where conservation is integrated with activities such as farm forestry and nature tourism, which can give, added economic benefits to primary production. Although not the primary motivator, many landholders showed a high degree of interest in the fauna and flora surveys and subsequent property management planning exercises.

Visual-Auditory learning

Even though it is recognized that landholders have identified field days, demonstrations, case studies as being their primary communication and learning strategies (Siepen and Stone 1997) many rely on reinforcing the information learned at these events through reading and by visual presentations. This "reflector" learning activity is catered for by the production of a resource manual (Dorricott), video and handbook (Siepen). Information from these source helps landholders understand the ecology of their properties, corroborates information gained at field days, demonstrations and in case studies, gives suggestions for integrating nature conservation with primary production and validates information given by peers.

The value of appropriate video and booklet packages has been shown by the success of two previous video-booklet combinations, which had been produced with the financial assistance of BHP Wire Products. They centred on landcare generally and catchment management as a means of better management of the natural resources. Both videos and booklets were distributed to all landcare groups in Australia and complemented national sustainable resource management and catchment planning programmes.

The effectiveness of video and booklet packages lies in their suitability for use at landcare and conservation group meetings and field days by individual landholders. Consideration was given to

- the length of the video
- the use of a known presenter
- the use of case studies
- testimonies of landholders who are successfully managing their resources in an ecologically and economically manner, and
- provision of ecological information to verify the case studies.

As a result, the video is 25 minutes long, has presenter Jim Downes (ex ABC *Countrywide* television programme), and contains case studies from various places in Australia presented by the landholders.

CONCLUSION

Kinesthetic learning styles (e.g., case studies, field days and demonstrations) have been identified by landholders in Queensland as the major means of learning about factors that affect their primary economic production and property management generally. These forms of communication are complemented by written materials and visual presentations (visual-auditory styles) to give landholders a full picture when making management decisions (Siepen and Stone 1997).

Understanding the learning styles and preferred learning activities of prospective target audiences are crucial to delivering effective extension and education programmes. Using extension and communication methods that allows the target audiences to set the agenda increases the likelihood of adoption of new technologies and management strategies.

There is a need by nature conservation agencies to recognize that nature conservation is not the fundamental motivator for the majority of landholder producers. Conservation information presented in the context of multiple natural resource use is likely to be accepted more readily.

REFERENCES

Campbell, C. A. and Siepen, G. L., 1994. Landcare: Communities Shaping the Land and the Future. Allen and Unwin: Sydney, New South Wales.

Dorricott, K. E., Lawrie, B. C. and Voller, P. J., 1997. Balancing Productivity and Conservation in southern inland Queensland: A Case Study at the Grass-roots. Pp. 233–40 *in* Conservation Outside Nature Reserves ed by P. Hale and D. Lamb. Centre for Conservation Biology: The University of Queensland, St Lucia, Queensland.

Garnett, S. and Crowley, G., 1997. The golden-shouldered parrot of Cape York Peninsular: the importance of cups of tea to effective conservation. Pp. 201–05 *in* Conservation Outside Nature Reserve ed by P. Hale and D. Lamb. Centre for Conservation Biology: The University of Queensland, St Lucia, Queensland.

Honey, P. and Mumford, A., 1986. Using Your Learning Styles. 3rd edition. Honey: Maidenhead, UK.

Siepen, G. L. and Stone, G., 1997. Identification of Rural Landholders' Requirements for Nature Conservation Information in Queensland. Pp. 226–32 *in* Conservation Outside Nature Reserves ed by P. Hale and D. Lamb. Centre for Conservation Biology: The University of Queensland, St Lucia, Queensland.

Siepen, Greg, n.d. Partnerships with nature: managing your farm in an ecologically sustainable manner. Landcare Australia Limited, Chatswood, New South Wales.

Vanclay, F. and Lawrence, G., 1995. The Environmental Imperative: Eco-social Concerns for Australian Agriculture. Central Queensland University Press: Rockhampton, Queensland.

Woodhill, J., 1992. Landcare in NSW: Taking the First Step. Centre for Extension and Rural Development: University of Western Sydney, Hawkesbury, Richmond, New South Wales.

CHAPTER 51

The private lives of protected areas: private land in the National Reserve System

Gillian Lee[1] and Steve Szabo[1]

Australia is committed to developing a comprehensive, adequate and representative National Reserve System. The land area of Australia proclaimed by State, Territory and Commonwealth Governments as protected areas is limited to about 7.6%.

In accordance with the Australian Constitution, the States and Territories have the main responsibility for land management. The Commonwealth's National Reserve System Programme assists Australia's States and Territories to strategically increase the reserve system through jointly funding acquisitions for national parks and reserves, and other strategic projects.

Given the high cost of land acquisition and the restricted range of land type on the property market, the Commonwealth Government has investigated alternatives to the traditional government-run, formally-gazetted nature conservation reserve system. Two new National Reserve System Programme components, Protected Areas on Private Land and Indigenous Protected Areas, have been developed to enable private and leasehold land to be added to the National Reserve System through partnerships between the government and landholders.

This chapter discusses the progress with the Protected Areas on Private Land and Indigenous Protected Areas components to date. It discusses the perceived overlap and complementarity between the National Reserve System Programme and other government programmes designed to encourage protection of remnant vegetation. It identifies key problem areas for managers to address. In particular it discusses the need for private landholders, including indigenous people, to secure government support to establish and manage such protected areas. The chapter argues that voluntary agreements are the best way for government nature conservation agencies and private landholders to achieve a co-operative approach to biodiversity conservation. It presents two Australian approaches which are key elements for achieving a comprehensive, adequate and representative National Reserve System.

INTRODUCTION

AUSTRALIA is committed to developing a comprehensive, adequate and representative National Reserve System. The land area of Australia proclaimed by governments as protected areas is limited to about 7.6%. Due to a range of factors the protected area system was largely established without a systematic approach to ecosystem representation and, as a result, is by no means representative of the full range of Australian ecosystems.

The National Reserve System Programme is one of the programmes of the Commonwealth Government's Natural Heritage Trust. Between 1997 and 2001 it will receive $80m from the Trust towards improving Australia's system of protected areas. Proposals are invited annually under the Natural Heritage Trust Guide to Applications and are assessed under a multi-level assessment process before a recommendation for approval is submitted to the Commonwealth Minister for the Environment. Projects are generally funded on a cost sharing basis with up to two-thirds of the cost of the acquisition and associated legal costs being provided by the Commonwealth as a once-off payment with the proponent being responsible for the remaining

[1]Biodiversity Group, Environment Australia, P.O. Box 787, Canberra, Australian Capital Territory, Australia 2601.

Pages 502–10 *in* NATURE CONSERVATION 5: NATURE CONSERVATION IN PRODUCTION ENVIRONMENTS: MANAGING THE MATRIX ed by J. L. Craig, N. Mitchell and D. A. Saunders. Surrey Beatty & Sons, 2000.

one-third and all the costs associated with ongoing management.

Programme managers in consultation with the States and Territories have developed a sound scientific basis for prioritising additions to the National Reserve System. Acquisition priorities are focused on national gaps in the coverage of ecosystems within protected areas, and ecosystem degradation or vulnerability to threatening processes. However, some acquisitions remain necessary within otherwise well-conserved regions when particular priority ecosystems are poorly reserved or significantly threatened.

The scientific basis for the establishment of the National Reserve System consists of a series of mechanisms for identifying gaps in the National Reserve System and setting priorities to fill these gaps using the resources available under the National Reserve System Programme.

All acquisition proposals are assessed against the agreed processes and reporting requirements. Three key elements underpin this scientific framework. They are:

1. The Interim Biogeographic Regionalization for Australia which is an agreed planning framework based on broad ecosystems;
2. The Interim Biogeographic Regionalization for Australia priorities map which shows an agreed set of national priorities for allocation of resources to each jurisdiction according to needs; and
3. The Interim Scientific Guidelines for Establishing the National Reserve System which give a nationally consistent process for identifying and selecting potential areas for inclusion in the National Reserve System.

Programme managers visualize the range of Australian native ecosystems maintained through the establishment of an innovative and diverse system of protected areas managed for conservation across all land tenure types.

The International Union for the Conservation of Nature definition of a protected area is "an area of land and/or sea especially dedicated to the protection and maintenance of biological diversity, and of natural and associated cultural resources, and managed through legal or other effective means".

This definition of protected areas is used in the National Reserve System Programme and programme managers have been innovative in seeking out options for management by "legal or other effective means". One solution is the establishment of protected areas on private land which is often in a production environment. Another solution is establishing protected areas on indigenous-owned land which, by its unique tenure arrangements will never come on to the property market. As a result the National Reserve System will consist of a system of core statutory reserves complemented by protected areas on indigenous owned and managed lands, and on private and leasehold lands.

Regardless of who owns or buys it, land acquisition will remain the primary tool used to fill the major gaps in the current reserve system. Of the $80 million allocated to the National Reserve System Programme over the four years 1997–2001, the majority will be allocated for land acquisition. The majority of property acquisitions for addition to the National Reserve System are expected to continue to be made by State and Territory Governments in the traditional manner through their nature conservation agencies.

THE CONCEPT OF PRIVATE PROTECTED AREAS

In considering the International Union for the Conservation of Nature definition of a protected area it is significant that there is no recommendation on who the manager of the land should be. Protection must be "legal or otherwise effective", and the area must be "particularly or chiefly dedicated to protection and maintenance of biological diversity, and of natural and associated cultural resources".

Investigations and analysis of Commonwealth, State and Territory legislation have concluded that there are no legal impediments to private landholders of any type declaring their land as Protected Areas according to the International Union for the Conservation of Nature protected areas categories in any of the Australian jurisdictions.

The International Union for the Conservation of Nature (1994) has identified and defined six categories for protected area management. These are:

Category I	Protected areas managed mainly for science or wilderness protection
Category II	Protected area managed mainly for ecosystem protection and recreation
Category III	Protected area managed mainly for conservation of specific natural features
Category IV	Protected area managed mainly for conservation through management intervention
Category V	Protected area managed mainly for landscape/seascape conservation and recreation
Category VI	Protected area managed mainly for sustainable use of natural ecosystems.

Using these categories, an extensive range of land ownership and management options could meet the definition of protected area.

Defining private land is a key issue. For National Reserve System purposes "private land" is defined as land owned by or leased by anyone other than Commonwealth, State or Territory nature conservation agencies. Private land includes local government land, Indigenous-owned land, land owned or leased by private individuals and companies, and land owned by State, Territory and Commonwealth

Governments other than nature conservation agencies. Land owned by State and Territory nature conservation agencies whether gazetted as national park or not, is termed public land and is not discussed in this chapter.

Combining these definitions, it follows that a **private protected area** is a protected area other than a formally-gazetted statutory protected area, managed for nature conservation and protected by legal or other effective means.

There are many models of organizations which own and manage protected areas which fit this definition. The United States of America-based non-government organization, The Nature Conservancy, has a mission "to preserve plants, animals and natural communities that represent the diversity of life on Earth by protecting the lands and waters they need to survive"[1]. It has been outstandingly successful, having 900 000 members in 1998 and having protected 10 million acres in the United States of America since 1953. The Nature Conservancy also manages 1 556 000 acres which the organization either owns or has under conservation easement. This is the largest private system of protected areas in the world.

The United Kingdom-based Wildlife Trusts are a network of 46 Wildlife Trusts and 50 Urban Wildlife Groups, which began, in 1912, to protect wildlife and habitat locally. The network now has 2 000 nature reserves managed by volunteers, and 220 000 members. The acquisition, management and associated works of the Wildlife Trusts are funded through memberships, fundraising, events, sponsorships, legacies, donations and grants.

Similarly, models of private protected area networks exist in Australia but they are smaller and not yet formally contributing to the National Reserve System. The Australian Bush Heritage Fund and the Trust for Nature (Victoria) are well established examples.

There are also many very good examples of Australian private protected areas. For example:

- Birds Australia has purchased Gluepot Station for protection of habitat of a small population of the endangered black-eared miner bird species.
- Australian Bush Heritage Fund has purchased Erith Island in Bass Strait, bushland on the Brogo River in the Bega Valley, New South Wales, Daintree Fan Palm Forest, North Queensland, and Wandoo Woodland, Kojonup, Western Australia.
- Oxygen Farm Association has purchased a property on the mid-north coast of New South Wales.
- Friends of Durras (with the NSW National Parks and Wildlife Service) have purchased bushland on Durras Lake near Bateman's Bay.

The National Reserve System first dabbled in private land with Anglesea Heathlands in Victoria. The Anglesea heathland, which was in private ownership in several contiguous blocks, was identified as an area of substantial value for nature conservation. Accordingly the area was of interest to the National Reserve System, but the value of the land (about $6m) was too great for acquisition. A novel approach developed by the State Planning Department enabled protection of the biodiversity values of the area with only a small government financial input.

A negotiated settlement resulted in the landowner gaining permission to develop 10 hectares of the 225 hectare property with controlled housing development (48 house lots and six bushland lots) to be confined to part of the perimeter of that block. In exchange, the Trust for Nature (Victoria) was to purchase the balance of that property, being 215 hectares, for $500,000. In late 1994, the Federal Government and State Government each contributed $250,000 to the Trust for Nature (Victoria) to purchase the property on the condition that the land to be added to the adjoining Angahook-Lorne State Park. Thus, 95% of the property was protected for 20% of the value of the land.

Although the land did not remain in private ownership as a private protected area, it did demonstrate to programme managers that innovative ways of delivering the National Reserve System were possible and, consequently, it led to the development of the components of the programme described in this chapter. Individuals, community groups and industry are all, as owners or leaseholders of parts of Australia, possible partners in biodiversity conservation. Approaches available for biodiversity conservation on private land vary in each State and Territory according to existing State and Territory mechanisms and legislation. Most will be able to meet National Reserve System requirements.

Programmes running in various States and Territories including the Trust for Nature (Victoria) Revolving Fund, and the South Australian Heritage Agreements reflect a broadening community awareness and willingness to protect some of Australia's natural assets on private land. They are the longest established initiatives on private land in Australia. Both are established under legislation; the Trust as a separate semi-government organization operating a purchase, covenant, resale system, and the Heritage Agreements as an effective mechanism for reducing land clearing.

The challenge has been to determine whether or not the outcomes of valuable conservation programmes such as these can form part of the National Reserve System and whether new approaches developed by building on and improving these models might be useful.

An area of private land will be considered within the scope of the National Reserve System only if it meets the International Union for the Conservation of Nature definition of a protected area. Examples of

[1]Page iii, The Year in Conservation. Annual Report Fiscal Year 1992. The Nature Conservancy.

legal and other effective means which could be used for management of such an area include:

- State and Territory legislative powers such as Nature Conservation Covenants (Victoria), Heritage Agreements (South Australia), Voluntary Conservation Agreements (New South Wales) etc.;
- Legally-binding management agreements;
- Easements in gross;
- Planning instruments, including those applied by local governments;
- In some cases, memorandums of understanding and leases; or
- Any combination of these.

It is these mechanisms the Commonwealth Government's National Reserve System Programme accesses to formalize and implement partnerships with private landholders to create protected areas on private land.

State, Territory and Commonwealth Governments have agreed that their preference is for funding for acquisitions to be channelled towards opportunities offering the highest levels of protection under International Union for the Conservation of Nature categories I to IV. However, voluntary arrangements on private and indigenous-owned land may accommodate other uses, as envisaged in other International Union for the Conservation of Nature categories, provided that biodiversity conservation is a primary objective of management.

Beginning in 1998/99 the Commonwealth Government invited the community to actively contribute to the enhancement of the National Reserve System and a proportion of the National Reserve System funding has been set aside specifically for the purpose. Some private land projects from previous years are also being funded through the States and Territories. Under the Community Groups Component of the National Reserve System Programme, programme managers aim to increase the representativeness of the National Reserve System through the contributions of community groups, industry bodies and local governments to land purchase or protected area declaration over currently-owned land.

Consultations have been carried out with those States and Territories which had an interest in establishing some kind of private protected area system:

- Western Australia has some existing agreements with mining companies regarding biodiversity conservation;
- South Australia has a successful programme of establishing Heritage Agreements;
- Victoria's Trust for Nature has been establishing Nature Conservation Covenants and a Revolving Fund, and the Victorian Government has several nature conservation agreements with mining companies;
- Tasmania has an existing National Reserve System project aiming to establish covenants in the midland areas;
- Researchers in New South Wales have joined forces with the NSW National Parks and Wildlife Service taking special interest in conserving the remaining grassy white box woodland and the National Parks and Wildlife Service has an active Voluntary Conservation Agreement Scheme; and
- The Australian Capital Territory Government is particularly concerned about protecting valuable biodiversity on Commonwealth-owned land.

Funding provided to successful applicants from the National Reserve System Programme Community Groups Component may be applied towards meeting up to two thirds of the costs associated with acquisition of land. The direct costs of purchase of the property, legal costs associated with the purchase and costs associated directly with formal protection such as those incurred as a result of gazettal of the property or entering into a nature conservation covenant agreement are all eligible for cost sharing. The proponent must supply the remaining costs of the acquisition and cover the costs of all ongoing management.

Any other proponent such as a group, individual, company or local government which already owns a property with significant conservation value which it would like nominally included as part of the reserve system may also apply for assistance. Funding from the programme can be applied towards establishing a formal protected area, assisting to meet such costs as those incurred as a result of gazettal of the property or entering into a nature conservation covenant agreement.

Establishment of a protected area as part of the National Reserve System requires a perpetual commitment to protect and manage the area according to the International Union for the Conservation of Nature definition of a protected area.

For projects involving land purchase, proponents must also demonstrate they meet the programme's eligibility criteria. They must:

- Be an incorporated body;
- Have demonstrated a capacity to make the purchase, including the ability to provide at least one-third of any acquisition costs;
- Have demonstrated a commitment to, and resourcing for, effective long-term management of the protected area for biodiversity conservation;
- Have demonstrated an ability to prepare and implement a plan of management for nature conservation;
- Provide a certified copy of the relevant documents of incorporation or founding document to check the principal purpose of the group and to check that no money goes to members. The documents must be certified by the Registrar of the Titles Office or at least by the Secretary beforehand;

- Be a Registered Environmental Organization[2] or an organization eligible to receive tax deductible donations under relevant legislation.

Proponents interested in establishing a protected area on their own land need not meet the criteria for eligibility described above.

Proponents are encouraged to contact their State or Territory nature conservation agency and to discuss their proposal. This is aimed at overcoming the problem of adversely affecting property values by disseminating information about priorities for land purchase.

The limited funds for the programme are applied to the highest priorities to ensure that the reserve system is expanded as effectively as possible. Priority is given to land purchases which:

- Secure ecosystems that are under-represented or threatened, particularly ecosystems that contain endangered ecological communities;
- Conserve ecosystems or ecological communities which contain nationally-listed threatened species or which allow for introduction of nationally-listed threatened species; and
- Are large enough to be viable in the long term either because of their size or their proximity to other areas managed for nature conservation.

Proposals can be in a priority bioregion[3] or an ecosystem under threat in any bioregion. Proposals that rate most highly are for ecosystems that are not currently well-represented in the National Reserve System and are under immediate threat. This may include threatened ecological communities.

All National Reserve System proposals are required to meet the requirements of the Interim Scientific Guidelines for Establishing the National Reserve System prepared to help prioritise additions to the National Reserve System which entails assessment of the site against a series of criteria. In addition to these guidelines, proponents are requested to address the following issues:

- Whether the proposal is compatible with regional land use strategies such as catchment management plans or bioregional conservation plans for the region;
- If the area does not have a regional plan, whether the proposal complements other reserves or places managed for conservation of natural values within the region;
- How management (e.g., for fire, ferals and weeds) fits in with other local management strategies; and
- How the purchase contributes to conserving ecological processes within the region.

A condition of funding for the establishment of a private protected area under the National Reserve System is that the property must become a formal protected area by conservation covenant, or a conservation reserve under relevant State or Territory legislation. In addition a Private Protected Area Establishment Agreement has been developed as a formal mechanism between the proponent and the Commonwealth Government where land is being purchased and the title will be held by the proponent. Its main purpose is to provide an additional measure of protection to the nature conservation covenant by ensuring active management of the values so that protected areas are managed to a standard similar to that of statutory national parks and reserves, and to provide for appropriate disposal options.

In cases where the National Reserve System is supporting establishment and management by proponents other than the owner of the property, a Deed of Agreement will be required with the landholder to ensure the National Reserve System interest in the property's values are similarly protected in the long term.

THE CONCEPT OF INDIGENOUS PROTECTED AREAS

About 15% of the land area of Australia is owned or leased by Aboriginal and Torres Strait Islander groups. Much of this land is in rural and remote parts of Australia and generally where contemporary environmental impacts have been relatively low.

From the early stages of the development of the National Reserve System it has been obvious that it will be difficult to establish a truly comprehensive, adequate and representative reserve system without the inclusion of indigenous land. Land owned and managed by indigenous people covers large areas of Australia and in some cases includes complete Interim Biogeographic Regionalization for Australia regions which are a priority for protected area management. National Reserve System programme managers, in consultation with these groups, have been investigating the potential for development of voluntary conservation agreements over land owned or leased by indigenous groups to expand and enhance the National Reserve System.

The Indigenous Protected Areas concept has attracted considerable support from indigenous people involved in consultations. Many groups see the concept as compatible with indigenous notions of stewardship of lands and seas and with their obligation to care for and maintain the estates for which they have traditional responsibilities. The Indigenous Protected Areas concept also provides a mechanism to address important social, cultural and economic issues for indigenous landholders.

The Indigenous Protected Areas component encompasses both the establishment and management

[2]The Register of Environmental Organizations is a register of approved organizations to which donations of money or property are tax deductible. The Register is listed under item 6.1.1 of table 6 in subsection 78(4) of the *Income Tax Assessment Act 1936*. Eligibility criteria for admission to the Register and the administrative requirements for the Register are prescribed in section 78AB of the Act.

[3]Interim Biogeographic Regionalization for Australia divides Australia into a series of biogeographic regions which are used as part of the basis for identifying deficiencies in the existing reserve system on a regional scale.

of protected areas on indigenous estates as well as the development of models for joint and co-operative management of publicly-owned protected areas of significance to indigenous peoples. Twelve pilot Indigenous Protected Area projects were initially funded nation-wide.

In seeking to develop an approach to establishment of protected areas on indigenous lands, noting that indigenous lands were not generally for sale, the International Union for the Conservation of Nature Category VI (Managed Resource Protected Area) was explored to determine whether it could be the basis for negotiating voluntary conservation agreements between governments and indigenous land owners. Category VI was initially considered most appropriate because it appeared to best conform to the needs of Aboriginal and Torres Strait Islander land holders and customary approaches to land use and management. That is, it was considered to give specific recognition to largely natural areas to ensure the supply of a sustainable flow of natural resources to local indigenous groups.

The development of the Indigenous Protected Areas concept was undertaken in three phases which have been completed.

Phase 1: Assess the legal feasibility of Indigenous Protected Areas. The first phase involved assessing the legal feasibility of establishing protected areas on indigenous lands within each jurisdiction. Our consultant concluded that all State and Territory Governments have the capacity to enter into agreements with indigenous land holders to manage such lands for conservation as a protected area (Sutherland and Smyth 1995).

Phase 2: Consult with indigenous peoples and nature conservation agencies. Next, selective consultations with Aboriginal and Torres Strait Islander organizations were undertaken with major representative indigenous organizations in each State and Territory with experience and interests in environmental management issues.

Two approaches were adopted at the beginning of the project to involve Aboriginal and Torres Strait Islander peoples in the exploration of the Indigenous Protected Areas concept:

- Meetings with selected land councils and other indigenous organizations;
- A National Working Group meeting of representatives of indigenous organizations and conservation agencies in Alice Springs (June 1995).

The National Working Group meeting had four outcomes:

1. Indigenous Protected Areas offered a worthwhile land use option which recognized and supported the land management capacity and authority of indigenous land holders, and that the International Union for the Conservation of Nature categories did not place undue constraints on continued customary use of their lands.
2. Ongoing resourcing and support were identified as the primary requirement.
3. An Indigenous Taskforce on Protected Areas was established to advise programme managers.
4. The Indigenous Protected Areas concept was broadened to provide resources to assist indigenous groups in negotiating co-operative management arrangements over publicly-owned protected areas.

The National Working Group meeting also developed a draft definition of an Indigenous Protected Area; "An Indigenous Protected Area is governed by the continuing responsibilities of Aboriginal and Torres Strait Islander peoples to care for and protect lands and waters for present and future generations."

Phase 3: Test the Indigenous Protected Areas concept using regional pilot projects. Twelve pilot projects were strategically identified on indigenous lands, on State/Territory owned and managed protected areas, and one project with a marine focus in the Torres Strait.

The concept of Indigenous Protected Areas would involve indigenous land owners announcing their intention to manage their land primarily for the protection of its natural and cultural values in accordance with International Union for the Conservation of Nature guidelines for Protected Area management. Management would be by local and resident Indigenous people with government support through resourcing, training and advice. Government assistance may include devolution of enforcement powers as applicable under various legislation and may be formalized through Memorandums of Understanding, legal agreements and partnerships (Thackway and Cresswell 1995).

There is no particular model being promoted as the ideal for an Indigenous Protected Area. It is clear from consultations that the best results will occur where initiatives are community-led and management structures are developed which are sensitive to local resources, needs, history and attitudes.

The Indigenous Protected Areas component is now established within the National Reserve System Programme. Funding has been allocated through the Government's Natural Heritage Trust with approximately 10% (i.e., $8 million over 1997–2001) of the National Reserve System Programme being allocated to Indigenous Protected Areas.

From the initial round of twelve Indigenous Protected Areas pilot projects the Indigenous Protected Areas concept has gained considerable support and interest from indigenous landholders and state and territory nature conservation agencies. Several communities involved in this round including Nantawarrina have since gone on to draft plans of management and are committed to establishing Indigenous Protected Areas.

With the management plans in place and approved by indigenous communities, State or

Territory nature conservation agencies and Environment Australia, a contract between these parties formalizes the inclusion of the Indigenous Protected Areas into the National Reserve System.

Indigenous Protected Areas component funding for the 1998/99 period will see the establishment of the first four Indigenous Protected Areas at Nantawarrina in South Australia, Oyster and Risdon Coves in Tasmania and Deen Maar (Framlingham) in Victoria. Other Indigenous Protected Areas which are close to declaration include Yalata in South Australia and Deliverance and Pulu Islands in the Torres Strait. The 1998/99 round of funding will also resource further projects to Indigenous Protected Areas component goals, including the development of joint management arrangements and further investigations to establish Indigenous Protected Areas.

BIODIVERSITY CONSERVATION BEYOND THE NATIONAL RESERVE SYSTEM

The Bushcare native vegetation protection programme is another of the Commonwealth Government's nature conservation programmes which protects biodiversity on private land and does so using the same State and Territory protection mechanisms. Bushcare works in partnership with the State and Territory governments to use funding to encourage landholders to access the same range of State and Territory mechanisms to ensure legal and effective protection of the biodiversity values on private land. The Bushcare and National Reserve System programmes are often considered to overlap. However, many areas protected under Bushcare are not within the scope of the National Reserve System.

The differences between the programmes are greatest in the planning phase (National Reserve System properties will meet National Reserve System scientific requirements), public interest (the ecological associations protected will be required in the long term as examples of Australia's biodiversity) and reporting (National Reserve System properties will be reported as part of Australia's protected area network). Figure 1 has been adapted from Carl Binning's "Motivating People"[4] report and attempts to demonstrate the complementary relationship between the programmes.

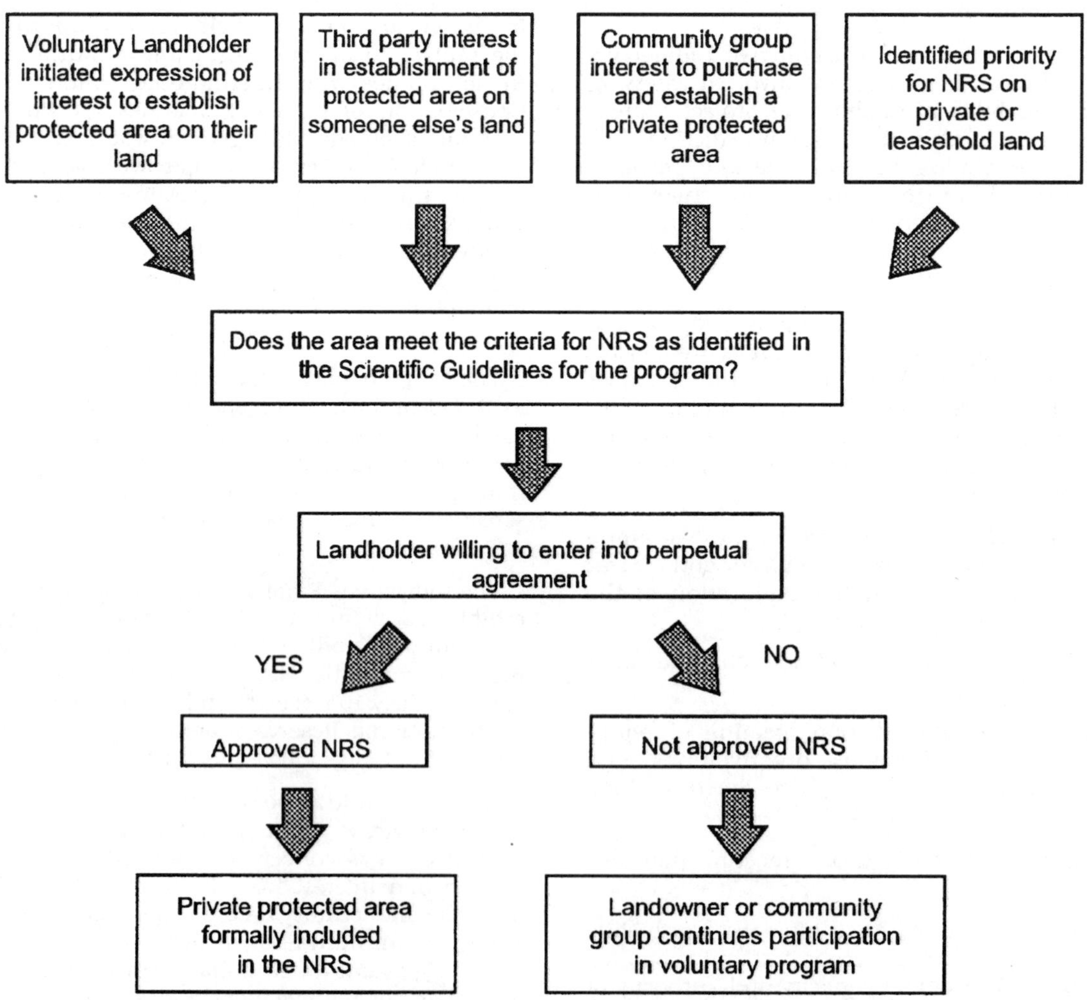

Fig. 1. Bushcare and the National Reserve System. Adapted from Binning and Young 1997.

[4]Page 62, Motivating People: using management agreements to conserve remnant vegetation by Carl Binning and Mike Young, CSIRO, 1997.

CURRENT ISSUES AND CONCLUSION

Both the Protected Areas on Private Land and the Indigenous Protected Areas components of the National Reserve System Programme are now integrated into normal programme delivery arrangements. However, outstanding issues for both programme components remain.

Common issues for adding private land to the National Reserve System include:

- Identification of priorities;
- Identification of properties with the attributes to meet those priorities;
- Willingness of the landowner or leasehold to enter into a long-term binding agreement;
- Ongoing funding;
- Maintenance of biodiversity values;
- Management;
- Monitoring; and
- Enforcement.

Establishing priorities at the finer scale for the programme generally is a key issue. While the Interim Biogeographic Regionalization for Australia provides a broad context for priority setting, finer analysis of ecosystem priorities is needed. Without clear and well publicised ecosystem priorities, effort preparing applications will be misdirected. Where the priority is clear, assessment of competing proposals can focus on values such as quality of management and other long-term issues. Establishing clear priorities assists in ensuring the transparency of programme processes. Conversely, providing too much detail on priorities too broadly may affect land values by artificially inflating real estate prices.

Currently State and Territory nature conservation agencies are the custodians for environmental data within their jurisdiction. Databases enable proposals to be assessed for their value to the National Reserve System. However, these systems are not accessible to the general public. This places a demand on the State and Territory agencies to provide a consultative role. When applying for assistance to establish Protected Areas on Private Land, National Reserve System Programme guidelines require each proponent to contact the State or Territory to determine the priority of the proposal before proceeding. Whilst this ensures States and Territories have the opportunity to endorse the priority of the proposals' environmental value, it does not accommodate the community's desire to have input into the development of National Reserve System priorities.

A Private Protected Area Establishment Agreement has been drafted in consultation with legal experts to implement acquisitions by the community. In drafting the agreement, eligibility requirements and the following issues were considered:

- Capacity to make the purchase;
- Commitment to and resourcing for effective long-term management;
- Implementation of a plan of management for nature conservation;
- Incorporation;
- Requirements for rules of incorporation or founding document;
- Registered Environmental Organizations or organizations eligible to receive tax deductible donations under appropriate legislation.

The agreement goes some way to ensuring that private protected areas are managed to a standard similar to that of State- and Territory-owned protected areas. It includes a basic set of standards for management and establishes a course of action designed to ensure long-term certainty and operational capacity even in the event the group is dismantled or land sold. The required standard for management plans is detailed and arrangements for endorsement of the plans and any alterations are catered for. Although public input into the plan of management is not essential as it is with the public land estate, the process of developing the plan and its implementation is open to scrutiny through the involvement of the Commonwealth or its nominated representative. Public access to the protected area may be restricted by permit to allow the managers greater control and to ensure that bonafide interests can be accommodated.

Each State and Territory has a suite of laws and policies governing a range of activities such as mining, communications infrastructure and grazing etc. which frequently over-ride nature conservation covenants and other arrangements on private land. While in many cases, gazetted national parks and other State- and Territory-owned protected areas are subject to the same laws and policies, the perception is that environmental values on private land cannot be as well protected.

Environment Australia supports the development of stewardship agreements as the most effective and flexible approach to the establishment of Indigenous Protected Areas and in some cases other Private Protected Areas.

Stewardship is a partnership between a management body and landholders, in which the responsibilities and obligations are identified for both parties. Given the nature of funding for Commonwealth programmes which is based on an annual cycle with programme funding generally indicated no more than four years at a time, stewardship presents a challenge for governments and communities committed to the long-term protection of important example of biodiversity on private and indigenous lands.

The future of these programme components looks promising. From proposals received in 1997/98, one proposal to purchase land for the establishment of a private protected area was funded. A significant increase in programme interest and acceptance is already apparent in 1998/99 as nine community land purchase proposals and five private protected area establishment proposals are to be supported. In 1997/98, 18 Indigenous Protected Area projects were funded. In 1998/99 it is anticipated that at least another 15 projects will be funded. Also in 1998/99 approximately 12–14 of the 1997/98 projects will progress to the next stage of development.

ACKNOWLEDGEMENTS

The assistance of the members of the Interstate Management Agreement Network is gratefully acknowledged in the development of the Protected Areas on Private Land component of the National Reserve System Programme, in particular the assistance of Dr Brian Whelan of the Trust for Nature (Victoria), and Carl Binning of CSIRO Wildlife and Ecology.

REFERENCES

Binning, C. and Young, M., July 1997. Motivating People. Using Management Agreements to Conserve Remnant Vegetation. CSIRO Wildlife and Ecology Resource Futures Program: Lynham, Australian Capital Territory.

Brunckhorst, D. J., Thackway, R., Coyne, P. and Cresswell, I. D., 1998. Australian protected areas — toward a representative system. *Nat. Areas J.* **18**(3): 255–61.

Cresswell, I. D. and Thomas, G. M. (eds), 1997. Terrestrial and Marine Protected Areas in Australia (1997). Environment Australia, Biodiversity Group: Canberra, Australian Capital Territory.

Environment Australia, 1997. Interim Scientific Guidelines for Establishing the National Reserve System [Online], http://www.biodiversity.environment.gov.au/protecte/nrs/sciguide.htm, 21 August.

International Union for the Conservation of Nature, 1994. Guidelines for Protected Area Management Categories. CNPPA with the assistance of WCMC. IUCN: Gland, Switzerland and Cambridge, United Kingdom.

Natural Heritage Trust Funding. A better environment for Australia in the 21st Century. Guide to new applications 1998–99. Commonwealth of Australia.

Thackway, R. and Cresswell, I. D. (eds), 1995. An interim biogeographic regionalisation for Australia: a framework for establishing the national system of reserves, Version 4.0. Australian Nature Conservation Agency: Canberra, Australian Capital Territory.

Smyth, D. and Sutherland, J., 1996. Indigenous Protected Areas. Conservation Partnerships with Indigenous Landholders. Environment Australia: Canberra, Australian Capital Territory.

The Nature Conservancy, 1992. Annual Report Fiscal Year 1992. Virginia, United States of America.

The Nature Conservancy, 1998. Wired for Conservation [Online], http://www.tnc.org/.

Wildlife Trusts, 1998. The Wildlife Trusts [Online], http://www.wildlifetrust.org.uk/index.htm, 16 September.

CHAPTER 52

Native grasslands and grassy woodlands: landholders' attitudes and incentives for conservation on private property

Kate Stothers[1]

Since European settlement of Victoria, Australia, 95% of the state's native grassland communities have been destroyed or drastically modified by the introduction of ungulates, introduced pests, exotic plant species and new fire regimes. This study, conducted in collaboration with Trust for Nature (Victoria), interviewed 52 landholders in western Victoria regarding their attitudes to native grasslands and grassy woodlands on private property. The participants surveyed were 33 people known to have an area of remnant native vegetation (Informed participants) and 19 landholders chosen randomly from the electoral role (Random).

Participants from the Informed group had a more sympathetic environmental attitude towards native grasslands and grassy woodlands than participants from the Random group, while there was no difference in agricultural attitude. For conserving native vegetation on private land, participants on the whole were cautious about the use of conservation covenants, but favoured tax or rate compensation as an incentive for refraining from deleterious development.

Key words: Landholder, Attitudes, Questionnaire, Incentives, Native grassland, Grassy woodland.

INTRODUCTION

NATIVE grasslands have suffered a serious decline in extent all around the world and this decline continues. Since European settlement of Australia, 95% of the State of Victoria's two million hectares of native lowland grassland communities have been lost (Department of Conservation and Environment 1990). Furthermore, all remaining native grassland sites on the western plains of Victoria, and probably throughout Victoria generally, have been altered to some degree (Stuwe 1986). Native grassland areas were selected at the outset of settlement as most suited for agricultural development; today, nearly twice as many threatened species occur in native grasslands and grassy woodlands than any other habitat (Kirkpatrick *et al.* 1995).

Prior to European settlement, grasslands in Victoria were shaped by macropod grazing and natural and Aboriginal burning regimes (Stuwe 1986). With the arrival of Europeans came not only extensive clearing, the introduction of exotic pasture plants and a change in burning frequency, but also a change in grazing pressure and type. Grazing pressure was altered by an increase in the number of native herbivores (i.e., large macropods) and the introduction of non-native animals, including domestic ungulates and the European Rabbit *Oryctolagus cuniculus* (Hobbs and Hopkins 1990). Native grasslands exist only as remnants today (Morgan 1996). These remnants persist because of the unsuitability of the land for agriculture, their reservation for public purposes, such as roads, railways, water and conservation reserves, or through the deliberate choice of private landholders (Main 1987).

More than half of the many rare or threatened plant species found in the lowland grasslands of southeastern Australia occur on private land (Kirkpatrick *et al.* 1995). In Victoria, 13% of agricultural land (1.7 million hectares) is classified

[1]Greening Australia Victoria, P.O. Box 525, Heidelberg, Victoria, Australia 3084.

Pages 511–23 *in* NATURE CONSERVATION 5: NATURE CONSERVATION IN PRODUCTION ENVIRONMENTS: MANAGING THE MATRIX ed by J. L. Craig, N. Mitchell and D. A. Saunders. Surrey Beatty & Sons, 2000.

as native grass pasture (Mitchell 1990). The overall goal of conservation on private land is to maintain a high level of indigenous biological diversity over an extended time scale in a way that is consistent with sustained production and land use demands (Williams *et al.* 1991). Native grassland and grassy woodland remnants have been unintentionally exposed to degrading processes by landholders who do not recognize these native areas and instead view them as degraded pastures (Mitchell 1990).

There have been conflicting views as to the role that landholders can play in scientific research. Chambers *et al.* (1989) suggest, for example, that farmers' knowledge, inventiveness and experimentation have long been undervalued, and that farmers and scientists can and should be partners. Thompson and Scoones (1994), however, argue that the knowledge of rural people is always fragmented, partial, and provisional in nature, never being fully unified into cultural logic or classification. Nevertheless, there is an increasing move towards collating and communicating the knowledge of farmers and their experience of establishing and managing native and introduced grasses (Millar and Curtis 1995). In order to identify information gaps in landholders' understanding of native grassland, it has become important to determine their experience and their knowledge of this ecosystem. Studies of the management of remnant areas on private land provide valuable information on maintaining native grasslands and grassy woodlands for the purposes of conservation. Landholders also have a special significance as managers and potential conservers of this land (Platt and Ahern 1995).

The initial idea for the project reported here came through an initiative of Trust for Nature (Victoria). The Trust is an independent body established by an Act of Parliament (*Victorian Conservation Trust Act 1972,* amendments 1978 and 1986), which is dedicated to nature conservation particularly on private land (Safstrom 1995). Economic, geographical, social and other factors may play important roles in determining the conservation of such areas. The Trust was interested in determining why remnants of native vegetation have persisted on private land, how they have been managed in the past and incentives needed to conserve these areas.

This chapter will cover three major themes in relation to a selected group of landholders' responses to a given questionnaire. The first theme is landholder knowledge of native grasslands and grassy woodlands, i.e., how they identify such vegetation, why they believe it has persisted on private land and what processes they believe have led to its decline. The second is landholder attitudes towards native grasslands and grassy woodlands, i.e., the advantages and disadvantages of native grasslands and grassy woodlands on private property, as well as their attitudes to conservation considerations and agricultural value of these remnants. The third theme is techniques for conservation of native grasslands and grassy woodlands on private land i.e., how government money should be spent, the role of conservation covenants, voluntary, permanent binding legal agreements between landholders and third parties (Safstrom 1995), the role of the administering body, Trust for Nature (Victoria) and finally, incentives preferred by landholders to conserve this endangered ecosystem.

STUDY AREA AND METHODS

Trust for Nature (Victoria) has an established base of landholders interested in native vegetation conservation within western Victoria. The 52 landholders surveyed for this survey were centred around the towns of Stawell, Horsham and Warracknabeal (Fig. 1). The regions concerned rely heavily on agriculture as the main source of income. Warracknabeal is located within a large cropping district, while landholders within both the Horsham and Stawell areas are largely graziers.

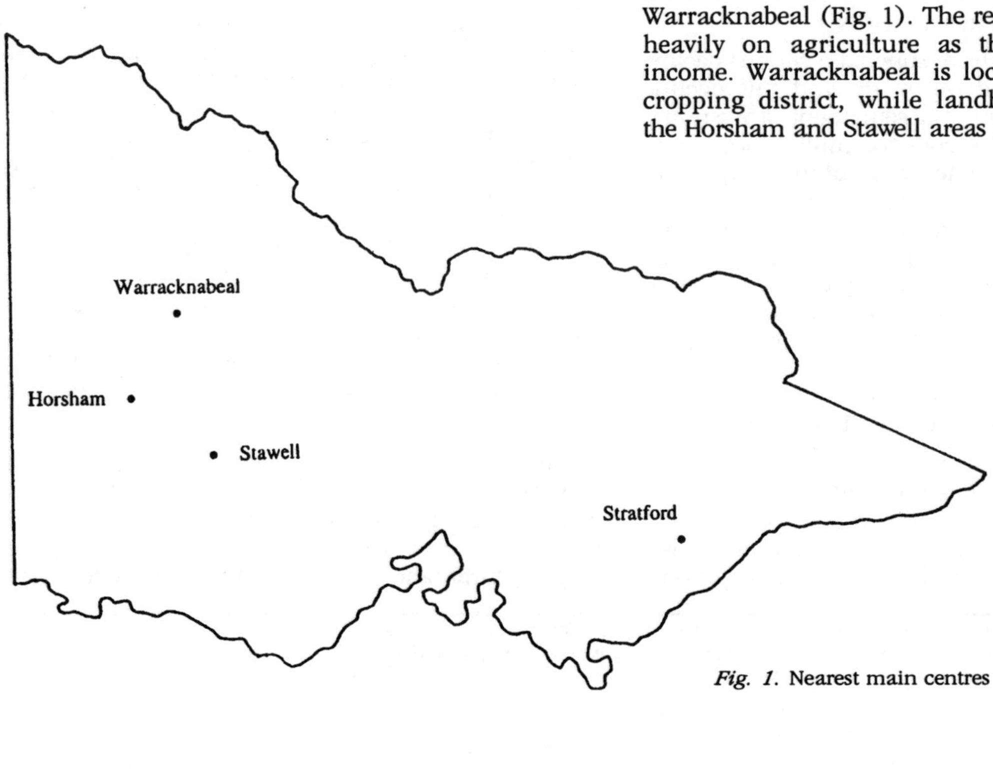

Fig. 1. Nearest main centres to surveyed landholders.

Trust for Nature had a group of landholders that were interested in having a botanical survey completed on their native vegetation remnants. This group became the "Informed" group and totalled 33 landholders. The remaining participants, a "Random" group, was chosen randomly from the electoral rolls in the district. Each participant from both the Informed and Random lists was sent an introductory letter to inform them of the aims of this project and their role. They were then contacted by telephone and a suitable time was arranged for an interview. Twenty-two landholders from the Random group were contacted by phone. Three were either not interested or too busy to take part, so this group totalled 19 landholders. All the properties of the Random participants had some form of native vegetation, ranging from a small remnant on a rocky hillside to an entire property that was maintained as a wildlife sanctuary.

To evaluate landholders' attitudes to native vegetation for this project, a questionnaire was developed for use in a semi-structured interview. Here, interviews are conducted informally with the questionnaire acting as a prompt, allowing the respondent scope in replying (de Vaus 1985). Millar and Curtis (1995) found that the technique of semi-structured interviews, where opinions can be expressed openly from the questions asked, maximizes the opportunity for respondents to describe their experiences, perceptions and understandings of the role and management of (perennial) grasses within their farming systems.

Some of the questions were adapted from previous questionnaires aiming to ascertain information similar to that which was sought for this project (see Gilfedder and Kirkpatrick 1995; Hamilton *et al.* 1997). In particular, questions from Hamilton *et al.* (1997) were used that covered incentives and use of available funds for native vegetation conservation on private property.

The questionnaire included both "extended response" questions, which allowed a free-flowing response, and "restricted response" questions, that came with a set range of responses for the respondant to scale. The extent of support given to these statements is displayed in the Figures as bar graphs with percentages calculated as a total of each group (i.e., n = 33 for the Informed and n = 19 for the Random) so that columns can be compared across the graphs. The answers to the extended response questions were categorized into similar themes and tallied.

Questions including the advantages and disadvantages of native grasslands and grassy woodlands on private land; why they still persist; how native vegetation is recognized; and the role of covenants and Trust for Nature in conserving these remnants, were all extended response questions. Restricted response questions included processes that have led to the decline of native grasslands and grassy woodlands, preferences for government funding, and preferred incentives. After each of these restricted response questions, there was an extended response question allowing the respondent to give their views about each question other than that stated in the set response.

Agricultural attitudes and environmental attitudes were investigated from the utilitarian values and non-utilitarian value placed on native grasslands and grassy woodlands by summing the level of importance (one to five) assigned to the responses to the statements shown in Table 1. An ***agricultural attitude*** scale was formed from the responses to a set of statements relating to agricultural values, while an ***environmental attitude*** scale was created the same way using numerical responses to statements with a more environmental theme. The assumption was made that the higher the rating that each respondent scored, after totalling each set of responses, the more importance they placed on environmental or agricultural values of native grasslands and grassy woodlands.

The original questionnaire was initially tried in Gippsland, southeastern Victoria, where remnant areas of native grassland and grassy woodland still occur (Lunt 1991). Six landholders from around the town of Stratford (Fig. 1) participated in this test. Three of these properties had native grassland and grassy woodland. The initial test was run to assess the clarity, appropriateness and validity of the questions asked. Following this testing, several modifications were made to the questionnaire before it became a standard form to use in western Victoria.

Associations between the two samples of participants and their attributes were compared using

Table 1. Characteristics used to determine an Agricultural and Environmental Attitude Scale in response to the prompt "Native Grasslands and Grassy Woodlands are important".

Agricultural: (Utilitarian)	• because a low level of input is required • in acting as a wind break, for example providing shelter at lambing time • because they can tolerate droughts • as high productive pasture • because they are less fire prone • to stabilize the soil • in controlling soil acidity • in helping combat land degradation • in adding capital value to the property
Environmental: (Non-utilitarian)	• in increasing the biodiversity, that is the amount of species living in the environment • as a wildlife sanctuary • for their natural beauty or aesthetic value • simply for its natural right to exist or belong

a 2 × 2 contingency table (χ^2), non-parametric Mann-Whitney U-test (U), and non-parametric t-tests with significance to the 5% level.

All properties selected as part of the Random group contained some form of native grassland or grassy woodland. This was unexpected, and questions on the amount of native vegetation occurring on the property had not been incorporated into the questionnaire for the Random group. No statistical comparison could therefore be carried out between the two groups concerning the amount of native vegetation existing on each property and whether this influenced their attitudes. Wilson (1992) found relative and absolute amounts of native forest present on individual holdings in southern New Zealand to be only weakly associated with attitudes of landholders. This, however, is in contrast with the findings of McDowell and Sparks (1989), who, in a study of farmers' conservation behaviour in South Africa, found that the extent of natural habitat on land influenced conservation attitudes.

RESULTS

Respondents' demographics

The participants from the Informed group and the Random group were similar in a number of ways. There was no significant difference (t = 1.28, df = 50, p = 0.21) between the mean size of the properties for the Informed group of 821 ha (st. dev. = 946) and the Random group of 465 ha (st. dev. = 344). There was also no statistical difference (t = 0.39, df = 50, p = 0.68) between the average year level of formal education of participants in the Random group (10.6) to that of the participants in the Informed group (10.8). Age of participants also did not significantly differ between the two groups (U = 307, p = 0.90). Other similarities were that both groups of participants tended to have off-farm income [60.6% of the Informed group and 52.6% of the Random group, with no significant difference between them (χ^2 = 0.07, p > 0.05)]. The majority of participants (66.7% of the Informed and 63.2% of the Random) worked their farm full-time.

KNOWLEDGE OF NATIVE GRASSLANDS AND GRASSY WOODLANDS

Identification

In response to an extended response question about how landholders recognize native grasslands and grassy woodlands, many participants (37%) thought they were simply "unimproved land in its original condition" (Table 2). Another common response (19%) was "it's simply my timber block," an association here with the remnant and wood production. A proportion however, (25%) were able to name particular native plant species, with the large majority of these respondants being for the Informed group.

Persistence

The most common response as to why remnants still remain on private land (Table 3) was that the remnants exist on very poor land that is not worth clearing (23% of respondents). Others (17%) said that they exist only on inaccessible areas, and 15% of respondents said that it was simply because it has not been developed, cultivated or cleared.

Decline

When given a set of possible causes of native grassland decline on private property, the respondents from both groups, on the whole, did not fully associate the decline with one particular cause (Fig. 2). Respondents from both the Informed and Random group rated highest "cultivation of the land for agricultural purposes" as contributing "a lot" to native grassland and grassy woodland decline. Respondents also rated "clearing of the land for industrial purposes" high in the "a lot" category. The process that had the greatest distribution across the ratings was "the alteration of previous fire regimes", where the majority of respondents did not rate it as contributing "nothing at all" or "a lot". The respondents also replied to an extended

Table 2. Definitions of native grasses or grassy woodland given by the Informed group and the Random group.

Definitions	Informed No. (%) n = 33	Random No. (%) n = 19	Total No. (%) n = 52
Unimproved land in original condition	11(33)	8(42)	19(37)
Named particular species	11(33)	2(11)	13(25)
Timber block	5(15)	5(26)	10(19)
It looks different — tussocky, seed ends	5(15)	4(21)	9(17)
Always had an interest	5(15)	1(5)	6(12)
Still green in summer	4(12)	2(11)	6(12)
Found in swamp areas	0(0)	5(26)	5(10)
How much I do not know	2(6)	2(11)	4(8)
Named introduced species	0(0)	3(16)	3(6)
Seeds stick in your socks	1(3)	2(11)	3(6)
Found in inaccessible areas	1(3)	2(11)	3(6)
Do not know	1(3)	2(11)	3(6)
Other[a]	7(21)	1(5)	8(15)

[a]The "Other" definitions mentioned included: an aesthetic recognition, what comes back after the land is left, what is locked up for regeneration and conservation.

Table 3. Reasons given by both the Informed group and Random group as to why native grasslands and grassy woodlands have persisted on private land.

Why native grasslands and grassy woodlands have persisted on private land	Informed No. (%) n = 33	Random No. (%) n = 19	Total No. (%) n = 52
Not worth clearing it, poor land	10(30)	2(11)	12(23)
Inaccessible area	6(18)	3(16)	9(17)
Have not developed, cultivated or cleared it	6(18)	2(11)	8(15)
Owners care, have taken an interest in it	2(6)	4(21)	6(12)
Family decision	4(12)	1(5)	5(10)
It has been fenced off	3(9)	2(11)	5(10)
Kept for shelter	4(12)	0(0)	4(8)
No fertilizer applied to that particular area	4(12)	0(0)	4(8)
Through conservative farming	3(9)	1(5)	4(8)
Luck	3(9)	1(5)	4(8)
Kept because it prevents erosion	1(3)	3(16)	4(8)
Do not know	2(6)	1(5)	3(6)
Through the viability of introduced species	0(0)	2(11)	2(4)
Other[a]	1(3)	3(16)	4(8)

[a]The "Other" category includes: one person from the Informed group thinking that it was just there, and three people from the Random group each commenting: because they have tough survival rates, because they are only found in salty areas, and because they are only found on public land.

answer question about what other processes may be the cause of native grassland decline. The majority of respondents (17%) thought that overstocking has contributed to the decline, while human activities (13% of respondents) and fertilizer (10% of respondents) were also processes mentioned.

ATTITUDES TOWARDS NATIVE GRASSLANDS AND GRASSY WOODLANDS

Advantages

The advantages of native grasslands and grassy woodlands identified by participants in an extended response question were largely related to the grazing value of native grasses (Table 4). A quarter of participants believed that the ability of native grasses to stay green throughout summer was an advantage. Other frequently mentioned advantages included the ability of native grasses to survive harsh conditions (19%), and their fast growth response to any amount of rain (10%).

Disadvantages

The most common disadvantage of native grasslands and grassy woodlands on private property (46% of participants) was that native vegetation areas, often being fenced off or inaccessible, reduced the productivity of the land. Amongst other perceptions native vegetation remnants were seen as areas that harbour vermin (27%) and as a fire risk (17%, Table 5).

Attitude scales

There was a significant difference ($t = 2.54$, $df = 50$, $p = 0.01$) for environmental attitudes (Fig. 3) between the means of the Informed group (0.87, st. dev. = 0.17) and the Random group (0.74, st. dev. = 0.21). The Informed group, on the whole, considered native grasslands and grassy woodlands to have a higher environmental value than the Random group. There was no significant difference, however, between the means of these groups for agricultural attitudes (Informed = 0.7, st. dev. = 0.2, Random = 0.6, st. dev. = 0.2; $t = 0.84$, $df = 50$, $p = 4.04$).

The environmental attitude and agricultural attitudes of both were examined in relation to property size, gender, education, income and age. The agricultural attitude of the Random group was significantly related to education ($r = 0.6173$, $p = 0.005$) and environmental attitude ($r = 0.5547$, $p = 0.014$). Therefore, if participants rated high on one scale, they were more likely to rate high on the other, and also the more highly educated participants from the Random group were, the higher they rated native grasslands and grassy woodlands. However, property size ($r = -0.4746$, $p = 0.005$) and gender ($r = 0.4444$, $p = 0.010$) were significant factors for participants from the Informed group. Therefore, the smaller the property, the higher participants from the Informed group rated native grasslands and grassy woodlands agriculturally, while female participants from this group were more inclined to have a higher agricultural attitude towards native grasslands and grassy woodlands. There were no significant correlations between the chosen factors and the participants environmental attitude rating from the Informed group. However, education was once again a significant factor for members of the Random group ($r = 0.5479$, $p = 0.015$). Participants from the Random group that were more highly educated, tended to rate native grasslands and grassy woodlands high from an environmental point of view than participants with a lower education.

TECHNIQUES TO CONSERVE NATIVE GRASSLANDS AND GRASSY WOODLAND ON PRIVATE PROPERTY

Uses of funding

Preferences towards the uses of government money were not strongly directed towards any specific one

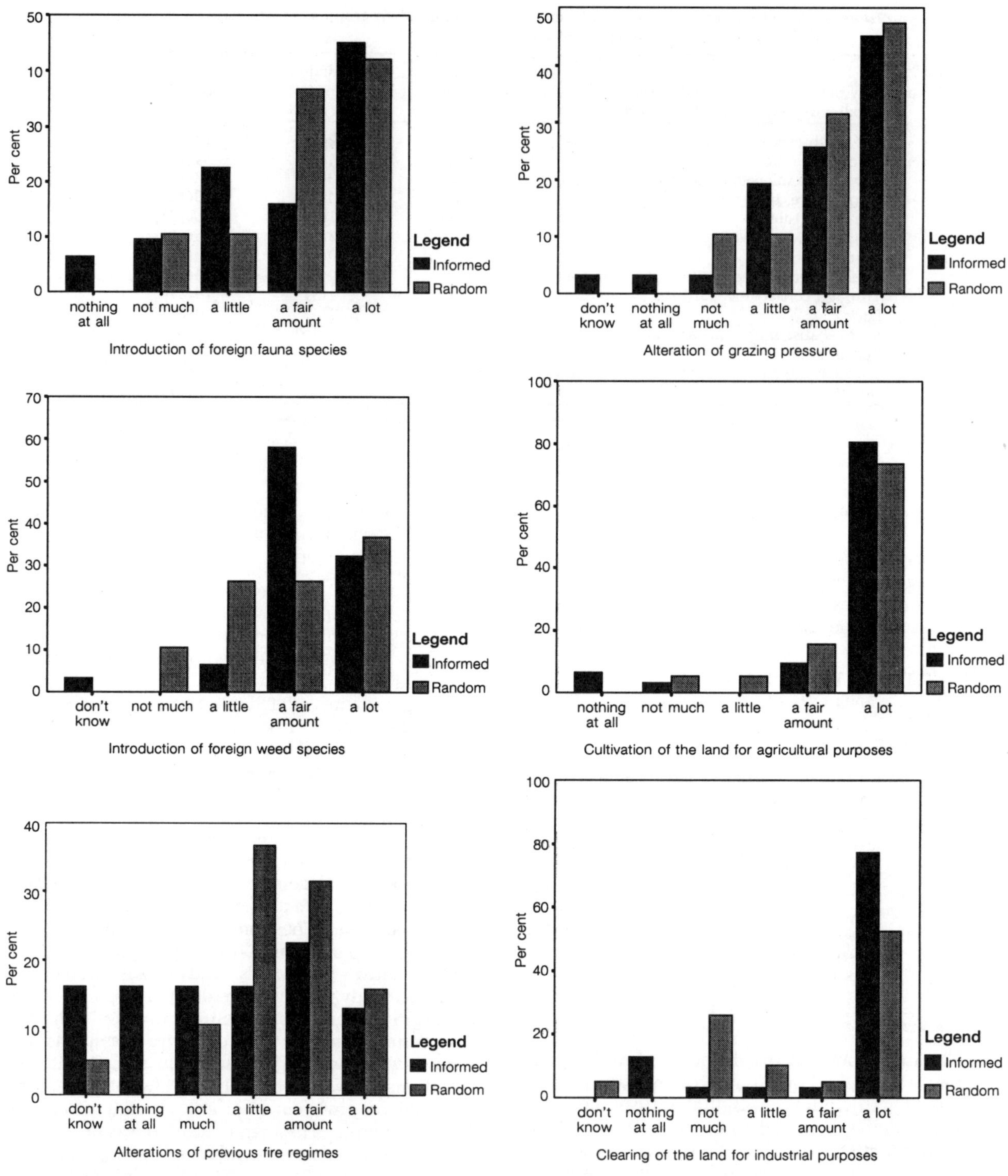

Fig. 2. Statements rated by the respondents as to how much the given processes contributed to the decline of native grasslands (Informed n = 33, Random n = 19).

of the given uses in a restricted response question (Fig. 4), although, the two given statements "Paying landholders to reduce stocking rates and encourage natural regeneration" and "Controlling weeds and vermin", resulted in strong feelings towards the more extreme responses. "Controlling weeds and vermin" was rated as the highest for "excellent use" of money by both groups, while "Paying landholders to reduce stocking rates and encourage natural regeneration", was rated the highest for "poor use" of money. Respondents within the Informed group rated highly "Establishing demonstration sites . . ." and "Paying part of fencing material costs" as excellent uses of money. The majority of respondents from the Random group favoured "Employing people to provide technical advice" and ". . . volunteer labour . . ." as "good uses of money".

In responding to an extended response question about other uses of government money, 13% of

Table 4. Responses given for advantages of native grasses and grassy woodland on private property by the 52 participants.

Advantages of Native Grasslands and Grassy Woodlands	Informed No. (%) n = 33	Random No. (%) n = 19	Total No. (%) n = 52
Summer green	10(30)	3(16)	13(25)
Survive harsh conditions	9(27)	1(5)	10(19)
Response to rain	4(21)	1(5)	5(10)
Windbreak	3(9)	2(11)	5(10)
Not much fertilizer needed	4(21)	0(0)	4(8)
Aesthetic	4(21)	0(0)	4(8)
Stabilize the soil	4(21)	0(0)	4(8)
Protecting native species	1(3)	3(16)	4(8)
Coverage on inaccessible areas	2(6)	1(5)	3(6)
No worming[a]	2(6)	0(0)	2(4)
Hope to collect seeds for sale	2(6)	0(0)	2(4)
Better protein[b]	1(3)	1(5)	2(4)
Other[c]	2(6)	1(5)	3(6)
None	5(15)	8(42)	13(25)

[a]Stock grazed on native pasture do not require drenching for worms.
[b]Native grasses have a better protein value for stock than introduced pasture.
[c]The "Other" category of advantages of native grasses included: there is less maintenance, they are more salt tolerant, they are a fire retardant, they maintain biodiversity and they are more economical.

Table 5. Summarized open-ended responses given by both the Informed and Random groups regarding disadvantages of native grasses and grassy woodlands on private property.

Disadvantages of Native Grasses and Grassy Woodlands	Informed No. (%) n = 33	Random No. (%) n = 19	Total No. (%) n = 52
Reduces productivity	14(42)	10(53)	24(46)
Vermin harbour	7(21)	7(37)	14(27)
Fire risk	4(12)	5(26)	9(17)
Weed control	2(6)	3(16)	5(10)
Kangaroo haven	2(6)	1(5)	3(6)
Lack of compensation	1(3)	2(11)	3(6)
Lot of work for no return	1(3)	2(11)	3(6)
Trespassers	1(3)	1(5)	2(4)
Low evaluation of land	1(3)	1(5)	2(4)
Other[a]	3(9)	1(5)	4(8)
None	7(21)	3(16)	10(19)

[a]The "Other" category of disadvantages included: native grasses are hard to revive after heavy grazing, a fear of government rules and regulations, sheep will not eat it unless they have to, and people like you annoying me.

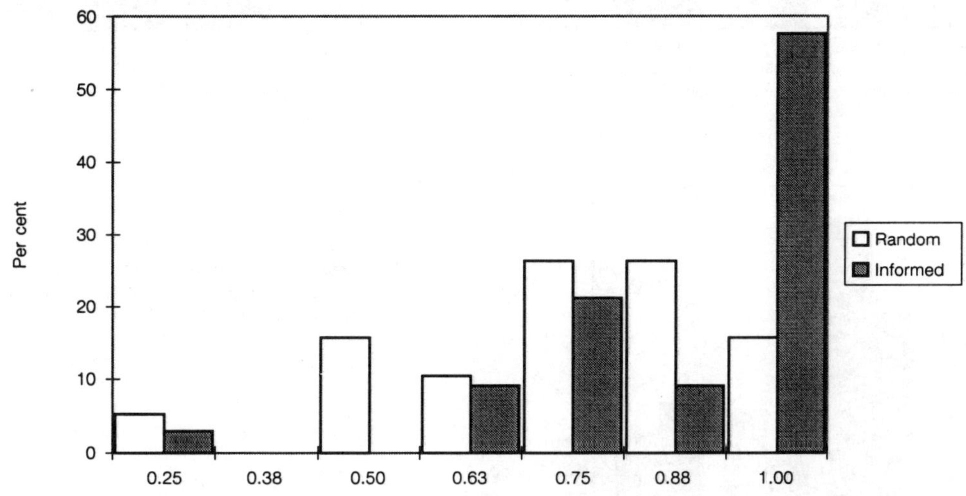

Fig. 3. Frequency distribution of environmental attitudes towards native grasslands and grassy woodlands where 0 = highly negative environmental attitude and 1 = highly positive.

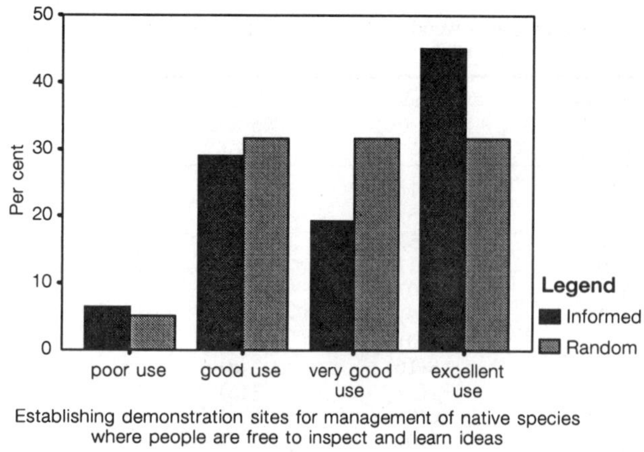

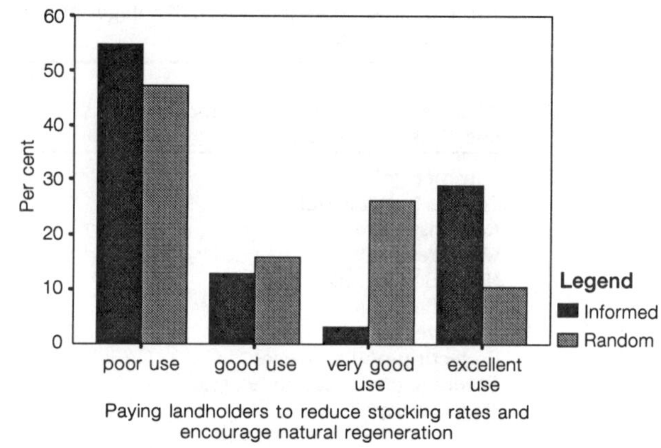

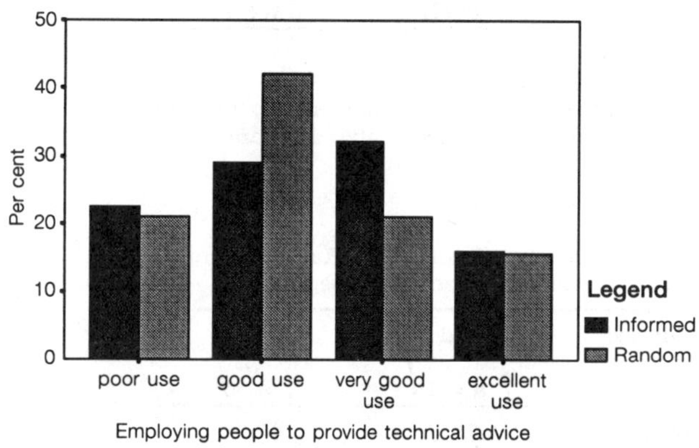

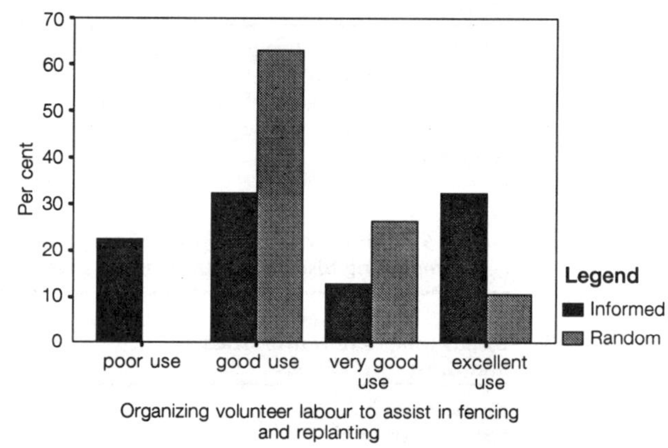

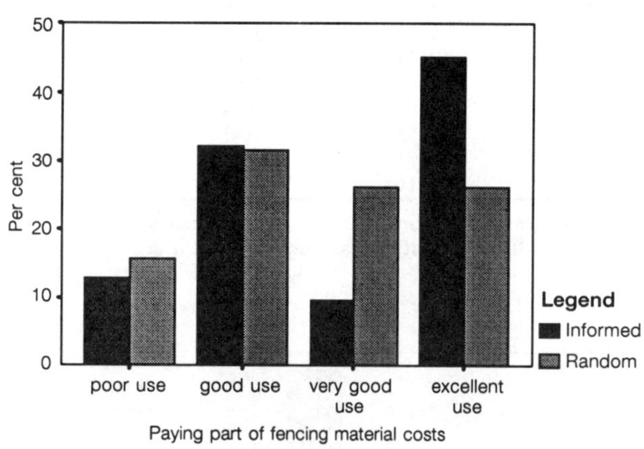

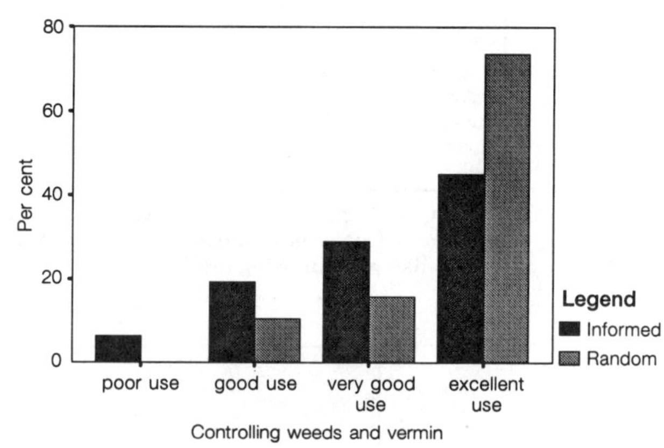

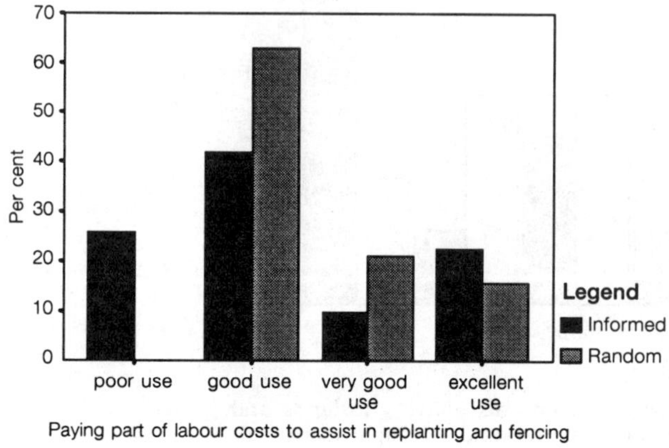

Fig. 4. Preferred uses of Government funding rated by the landholders to the given statements (Informed n = 33, Random n = 19).

Table 6. Other open-ended responses stated by participants for uses of government money in conserving native vegetation on private property.

Other uses of money	Informed No. (%) n = 33	Random No. (%) n = 19	Total No. (%) n = 52
Increase public awareness	4(12)	3(16)	7(13)
Subsidize or reduce rates	3(9)	3(16)	6(12)
Control vermin	5(15)	1(5)	6(12)
Control weeds	4(12)	2(11)	6(12)
More research into benefits	4(12)	1(5)	5(10)
Increase education	4(12)	0(0)	4(8)
Demonstration plots	3(9)	0(0)	3(6)
No other ideas	7(21)	3(16)	10(19)
Other[a]	3(9)	7(37)	10(19)

[a]The "Other" category were singular answers which included: more appropriate water supply, employing a grassland manager, government buying the land to maintain, mapping of grassland areas for display so people actually know where it is, soil erosion control, and tidying areas up such as clearing wood.

participants suggested to increase public awareness through media networks, schools and open days (Table 6). Other responses, such as controlling vermin (12%) and controlling weeds (12%), were reiterating ideas that were previously given as choices. The idea of rate subsidies or reduction was given by 12% of participants.

Incentives

Participants' responses to a restricted response question, where support given to a set of incentives to conserve native grasslands and grassy woodlands on private property was rated, showed a preference for economical incentives (Fig. 5). "Tax concessions" showed the majority of participants from both groups giving it full support. Landholders maintaining their area for purely ecological reasons however, expressed dissatisfaction at not being able to claim tax rebates on the required herbicides and equipment they purchase. Similarly, farmers were dissatisfied at having to pay the same amount of rates per ha for their farming land, where they can potentially earn $AUD200 per ha, as they did for their native vegetation remnant, which was much less, and in some cases not at all, productive. Fencing was also a strong choice with both "assistance with fencing" and "payment for fencing costs" given full, high or medium support by the majority of both groups. Results to an extended response question of other incentives they would support, showed that the idea of tax or rate compensation for locking remnant areas away was once again reiterated (17% of respondents). Twelve per cent supported education as an incentive.

Covenants

Participants, on the whole, were slightly wary of covenants and committing themselves to a precise answer when replying to an extended response question on what their general attitudes was to covenants (Table 7). The most common response (by 23% of participants) was that covenants were too legally binding and that farmers in particular, need more flexibility. This response was given only by members of the Informed group. Seventeen per cent thought it was a good idea, but were not willing to commit themselves with any further comment. The same proportion, 17%, saw covenants as a loss of productivity and largely believed that the land is simply locked up and left. A number of respondents (12%) had already entered into a conservation covenant. The main concern for these respondants was that the legislation agreement could always be overturned by the governing body at the time.

Trust for Nature input

Suggestions about how an outside body such as Trust for Nature can help to conserve native vegetation on private land were reasonably varied:

> *"I'd like to see people from that organization who are experienced, socially competent, visit farms to try and generate a bridge between the greenie element and the farming element. Ignorance is the greatest source of prejudice."*

> *"They've got to stay in their suits and not get their hands dirty. They must stay as an administrative management situation. Get the money from the city grants that is available and channel it to the workers. They haven't got enough resources to get involved at that sort of scale."*

> *"The problem with most of these bodies is the bureaucracy — the amount of money chewed up. An efficient way of funding the situation is to aim it directly where it's targeted."*

> *"All that has been taken care of by Landcare. Landcare is a good organization in that it is looking after all those things and providing programmes for us to take care of it ourselves."*

Overall, Trust for Nature was seen as a way of increasing awareness of the importance of native grasslands and grassy woodlands, not only for conserving them, but also by publishing and disseminating the agricultural benefits of such areas to the wider public (Table 8).

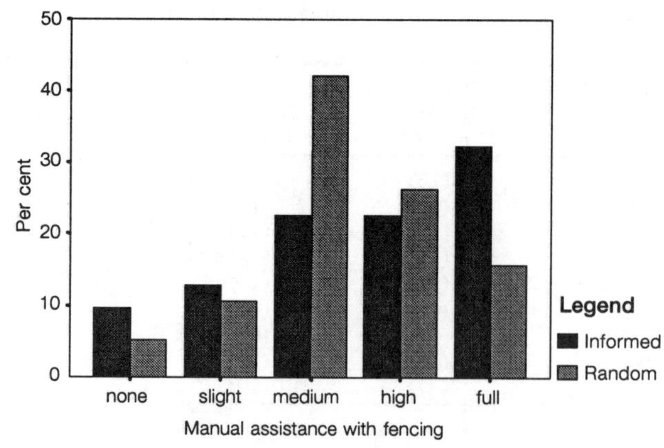

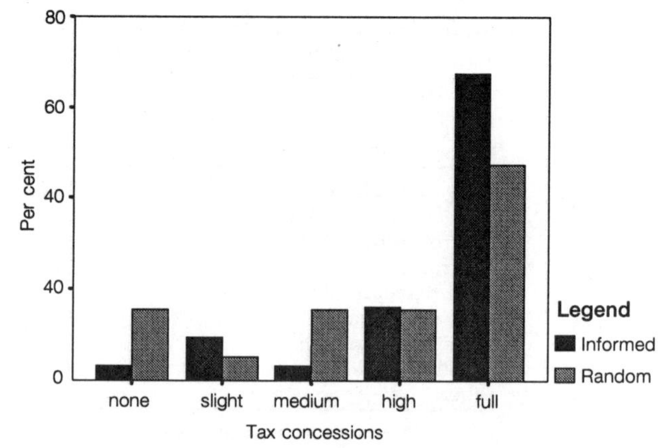

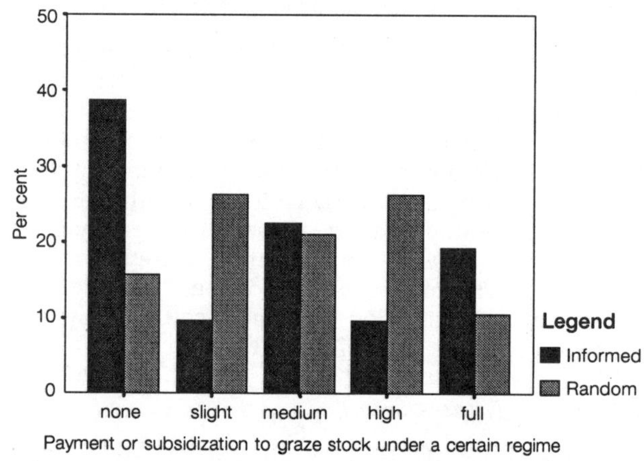

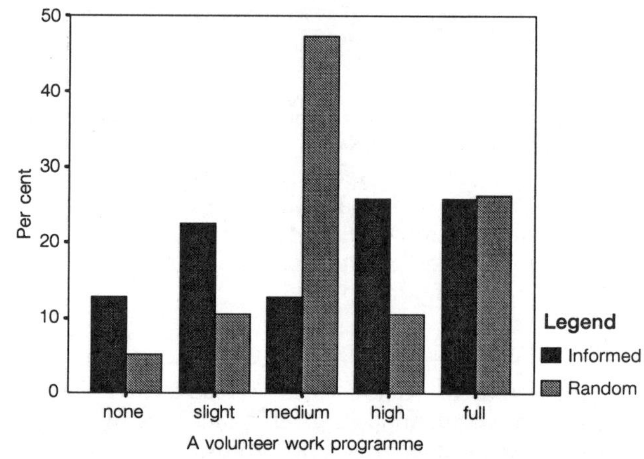

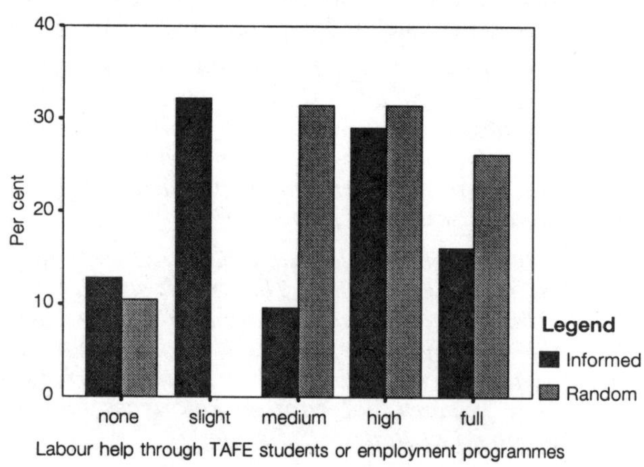

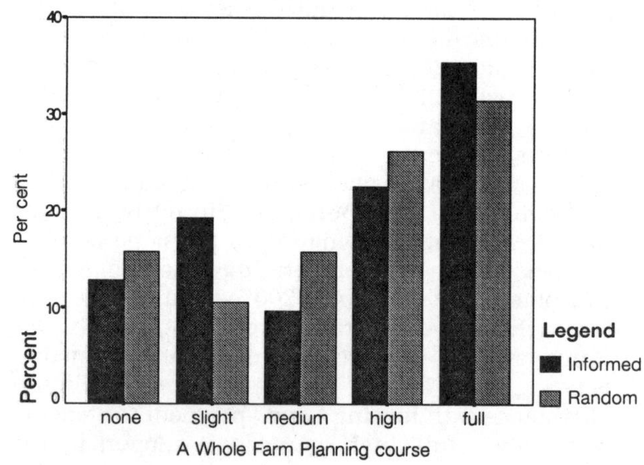

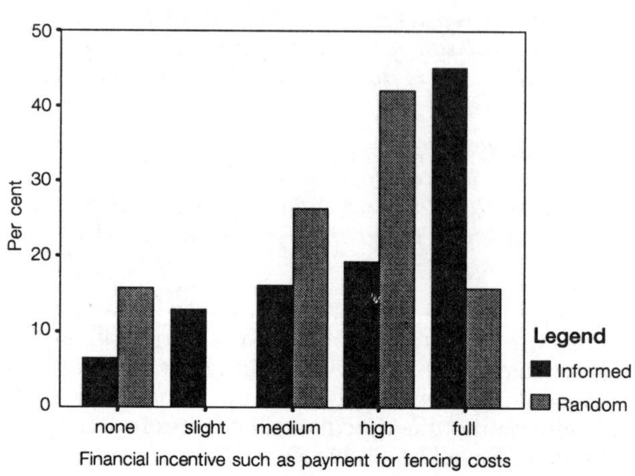

Fig. 5. Respondents support given to incentives to conserve native grasslands and grassy woodlands (Informed n = 33, Random n = 19).

Table 7. Attitudes had by the Informed and Random groups towards the protection of native grasslands and grassy woodlands on private land through covenants.

Attitudes towards covenants	Informed No. (%) n = 33	Random No. (%) n = 19	Total No. (%) n = 52
Too legally binding, want more flexibility	12(36)	0(0)	12(23)
Good Idea	6(18)	3(16)	9(17)
Loss of Productivity	4(12)	5(26)	9(17)
Definitely no	4(12)	3(16)	7(13)
Do not think so	3(9)	3(16)	6(12)
Have entered into one or are going to	5(15)	1(5)	6(12)
Definitely Yes	3(9)	2(11)	5(10)
Do not want to restrict others in the future	2(6)	3(16)	5(10)
Good idea but worried can get out of it	3(9)	1(5)	4(8)
Do not know	2(6)	2(11)	4(8)
Never heard of them before	0(0)	3(16)	3(6)
Do not want to be forced	1(3)	2(11)	3(6)
It would do no harm	1(3)	2(11)	3(6)
Other[a]	6(18)	3(16)	9(17)

[a]The "Other" category were responses from the Informed group only who offered answers such as: they are against it as it tells them how to run their land (3.8%), some offered suggestions such as education and ***Land for Wildlife*** (State government voluntary conservation scheme) as better alternatives (3.8%), 3.8% are "not yet" willing to obtain a covenant, one person was not willing to conform to a covenant as her son did not have the same feelings, and one person thought covenants were more applicable to trees only.

Table 8. Suggestions by participants about how an outside body such as Trust for Nature can help to conserve grasslands and grassy woodlands on private property.

Suggestions	Informed No. (%) n = 33	Random No. (%) n = 19	Total No. (%) n = 52
Increase awareness	6(18)	4(21)	10(19)
Education	5(15)	2(11)	7(13)
Advice/encouragement	4(12)	2(11)	6(12)
Control weeds and vermin	3(9)	1(5)	4(8)
Covenants	2(6)	0	2(4)
Demonstrations/field days	2(6)	0	2(4)
Other[a]	4(12)	4(21)	8(15)

[a]The "Other" category included single statements such as: portray more agricultural information, monitor the decline so that there is a benchmark, continue research, and make the government aware of the issues involved.

DISCUSSION

The reasons for native grasslands and grassy woodlands remaining on private land across the two groups were varied. The reasons ranged from a pure conservation awareness, through to utilitarian reasons ("stays green in summer") through to a perceived association of low productivity ("too poor to clear") through to difficulties in working the area (for example, remnant situated in a depression or rocky slope). These reasons highlight the fact that "luck" has played a significant role in remnants of native vegetation persisting. On a national and international basis, luck can not be relied on in the long term for the future conservation of these remnants. A more active role has to played by conservation and land management agencies and concerned individuals in the long-term conservation of a large proportion of native vegetation remnants.

Attitudes

In general, attitudes of landholders to native vegetation, do not appear to always have a consistent foundation. These varied attitudes are derived from a number of sources, e.g., the unquestioning adoption of previous land practises, economic pressures, aesthetic values, or a combination of all of these. McDowell and Sparks (1989), in a study on attitudes of landholders to natural ecosystems on farms in South Africa, concluded that for owners who have to make a living from the land, the economic role of that land tends to take precedence over its aesthetic values. Hamilton *et al.* (1997) found that the relationship between perceptions and management of Box-Ironbark remnants in central Victoria was largely dependant on the size of the property. Knowledge concerning the origins of these attitudes is useful in dealing with situations on a case by case basis.

This is an issue also raised by other similar surveys (Gilfedder and Kirkpatrick 1997; Hamilton *et al.* 1997) where it is stressed that policies and programmes need to address a wide range of opinions and situations.

Raising awareness

The interview process itself appeared to act as a means of increasing most landholders' knowledge of indigenous vegetation. Personal contact with these landholders and discussion of issues specific to their property, were often effective ways of increasing their awareness and appreciation of native grasslands and grassy woodlands. Many participants commented that the "one on one" situation created by the interview was of greater value to them as a learning tool in relating to situations on their own property than, for example, "walking around seeing what other people had done" (i.e., open days, field days). Also, it was often remarked that through answering the questions, they had started thinking about more profound issues involving native grasslands and grassy woodlands. Communication needs to occur in group and individual levels as is appropriate to the situation and target audience.

Avenues to effective conservation — education and incentives

Mirroring the variety of reasons for the persistence of native grasslands and grassy woodlands on private land, the nature of education and types of incentives needed must be able to target specific landholder situations. Feedback gained during the interview process strongly indicated that education themes aimed at the practical benefits of native grasslands and grassy woodlands will have the greatest conservation effect on the majority of landholders. The property must be looked at from a holistic point of view including how the native grassland or grassy woodland remnant will have real agricultural benefits, as well as achieving conservation outcomes. Research needs to be ongoing to investigate the practical benefits of conserving native grassland and grassy woodland remnants in a production environment. Examples of such benefits include: the value of remnants as wind breaks during lambing time; supporting insectivorous birds as pest controls for crops; and productivity of native grasses without fertilization application. These associations should form the basis for any efforts to raise awareness of native grasslands and grassy woodlands on private land. Education programmes on these themes would also promote understanding of ecological processes.

Grassy ecosystems have particular management requirements that must be taken into consideration when formulating incentives. For example, periodic grazing may be required as an ongoing management tool to maintain biodiversity values, however, the results in this chapter suggest that paying landholders to graze at reduced levels was particularly disfavoured. There is a gap in scientific information as to what effect the current management regimes have had on the structure and floristics of the existing grassy remnants. In the absence of this information, maintaining the *status quo* may be the most sensible approach. In order to achieve this, tax concessions and rate compensation for the area of native grassland or grassy woodland could be offered to the landholder. These incentives were the most favoured among the respondents. Landholders would rather believe they are doing something worthy with the land, whether it be for economic gain or increasing conservation value, rather than be paid to simply set land aside or engage in management practices that apparently decreases agricultural productivity.

Conservation covenants, mostly as they are perceived, will only be appealing to a certain number of landholders. Awareness is needed to increase the knowledge about the scope that covenants can offer, if they are to be more appealing to the majority of the community. As long as conservation values of the area are not compromised, covenants can be designed to suit the particular property (Todd 1997). The strong resistance to the idea of covenants and altered stocking rates on native grasslands or grassy woodlands by the respondents highlights the need for education and incentives to be often applied in tandem.

ACKNOWLEDGEMENTS

This project was funded by the generous support of the Grassy Ecosystem Reference Group. I would like to thank all the participants, as well as Graeme Coulson, Neil Marriott, John Van Brahn, James Todd, Barry Traill, Jim Crosthwaite, Mark Burgman, Gary and Helga Saunders and Lance Williams.

REFERENCES

Chambers, R., Pacey, A. and Thrupp, L., 1989. Farmer first — Farmer innovation and agricultural research. Intermediate Technology Publications: London.

Department of Conservation and Environment, 1990. Remnant native grasslands and grassy woodlands of the Melbourne area — an action plan for conservation based on biological values. Department of Conservation and Environment: East Melbourne, Victoria.

Gilfedder, L. and Kirkpatrick, J. B., 1995. A survey of landholder attitudes and intentions towards the longterm conservation of native lowland grasslands. A Report to the Grassland Ecology Unit, Australian Nature Conservation Agency: University of Tasmania, Hobart.

Gilfedder, L. and Kirkpatrick, J. B., 1997. Remnant bush conservation in Tasmania. Pp. 217–25 *in* Conservation Outside Nature Reserves ed by P. Hale and D. Lamb. Centre for Conservation Biology: University of Queensland, Queensland.

Hamiliton, S. D., Dettmann, P. D. and Curtis, A. L., 1997. Perceptions of remaining native bush on private land in the Box-Ironbark region — Landholder survey. Land and Water Resources Research and Development Corporation: Melbourne University — Dookie College, Charles Stuart University, Victoria.

Hobbs, R. J. and Hopkins, A. J. M., 1990. From frontier to fragments: Europeans impact on Australia's vegetation. *Proc. Ecol. Soc. Aust.* **16:** 93–114.

Kirkpatrick, J., McDougall, K. and Hyde, M., 1995. Australia's most threatened ecosystems — the southeastern lowland native grasslands. Surrey Beatty & Sons: Chipping Norton, New South Wales.

Lunt, I., 1991. Management of remnant, lowland grasslands and grassy woodlands for nature conservation: a review. *Vic. Nat.* **108:** 56–66.

Main, A., 1987. Management of remnants of native vegetation — a review of the problems and the development of an approach with reference to the Wheatbelt of Western Australia. Pp. 1–13 *in* Nature Conservation: the Role of Remnants of Native Vegetation ed by D. A. Saunders, G. W. Arnold, A. A. Burbidge and A. J. M. Hopkins. Surrey Beatty & Sons: Chipping Norton, New South Wales.

McDowell, C. and Sparks, R., 1989. The multivariate modelling and prediction of farmers' conservation behaviour towards natural ecosystems. *J. Env. Man.* **28:** 185–210.

Millar, J. and Curtis, A., 1995. Farmer Knowledge and Experience with Perennial Grasses. Report No. 41. Johnstone Centre of Parks, Recreation and Heritage. Charles Sturt University: Albury, New South Wales.

Mitchell, M., 1990. The status of native grasses in Victoria. Pp. 105–10 *in* Native Grass Workshop Proceedings ed by P. Dowling and D. Garden. Research and Development Group, Australian Wool Corporation: Dubbo, New South Wales.

Morgan, J., 1996. Secondary juvenile period and community recovery following late-spring burning of a Kangaroo Grass *Themeda triandra* grassland. *Vic. Nat.* **113**(2): 47–57.

Platt, S. J. and Ahern, L. D., 1995. Nature conservation on private land in Victoria, Australia — the role of Land for Wildlife. Pp. 300–11 *in* Nature Conservation 4: The Role of Networks ed by D. A. Saunders, J. L. Craig and E. M. Mattiske. Surrey Beatty & Sons: Chipping Norton, New South Wales.

Safstrom, R., 1995. In trust; working with people to achieve conservation on private land in Victoria. Pp. 293–99 *in* Nature Conservation 4: The Role of Networks ed by D. A. Saunders, J. L. Craig and E. M. Mattiske. Surrey Beatty & Sons: Chipping Norton, New South Wales.

Stuwe, J., 1986. An assessment of the conservation status of native grasslands on the western plains, Victoria and sites of botanical significance. Arthur Rylah Institute Technical Report Series No. 48. Arthur Rylah Institute for Environmental Research, Heidelberg, Victoria.

Thompson, J. and Scoones, I., 1994. Challenging the populist perspective: Rural people's knowledge, agricultural research, and extension practice. *Agric. Human Values* **11**(2/3): 58–76.

Todd, J. A., 1997. Victoria's Conservation covenant program: How effective has it been at achieving private land conservation? Pp. 173–75 *in* Conservation Outside Nature Reserves ed by P. Hale and D. Lamb. Centre for Conservation Biology: University of Queensland, Queensland.

de Vaus, D., 1985. Surveys in Social Research. Allen and Unwin: Sydney, New South Wales.

Williams, D., Georges, A., Dearn, J. and Sharp, S., 1991. Information requirements for native grassland conservation in the A.C.T. Applied Ecology Research Group: University of Canberra, Australian Capital Territory.

Wilson, G. A., 1992. A survey on attitudes of landholders to native forest on farmland. *J. Env. Man.* **34:** 117–36.

CHAPTER 53

Reconciling conservation and production values: a case study of the role of Landcare groups in the Waikato region, New Zealand

Helen Ritchie[1]

"Landcare" refers to local residents' groups who take action to enhance their environment. The primacy afforded to production values of land in New Zealand's agricultural landscapes is increasingly challenged by those promoting ecological, landscape, recreational and indigenous spiritual/cultural concerns. Questions then arise as to whether local approaches like Landcare reflect the many diverse environmental values held by those who may or may not reside locally or own land. A study was conducted to investigate this question, and to examine the appropriate role of government in enhancing the democratic use of land and resources. An action research approach was used to work with Landcare groups and regional government staff in examining and improving Landcare practice in Waikato, New Zealand. A range of qualitative research techniques were employed to make a critical assessment of the planning and action undertaken by groups. Results showed that landholders did take action on issues which had wider benefits, particularly water quality, though their actions to enhance environmental values were often limited by economic constraints. Farmers were initially resistant to the participation of others in Landcare groups. This resistance lessened when they were provided with information and exposed to different perspectives at interactive field days where a range of environmental values were discussed. A call is therefore made for government resources to be committed to supporting the information needs of Landcare groups, and to fostering dialogue and participatory planning. Economic policies which favour production over environmental values may also need review. If governments in New Zealand and elsewhere take up these concerns, Landcare is a model which holds considerable potential for reconciling nature conservation and production in rural environments.

Key words: Landcare, Voluntary action, Resource management, Role of government, Action research.

INTRODUCTION

LOCAL groups concerned with managing rural environments, called Landcare groups, have arisen in Australia and more recently in New Zealand. These are voluntary, community-based groups aiming to tackle some aspect of land management which they have in common. The groups identify local issues and take action to address them. Landcare emerged in Australia in the early 1980s in response to evidence of land degradation problems, and pressure from conservation groups on landholders and government to do something about these issues. Since then, about 2 600 groups are estimated to have formed in Australia (Martin and Halpin 1998), far outstripping expected rates of growth (Mues 1997).

In New Zealand, government policy documents on environmental management propose that voluntary action, including through Landcare groups, could be an important means to manage environmental

[1]School of Agriculture and Rural Development, University of Western Sydney, Hawkesbury, Richmond, New South Wales, Australia 2753.

Pages 524–33 *in* NATURE CONSERVATION 5: NATURE CONSERVATION IN PRODUCTION ENVIRONMENTS: MANAGING THE MATRIX ed by J. L. Craig, N. Mitchell and D. A. Saunders. Surrey Beatty & Sons, 2000.

degradation (Ministry for the Environment 1995, 1996, 1997). Such groups are seen as particularly useful for managing issues on private farmland, where regulation is difficult to enforce and unpopular with landholders.

The benefits of involving local people in learning and action to manage their environments better are also recognized in international development practice. Prominent examples are the concepts of People-Centered Development (Korten 1992), and Farmer First Research (Chambers *et al.* 1987). Agenda 21, produced in 1992 from the United Nations "Earth Summit" conference in Rio de Janeiro, affirms the centrality of local action in the management of resources and the environment. Much of the writing on People-Centered Development and Farmer First Research arises from situations in developing countries. In these cases rural development is often village-centred, and through involving the poor and powerless residents of these villages the goal of empowerment is pursued (Chambers 1983). Landcare also seeks to empower rural people. However, in a New Zealand context, agricultural environments are largely owned and managed by farm families, and while many of these farming families struggle to meet debt repayments, they do not fit comfortably in a class typology as "the poorest of the poor". Indeed, for those who already own their farm, some comfortable livings are made. In addition, as New Zealand has always relied heavily on agricultural exports for foreign exchange, the farming industry has in the past enjoyed a disproportionate degree of political influence (Clough and Hicks 1992).

More recent political trends in New Zealand have removed much of the policy privileging the farming sector such as subsidies and price support (Horesh 1995). However, the agrarian beliefs in farming as a superior lifestyle and in the right of land owners to do what they wish on their own land persist, and may often be heard during debate over managing rural environments (for example in recent television and radio coverage of farmers fighting district plan provisions). Such values could potentially act to exclude the interests of others in rural resource management if voluntary action is relied upon. According to Campbell (1989), the key feature of Australian Landcare groups which set them apart from the previous formal advisory structures was their freedom to determine their own direction and agenda. Much of the attraction of Landcare for landholders lies in this autonomy from government control, and in the opportunity for action without undue bureaucratic hindrance (Alexander 1995). In this light, it is interesting to examine whether the Landcare model of empowering local farmers to act in small groups will reflect the range of values associated with natural resources.

As a voluntary activity, Landcare groups can be seen as a form of self-regulation in environmental management (Martin 1997). A move towards greater self-regulation in agriculture is consistent with other policies adopted by recent administrations in New Zealand (Ritchie 1996) and in Australia (Martin and Ritchie 1999), such as the removal of subsidies and disaster relief measures. These measures are part of international trends in governance which stress individual responsibility, fiscal restraint on the part of government and deregulation of business conditions (Tickell and Peck 1995).

Yet, in spite of government confidence that voluntary action will provide adequate self-regulation of environmental effects, there is little documented evidence to confirm that local rural groups will integrate nature conservation into farm management in New Zealand. Conservation lobbyists ask if Landcare groups will be sufficiently accountable to a wider public. When the New Zealand Landcare Trust was launched in 1996, one such group, the Maruia Society, issued a press release entitled "Maruia Society Welcomes Moves to Clean Up Agriculture". In it, they expressed support for the Landcare Trust, but cautioned that the promotion of Landcare by government and landholders could become a "cosy but ineffectual exercise in greenwashing". Part of this doubt centres on whether self-regulation and private land management will be able to reconcile conservation and production values regarding the use of rural environments.

VALUES ASSOCIATED WITH RURAL ENVIRONMENTS IN NEW ZEALAND

Outdoor environments hold an important place in the psyche of the New Zealand public. Fishing, hunting and tramping, or bushwalking, are enjoyed by a broad cross section of the population (Palmer 1990). These recreational activities are affected by the decisions of private landowners as private landowners provide access to some public lands, such as reserve areas or state parks. Their actions can also impact upon the habitat of game species, both terrestrial and aquatic. For example, the quality of habitat for freshwater fish is determined to no small extent by land use in the catchment of the waterway (Rowe 1996; Quinn *et al.*, this volume). Practices which impact on habitat values include land clearance, stocking regimes and the construction of culverts and farm tracks.

Conservation values are also affected by farm management. For example, the protection of indigenous forest on private land is enhanced when farmers exclude stock and conduct pest control. Conservation groups increasingly call into question the sustainability of farming practices such as clearing native forests or burning off high country tussock land, due to the impacts on vegetation and fauna (Nield 1993; Hewison 1995).

There are also spiritual values held by Maori regarding environmental management. For example, water is viewed by Maori as having a spiritual life which is compromised or damaged by pollutants (Kaipo 1997). In addition, the concept of "kaitiakitanga" refers to a guardianship role, borne by specific people in a tribe, pertaining to a specific resource or site, which is not extinguished when ownership of the property is not held by that person or tribe (Ritchie 1992). Maori therefore maintain an interest in resources which is not related to ownership in the legal sense. Such concerns are recognized in New Zealand environmental law (under the *Resource Management Act of 1991*).

All of these are examples of interests in land and water management which are not limited to private use for the purposes of production. They serve to illustrate that agricultural, spiritual, recreational, and conservation values exist side-by-side within production environments, and may conflict with one another.

Such diverse public interests often sit uneasily with the farming fraternity, who have long maintained that they are the adequate and rightful guardians of the land, and that their rights to determine what practices occur on their private property are sacrosanct and should not be restricted by government intervention (Reeve 1997). While in practice, private property rights are limited by various public interest laws, their primacy in farming culture is upheld. As Landcare relies on private action for environmental management, it may potentially reinforce what Reeve terms "presumed property rights". A study of the issues which farmer groups choose to address, and the processes they use to plan their activities, can reveal how well their private actions incorporate the range of interests described above.

Uphoff (1992) suggests that the nature of costs and benefits from local management of natural resources are important in determining which issues groups will choose to address. "Getting local people to take responsibility for natural resource management will be easier where the benefits accrue more quickly, locally, visibly, and to those who bear the cost. In the opposite situation, where benefits are delayed, remote, hard to identify, and do not go to those who invested effort, money or foregone use, local management and maintenance of natural resources would be most unpromising". These comments provide an indication of the issues we may expect local farmer groups to pursue, and therefore the environmental interests which they will seek to protect. It could be predicted that farmers would be most interested in undertaking activities which have on-site results, and which produce immediate benefit to them as farmers. It might also be assumed that this would involve increasing their income in some way.

The evidence from Australia tends to support this view. Martin and Woodhill (1995) have examined the Australian Landcare and Total Catchment Management approaches and concluded that in order for the perspectives of diverse groups to be integrated into resource management, a level of co-ordination, monitoring and perhaps regulation is required to complement local action. They cite the example of biodiversity and conservation of native vegetation as an issue which is largely bypassed or given low priority by Landcare groups in Australia because they focus on soil fertility and salinity — issues which directly affect productivity. Some groups, such as Bushcare and Dunecare, have formed in Australia to take action on other issues. However, Martin and Woodhill (1995) maintain that the principal concern of rural Landcare groups formed by farmers is to address land degradation affecting production. This analysis suggests that nature conservation will be of only secondary interest to private landholders, and to Landcare groups.

This brief review of the values held by different groups confirms the existence of a range of interests in natural resources. Recent reforms in environmental legislation in New Zealand, culminating in the *Resource Management Act in 1991,* uphold that public interests in resources, (even those resources held in private ownership) are legitimate, and must be considered. To what extent, then, will local, voluntary groups, made up principally of private landowners, reflect these wider values? In order to answer this question, this chapter examines the choice of activities and the decision making processes within Landcare groups in the Waikato region of New Zealand. It assesses whether Landcare compromised or facilitated the inclusion of the broad range of values associated with rural environments.

THE RESEARCH PROJECT

The research project which is the subject of this chapter studied Waikato Landcare groups, which were working to enhance their local environment. While urban and coastal groups are active in the region, this chapter is concerned with the integration of nature conservation and production, and therefore focuses on the activities of rural farmer groups. The Waikato region is intensively farmed for dairy and dry stock production, which can create problems with soil erosion and the water quality in streams and rivers. These issues may be addressed through soil conservation plantings, careful stock and fertilizer management, and through better riparian (streamside) protection. This entails fencing waterways to exclude stock, and planting the fenced areas with trees. Other nature conservation issues in the region include the protection and enhancement of habitat such as forests or wetlands.

The project studied the actions undertaken by Landcare groups to address environmental issues. It also aimed to examine and enhance the interface and communication between Landcare groups and the regional government (known as the regional council). Regional councils are in the middle of a three-tier system of government in New Zealand. They have responsibility for promoting the sustainable management of resources (in particular soil, water and air) in the region. The Waikato Regional Council (which has adopted the name Environment Waikato) seeks to enlist local people in managing resources, and has a policy of supporting Landcare groups by providing facilitation, information, networks, and training.

This research took place over a two-year field work period, beginning in July 1995. During that time I was employed at the regional council, under the programme which provides support to Landcare groups. The first Landcare group had formed in the Waikato region in 1993. In 1995 there were few active groups, but several were in their establishment phase, and many others had been proposed but not yet formed.

The study adopted an action research approach (i.e., one which aims to generate both new knowledge and change (Lewin 1946)). Members of Landcare groups and staff from regional government were invited to participate in the research. Staff from a range of programmes within the regional government, and members from eight rural Landcare groups (established and establishing) volunteered to take part. In general, they were the more active members of the groups. These participants were encouraged, through informal discussions, to reflect on their current situation, and identify changes which would increase the effectiveness of Landcare groups. Steps were then taken to implement such changes wherever feasible. Ideas for supporting Landcare groups which were pursued under the project included the provision of training in the form of workshops on collecting seed and growing native plants, and seminars on seeking funding for group projects. Information was provided on environmental issues and the state of local resources, and a series of interactive field days were held to meet information needs on managing agricultural land to protect and enhance waterways.

Within the regional council, a steering committee of staff was established to incorporate the findings of the research project into the support activities of the council. The context of the project was discussed, issues were examined as they arose, and appropriate actions for council staff were identified.

Ethnographic techniques such as participant observation and interviews were used throughout the research, in order to gather qualitative data regarding the interaction of forces and influences as they affected both landowner and government decisions regarding resource management. Based on these data, a critical assessment was made of the changes in environmental management which Landcare groups were able to effect, and of the factors limiting effective action. The same data collection techniques were used to assess the environmental values considered in the decision making and planning processes of local groups and of individual landholders.

Evaluations of the project activity were conducted through interviews and workshops after one year, and again at the conclusion of the project. These evaluations sought input from a range of participants, including Landcare group members, regional council staff, Maori, scientists, and conservation and recreation advocates. At the workshops, interpretations of the data collected through interviews and participant observation were presented back to participants for confirmation, critique, or alternative interpretations, and future action was planned.

The research addressed a range of questions, including the strengths and limitations of local groups in achieving sustainable management of rural environments, the learning which occurred in the groups, the range and effectiveness of actions which resulted, and the impact of the action research approach on group activity and focus. This chapter discusses the broader questions surrounding participation in decision making about rural environments, and the extent to which different perspectives and values were embraced by Landcare groups when undertaking local action. The appropriate role of government in protecting wider public interests in resource management is a further consideration.

RANGE OF ACTIONS TAKEN BY LOCAL GROUPS

Much of the action taken by Landcare groups was directed towards fencing and planting riparian (streamside) areas. The aim of most groups in doing this was to minimize bank erosion. Their primary interest in stabilizing banks was to protect farmland, and to reduce threats to fences, crossings or bridges. Habitat for fish and water quality enhancement for amenity reasons were important secondary goals. The environmental activity of group members was generally limited to riparian strips at the edge of productive farmland, and did not constitute a review of whole-farm management. Farmers expressed some resistance to extending these riparian strips further into production land. As one farmer said: "You just can't expect cockies to give up that much turf".

There are indications that other considerations grew out of the raised awareness of waterways which resulted from riparian management activity. These included concern over fertilizer use and stock management causing run-off into streams or leaching into groundwater, sediment loss from farm tracks, loss of biodiversity and compaction or loss of soil on hillsides. Numerous Landcare group members interviewed at the end of the field work identified alterations they had made to their own farm practices for environmental reasons, as well as changes others had made. These included altering the type of application method of fertilizer (trucking rather than flying on the fertilizer to avoid applying it to waterways), more careful stock management during wet weather, and designing the construction and placement of tracks and stock pads with consideration of run-off. While such wider land management considerations were still peripheral to the groups' focus, evaluations at the conclusion of the field work showed that group members' awareness of a range of issues in sustainable management of land and water was higher than at the beginning of the project. "We've swung from bank protection to being more aware of fencing and planting to prevent run-off, shading for fishlife and other forms of life. And then looking at nitrogen use and pugging. It's not 100% yet — the majority are still bank-protection oriented but they are changing over time — they need the information." — Landcare group member. It can be concluded that broad concerns with sustainable land management issues had been raised within these groups.

These findings suggest that private landholders will take action to integrate nature conservation into aspects of farm management, and will do so for reasons other than immediate economic benefits to them. This contradicts the predictions made earlier, based on Uphoff's (1992) analysis of local institutions for resource management, which suggested that such groups are most likely to focus on issues of immediate benefit to local residents. The role of Landcare in influencing farm management decisions to consider nature conservation values therefore merits further discussion.

HOW LANDCARE INFLUENCED THE CONSIDERATION OF VALUES

Landcare encouraged the inclusion of environmental values in the farm management decisions of private landholders in two important ways. Firstly, Landcare provided an opportunity for information exchange, and exposure to a range of perspectives on resources. Secondly, Landcare functioned as a means to change local norms and expectations of farm managers subtly, so that environmental considerations became more acceptable criteria for decision making.

Learning opportunities provided by Landcare

Through involvement in local Landcare groups, farmers reported that their understanding and awareness of a range of environmental issues in land management had increased. "We started off fencing and planting [streams] from a fishing point of view but the more you read and talk and go to field days and so on you wonder how much good you are doing if adjacent land practices like pugging, strip grazing and bad practices aren't addressed . . . That kind of thing has really come home to us since the group got going" — Landcare group member.

The idea that small groups provide an effective learning environment is consistent with the work of Vanclay (1992) in Australia. He found that Landcare groups offered advantages over one-on-one advice given by government agricultural extension agents, because they allowed ideas to be discussed, applied to local situations and appropriated by landholders as valid knowledge. In New Zealand, the government no longer funds agricultural extension agents. However, in this case, learning about natural resource issues was facilitated by the provision of information to groups, and by field days and workshops which were organized as part of the research project.

Evaluations indicated that participants gained new understandings from these events. In one case, a field day held on a Landcare group member's farm created new interest in the river as habitat for fish, as the farmer reported afterwards: "That guy who was here from that other stream care group, he reckoned our river'd support brown trout — he reckoned they'd like it in there. But they just can't get up here from downstream, because that darker water down there is too hot for them. Maybe we could introduce brown trout here though . . . Should be cool enough for 'em here with the trees here and the new plantings we're doing further up."

Importantly, these field days and workshops also allowed Landcare group members to interact with those from other Landcare groups, and with people from diverse government agencies, scientific institutions and non-governmental organizations with relevant expertise. This interaction was valued by participants perhaps as much as the skills and information presented by designated "experts" during the field days. When asked to evaluate the best aspects of these events, participants identified "opportunities for discussion", "chance to question the scientists", "free flow of ideas", "learning what others have done" and "range of groups and people" as highlights. These findings support the conclusions of Vanclay (1992), that opportunities for farmers to discuss and apply new knowledge to the local situation can facilitate learning. Given appropriate support, the Landcare model provides these learning opportunities.

Landcare as an influence on local norms

Farmers participating in this study stated that high production has traditionally defined "good farm management" in New Zealand farming culture, and that farmers undertaking environmental activity have sometimes been ridiculed or isolated by their neighbours. "You know really, in the past, a good farmer was one that topped the production statistics in the district. And to a certain extent that is still the case . . . Anyone doing something really different in an environmental way gets called a 'greenie', which is to say they aren't a real farmer, they are a bit weird or something."

They reported that often, farmers (including themselves) sought gains in production even when high costs were incurred in achieving them. For example, they sometimes applied fertilizer to marginal hillsides in an effort to increase grazing even when the cost of that fertilizer exceeded the value of the extra grazing. Similarly, they continued to allow grazing of steep gullies, rather than retiring them, even though stock losses through falls in these sites outweighed the grazing advantage. In such cases, pasture production was favoured above forestry or retirement options without any clear economic or practical rationale, reflecting instead a cultural preference for pastoralism.

Farmers in Landcare groups considered that the group had helped to make environmental and community concerns more acceptable criteria for decision making for farm management. In spite of the pre-eminence of agricultural production goals in New Zealand's commercial farm sector, other factors have been and continue to be considered by land managers, and Landcare can serve to reinforce these alternative values. The concept of a neighbourhood effort to improve local conditions draws on a history for rural areas of self-help in services such as fire fighting, sports clubs and school administration (Ritchie 1996). Rural people are well represented among those who pursue recreational activity in the natural environment. In addition, the role of stewards of water and land resources is one which farmers have long claimed for themselves. It is by bringing these community, recreational and stewardship values to the fore that local groups can influence resource management practices. Thus, Landcare groups can build upon existing rural cultural values and still bring about changes which protect some of the environmental values held by others.

In many small communities, maintaining social relations is important, and changes in neighbours' expectations and actions can apply subtle yet powerful pressure to conform (Uphoff 1992). This concept is supported by the work of Vanclay (1997) in Australia, who states that farmers do things which are consistent with what is locally considered "good farm management", and that this is largely determined by the local farming subculture. Vanclay's previous (1992) work in Australia found that Landcare can effect changes in the local farming subculture, and this has also been shown in New Zealand (Ritchie 1995). What is deemed "acceptable" or "good land management" or even "being a good neighbour" is gradually altered through the establishment of new precedents and a subsequent change in local expectations and norms. These effects were confirmed by participants in the research reported in this chapter. As one said, "Without realizing it we changed people's goals through communicating . . . Goals changed subtly from meeting to meeting". Another reported: "You sort of all know what each other have going on and you sort of say such-and-such next door, you know, he should be doing this and he should be doing that. And so it makes you realize well you better make sure your own back yard's clean too." Group members also said that undertaking action as a group removed the anxiety and social risk of being the first to institute a change in land management practices. "I think that made it easier for people — all of us doing something together. Nobody had to go out on a limb." — Landcare group member.

Landcare groups can therefore encourage individuals to consider a range of values in environmental management when making their farm management decisions. Landcare can provide useful learning opportunities, and it can be particularly powerful in bringing to the fore underlying values which already exist in the local rural culture. But what opportunities do these groups offer outside interests to make their own values known? Does Landcare allow non-landholders to participate directly in decision making, planning and action regarding production environments?

DIRECT PARTICIPATION IN LOCAL GROUPS' DECISION MAKING

Vanderplaat (1995) has suggested that an important criterion for evaluating participatory projects is the extent to which they "create space" for communicative planning — the interchange of ideas and presentation of interests, so that people have an opportunity to put forward arguments for consideration by others. This contrasts with the predominant model for planning in New Zealand, which invites public input through a formalized process of submissions and hearings. Decisions are then recommended by planning staff to the elected regional or district councillors. Because resource management legislation requires councils to prepare plans for the region or district, few resources are currently committed to more communicative planning processes at a catchment or community level. The potential for Landcare to provide for such processes is therefore of interest.

Landcare groups in this study did not always create a communicative space where an open interchange of opinions could occur regarding plans for local action in production environments. When farmers considered forming a new Landcare group, they frequently asserted that such a group should consist purely of landowners, to avoid the possibility of conservationist or "greenie" challenges to farmer control. There was a strong preference for excluding "outsiders" from planning action for environmental enhancement. This was related to a mistrust of government or outside interests "telling farmers what to do on their land". While it is regional council policy to encourage broad public participation in Landcare groups, often in the initial stages of building trust with a group, the first meetings are held with landholders only. For advocates of democratic decision making, this might pose some concerns. However, dialogue with Landcare group members during this action research project indicated a growing willingness to involve others. Partly, this was because groups gained confidence in themselves and a greater sense of practising environmental stewardship. "People are doing more things — collecting seeds, growing things, even just planting — and they realize they can do things themselves. And they are feeling better about doing it. They don't feel like farmers are "under siege" any more — they are doing it themselves." — Landcare group member. Forming a Care group had allowed landholders to feel they were "on the front foot" with respect to environmental care, which permitted a less defensive attitude over time to input from other perspectives regarding local resources.

Through the action research approach used in this project, further communicative space was created at field days on managing agricultural waterways, which attracted farmers, conservationists, scientists, and recreational users of streams. Lively interchange occurred during which different perspectives were articulated.

Another opportunity for landowners to gain greater appreciation of the claims and concerns of others regarding shared resources occurred when local interest groups collaborated in action. An example of this arose when a conservation group provided trees for riparian planting to Landcare farmers and assisted with planting them, along with the local school. A landowner remarked afterwards, referring to the conservation group: "I really had no idea before what they were all about, but it was really great that they showed up to help out with the planting."

These examples indicate that while no formal participatory decision making forum was instituted to plan and direct Landcare action, Landcare participants became more open to the participation of others in their planned activities (and especially if those "others" were prepared to assist in some way). Indeed, if the same action could meet the needs of aquatic life, recreational anglers, conservationists and Maori, as well as enhancing farmers' own goals, then the satisfaction associated with that action was multiplied. Some conflicts may occur between these perspectives, but can often be accommodated with compromises. On-site action planning was arranged through this research project which allowed for the identification of such compromise solutions. A relevant example was the planning of a riparian planting, where less dense plantings were designed to filter run-off by trapping sediment in the long grass growing between trees, and also to allow access for anglers, while still providing adequate wildlife corridors and in-stream benefits.

It would be naïve or overly optimistic to assume that landowners will always act in the public interest, particularly where this significantly compromises their production. As previously noted, the retiring of adequate land to ensure effective nutrient removal is seen by landholders in Landcare groups as "asking a farmer to give up too much turf". This constraint is particularly important when sheep and beef farmers are struggling to maintain financial viability in current market conditions. Voluntary action in this respect will in most cases compromise conservation interests in order to retain grazing area and avoid associated costs such as reticulating stock water once animals are excluded from streams. These results serve to highlight the very real conflicts of interest in rural environments, where economic forces have been deemed to best determine land use. However, potential exists for gradual change in rural community members' expectations of each other.

The model of supplying voluntary groups with information and sites for communication of various interests does not address power issues. In the final analysis, this model reinforces a concept of private property rights and local autonomy whereby property owners remain in control of decisions. In this study, there were no deeply entrenched conflicting positions over resource use, as there are in the high country of the South Island, where the ecological viability of pastoral farming is seriously questioned by many (Neild 1993). The effectiveness of local voluntary groups to resolve contentious issues such as these remains uncertain.

Furthermore, the process of allowing space for the articulation of interests may effectively preclude the participation of some. For example, Maori perspectives have been sought in organizing events for Landcare groups, and strong support has been expressed by Maori for any activities aimed at cleaning up waterways. However, the multiple activities that tribal structures assume (including support and services for the welfare of their people and inputs into formal resource management proceedings such as plans and consents for development) create enormous demands on their time and energy. The model of voluntary communicative exchange provides neither alleviation nor compensation for this. Historical factors have largely removed from Maori the economic base they once enjoyed. Until settlement of grievances over resource loss occurs, their participation in managing local environments may be precluded by other demands on their time and finances.

The question then becomes how government might best intervene to address these power differentials and to facilitate the consideration of the multiple values associated with rural environments.

A ROLE FOR GOVERNMENT IN FACILITATING DEMOCRATIC MANAGEMENT OF RURAL ENVIRONMENTS

This research shows that given adequate information, local action groups are often willing to incorporate the concerns of others when planning their projects. In many cases a facilitator, such as a local government staff member, may be able to introduce matters of public concern to Landcare group members, or alternatively organize an information forum which explores various perspectives on an issue.

Wider co-ordinating and integrative mechanisms may be required to complement local action, in particular for deeply conflicting issues or complex situations where many players are involved and power differentials exist. This will be the case in larger scale planning, such as total catchment or harbour management. Similarly, if redistribution of financial resources is required for action in some areas (for example, where farmers cannot afford to fence or provide alternative stock watering sources), then more formalized processes to examine issues collectively and determine community priorities may be justified. Where the issues are sufficiently contentious that communities are expressing concerns about them, mechanisms need to be provided to discuss, learn about and plan action to address the issues, taking into account the diverse values associated with resources. Some intervention prior to communicative planning exercises may be required to achieve a similar capacity to participate for all parties. Such interventions could support groups to participate equally through training, and provide information to individuals so they could better decide how different management options impact upon their interests. This support and capacity-building might also compensate for the time demands on overburdened participants.

Local action will never supplant the need for governance at the regional or national level. An overview of cumulative effects is needed which identifies where voluntary mechanisms will not be sufficient to protect the environment equitably for current and future generations. Central and regional governments in New Zealand (and elsewhere) will need to monitor the adequacy of relying on voluntary action and market forces for environmental management.

Economic factors are important influences on farm management decisions. Some observers of market reform argue that the removal of subsidy programmes for land development, inputs and price support in agriculture has increased the sustainability of New Zealand farming (Reynolds *et al.* 1993). However, the government itself recognizes that investment in conservation works has declined dramatically as farm incomes have fallen and subsidies have been withdrawn (Ministry of Agriculture and Fisheries 1993). Attention has been drawn to the positive influence which may be manifested as New Zealand farmers respond to overseas consumer preference for sustainably produced goods (Williams 1995). Government's role has been conceived as "primarily that of encouraging market-led adjustment to sustainable practices" (Ministry of Agriculture and Fisheries 1993). The evidence to date that greater deregulation has produced more sustainable land management is equivocal (Bradshaw and Smit 1996). Even if it is accepted that future markets may produce incentives for the protection of conservation values in production environments, this raises a further question: Is it desirable for the shape of the local environment and the nature of resource use to be defined according to the preference and impulse of geographically distant consumers, who may have little understanding of the cultural and environmental idiosyncrasies and values associated with the local landscape? This concern suggests a role for government in planning and fostering rural development with provision for the protection of conservation values. Such a regional development policy could incorporate the views of a range of interested parties through participatory planning processes.

A call for greater use of participatory planning should not obscure the fact that it is the rapid achievement of results and the ensuing feeling of group effectiveness and power which is the strength and the beauty of local action. It is inappropriate, in many cases, to hold up such action until a full community consensus has been reached as to what the issues are, and what perspectives should be accommodated in any action. It is fruitless to try to drag busy people into a planning process in which they are not especially interested. Rather, it may be more appropriate to continue the process of informally discussing various perspectives on resources with groups who are interested in carrying out action. There is also a need to provide for learning events where views can be presented and debated, scientists alongside interest groups, government information alongside local knowledge, so that all can go away with a broader understanding of an issue.

Communicative planning can take place informally through the interchange of ideas, provided entrenched power and interests are not at play. In the case of Landcare groups, who have indicated a willingness to take on board public values associated with the streams on their land, staff members who are aware of various concerns with water management can either represent these concerns themselves or bring in the appropriate person to do so. Staff need to seek opportunities to widen participation while still building on local institutions. This should be a recognized and explicit part of their role. If further experience indicates that such an approach is failing to allow certain interests to be represented, more intensive democratic planning exercises may be required.

It seems important that research and support activity carried out with landholders aimed at improving environmental management be conducted with considerations of equity and democracy in mind. This implies a need for participatory reflection and evaluation to review critically which interests and issues are being addressed or left out, and whose voice is being heard. It is not enough simply to engage in "empowering" local people, without examining the broader claims on the resources they are managing.

CONCLUSIONS

Without enquiring into the power structures surrounding a resource, and the adequacy of local knowledge to manage it sustainably, local action can be counterproductive and disempowering of other interests. However, this New Zealand study indicates that Landcare can be a powerful force for change based on the values held by landholders, reinforced by the learning and norming effects of local groups. If appropriate support services are provided and economic constraints on landholders alleviated, Landcare groups could prove a useful model for the reconciliation of nature conservation and production in rural environments elsewhere.

ACKNOWLEDGEMENTS

Certain ideas in this chapter were presented at The World Congresses: 4th on Action Research, Action Learning and Process Management, 8th on Participatory Action Research, in Cartagena, Colombia, in June, 1997. I am grateful to participants at that conference and to other colleagues in New Zealand for their helpful comments. I am also appreciative of the comments provided by referees on an earlier draft.

This research was jointly funded by Environment Waikato (Regional Council), and Ministry for the Environment (Sustainable Management Fund). The conclusions presented in this chapter are the authors and should not be interpreted as policy of the funding agencies.

REFERENCES

Alexander, H., 1995. A Framework for Change: the State of the Community Landcare Movement in Australia. National Landcare Facilitator Annual Report. National Landcare Program: Canberra, Australian Capital Territory.

Bradshaw, B. and Smit, B., 1996. Resource management implications arising from the reform of state intervention and support in Canadian agriculture: Insights from New Zealand. Pp. 22–36 *in* Proceedings, Resource Management: Issues, Visions, Practice. A Symposium. Centre for Resource Management: Lincoln, New Zealand.

Campbell, A., 1989. Landcare in Australia — an overview. *Aust. J. Soil Water Cons.* **2**(4): 18–20.

Chambers, R., 1983. Rural Development: Putting the Last First. Longman: London.

Chambers, R., Pacey, A. and Thrupps, L., 1987. Farmer First: Farmer Innovation and Agricultural Research. Intermediate Technology Publications: London.

Clough, P. and Hicks, D., 1992. Soil Conservation and the RMA. Report to the Ministry of Agriculture and Fisheries. New Zealand Institute of Economic Research Working Paper 93/31. New Zealand Institute of Economic Research: Wellington, New Zealand.

Hewison, G., 1995. Reconciling Trade and the Environment: Issues for New Zealand. Institute of Policy Studies: Wellington, New Zealand.

Horesh, R., 1995. Farming for the Market: the New Zealand Experience. Ministry of Agriculture and Fisheries: Policy: Wellington, New Zealand.

Kaipo, S., 1997. Te Tangi a te Iwi. Why Our People Cried. Managakahia Irrigation from a Tangata Whenua Perspective. Occasional Publication 35. Department of Geography, University of Auckland: Auckland, New Zealand.

Korten, D., 1992. A deeper look at sustainable development. *World Policy J.* **Winter 1991–2:** 157–90.

Lewin, K., 1946. Action research and minority problems. *J. Social Issues* **2:** 34–46.

Martin, P., 1997. The constitution of power in Landcare: a post-structuralist perspective with modernist undertones Pp. 45–56 *in* Critical Landcare ed by S. Lockie and F. Vanclay. Centre for Rural Social Research: Charles Sturt University, Wagga Wagga, New South Wales.

Martin, P. and Halpin, D., 1998. Landcare as a politically relevant New Social Movement. *J. Rural Studies* **14**(4): 445–57.

Martin, P. and Ritchie, H., 1999. Logics of participation: rural environmental governance under Neoliberalism in Australia. *Environ. Politics* **8**(2): 145–67.

Martin, P. and Woodhill, J., 1995. Landcare in the balance: roles of government and policy directions for degrading rural environments. *Aust. J. Environ. Man.* **2**(3): 173–83.

Ministry for the Environment, 1995. Environment 2010 Strategy. A Statement of the Government's Strategy on the Environment. Ministry for the Environment: Wellington, New Zealand.

Ministry for the Environment, 1996. Sustainable Land Management. A Strategy for New Zealand. Ministry for the Environment: Wellington, New Zealand.

Ministry for the Environment, 1997. Reducing the Impacts of Agricultural Runoff on Water Quality. A Discussion of Policy Approaches. Ministry for the Environment: Wellington, New Zealand.

Ministry of Agriculture and Fisheries, 1993. Sustainable Agriculture. Policy Position Paper 1. Ministry of Agriculture and Fisheries: Policy: Wellington, New Zealand.

Mues, C., 1997. Landcare's role in promoting sustainable farm management. Pp. 99–100 *in* Landcare Changing Australia National Conference, 16–19 September. Conference Proceedings, Volume 1, Adelaide, South Australia.

Neild, J., 1993. Synopsis of Submissions on Ministry of Agriculture and Fisheries Policy Paper 106: Sustainable Agriculture — A Policy Proposal. Ministry of Agriculture and Fisheries: Policy Public Information Paper 4. Ministry of Agriculture and Fisheries: Policy: Wellington, New Zealand.

Palmer, G., 1990. Environmental Politics. A Greenprint for New Zealand. McIndoe: Dunedin, New Zealand.

Reeve, I., 1997. Land resource management: management or negotiation? Pp. 83–96 *in* Critical Landcare ed by S. Lockie and F. Vanclay. Centre for Rural Social Research: Charles Sturt University: Wagga Wagga, New South Wales.

Reynolds, R., Moore, W., Arthur-Worsop, M. and Warmington, B., 1993. Impacts on the Environment of Reduced Agricultural Subsidies: A case study of New Zealand. Ministry of Agriculture and Fisheries Policy Technical Paper 93/12. Ministry of Agriculture and Fisheries: Policy: Wellington, New Zealand.

Ritchie, H., 1995. Landcare in Rural Waikato. Report on a Study for Environment Waikato. Environment Waikato: Hamilton, New Zealand.

Ritchie, H., 1996. Voluntary and participative approaches to rural environmental management: the case of Landcare groups in Waikato. Pp. 223–32 *in* Proceedings: Resource Management: Issues, Visions, Practice. A Symposium. Centre for Resource Management: Lincoln, New Zealand.

Ritchie, J., 1992. Becoming Bicultural. Huia Publishers/ Daphne Brasell Associates Press: Wellington, New Zealand.

Rowe, D., 1996. Interim Report on Studies Related to the Effects of Land-Use Changes on Native Fish. National Institute for Water and Atmospheric Research: Hamilton, New Zealand.

Tickell, A. and Peck, J., 1995. Social regulation after Fordism: regulation theory, neo-liberalism and the global-local nexus. *Econ. Soc.* **24**(3): 357–86.

Uphoff, N., 1992. Local Institutions and Participation for Sustainable Development. International Institute for Environment and Development Gatekeeper Series No 31. International Institute for Environment and Development: London.

Vanclay, F., 1992. The social context of farmers' adoption of environmentally sound farming practices. Pp. 94–121 *in* Agriculture, Environment and Society ed by G. Lawrence, F. Vanclay and B. Furze. MacMillan: Melbourne, Victoria.

Vanclay, F., 1997. The social basis of environmental management. Pp. 9–27 *in* Critical Landcare ed by S. Lockie and F. Vanclay. Centre for Rural Social Research: Charles Sturt University: Wagga Wagga, New South Wales.

Vanderplaat, M., 1995. Beyond technique. Issues in evaluating for empowerment. *Evaluation* **1**(1): 81–96.

Williams, J., 1995. Searching for the Ingredients of More Sustainable Landuses: A study tour report. Ministry of Agriculture and Fisheries Policy: Christchurch, New Zealand.

CHAPTER 54

Nature conservation: actions, perceptions, experience and requirements of rural groups

L. Stocker[1]

Many rural community groups in the Central Wheatbelt and Great Southern Region of Western Australia are undertaking nature conservation projects in a bid to retain or even regenerate native bushland and the wildlife that inhabits it. The initial aim of this project was to improve understanding of nature conservation in rural areas of Western Australia, and thereby to improve the practice of it and policy towards it. A team conducted workshops with nine rural nature conservation groups in south-west Western Australia, using a tripartite methodology: questionnaire, field-trip and structured discussion. This approach allowed the provision as well as the collection of information in a field-based environment. The findings of this project offer some insights into rural groups' understandings and practices of nature conservation in Western Australia, whereas previously the emphasis of similar projects has been on landcare. The findings show that nature conservation is among the top three decision making factors among rural conservation group members. It can be seen that the culture of nature conservation in many rural groups is complex and sophisticated, and the environmental ethic is ecocentric. These results suggest that existing conservation programmes aimed at awareness-raising in rural areas using broad-brush approaches, on one hand, and linear extension practices of knowledge delivery, on the other, could both be replaced in some cases by interactive learning programmes designed to resource rural groups so that they can educate their own communities, and act on their existing levels of awareness.

INTRODUCTION

THE analysis and discussion presented in this chapter form a part of a larger project. The broad aims of this larger project were to enhance the understanding and practice of nature conservation in rural Western Australia. These aims were pursued by working with rural groups as so much nature conservation work is group-oriented. The specific objectives of this project were: to improve government and conservation agencies' understandings of nature conservation by rural groups in Western Australia, and to improve their policy and practice towards it; to improve rural groups' understanding and practice of nature conservation generally, and of the issues of particular concern to them; and to identify further strategies that rural groups and their local communities require to pursue their nature conservation understanding and practice.

Factors considered most likely to affect the rural group members' level of action in nature conservation related first to their perception of nature conservation issues, and second to their appreciation of nature itself. Third, the capacity for groups to undertake nature conservation may depend on the capacity of government conservation programmes to facilitate learning.

In the agricultural sector throughout the world there has been a shift away from the conventional linear extension model of knowledge delivery or transfer of technology towards a farmer first approach (Chambers *et al.* 1989) that involves interactive learning environments, institutional support and context, and participatory approaches and methods (Pretty and Chambers 1994). This shift has been associated with an awareness that tacit, local or indigenous rural knowledge forms a fundamental interactive component of rural development (Scoones and Thompson 1994; Stocker 1995; Carr 1997). This collaborative approach to knowledge has also been called community science (Stocker 1995). In many cases this shift towards

[1]Institute for Sustainability and Technology Policy, Murdoch University, Western Australia, Australia 6150.

Pages 534–41 *in* NATURE CONSERVATION 5: NATURE CONSERVATION IN PRODUCTION ENVIRONMENTS: MANAGING THE MATRIX ed by J. L. Craig, N. Mitchell and D. A. Saunders. Surrey Beatty & Sons, 2000.

farmer first has also gone hand in hand with an increased focus on ecological sustainability. In Australia these changes are exemplified by the landcare movement (Chamala and Mortiss 1990; Curtis and De Lacy 1994; Lockie and Vanclay 1997).

In nature conservation, too, there has been a building of networks both within the community, and between agencies and the community (Saunders *et al.* 1995). An interactive learning model has not, however, been adopted on a systematic basis for nature conservation. Rather, the posture of conservation biology towards the rural community in Australia is still more akin to a linear extension model. Indeed it is only in the last decade that many ecologists have begun to communicate their findings outside of scholarly conferences and journals.

But what do we really know about levels of awareness and practice of nature conservation in our rural communities? In Western Australia we do know that there are few individuals who are extraordinarily interested in, and knowledgeable about, nature conservation; and there are others who are hostile to it. We have few published accounts, however, of the general level of awareness about nature conservation within the rural community. We do not even know what nature conservation means for rural people. We certainly do not know what sort of environmental ethic rural people espouse.

This chapter therefore aims to enhance our understanding of how members of rural groups in Western Australia perceive nature conservation and how they relate to nature itself. It also aims to identify how nature conservation learning and communication might best be achieved in order to promote greater action.

In the first part of this chapter I describe the methodology in detail because it was significant to the process of this project. I then present the findings of surveys and structured discussions.

Research questions addressed in this chapter are as follows. In what kinds of nature conservation activities are rural groups involved? How do rural groups perceive nature conservation? Are rural groups' perceptions of nature conservation problems linked to their nature conservation activities? How do rural groups perceive nature? Are rural groups' perceptions of nature linked to their nature conservation activities? How is information on nature conservation best exchanged with, and within, rural communities? What are the implications of the current project's findings to policy and funding for improving the communication of nature conservation?

METHODOLOGY

Choosing the groups and establishing contact

The project targeted a range of rural conservation groups, including those with a landcare focus, such as catchment groups and Land Conservation District Committees, and those with a natural history focus, such as herbarium groups. Groups were sought that were representative of the Great Southern Region and the Wheatbelt. These were: Wilson Inlet Catchment Group, Denmark Wetlands Group and Young River Catchment Group in the Great Southern Region; and Wyalkatchem Land Conservation District Committee, Elashgin Catchment Group, Wyalkatchem Herbarium Group, Cunjardine Catchment Group, Ucarty Catchment Group and Corrigin Herbarium Group in the Central Wheatbelt. In addition, questionnaires were handed out at a Wheatbelt Group workshop (titled "Woodland, Wire and Weeds") in Merredin in the Central Wheatbelt.

Approach to workshops

Workshops were divided into three sectors: a field trip, a questionnaire and a structured discussion. The field trip to local bushland was made with a specialist speaker. Each speaker was chosen on the basis of the expressed information requirements of each rural group. During the field trip, the group described to the research team the nature conservation issues and/or achievements relevant to the bushland; and the speaker answered questions and provided some specialist input. On returning to the house or hall, the group completed a questionnaire. A team member then facilitated a structured discussion on resources and strategies required to promote nature conservation. Because the groups' requirements and availability varied, the total time taken for a workshop varied between two and half and four hours.

The tripartite participatory approach was used for several reasons. It enabled the exchange and development of knowledge. It ensured that participants came away with information useful to their nature conservation goals in the short term, and that the research team came away with information useful to the project. It provided three different angles that either confirmed a particular perspective or suggested contradictions in the way participants thought about an issue. It allowed participants who were not comfortable with a particular medium, say reading and writing, to express themselves verbally on a one to one basis in the field.

Presentation of findings

In the following sections, the profiles of the workshops are provided first. The subsequent descriptions of the groups' conservation projects are based on observations and discussions during the field trip. The next sections, dealing with respondents' nature conservation activities, their perceptions of nature conservation, and their appreciation and experience of nature, are based on the results of the questionnaires. Lastly, the section on strategies for communicating conservation is a summary of the structured discussions held with each group. Ideas presented are those of the workshop participants.

FINDINGS

Workshop profiles

Numbers attending the workshops ranged from 4 to 10. A total of 72 respondents were surveyed. This included the Merredin workshop which was organized by the Wheatbelt Group as an all-day seminar. Questionnaires were handed at this workshop, but the group was not involved in other activities of this project.

Overall, 39 women and 33 men completed the surveys; of these, 36 women and 28 men attended the workshops. Merredin respondents did not attend a project workshop. The two herbarium groups and the Wilson Inlet Catchment Group were conspicuous for their relatively high percentage of women (Corrigin Herbarium 77%; Wyalkatchem Herbarium 75%; Wilson Inlet Catchment Group 75%).

Ages of people attending the workshops varied from 12 to 71, with the average age overall being 43. Not all respondents provided their ages.

Most participants were farmers, or employed in land or nature conservation in some way.

Group conservation projects

During the field trips, rural groups explained their projects to the research team. A brief summary of each project is set out below. In Denmark, the Wetlands Group is aiming to protect the wetlands that are the home of two species of Ibis (*Threskiornis molucca* and *T. spinicollis*). The group is helping to regenerate vegetation around ponds and to build islands to increase the habitat of the ibises.

Close by, the Wilson Inlet Catchment group, has just finished compiling a directory of landcare services in their local area.

Members of the Young River Catchment group, just out of Mt Barker, have made substantial efforts at two very different sites on private property: fencing off and mapping a floristically diverse 176 hectare remnant; and revegetating a salty creek line.

Based further north in the central wheatbelt, the Corrigin Herbarium Group aims to produce good quality pressings as a local resource for identification, and is involved with the local Land Conservation District Committee in monitoring and rehabilitating a 162 hectare patch of remnant bush which contains extensive Salmon Gum *Eucalyptus salmonophloia* stands.

Three groups in the Wyalkatchem area are also interested in the protection of Salmon and York Gum *Eucalyptus loxophleba* remnants. The Elashgin Creek Group has been active in rabbit and fox shooting, water monitoring, tree planting, soil testing, mapping, and fencing remnant bush. The Wyalkatchem Land Conservation District Committee has similar goals. The herbarium group in Wyalkatchem has been involved in flora and fauna identification and protection; controlling water with banks and dams; redirecting water to control erosion; tree planting; mapping; photographing trees planted on various sites; promotion and education in local school.

The Ucarty Group has been active in both the planning and practical dimensions of land conservation. This group is trying to improve the ecological value of this and other remnants in the area.

The Cunjardine group has been involved in both catchment and farm planning, flora photography, soil testing, fencing, education and tree planting. Among them they have impressive taxonomic knowledge and photographic records of the flora of the area.

Action towards nature conservation

RURAL NATURE CONSERVATION ACTIVITIES BY GROUPS

Respondents were asked to indicate the number of days per year spent as a group on a range of nature conservation activities. A total of 85 days were spent on group-based nature conservation activities (all groups combined). This figure does not include the days spent by individuals in the groups doing non-group based nature conservation activities. By far the most common nature conservation activities in which groups were involved were tree planting (19 days per year or 22% of total conservation days) and remnant bush protection (13 days per year, or 15% of total conservation days), including rehabilitation and regeneration. Mapping (9 days per year, or 11% of total conservation days) and education (9 days per year, or 11% of total conservation days) were also common group activities.

INTERACTION AMONG GROUPS ABOUT RURAL NATURE CONSERVATION

Respondents were asked whether they belonged to other community groups and, if so, how often they talked about nature conservation in those groups. Around half the people surveyed belonged to community groups other than the group they were a part of for the purposes of the workshop. The most common community group to which respondents belonged was a sports club. Nature conservation tended to be discussed more often in groups whose purpose was associated with the land in some way. Eleven respondents out of the 13 who belonged to landcare consultancy groups (85%) discussed nature conservation regularly or always in their groups. Similarly, 8 out of 10 (80%) of gardening club members; 8 out of 11 wildflower society members (73%) and 7 out of 10 (70%) Federated Farmers or Pastoralist and Graziers Association members discussed nature conservation regularly or always in their groups. Craft and hobby groups also talked regularly about nature conservation (8 out of 11 members). By contrast, those groups not directly involved in land-based activities groups rarely or never talked about nature conservation. Thus, church groups (6 out of 8 members), Rotary, Apex and Lions groups (3 out of 3 members), bridge groups (2 out of 2 members), Parent and Community groups (6 out of 7 members) and sports groups (14 out of 17 members) rarely or never talked about nature conservation in their groups.

REASONS FOR INVOLVEMENT IN RURAL NATURE CONSERVATION

Respondents were asked why they became involved in nature conservation, and the reasons for which they remained involved. The majority of the 64 respondents answering this question became involved in rural nature conservation because they see it an integral part of their livelihood. Many did not see nature conservation as something separate from farming; it was a part of their farm work, and a way of life that they had known for a long time or had inherited from their parents. Many others became involved because it was satisfying in some way: they enjoy it, find it relaxing, challenging, rewarding or interesting. "It's my work and I enjoy it!" was a common response. For many that became involved in nature conservation because it was part of their livelihood in the first place, remaining involved in nature conservation has been a function of enjoyment. Other reasons for involvement in nature conservation included: repairing damage, wanting to continue to exist in harmony with nature, being involved in a process of improvement.

Perceptions of rural nature conservation

WHAT IS RURAL NATURE CONSERVATION?

Respondents were asked about the relationship between land conservation and nature conservation. For nearly two-thirds (62%) of the 72 respondents to the questionnaire, nature conservation goals were seen as separate from land conservation goals but related to them. One-third of respondents (33%) thought that nature conservation and land conservation goals were the same. Only two people saw nature and land conservation goals as different.

Respondents were asked to tick boxes to indicate which of a range of activities they considered to be nature conservation. They were also given space to amplify the given activities with their own suggestions. The kinds of issues or activities that were cited as "nature conservation" were broadly inclusive. Management of bushland, management of wildlife, and monitoring soil, water and bush quality were the three sets of activities most commonly cited as "nature conservation". For each of these activities, 70 or more out of the 72 respondents cited them as nature conservation.

The term "management of bushland" was amplified by respondents with suggestions such as the protection, fencing and regeneration of remnant vegetation, prohibition of grazing on reserves, reforestation, establishing and protecting understorey. The term "management of wildlife" was amplified by respondents with suggestions such as protection of endangered species. The term "monitoring" included responses that related to the monitoring of soil, water, wildlife and bush.

Management of wetland habitats, forest habitats, native plants, and land conservation generally were also commonly viewed as "nature conservation".

A few respondents considered the management of rose gardens and sports ovals to be nature conservation.

IMPORTANT RURAL NATURE CONSERVATION ISSUES

Respondents were asked to nominate the three nature conservation issues most important to them. Remnant bushland protection (including fencing, restoration, and rehabilitation) was the issue considered important by the largest number of people (40 out of 72 respondents). Soil and land conservation was the issue considered important by second largest number of people (32 respondents). Other factors considered important by some people were wildlife protection (22 respondents), feral animal control (14 respondents) and water quality maintenance (12 respondents).

RELATIVE IMPORTANCE OF RURAL NATURE CONSERVATION

Respondents were asked to rank the three factors (from a list provided) they considered most important to decision making in the farm business or livelihood. Farm productivity (53 respondents), family well-being (38 respondents) and nature conservation (37 respondents), in that order, were the three factors most commonly given a ranking of one, two or three. Commodity prices (30 respondents) and government regulations (15 respondents) were also commonly given a ranking of one, two or three.

IMPACTS OF RURAL NATURE CONSERVATION

Respondents were asked whether nature conservation increased productivity, decreased productivity or had no effect. Over 70% of the 72 respondents believed that nature conservation increased farm productivity, while the remainder believed it decreased farm productivity (25 %), had no effect (3%), or they did not know (1%). People who thought that nature conservation on farms increased productivity were involved in nature conservation activities on a greater number of days per year than those who thought nature conservation decreased productivity ($r = -0.2735$, $p = 0.041$).

RESPONSIBILITY FOR ACTION IN RURAL NATURE CONSERVATION

Respondents were asked to indicate by ascribing percentages how responsibility for nature conservation should be shared among the various stakeholders listed. On average, respondents perceived that 30% of the responsibility for nature conservation action lay with the farmers themselves; 16% lay with federal government; 15% lay with state government; and 12% lay with local communities. Other sectors to which responsibility was apportioned included: environment groups (9%), the urban community (6%), corporate business (4%) and banks (4%).

Appreciation of nature

VISIONS OF NATURE

Respondents were asked three questions that related to visions of nature. How do you picture your ideal patch of nature? Within this patch of nature what plants and animals would you like to see? How should this patch of nature relate to the rest of the landscape? The questions were thus qualitative and open ended. Responses were grouped according to common themes.

Most of the 72 respondents pictured their ideal patch of nature as a native environment (63 respondents), usually bushland (53 respondents). A few people (3 respondents) specified a synthetic (mixed native and exotic) environment as their ideal patch of nature, such as their own farm with both bushland and agriculture on it. Notions of a pristine or untouched natural environment (10 respondents) with a diversity of plants and animals were commonly mentioned (12 respondents). It was also common for respondents to include rock or soil types (9 respondents), streams (11 respondents), and wetlands or lakes (3 respondents) in their vision of a patch of nature. For a few respondents (4), their ideal patch of nature was a particular place, usually a favourite place.

Most people wanted to see native animals within their ideal patch of nature (54 respondents). Some people provided considerable detail about the kinds of native animals they wanted. The most common kinds of animals specified were birds (26 respondents) such as parrots, wedge-tail eagles *Aquila audax*, wrens; and mammals (20 respondents) such as wallabies, kangaroos, numbats *Myrmecobius fasciatus*, bandicoots, possums and echidnas *Tachyglossus aculeatus*. Reptiles including lizards, geckos, tortoise, and snakes were also commonly mentioned (9 respondents). Fish (1 respondent), insects (6 respondents) and frogs (5 respondents)were also mentioned. A few people specified such things as "at least 80 species of aquatic invertebrates". In some cases this was in the context of ensuring ecological health or ecosystem function, as discussed below. Many people also specified the exclusion of feral animals or native pests (14 respondents).

Most people (51 respondents) indicated that they would like to see native plants in their patch of nature. Many respondents (31) considered that local provenance species were most appropriate for their patch of nature. The most common kinds of plant that people would like to see within their ideal patch of nature were native wildflowers (16 respondents) and trees (21 respondents). A number of people (12) were quite specific about the plants they wanted to see in their patch of nature. Wildflowers included spider and enamel orchids, and everlasting daisies. Specific trees included the York Gum, Salmon Gum, Wandoo *Eucalyptus capillosa*, Sheoaks *Casuarina* and *Allocasuarina* spp., and Paperbarks *Melaleuca* spp. Many people (14 respondents) also specified ground covers and a healthy understorey for their patch of nature, nominating *Hakea* spp., *Callistemon* spp., *Grevillea* spp., and *Acacia* spp. Others (11 respondents) specified the exclusion of weeds. Again, these responses indicated an interest in, and knowledge of, the ecology of bush remnants. Only one person specifically mentioned the inclusion of rare and endangered species.

Respondents had diverse views on how their ideal patch of nature should relate to the rest of the landscape, and, by implication, what role it should perform.

Respondents' comments confirmed that the majority of people do not see bushlands or wetlands as isolated patches but as functionally and structurally connected to the wider landscape. In terms of the structure of the patch in relation to the rest of the landscape, the most common perspective (40 respondents) was that the patch of nature should complement, be integrated with, or be compatible with, the surrounding landscape. Many people (16 respondents) suggested that integration and nature conservation could be enhanced through the development of corridors. It was recognized by some people (8 respondents) that "integration" could only be achieved where patches of nature, such as bushland, were in fact protected from the rest of the landscape, typically by fencing.

In terms of the functioning of a patch, the concept of a "patch of nature" as an ecosystem unto itself was also prevalent. As suggested above, some respondents indicated through their choice of plants and animals their understanding of, and preference for, an ecosystem. Many respondents, however, specified "ecosystem" explicitly (22 respondents), or else implied it through the use of allied concepts (6 respondents), such as "self-sustaining", "balanced", "regenerating", and "self-regulating". A few people conveyed a vision of a patch of nature as an "oasis" (6 respondents), bringing life to the surrounding "desert". Some respondents (8) felt that a "patch" implied an unacceptably small representation of "nature", and that it should be seen as an exemplar, or the rule rather than the exception. "A greater percentage of it than there is now. A decrease in land available to farming . . ." "An example of how the rest of the landscape could be." "These areas all linked together only interrupted by the occasional paddock block."

In terms of the management of patches of nature, some respondents (6) believed that the patches should be left alone altogether by humans, but more people (10 respondents) identified the need for a management plan at a district scale to ensure continued protection.

A few people commented explicitly on the use value of the patch of bush to humans: one person emphasized the productivity aspect of integration by specifying that stock access to bushland was important; others (3 respondents) indicated recreational and aesthetic use.

The large majority of people who provided examples of particular species to illustrate the plants and animals of their patch of nature were women (10 out of 12).

Experience of nature

Many people become interested in nature or involved in nature conservation as the result of one or more transformative experiences. Respondents were asked whether they had ever had an experience that suddenly transformed how they felt about nature or their views on nature conservation. Of 72 respondents, nearly half (49%) reported having had such transformative experiences. Many provided details of their experiences, which fell into several categories: education/awareness; witnessing the decline of natural forests; enjoying the pleasures of nature observation and nature care; and wildlife; observing specific acts of human destructiveness (such as "the day the bulldozers rolled [through the wetland]"); observing land degradation; and witnessing destructive "natural" events. The most powerful transformative experience was the observation of, and care for, nature.

Quotes revealed a spiritual and emotional delight in the miracles of nature as well as a more intellectual appreciation. In terms of education and awareness, people were transformed through reading, formal education, television, personal contact and family values. Concern over environmental degradation through human activity, whether direct or indirect also affected people forcefully.

RELATIONSHIP WITH NATURE

People find different values in nature, some negative and some positive. Respondents were asked what local bushland provides them with. The value respondents most commonly attached to bushland near them, was "beauty" (50 out of 72 respondents). "Improved health of the land" (45 respondents), "inspiration" (38 respondents), enhanced productivity (20 respondents), a spiritual experience (20 respondents) were also important values that people attached to bushland. Negative values such as "a management problem" (18 respondents), "a fire hazard" (15 respondents), "a haven for weeds" (11 respondents), "an expense" (7 respondents), and "reduced productivity" (2 respondents) were less commonly attributed to bushland.

People relate to nature in a variety of ways. They may feel separate from nature, even in opposition to it, or at the other extreme they may feel an integral part of nature. Respondents were asked how saw their relationship to nature. The most common relationship with nature experienced among the 72 respondents was being one creature in nature's web (22 respondents). This position corresponds to an ecocentric environmental ethic. By contrast, no respondents saw themselves as an "opponent of nature in a difficult struggle" and only 2 saw themselves as "using nature's goods and services". This position corresponds to an utilitarian anthropocentric environmental ethic. An ethic somewhere between the two poles of ecocentrism and utiliatarian anthropocentrism was suggested by a number of respondents who reported their relationship with nature as "living with nature" (17 respondents), "being a steward of nature for mutual benefit" (19 respondents), or "being a good neighbour of nature" (10 respondents).

There was a correlation between the felt depth of relationship to nature and number of days per annum involved in nature conservation activities ($p = 0.019$). People who reported a deeper relationship with nature were more involved in nature conservation.

In addition, there was a negative correlation between the felt depth of relationship to nature and age ($r = -0.281$, $p = 0.023$). This result suggested that younger people tended to see themselves as more a part of nature than did older people.

FEELINGS ABOUT NATURE

Respondents were asked to consider the piece of bushland closest to their house and to indicate how they felt about it. People's feelings were both positive (49 out of 72 respondents), such as pride, empowerment, gladness, happiness and satisfaction; and negative (33 out of 72 respondents), such as anger, embarrassment, frustration, overwhelmedness and sadness.

People who had positive feelings tended to be involved in naturalist/environment groups ($r = 0.217$, $p = 0.009$), bushwalking ($r = 0.265$, $p = 0.031$), lobbying government ($r = 0.290$, $p = 0.018$), running courses ($r = 0.272$, $p = 0.025$), and education ($r = 0.320$, $p = 0.011$).

People who had negative feelings tended to be involved in discussions with friends and colleagues ($r = 0.405$, $p = 0.001$), on-farm monitoring of soil and water ($r = 0.328$, $p = 0.008$), and on-farm monitoring of bush and wildlife ($r = 0.435$, $p = 0.000$).

Strategies for communicating conservation

In the structured discussions with the rural groups, participants raised issues that were surprisingly consistent across groups. The main points relevant to this chapter are listed below.

RURAL COMMUNITIES SHOULD BE MADE AWARE OF NATURE CONSERVATION ISSUES

An improved image of landcare and conservation should be communicated to local farmers. Many rural people are not aware of the benefits that a healthy piece of vegetation can offer. This is especially true of first generation farmers who may have come from somewhere else. Awareness of the bush and the value they place on bush needs to be increased. A landcare ethic, an appreciation of natural heritage, a sense of belonging, and a sense of place also need to be fostered in rural communities.

Local shires need to be more involved. Shire officers and councillors need education about conservation. Shires should also be resourced to employ a dedicated landcare/conservation officer. In many cases in Western Australia this is already happening with the employment of Landcare Co-ordinators partly funded by shires and partly by National Heritage Trust funds.

Local schools should be more involved. They should be resourced to teach and undertake nature conservation practices, and to disseminate nature conservation information. Schools could, for example, produce nature conservation newsletters.

RURAL PEOPLE THEMSELVES SHOULD BE USED TO DISSEMINATE CONSERVATION UNDERSTANDING AND PRACTICE

Farmers enjoy interacting with each other and learn much this way. They like to talk, and to compare notes and experiences. They are practical people who enjoy field days, practical demonstrations and trials.

The dissemination of information coming from grass roots should be increased. Local residents who are forward thinking, who gain results, and who will encourage others by example are ideal disseminators.

INFORMATION AND HELP IS IDEALLY EXCHANGED IN A PERSONAL ATMOSPHERE WITH SPECIALIST PERSONNEL AND RURAL GROUPS WORKING TOGETHER

Farmers enjoy interacting with specialists in small groups out in the field, and learn much this way. Specialists are needed to help with fauna and flora identification, salination control, wind erosion control, and establishing conservation trials.

INFORMATION, UNDERSTANDING AND HELP SHOULD TO BE LOCALLY RELEVANT AND LOCALLY APPLIED

Many nature conservation issues are so site specific that farmers need information that relates to their immediate locale rather than to their region.

Information needs to be in ordinary language and easily accessible. Information in books, workshops and seminars should not be too scientific, technical or heavy, nor should farmers have to go to far to attend seminars.

DISCUSSION

This chapter points clearly to some facts which may be surprising to some urban environmentalists and conservation biologists. First, the culture of nature conservation in some rural groups is quite sophisticated. In many cases it is underpinned by a strongly ecocentric ethic and a sense of stewardship or duty of care for the land. Second, this culture is often complemented by good levels of ecological and taxonomic knowledge. Awareness is high of issues such as self regeneration, sustainability, ecological webs, weeds, and of the importance of species diversity and presence of understorey vegetation. The conservation biology concept of corridors is still alive and well in many places, even though ecologists themselves may have moved away from some of these concepts. In many cases knowledge of wildflowers and vegetation is extensive. Third, members of rural groups trust information most that comes from other locals; but they also highly value information from specialists provided in an informal, field-day environment. These findings suggest that the uptake of nature conservation ideas and practices in rural areas may be enhanced by a greater acceptance of their culture, knowledge and information needs on the part of conservation biologists.

Although there are many differences between science and technology for agricultural extension purposes and for nature conservation purposes, there are many similarities. Conservation biologists should be aware that whereas they may have many hard-won answers from a scientific perspective, convincing rural people to take up their suggestions may be even harder.

The history of approaches to rural agricultural change throughout the world have shown several patterns. First, where a "transfer of technology" or linear extension approach is taken, i.e., innovation is unidirectionally diffused from scientists to farmers, the uptake of new approaches is likely to be low initially and then declines quickly after the scientists leave or the programme finishes (Pretty 1995). Second, where financial subsidies or other incentives are offered, initial uptake can be very high but declines quickly when the incentives are withdrawn (Pretty 1995). Third, where farmers' knowledge is incorporated into the design of the approach and where they are encouraged to adapt and modify this on an ongoing basis according to their own experience, then there may still be an initial decline in uptake after the programme finishes but uptake will swing up again as modifications make the approach more locally workable (Pretty 1995).

The approach to the diffusion of nature conservation understanding and action is still in many quarters similar to the old agricultural extension model of transfer of technology. There is still an assumption of "scientist knows best" within the scientific profession. Certainly scientists do know a good deal and many rural people rightly value this specialist knowledge. Indeed, many farmers are keen to have scientists provide them with unambiguous answers about conservation and revegetation so they can act on the advice to achieve efficient and effective conservation outcomes. However, professional scientists' knowledge is not always practical enough nor locally relevant to farmers' nature conservation needs. "Information about the whole region is not useful to us; we need information about our local area!" is a typical plea from farmers. Other farmers may have particular conservation interests such as favourite wildflower species or animals which they would like to protect. These farmers may be less amenable to being told the correct approach to conservation, especially given that scientists have been known to change their minds about what needs to be done, anyway.

A model for interactive learning for conservation biologists and farmers is needed that is more effective than the conventional model of nature conservation extension or transfer of technology. The results of this current project suggest a model which helps to convey professional scientists' knowledge in an efficient, effective, geographically relevant and socio-culturally appropriate manner, in response to local concerns. In the first phase of the proposed model, field-based workshops would be run which would each involve 1–3 scientists and a rural conservation group, with a typical size of 5–15. In this forum, interactive learning can occur: information can be exchanged, goals compared, and needs expressed. I suggest here that a useful approach on the part of scientists is to start where the enthusiasm and energy of the rural conservation group lie (for example, a passion for a particular species of orchid). Once a process of working together begins, relationships develop, and trust is established, other scientific ideas can be introduced to build on the initial work. This recognition of the knowledge and socio-cultural needs of rural people is a part of the collaborative approach of community science. This phase clearly has substantial resource implications and would require resources to be redirected from other efforts.

In the second phase of the proposed model, rural groups would interact with their local schools, shires, other community groups and individual farmers to disseminate conservation information and promote action. Our findings show that rural groups see the need for greater education of their own communities about nature conservation and believe that this is best done by farmers or other local residents. Again, the willingness of rural groups to undertake this educational role depends on resources from funding agencies being allocated to the task.

Certainly, the elements of this model are not new; indeed the "Nature Conservation: Role of Networks" conference (Saunders *et al.* 1995) highlighted some excellent community science by conservation groups some of which involved collaboration with professional scientists. I am suggesting here, however, that the culture of rural conservation groups is such that a systematic collaborative approach of working with groups, similar to the Landcare program (Roberts 1992) could be introduced for fostering nature conservation.

ACKNOWLEDGEMENTS

I would like to acknowledge the enthusiasm and extensive knowledge base of the rural group members who participated in this project. I also thank members of the project team, including Fionnuala Frost, Peter Cale, Colin Yates, Patrick Piggott, Joanna Seabrook and Wendy Dymond for their involvement in the planning and presentation of various aspects of the workshops.

REFERENCES

Carr, A., 1997. Innovation of diffusion: landcare and information exchange. Pp. 201–16 *in* Critical Landcare ed by S. Lockie and F. Vanclay. Centre for Rural Social Research: Charles Sturt University, Wagga Wagga, New South Wales.

Chamala, S. and Mortiss, P. D., 1990. Working Together for Landcare. Australian Academic Press: Brisbane, Queensland.

Chambers, R., Pacey A. and Thrupp, L. A., 1989. Farmer First. Intermediate Technology Publications: London.

Curtis, A. and De Lacy, T., 1994. Landcare: Does it Make a Difference? Centre of Parks, Recreation and Heritage: Charles Sturt University, Albury, New South Wales.

Lockie, S. and Vanclay, F., 1997. Critical landcare. Pp. 1–8 *in* Critical Landcare ed by S. Lockie and F. Vanclay. Centre for Rural Social Research: Charles Sturt University, Wagga Wagga, New South Wales.

Pretty, J. N., 1995. Regenerating agriculture: policies and practices for sustainability and self-reliance. Earthscan Publications: London.

Pretty J. and Chambers, R., 1994. Towards a learning paradigm: new professionalism and institutions for agriculture. Pp. 182–202 *in* Beyond Farmer First ed by I. Scoones and J. Thompson. Intermediate Technology Publications: London.

Roberts, B., 1992. Landcare manual. New South Wales University Press: Kensington, New South Wales.

Saunders, D. A., Craig, J. L. and Mattiske, E. M. (eds), 1995. Nature Conservation 4: The Role of Networks. Surrey Beatty & Sons: Chipping Norton, New South Wales.

Scoones, I. and Thomson, J., 1994. Knowledge, power and agriculture — towards a theoretical understanding. Pp. 16–31 *in* Beyond Farmer First ed by I. Scoones and J. Thompson. Intermediate Technology Publications: London.

Stocker, L. J., 1995. Community science and community scientists: their role in conservation. Pp. 548–55 *in* Nature Conservation 4: The Role of Networks ed by D. A. Saunders, J. L. Craig and E. M. Mattiske. Surrey Beatty & Sons: Chipping Norton, New South Wales.

CHAPTER 55

Decision support for land restoration, rehabilitation and soil conservation

Paul G. Luckman,[1] Colin D. Meurk, [1] Ramana R. Chamarti[1] and Mark C. Smale[1]

Local governments and communities in New Zealand are increasingly seeking to achieve a range of ambitious environmental management goals by manipulating vegetation cover. However, the knowledge-base to support these efforts is highly fragmented. The *Restoration and Rehabilitation PlantFinder* represents the first step towards computerized support for vegetation management decisions with environmental improvement objectives. It assists users to select plants for ecological restoration and rehabilitation or soil conservation projects, based on a match between site or user requirements and plant characteristics, and on a reconstruction of the original indigenous vegetation cover. The *PlantFinder* is complemented by the *Soil Conservation Planner* (under development), which aims to support vegetation management decisions in erodible hill country, and the *Land Management Simulator*, a tool for predicting the on- and off-site environmental effects of progressive changes in land use and vegetation cover. Together, these tools will provide a robust foundation upon which to design more sustainable land management mosaics at farm, catchment and regional scales.

INTRODUCTION

MANAGEMENT of vegetation is critical to achieving an array of socially desirable on- and off-site environmental and economic outcomes. Many of these outcomes are now appearing as the goals of a new generation of management plans being prepared under New Zealand's *Resource Management Act (1991)*. While these goals continue to relate to soil conservation, local ecological restoration and land rehabilitation, there is emerging interest among land managers, policy-makers, scientists and community groups in co-ordinating vegetation management to achieve landscape or catchment-level goals. Consequently, vegetation management decisions are becoming increasingly viewed in terms of land cover mosaics, landscape ecology, catchment ecosystem health, and community well-being (Lee *et al.* 1992; Mallenson 1993; Naiman *et al.* 1993; Lineham *et al.* 1995; Ndubisi *et al.* 1995).

In turn, this interest is producing a growing demand for information on how vegetation can be manipulated in a variety of local environments to achieve a range of ambitious landscape-level goals. However, precedents are limited for many of these projects. This can be because site conditions are demanding or uncertain, or because there is little documented experience with the full range of plant species concerned. The resulting mismatch between community interest and commitment, on the one hand, and lack of accessible information on the other, is especially pronounced for vegetation strategies involving indigenous plants. Consequent poor planning can lead to failure of the restoration strategy, failure to realize project goals, wasted resources, and reduced commitment to environmental improvement (Atkinson 1994).

Information required to support environmental improvement projects varies from advice on appropriate species and establishment and maintenance requirements, to measures of economic benefits and environmental effects on plant or animal biodiversity, soil health, erosion and sediment delivery, water resources, water quality or aquatic habitat. In this chapter we first describe a decision support system, the *Restoration and Rehabilitation PlantFinder*, to aid selection of plants for a variety of soil conservation, land restoration and rehabilitation projects. We then discuss a modelling

[1]Landcare Research, Private Bag 11 052, Palmerston North, New Zealand.

Pages 542–50 *in* NATURE CONSERVATION 5: NATURE CONSERVATION IN PRODUCTION ENVIRONMENTS: MANAGING THE MATRIX ed by J. L. Craig, N. Mitchell and D. A. Saunders. Surrey Beatty & Sons, 2000.

tool designed to help communities set goals for catchments or landscapes (*Land Management Simulator*), and finally describe progress towards a system to help farmers, land management advisors and agricultural consultants plan vegetation change projects in erodible hill country (the *Soil Conservation Planner*).

SELECTING SPECIES: *PLANTFINDER*

Context

Land managers aiming to rehabilitate, stabilize or restore vegetation in disturbed or biologically depleted sites have relied on a variety of publications (e.g., Pollock 1986; Van Kraayenoord and Hathaway 1986a,b; Metcalfe 1991; Porteous 1993; Lucas *et al.* 1995–97) for information on species suitability to sites and objectives. However, many of these publications have been oriented toward use of exotic species and become outdated when new varieties bred for disease resistance, greater vigour, or timber potential become available, or when species become recognized as having weed potential. In addition, community interest in using indigenous plants to improve a depleted biodiversity is revealing the need for a more comprehensive information source on these plants. Finally, there has been interest among resource management agencies in moving beyond publications as an information transfer mechanism, and utilizing some of the newer computer technologies to provide greater and more effective access to the expertise needed to ensure community environmental projects achieve success. One possibility is to link plant lists to geographic information systems to help select plants for particular sites, employing available spatial information on climate, geomorphology, geology, soils, aspect, and pre-existing indigenous vegetation. Together, these factors have led to development of a PC-based decision support system, the *Restoration and Rehabilitation PlantFinder*, as a nucleus for a range of information products to meet the needs of land managers.

Design

The *PlantFinder* functions as a selection tool (Fig. 1). It helps users construct a list of plant species which meet user-specified criteria. Separate modules address exotic and indigenous species. Given that end-users are usually sceptical of recipes, the plant list is generated interactively by adding or deleting selection criteria. The user specifies conditions in order of importance, progressively eliminating unsuitable species. Other attributes of the selected plants are then browsed to determine their suitability for the application. Experiments with managed dialogues (in which the user is encouraged to specify requirements through a sequence of questions) commonly resulted in over-specification of requirements and empty plant lists, while the sequencing of questions required assumptions about user priorities. Although they are interactive, managed dialogues give the user less control, and are best employed as an accessory technique. We concluded they would be effective if coupled with fuzzy (probabilistic) rather than sharp selection criteria.

Attributes available to assist plant selection are shown in Figure 2 (exotic species) and Figure 3 (indigenous species). While the user must specify requirements in terms of these attributes, future links to geographic information systems will allow some of this information to be extracted automatically, using on-screen graphics and maps. There are, however, two fundamentally different approaches to plant selection within *PlantFinder*.

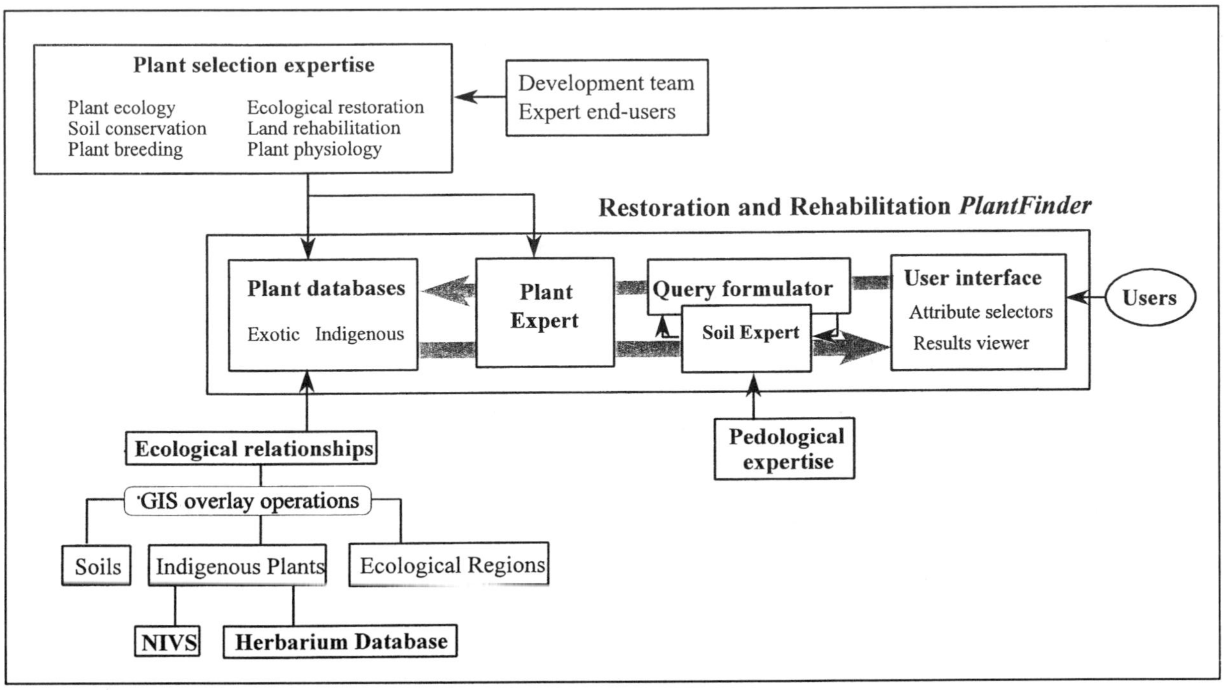

Fig. 1. Knowledge sources and design of the *Restoration and Rehabilitation PlantFinder*.

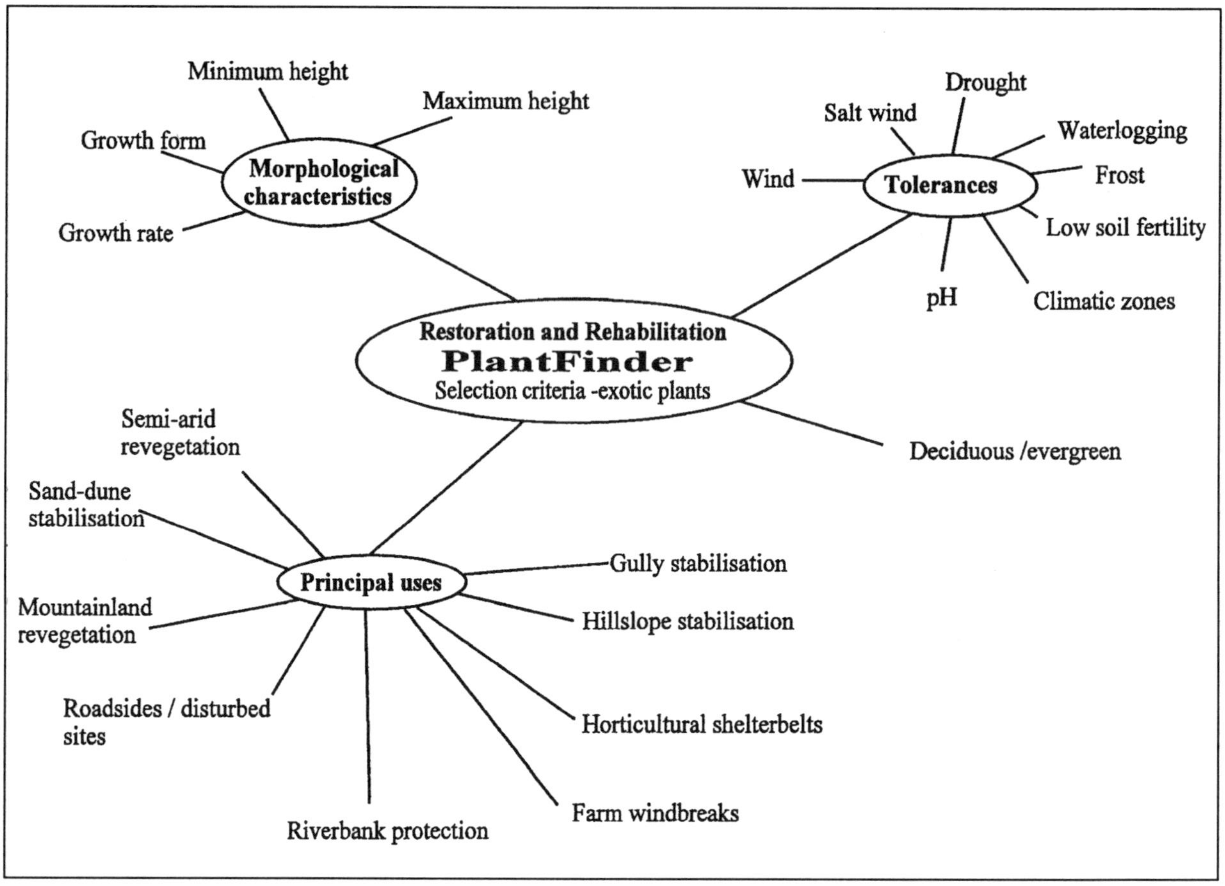

Fig. 2. Attributes for selecting exotic plants.

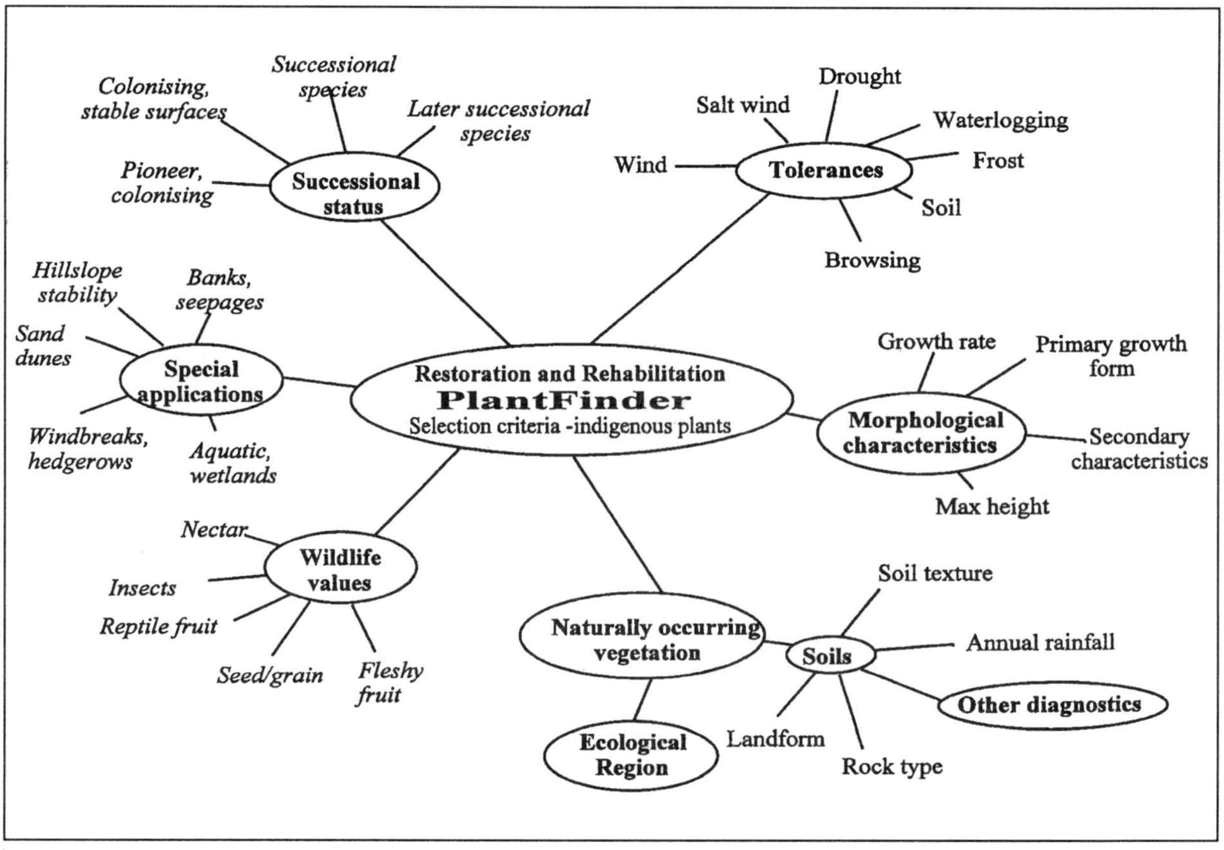

Fig. 3. Attributes for selecting indigenous plants.

The first approach is physiologically- and functionally-based, and is especially suited to exotic species, but is also used with indigenous species. Plants are selected if they can grow satisfactorily in given site conditions, can serve particular purposes, and perhaps also meet certain morphological requirements. This approach to plant species selection requires the user to have a knowledge of attributes of the plant, site, or application.

The second approach, appropriate only for indigenous plants, aims to determine the naturally occurring plant species of a site by specifying the location, in terms of the New Zealand Department of Conservation's Ecological Regions (McEwen 1987), and the local soil. These two attributes provide a simple but powerful means for determining the structural elements of the indigenous vegetation which preceded that modified by Maori and European land use activities. Ecological Regions have been defined as regions with distinctive ecological integrity in terms of climate, geology, soils, vegetation and land use. As New Zealand's plant communities and soils have evolved together, the soil encapsulates many of the environmental factors which also determined the composition of the indigenous plant communities.

A third approach also has potential. This is the "case-based" approach (Hammond 1986; Rich and Knight 1991), in which a vegetation management strategy is developed by importing and adapting a strategy from a "similar" context. Implementing the approach requires a database of case studies and robust measures of site similarity. It is applicable when strong local precedents can be used to compensate for uncertainty about the physiological or functional suitability of plant species.

Knowledge is inevitably incomplete and uncertain. In *PlantFinder* this is recognized first in the definition of the attributes, which may be qualitative or specified in broad ranges. Second, plant attributes may be multi-valued, reflecting either known variability or uncertainty. A plant with frost tolerance specified as "medium-high" (defined in terms of minimum temperatures tolerated) will be conservatively passed over if the user requires high frost tolerance, but will be included in the selection list if the user relaxes the condition to medium tolerance. Uncertainty about the tolerance can be taken into account when the user browses attributes of selected plants. For indigenous plants, the importance of this issue has been reduced by providing for plant selection through analysis of extant or prior natural vegetation. However, this approach requires knowledge of the soil at the site, and therefore is also subject to uncertainty, since even very detailed maps cannot capture the microtopographic variability and sharp gradients within some soil-landscape patterns. As an additional mechanism to improve accuracy or defensibility of *PlantFinder* conclusions, the databases of plant attributes can be browsed, edited and extended by expert users to reflect local knowledge, as well as to add plants to the database.

Technical systems such as *PlantFinder* inevitably bring the user into contact with unfamiliar subject areas. In *PlantFinder* this has been addressed through textual help files providing attribute definitions and, in the case of soils, an embedded expert system to aid soil identification. Soils are characterized in terms of the Soil Order (Hewitt 1992), which is the broadest soil unit recognized and can be inferred from landform, parent rock type, soil texture, annual rainfall and a small additional set of diagnostic characteristics. The expert system interprets the information which the user supplies about these attributes, to reduce the number of possible Soil Orders from the maximum of 15 to a few or, in the case of a positive identification, just one. This process relies on a set of rules which either positively identify a Soil Order, or exclude one or more from consideration. The short list of Soil Orders is then used by the *PlantFinder* as one of the constraints. The user can override the results of the *PlantFinder*'s soil identification process after accessing and browsing descriptive paragraphs on each of the Soil Orders.

Knowledge sources and errors

Tolerances and suitabilities of 197 introduced plant species have been obtained primarily from Van Kraayenoord and Hathaway (1986a), who based assessments on the results of field trials, climate laboratory studies, and direct experience. This information base has been supplemented to a limited extent by data on recent plant releases (A. G. Wilkinson, pers. comm.). The expense of field trials has ensured that heavy reliance is placed on a synthesis of experience and experimentation.

Information on indigenous plants has been obtained from three sources. First, the initial database of plant species tolerances and suitabilities was derived from data for 80 native species (trees, shrubs, grasses and herbs) compiled by Pollock (1986). Second, this database has been expanded to cover 320 indigenous plant species considered to have practical value in restoration projects, with plant tolerances and suitabilities evaluated by plant ecologists. Third, distribution of species in relation to Ecological Regions and Soil Orders has been obtained initially by analysis of data held in national databases. Indigenous plant databases employed are National Indigenous Vegetation Survey and the Landcare Research Herbarium database.

The National Indigenous Vegetation Survey database contains information on plant species composition for more than 50 000 plots in lowland and high country areas (Fig. 4). The Herbarium database contains over 75 000 grid-referenced records of occurrences of some 2 000 indigenous plant species. For the 320 plant species of interest, we derived species distribution maps from both the National Indigenous Vegetation Survey and Herbarium databases. Inaccuracies in recording grid references for some records resulted in some loss of data in this process. These distributions were then

Fig. 4. Extent of coverage of the National Indigenous Vegetation Survey indigenous plants database.

digitally overlaid onto maps of Ecological Regions and Soil Orders, the latter having been generated by interpretation of the soils code in the New Zealand Land Resources Inventory (Fletcher 1988). This process of matching distributions and geographical attributes resulted in lists of Ecological Regions and Soil Orders for each of the 320 indigenous species. Errors in these lists are considered to arise principally from nonuniform sampling of plant distributions. Other sources of error include incomplete grid references, and errors in predicting Soil Order from maps intended for use at 1:50 000 to 1:250 000. Consequently, the lists have been edited by plant ecologists. While this has reduced errors, the accuracy of species distribution data will be further improved by research aimed at better defining the species composition of vegetation classes mapped in surveys such as the New Zealand Land Resources Inventory (E. R. Nicol, pers. comm.) and improving understanding of the environmental factors governing indigenous plant species distribution (Leathwick *et al.* 1996).

Knowledge for the soil identification expert system was elicited directly from soil scientists, focusing first on positive identification rules. Individual attributes were then interpreted separately in a series of rules which excluded plants from selection. These rules can then be applied in combination with the positive identification rules. The soil identification expert system has been validated by soil scientists "mentally touring" through New Zealand, generating test combinations of diagnostic criteria and known soils. While the knowledge-base has been tuned to provide accurate identification for these sites, identification errors will occur, and consequently it is helpful for the user to browse textual information about the soils to check the identification.

Because *PlantFinder* is a general tool with national application, soil identification is not approached in terms of landscape models. While such an approach is potentially more powerful and easier to use, the many landscape models which would be required are not available. Landscape models could be included in regional enhancements of the software.

It will always be possible to improve the accuracy of the knowledge contained within the *PlantFinder*. While it provides assistance in selecting plants for land management and restoration projects, the final choice must be made only after scrutiny of all the information available on the candidate species. The software package will assist preparation of specifications, but requires the user to have access to some biological, resource management or horticultural expertise.

MANAGING THE MATRIX: THE *LAND MANAGEMENT SIMULATOR*

Development of policies and plans for vegetation management requires a means for assessing the spatially and temporally cumulative effects of vegetation management decisions. Numerous modelling efforts in New Zealand and overseas aim to provide improved predictions of environmental and economic effects of changing land management. However, there have been few attempts in New Zealand (e.g., Lee 1996) to model the *process* of achieving environmental goals and objectives, which involves modelling spatial land management decisions as well as predictions of their effects as they unfold through time. This allows estimation of the timing of desirable or undesirable outcomes, which can be critical to effective goal setting.

In conjunction with catchment-based studies of the effects of land management on rates of erosion and sedimentation, a workstation-based spatio-temporal modelling tool, the *Land Management Simulator* has been constructed (Luckman *et al.* 1997; Fig. 5). A core feature of *Land Management Simulator* is a simulation of land use and land cover patterns through time and prediction of their consequent effects. *Land Management Simulator* has been applied to analysis of lake sedimentation risk, and is currently being used to investigate the impact of changes in land management on erosion rates and sediment delivery from the Waipaoa catchment, in the East Coast of New Zealand.

Changes in land use are modelled using user-defined land use transition "rules". These rules specify when and where land use change will occur, and the nature of the land use transition. The land use change conditions in these rules explicitly take into account terrain class, position in the

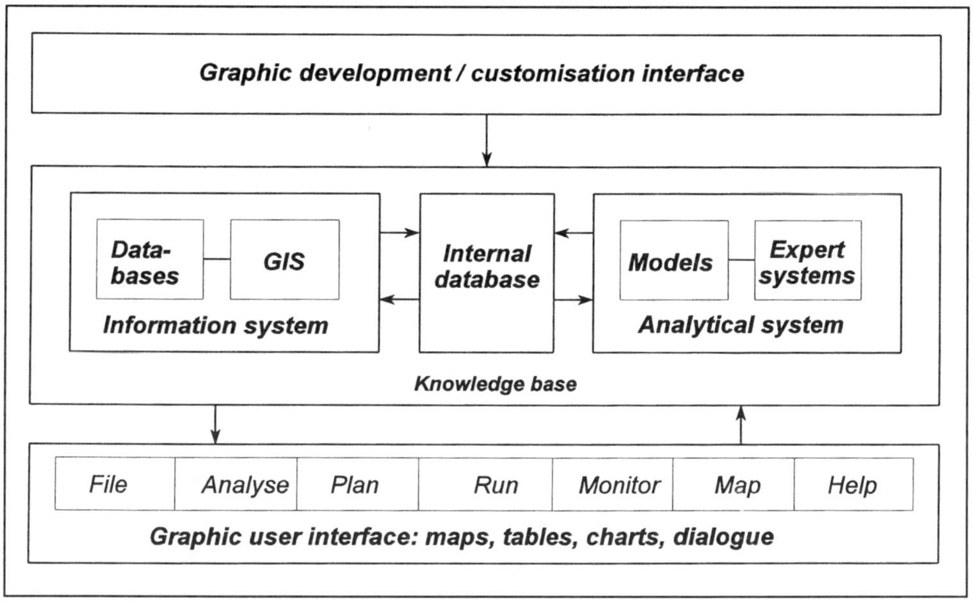

Fig. 5. Architecture of the *Land Management Simulator*.

catchment, ownership, and existing land use. Factors which motivate the land use change, such as the state of neighbouring properties, economic indicators, or whether there has recently been a major erosion event, are therefore treated implicitly. Rules which consider these underlying factors can be created using the graphic expert system development interface directly. Collectively, groups of land use transition rules specify land use scenarios.

Temporal changes in the nature of land cover, which occur as a result of growth and succession of vegetation, are modelled using the concept of vegetation phase. Land not in productive use follows different succession patterns (sequences of phases) under different conditions. A specified phase sequence is associated with each area which is removed from productive use. These phase sequences can be determined empirically, or through use of simulation models. Recent developments in simulating regenerating native vegetation and landscape interactions suggest the possibility of explicitly taking account of factors such as nearby seed sources, availability of vectors, and wind directions in these succession models (Meurk *et al.* 1997).

Land Management Simulator has been used to determine how land use changes (retirement, revegetation, conversion of pasture to commercial forestry) within the Tutira catchment in northern Hawke's Bay affect the risk of sedimentation within Lake Tutira (Luckman *et al.,* in review). Risk is computed as the probability that more than a given depth of sediment is deposited in the lake as a result of a storm of given magnitude. This probability is obtained by adjusting the (empirically-determined) probability for the current vegetation regime, to reflect likely effects of changes in land management on erosion and sedimentation processes. Running the model for 20–50 years enables the user to determine when sedimentation risk becomes acceptable, thus providing an idea of the timeframe required to achieve goals under different management regimes (Fig. 6).

PROJECT PLANNING: THE *SOIL CONSERVATION PLANNER*

While the *PlantFinder* and *Land Management Simulator* systems will continue to evolve, our next step in providing decision support for land management and ecological restoration applications is to develop a tool to aid project planning. The *Soil Conservation Planner* will support vegetation management decisions in New Zealand's North Island hill country. Work to date on the *Planner* has focused on analysis and prioritization of needs associated with this planning process. The users are those farmers in erodible hill country who want to change their vegetation management, and are seeking advice from regional council land management advisors or agricultural consultants. We assume that the farmer's goals have yet to crystallize, so a major task is to help determine what can be achieved through the change in management, recognizing the limitations of an analysis which focuses mostly on sites or paddocks.

Planning is an iterative process in which candidate plans are hypothesized, tested and refined. The process involves detailed consideration of site factors and high level testing against wider goals and constraints, including aesthetics. Our model of the process involves (i) rapid development of a *concept plan* for addressing vegetation management issues at selected parts of the farm, (ii) revising and refining this plan by considering the details at each site, leading finally to (iii) commitment to an *action plan*, which focuses especially on the first five years. Unlike whole farm planning, which can be approached as an economic allocation and optimization process, the process for decisions involving vegetation management in erodible hill

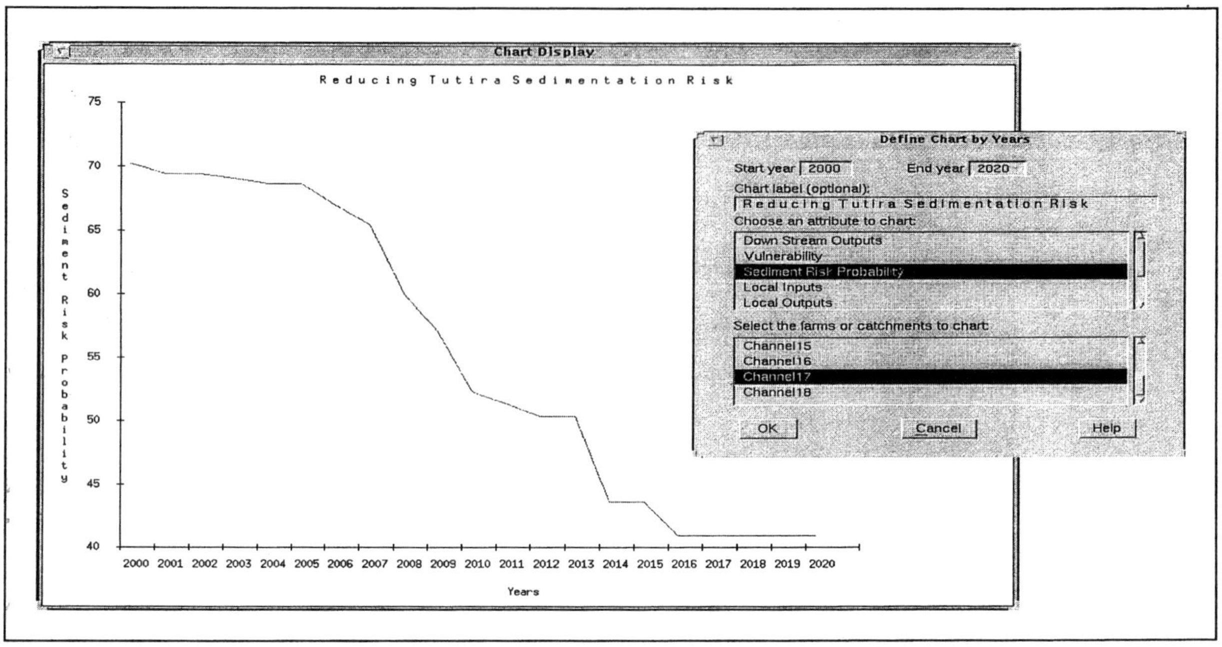

Fig. 6. Land Management Simulator chart showing progressive reduction in sedimentation risk resulting from land use changes in the Tutira catchment, Hawkes Bay, New Zealand. Risk is measured for this example as the probability that more than 10 mm of sediment accumulates in the lake as a result of a 350 mm rainstorm.

country appears most commonly to involve iterative refinement of plans for collections of sites. We distinguish between simple sites and complex sites. A simple site consists of one land class or landform, and would be expected to involve a single vegetation management option. Complex sites, which could be paddocks, can involve entire hillsides or small catchments, and consequently vegetation management should vary spatially within these to reflect variability in site conditions.

Desirable features of an ideal planning aid which emerged from a survey of present planning processes are illustrated in a non-prioritized manner in Figure 7. These features collectively include information on site conditions, options, models for biophysical and financial analysis and prediction, information on case studies, spatial site planning aids, and a variety of diagnostic tools to provide an expert overview of the plan to check for farm impacts, consistency with regional land management policies, and weed and pest issues. The development process for the *Soil Conservation Planner* will progressively deliver many of these capabilities, focusing first on components which assist clarification of site conditions.

CONCLUSION

Computer based decision support systems have an important role to play in achieving more sustainable futures and land cover mosaics, particularly through adding rigour to the planning process. This added rigour results from use of a systematic process for constructing and comparing alternatives, access to information and models to ensure reliable predictions, and inclusion of mechanisms to ensure that important factors are not omitted from consideration.

The role of these decision support systems is most obvious at the site level, in products such as the *PlantFinder* and *Soil Conservation Planner*. For site level decisions, the focus is on questions relating to how objectives are to be achieved, a relatively small number of people are involved in the decision, and decisions are made more frequently. Site level decisions, however, are made in the context of strategies and plans developed at higher levels. While these strategies and plans emerge from more complex decision processes, there is a need for systems such as the *Land Management Simulator* which can show what environmental improvements can be achieved through changes in land cover, as well as how risk of further degradation can be reduced. The systems will also reveal when measurable improvements will occur and, at a broad level, the changes in land cover required to achieve them. Used in conjunction with planning exercises such as visioning, these tools can lead to clearly defined and achievable goals, such as specific changes to sedimentation risk probabilities, which are important to maintaining commitment and cohesion of the communities who work to achieve them.

Significant progress is being made towards an integrated collection of tools, designed specifically for New Zealand conditions and issues, to support decisions at different scales. While the structures and knowledge requirements for these tools are now emerging, opportunities exist to exploit existing spatial databases further, develop regional knowledge-bases, and integrate improved models of terrestrial ecosystems into landscape scale tools. There is also a need for improved methods for handling uncertainty, so that risks associated with decisions based on incomplete information

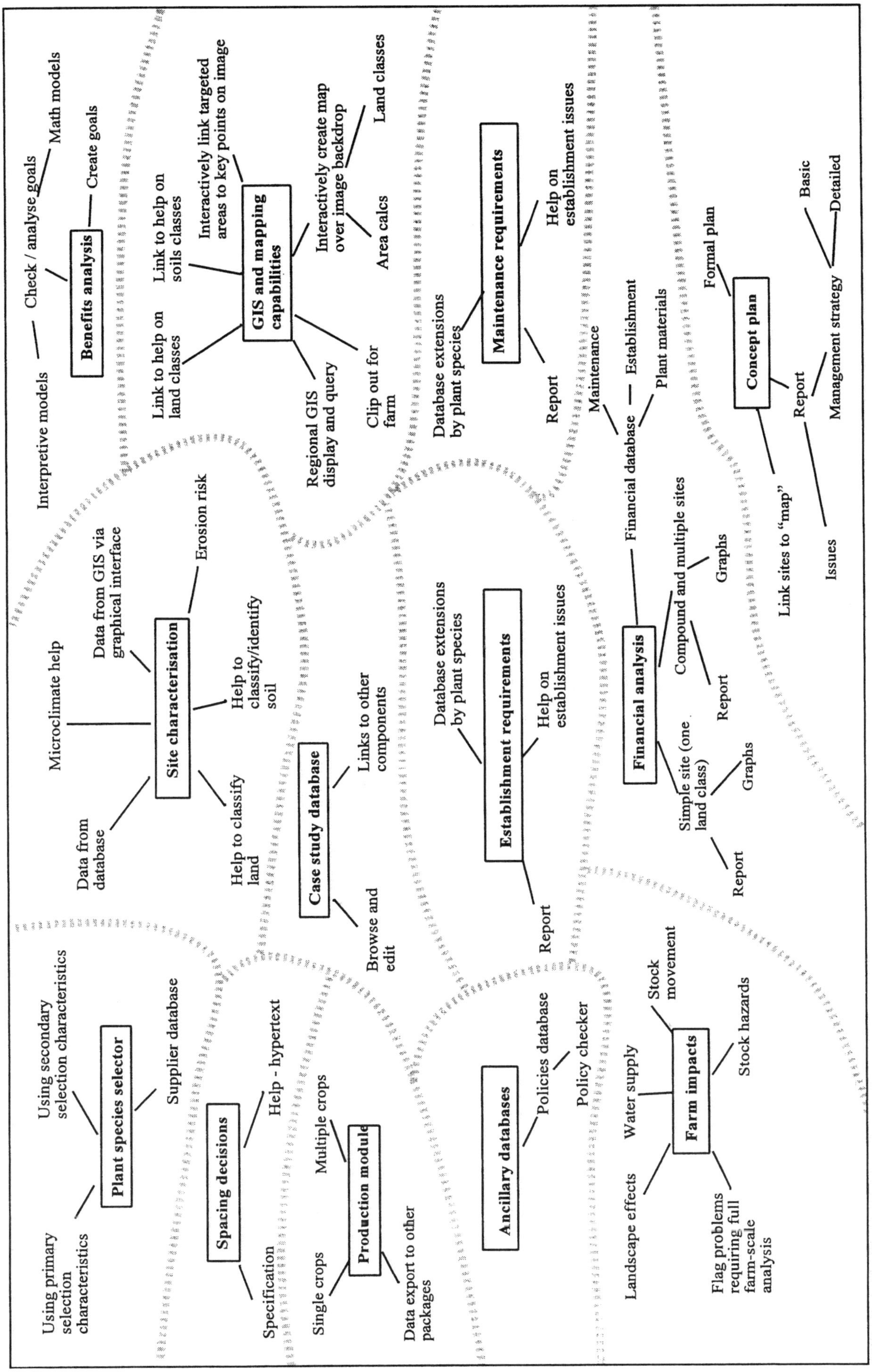

Fig. 7. Soil Conservation Planner features suggested by analysis of planning processes.

are understood. Central to future development of knowledge bases, and to robust methods for dealing with uncertainty, will be to develop the knowledge bases in conjunction with communities actively engaged in redesigning their own land cover mosaics to achieve goals they have set themselves.

ACKNOWLEDGEMENTS

Funding for this research has been provided by the Foundation for Research Science and Technology under contracts C09622, C09641, and C09612, and through technology transfer grants from the Regional Council Plant Materials Collective, in 1996–97, and 1997–98. GIS coverages for the spatial analysis were prepared by Graeme Hall, Euan Nicol and Robert Gibb. Soil expertise was elicited from Allan Hewitt and Hugh Wilde, who also evaluated the soil expert system. Figure 5 is an interpretation of data on user needs collected by Terry Parminter and Annie Perkins (of AgResearch, Ruakura), and Roger Wilkinson. Additional programming support was provided by Rhys Gibson and Chris Scott. Helpful comments on this chapter were provided by Robert Gibb, Leslie Reid, Robyn Simcock, Peter Stephens and Hugh Wilde.

REFERENCES

Atkinson, I. A. E., 1994. Guidelines to development and monitoring of ecological restoration programmes. *New Zealand Department of Conservation Technical Series* No. 7.

Fletcher, J. R., 1988. The New Zealand Land Resource Inventory — a national evaluation of land resources. *Soil Surv. Land Eval.* **8:** 129–37.

Hammond, K., 1986. *CHEF*: A model of case-based planning. *Proceedings AAAI-86* (American Association of Artificial Intelligence).

Hewitt, A. E., 1992. New Zealand Soil Classification. *DSIR Land Resources Scientific Report* No. 19. 133 Pp.

Leathwick, J. R., Whitehead, D. and McLeod, M., 1996. Predicting changes in the composition of New Zealand's indigenous forests in response to global warming: a modelling approach. *Environ. Software* **11:** 81–90.

Lee, A., 1996. Simulating impacts of regional land use policy on spatial dynamics of land use change in New Zealand. M.Sc. thesis. Department of Geography, University of Auckland, New Zealand.

Lee, R. G., Flamm, R., Turner, M., Bledsoe, C., Chandler, P., DeFerrari, C., Gottfried, R., Naiman, R. J., Schumaker, N. and Wear, D., 1992. Integrating sustainable development and environmental vitality: a landscape ecology approach. Pp. 499–521 *in* Watershed Management — Balancing Sustainability and Environmental Change ed by R. Naiman. Springer Verlag.

Lineham, J., Gross, M. and Finn, J., 1995. Greenway planning: developing a landscape ecological network approach. *Lands. Urb. Plan.* **33:** 179–93.

Lucas, D., Meurk, C. D., Head, J. and Lynn, I., 1995–97. *Indigenous ecosystems of Otautahi Christchurch, Sets 1–4.* Christchurch-Otautahi Agenda 21 Forum, Christchurch, New Zealand.

Luckman, P. G., Gibson, R. D. and Chamarti, R. R., 1997. A hybrid rule-object spatial modelling tool for catchment analysis. Pp. 13–21 *in* Proceedings of 2nd International Conference on GeoComputation (GeoComputation 97/SIRC 97), 26–29 August, 1997, Dunedin. Ed by R. T. Pascoe.

Luckman, P. G., Gibson, R. D., Scott, C. and Trustrum, N. A. (in review). Evaluating environmental risks for catchments undergoing change. *J. Environ. Manage.*

Mallenson, G. P., 1993. Riparian landscapes. Cambridge University Press. 286 Pp.

McEwen, W. M., 1987. Ecological Regions and Districts of New Zealand. Third revised edition in four 1:500 000 maps. *New Zealand Department of Conservation Publication* No. 5.

Metcalfe, L. J., 1991. The Cultivation of New Zealand Trees and Shrubs. Reed: Auckland, New Zealand.

Meurk, C. D., Hall, G. M. J., Burrows, L. E. and Smale, M. C., 1997. Towards a dynamic model of indigenous landscape elements. *Proc. NZ Ecol Soc. Conference.* (Abstract).

Naiman, R. J., Decamps, H. and Pollock, M., 1993. The role of riparian corridors in maintaining regional biodiversity. *Ecol. Appl.* **3:** 209–12.

Ndubisi, F., DeMeo, T. and Ditto, N. D., 1995. Environmentally sensitive areas: a template for developing greenway corridors. *Lands. Urb. Plan.* **33:** 159–77.

Porteous, T., 1993. Native forest restoration. Queen Elizabeth II National Trust: Wellington, New Zealand.

Pollock, K. M., 1986. Plant Materials Handbook for Soil Conservation. Volume 3: Principles and practices. *Water and Soil Miscellaneous Publication* 95. 66 Pp.

Rich, E. and Knight, K., 1991. *Artificial Intelligence.* 2nd edition. McGraw-Hill, Singapore. 620 Pp.

Van Kraayenoord, C. W. S. and Hathaway, R. L., 1986a. Plant Materials Handbook for Soil Conservation. Volume 1: Principles and practices. *Water and Soil Miscellaneous Publication* 93. 296 Pp. Ministry of Works and Development: Wellington, New Zealand.

Van Kraayenoord, C. W. S. and Hathaway, R. L., 1986b. Plant Materials Handbook for Soil Conservation. Volume 2: Introduced plants. *Water and Soil Miscellaneous Publication* 94. 299 Pp.

CHAPTER 56

Extension efforts for rural nature conservation on the New England Tablelands, New South Wales, Australia — a case study

R. Trémont,[1] S. Falconer,[1] R. Morsley[1] and R. Nias[2]

The New England Tablelands of New South Wales, Australia have been developed for sheep and cattle grazing. However, "dieback" of many native trees across the region has been a sequel to pasture improvement. A network of local people, concerned about the tree and native vegetation decline, has generated and disseminated information to help change the behaviour of land managers, so that tree regeneration and nature conservation are recognized and assisted. As part of this network the World Wide Fund for Nature Australia (WWF) has developed a regional plan for nature conservation on the New England Tablelands and investigated ways in which native birds and vegetation can be conserved.

A current WWF project is now encouraging and supporting individual land managers on the New England Tablelands to adopt management practices which include nature conservation. The success of this rural nature conservation project is largely due to close personal interaction between the project's Field Officers and people responsible for day-to-day land management. WWF has enhanced the success of this project by employing local people with good understandings of the social and biophysical processes operating in the region. By communicating effectively and providing continuity these officers have engendered trust among local land managers. The provision of specific but adaptive technical information and advice, relevant on-ground assistance and follow-up support have also helped land managers to actively include nature conservation in their rural enterprises. This chapter confirms that partnerships (not client-provider relationships) and the facilitation of self-directed learning are essential for successful rural nature conservation.

Key words: Rural nature conservation, Rural extension, Extension methods.

INTRODUCTION

DURING the 1970s many of the native trees in the open eucalypt woodlands of the productive "improved" pastures on the New England Tablelands of New South Wales, Australia became sick and died. This highly visible "New England Dieback" was the stimulus for local people to begin to develop an appreciation of native trees on farms and the value of agricultural ecology and rural nature conservation. Researchers, graziers, extension officers, tree planting teams, government agencies, retired citizens, community organizations and school children became interested and active in learning about and addressing rural tree and vegetation decline, as well as other land degradation issues. Today, within two to three decades, this involvement has begun to restore some balance between nature and production within the pastoral landscape and culture of the New England region. The development of a network of people and groups, in and around which information has been generated and communicated, has been essential for achieving change in peoples' attitudes and behaviour. This network is dynamic and the result of many influences (see Curtis *et al.* 1995) but its

[1]World Wide Fund For Nature Australia, P.O. Box 949, Armidale, New South Wales, Australia 2350.
[2]World Wide Fund For Nature Australia, G.P.O. Box 528, Sydney, New South Wales, Australia 2001.

Pages 551–56 *in* NATURE CONSERVATION 5: NATURE CONSERVATION IN PRODUCTION ENVIRONMENTS: MANAGING THE MATRIX ed by J. L. Craig, N. Mitchell and D. A. Saunders. Surrey Beatty & Sons, 2000.

success has depended on the development of sound technical information and the presence of people who communicate this information so that it is widely understood and applied.

The history behind the development of the network of people which grew in response to the rural tree decline on the New England Tablelands and the changes this network has facilitated have been presented by Curtis *et al.* (1995). The network that now exists among land managers, government agencies, researchers, non-government organizations, community groups, retired citizens and interested individuals is strong and likely to be vital to the continuing and active integration of nature conservation and pastoral production in the region. However, there is room to improve the flow of information and level of trust within this network and encourage land managers to take on new ethics and actions. In addition, a challenge of the moment is to mesh the needs of individual farmers, graziers and other land managers with the emerging legal requirements for nature conservation.

In this chapter a small rural nature conservation extension project run by a non-government organization is briefly described. This project has focused on conserving native habitat and rare species, educating and empowering rural landholders to change their approaches to land management, and building partnerships to address native vegetation decline. The main purpose of this chapter is to present the methods and approaches the authors have found most useful for changing the behaviour of local land managers and helping to implement land management practices which include nature conservation.

SUCCESSFUL EXTENSION FOR RURAL NATURE CONSERVATION — AN EXAMPLE

Introducing the WWF Rural Nature Conservation (New England Tablelands) Project

One of the key outcomes of "dieback" research and early tree planting efforts on the New England Tablelands was the realization that the conservation of remnant vegetation and the encouragement of natural regeneration are essential to restoring biological and ecological function and supporting long-term pastoralism.

This outcome provided an opening for the World Wide Fund for Nature Australia (WWF) to contribute to rural nature conservation efforts on the New England Tablelands. The first project to be supported by WWF in this region established basic principles and methods for the development of a regional plan for land management and nature conservation. At its conclusion, this project provided a systematic description of the biophysical attributes of the New England Tablelands and identified areas of high nature conservation value. Recommendations were also made for: establishing land management priorities; retaining multiple use zones of undeveloped native vegetation within areas of more intensive land use; developing a framework for vegetation retention, tree planting and natural regeneration; and demonstrated a method of land assessment and land use planning for nature conservation (Morgan and Terrey 1999).

In 1989, WWF began another regional nature conservation study on the New England Tablelands. This second project investigated some of the conservation needs of native birds and their woodland and forest habitats on the New England Tablelands. A regional plan was developed to conserve areas of native vegetation and the particular bird species and communities that these habitats support. The project identified groups of birds with varying tolerances of habitat fragmentation and/or degradation and recommended vegetation management approaches which could be taken to conserve each group of increasingly vulnerable birds. In addition, land managers likely to be responsible for conserving each bird group and habitat type were identified (Ford *et al.* 1995).

Both these WWF projects contributed significantly to the information resources for rural nature conservation in south-east Australia. In particular, the outcomes of both projects reinforced the message that tree clearing and woodland and forest fragmentation are detrimental to both nature conservation and agricultural and pastoral production.

However, both these WWF projects took a "top down" non-collaborative approach to changing land management methods. WWF quickly identified that there was a need for more effective dissemination of scientific information to the general rural community and to land managers in particular. Land use methods favouring nature conservation needed to be promoted to as wide a group of land managers as possible across the New England Tablelands. WWF also realized that such promotion could only be effective as a result of close personal interaction and trust.

Thus, in March 1993, the WWF Rural Nature Conservation (New England Tablelands) Project began. The initial aim of this third WWF project was to increase awareness of nature conservation issues and provide management support for individual land managers responsible for "core conservation areas" identified for the region in the first WWF project report (Morgan and Terrey 1999).

In late 1997, this third WWF project was still running and was set to continue. During its first two years the project assisted with the development of management options favouring rural nature conservation at about 50 sites on the New England Tablelands, helped protect and manage 635 ha of high conservation value land and located several new populations of rare or significant species. In addition, a joint WWF-New South Wales Farmers' Association-Total Environment Centre-Greening Australia proposal for a New South Wales Government funded fencing subsidy scheme was made and subsequently considered as part of the legislative package for the conservation of native vegetation in New South Wales. The Dumaresq and Uralla Shires' Roadside Management Project on the New England Tablelands was initiated by this WWF project and a plan for managing roadside vegetation and rare plant species was subsequently developed and implemented. Shortly after its implementation this Roadside Management Plan was used as a model for roadside management planning across New South Wales (New South Wales Roadside Environment Committee 1996).

Since 1995 this WWF project has fenced 800 ha of remnant vegetation on private properties and provided management advice about nature conservation to many land managers. A description of a newly discovered plant species known from only one property in the New England region has been prepared and WWF Field Officers have nominated this species as a threatened Australian plant and developed a plan for the plant's long-term protection and conservation. The publication of brochures and a booklet on birds and habitats of regional conservation significance is occurring; while information is being prepared and advice extended on monitoring changes in bushland remnants and implementing adaptive management approaches for on-farm nature conservation.

Integral to the success of these on-ground actions have been the extension methods adopted by this WWF project, as well as the complementary and co-operative liaison between the WWF Field Officers and other people within the region's network of contacts.

Extension methods adopted by the WWF Rural Nature Conservation (New England Tablelands) Project

A simplistic interpretation of the role of rural nature conservation extension is that it assists in the evolution of land management practices which are more sympathetic to the needs of the natural environment. However, the processes involved in this evolution are complex, occur at many levels, use a variety of techniques and are reflected in areas as diverse as government legislation, school classrooms, industrial codes of practice and the day-to-day activities of farmers and other land managers. At a local level the authors have found that the consideration and use of a particular selection of communication methods has led to success in achieving rural nature conservation.

Continuity and Trust

The value of establishing trust and credibility within a rural community cannot be overestimated. The authors have observed that it can take many years to gain a substantial level of trust and respect, especially with farmers. It is also important that consistency and continuity are features of extension work. Inconsistency, lack of continuity, as well as low credibility have negative impacts on the learning patterns of farmers and hinder the development of trust in and commitment to any changes in behavioural practices by rural people. It is necessary for extension personnel to share difficulties and challenges with farmers, by working through new information and alternative approaches, then appraising outcomes and revising or adapting management over a period of at least five to ten years. It is essential that the land manager's perspective and constraints, which are usually different for every case, are taken into account.

Given these requirements for effective extension, it has been invaluable that the WWF Field Officers are local people with sound appreciations of local attitudes, local land management practices and the rural economy. Equally important is the fact that these officers have their own long-term commitments to rural nature conservation in local and regional production settings.

In contrast with this WWF project, funding for extension officers now tends to be short-lived, lasting a maximum of three years. The public are consequently resigning themselves to the expectation that extension personnel are transient. Such transience is unfortunate, as it is frequently incompatible with the needs of farmers. In addition, the attitudes and commitment of extension officers may be compromised, particularly for those who are beginning their careers or who have been transferred to a new location for a short-term project. Such situations do not favour the development of continuity and trust.

Close Personal Interaction

Personalized consultation is considered by many organizations to be a poor use of resources. In some instances this holds true, but the authors feel that if performance indicators for extension staff were adjusted toward outcomes rather than numbers or processes the cost-benefit of interaction on an individual basis would be viewed more favourably and even considered an integral part of rural nature conservation extension.

For the present WWF project, close personal interaction usually occurs as discussion with a single farmer or grazier on their property. Sometimes a partner or trusted friend of the land manager is also involved. Close interaction at a familiar location creates a relatively comfortable atmosphere in which an individual's knowledge, aspirations and needs can be exposed and explored. In the authors' experience working closely with a farmer to devise a clear and mutually agreeable course of action, suited to the farmer's unique situation, combined with discussion and justification of these actions is more likely to achieve results on the ground than the landholder's attendance at a group workshop. Indeed, the authors' observations suggest that many rural people find groups intimidating and are frustrated by the limited opportunities to apply or adapt general information to their own particular situations.

Another advantage of providing such close interaction is that it encourages farmers who are not part of the Landcare (see Campbell and Seipen 1994) or field day fraternity to become involved in nature conservation in their own way. These farmers contact the WWF Field Officers as a result of word-of-mouth or in response to the project's media presence. No prerequisites are demanded of any land managers and as the WWF Field Officers are independent consultants they are rarely seen as intimidating or threatening.

Farmers know their properties intimately and there is probably no better way of increasing their knowledge and awareness than to share with them, at a personal level, a new way of viewing something that they have been looking at for years. Many farmers have sound ideas and answers of their own, or already know what should be done to solve particular problems. However, many need reassurance, some ask for guidance and almost all require financial support (at least to take that first step). Once change has begun many land managers then follow up outcomes and formulate their next steps on their own.

Close personal interaction requires relevant and well developed communication skills if it is to be effective. Talking must be balanced with (or often outweighed by) listening; understanding often needs to be reinforced by "seeing and doing"; knowledge and confidence must be created to enable action; and positive reinforcement, not judgement, is necessary. In addition, outcomes must be appraised and the development of adaptive management options facilitated. In these ways managers can be empowered and partnerships, not client-provider relationships, can be built.

Effective rural nature conservation extension personnel must also have a sound knowledge of the scientific bases of ecology, biology and agriculture, as well as good interpersonal skills and an appreciation of social processes. Such personnel must be able to speak and understand the "languages" of farmers, researchers, government agency staff, policy makers, planners, local organizations and private citizens. Good extension people know the backgrounds and priorities of these groups and can walk the "bridges" between them, effectively transferring information and skills in a way that builds the confidence, knowledge and experience of individuals and groups. People having all these talents are relatively rare (particularly in rural areas) and should not be underrated. The contributions and energies that a stable, committed network involving these sorts of people may provide at personal, local and regional levels can be far-reaching.

The technical focus adopted by the WWF Rural Nature Conservation (New England Tablelands) Project

Rare Species and Habitat Management

Having rare species as a focus has helped the current WWF project to establish good relationships with land managers who are not particularly conservation-minded. The increasing recognition of the need to manage rare native species in agricultural and pastoral production environments has provided a reason for the WWF project to help develop appropriate management practices at the level of individual properties, as well as with organizations such as Rural Lands Protection Boards, local governments and educational facilities. This role has become more important during the last two or three years, as recently developed State legislation and government policy expects land managers to consider the importance of native vegetation and threatened species when planning and undertaking routine land use activities. In addition, however, the authors have confirmed that if appropriate efforts are made to protect a variety of native habitats overall, then rare species and communities are automatically assisted (and frequently noticed).

It is important that the processes involved in the identification and long-term management of rare species and habitats in rural settings are transparent and that farmers are involved from the start. In the authors' experience, government agencies and researchers often seem reluctant to reveal the locations of rare species to landholders because the former believe the security of these species will be jeopardized. It is true that in some instances landholders intentionally destroy precious species due to a fear of having their land or rights taken away from them. However, it is naïve to think that populations of rare species will remain secret and, indeed, rare species are being lost simply because land managers are unaware that the species are present or valuable. In addition, opportunities to investigate the character of rare species and communities are being missed because some landholders hesitate to reveal their knowledge or custody of such entities.

Poor communication inevitably causes confusion and distrust. Ironically for the present WWF project, the distrust some landholders have toward a number of conservation organizations has enabled the project to expand its network and to work co-operatively with land managers to help them interpret conservation-based legislation, avoid penalty and to integrate habitat and rare species conservation with production activities. In the authors' experience, many farmers and other land managers are interested in and even proud of having rare species or communities on their properties. Furthermore, once confusion and mis-information are appropriately resolved, many land managers are keen to co-operate with a whole range of conservation initiatives.

Specific but Adaptive Management Advice

It is important to draw the distinction between general technical information and specific technical advice. Many government and semi-government extension programmes focus on providing farmers with general technical information about things such as the value of trees and bushland, or managing native pastures, soils and riparian zones.

The aim of these general programmes is to educate, to a point at which farmers are aware of and interested in exploring alternatives and applying different management actions. This is an important and continuing process, particularly considering the variety of attitudes and levels of progress within different regions. However, the time is ripe to capitalize on the substantial progress that these general approaches have made. Furthermore, the authors are seeing indications that farmers (at least on the New England Tablelands) are becoming frustrated with the plethora of activities centred on talking and planning, as these activities rarely carry through to implementation.

Specific technical advice should provide the information and definitive guidelines needed to support action. It is a tool for achieving outcomes; not, as has sometimes been thought, an outcome in itself. Good technical advice adapts general information, redefines it and makes it relevant to individual land managers. In this context, personnel working on the current WWF project only provide individual land managers with the specific information requested (as opposed to loading people with unsolicited generic information and publications) and by giving advice on an individual basis. This approach also addresses the fact that there are no simple recipes available for farmers wanting to integrate nature conservation with other management activities on their land. By providing useful technical advice in an integrated and adaptive "best bet" style, based on personal experiences and scientific information, information is made relevant to land managers and they put theory into practice.

ON-GROUND ASSISTANCE

Effective on-ground assistance is often required, in conjunction with specific technical advice, if nature conservation is to be facilitated in private off-reserve settings. On-ground assistance is usually a catalyst for action and enables farmers who appreciate at least some of the ecological and technical information to take their first step in identifying and then specifically managing their conservation areas. Beyond this step is the opportunity (and frequently the desire) for integrating conservation and production in an environmentally and economically viable way, on individual farms as well as across numerous land holdings.

In the authors' experience, changes in livestock management which favour rural nature conservation usually require the construction of stock-proof fencing. However, from a farmer's perspective, investing in infrastructure that will provide no short-term financial benefit (and, in the eyes of some, is of no proven long-term benefit) is hard to justify.

Supplying fencing materials to farmers not only enables them to change and improve the management of their bushland but it also sends a clear message that such action is endorsed by the wider community.

Furthermore, the capacity to provide tangible assistance such as fencing materials, even if only intermittently, supports the continuing process of learning by observation and action. For example, land managers can see, on a regular and continuing basis, the impact that the removal of livestock (and feral animals) has on a site. As a result of this opportunity land managers are often inspired to learn more about the new things they observe and the ecological roles these things play. The outcome is a more aware land manager who is better able to incorporate actions to maintain some ecological balance amongst their production enterprises. Such change frequently follows the simple act of providing fencing materials. In addition, putting nature conservation concepts into the land manager's own context and language is important.

During the current WWF project, opportunities to offer material incentives have sometimes also provided a positive means of establishing dialogue with farmers who are outside the Landcare circle (see Campbell and Seipen 1994). The occurrence of such dialogue has, however, required that the project have an effective media presence and referral network across the region. In this instance, a referral network that has benefited from and trusts the Field Officers associated with this project has been invaluable.

SOME IMPLICATIONS FOR RURAL NATURE CONSERVATION EXTENSION

The growing success of rural nature conservation efforts, as well as the need for more moderate and flexible approaches to rural land management are encouraging many Australian farmers and other land managers to make new types of decisions about how they manage their properties. The decision-making processes of land management are complicated and influenced by a host of factors, the most immediate being financial. This complexity and the effects of past attitudes and experiences on rural communities and their forebears should not be underestimated by extension personnel.

Efforts to conserve nature in the face of market-driven rural production must address human development needs, technical issues and the desire for quick fixes. While some people suggest that a consideration of human development needs is only relevant to developing parts of the world, it appears that these considerations are also important in developed countries. Furthermore, all around the world, technology-transfer models of extension are being superseded. Individuals and local and regional communities are starting to ask their own questions, set their own priorities and direct their own learning processes (van Den Ban and Hawkins 1985; Siepen and Dorricott, this volume).

Contemporary extension processes, however, have yet to evolve fully from the client-provider technology-transfer (magic bullet) model to a network model built on partnerships and experiential participatory learning (Campbell and

Siepen 1994; Siepen and Stone 1997). The key role for extension personnel dealing with rural nature conservation is to be effective conduits for information flows among land managers and between land managers, researchers, policy makers and planners. In addition, extension personnel must support self-directed learning by land managers, facilitate human development and utilize appropriate social processes (Miller and Prior 1992; Campbell and Siepen 1994) in order to promote rural nature conservation in production settings.

Unfortunately, government and semi-government extension services in many rural areas of Australia (e.g., the New England Tablelands of New South Wales) are increasingly focusing on assisting groups of people, rather than individuals. Yet, farmers and other land managers form an heterogeneous group. In addition, the authors have observed that new legislative requirements, as well as recent changes to government policies and services, are not promoting trust or continuity within rural communities served by rural nature conservation extension programmes. Thus the demand for long-term extension services which offer consistent and close personal interaction, together with specific but adaptive approaches to nature conservation continues and is probably increasing. The authors feel that the local focus and personalized methods of the current WWF project have made significant contributions to the adoption of land management practices favouring native vegetation and rare species conservation by individual land managers on the New England Tablelands of New South Wales, Australia.

ACKNOWLEDGEMENTS

We are grateful to TransGrid (previously Pacific Power) for the funding that has made the WWF Rural Nature Conservation (New England Tablelands) Project possible. In addition, the interest and support provided by Bob Boland (of TransGrid) have been invaluable in maintaining the continuity of the project. All members of the project's Steering Committee have also made considerable contributions to the success of the Field Officers' activities. Staff of WWF's Sydney office have injected many useful ideas during the course of the project and have seen to much of the administrative work for the project. The help of WWF staff in making it possible for one of us (Ruth Trémont) to take part in the "Nature Conservation in Production Environments: Managing the Matrix" Conference is also gratefully acknowledged.

We thank staff of the Dumaresq Shire Council, Armidale for assistance with clerical and financial matters. Hugh Ford and members of the technical staff of the Zoology Department, University of New England, Armidale are also thanked for their commitment to and assistance with the project.

Collation of this chapter would have been impossible without the participation of others involved in the New England Tablelands' network of rural nature conservationists. Specifically, we thank Margie Crowther (Southern New England Landcare Co-ordinating Committee), David Curtis (Greening Australia), Chris Nadolny (NSW Department of Land and Water Conservation), Jamie Pittock (World Wide Fund for Nature Australia), Julian Prior (NSW Department of Land and Water Conservation) and Kath Wray (Citizens Wildlife Corridors Inc.) for their contributions to a Rural Nature Conservation Workshop on 7th August, 1997. In addition, we are grateful to Julian Prior, Chris Nadolny and Hugh Ford for their comments and suggestions regarding drafts of this chapter.

Most importantly, we acknowledge all the individuals across the New England Tablelands who have contributed in their own ways to restoring the natural environment of the region and promoting a nature conservation ethic out there.

REFERENCES

Campbell, A. and Seipen, G., 1994. Landcare: Communities Shaping the Land and the Future. Allen and Unwin: Sydney, New South Wales.

Curtis, D., Nadolny, C. and Ford, H. A., 1995. Lessons from a network of dieback repairers. Pp. 175–86 *in* Nature Conservation 4: The Role of Networks ed by D. A. Saunders, J. L. Craig and E. M. Mattiske. Surrey Beatty & Sons: Chipping Norton, New South Wales.

Ford, H., Barrett, G. and Recher, H., 1995. Birds in a degraded landscape — safety nets for capturing regional biodiversity. Pp. 43–50 *in* Nature Conservation 4: The Role of Networks ed by D. A. Saunders, J. L. Craig and E. M. Mattiske. Surrey Beatty & Sons: Chipping Norton, New South Wales.

Miller, D. and Prior, J., 1992. The agricultural facilitator: a sensitive new age advisory officer. Pp. 133–37 *in* Future Directions for Agricultural Education and Training. Proceedings of a Conference and Workshop held at the University of Western Sydney, Hawkesbury, New South Wales, September 30–October 2, 1992. Centre for Extension and Rural Development, University of Western Sydney: Hawkesbury, Richmond, New South Wales.

Morgan, G. and Terrey, J., 1999. The New England Tableland: A Bioregional Strategy. Greening Australia NSW Inc.: Armidale (revised edition).

New South Wales Roadside Environment Committee, 1996. Managing Roadsides. An information package of handbooks, information sheets and a video. New South Wales Roadside Environment Committee, GPO Box 3482, Sydney, 2001.

Siepen, G. L. and Stone, G., 1997. Identification of rural landholders' requirements for nature conservation information. Pp. 226–32 *in* Conservation Outside Nature Reserves ed by P. Hale and D. Lamb. Centre for Conservation Biology, University of Queensland: St. Lucia, Queensland.

van Den Ban, A. W. and Hawkins, H. S., 1985. Agricultural Extension. Longman Scientific and Technical: Harlow, England.

CHAPTER 57

Evaluating flagship species for ecosystem restoration; a case study involving freshwater crayfish and a river in southwestern Australia

Rachael Nickoll[1] and Pierre Horwitz[1]

Using flagship species to provide a tangible endpoint, and to make clear the local or regional benefits from the restoration work, should be considered integral to restoration planning. In this chapter, criteria are assembled for evaluating potential flagship species in restoration projects. Such criteria include the ability of the species to respond to restoration activities and/or recover from endangerment, and the degree to which the species provides benefits to local or regional communities. A case study using the Blackwood River system in southwestern Australia, and the potential use of a freshwater crayfish (marron; *Cherax tenuimanus*) as a flagship for river restoration, demonstrates how a species can be tested against the assembled criteria. Through a review of literature and other sources of ecological information, and a questionnaire, marron were shown to fulfil most criteria. The questionnaire (within shires along the river) revealed that perceptions of river restoration were widespread. The most popular endpoint statement for river restoration was that river animals which were once common have returned. The public appeal and utility of marron was demonstrated, and marron were deemed the most appropriate biotic symbol for the Blackwood River. The decline of marron has engendered sympathy towards its plight, and ensured a linkage between the need to restore the river and the alleviation of threat to the species. Using the recovery of marron as an endpoint for the restoration of the middle catchment of the Blackwood River should enhance public understanding and acceptance of proposed restoration activities.

Key words: River restoration, Flagship species, Indicator species, Blackwood River, Freshwater crayfish, Distribution, Southwestern Australia.

INTRODUCTION

RESTORATION aims to return a degraded system to a condition which is protective, productive, aesthetically pleasing, or valuable in a conservation sense, and aims to develop a system which is sustainable in the long term (Hobbs and Norton 1996). The characteristics of a restored ecosystem can be bound by two significantly different perceptions: the public perceived restoration to "usefulness" and the scientifically documented restoration to original structure and/or function (Cairns *et al.* 1977). Consequently, the difficulties created by differences in social perception and scientific documentation require restoration programmes integrating or linking these diverse views.

Notions of recreating a condition which "originally" occurred in the historical past raise questions concerning the actual nature of the past condition, whether it was stable and static, whether it was sufficiently well known to enable reconstruction, or indeed whether it is socially desired. Restoration programmes with such aims are hampered by ambiguous goals and criteria for success (Hobbs and Norton 1996). Successful restoration programmes, it might be argued, will be those with endpoints which are clearly enunciated

[1]Centre for Ecosystem Management, Edith Cowan University, 100 Joondalup Drive, Joondalup, Western Australia, Australia 6027.

Pages 557–64 *in* NATURE CONSERVATION 5: NATURE CONSERVATION IN PRODUCTION ENVIRONMENTS: MANAGING THE MATRIX ed by J. L. Craig, N. Mitchell and D. A. Saunders. Surrey Beatty & Sons, 2000.

(Kondolf 1995), popular (to, for instance, the developer, manager, polluter and general public), and measurable (Milner 1996). Referring to the management of rivers recovering from disturbances, Milner (1996) commented that the establishment of acceptable endpoints was "a complex issue and a criterion of immense importance" where common endpoints used in assessing recovery of rivers ranged from habitat and water quality parameters, to structural and functional biotic parameters.

River restoration cannot be successful, let alone initiated, without the understanding and commitment of people who live around the river. There is an inherent difficulty in convincing local people, for example, that a 10% reduction in salt load over a 10 year period is a valuable endpoint for river restoration. Such measures of restoration are probably neither tangible nor valuable to local communities. The restoration process, we believe, will be greatly aided by an icon/flagship species to convey understanding in the community.

The flagship species concept is another priority conservation grouping, relatively new compared to the keystone, umbrella and indicator species or group. A flagship species, as outlined by New (1995), gains public and political sympathy based on charismatic appeal, serving to increase public awareness. Flagship species are becoming increasingly important with the growing need for community awareness and involvement in conservation and restoration programmes.

Restoration programmes and conservation of flagship species are logically good for one another. An ideal flagship will have a high public profile and will require conservation. The latter may come in the form of reservation of critical habitat and use of protective legislative provisions; additionally more appropriate management of the degraded habitat (some restorative action) will be required to alleviate the threat. The high public profile lends considerable support to the restoration process.

A list of criteria for invertebrate flagships, adapted from New (1995), Yen and Butcher (1997) and Kellert (1993) has been assembled. The criteria can be grouped according to ecological knowledge of the species, their indicator properties, restorative value, cultural, symbolic or historical value, aesthetic value, and utilitarian value. Similar sets of criteria have been compiled, for instance, for indicator taxa and umbrella taxa (see Yen and Butcher 1997).

CASE STUDY

The aim of this study was to establish whether the use of a high profile riverine species, by evaluating it against the criteria outlined above, could be effective in providing a community with a tangible understanding of river restoration, its objectives and endpoints.

The study was based on two assumptions: the first was that marron (*Cherax tenuimanus*), a freshwater crayfish species endemic to rivers in southwestern Australia, might represent a charismatic icon. The second was that the Blackwood River in southwestern Australia is in need of restoration, where the Blackwood Basin Group (an organized community group) is established to enable a successful restoration programme to be planned and implemented. Marron could, as a flagship, provide direction for the Blackwood Basin Group, and provide a focal point for research, Landcare activities, and foreshore restoration along the river.

The Blackwood River and its catchment represent an example of the changes in aquatic environments of southwestern Australia which have occurred over the last hundred years or so. Widespread clearing of vegetation has indirectly raised the groundwater table, leading to waterlogging, induced salinity problems for both soil and rivers (Schofield *et al.* 1988; McFarlane *et al.* 1993), increased erosion and silting of rivers, and promoted river training, desnagging and channelization (Olsen and Skitmore 1991). Riverine riparian vegetation has been damaged by clearing and grazing so that agricultural pollutants such as pesticides, herbicides, fertilizers and organic debris can more readily flow into river systems (Olsen and Skitmore 1991). Many of the Blackwood River's summer pools no longer function as high quality seasonal refuges for a diversity of aquatic biota, or as a source of water for domestic and agricultural use (Western Australian Water Resources Council 1992). Attempts are being made to rehabilitate river banks and restore the riparian zone function along the Blackwood River.

METHODOLOGY

Details of methodologies used in the study are presented in the context of each criterion.

Ecological knowledge and indicator properties

Relevant information from the literature was sought on the taxonomy of marron, its biology, tolerances to disturbances and distribution. Additional data on the distribution of marron were obtained from records held by government departments, recollections published within shire histories, and oral histories as per Nickoll (1996). In addition, the distribution, population and water quality aspects of Morrissy's work (1978b) were replicated by Nickoll (1996).

Restorative, aesthetic and cultural, symbolic and historical values: The questionnaire

A face to face questionnaire sampling the Middle Blackwood Catchment (see Fig. 1) community was used to: determine the community value and awareness of, and preferred future for the river environment; identify community perception of river restoration, its endpoint, and support for a restoration programme; identify potential flagship species to direct river restoration; and to test the marron's (comparative) ability to evoke public sympathy and pride, and assess marron popularity and appeal locally. The questionnaire was also aimed at identifying whether a relationship already existed between the community understanding of river restoration and high profile riverine species.

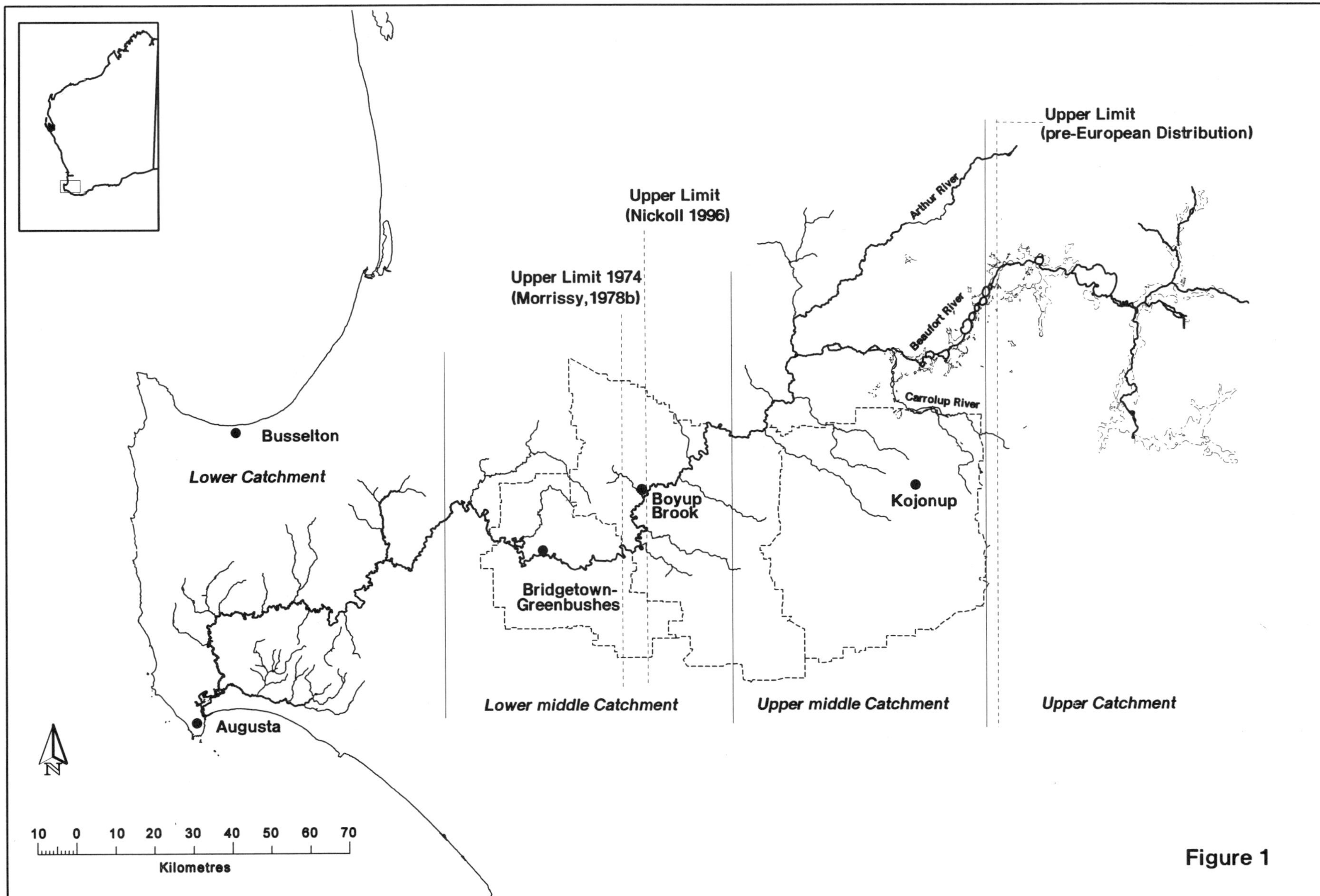

Fig. 1. The Blackwood River basin system, showing the shires sampled under the questionnaire, the Lower, Middle (Lower and Upper), and Upper Catchment divisions of the basin, and the upper limits of the distribution of marron (pre-European, and "present-day" identified by Morrissy 1978b and Nickoll 1996).

The questionnaire consisted of 36 closed questions which tested behaviour, beliefs, attitudes and provided demographic information. Visual aids were used along with rating scales, lists and semantic differential questions. The questionnaire took between 15 to 20 minutes to complete. A single sample was conducted at strategic locations within the three main towns (Bridgetown, Boyup Brook and Kojonup; Fig. 1) making up the (lower and upper parts of the) Middle Catchment. Three interviewers were assigned to each town, sampling during the Saturday morning shopping period from 8.30 a.m. to 12 noon. The 71 respondents represented a cross-section of the Middle Catchment community, and were permanent adult residents of the catchment. Analysis of the responses involved both descriptive (frequency distribution and cross-tab analysis) and inferential statistics (Chi-square and Spearman Rank Correlation) conducted on the statistical package SPSS-for-Windows Version 6.

RESULTS

Performance of marron against each of the criteria is shown in Table 1, and discussed below.

Ecological knowledge criteria

Examination of the literature confirms that sufficient information exists to characterize the taxonomy and distribution of marron. Marron belong to a relatively diverse and widespread genus in a southern hemisphere family (see Riek 1967; Austin 1986), arguably the most easily identified member of the genus (Riek 1967; Austin 1986). Austin and Knott (1996) demonstrated that some geographical differentiation could be detected both morphologically and genetically within the range of the species. Such differentiation becomes of managerial concern if restoration requires the reintroduction of marron (if the threat of genetic pollution is to be avoided; Horwitz 1995). Marron have a known distribution which was localized prior to the spread of the species into artificial and natural waters, and contraction of the species' range in river systems (see Shipway 1951; Riek 1967; Morrissy 1978b; Horwitz 1990). Figure 1 shows past and present upper limits of the species in the Blackwood River, showing a reduction of approximately 100 km from the pre-European distribution (Morrissy 1978b).

The functional importance of a flagship species for river restoration could be demonstrated if it were known that other elements of the riverine biota were dependent on that species, so that any restorative action targeting the flagship would have wider benefits. The size and density of marron suggest they represent a significant component of biomass in inland aquatic systems of southwestern Australia. In addition, marron are similar to other freshwater crayfish functioning as host and habitat to a diversity of symbiotic biota; conserving the host has been identified as an important means for conserving their symbionts (see Cannon and Sewell 1994).

A suitable flagship species should be sufficiently well known biologically to establish baseline information for the definition of habitat requirements and of the species' likely response to environmental change. Elementary descriptive data have shown that marron occupy broad and deep permanent water bodies, and are dependent on the provision of shelter and permanent water (see Austin and Knott 1996). Aspects of the breeding

Table 1. Criteria for evaluating the utility of species as flagships for restoration activity, and observed evaluation of marron for these criteria. Criteria are ranked according to the degree to which that characteristic is required for the species to be considered suitable as a flagship for restoration, *viz* +, ++, +++ = useful, important, or essential (adapted from New 1995, Yen and Butcher 1997). Essential criteria are best regarded as absolute; others add to the potential utility of the species for the purpose. Observed characteristics are shown according to the estimated degree to which they satisfied the criterion *viz* ✓✓✓, ✓✓, ✓ = easily achieved, probably sufficient, doubtful. Supporting details are provided in the text.

	Required	Observed
Ecological knowledge criteria		
1. Diverse and widespread grouping	++	✓✓✓
2. Known, stable and easily interpreted taxonomy	+++	✓✓✓
3. Known and localized distribution	+++	✓✓✓
4. Functionally important (i.e., keystone species)	+	✓✓
5. Biology sufficiently known	+++	✓
Indicator properties		
6. Relevant lifestyle (sedentary)	++	✓✓
7. Indicator of habitat change	++	✓✓
Restorative value		
8. Ability to link habitat degradation with restorative action	+++	✓✓✓
9. Ability to recover from threatening processes	+++	✓✓
Cultural, symbolic or historical value		
13. Ability to evoke public sympathy	+++	✓✓✓
14. Ability to evoke community pride	+++	✓✓✓
Aesthetic value		
10. The species may need to be aesthetically appealing	+	✓✓
Utilitarian value		
11. Direct monetary value	+	✓✓
12. Indirect monetary value	+	✓✓✓

biology have been established, however, physico-chemical conditions, ecological functioning and structural characteristics of their habitats, and physiological tolerances to changes in these *in situ*, have been poorly treated in the literature.

Indicator properties

Suitable indicator properties of flagship species include the ability to be sampled and monitored easily, and to reflect the condition of local environments. Being non-burrowing, relatively sedentary benthic dwellers, easily trapped, marked and spotted, marron have these characteristics, although more data on movement of marron along and between watercourses would be useful.

Ideally the flagship species should show a predictable, rapid, analysable and linear response to environmental change which enables cause and effect to be determined (see Yen and Butcher 1997); the species should show an identifiable response to recovery action (see criterion 9). Research on the environmental tolerances of marron yields information for salinity (Morrissy 1978b; Morrissy *et al.* 1990), heavy metal pollution for instance cadmium (Chambers 1994); temperature and dissolved oxygen levels (Morrissy *et al.* 1990), and ammonia (Constantine 1988). Aquacultural (Morrissy 1980) and anecdotal information suggests a strong negative relationship between marron and organic pollutants and pesticides. However, most scientific knowledge on marron biology and environmental tolerance have been based on aquacultural research and toxicological testing within a laboratory, potentially limiting its utility as a reliable indicator. This becomes critical when determining likely causes for its decline, and potential restorative action.

At least three possibilities for the shrinkage in marron distribution from the Upper Catchment of the Blackwood River have been considered: over fishing, rising river salinity and oxygen depletion. Of these, Morrissy (1978b) considered that oxygen depletion in the bottom water of river pools during the summer months was the most likely cause. A reasonable case can still be made to link the combined effects of oxygen depletion in the water column (see Nickoll 1996) with removal of riparian vegetation, organic enrichment, and summer storms, to result in marron mortalities. The role of rising river salinity as a principal cause for the decline needs further investigation, particularly given an absence of reliable data on the tolerance of eggs and adult marron to increasing salinity, despite levels remaining within the predicted salinity tolerance range (see Morrissy 1978b) for marron over parts of its range from which it has disappeared.

Restorative value

Ideally a flagship's ability to recover from threatening processes enables objectives or an endpoint to be identified for conservation and restoration programmes. If Morrissy's hypothesis is correct, this means that marron must respond to attempts to address the water column degradation (O_2 depletion). Aquacultural research has shown that the threat from such a condition can be ameliorated with daily or seasonal mixing of the water column by cooling air temperatures, mechanical mixing of the wind and the flushing effect of winter rain. Some evidence for the restorative potential comes from Western Australian Fisheries Department catch return data for the marron fishery (unpubl.) demonstrating a trend in catch rate which is positively correlated with rainfall. These results suggest that the diluting and flushing effect associated with high rainfall improve environmental conditions for marron catchability and/or survival.

Consequently a restoration programme designed for the Blackwood River may justifiably involve the following actions to enhance the rate of recovery of marron and water quality: artificial aeration of certain pools during summer; earthen or mesh contour banks trapping organic runoff; restoration of the riparian zone; mechanical removal of anoxic sediments and artificial flushing of pools. As restoration proceeds, it is hypothesized that marron numbers would increase and their range would be extended upstream. In so doing, marron can provide an indicator of river health, allowing the restoration process to be mapped and monitored, and enhancing community education and satisfaction. If marron were once again caught throughout the Middle Catchment of the Blackwood River, the restoration process could be deemed to be working, ensuring continued community support. If marron did not respond, then assumptions concerning habitat degradation and oxygen depletion as the causal mechanisms for the decline may need to be re-assessed. A realistic restoration programme would have such a reflexive, adaptive approach.

Nevertheless, restoration of the river environment using marron as a flagship species will not occur unless the community desire an improvement of their river and value marron enough to be sympathetic to the plight of the species within the Blackwood River.

Investigation of the questionnaire sample population revealed a level of awareness and interest in river management issues, and an importance placed on the river environment and its associated native vegetation. These were significantly related to a respondent's level of contact with the river environment. People who lived near the river (X^2 = 13.9, df = 6, p = 0.02), and those who generally used the river (X^2 = 6.8, df = 3, p = 0.007) had a higher level of awareness and placed greater importance on the river.

A significant difference (X^2 = 10.37, df = 2, p = 0.005) was recorded in river use between shires, where river use declined as sampling moved up the river from the shire of Bridgetown-Greenbushes (82%, n = 27 used the river) to Kojonup (43%, n = 9 used the river). This result reflects a trend of deteriorating water quality from the Lower to the Upper Blackwood Catchment. Of those who had an opinion on river water quality, a majority of the respondents identified salinity, followed by paddock runoff as the major threats to river health.

When asked for preferred "potential future river use activities" respondents recognized the importance of river health. Activities deemed detrimental to the river environment, such as farming and domestic/household use of the river, were generally rejected as a preferred use. Recreational and tourist activities were strongly favoured, with primary contact activities such as swimming and fishing having the highest response.

In terms of community perception of restoration and the level of support for a restoration programme, all but four people sampled indicated that the river required restoration to some acceptable level. A similar finding was reported by Butterworth and Carr (1996).

Thirteen restoration endpoint statements were ranked by each respondent in the sample population in terms of their importance/necessity in achieving "successful" restoration (see Nickoll 1996). The lowest priorities were for complete fencing of the river foreshore, and agricultural and domestic use of the water in the river. The need to establish a river environment similar to that experienced prior to clearing was perceived as having moderate importance. Of the statements ranked more highly, aesthetic, educational, human health and safety, and recreational elements were predominant, or could be inferred. Perhaps the most relevant in terms of the context of this investigation was that according to almost 80% of those sampled, it was deemed most important that riverine species once common to the Blackwood River return to the successfully restored river environment.

This final statement represents the most strongly accepted endpoint for restoration of the Blackwood River (although it is acknowledged that other endpoints are highly acceptable too). It indicates that a relationship exists between the community perception of restoration and high profile riverine species.

Cultural, symbolic or historical value

From a list of animal and plant species relevant to the Blackwood River (or species nominated by the respondent themselves), marron received the highest response rate from the sample population when asked to select the most appealing species (38%, $n = 27$), the species which evoked the greatest sense of pride (48%, $n = 34$) and the most appropriate symbol for the Blackwood River (72%, $n = 51$, respondents were given three choices). Together all three questions were designed to elucidate appropriate flagship species from the community. The River Gum *Eucalyptus rudis* was the second most popular species (with a response rate of 28%, $n = 20$; 21%, $n = 15$; and 60%, $n = 43$, respectively). Both the marron and the River Gum had a response rate over three times that of all other species identified as most appealing and promoting most community pride.

Overall the sample population's response to the three questions revealed the frog, tortoise and paper bark as other popular species, and potential flagships. Moderate support was given for cobbler, duck, koonac, water lily and water rat. Very little support was given for freshwater mussel, gilgie, perch, cormorant, minnow, phascogale and leech. Marron, koonac and gilgie are different types of freshwater crayfish; river gum and paper bark are riparian trees; cobbler, perch and minnow are types of fish; and a phascogale is a small marsupial.

Cross-tab analysis revealed that those respondents most likely to select the River Gum over the marron were generally female, identified riparian vegetation as "very" important, had lived in the catchment for less than five years, and lived in the shire of Bridgetown-Greenbushes.

Differences in attitude towards, and knowledge of, marron were identified between the different shires, genders, and those who did and did not fish for marron. For example, females generally showed a lower level of knowledge and appreciation for marron compared to males. Participants from shires in the Lower Middle Catchment (Bridgetown-Greenbushes and Boyup Brook) showed a lower appreciation of marron compared to those sampled from Kojonup. As sampling moved up the river from Bridgetown-Greenbushes to Kojonup, significantly more people agreed that marron previously occurred in the Upper Blackwood ($X^2 = 10.77$, $df = 4$, $p = 0.029$), were endemic to the south-west, and could correctly distinguish between photographs of marron and other freshwater crayfish species. All respondents, except those who "didn't know", were of the opinion that marron numbers had declined within the Blackwood River over the last 40 years. Seventy per cent of the sample population identified either increasing salinity or fertilizer input into the river as the major threat to marron survival in the Blackwood.

Aesthetic appeal

The results from three semantic differential questions (see Nickoll 1996), designed to identify the level of appeal associated with marron within the Blackwood community, revealed that the sample population generally perceived marron to have moderate to quite high "public appeal" (70%, $n = 50$) in comparison to the extremes reflected by the spider ("low" appeal) and koala ("high" appeal). Marron were considered to be moderate to very beautiful (85%, $n = 60$) on the ugly-beautiful scale, and on average the sample found marron to be quite to very "interesting" (80%, $n = 57$).

Respondents who indicated marron were "boring" creatures were significantly ($r = 0.5$, $df = 70$, $p < 0.001$) likely to state that marron were "ugly" too, and vice versa. As before, individuals who lived higher in the catchment (Kojonup) perceived more interest and greater attractiveness of marron.

Much of the sample's perception of marron and linkage to river management issues might be explained by any individual's direct encounters with marron in the past. Analysis revealed that of the 71 participants, 60 indicated that they or their immediate family had fished for marron in the last 25 years. Of these, thirty-nine were active fishers, fishing in both the Blackwood River and from farm dams. When fishing from the river, 85% of respondents fished below Boyup Brook, which complements the study's findings on the current distribution of marron in the Blackwood. Since fishing is a river use, it has a significant influence on an individual's level of awareness and value for the river environment. The act of marron fishing therefore can promote a high level of value for the species and the river environment. This link is fundamental for using marron as a flagship for restoration of the Blackwood River.

Utilitarian values

Marron have economic and cultural value in that the species supports both a world class amateur fishery (a leisure activity typically south-west Australian) (Morrissy 1978a), and a commercial aquaculture industry (Morrissy *et al.* 1990). These economic and cultural incentives have raised the number of prospective marron farmers and have prompted concerted government, university and industrial-based aquacultural research focused on improving marron farming technologies (Evans and Fotedar 1996).

DISCUSSION

As outlined in Table 1, marron generally meet all flagship selection criteria. The only criterion not fulfilled is whether the biology of marron is sufficiently well known to make, with absolute confidence, the assumptions and assessments required elsewhere in the Table.

Provided this limitation is overcome, it is concluded that marron are an appropriate flagship for the restoration of the Blackwood River; returning marron to the upper parts of the Middle Catchment of the Blackwood River to set targets could be defined as an endpoint of the river restoration process, by providing direction for the improvement of river water quality, riparian zone function and the implementation of sustainable land use practices, all of which could be defined as restoration objectives. Marron have the potential to provide direction for restoration action and milestones for a restoration programme tangible to most if not all stakeholders within the Blackwood River Catchment.

If marron is to be raised to flagship status then efforts could be directed towards educating the minority within the local community who were unaware of the species' endemic nature and the extent of its decline. It may be beneficial to utilize the popularity for marron in conjunction with that of the River Gum to provide widespread effective community understanding and involvement in a restoration programme for the Blackwood River.

We suggest that the Fisheries Department of Western Australia, in whose jurisdictional area marron lie, need to divert some attention away from creating the bigger, better aquacultural product, to providing baseline data on the population structure, feeding biology, reproductive and life history of marron in the wild. The knowledge on wild stock is poor considering that marron has been officially recognized as important in Western Australia for over 30 years. Similarly the marron aquaculture industry must adopt part ownership in the welfare of marron, in return for the utilization of its genetic material. It is recommended that the industry promote marron and become involved in the recovery of marron populations by supplying ecological information and lobbying for effective funding.

ACKNOWLEDGEMENTS

The authors are grateful to Dr Noel Morrissy for sampling advice and use of unpublished data from catch returns, and Dr James Butterworth for advice on questionnaire construction. Ms Bea Sommer provided comments on an earlier draft of the manuscript and Mr Craig Jaques drafted Figure 1. Ms Susan Masterson showed constant enthusiasm for the project. Editorial and reviewer comments improved the chapter.

REFERENCES

Austin, C. M., 1996. Electrophoretic and morphological systematic studies of the genus *Cherax* (Decapoda: Parastacidae) in Australia. Ph.D. thesis, Zoology Department, University of Western Australia.

Austin, C. M. and Knott, B., 1996. Systematics of the freshwater crayfish genus *Cherax* Erichson (Decapoda: Parastacidae) in south-western Australia: electrophoretic, morphological and habitat variation. *Aust. J. Zool.* **44:** 223–58.

Butterworth, J. E. and Carr, A. J. L., 1996. Riparian zone management in the Middle Blackwood Catchment: a baseline study of landholders' views. CSIRO Division of Water Resources Consultancy Report No. 96. CSIRO: Perth, Western Australia.

Cairns, J., Dickenson, K. L. and Herricks, E. E., 1977. A challenge for action: symposium analysis. Pp. 522–31 *in* Proceedings of the International Symposium on the Recovery and Restoration of Damaged Ecosystems. University Press of Virginia: Charlottesville.

Cannon, L. R. G. and Sewell, K. B., 1994. Symbionts and biodiversity. *Mem. Qld Mus.* **36:** 33–40.

Chambers, M., 1994. Cadmium toxicity studies in *Cherax tenuimanus*. Honours thesis, Curtin University of Technology, Western Australia.

Constantine, C., 1988. *Cherax tenuimanus*. Ammonia toxicity of marron. Honours thesis, University of Western Australia.

Evans, L. H. and Fotedar, R. F., 1996. Current status of marron farming in Western Australia. Pp. 1–8 *in* Proceedings of the Marron Growers Association of Western Australia, Open Seminar, May 1996. Aquatic Science Research Unit: Curtin University of Technology, Perth, Australia.

Hobbs, R. J. and Norton, D. A., 1996. Towards a conceptual framework for restoration ecology. *Rest. Ecol.* **4:** 93–110.

Horwitz, P., 1990. Translocations of Australian freshwater crayfish; potential impact and the need for control. *Biol. Conserv.* **54:** 291–305.

Horwitz, P., 1995. An environmental critique of some freshwater captive breeding and reintroduction programmes in Australia. Pp. 75–80 *in* Reintroduction Biology of Australasian Fauna ed by M. Serena. Surrey Beatty & Sons: Chipping Norton, New South Wales.

Kellert, S. R., 1993. Values and perceptions of invertebrates. *Conserv. Biol.* **7:** 845–55.

Kondolf, G. M., 1995. Five elements for effective evaluation of stream restoration. *Rest. Ecol.* **3:** 133–36.

McFarlane, D. J., George, R. J. and Farrington, P., 1993. Changes in the hydrologic cycle. Pp. 146–89 *in* Reintegrating Fragmented Landscapes: Towards Sustainable Production and Nature Conservation ed by R. J. Hobbs and D. A. Saunders. Springer Verlag: New York.

Milner, A. M., 1996. System Recovery. Pp. 205–26 *in* River Restoration ed by G. Petts and P. Calow. Blackwell Science: Oxford.

Morrissy, N., 1978a. The amateur marron fishery in south western Australia. *Fish. Res. Bull. West. Aust.* **21:** 1–44.

Morrissy, N., 1978b. The past and present distribution of marron *Cherax tenuimanus* in Western Australia. *Fish. Res. Bull. West. Aust.* **22:** 1–38.

Morrissy, N., 1980. Production of marron in Western Australian wheatbelt dams. *Fish. Res. Bull. West. Aust.* **24:** 1–79.

Morrissy, N. M., Evans, L. H. and Huner, J. V., 1990. Australian freshwater crayfish: aquaculture species. *World Aquacul.* **21:** 113–22.

New, T. R., 1995. Introduction to Invertebrate Conservation Biology. Oxford University Press: New York.

Nickoll, R., 1996. An investigation into the use of the freshwater crayfish Marron (*Cherax tenuimanus*) as a flagship for the restoration of the Blackwood River. Honours thesis, Edith Cowan University, Perth, Western Australia.

Olsen, G. and Skitmore, E., 1991. State of the rivers of the south west drainage division. Water Resources Perspectives. Western Australia Water Resources Council Publication No. WRC 2/91. 171 Pp. Perth, Western Australia.

Riek, E. F., 1967. The freshwater crayfish of Western Australia (Decapoda: Parastacidae). *Aust. J. Zool.* **15:** 103–21.

Schofield, N. J., Ruprecht, J. K. and Loh, I. C., 1988. The impact of agricultural development on the salinity of surface water resources of south western Western Australia. Water Authority of Western Australia, Report WS27, Perth, Western Australia.

Shipway, B., 1951. The natural history of the marron and other freshwater crayfishes of south-western Australia. *West. Aust. Nat.* **3:** 7–12 and 27–34.

Western Australian Water Resources Council, 1992. The state of the rivers in the south west. Water Resources Perspective. Western Australian Waters Resources Council. Publication No. WRC 2/92. Perth, Western Australia.

Yen, A. L. and Butcher, R. J., 1997. An Overview of the Conservation Status of Non-marine Invertebrates in Australia. Environment Australia: Canberra.

CHAPTER 58

Assessing the ecological sustainability of strategic planning for the conservation of biological diversity

Susan F. Jennings[1] and Susan A. Moore[1]

Integrated management of production and conservation landscapes requires recognition and inclusion of diverse interests and values. Such management in the longer term also depends on decision making and decisions that are "ecologically sustainable". A key component of sustainability is the conservation of biological diversity. Such integrated, long-term outcomes are increasingly being sought via strategic planning.

This chapter assesses several Australian strategies, for the conservation of biological diversity, for their ecological sustainability. Strategies were selected to investigate three scales — national, state and landscape — and included the National Strategy for the Conservation of Australia's Biological Diversity (Commonwealth of Australia 1996), Draft Nature Conservation Strategy for Western Australia (Department of Conservation and Land Management 1992), and Local Greening Plans (Greening Australia 1995) respectively. The ecological sustainability of the decision making approaches and on-the-ground actions (i.e., decisions) recommended in each strategy were assessed.

The strength of the National and State Strategies was their emphasis, through recommended management actions, on ecological integrity. When opportunities for adaptability and community consultation were assessed, these strategies fared less well. Local Greening Plans, on the other hand, not only provided for dynamic, adaptive approaches to ongoing decision making, their recommended management actions also contributed to ecological sustainability. The areas of least strength in all three strategies were satisfying equity and justice concerns and facilitating ongoing goal-seeking approaches. These results indicate that currently, for decision making directed toward conserving biological diversity, landscape level planning provides the greatest potential for satisfying ecological sustainability criteria.

INTRODUCTION

INTEGRATED management of a matrix landscape of production and conservation areas is a complex task. Such management, especially over the longer term, requires decision making and decisions that are "ecologically sustainable". This means decision making resulting in outcomes that can be implemented as well as including affected interests, both present and future (Moore *et al.* 1998). The involvement of stakeholders, individuals and communities in such decision making is increasingly being recognized as integral to sustainability (Natural Resources Canada 1998).

Decisions for the sustainable management of natural resources, such as biological diversity, are made at a number of scales of human organization, including national, state, regional, landscape and local levels. These decisions, particularly at higher scales, are developed and produced as strategic plans or policies. Such plans may or may not have a legislative base, although biological diversity activities at the national and state levels are generally guided by the *Endangered Species Protection Act 1992* (Commonwealth), and in Western Australia, for example, by the *Conservation and Land Management Act 1984* (Western Australia) and the *Wildlife Conservation Act 1950* (Western Australia).

[1]School of Environmental Science, Murdoch University, Western Australia, Australia 6150.

Pages 565–73 *in* NATURE CONSERVATION 5: NATURE CONSERVATION IN PRODUCTION ENVIRONMENTS: MANAGING THE MATRIX ed by J. L. Craig, N. Mitchell and D. A. Saunders. Surrey Beatty & Sons, 2000.

The scale at which decisions should be made for sustainability has been the subject of continued debate (Lee *et al.* 1992; McDonald 1996). Some researchers advocate state or regional control as local decisions are seen as limiting effectiveness (McDonald 1996). The watershed or regional level is considered by Lee *et al.* (1992) to be the best management scale for sustainable landscapes, as changes at this scale significantly influence decisions at smaller scales. As ecological and socioeconomic systems are linked and have different boundaries, it is unlikely there is a single "best" scale of operation where decision making for ecological sustainability can adequately cover and co-ordinate the management of biological diversity in conservation and production areas (Dovers 1995).

This chapter assesses strategic planning for conserving biological diversity at three scales — national, state and landscape — to determine the ecological sustainability of each strategy. Strategies were selected as representative of planning at each of these three scales. Included were the National Strategy for the Conservation of Australia's Biological Diversity (Commonwealth of Australia 1996), Draft Nature Conservation Strategy for Western Australia (Department of Conservation and Land Management 1992) and Local Greening Plans (Greening Australia 1995).

The study focus was the ecological sustainability of the decision making approaches and on-the-ground actions (i.e., decisions) recommended in each strategy. Criteria developed by Gardner (1989), for assessing the ecological sustainability of environmental decision making and decisions in Canada, provided the analytic framework for this study. These criteria, described below, were used to determine the strengths and weaknesses of each strategy in achieving ecologically sustainable decision making and decisions. This analysis allowed broad inferences about the effectiveness of strategic planning at various scales, in terms of achieving ecological sustainability, to be drawn.

Sustainable development and ecologically sustainable development must be two of the most frequently defined, interchanged and contested terms in natural resource management today. The most widely-used definition of sustainable development is "development that meets the needs of the present without compromising the ability of future generations to meet their own needs" (World Commission on Environment and Development 1987). In Australia, a national strategy for ecologically sustainable development was produced by the Commonwealth Government in 1992 to provide "a broad strategic and policy framework under which governments will co-operatively make decisions and take actions to pursue ecologically sustainable development in Australia" (Beder 1996). It aims to: integrate environmental and economic goals in policies and activities by ensuring that environmental assets are appropriately valued; provide equity within and between generations; deal cautiously with risk and irreversibility; and recognize global dimensions of policies and decisions (Beder 1996).

THE PLANNING STRATEGIES

As mentioned previously, three strategies were selected to represent planning at different scales. The National Strategy for the Conservation of Australia's Biological Diversity (Commonwealth of Australia 1996) was prepared as part of implementation of the International Convention on Biological Diversity, one of four main agreements reached at the Earth Summit in Rio de Janiero and ratified by Australia in June 1993. The strategy acknowledges the core objectives and accepts the guiding principles of the National Strategy for Ecologically Sustainable Development (1992). Widely endorsed by all levels of Australian Government, collective responsibility for the national biodiversity strategy rests with Commonwealth and State Governments. The strategy's goal is protecting biological diversity and maintaining ecological processes and systems by a comprehensive approach to identifying, conserving and managing Australia's biological diversity. Key components include: a bioregional approach to planning and management; managing for conservation; establishing protected areas; improving conservation off-reserves; and recognizing indigenous peoples' contribution to conserving biological diversity.

The strategy also aims to develop and improve techniques for integrated management of ecologically-viable protected areas within agricultural and resource production systems. It promotes action on a regional basis to address biological diversity loss resulting in part from incremental decision making. This regional action is intended to achieve integration across all levels of government and community activity. The strategy also calls for the consideration of biological diversity conservation in decision making by all levels of government, in accordance with principles of the National Strategy for Ecologically Sustainable Development. These guiding principles require decision making to: integrate both short- and long-term economic, environmental, social and equity factors; promote the precautionary principle; recognize global dimensions of actions; and culminate in decisions providing for broad community involvement where necessary (Commonwealth Government 1992).

The second strategy was a state-level plan, the Nature Conservation Strategy for Western Australia: A Draft for Public Comment, published by the Western Australian Department of Conservation and Land Management in 1992. It has not been produced in final form, however it remains the main strategic guide to biological diversity management in Western Australia. Its objective is "to conserve in perpetuity the widest possible diversity of indigenous landscapes, ecosystems and species (including their genetic variability), in natural habitats in Western Australia" (Department of Conservation and Land

Management 1992). Associated actions are directed towards integrating nature conservation with a range of natural resource management activities. The strategy focuses on the direct responsibilities of the Department of Conservation and Land Management – management of Department of Conservation and Land Management lands and waters, including national parks, marine reserves and state forests, and management of biota on lands and waters of all tenure. The document includes sections on: the need for and current status of biological diversity conservation; placing the strategy in a global, national and state; existing nature conservation reserve systems and their future planning and management; and co-ordinating agency and community involvement. It emphasizes the need for partnerships of mutual benefit between the Department of Conservation and Land Management, resource users and the wider community to enable integration of activities and proper management of lands and waters for nature conservation.

The landscape-level strategy assessed was the Local Greening Plans: A Guide for Vegetation and Biodiversity Management (Greening Australia 1995). The guide was developed by the non-government organization Greening Australia with support from the Commonwealth Department of the Environment, Sport and Territories. It was produced to assist local governments in making decisions about sustainable use of private and public lands. The document provides guidelines on how to produce a "greening plan" which is "a structured and systematic approach to managing vegetation and to helping maintain biodiversity within a local government area for the long term" (Greening Australia 1995). Greening plans are expected to help local government assess development and land use change proposals. This strategy was a response to an increasing awareness of the effects of local government decisions, projects, plans and programmes on plant communities, wildlife and their habitats, and pests. The Local Greening Plan Strategy includes: an explanation of the steps in creating a greening plan, including partnership building, resource and management assembly and vegetation inventory; actions needed to implement the greening plan on-the-ground, especially details on local government and community action; a national perspective on greening plans; and methods for keeping government and community working together at the grassroots.

THE ASSESSMENT CRITERIA

Criteria derived from the work of Gardner (1989) were used to assess the strategies. Gardner (1989) originally developed her criteria to assess the ecological sustainability of decision making and decisions/outcomes in Canadian environmental assessment, planning and management. This was the first attempt to comprehensively describe and assess the attributes of decision making contributing to ecological sustainability. Subsequently Slocombe (1993) used a similar approach to assess project-based, environmental, regional planning. Sutton (1995) used Gardner's criteria to assess the ecological sustainability of decision making and an on-the-ground decision regarding management of the Wilson Inlet on the south coast of Western Australia.

Gardner (1989) and subsequently Sutton (1995) used four criteria to assess the ecological sustainability of decision making. The first of these was goal-seeking. If decision making is goal-seeking, then goals are clearly stated early in decision making, they are consistent with the principles of sustainability, a wide range of options are proactively considered, and both individual and societal interests are reflected. The second criterion is a systems approach to decision making. Such an approach is characterized by recognition of the connections within and between socioeconomic (human) and biophysical systems and the spatial and temporal context of decision making. The third criterion for assessing decision making is adaptability. An adaptable decision making process is dynamic, responsive, evolutionary, and iterative, facilitating experimentation and learning by feedback and self-reflection. The last criterion is interactive organizations. Characteristics include transdisciplinary collaboration between experts and integration within and across sectors.

The criteria for assessing decisions for their contributions to ecological sustainability are satisfying human needs, maintaining ecological integrity, equity and social justice and providing for self-determination and cultural diversity (Gardner 1989; Sutton 1995). Maintaining ecological integrity is characterized by recognizing and adhering to the limits to ecological carrying capacity, as well as achieving environmental rehabilitation and mitigating impacts. Satisfying human needs is based on the precept that ecosystem conservation depends on sustaining human culture. This criterion is met if human culture is sustained. Equity and social justice can be achieved if decision making power is equitably distributed and intergenerational equity addressed. Intergenerational equity relies on "protecting" future generations from current ones and retaining diverse, future opportunities. The attributes of self-determination and cultural diversity are equitable distribution of decision making power, self-reliance, and individual development and fulfilment.

ECOLOGICAL SUSTAINABILITY OF THE STRATEGIES

The three strategies were assessed using the above criteria, applied as two discrete sets, one addressing sustainability of the decision making processes advocated in each strategy and the other set assessing the recommended actions or decisions. This assessment concludes with a comparison between the strategies.

The strength of the National Strategy lies in its decisions rather than its provision for sustainable decision making (Table 1). The decisions are primarily directed toward maintaining ecological integrity, the first criterion for assessing the

Table 1. National Strategy strengths and weaknesses identified through criteria analysis.

Criteria	Strengths	Weaknesses
A. DECISION MAKING CRITERIA A1. Goal-seeking approach	—	• Lacks early goal identification, proactive goal seeking approaches and integration of decision making with monitoring and management systems and within and between management systems.
A2. Systems-oriented approach	• Adopts comprehensive and interdisciplinary approach using farm management and bioregional and catchment planning. • Emphasizes importance of improved knowledge in identifying species and communities of value to biophysical and human systems.	• Lacks systems approach and use of spatial and temporal boundaries in decision making. • No acknowledgment of human systems and interaction with biophysical systems.
A3. Adaptive approach	• Provides mechanisms for continued review of impacts on biodiversity from resource production management and gathering and responding to new information. • Recommends a suite of actions — reserve systems, off-reserve conservation, bioregional planning and ecologically sustainable use in production environments.	• Lacks means for managing risk through anticipation and prevention and for accounting for changes in physical, social and ecological systems. • No promotion of modelling, monitoring, feedback and self-reflection necessary for adaptation.
A4. Interactive organizations	• Recommends transdisciplinary collaboration by stakeholders through advisory councils.	• Lacks integration of decision making processes with monitoring and management systems, organizational means for transdisciplinary collaboration between experts and community, participatory processes integrating socio-cultural, technical and institutional factors, and ways of transferring information between scales and groups.
B. OUTCOME/DECISION CRITERIA B1. Maintaining ecological integrity	• Provides for genetic diversity preservation (by in-situ and ex-situ conservation), long-term ecological monitoring, and a shift from maximum yield to sustained yield. • Recommends focusing research and management on identifying valued ecosystem components. • Promotes attitudinal change in community and industry.	• Lacks spatial and temporal boundaries for management, recommendations for modelling and monitoring, and explicit means for maintaining essential ecological processes and life support systems.
B2. Satisfying human needs	• Values community needs (cultural, social, medicinal) as part of biodiversity conservation. • Identifies multiple values and integral relationship society has with resource base.	• Lacks recognition of economic development as part of human needs.
B3. Achieving equity and social justice	—	• Lacks recommendations for dealing with conflict. • Lacks means of dealing with power imbalances in decision making and equity concerns within and between generations.
B4. Providing for self-determination and cultural diversity	• Recognizes importance of culturally appropriate forms of development by inclusion of cultural group representation.	• Lacks recognition of community self-reliance, individual development and fulfilment. • No recommendations for community participation in decision making.

ecological sustainability of decisions, through recommendations for genetic diversity preservation and long-term monitoring. An important attribute of this criterion — providing temporal and spatial boundaries — is absent. Satisfying human needs, another outcome-directed assessment criterion, is met by this Strategy through recognizing community needs as part of biodiversity conservation. On the downside, the National Strategy's decisions, in other words, its recommendations for action, do not actively address equity and social justice, or self-determination (Table 1). Particularly problematic is the lack of recommendations for dealing with conflict, with this shortcoming related then to difficulties in achieving equity and social justice. Also potentially problematic is the bioregional focus that may preclude local-level planning and associated self-determination.

The decision making processes advocated do not fare well against the sustainability criteria (Table 1). Although the strategy provides for an interdisciplinary approach and ongoing review (Table 1, Criterion A2 and A3), there is no overt acknowledgement of the interaction between human and biophysical systems. And, in terms of

Table 2. State Strategy strengths and weaknesses identified through criteria analysis.

Criteria	Strengths	Weaknesses
A. DECISION MAKING CRITERIA A1. Goal-seeking approach	—	• Lacks early goal identification, proactive goal seeking approaches and explicit consideration of societal needs in decisions.
A2. Systems-oriented approach	• Identifies potential threats to ecosystem components.	• Lacks systems and interdisciplinary approaches to analysis. • No acknowledgement of human systems and interaction with biophysical systems.
A3. Adaptive approach	• Emphasizes conserving multiple options in human and biophysical systems and improving knowledge acquisition through several means to provide an information base for future decisions.	• Lacks means for managing risk through anticipation and prevention and for accounting for changes in physical, social and ecological systems.
A4. Interactive organizations	• Promotes free exchange of views and information between involved parties through liaison, public access to information and communication programmes.	• Lacks integration of decision making processes with monitoring and management systems, organizational means for transdisciplinary collaboration between experts and community, participatory processes integrating socio-cultural, technical and institutional factors, and ways of transferring information between scales and groups.
B. OUTCOME/DECISION CRITERIA B1. Maintaining ecological integrity	• Provides for genetic diversity preservation, and ecological modelling and monitoring. • Recommends basing use of resources and ecosystem restoration on ecosystem limits, plus expanding the information on which future decisions depend. • Recognizes the importance of processes and systems for future biodiversity and human survival.	• Lacks explicit means for identifying valued ecosystem components.
B2. Satisfying human needs	• Recognizes the economic value of conservation.	• Lacks recognition of relationship between society and natural resource management and including management recommendations directed towards meeting human needs.
B3. Achieving equity and social justice	• Provides for equity within and between generations and sustainable use of the resource base.	• Lacks recommendations for dealing with conflict and for community participation in decision making.
B4. Providing for self-determination and cultural diversity	• Recognizes cultural diversity through recommendations for participation and recognizes cultural values in decision making.	• Lacks recognition of community self-reliance, individual development and fulfilment. • No recommendations for broad community participation in decision making.

adaptability, the third decision making criterion, there is no means of informing adaptation through modelling, monitoring or feedback. A goal-seeking approach, the first criterion given for assessing decision making, is not apparent. The last criterion, interactive organizations, is partially met by recommendations for advisory councils as conduits for information exchange and collaboration.

For the State Strategy, similarly to the National Strategy, its strength lies in the sustainability of its decisions rather than its recommended decision making processes (Table 2). Its decisions, similarly to the National Strategy, are directed toward ecological integrity with an emphasis on preserving genetic diversity, ecological processes and systems, and satisfying human needs. The economic element of human needs is clearly identified and addressed. Again similarly to the National Strategy, social justice and self-determination are not actively addressed. And, methods for resolving conflict are not obvious. In contrast to the National Strategy, the State Strategy considers equity by advocating sustainable resource use and avoiding future costs through planning and management.

The State Strategy did not fare well against the sustainability criteria for future decision making (Table 2). Although the strategy generally met the adaptability criterion, through its emphasis on knowledge acquisition and use for future decisions, it only partially met the system-oriented criterion. Limited recognition of a systems-oriented approach is evidenced by little acknowledgement of human systems and their interactions with biophysical ones. In common with the National Strategy, this strategy provides limited guidance as to how interactions between organizations across spatial scales are to be achieved. This criterion of interactive organizations is, however, partially met by recommending promotion and open exchange of views and information by involved parties.

Table 3. Landscape-level Strategy strengths and weaknesses identified through criteria analysis.

Criteria	Strengths	Weaknesses
A. DECISION MAKING CRITERIA A1. Goal-seeking approach	• Provides guidelines for land use planning with an environmental emphasis. • Facilitates forward planning through recommending strategic planning and development control plans.	• Lacks reference to the need to develop economic, social, cultural and political goals early in the planning process.
A2. Systems-oriented approach	• Emphasizes a comprehensive and interdisciplinary approach and analysis using community and environmental knowledge. • Addresses human systems and their interactions with biophysical systems. • Identifies potential threats to ecosystem components.	—
A3. Adaptive approach	• Emphasizes monitoring using key sources of information, flexible planning, updating planning as-needed, taking multiple actions and adapting implementation.	• Lacks means for managing risk through anticipation and prevention.
A4. Interactive organizations	• Recommends working groups to build community support, explicit processes for seeking community agreement, transferring information within and across scales to build on existing knowledge and interaction between organizations and between human and biophysical systems.	• Lacks recommendations for integrating sociocultural, technical and institutional factors.
B. OUTCOME/DECISION CRITERIA B1. Maintaining ecological integrity	• Links local plans with other adjacent greening plans, regional and catchment plans. • Integrates anthropocentric and biocentric principles.	• Lacks spatial and temporal boundaries for management, recommendations for modelling and monitoring, explicit means for maintaining essential ecological processes and life support systems and recommendations for preserving genetic diversity.
B2. Satisfying human needs	• Recognizes the economic costs of losing native vegetation and economic benefits of better management and the broad range of factors required to fulfil human needs.	• Lacks mention of need to explicitly consider human needs in developing management objectives.
B3. Achieving equity and social justice	• Provides opportunities for bargaining and negotiating and for resolving conflicts between social and ecological objectives.	• Lacks consideration of need to include equity in decision making.
B4. Providing for self-determination and cultural diversity	• Recognizes value of landscape-level planning for expression of cultural needs and wide community involvement.	• Lacks recognition of community self-reliance, individual development and fulfilment.

The strength of the Landscape Strategy rests on the apparent contribution of its decision making processes to ecological sustainability (Table 3). This is in contrast to the previous two strategies that performed better in terms of the sustainability of their decisions, rather than their recommended decision making processes. Notable strengths of the decision making processes in the Landscape Strategy are its systems-oriented and adaptive approaches, and interactive organizations. For example, a collaborative approach to decision making is advocated, with local governments working with communities to create and implement local plans.

The decisions encapsulated in the Landscape Strategy met the sustainability criteria of satisfying human needs, equity and social justice, and social self-determination and cultural diversity (Table 3). Human needs were satisfied through recognizing the economic benefits of conservation. Actions in the Strategy directed towards community participation, empowerment, ownership and self-determination helped meet the remainder of these criteria. Unlike the other two strategies, this one provided opportunities for bargaining and negotiating, contributing towards equity and social justice. Transferable development rights between landholders and councils is an example of a bargaining framework. Resolving conflict is a fundamental objective of Local Greening Plans. Unlike the other two Strategies, ecological integrity was not the best-met criterion for decisions and preserving genetic diversity was not addressed.

Comparing the three strategies provides insights into general strengths and weaknesses of current strategic planning, in terms of both the contributions of recommended decision making processes and the decisions embodied in the plans, for achieving ecological sustainability. Care should be taken in any conclusions drawn from such a comparison given the very different reasons and audiences for each strategy. However, useful inferences can be made about the extent to which the recommended decision making processes and outcomes in current strategic planning contribute to ecological sustainability and the associated potential benefits of planning at different scales.

The greatest strength of both the National and State Strategies was their recommendations for maintaining ecological integrity. Managing for ecological integrity is the "core business" of the government departments who produced these strategies. These departments have legislative and jurisdictional responsibility for selected ecological elements of the environment. Also, the public expects them to be responsible for managing these elements. The ability of these two strategies to satisfy human needs, another of their strengths, reflects recognition of the economic benefits of nature conservation and perhaps more realistically, recognition of the importance of couching these discussions in economic terms. Lack of decisions directed towards maintaining ecological integrity was the main weakness of the Landscape Strategy. The absence of detail on maintaining ecological processes and preserving genetic diversity may reflect one or more of the following: belief that such matters should be addressed at other strategic levels; lack of expertise to include such details; and concern that jurisdictional responsibility lies elsewhere.

Ecological integrity aside, the Landscape Strategy performed well against both the decision and decision making sustainability criteria. Particular strengths were its systems-oriented and adaptive approach complemented by means for interactions between organizations. This explicit, detailed attention to ongoing adaptation and decision making reflects the origins and purpose of the strategy. It was produced by a non-government organization, Greening Australia, with no responsibilities for implementing decisions and no means of doing so. Its purpose was to provide a process for others to use.

There are some weaknesses common to the National and State, and all three strategies. The decisions in the National and State Strategies did not perform well against the sustainability criteria of equity and social justice, and self-determination and cultural diversity (Tables 1 and 2, Criterion B3 and B4). Part of this is a jurisdictional problem, in that the lead departments are responsible for the biota but not for the publics associated with it. However, more than this, this lack of emphasis reflects the history, culture and training of those involved in nature conservation management. As such, equity is still afforded low priority in final decisions in conservation and traditional production systems (Payne 1996). Ways of ensuring interactions between organizations was another area receiving limited attention in the National and State Strategies. Lack of interaction between organizations and disciplines, and subsequent problems integrating sociocultural, technical and institutional factors, has been flagged by many researchers as problematic in natural resource management (Dovers 1995; Cortner *et al.* 1996). Inattention to this area in the National and State Strategies is more likely to reflect jurisdictional constraints than lack of awareness of its importance.

Several weaknesses were common to all three strategies. None overtly included ongoing goal seeking as part of their recommendations for future decision making. For the National and State Strategies, goals would have been set either by legislative or administrative direction or subsequently as part of development of the strategy itself. As such, the strategy authors would be unlikely to see any need to allow for subsequent changes. For the Landscape Strategy, such goal setting may not have been considered necessary at this scale. Another weakness common to all three strategies was lack of reference to spatial and temporal boundaries. Perhaps these were assumed to be widely known and were therefore not acknowledged. The last weakness and warranting special mention, was the lack of attention to self-determination and the accompanying concept of self-reliance. Gardner (1989) suggested self-determination is thwarted by centralized control stifling public initiative. This, however, does not explain its absence from the Landscape Strategy. Rather, it may reflect a lack of awareness among those involved in planning for biological diversity that sustainability depends on local initiative and the well-being of communities as well as a myriad of more biologically-directed factors.

RECOMMENDATIONS — STRATEGIC PLANNING FOR ECOLOGICAL SUSTAINABILITY

Inferences drawn from this work provided the basis for the following recommendations, all directed towards improving the contribution by strategic planning to ecological sustainability. The decision making processes advocated in the Landscape Strategy best met the sustainability criteria. The obvious recommendation is planning at this scale to achieve these benefits. More importantly, however, is making provision no matter what the scale for meeting the criteria listed against sustainable decision making. This means including elements in strategic plans directed toward ongoing and flexible goal-seeking and adaptive approaches to management (Haney and Power 1996). Also important is using systems-approaches in developing strategies and providing for ongoing interactions between organizations, recognizing that moving across jurisdictions, disciplines and traditional activities will not be easy (McDonald 1996). From this analysis and current research by the first author it seems that flexibility and interaction between organizations are most easily achieved at smaller, local scales.

The State and National Strategies are more "static" than the Landscape Strategy. The former two provide numerous recommendations for actions but few for negotiations regarding future directions. In contrast, the Landscape Strategy is predominantly a guide for developing future directions. The emphasis of the government strategies on outcomes with little attention to further, strategic, decision making processes is understandable. Both documents probably reflect carefully-won consensus between numerous interests; each establishes and refines responsibilities between competing interests; and substantial resources were most likely dedicated to each document's development. To add dynamism to such strategies, including some negotiable elements, and the means for doing so and for dealing with conflict, seems essential. Mechanisms for facilitating consultation and dispute resolution are an important consideration (Fowler 1992; Sandford 1992; Dovers 1995). If such changes were made, these strategies would fare better when assessed against the sustainability criteria of adaptability and achieving equity and social justice.

Loss of biological diversity is largely a result of socioeconomic causes. Thus, solutions must be both ecologically and socially derived (Machlis 1992). Thus, strategic planning must address human as well as ecological needs. This is new territory for many natural resource managers, although the evidence from this analysis is that satisfying human needs has been embraced by strategic planners addressing biological diversity issues. Recognition of human needs and concerns in strategic planning needs to go much further to meet the sustainability criteria given here. In particular, providing for self-determination is important. This means ongoing, meaningful inclusion of communities in decision making and allowing full expression of their cultural needs in such processes. Local-level approaches are usually advocated as the most effective and cost-efficient (Moore and Bache 1997).

Is there one best scale for strategic planning? The evidence from this study is "no". Maintaining ecological integrity, and the associated issue of preserving genetic diversity, appears easier to consider at larger scales, as in the State and National Strategies. On the other hand, local-level planning such as the Landscape Strategy appears the best way to gain public participation and facilitate community self-determination, while addressing equity and social justice concerns. Ideally, it seems strategic planning should occur at several scales, to make the most of the strengths of each scale, with links between scales. Although this study did not address or explore these links, it seems imperative to link actions across scales to allow for adaptability, learning and complementary activities.

ACKNOWLEDGEMENTS

Comments by Ken Wallace, Keiran McNamara and Stephen Renton contributed significantly to this chapter.

Susan Jennings acknowledges the support of the Land and Water Resources Research and Development Corporation.

REFERENCES

Beder, S., 1996. The Nature of Sustainable Development. Scribe: Newham, Australia. 2nd Ed.

Commonwealth Government, 1992. National Strategy for Ecologically Sustainable Development. AGPS: Canberra.

Commonwealth of Australia, 1996. The National Strategy for the Conservation of Australia's Biological Diversity. Department of the Environment, Sport and Territories: Canberra.

Cortner, H. J., Shannon, M. A., Wallace, M. G., Burke, S. and Moote, M. A., 1996. Institutional Barriers and Incentives for Ecosystem Management: A Problem Analysis. General Technical Report PNW-GTR-354 February 1996. United Stated Department of Agriculture, Pacific Northwest Research Station: Portland, Oregon.

Department of Conservation and Land Management, 1992. A Nature Conservation Strategy for Western Australia. A Draft for Public Comment. Department of Conservation and Land Management: Perth, Western Australia.

Dovers, S., 1995. Information, sustainability and policy. *Aust. J. Environ. Man.* **2:** 142–56.

Fowler, R. J., 1992. Environmental dispute resolution techniques — What role in Australia? *Environ. Plan. Law J.* **9:** 122–31.

Gardner, J. E., 1989. Decision Making for Sustainable Development: Selected Approaches to Environmental Assessment and Management. *Environ. Impact Assess. Rev.* **9:** 337–66.

Greening Australia, 1995. Local Greening Plans. A Guide for Vegetation and Biodiversity Management. Greening Australia Ltd: Canberra.

Haney, A. and Power, R. L., 1996. Adaptive management for sound ecosystem management. *Environ. Man.* **20:** 879–86.

Lee, R. G., Flamm, R., Turner, M. G., Bledsoe, C., Chandler, P., DeFarrari, C., Gottfried, R., Naiman, R. J., Schumaker, N. and Wear, D., 1992. Integrating Sustainable Development and Environmental Vitality: A Landscape Ecology Approach. Pp. 499–522 *in* Watershed Management: Balancing Sustainability and Environmental Change ed by R. J. Naiman. Springer-Verlag New York Inc: New York.

Machlis, G. E., 1992. The contribution of sociology to biodiversity research and management. *Biol. Cons.* **62:** 161–70.

McDonald, G. T., 1996. Planning as sustainable development. *J. Plan. Educa. Res.* **15:** 224–36.

Moore, S. A. and Bache, S. J., 1997. Spatial scale and environmental justice in Australian environmental dispute resolution: does it matter and to whom? Environmental Justice: Global Ethics for the 21st Century Conference, The University of Melbourne, Victoria, Australia, 1–3 October 1997.

Moore, S. A., Jennings, S. F. and Tacey, W. T., 1998. A crisis in commitment: re-thinking environmental decision making for ecologically sustainable development. Paper presented at Managing for Ecological Sustainability Conference, The University of Queensland, Queensland, Australia, 22–24 September 1998.

Natural Resources Canada, 1998. www.nrcan.gc.ca/mms/pubs/monoe4.htm, accessed 1 September, 1998.

Payne, R. J., 1996. Values in Integrated Resource Management: Timber Management on Crown Land in Ontario. Pp. 9. The Sixth International Symposium on Society and Resource Management: Social Behavior, Natural Resources, and the Environment. May 18–23, 1996. The Pennsylvania State University, University Park, Pennsylvania.

Sandford, R., 1992. Conflict management, dispute resolution, and ecologically sustainable development. *Environ. Impact Assess. Rev.* **12:** 181–83.

Slocombe, D. S., 1993. Environmental planning, ecosystem science, and ecosystem approaches for integrating environment and development. *Environ. Man.* **17:** 289–303.

Standing Committee on Agriculture, 1991. Sustainable Agriculture. Report of the Working Group on Sustainable Agriculture. SCA Technical Report Series No 36. CSIRO: Victoria.

Sutton, A., 1995. Decision making for ecologically sustainable development: Wilson Inlet a case study. Master of Science (Environmental Science) Dissertation. Murdoch University: Murdoch, Western Australia.

World Commission on Environment and Development, 1987. Our Common Future. Oxford University Press: Oxford.

CHAPTER 59

Planning the move to mosaics

An assessment, planning and practical phased approach to protecting and restoring indigenous ecosystems in production focused landscape

Grant B. Rosoman[1] and Diane Lucas[2]

The rationale and context for why farmers, foresters, and developers should be concerned about protecting and restoring indigenous ecosystems is discussed. This includes, society values and expectations, productivity benefits, social and ecological benefits, legislative requirements, local action on meeting New Zealand's international responsibilities under the Biodiversity Convention, and the impact and potential of "green" markets. The need for identifying ecosystems is introduced, and a process outlined for Landscape Assessment that identifies underlying ecosystems and their current status. It is proposed that land use can maintain a productivity focus and incorporate biodiversity restoration and conservation back into the fabric of the production dominated landscape. Nature conservation and conservation biology principles are drawn on in a planning approach that involves step by step phases of protection and restoration. Case study examples are used: of a landscape assessment and planning technique that reads the "stories in the land" and uses a participatory approach for the "community of interest", of a plantation forestry development to show what can be achieved if a commitment is made and phases through tree crop rotations used to implement ecosystem restoration and protection; and an urban development that has involved protection and restoration. It is concluded that the step by step process within a restoration planning framework allows economic activity to continue and provides a realistic pathway to ecosystem restoration.

Key words: Biodiversity, Landscape, Ecosystems, Restoration, Assessment, Planning, Mosaics, Nature conservation.

INTRODUCTION — THE TASK OF CONSERVING BIODIVERSITY

"*THE last word in ignorance is the man who says of an animal or plant: 'What good is it?' If the land mechanism as a whole is good, then every part is good, whether we understand it or not. If the biota in the course of aeons, has built something we like but do not understand, then who but a fool would discard seemingly useless parts? To keep every cog and wheel is the first precaution of intelligent tinkering.*" (Aldo Leopold 1953)

The urgent need

Around the middle of this century the human race reached a profound point — we humans had developed the power to dominate, alter and manipulate all life on Earth. The degradation of unique landscapes and ecosystems, the extinction of species and loss of genetic material has occurred like no other period in the history.

Aotearoa's (New Zealand's) biota is unique. We have a high level of endemism, matched only by islands like Hawaii and Madagascar, with more than three quarters of higher plants, and all reptiles, amphibians and bats endemic to New Zealand. We have a number of taxa that are "missing" such as terrestrial mammals, snakes, turtles and salamanders (Diamond 1990). We also have animals that have evolved to fill the niches of those missing, such as large ground herbivorous birds (moa, takahe etc.), and giant wetas behaving like mice. The inherent vulnerability of Aotearoa's biodiversity is illustrated by 43 terrestrial birds becoming extinct since the arrival of humans.

We may still have approximately 24% indigenous forest cover (Ministry for the Environment 1997) but this statistic tells us nothing of either its

[1]Greenpeace New Zealand, Private Bag 92507, Auckland, New Zealand.
[2]Lucas Associates, Marokapara, 351 Manchester Street, Christchurch, New Zealand.
Pages 574–85 *in* NATURE CONSERVATION 5: NATURE CONSERVATION IN PRODUCTION ENVIRONMENTS: MANAGING THE MATRIX ed by J. L. Craig, N. Mitchell and D. A. Saunders. Surrey Beatty & Sons, 2000.

representativeness, distribution, or its condition. It is largely the areas of land that were left as they had the lowest economic value for agriculture. Lowlands forest systems are only around 10% of their original area and wetlands now less than 5%. Areas of our native grasslands such as red tussock are now below 5% of their original range. Under the Australian national reserves system criteria any ecosystem occupying less than 10% of its former range or area, is considered endangered (Joint Australian and New Zealand Environment and Conservation Council/ Ministerial Council on Forestry, Fisheries and Aquaculture National Forest Policy Statement Implementation Subcommittee 1997).

Although the voluntary non-governmental organization/industry agreement, the 1991 New Zealand Forest Accord, has stemmed much of the conversion of native forest to plantations, significant areas (thousands of hectares, including regenerating closed-canopy manuka/kanuka that may be required for protection) are still being cleared for plantation establishment. This is particularly in Northland, and the East Coast including 530 hectares under the government subsidized East Coast Forestry Scheme (Minister of Forestry 1995). The collapse of thousands of hectares of forest in the Ruahine Range due to possum and deer browsing is the tip of the iceberg in terms of ecosystem degradation due to exotic pest impacts.

Focus on the lived in lowlands

In Aotearoa the lowland systems in particular were cleared for farming, to the point where most are now represented only by small remnants. The destruction of the great kahikatea plains' forests or Nga Uruora (groves of life) has been well documented by Park (1995). A "Gap" type analysis carried out by Harding (1997) showed that for the rich dense lowland podocarp ecosystems within the Waikato Department of Conservation Conservancy (excluding the volcanic plateau regions), of the estimated original extent of 349 180 ha, only 2.1% remains, and 0.15% is protected. Lowland wetlands in general have fared even worse, where society has favoured draining them to farm and live on.

Given the high biodiversity values of lowland systems and the degree to which they have been modified, it is here that the task of conservation of biodiversity will need to focus. It is only through protecting, expanding and restoring remnants in lived in landscapes that a representative network of protected ecosystems can be completed. Conservation through protecting and creating publicly-owned natural areas will be insufficient. A process and mechanisms for working within private ownership will be key for achieving this protection and restoration in the lived in landscape.

Society is now in a position where it can make a choice on whether to try to go on living in the land with an impoverished biodiversity, or actively seek a path of restoring former systems. Humans have the power to destroy and alter, but with it also carries a mantle of responsibility to protect, restore and renew.

This chapter sets out some of the rationale for protection of the ecosystems of Aotearoa, an approach for assessment, identification, design, and planned phases to achieve adequate biodiversity protection in the lived in landscapes.

SUMMARY OF VALUES ATTACHED TO BIODIVERSITY[1]

Underpinning the protection and restoration of indigenous communities lie values that are part of the rationale for nature conservation in production focused landscapes. These values will vary from the purely mechanistic view of utilitarian values, where the language of "resources" is expressed in terms of the "ecomachine", to appreciating nature for its own sake, including herein the thoughts of intrinsic value (Jackson 1992).

Direct use values

People rely on plants and animals for much of their food, medicine, clothing, and in some parts of the world almost entirely for fuel and building materials. There is hardly a dwelling in New Zealand that does not contain some native timber. Wild genes have been used repeatedly to provide particular characteristics or disease resistance for the main food plants such as rice, potatoes and corn. The potential for new food crops and improving the viability of current crops is huge.

Medicinal value also provides a strong argument for the protection of flora and fauna. Our native biota is a source of compounds and cures to cancer and other ills. Although indigenous peoples have knowledge of the medicinal uses of most plants, it has been estimated that only about 2% of vascular plants have been screened for their chemical compounds (Worldwide Fund for Nature 1991). A simple New Zealand example is Tea tree or manuka *Leptospermum scoparium*, which has recently been "discovered" as a valuable medicinal oil and honey producing plant in parts of New Zealand. Previously manuka was largely viewed as a woody weed only good for firewood.

Tourism and the recreation sector is now a major part of the New Zealand economy. Behind this is New Zealand's natural heritage. The active use and enjoyment of the natural environment by "paying users" is increasing, in recognition of Aotearoa's unique natural features.

Ecosystem services

The evolution of thinking in terms of the construct "ecosystem" has allowed us to encompass the broader values that biodiversity contain: to see

[1]From the Convention on Biological Diversity, this is defined as "the variability among living organisms from all courses including, *inter alia*, terrestrial, marine and other aquatic ecosystems and the ecological complexes of which they are part; this includes diversity within species, between species and of ecosystems."

interconnections, processes, landform as well as the individual parts or species, and even former functions in degraded areas. Biodiversity and natural ecosystems provide a range of indirect uses for society, or as Ehrlich and Ehrlich (1981) termed them "ecosystem services". It has been found that biodiverse systems are more efficient at harnessing sunlight, consuming CO_2 and are more productive than simple systems with few species (Cherfas 1994). Other ecosystem services include; maintaining air quality and gas ratios, stimulating local and regional rainfall, regulating water quality and quantity, cycling of nutrients and absorbing waste, and pest and disease control processes. These largely "hidden" services are little known but are none-the-less equally important as any direct or economic benefits. It may be that many of these services can be performed by human-created ecosystems (Noss and Cooperrider 1994). However, for whatever reason natural ecosystems have evolved into their current form and function over millions of years. Once ecosystems or species are destroyed they are impossible to re-create. The opportunity cost of this loss through extinction is inestimably high (Ehrlich and Ehrlich 1981).

Recreational and aesthetic

Recreational and aesthetic values are based around people's experiences of biodiversity, with an appreciation and valuing following this. These experiences can be passive or active, and may relate specifically to a component, such as being indigenous. However, areas set aside for recreational and aesthetic objectives may not necessarily meet biodiversity conservation goals (Noss and Cooperrider 1994).

Cultural importance of biodiversity

For some people biodiversity may have immense significance in their lives. For Maori, notable landscape features are viewed as ancestors and have living form. Maori recognized that their existence depended on the diversity of species and the sustainability of these species (Te Puni Kokiri 1994). The Treaty of Waitangi lays the foundation for both Pakeha and Maori cultures with respect to biodiversity through the concept of the kaitiakitanga (stewardship) being explicitly stated in the Treaty. For many New Zealanders some species such as kiwi, tuatara and Hectors Dolphin, and ecosystems such as kauri forest, harakeke (flax) and ti kouka (cabbage tree) wetlands, are icons that are identified with strongly.

Ethical, moral, spiritual and intrinsic values

With the capability for control and modification comes the moral responsibility for conservation. Our ethical obligations must encompass more than to our fellow human beings, as was expressed by Leopold (1949) in his essay on the land ethic (Noss and Cooperrider 1994). The concept of species or ecosystem right is debatable but refers more to the modern idea of intrinsic value, where "... the species which evolved here have a basic right to be here, whether we need them or not." (Ministry for the Environment 1997).

GOING BEYOND GOVERNMENT TO CONSERVE BIODIVERSITY

Legal requirements and the Resource Management Act

While there are a number of Acts in New Zealand that cover conservation of biodiversity, such as the *Reserves Act 1977, Conservation Act 1987, National Parks Act 1980,* and *Wildlife Act 1953,* these are primarily instruments used by the Crown for public lands. With the key issue of protecting and restoring remnants in the production focused land matrix and private land, the primary piece of legislation is the *Resource Management Act 1991.*

The *Resource Management Act's* purpose (Section 5) is *"to promote the sustainable management of natural and physical resources".* "Sustainable management" includes the *protection* of natural and physical resources as well as their "... *potential ... to meet the reasonably foreseeable needs of future generations* (5.2.a); *and safeguarding the life supporting capacity of air, water, soil and ecosystems* (5.2.b)." The definition of "Protection" comes from the *Conservational Act 1987,* and includes *"its restoration to some former state".*

Section 6 concerns Matters of National Importance, including "a) *The preservation of the natural character of the coastal environment, wetlands and lakes and rivers ...",* "b) *The protection of outstanding natural features and landscapes",* and "c) *The protection of areas of significant indigenous vegetation and significant habitats of indigenous fauna."* The preservation of natural character in (a) requires the maintenance of the viable functioning of natural processes and systems together with the visual or perceived attributes of "naturalness" (Lucas 1994).

International responsibilities under the Convention for Biological Diversity

The Convention on Biodiversity was signed by New Zealand in 1992 and since ratified by the New Zealand Government. It clearly recognizes the importance to society of protecting and restoring biodiversity. More than 140 other nations have signed the Convention on Biodiversity.

As part of the New Zealand Government's obligations under the Convention on Biodiversity a *New Zealand Biodiversity Strategy* is under development. It will need to lay a framework for a protected area system that adequately protects and restores Aotearoa's biodiversity. The challenge will be in finding mechanisms and a process for dealing with completely modified lowland ecosystems that are in production focused landscapes on privately owned lands. As yet, there is no broad strategy for enhancing native biodiversity in lowland areas (Ministry for the Environment 1997).

EXISTING MECHANISMS FOR VOLUNTARY PROTECTION AND RESTORATION ON PRIVATE LANDS IN NEW ZEALAND

There are a number of existing mechanisms for landowners to protect biodiversity on their land. They are primarily to address the costs associated with setting aside areas for conservation and they have made a substantial contribution to the national conservation estate. However, they are focused on protection of existing areas rather than restoration of threatened or critically under-represented ecosystems. Nearly 50 years ago Aldo Leopold raised these issues in his essay on "the land ethic", and they are equally relevant today.

"When the private landowner is asked to perform some unprofitable act for the good of the community, he today assents only with outstretched palm. If the act costs him cash this is fair and proper, but when it costs only forethought, open-mindedness, or time, the issue is at least debatable." (Leopold 1949)

Forest Heritage Fund and Nga Whenua Rahui

Both are government contestable funds to assist protection of indigenous forests and associated ecosystems that represent the full range of natural diversity originally present in the New Zealand landscape. Since 1990 the Forest Heritage Fund has protected over 100 000 hectares through either direct purchase, legal covenanting or assistance with fencing. Nga Whenua Rahui is dedicated to assisting Maori owners protect their indigenous forest, largely through covenants, and assisting with costs of protection. More than 50 000 hectares of forest have been protected.

Queen Elizabeth II National Trust

The Queen Elizabeth II National Trust provides assistance for landowners in establishing open space covenants to protect particular natural features in the landscape. Over 1 000 landowners have entered into Queen Elizabeth II National Trust open space covenants protecting some 45 000 hectares, predominantly indigenous forest remnants (Porteous, this volume).

New Zealand Forest Accord

The New Zealand Forest Accord is an agreement signed in 1991 between the plantation forest industry and environmental non-governmental organizations restricting the clearance of native forests for plantations. The forest industry has largely followed this although considerable areas of regenerating indigenous forest continue to be cleared.

Local and regional council initiatives

A recent review of local councils found that one-third intended to offer rate relief on application for protected areas, and one-quarter some form of financial assistance for conservation such as legal protection and fencing (Froude 1997). Several councils propose or offer development privilege incentives in return for landowner conservation actions. Some regional councils offer assistance also, such as Environment Waikato Environmental Initiatives Fund, and Environment Bay of Plenty's Environmental Farm Plans.

Other

There are many other initiatives that in particular support restoration of indigenous ecosystems, such as the New Zealand Native Forest Restoration Trust, many community and conservation organization restoration and replanting projects, and the Lottery Environment and Heritage Fund.

CERTIFICATION AND MARKET INCENTIVES

The market is one channel for society to express preferences. Forest management certification and performance standards cover biodiversity protection and restoration in response to consumer concerns. Agricultural certification is moving towards recognizing biodiversity e.g., Project 98 (Salmon, in litt.). Also, in many cases there is considerable market benefits through increased prices and demand for semi-rural subdivisions that have a component of native area protection and restoration.

Forest management certification

Most international non-governmental organizations working on forests are supporting certification as a way of providing credible product information for the buyer on the environmental and social performance standards of the forest operation and the chain of custody of the product. Non-governmental organizations view certification as an important market tool that can reward producers who address environmental and social concerns, including protection of indigenous ecosystems. The demand for certified wood is on the increase.

Over the past five years international non-governmental organizations have developed some clear requirements for certification. In New Zealand, environmental non-governmental organization's view that, at the moment, certification is only appropriate for plantations. Certification must be based on clear environmental and social performance standards, which;

- have explicit support from ecological and social groups,
- are credible and understandable to consumers,
- are based on objective and measurable criteria,
- have been independently assessed,
- are transparent and not set by parties with vested interests (e.g., industry-set standards),
- are implemented at the forest management unit level, and
- are linked to a chain of custody and product lifecycle verification.

Greenpeace is supporting credible certification around the world, such as in Germany, Canada, Austria, Switzerland and Solomon Islands. The first New Zealand certification assessment under the Forest Stewardship Council for plantations in Southland is nearing completion and includes standards for biodiversity protection and restoration.

The Forest Stewardship Council is currently the only credible international certification framework. It is supported by the major international non-governmental organizations working on forests including Greenpeace, World Wide Fund For Nature and Friends of the Earth, but also the timber trade such as a United Kingdom buyer's group of 70 retailers with a market of over NZ$15 billion who have committed to Forest Stewardship Council certified products. The Forest Stewardship Council has accredited five certifying organizations that are now around the world certifying forests to the agreed set of Principles and Criteria for Forest Management.

Examples of Forest Stewardship Council Criteria for Forest Management that refer to biodiversity:

6.1 ". . . Assessments shall include landscape level considerations . . ."

6.2 ". . . Conservation zones and protection areas shall be established . . ."

6.4 "Representative samples of existing ecosystems within landscapes shall be protected in their natural state . . ."

10.2 ". . . Wildlife corridors, streamside zones and a mosaic of stands of different ages and rotation periods, shall be used in the layout of plantations, . . ."

10.5 "A proportion of the overall forest management area [plantation] . . . shall be managed so as to restore the site to a natural forest cover."

(Forest Stewardship Council 1996).

CASE STUDY OF A LANDSCAPE ASSESSMENT AND PLANNING TECHNIQUE

The landscape provides the context for forest and land use planning. The landscape that exists and potential landscape change from land development, management and harvest must all be addressed. For forest or land management, ecosystem assessments need to be undertaken at a landscape scale. Landscape assessments are integrative of the social elements in the landscape and act as a process for land managing communities to learn more of the rationale behind biodiversity protection, to participate in it, and create their own version of a network of protected systems.

Reading the stories in the land

To plan for change people need to understand a landscape to assess it. Assessment needs to consider the physical land as well as people's relationship with it. It is important to read the site together with its context, particularly in relation to the adjoining lands and the rest of the catchment. Assessing the landscape thus involves attempting to understand something of the layers of nature and layers of culture that together make that place. Ecosystem and landscape assessments (vas. species inventories) are an efficient (fast) tool for designing protected ecosystem networks as it is easier to deal with dozens or hundreds of ecosystems rather than thousands or millions of species (Noss and Cooperrider 1994).

It is limiting to just read the land surface and the vegetative cover that currently exists. With so much of the production and lived in landscapes of Aotearoa simplified to just pastoral, plantation or scrub cover, the underlying natural diversity and potential can easily be overlooked. Also, merely identifying the remnants of nature, the remnants from fire, browse and clearing may be inadequate as they may not be typical of the vegetation of the past.

To plan the move to a richer landscape, one planning technique involves firstly seeking to understand "what nature intended" for that place — the type of forest or grassland ecosystem that would occur there naturally if there was not intervention. Getting an idea of the nature of the past, both pre-people and pre-European, will help understand something of the natural potential of the place. Knowing what types of vegetation probably existed, and what caused the changes that lead to the vegetation of today, provides useful information for exploring options for future vegetation. Provide easily understood data to enable people to understand the underlying nature of their place.

By identifying the underlying ecosystems, and something of their dynamics — the seral stages, the conditions for establishment, the vulnerability — will provide information for the community of interest to plan for potential restoration opportunities.

Thus putting together an assessment involves mapping the lands on aerial and oblique photographs and topographic maps to identify site diversity and patterning in terms of landform, soils, past and present vegetation, hydrology/drainage, and also for land use, productivity and amenity. Not just opportunities, but production limitations also need to be identified and mapped in an assessment.

Assessment and plan development guidelines

Identify and map catchments as possible planning units. Habitat values both on-site and beyond, and any indication of trends in their expansion or reduction, need to be identified. Special niches, linkages and corridors need to be noted.

Provide supporting information, but encourage the community of interest to input their knowledge and understanding. The community of interest includes the main stakeholders of landholders, managers, users and neighbours, as well as those with an interest in the area, such as community groups, scientists, government agencies, and local

Table 1. Factors influencing protected ecosystem design requirements. Based on Noss and Cooperrider (1994). They range from the purer ecological to the socio-political, and their impact on protected area requirements is noted.

Factor	Influence
1. Scale of the landscape	Larger landscapes such as plains, may require less area than smaller "micro" land scapes with uncommon or unique features.
2. Degree of homogeneity	Landscapes with considerable diversity of species, habitat and ecosystems will require larger areas and range of sites.
3. Proportion of Indigenous Ecosystems in a Landscape	In landscapes that have large indigenous areas remaining, it will be easier to protect larger areas. In depleted and degraded landscapes any remnant will be significant. However, in production dominated landscapes with high degrees of private ownership, protection and restoration will face more barriers and take longer.
4. Connectivity	In fragmented landscapes with low connectivity, area increase required (see Saunders *et al.* 1991), if remnants well connected then less. Position in the landscape may be important. A few large areas are better than many small remnants. It may be important to maintain mainland "islands" for ease of predator control, and they may not need land corridor to protect key species and elements.
5. Natural Disturbances	Such as drought, catastrophic storms, volcanic eruptions: if the risk is high then area and range of sites higher.
6. Distribution of Species or Ecosystem	If widespread then lesser area, if rare or local endemic (due to human influence) then greater proportion required. Species that have a narrow range of movement will require less, whereas long-ranging species such as kaka *Nester meridionalis* and Karearea *Falco novaeseelandiae* will require a larger range.
7. Replication	Number of sites to provide an adequate and representative system (e.g., Joint Australian and New Zealand Environment and Conservation Council/Ministerial Council on Forestry, Fisheries and Aquaculture National Forest Policy Statement Implementation Subcommittee 1997). Will depend on site viability and risks.
8. Quality of Habitat	Linked to predator and exotic weed/pest influences in New Zealand e.g., for kokako *Callaeas cinerea* and kereru *Hemiphaga novaeseelandiae* densities, and area required may depend on the degree of management intervention.
9. Population Viability Criteria	Time scales and probability of persistence will influence target population size and area required. Estimates will vary depending on assumptions and parameters (Noss and Cooperrider 1994), especially whether evolutionary potential is included.
10. Degree of "Unnatural" disturbance	Level of human use such as grazing or recreation, or pest species density will effect the quality of the habitat. If effect high then larger areas may be required.
11. Adjacent Land Use	If management in areas adjacent to proposed protected areas has a negative impact on biodiversity, such as edge effects, spray drift, nutrient or sediment influx, then areas will need to be larger. If adjacent areas are managed to maintain and mimic indigenous components and structure, then the protected area requirement will be less, e.g., light grazing of short tussock grasslands, long-rotation plantations, native tree plantations.
12. Cultural Values	If indigenous ecosystems prioritized for production by Maori, then pressure will be for smaller protected and restored areas. If landscapes contain many areas that are of significance to Maori and/or Pakeha then there may be a need for larger areas/degree of protection.
13. Public Support	If society or local population value indigenous areas as a key part of the matrix in a sustainable landscape, then areas likely to be larger and more varied, compared with areas where there is little support and will therefore put pressure on area reduction.
14. Political Will	Generally lacking and erring on the side of compromising ecological priorities, so will in most instances put pressure for a reduced area.
15. Level of Support for Protection on Private Lands	If mechanisms and financial support are easily available (such as via local councils) then this will push area up, if not then there is a reduced incentive for private land holders to protect areas for the public good.

individuals, as well as people with an historic or cultural relationship with the area.

Identify what is valued

Rather than planning by "experts" alone:

- Encourage the community of interest to articulate what is valued, what is important and special about a place — perhaps particular wildlife, vegetation, outcrops or waterfalls.
- Identify the physical, the utility, the historic, the spatial and other amenity values such as key views and vistas.
- Seek to identify consensus on what is important, whilst recognizing that values will have different weightings for everyone and will change.

Identify the issues

- Have the community of interest articulate the issues as they perceive them — e.g., that stream health may be threatened; a visually dominant monoculture may emerge; indigenousness may be scarcely evident.
- Identify the common ground, a consensus on what is of real concern — but do not ignore minority concerns.

- Investigate the extent, triggers, implications and seriousness of these issues for resource sustainability, including for sustaining utility, landscape and amenity values.
- Where concerns appear unfounded, provide additional information and work through with the community perceived concerns and seek resolution.

Articulate their vision

- Have the community of interest explore opportunities for the land — perhaps, to sustain particular identified biota in the catchment; to have highly indigenous stream corridors; to maintain shrubland feeding areas for birds.
- Ensure technical and design assistance is available to encourage constructive and creative exploration of solutions and opportunities.
- Encourage the community to articulate their conceptual vision and seek a consensus on directions.

Concept plan development

- Pool the information.
- Provide technical and design input to develop practical development and management plans that recognize the nature and culture of the place.
- Recognize what is valued.
- Address the issues.
- Pool the information gathered from landholders, land managers, scientists, tangata whenua, public groups, and others recognized as the community of interest for the area.
- Explore the visions, the potential core areas and linkages for nature. Identify the potential as well as the problems for restoration.
- Ensure a diversity of natural values will be protected and/or restored — not all steep dark faces, that have the thinner soils, limited species ranges and of little appeal to birds in winter!
- Provide a concept (quickly) that is friendly, easy to read, use and identify with.
- Seek an interactive and/or iterative process to develop consensus in order to produce the community of interest's plan, not just the expert's plan.
- Identify opportunities for trialing and staging new management styles.
- Ensure the style of plan responds to local nuances, language, landscape, etc.
- Develop an action plan or programme and consider allocating tasks.
- Provide further guidance for developing and managing areas and stages over time, or for reviewing directions.
- Seek to ensure their concept plan is readily available and regularly referred to — a living document.

PHASED PLANNING FOR PROTECTION AND RESTORATION — MOVING TO MOSAICS

Scientific criteria for designing a protected ecosystem network

There is a reasonably good consensus on the criteria used for selection of areas for nature conservation. Frameworks have emerged to both theoretically and practically weigh up the factors for protected area decisions. These include; New Zealand Protected Natural Area programme, ecological evaluation for nature conservation such as in O'Connor *et al.* (1990), Harding (1994) and Spellerberg (1992), Man and Biosphere Programme for designing integrated protected areas, Gap Analysis projects (Scott 1996) and Sanjayan and Soulé (1997), The Wildlands Project (Noss 1992), Joint Australian and New Zealand Environment and Conservation Council/Ministerial Council on Forestry, Fisheries and Aquaculture National Forest Policy Statement Implementation Subcommittee (1997), and a Focal/Multi-species approach (Lambeck 1997).

For landscapes that have sufficient indigenous ecosystems remaining to be able to "select" areas (mostly publicly owned lands), then the above criteria are adequate. However, for most New Zealand landscapes, and in particular the regions where native communities are most threatened, there are no intact areas to *select* from but rather fragmented remnants from which to restore and rehabilitate a network. Therefore the evaluation and selection for restoration, and *design* of a protected network becomes the key focus. This is particularly important when integrating restoration with other land uses, and working within the dynamics of fragmentation (e.g., Saunders *et al.* 1991).

Evaluation and design criteria for restoration and protection

The primary goal could be seen as to "viably represent all elements of biodiversity within a network of protected ecosystems in the landscape". But just how much of all the different ecosystems is required to protect them is a key question that has no simple answer. The 10% figure (based on island biogeography (e.g., MacArthur and Wilson 1967)) as a goal for protection has been used widely from the International Union for the Conservation of Nature Bali Plan of Action in 1982, recommended here in New Zealand (Molloy *et al.* 1980), to the slightly higher 12% recommended by the Brundtland Report in 1987 (Sanjayan and Soulé 1997) and 15% for Australia (Joint Australian and New Zealand Environment and Conservation Council/Ministerial Council on Forestry, Fisheries and Aquaculture National Forest Policy Statement Implementation Subcommittee 1997). It is apparent that these figures have been chosen primarily for political, not ecological reasons. A recent survey of conservation scientist views confirms this and that these low percentages would not "maintain viable populations of species and adequately represent ecosystems in biologically heterogeneous areas" (Sanjayan and Soulé 1997).

Given that 10 to 15% figure is seen as too low, then where between there and 100% is adequate? The percentage chosen will involve weighing up many key conservation biology factors against the social, political and economic realities. Analysis by Sanjayan and Soulé (1997) found that the proportion of regions actually recommended for protection ranged between 25% and 99.7%, with most falling in the 40% to 70% range. They concluded that:

- some degree of protection for about half of the terrestrial land and fresh water would be required to protect biodiversity,
- the proportion might be lower in temperate areas and higher in regions with high endemism and habitat heterogeneity, and
- not all of the % need be in strict reserves but would need to be managed with the objective of protecting ecological integrity and species diversity.

It should be emphasized that these percentages were determined largely on ecological criteria, without the compromise that socio-political aspects bring. They are useful as they give an indication of what sort of commitment it will take to maintain ecological integrity in the landscape in the long term, that it will vary considerably between landscapes, and that for most landscapes in Aotearoa the focus moves to protection and restoration of production areas.

Considering the studies above and New Zealand's temperate ecosystems, a generalized vision of the components of the mosaic can determine a draft proportion of the landscape to be in indigenous systems and under some form of protection. This is the planning goal. There is no "right" figure that can be used for any landscape but it acts as a starting point for adjustment relevant to different factors. We know that if ecosystem integrity and biodiversity is to be taken seriously, then protected and restored areas will have to transcend the politically constructed lower and poorly representative percentages. This poses a significant challenge and effectively a land use transition for many landscapes. This will be especially so in the production dominated landscapes of New Zealand such as the Waikato, Taranaki and Southland lowland areas, and for ecosystems such as kahikatea flood plains, most wetlands, and coastal forest systems.

To guide the process of transforming the current land use matrix to include a restored network of protected ecosystems, adjustments can be made to the draft plan to consider key factors. These factors will influence the proportion of a landscape to be protected and restored, the design and the management of protected ecosystems.

Case study of phased plantation management for restoration and protection

The reasons for the protection and restoration biodiversity have been given above, as has the rationale behind the selection and design of a protected ecosystem network. Given that the fundamental purpose of plantation forestry is to make a profit, then meeting the biodiversity protection goals in a plantation dominated landscape will necessarily involve temporal and spatial steps that are economically viable. Most plantation managers believe that their commitment to biodiversity protection is fulfilled through meeting the New Zealand Forest Accord requirements of not clearing native areas and the conservation of threatened species that are known to occur in their plantations (Dyck 1997). It is very unlikely that plantation management will be able to "make the leap" to an adequate and viable protected ecosystem network over one tree crop rotation without considerable subsidization from outside sources in recognition of the wider public benefits. Achieving certification such as the Forest Stewardship Council, would allow some compensation for restored production areas.

An achievable pathway to establishing an adequate and viable protected ecosystem network is using phases of protection and restoration (Rosoman 1996). Following on from an Ecological Landscape Assessment a Protected Ecosystem Network can be overlayed on the production landscape matrix to identify required protection and restoration areas. An implementation plan would create the transition pathways from production to protection areas over future rotations, and any management and renewal work such as pest control. It could be carried out as bands of restoration mimicking seral stages of the core area, gradually moving the edge out to increase the viability of the core and if necessary establishing greater linkages in the landscape. If this planning approach was carried out and its implementation begun, then baseline indigenous biodiversity protection would be met.

Phasing to meet financial constraints and make restoration more adaptive and practically achievable could also be used in the farming situation. The trend in the pastoral landscape is for woodlot planting. These could be combined with mobile electric fences that move out with the restoration phase. We have the obligation of restoring ecosystems that land management in the past has degraded. Plantation growers and farmers must take some responsibility for past ecological mistakes and restore indigenous ecosystems back in the landscape.

Additional biodiversity benefits would be obtained from plantation management by planting a range of production species, including natives, by encouraging and retaining an indigenous understorey, using non-chemical methods of weed control (e.g., Kintop 1992; MacDonald and Fiddler

A Case Study of a Plantation Forestry Development Proposal

(Protected ecosystem restoration in a real, largely deforested landscape in Nth Canterbury, New Zealand, that is proposed for plantation establishment)

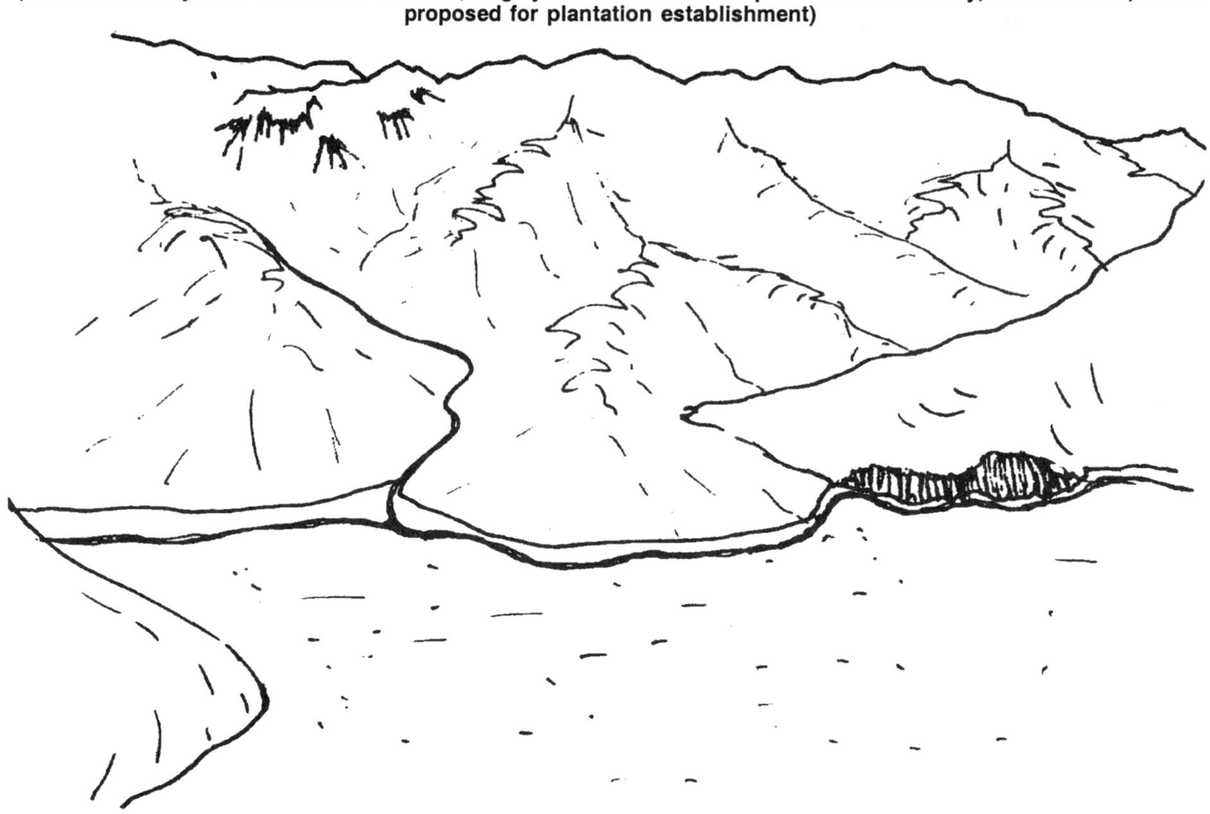

Phase 1: Landscape Assessment to identify areas required for protection and restoration: Protected Ecosystem Network selection and design.

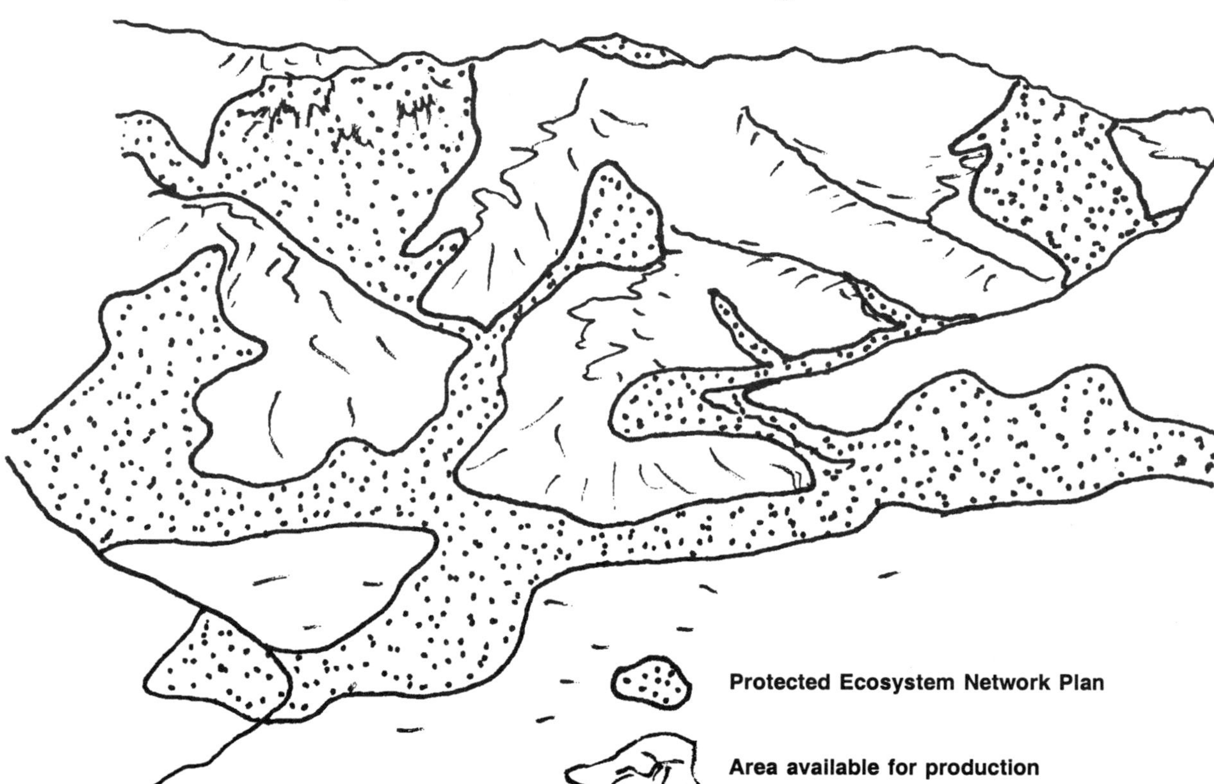

Fig. 1 (Above and opposite). Phases of plantation restoration.

Phase 2: **Protection and Restoration Planning and Establishment, working initially from remnant areas, streamside zones, sensitive areas, and special sites.**

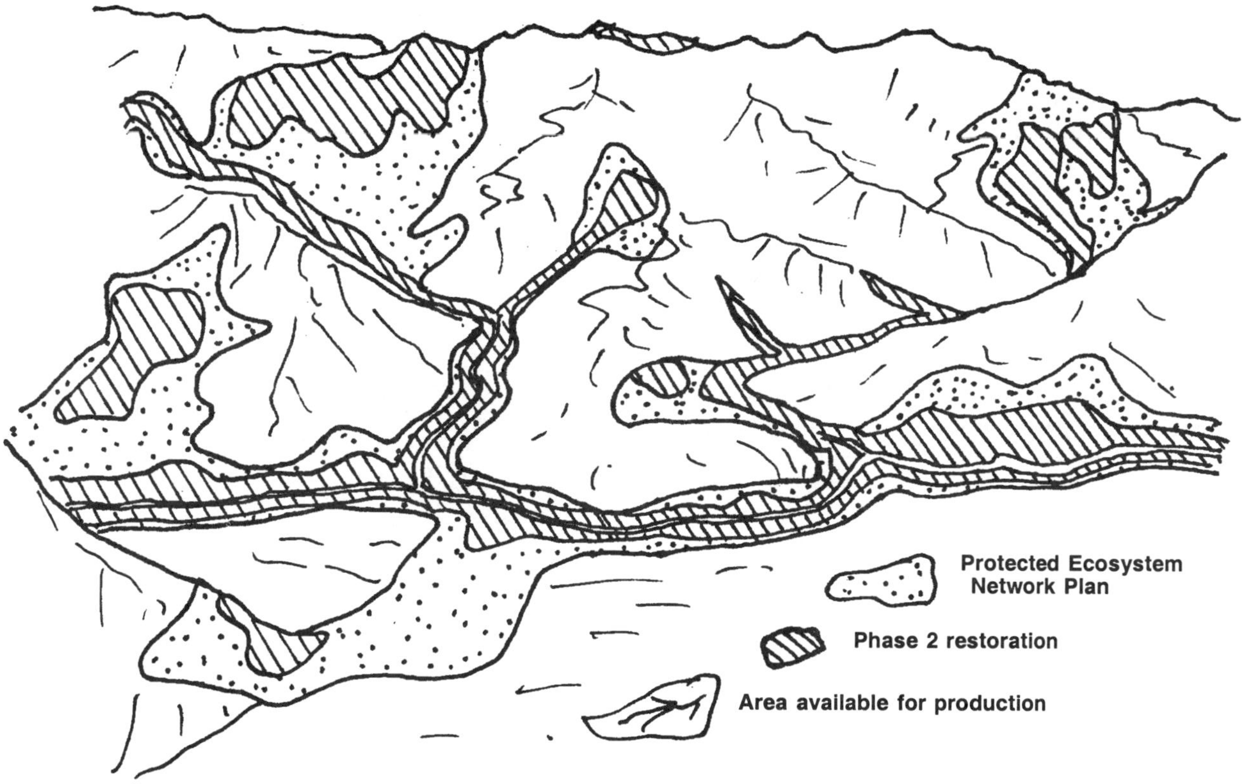

Phase 3, 4 and on: **Restoration over time or rotations e.g., using edge expansion, cover crops to assist structural complexity, towards a goal of completing the Protected Ecosystem Network Plan.**

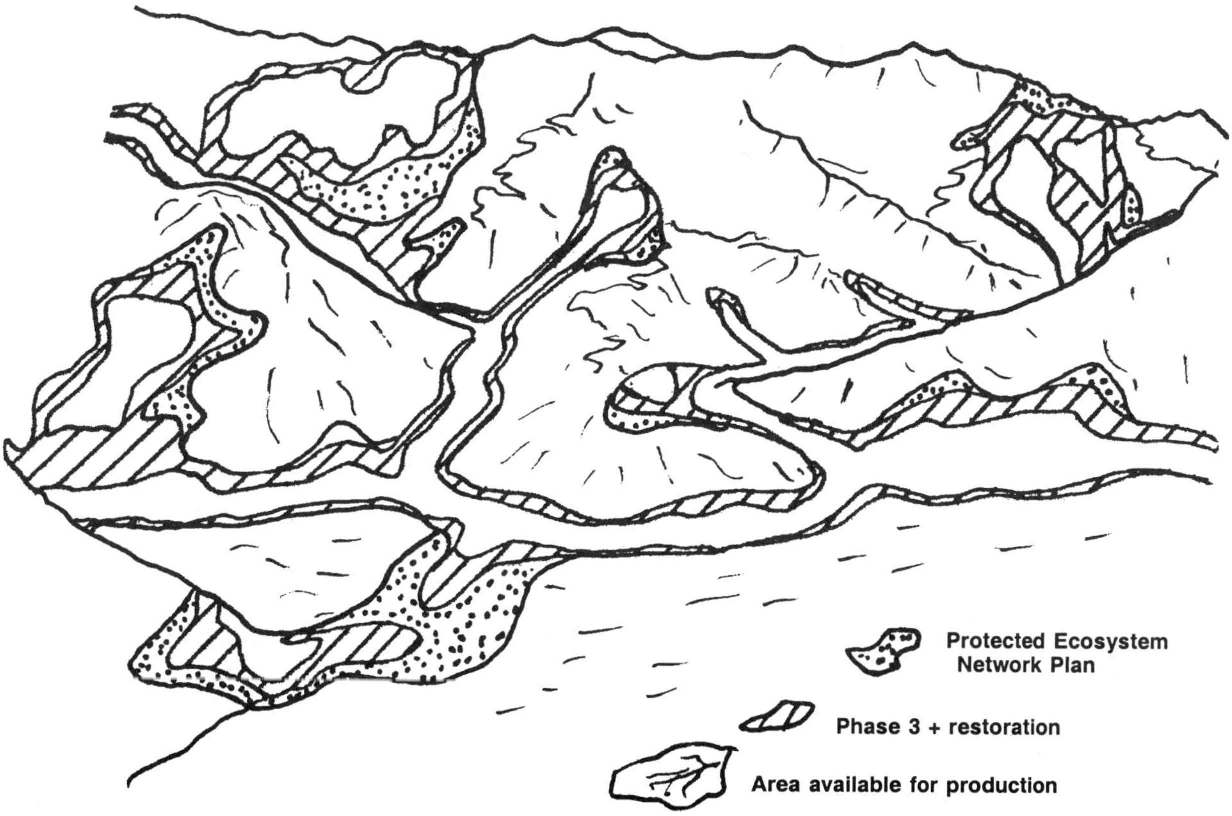

1992), by increasing rotation length, by having full age range compartments and reducing stand size, converting monocultures into mosaics, and by carrying out pest control. Planning that takes into account landform and the indigenous component, can utilize patch planting, a range of species and their different growth forms to harmonize plantations into the visual landscape. Harvesting a mosaic of patches avoids the spectacle of large bare areas, and retains more forest structure and function in the land.

Case study of rural residential restoration on Waiheke, New Zealand

This case study looks at the nature conservation benefits achieved by a shift to a higher value land use within a local planning environment that is supportive of protection and restoration. There has been considerable criticism of regulatory controls in district plans but some rules and incentives in parts of the Hauraki Gulf Islands Section of the Auckland City District Plan have been working well for conservation on Waiheke Island. The Hauraki Gulf Islands Section (specifically for Land Unit 22) emerged following a community visioning process more than 10 years ago that identified community preferences for the western end of Waiheke Island (Scott 1989). This laid the framework for subdivision development that has involved nearly 80% restoration of the landscape or approximately 450 ha.

Key components of the District Plan rules that have facilitated the protection and restoration are;

- *Dwelling Density*: ratio of one lot per 5 ha of gross land area, with a minimum size lot of 1.5 ha
- *Bonus Density Provisions:* up to a maximum of 1 lot per 3.5 ha per gross land area, for each additional lot created 2 ha must be added to public open space, reserve, protected area or protected significant natural feature.
- *Subdivision Financial Contributions*: either cash equivalent to 10% of value of lots for reserve purchase by the Council, or between 2.5% and 5% gifting of land outside other reserve contributions.

For this case study we are largely reporting on the work of Dennis Scott of Auckland, who has carried out the landscape planning for the subdivisions (Scott 1989, 1991 and 1996).

Scott used a land typing and catchment focus for assessment of the land areas, including topography, archaeology, geology, soils, hydrology, slope, and biotic/ecologic components. His analysis looked at "critical landscape elements" to determine appropriate land use. These focused around coastal margin, riparian areas, cultural sites, steep and sensitive sites, and existing remnant indigenous ecosystem areas. The design concentrated on representation to mimic former indigenous system patterns, including making property boundaries around natural landform patterns to create integrated units. As well, key social criteria were used such as access, visibility and shelter from winds, public access and walkways, and suitable house sites. A significant outcome of the assessment and design process is a learning and appreciation for those involved, especially the developers and Council staff, on the importance of the protection and restoration of the indigenous systems. This is carried through by further management and interactive work with the new land owners including the development of a management booklet, weed and pest control, and follow-up successional planting.

CONCLUSIONS

There is an urgent need for the protection and restoration of a viable and adequate network of protected ecosystems in New Zealand. To complete this the focus for nature conservation must move beyond dedicated public protected areas to production focused landscapes and on private land. Society desires mosaics of indigenous biodiversity in production landscapes.

Biological systems are complex and there are major information gaps in both the current status of biodiversity and which path forward will achieve the desired result. A conservation science basis exists for the selection and design of, and planning for restored reserve networks. Considerable areas within many production dominated landscapes are required to be restored if the indigenous communities represented as small fragmented remnants in those landscapes are to survive. Landscape assessments are required to identify the underlying ecosystems, identify the community of interests' values, concerns and opportunities, and to develop multi-dimensional concept plans.

To reduce the financial burden of loss of production from restored areas, following a concept plan, a phased approach to restoration of an adequate network of indigenous ecosystems within plantation forestry management and agriculture offers a way forward. This could be supported by eco-certification market-based incentives, and local and national government financial incentives. The market for eco-subdivisions within a supportive local planning environment, offers a viable restoration pathway for rural residential areas.

ACKNOWLEDGEMENTS

Thanks to emeritus Prof. Kevin O'Connor and Dr Ian Spellerberg for comments on our draft, and to Dennis Scott for detailed information on the rural residential case study.

REFERENCES

Cherfas, J., 1994. How many species do we need? *New Scientist* **38.**

Diamond, J. M., 1990. New Zealand as an archipelago: An international perspective. Pp. 3–8 *in* Ecological Restoration of New Zealand Islands ed by D. R. Towns, C. H. Daugherty and I. A. E. Atkinson. Department of Conservation: Wellington, New Zealand.

Dyck, W. J., 1997. Biodiversity in New Zealand Plantation Forestry — an Industry Perspective. *NZ Forestry,* **November 6–8.**

Ehrlich, P. R and Ehrlich, A. H., 1981. Extinction: The causes and consequences of the Disappearances of Species. Random House: New York.

Froude, V., 1997. Implementing the Biodiversity Protection Provisions of the Resource Management Act — a review of council progress to date. Pacific Ecologic Resource Management Associates: Wellington, New Zealand.

Harding, M., 1997. Waikato Protection Strategy: a report to the Forest Heritage Fund Committee. Forest Heritage Fund: Wellington, New Zealand.

Harding, Mike, 1994: Implementing Biodiversity Conservation — an assessment of the strategic direction of the Forest Heritage Fund. Forest Heritage Fund: Wellington, New Zealand.

Jackson, L. L., 1992. The Role of Ecological Restoration in Conservation Biology. Pp. 434–51 *in* Conservation Biology: The Theory and Practice of Nature Conservation, Preservation and Management ed by P. L. Fielder and S. K. Jain. Chapman and Hall.

Forest Stewardship Council, 1999. Principles and Criteria For Forest Stewardship. Document 1.2. Forest Stewardship Council: Mexico.

Joint Australian and New Zealand Environment and Conservation Council/Ministerial Council on Forestry, Fisheries and Aquaculture National Forest Policy Statement Implementation Subcommittee, 1997. Nationally Agreed Criteria for the Establishment of a Comprehensive, Adequate and Representative Reserve System for Forests in Australia. A report by the Joint Australian and New Zealand Environment and Conservation Council/Ministerial Council on Forestry, Fisheries and Aquaculture National Forest Policy Statement Implementation Sub-committee.

Kintop, C. L., 1992. Paper mulches for Herbaceous Weed Control in Conifer Plantations of the Roseburg BLM District. Proceedings of a presentation at the workshop on Forest Vegetation Management without herbicides, Oregan State University: Corvallis.

Lambeck, R. J., 1997. Focal Species: A multi-species umbrella for nature conservation. *Conserv. Biol.* **11:** 849–56.

Leopold, A., 1953. Round River. Oxford University Press: New York.

Leopold, A., 1949. A Sand County Almanac: and sketches here and there. Oxford University Press: New York.

Lucas, D., 1994. Landscape and Resource Management. Unpublished paper to Royal Forest and Bird Protection Society AGM. Wellington, New Zealand.

MacArthur, R. H. and Wilson, E. O., 1967. The Theory of Island Biogeography. Princeton University Press: Princeton, New Jersey.

MacDonald, P. M. and Fiddler, G. O., 1992. Mulching and Grubbing Techniques in Forest Vegetation Management: A research synthesis. Proceedings of a presentation at the workshop on Forest Vegetation Management without herbicides, Oregan State University: Corvallis.

Ministry for the Environment, 1997. State of the Environment Report. Wellington, New Zealand.

Minister of Forestry, 1995. Letter to M. Mackey dated 1st December.

Molloy, L. F. in association with Forde, B. F. *et al.* (Compilers), 1980. Land Alone Endures: Land Use and the Role of Research. New Zealand DSIR Discussion Paper No. 3. Wellington, New Zealand.

Noss, R. F., 1992: The Wildlands Project. land conservation strategy. *Wild Earth* **(Special Issue):** 10–25.

Noss, R. F. and Cooperrider, A. Y., 1994. Saving Nature's Legacy: Protecting and Restoring Biodiversity. Island Press: Washington DC.

O'Connor, K. F., Overmars, F. B. and Ralston, M. M., 1990. Land Evaluation for Nature Conservation — a scientific review compiled for application in New Zealand. Conservation Sciences Publication No. 3. Dept. of Conservation: Wellington, New Zealand.

O'Connor, K. F., Batchelor, G. W. and Davison, J. J., 1982: Mavora: development of a planning process for reconciliation of interests in wilderness. Tussock Grasslands and Mountain Lands Institute, Lincoln College, Lincoln. Papers in Resource Management No. 4.

Park, G., 1995. Nga Uruora — the groves of life. Ecology and History in a New Zealand Landscape. Victoria University Press: Wellington, New Zealand.

Rosoman, G. B., 1996. Beyond the Monoculture: An environmental and social perspective on alternative approaches to forestry. Proceedings of the New Zealand Institute of Forestry Conference "Alternative Approaches to Forestry — time for a review", Invercargill.

Sanjayan, M. A. and Soulé, M. E., 1997. Moving Beyond Brundtland: The conservation value of British Colombia's 12% protected areas strategy — A Preliminary Report. Unpublished report to Greenpeace.

Saunders, D. A., Hobbs, R. J. and Margules, C. R., 1991. Biological Consequences of Ecosystem Fragmentation : A Review. *Conserv. Biol.* **5:** 18–32.

Scott, D., 1989. The Waiheke Landscape. Waiheke County Council: Auckland, New Zealand.

Scott, D., 1991. Church Bay Resource Consents Documents. Unpublished contract reports for N. and A. Johnstone.

Scott, D., 1996. Owhanake — Assessment of Environmental Impact. D. J. Scott Associates. Unpublished report.

Spellerberg, I. F., 1992: Evaluation and Assessment for Nature Conservation. Chapman and Hall: London.

Te Puni Kokiri, 1994. Te Ara o Te Ao Turoa: Biodiversity and Maori. Te Puni Kokiri (Ministry of Maori Affairs): Wellington, New Zealand.

World Wide Fund for Nature, 1991. The Importance of Biological Diversity. Yale Press, New Haven, CT, *in* Noss, R. F. and Cooperrider, A. Y., 1994. Saving Nature's Legacy: Protecting and Restoring Biodiversity. Island Press: Washington DC.

CHAPTER 60

Legislation and practice in the protection of indigenous biodiversity in production landscapes

Mairi Jay[1]

Within New Zealand, almost all land in production is subject to the *Resource Management Act*. The *Resource Management Act* provides a strong legislative mandate for the protection of indigenous vegetation and habitat of indigenous species in developed landscapes. Despite the strongly worded legislative requirements, there are many practical difficulties in the way of achieving such protection. These difficulties relate not only to the attitudes and priorities of land users, but to "institutional drag" represented by the people and practices within the local government agencies that are involved in land management. This chapter will detail some of these problems, and suggest some non-statutory measures which may be required to achieve long-term conservation of indigenous biodiversity in developed landscapes.

Key words: Biodiversity conservation, Production landscapes, *Resource Management Act, Local Government Act.*

INTRODUCTION

In New Zealand, local government is the key to nature conservation in production environments. The knowledge, attitudes and skill of local government staff and councillors, and the organizational capacity of district and regional councils will determine to a considerable degree, the extent to which New Zealand's natural heritage will be conserved for future generations.

The aim of this chapter is to explain how and why local government is so important for nature conservation in production environments; how and why it frequently fails; and how and why it might improve.

"Indigenous biodiversity" or simply "biodiversity" refers to species that are indigenous to New Zealand and to ecosystems and landscapes that are still predominantly indigenous in their species composition or structure. This excludes biodiversity that results from the invasion and naturalization of introduced plants, animals and diseases. "Local government" refers to the system of government that operates at sub-national level. It comprises district councils, which include cities and counties; and regional councils, which overarch districts.

Information for this chapter has come from four sources: personal experience as an environmental advocate in the context of local government; discussions with planners around the country, a survey of relevant literature, particularly papers presented to the Significant Natural Areas Workshop, held in Wellington on 26th September, 1997 by the Local Government Association, in conjunction with the Ministry for the Environment; and preliminary results from a survey of district council planning departments in the country, undertaken by myself and Rachael Carruthers from the University of Waikato.

NEW ZEALAND'S LEGAL AND ADMINISTRATIVE FRAMEWORK FOR NATURE CONSERVATION

For New Zealand's remaining natural landscapes, there are two different types of administrative and legal framework that apply: that which applies to public land, and that which applies to private land.

The bulk of conservation land in public ownership is administered by the Department of Conservation under "protected areas" legislation. This includes

[1]Department of Geography, University of Waikato, Private Bag 3105, Hamilton, New Zealand.

Pages 586–93 *in* NATURE CONSERVATION 5: NATURE CONSERVATION IN PRODUCTION ENVIRONMENTS: MANAGING THE MATRIX ed by J. L. Craig, N. Mitchell and D. A. Saunders. Surrey Beatty & Sons, 2000.

national and forest parks administered under the *National Parks Act 1980,* the *Conservation Act 1987,* the *Reserves Act 1977,* and a number of others. Most of this land covers wilderness areas outside "production environments". Only a very small proportion of the areas of conservation value within production environments come under the legal and administrative framework of protected areas legislation.

Virtually all land under production, whether agriculture or forestry, is private land, or multiply owned Maori land. As such, it is subject to a suite of legislation which includes the *Resource Management Act 1991*, the *Forests Amendment Act 1993,* the *Biosecurity Act 1993,* and, indirectly, the *Local Government Act 1974.* (*The Wild Animal Control Act 1977* and the *Wildlife Act 1953,* also apply to private land but they are of little significance to the way private landowners manage their land). The *Resource Management Act* is the most powerful and far-reaching in its immediate impact, but the Local Government and Biosecurity Acts may well be of most long-term value as instruments for continued ecological management and for bringing about changes in public attitude and behaviour.

The purpose of the *Resource Management Act* is,

> "to promote the sustainable management of natural and physical resources by managing the use, development, and protection of natural and physical resources in a way, or at a rate, which enables people and communities to provide for their social, economic, and cultural well being and for their health and safety while:—
>
> (a) Sustaining the potential of natural and physical resources (excluding minerals) to meet the reasonably foreseeable needs of future generations; and
>
> (b) Safeguarding the life-supporting capacity of air, water, soil, and ecosystems; and
>
> (c) Avoiding, remedying, or mitigating any adverse effects of activities on the environment."

Section 6 of the Act states that, as a matter of national importance, all persons exercising functions and powers under the Act "shall recognize and provide for . . . (a) The preservation of the natural character of the coastal environment (including the coastal marine area), wetlands, and lakes and rivers and their margins, . . . (b) The protection of outstanding natural features and landscapes . . . ; and (c) The protection of areas of significant indigenous vegetation and significant habitats of indigenous fauna". Section 7 (d) also states that persons exercising functions and powers under the Act shall have particular regard to the "intrinsic values of ecosystems". In short, the provisions of the Act create a strong legal mandate for conservation of natural resources in production landscapes.

The key themes of the *Resource Management Act* are sustainable management of natural and physical resources, their integrated management under one piece of legislation, and control of the adverse effects of activities on the environment (Williams 1997:68). In principle, therefore, the Act provides a powerful instrument for the sustainable management of biodiversity "resources".

The *Resource Management Act* divides responsibility for resource management between central, regional and district levels of government (Williams 1997:70). Central government involvement has been largely confined to a monitoring role, apart from the involvement of the Department of Conservation in management of the coastal environment. On the ground, the main agents of *Resource Management Act* responsibility are regional and district councils. Broadly speaking, regions are mainly responsible for management of water, soil, geothermal resources, pollution control, natural hazard mitigation, lake and river-bed management, the coastal marine area (Milne 1992:49) and policies for the integrated management of the resources of the region (including biodiversity resources).

Districts are largely responsible for land use management (including designations and heritage orders), land subdivision, noise control, and hazard mitigation (Milne 1992:54). Districts are the level of government most immediately responsible for the retention or creation of ecological corridors, or the creation or protection of landscape mosaics, including protection of riparian margins. However, regional councils are responsible for ensuring integrated management of the resources of the region by the preparation of regional policy statements which articulate policies for the integrated management of all natural and physical resources of a region. Insofar as "biodiversity" may be viewed as a natural resource, and to the extent that biodiversity maintenance requires the maintenance or restoration of indigenous vegetation and habitat of indigenous species, regional councils may include relevant policies for this purpose within the regional policy statement. According to the wording of the *Resource Management Act* "district plans shall not be inconsistent with" regional plans and policy statements.

All development of natural and physical resources is, in theory, subject to the policies of district and regional councils. Resource management policies are developed and codified in district and regional plans and policy statements through public consultation procedures, and then implemented by the application of rules, incentives, education, and development control (resource consent applications). Districts and regional councils are encouraged to work together, and to hold joint hearings where development proposals relate to issues that involve mixed responsibility.

In theory it is possible under the *Resource Management Act* for regions and districts to work together in an integrated way, to protect natural resources such as ecosystems and species.

The *Biosecurity Act 1993* is important because it applies to the control and management of pests (plants, animals and diseases). In New Zealand, invasion of indigenous habitat by introduced animals, weeds and diseases is possibly the single most important cause of loss of indigenous biodiversity. The *Biosecurity Act* gives regional councils the power to prepare and implement regional pest management strategies to control pests in individual regions.

Finally, another piece of legislation that is indirectly relevant to nature conservation, is the *Local Government Act 1974* and its amendments. This Act sets out the powers and responsibilities of local government. Major amendments of the *Local Government Act* in 1989 created the current structure of local government into district and regional councils (from a previous host of territorial and ad hoc agencies) that was specifically and intentionally related to responsibilities and powers of resource management now contained in the *Resource Management Act*.

The *Local Government Act* also gives councils broad freedom to choose the activities in which they become engaged so long as their activities relate to objectives stated in an annual plan and long-term financial strategy. In terms of nature conservation, they allow councils to plan and implement projects that cannot be implemented under the *Resource Management Act*. For example, while the *Resource Management Act* does not directly empower councils to implement landcare projects, environmental education, ecological restoration, or on-going ecosystem management, these activities are enabled by the financial provisions of the *Local Government Act* if they come within a long-term strategic plan.

THE IMPORTANCE OF LOCAL GOVERNMENT FOR BIODIVERSITY CONSERVATION

So, how and why is local government important for nature conservation in production environments?

It is important in at least three ways: it has legislative power and responsibility for environmental issues at local and regional level; it is accountable to individuals and communities for environmental conditions within their local area, and can harness their energies and commitment for environmental action; and it is potentially the level of government that can provide the ongoing care that is necessary for long-term ecological protection and restoration.

It has been argued for New Zealand (e.g., James 1993:10; Bayfield 1997; Froude 1997:17–20) that landowners tend to react negatively to regulatory mechanisms of conservation and to prefer positive approaches such as incentives and provision of information. Landcare groups have been formed by some councils as a means to promote environmental education and individual motivation (Environment Waikato 1997:40). Regional and district councils are able to target such education and incentives to groups within communities that are likely to be most receptive.

Ecologically, local government has the power (seldom realized to date) to develop and implement long-term strategies for ecological restoration and ongoing management of ecosystem processes for important types of indigenous ecosystems. For example, regional councils have responsibility for deciding pest management strategies that involve long-term control of pest animals such as possums, goats and deer and regions and districts can influence drainage of wetlands by farmers. Regional councils can, in theory, work with districts to target areas of remnant forest for pest control programmes and long-term protection from development. Regional councils can also, in theory, work with district councils to determine critical river or stream catchments and to provide funding for restoration or protection of riparian margins.

In short, regional and district councils are important for biodiversity conservation because they administer the legislation that most directly influences the actions of local landowners and because they are accountable to people for environmental conditions and activities within their district or region.

THE LIMITATIONS OF LOCAL GOVERNMENT ADMINISTRATION FOR BIODIVERSITY CONSERVATION

Although local government is important for nature conservation, it is subject to a range of responsibilities and duties beyond those of environmental management. This broader structure of responsibilities shapes the way that specific local government agencies perceive and respond to their responsibilities for nature. Frequently, it constrains the resources they put toward environmental management and nature conservation. Some of the main constraints can be identified as follows:

Conflicts between administrative and ecological organization and structure

At least three different types of conflict arise: these are, boundary conflicts, process conflicts, and conflicts between different levels of government in relation to their functional responsibilities.

In terms of boundaries, council boundaries are fixed and generally follow cadastral and property boundaries. Ecosystems seldom have fixed and determined boundaries but may have permeable, intermittent, indeterminate boundaries, or ones that change over time. Species tend to have distributions rather than boundaries, but also may have critical sites for particular parts of their lifecycle (for example, native fish which require upstream and downstream passage to estuarine spawning areas, waterfowl which require separate feeding and loafing areas, and migratory species which require flyway resting and feeding sites). District and regional councils seldom look beyond their boundaries to consider the health and welfare of species or ecosystems as a whole. Within the Waikato region, for example, peatlands and swamps are still significant ecosystem types that many people view as a nuisance. Because most people do not look beyond their own immediate local district or region, they do notice that wetlands and wetland species have become rare or endangered ecosystems and species within New Zealand as a whole. There is little sense that wetlands of the Waikato need to be protected because they represent a significant proportion of New Zealand's total remaining wetlands (certainly this is true in relation to restiad bogs, but also in relation to mineralized wetlands).

An important aspect of biodiversity maintenance is the fact that ecosystem function involves processes (of nutrient cycling, reproduction etc.) which may extend over a range of habitats and sites within a region or district. Biodiversity retention may thus be not so much a function of protecting this or that specific area of bush or wetland, but of retaining connections and relationships between features across a landscape. Because district councils are confined legally (and often mentally) within their territorial boundaries, management of ecosystem functions across district territorial boundaries is a major long-term problem for conservation of indigenous biodiversity.

Process conflicts relate to the fact that council timescales are influenced strongly by the 3-year election cycle, and then by the comparatively short 10-year life span of most statutory plans. Ecological cycles by contrast are measured in decades and centuries. Few people in today's day and age seem prepared to work on conservation or restoration projects which may not show definitive results for decades. Equally true, many councils hold back at the thought that they might be required to commit pest control funds indefinitely in order to protect or restore an area of remnant forest.

District and regional levels of government have different statutory responsibilities and different, frequently overlapping boundaries. In simplified terms, district councils control the management of land and land resources (including natural habitat, but not soil), and regional councils are responsible for water and soil conservation. These different responsibilities mean that there can be inconsistencies on the ground that result from conflicting priorities of the government agencies concerned. There are 12 regional councils, 74 district councils (which include cities such as Auckland, Wellington and Christchurch), and four unitary authorities (which combine the functions of both regions and districts). Although this division of responsibilities between regional and district councils makes sense if land and water are regarded as isolated and separate resources, it is a major weakness in relation to the management of ecosystems. For ecosystem management, the interaction of land and water is usually a critical feature, whether one is talking about terrestrial or aquatic ecosystems. The same is true of the separation between land management and pest control: although pest control on a regional basis is logical where the main focus is a pest species (e.g., control of possums, nodding thistle or ragwort), it makes less sense where the focus is ecosystem management or the protection of a particular biotic community.

Competing local government responsibilities and priorities

It is equally important to recognize that the *Local Government Act* imposes on councils a wide range of statutory duties and responsibilities many of which, one way or another, they **must** perform. These include provision for rubbish, roads, reserves and recreation (sports fields etc.), water and sewage waste, control of building and development, animal control (e.g., wandering dogs and stock), noise control, libraries and community services, environmental health, and many more. Many of those responsibilities may be urgent or impossible to delay, because of their intrinsic nature (e.g., environmental health, dog and livestock control) or because of statutory deadlines (e.g., resource consents).

It is frequently the case that urgent or politically significant issues take precedence over issues such as environmental monitoring and protection. These latter are, for the most part, long-term and politically invisible (until there is an environmental emergency which, likely as not, has its root causes in the long-term cumulative effects of environmental mismanagement). Construction and maintenance of roads, water mains, and sewer mains tend to take precedence over the protection or restoration of an area of remnant bush. For a district council such as Thames-Coromandel, Waikato or Waitomo, community playing fields for rugby, soccer, netball or bowls tend to be viewed as more necessary than protection of the margins of a small rural stream. They benefit more people in the community and they provide more immediate social value.

The conflict between environmental management and other responsibilities is particularly significant in the case of district councils. District councils tend to have a wider range of responsibilities than regional councils, and their other responsibilities (roads, rubbish, water supply, parks and reserves, and the like) tend to outweigh their resource management responsibilities as matters of political importance. Regional council responsibilities, by contrast, are more directed toward regional resource management but they are spread over larger areas, so, consequently, may be spread more "thinly".

Rural decline over the past few decades means that geographically widespread rural local authorities which depend on a small rating base to fund an expensive network of rural roads and other services are especially disinclined to incur extra costs for conservation of indigenous biodiversity. In New Zealand, many of the councils most likely to experience this conflict are ones which include areas of high nature conservation value (e.g., Kaipara, Coromandel, Waitomo, Opotiki, South Wairarapa, Gisborne, and most of rural South Island).

Table 1 provides a comparison of selected local government statistics. They show wide variation between local governments in terms of population base, population growth (and consequent growth pressures), number of occupied dwellings (which provide the main bulk of the rates base), and the number of full-time staff employed by councils. It is interesting to note that some regional councils have a smaller population base than some districts (e.g., West Coast Regional Council with 34 000 people compared with Rotorua District with 65 000 people).

The number of full-time staff shown on Table 1 is not an accurate representation of a council's ability or willingness to fulfil its nature conservation responsibilities, but it suggests that the capacity of some councils to promote conservation may be affected by staff work loads and the availability of in-house expertise. Councils that have a large staff/population ratio and resource base (e.g., Waitakere City) are more likely to include staff who have the appropriate skills for nature conservation than councils with a small staff/population ratio and limited resource base. The availability of knowledgeable staff does not necessarily guarantee that such staff will be given the time and responsibility to attend to issues of nature conservation, but they can at least provide the knowledge if called upon.

The capacity of local government to conserve biodiversity depends greatly on the knowledge, commitment and skill of staff, and on their relationships with developers, landowners, and members of the public. As noted by Roper-Lindsay (1997), the *Resource Management Act* requires local authorities to deal with subject areas that were, prior to the implementation of that Act, only peripheral to their core concerns. Even today, six years after the Act came into power, many local authority staff have little understanding of ecosystem structures and processes, or the habitat requirements of and threats to indigenous species (see also, Froude 1997:20).

Within every organization, there is "institutional drag" created by past history of knowledge and ways of doing things. Staff involved in land management build up relationships within the organization and with individuals and interest groups outside. Sometimes these relationships encourage conservation attitudes and sometimes they can operate against such attitudes. For example, in many councils there is a division between staff involved with *Resource Management Act* procedures, and staff involved with operations (streets, reserves, water and sewage, flood control, etc.). While those involved directly with the *Resource Management Act* usually have a reasonable understanding of the purpose and principles of the Act (although they may not know what is an ecosystem or what comprises indigenous vegetation), staff involved in other areas of council seldom have such knowledge. They may feel opposed to or impatient of the *Resource Management Act* requirements.

Table 1. Statistical comparison of selected councils. F/T staff = number of full-time staff; T.O.D. = Total Occupied Dwellings.

Regional Council	F/T staff	1991 Pop.	1996 Pop.	% change	T.O.D. '96
Northland	80	131 620	141 865	8%	48 665
Auckland	286	953 980	1 077 205	13%	356 648
Waikato	248	338 883	357 294	5.00%	122 415
B. of Plenty	107	208 239	230 465	11%	80 189
Gisborne	340	44 387	46 089	4%	15 367
Hawkes Bay	160	139 483	144 292	3%	51 456
Taranaki	73	107 222	106 570	−1%	38 980
Manawatu-Wanganui	180	226 612	229 989	1%	81 612
Wellington	338	402 892	416 019	3%	150 139
West Coast		33 961	35 671	5%	12 494
Canterbury		446 114	478 912	7%	176 255
Otago	127	186 067	193 132	8%	69 410
Southland	49	103 442	100 758	−3%	35 818
Tasman	176	36 418	40 036	10%	14 374
Nelson	304	37 995	42 073	11%	15 394
Marlborough District	165	36 771	40 242	9%	14 500
Councils					
Thames-Coromandel	230	25 032	27 715	11%	10 431
Franklin	200	42 193	48 285	14%	15 837
Hauraki	121	16 921	17 263	2%	6 172
Waikato	133	37 556	39 105	4%	12 807
Matamata-Piako	137	29 408	29 655	1%	10 562
Hamilton	685	101 448	109 043	7%	37 775
Waipa	196	37 031	39 621	7%	13 340
Otorohanga	35	9 231	9 822	6%	3 043
South Waikato	130	26 186	24 671	−6.00%	8 326
Waitomo	80	10 074	9 991	−1%	3 446
Taupo		30 721	33 778	10%	11 036
Rotorua	396	65 096	68 990	6%	22,028

Thus, despite the legislation, we have officials and communities that simply do not acknowledge that remnant areas are significant, subdivision consent officers who ignore the principles of Section 6(c); resource consent administrators who do things the way they have always been done. There are the people who are opposed to the purposes of the Act on principle and actively find ways to get around it. There are relationships and commitments between developers (e.g., surveyors, builders) and council officers. There is the pressure to provide a service, be client driven, not to obstruct or ask for additional information. There is confusion and controversy about criteria for "significant indigenous vegetation" and "significant habitat" and a host of associated issues related to lack of knowledge about the presence or absence of species, the dynamics of ecosystems, cumulative effects of change, threshold limits etc.

LOCAL GOVERNMENT SUCCESSES

Despite a lack of ecologically qualified staff, there are examples of local authorities that have taken their *Resource Management Act* responsibilities seriously and sought to find effective ways to encourage conservation of our remaining indigenous ecosystems and species.

At the level of regional councils, and in relation to water quality rather than biodiversity protection, Environment Waikato has achieved significant successes in relation to reduction of dairyshed effluent into streams. The Dairy Farm Effluent Management provisions of the regional transitional plan make it easy for farmers to undertake land-based disposal of dairyshed waste, and difficult (and expensive) to dispose of such waste by other methods. Between 1992/93 and 1995/96, the proportion of farms using land-based effluent systems increased from approximately 35% to 65% while the number of farms using other means (mostly pond-based systems) reduced from 67% to 40%. Much of this turn-around is related to the provisions of the regional plan, but they are also influenced by one-to-one consultations between individual farmers and council staff (Don Scarlet, pers. comm.).

At the district council level, Waitakere City Council is a widely acknowledged front-runner in the field. The Waitakere District Plan includes a complex series of policies and rules that are designed to protect and restore areas of native vegetation and fauna habitat. All areas of "significant and outstanding native fauna habitat, and those areas of significant native vegetation that contain greater than 50% native vegetation cover and are greater than 300 m² in area" (including some areas of housing) are covered by a type of "overlay" zone termed "Managed Natural Area" (Waitakere District Plan, Policy Section, The response, objectives, policies and methods, P. 2). Within this zone cutting and clearance of indigenous vegetation must conform with rules laid out in the Plan. The provisions of the Waitakere City plan have been developed in a process of prolonged and intensive consultation with the Auckland Regional Council, the Department of Conservation, and the public. This consultation process is important both because it means that the provisions rest on a broader knowledge base of regional expertise, and because it helps to generate understanding and support for the provisions from the public at large. Support from the public means that homeowners and developers may be more proactive in helping to protect native vegetation as well as less resistant to the restrictions on their rights of private property.

Within the Waikato Region, most of the district councils have used a combination of consultation with the Department of Conservation and consultation with the public to identify and develop conservation measures for areas of high conservation value. They have also involved patient and lengthy negotiations between council officers, councillors and submittors who have made submissions to district plans and the Waikato Regional Policy Statement. By using such procedures, council officials in the region have been able to develop provisions for protection of local ecosystems and remnant vegetation without causing enormous resistance from landowners. Thus Waipa District Council for example, not only has protection for its ecologically unusual peat lakes but has established a sizeable fund for protection of the margins of its lakes. Ruapehu District Council, through careful and patient discussion has achieved acceptance by farmers and stakeholder groups of a number of significant principles for conservation of remaining indigenous forest in a district where remnant forests are important from both the viewpoint of nature conservation and making extra money for hard-pressed hill country farmers.

Although it is not possible to generalize that extensive consultation with the public and with other agencies (e.g., other levels of government) is a prerequisite for successful introduction of nature conservation policies for private land, it does appear the case that **lack** of consultation is likely to result in opposition and resistance. Such consultation is particularly necessary in relation to biodiversity conservation because protection of native vegetation and fauna habitat often means a loss of production potential for the land user. In addition, land user management support or co-operation (e.g., in the maintenance of stock-proof fencing, retention of the water-table) is often necessary for the long-term protection of the vegetation against invasion of weeds and animal pests.

While some of the consultation may involve public meetings or meetings between council staff and particular interest groups, one-to-one consultation with landowners or with small groups of landowners appears to be equally or more important at least in rural areas.

In short, examples of effective biodiversity conservation by local government seem to share the following common features:

- supportive policy statements in strategic documents such as a district or regional plan.

Such statements draw attention to the issue and can provide justification for enforcement of protective measures in the face of individual or community opposition.

- personal commitment by one or two key mid-level and/or senior staff;
- a willingness to engage in thoroughgoing consultation with stakeholders, particularly the landowners themselves;
- linkages between members of the community and council staff (i.e., some history of trustful interaction between council and the public);
- incentives (some means of reward for desired behaviour. As with the example of the Waikato Regional Council's dairy shed effluent regulations, this can be the promise of fewer administrative hurdles and reduced expense associated with the preferred option);
- disincentives (some regulatory prescription or negative feature associated with the less preferred option, such as fines, greater administrative difficulty or expense).

CONCLUSION

District and regional councils are of particular importance for conservation of New Zealand's indigenous biodiversity in production landscapes because they administer the legislative instruments that are most relevant for conservation on private land.

The legal and administrative structure of district and regional government offers both opportunities and constraints for conservation of biodiversity. Strengths offered by the system include the clear mandate of the *Resource Management Act* for sustainable management of natural and physical resources, including indigenous vegetation and habitat of indigenous species; legislative support from the *Biosecurity Act,* which provides for a clear and systematic way of determining regional priorities for the control of introduced species; and the provisions of the *Local Government Act,* which enable district and regional council to develop policies and strategies for the implementation of community objectives, including objectives which apply to conservation.

The weaknesses or constraints to biodiversity conservation by local government include conflict between the administrative and legal structures on the one hand with ecological structures and processes on the other; divisions and conflict between the responsibilities of different levels of government, especially in relation to the interaction of land (a district responsibility), and water (a regional responsibility), and between districts and regions in relation to pest control; the fact that environmental issues and biodiversity conservation must compete with other responsibilities and priorities of local government and with competition for funds; and with the patchy and sometimes inadequate capacity of local government to provide the staff skills and resource commitments for effective conservation. This is particularly true for rural district councils, which may suffer a low funding base (due to small numbers of ratepayers) and high service costs such as roading.

Despite difficulties and constraints, progress toward greater awareness of conservation responsibilities is being made by many regions and district councils. Where progress occurs, it is often due to co-operation between individuals and organizations (such as councils and the Department of Conservation, or between regional and district councils), and to patient and hard working involvement of non-governmental organizations such as Forest and Bird into *Resource Management Act* processes. It is also due to effective consultation between councils and private landowners. In the end, it is the landowner who can cut down the last remaining habitat of a local population of native bird; it is the local landowner who must keep his stock from grazing the remnant forest or trampling the edges of his local stream. It is landowners of the catchment, organized into groups, who will protect or recreate the corridors and patches of a future biodiverse landscape. It is local government which can most effectively help or hinder the process.

ACKNOWLEDGEMENTS

I would particularly like to thank the following people for their ideas and information: Jeff Flavell, Department of Conservation; Vicky Froude, Pacific Ecologic Resource Management Associates; Jenny Fuller, Waitakere City Council; Michael Harte, Ministry for the Environment; Doris Johnston, Department of Conservation; Gerry Kessels, Te Pahu, Hamilton; Rebecca Maplesden, Ministry for the Environment; Shona Myers, Auckland Regional Council, Rachael Carruthers, Department of Geography, University of Waikato.

REFERENCES

Bayfield, Maggie, 1997. Significant Natural Areas. Unpublished paper presented to Local Government Association/Ministry for the Environment Workshop on 26 September, 1997.

Department of Statistics, 1997. Summary of regional 1996 census statistics. New Zealand Government: Wellington, New Zealand.

Environment Waikato, 1997. 1997–98 Annual Plan. Environment Waikato Regional Council: Hamilton, New Zealand.

Environment Waikato, 1996. Operative Regional Pest Management Strategy for the Waikato Region. Environment Waikato: Hamilton, New Zealand.

Froude, V., 1997. Biodiverstiy Protection Provisions, Council Implementation of the Biodiversity Protection Provisions in the Resource Management Act. *Plan. Quart.* 126, September.

James, B., 1993. Environmental Issues, Values and Behaviour. Department of Conservation: Wellington, New Zealand.

Milne, C. D. A., 1992. Handbook of Environmental Law. Royal Forest and Bird Protection Society: Wellington, New Zealand.

New Zealand Government, 1974–1996. *Local Government Act, 1974* plus amendments. Government Printer: Wellington, New Zealand.

New Zealand Government, 1994. *Resource Management Act 1991,* reprinted as on 1 March 1994. Government Printer: Wellington, New Zealand.

Roper-Lindsay, J., 1997. Ecology, Significance and Section 6(c). *Res. Manage. News* **5**(2): 18–23.

Waitakere City Council, 1995. Our Place, Our Future, Waitakere City Proposed District Plan 1995.

Williams, D. A. R. (editor in chief), 2nd edn. 1997. Environmental and resource management law. Butterworths: Wellington, New Zealand.

Works Consultancy Services, 1995. Ruapehu Proposed District Plan: Audit of Indigenous Vegetation Provisions. Ruapehu District Council: Taumaranui, New Zealand.

CHAPTER 61

Caring for country today: towards an ecopolitics of sustainable indigenous land management regimes

J. Birckhead,[1] J. Collins,[1] H. Sakulas[2] and J. Bauer[1]

Until the 1970s, most protected areas were founded on premises of strict nature protection, thus, often resulting in the eviction of indigenous peoples from their homelands in the name of conservation — lands which for them constituted *production environments* in which they managed and utilized resources. However, with a growing realization that most biological diversity is in landscapes and seascapes inhabited and used by local peoples, indigenous "involvement" in conservation and protected area management has emerged as a much lauded, but highly contested domain of policy and practice in the 1990s. While New Age discourses have valorized "ecosystem people", their knowledge, culture, and cosmic connection with Gaia, conservation agencies in a number of countries have sought to "involve" "first" people in the management of protected areas and to incorporate the "wisdom of the elders" (traditional ecological knowledge) into mainstream conservation agendas. Ideally, such arrangements empower indigenous people by promoting local level control and management of resources, and by contributing to their development through sustainable economic and cultural production enterprises. Yet, all too often agencies construct indigenous people and their knowledge as abstract, fragmented, and disembodied; not grounded in concrete social relations or real economic options.

We critically examine regimes that attempt to reconcile nature conservation with indigenous development and production, drawing on case material from Australia, Papua New Guinea, and Nepal. Key issues include; how agencies conceptualize and structure indigenous "involvement", local level decision making and power relations, control of resources, implications for conservation, and underlying contradictions.

INTRODUCTION

THIS chapter considers the fundamental issue of indigenous people and the environment today; who are "first peoples", what roles do/should they assume in the nature conservation policies and regimes of various nation-states, and how do they fit into the production environment matrix? These, of course, are not new questions, but ones that are being asked increasingly as we approach the millennium. To date, attempts to answer such questions and to "involve" indigenous people in conservation agendas have produced a variety of international covenants, policy options and management structures in a number of countries, and have generated a growing plethora of literature and popular discourse.

The 5 000 indigenous groups with a total population of more than 600 million comprising some 5 to 10% of the world's population (Stevens 1997) are emerging as an important force in world conservation. They today inhabit, and claim as traditional homelands, at least 20% of the earth's surface and perhaps as much as 30%, four to six times more territory than is encompassed within the

[1]Johnstone Centre, Charles Sturt University, P.O. Box 789, Albury, New South Wales, Australia 2640.
[2]Wau Ecology Institute/Johnstone Centre, Charles Sturt University, P.O. Box 789, Albury, New South Wales, Australia 2640.

Pages 594–605 *in* NATURE CONSERVATION 5: NATURE CONSERVATION IN PRODUCTION ENVIRONMENTS: MANAGING THE MATRIX ed by J. L. Craig, N. Mitchell and D. A. Saunders. Surrey Beatty & Sons, 2000.

entire global protected area system, and with native title claims in a number of countries, may in the future control significantly more land than they do at present (Stevens 1997). Many of these environments are fragile or under threat from development and are characterized by high levels of biodiversity and therefore significant to global conservation. As well, Andrew Gray (1991) argues that they account for as much as 90 to 95% of the world's cultural diversity. "First Peoples" then, "are not marginal peoples who are disappearing but rather marginalized peoples who are seeking accommodation with the states in which they live" (Maybury-Lewis 1997).

Yet, despite the apparent centrality of indigenous people in popular and policy discourses, indigenous people and their issues continue to be marginalized in mainstream ecology and conservation fora. Fiona Walsh (1995), for example, commented on the minimal Aboriginal participation at the 1994 "The Role of Networks" conference at Geraldton, Western Australia, noting that "this was indicative of the progress that still needs to be made in facilitating increased participation of Aboriginal people in nature conservation". Similarly, the 1997 "Conservation in Production Environments — Managing the Matrix" conference at Lake Taupo would have benefited from increased indigenous participation and themes, or even motions or recommendations regarding indigenous people as an essential component in the elusive matrix. This is a reflection of the distance that still exists between indigenous people and conservationists. Given this absence of indigenous voices, one could derive the impression that these are issues on the periphery, not central to conservation and production environments in countries such as Australia and New Zealand (Aotearoa).

This latter conference occurred, ironically, at a time when some New Zealand farmers were driving their tractors to Parliament House in Wellington to protest the Maori Reserved Lands Bill. New Zealand papers (e.g., *New Zealand Herald,* 5 December 1997:3) explained how farmers were worried by the loss of equity because of the uncertainty attached to the leasehold issue. It further reported that in response to a farmer who said "the land had been given to Pakehas as a lease in perpetuity, Mr Delamere (the Associate Treasurer) said it had been stolen from Maori. 'All the Maori owners want is a fair deal — when you have lived off the pig's back for 110 years'."

At the same time in Australia there was considerable media debate over the proposed amendments to the Wik High Court Decision regarding the rights of Aboriginal native title holders to negotiate over the uses of pastoral leases and on freehold land. Underlying this debate, is the real question of the status and place of Aboriginal people and Torres Strait Islanders in Australian society and their power to control resources on public lands and in production environments. Indeed, the fear that an indigenous group can veto development and limit mine, forest, marine, and agricultural production is a root cause of much of the panic and apprehension in Australia, Aotearoa, and in other countries. Underlying such fears often lies the assumption that corporate interests are synonymous with the "national interest" whilst indigenous interests are perceived as being external and antagonistic to this expression of nationhood (Howitt *et al.* 1996). Thus, we agree with Jeffrey McNeely's (1993) contention that "conservation is far more a social challenge than a biological one". It is for this reason that we emphasize "ecopolitics": the linking of politics and other social forces with ecology and issues of sustainability and biodiversity. In other words, indigenous land management issues are as much matters of cultural politics, political history, ideology and discourse as they are of ecology, natural history and protected area management (Birckhead 1996).

We now explore further some questions of indigenous participation in "nature conservation" that derives from our long-term interdisciplinary research through the Johnstone Centre at Charles Sturt University in three diverse settings: Australia (Birckhead *et al.* 1992, 1996; Birckhead 1996, 1999; Collins *et al.* 1996, 1997; Roberts *et al.* 1996), Papua New Guinea (Sakulas *et al.,* this volume), and Nepal (Bauer *et al.* 1994; Furze *et al.* 1996). Our overall goal in these projects is to combine ecological perspectives with social science methods and theory to develop a framework for linking "conservation" and community and cultural development at the local level, both inside and outside protected areas, through participatory approaches (see Furze *et al.* 1996). While our Australian work comprises the core of this discussion, perspectives from our work in Papua New Guinea and Nepal are brought in to frame key questions and to provide points of comparison and contrast. But, first we briefly review the emerging role of indigenous people in nature conservation agendas world-wide.

"FIRST PEOPLES" — FROM PERIPHERAL PRIMITIVES TO WISE ELDERS

Indigenous or "first peoples" literally "were the original inhabitants of lands, since colonized by foreigners" (Burger 1990). A more inclusive definition sees such people as "the 'original' or oldest surviving inhabitants of an area" who "have typically lived in a traditional homeland for many generations — usually for many centuries" (Stevens 1997). While exact definitions of indigenous people vary, they all have one thing in common; their subsistence practices (now or until relatively recently at least) heavily relied on the use of local resources and ecosystems (Stevens 1997). However, indigenous identity is not solely dependent upon or constructed within the constraints of ancient links to place, but is a dynamic contemporary process reflected in, and reflective of, contemporary geopolitics (Cobo in Howitt *et al.* 1996). Indeed, the term "indigenous" has gone beyond the academic and self-definition to assume importance as a category in the practical and political agendas of not only nation states but mainstream international organizations such as the World Bank (Howitt *et al.* 1996).

Indigenous peoples until relatively recently occupied the "savage slot" (Trouillot 1991) in popular and scientific discourse and have suffered throughout the world from the appropriation, colonization and development of their land by others (see Kemf 1993; Furze *et al.* 1996). In fact, indigenous peoples have often been victims of state sponsored, or tolerated, "ethnocide", or even genocide, where indigenous populations are simply considered to be "in the way" of national destiny, development, resource extraction, dam building or anything else (Maybury-Lewis 1997). Whilst being pushed into increasingly marginal habitats by the onslaught of developers, indigenous interests in even these lands have often been frustrated by the imposition of conservation declarations. In Australia, for example, the declaration of Archer Bend National Park on Cape York Peninsula in the 1970s was used by the Queensland Government to thwart the land acquisition aspirations of clans of the Wik Nation (Dale 1997).

Seen as "primitive" remnants of societies passed over by progress at the margins of "civilization", it was seldom considered that such peoples would have anything to teach scientifically "advanced" urban-industrial societies about caring for nature. However, times are changing. Although indigenous rights are far from secure in many parts of the world, indigenous people are increasingly active on the world stage, fighting for rights to land and self-determination[1], and the protection of the environment (Burger 1990). Increasingly, indigenous people, their knowledge and wisdom, are being looked to by both scientists and the general public to provide a corrective to our alienated, rapacious use of the environment. Indeed, a growing body of popular and scientific literature, extols the virtues of indigenous "earth honouring" (Jacobs 1994), "caring for country" (Young *et al.* 1991), "tribal wisdom in the modern world" (Maybury-Lewis 1992), "wisdom of the elders" (Knutson and Suzuki 1992), "traditional ecological knowledge" (Williams and Baines 1993), and "conservation through cultural survival" (Stevens 1997).

Indigenous people are quite often associated with parks and protected areas. This reflects the current fashion of treating environmental issues as global problems that has led to greater state and international intervention in land and natural resource management issues (Colchester 1994), particularly in the establishment of protected areas. Defined as "an area of land or sea especially dedicated to the protection and maintenance of biological diversity, and of natural and associated cultural resources, and managed through legal or other effective means" (International Union for the Conservation of Nature 1994), national parks and protected areas have become a major global phenomenon during the last 130 years and are a major tool in preserving endangered species, habitats, and ecosystems (Stevens 1997). Until the 1970s, most protected areas were founded on premises of strict nature protection, thus often resulting in human rights abuses with the eviction of indigenous peoples from their homelands, lands which for them constituted production environments in which they managed and utilized resources. However, in recent decades there has been increasing recognition of indigenous rights within protected areas globally and some indigenous participation in protected area management (Birckhead *et al.* 1992; Stevens 1997). Such changes have been facilitated by a growing international and national recognition of the rights of indigenous peoples and the realization that the conservation of biodiversity is unlikely to succeed without the support of local communities. Indeed biological and cultural diversity appear to be interdependent. As Neitschmann (in Stevens 1997) notes, most remaining biological diversity world-wide is intimately linked to areas used by local, and especially indigenous, peoples and regulated by their cultural norms and land use practices. In the case of Australian Aboriginal peoples, their use of fire for millennia has had a significant impact on the composition of the flora and fauna of Australia (Flannery 1994).

A number of international resolutions and protocols through the United Nations and many other organizations have articulated general principles concerning the rights of indigenous peoples and how these include environmental protection. One of the most frequently cited of these asserts:

> "Indigenous people and their communities, and other local communities, have a vital role in environmental management and development because of their knowledge and traditional practices. States should recognize and duly support their identity, culture and interests and enable their effective participation in the achievement of sustainable development" (United Nations Conference on Environment and Development, Rio Declaration, Principle 22, 1992, in Webb 1996:98).

More specifically, the International Union for the Conservation of Nature "Caracas Declaration — Protected Areas and the Human Future", 1992, endorsed six principles, which delineated the basic rights of indigenous and local peoples with respect to their "involvement" in nature conservation

[1] "Self-determination" is a term that is used frequently in this chapter in connection with the rights of indigenous peoples. The United Nations Draft Declaration on Indigenous Rights states: "Indigenous peoples have the right of self-determination. By virtue of that right, they freely determine their political status and freely pursue their economic, social and cultural development" (Anaya 1997). In 1991, an Australian Federal Labor Government acknowledged a more inclusive definition of "self-determination" and the legitimacy of indigenous aspirations. However, the current Australian Federal Coalition Government opposes references to: self-determination", believing phrases such as "self-management" or "self-empowerment" better reflect Australia's national interest. If the Australian Government gains support at the United Nations, this would diminish a principle fundamental to the recognition of indigenous rights.

arrangements with governments. Such rights include, the recognition of indigenous territorial rights, indigenous environmental knowledge and "traditional management systems" and their contemporary applications to environmental management, the role of women's environmental knowledge in such systems, and intellectual property rights and proper compensation for use by agencies or commercial interests (International Union for the Conservation of Nature 1993).

The first national parks, modelled on Yellowstone National Park, sought to preserve natural resources from destruction and excessive development while at the same time providing recreation for tourists. By contrast *Parks for Life* (International Union for the Conservation of Nature 1993) aims to ensure that protected areas can continue to be a mechanism for conserving biodiversity as well as becoming models for sustainable development by local communities. Furthermore, the International Union for the Conservation of Nature Commission on National Parks and Protected Areas (1994) now advocates that taking into account the needs of indigenous peoples, including subsistence resource use is a *specific objective* of the creation of national parks (our emphasis).

While the above principles provide useful ideals for the participation of indigenous people in state conservation agendas, the question arises, what actually happens on the ground in various settings, when ideal principles are translated into policy and everyday management structures? We now briefly review examples from Australia, Papua New Guinea, and Nepal to illustrate key issues in the participation of indigenous people in the local level management and control of natural and cultural resources. Of specific interest, are questions of empowerment and decision making, sustainable use of resources and cultural and community development, relations with the state and other bureaucratic structures, and the "incommensurability" between indigenous and non-indigenous cosmogonies (Roberts *et al.* 1995).

AUSTRALIA — SHARING COUNTRY?

Although the first peoples of Australia can justifiably claim to be the world's oldest continuous culture, spanning at least 60 000 years (Flannery 1994), their place in Australian society is much less assured or secure than their long occupation of the continent might imply. Representing less than 2.5% of the population, Aboriginal and Torres Strait Islander people have suffered the effects of dispossession and marginalization at the hands of the settler populations. Decimated and dispersed by the British who took their lands under the fiction of "terra nullius", and a succession of assimilationist government policies, the indigenous people here are very much fighting for their identity and proper place in a contemporary Australia (see Birckhead 1996). At present, native title issues, especially the right recognized by the Wik High Court Decision for Aboriginal people to negotiate with farmers over the use of pastoral leasehold land and the subsequent legislative response of the Federal Government, (which are proving to be quite contentious at a national level) (Chalk 1997), reflect the paramount importance of environmental and land issues to Aboriginal people today. Also, there is the real potential for conflict when indigenous people claim or desire to control land in production environments (see Ross 1995 and above). More pertinently for the purposes of this chapter, there is a reluctance to perceive that many of our protected areas constitute cultural landscapes and hence "production environments" for indigenous peoples who have lived off and managed the land for millennia. Yet, the popular definition of "wilderness" persists, along with a desire to exclude or limit all human interaction within an allegedly pristine area. Where Aboriginal people have been brought to the brink of annihilation, their territories are subsequently recast as "wilderness" (Langton 1996). An indigenous presence is accepted if it is consistent with the image of the "noble savage", being at one with the natural environment "frozen in time", not part of a global economy with aspirations to share in the fruits of Western development.

Aboriginal people hold land tenure over approximately 15% of the land mass of Australia (Thackway *et al.* 1997). While this may sound like a large chunk of the continent, it must be noted that Aboriginal people have been dispossessed from the most productive land in the country. Most of the land that they have retained or reclaimed has been deemed "unproductive" and has often been badly degraded by inappropriate non-indigenous land use (although, paradoxically, much of this land has subsequently been found to be rich in mineral resources). This inequitable distribution of productive land obviously presents obstacles to the empowerment and self-determination of Aboriginal people. However, for the purposes of this chapter we will focus on the fact that ecotourism is often perceived to be the most productive and culturally appropriate land management regime for the marginal land currently in Aboriginal possession. Furthermore, a portion of the conservation estate, which also occupies marginal and non-productive land, may also in the future be owned by Aboriginal people, facilitated by Native Title and other state legislation. It is vitally important, therefore, that conservation practices are linked to the interests of Aboriginal people (Walsh 1995).

With respect to national parks, Australia is well known and acclaimed for its Aboriginal ownership and joint management of Uluru-Kata Tjuta, Kakadu and Booderee national parks. As well, other states and territories in Australia are developing their own models of indigenous involvement in national park management with varying degrees of empowerment for indigenous participants (Young *et al.* 1991; Birckhead *et al.* 1992; Cordell 1993; DeLacy 1994; Woenne-Green *et al.* 1994; Larritt 1995; Webb 1996; Collins *et al.* 1997; DeLacy and Lawson 1997; Ellwood and Andrews 1997; Lawrence 1997; Smyth and Sutherland 1997). However, although indigenous landholders have a major say in the management of jointly managed national parks in Australia, they did not have a choice as to whether the national park was declared or not (Thackway *et al.* 1997). Therefore it cannot be said that jointly managed national parks are an expression of indigenous self-determination. They are simply a mechanism to accommodate Aboriginal interests in the management of a pre-determined conservation framework. Attempts to persuade indigenous landholders to voluntarily enter into joint management arrangements for land they already own have not been successful. Furthermore, such lease back arrangements are subject to the vagaries of the government of the day. The current Federal Coalition Government has recently proposed changes to environmental legislation which may undermine the joint management structure of Uluru, Kakadu and Booderee national parks, reserving greater powers for government. While both the House of Representatives Standing Committee on Environment, Recreation and Arts (1993) and the Australian Law Reform Commission (1986) recognized indigenous rights to land and resources, they considered that the interests of conservation represented a legitimate limitation on the rights of Aboriginal and Torres Strait Islander peoples to hunt and fish. Therefore, indigenous resource use is still controlled by the management agency and resource managers' perceptions of appropriate land use (Stevens 1986, as cited in Lawrence 1997).

There are many issues relating to indigenous involvement in the conservation estate. However, probably the most fundamental and controversial for policy makers is that of respect for indigenous settlement and resource use. Concern for indigenous rights requires the fullest possible recognition of indigenous peoples' land tenure and subsistence practices within protected areas (Stevens 1997). Restrictions by outside authorities violates the basic human right of indigenous peoples to self-determination as articulated by the United Nations Draft Declaration on Indigenous Rights. Many conservationists, however, are uneasy with broad endorsements of resource use within protected areas or for indigenous control of agricultural lands or of sea country. A blanket rejection of hunting and gathering as unsustainable not only ignores the rights and interests of indigenous peoples but also the changing approach to the management of biodiversity. It is now accepted that indigenous knowledge is beneficial to biodiversity conservation. Therefore, to reject hunting and gathering rights is hypocritical, perpetuating Australia's colonial legacy of indigenous dispossession (English 1997). Perhaps justifiably, concern is expressed regarding changes in resource use patterns due to more sedentary settlement patterns, population growth, cultural change leading to a decline in conservation values and the use of modern technology, in particular, guns, motor vehicles and boats. All this in a world where species may already be threatened by non-indigenous land use, such as habitat destruction and degradation and the introduction of feral species (Collins *et al.* 1996). Such concerns have led to the development of several forms of conditional recognition of indigenous settlement and land use within protected areas. The approach most often adhered to in the Australian context, has been to authorize only "traditional" land uses. However, traditional use policies when implemented from the outside can be paternalistic and coercive, creating "cultural zoos" (Stevens 1997).

Questions of authenticity also arise as to who has the rights to determine what is "traditional" in diverse Aboriginal societies (see Birckhead 1992). The Australian Law Reform Commission (1986) has maintained that in determining if an activity is traditional or not, the purpose of the activity should be considered and not the technology involved. As a result, most authorities have not precluded the use of firearms and vehicles in the pursuit of a "traditional" activity (see Lewis 1992). Nevertheless, restricting land use to only long-standing traditional practices raises significant human rights issues by sharply limiting self-determination and avenues of development (Stevens 1997). Some have argued that because the system of barter is a long-standing tradition in Aboriginal societies then this should facilitate the recognition of commercial exploitation as a legitimate activity (Sweeney 1993). However, this argument for the most part has been ignored. It is difficult to argue that indigenous rights to wildlife should be limited to utilization for domestic and cultural purposes only when evidence would suggest that the commercial use of wildlife can assist conservation and bring benefits to local communities (Wilson *et al.* 1992; English 1997). While community rights to the ownership of wildlife resources have been advocated internationally, a lack of empirical data regarding the sustainability of community wildlife management has led to an increased emphasis on monitoring indigenous wildlife use (Davies and Higginbottom 1997). Self-monitoring is encouraged, but a trend towards "monitored lives" implies unequal power relations. Undoubtedly, concepts of "sustainable use" will become increasingly important in Australia and globally as governments attempt to incorporate more land in a protected area system in an attempt to conserve biological diversity. Currently however, "sustainable use" can be a vague goal (Stevens 1997).

The World Commission On Environment and Development (1987) defined "sustainable development" as "meeting the needs of the present generation without compromising the needs of future generations". While this is an important term and development concept, it is not very precise (Furze *et al.* 1996). Opinions may vary as to the sustainability of various land use activities. Practices designed to facilitate the sustainable use of one species may have unforseen impacts on other species and entire ecosystems. Many common "scientific" sustainable systems, for example, forestry practices, often radically alter the characteristics of forests. It is therefore important that notions of sustainable development need to be coupled with explicit biodiversity protection goals. Furthermore, the adoption of new land use activities, even when environmentally friendly and sustainable, may mean an abandonment of traditional land use regimes, something which can easily lead to the loss of a cultural landscape and ecosystem which has been shaped by a specific human intervention. What does seem clear both in Australia and overseas is that the conservation of both biological and cultural diversity are intertwined. One will not survive without the other. It is also clear that whatever form regulation of resource use takes indigenous stakeholders should be instrumental in developing management plans and specific regulations. It may be that in promoting the long-term viability of a protected area, indigenous rights should be allowed occasionally to compromise biodiversity objectives (Stevens 1997).

While few would disagree that indigenous people should be entitled to some economic return from their land, what is deemed appropriate in the context of national parks is not always easy to define. Revenue from entrance fees and rent for the use of the land by a national park authority is possible, but such sums may be limited in application; for example, all rental income to be paid by the NSW National Parks and Wildlife Service to Aboriginal landowners under the new *National Parks and Wildlife (Aboriginal Ownership) Amendment Act 1996* can only be used in the management of the protected area. What constitutes management expenditure is broad and flexible, but this arrangement does not facilitate economic independence. Involvement in the day to day running of the park, interpretive tours and artefact production for tourists are usually seen as the most appropriate non-consumptive activities for indigenous participants in national parks. An approach to protected area management less likely to violate human right to self-determination is to support the freedom of indigenous peoples to adopt whatever form of land use they choose so long as they do not significantly compromise the conservation of biological diversity and cultural landscapes within the protected area. This makes principles such as environmental sustainability and ecological sensitivity paramount as guidelines for management planning whilst at the same time promoting the social, cultural and economic sustainability of traditional owners.

In recent years the concept of Indigenous Protected Areas has been developed, following a realization that the incorporation of indigenous lands into the reserve system was essential if all bioregions of Australia are to be conserved for the future. The Indigenous Protected Areas programme allows indigenous landholders with land of high conservation value to enter into conservation partnerships with Environment Australia (Biodiversity Group) Such protected areas are classified as category VI of the new International Union for the Conservation of Nature guidelines, a "Managed Resource Protected Area", which allows a sustainable flow of goods and services to local indigenous groups (Thackway *et al.* 1997). Projects being funded under the Indigenous Protected Areas scheme are located in all states of Australia and the Northern Territory. An issue of major concern to indigenous organizations is the extent to which the establishment of protected areas may compromise their existing rights to use, control and manage their land and resources. Ideally, the schemes allow indigenous groups to retain full management control of their lands but offers them financial support to do so in a sustainable fashion, promoting such activities as ecotourism rather than pastoralism. The success of Indigenous Protected Areas will require conservation agencies to display a greater level of trust in indigenous land management than has been shown in the past (see Moore 1995). Importantly, standards should not be imposed on Indigenous Protected Areas which are currently not imposed on the government's own protected areas. There will need to be far greater sharing of power, knowledge and resources and a flexible collaborative approach. For their part, indigenous groups may have to be prepared to make legally binding commitments to sustainable management objectives. While much may be made of the principle of empowerment for indigenous people in relation to land and resources very little is ever said of how the consequent disempowerment of state resource management agencies will be achieved (Walsh 1995).

The fact that indigenous lands have remained areas of high conservation value is not simply a historical legacy (Dale 1997). Indigenous peoples have continued to manage their lands sustainably outside the protected area system. Many communities in Northern Australia support community ranger programmes to promote the sustainable management of their lands. Kowanyama, on Cape York, is often cited as an example of an indigenous community which has actively campaigned for self-management of its land and resources, operating its own natural resource management office using indigenous ecological knowledge and Western science to promote sustainable management of its resources (Sinammon 1997). The Kowanyama community have also initiated total catchment management of the Mitchell River Delta, entering into co-operative management arrangements with commercial fisherman and other stakeholders. The Bawinanga Aboriginal Corporation at Maningrida in north-central Arnhem Land has also demonstrated a strong commitment to managing their lands in an ecologically sustainable fashion. The community ranger programme there is involved in feral weed and animal control and many ecological surveys to determine the distribution of rare indigenous species (Carter and Yibarbuk 1997). The community has also involved the Northern Territory University in assessing the viability of a commercial trepang fishery. Such initiatives have been very much community driven with limited support and funding from state agencies or other conservation agencies. While there has been financial support for the community ranger programme, this is inadequate and constrained by state mandated bureaucratic structures. In general, decision making and leadership through government established community councils and land councils, are often at variance with the customary indigenous leadership of traditional owners (Birckhead and Wallis 1994; Furze *et al.* 1996; Birckhead 1999).

While many look to north and central Australia as repositories of traditional ecological knowledge, communities in southern Australia have also established land management initiatives which maximize both cultural and biological diversity. At Corindi on the north coast of New South Wales the Gumbaingirr people successfully established themselves on land outside of the reserve system and thus outside of government control (Anon. 1997). The community resettled traditional camping places located on public land or "no mans land" along Corindi Creek and Corindi Lake. The community is actively promoting the cultural and biological diversity of these bushlands through the establishment of a cultural heritage ecotourism venture, utilizing western expertise when appropriate. These are just a few examples. There are many more throughout Australia. Many such landholders may prefer to remain outside the protected area system, fearing it entails an unacceptable level of government interference (Stevens 1997). However, it would be desirable if such unofficial protected areas should be recognized by governments, the International Union for the Conservation of Nature, and international conservation agencies so that they are afforded the same funding opportunities and the same protection from encroachment by development as state managed protected areas.

Aboriginal land management initiatives such as those described at Kowanyama, demonstrate the capabilities and aspirations of indigenous people in relation to land management. However, many indigenous groups in Australia, particularly in Western Australia, are largely preoccupied with acquiring secure title to their land and adequate water, housing, food and education. Until these basic rights are secured, land management cannot be a central focus of Aboriginal organizations. As an Aboriginal representative to the Ecopolitics IX Conference (Perspectives on Indigenous Peoples Management of Environment Resources) said, "Social justice first!, then we can talk about sustainable land management" (pers. obs.).

INDIGENOUS LAND MANAGEMENT IN PAPUA NEW GUINEA

Land Management in Papua New Guinea is very much an indigenous issue, after all the great majority (98%) of its inhabitants are indigenous and traditional landowners. On the other hand, development and conservation are mostly imposed by foreign investors, non-government organizations or International Agencies. Papua New Guinea's great mining and timber resources have encouraged a kind of development which has, while promoting the central government in Port Moresby, challenged and often ignored traditional landownership. While there have been recent successes of landowners to prevent this type of unwanted development or to ask for compensation for environmental damage (e.g., Ok Tedi), the overwhelming desire of remote communities to become part of the bigger world and to obtain at least some basic infrastructure seems to be most easily realized in the wake of mine sites and logging camps.

During this process, however, communities lose control over their natural resources. They also lose control of their social structures; indeed, over their lives in general. The benefits they obtain such as roads, better housing, a school, perhaps even a hospital, come in exchange for erosion, displacement by labour immigration, alien diseases, social destabilization, and often unwanted knowledge. When the exchange is realized, there is no going back and communities are overwhelmed within a few years by the consequences of short-term benefits and a lack of long-term vision. This is the general outcome of logging and mining operations.

At present the situation in Papua New Guinea is characterized by battles for control. Battles between communities and Port Moresby over control of their natural resources. Battles between traditional elders and rebellious youth who gain power by having better access to Western thinking. The evolution of a new world order which has little regard for traditional values, beliefs, knowledge and order. The new order might be Western or Eastern, in any case alien models which find traditional landownership cumbersome and unproductive. Will this be to the benefit of Papua New Guinea? It will be beneficial for an elite in Port Moresby and for a group of people from communities who have used the new order to their advantage. In this world indigeneity will have ceased to be a word with meaning. It will be used to attract foreign ecotourists into indigenous Disneylands and perhaps later on as a vision for a lost national identity. For Papua New Guinea this would be tragic, as it still remains one of the few places on earth where truly indigenous people can shape their own destiny and are not a disempowered minority which has to accept Western (or Eastern) way of thinking.

What can be the model for indigenous land management in Papua New Guinea, with its obvious differences to the Australian situation? How can conservation and development be reconciled in this rapidly changing social, economic and ecological environment? In lots of ways, indigenous management in Papua New Guinea rests on community and village control of resources. Sakulas *et al.* (this volume) in their critique of the Integrated Conservation and Development process in Papua New Guinea suggest land tenure, education and social integration as one of the key processes. Their analysis also suggests that conservation can only compete with degrading development (such as unsustainable logging and unmitigated mining) if it can match the benefits offered by these schemes or offer different ones. The history and long-term success of Wau Ecology Institute (see Sakulas *et al.*, this volume and Bauer *et al.,* in press) demonstrates that Papua New Guinea nationals have succeeded in developing their own, decentralized, community-driven and indigenous model for the Integrated Conservation and Development process. The lack of support, however, this indigenous and decentralized institution, which indeed incorporates many features of the Integrated Conservation and Development process, receives from its own government and from international conservation seems to suggest to us that the control of the conservation and development process is not being given up willingly. Western-driven conservation has to be sensitive to the increasing local ability in a particular country, to manage their own resources. There is definitely a time, when the control of processes driven by foreign United Nations-based, or international non-government organization's has to be handed over to locals. At such a time it is beneficial to connect institutions (educational, protected area management, minesite management etc.) of the newly developed capacities in Papua New Guinea with similar institutions overseas. The billion-dollar consulting industry world-wide appears to resist this process, appearing more interested in its own sustainability than sustainable local development.

INDIGENEITY AND CONSERVATION IN NEPAL'S MULTICULTURAL CONTEXT

The concept of "Indigeneity" in Nepal has to be applied to a society of many different cultures and languages merged from people who have arrived during various phases of its history and who have had hundreds, indeed thousands of years to coevolve into one society. Integration of a continuous stream of newcomers from the north (the Sherpas, originally a Tibetan tribe arriving 800–1 000 years ago, others as recently as 400 years ago, and many Tibetan refugees recently) with Indian Tribes (e.g., the Tharu) or eastern colonizers (the Rai and Magar) have formed a culture, where Nepal's dominant and oldest group, the Nehwari, originally centred in Kathmandu valley, have merged into one superculture as unique in its diversity as its natural heritage.

As of today Nepal is known as a country where this intensive mixing has led to a multi-cultural society which has incorporated many aspects of this diversity in a fascinating and apparently well balanced blend of mutual tolerance. It is not surprising that this has not only acted as a magnet to an increasing number of visitors but to an almost ludicrous plethora of aid and conservation projects from many parts of the world, each of which wishes to implement their own vision of a sustainable future for Nepal.

While at a first glance it would seem that the protected area proponents have had major gains in Nepal — unlike to the situation we describe in Papua New Guinea — the past ten years have seen an increasing number of concessions to local communities — which have, at least in theory, returned more control to locals. National Parks, while initially modelled after the African experience to exclude local people (in the early days local people had been "removed" e.g., from Lake Rara National Park and settled in the Terai region, where many old people succumbed to Malaria) have now accepted their human inhabitants, at least generally. Sherpa culture in Sagarmatha National Park has not only become the very embodiment of what constitutes the natural Mount Everest environment, they are also a main "attraction" for the visiting tourist. Nepal's most successful and by now best known protected area, Annapurna Conservation Area, is similarly built on the cultural environment and attempts to explore the concept of conservation not by protection alone but by wise and local use (see Furze *et al.* 1996).

In other areas of Nepal, this mix of natural and indigenous identity is less successful. While mountain cultures have, in Nepal as anywhere else, tended to be protected (or excluded) from many forms of development and change by remoteness, lack of access and low human carrying capacity, this situation is dramatically different in its southern protected regions in the Terai. Here, protected areas such as Royal Chitwan and Bardia National Park and Kosi Tappu, Parsa and Suklaphanta Wildlife Reserves have become subject to an ever increasing pressure from agricultural development, descending hill people, Indian migrants, and hydro development (e.g., Bauer *et al.* 1994, 1996a,b). Conversely, the indigenous people of this region, the Tharu, have similarly not only lost many of their original foraging, hunting and fishing grounds but are being swamped by migrants from other parts of Nepal. In some parts of Nepal (Western Terai) this annual immigration amounts to more than a quarter of the local populace (Bauer *et al.* 1996a,b).

Western type conservation efforts in Nepal have evolved from mountaineering expeditions in the Himalayas and early protection efforts of the Royal hunting grounds in the Terai boosted with an increasing drive of the central government in Kathmandu to exert more control in formerly remote areas. Conservation in Nepal has a history where disempowerment of local communities has led to devastating impacts on the environment. In 1950, the central government in Kathmandu removed community control from forests and replaced it with the centralized Panchayat system. This move is generally held responsible for the subsequent rapid destruction of forest resources and continued to operate until 1993 when community control was at least partly reinstated to villages (see Rustomji and Ramble 1990; Fisher 1999).

This situation is not confined to forestry. The establishment of the National Park system mirrors the effects of local disempowerment. When Nepal's protected areas were established twenty-five years ago a further element of government control by a bureaucracy in Kathmandu has continued this trend. Over the years this situation has become more endemic as more and more national experts, trained in Western ways of thinking in conservation, have taken over functions formerly introduced by foreign experts and institutionalized Western thoughts.

It would appear to us that Nepal is now in the position where many indigenous values, religious beliefs and attitudes are constantly challenged within an environment where development and conservation is driven by a multitude of foreign organizations, many of them competing for funds and working under different paradigms and ideologies. Many Nepali experts, most of them trained overseas, have to find their place in this confusing environment which constantly forces them to realign their ideologies with new projects and discard their own experiences. Within a powerful bureaucracy, which in particular during the past five years was entirely unpredictable and driven to a large extent by foreign projects (with new expatriate project leaders seeking their new place every few years, often coming from vastly different background), many committed Nepali scientists, managers and field staff face a constant struggle for self expression. In this world Nepali scientists who align themselves with Western development projects and Western thought are the winners. Their colleagues who have not done so are now actually competing with Western experts and Western jobs and tend to miss out because of their lack of alignment. In this way a considerable number of Nepali conservation scientists have been marginalized by the international community.

In many ways the overproportional commitment of the Western world to aid Nepal in conserving its natural and cultural diversity and grandeur has led to a situation where positive aspects are constantly challenged by negative implications and where project gains are offset by undesirable shifts of power. Nepali and foreign people have not only initiated many interesting and resourceful approaches to reconcile development with conservation and finding the right answer for Nepal, but they have also created a situation where competition, opportunism, conflicts between key players and political considerations are constantly thwarting and reducing the otherwise remarkable success stories. The present day reality of conservation in Nepal often is to find (with difficulties) something which has not been done by others yet, something which is funded (generally by the World Bank), something which does not antagonize the many key players in the country and something which corresponds to the latest conservation ideologies and fashions from New York, Washington, Switzerland or United Kingdom. Perhaps not surprisingly Nepal's own remarkable identity, its remarkable people and its unique cultural and biological heritage are not getting the best deal from all of this.

CONCLUSIONS

The above examples demonstrate ably how conservation is interacting with indigenous people and how their care for their country, homeland and village is affected by Western concepts. It seems obvious that in this comparison Aboriginal people in Australia are getting by far the worst deal in terms of having least access to land, more constraints on land use and least control, while in Papua New Guinea where indigenous people have never been subdued, Wau Ecology Institute is a clear demonstration of the ease with which prominent village culture and community can accommodate the goals of biodiversity conservation. Wau Ecology Institute seems to us to demonstrate that the indigenous approach can integrate the conservation of biodiversity into the cultural landscape, while in Australia, reconciling goals of cultural survival through self-determination with those of state conservation agencies, which are tightly controlled economically, ideologically and bureaucratically, is more problematic. In particular, while security of land tenure remains uncertain for many Aboriginal groups, sustainable land management will not be a primary goal and declarations of protected area status will be seen as just another means of colonizing Aboriginal land and dispossessing Aboriginal people. Nepal in contrast is in a unique position which seems to be characterized by almost chaotic and frantic experimentalism, still largely controlled by central and international agencies. While none of these systems seem to be entirely comparable we believe each of them holds important lessons for the other.

There is one more important point to be made. Where do these observations leave our production landscape? We have realized that this concept of distinguishing between production and protection landscape is also a very Western one which marginalizes indigenous people. If we accept the existence of indigenous people within protected areas and their rights to use "protected" natural resources, we must challenge the notion of contrasting protected areas with production lands. Perhaps it tells us that again we have been overly simplistic. We would even like to go further. If we accept tourism as a form of land use, and a land use which is centred in protected areas, this again makes protected areas, pristine areas and wilderness areas production landscapes. If we do this however, we realize that by now there are no "protected areas" left in the world. They are just utilized for different purposes.

Nevertheless, in Australia and many other countries, current models for protected areas still place emphasis on developing acceptable patterns for using the physical environment, often at the expense of recognizing the social and spiritual value of land to indigenous peoples (Lawrence 1997). It is imperative, therefore, that management agencies seeking to equitably conserve both cultural and biological diversity within a protected area system, need to have not only scientific expertise in conserving ecosystems but also the expertise of a social scientist, and a local person (or people), in understanding local economies, cultures, histories, political structures and the needs and aspirations of traditional owners. Most importantly, the management of protected areas should involve traditional owners in an empowering process, ultimately promoting self-determination for indigenous land managers.

It has been argued that the second thing future generations will most blame us for after the loss of biological diversity is the loss of cultural diversity (Wilson 1988). Until indigenous people come to play a central role in global conservation, the future of both global biological and cultural diversity is at risk (Stevens 1997).

REFERENCES

Anaya, S. J., 1997. Indigenous self-determination. Plenary Session 2. Human Rights and Indigenous Australians. Proceedings of the Australian Reconciliation Convention. *Book* **3:** 13–18.

Anon., 1997. The Cultural Heritage of the Gumbaingirr People. (Phamphlet). Yarrawarra Aboriginal Cultural Centre: Corindi, New South Wales.

Australian Law Reform Commission, 1986. The recognition of traditional hunting, fishing and gathering rights. Pp. 75–81 *in* Report No. 31, The Recognition of Aboriginal Customary Law, Vol. 1. Australian Government Publishing Service: Canberra, Australian Capital Territory.

Bauer, J. J., Maskey, T. and Rast, G., 1994. River systems, hydrodevelopment and the species crisis in the Terai. Pp. 137–145 *in* Safeguarding Wetlands in Nepal ed by B. Bhandari, T. B. Shresta and J. McEachem, IUCN, The World Conservation Union: Gland, Switzerland.

Bauer, J. J., Maskey, T., Rast, G., DeLacy, T. and Furze, B., 1996a. Hydro-development, Agriculture and Conservation in Nepal's Terai — A Case Study on River-Basin Management from the Upper Ganges System. UNEP/AVA: Kuala Lumpur, Nairobi, Kenya, 85 Pp.

Bauer, J. J., Maskey, T. and Rast, G., 1996b. The environmental costs of river regulation in Nepal — present evidence and scenarios for the future. Pp. 8–14 *in* Proceedings Intenational Conference on Wetlands and Development, Selangor, Malaysia.

Bauer, J. J., Birckhead, J., Sakulas, H., DeLacy, T., Thwaites, R. and Iland, J., in press. Biosphere reserves in Papua New Guinea — a model for the future? *In* Wau Ecology Institute, 25 Years: A Perspective ed by A. Allison. Bishop Museum: University of Hawaii, Honolulu.

Birckhead, J., 1992. "Traditional Aboriginal land management practices" at CSU — the cultural politics of a curriculum innovation. Pp. 297–306 *in* Aboriginal Involvement in Parks and Protected Areas ed by J. Birckhead, T. DeLacy and L. Smith. Aboriginal Studies Press: Canberra, Australian Capital Territory.

Birckhead, J., 1996. "Dreaming" down under: the cultural politics of people and "country". Pp. 205–14 *in* Nature and the Human Spirit — Toward An Expanded Land Management Ethic ed by B. L. Driver, D. Dustin, T. Baltic, G. Elsner and G. Peterson. Venture: State College, Pennsylvania.

Birckhead, J., 1999. Brief encounters: doing "rapid ethnography" in Aboriginal Australia. Pp. 195–228 *in* Applied Anthropology in Australasia ed by S. Toussaint and J. Taylor. University of Western Australia Press: Nedlands, Western Australia.

Birckhead, J. and Wallis, A., 1994. An Evaluation of the Cairns College of TAFE 'Community Ranger' Program — Final Report. The Johnstone Centre: Albury, New South Wales.

Birckhead, J., DeLacy, T. and Smith, L. (eds), 1992. Aboriginal Involvement in Parks and Protected Areas. Aboriginal Studies Press: Canberra, Australian Capital Territory.

Birckhead, J., Klomp, N. and Roberts, A., 1996. Protocols for the participation of Aboriginal communities in research and monitoring. Pp. 126–35 *in* Sustainable Use of Wildlife by Aboriginal Peoples and Torres Strait Islanders ed by M. Bomford and J. Caughley. Australian Government Publishing Service: Canberra, Australian Capital Territory.

Burger, J., 1990. The Gaia Atlas of First Peoples — A Future for the Indigenous World. Penguin: London.

Carter, J. L. and Yibarbuk, D., 1997. A cooperative research plan for the commercial harvest and management of trepang by Aboriginal communities in the Top End. Pp. 290–92 *in* Conservation Outside Nature Reserves ed by P. Hale and D. Lamb. Centre for Conservation Biology: The University of Queensland, St. Lucia, Queensland.

Chalk, A., 1997. The Wik Decision and the 10 Point Plan — implications for pastoralists. Paper delivered to LEXPO Forum, Centrepoint Sydney, 22 October 1997.

Colchester, M., 1994. Salvaging Nature: Indigenous Peoples, Protected Areas and Biodiversity Conservation. United Nations Research Institute for Social Development Discussion Paper; DP 55, Geneva.

Collins, J., Klomp, N. and Birckhead, J., 1996. Aboriginal use of wildlife: past, present, and future. Pp. 14–36 *in* Sustainable Use of Wildlife by Aboriginal Peoples and Torres Strait Islanders ed by M. Bomford and J. Caughley. Australian Government Publishing Service: Canberra, Australian Capital Territory.

Collins, J., Klomp, N. and Birckhead, J., 1997. The Recognition of Aboriginal and Torres Strait Islander Hunting, Fishing and Gathering Rights in Australia: A Review of Legislation, Policy, International and Common Law. The Johnstone Centre, Charles Sturt University: Albury, New South Wales.

Cordell, J., 1993. Who owns the land? indigenous involvement in Australian protected areas. Pp. 104–13 *in* The Law of the Mother: Protecting Indigenous Peoples in Protected Areas ed by E. Kemf. Sierra Club Books: San Francisco.

Dale, A., 1997. Principles for negotiating conservation on indigenous lands. Pp. 74–79 *in* Conservation Outside Nature Reserves ed by P. Hale and D. Lamb. Centre for Conservation Biology: University of Queensland, St. Lucia, Queensland.

Davies, J. and Higginbottom, K., 1997. Evaluating indigenous community wildlife management: work in progress. Paper presented at the Institute of Australian Geographers and New Zealand Geogaphical Society Second Joint Conference, Hobart, Tasmania, 28–31 January 1997.

DeLacy, T. and Lawson, B., 1997. The Uluru/Kakadu model: joint management of Aboriginal-owned national parks in Australia. Pp. 155–87 *in* Conservation Through Cultural Survival: Indigenous Peoples and Protected Areas ed by S. Stevens. Island Press: Washington, DC.

DeLacy, T., 1994. The Uluru model — Anangu Tjukurrpa: 50,000 years of Aboriginal law and land management changing the concept of national parks in Australia. *Soc. Nat. Res.* **7:** 479–98.

Ellwood, W. and Andrews, G., 1997. Aboriginal Ownership of National Parks: Explanatory Notes. NSW National Parks and Wildlife Service: Sydney, New South Wales.

English, A. J., 1997. Terrestrial hunting and gathering by Aboriginal people in New South Wales: an assessment of law and policy. *Environ. Plan. Law J.* **14**(6): 437–55.

Fisher, R. J., 1999. Anthropology and community forestry in Nepal. Pp. 63–87 *in* Applied Anthropology in Australasia ed by S. Toussaint and J. Taylor. University of Western Australia Press: Nedlands, Western Australia.

Flannery, T., 1994. The Future Eaters: An Ecological History of the Australasian Lands and People. Reed: Sydney, New South Wales.

Furze, B., DeLacy, T. and Birckhead, J., 1996. Culture, Conservation and Biodiversity — The Social Dimension of Linking Local Level Development and Conservation Through Protected Areas. John Wiley and Sons: Chichester.

Gray, A., 1991. Between the Spice of Life and the Melting Pot: Biodiversity Conservation and Its Impact on Indigenous Peoples. (IWGIA Document 70-). International Working Group for Indigenous Affairs, Copenhagen.

House of Representatives Standing Committee on Environment, Recreation and Arts, 1993. Biodiversity, the role of potected aeas. Australian Government Publishing Service: Canberra, Australian Capital Territory.

Howitt, R., Connell, J. and Hirsch, P., 1996. Resources, Nations and Indigenous Peoples. Oxford University Press: Oxford.

International Union for the Conservation of Nature, 1993. Parks for Life: Report of the IVth World Congress on National Parks and Protected Areas. International Union for the Conservation of Nature: Gland, Switzerland.

International Union for the Conservation of Nature, 1994. Guidelines for Protected Areas Management Categories. International Union for the Conservation of Nature: Gland, Switzerland.

Jacobs, J., 1994. Earth honouring — Westem desires and indigenous knowledges. *Meanjin* **53**(2): 305–14.

Kemf, E. (ed), 1993. The Law of the Mother — Protecting Indigenous Peoples in Protected Areas. Sierra Club Books: San Francisco.

Knudtson, P. and Suzuki, D., 1992. Wisdom of the Elders — Honouring Sacred Native Visions of Nature. Allen and Unwin: Sydney, New South Wales.

Langton, M., 1996. Art, wilderness and *terra nullius*. Pp. 11–25 *in* Ecopolitics IX Conference papers and resolutions. Northern Land Council: Darwin, Northern Territory.

Larritt, C. S., 1995. Taking part in Mutawintji: Aboriginal involvement in Mootwingee National Park. *Aust. Geog. Stud.* **33**(2): 242–56.

Lawrence, D., 1997. Managing Park/Managing "Country": Joint Management of Aboriginal Owned Protected Areas in Australia. Parliamentary Research Service, URI:httl2:/www.aph.gov.au/prs/rp/97rp2.htm

Lewis, H. T., 1992. The technology and ecology of nature's custodians: anthropological perspectives on Aborigines and national parks, Pp. 15–28 *in* Aboriginal Involvement In Parks and Protected Areas ed by J. Birckhead, T. DeLacy and L. Smith. Aboriginal Studies Press: Canberra, Australian Capital Territory.

Maybury-Lewis, D., 1992. Millennium — Tribal Wisdom and the Modem World. Viking: New York.

Maybury-Lewis, D., 1997. Indigenous Peoples, Ethnic Groups, and the State. Allyn and Bacon: Boston.

McNeely, J., 1993. A new future for biosphere reserves? how to adapt to the 21st century. Paper presented at International Conference on Biosphere Reserves, Sevilla, Spain, 22 March 1995.

Moore, S. A., 1995. The role of trust in social networks: formation, function and fragility. Pp. 148–54 *in* Nature Conservation 4: The Role of Networks ed by D. A. Saunders, J. L. Craig and E. M. Mattiske. Surrey Beatty & Sons: Chipping Norton, New South Wales.

New Zealand Herald, 1997. Land bill passes, lessees vow fight', 5 December, P. 3.

Roberts, A., Klomp, N. and Birckhead, J., 1996. Monitoring marine and terrestrial hunting in an Aboriginal community in north Queensland. Pp. 152–66 *in* Sustainable Use of Wildlife by Aboriginal Peoples and Torres Strait Islanders ed by M. Bomford and J. Caughley. Australian Government Publishing Service: Canberra, Australian Capital Territory.

Roberts, M., Norman, W., Minhinnick, N., Wihongi, D. and Kirkwood, C., 1995. Kaitiakitanga: Maori perspectives on conservation. *Pac. Conserv. Biol.* **2**: 7–20.

Ross, H., 1995. Aboriginal Australians' cultural norms for negotiating natural resources. *Cultur. Surv. Quart.* **Fall**: 33–38.

Rustomji, N. K. and Ramble, C. (eds), 1990. Himalayan Environment and Culture. Indian Institute of Advanced Study and Indus Publishing Company: New Delhi.

Sinammon, V., 1997. Co-management and self-govemance: contemporary indigenous natural resource management. Pp. 10–16 *in* Conservation Outside Nature Reserves ed by P. Hale and D. Lamb. Centre for Conservation Biology: University of Queensland, St. Lucia, Queensland.

Stevens, S., 1997. The legacy of Yellowstone Pp. 13–32 *in* Conservation Through Cultural Survival — Indigenous Peoples and Protected Areas ed by S. Stevens. Island Press: Washington, DC.

Smyth, D. and Sutherland, J., 1997. Indigenous Protected Areas: Conservation Partnerships with Indigenous Landholders. Indigenous Protected Areas Unit, Biodiversity Group, Environment Australia: Canberra, Australian Capital Territory.

Sweeney, D., 1993. Fishing, hunting and gathering rights of Aboriginal peoples in Australia. *Uni. NSW Law J.* **16**(1): 97–160.

Thackway, R., Szabo, S., Smyth, D., 1997. Indigenous protected areas: new opportunities for the protection of biodiversity. Pp. 62–73 *in* Conservation Outside Nature Reserves ed by P. Hale and D. Lamb. Centre for Conservation Biology: University of Queensland, St. Lucia, Queensland.

Trouillot, M., 1991. Anthropology and the savage slot: the poetics and politics of otherness. Pp. 17–44 *in* Recapturing Anthropology: Working in the Present ed by R. Fox. School of American Research Press: Santa Fe, New Mexico.

Walsh, F., 1995. Interactions between land management agencies and Australian Aboriginal people: rationale, problems and some lessons. Pp. 88–106 *in* Nature Conservation 4: The Role of Networks ed by D. A. Saunders, J. L. Craig and E. M. Mattiske. Surrey Beatty & Sons: Chipping Norton, New South Wales.

Webb, T., 1996. Co-management of wildlife resources by indigenous peoples. Pp. 97–109 *in* Sustainable Use of Wildlife by Aboriginal Peoples and Torres Strait Islanders ed by M. Bomford and J. Caughley. Australian Government Publishing Service: Canberra, Australian Capital Territory.

Williams, N. M. and Baines, G. (eds), 1993. Traditional Ecological Knowledge: Wisdom — for Sustainable Development. Centre for Resource and Environmental Studies: Australian National University, Canberra, Australian Capital Territory.

Wilson, E. O., 1988. The current state of biological diversity. Pp. 3–18 *in* Biodiversity ed by E. O. Wilson. National Academy Press: Washington, DC.

Wilson, G., McNee, A. and Platts, P., 1992. Wild Animal Resources: Their Use by Aboriginal Communities. Bureau of Rural Resources and Australian Government Publishing Service: Canberra, Australian Capital Territory.

Woenne-Green, S., Johnston, R., Sultan, R. and Wallis, A., 1994. Competing Interests: Aboriginal Participation in National Parks and Conservation Reserves in Australia: A Review. The Australian Conservation Foundation: Fitzroy, Victoria.

World Commission on Environment and Development, 1987. Our Common Future. Oxford University Press: Oxford.

Young, E., Ross, H., Johnson, J. and Kesteven, J., 1991. Caring for Country: Aborigines and Land Management. Australian National Parks and Wildlife Service: Canberra, Australian Capital Territory.

CHAPTER 62

The importance of personality to ecologically sustainable rural management policy

Marilyn Shrapnel,[1] Jim Davie,[1] Alan Freed[2] and Bruce Frank[1]

The research presented in this chapter is based on the observation that people vary in their capacity to identify problems, to admit to those problems and to take planned action to seek a solution. Where stresses exist, the capacity of people to respond actively to a planned problem solution diminishes. The psychological profiles of people tend to fall into three groups — those who can generally cope; those who can usually cope but undergo dysfunctional episodes under acute stresses, and those who generally cannot cope. Rural land holders in many parts of Queensland are undergoing an extended period of economic hardship which is producing a range of social and ecological problems which limit the capacity of the region to recover and to perform at economic levels which will sustain settlement. Government attempts at defining solutions assume that all people are potentially able to respond logically to the problems with which they are faced and to implement, through subsidy or other expert technical assistance, restorative land management. That the emotional rather than the logical dimension of decision making is dominant in many circumstances is not acknowledged in the suite of government policies and options being explored. The present research is endeavouring to describe the range of emotional circumstances which may characterize a rural community under stress and to determine alternative policy options.

Key words: Personality, Psychiatry, Psychological functioning, Ecologically sustainable managment, Rural policy.

INTRODUCTION

As a consequence of our history of land use, large scale ecological processes have been set in motion in inland Australia, which have led to serious land degradation. Intervention is essential to halt the progression of these processes. A successful outcome will only be achieved by a population with the resources to employ ecologically sustainable land management practices.

Over the past 20 years, a decline in the rural population has occurred with a demographic shift to urban and coastal areas (Salt 1996). In response to the change, government policy has favoured the withdrawal of services and resources, resulting in an acceleration of the process of depopulation (Campbell 1994). Reduced government service delivery makes it more imperative that people remaining on the land are able to be effective resource managers. Potential effectiveness is typically assessed by criteria which emphasize economic measures. Those who stay on the land are usually perceived as being the most economically viable despite accumulating evidence of poverty and debt.

However, economic capability represents only one dimension of the human profile required to successfully deal with the rapidly changing face of rural Australia. In 1991 the Ecologically Sustainable Development Working Group on Agriculture defined three key ingredients for the development of sustainable agricultural practices (Prevett *et al.* 1995):

1. farmers must want to change,
2. they must know how to change, and
3. they must have the necessary material resources to bring change about.

We would add to this list the fourth ingredient that *farmers must have the necessary psychological resources to bring change about*. The existence

[1]School of Natural and Rural Systems Management, The University of Queensland, Lawes, Queensland, Australia 4343.
[2]Antenor House, 16 South Street, Ipswich, Queensland, Australia 4305.
Pages 606–11 *in* NATURE CONSERVATION 5: NATURE CONSERVATION IN PRODUCTION ENVIRONMENTS: MANAGING THE MATRIX ed by J. L. Craig, N. Mitchell and D. A. Saunders. Surrey Beatty & Sons, 2000.

of this fourth factor, related to personality, is occasionally noted but little researched and, where interpretations have been attempted, they are invariably done within a broader cultural and sociological context (e.g., McAllister 1991; Stern 1992; Curtis *et al.* 1995).

The absence of research into the role of personality has had two important implications for rural policy. The first is that State and Federal Governments have had to assume that the rural community is an homogeneous group to which a common set of assistance practices can be applied. The second is that in implementing policies it has been important to assume that all individuals within a community are capable of changing their behaviour given the right circumstances. The challenges for policy makers of confronting these two assumptions have also been very serious. However, there is now a growing recognition that there needs to be greater individual assessment within rural communities in order to better target government assistance. For example, social security concerns have now been considered in the operation of the Rural Adjustment Scheme (Professor Beth Woods, pers. comm.), and this lends urgency to the present research.

Our research is designed to explore the way the personality of individual farmers influences their behaviour towards the land, as exhibited by their land management practices. We are working with a voluntary group of farmers who are presently exposed to an array of severe environmental problems and economic stresses. The primary data are being collected through semi-structured interviews which explore individual personality profiles on the basis of the landholders' perceptions of the problems facing rural communities, and their views on solutions. Through the process of interview it is possible for a trained psychiatrist to determine the presence of personality factors associated with an individual's capacity to cope with change. Although largely qualitative, the results of these investigations provide a tool for understanding differences in individual responses to land resource management. It is our expectation that the insights gained from the research will suggest options for ecologically sustainable management within the context of the apparent psychological resources which characterize this sample of the farming community.

As the research is still in its early stages, detailed results from rural properties are not yet available. Rather, this chapter presents a brief review of what we believe are the most critical underlying theoretical concepts drawn from contemporary psychiatry. It then discusses how this approach to an analysis of land management might be used to formulate effective management interventions.

BEHAVIOURAL ASPECTS OF LAND MANAGEMENT

If assistance is to be provided to landholders to manage rural environments in an ecologically sustainable manner, all the factors which may influence farmers' practices need to be considered. Those factors which prevent an ecologically sustainable approach may then be modified or eliminated. Vanclay and Lawrence (1995) list educational factors (e.g., amount and adequacy of information); perceptions of the problems (including the influences of public media); economic and financial considerations; socio-cultural factors, and situational factors (e.g., commodity prices, amount of time, stage in the life of the farm) as being particularly important. These views have clearly figured strongly in the range of policy initiatives which have been taken by governments in an attempt to enhance rural industry and to put farming practice onto an ecologically sustainable management footing (Fig. 1).

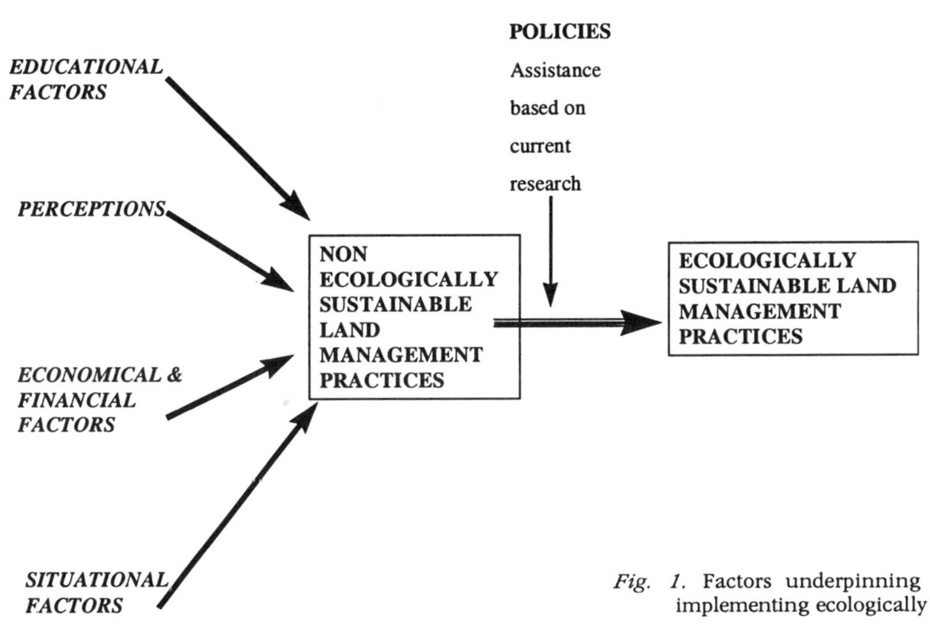

Fig. 1. Factors underpinning current policy approaches to implementing ecologically sustainable land management.

Research to date has indicated that the solutions are complex. Nevertheless, provision of correct information and education or extension regarding good land management practices — even when coupled with the provision of adequate financial assistance — does not automatically result in adoption of ecologically sustainable land management.

Much research on environmental attitudes and perceptions has sought to explain and predict conservation behaviour. For example, the theory of post-materialist values, based on Maslow's hierarchy of needs theory, attempts to explain the rise in environmental concerns associated with the satisfaction of basic needs and the concept of a conservation ethic (Curtis *et al.* 1995). However, as noted by many researchers, attitudes and perceptions are not accurate predictors of behaviour. Perception of environmental problems by farmers does not necessarily relate to subsequent adoption of solutions (Vanclay 1992). Many theories have been advanced to explain these discrepancies. Most are based on social theory and postulated on the basis of studies of populations, rather than individuals. Two examples are the classification of farmers into groups based on rationality (Vanclay and Lawrence 1995); and the concept of the farming subculture, which emphasizes the influence of peer group pressure on behaviour (Vanclay 1992).

There has been little research on the psychological dimensions of individual farmers and its influence on their behaviour. Traditionally, the psychological aspects of conservation behaviour have been studied and analysed in learning theory terms. This approach regards behaviour as being primarily learned and equates change with learning. The underlying assumption is that undesirable behaviour has been shaped by a certain set of external factors and can be changed by applying a more positive set of stimuli, once these factors are determined. Such a behavioural interpretation fails to address the differences which clearly exist between individual personality styles and which determine the capacity of people to change. The two most frequently discussed psychological aspects of farmers' behaviour are locus of control and self-efficacy.

Locus of control

Locus of control is best conceptualized as a bipolar continuum, ranging from internal locus of control to external locus of control. Those with an internal locus of control believe their success is controlled by their own efforts, while those with an external locus of control perceive themselves to be controlled by factors outside themselves. Studies of locus of control indicate that those with internal locus of control are more *responsible and resourceful problem solvers,* whereas others are *more alienated and apathetic, tending to blame other people and chance circumstances for problems* (Cloninger *et al.* 1993). Research also indicates that locus of control has implications for the use of information. Those with an internal locus of control are more likely to seek out information, and are more likely to make use of that information than those with an external locus of control. Locus of control is a concept that has been used to explain the variation of individual effectiveness in addressing various conservation issues (Ingold 1989).

Prevett *et al.* (1995) have researched psychological variables in property management (adoption of rabbit control practices) and are of the opinion that locus of control is important in determining behaviour because of its influence on an individual's perception of him/herself as being capable of successfully managing the problem (belief in abilities).

Self-efficacy

Self-efficacy is the capacity of an individual to adapt to changes and is determined by the combination of an individual's knowledge and confidence in his/her ability to act on that knowledge. O'Brien (1995) considers self-efficacy to have an important influence on conservation behaviour and postulates that successful experiences improve self confidence and thereby promote self-efficacy.

The assumption from the research on these two psychological variables is that appropriate intervention can move an individual from one point on the behavioural continuum to another, so that change can be effected. In other words, that people can learn. We believe that the application of learning theory to explain and predict behaviour overlooks the powerful influences of the underlying personality structure of individuals.

APPLICATION OF PSYCHIATRIC RESEARCH TO LAND MANAGEMENT PRACTICES

The concept of personality is an important one to appreciate, since an individual's personality colours a person's perceptions, the way they see the world, what is important to them, how they relate to others and how they behave. Personality describes an individual's characteristic pattern of thinking, feeling and behaving. It also describes an individual's coping mechanisms. It is a *person's distinctive pattern of psychological functioning* (Oldham and Morris 1995), and it is unique, much the same as a fingerprint. It is thought to be well established from late adolescence. Personality has predictive value, pointing to the way an individual may meet future life challenges.

Recent advances in the theoretical underpinning of psychiatry suggest that individual personalities can be understood in terms of seven dimensions, which fall into character (three dimensions) and temperament (four dimensions) (Cloninger 1996). Temperament (emotionality) refers to the automatic emotional responses to stimuli. It is thought to be inherited and involves the four dimensions of harm avoidance, novelty seeking, reward dependence and persistence. Character, on the other hand, is what we make of ourselves. It involves three aspects of self concept: self directedness (self reliance), co-operativeness (as part of society) and self transcendence (as part of the universe). The two most important dimensions of personality in terms of assessing capacity to change and meet challenges are self directedness and co-operativeness.

Self directedness

Self directedness refers to self determination, or will power. It is *the ability of an individual to control, regulate and adapt behaviour to fit the situation in accord with individually chosen goals and values* (Cloninger *et al.* 1993). To determine self directedness, the following are examples of characteristics to be considered

- Responsibility *vs* blaming.
- Purposefulness *vs* lack of goal direction.
- Resourcefulness *vs* inertia.
- Congruent second nature *vs* incongruent habits.

Those with low self directedness have difficulty accepting responsibility, setting meaningful goals and resourcefully meeting challenges. They tend to be blaming and reactive to external events. Self directedness has some similarities to the concepts of locus of control and self-efficacy.

Co-operativeness

This aspect of personality *accounts for individual differences in identification with, and acceptance of other people.* It measures the *different aspects of character related to agreeability* vs *self-centred aggression and hostility* (Cloninger *et al.* 1993). Co-operative individuals will be socially tolerant, empathetic, helpful and compassionate, concerned with the rights of others, compared with unco-operative individuals who will be socially intolerant, disinterested in other people, unhelpful and revengeful.

Empirically we have found that self-directedness and co-operativeness tend to occur together in individuals. Psychiatric research indicates that a low measure of these two factors is associated with an immature character (i.e., a personality disorder). Those who are low on self directedness and co-operativeness would find significant long-term change extremely difficult (Cloninger 1996).

Personality disorder

Most people's personality styles allow them enough flexibility to cope with life's difficulties, which they do in their own characteristic fashion and within the normal boundaries of social interaction. They adapt to changes and grow wiser with life's experiences. However, when personality traits are inflexible and maladaptive, they interfere with an individual's capacity to function appropriately or adequately. Individuals with this condition are then said to be suffering from a personality disorder. People with a personality disorder show deeply ingrained, inflexible and maladaptive patterns of relating to and perceiving both the environment and themselves.

The difference between a personality style (functioning) and a personality disorder (malfunctioning) is really a matter of degree and quantity, rather than quality. Oldham and Morris (1995) liken personality disorder to high blood pressure, in that it represents extremes of normal human patterns. People are said to be *suffering* from a personality disorder. However, while causing functional impairment, there may be no subjective distress on the part of the *sufferer*. They are far more likely to deny their problems and do not routinely experience anxiety about or acknowledge pain from what society perceives as their symptoms.

Studies indicate that personality disorders occur in 10—14% of the total population, although it is likely that the true frequency may be higher than that (Oldham and Morris 1995). We consider that acknowledgement of the existence of personality disorder is extremely important, as psychiatric research indicates that those individuals with personality disorders are unlikely to change their behaviour, regardless of the type of approach undertaken.

DISCUSSION

The empirical evidence from psychiatric practice suggests that people tend to fall into three functional groups depending on their psychological resources.

- Group 1 consists of those who are able to cope despite adversity, because of underlying personality structure. They are self aware and confident and possess a high measure of self directedness and co-operativeness. For this group change is a challenge which they are normally able to embrace effectively because they are predisposed to adapting and learning new ways.
- Group 2 have similar robust personality structures to Group 1 but, while normally able to cope, are temporarily emotionally incapacitated by virtue of a combination of severe stresses, e.g., illness, crop failure or death in the family. Change for this group is an extra stress which they find difficult to face up to in their vulnerable condition.
- Group 3 represent those whose inherent psychological capacity is limited when it comes to coping with problems or change. They have a low measure of self directedness and co-operativeness and find it extremely difficult to effectively deal with other than routine challenges. Typically they are unable or unwilling to see the broader picture and tend not to be open to learning new ways of approaching problems for which they assume no personal responsibility.

Conservation of the environment, as the basis for ecologically sustainable production, is determined by people's behaviour towards it. Personality characteristics and other psychological factors alter an individual's relationship with the environment. Effective management of the environment involves recognition of the quality of this relationship. If we consider the characteristics of the landowners in relation to the three groups above, there are clear

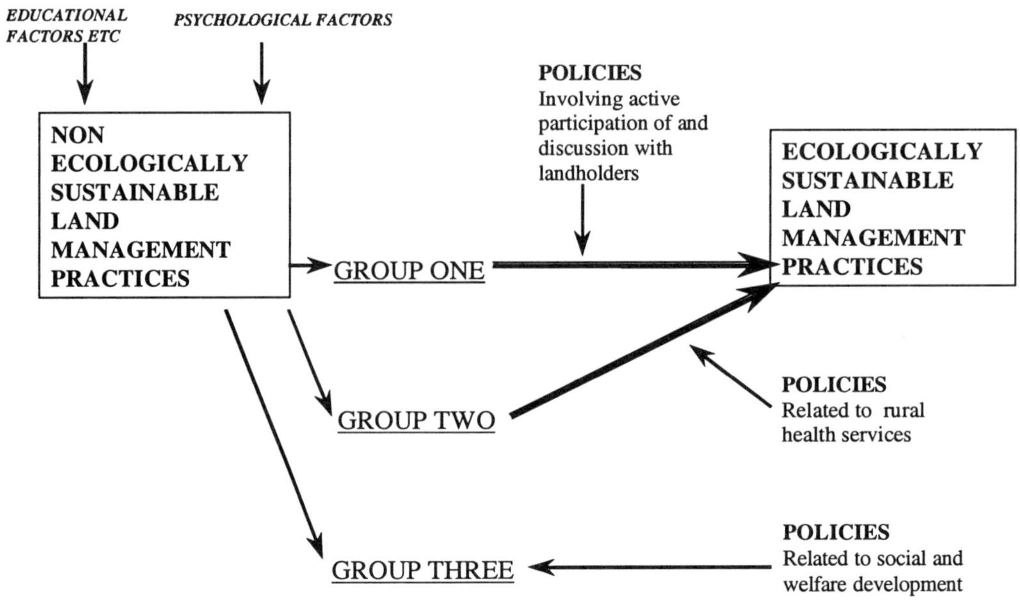

Fig. 2. Policy pathways to ecologically sustainable land management which are sensitive to individual psychological profiles.

implications for government and community based management approaches. These implications are that different policy approaches may be necessary in order for government to successfully assist the transition to sustainable land management practices (Fig. 2).

As Figure 2 shows, members of group one are likely to be actively involved in seeking out information about the problems and their effective solutions, behaviour indicative of high self directedness. They will be the ones who organize themselves into Landcare groups or other self-help organizations — indicative of high co-operativeness. They will also be the group who will be looking for planning and strategic opportunities for arresting environmental degradation and shifting or broadening the economic base of their properties. Members of this group will typically require only minimal supervision and may become frustrated if they do not have the opportunity to be actively involved in decision making. These characteristics mean that members of this group will generally be able to move directly through the changes necessary for ecologically sustainable management practices and will benefit from policies which facilitate active interaction with government and other outside agencies.

Members of group two are those temporarily distracted and immobilized in their capacity to respond to sources of assistance. They may require medical/psychiatric treatment in order to restore them to their fully functioning selves. There is a higher incidence of physical and emotional health problems in rural areas (Clarke 1989) and yet, rural communities are greatly disadvantaged in respect to access to these vital services. Figure 2 emphasizes that solutions for this group will involve consideration of policies related to the issues of rural health, its structure and services.

Members in group three tend not to take responsibility for their difficulties and, instead, willingly attribute blame on external factors, such as drought and lack of government financial support. This is behaviour indicative of low self-directedness. They also tend to be distrustful of community organizations such as Landcare, show minimal interest in other members of the community, and exhibit little in the way of land stewardship — behaviours indicative of low co-operativeness. They have limited capacity for change regardless of the approach taken. Members in this group may require social development programmes. Some may also benefit from access to psychiatric and/or counselling services. Figure 2 emphasizes that progress with environmental problems, with which members of this group are confronted, will require major supervision, driven through policies related to social development.

CONCLUSIONS

Interpretation of personality types suggests that it is necessary to re-examine the assumptions that form the basis of the current policies. Policy makers cannot approach rural populations as an homogenous group. Each community consists of individuals who possess different psychological resources. Some individuals are limited in their capacity to respond to the present challenges and to change their behaviour, regardless of the assistance practices applied.

We believe that psychiatry, by virtue of its focus on individual personality, vulnerability and functioning can make a major contribution to the understanding of the behaviour of farmers towards their land and can thereby complement the more usual analyses by natural scientists, psychologists and rural sociologists.

REFERENCES

Campbell, A., 1994. Landcare. Communities Shaping The Land and the Future. Allen and Unwin: Sydney, New South Wales.

Clarke, L., 1989. An Analysis of the Health Status of Rural People. Pp. 132–48 *in* Crisis in Rural Health and Welfare: Myth or Malady. Proc. 10th National Student Initiatives in Community Health Conference 26–28 September 1989 ed by T. Cullen, P. Dunn and G. Lawrence. Centre for Rural Welfare Research: Charles Sturt University, Albury, New South Wales.

Cloninger, C. R., 1996. A psychobiological model of temperament and character: fundamental findings for use in clinical practice. Pp. 975–90 *in* New Research in Psychiatry ed by H. Hafner and E. Wolpert. Hugrefe and Huber Publishers: Gottingen.

Cloninger, C. R., Svrakic, D. M. and Przbeck, T. R., 1993. Psychobiological model of temperament and character. *Arch. General Psychiatry* **50:** 975–90.

Curtis, D., Nadolny, C., Ford, H. A., 1995. Lessons from a network of dieback repairers. Pp. 175–86 *in* Nature Conservation 4: The Role of Networks ed by D. A. Saunders, J. L. Craig and E. M. Mattiske. Surrey Beatty & Sons: Chipping Norton, New South Wales.

Ingold, C., 1989. Locus of control and use of public information. *Psychol. Rept.* **64:** 603–07.

McAllister, I., 1991. Community attitudes to the environment, forests and forest management in Australia. Forest and Timber Inquiry, Consultancy Series, No. FTC91/09. Resource Assessment Commission: Canberra, Australian Capital Territory.

O'Brien, M., 1995. Community perspectives of riparian management. Pp. 228–33 *in* Nature Conservation 4: The Role of Networks ed by D. A. Saunders, J. L. Craig and E. M. Mattiske. Surrey Beatty & Sons: Chipping Norton, New South Wales.

Oldham, J. M. and Morris, L. B., 1995. The New Personality Self Portrait. Why You Think, Work, Love and Act The Way You Do. Bantam Books: New York.

Prevett, P. T., Murphy, M. and Smithyman, S., 1995. An investigation of the human factors relevant to rabbit pest management in Victoria. Pp. 220–27 *in* Nature Conservation 4: The Role of Networks ed by D. A. Saunders, J. L. Craig and E. M. Mattiske. Surrey Beatty & Sons: Chipping Norton, New South Wales.

Salt, B., 1996. Demographic and Cultural Change in the Bush. Pp. 119–21 *in* Proc. National Agricultural and Resources Outlook Conference, Canberra 4–6 1997. ABARE: Canberra, Australian Capital Territory.

Stern, P. C., 1992. Psychological dimensions of global environmental change. *Ann. Rev. Psychol.* **43:** 269–302.

Vanclay, F., 1992. The social context of farmers' adoption of environmentally sound farming practices. Pp. 94–119 *in* Agriculture, Environment and Society: Contemporary Issues for Australia ed by G. Lawrence, F. M. Vanclay and B. Furze. Macmillan: Melbourne, Victoria.

Vanclay, F. and Lawrence, G., 1995. The Environmental Imperative. Eco-Social Concerns For Australian Agriculture. Central Queensland University Press: Rockhampton, Queensland.

CHAPTER 63

Community consultation and the New South Wales Biodiversity Strategy: embarking on a journey of action

Lynn Webber[1] and Raymond Fowke[2]

The National Parks and Wildlife Service has the responsibility to facilitate the development and implementation of the New South Wales Biodiversity Strategy as a whole-of-government and community process. This strategy complements the National Strategy for the Conservation of Australia's Biological Diversity *and* contributes to Australia's commitment to the International Convention on Biological Diversity.

A state-wide process of consultation has been undertaken, which recognizes the critical importance of community participation in decision making about biodiversity conservation across landscapes. This chapter discusses the design, implementation and evaluation of the community consultation process undertaken to develop the strategy.

Outcomes of the consultation process highlighted the importance of community participation in decision making, the need for clarity of objectives and mode of participation *and* the value of consultation as a pro-active investment in future action.

Important principles underpinning effective design of community consultation processes were revealed to include the recognition of existing agency and community efforts, seeking advice from stakeholders in designing processes, ensuring openness and transparency, having clarity of purpose, offering participation options, providing co-learning opportunities, establishing information relevance, engaging in active listening, focusing on the quality of participation, documenting outcomes without alteration or editing, using participant language and organizing frameworks, providing opportunities to build ongoing positive relationships, encouraging and extending communication networks *and* feeding back information and updates.

Key drivers for effective community consultation from an agency perspective included legislative recognition in the importance of consultation, adequate staff and financial resource investment, commitment to learning and development of staff, focus on continuous improvement *and* forging community partnerships.

Key words: Participation, Community consultation, Process management, Policy, Biodiversity.

INTRODUCTION

CONCERTED efforts to protect New South Wales' native biodiversity are reflected in the extensive protected area system of national parks and nature reserves under the management of the state's leading conservation agency, the National Parks and Wildlife Service. However, conservation of native biodiversity has tended to be seen as a government responsibility, rather than as a goal permeating the matrix of land uses. A lack of meaningful participation in decision making by all land managers in the formulation of strategies for biodiversity conservation has contributed to responsibility for native biodiversity conservation remaining largely with government agencies.

The *Threatened Species Conservation Act 1995* heralds a new era of biodiversity conservation in New South Wales. This Act establishes biodiversity as

[1]Community Relations NSW National Parks & Wildlife Service, P.O. Box 1967, Hurstville, New South Wales, Australia 2220.
[2]Environmental Policy NSW National Parks & Wildlife Service, P.O. Box 1967, Hurstville, New South Wales, Australia 2220.

Pages 612–22 *in* NATURE CONSERVATION 5: NATURE CONSERVATION IN PRODUCTION ENVIRONMENTS: MANAGING THE MATRIX ed by J. L. Craig, N. Mitchell and D. A. Saunders. Surrey Beatty & Sons, 2000.

an umbrella for ensuring the survival and evolutionary development in nature of all species, populations and communities of plants and animals (including management of threats and recovery of threatened species); recognizing the importance of ecologically sustainable development (including the integration of conservation and natural resource management); and identifying the need for research and education. Significantly, the Act establishes community participation as central to achieving its aims.

The impetus for the development of the New South Wales Biodiversity Strategy is the continuing loss and decline of native biodiversity. This concern resonates in the National Strategy for the Conservation of Australia's Biological Diversity (Commonwealth of Australia 1996a). The first national State of the Environment Report for Australia declares that the loss of biodiversity is our most serious environmental problem (Commonwealth of Australia 1996b). Ecologically Sustainable Development has also been a focus of discussion over the past two decades in Australia. The agreed national approach has three core objectives: to enhance individual community well-being and welfare by following a path of economic development that safeguards the welfare of future generations; to provide equity within and between generations; and, to protect biological diversity and maintain essential ecological processes and life support systems.

Under the *Threatened Species Conservation Act 1995*, the National Parks and Wildlife Service is charged with the key role in facilitating the development and implementation of the New South Wales Biodiversity Strategy across government agencies and with the community. This chapter discusses the design, implementation and evaluation of a process of agency and community participation in the development of the New South Wales Biodiversity Strategy as a basis for creating ownership of strategies that are meaningful to those who are involved in its implementation and which invites responsibility for collective action.

Defining participation

Participatory strategies emerged in the early 1970s in Latin America, Africa and Asia, initiated by proponents of development reacting to difficulties with conventional extension (technology transfer) approaches.

The meaning of participation has changed over the years (Bryant and Whyte 1982). The 1980s saw participatory approaches used in a variety of ways to involve people. The nature of participation was revealed through action rather than explicit discussion of theory or concepts. Defining participation as a process of active involvement reveals its dynamic nature but tends to conceal power relations established in the range of mechanisms available. It is necessary to question the use of participation as an organizing principle because it is not specific enough to assess the particular power relations that are embodied in different participatory approaches (Tripp 1989).

Participation provides a structure which controls the different ways that people can act (Foucault 1971). A delineation of the different modes of participation can clarify the nature of power relationships: *compliance* (legal structures establish possible actions and punish transgression), *co-operation* (agreement to comply), *consultation* (actively seeking issues and ideas to be considered in decision making), *collaboration* (partnership in decision making process) and *self-directed action* (initiate decision making) (Biggs 1989). Power relationships which also describe modes of participation include voting, consumer choice, negotiation and third party (usually legal) determinations. These modes of participation may be more or less desirable in application. However, they clarify power relationships for all participants.

Attempts to classify power relations in participation have been based mainly on the nature of responsibility and degree of involvement in the process. This can encourage an assumption that participation is inherently good (Webber 1995). The objectives established to achieve a particular goal offer a useful guide in determining the most appropriate mode(s) of participation in designing a particular process (L. Webber, pers. comm.).

Consultation process

The objectives established when developing the draft strategy emphasized the importance of participation: "to build on the success of existing initiatives to develop co-ordinated and cost-effective biodiversity conservation programme involving the community, industry and all levels of government and ensure that the rights, knowledge and values of local and Aboriginal communities are properly recognized and reflected; and, to strengthen actions to inform, motivate and achieve the support of the community, including local and Aboriginal communities, industry, state government agencies and local government, in conserving biodiversity" (New South Wales Government 1997).

Participation by government agencies and the community in the development of the strategy was seen as an opportunity to forge partnerships for mutually beneficial action. Consultation was established as an appropriate mode of participation for the following reasons:

Consultation means "to take counsel — to seek information or advice from person(s)" (Concise Oxford Dictionary 1976). This describes a relationship involving a process of actively listening to issues of interest and concern, genuinely considering contributions to address these issues and adopting a course of action which recognizes a government's democratically elected right to decide.

Consultation focuses on the quality of participation and contributions to the process, rather than establishing a process which seeks to achieve absolute consensus or majority voting outcomes.

Consultation can provide an opportunity to build positive relationships through recognizing, valuing and acknowledging input from participants to the process. Over time, working together has the potential to bring government agencies and the community closer together in achieving mutual goals. At its most unethical, consultation can be used as a token exercise to placate key stakeholders before proceeding with a predetermined strategy. Community response to such consultation can be difficult to predict, but risks community outrage and denigrates other genuine efforts to consult. A critical element is trust, which is built on honesty, openness and a proven track record based on previous experiences (Sandman 1993). Trust by the community is earned through accountability and willingness to collaborate in action. Engaging in unethical consultation risks a loss of confidence by the community which can take years to regain.

Consultation can be a creative process. Consultation as an "event" is commonly characterized in the development of draft strategies, developed without formal consultation and placed on public exhibition or circulated to invite "responses". Opportunities for the community to participate and benefit from creative involvement are limited. This approach has the potential to evoke a "reaction" rather than a "response". With attention to process design, consultation can offer meaningful and creative participation in decision making frameworks.

Consultation can be a formal process. Inviting participation in a formal process can recognize and address equity and access issues such as nepotism, geographic remoteness, literacy, languages other than English, disability, time frames for consultation, opportunities for public and private submissions, individual and group participation and intergenerational equity. It provides an opportunity to provide key information on the basis for consultation, time frames and background information required for meaningful participation which minimizes the risk of misunderstanding intention or attenuating the decision making process unnecessarily.

Consultation has an inherent capacity to value local and indigenous peoples' knowledge and experience in its own right. This lies in the creation of space for a range of perspectives on an issue to emerge and be acknowledged, valued and accepted within the process. As the dominant form of knowledge in western society, science tends to overpower other forms of knowledge, such as local knowledge (Webber 1996) or indigenous knowledge (Young and Ross 1994) to such an extent that they are excluded from discussions in many situations. Determination of relevance is undertaken by those in power, who judge the validity of knowledges (Mosse 1993). "[There are] a whole set of knowledges that have been disqualified as inadequate to their task or sufficiently elaborated: naive knowledges, located down on the hierarchy, beneath the required level of cognition or scientificity" (Foucault 1980 P 82).

Consultation can encourage community ownership and acceptance. The planning phase of a strategy is essentially the stage where decisions are made or influenced and agreement to the process is established. The planning process provides an opportunity to build community ownership and responsibility for proposed actions and priorities and to create a basis for sharing responsibility for outcomes and benefits accrued. In many instances approaches by government agencies to consultation have been confined to inviting community "response" to a draft version of a plan, rather than inviting the community to take an active part in its creation.

Consultation has the potential to trigger other modes of participation in a continuing process, such as co-operative or collaborative action. Often consultation is seen as an end in itself — an "event" which elicits information to aid decisions rather than as a mode of participation which establishes a basis for future action. A lack of recognition of other knowledges besides western scientific knowledge has contributed to a notion that there is some divine separation of mental (the thinkers) from manual labour (the do-ers). Community participation in biodiversity conservation ultimately involves taking action. It is important to avoid a paternalistic model where those who are meant "to do" will be advised or told what to do because it is assumed they are not capable of understanding their own situation and need enlightenment.

Consultation can provide a co-learning opportunity for all participants, where mutual respect for contributions to the process can raise awareness of different perspectives on issues and in some cases, change understandings (Maturana 1988). This moves beyond traditional notions of "educating the community" towards "learning together". Traditional notions of "education" establish a tenet that "those with knowledge" teach "those without knowledge" (Chambers and Ghildyal 1985). Any opportunities to discuss issues from a different perspective seek to achieve a convergence or match with the "right" understandings constructed by those in power (Scoones and Thompson 1994).

Consultation can establish a relevance for information seeking to support participation. Information provided to support consultation does not have to create an "information washover" effect, but can offer information targeted at the level of interest of audiences. For those who may be browsers (have a general interest in the process), attentives (those with a peripheral interest but may later become involved) or actively engaged, different levels of information will need to be provided to support their participation (Hance *et al.* 1990).

Consultation can provide a framework for participant aspirations and expectations. Experiencing the complexity of the issue and the diversity of views offer participants an appreciation of the challenges of decision making, time involved, legislative requirements and the resources available.

PROCESS DESIGN AND IMPLEMENTATION

In order to achieve effective native biodiversity conservation across landscapes, there needed to be ownership of the New South Wales Biodiversity Strategy and responsibility for its implementation held by the community and other government agencies. In designing the consultative process, it was recognized that consultation can act to limit participation by some groups and individuals in the community (Foucault 1980) through the selection of time frames, participants and who may speak. The facilitator is in a position of power to limit and govern fields of action in both design and conduct of processes (Webber 1995).

Establishing roles

As the government agency responsible for leading the development and implementation of the biodiversity strategy, the National Parks and Wildlife Service was placed in the role of "lead agency" in a whole-of-government initiative. In terms of the development of the strategy, this role was interpreted as "facilitator" of the process.

The National Parks and Wildlife Service also had an interest in the content of the strategy. An important distinction was made between facilitating the process and participating in the development of strategy content. A separation of these different roles established an ethical framework for participation. National Parks and Wildlife Service staff not engaged in facilitating the process were able to participate in the development of the strategy, along with other government agency staff and interested groups and individuals in the community.

Many government agencies support advisory committees and councils established under legislation. These mechanisms can be described as "vertical" forms of participation (Bryant and Whyte 1982). The *Threatened Species Conservation Act 1995* makes provision for a Biological Diversity Advisory Council, which includes representatives from a broad cross-section of the community: scientific organizations, peak environmental groups, local government, industry and the Aboriginal community. The role of the Biological Diversity Advisory Council is to provide an independent source of expertise and advice on biodiversity issues and has a specific function to advise the Minister for the Environment and Director-General of National Parks and Wildlife Service on the development of the biodiversity strategy and a comprehensive consultation process.

The Biological Diversity Advisory Council played an important role in identifying the most appropriate ways to invite community participation in the development of the strategy. As representatives of a wide range of community interests, members of the Biological Diversity Advisory Council provided vital advice about the best ways to raise awareness about the strategy, the most accessible forms of consultation in terms of methods, locations and timings and the strategies for inviting maximum participation in the development of the strategy. Members of the Biological Diversity Advisory Council were also part of the regional workshop process, giving presentations about their role and discussing issues with participants during workshop session breaks.

Establishing consultation objectives

The consultation objectives reflected both rational and experiential dimensions of the process. Rational objectives relate to the physical product of the process and experiential objectives relate to the way participants feel about their involvement in the process (Canadian Institute of Cultural Affairs 1989).

The rational objectives aimed to raise awareness about biodiversity in the general community, enhance understanding about biodiversity among interested individuals and groups in the community; establish the important areas which needed to be addressed in the New South Wales Biodiversity Strategy, formulate the options for action and develop a New South Wales Biodiversity Strategy taking into account all participant contributions. The experiential objectives focused on encouraging active involvement by participants, valuing contributions and having participants enthusiastic about taking action on a plan they helped to develop.

Intra-agency support

The success of the process hinged on the whole agency working together to fulfill a range of roles. This required flexibility to deploy people and financial resources in a way that was efficient and effective to deliver on a major initiative. A financial costing on the consultation process design was prepared and endorsed by Biological Diversity Advisory Council. Community relations staff within National Parks and Wildlife Service provided the communication specialists to edit and design the community information and design and co-ordinate the consultation process across the state. National Parks and Wildlife Service policy staff were responsible for drawing together all the content in drafting the strategy and community information, providing logistical support to Biological Diversity Advisory Council and the workshop teams and managing the submissions and analysis phase. National Parks and Wildlife Service District staff across the state supported the Public Exhibition phase through their local offices. Specialist technical staff were also involved, acting as "resource people" for workshop participants.

National Parks and Wildlife Service has been implementing a comprehensive workshop facilitator programme over the past two years. This investment in people provided a critical pool of 70 staff, at all levels and in all directorates of the agency, who could organize and facilitate workshops. In addition, a small team of communication specialists with extensive skills and experience in a range of communication areas were a key design and co-ordinating group for the community workshop phase of the process.

The decision to use government agency staff to design, implement and evaluate the consultative process was based on the potential benefits accrued in building positive government agency and community relationships through working together to develop the strategy. Specifically, the community should experience a public service working directly with them and demonstrating an ability to confidently engage on issues of mutual interest and concern.

Consultative process

A common government agency approach to consultation is to establish a formal written submission process as the sole mechanism for community consultation. This subscribes an electronic transmission model of communication, where there is a movement of a product (information) from a source (government agency) to receivers (community) with emphasis placed on the message (content of the draft document) with feedback mechanisms (the submission from the community and the report on the analysis of submissions from the agency) to create a two-way flow of information (Shannon and Weaver 1949).

Inviting written submissions provides an opportunity for participants to contribute thoughtful submissions of a variety ranging from a straightforward letter to documents of great detail and complexity and to maintain confidentiality. Many approaches to consultation require public and group activities, which creates a different context for participation to that which is private (Mosse 1993). However, it also creates a context for participation that excludes those people who have low levels of literacy and excludes knowledge that cannot be expressed in written form. In the case of this process, it was important to involve rural communities and Aboriginal communities, whose knowledge and experience is predominantly verbal and even non-linguistic (McCorkle 1989; Bloch 1991).

Encouraging ownership of the strategy recognized the need to move beyond merely transmitting information. Sole reliance on a submission approach to consultation does not provide an opportunity for personal interaction among other interested members of the community and may also contribute to community perceptions of government agencies being remote. Consultation approaches which involve personal interaction by interested groups and individuals offered a different context for participation. These approaches subscribe to a constructivist model of communication (Kelly 1955; Maturana and Varela 1992; Webber 1995). Formal interactive group approaches focus on encouraging communication from a range of perspectives, tap the synergy of people working as a group and offering an inclusive framework for verbal and non-verbal communication to occur.

The design of the process sought to capture the features of consultation and address power issues related to the design framework. The design included intra-agency communication and support, initial consultation for information gathering, inviting interagency partnership, preparation of community information, finding participants, public exhibition phase, community workshops phase, submission phase, analysis phase, community reporting phase and finalization of the strategy.

A framework outlining the key components of the consultative process and the associated time frames was developed with advice from the Biological Diversity Advisory Council about the most effective ways of consulting with the broad range of interested groups they represented. This assisted in setting intermediate targets within the longer term process and providing information to government agencies and the community when opportunities to participate were scheduled, closing dates for submissions and when outcomes would be available. The consultation process did not proceed in a linear fashion — at one stage public exhibition, workshops and submission processes were concurrent.

Information gathering. The draft strategy explicitly recognizes that many agencies, community groups, councils and individual landholders are already making major contributions to the conservation of biodiversity. The intention is to invite an opportunity for co-ordinated and collaborative action — a synergy rather than a reinvention or duplication. In order to make these important linkages, government agencies were consulted about activities in progress which supported the goal of conserving native biodiversity across landscapes. Building existing government agency and community initiatives into the strategy sought to recognize and acknowledge efforts and value the important work already being undertaken. This provided an opportunity to establish which dimensions of a biodiversity strategy may require enhancement or adaptation of action already being taken, or new initiatives, which supported the goal.

Inter-agency partnerships. An outcome of the initial consultative process with other government agencies was not only a picture of what actions were already being undertaken and proposed initiatives, but also the opportunity to invite partnership in the consultation process. The biodiversity strategy is a whole-of-government initiative and collaborating in the consultative process offered an additional dimension to that of participating in developing the content of the strategy.

The Director-General invited other agency heads to work with the National Parks and Wildlife Service in a whole-of-government approach to the consultative process. Other government agencies responded positively. Their staff established areas for public exhibition in their local offices across the state to enhance community access to information about the strategy beyond what National Parks and Wildlife Service could provide alone. Assistance was given by other government agency staff in facilitating workshops in regional centres across the state. This contributed to a view that biodiversity was a whole-of-government issue and that agencies were working together with the community to achieve the ultimate goal.

Preparation of community information. The development of community information to support the consultative process followed a model of using three broad categories of community audiences: actives, attentives and browsers (Hance *et al.* 1990).

For the browsers, a DL full colour brochure was developed, discussing the meaning and importance of biodiversity and the development of the strategy. For the attentives, an A4 full colour brochure was developed which discussed the meaning and importance of biodiversity and a summary of the key areas of the draft strategy to be discussed in the consultative process. For the actives, a black and white publication of the draft strategy provided a comprehensive background and detail of the information gathered and structured into a draft strategy framework.

Perceptions of *fait accompli* are heightened by the publication standards of "glossy" draft versions and use of directive language such as "will" to describe actions in a strategy which at this stage have the status of proposals. The major document containing the draft strategy framework was produced with a low sheen black and white cover and content of black and white text without pictures. In the editing stages attention was paid to the use of words such as "encourage", "support" and "strengthen" to describe possible actions.

It is recognized that language can act to exclude participation in processes (Foucault 1986). The development of community information in written form focused on the use of plain English and where complex words had to be used to maintain the integrity of meaning, a glossary was provided to explain them. An image and branding was designed to convey a visual meaning of biodiversity complemented by a short catch-phrase.

A non-linguistic form of communication has an inherent ability to be inclusive through transcending language. Many participatory processes are dependent on linguistic and sometimes spatial or diagrammatic forms of expression which exclude non-linguistic and practical expressions of knowledge (Mosse 1993 P11, P24). The use of an image also provided consistency and cohesion between phases of the process for participants by encouraging visual association between the image and the biodiversity strategy.

Inviting participation. It was recognized that "the community" is not homogenous, but composed of a rich tapestry of people with diverse interests and experiences and a wide range of perspectives on issues. Within the community, a number of groups exist, made up of people who identify with each other through their common interests.

In identifying individuals and groups to invite for participation in the consultative process, attention was given to finding those individuals and groups in the community who were known to have an interest or likely to have an interest in the biodiversity strategy. A full range of key stakeholder individuals and groups were contacted by letter using lists built up in ongoing communication by the National Parks and Wildlife Service and other government agencies. This included attention to ensuring a diverse range of groups were contacted.

Sole reliance on participation via representation assumes strong communication mechanisms within organizations and therefore a representative is articulating the consensus or majority view of that constituency. Representatives of particular interest groups are more easily identifiable to invite their participation and are important in gaining acceptance of the outcomes of a consultative process. However, in keeping with the principle of an open process, any member of the community with an interest in the biodiversity strategy was invited to participate in its development. It is not uncommon for this dimension of inviting participation to be overlooked, particularly as a way to exclude people who may hold a divergent view.

The draft document on the biodiversity strategy was launched in Sydney by the Minister for the Environment. There was a strong media presence at the launch and journalists in attendance were supplied with kits containing press releases, information on biodiversity and information about the strategy and consultation process. Coverage in metropolitan and regional media followed, which assisted in raising awareness about the biodiversity strategy and its development. Advertisements about the public exhibition of the draft strategy were placed in metropolitan and regional newspapers and other media opportunities (particularly radio in regional areas) was used to raise awareness of biodiversity and the strategy and provide information about how people could become involved.

In the lead up to the public workshop phase, advertisements were placed in metropolitan and regional newspapers, outlining the dates, times and locations of the workshops, with an invitation to register interest in attending by phone, fax or mail. Again regional radio opportunities complemented the efforts.

Public exhibition. The public exhibition phase was advertised in metropolitan and regional newspapers, outlining reason for public exhibition, details for accessing information and the other key phases of the process (community workshops and submissions). The duration of public exhibition phase was initially three months, extended for an additional month after community feedback found that more time was needed.

A wide range of accessible locations were available to the community, with the 27 National Parks and Wildlife Service District Offices across the state and the partnering of Local Government Councils, Taronga Zoo, Australian Museum, Royal Botanic Gardens. Catchment Management Committee representatives and Regional Landcare co-ordinators also assisted in distribution of information. Displays included a poster, the three levels of information (the DL brochure and the A4 booklet were on the display counter and the comprehensive document was provided on request) and information about the community workshops and the submission process.

Counter staff were available to answer any questions relating to the consultative process and a telephone service also provided a mail-out option across the state.

The Community Biodiversity Network, a non-government organization with a key interest in biodiversity conservation, provided an opportunity to place the draft document for the biodiversity strategy on to the world-wide web to provide an internet access point. This opportunity provided an important new way to provide community access to information and raise awareness about the strategy, particularly from remote rural areas.

Community workshops. A series of community workshops were scheduled two months into the three month public exhibition phase. The inclusion of community workshops in the consultation process recognized the benefits of personal interaction with people in the community with a mutual interest in native biodiversity conservation. This sought to maximize the opportunity for forging new links between networks of people and tapping the synergy of working as a group to generate new ideas.

Community workshops were conducted in six regional centres throughout New South Wales in an effort to make the workshops geographically accessible for rural communities. These were located in Coffs Harbour (north coast), Armidale (northern tablelands), Dubbo (western area), Wagga Wagga (southwestern area), Queanbeyan (southern area) and Sydney (metropolitan). An additional workshop was conducted in Cobar (outer west) in response to rural community concerns about participants in more remote areas of western New South Wales travelling to Dubbo. Timings for the workshops were set with a conscious effort to accommodate participant needs.

The workshop was designed to provide a consistent framework statewide and to allow all workshop outcomes to be documented and analysed as submissions. In line with the rational and experiential objectives of the consultation process, the workshop design aimed to achieve the following.

1. Introduce members of Biological Diversity Advisory Council, the National Parks and Wildlife Service and other government agencies.
2. Provide opportunities to work with people who have a common interest about biodiversity.
3. Provide a basis for encouraging future participation in implementing action strategies for native biodiversity conservation.
4. Provide a "contexting" for participation in the workshop recognising that not all participants have read, are able to read, or have been able to access or understand the information available.
5. Provide an opportunity for participants formulate ideas, issues of interest or concern in relation to the development of the biodiversity strategy.
6. Provide an opportunity for participants to formulate options to build actions into the strategy.
7. Actively value participants knowledge and experience in its own right, including the use of their "language" in accurately recording contributions without alteration or editing.
8. Provide an opportunity for participants to extend and link networks.
9. Provide maximum opportunity to contribute during the workshop and leave the workshop feeling they have done so.
10. Inform participants about the other phases of the consultation process and the opportunities to further contribute.
11. Inform participants about action on the workshop outcomes and "next steps" in the consultation process.

In achieving these aims, the Community Workshop Programme was designed as a 10 stage process.

1. Registration of name and contact details, issue of name tag and allocation to group based on principle of diverse backgrounds.
2. Welcome to participants, introduction of facilitators and programme outline.
3. Presentation by member of Biological Diversity Advisory Council outlining role and work to date.
4. Overview of the development of the draft strategy (including question time) to set the context.
5. Introduction to small group session, outline and clarify focus questions, group area and facilitator allocations.
6. Small group session 1 — exploring important areas that need to be addressed in the strategy using a facilitated process of focus question, brainstorming ideas, grouping ideas and naming each group.
7. Break — informal networking and discussion, examining other group's work.
8. Small group session 2 — formulating how the areas identified in session 1 could be built into the strategy using a facilitated process of discussion and documentation.
9. Large group session — to report back to whole group on small group outcomes, including questions.
10. Summing up — recapping workshop achievements, feedback arrangements, next steps, thank you notes and evaluations.

Over 450 people participated in the regional community workshops from a wide range of government agencies, non-government agencies, community groups, Landcare, Total Catchment Management, universities, Aboriginal communities and representative bodies, farmer representative organizations, rural women's organizations, conservation organizations, individual rural landholders and interested citizens. Table 1 summarizes the key areas formulated by workshop participants to be included in the biodiversity strategy.

An area given little consideration is how the content and structure of a strategy reflects the language and the ways participants organize their ideas. In this workshop, recording of participant contributions was integral to the process design. All contributions were recorded in an open and transparent process of writing statements and ideas exactly as they were formulated by individuals and the group on A5 and poster size paper so everyone could see what was being written.

Workshop outcomes were transposed from the papers and posters into a workshops document, which provided participants with a written record of their own workshop outcomes with their group and their own ideas recorded without editing or alteration and the outcomes of the other workshops across the state which provided a broader view of diversity of issues and ideas emerging in other areas. The document was posted directly to all participants using the registration lists from the workshops within four weeks. The workshops document was included as a submission for analysis along with all other written submissions.

PROCESS EVALUATION

Workshop Evaluation. Workshop participants were invited to evaluate the workshop process through a written feedback survey containing both forced choice and open-ended questions. These could be completed on exit, faxed or mailed in, anonymously if desired. This method of endogenous evaluation (Webber and Ison 1995) offered some insight into the experiences of the participants and provided a basis for analysis of rational and experiential objectives established for the consultation process and the workshop.

Workshop evaluation sought participant perspectives on what they considered to be the most important parts of the workshop for them, its strengths and weaknesses and suggestions for improvement of future workshops. In measuring an increased awareness and understanding about the biodiversity strategy, a question was asked of participants which sought insight into if and how, the process helped them understand more about the biodiversity strategy. Provision was also made for additional comments to be made.

Ninety-three participants completed an evaluation survey, representing 21% of the total number of participants. The analysis of the survey followed a content analysis methodology, with a grouping of similar ideas or issues into key themes for each question.

The most important features of the workshop process for participants included the opportunity to provide feedback, information about the process of getting the strategy up and understanding that process, meeting new people with an interest in biodiversity, the opportunity to have input through small group discussions and the opportunity to hear a diversity of views.

The strengths of the workshop cited by respondents included the opportunity for the community to have input into the strategy, the educational experience, small groups, discussion, the opportunity to participate and have input no matter what your view the 70% listening/30% talking approach. Weakness of the workshop identified by respondents included a concern that some groups were poorly represented (for example Aboriginal communities were not represented at one workshop), more topics for discussion was needed, the different levels of understanding on biodiversity among the groups, not enough time to discuss everything, the experience was one step in a long process so there was a risk of input being clouded and lose its relevance and some groups went off the topic a bit. Suggestions on ways to improve future workshops included starting an hour earlier than this time so that it is not too late for farmers getting home, greater input by local government and landholders, more time for large and small group sessions, send out focus questions before the workshop and ask participants to read before the workshop.

In terms of understanding more about the biodiversity strategy, respondents indicated that the workshop introduced the basic elements to them, that issues raised in the small groups helped put the whole picture together and that the experience reinforced their understanding of the biodiversity strategy. Participants added that some good ideas emerged from the group discussions and the workshop made players more aware of the complexity of the issue where there is no simple answer. The experience of the workshop also enabled input without feeling intimidated.

Evaluation revealed that many of the strengths of the workshop reflected the features of consultation and the principles underpinning the workshop design. In terms of the rational and experiential objectives, there was a strong linkage between established objectives and participant experiences revealed in the feedback survey, such as meeting new people with a mutual interest in biodiversity, opportunities for ideas formulation, valuing participant contributions, providing information about other phases of the consultation process and next steps.

Areas for improvement suggested by participants were considered in further work on the development of the strategy. In particular, discussions regarding Aboriginal community participation in the development of the Strategy were followed up with the New South Wales Aboriginal Land Council as the Aboriginal community representative on Biological Diversity Advisory Council.

Analysis of submissions and *reporting*. There were more than 170 submissions contributing to the development of the biodiversity strategy, in addition to the workshop outcomes reports from the seven workshops conducted. A content analysis methodology was used in analysing the submissions, establishing key themes and associated issues (Table 1). Submissions reflected the multiple goals in natural resource management of an economic, environmental and social nature across a broad range of issues, as well as highlighting the effects of global influences on native biodiversity.

A report on submissions document was prepared for consideration by the Biologically Diversity Advisory Council and used as a basis for incorporating contributions into the Strategy. This established where areas were consistent or had common threads, where changes were needed, where enhancements or improvements were needed, where inclusions needed to be made and where priority action lay.

The key themes to emerge from the consultative process have formed the basis of the final strategy document and guidance for priority actions revealed as community involvement, incentives for community involvement, education and information, a bioregional approach to conservation and research and monitoring.

Finalization of living document. The Biodiversity Strategy is seen as a living document, providing a framework for action which will be revisited on a regular basis with other government agencies and with the community to monitor progress, celebrate achievements, enhance co-ordination and make improvements to the strategy. The document developed will be disseminated to all participants in

Table 1. Key areas to be addressed in the development of the New South Wales Biodiversity Strategy: summary of workshop outcomes.

Key area	Components
Philosophy	Why we need to conserve biodiversity, the cost of failure, Aboriginal statement about the earth, imbue a conservation culture, the bottom line, ethos, address cultural/aesthetic values.
Forward look	Vision, attitudes now, the future.
Guiding principles	
Realistic goals	
Balance	
Global perspective	Environmental issues, human population and pressures, big picture ecology, global warming and its effect.
Consultation/ Involvement	Land managers and users, ground up approach, process, human focus, landowner connection, landowner support, rights of individuals, shared responsibility, key areas — local government, rural areas, schools, community ownership and leadership, use existing networks, regional basis, stakeholder partnership, whole community responsibility.
Awareness	Promotion.
Education and knowledge sharing	Knowledge, understanding, land management that is a plus for biodiversity, development and support recognition of on-ground practitioners, appreciation of biodiversity, attitudes.
Information	Availability of useful information, management guidelines, meeting information needs, management tools.
Communication	
Role of government	Co-ordination; integration, administration, policy, long term political commitment — bipartisan; local government involvement.
Data management	Factual information to back credibility, decision making justification.
Biodiversity planning	Holistic approach, bioregional approach, definition and scale of assessment, conservation reserves, exploration of natural resources, compensatory habitats, current land management approaches, protection of biological values and habitat, industry, ecological communities, Ecologically Sustainable Development — biodiversity and productivity, protection of key areas, effects on current activity, identification of critical habitat, economics, cost-benefit, social impacts, social and community expectations, political process.
Threats	Ferals (including carp); threatening processes; exotics; kangaroo management; woody weed control; Medicare equivalent for desexing dogs and cats.
Investigating and working with existing biodiversity initiatives	
Implementation	Looking after, reversing, restoration, repair, improving, rivers and riparian vegetation, roadside vegetation and remnants, aquatic biodiversity, practical implementation, off reserve conservation, conservation of common as well as threatened species, cross tenure — across landscapes, different land uses, oil spill management — pollution incidents, wildlife corridors.
Regulations	Too much legislation, compliance, need for enforcement as last resort.
Documentation	
Research and development	Monitoring, investigation research, interpretation, get the science right — benchmarking and recognizing roles in landscape, science.
Resources	Money; Who pays?; Who bears the cost; value for money strategy; incentives; commitment; $ driven.
Priorities	
Insurance	Gene banks.
Performance accountability	Success/achievements.

the process to date and encourage further distribution of additional copies to expand the network. Through this process, the network of interested groups and individuals can grow. Encouraging the sharing of different knowledges and experiences can contribute to enhanced co-ordination and communication among all who are interested in biodiversity. The initial Biodiversity Strategy has also provided a basis for the formulation of a resourcing bid and National Parks and Wildlife Service is working in collaboration with Biological Diversity Advisory Council and an interagency working group for submission in early 1998.

CONCLUSIONS AND FUTURE DIRECTIONS

Considerable effort has been made to invite other government agencies and the community to participate in a process which seeks to recognize and acknowledge the important achievements and work underway in many areas. This provided an opportunity to develop a co-ordinating umbrella under which to encourage synergy and integration to maximize effectiveness of a whole-of-government and community quest for native biodiversity conservation across landscapes.

The key findings from the design, implementation and evaluation of this community consultation process highlighted the importance of establishing a clear rationale for participation and setting objectives with a view to the desired outcomes of the process in both rational (product, output) and experiential terms.

The important design features which raised community awareness about biodiversity included interagency partnerships, media launch of the draft strategy and consultation process, advertising the public exhibition and workshops and use of local radio to promote the strategy. Attention to the preparation of appropriate community information which catered for a range of interest levels, attention to use of plain English, employing images, brands and slogans to achieve consistency of message and providing opportunities for people to come together as a group enhanced participant understanding of the biodiversity strategy.

The process features which secured meaningful contribution by participants included the accessibility of appropriate supporting information, community workshops in regional locations at appropriate times, opportunities for personal interaction during the process, use of small groups to maximize synergy, productivity and mutual valuing of different world views. The engagement of skilled facilitators to work with small groups during the workshops ensured participants had an equitable opportunity to establish the important areas which need to be addressed in the strategy, the options to build them into the plan and achieve a tangible product which was accurately documented. Enthusiasm for taking action on a plan was observable among participants and the degree to which involvement in the planning process catalyses future action will be of interest in monitoring the implementation phase of the strategy cycle.

The consultation process highlighted some important policy principles. Citizens should have an opportunity to participate in decision making about issues which affect them, objectives for inviting community participation should be clear, the mode of participation to be utilized should be clear and government agencies are able to facilitate processes which invite community ownership and responsibility for planning and action without abrogating their own responsibilities. Policy development can be pro-active and brought alive. Creative community participation in the development of strategy should be seen as an investment in future action.

Important design principles were revealed and reinforced through this process: recognize important work already being undertaken in agencies and the community which contributes to the outcome, ensure appropriate design of processes involves seeking advice from stakeholders, commit to an open and transparent process, provide options as to how people may participate, create space for active listening, value all contributions (scientific, indigenous and local knowledge), focus on the quality of participation rather than numbers, engage in opportunities to build positive relationships, encourage and extend networking among people with common interests and provide a meaningful and creative opportunity to contribute issues and ideas.

The importance of establishing formal processes was highlighted in attending to equity and access issues, policy paralysis, agency and community understanding of opportunities and limits, building-in flexibility to be responsive to community needs or concerns, providing a co-learning opportunities for all participants, establishing a relevance for participants to seek information which enhances meaningful participation, documenting outcomes without alteration or editing, using participant language and organizing frameworks wherever possible with attention to plain language or use of glossaries and giving back information and updating progress at key stages to participants.

From an agency perspective, the commitment to facilitating the development of the strategy and the future implementation work in partnership with other government agencies and with the community is supported through legislation which recognizes the importance of community participation in decision making. This requires commitment of people and financial resources and commitment to developing skills and experience in agency staff. Forging agency and community partnerships wherever possible in the process is critical to engendering ownership and responsibility for outcomes. Ongoing dialogue with agencies and the community about the strategy will encourage a further building of networks.

The development of the New South Wales Biodiversity Strategy has been a whole-of-government process and indeed, a whole-of-community process. The Strategy identifies priorities and common objectives for biodiversity conservation in New South Wales. The primary focus for initial implementation will be placed on actions that are achievable in the first few years in order to establish momentum and build community support. Community participation in implementing the Strategy will be ongoing and will now shift from planning to implementation and then review, in an ongoing adaptation to changing circumstances, new understandings or insights and different priorities. The experiences in developing the New South Wales Biodiversity Strategy provides considerable insights for other governments who are seeking to develop comprehensive approaches to biodiversity conservation across landscapes and ensure the community has a meaningful role in strategic policy development.

ACKNOWLEDGEMENTS

We wish to thank the following people: all community groups and individuals who have been involved in the development of the strategy to date. The Biological Diversity Advisory Council for their advice and assistance in design and implementation of the consultation process. The Community Biodiversity Network for establishing Internet access for the consultation process. Staff in government agencies who partnered National Parks and Wildlife Service by providing public exhibition display sites and workshop facilitators. Staff of the National Parks and Wildlife Service who supported the consultation process, especially Annette Bonnefin, Stuart Cohen, Lawrence Orel and Jacqui Payne — the Community Programme Officers who co-ordinated the regional community workshops.

REFERENCES

Biggs, S. D., 1989. Farmer Participation in Research: a synthesis of experiences from nine agricultural research systems. *OFCOR Comparative Study Paper No. 3*, International Service for Agricultural Research.

Bloch, M., 1991. Language, anthropology and cognitive science. *Man* **26**(2): 183–98.

Bryant, C. and Whyte, L. G., 1982. Managing participation. *In* Managing Development in the Third World. Westview Press: Boulder.

Canadian Institute of Cultural Affairs, 1989. The LENS Method. Monograph. Canadian Institute of Cultural Affairs: Canada.

Chambers, R. and Ghildyal, B. P., 1985. Agricultural research for resource poor farmers: the farmer first and last model. *Agricul. Admin.* **20:** 1–30.

Commonwealth of Australia, 1996a. The National Strategy for the Conservation of Australia's Biological Diversity. Department of Environment, Sport and Territories: Canberra, Australian Capital Territory.

Commonwealth of Australia, 1996b. Australia: State of the Environment. CSIRO Publishing: Melbourne, Victoria.

Concise Oxford Dictionary, 1976. The Concise Oxford Dictionary of Current English ed by J. Sykes. Oxford University Press.

Foucault, M., 1971. The order of discourse. *In* Language and Politics ed by M. Shapiro. Basil Blackwell.

Foucault, M., 1980. Power/Knowledge: selected interviews and other writings 1972–1977 ed by C. Gordon. Pantheon Books: New York.

Hance, B. J., Chess, C. and Sandman, P., 1990. Industry Risk Communication Manual: improving dialogue with communities. Lewis Publishers: Florida.

Kelly, G. A., 1955. The Psychology of Personal Constructs. Norton and Co.: New York.

Maturana, H. R., 1988. Reality: the search for objectivity of the quest for a compelling argument. *Irish J. Psychol.* **9**(1): 25–82.

Maturana, H. R. and Varela, F. J., 1992. The Tree of Knowledge: the biological roots of human understanding. Revised ed. Shambhala Publications: Boston.

McCorkle, C. M., 1989. Toward a knowledge of local knowledge and its importance for agricultural RD&E. *Agricult. Human Values.* Vol. 6, No. 3.

Mosse, D., 1993. Authority, gender and knowledge: theoretical reflections on the practice of participatory rural appraisal. *Overseas Development Institute, Agricultural Administration (Research* and *Extension) Network Paper* **44:** December.

New South Wales Government, 1997. Draft NSW Biodiversity Strategy. Monograph. NSW National Parks and Wildlife Service.

Sandman, P., 1993. Responding to Community Outrage: strategies for effective risk communication. American Industrial Hygiene Association.

Scoones, I. and Thompson, J. (eds), 1994. Beyond Farmer First: rural peoples knowledge, agricultural research and extension practice. Intermediate Technology Publications: London.

Shannon, C. and Weaver, W., 1949. The Mathematical Theory of Communication. University of Illinois Press: Urbana.

Tripp, R., 1989. Data collection, site selection and farmer participation in on-farm experimentation. *CIMMYT Working Paper* **82:** 1

Webber, L. M., 1995. A Participative Communication Process for Agricultural R&D Prioritisation. Unpublished PhD Thesis. Macquarie University.

Webber, L. M., 1996. Managers' perspectives of weed management within rangelands management systems. Pp. 214–22 *in* Proceedings of the 11th Australian Weeds Conference. Council of Australian Weed Science Societies: September.

Webber, L. M. and Ison R. L., 1995. Participatory Rural Appraisal: conceptual and process issues. *Agricul. Syst.* **47:** 107–31.

Young, E. and Ross, H., 1994. Using the Aboriginal rangelands: insider "realities" and "outsider" perceptions. *Rangel. J.* **16:** 184–97.

Threatened Species Conservation Act 1995. New South Wales Government Printer: Sydney, New South Wales.

CHAPTER 64

On the future for birds in a multi-use landscape

Andrew M. Fisher[1]

The landscape around Bathurst, in the central tablelands of New South Wales, Australia, comprises a range of land uses, including farming, nature conservation, forestry and urban. Within this multiple-use landscape, the avifauna was surveyed at 39 sites twice per season over two-and-a-half years. A total of 116 bird species was recorded, of which 60.3% were detected at ten sites or less and 24 species at one site. No individual species was recorded at all study sites. Seven main species groups were identified by Hierarchical Cluster Analysis: species associated with riparian strips through grazing land, species of degraded remnants, open woodland species, species sparsely distributed or with narrow habitat requirements, species of densely forested areas, widespread species and species associated with well-developed understorey.

The future regional trend was determined for each species based upon the number of sites at which the species was recorded during this study, the abundance of the species, distribution of the species across the study area, seasonal occurrence and response to habitat change. Thirty-nine bird species are considered to be declining within the Bathurst landscape, 18 species are increasing and 35 species are regionally stable. A further 24 species were recorded as vagrants.

This study indicates that bird species from the open woodland group and those associated with riparian strips through grazing land are most at risk in the Bathurst landscape. Key land management actions needed to conserve and/or enhance appropriate habitats include: fencing native vegetation, especially box woodland remnants and watercourses on private land, to restrict stock access and to encourage regeneration; enhancement of riparian strips of native forest within softwood plantations; and the establishment of structurally diverse native vegetation in urban areas. Birds are presented as a tool whereby ecologists can effectively communicate with land managers whose decisions are likely to impact upon the sustainability of this and similar landscapes.

Key words: Birds, Avifauna, Habitat, Landscape ecology, Multiple-use landscapes, Land management, Conservation.

INTRODUCTION

FEW landscapes comprise single-use enterprises; most embrace a range of seemingly competing management regimes. As the inability of nature reserve systems to conserve species, communities and ecosystems adequately has become apparent, increased attention has been given to the management of whole landscapes and the role of component land uses in conservation (e.g., Bennett *et al.* 1995; Hale and Lamb 1997). Given that viable agricultural and conservation systems require the maintenance of ecosystem processes, it is clear that an integrated approach to landscape management is required, in which all stakeholders (e.g., nature conservation agencies, forestry, local governments and farmers) understand their role and influence on other stakeholders and the system as a whole (Saunders *et al.* 1995).

This study forms part of an extensive socio-ecology-extension programme underway in the central western region of New South Wales, Australia (Goldney *et al.* 1995a; Goldney, Wakefield, Watson, Hodgkins and Baldwin, chapter 22, this volume). Components of this programme cover a range of spatial scales and study systems, including tree regeneration across the regional landscape

[1]Environmental Studies Unit and the Johnstone Centre, Charles Sturt University, Panorama Avenue, Bathurst, New South Wales, Australia 2795.

Pages 623–32 *in* NATURE CONSERVATION 5: NATURE CONSERVATION IN PRODUCTION ENVIRONMENTS: MANAGING THE MATRIX ed by J. L. Craig, N. Mitchell and D. A. Saunders. Surrey Beatty & Sons, 2000.

(Windsor 1998), farm restoration (Bauer *et al.*, unpubl. data) and conservation of endangered species (Date *et al.*, chapter 14, this volume). This chapter presents some findings from a recent PhD project which investigated the effects of land management practices on the avifauna of the Bathurst landscape, located in the eastern portion of the central western region (Fisher 1997). The aims of this chapter are to:

1. provide an overview of the avifauna of the Bathurst landscape;
2. group bird species based on broad habitat requirements within the Bathurst landscape;
3. assess the status of individual species in the Bathurst landscape in order to determine future regional trends; and
4. explore the implications of this work for conservation and land management in the Bathurst landscape.

STUDY AREA

The study area is located in the central tablelands of New South Wales (Fig. 1) and extends from the western outskirts of Bathurst (33°26'S, 149°34'E) to the Great Dividing Range at Sunny Corner (33°23'S, 149°43'E). This area is a subset of the Bathurst 1:100 000 topographic map sheet, representing a 35 km east-west landscape transect. Altitude ranges from 590 m in the west to 1 282 m on the Great Dividing Range.

The climate of the study area is characterized by marked seasonal temperature changes, with relatively hot summers and cold winters. Temperatures range from a summer maximum of 27°C (min. 13°C) in Bathurst to a winter maximum of 8°C (min. 0°C) at Sunny Corner. Frosts occur from May to September while snow falls from 5–15 times per year. Mean annual rainfall declines along an east-west gradient from 935 mm at Sunny Corner to 630 mm at Bathurst (Forestry Commission of New South Wales 1987). Rainfall is generally highest in the spring and summer months. During the period of this study (September 1993 to January 1996) the area experienced drought conditions, with rainfall in 1994 and 1995 below average.

Five vegetation alliances are recognized in the study area (Bower and Semple 1993). These are:

1. *Eucalyptus melliodora–E. blakelyi* (Yellow Box–Blakely's Red Gum) woodland (occurs on well-drained soils of granitic origin, in agricultural lands in the western half of the study area below 700 m);
2. *E. albens* (White Box) woodland (on undulating country, occurs higher on the slope than *E. melliodora–E. blakelyi*; associated with *Brachychiton populneus* (Kurrajong) on rockier sites);
3. *Casuarina cunninghamiana* (River She-oak) open forest (occurs mainly along rocky watercourses in alluvial soil, generally at lower altitudes);
4. *E. macrorhyncha–E. rossii* (Red Stringybark–Inland Scribbly Gum) open forest and woodland (occurs on the rocky, skeletal soils of the study area, from about 700–1 000 m); and
5. *E. fastigata–E. viminalis* (Brown Barrel–White Gum) open forest (occurs in the east of the study area above 900 m; extensively cleared for *Pinus radiata* (Radiata Pine) plantations).

METHODS

Site selection

Thirty-nine sites were selected using a modified gradsect technique (Gillison and Brewer 1985; Austin and Heyligers 1989, 1991) to ensure that the range of vegetation alliances and land management practices in the region were sampled (McGarigal and McComb 1995). Where possible the management regimes recognized within the vegetation alliances were represented at least in duplicate. All sites selected were within walking distance of access roads.

Bird surveys

The methodology used in this study for surveying birds has been described elsewhere (Fisher and Goldney 1997). Briefly, a 20-minute count was used to survey birds along a 200 m × 50 m strip transect at each site, twice per season from spring 1993 to summer 1996.

Data analysis

(i) OVERVIEW

An Occurrence Index was derived for each species at each site. This was the number of times a given species was recorded at a particular site divided by the number of visits to that site. This resulted in the value for the Occurrence Index ranging from 0 (not recorded at that site during this study) to 1 (recorded on every visit to the site). The number of sites at which a species was detected was also determined.

(ii) GROUPING OF SPECIES BASED ON OCCURRENCES ACROSS SITES

Hierarchical Cluster Analysis was used to investigate similarities between occurrences of individual bird species across the study sites, in order to broadly categorize species into groups with similar habitat/landscape requirements. Clusters based on the Occurrence Index data for each species at each site were formed using Ward's method (Norusis 1994), in which the means for all variables are calculated for each cluster, then the squared euclidean distance to the cluster means is calculated and summed for all cases. At each step those clusters which result in the smallest increase in the overall sum of the squared within-cluster distances are the two that are joined (Norusis 1994). The chi-square measure was used for this frequency data (Norusis 1994). All Hierarchical Cluster Analysis were undertaken using SPSS 6.1.2. All clustering methods available produced similar patterns, however, the method

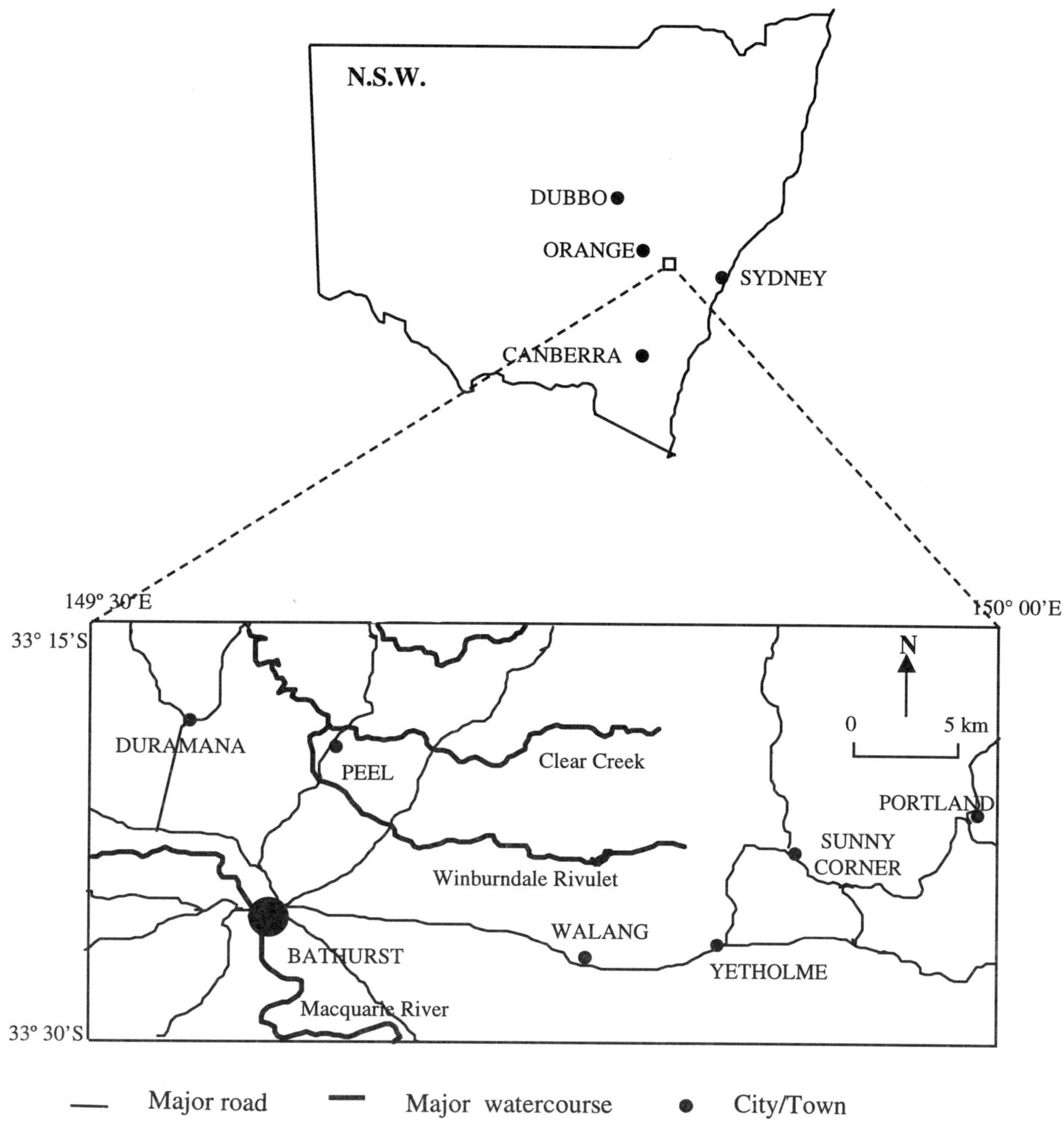

Fig. 1. Location of study area.

outlined here gave the clearest representation of the species groupings and thus was selected to facilitate data presentation and interpretation. Species recorded at only one site were excluded from this analysis in order to eliminate the formation of spurious groupings of species.

(iii) PREDICTED CHANGES IN THE AVIFAUNA OF THE BATHURST LANDSCAPE

The future regional trend for each species (increasing, decreasing or stable) was determined, based upon:

1. the number of sites at which the species was recorded during this study;
2. the abundance of the species (as measured by the mean Occurrence Index);
3. the distribution of the species across the study area (extensively cleared low altitude landscape, well-timbered high altitude landscape, or both);
4. the seasonal status (migrants will have lower mean Occurrence Index); and
5. the response to habitat change. Bird species in the study area exhibit four main responses to habitat alteration (Fig. 2). High sensitivity decreasers are highly sensitive to habitat loss and fragmentation. Species tolerant of habitat loss and fragmentation but sensitive to moderate

levels of habitat degradation are termed medium sensitivity decreasers. Species with a low sensitivity to habitat degradation are designated low sensitivity decreasers. The remainder of species are those which benefit from habitat simplification. There are two types of such increaser species, specialist increasers, for which the availability of suitable habitat will increase, and generalist increasers which are able to utilize a range of habitats. These response categories have been derived from observations within the central tablelands along with reference to Barrett *et al.* (1994) and Ford *et al.* (1995).

RESULTS

(i) Overview

A total of 116 bird species was recorded during this study from spring 1993 to summer 1996 (Table 1, which also provides the common and scientific names of all bird species recorded during this study). Twenty-four (20.7%) of these species were only recorded at one site over this two-and-a-half year period (Fig. 3). Seventy species (60.3%) were detected at ten sites or fewer. No species were found at all sites. The Crimson Rosella was recorded at all but one of the study sites, while the Red Wattlebird was recorded at 37 of the 39 sites.

(ii) Grouping of species based on occurrences across sites

Seven main clusters were formed by the Hierarchical Cluster Analysis of bird species based on Occurrence Index data collected from the 39 study sites surveyed from spring 1993 to summer 1996 (Table 1). These were species associated with riparian strips through agricultural land (16 species), species of degraded woodland (17 species), open woodland species (12 species), species distributed sparsely or with narrow habitat requirements (12 species), species associated with dense forest (10 species), widespread species (8 species) and species associated with well-developed understorey vegetation (17 species). The remaining 24 species were regarded as vagrants, being recorded at a single site only over the period of this study.

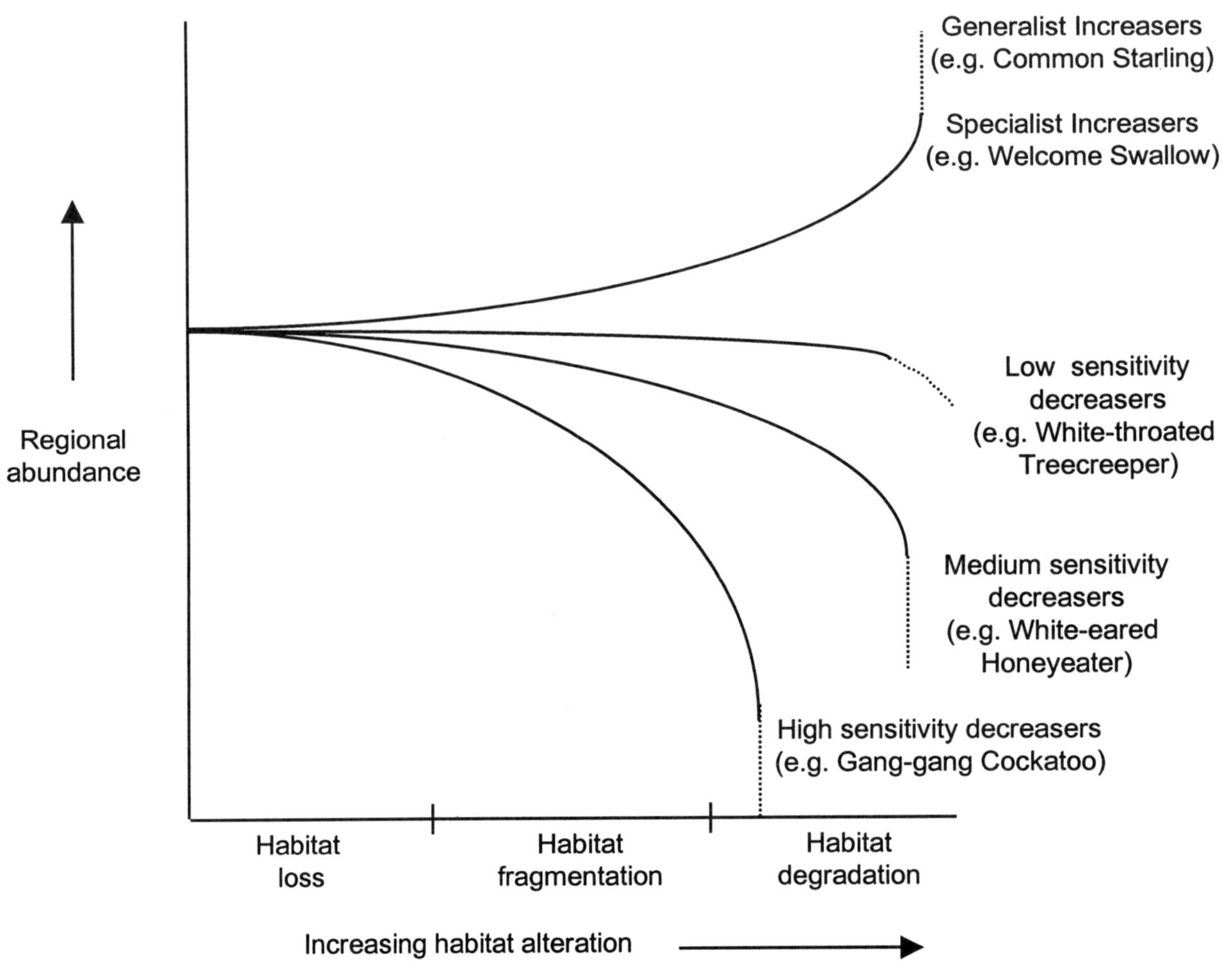

Fig. 2. Generalized responses by bird species to habitat change within the Bathurst landscape.

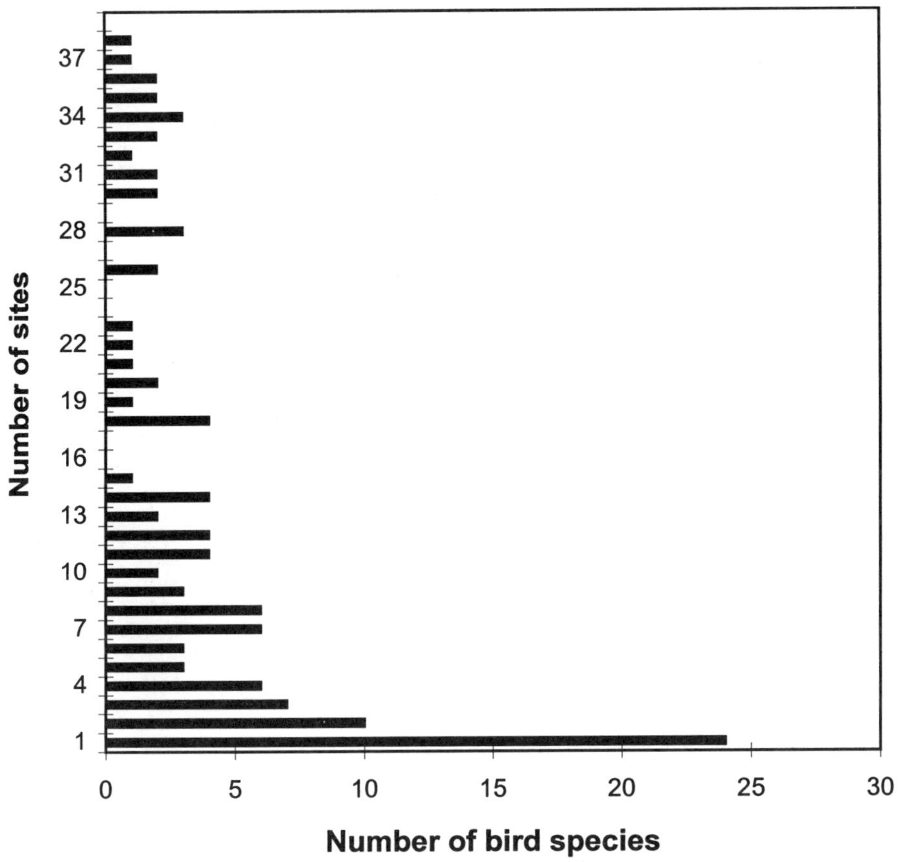

Fig. 3. The distribution of bird species as a function of the number of sites at which each species was recorded.

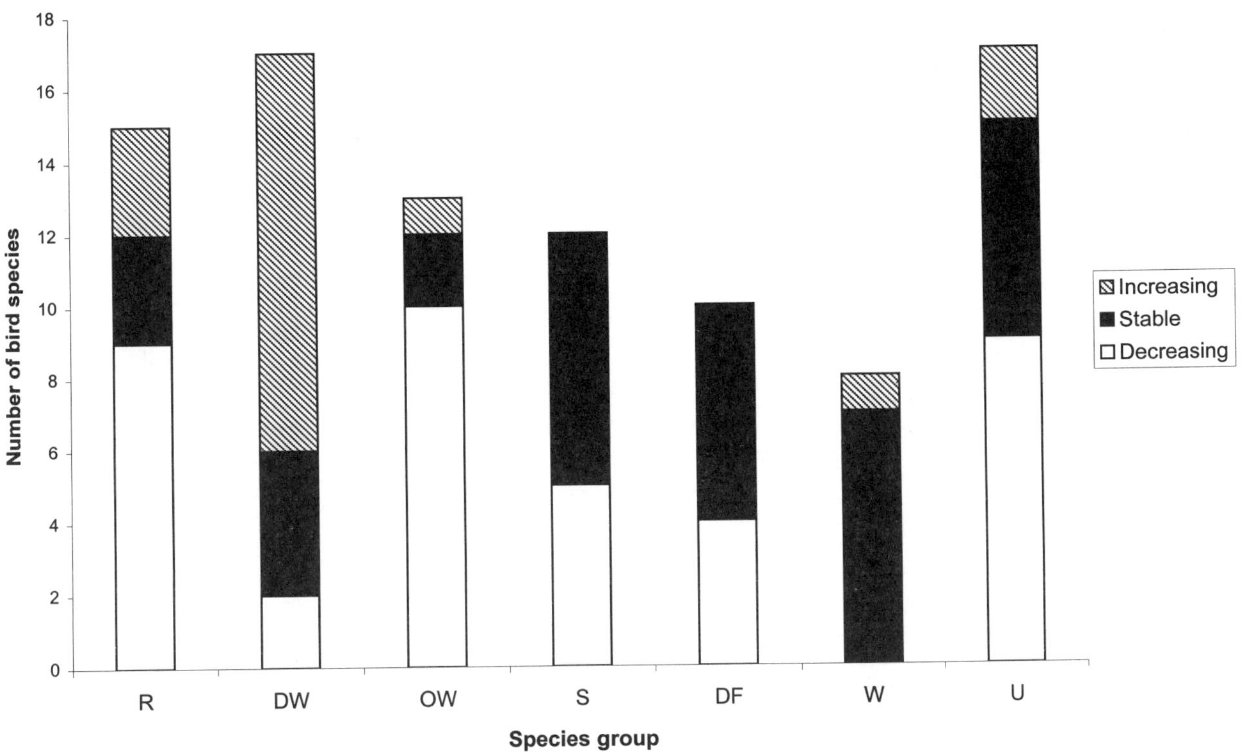

Fig. 4. Regional trends of bird species within designated groups in the Bathurst landscape. R = species associated with riparian strips through grazing land; DW = species associated with degraded woodland; OW = open woodland species; S = species sparsely distributed or with narrow habitat requirements; DF = species associated with densely forested areas; W = widespread species; U = species associated with well-developed understorey.

Table 1. Classification of bird species (n = 116) recorded during this study. Species groupings are according to the Hierarchical Cluster Analysis output. Nomenclature follows Christidis and Boles (1994). The symbol * denotes introduced species. A total of 39 sites were surveyed. The mean Occurrence Index is calculated from the sites at which each species was recorded (i.e., zero values excluded). The Bathurst landscape was divided into low altitude (<900 m, extensively cleared) and high altitude (>900 m, relatively intact forest cover). Based on this study, bird species were designated as occurring in the low altitude (Low) or high altitude (High) landscape, or both if records were from throughout the study area (Both). Seasonal status was categorized as R = resident, M = migrant or V = vagrant. The generalized response of each species to habitat alteration was classified as LSD = Low sensitivity decreaser, MSD = Medium sensitivity decreaser, HSD = High sensitivity decreaser, GI = generalist increaser or SI = specialist increaser. Projected regional trends for each species are given as ↑ (increasing), ↓ (decreasing) or • (stable).

Bird species	Scientific name	No. of sites	Mean OI (± SD)	Seasonal status	Landscape	Response to habitat change	Regional trend
Riparian Strips							
Australian Hobby	*Falco longipennis*	2	0.05 (± 0.00)	R	Low	LSD	•
Peaceful Dove	*Geopelia striata*	5	0.07 (± 0.04)	R	Low	MSD	↓
White-plumed Honeyeater	*Lichenostomus penicillatus*	8	0.44 (± 0.41)	R	Low	MSD	↓
Willie Wagtail	*Rhipidura leucophrys*	13	0.27 (± 0.25)	R	Low	GI	↑
Dollarbird	*Eurostomus orientalis*	4	0.13 (± 0.10)	M	Low	MSD	↓
Pacific Black Duck	*Anas superciliosa*	9	0.22 (± 0.17)	R	Low	LSD	•
Double-barred Finch	*Taeniopygia bichenovii*	5	0.09 (± 0.07)	R	Low	HSD	↓
Rainbow Bee-eater	*Merops ornatus*	3	0.18 (± 0.10)	M	Low	HSD	↓
Restless Flycatcher	*Myiagra inquieta*	3	0.13 (± 0.10)	M	Low	MSD	↓
Rufous Songlark	*Cincloramphus mathewsi*	3	0.10 (± 0.05)	M	Low	SI	↑
Dusky Woodswallow	*Artamus cyanopterus*	3	0.27 (± 0.23)	M	Low	MSD	↓
Black-fronted Dotterel	*Elseyornis melanops*	2	0.18 (± 0.18)	M	Low	HSD	↓
White-faced Heron	*Egretta novaenollandiae*	3	0.13 (± 0.14)	R	Low	LSD	↓
Welcome Swallow	*Hirundo neoxena*	4	0.13 (± 0.10)	M	Low	SI	↑
Red-browed Finch	*Neochmia temporalis*	15	0.16 (± 0.11)	R	Both	LSD	↓
Superb Fairy-Wren	*Malurus cyaneus*	23	0.38 (± 0.25)	R	Both	LSD	•
Degraded Woodland							
Common Starling*	*Sturnus vulgaris*	11	0.34 (± 0.31)	R	Both	GI	↑
Red-rumped Parrot	*Psephotus haematonotus*	8	0.49 (± 0.36)	R	Low	SI	↑
Crested Pigeon	*Ocyphaps lophotes*	12	0.29 (± 0.21)	R	Low	GI	↑
Noisy Miner	*Manorina melanocephala*	12	0.55 (± 0.43)	R	Low	SI	↑
Magpie-lark	*Grallina cyanoleuca*	12	0.43 (± 0.27)	R	Low	GI	↑
Grey Butcherbird	*Cracticus torquatus*	12	0.21 (± 0.18)	R	Both	SI	•
Pied Butcherbird	*Cracticus nigrogularis*	8	0.09 (± 0.06)	R	Low	SI	•
Eastern Rosella	*Platycercus eximius*	26	0.34 (± 0.32)	R	Both	GI	↑
Pallid Cuckoo	*Cuculus pallidus*	8	0.06 (± 0.02)	M	Low	SI	•
Yellow Thornbill	*Acanthiza nana*	4	0.05 (± 0.00)	R	Low	LSD	↓
Musk Lorikeet	*Glossopsitta concinna*	2	0.05 (± 0.00)	M	Low	MSD	↓
Nankeen Kestrel	*Falco cenchroides*	2	0.05 (± 0.00)	R	Low	SI	↑
Brown Goshawk	*Accipiter fasciatus*	7	0.13 (± 0.11)	R	Low	SI	↑
Australian Wood Duck	*Chenonetta jubata*	9	0.19 (± 0.16)	R	Both	LSD	•
Galah	*Cacatua roseicapilla*	19	0.18 (± 0.22)	R	Both	GI	↑
Australian Raven	*Corvus coronoides*	28	0.80 (± 0.14)	R	Both	GI	↑
Australian Magpie	*Gymnorhina tibicen*	31	0.49 (± 0.25)	R	Both	GI	↑
Open Woodland							
Speckled Warbler	*Chthonicola sagittata*	7	0.10 (± 0.06)	M	Low	MSD	↓
Yellow-rumped Thornbill	*Acanthiza chrysorrhoa*	7	0.16 (± 0.09)	R	Low	SI	↑
White-throated Gerygone	*Gerygone olivacea*	14	0.22 (± 0.14)	M	Low	MSD	↓
Mistletoebird	*Dicaeum hirundinaceum*	14	0.17 (± 0.07)	M	Both	LSD	•
Noisy Friarbird	*Philemon corniculatus*	26	0.17 (± 0.11)	M	Both	LSD	↓
White-winged Chough	*Corcorax melanorhamphos*	28	0.24 (± 0.21)	R	Both	MSD	↓
Brown-headed Honeyeater	*Melithreptus brevirostris*	8	0.13 (± 0.11)	M	Both	MSD	↓
Olive-backed Oriole	*Oriolus sagittatus*	11	0.11 (± 0.07)	M	Both	MSD	↓
Common Bronzewing	*Phaps chalcoptera*	7	0.14 (± 0.08)	R	Low	MSD	↓
Weebill	*Smicrornis brevirostris*	10	0.11 (± 0.09)	R	Low	MSD	↓
Brown Treecreeper	*Climactris picumnus*	4	0.18 (± 0.25)	R	Low	MSD	↓
Buff-rumped Thornbill	*Acanthiza reguloides*	30	0.30 (± 0.24)	R	Both	LSD	•
Sparsely Distributed/Narrow Habitat							
Crescent Honeyeater	*Phylidonyris pyrrhoptera*	2	0.15 (± 0.07)	M	High	MSD	•
Flame Robin	*Petroica phoenicea*	8	0.11 (± 0.10)	M	High	LSD	•
Rufous Fantail	*Rhipidura rufifrons*	6	0.10 (± 0.08)	M	High	MSD	•
Diamond Firetail	*Stagonopleura guttata*	2	0.10 (± 0.07)	M	Low	HSD	↓
Jacky Winter	*Microeca fascinans*	3	0.10 (± 0.05)	M	Low	LSD	↓
Horsefield's Bronze Cuckoo	*Chrysococcyx basalis*	2	0.05 (± 0.00)	M	Both	LSD	•
Varied Sittella	*Daphoenositta chrysoptera*	2	0.05 (± 0.00)	R	High	LSD	•
Brown Quail	*Coturnix ypsilophora*	2	0.05 (± 0.00)	R	Low	MSD	↓
Spotted Quail-thrush	*Cinclosoma punctatum*	2	0.13 (± 0.11)	R	High	MSD	•
Leaden Flycatcher	*Myiagra rubecula*	4	0.06 (± 0.03)	M	Low	MSD	↓

Table 1 — *continued*

Bird species	Scientific name	No. of sites	Mean OI (± SD)	Seasonal status	Landscape	Response to habitat change	Regional trend
Sparsely Distributed/Narrow Habitat — *continued*							
Superb Lyrebird	*Menura novaehollandiae*	3	0.10 (± 0.05)	R	High	MSD	•
Sacred Kingfisher	*Todiramphus sanctus*	7	0.06 (± 0.02)	M	Both	MSD	↓
Dense Forest							
Sulphur-crested Cockatoo	*Cacatua galerita*	11	0.10 (± 0.09)	R	Both	LSD	•
White-naped Honeyeater	*Melithreptus lunatus*	14	0.12 (± 0.19)	M	High	LSD	•
Satin Flycatcher	*Myiagra cyanoleuca*	18	0.14 (± 0.12)	M	Both	MSD	↓
Brush Cuckoo	*Cacomantis variolosus*	6	0.07 (± 0.03)	M	High	LSD	•
Yellow-tailed Black Cockatoo	*Calyptorhynchus funereus*	9	0.11 (± 0.07)	R	High	HSD	↓
Australian King-Parrot	*Alisterus scapularis*	9	0.09 (± 0.05)	R	High	MSD	•
Red-browed Treecreeper	*Climacteris erythrops*	11	0.10 (± 0.06)	R	High	HSD	↓
Grey Currawong	*Strepera versicolor*	11	0.10 (± 0.04)	R	High	LSD	•
Fan-tailed Cuckoo	*Cacomantis flabelliformis*	18	0.08 (± 0.04)	M	Both	LSD	•
Gang-gang Cockatoo	*Callocephalon fibriatum*	20	0.24 (± 0.21)	R	High	HSD	↓
Widespread Species							
Spotted Pardalote	*Pardalotus punctatus*	34	0.36 (± 0.18)	R	Both	LSD	•
White-throated Treecreeper	*Cormobates leucophaeus*	36	0.67 (± 0.28)	R	Both	LSD	•
Crimson Rosella	*Platycercus elegans*	38	0.63 (± 0.23)	R	Both	LSD	•
Laughing Kookaburra	*Dacelo novaeguineae*	31	0.23 (± 0.18)	R	Both	LSD	•
Red Wattlebird	*Anthochaera carunculata*	37	0.32 (± 0.23)	R	Both	LSD	•
Pied Currawong	*Strepera graculina*	36	0.29 (± 0.14)	R	Both	GI	↑
Black-faced Cuckoo-shrike	*Coracina novaehollandiae*	33	0.23 (± 0.13)	M	Both	LSD	•
Striated Pardalote	*Pardalotus striatus*	34	0.41 (± 0.27)	R	Both	LSD	•
Dense Understorey							
Brown Thornbill	*Acanthiza pusilla*	33	0.30 (± 0.24)	R	Both	LSD	•
Yellow-faced Honeyeater	*Lichenostomus chrysops*	35	0.48 (± 0.28)	R	Both	LSD	•
Grey Fantail	*Rhipidura fuliginosa*	35	0.47 (± 0.21)	M	Both	MSD	↓
Striated Thornbill	*Acanthiza lineata*	32	0.20 (± 0.16)	R	Both	LSD	•
White-eared Honeyeater	*Lichenostomus leucotis*	28	0.26 (± 0.14)	R	Both	MSD	↓
Bassian Thrush	*Zoothera lunulata*	6	0.05 (± 0.00)	R	High	MSD	•
Crested Shrike-tit	*Falcunculus frontatus*	5	0.05 (± 0.00)	R	High	MSD	•
European Goldfinch*	*Carduelis carduelis*	7	0.09 (± 0.06)	R	Both	SI	↑
Silvereye	*Zosterops lateralis*	13	0.10 (± 0.06)	R	Both	LSD	•
Golden Whistler	*Pachycephala pectoralis*	18	0.09 (± 0.06)	M	Both	MSD	↓
Eastern Yellow Robin	*Eopsaltria australis*	18	0.10 (± 0.07)	R	Both	MSD	↓
Eastern Spinebill	*Acanthorhynchus tenuirostris*	20	0.14 (± 0.11)	M	Both	MSD	↓
Common Blackbird*	*Turdus merula*	14	0.15 (± 0.16)	R	Both	GI	↑
Rufous Whistler	*Pachycephala rufiventris*	30	0.17 (± 0.12)	M	Both	MSD	↓
Scarlet Robin	*Petroica multicolor*	21	0.11 (± 0.07)	R	Both	MSD	↓
Grey Shrike-thrush	*Colluricincla harmonica*	35	0.38 (± 0.23)	R	Both	MSD	↓
White-browed Scrubwren	*Sericornis frontalis*	22	0.46 (± 0.29)	R	Both	LSD	↓
Vagrant							
Little Pied Cormorant	*Phalacrocorax melanoleucos*	1	0.05	V			
Great Cormorant	*Phalacrocorax carbo*	1	0.05	V			
Nankeen Night Heron	*Nycticorax caledonicus*	1	0.05	V			
Straw-necked Ibis	*Threskiornis spinicollis*	1	0.05	V			
Little Eagle	*Hieraatus morphnoides*	1	0.05	V			
Grey Falcon	*Falco hypoleucos*	1	0.05	V			
Masked Lapwing	*Vanellus miles*	1	0.25	V			
Shining Bronze Cuckoo	*Chrysococcyx lucidus*	1	0.05	V			
Powerful Owl	*Ninox strenua*	1	0.05	V			
Southern Boobook	*Ninox novaeseelandiae*	1	0.05	V			
White-throated Nightjar	*Eurostopodus mystacalis*	1	0.10	V			
Australian Owlet-Nightjar	*Aegotheles cristatus*	1	0.05	V			
Azure Kingfisher	*Alcedo azurea*	1	0.10	V			
New Holland Honeyeater	*Phylidonyris novaehollandiae*	1	0.10	V			
Red-capped Robin	*Petroica goodenovii*	1	0.05	V			
Rose Robin	*Petroica rosea*	1	0.05	V			
Hooded Robin	*Melanodryas cucullata*	1	0.05	V			
Eastern Whipbird	*Psophodes olivaceus*	1	0.20	V			
Masked Woodswallow	*Artamus personatus*	1	0.05	V			
White-browed Woodswallow	*Artamus superciliosis*	1	0.05	V			
Black-faced Woodswallow	*Artamus cinereus*	1	0.10	V			
Skylark*	*Alauda arvensis*	1	0.10	V			
Richard's Pipit	*Anthus novaeseelandiae*	1	0.05	V			
Fairy Martin	*Hirundo aerial*	1	0.05	V			

(iii) Predicted changes in the avifauna of the Bathurst landscape

Of the 92 bird species recorded at two or more sites in the course of this study, 39 (42.4%) are in regional decline (Table 1). Eighteen species (19.6%) are increasing in the Bathurst landscape while the remaining 35 species (38%) are regionally stable.

The combined regional trends of the species within the groups defined by the Hierarchical Cluster Analysis are presented in Figure 4. The greatest decline in bird species in terms of distribution and abundance in the Bathurst landscape is in the open woodland group, in which ten of the 13 species are considered to be decreasing in the study area. Nine of the fifteen species associated with riparian strips are decreasing in regional distribution and abundance. Nine of the seventeen understorey species and ten of the twenty-four vagrants are declining in the Bathurst landscape. The highest proportion of increasing species is within the degraded woodland group (11 of 17 species). Seven of the eight widespread species are considered to be regionally stable, while six of ten dense forest species and seven of the twelve species sparsely distributed or with narrow habitat requirements are considered to be relatively stable in terms of their distribution and abundance in the study area.

DISCUSSION

The 116 species recorded during this study represent 62.4% of the species previously recorded or predicted to occur in the Bathurst 1:100 000 topographic map sheet (Goldney 1987). As this study did not include wetlands or intentional surveys for raptors or nocturnal species, the diurnal terrestrial avifauna occurring in the Bathurst landscape was effectively sampled.

In a result similar to that presented by Recher *et al.* (1991) for the south-east forests, a high proportion of species was recorded infrequently in this study. Neave *et al.* (1996) also found that less than 20% of bird species were detected at more than half of the 129 sites they surveyed in southeastern Australia.

The grouping of bird species by cluster analysis based on the Occurrence Index data provides results similar to those presented by Neave *et al.* (1996), but species groups are more clearly delineated in this study by virtue of the systematic sampling of a range of habitat and landscape situations represented within the Bathurst landscape.

The assessment of the regional status of bird species indicates that 39 bird species are declining in distribution and abundance, and may be lost from the Bathurst landscape within 30 years if the present rate of habitat degradation continues (Goldney *et al.* 1997; Bauer and Goldney, in press). In addition there were 24 species recorded as vagrants during the present study, many of which may have this status as a consequence of regional declines (e.g., Hooded Robin). Most at risk are those species associated with open woodland and riparian woodland habitats. The increase in suitable habitat for species of degraded woodland suggests that this group will become dominant within the study area in the near future particularly below elevations of about 800 m. However, as trees on farmland, particularly box species, are dying and not being replaced, bird species which require degraded woodland for breeding and/or feeding (such as the Noisy Miner) will subsequently decline. Widespread species, which tend to be habitat generalists, will form an increasingly large component of the avifauna of the Bathurst landscape. The tale of woodland bird decline presented here concurs with that of Saunders (1989), Barrett *et al.* (1994), Ford *et al.* (1995), Saunders and Ingram (1995) and Robinson and Traill (1996), illustrating this to be a widespread phenomenon throughout the temperate woodlands of southern Australia. Mac Nally and Bennett (1997) sought to forecast the proneness to local extinction of bird species in the box-ironbark forests of central Victoria. Their modelling approach, based upon the mobility, population density and habitat specificity of each species, has been subsequently invalidated by field data, with indications that spatial and temporal variability play an important part in this system (Bennett, pers. comm.). This suggests that the predictions made in this chapter should be treated carefully, with a need to determine precisely the characteristics of bird species likely to decrease across temperate Australia if land management practices do not radically change.

MANAGEMENT IMPLICATIONS

The Bathurst landscape embraces a range of vegetation alliances and associated management regimes. While all of the land tenures within this multi-use landscape have a role to play in the long-term conservation of the avifauna, of critical importance is the privately-owned farming land, which includes the majority of the *E. melliodora–E. blakelyi* and *E. albens* open woodlands and a large proportion of the *Casuarina cunninghamiana* fringing forest within the study area. These areas have been severely impacted by overgrazing, and in the case of the riparian strips, excessive bank erosion by the continual accessing of the water by stock and a lack of stability during flood events (Fisher and Goldney 1997). A principal action for bird conservation involves fencing woodland remnants and riparian strips to restrict stock access. This will facilitate the regeneration of native vegetation and thereby provide habitat for birds. Some active management may be also be needed (e.g., direct seeding or tubestock planting; weed control — see Wakefield 1997).

The other main land uses within the study area are nature reserves, pine plantations and urban development. The nature reserves (mostly managed by the New South Wales National Parks and Wildlife Service) tend to be on the low fertility soils and rocky ridges not historically cleared for agriculture, consisting primarily of *E. macrorhyncha–E. rossii* open forest with some *E. fastigata–E. viminalis* tall open forest. A sizeable proportion of the species distributed sparsely or with narrow habitat requirements are found in these nature reserves, and given that these reserves are unlikely to be adversely affected by current management practices, these species are reasonably secure in the Bathurst landscape.

State Forests of New South Wales manage the majority of the high altitude *E. fastigata–E. viminalis* tall open forest in the eastern portion of the study area. Much of the native vegetation in this area has been cleared and replaced by exotic *Pinus radiata* plantations. The fragments of native forest retained within the State Forests provide important regional habitat for avifauna (Fisher and Goldney 1998). With tight controls on the management of remnant vegetation within State Forests (State Forests of New South Wales 1997) six of the ten species associated with dense forest are considered to be regionally secure, while the remainder are at risk because they require large areas of native forest (e.g., Red-browed Treecreeper, Yellow-tailed Black-Cockatoo). In order to increase the habitat value for birds of native vegetation within extensive plantations, it is suggested that riparian strips be extended and replanted, with the current 20 m buffer increased to at least 50 m and preferably 100 m (Fisher and Goldney 1998).

The urban population within the study area is steadily growing, with the population of the largest centre of Bathurst at just over 30 000 people. Among the most common bird species in Bathurst City are the introduced Common Starling and Common Blackbird (pers. obs.). Ideal habitat is provided for these species in the form of parks and gardens of exotic berry-producing trees and shrubs throughout the city. One native species, the Pied Currawong, has similarly learnt to exploit this winter food source offered by exotic plants. This opportunistic omnivore congregates in large numbers in urban areas during winter (Farrell 1995) and many remain to nest in spring close to habitation, feeding on the eggs and nestlings of native species (Bass 1995; Major *et al.* 1996). The establishment of structurally diverse native vegetation should be encouraged in new housing developments and the benefits of native plants promoted to owners of older-style gardens. Increasing the amount of native vegetation cover at the expense of introduced plants should provide less suitable habitat for introduced bird species and predators such as the Pied Currawong and more resources for a range of native species.

TOWARDS AN INTEGRATED LANDSCAPE APPROACH: THE ROLE OF BIRDS

While most people are able to recognize cleared or degraded land, the majority fail to see their actions as impacting upon nature conservation. It has been demonstrated that while landholders in the central western region have a good understanding of global environmental issues, they fail to translate this into local activity (Goldney *et al.* 1995a,b; Hodgkins *et al.,* in press). Many bird species perceive the environment at a landscape-scale and do not recognize the boundaries that humans use (Recher 1993), such as that between a nature reserve and private property. Given the large proportion of bird species which move across the Bathurst landscape, either as summer migrants, non-breeding residents or vagrants, an integrated landscape approach to the conservation of the avifauna is essential. The reality of ecological fluxes and the importance of landscape connectivity for ecological processes and bird movements underscores the need for a landscape approach. As many land managers (farmers, foresters, park rangers, council employees) are able to recognize many of the bird species occurring in the Bathurst landscape (pers. obs.), it is possible that the avifauna may provide a medium for communicating with those whose decisions are likely to impact on the ecology of the region. The simple observation that many birds move across a landscape can assist in explaining the way in which the various components of the landscape are inter-related. The challenge for conservation biologists is to demonstrate unequivocally that the movement of birds across a landscape is linked to the long-term sustainability of that landscape.

ACKNOWLEDGEMENTS

Thanks to John Suttor, Richard Dutton, Bernard Lenahan, Terry Robinson, Peter Style, Vern and Joyce Moffit and Mrs. Charman who allowed access to their properties.

Meteorological data for Bathurst were supplied by Noel Mitchell (Bathurst Agricultural Research Station). Arthur Henry (State Forests, Bathurst) provided climatic data for Sunny Corner and gave useful insights into the management of pine plantations.

Thanks to Steve Woodhall and Ian McCartney for facilitating access to the Winburndale and Wambool Nature Reserves. Permission to conduct surveys within nature reserves was given by the NSW National Parks and Wildlife Service (Authority No. A 1355) and within state forests by State Forests of NSW (Special Purposes Permit Nos. 04827 and 04948). Debi Fisher patiently helped with data collation. This study contributes to the Remnant Woodland Ecology Program and was partly funded by WWF Australia and the Johnstone Centre at Charles Sturt University. David Goldney helped with the design of the project and provided useful comments on a previous version of this chapter, as did Donna Windsor. Peter Cale, Denis Saunders and an anonymous referee helped to distil the main message from an earlier manuscript.

REFERENCES

Austin, M. P. and Heyligers, P. C., 1989. Vegetation survey design for conservation: Gradsect sampling of forests in north-eastern New South Wales. *Biol. Conserv.* **50:** 13–32.

Austin, M. P. and Heyligers, P. C., 1991. New approach to vegetation survey design: Gradsect sampling. Pp. 33–36 *in* Nature Conservation: Cost Effective Biological Surveys and Data Analysis ed by C. R. Margules and M. P. Austin. CSIRO: Australia.

Barrett, G. W., Ford, H. A. and Recher, H. F., 1994. Conservation of woodland birds in a fragmented rural landscape. *Pac. Cons. Biol.* **1:** 245–56.

Bass, D. A., 1995. Contribution of introduced fruits to the winter diet of Pied Currawongs in Armidale, New South Wales. *Corella* **19:** 127–32.

Bauer, J. J. and Goldney, D., in press. Extinction and degradation in complex landscape systems. *In* Temperate Woodlands in Australia: Biology, Conservation, Management and Restoration ed by R. J. Hobbs and C. J. Yates. Surrey Beatty & Sons: Chipping Norton.

Bennett, A., Backhouse, G. and Clark, T. (eds), 1995. People and Nature Conservation: Perspectives on Private Land Use and Endangered Species Recovery. Royal Zoological Society of NSW: Mosman.

Bower, C. C. and Semple, W. S., 1993. A guide to the eucalypts of the central west of NSW. Department of Conservation and Land Management Technical Report No. 30. Department of Conservation and Land Management: Sydney.

Christidis, L. and Boles, W. E., 1994. The Taxonomy and Species of Birds of Australia and its Territories. Royal Australasian Ornithologists Union Monograph 2, Royal Australasian Ornithologists Union: Melbourne.

Farrell, J. R., 1995. Movement patterns of Pied Currawongs *Strepera graculina* in central western New South Wales. *Corella* **19:** 95–102.

Fisher, A. M., 1997. The distribution and abundance of avifauna in the Bathurst landscape: implications for conservation and land management. Ph.D. thesis, Charles Sturt University, Bathurst, New South Wales.

Fisher, A. M. and Goldney, D. C., 1997. Use by birds of riparian vegetation in an extensively fragmented landscape. *Pac. Cons. Biol.* **3:** 275–88.

Fisher, A. M. and Goldney, D. C., 1998. Native forest fragments as critical bird habitat in a softwood forest landscape. *Aust. For.* **61:** 287–95.

Ford, H., Barrett, G. and Recher, H., 1995. Birds in a degraded landscape — safety nets for capturing regional biodiversity. Pp. 43–50 *in* Nature Conservation 4: The Role of Networks ed by D. A. Saunders, J. L. Craig and E. M. Mattiske. Surrey Beatty & Sons: Chipping Norton.

Forestry Commission of New South Wales, 1987. Management Plan for the Bathurst Management Area. Forestry Commission of New South Wales: Sydney.

Gillison, A. N. and Brewer, K. R. W., 1985. The use of gradient directed transects or gradsects in natural resource surveys. *J. Environ. Manage.* **20:** 103–27.

Goldney, D. C., 1987. The vertebrate fauna of the central western region. Pp. 136–60 *in* The National Trust of Australia (NSW) Scenic and Scientific Survey of the Central Western Region. A report to the Australian Heritage Commission. Volume 1 ed by D. C. Goldney and I. J. S. Bowie. Mitchell College of Advanced Education: Bathurst, New South Wales.

Goldney, D., Bauer, J., Bryant, H., Hodgkins, D. and Watson, G., 1995a. Winning battles but losing the war: The education marketing imperative. Pp. 574–88 *in* Nature Conservation 4: The Role of Networks ed by D. A. Saunders, J. L. Craig and E. M. Mattiske. Surrey Beatty & Sons: Chipping Norton.

Goldney, D., Bryant, H., Tyson, G. and Stone, B., 1995b. Burrendong Catchment Land Management Survey. Environmental Studies Unit, Charles Sturt University: Bathurst, New South Wales.

Goldney, D., Stone, B. and Croft, M., 1997. Integrating satellite imagery with other data sources to better manage bushland in the Central Western Region of New South Wales. Pp. 98–113 *in* Ecology at the Cutting Edge: Information Technologies for Managing Biodiversity and Ecological Processes ed by B. Diekman, E. Higginson, F. Sutton and H. Webb. Nature Conservation Council of New South Wales: Sydney.

Hale, P. and Lamb, D. (eds), 1997. Conservation Outside Nature Reserves. Centre for Conservation Biology, University of Queensland: Brisbane.

Hodgkins, D., Watson, G., Goldney, D. and Tyson, G., in press. The attitudes of landholders to a range of environmental issues, including the values of remnant bushland in the Central Western Region of New South Wales. *In* Temperate Woodlands in Australia: Biology, Conservation, Management and Restoration ed by R. J. Hobbs and C. J. Yates. Surrey Beatty & Sons: Chipping Norton.

Mac Nally, R. and Bennett, A. F., 1997. Species-specific predictions of the impact of habitat fragmentation: Local extinction of birds in the box-ironbark forests of central Victoria, Australia. *Biol. Cons.* **82:** 147–55.

Major, R. E., Gowing, G. and Kendal, C. E., 1996. Nest predation in Australian urban environments and the role of the Pied Currawong *Strepera graculina*. *Aust. J. Ecol.* **21:** 399–409.

McGarigal, K. and McComb, W. C., 1995. Relationships between landscape structure and breeding birds in the Oregon Coast Range. *Ecol. Monogr.* **65**: 235–60.

Neave, H. M., Norton, T. W. and Nix, H. A., 1996. Biological inventory for conservation evaluation II. Composition, functional relationships and spatial prediction of bird assemblages in southern Australia. *For. Ecol. Manage.* **85:** 123–48.

Norusis, M. J., 1994. SPSS Professional Statistics™ 6.1. SPSS Inc.: Chicago, Illinois.

Recher, H. F., 1993. Why conservation biology? An Australian perspective. Pp. 1–15 *in* Conservation Biology in Australia and Oceania ed by C. Moritz and J. Kikkawa. Surrey Beatty & Sons: Chipping Norton.

Recher, H. F., Kavanagh, R. P., Shields, J. M. and Lind, P., 1991. Ecological associations of habitats and bird species during the breeding season in southeastern Australia. *Aust. J. Ecol.* **16:** 337–52.

Robinson, D. and Traill, B. J., 1996. Conserving woodland birds in the wheat and sheep belts of southern Australia. Royal Australasian Ornithologists Union Conservation Statement No. 10. Supplement to *Wingspan* 6(2) June 1996.

Saunders, D. A., 1989. Changes in the avifauna of a region, district and remnant as a result of fragmentation of native vegetation: The wheatbelt of Western Australia. A case study. *Biol. Conserv.* **50:** 99–135.

Saunders, D. A. and Ingram, J. A., 1995. Birds of Southwestern Australia. An Atlas of Changes in Distribution and Abundance of the Wheatbelt Fauna. Surrey Beatty & Sons: Chipping Norton.

Saunders, D. A., Craig, J. L. and Mattiske, E. M. (eds), 1995. Nature Conservation 4: The Role of Networks. Surrey Beatty & Sons: Chipping Norton.

State Forests of New South Wales, 1997. Forest Practices Code. Part Three. Plantation establishment and maintenance. State Forests of New South Wales: Pennant Hills.

Wakefield, S. 1997. Managing farm bushland, Kit 2. *In* Save the Bush Toolkit ed by S. Wakefield and D. Goldney. Charles Sturt University: Bathurst, New South Wales.

Windsor, D. M., 1998. A landscape approach to optimise recruitment of woodland species in an intensive agricultural environment in the central tablelands of NSW. Ph.D. thesis, Charles Sturt University, Bathurst, New South Wales.

CHAPTER 65

Arguing for conservation of wetlands in landscapes with intensifying land use pressures: *Melaleuca quinquenervia* south-east Queensland

Toivo Zoete[1] and Jim Davie[2]

Traditional site and species based approaches to conservation have failed to prevent the large scale conversion of wetlands for rural and residential purposes. Today many types of wetlands are seriously endangered. This is the situation in the forested wetlands dominated by *Melaleuca quinquenervia* in south-east Queensland where conversions since 1974 have resulted in large losses of the original area across the region. Continuing high rates of urban expansion in the region mean that there is continuing high pressure on the remaining wetlands. The retention of these areas in a condition which conserves their contribution to local and regional biodiversity, as well as to a range of ecological, physical and social functions, requires reframing arguments to planners and other land use decision makers. Developments in landscape ecology and ecological economics are providing opportunities to widen the scope and improve the basis for nature conservation by providing a greater emphasis on ecological functions, such as primary production and nutrient removal and transformation, the regulation of water flows and flooding. In this chapter the range of wetland ecosystems defined by *M. quinquenervia* is described and related to their significance for a range of ecological landscape functions, independently assessed using Wetland Evaluation Technique (Adamus 1987) and a Delphi analysis by an expert panel. It is argued that the strong correlation between apparent ecological functions, which have economic implications, and the structure and composition of the vegetation component of the wetland ecosystems, will enhance opportunites for conservation and restoration of otherwise endangered sites.

Key words: Conservation, *Melaleuca quinquenervia*, Wetlands, Ecological functions, Economic values, South-east Queensland.

INTRODUCTION

TRADITIONAL site and species based approaches to conservation have failed to prevent the conversion of wetlands for development purposes. Species diversity in wetlands is often low, species are usually common, and the pervading influence of water implies that conservation of part of a wetland has no meaning without sensitive management of the other parts of the wetland and, indeed, its catchment. Today many wetlands are seriously endangered (Dugan 1993).

This is the situation in the forested wetlands dominated by *Melaleuca quinquenervia* in south-east Queensland. Although in the past agricultural conversion has resulted in the loss of much of the originally extensive areas of *M. quinquenervia* which occurred in coastal south Queensland, there is now a renewed threat to remaining stands as the region struggles to adjust to the pressure for settlement of internal migrants, which have been expanding the population of south Queensland, at unprecedented rates, since the late 1980s (Regional Planning and Advisory Group 1993).

[1]Australian Housing and Urban Research Institute, Queensland University of Technology, G.P.O. Box 2434, Brisbane, Queensland, Australia 4001.
[2]School of Natural and Rural Systems Management, The University of Queensland, Gatton Campus, Lawes, Queensland, Australia 4343.
Pages 633–43 *in* NATURE CONSERVATION 5: NATURE CONSERVATION IN PRODUCTION ENVIRONMENTS: MANAGING THE MATRIX ed by J. L. Craig, N. Mitchell and D. A. Saunders. Surrey Beatty & Sons, 2000.

One reason for the ease with which planning decisions for conversion have been granted in the past has been the perception that the *M. quinquenervia* wetlands are species poor. Indeed the canopy stratum is dominated by *M. quinquenervia*, which is frequently the only tree species. It is common for all the *M. quinquenervia* wetlands to be lumped as a single ecosystem for conservation planning purposes (e.g., Young and Cotterell 1993) although vegetation maps showing the extent of *M. quinquenervia* forested wetlands in south-east Queensland (Dowling and McDonald 1976; Elsol and Dowling 1978; Elsol and Sattler 1979) describe a number of different plant associations.

Two recent developments are providing opportunities to broaden the scope and improve the basis of traditional conservation. Firstly, in what has become the field of landscape ecology, there has been an increased recognition of the importance of the ecosystem and landscape levels of ecological organization for conservation purposes (e.g., Naveh and Liebermann 1984; Noss 1990). Interest in these levels has resulted from the recognition that species and populations require healthy, functioning systems which in turn are dependent on processes through which adjacent ecosystems interact within the wider landscape (Davie 1997). Such processes include biogeochemical cycles, primary and secondary production, mineralization of organic matter in the soils and sediments, storage and transport of minerals and biomass, and regulation of these processes (Ricklefs *et al.* 1984). By concentrating on the maintenance of ecological processes at the ecosystem and landscape levels, the traditional economic and instrumental species as well as bacteria, fungi, and unknown species are looked after (McNaughton 1989).

Secondly, the increasing faith in markets and the concomitant push for the extension of private property (which gained greater emphasis during the 1980s) has not bypassed conservation. In fact, much has recently been made of the natural linkage between ecology and economics and the benefits to conservation objectives of economic arguments (e.g., Pearce *et al.* 1989; Tisdell 1991; Barbier 1993). As a result of these discussions, ecosystems have begun to be regarded as providers of goods and services. By valuing and summing the goods and services that flow from an ecosystem in its natural state versus those that would flow from the system in some modified state, it is proposed that decisions regarding modification for some development or preservation can be made on an economic basis (de Groot 1992).

However, for either of these alternative approaches to gain credibility there needs to be a substantial body of information about the target ecosystems. The fact is that many systems are still insufficiently known to determine what ecological functions they perform or what marginal economic value might be placed upon the loss of part or all of them. At the outset of the research reported in this chapter, less than a dozen significant references existed on the biology and ecology of *M. quinquenervia*. Most of the references were from Florida, United States of America, where *M. quinquenervia* is an important invasive species and most of these focused on aspects of the population biology of *M. quinquenervia* from the perspective of control.

The project on which this chapter reports had two objectives (Zoete 1997). The first was to find out more about the composition, structure, and ecological relationships of the *M. quinquenervia* wetlands in south-east Queensland. The second sought to assess the adequacy of scientific arguments for the conservation of *M. quinquenervia* dominated wetlands. The research addressed two gaps in the information base on these wetlands. Firstly, appealing to traditional conservation concerns, there was a need to investigate the composition and structure of the whole vegetation, including vegetation strata other than the canopy. A second strand of investigation represented an assessment of the likely performance by these wetlands of a range of ecological functions which have values to which the surrounding community could relate.

In this chapter we first describe some of the variety of *M. quinquenervia* ecosystems that may be found and dispel the myth about their low species richness. We then present an analysis of the range of ecological functions which are served by different *M. quinquenervia* ecosystems before discussing how these insights might be used in arguing more comprehensively for their conservation or restoration in otherwise endangered sites.

THE STUDY AREA

The study area is located in coastal southern Queensland to the east of the Great Dividing Range and between Noosa, on the "Sunshine Coast", to the New South Wales-Queensland border (Fig. 1). The coastline is dominated by Moreton Bay, located centrally within this area and the region is thus often called the "Moreton Region". The City of Brisbane is located near the mouth of the Brisbane River and with other major urban areas such as the City of the Gold Coast, the cities of Redcliffe, Caloundra, Maroochydore and Noosa and urban areas in surrounding shires, it is the focus of a conurbation which extends almost without break along the Moreton coastline. The larger south-east Queensland region, of which the Moreton Region is part, contained 1 849 000 inhabitants in 1991 (Regional Planning and Advisory Group 1993) and that number is projected to grow to 3 050 000 by 2011. Most of the growth is likely to occur in the coastal zone (Regional Planning and Advisory Group 1993), where *M. quinquenervia* wetlands provide some of the most available and economically most affordable sites for development.

The Moreton Region can be subdivided into three topographic units (Beckmann 1967; Beckmann *et al.* 1987; Capelin 1987). Inland are undulating to low hilly lands; with steep hilly and mountainous lands associated with a number of elongated subcoastal ranges in the northern and southern parts.

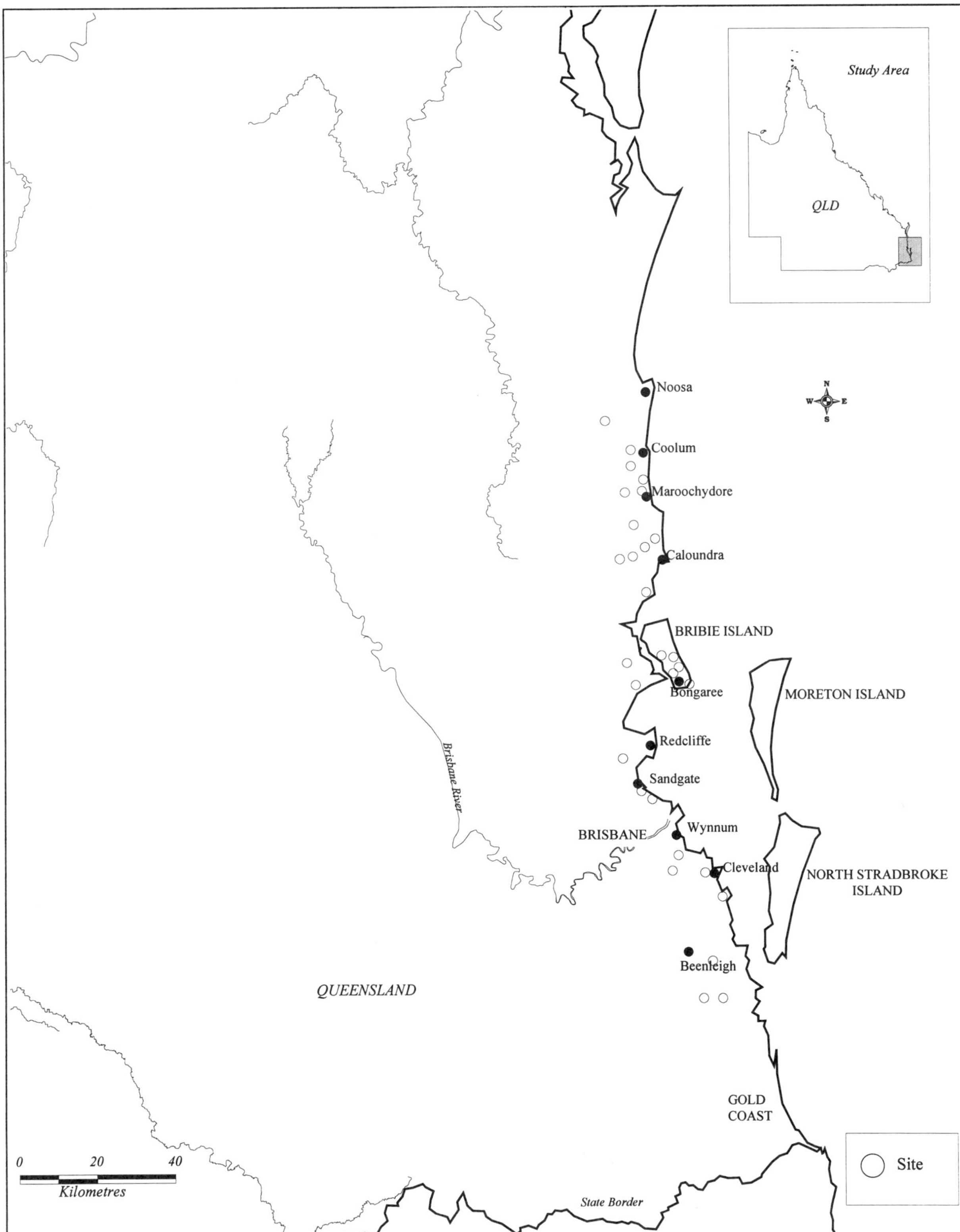

Fig. 1. Map of south-east Queensland showing location of sites (small open circles).

The third unit, the coastal plain, consists of low lying, gently undulating land generally less than 10 m above sea level. Large parts of the coastal plain consist of marine deposited sandy substrates, particularly on the Sunshine Coast. Finer materials of fluvial origin prevail in the Brisbane and southern areas.

As the region has been settled since the early 1800s, large tracts of the coastal plain were cleared for rural pursuits long ago. The flood prone nature of the *M. quinquenervia* wetlands ensured that much of this system was spared from conversion until relatively recently. However, as the supply of more suitable land has run out, pressures to use the *M. quinquenervia* system to provide land for pine plantations, urban developments (Queensland Department of Environment and Heritage 1992), sugarcane farming (Blake 1968; Catterall 1991), and grazing (Stick and Legessner 1969) have increased substantially.

By comparing the most recent aerial photos with vegetation maps based on photography from 1972 and 1974 (Dowling and McDonald 1976; Elsol and McDonald 1978; Elsol and Sattler 1979), Davie (1991) showed that between 22% and 77% of the *M. quinquenervia* wetlands of south-east Queensland had been converted for either farming or urban purposes during the 15 years covering this period (Table 1). The highest rates were in the predominantly rural southern part of the region (the Beenleigh Sheet) where pressure to increase the acreage under sugar cane has been intense. The lowest rates were experienced in the north (the Caloundra Sheet), where the sandy substrate was in less demand for agriculture. However, the rate of wetland conversion on the Sunshine Coast has increased sharply since the 1980s due to an urban expansion fueled by high rates of immigration into the area. Catterall (1991) considered that 95% of the area in the Brisbane Region originally covered in *M. quinquenervia* wetlands was lost by 1991. As a result of these losses, *M. quinquenervia* forests have been identified as a conservation priority for south-east Queensland (Regional Planning and Advisory Group 1993; Young and Cotterell 1993).

Despite the recent and continued contraction of *M. quinquenervia* wetlands, large areas still remain (Table 1). Much of the remaining resource is, however, degraded by changes in the hydrology and water quality of the catchment, an influx of weeds and other edge effects, and an increased incidence of fires. Most of the approximately 150 remaining stands over 1 ha (Davie 1991) are isolated small patches, often bisected by roads and other infrastructure. However, some large and relatively intact stands still occur, particularly in the northern part of the study area.

THE *MELALEUCA* ECOSYSTEMS

It has become common in Australia to identify ecosystems as plant communities defined as Plant Associations (Specht 1972). Plant Associations share the same Structural Formation and the same or related species in the canopy stratum. They therefore share similar biomass (estimated as the height of the canopy stratum) and energy assimilative capacity (estimated as Foliage Projective Cover). Plant Associations have become the basis for the classification of plant communities in Australia (Specht 1972; Specht *et al.* 1974; Specht 1981) and have been widely used for vegetation mapping.

Mapping the *Melaleuca quinquenervia* plant associations was done as part of a regional mapping project, at a scale of 1:100 000 in the Moreton Region (Dowling and McDonald 1976; Elsol and McDonald 1978; Elsol and Sattler 1979). The exercise indicated the range of environmental conditions under which this forested wetland species could be found, and by implication the range of ecosystem energy relationships which could be expected. However, in our preliminary field surveys it soon became clear that the range of significantly different environmental conditions across which the species could be found as the dominant, was far more finely differentiated by the vegetation than shown on the maps. Differences extended beyond structural characteristics of the canopy layer to the structure of the understorey strata and their species composition. It was decided that a finer level of ecosystem resolution had to be sought to more accurately reflect the differences in the ecological processes acting on and in these ecosystems.

The approach which we took draws upon the pioneering work of Gams (1918) in Austria, which subsequently evolved into the vegetation ecology school known as the *Northern Tradition,* in recognition of its development in Scandanavian plant communities (Whittaker 1978). The characteristics of these plant communities are a tree stratum of only one or two species differing little over environmental gradients. These gradients do, however, strongly influence the rich and site specific understorey stratum. The *Northern Tradition* held the view that environmental relations in these situations are better revealed by the lower strata than by the canopy (Barkman 1978). The *Finnish School* believed that the composition of the forest understorey was the most appropriate basis for vegetation classification, and used it as an indicator of site quality for the growth and establishment of trees (Cajander 1909, 1949).

On subtle environmental gradients, the tree, shrub, and ground strata cannot be expected to respond similarly to environmental parameters (Whittaker 1960). In these situations, the boundaries between changes in the tree stratum do not necessarily coincide with those of the understorey (Bratton 1975; Del Moral and Watson 1978; McCune and Antos 1981; Rheinhardt 1992; Gilliam *et al.* 1995), or the correlation between the strata may be very weak (Bradfield and Scagel 1984). Equal treatment of all species in a stand of forest therefore

Table 1. Losses of *Melaleuca quinquenervia* dominated forested wetlands in coastal mainland south-east Queensland 1972–1990 (Davie 1991).

Map sheet	Starting total area (km^2) (year of photography)	Total area lost (km^2)	Percentage loss
Caloundra (north)	159.0 (1972)	35.0 (1988)	22
Brisbane (central)	78.0 (1972)	32.5 (1987)	42
Beenleigh (south)	19.5 (1974)	15.0 (1990)	77

obscures the different responses of the various vegetation strata to environmental variables.

As *M. quinquenervia* forested wetlands appear to have strong structural similarities to high latitude plant communities, the vegetation was analysed in terms of its canopy, shrub layer, and ground layer, each of which is readily recognized as a separate stratum in the field. These strata were characterized by the foliage projective cover of each plant species occurring within the respective strata. Tree and shrub stems with diameters over 2.5 cm at breast height were included as a separate stratum in order to illustrate the pleiomorphic nature of the *M. quinquenervia* canopy within the study area.

Thirty wetland sites were selected for closer investigation on the basis of geographical spread, degree of disturbance, and the composition and structure of the vegetation (canopy, shrub layer, ground layer, and stems) (Fig. 1).

The data for each stratum were subjected to a variety of analyses, including numerical classification and ordination in order to reveal similar sites which could be combined to define vegetation types (Zoete 1997). Table 2 presents a classification of the floristic, structural and environmental characteristics recognized in each of four vegetation strata. By combining the appropriate descriptors for each stratum, the vegetation type at any site can be defined. However, since for any site, some descriptors may be more defining than others, communities may often be typed without referring to all strata. For example, stem dimensions, with or without associated canopy characters, distinguish several wetland environments. This is seen in the tree stratum where vegetation types with rainforest elements (primarily closed forest with few, tall and thick stems — the community D6/C4 — but also more open forest, with more moderate sized stems (D1/C2 or D2/C2), contained 25–30 m tall trees with basal areas of 70–96 m^2/ha, and foliage projective cover over 50%. By comparison, trees growing on sandy substrates were only 15 m tall, with basal areas of approximately 20 m^2/ha and foliage projective cover's of 10–30% (D4/C1). In one site immediately behind dunes, *M. quinquenervia* occurred only as a low heath of less than 1 m height (D7/C6).

Except for the two sites where the canopy was dominated by *Eucalyptus robusta* (the community D3/C5), *M. quinquenervia* dominated all sites, and approximately half the sites consisted of pure *M. quinequenervia*. Only one site (the community D6/C4) had a substantially larger number of species in the canopy. The community at this site was influenced by rainforest elements and contained nine of the total of 18 species encountered within the tree layer. Thus, while almost all of the communities were strongly dominated by *M. quinquenervia* and the species diversity was low, the structure of the communities varied

Table 2. Classification of the various vegetation strata of *Melaleuca quinquenervia* dominated forested wetlands in south-east Queensland, including stems (D), canopy foliage (C), shrub stratum (S), and ground stratum (G).

Type	Charcteristics
STEMS (TREES AND TALL SHRUBS)	
D1	few, thick, tall stems on moderately moist, clayey substrates
D2	few, relatively thick stems with *Glochidion ferdinandi*, *Lophostemon suaveolens* on moderately moist, clayey substrates
D3	moderate sized stems, containing *Eucalyptus robusta* on drier substrates
D4	moderate sized stems, containing *Eucalyptus* spp. and/or medium sized *Melaleuca quinquenervia* on drier substrates
D5	many, thin *M. quinquenervia* on regularly inundated, clayey substrates
D6	few, thick, tall stems, containing rainforest spp. on moderately moist, clayey substrates
D7	no stems >2.5 cm dbh on a deep sand substrate
CANOPY LAYER	
C1	woodland, foliage projective cover less than 22%, *M. quinquenervia* often on sandy substrates
C2	open forest, foliage projective cover more than 25%, *M. quinquenervia* on moist to regularly indundated substrates
C3	open forest, foliage projective cover >25%, dominated by *Casuarina glauca* on regularly inundated saline substrate
C4	closed forest, foliage projective cover >70%, with many rainforest species on moderately moist, clayey substrates
C5	open forest, dominated by *E. robusta* on drier substrates
C6	no tree canopy on a deep sandy substrate
SHRUB LAYER	
S1	no shrub stratum on regularly inundated substrates
S2	contains sparse to dense shrub stratum with *M. quinquenervia* and *L. suaveolens* on drier, often sandy substrates.
S3	contains sparse to dense shrub stratum with *C. glauca* and *Baccharis halimifolia* on tidally influenced substrates
S4	moderately dense shrub stratum with *Livistona australis* on moderately moist, clayey substrates
S5	moderately dense shrub stratum with rainforest species on moderately moist, clayey substrates
S6	sparse shrub stratum with *Ficus virens* on moderately moist, clayey substrates
S7	sparse shrub stratum with *E. robusta* on drier substrates
GROUND LAYER	
G1	dense ground stratum dominated by grasses on drier, often tidally influenced, substrates
G2	dense ground stratum dominated by sedges and forest herbs on drier substrates
G3	dense ground stratum dominated by wallum heath on sand
G4	sparse ground stratum containing salt tolerant species on tidally influenced substrates
G5	sparse ground stratum containing aquatic species on regularly inundated, tidally influenced substrates
G6	dense ground stratum dominated by *Blechnum indicum* on substrates regularly inundated by fresh water
G7	sparse ground stratum containing rainforest canopy regeneration on moderately moist, clayey substrates

substantially from site to site, and from environment to environment.

Variation in the structure of the understorey allowed a similar differentiation between wetland environments. While some sites with a shrub stratum consisting of *M. quinquenervia*, *Lophostemon suaveolens*, *Baccharis halimifolia*, and *Casuarina glauca* (communities S2 and S3) had foliage projective cover of 40 to 60%, almost half the sites had no shrub stratum at all. With regards to the ground layer, sites with a dense grass-dominated ground stratum were found on dry, often saline substrates (communities G1), whereas those with a dense fern ground layer (G6) occurred on sites regularly inundated by fresh water. Those communities with a sparse (<20%) ground layer of aquatic or salt tolerant species (G4 and G5), occurred in regularly inundated, but tidally influenced wetland environments.

In contrast to the overstorey, the understorey also varied considerably in terms of species composition. For example, apart from *Baccharis halimifolia* in communities S2 and S3, there were no shared species within the shrub layer among any of the vegetation types. Although the shrub stratum was absent from almost half the sites, a total of 26 species were encountered within this layer.

The number of species encountered within the ground stratum was higher still. The 236 species encountered here represents 10% of all species occurring in the Moreton Region (McDonald and Elsol 1984). As a consequence, there was considerable sharing of species among the vegetation types. Nevertheless, the first axis of the detrended correspondence analysis of sites had a length of 9.6 average standard deviations of species. Assuming that a species appears, rises to its mode, and disappears over a span of about 4 standard deviations (Gauch 1982), a length of 9.6 standard deviations represents more than two full species turnovers along the gradient represented by that axis. Thus, although the species richness of individual sites (α diversity) was not very high (ranging from a minimum of six species to a maximum of 35 per 20 m^2 sampling area), the diversity among sites (β diversity) was much higher and indicative of the large ecological amplitude of *M. quinquenervia*.

Non-parametric analyses of variance among vegetation types and environmental attributes (soils, water levels) show that, of all vegetation strata, the ground stratum and then the density and size of the stems, appear to respond most strongly to most environmental variables, while the shrub stratum and the canopy stratum respond least (Zoete 1997). These analyses suggest that while environmental conditions are indicated by subtle variations in the tree layer, such as degree of abundance of very few species, particularly in the dimensions and number of stems, the presence of particular species assemblages in the ground stratum is a more precise and convenient indicator of site conditions.

WETLAND GOODS AND SERVICES

Over the past two decades attitudes towards tidal wetlands have changed as decision makers and the public have accepted that these ecosytems provide a range of goods and services free of charge, such as fish habitats and coastal protection. The economic implications of mangrove destruction have been appreciated by linking these functions to the value of the fisheries industry and the engineering costs of artificial erosion controls. Freshwater wetlands also provide a range of goods and services which are at least as valuable in economic terms (Dugan 1990). However, public awareness of these values has not yet reached such a level as to prevent continuing widespread conversion. One reason for this is that the link between polluted water, flooding, saline intrusion into groundwater, or the diversity of songbirds in suburban gardens; and the number, distribution and type of wetlands is, for many, quite subtle.

Nevertheless it is clear to us that the success of nature conservation ultimately depends on how people value natural environments. While the intrinsic value of native species and native ecosystems may be sufficient for some people, for many hard planning decisions a broader range of values, perhaps with clear economic significance, will be required. The difficulty of determining these values is widely recognized. In North America Adamus (1987) has developed a rapid assessment method termed the Wetland Evaluation Technique which is now commonly adopted by public land management agencies (Leitch and Shabman 1988).

In this research we have adapted The Wetland Evaluation Technique as the method to evaluate a range of ecological functions, which we believe also occur within the *M. quinquenervia* forested wetlands of south-east Queensland. The transferability of the technique to south-east Queensland was evaluated by a Delphi questionnaire (Linstone and Turoff 1975) distributed to an expert panel of local scientists. This analysis is detailed in Zoete (1997). Apart from findings pertaining to aquatic bird habitat functions — expected to be region specific — there was broad overlap between the results from the Wetland Evaluation Technique and the Delphi survey.

Wetland Evaluation Technique was applied to the 30 sample sites shown in Figure 1. Table 3 shows the Wetland Evaluation Technique ratings of the probability that particular ecological functions are operating in the various melaleuca wetland types. *M. quinquenervia* wetlands, in contrast to other wetlands do not appear to contribute significantly to the *provision of aquatic diversity*, as they are generally too dry (at least periodically) to provide habitat for aquatic organisms. Non-parametric analysis of variance, as measured by Kruskal-Wallis values, shows that most of the functions shown in Table 3 are significantly linked to variation in three or four strata of the vegetation. The only functions not strongly reflected in the analysis are

Table 3. Wetland Evaluation Technique relative probability ratings showing the environmental goods and services provided by *M. quinquenervia* wetlands (ns = not statistically significant, $p \geq 0.05$) (derived from analyses in Zoete 1997).

Group	Groundwater recharge	Groundwater discharge	Floodflow alteration	Sediment stabilization	Sediment and toxicant retention	Nutrient removal and transformation	Production export
GROUND LAYER							
G 1	low	low	moderate	low	ns	moderate	low
G 2	low	low	moderate	low	ns	low	low
G 3	high	low	moderate	low	ns	low	low
G 4	moderate	moderate	moderate	low	ns	high	low
G 5	low	high	low	high	ns	moderate	moderate
G 6	low	moderate	moderate	low	ns	moderate	low
G 7	low	moderate	low	high	ns	moderate	moderate
SHRUB LAYER							
S 1	ns	moderate	modertate	low	moderate	moderate	low
S 2	ns	low	moderate	low	moderate	low	low
S 3	ns	moderate	low	low	moderate	high	low
S 4	ns	low	moderate	low	high	high	low
S 5	ns	moderate	low	high	high	moderate	moderate
S 6	ns	low	low	low	low	high	low
S 7	ns	moderate	moderate	low	high	low	low
CANOPY							
C 1	ns	low	moderate	low	moderate	moderate	low
C 2	ns	moderate	moderate	low	moderate	moderate	low
C 3	ns	moderate	low	low	high	high	low
C 4	ns	moderate	low	high	high	moderate	moderate
C 5	ns	moderate	moderate	low	moderate	low	low
C 6	ns	low	moderate	low	high	low	low
STEMS							
D 1	ns	low	moderate	low	moderate	moderate	low
D 2	ns	moderate	high	moderate	moderate	moderate	moderate
D 3	ns	moderate	moderate	low	moderate	low	low
D 4	ns	low	moderate	low	moderate	moderate	low
D 5	ns	moderate	low	moderate	moderate	moderate	low
D 6	ns	moderate	low	high	high	moderate	moderate
D 7	ns	low	moderate	low	high	low	low

those related to groundwater recharge, and sediment and toxicant retention.

The results clearly show that there is variation in the functioning of different types of *M. quinquenervia* dominated communities, much of which parallels the variation in environmental conditions. Vegetation groups from the wetter sites (e.g., sites with high *M. quinquenervia* basal areas (D2, D5 or D6), dense *M. quinquenervia* canopy foliage projective cover (C2 or C3), rainforest (D6/C4), or very little ground cover (G4, G5 or G7)) tended to score higher means for most functions than those containing the drier sites (e.g., sites with *Eucalyptus intermedia* or *E. signata* (D4), low *M. quinquenervia* foliage projective cover (C1), grasses (G1), sedges (G2), or heath (G3)), except for groundwater recharge and floodwater alteration.

Only a few functions were moderately or highly likely to be performed by vegetation groups occurring on sandy substrates (e.g., sites with *M. quinquenervia* heath and other heath and sedge species (D7, C6, G2, or G3), reflecting the inability of these substrates to hold water or nutrients relative to clay soils. However, sites with sandy substrates are more likely to recharge groundwater, than sites with clay substrates.

Sites with rainforest (e.g., D6/C4, S5 or G7) tended to be more likely to perform more functions than other vegetation groups, possibly a reflection of the optimal moisture conditions and associated clay soils. Vegetation groups occurring on substrates affected by salinity (e.g., sites with many thin stems (D5), containing *C. glauca* in the canopy (C3) or shrub stratum (S3) were more likely to perform the sediment and toxicant retention and nutrient removal and transformation functions than most vegetation groups on non-saline soils, reflecting the position of these saline sites lower in the catchment.

Each function was significant at the 95% level in at least one vegetation stratum. Most indicative were the ground and stem strata. The mean significance level of the Kruskal-Wallis values of these strata was the lowest of all strata. Whereas the mean significance level of the ground stratum and stem strata was 0.04 and 0.03 respectively, the levels for the canopy and the shrub strata were 0.15 and 0.16, respectively (Zoete 1997). This parallels the results obtained in the ecological study, in which the species composition and structure of the ground stratum and stem dimensions were found to be more responsive to environmental data than the composition and structure of the shrub and canopy layers. As the species composition and structure of the vegetation in the ground stratum is relatively easily assessed compared to the dimensions of stems, it suggests that the ground stratum could be used as a convenient indicator of site conditions and wetland functioning. This requires a shift in the current emphasis on canopy cover data.

DISCUSSION

In order to make more informed decisions on the use of *M. quinquenervia* wetlands, the goods and services provided by the wetland need to be known. The widespread realization that wetlands perform a range of economically valuable ecological functions is comparatively recent, attesting to the difficulty many people have in observing non-market economic benefits. By their very nature ecosystems are not simple, since they are a manifestation of the interaction of many factors. Site specific data on functions are, therefore, expensive and time consuming to obtain from the field and are usually not available. Even if comprehensive data were available, experience suggests that it is rarely in a form easily accessible to decision makers.

Environmental Impact Assessment has become the principal administrative instrument through which ecological insights are introduced to decision makers. However, the usual constraints of time, money and expertise invariably limit Environmental Impact Assessment to a selected range of *important* attributes (e.g., United States Atomic Energy Commission 1973). Typically there is an emphasis on biological species and their habitats. Beanlands (1988) notes: *"The ecological implications of a proposed development usually get translated into effects on physical and biotic resources valued by man for commercial, recreational, or aesthetic purposes. From the perspective of an ecologist, more profound changes to the intrinsic structure and function of natural systems may be involved, but their significance probably will be evaluated by the public in terms of the implications for such resources"*. Since the linkage between ecological implication and cause is frequently complex, spatially wide ranging, and obscure, Environmental Impact Assessment is often distinctly *satisficing* in character (satisfactorily sufficing — March and Simon 1958) and secondary effects are neglected (de Groot 1992).

Recent research on wetlands has shown that the value of many environmental functions and hazards can be substantial in terms of the level of costs and benefits involved, (e.g., Maltby 1986), suggesting that the scope of Environmental Impact Assessment should be expanded to take these into account. There is a need to introduce ecological insights into the decision making mechanism in a simplified form and at a minimum cost. Whenever land use decisions are taken which interfere with ecological functions, there are costs to society, as goods and services are lost.

The research reported in this chapter suggests that wetland functions can be simply assessed (at least to a level justifying the application of the *precautionary principle*) by using vegetation indicators. Since plants are sedentary, rich in species with different niches, and are easily detected in the field, they represent good indicators of site conditions. Plants have been used to indicate levels of pollution (Hellawell 1986; Bohác and Ruzicka 1989; Weinert 1991); changes in climate (Retuerto and Carballeira 1991); levels of recreational use (Rysin 1989); agricultural and forestry potential (Clements 1920; Cajander 1949); and the suitability of land uses in general (Ignatyev 1968; Naveh and Liebermann 1984).

The use of vegetation to indicate the presence or absence of wetland functions has been surprisingly limited. This is despite a widespread recognition that the essential ecological processes occurring within wetlands, such as hydrology, geochemical processes in soils, and biological processes, are reflected in the vegetation (e.g., Gosselink and Turner 1978; Theriot 1988; Breen 1990; Conner *et al.* 1990; Malanson 1993; Murphy *et al.* 1994). In the United States, for example, hydrophytic vegetation is used to delineate wetlands under Section 404 of the *Clean Water Act* (Tiner 1989, 1993; Tammi 1994). Firth and Hooker (1989) showed that artificial changes in the hydrological regime of the Steel Creek wetlands (South Carolina, United States) caused changes in vegetation from a high diversity of tree and vine taxa in undisturbed sites to a high diversity of shrubs and herbs in the disturbed sites. These authors noted *"the functional effectiveness of wetlands is linked to their plant community structure. With the community structure rapidly shifting at the disturbed sites, it is likely that animal habitat, flood control, sediment stabilization, and pollutant filtration were all adversely affected in the Steel Creek corridor"*.

A functional evaluation approach such as the one outlined in this chapter, which uses characteristics of the vegetation to indicate wetland functioning, allows decision makers to direct the course of data gathering and other research in a flexible manner. The priority for data collection is a quantitative inventory of the species present in the ground stratum, as it is this component of the community which appears to be most responsive to environmental conditions. Such an indicator approach can be used as a framework for scoping in Environmental Impact Assessment, as a very cheap method to provide input information to a partial cost-benefit analysis, or as a preliminary filter for further quantitative studies for a more comprehensive analysis.

Use of plant indicators alone for predicting wetland functions will not substitute for more intensive and broader ranging studies necessary for quantifying functions or refining intervention designs. There is always a trade-off between reliability and cost. However, the results shown in Table 3 demonstrate that at a 95% probability level vegetation indicators do appear to offer a rapid, qualitative method of assessment of the functional values of *M. quinquenervia* wetlands, which is reliable.

Any method which avoids the need for time-consuming, intensive site investigations every time a proposal for wetland conversion is made, reduces the proponent's risk of over-investing in an endeavour which may be refused planning permission. An indicator approach, which identifies what ecological functions are likely to be present, offers an efficient means of scoping the potential environmental impacts which will benefit both the proponent and the decision making body.

The potential economic value of the goods and services which all types of *M. quinquenervia* wetlands appear to provide, suggests that economic and social returns to the people of south-east Queensland would need to be substantial in order to justify the losses which continued conversion of these wetlands will produce.

CONCLUSION

The research has shown that, even though the canopy of forested wetlands may be dominated by a single species, often to the exclusion of all other species, diversity in growth form of canopy trees, and in the species composition and structure of the groundlayer, is closely related to variation in environmental site conditions and economically valuable ecological functions.

M. quinquenervia dominated forested wetlands are not a uniform ecosystem. To treat them in this way in land use planning is to ignore ecological and economic values which may eventually prove to be very expensive to restore. In traditional conservation terms, the substantial β diversity of the groundlayer would alone represent a compelling argument for protection of a larger range of wetlands than is presently the case. Conversely, there needs to be a greater awareness in wetland restoration that some environmental site conditions support much more complex species associations than others.

From the perspective of conservation of ecological processes, the fact that each of the vegetation types was economically valuable for at least one ecological function, indicates that compelling arguments for conservation can be put to land use decision makers that are based on tangible goods and services which these wetlands provide directly to people. While this study did not have the resources to put a financial value on the goods and services, experience of other wetland systems — most notably mangroves — has shown that ecological functions are an effective argument for conservation (Hamilton and Snedaker 1984). As a convenient indicator of the presence of ecological functions, the ground stratum should therefore be afforded more attention than often has been the case in the past.

ACKNOWLEDGEMENTS

This chapter forms part of a study into the conservation of the *Melaleuca* forested wetlands in south Queensland, which was conducted between 1992 and 1996. The support of Australian Housing and Urban Research Institute and the Faculty of Built Environment and Engineering at the Queensland University of Technology, the School of Natural and Rural Systems Management at The University of Queensland and WBM Australia, is gratefully acknowledged. Comments offered by the reviewers John Craig, Mark Bellingham and Graham Ussher were also appreciated.

REFERENCES

Adamus, P. R., Clairain, E. J., Smith, R. D. and Young, R. E., 1987. Wetland Evaluation Technique (WET), Volume 2: Methodology. Waterways Experiment Station: Corps of Engineers, Department of the Army, Vicksburg, Mississippi, USA.

Barbier, E., 1993. Economics and Ecology: New Frontiers and Sustainable Development. Chapman and Hall: London, UK.

Barkman, J. J., 1978. Synusial Approaches to Classification. Pp. 111–66 *in* Classification of Plant Communities ed by R. H. Whittaker. Junk Publishers: The Hague, The Netherlands.

Beanlands, G., 1988. Scoping Methods and Baseline Studies. Pp. 33–46 *in* Environmental Impact Assessment: Theory and Practice ed by P. Wathern. Routledge: London, UK.

Beckmann, G. G., Hubble, G. D. and Thompson, C. H., 1987. The Soil Landscapes of Brisbane and South-eastern Environs. Soil and Land Use Series No. 60. CSIRO: Melbourne, Victoria.

Beckmann, G. G., 1967. Soils and Land Use in the Beenleigh-Brisbane Area, South-Eastern Queensland. Soils and Land Use Series No. 50. CSIRO: Melbourne, Victoria.

Blake, S. T., 1968. A Revision of *Melaleuca leucadendron* and its Allies (Myrtaceae). Contributions from the Queensland Herbarium No.1. Queensland Herbarium, Department of Primary Industries: Brisbane, Queensland.

Bohác, J. and Ruzicka, V. (eds), 1989. Bioindicatores Deteriorisationis Regionis: Proceedings of the 5th International Conference. South Bohemian Biological Centre, Ceské Budcjovice, Czechoslovakia.

Bradfield, G. E. and Scagel, A., 1984. Correlations among Vegetation Strata and Environmental Variables in Subalpine Spruce-Fir Forests, Southeastern British Columbia. *Vegetatio* **55**: 105–14.

Bratton, S. P., 1975. A Comparison of the Beta Diversity Functions of the Overstory and Herbaceous Understory of a Decideous Forest. *Bull. Torrey Bot. Club* **102**(2): 55–60.

Breen, P. F., 1990. Hydrology, Structure and Function of Natural Wetlands. *In* Wetlands: Their Ecology, Function, Restoration, and Management, Proceedings of the Applied Ecology and Conservation Seminar Series Wildlife Reserves, La Trobe University October — December 1989. Wildlife Research: La Trobe University, Bundoora, Victoria.

Cajander, A. K., 1909. Über Waldtypen. *Acta For. Fenn.* **1**(1): 1–75.

Cajander, A. K., 1949. Forest Types and their Significance. *Acta For. Fenn* **56**(4): 1–71.

Capelin, M. A., 1987. Horticulture Land Suitability Study, Sunshine Coast, Southeast Queensland. Land Res. Bull. QV87001, Queensland Department of Primary Industries: Brisbane, Queensland.

Catterall, C. P., 1991. Briefing Notes on the Regional Ecological Significance of the Dèagon Melaleuca Woodlands. Paper prepared for the Queensland Conservation Council, Brisbane, Queensland.

Clements, F. E., 1920. Plant Indicators: The Relation of Plant Communities to Process and Practice. Carnegie Institution of Washington: Washington, USA.

Conner, W. H., 1990. Composition and Productivity in Bottomland Hardwood Forest Ecosystems: The Report of the Vegetation Workgroup. Pp. 455–79 *in* Ecological Processes and Cumulative Impacts, Illustrated by Bottomland Hardwood Wetland Ecosystems ed by J. G. Gosselink, L. C. Lee and T. A. Muir. Lewis Publishers: Chelsea, USA.

Davie, J. D. S., 1991. The Ecological Significance of the Deagon Wetlands to the Brisbane and South-East Queensland Region. Paper presented to the Land and Environment Court for the Queensland Conservation Council. Queensland University of Technology: Brisbane, Queensland.

Davie, J., 1997. Is biodiversity really the link between conservation and ecologically sustainable management? A reflection on paradigm and practice. *Pac. Conserv. Biol.* **3:** 83–90.

De Groot, R. S., 1992. Functions of Nature: Evaluation of Nature in Environmental Planning, Management and Decision Making. Wolters — Noordhoff: Groningen, The Netherlands.

Del Moral, R. and Watson, A. F., 1978. Gradient Structure of Forest Vegetation in the Central Washington Cascades. *Vegetatio* **38**(1): 29–48.

Dowling, R. M. and McDonald, W. J. F., 1976. Moreton Region Vegetation Map Series: Brisbane. Queensland Department of Primary Industries: Brisbane, Queensland.

Dugan, P. (ed), 1993. Wetlands in Danger. IUCN/Mitchell Beazley: London, UK.

Elsol, J. A. and Dowling, R. M., 1978. Moreton Region Vegetation Map Series: Beenleigh. Queensland Department of Primary Industries: Brisbane, Queensland.

Elsol, J. A. and Satler, P. S., 1979. Moreton Region Vegetation Map Series: Caloundra. Queensland Department of Primary Industries: Brisbane, Queensland.

Firth, P. L. and Hooker, K. L., 1989. Plant Community Structure in Disturbed and Undisturbed Forested Wetlands. Pp. 101–14 *in* Wetlands: Concerns and Successes ed by D. W. Fisk. American Water Resources Association: Bethesda, Maryland, USA.

Gams, H., 1918. Prinzipienfragen der Vegetationsforschung: Ein Beitrag zur Begriffserklärung und Methodik der Biocoenologie. *Vjschr. naturf. Ges. Zürich* **63:** 293–493.

Gauch, H. G., 1982. Multivariate Analysis in Community Ecology. Cambridge Studies in Ecology, Cambridge University Press: Cambridge, UK.

Gilliam, F. S., Turill, N. L. and Adams, M. B., 1995. Herbaceous-Layer and Overstory Species in Clear-Cut and Mature Central Appalachian Hardwood Forests. *Ecol. Appl.* **5**(4): 947–55.

Gosselink, J. G. and Turner, R. E., 1978. The Role of Hydrology in Freshwater Wetland Ecosystems. Pp. 63–78 *in* Freshwater Wetlands: Ecological Processes and Management Potential ed by R. E. Good, D. F. Whigham and R. L. Simpson. Academic Press: New York, USA.

Hamilton, L. S. and Snedaker, S. D. (eds), 1984. Mangrove Management Handbook. IUCN/ EWC/ UNEP/UNESCO: Honolulu, Hawai, USA.

Hellawell, J. M., 1986. Biological Indicators of Freshwater Pollution and Environmental Management. Elsevier Applied Science Publishers: London, UK.

Ignatyev, G. M., 1968. Classification of Cultural and Natural Vegetation Sites as a Basis for Land Evaluation. Pp. 104–11 *in* Land Evaluation: Papers of a CSIRO Symposium Organized in Cooperation with UNESCO 26–31 August 1968 ed by G. A. Stewart. MacMillan of Australia: South Melbourne, Victoria.

Leitch, J. A. and Shabman, L. A., 1988. Overview of Economic Assessment Methods Relevant to Wetland Evaluation. Pp. 95–102 *in* The Ecology and Management of Wetlands, Volume 1: Ecology of Wetlands ed by D. D. Hook, W. H. McKee, H. K. Smith, J. Gregory, V. G. Burrell, M. R. DeVoe, R. E. Sojka, S. Gilbert, R. Banks, L. H. Stolzy, C. Brooks, T. D. Matthews and T. H. Shear. Croom Helm Ltd: Beckenham, UK.

Linacre, E. and Hobbs, J., 1977. The Australian Climatic Environment. John Wiley and Sons: Milton, Queensland.

Linstone, H. A. and Turoff, M. (eds), 1975. The Delphi Method: Techniques and Applications. Addison-Wesley: Reading, Massachusetts, USA.

Malanson, G. P., 1993. Riparian Landscapes. Cambridge University Press: Cambridge, UK.

Maltby, E., 1986. Waterlogged Wealth: Why Waste the World's Wet Places? International Institute for Environment and Development: London, UK.

March, J. and Simon, H., 1958. Organizations. Wiley and Sons: New York, USA.

McCune, B. and Antos, J. A., 1981. Correlations Between Forest Strata in the Swan Valley, Montana. *Ecology* **62**(5): 1196–204.

McDonald, W. J. F. and Elsol, J. A., 1984. Moreton Region Vegetation Map Series, Summary Report and Species Checklist for Caloundra, Brisbane, Beenleigh and Murwillumbah Sheets. Queensland Department of Primary Industries: Brisbane, Queensland.

McNaughton, S. J., 1989. Ecosystems and Conservation in the Twenty-First Century. Pp. 109–20 *in* Conservation for the Twenty-first Century ed by D. Western and M. Pearl. Oxford University Press: Oxford, UK.

Murphy, K. J., Castella, E., Clément, B., Hills, J. M., Obrdlik, P., Pulford, I. D., Schneider, E. and Speight, M. C. D., 1994. Biotic Indicators of Riverine Wetland Ecosystem Functioning. Pp. 659–82 *in* Global Wetlands: Old and New ed by W. J. Mitsch. Elsevier Science: B.V., The Netherlands.

Naveh, Z. and Lieberman, A. S., 1994. Landscape Ecology: Theory and Application. Second Edition. Springer-Verlag: New York, USA.

Nguyen, L. M., Cook, J. G. and McBride, G. B., 1995. Phosphorous retention and release characteristics of sewage-impacted wetland sediments receiving reactants from natural wetlands. Wetlands for Water Quality: Conference Papers. James Cook University, Townsville, 25–29 September 1995. Queensland Department of Primary Industries: Brisbane, Queensland.

Noss, R. F., 1990. Indicators for Monitoring Biodiversity: A Hierarchical Approach. *Conserv. Biol.* **4:** 355–64.

Pearce, D., Markandya, A. and Barbier, E. B., 1989. Blueprint for a Green Economy. Earthscan: London, UK.

Queensland Department of Environment and Heritage, 1992. Management Planning for National, Environmental and Marine Parks of Bribie Island and Pumicestone Passage. Brisbane, Queensland.

Retuerto, R. and Carballeira, A., 1991. Numerical Estimation of Climate from Indicator Plants. Pp. 191–206 *in* Bioindicators and Environmental Management ed by D. W. Jeffrey and B. Madden. Academic Press: London, UK.

Rheinhardt, R. D., 1992. Disparate Distribution Patterns between Canopy and Subcanopy Life-Forms in Two Temperate North American Forests. *Vegetatio* **103:** 67–77.

Ricklefs, R. E., Naveh, Z. and Turner, R. E., 1984. Conservation of Ecological Processes. Commission of Ecology Papers Number 8. IUCN: Gland, Switzerland.

Regional Planning and Advisory Group, 1993. Creating our Future: Towrds a Framework for Growth Managment in South East Queensland. Regional Planning and Advisory Group: Brisbane, Queensland.

Rysin, L. P., 1989. The Resistance of Forest Plants and Communities to the Recreation. Pp. 65–66 *in* Bioindicatores Deteriorisationis Regionis: Proceedings of the 5th International Conference, Volume 1 ed by J. Bohác and V. Ruzicka. South Bohemian Biological Centre: Ceské Budcjovice, Czechoslovakia.

Stick, M. M. and Lergessner, D. A., 1969. Conservation of Plant Communities on the Near North Coast, Queensland. *Geography IVH:* 13–19.

Tammi, C. E., 1994. Onsite Identification and Delineation of Wetlands. Pp. 35–54 *in* Applied Wetlands Science and Technology ed by D. M. Kent. CRC Press: Boca Raton, Florida, USA.

Theriot, R. F., 1988. Flood Tolerance Indices for Palustrine Forest Species. Pp. 477–88 *in* The Ecology and Management of Wetlands, Volume 1: Ecology of Wetlands ed by D. D. Hook, W. H. McKee, H. K. Smith, J. Gregory, V. G. Burrell, M. R. DeVoe, R. E. Sojka, S. Gilbert, R. Banks, L. H. Stolzy, C. Brooks, T. D. Matthews and T. H. Shear. Croom Helm Ltd: Beckenham, UK.

Tiner, R. W., 1993. Field Recognition and Delineation of Wetlands. Pp. 153–98 *in* Wetlands: Guide to Science, Law, and Technology ed by M. S. Dennison and J. F. Berry. Noyes Publications: Park Ridge, New Jersey, USA.

Tiner, R. W., 1989. An Update of Federal Wetland Delineation Technique. Pp. 13–24 *in* Wetlands: Concerns and Successes ed by D. W. Fisk. American Water Resources Association: Bethesda, Maryland, USA.

Tisdell, C., 1991. Economics of Environmental Conservation. Developments in Environmental Economics, Volume 1, Elsevier: Amsterdam, The Netherlands.

United States Atomic Energy Commission, 1973. Preparation of Environmental Reports for Nuclear Power Plants. Regulatory Guide 4.2, United States Atomic Energy Commission Regulatory Guide Series. United States Atomic Energy Commission Directorate of Regulatory Standards: Washington DC, USA.

Weinert, E., 1991. Biomonitoring of Environmental Change Using Plant Distribution Patterns. Pp. 179–90 *in* Bioindicators and Environmental Management ed by D. W. Jeffrey and B. Madden. Academic Press: London, UK.

Whittaker, R. H., 1960. Vegetation of the Siskiyou Mountains, Oregon and California. *Ecol. Monogr.* **30:** 279–338.

Whittaker, R. H. (ed.), 1978. Ordination of Plant Communities. Junk Publishers: The Hague, The Netherlands.

Young, P. A. R. and Cotterell, M. A., 1993. Conservation Assessment, SEQ 2001 Region. Department of Environment and Heritage: Brisbane, Queensland.

Zoete, T., 1997. Conservation in the face of rapid change: *Melaleuca quinquenervia* wetlands in SE Queensland. Ph.D Thesis, Queensland University of Technology, Brisbane, Queensland.

CHAPTER 66

Sustainable development and biodiversity conservation in Wuyishan Biosphere Reserve, Fujian Province, China

Ren Zhuge,[1] Terry De Lacy,[2] Johannes Bauer[3] and Haohang Wu[4]

Conflicts between traditional forest resource management systems and demands for modern commodity production frequently result in degradation and loss of biodiversity. This is particularly true where traditional multiple use systems are being replaced by single species systems. In Wuyishan Biosphere Reserve, China, the reallocation of property rights from the state to individual households has encouraged thriving micro-industries in bamboo which provides the opportunity for householders and reserve administration to collaborate in managing the resources of the reserve to promote sustainable local development and conservation of biodiversity. However, this move to bamboo monoculture is posing a threat to the region's highly diverse flora and fauna. This research investigates the interactions between national policy reform, community institutions, and the local, regional and national economy, in Wuyishan biosphere reserve. A regional joint-stock company with more extensive collaboration is proposed to reduce negative impacts of bamboo monoculture on biodiversity conservation. An ecological environmental compensation system to reduce the direct consumption of natural resources and disturbance to the natural system in the protected areas is also proposed.

INTRODUCTION

PROTECTED area management is undergoing a significant redefinition in the developing world. Traditional concepts of biodiversity conservation that prohibited use of protected areas, have resulted in frequent conflicts between protected area management authorities and local communities. Experience has shown that the closed protection model without involving the local community is not successful (Furze, De Lacy and Birckhead 1996). This is especially so in regions with high population density and immense pressures for human development. The United Nations Education, Scientific and Cultural Organization Man and the Biosphere reserve model which deliberately seeks to integrate conservation with local development is seen by some as a more appropriate protected area model in such regions, than the traditional closed United States national park model. During the second International Conference on Biosphere Reserves held in Seville in March 1995, two important documents, the Seville Strategy for Biosphere Reserves and Statutes of the World Network of Biosphere Reserves, were developed as the basis for biosphere reserves in the 21st century (Zhao 1996). The biosphere reserve model is increasingly seen as an important means of resolving conflicts over resource use and integrating sustainable development of local communities with biodiversity conservation.

China has drawn up "China's Agenda 21" and "China Biodiversity Conservation Action Plan". More than 1 000 nature reserves, national parks and forest parks have been set aside covering a total land area of 8% (National Environmental Protection Agency 1996). However, like all other developing countries, China's biodiversity conservation has to face handicaps of pressure on the environment from economic development, shortage of natural

[1]School of Natural and Rural Systems Management, The University of Queensland, Queensland, Australia 4072.
[2]CRC for Sustainable Tourism Pty Ltd, Australia.
[3]International Conservation Program, Johnstone Centre, Charles Sturt University, Australia.
[4]Global Environment Facility Program of Wuyishan Nature Reserve, Fujian, China.

Pages 644–52 *in* NATURE CONSERVATION 5: NATURE CONSERVATION IN PRODUCTION ENVIRONMENTS: MANAGING THE MATRIX ed by J. L. Craig, N. Mitchell and D. A. Saunders. Surrey Beatty & Sons, 2000.

resources, insufficient financial support from government, and lack of incentives for the participation of local people in surrounding regions. As an acceptable and effective model the biosphere reserve concept, which was first introduced to China's nature reserve system in 1979, has been accepted more and more widely within the country. Currently United Nations Education, Scientific and Cultural Organization has designated 12 of China's nature reserves as biosphere reserves. One of them is Wuyishan Nature Reserve in Fujian Province.

In recent years Wuyishan Nature Reserve has actively taken the opportunity provided by the market to develop the local economy and to make efforts for sustainable use of natural resources. The reserve has practised an efficient policy to support a village based bamboo harvesting and micro manufacturing industry. Since the middle of the 1980s remarkable achievements have been obtained in developing the local economy, but the fundamental question of whether community development is compatible with the mission of nature conservation is still not clear. This chapter reports on research into integrated conservation and development in Wuyishan Nature Reserve. It describes changes in the bamboo industry of Wuyishan Nature Reserve and suggests implications for sustainable development and biodiversity conservation through analysis of community decision making systems, property rights, ecosystem change and ecotourism.

BACKGROUND

Wuyishan Nature Reserve is located in the northwestern part of Fujian Province (Fig. 1). Its total area is 56 527.4 ha, of which 34 771.3 ha are designed as core protected zone, 911.3 ha for scientific and educational purposes and 20 844.8 ha is an "experimental zone" similar to a multi-use zone in protected areas in other parts of the world. The highest peak, Huanggangshan, is 2 158 m above sea level and the lowest place is only 350 m in altitude. The average altitude in the reserve is 1 200 m. Most areas in the reserve have steep slopes over 30–40 degrees with strong gorges along river valleys. Some are 500–1 000 m in depth. Jiuquxi River, Mayanxi River, Beixi River and Qianshanhe River originate from this region.

Subtropical monsoon effects the whole nature reserve. Evergreen broad-leaf forest covers 95.3% of the land with 2 466 species of higher plants recorded, belonging to 267 families and 1 028 genera in this region. The forest is distributed in the hilly areas between 350–1 400 m with the biggest biomass in the reserve. There is a described fauna of 5 614 species, belonging to 66 orders and 448 families, including 71 mammals, 256 birds, 73 reptiles, 35 amphibians, 40 fishes. *Neofelis nebulosa* Clouded Leopard, *Panthera pardus* Leopard, *Panthera tigris amoyensis* South Chinese Tiger, *Muntiacus crinifrons* Black muntjac, *Ciconia nigra* Back Stork, *Mergus squamatus* Chinese Merganser, *Tragopan caboti* (a rare pheasant), *Syrmaticus ewllioti* Long-tail Pheasant and *Teinopalpus aureus* (a rare butterfly) are Class A endangered species. Another 48 protected species including some well-known animals such as *Macaca thibetana* Stump-tailed Macaque, *Capricornis sumatraensis* Mainland Serow, *Lophura nycthemera* Silver Pheasant and

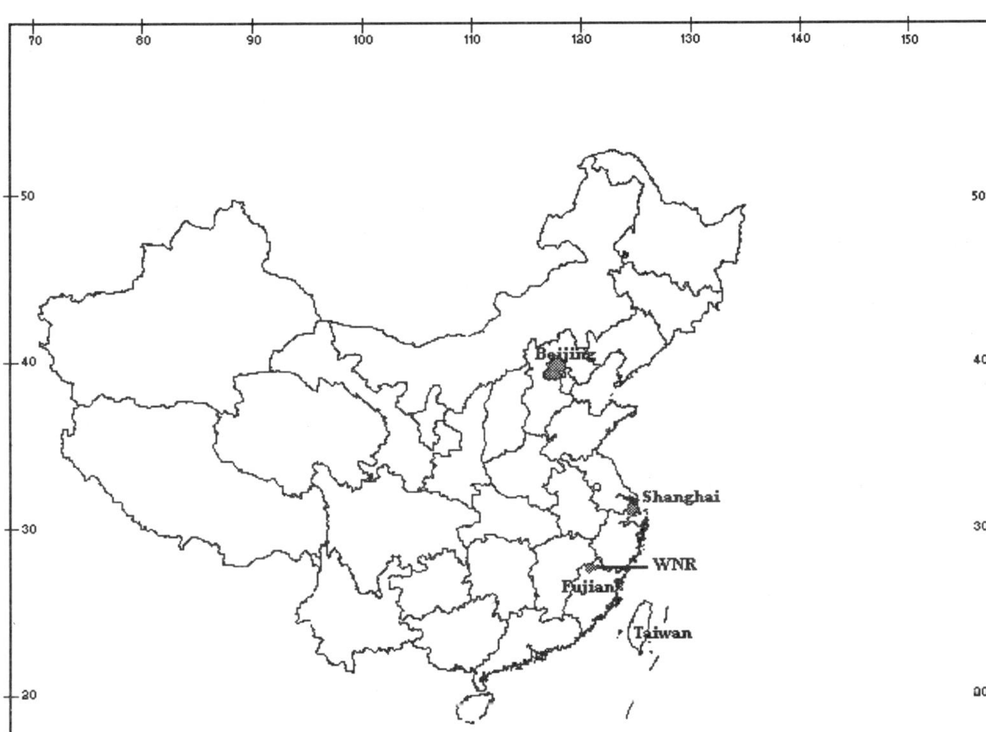

Fig. 1. The location of the Wuyishan Nature Reserve, Fujian Province, China.

Selenarctos thibetanus Asiatic Black Bear occur. Totally, there are 57 animals listed as national protected species. In addition, there are 81 species of migratory birds belonging to the China-Japan migratory bird protection net and 20 species of migratory birds in the China-Australia net (Wuyishan Nature Reserve Administration 1997a).

In the 1950s, this region was managed by the Nanping Prefecture of Fujian Province and, as with most sparsely populated forestlands, was regarded as restricted no-felling areas. Timber production activities on a large scale happened in 1960s—1970s. During that period, there were four state-run timber stations and several local community-owned timber and bamboo stations in operation. It was not until the establishment of the nature reserve in 1979 that all those stations were dismantled according to the order No. 169 of the State Council. Since then scientific survey and planning has taken place. The Fujian provincial government promulgated a management regulation for the reserve and an administrative organization at a county level began to operate. There are 42 natural villages with 6 000 people living in the reserve (Wuyishan Nature Reserve Administration 1997a).

The reserve was accepted as the fifth Biosphere Reserve in China by United Nations Education, Scientific and Cultural Organization's Man and the Biosphere Program in 1987 (Wuyishan Nature Reserve Administration 1997a). It also became one of the five nature reserves funded by the Global Environment Facility of the World Bank in 1995. A grant of US$ one million has been put in the reserve for developing a management plan, improving management conditions, training staff, establishing a Geographic Information System and promoting joint management with local communities. Although basic surveys of the natural resources and environment have been conducted, there has been very little work carried out on the distribution, status or conservation biology of individual species, the integration of local development with biodiversity conservation and the participatory management with local communities in the reserve. A joint research project between University of Queensland and Charles Sturt University in Australia and the Chinese Ministry of Forestry and Wuyishan Nature Reserve, funded by the Australian Research Council and Global Environment Facility, is investigating these issues (De Lacy and Bauer 1997).

METHODS

A Rapid Rural Appraisal was undertaken by the authors in May and June 1997. Secondary information was gathered and analysed; direct observation of forestry, manufacturing, agriculture and conservation activities were made; semi-structure interviews were held with village leaders and individual householders; and a meeting with senior officials of the reserve management authority was conducted.

Ecological surveys of medium and large mammals' status, habitat and distribution in Wuyishan Nature Reserve based on gradsect survey technique were undertaken. Census techniques used were forest transacts, stream transacts, observation points, villager interviews, spotlighting and air tubes. Of the 70 species of mammals recorded in Wuyishan Nature Reserve, 34 medium and large species were chosen for the study. The details of this study are reported separately (Bauer 1998)

A National Conference of China-Man and the Biosphere Network including a review of Wuyishan Biosphere Reserve was held on September 16—22, 1997 in Wuyishan Nature Reserve. During the conference, a workshop on evaluating the joint management of local communities in Wuyishan Nature Reserve was conducted. Management experts from national authorities and scientific institutions and managers from other nature reserves nationwide attended this meeting. One of the research team, Ren Zhuge, participated in the meeting.

DECISION MAKING STRUCTURE AND MANAGEMENT OF BAMBOO PRODUCTION

China has had over two thousand years' history of centralized state power. Hierarchy that is reflected in the nature reserve system of China identified at national level, provincial level, prefecture (municipal) level and county level. Wuyishan Nature Reserve is a national level nature reserve. Its administration is by a county level institution directly subordinate to the Fujian Provincial Department of Forestry and supervised by the National Ministry of Forestry in aspects of policies and management plans. The Ministry of Forestry approves the management plan for national level nature reserves and supervises the implementation of related policies based on national legislation and regulations. The provincial forestry department sets out the staffing quota for the reserve administration, appoints the director and the vice-director of the reserve, and assigns an annual budget and production quotas for timber and bamboo.

Most of China's nature reserves are located in poor areas with limited opportunities for economic development. People in these regions depend heavily on natural resources for subsistence and cash generating activities. Increasing population and economic growth lead to over-exploitation of the land and degradation of the resource base on which the people depend. Pressures then increase to use resources in nature reserves, especially after the collective land use right has been allocated to individual households, and ecological damages of surrounding areas threaten the function of protected ecosystems (National Environmental Protection Agency 1994).

At the beginning of Wuyishan Nature Reserve, the managers followed the related regulations from national level and provincial level authorities to prohibit most productive activities in the reserve, i.e., emphasizing strict protection of nature and excluding local people's demands from nature (Zhao and Han 1996). Unfortunately, it resulted in serious antagonism between the reserve administration and local communities even between local governments of county level and township level (Wuyishan Nature Reserve Administration 1997b). In this context, in 1985 the reserve managers readjusted their policy to allow local people to use forest resources in the experimental zone, and subsequently, bamboo processing in the reserve was permitted in 1989.

Wuyishan Nature Reserve administration has 99 permanent staff members and consists of an administrative office and six sections, namely: protection, scientific management, community management, production management, fire prevention and forest police. Four management branches in charge of checkpoints and watchtowers are located in different villages. In addition, the reserve has its own company for running bamboo processing factories and the ecotourism industry.

The production management section directly controls all the production activities of bamboo such as harvest, processing, transport and trade. The managers in this section give suggestions to the local producers and supervise the operation of bamboo based on the reserve's rules and regulations. The protection section checks annual production plans of bamboo from local villages and the bureau according to the annual change of natural resources in the reserve and the demands of local communities. The community management section is a new section in the bureau. Its role is serving local communities, leading local people to engage in diversified economic activity, helping village leaders to resolve difficulties in bamboo production, and usually improving the life of local people.

Annually reserve managers establish bamboo management plans setting permitted harvest output and locations of quotas of bamboo for processing in the reserve. During the production period, managers have to make an inspection tour to check that production processes are following operation rules. For example, according to the rules, villages must keep 130–140 bamboo living trees in one mu (= 1/15 ha) of bamboo forest and all the remnants of bamboo processing should be burnt in specially designed cremators to avoid potential forest fire and reduce environmental pollution. Furthermore, the managers also have responsibilities to guide safe production and to promote biodiversity conservation through regular meetings with the local community.

Administrative village councils are the lowest administrative organization in rural China. Usually an administrative village governs a few or several tens of natural villages. It depends on the population of those natural villages. A village group is a basic cell that is subordinate to the administrative village. It consists of a big natural village or a few small natural villages. In the reserve region, there are three administrative villages supervised by the Wuyishan Nature Reserve administration in aspects of production and livelihood, but executively, these three villages are subordinate to township governments. For example, Tongmu Administrative Village, the biggest village in the reserve, is subordinate to Xingcun Township. Township government has power to appoint and dismiss a village head. The village council has the responsibility to implement policies and to carry out orders from township government such as birth control, conscription, fiscal levy and so on. The reserve administration concentrates its efforts on developing the economy of local communities and the protection of biodiversity in the reserve.

Under the annual quota control of the Wuyishan Nature Reserve, each village should submit the production plan of bamboo for the coming year to the Wuyishan Nature Reserve administration. When the approved plan is returned to the village, the village council breaks up production quotas to each village group, and then the quotas are allocated to every household in the village group. The village-run bamboo company has priority in getting some production quota, and controls the marketing of bamboo products in the administrative village. According to the stipulation of village council, all bamboo products must be sold to the village-run bamboo company. Individual households have no right to sell their bamboo products to outside companies without the permission of the village council unless they have applied for an independent licence for production of bamboo. For example, in Aotou Village there are two independent private bamboo factories. The owners sell their bamboo products to outside companies directly.

The reserve administration, and local governments at township, county or prefecture levels usually have different objectives for reserve management. For example, local governments assign economic targets to the village according to a gradually increasing basis. They expect the local communities to obtain increasing bamboo benefits through increasing production targets. The reserve administration, which distributes production quotas to village, has to base these quotas on potential impact on conservation. Local governments want the villagers to become rich as quickly as possible and pursue a high achievement (usually measured by annual average per capita income). They may often ignore the conservation objectives of the reserve. They may often stand by the local people in conflicts with the reserve administration, the "outside encroacher". On the other side, the managers have to control the productive activities following regulations and administrative orders from higher authorities. The only way for the reserve managers to relax the tension with local communities is by means of preferential policies to develop the local economy, or return some benefits to local people by supporting public construction such as roads, bridges, telephone line, power stations and so on. But they often wonder whether they have gone too far by encouraging local economic development or acting in excess of authority. The administrative villages have to balance the two decision making systems, management bureau and township.

The economic welfare in the villages is dependent on decisions made by the reserve, but their township governments based on the common vote of villagers appoint the heads of the villages. As a result, the leaders of the villages have to strive for more preferential policies from the reserve authority and pursue higher economic development that may sometimes run counter to the purpose of the reserve.

PROPERTY RIGHTS AND EQUAL USE OF THE NATURAL RESOURCES

According to China's Land Management Law (1987) and Forest Law (1985), which are based on the Constitution of the People's Republic of China, state ownership and collective ownership are the only two existing legitimate forms of land and forest ownership in China. Individuals can only obtain forest tree ownership or land use rights and can contract forest or forestland from both collective owned or state owned lands (Tang and Du 1995). Since 1982, use rights and management rights in most collective lands have been transferred to individual households (the household responsibility system). Of course, the use and management rights of forestlands are limited by the regulations within nature reserves.

The total land area of Wuyishan Nature Reserve is 56 527.4 ha, of which 59.7% belongs to local communities: so called "collective land" as compared to "state-owned land". In the 1980s, most people did not take the new land policy seriously because they had experienced frequent policy changes. When the "household responsibility system" was put into practice, many people in the reserve were sceptical about how long forest resource use would be allocated to them. The collective land was separated into small pieces in a simple way by rough estimation of land area. Some people were unlucky because they were allocated land with infertile soil or fewer trees, but they accepted it without strong complaint since they believed that the policy would change in a short time. As time goes on, those people recognize that they have lost a lot since their welfare depends on the natural resources of the land over which they have been allocated use rights.

At the time we carried out our rural appraisal, bamboo resource allocation was undergoing a review. The bamboo resource is now the focal point of management. The Tongmu Village Council has put forward a preliminary plan based on discussions with village people. The new idea for resource allocation is to calculate the numbers of bamboo shoots in the collective land and to divide the land into equal parcels depending on the numbers of bamboo shoots. It will then be allocated to householders based on the number of people in each household. It is said that this method will reflect more closely the economic value of land. But, the conflicts on benefits distribution between different communities and households will continue unless an effective mechanism for equitable use of the natural resources is established. Just as important is the time frame over which the resource use rights are allocated. If householders are to make resource use decisions based on sustainable criteria, then they need to believe they will have long-term rights to the resource.

Recently, Aotou and Dabo villages have changed the ownership system in the bamboo processing industry and opened their own bamboo joint-stock companies. Local villagers become the stockholders by joining the company and contributing their annual supply of raw materials for processing, together with their labour. This may reduce the conflicts on benefit distribution between administrative village and households and on management of natural resources between conservation and development. Joint-stock company seems to be an effective model for incorporating small-scale production units into a bigger company so as to produce greater efficiency, a broader economy and increase the utilization of natural resources. Furthermore, the joint-stock company can be used to develop a diversified economy based on traditional activities of tea growing, timber harvesting, gathering plants for medicines and so on. This seems to be an effective approach for reducing negative impacts of bamboo monoculture on biodiversity conservation.

LOCAL ECONOMY BASED ON MONOCULTURE OF BAMBOO PRODUCTION

Historically tea, timber and grain products provided essential sources for the life of local people in the Wuyishan area. Before the establishment of the reserve in 1979, this region produced some 60 thousand kg of tea, 440 thousand kg of grain, 31 thousand cubic meters of timber and one million bamboo trees. Approximately 30% of local income came from timber and bamboo production, 45% from tea products, and 25% from grain and others. Average annual income in the three villages (i.e., Tongmu, Aotou, and Dabo) was only 204 yuan (100 Chinese yuan is about 83 US$) per head. The situation has changed quickly since the founding of the reserve. Figure 2 illustrates what has happened to average income in the local communities in the past 15 years.

However, the economic achievements of local communities is not based on the demands of more consumption of natural resources. According to statistics, the annual average harvests of timber and bamboo have dropped down to 2 000 cubic meters of timber and 300 000 bamboo trees from 31 000 cubic meters of timber and 1 million bamboo trees, accounting for 6.4% of timber and 30% of bamboo harvested in 1979 (Wuyishan Nature Reserve Administration 1997b).

The local bamboo industry includes products of original bamboo (Maozhu) tree harvest as well as processing. Before 1989 there was no bamboo processing industry in this area. Local people sold their bamboo trees. In 1989, the reserve administration issued temporary provisions for processing, transporting and marketing of bamboo products in the experimental zone of the reserve. Since then the bamboo resource has played a much more

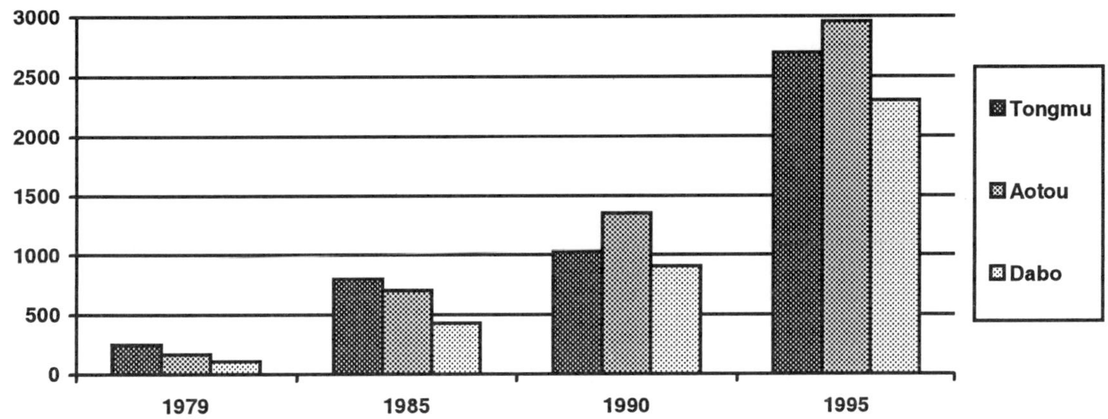

Fig. 2. Average annual per head income in the three villages (Data Source: *Annals of the Wuyishan National Nature Reserve* 1997).

important part in the local economy. For example, Tongmu, the biggest village in the reserve has developed several bamboo processing factories, a tea refinery and flagstone factory. Over 400 people work in these enterprises. The income from processing industry in 1993 was 6.5 million yuan, accounting for 60.2% of total income in this year. Aotou Village developed their bamboo processing factory in 1989 to produce bamboo ware, bamboo chopsticks and bamboo skewers. More than one hundred people become workers of the factory and 44.9% of the annual total income was from the processing industry in 1993 (Fig. 3)

In the income distribution pies, timber and bamboo incomes are from original products. "Processing" includes incomes from processing mostly based on bamboo and a few other raw materials. "Sideline" means incomes from some small scale plantations other than grain and tea plantations such as fruit, oil crops, palm fibre, bamboo shoots and honey collection, animal husbandry and so on (Wuyishan Nature Reserve Administration 1997a). In 1993 bamboo processing incomes accounted for 60.2% and 44.9% of the annual incomes of Tongmu and Aotou respectively. Although bamboo production in Dabo Village was not fully exploited in 1993, the bamboo processing also occupied 1/3 of total annual income.

The thriving development of bamboo processing in the reserve makes an important contribution to the local economy and the living standard of local people has been dramatically improved since the establishment of the Wuyishan Nature Reserve. Their annual average per head incomes in the reserve's villages in 1995 was 2 700 yuan in Tongmu, 2 954 yuan in Aotou and 2 299 yuan in Dabo compared with 249 yuan, 165 yuan and 107 yuan in 1979. There now exists a sound relationship between local people and the reserve authority, after undergoing a severe period of conflicts in the past.

Besides the above mentioned three villages, there are eight other villages with their forestlands in the reserve, but the production and livelihood of these villages are all under their township's governments. In comparison with the three villages in the reserve, these villages have smaller pieces of forests in the reserve and their traditional production areas are mostly outside the reserve. Due to the preferential policies from the reserve, such as low interest loans, low taxes and strong technical support, there exist differences in community economy between reserve-managing villages and local government-managing villages (Fig. 4). However, the poverty in neighbouring communities and the exhaustion of natural resources around the reserve will increase the pressures on biodiversity conservation. That means

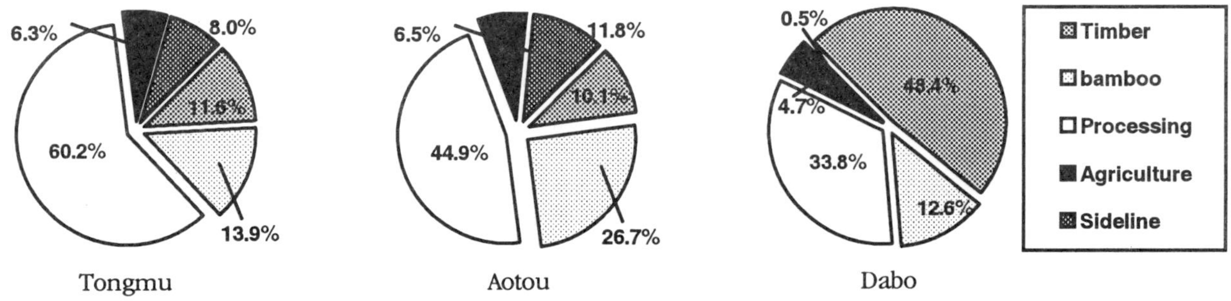

Fig. 3. Distribution of annual incomes in the three villages in 1993 (Data Source: *Annals of the Wuyishan National Nature Reserve* 1997).

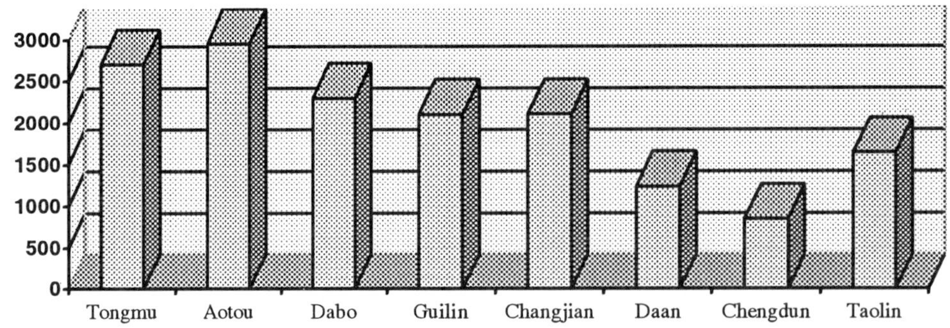

Fig. 4. Average per head incomes of the three reserve-managing villages and other local government-managing villages in 1995 (Data Source: copied from above village councils' *Annual reports on 1995*).

the reserve could become an isolated island ecologically.

THE ENVIRONMENTAL IMPACTS OF BAMBOO MONOCULTURE

Bamboo, taxonomically belonging to bamboo sub-family of Gramineae, is a very special group of higher plants. It is dominantly distributed in the rainy tropics and subtropics in Asia, Africa, America and Australia. However, China is a country with the richest bamboo plants, including about half of the 1 000 bamboo species in the world. Bamboo's rapid growth and non-seed sprouting nature keep green cover over a harvested bamboo forestland that causes less erosion of soil than other forestlands. Usually, from a bamboo shoot to a bamboo tree needs only a few months and about 5–7 years' growth for harvest. Bamboo has almost no seeds in its 60–100 years' growing term until one day it blossoms into death. Bamboo shoots take shape from underground "roots" every year and rapidly expand. Besides its well-known ornamental and aesthetic values in China's history and culture, bamboo is widely used for food and wood.

Since the incentive of profits from bamboo processing make local people take care of bamboo trees in their "responsibility land"; the scattered bamboo trees and mixed forests alongside the villages and by the roads have been gradually changed into bamboo-dominant or pure bamboo forests. Bamboo forests have become one of the most important vegetation types in Wuyishan Nature Reserve, accounting for one seventh of the total reserve area. There are 40 species, including 5 varieties and 2 cultivated varieties, belonging to 11 genera, which consists of 10 types of bamboo forests. In these types of bamboo forests, Maozhu *Phyllostachys pubescens* forest undoubtedly plays a crucial role in the development of local economy, not only for its heavy stock in the reserve but also for its wide uses. Maozhu forest has an area of 7 938 ha with 21.4 million of trees in the reserve, accounting for 15.4% of total forestland. It is mostly distributed in the traditional production zones of local communities. Table 1 shows the distribution of Maozhu resources in the reserve.

The survey conducted by the reserve authority in 1990 indicated an increase of bamboo forests and a decrease of scattered bamboo trees during 1980–1990 (Table 1).

According to He *et al.* (1994), in a mixed bamboo forest (*Phyllostachys pubescens* + *Castanopsis eyrei* + *Acer sinense*), there are 11 tree species in the level of forest crown, 14 species of shrubs in the bush level, and a few vines and herbs. In comparison a secondary bamboo forest has only one species (*Phyllostachys pubescens*) in the level of forest crown, 13 species in the level of bush and two species of herbs. There is no other tree or shrub under the bamboo trees in a carefully tending bamboo forest. Only *Rhizomys sinensis davidi, Sus scrofa chirodonts, Muntiacus reevesi, Muntiacus*

Table 1. Distribution and Changes of bamboo forests and trees between 1980 and 1990. Data Source: *Wuyishan Research Series* 1994.

Category	1980	1990	Results	Change (%)
Scattered bamboo trees	**6 117**	**5 114**	**−1 003**	**−16.4**
Core Zone		2 323		
Experimental Zone		2 791		
Scientific Zone				
Bamboo forest area (ha)	**6 477**	**7 938**	**+1 461**	**+22.6**
Core Zone		1 613		
Experimental Zone		6 119		
Scientific Zone		206		
Bamboo trees in forest	**147 828**	**209 262**	**+61 434**	**+41.6**
Core Zone		30 327		
Experimental Zone		174 097		
Scientific Zone		4 838		

crinifrons, Elaphodus cephalophus and *Capricornis sumatraensis* feed or den in pure bamboo forests. A field survey of 34 medium and large mammals in Wuyishan Nature Reserve was undertaken during November 1996 and February 1997. It indicated that a total of 20 species were adversely effected by bamboo monoculture (Bauer 1998). Furthermore, poorer soil fertility and more soil erosion may occur in a pure bamboo forest than in a mixed forest (Li 1994). Consequently the dependence of the local community on the bamboo industry is gradually decreasing the biodiversity of Wuyishan Nature Reserve.

ECOTOURISM

Tourism is booming throughout China and indeed internationally. The Chinese government has regarded tourism as one of the major currency earners and a development strategy for tourism forms a major part of both the Eighth (1990–1995) and the Ninth (1996–2000) Five Year Plan (Wen and Tisdell 1997). Ecotourism, a non-consumptive use of resources, appears to be a sound option for both biodiversity conservation and local development in protected areas. General shortage of budget for the protection of biodiversity and operating expenses of management authorities encourages the nature reserves of China to develop ecotourism. In Wuyishan Nature Reserve, we estimated that there are about 30 000 tourists expending some 500 thousands yuan annually on entrance fees. We did not estimate expenditure in local villages. The current strategies are to develop tourism in the reserve by building facilities such as hotels, restaurants, an entertainment centre, bird's-eye view towers and pathways to access beautiful scenic sites. A special agreement between the reserve and tourist agencies let tourist guides who introduce travellers to visit the reserve to earn ⅓ entrance ticket fee for each visitor. The local restaurants also pay guides for patrons.

Local villages such as Tongmu and Aotou are also enthusiastic to join the ecotourism activity. Tongmu, the biggest village with 324 households and 1 495 persons and situated in the location of the reserve authority, has developed about 30 family-based shops, restaurants and hotels. Two of the twelve households supply a passenger transport network between Tongmu and Wuyishan City (Cai 1997). Aotou, a village with 82 households and 373 persons, started up an ambitious development plan, involving family-based summer hotels, for tourism in 1996. An eye-catching slogan standing at the entrance of the village says: *to protect forests for benefiting village and to develop tourism for promoting economy*. Up to now, the family-based summer hotel plan has involved four households, and an estimated total income of 60 000 yuan was earned in 1996.

Ecotourism development brings great benefits to the local economy but some potential threats to the biodiversity. Most commonly cited threats are development of cluster-style tourism facilities involving intensive construction of buildings in limited land areas near natural and cultural attractions; discharge of effluence from tourism facilities; souvenir collection by tourists; and noise, water and air pollution (World Tourism Organization 1993). Resolving these potential problems depends on a sound planning. But up to now, there is no comprehensive plan made for the rational distribution and management of ecotourism projects. Most local facilities designed for tourism are too modern or town-style to be compatible with local natural landscape. Since the enthusiasm of the local communities runs high, the reserve managers have found they are facing strong pressure from local people. In other words, local ecotourism exploitation may go out of control and threaten the biodiversity conversation objectives while individual units in the reserve pursue their own economic interests and follow different decision making systems.

SUSTAINABLE MANAGEMENT OF WUYISHAN NATURE RESERVE

Wuyishan Biosphere Reserve has experienced great changes in its management since its establishment in 1979. Economic activity based on bamboo processing has provided the key incentive to participation of local people in reserve management. As a result the initial conflict between the local community and reserve administration has been replaced by a good working relationship. Although there has been some loss of biodiversity due to increasing bamboo monoculture, the co-operation between stakeholders has enabled effective conservation measures to be implemented. This makes Wuyishan Nature Reserve a significant model of participatory management that could be used in other Chinese nature reserves.

Due to the existence of different decision making systems and mixed property rights in the protected area, it is hard to co-ordinate the relationship among varied parts. Regional co-ordinating organizations, such as the joint protection committee composed of local agencies and communities, can reconcile some problems and conflicts, but have no authority to control economic activity. However, the joint-stock company model, which has been put into practice in some parts of the reserve, may be a way for incorporating individual interests in the new model of property rights. A regional joint-stock company with more extensive collaboration could form a basis to reduce negative impacts of bamboo monoculture on biodiversity conservation through better planning and increasing yield from varied natural resources.

Although the bamboo industry has produced dynamic local economic development, so important in the context of China, the resulting progression towards a monoculture has resulted in some reduction of biodiversity. Ways of reducing reliance on a monoculture economy are an urgent task facing to the reserve managers. Ecotourism development is one suggested approach to make up the deficiency of a monoculture economy. The current situation of unplanned ecotourism exploitation in the reserve needs prompt attention by developing an effective ecotourism plan.

Finally, sustainable development in protected areas should not depend on overly increasing the direct consumption of natural resources. This is a difficult balancing act in the context of China, which must encourage the development of local economies, which can often only be done by relying on further exploitation of natural resources. An integration of local development with biodiversity conservation in nature reserves needs to establish a sound compensation system in which biodiversity and ecological benefits can be well evaluated. Companies and individuals that earn benefits from using the natural resources must contribute some money for the costs of the resource and environment protection. For example, 47 km away from Wuyishan Nature Reserve there is the well-known Wuyishan National Scenic Spot. Annually, this area is always crowded in spring, summer and autumn by visitors attracted by the wonderful experience of drifting down the Jiuquxi River through magnificent scenic gorges. Nevertheless, this river almost dried up due to the logging in the upper reaches before the establishment of Wuyishan Nature Reserve. Now water sources in the well-protected forests in Wuyishan Nature Reserve become an inexhaustible supply and always available for use by tourists. This will certainly benefit the tourism of lower reaches but the reserve management and its communities get no financial return from this tourism activity.

It is necessary for an ecological environmental compensation system to operate so as to reduce the direct consumption of natural resources and disturbance to the natural system in the protected areas. Society, companies and individuals that earn benefits from the protection of biodiversity and natural environment should contribute to the value created by the reserve enterprises. However, establishment of this system to a great extent depends on the estimation of biodiversity values, environmental education and consciousness of the whole people and the foresight and sagacity of governments and policymakers.

REFERENCES

Administration of Wuyishan National Nature Reserve (Wuyishan Nature Reserve Administration), 1997a. Annals of the Wuyishan National Nature Reserve (in press).

Administration of Wuyishan National Nature Reserve, 1997b. A report on the joint protection and sound development of local communities in Wuyishan Nature Reserve. National Conference of China's Man and the Biosphere Network and Review of Wuyishan Biosphere Reserve, Fujian Province, China on 16–22 September 1997.

Bauer, J. J., 1998. Bamboo vs. Tiger. A preliminary evaluation of the status and ecology of the mammal fauna of the Wuyishan Biosphere Reserve. Johnstone Centre: Charles Sturt University, Bathurst, New South Wales.

Cai Dingkui, 1997. A report on the community development in Tongmu Village. National Conference of China's Man and the Biosphere Network and Review of Wuyishan Biosphere Reserve, Fujian Province, China on 16–22 September 1997.

De Lacy, T. and Bauer, J. J., 1997. A Project Proposal for Sustainable Development and Biodiversity Conservation in Wuyishan Biosphere Reserve, Fujian Province, China. ARC grant application, Canberra, Australian Capital Territory.

Furze, B., De Lacy, T. and Birckhead, J., 1996. Culture, Conservation and Biodiversity: the Social Dimensions of Linking Local Level Development and Conservation Through Protected Areas. John Wiley: Brisbane, Queensland.

He Jianyuan (ed), 1994. Wuyishan Research Series (Natural Resources). Xiamen University Press: Xiamen.

He Jianyuan, Xu Dingcheng, Liu Chuntian, Fan Yanhong and Chen Renhua, 1994. The forest resources in the Wuyishan National Nature Reserve. *In* Wuyishan Research Series (Natural Resources) ed by He. Xiamen University Press: Xiamen.

Li, linghao, 1994. An outline of the physical geography in Wuyishan Nature Reserve. *In* Wuyishan Research Series (Natural Resources) ed by He. Xiamen University Press, Xiamen.

National Environmental Protection Agency, 1994. China: Biodiversity Conservation Action Plan. National Environmental Protection Agency: Beijing.

National Environmental Protection Agency, 1996. National Report on China's Biodiversity. National Environmental Protection Agency: Beijing.

Tang Hongqian and Du Shouhu (eds), 1995. Forest Land and Tree Tenure and Social Forestry. Chengdu University of Science and Technology Press: Chengdu.

Wen, J. and Tisdell, C. A., 1997. Regional inequality and decentralisation of China's tourism industry. *In* China's Economic Growth and Transition: Including Regional, Environmental, Technological and Other Dimensions ed by C. A. Tisdell and J. C. Chai. Department of Economics, The University of Queensland: Brisbane, Queensland.

World Tourism Organization, 1993. Yearbook of tourism statistics. World Tourism Organization, Madrid.

Zhao Xianying, 1996. Preface. China's Biosphere Reserve Special Issue 2.

Zhao Xianying and Han, Nianyong, 1996. Open management strategy: Biosphere Reserves in China. China's Biosphere Reserve Special Issue: 4–6.

CHAPTER 67

Improving the assessment of risk of extinction and management requirements of flora taxa in Tasmania

A. J. J. Lynch[1]

The National Forest Policy Statement, agreed to by the Commonwealth, State and Territory Governments of Australia, established that an extensive and permanent native forest estate would be maintained and managed in an ecologically sustainable manner with parallel development of internationally competitive and ecologically sustainable forest-based industries. Joint Commonwealth-State Comprehensive Regional Assessments of the environmental, heritage, social and economic values of the forests would be undertaken to develop a comprehensive, adequate and representative reserve system as part of a Regional Forest Agreement. In developing the reserve system, the vegetation community is the surrogate used for biodiversity and accommodates most of the common and widespread flora species. However, rare and threatened species tend to occur in localized or specialized habitats and, therefore, their conservation needs to be specifically addressed.

Previous to the Tasmanian Regional Forest Agreement, the assessment of the conservation and reservation status of flora in Tasmania had been conducted predominantly by expert opinion based on the distribution of the taxa and their presence within areas of secure reserve tenure. As part of the Tasmanian Comprehensive Regional Assessment, an explicit analysis of priority rare or threatened flora was developed with three main assessment phases: intrinsic risk of extinction, the effects of disturbance on the taxa, and the adequacy of current management prescriptions. The management requirements of the taxa were addressed in terms of protection of populations in dedicated and/or informal reserves as well as management prescriptions of these and additional populations. The reservation levels of the priority taxa under the finalized comprehensive, adequate and representative reserve system are presented and compared to their former reservation levels.

Key words: Extinction, Management, Rare, Threatened, Flora, Tasmania.

INTRODUCTION

ASSESSMENTS of the conservation status of flora taxa in Australia have generally been conducted in order to identify classes of taxa at risk of extinction for incorporation into schedules of threatened taxa. In contrast, the determination of the conservation and long-term management requirements of flora taxa on- and off-reserve is required as part of the Commonwealth-State Comprehensive Regional Assessments being conducted of forested environments in Australia. During the Tasmanian Comprehensive Regional Assessment, a methodology was developed to rapidly assess these requirements for flora based on a combination of available literature and local expert opinion. Protection targets were identified for all priority taxa, and this information formed one of the environment and heritage data layers integrated with socio-economic data in the development of a Tasmanian comprehensive, adequate and representative reserve system. This chapter outlines the process used to assess the conservation and management requirements of priority flora in the context of the Tasmanian Regional Forest Agreement.

[1]Environment Forests Taskforce, Environment Australia, G.P.O. Box 787, Canberra, Australian Capital Territory, Australia 2601.

Pages 653–61 *in* NATURE CONSERVATION 5: NATURE CONSERVATION IN PRODUCTION ENVIRONMENTS: MANAGING THE MATRIX ed by J. L. Craig, N. Mitchell and D. A. Saunders. Surrey Beatty & Sons, 2000.

FLORA CONSERVATION ASSESSMENTS IN AUSTRALIA

Conservation assessments of flora in Australia have generally been based on attributes and thresholds relating to the demography, geography and probability of extinction of taxa. These attributes enable the clustering of taxa according to their perceived current risk of extinction. Several classifications have resulted, including: the Rare or Threatened Australian Plants system (Hartley and Leigh 1979; Leigh, Briggs and Hartley 1981; Briggs and Leigh 1988; Briggs and Leigh 1996) which is based on the national distributions of taxa and used at a national level; variants of the Rare or Threatened Australian Plants system applied at a state level (e.g., Kirkpatrick *et al.* 1988); and modified versions of the International Union for the Conservation of Nature Red List Criteria also applied at a state level in Western Australia (K. Atkins, pers. comm.) and New South Wales (Keith 1998).

In Tasmania, prior to the Comprehensive Regional Assessment, the conservation status of flora was assessed using similar categories as the Rare or Threatened Australian Plants system but with more explicit subcategories of rare defined at the statewide level (Kirkpatrick *et al.* 1988; Kirkpatrick *et al.* 1991; Flora Advisory Committee 1994). The taxa were evaluated by a panel of experts using local knowledge of the distribution of the taxa and their presence within the formal reserve system. Also, Tasmania's geographic isolation as an island at the southeastern extremity of Australia has resulted in many of the non-endemic taxa being at the edge of their range. This isolation creates an additional factor contributing to the conservation value of Tasmanian flora. The conservation status of the species was assigned on an intuitive basis by expert opinion as up-to-date information was not systematically collated or databased. Taxa were assigned a reservation status based on the number of locations recorded for each taxon within an area of secure reserve tenure. Endemic taxa were considered reserved if two populations of the taxon occurred within secure reserves, non-endemics if one population was within a secure reserve.

THE COMPREHENSIVE, ADEQUATE AND REPRESENTATIVE RESERVE SYSTEM REQUIREMENTS

The Tasmanian Comprehensive Regional Assessment formed part of the implementation of the National Forest Policy Statement (Commonwealth of Australia 1992). The National Forest Policy Statement identified that an extensive and permanent native forest estate should be maintained and managed in an ecologically sustainable manner with parallel development of internationally competitive and ecologically sustainable forest-based industries under a Regional Forest Agreement (Commonwealth of Australia 1992). A vital element of the Statement was that joint Commonwealth-State Comprehensive Regional Assessments of the environmental, heritage, social and economic values of the forests would be undertaken to develop a comprehensive, adequate and representative reserve system. The environment and heritage assessments identify biodiversity and conservation values requiring protection while a parallel socio-economic assessment analyses the regional industry and social impacts of proposed conservation options.

The nationally agreed comprehensive, adequate and representative criteria state that the "reserve system should seek to maximize the area of high quality habitat for all known elements of biodiversity wherever practicable" including the protection of 15% of the pre-1750 (i.e., pre-European settlement) extent of forest communities and the maintenance of viable populations of native forest species throughout their natural ranges (Joint Australian and New Zealand Environment and Conservation Council/Ministerial Council on Forestry, Fisheries and Aquaculture National Forest Policy Statement Implementation Subcommittee 1997). The categories of species that should be represented in the reserve system include species which are threatened, rare, declining, migratory, dependent on old growth forest, phylogenetically distinct, bioregional endemics, disjunct, at the edge of their range, indicators, and functionally, economically or culturally important (Joint Australian and New Zealand Environment and Conservation Council/Ministerial Council on Forestry, Fisheries and Aquaculture National Forest Policy Statement Implementation Subcommittee 1997). While the community level reservation acts as a surrogate for biodiversity and accommodates most of the common and widespread species, rare and threatened species tend to occur in localized or specialized habitats and, therefore, need to be conserved specifically.

Assessment procedures need, under the National Forest Policy Statement, to be transparent in process, inputs and outputs. Prioritization of conservation actions also needs to be based on objective and repeatable methods, the resolution and detail of which are commensurate with well defined aims and consistent with data reliability (Burgman *et al.* 1997). Explicit, quantitative and systematic methodologies, such as rulesets, ensure a level of precision, reliability and repeatability that is necessary to defend the outputs when subject to challenge (Keith 1998). Explicit schemes allow for the inclusion of a wider group of experts, especially those with data on a limited number of taxa or only proportions of the range of a taxon, and also for consistency between different assessors that is difficult to achieve under intuitively-based assessments (Mace and Lande 1991). Experts may not necessarily prove to be reliable or consistent (Lunney *et al.* 1996) and even consistency is only weak evidence of reliability (Burgman *et al.* 1997). Intuitive assessments also may obscure sources of bias (Burgman *et al.* 1997).

DEVELOPMENT OF THE FLORA CONSERVATION ASSESSMENT METHODOLOGY

The determination of the conservation and long-term management requirements of flora taxa on- and off-reserve require that protection targets be identified for all taxa to be conserved through reservation and complementary off-reserve management action. Established prescriptions need to be collated and assessed for their adequacy and appropriateness for sustaining the taxa long term.

During the Tasmanian Comprehensive Regional Assessment, a methodology was developed to identify the habitat and target areas required to maintain populations of priority flora. The methodology was developed iteratively as explicit rulesets in conjunction with a local expert panel; representatives included officers from the Tasmanian Parks and Wildlife Service, Forestry Tasmania, the Queen Victoria Museum and botanical consultants. The data for the assessment were compiled from literature, scientific reports (published and unpublished) and expert opinion. The assessment system operated as a rule-based system rather than a scoring procedure. A scoring system was seen to be inappropriate to relate the varying types of data available, and to be less applicable to data varying in consistency of precision and availability between taxa. The data were entered into two databases set up for the Comprehensive Regional Assessment (the "Response to Disturbance" and "Management Prescriptions" databases). Establishment of the databases not only made apparent which data were available but simultaneously enabled the identification of knowledge gaps.

The priority taxa were agreed to be those classified as rare or threatened. Rare species were presumed to be those of limited range and/or abundance, but whose populations are currently stable and not threatened by extinction, whereas the threatened categories, of critically endangered, endangered and vulnerable, were presumed to include species which are both rare and threatened by extinction in descending degree of risk.

The prioritization system comprised three main assessment stages of the priority taxa leading to conservation recommendations (Fig. 1):

(i) intrinsic risk of extinction;

(ii) the effects of various disturbances or management practices on the taxa, and

(iii) compilation and assessment of the adequacy of current management and prescriptions.

The first stage (Fig. 2, Appendix 1) was based on the rulesets in the International Union for the Conservation of Nature Red List Criteria (International Union for Conservation of Nature and Natural Resources 1994b). The rulesets were simplified in accordance with the data available for most of the priority taxa and the thresholds were lowered as suggested by Keith (1998) to be more appropriate for sessile organisms such as flora. Different thresholds were applied in the Tasmanian assessment than those suggested by Keith (1998), however, due to expert advice and the coarseness of the distribution data (Appendix 1).

The taxa were separated using the rulesets into one of five categories using data on the taxon's intrinsic risk of extinction, that is, on its total population size, geographic range, number and size of populations, and whether continuing decline or extreme fluctuation could be demonstrated for any of the attributes (Step 1 (Fig. 2)). The initial designation of taxa to categories 1, 2 and 3 equated approximately to the International Union for the Conservation of Nature Red List categories of critically endangered, endangered and vulnerable. Category 4 essentially related to taxa considered to be rare but not threatened, while category 5 included taxa for which there was insufficient knowledge to categorize as either rare or common, were taxonomically uncertain or presumed extinct. The priority taxa were generally required to be of a total population size of less than or equal to 10 000 reproductively mature individuals.

Step 2 (Fig. 2, Appendix 1) revised the taxon's initial category placement according to data on the extrinsic threats to the taxon. These data incorporated information on known disturbance regimes or threatening processes, the intensity of any adverse impacts, and the proportion of the taxon's range affected. Data were also compiled on whether the impacts were beneficial or adverse to the taxon's survival or did not affect the stability of the taxon. The experts also estimated the proportion of the taxon's range potentially affected by each adverse impact and its intensity, as well as the total proportion of the range affected by any adverse impact and to what degree (severe, moderate, minimal). A taxon could be upgraded to a higher degree of threat if it was disturbance sensitive, e.g., a large proportion of the taxon's range was severely affected by adverse impacts, or downgraded if the taxon was disturbance tolerant.

Management requirements were finally assigned to the taxa by their revised category (Step 3 (Fig. 2)), with management, incorporating both reservation and additional areas for protection through management prescription, in accordance with Joint Australian and New Zealand Environment and Conservation Council/Ministerial Council on Forestry, Fisheries and Aquaculture National Forest Policy Statement Implementation Subcommittee (1997). The explicit recognition of the threatening processes affecting species was important for determining the ongoing management requirements of the taxa, however, the conservation strategies were still largely determined by the main intrinsic risk attributes (geographic range, total population size, number of populations). The Tasmanian flora expert panel assessed whether the assigned management requirements for taxa of each category were appropriate. Some flexibility was required where populations were very small and management was not considered to be feasible. All decisions by the expert panel were documented to justify and make transparent the management recommendations.

APPLICATION OF THE METHODOLOGY TO THE PRIORITY TASMANIAN FLORA

The 171 taxa initially selected for the Comprehensive Regional Assessment flora assessment

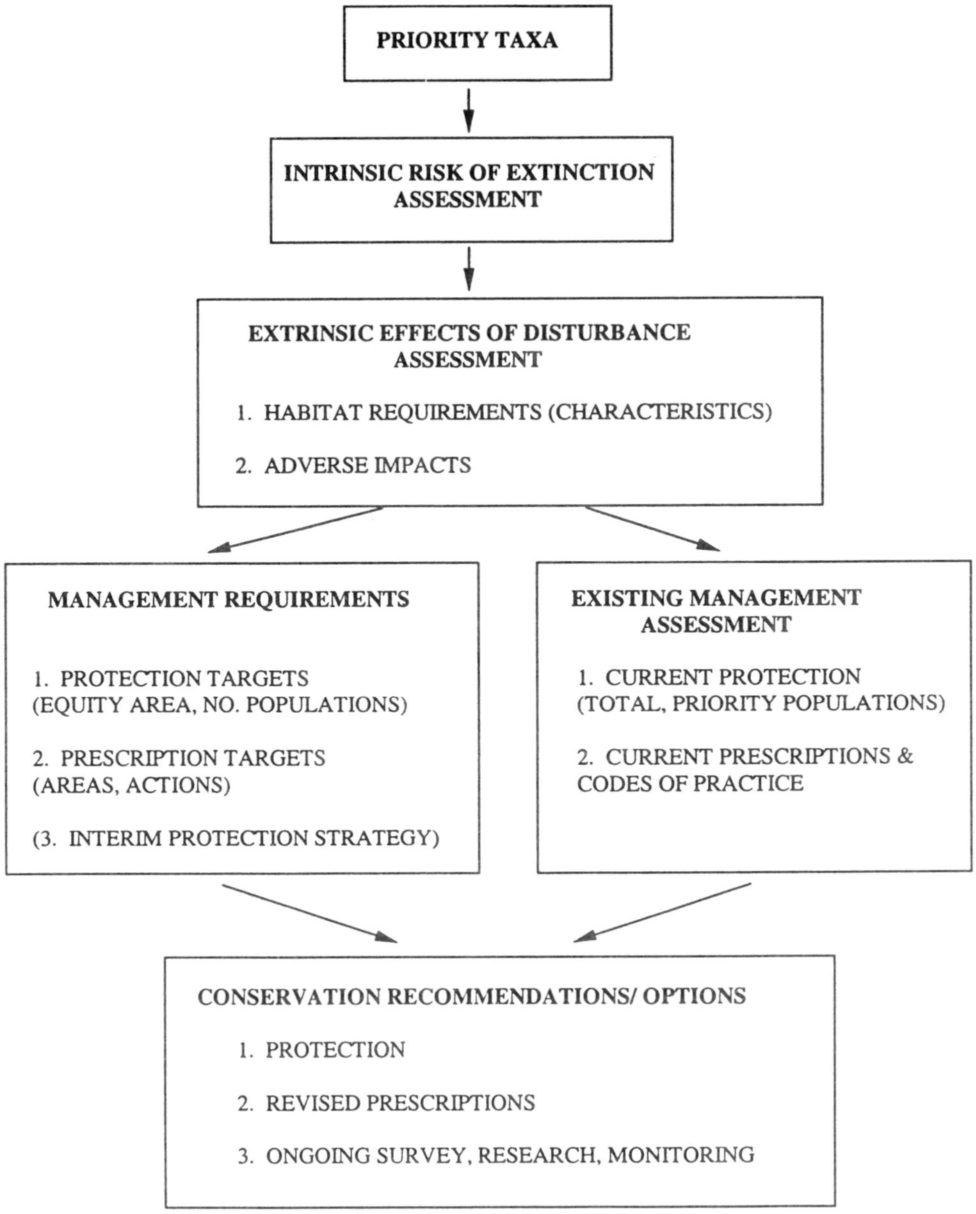

Fig. 1. Stages in the conservation assessment of the Tasmanian flora.

consisted of all forest dependant flora taxa which were both listed as rare or threatened in Tasmania (Flora Advisory Committee 1994) and also categorized as the highest priority (i.e., priority one) for forest management in the Forest Botany Manuals of Forestry Tasmania's Forest Practices Board Forest Botany Manuals (e.g., Duncan 1991; Duncan and Johnson 1995). The preliminary list included species, subspecies, varieties or types of species, identified as forest dependant, that is, occurring in forests or forested catchments and potentially affected by forest management processes. During the iterative analyses of the taxa, the expert panel revised the priority list to 168 taxa distributed across Tasmania. A subset of 67 of the priority flora, most of them listed as nationally threatened, were researched and surveyed during the Comprehensive Regional Assessment to improve knowledge on their distribution and threats. These and other recently studied taxa contributed the main body of data to the assessment and therefore could be evaluated the most adequately. The precautionary principle was applied where wide estimates were supplied on the values of attributes in the rulesets. As recommended by the International Union for the Conservation of Nature (1994b), the estimate (providing it was credible) was used that led to the taxa being listed in the highest category of risk.

DEVELOPMENT OF THE COMPREHENSIVE, ADEQUATE AND REPRESENTATIVE RESERVE SYSTEM

During the integration phase of the Comprehensive Regional Assessment, the data layers from each of the major environment and heritage assessments (forest communities, old growth forest,

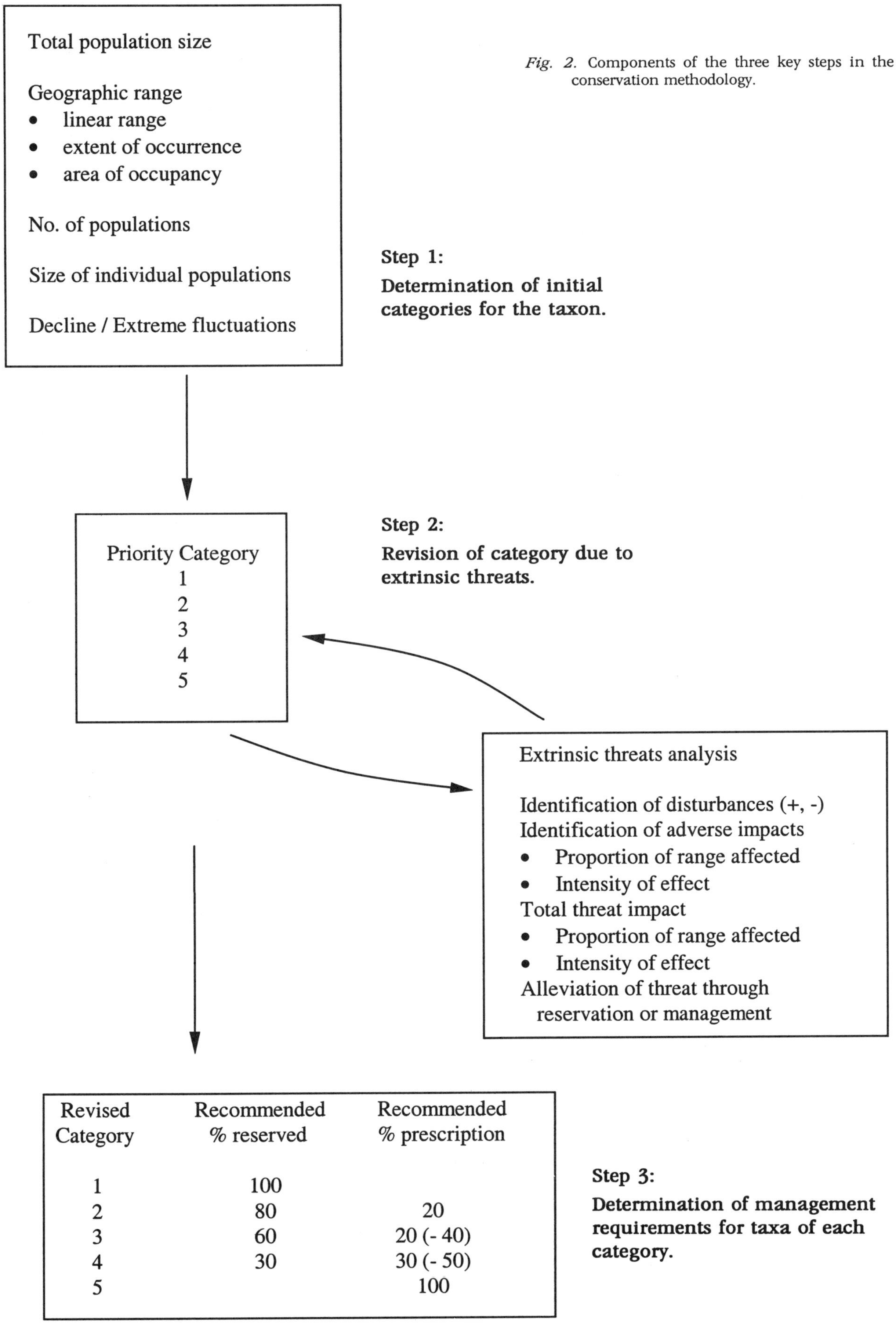

Fig. 2. Components of the three key steps in the conservation methodology.

priority flora and fauna species, National Estate, World Heritage, wilderness) were used to design the comprehensive, adequate and representative reserve system. Over a period of several months, various reserve scenarios were devised to maximize the environment and heritage components under the Joint Australian and New Zealand Environment and Conservation Council/Ministerial Council on Forestry, Fisheries and Aquaculture National Forest Policy Statement Implementation Subcommittee (1997) criteria while minimizing the regional socio-economic impacts. The comprehensive, adequate and representative reserve system was finalized with the signing of the Tasmanian Regional Forest Agreement by the Commonwealth of Australia and the State of Tasmania on 8th November, 1997.

The comprehensive, adequate and representative additions to the Tasmanian reserve system improved the reservation status of a significant proportion of the priority taxa (Table 1). These taxa either had additional sites reserved or some of their sites upgraded in status to a more secure tenure for conservation purposes, e.g., from an informal to a formal reserve. Thirty-six per cent of the taxa were reserved to their full extent on public land. Overall, 20% of the taxa were reserved up to or exceeding their reservation target; a value of 23% if only priority categories 1–4 are considered. Category 5 taxa are recommended for management wholly through prescription until the taxa can be reclassified as Category 1–4 or delisted. Eleven per cent of taxa also met their total protection target (including both reservation and prescription area targets).

In achieving the management targets for the flora species, those areas which form the Dedicated Reserve component of the comprehensive, adequate and representative reserve system are equivalent to Categories I, II, III or IV as defined by the International Union for the Conservation of Nature Commission for National Parks and Protected Areas (International Union for Conservation of Nature and Natural Resources 1994a) and in accordance with Joint Australian and New Zealand Environment and Conservation Council/Ministerial Council on Forestry, Fisheries and Aquaculture National Forest Policy Statement Implementation Subcommittee (1997). In situations where it is not possible or practicable to include conservation values (in this case, some species' populations) into Dedicated Reserves, areas are protected under other secure tenure or management arrangements as informal reserves set aside specifically for conservation purposes and meeting the principles set out in Joint Australian and New Zealand Environment and Conservation Council/Ministerial Council on Forestry, Fisheries and Aquaculture National Forest Policy Statement Implementation Subcommittee (1997).

Where neither a Dedicated nor an Informal reserve is practicable (e.g., a small population on a roadside), then protection will be prescribed through Codes of Practice or Management Plans (Joint Australian and New Zealand Environment and Conservation Council/Ministerial Council on Forestry, Fisheries and Aquaculture National Forest Policy Statement Implementation Subcommittee 1997). If, however, a population was considered by the Tasmanian flora expert panel to be too small for management (protection and/or prescriptions) to be practicable, then such populations were excluded from consideration for protection and management assessment, unless those individuals made up a substantial proportion of the total known population of that taxon. Species for which there were few or no known viable populations listed in the "Rare and Threatened Species" database and/or defined by the Tasmanian flora expert panel may require rehabilitation of populations to ensure persistence of the species.

Taxa of category five, i.e., those which are insufficiently known to categorize as rare or common, will be managed under an interim management strategy to be implemented on a case by case basis. This strategy will aim to manage all core populations by protection and/or prescription until the taxon can be adequately assessed. Where data are inadequate to identify the risk of extinction and disturbances adversely affecting taxa, additional research, survey and/or rehabilitation may be identified as ongoing actions for the post-Regional Forest Agreement and review period.

Management prescriptions will not only apply to species or populations outside the Dedicated or Informal reserve system, but will also be required for all species for which adverse disturbance impacts have been identified which are alleviated by prescription. Management requirements will be determined for the priority taxa prior to the first (of four) five yearly reviews of the Regional Forest Agreement. The databases set up for the Comprehensive Regional Assessment will be maintained and updated as a basis for updating relevant state management documents, and made available periodically for public comment. Forest management will be further improved across all agencies and land tenures through development and implementation of a Threatened Species Protection

Table 1. Reservation and protection levels of priority flora taxa.

	Reservation target met	Protection target met	All public land protected	Reservation increased by reserve system
All taxa (total 168)	20% (34)	11% (18)	36% (60)	41% (69)
Category 1–4 (148)	23% (34)			
Category 5 (20)	not applicable			

Strategy and a Tasmanian Biodiversity Strategy (Tasmanian Public Land Use Commission 1997).

A programme to protect comprehensive, adequate and representative values on private land by the voluntary participation of private landowners in the comprehensive, adequate and representative reserve system is being established, including mechanisms of compensation, stewardship agreements and voluntary sale. Also, timber harvesting plans on private land will include conditions to ensure the application of appropriate management prescriptions to these taxa (Tasmanian Public Land Use Commission 1997).

DISCUSSION

The need for a species level assessment as well as forest community and old growth forest assessments to conserve biodiversity was highlighted by the Tasmanian Comprehensive Regional Assessment. The higher level community assessments did not perform as adequate surrogates for biodiversity of flora species in Tasmania as most of the priority taxa did not overlap with the priority forest types, that is, those significantly depleted since European settlement of the state, or with areas of old growth forest.

The management of biodiversity on private land was also identified as an important issue. A significant proportion of the priority taxa occurred in the dry sclerophyll forests of the Midlands area of central Tasmania; an area predominantly of freehold tenure. Approximately 80% of taxa would need some protection of populations on private land to meet their recommended protection targets.

Another limitation to achieving the protection targets was that the priority flora generally occurred as localized populations with clustered individuals. Such small, isolated sites could not easily be incorporated through formal reservation into the comprehensive, adequate and representative reserve system as they would create significant management difficulties. Necessarily, management of such populations will be through management prescription, including Management Decision Classification zoning with an appropriate buffer area for sites in State Forest.

The Comprehensive Regional Assessment has enabled the development of an explicit and transparent conservation assessment methodology that can be systematically applied to a large number of taxa, as well as the compilation and documentation of the supporting data. The establishment of this broad coverage of data in centralized databases creates a reference for the current status of the taxa which can form the basis for longer term assessment and monitoring of the taxa. It can also be used to incorporate quantitative data, such as estimates of declines of taxa, and to distinguish known declines or inferred declines (past adverse processes) from projected or potential declines (future continuation of adverse processes). The Management Prescriptions database enables the assessment of the adequacy of current protection and prescriptions to address the processes threatening taxa as identified during the Comprehensive Regional Assessment. The proportion of the total populations of each taxon requiring active management can also be assessed. Criteria have been developed to guide such assessments.

The flora assessment has established priorities for conservation and management in Tasmania, and its emphasis on both intrinsic risk of extinction or level of endangerment and the extrinsic effects of threatening processes. It achieved its aims of assessing the conservation and management requirements of the taxa through determination of their habitat, estimation of long-term population requirements, and identification of threats. A framework has been established to enable ongoing monitoring of proposed management levels and actions to ensure their appropriateness for maintaining long-term perpetuation of the taxa. Management prescriptions will continue to be developed to mitigate or alleviate disturbances causing adverse impacts, either directly on the taxon or indirectly on its habitat.

The inclusive conferral and liaison of local experts encouraged during the Comprehensive Regional Assessment has enabled spatial and temporal cross-referencing of data on the taxa and conservation and management activities. Continued co-operation between agencies and experts is a desirable outcome of the assessment, not only in relation to management of rare and threatened species but also on broader conservation issues.

The Tasmanian flora assessment methodology has provided a starting point for addressing flora conservation in other Comprehensive Regional Assessment regions. Subsequent Comprehensive Regional Assessments are currently being conducted in New South Wales, Queensland, Western Australia and Victoria. The methodology is being revised and customized in those regions.

The framework also has wider application to other programmes, for example, for species recovery planning on the Greater Bilby *Macrotis lagotis*, emphasizing the need for management of the species through knowledge of its threats and determination of required ameliorative and/or preventative management actions. Population Viability Analysis is time and resource intensive and recovery planning must as a necessity focus on prioritizing recovery plan actions with funding goals and critical paths to achieve the objectives of the plan with minimum expenditure of time and resources (Rounsevell 1997).

The flora assessment methodology is, however, currently being extended to incorporate experience from Population Viability Analysis studies in the framework of intrinsic and extrinsic extinction risk (Burgman *et al.*, in press). While the protection targets identified for the priority flora in Tasmania initially seemed much higher than previous estimates of the conservation requirements of the taxa, the conservation of flora needs to be perceived as of importance commensurate with other measures of biodiversity (e.g., fauna, vegetation communities, old growth forest) and also to be considered in the perspective of their long-term viability.

ACKNOWLEDGEMENTS

The development and successful application of the flora assessment as part of the Tasmanian Comprehensive Regional Assessment involved the contributions of a large number of people from the forest management agencies as well as expert scientists. The contributions of Mick Brown, Phil Barker and Andrew Taplin were especially valuable. The assistance of Maria Moore in provision of data on the reservation status of the taxa and of Andrew Tytherleigh was also appreciated.

APPENDIX 1

Step 1. Determination of initial categories for the taxon

Category 1

(a) Total population size of 1–50 mature individuals; **or**

(b) Geographic range is ≤10 km (linear range) **or** ≤100 km² (extent of occurrence) **or** ≤10 km² (area of occupancy; 1 x 5 km by 5 km grid cells) **and** total population size is ≤10 000 mature individuals **and either:**

- No. populations = 1; **or**
- extreme fluctuations are recorded in any attribute; **or**

(c) Total population size of 50–250 mature individuals **and either:**

- No. populations = 1; **or**
- size of individual populations are all ≤50 mature individuals; **or**
- extreme fluctuations are recorded in any attribute.

Category 2

(a) Total population size of 50–250 mature individuals; **or**

(b) Geographic range is ≤80 km (linear range) **or** ≤5 000 km² (extent of occurrence) **or** ≤250 km² (area of occupancy; 10 x 5 km by 5 km grid cells) **and** total population size is ≤10 000 mature individuals; **and either:**

- No. populations = 1–5; **or**
- extreme fluctuations are recorded in any attribute; **or**

(c) Total population size of 250–2 500 mature individuals **and either:**

- No. populations = 1; **or**
- size of individual populations all ≤250 mature individuals; **or**
- extreme fluctuations are recorded in any attribute.

Category 3

(a) **Either:**

- total population size is ≤1 000 mature individuals; **or**
- area of occupancy is ≤250 km² (10 x 5 km by 5 km grid cells); **or**
- No. populations = 1–5; **or**

(b) Geographic range is ≤160 km (linear range) **or** ≤20 000 km² (extent of occurrence) **or** ≤1 000 km² (area of occupancy; 40 x 5 km by 5 km grid cells) **and**

- No. populations = 6–10 **and**
- total population size is ≤10 000 mature individuals; **or**

(c) Total population size is ≤10 000 mature individuals **and either:**

- No. populations = 1; **or**
- size of individual populations are all ≤1 000 mature individuals.

Category 4

(a) **Either:**

- total population size is ≤10 000 mature individuals; **or**
- geographic range is ≤160 km (linear range) **or** ≤20 000 km² (extent of occurrence) **or** ≤1 000 km² (area of occupancy; 40 x 5 km by 5 km grid cells); **or**

(b) Taxon is currently listed as a rare or threatened forest taxon in Tasmania **and** does not meet the above criteria.

Category 5

(a) Taxon is insufficiently known; **or**

(b) Taxon is taxonomically uncertain; **or**

(c) Taxon is presumed extinct.

Step 2. Revision of assigned categories due to extrinsic threats to taxa

The placement of taxa in categories was revised according to the following guidelines and based on the amount of information available for the disturbances affecting the taxa. The information collated in the database was used to compile two tables:

(i) lists of the disturbances known to affect each of the taxa and whether the effect is beneficial, adverse or the taxon persists but does not increase or decrease; and

(ii) the estimation by experts of the proportion of the taxon's range potentially affected and the intensity of each adverse impact affecting a taxon; and estimates of the total proportion of the range affected by any adverse impact and at what level (severe, moderate, minimal) across the taxon's range.

(a) For taxa with both disturbance data tables assessed (adverse impacts identified and their intensity/ range of taxon affected estimated):

- If ≥50% of the taxon's area of occupancy is affected by severe adverse impacts, then the taxon should be raised to the next highest category;
- If ≥80% of the taxon's area of occupancy is affected by moderate adverse impacts, then the taxon should be raised to the next highest category;
- If the taxon is disturbance tolerant (i.e., ≥75% of the disturbances identified for the taxon are listed as the taxon "persists" or the disturbance is a "beneficial impact" **and** the taxon has not been identified as continuing to decline, then the taxon should be lowered to the next lowest category.

Known, inferred or projected declines from literature or expert opinion were used in the Tasmanian analysis, however, the "Rare and Threatened Species" database will be used in future to document explicit quantitative estimates of declines of taxa and to distinguish known declines or inferred declines (past adverse processes) from projected or potential declines (future continuation of adverse processes).

*(b) For taxa without **BOTH** disturbance data tables assessed (adverse impacts identified and their intensity/ range of taxon affected estimated):*

The table categorizing the adverse impacts for a taxon by those alleviated by protection and those alleviated by management prescription was used to indicate:

- If all adverse impacts can be alleviated by protection, then the taxon remains in its designated category;
- If all adverse impacts can be alleviated by prescription, then the taxon should be lowered to the

next lowest category, but the percentage of populations for which prescriptions are required should be increased to a level commensurate with the taxon's original category (e.g., a taxon of category 2 with 80% protection and 20% prescription required would go to category 3 but with 60% protection and 40% management prescription);

- If a mixture of protection and management prescription is required to alleviate all adverse impacts, then the taxon remains in the same category.

REFERENCES

Briggs, J. D. and Leigh, J. H., 1988. Rare or Threatened Australian Plants. Australian National Parks and Wildlife Service, special publication No. 14. Commonwealth of Australia: Canberra, Australian Capital Territory.

Briggs, J. D. and Leigh, J. H., 1996. Rare or Threatened Australian Plants. CSIRO: Melbourne, Victoria.

Buchanan, A. M., 1995. A Census of the Vascular Plants of Tasmania. Tasmanian Herbarium occasional publication No. 5. Tasmanian Museum and Art Gallery: Hobart, Tasmania.

Burgman, M. A., Keith, D. A., Rohlf, F. J. and Todd, C., 1997. Probabilistic classification rules for setting conservation priorities. Pp. 43–55 *in* Assessing the Status of Threatened Plants: A New Methodology and an Application to the Vascular Flora of New South Wales ed by D. A. Keith, J. M. Chalson and T. D. Auld. Project No. 450, Endangered Species Program, Biodiversity Group, Environment Australia: Canberra, Australian Capital Territory.

Burgman, M. A., Possingham, H. P., Lynch, A. J. J., Keith, D. A., McCarthy, M. A., Hopper, S. D., Drury, W. L., Passioura, J. and DeVries, R. J., in press. A decision support protocol for setting plant conservation target areas. *Conserv. Biol.*

Commonwealth of Australia, 1992. National Forest Policy Statement. A New Focus for Australia's Forests. Commonwealth of Australia: Canberra, Australian Capital Territory.

Duncan, F., 1991. Forest Practices Forest Botany Manual — Nature Conservation Region 7. Forestry Commission: Hobart, Tasmania.

Duncan, F. and Johnson, K., 1995. Forest Practices Forest Botany Manual — Nature Conservation Region 4. Forestry Tasmania: Hobart, Tasmania.

Flora Advisory Committee, 1994. Native Higher Plant Taxa which are Rare or Threatened in Tasmania. Edition 1, Species at Risk, Tasmania — Flora. Parks and Wildlife Service: Hobart, Tasmania.

Hartley, W. and Leigh, J. H., 1979. Plants At Risk. Australian National Parks and Wildlife Service, special publication No. 3: Canberra, Australian Capital Territory.

International Union for Conservation of Nature and Natural Resources, 1994a. Guidelines for Protected Area Management Categories. International Union for the Conservation of Nature Commission on National Parks and Protected Areas and World Conservation Monitoring Centre: Gland, Switzerland and Cambridge, United Kingdom.

International Union for Conservation of Nature and Natural Resources, 1994b. International Union for the Conservation of Nature Red List Categories. International Union for the Conservation of Nature: Gland, Switzerland.

Joint Australian and New Zealand Environment and Conservation Council/Ministerial Council on Forestry, Fisheries and Aquaculture National Forest Policy Statement Implementation Subcommittee, 1997. Nationally Agreed Criteria for the Establishment of a Comprehensive, Adequate and Representative Reserve System for Forests in Australia. Commonwealth of Australia: Canberra, Australian Capital Territory.

Keith, D. A., 1998. An evaluation and modification of World Conservation Union Red List Criteria for classification of extinction risk in vascular plants. *Conserv. Biol.* **12:** 1076–090.

Kirkpatrick, J. B., Gilfedder, L. A. and Fensham, R. J., 1988. City Parks and Cemeteries: Tasmania's Remnant Grasslands and Grassy Woodlands. Tasmanian Conservation Trust: Hobart, Tasmania.

Kirkpatrick, J. B., Gilfedder, L., Hickie, J. and Harris, S., 1991. Reservation and Conservation Status of Tasmanian Native Higher Plants. Wildlife Division scientific report No. 91/2. Department of Parks, Wildlife and Heritage: Hobart, Tasmania.

Leigh, J. H., Briggs, J. D. and Hartley, W., 1981. Rare or Threatened Australian Plants. Australian National Parks and Wildlife Service, special publication No. 7: Canberra, Australian Capital Territory.

Lunney, D., Curtin, A., Ayers, D., Cogger, H. G. and Dickman, C. R., 1996. An ecological approach to identifying the endangered fauna of New South Wales. *Pac. Conserv. Biol.* **2:** 212–31.

Mace, G. M. and Lande, R., 1991. Assessing extinction threats: toward reevaluation of IUCN threatened species categories. *Conserv. Biol.* **5:** 148–57.

Rounsevell, D., 1997. Endangered Species Program (TSCS) Project Review. Reviewer's Comments, ESP Project 185, Greater Bilby *(Macrotis lagotis)* Recovery Plan, Queensland. Unpublished report.

Tasmanian Public Land Use Commission, 1997. Tasmanian Regional Forest Agreement between the Commonwealth of Australia and the State of Tasmania, 8 November 1997, Tasmanian RFA Information kit. Tasmanian Public Land Use Commission: Hobart, Tasmania.

CHAPTER 68

Retaining native vegetation on farms: understanding its private value

J. Crosthwaite[1] and N. D. MacLeod[2]

Conflicts commonly emerge in farm businesses between the short to medium term private imperative of generating income and public goals that relate to conservation. A common outcome is that native vegetation is replaced or irreversibly degraded. The private value of land with native vegetation depends on its use within the context of the farm system. This value will change if the land in question is managed differently, and may change if there are changes elsewhere in the farm system. It is these values that principally drive decisions concerning exploitation of native vegetation. Accordingly, their identification can help to clarify the nature of the conflicts and positively contribute to their resolution. In particular, it may lead to a shifting emphasis from exploitation of the native vegetation to alternative ways to increase income on the farm.

Time and scarce resources preclude comprehensive assessments of the native vegetation or the farming system each time potential loss of remnant vegetation comes to public attention. A rapid appraisal approach for responding quickly to potential loss is illustrated using a case study property where loss of conservation values is imminent. The approach establishes the current status of the native grassland and its place in the farming system. The expected benefits to the owners of changing management are estimated, and alternative means of increasing income considered. For the case study farm, it is found that there are some alternatives, but the expected benefits of the change are small. Nevertheless, they are important for this farm and the owners' present needs. Such information is the basis for determining the most appropriate policy interventions. In this case, further investigations into alternative farm systems, uses of the native vegetation and changing property rights are being pursued in conjunction with the owners.

Key words: Remnant vegetation, Native grassland, Farming system, Economics, Grazing, Southwestern Victoria.

INTRODUCTION

NATIVE grasslands represent one of the most threatened ecosystems in southeastern Australia. The remaining area is disproportionately small in comparison to its pre-European distribution, and most of the remnants are located on private grazing land (Kirkpatrick *et al.* 1995). Native grasslands are now recognized under both national and state biodiversity strategies (Department of Environment, Sport and Territories 1996; Department of Natural Resources and Environment 1997), as well as in regional catchment management strategies, and to a lesser extent in native vegetation retention regulations and property management planning guidelines. Nevertheless, despite these various public policy initiatives, the number and area of relatively intact grassland remnants on the Monaro, the riverine plains of New South Wales, the basalt plains of southwestern Victoria and elsewhere continues to decline. Principal causes of this decline are conversion to cropping and sown pastures, as well as through ongoing degradation from over-grazing and fertilizer use.

In this chapter the problem, and a path towards finding possible solutions, is illustrated through a case study from the basalt plains of southwestern

[1]Institute of Food and Land Resources, University of Melbourne, Parkville, Victoria, Australia 3053.
[2]CSIRO Tropical Agriculture, St Lucia, Queensland, Australia 4067.
Pages 662–69 *in* NATURE CONSERVATION 5: NATURE CONSERVATION IN PRODUCTION ENVIRONMENTS: MANAGING THE MATRIX ed by J. L. Craig, N. Mitchell and D. A. Saunders. Surrey Beatty & Sons, 2000.

Victoria. These plains extend from the west of Melbourne almost to the South Australian border, and from Beaufort in the north to Colac in the south. They were favoured by early European settlers and agricultural activities in the region are now dominated by wool, fat lamb and beef production with an increasing level of cropping. The area of native grassland has contracted dramatically and is now restricted to less than 1% of its previous range (Department of Natural Resources and Environment 1997). There are few intact native grasslands in public reserves, and these are mostly located on roadsides and in cemeteries. The protection of the native grasslands that remain on private land is a priority in the relevant catchment management strategy (Glenelg Regional Catchment and Land Protection Board 1997) and the *Western (basalt) Plains Grassland Community* Action Statement (Department of Conservation and Natural Resources 1994).

Actions by conservation agencies to deal with immediate threats are (correctly) primarily directed at native grasslands with the highest conservation ranking. As resources provided by governments to achieve native grassland conservation are limited, these agencies are relatively powerless to halt the losses of less significant grasslands, particularly those on private farming land. Contributing factors include the aforementioned lack of resources, perceived limitations of regulatory approaches but also the powerful economic imperatives driving most farm businesses and a serious shortage of experienced extension officers on the ground.

Appropriate ways of dealing with immediate threats to those native grasslands that are not able to attract scarce resources for their protection is the subject of this chapter. On the selected case study farm, the owners' land development plans for the coming season will seriously compromise conservation values of a tract of remnant native grassland and there is little time for detailed analysis — practical solutions have to be found quickly.

The key elements of the recommended approach to finding a solution include: (a) organizing relevant expertise for a rapid appraisal of the situation; (b) considering the conservation area in the whole farm context (as in whole farm planning); and (c) establishing the net gain/loss to the owners from the proposed change and alternative investments across the farm. These steps are necessary preliminaries to negotiating a solution that protects the native vegetation. For the case study farm, further investigations into alternative farm systems, uses of the native vegetation and changing property rights are being pursued in conjunction with the owners. The lessons from this case study can be applied to other threatened ecosystems on private land, such as wetlands and grassy woodlands.

CONSERVATION AND PRODUCTION

On-going causes of the loss of native grasslands

Farming activities involve intervention in, and management of, ecosystem processes (Lefroy *et al.* 1993; Swanson 1996) to produce a sufficient quantity of saleable commodities at low enough cost to satisfy the need for profit, debt servicing and a livelihood. This usually requires some form of intensification of activities which typically involves manipulation of ecosystem processes with fertilizer, pesticides and other agricultural inputs. While conservation requires low fertility, intensive agriculture requires fertility build-up and reduced competition from other species (Wibberley 1982). As the ecosystem inevitably changes, higher levels of input may be required simply to maintain production (Swanson 1996) e.g., to combat pesticide resistance or increasing soil acidity.

The cause of the loss of native grasslands can also be related directly to farm size, the cost-price squeeze and farmer knowledge. Most farms are too small to generate sufficient income using "low input" management. The cost-price squeeze, whereby prices received are static or falling while costs increase, means that farmers have to adopt new techniques and intensify use of their land to stay ahead. The limited ecological knowledge of most managers means that they will inadvertently, or otherwise, contribute to the degradation of remnant vegetation.

Tracts of remnant vegetation will usually only be conserved if this action is consistent with overall farm business goals. This, in turn, will require future investments that are necessary to maintain or increase income to be concentrated on other areas of the farm. However, in the face of an on-going struggle for most farmers to make a living, the future of native vegetation is not assured in absence of public policy initiatives which anticipate possible changes to their management (Bowers 1999; Young *et al.* 1996). Therefore, remnant native vegetation will be under threat on most farms today, if not tomorrow. Analysing the conservation problem at the right scale is important (Lefroy *et al.* 1993; Brown and MacLeod 1996). In the context of grazing properties, the critical scale for *private* decision-making on remnant vegetation use is that of the whole farm business.

Benefits of remnant vegetation

Remnant vegetation will usually generate at least some economic benefits to farmers. Therefore, a critical issue for those farmers who are seeking to increase their incomes is how significant these benefits are, or could be, in comparison to alternative uses of their land. It will also depend on whether the vegetation can play a complementary role in the whole farm system (Mitchell 1994; Crosthwaite 1997). Examples include: animal health gains through spelling improved pastures; year round production of green forage with reduced need to supplementary feed; harbour for beneficial organisms that promote pasture health; and native seed harvesting.

The potential economic gain from changing management of remnant vegetation has to be assessed relative to other opportunities to increase income on the farm, including restructuring the farming system to gain more from potential complementarities. The income foregone by the owners in continuing current management can then be determined. This information is vital for determining feasible actions to conserve remnant vegetation that is under threat.

Evaluating the contribution of remnant vegetation to the farming system

The contribution of remnant vegetation will depend on its current and prospective roles within the farming system. Evaluating these roles requires an accurate assessment of relevant bio-physical and technical factors across the farm, and linked to this an economic evaluation of the current farming system and possible alternatives. On the technical side, soil conditions need to be systematically related to the botanical and agronomic characteristics of each pasture type on the property. This information can then be linked to livestock enterprises through estimates of the utilization of forage supplied from various pasture types. Whole farm forage and financial budgets can then be estimated to determine the productive value of the remnant in its current use.

To determine the future contribution will require the identification of technically feasible options for changing the farm system to meet the owners' goals. These should include options for other areas of the farm, as well as those that may directly threaten the remnant vegetation. They should also account for any inter-dependencies between the various areas. Short-term profitability and cash flow effects of the "before change/after change" situations can then be compared. The long-term effects of changing management are difficult to quantify, and also uncertain given the possibility of an ecosystem irreversibly switching to a less desirable state (Perrings and Walker 1997). However, enough is generally known about the effects of different land management practices to make reasonable judgements about likely outcomes.

To summarize, a whole farm analysis in the form of a case study will help to quantify the on-farm contribution of native vegetation, to highlight options and allow for more informed decision-making. The next sections show how an evaluation can be performed if time and resources are limited.

RAPID APPRAISAL OF THE CASE FARM

Selection of the case study farm

The Trust for Nature initiated a survey in 1995 to locate and characterize remaining native grasslands in southwestern Victoria (Barlow 1996). The owners of the case farm responded to an invitation seeking the involvement of farmers with native grasslands. Two native grasslands on the property were surveyed at this time. The owners contacted conservation agency officers and the University of Melbourne in late 1997 seeking advice on how to increase production from the native grasslands to help meet their short-term cash flow problems. Specific information was sought on the most appropriate fertilizer application strategy to increase production, while retaining the native grass base of the pasture.

A team approach to assessment

A small specialist team was assembled to visit the property in November 1997 with the following objectives: (a) to evaluate the income and conservation effects of altering management of the grasslands, and (b) to explore alternative strategies for increasing income on the farm while leaving the grassland intact.

Participants included a farm management consultant, an economist, and conservation officers from two agencies. The duration of the visit was one day.

The programme included:

1. clarifying objectives and deciding which paddocks to inspect;
2. inspecting two native grasslands and (briefly) three other pasture types;
3. listing species by abundance, estimating the percentage contribution to dry matter of each vegetation class, noting current stocking and estimating carrying capacity, noting management issues for each paddock;
4. obtaining basic economic and financial data;
5. discussing options at the farm level, and ranking options in terms of income potential; and
6. discussing management possibilities for the native grassland.

This approach was designed to capture all of the necessary information to answer the following three questions in a single visit.

1. What are the options and likely economic outcomes for seeking to raise income from the native grasslands? This is a question of immediate interest to the property owners.
2. Given that the native grasslands have high conservation values, can farm business income be lifted in other ways, while managing for the conservation values? This is the question of interest from a public conservation viewpoint.
3. If there are no alternative opportunities on-farm, what options are available to meet the owners' income needs while protecting the public conservation values?

The following sub-sections provide an overview of (a) the conservation values of the native grasslands, (b) the existing farming system and (c) how the native grassland fits into that farming system.

Conservation overview

The Trust for Nature survey (Barlow 1996) has described the two native grassland paddocks (Sandy paddock and Rocky paddock) as *Microlaena/ Danthonia* grassland. The two paddocks both have a very high diversity with in excess of 70 species of grasses, lilies, daisies and orchids (Barlow 1996). The nearest sites with significant conservation values are on adjoining roadsides, some parts of which have a very rich species diversity. There are several other isolated sites in the district which are also located on private farmland (Barlow 1996).

Sandy paddock: The grassland in this 32 ha paddock was found to have a total of 70 species which included the nationally vulnerable *Glycine latrobeana* (Glycine Clover), as well as 11 species of regional significance. There were also 15 exotic species which comprised an estimated 30% of total cover at the time of survey.

Rocky paddock: The grassland in this 25 ha paddock is of lesser conservation value with 76 species, of which 26 were introduced. Eight species are significant for conservation at a regional or higher level, with *Glycine latrobeana* also found here. It was found to have a higher proportion of non-native species, which at the time of the survey accounted for 50% of total cover. As this grassland is situated on relatively fertile volcanic soils, the weeds will be difficult to control while maintaining conservation values.

The high diversity of species remaining in these two paddocks is most likely related to a combination of relatively light stocking, a limited history of fertilizer use, and a general absence of cultivation. Given the historical loss of native grasslands over most of southwestern Victoria, it is reasonable to assume that changing management to more intensive land uses will quickly lead to a loss of conservation values in these two paddocks.

The current farm system

The case study farm is 400 ha and runs a flock of 1 700 superfine Merino ewes, 500 of which are joined to British breed rams to produce fat lambs. It is run by a family partnership which supports a young family and a retired parent. Because it is a relatively small farming operation, one of the owners does wool classing to earn additional income. However, this is not sufficient to meet the owners' aspirations and borrowing commitments, and so further increasing income is an over-riding goal for them. To this end, the actual size of the farm business has been expanding and the level of management has been intensifying. For example, there are 3 700 grown sheep presently run compared to 2 100 six years ago, while expenditure on superphosphate has risen from nil to $AUD6,000 per annum over this period. In production terms, the owners are seeking to increase further the flock from 3 700 to 6 000 adult sheep with wool production to rise from 30 kg to 50 kg per hectare.

The current farm financial position is summarized in Table 1. Beyond providing for family needs, the farm enterprise must generate sufficient income to service existing business debt. After cash costs have been deducted from cash income, there is only $AUD42,000 left over to cover depreciation of assets, debt servicing and family living expenses, and any development expenditure. Off-farm income of about $AUD15,000 per annum is both important and insufficient to meet the family and business goals.

Table 1. Whole farm budget for case study farm.

	$'000 (AUD)
Farm Income	
Wool sales	85
Livestock sales	15
Total Income	100
Farm Costs	
Sheep enterprise costs	25
Other variable costs	12
Overhead costs	19
Total Costs	56
Net Cash Flow	42

In summary, the farm is a small one, and family income is already being supplemented by off-farm work. The long-term viability of the operation is, therefore, likely to depend on significant intensification of operations, an expansion in area, an increase in off-farm income or addition of another more intensive activity such as cropping on part of the farm. If the present owners were forced to sell, the conservation problem would not necessarily disappear, as new owners are likely to also adopt an intensified farming system.

The place of the native grasslands within the farm system

The farm is comprised of two separate blocks. Both native grassland paddocks are situated on the same block which is several kilometres from the home farm. It is used to run wether sheep. This is the strategic value of the grassland to the whole farm enterprise. The block also has a cropping paddock and some saline areas that are planted to tall wheat grass. The purpose of fertilizing these paddocks is to increase carrying capacity of the pastures, and to reduce the amount of supplementary feeding required for the wethers in autumn and winter. While the grazing benefits from the native grasslands are recognized by the owners, there is insufficient green feed produced to carry the wethers through the year without supplementary feed.

There appear to be no significant interdependencies, whereby the rest of the farm benefits directly from the native grasslands. While these paddocks could be used for young sheep which might not scour as readily in winter as on lush introduced pasture, logistical considerations make this impractical. The native grasslands could only carry small flocks over winter, and the effort required to split up the weaners and drive them to the block is not justified. However, there are some complementarities with the small saline areas adjacent to the native grassland which have been sown to tall wheat grass. These areas are usually stocked from December through to late winter when production from the native grasslands is very low. Nevertheless, supplementary feeding of the wethers is still necessary in most years, whereas wethers on winter-growing pastures on the home block do not have to be fed.

Current stocking on Sandy paddock is estimated to be between 2.5 and 3.7 dry sheep equivalent ha^{-1}, excluding supplementary feeding. The soils are a poor sandy type. At the time of inspection native grasses (mainly *Microlaena stipoides, Poa* spp. and *Danthonia* spp.) were estimated to be contributing 40% of dry matter. Native forbs were contributing an estimated 2% dry matter. Few orchids, lilies and daisies were then visible due to the very dry season and the lateness of the inspection. Desirable exotic species contributed little to production — 5% dry matter was from introduced annual grasses and 2% from introduced legumes. The remainder was mostly undesirable exotic species.

On Rocky paddock, dry matter at the time of inspection was estimated at 60% from native grasses, 5% introduced legumes (mainly sub clover), and 35% other species (nearly all onion grass). Until very recently this paddock has been run as a single unit with Sandy paddock. Given the higher fertility of this paddock, it is likely that the effective stocking rate is between 3.7 and 5.0 dry sheep equivalent ha^{-1}. Five hundred and fifty wethers have been stocked on both these paddocks for most of the year, with access to 12 ha of tall wheat grass for 8–9 months and with supplementary feeding.

FERTILIZING THE GRASSLANDS

Conservation implications of fertilizing the native grasslands

When the property was purchased 15 years ago, the grasslands had received little or no fertilizer. After purchase, the grasslands were fertilized for about five years. As the intention is to apply fertilizer infrequently, the owners do not see much risk to the conservation values and have the perspective of "nothing ventured, nothing gained". By contrast, from a public policy viewpoint a precautionary approach is justified (Department of Natural Resources and Environment 1997). Ecologists would likely see increasing pasture productivity as involving considerable uncertainty. Given the low fertility adaptation of native grasslands in Australia, the consequences of fertilizer use and associated increase in stocking rates may be an irreversible loss of species diversity due to competition from exotic species and intolerance to fertilizer (Kirkpatrick *et al.* 1995).

The owners may also be made worse off by fertilizing the native grasslands if it is technically feasible to change the farm system to achieve more of the complementarities that have been identified in the earlier section (Mitchell 1994; Crosthwaite 1997). These complementarities are not easily measured by contrast to the gain from an increase in stocking rate. Further, at least some of the benefits will still be achievable with a pasture based on a few native grasses, compared with a grassland rich with other species.

Estimating the gain from fertilizing the native grassland

Fertilizing the grassland must pass three tests to be worthwhile to the owners: (a) it must be profitable; (b) it must be affordable; and (c) it must represent a better use of scarce resources than any alternative use.

The potential gain from fertilizing the 32 ha Sandy paddock in the first year is an estimated increase in stocking rate of 3.7 dry sheep equivalent/ha resulting in net income rising by $AUD1,394. This increases to $AUD2,154 in the second year when the stocking rate is expected to stabilize at 4.9 dry sheep equivalent/ha (Table 2). These results are based on a gross margin of $AUD19/dry

Table 2. Projected benefit from fertilizer application — Sandy paddock.

	Year 1	Year 2
INCOME		
Increase in stocking rate — dry sheep equivalents	3.7 dse ha^{-1}	4.9 dse ha^{-1}
Gross margin @ $19/dse	$2,280	$3,040
Total cash in	**$2,280**	**$3,040**
COSTS		
Superphosphate 70 kg/ha @$351/tonne	$796	$796
Spreading cost @ $40/tonne	$91	$91
Total cash out	**$886**	**$886**
Tax	$0	$0
NET CASH FLOW	**$1,394**	**$2,154**
ECONOMIC ANALYSIS		
1. Net cash flow after tax	$1,394	$2,154
2. Return to marginal capital		
Extra wethers purchased @ $25/head	$3,000	$4,000
Return to marginal capital	46%	54%
FINANCIAL ANALYSIS		
Cost of deferred payment on superphosphate	$18	$18
NET CASH FLOW AFTER INTEREST	$1,376	$2,136

sheep equivalent, a fertilizer cost of approximately $AUD900, and zero tax (because taxable income is negative due to high farm debt). While these are not large increases in income, they represent significant rates of return on the additional capital investment and would be important to the owners to reduce the present level of outstanding debt. Return to the additional capital invested is 46% in the first year and 54% thereafter (Table 2). These results are quite robust in face of changing assumptions on both wool prices and expected stocking rate increase. For example, using fertilizer would be profitable even if wool prices fall drastically. If the gross margin/dry sheep equivalent were to fall to $AUD11, the return to marginal capital is still an acceptable 15% even in the first year. Alternatively, if a stocking rate increase of only 1.0 dry sheep equivalent/ha were to be achieved and gross margin is at least $AUD15/dry sheep equivalent, the return to additional capital will exceed 15%.

The owners can easily afford the necessary outlays. Cash flow is positive in year one, and the only borrowing cost is a small charge for deferring payment for fertilizer until after payment for the wool is received. No direct outlays are required for extra livestock as they can be reared on the farm.

Finally, available resources are scarce. The owners cannot afford to apply as much fertilizer as they would like across the farm, so the gain from fertilizing the native grassland must be compared to using the fertilizer elsewhere. The net gain should be calculated as: net cash flow from fertilizing the native grassland *less* net cash flow from fertilizing a different area *equals* net gain/loss from fertilizing the native grassland.

If the first year increase in stocking rate on an alternative site is 3.0 dry sheep equivalent ha^{-1} compared to an increase of 3.7 dry sheep equivalent ha^{-1} on the native grassland, the actual net gain from fertilizing the native grassland is only $AUD744 (Table 3). If the net increase in stocking rate on the other site were reduced to only 2.5 dry sheep equivalent ha^{-1} the benefit would only be $AUD1,000. Naturally, there would be no net gain from fertilizing the native grassland if the increase in stocking rate on the alternative area is the same or higher.

The sum required to compensate the owners for not fertilizing

An annual "compensating payment" of $AUD2,154 (Table 2) would almost certainly convince the owners not to fertilize the Sandy paddock in later years when the owners have the resources to fertilize as much as is profitable across the farm. However, the preceding calculations suggest that in the first year, the foregone benefit is likely to be under $AUD1,000. It may even be nil if stocking rates cannot be raised higher on Sandy paddock than elsewhere on the farm.

Foregone income has been used to calculate the compensation that might be required to achieve grassland conservation outcomes in the short term. However, if alternative ways of increasing farm income are found, the need for income from the grassland will be less pressing and lower payments may be acceptable. Nevertheless, as long as conservation of the native grasslands risks medium-term farm business survival, payments equivalent to foregone income will be required. Some alternatives are now considered.

Alternative strategies immediately available

There are several apparent options for increasing income from the farm. An important question is firstly whether these options are both profitable and financially more feasible than fertilizing the native grassland. Secondly, it may be possible within cash flow constraints to both fertilize the native grassland and pursue the other option(s). The available options depends partly on how other areas of the farm are currently being used.

The home block carries ewes and lambs, as well as replacement ewes and some wethers. Most of the paddocks are sown to ryegrass and subterranean clover which are regularly fertilized and carry approximately 15 dry sheep equivalent ha^{-1}. Dry matter currently standing in the inspected paddocks was estimated at 20% introduced perennial grasses, 15% legumes, 15% introduced annuals and 50% onion grass. These paddocks are, in the opinion of the consultant, being generally well run. Sowing improved species into a cultivated seed-bed might increase production, but the gain would still be marginal. Given the tight cash situation, this option is not recommended by the consultant.

There are some other run-down paddocks on the home block that have not yet been brought into the present fertilizer programme. These paddocks were last cultivated prior to 1965 and fertilizer has not been applied for many years. The paddocks carry about 7.0 dry sheep equivalent ha^{-1}. Dry matter was estimated at 10% introduced perennial grasses, 5%

Table 3. Net income in Year 1 from fertilizing Sandy paddock compared to an alternative site.

	Sandy paddock	Other site	
Increased carrying capacity — dry sheep equivalents	3.7 dse ha^{-1}	3.0 dse ha^{-1}	2.5 dse ha^{-1}
Return to capital	46%	27%	20%
Net cash flow after interest	$1,376	$632	$376
Difference compared to fertilizing Sandy paddock		–$744	–$1,000

legumes (mainly suckling clover), 15% introduced annuals and 70% onion grass. There is a small amount of *Danthonia* spp. in the paddocks. The consultant believes that production can be increased to 15 dry sheep equivalent ha^{-1}. The owners will begin to regularly fertilize the remaining run-down and unfertilized paddocks this year. By promoting the growth of the clovers, this strategy would particularly increase the effective winter carrying capacity, more so than increasing total annual dry matter production. Pasture topping for weeds and direct drilling pasture seed are further management options once production has increased. The tight cash flow will not permit sowing into a cultivated seed-bed.

Apart from increasing the number of paddocks being regularly fertilized, cropping a larger area is the other feasible option. One of the possible cropping paddocks, a 10 ha paddock on a sandhill, was inspected. It currently carries 170 ewes and lambs from June to January. It was last cropped seven years ago and has received no fertilizer since that time. The consultant regards the sandhill as a very valuable paddock, given a shortage of well drained land in the district, and believes its current contribution to farm income can be further raised by a factor of 400–500%. Long-term options for the sandhill include growing grapes or olives, both of which would require substantial investment.

Winning a breathing space

The initial evaluation suggested that action to increase income on the home block would give a breathing space of one or two years to develop a whole farm plan which both increased income and protected the native grassland. Fertilizing the home block and additional cropping were established as priorities, because cash availability appeared to restrict the number of options that could be managed and these two options were more profitable than fertilizing the native grassland.

However, some crucial factors remain. The most important is that fertilizing the native grassland could be achieved within the tight financial constraints presently facing the owners. In drawing up their plans, the owners were able to count on obtaining fertilizer on credit and also on increasing stocking numbers on the fertilized native grassland with sheep from parts of the home block that are to be cropped. The farm consultant has also revised his estimate of the potential stocking rate of Sandy paddock upwards after seeing similar areas elsewhere on the property. Sandy paddock currently carries 2.5 and 3.7 dry sheep equivalent ha^{-1}, with the initial assessment of fertility suggesting it could carry at most 5.0 dry sheep equivalent ha^{-1} or a little higher. This has since been revised to approximately 7.5 dry sheep equivalent ha^{-1}.

The owners ordered a bulk load of fertilizer to be delivered in May 1998. The plan included fertilizing Sandy paddock, but not Rocky paddock. After discussion with a conservation agency officer, the plan was first revised to exclude the most valuable two hectares of native grassland in Sandy paddock. Later, it was agreed that the whole paddock would be excluded for one year. The owners chose to use the two tonnes of fertilizer planned for Sandy paddock on the native grassland in Rocky paddock.

THE FUTURE — ARE THERE ALTERNATIVE ROUTES TO CONSERVATION?

As well as agreeing to a breathing space of one year, the owners have agreed to take part in a further review of alternatives. It is clear, however, that the option of fertilizing Sandy paddock remains high on their agenda. The farm is a small one, the owners are relatively young with a family, and they are likely to be continually looking for ways to increase income for the next 10 years or more. There are three broad approaches which may increase farm income sufficiently to allow a negotiated solution which conserves the native grassland.

The first approach is to find a path for farm development which reduces dependence on the native grassland. The rapid appraisal identified some development opportunities on the farm, and may have missed others that should be explored. Discussions with the financing body may open up options, including possible restructure of farm debt, which allow other investments to occur sooner than otherwise.

The second approach is to clarify the economic value of the native grassland to the farm system in different uses. Fertilizer application may not be as valuable as expected. Setting up a trial will allow expected dry matter production and stocking increase to be tested, as well as indicate changes in botanical composition. Harvesting native flora and native grass seed may combine well with continued grazing to increase the economic value of the native grasslands. Consideration of the potential economic value of the unique genetic resources in this native grassland to Victorian and Australian agriculture in 50 years time, may lead to some form of "reimbursement" to the owners as custodians (Crosthwaite 1997).

The third approach is to purchase, in one form or another, the rights to development (Young *et al.* 1996) on the grassland which will effectively compensate the owners for the foregone income. Options to explore include seeking funding to support a long-term management agreement, a land swap or to buy the land and lease it back on concessional terms.

CONCLUSION

The "here and now" issues faced by both farmers and conservation agencies have been considered in this chapter. Given the frequency with which conservation agencies find important conservation sites on private land on the brink of destruction, practical and cheap methods of rapid appraisal are needed to assess both remnant vegetation and farming systems. Regional, state and national strategies relevant to conservation provide the strategic framework within which such appraisals should be conducted.

The case example presented in this chapter involved the formation of a small team with relevant ecological, agronomic and economic expertise to spend one day assessing a farm. The process helped educate all of the parties, not least the farm consultant who deals with many farmers each year. The method involves collecting the minimum of technical and economic information necessary to broadly characterize the native grassland and other pastures on the farm, to identify development opportunities open to the owners, and to quantify the private gain or loss from changing management of the native grassland. The whole farm approach that is recommended allows a focus on strategic planning and is likely to reveal options, not otherwise obvious, for influencing farmer behaviour.

On the case study farm, no feasible options have yet been found for changing the management system to increase farm business income while protecting the native grasslands. In fact while one grassland has been protected for an extra year, another has been fertilized. The income foregone by continuing current management of the remaining native grassland is small. The public conservation values do not have to be high to exceed the private costs. However, public values are not the critical factors influencing how farmers will act. The importance of the extra income to farms with short-term cash flow difficulties, like the case study farm, cannot be under-estimated.

The breathing space of one year that the owners have agreed to offers an opportunity to find business development alternatives that may reduce the pressure to exploit the native grassland. Alternatively, it may allow a package to be developed which compensates the farmer for income losses as part of a change to property rights. The challenge is a big one, as conservation agencies are unlikely to fund such a package given that there are native grasslands of higher conservation priority. Nevertheless, investigating these possibilities further is important before accepting the loss of the native grassland as a *fait accompli*.

While strategies to protect native grasslands and other remnant vegetation on private land have been adopted by Australian governments, losses of important conservation sites are continuing. The approach outlined in this chapter may assist in finding solutions.

ACKNOWLEDGEMENTS

This chapter derives from a project (UME 25) that is being undertaken within the National Remnant Vegetation Program being funded by the Land and Water Resources Research and Development Corporation and Environment Australia. The Department of Natural Resources and Environment (Victoria) has provided additional support for the work.

The authors have benefited from information provided by the case farm owners and farm consultant. The authors retain responsibility.

REFERENCES

Barlow, T., 1996. The Vegetation of 25 Grassland and Grassy Woodland Remnants on Farms in south-western Victoria. Trust for Nature (Victoria): Melbourne, Victoria.

Bowers, J., 1999. Policy instruments for the conservation of remnant vegetation on private land. *Biol. Conserv.* **87:** 327–39.

Brown, J. R. and MacLeod, N. D., 1996. Integrating ecology into natural resource management policy. *Environ. Manage.* **20:** 289–96.

Crosthwaite, J., 1997. Economic benefits of native grasslands on farms. A report to Environment Australia, Grassland Ecology Program Project Number GEP 017. Department of Natural Resources and Environment: Victoria.

Department of Conservation and Natural Resources, 1994. Western (basalt) Plains Grassland Community — Action Statement No. 53. Department of Conservation and Natural Resources: Melbourne, Victoria.

Department of Environment, Sport and Territories, 1996. The National Strategy for the Conservation of Australia's Biological Diversity. Department of Environment, Sport and Territories: Canberra, Australian Capital Territory.

Department of Natural Resources and Environment, 1997. Victoria's Biodiversity: Directions in Management. Department of Natural Resources and Environment: Melbourne, Victoria.

Glenelg Regional Catchment and Land Protection Board, 1997. Glenelg Regional Catchment Strategy. The Board: Hamilton, New Zealand.

Kirkpatrick, J., McDougall, K. and Hyde, M., 1995. Australia's Most Threatened Ecosystems — the Southeastern Lowland Native Grasslands. Surrey Beatty & Sons: Chipping Norton, New South Wales.

Lefroy, E. C., Salerian, J. and Hobbs, R. J., 1993. Integrating Economic and Ecological Considerations: A Theoretical Framework. Pp. 209–44 *in* Reintegrating Fragmented Landscapes. Towards Sustainable Production and Nature Conservation ed by R. J. Hobbs and D. A. Saunders. Springer-Verlag: New York.

Mitchell, M. (ed.), 1994. Australian native grasses. An untapped resource. Agriculture Victoria, Rutherglen and Department of Conservation and Natural Resources: Ballarat, Victoria.

Perrings, C. and Walker, B., 1997. Biodiversity, resilience and the control of ecological-economic systems: the case of fire-driven rangelands. *Ecol. Econ.* **22:** 73–83.

Swanson, T., 1996. Biodiversity as information. *Ecol. Econ.* **17:** 1–8.

Wibberley, G., 1982. Public pressures on farming: the conflict between national agricultural and conservation policies. *Farm Manage.* **4:** 373–79.

Young, M. D., Gunningham, N., Elix, J., Lambert, J., Howard, B., Grabosky, P. and McCrone, E., 1996. Reimbursing the Future. An evaluation of motivational, voluntary, price-based, property-right and regulatory incentives for the conservation of biodiversity. Biodiversity series, Paper No. 9. Biodiversity Unit, Department of Environment, Sport and Territories: Canberra, Australian Capital Territory.

CHAPTER 69

Conserving threatened fauna in production forests: the Tasmanian process

S. A. Munks[1] and R. J. Taylor[2]

The Tasmanian Forest Practices Code requires that threatened fauna species likely to be affected by forestry activities are taken into account in Timber Harvesting Plans. To satisfy this requirement a process was developed which ensures that Forest Practices Officers responsible for drawing up timber harvesting plans have access to all known locations of species, simple descriptions of their habitats and recommendations for their conservation. Success of the process depends on providing information in a "user-friendly" format, co-operation between participants, research to enable the development of meaningful management recommendations and monitoring of the outcomes of the process.

INTRODUCTION

REVISION of the Tasmanian Forest Practices Code in 1993 introduced a series of measures to cater for fauna conservation (Taylor 1991a and b), including the requirement that threatened species be taken into account in Timber Harvesting Plans on Crown land, and also on Private land if the species is inadequately reserved on Crown land. Threatened fauna species lists were subsequently developed (Invertebrate Advisory Committee 1994; Vertebrate Advisory Committee 1994; Taylor and Bryant 1997) and included in the schedules of the Tasmanian *Threatened Species Protection Act 1995*. This chapter outlines the process which was implemented to satisfy the intent of the Code in relation to management of the listed threatened species which occur in the production forest matrix. It also discusses ways in which the process is being further developed and formalized to meet the requirement of the Tasmanian Regional Forest Agreement (1997).

THE PROCESS AND ITS USE

The steps involved in the assessment and planning process for Threatened Fauna during the drawing up of Timber Harvesting Plans are based on consideration of range and likely presence of habitat (Fig. 1.) The "Threatened fauna manual for production forests in Tasmania" (Jackson and Taylor 1995) plays a central role in the process. This manual includes the 79 threatened invertebrate species and nine threatened vertebrate species listed in the Tasmanian *Threatened Species Act 1995* which occur in vegetation types or areas affected by forestry activities. The manual is divided into two parts and is designed to be "user friendly". The first part of the manual contains map based information on the known range of these threatened fauna species and their habitats to enable a determination to be made of the likelihood of the presence of threatened fauna species. Information on the biology, distribution and habitat requirements of each threatened species is presented in the second part of the manual. Selected Forestry Officers from each district and company also undergo training in the identification of threatened species habitat. Site surveys by specialists are undertaken in some cases.

Advice is sought by the Forest Practices Officer if the coupe or area to be harvested is identified as being within the range of a threatened species and contains habitat important for that species. The actions required are formulated by the Forest Practices zoologist in consultation with Parks and Wildlife specialists following the procedure outlined in the Tasmanian Forest Practices Code (Forestry Commission 1993). An example of logging recommendations for the protection of habitat important for two threatened species, the Grey

---

[1]Forest Practices Board, P.O. Box 180, Kings Meadows, Launceston, Tasmania, Australia 7249.
[2]Forestry Tasmania, G.P.O. Box 207B, Hobart, Tasmania, Australia 7001.

Pages 670–74 *in* NATURE CONSERVATION 5: NATURE CONSERVATION IN PRODUCTION ENVIRONMENTS: MANAGING THE MATRIX ed by J. L. Craig, N. Mitchell and D. A. Saunders. Surrey Beatty & Sons, 2000.

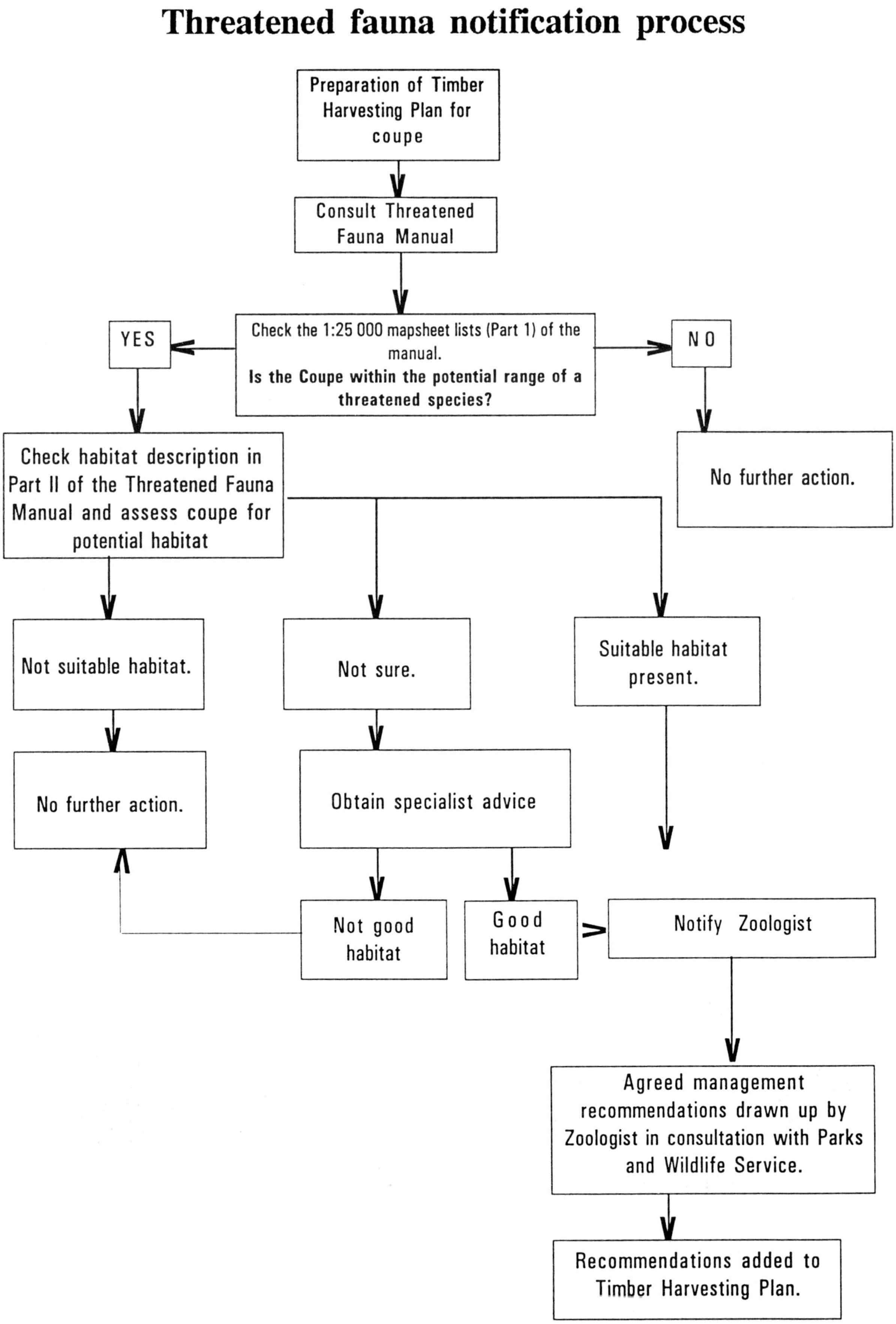

Fig. 1. Flow diagram illustrating stages in "Threatened Fauna Notification Process".

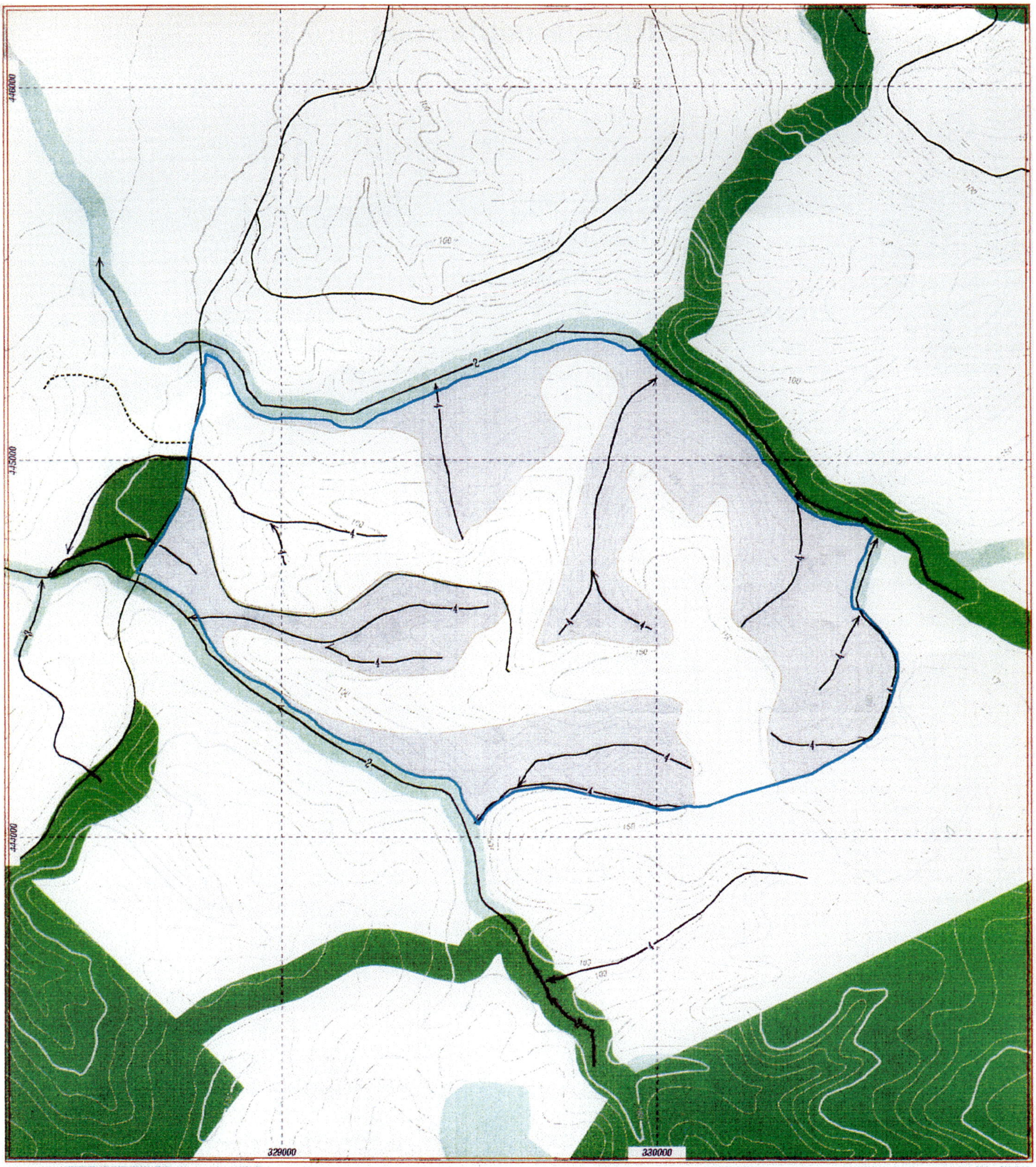

Fig. 2. Coupe map from the north-west of Tasmania illustrating implementation of actions recommended for the giant freshwater crayfish and grey goshawk via the "Threatened Fauna Notification Process". Shaded area = reserved area. Note exclusion of most riparian areas alongside small streams.

Goshawk *Accipiter novaehollandiae* and the Giant Freshwater Crayfish *Astacopsis gouldi* is given in Figure 2. The recommendations for both these species were based on available information on their habitat requirements (Mooney and Holdsworth 1988; Brereton and Mooney 1994; Lynch and Bludorn 1997) and involve additional reservation of riparian areas.

MONITORING

Monitoring of the Threatened Fauna Notification Process and implementation of the resultant recommendations is vital for subsequent refinement and to gauge the effectiveness of the process. There has been a significant increase in notifications between 1995/96 and 1996/97 (Table 1). Requests for advice for all 18 species increased almost four fold (from 116 to 428). This increase is partly due to the increased acceptance and utilization of the Threatened Fauna Notification Process but is also due to a significant increase in forestry activities within the range of certain listed species.

Monitoring of the application of recommendations in the field is undertaken as part of the annual auditing of forest practices carried out by the Forest Practices Board. The results of the 1996/97 audit, with respect to threatened fauna, are illustrated in Table 1. This audit samples at least 15% of the 800—1 000 Timber Harvesting Plans each year. Two of the nine fauna questions assessed by the monitors cover the threatened fauna process. One question aims to assess if the process has been followed correctly and the other checks if the recommended prescriptions have been implemented in the harvest area. The results for 1996/97 suggest that there is still room for improvement with respect to following the process but where it has been followed the prescriptions are generally implemented satisfactorily in the field.

To date, assessment of the success of management for ameliorating the effects of forestry operations on threatened species has only been carried out in detail for the endangered Wedge-Tailed Eagle *Aquila audax fleayi* (Mooney and Taylor 1996). Forestry activities threatening the long-term survival of this species include disturbance and loss of nests during logging operations. Monitoring of the use of protected and unprotected nests in disturbed areas showed that implementation of protection measures for nest sites markedly improved the breeding success of the eagle pair subject to the disturbance. Collection of such information is important to maintain confidence amongst forest managers in the actions recommended via the Threatened Species Notification Process.

DEVELOPING RECOMMENDATIONS — THE IMPORTANCE OF RESEARCH

The ultimate success of the process depends upon a thorough knowledge of the distribution and habitat of a species and a knowledge of the impacts of forestry operations. Since few of the species, particularly the invertebrates, are known well enough to provide such information, research is being undertaken into the requirements of threatened species as funds permit. The results of two recently completed research projects highlight the importance of such research to aid in the development of meaningful recommendations.

The Scottsdale Burrowing Crayfish *Engaeus orramakunna* was listed as Vulnerable in the *Tasmanian Threatened Species Act 1995* due to its restricted distribution (only three known localities in 1995) and presumed sensitivity to habitat disturbance. However, a recent study by Doran and Richards (1996) has extended the distribution of the species from the three known original sites to seventy-nine known sites which include several gazetted reserves. The study also suggests that the species is highly tolerant of disturbance due to forestry activities with healthy, reproductively active individuals found in high numbers in culverts, streams in pine plantations and small drainage areas in cleared and burnt plantations.

Another recent study investigated the distribution and habitat characteristics of a Stag Beetle *Hoplogonus simsoni* (Meggs 1997). This study confirmed the restricted distribution of the beetle and supported its listing as Vulnerable. Generalized linear modelling was used to identify the characteristics of habitat important to the species. The models produced were coupled with spatial information to produce Geographic Information System maps predicting the distribution and abundance of the beetle throughout its range. This information is currently being used to develop management recommendations, in collaboration with relevant experts and forest managers, for areas of importance to the species and the forest industry.

FUTURE DIRECTIONS

The success of the Threatened Fauna Notification Process relies heavily on co-operation between government agencies, forest companies, private forest consultants and private landowners. It also depends on the quality and regular updating of information provided in the Threatened Fauna Manual. The need to supply advice for an increasing number of areas and individuals, the lack of funds for additional zoological trained staff and the costs

Table 1. Outcome of recommended actions given for the conservation of threatened fauna species.

Land tenure	Number of coupes 95/96	Number of coupes 96/97	No changes required to operation 95/96	No changes required to operation 96/97	Recommendations given 95/96	Recommendations given 96/97	Part or all of the coupe excluded from logging 95/96	Part or all of the coupe excluded from logging 96/97
Private	72	150	67%	22%	30%	78%	3%	53%
State Forest	37	106	70%	7%	16%	92%	14%	69%

involved in regularly updating the Threatened Fauna Manual has lead to the need to develop a Decision Support System (Expert System). This captures current scientific and expert knowledge with respect to each species and enables the forest manager to arrive at the recommended action for some harvest areas. It is hoped that use of the programme by forest managers in the future will free resources for more pre-logging site assessments of problem areas and scientific monitoring projects.

ACKNOWLEDGEMENTS

The authors thank Kylie Dillon for her assistance in the preparation of figures for this manuscript and Circular Head District, Forestry Tasmania for the Timber Harvesting Plans coupe map (in particular Dot Drevenkar and Phil Donnelly). We also thank Graham Wilkinson for his valuable comments on the first draft of this manuscript.

REFERENCES

Brereton, R. and Mooney, N., 1994. Conservation of the nesting habitat of the grey goshawk, *Accipter novaehollandiae* in Tasmanian State Forests. *Tasforests* **6:** 79–91.

Doran, N. and Richards, K., 1996. Management requirements for rare and threatened burrowing crayfish in Tasmania. Report to the Tasmanian Regional Forest Agreement Environment and Heritage Technical Committee.

Forestry Commission, 1993. Forest Practices Code. Forestry Commission: Hobart, Tasmania.

Horwitz, P., 1994. Distribution and conservation status of the Tasmanian giant freshwater lobster, *Astacopsis gouldi* (Decapoda: Parastacidae). *Biol. Conserv.* **69:** 199–206.

Jackson, J. and Taylor, R., 1995. Threatened fauna manual for production forests in Tasmania. Forestry Tasmania and Forest Practices Board: Hobart, Tasmania.

Lynch, T. P. and Bludorn, D. R., 1997. Reservation assessment and habitat requirements of the giant Tasmanian freshwater lobster, *Astacopsis gouldi*. Report to the Tasmanian Regional Forest Agreement Environment and Heritage Technical Committee.

Meggs, J., 1997. Simsons stag beetle, *Hoplogonus simsoni*, in North-east Tasmania: Distribution, habitat characteristics and conservation requirements. Report to Forest Practices Board and Forestry Tasmania.

Mooney, N. and Holdsworth, M., 1988. Observations on the use of habitat by the grey goshawk in Tasmania. *Tas. Bird Report* **17:** 1–12.

Mooney, N. J. and Taylor, R. J., 1996. Value of nest site protection in ameliorating the effects of forestry operations on wedge-tailed eagles in Tasmania. Pp. 275–82 *in* Raptors in Human Landscapes — Adaptations to Built and Cultivated Environments ed by D. M. Bird, D. E. Varland and R. J. Jose. Academic Press: London.

Taylor, R. J., 1991a. Fauna management practices in state forests in Tasmania. Pp. 259–64 *in* Conservation of Australia's Forest Fauna ed by D. Lunney. Royal Zoological Society of New South Wales: Sydney, New South Wales.

Taylor, R. J., 1991b. The role of retained strips for fauna conservation in production forests in Tasmania. Pp. 265–70 *in* Conservation of Australia's Forest Fauna ed by D. Lunney. Royal Zoological Society of New South Wales: Sydney, New South Wales.

Taylor, R. J. and Bryant, S. L., 1997. Compilation of a list of threatened invertebrates: The Tasmanian experience. *Mem. Mus. Vict.* **56** (2): 605–09.

Vertebrate Advisory Committee, 1994. Native vertebrates which are rare or threatened in Tasmania. Edition 1. Species at Risk, Tasmania — Vertebrates. Parks and Wildlife Service: Tasmania.

CHAPTER 70

The policy challenge of conserving biological diversity in production environments

Michael Harte[1] and Glen Lauder[2]

The term biodiversity has come to replace terms like "wildlife" in the environmental policy lexicon because it better portrays the idea of ecological complexity and the interrelated nature of the biotic and abiotic aspects of the environment. In this chapter an approach for guiding biodiversity policy development in production environments is described. Two pervasive policy problems: uncertain ecological foundations for policy development; and land management goals inconsistent with indigenous biodiversity conservation are briefly explored. The chapter concludes with two suggested policy responses: integrating science and policy; and recasting biodiversity management as a social rather than scientific challenge.

INTRODUCTION

THE conservation and sustainable use of biodiversity is a contemporary international and national environmental policy issue. The term biodiversity has come to replace terms like "wildlife" in the environment policy lexicon because it better portrays the idea of ecological complexity and the interrelated nature of the biotic and abiotic aspects of the environment.

A wide range of actions intended to sustainably manage our biodiversity are being considered or undertaken by landholders, local government, research agencies, conservation groups, and central government departments. Indicative of this activity, the New Zealand Government is in the process of developing a Biodiversity Strategy for New Zealand. The goals of the Strategy reflect the need to manage biodiversity for all level of ecological organization — genetic, population, biological community, and landscape (Table 1).

Developing New Zealand's Biodiversity Strategy has demonstrated just how complex, and perhaps intractable the development of a single framework for biodiversity management is. Viewed scientifically, biodiversity has multiple levels — genetic, species, ecosystem and landscape. It is also characterized by a high degree of spatial and temporal variability and by uncertainty. From a social perspective, this complexity is exacerbated over the 70% of New Zealand dominated by production environments. These environments are shaped by management regimes that are often ambivalent to indigenous biodiversity, and at worst hostile to its existence.

In this chapter we sketch a framework for guiding biodiversity policy development in production environments. We briefly explore two pervasive policy problems: uncertain ecological foundations for policy development; and land management goals inconsistent with indigenous biodiversity conservation. We then suggest two policy responses: integrating science and policy; and recasting biodiversity management as a social rather than scientific challenge.

POLICY CHALLENGES

Uncertain ecological foundations

Although policies for managing biodiversity are often assumed to be underpinned by robust ecological principles and evidence, ecology is often

[1]New Zealand Seafood Industry Council PB 24-901 Wellington, New Zealand.
[2]Project Facilitator, New Zealand Biodiversity Strategy, P.O. Box 10-420, Wellington, New Zealand.
Pages 675–77 *in* NATURE CONSERVATION 5: NATURE CONSERVATION IN PRODUCTION ENVIRONMENTS: MANAGING THE MATRIX ed by J. L. Craig, N. Mitchell and D. A. Saunders. Surrey Beatty & Sons, 2000.

Table 1: Proposed goals of New Zealand's draft Biodiversity Strategy.

Goal 1 – Natural habitats and ecosystems

Maintain and restore a comprehensive and representative range of remaining natural habitats and ecosystems to a healthy functioning state, and sustain those features which support indigenous biodiversity in a range of more modified ecosystems (including those in primary production areas and urban environments).

Goal 2 – Indigenous species

Maintain and restore representative populations of all indigenous species in selected natural habitats where threats can be controlled or there are reasonable prospects for controlling them in the future.

Goal 3 – Genetic resources

Maintain the natural resilience and economic usefulness of our domesticated and cultivated species by conserving their genetic diversity.

unable to provide the degree of guidance and/or certainty sought by policy makers. Many of ecology's theoretical foundations have been challenged in recent times (see, for related discussions, Sagoff 1985; McIntosh 1987; Botkin 1990; Peters 1992; Pickett *et al.* 1992; Caldwell and Shrader-Frachette 1993; Demeritt 1994; Shrader-Frechette and McCoy 1994; Zimmerer 1994). Ecology has neither unambiguous definitions of basic concepts such species, niche, or ecosystems, nor a testable general theory capable of providing predictive or prescriptive guidance to environmental managers and analysts. Emerging ecological debates, both theoretical and empirical, also tend to undermine some of the ecological rationale for biodiversity management practices based around notions of stability and equilibrium.

It is becoming clear that there is no one state for biodiversity that can be defined as sustainable in production environments, or identified as the goal of sustainable land management. Take, for example, one of New Zealand's most problematic production environments — the South Island high country. The evidence of biodiversity recovery in the high country when grazing pressure is removed is mixed. Kerr and Treskanova (1993) point to the return of indigenous (and other exotic) vegetation species at the expense of *Hieracium* on one high country run once sheep were removed and rabbits controlled in 1991. O'Connor (1983), however, suggests that biodiversity conservation by exclusion of livestock and fire may be interesting ecologically but is often futile in conserving native flora and its dependent fauna. O'Connor points out that exotic shrub and tree species may invade grasslands under low grazing intensity in many localities and that other exotic elements already present may end up dominating.

Land management goals and indigenous biodiversity

Past government policies and associated social and institutional arrangements have reflected a preference for landscapes dominated by primary production and the economic, environmental and amenity services it offered. People were willing to accept the degradation of indigenous biodiversity to allow the expansion of this preferred landscape. The result is that:

> *Most of the New Zealand landscape is now ecologically hostile to many native species. Although nearly 30% of the land area is protected in the conservation estate, most of this is on steep and mountainous land* (Taylor and Smith 1997, P. 9.7).

Recent changes in attitudes towards indigenous biodiversity, together with changing economic incentives, regulatory frameworks, and a better understanding of ecological and economic realities, have forced a rethink of expectations for the nature of goods and services produced in our production environments, and we are now in the process of creating a new mix of land uses over much of New Zealand. The challenge for the Biodiversity Strategy is to ensure this new landscape is no longer hostile to indigenous biodiversity.

FINDING POLICY SOLUTIONS

Two key needs emerge from the development of the Biodiversity Strategy. These are to (1) establish a strong and consistent link between science and biodiversity policy development; and (2) to find ways of recognizing the social and cultural factors that determine both the content and implementation of biodiversity policy.

Integrating science and biodiversity policy

Recent developments in New Zealand's resource management framework have begun to expand the biodiversity policy horizon from a focus on species and populations to higher levels of ecological organization such as communities, ecosystems, and landscapes. Section 5(2)(b) of the *Resource Management Act 1991*, for example, refers to *safeguarding the life-supporting capacity of air, water, soil and ecosystems.* Putting such legislative intent into practice, however, requires that policy makers must have an understanding of controlling system processes; the context of system processes; the historical range of system perturbations; the evolutionary and physiological limits of system components; and the characteristics and effects of short and long-term phenomena.

The research community is increasingly well equipped to provide this information. A new ecology is emerging that highlights disequilibria, instability, and chaotic fluctuations in biophysical environments and blunts the distinction between natural and human systems (see, among others, McIntosh 1987; Allen and Starr 1988; Turner 1989; Noss 1990). Perhaps more importantly from a policy context, system processes, interactions and overall ecological context are increasingly emphasized, rather than the final state produced by these processes (Kay 1991a,b; Pickett *et al.* 1992; Walker 1992).

Better linkage of research and policy will help ensure that there is an appropriate synergy between policy and science. New Zealand's Biodiversity Strategy proposes to secure a lasting association between science and biodiversity policy by advocating initiatives that enhance the: identification of gaps in existing knowledge of biodiversity; establishment of inventories and monitoring biodiversity; co-ordination and prioritization of research; management and dissemination of knowledge; and building of knowledge through partnership with Maori.

Redefining biodiversity conservation as a social concept

While developing New Zealand's Biodiversity Strategy we deliberately cast biodiversity conservation as a social, rather than scientific issue. While acknowledging that biodiversity as a technical term in the biological sciences can be measured at a genetic, species, community, ecosystem or landscape scale, in the text of the strategy and its articulation in the public process focused on its human meanings: contributions to national and (bi-) cultural identity; economic and ecological utility; responsibilities to neighbours and to future generations.

A wider acceptance of policy makers of a social approach to environmental policy development in general may break a long tradition of issue polarization in the production environment by leading to a greater awareness of the need to understand the social context in which issues such as the conservation of biodiversity is debated. A social approach at its simplest means that policy makers actively seek to understand the assumptions tacitly shaping their own understanding of sustainable management; recognize other stakeholders' assumptions and values; and engage in the collective negotiation of meanings, definitions and assessments that form the basis for subsequent policy development.

Without an understanding of the social dimension of biodiversity management it is unlikely that biodiversity policy will be structured in a way that promotes solutions that are ecologically sound, socially acceptable and politically supported. For example, unconvinced by the economic and environmental arguments presented, many individuals and organizations still distrust the government's motives for calling for the sustainable management of biodiversity on private land and are reluctant to be involved in, and/or actively oppose the policy development process. Thus, if policy makers ignore the social context in which an environmental issue arises, an attempt to remedy one environmental problem may reduce the likelihood of achieving sustainable management due to an increased polarization of society's values and attitudes toward the environment, and government's role in its management.

REFERENCES

Allen, T. F. H. and Starr, T. B., 1988. Hierarchy: Perspectives for Ecological Complexity. University of Chicago Press: Chicago.

Botkin, D. B., 1990. Discordant Harmonies: A New Ecology for the Twenty-first Century. Oxford University Press: New York.

Caldwell, L. K. and Shrader-Frechette, K., 1993. Policy for Land: Law and Ethics. Rowan and Littlefield: Maryland.

Demeritt, D., 1994, Ecological, objectivity and critique in writings on nature and human societies. *J. Hist. Geog.* **10:** 22–37.

Kay, J. J., 1991a. The concept of ecological integrity, alternative theories of ecology and implications for decision-support indicators. Pp. 23–58 *in* Economic, Ecological, and Decision Theories: Indicators of Ecologically Sustainable Development ed by P. A. Victor, J. J. Kay and H. J. Ruitenbeek. Canadian Environmental Advisory Council: Ottawa.

Kay, J. J., 1991b. A Nonequilibrium Thermodynamic Framework for Discussing Ecosystem Integrity. *Environ. Manage.* **15:** 483–95.

Kerr, I. G. C. and Treskanova, M., 1993. Evaluation of the grasslands of Mount John for sustainable management. Unpublished Report to New Zealand Defence Forces.

McIntosh, R. P., 1987. Pluralism in Ecology. *Ann. Rev. Ecol. System.* **18:** 321–41.

Noss, R. F., 1990. Indicators for Monitoring Biodiversity: a Hierarchical Approach. *Conserv. Biol.* **4:** 355–64.

O'Connor, K. F., 1983. Land Use in the Hill and High Country. *In* Canterbury at the Cross-roads ed by R. D. Bedford and A. P. Sturman. Miscellaneous Series No 8. New Zealand Geographical Society Special Publication. Christchurch, New Zealand.

Peters, R. H., 1992. A Critique for Ecology. Cambridge University Press: Cambridge.

Pickett, S. T. A., Park, V. T. and Fiedler, P. L., 1992. The New Paradigm in Ecology: Implications for Conservation Biology above the Species Level. Pp. 66–68 *in* Conservation Biology: The Theory and Practice of Nature Conservation Preservation and Management ed by P. L. Fiedler and S. K. Jain. Chapman and Hall: London.

Sagoff, M., 1985. Fact and Value in Environmental Science. *Environ. Ethics* **7:** 99–116.

Shrader-Frechette, K. and McCoy, E. D., 1994. What Ecology Can Do for Environmental Management *J. Environ. Manage.* **41:** 293–307.

Taylor, R. and Smith, I., 1997. The State of New Zealand's Environment 1997. Ministry for the Environment and GP Publications: Wellington, New Zealand.

Turner, M. G., 1987. Landscape Ecology: the Effect of Pattern on Process. *Ann. Rev. Ecol. System.* **20:** 171–97.

Zimmerer, K. S., 1994. Human geography and the "new ecology": The prospect and promise of integration. *Annals Assoc. Ameri. Geog.* **84:** 108–25.

CHAPTER 71

Nature conservation in production environments: What do we know about managing the matrix?

Denis A. Saunders,[1] John L. Craig[2] and Neil Mitchell[2]

INTRODUCTION

THIS volume is the fifth in a series dealing with nature conservation. Each volume has been based on an international conference addressing a particular conservation theme. The first was on the role of remnants of native vegetation in conservation (Saunders *et al.* 1987). The aim of that volume was to look at methods of integrating research on, and management of, remnants in order to make more efficient use of the generally scarce resources allocated to nature conservation. The second volume developed directly from the first, but, it had a narrower focus; the role of corridors of vegetation in nature conservation (Saunders and Hobbs 1991). It addressed: inventory and assessment; the values of corridors; the movement of biota; and the management, establishment, maintenance and rehabilitation of corridors. The third volume had a broader focus addressing the reconstruction of fragmented ecosystems particularly those fragmented by agricultural use (Saunders *et al.* 1993a). The fourth volume concerned the role of networks in conservation (Saunders *et al.* 1995). It explored the themes that networks of conservation areas constitute our conservation resource and that networks of people constitute our conservation force. This volume concluded that effective conservation requires people who are networked for action (Craig *et al.* 1995).

Each volume has been based on the fact that loss of biological diversity is being recognized increasingly as <u>the</u> major national and international environmental issue (e.g., Commonwealth of Australia 1996a; Ministry of the Environment 1997) and on the need to do something about this biotic impoverishment and its ecosystem consequences. The effects of humans co-opting natural resources for their own use through agriculture, forestry, fisheries, pastoralism, urban development, mining, and a variety of other enterprizes have been well documented. The direct results have been that 39–50% of Earth's land area has been transformed or degraded (Vitousek *et al.* 1997). Human activities have impacted adversely on every terrestrial, aquatic and marine ecosystem (Ehrlich and Ehrlich 1991). Concern about these detrimental effects has reached the international political process and an international *Convention on Biological Diversity* was one of the major outcomes of the Earth Summit held in Rio de Janeiro in June 1992. One of the key aims of this convention is the conservation of biological diversity and the sustainable use of its components.

This Convention has formed the basis for the approach of some countries to develop strategies for the conservation and maintenance of biological diversity. For example, in 1996 all Australian States and Territories, together with the Federal Government endorsed *The National Strategy for the Conservation of Australia's Biological Diversity* (Commonwealth of Australia 1996b). Implementation of this strategy will require actions that affect all of Australia's terrestrial, aquatic and marine ecosystems, most of which will continue to be subject to a multiplicity of uses. Most of these uses depend on, or are impacting on, biological diversity. New Zealand is currently negotiating a national biodiversity strategy (Department of Conservation and Ministry of the Environment 1998).

In order to meet the objectives of the international Convention and strategies like the one drawn up for Australia and the one being drawn up for New Zealand, it will be necessary to manage for the conservation and maintenance of biological diversity in every terrestrial, aquatic and marine production landscape. This means developing an integrated and holistic approach to management rather that the more recent, narrow, production-focused approach in non-reserve areas plus a narrow species-focused approach in reserves.

Recognition of the need to develop a more holistic approach has now reached the point where politicians are trying to come to terms with

[1]CSIRO Wildlife and Ecology, G.P.O. Box 284, Canberra, Australian Capital Territory, Australia 2601.
[2]School of Environmental and Marine Sciences, University of Auckland, Private Bag 92019, Auckland, New Zealand.
Pages 678–89 *in* NATURE CONSERVATION 5: NATURE CONSERVATION IN PRODUCTION ENVIRONMENTS: MANAGING THE MATRIX ed by J. L. Craig, N. Mitchell and D. A. Saunders. Surrey Beatty & Sons, 2000.

the concept and in some cases are seeking a way forward. The New Zealand Minister for Conservation clearly articulated this in the speech he gave when opening the conference on which this volume is based (Smith, see Preface):

In the last century we have extensively destroyed the globe's forests. World-wide only 20% of our ancient forests are still standing. Seventy-six countries have cut down all their ancient forests. We have plundered Earth's fishing stocks, to the point where some fisheries have completely collapsed. We have poisoned our atmosphere with billions of tonnes of carbon and sulphur, and used chemicals that have put the ozone layer at risk. Human induced erosion, desertification and chemical contamination have made vast tracts of land unusable. With the loss of forests and wetlands we have caused the extinction of 20% of the world's species of birds. Over that same century our population has increased six fold. Only a fool would pretend we could take the same destructive practices with us from the 20th century into the 21st.

This conference is about providing hope and inspiration. It is about going beyond the cliches and into some of the serious grunty issues that we need to confront in the new century. The theme of conserving nature in productive environments is one of the more difficult, but relevant, issues of our future . . . The challenge at the core of the conference agenda is how to marry together those conservation and economic objectives to create a win-win situation.

Over the past 100 years, most countries have established agencies charged with the stewardship and management of plant and animal communities. In the recent past, this has been attempted through a network of reserves dedicated for nature conservation or some equivalent purpose. However, there is more land and sea outside such reserves than within. Moreover, the boundaries of such reserves are human constructs that are meaningless to the associated biota. It is now widely recognized that effective nature conservation will only occur by integrating actions across all environments.

The conference on which this volume is based, called for papers to explore these issues and had five objectives. The first was to provide an understanding of ways that conservation and production can be integrated, including the development of accounting links which value biological diversity, rather than discount it. The second was to understand how knowledge of the current functional and dysfunctional links between conservation and production could be used to enhance sustainability. The third was to provide methods of achieving both economic benefit and conservation as ways of enhancing the quality of life. The fourth was to understand how a landscape approach can integrate social, economic and conservation goals. The final objective was to develop performance criteria that integrate the matrix encompassed by production and conservation.

WHAT DID THE CONFERENCE ACHIEVE?

The conference drew together people from eight countries, many disciplines and affiliations and placed them in a small town in New Zealand to discuss conservation in production environments. Over five days, 93 spoken presentations and 18 poster papers covered subjects ranging from conservation policy, conceptual frameworks for guiding integration of conservation and production, research (including surveys of attitudes and behaviours), and accounts of actions being taken to address environmental problems. The conference closed with an open plenary discussion session to draw out the issues raised, and to explore future directions. While no one attempted to cover the final objective of the conference, participants did raise a number of issues relating to the theme of the conference and these issues are presented below.

IDENTIFICATION OF ENVIRONMENTAL PROBLEMS

There are international and national conventions or agreements, state of the environment reporting systems, many scientific publications, reports in the media, community action groups, and other avenues for presenting information and addressing the environmental problems we are now facing, particularly the loss of biological diversity and the consequences of this loss. However, there is still widespread ignorance about the extent of these environmental problems, their implications, the need for concerted action to address them and, by almost any measure, the situation is not improving.

Virtually all of the environmental problems we face stem from increasing human populations and their associated economic systems which take no account of natural assets. Consequently attempts to stem increasing demands on natural resources are seen as barriers to growth in so-called human welfare. While most papers presented at the conference stated the environmental problems being addressed were a direct consequence of loss of biological diversity, only one raised the issue of human population increase and the need to address that as part of a holistic approach to environmental management (Gunarso and Davie, this volume). A minority (see below) raised the issue of economic systems. There is little doubt that if the issue of human demands on the environment is not addressed, and these demands continue to rise as is currently happening, then any measures implemented for integrated land — water management will be as much use in the long term as a line in the sand for stemming an earthquake-induced tidal wave.

In the light of the importance of the human population issue and its degrading environmental consequences, experts proposing solutions to environmental problems should preface their publications or communications detailing the solutions with some disclaimer relating to the level of human populations and demand on natural resources within which their solutions would operate. This issue of human population needs to be constantly put forward as one part of any integrated solution.

Similarly, the overwhelming belief that natural systems and the biodiversity they contain sit outside economic considerations prevents conservation integrating with production. In much of the world, conservation is largely a government monopoly disconnected from normal costs such as land taxes and from income related to services provided for known customers. Some chapters (e.g., Davis and Cocklin; Mitchell and Craig, both this volume) touch on this whereas most papers offer solutions that assume that such irrational economics are possible everywhere.

While there is a section of the community who believe that too much time is spent stressing environmental problems, there are still many who are not aware of the extent of our environmental problems. For example, the eight New Zealand high school students who attended the conference, reported to the closing plenary session that while they did know of some of the problems raised, the conference made them realize that the problems were much worse than they thought. Accordingly, there is still need to raise community awareness and debate on environmental issues and potential solutions through every forum available.

The problem with communicating environmental issues is that the environment degrades incrementally with some components being adversely affected differentially. Environmentally there is no sudden threshold when ecosystems collapse and life becomes untenable. This makes it difficult to communicate the problems via media which want "a new angle" or "an environmental sensation" to sell the message that the environment needs urgent attention, instead of passing on the "same old doomsday messages."

In presenting material on environmental problems, agencies, scientists, environmentalists and others should be aware of the need to use the problems as the introduction to developing solutions that are practical and economically sustainable. There is also the need to offer people some hope that they can be successful in addressing the problems that affect their lives. It is important to recognize that social solutions to technical problems within production landscapes are also needed.

ACTION DOES NOT MATCH RHETORIC

The one message that came through strongly throughout the conference was that, in every case discussed, in terms of integrated solutions to landscape-scale problems we are long on rhetoric and short on practical actions. For example, to demonstrate this point Hobbs and Saunders (this volume) use the wheatbelt of Western Australia which has rapidly developing and widespread ecological problems that are having major impacts on the viability of agriculture and nature conservation. Despite considerable research on defining the ecological and social problems, suggestions for solutions, and political rhetoric stressing the need to act, significant areas of native vegetation continued to be cleared. Moreover, less than 10% of the amount of annual revegetation thought to be necessary over a decade to reverse land degradation and loss of biological diversity is being carried out.

The Government of Western Australia has taken action through the four main government agencies with responsibilities for natural resouce management in the wheatbelt. They have produced an action plan which uses specially selected catchments as foci for applying remedial actions. However, the four agencies have each selected separate catchments to carry out remedial actions; in no catchment is there is any attempt on behalf of all agencies to demonstrate an integrated approach to the ecological problems to prove that land degradation and loss of biological diversity can be arrested, and in some cases, reversed. In addition, most landholders seem to be willing to take the risk of doing nothing (the business as usual approach), rather than taking the risk of implementing best practice approaches to management (i.e., widespread revegetation to provide windbreaks, draw down rising water tables, provide more habitat for the biota). This is a clear example of widespread and well-publicized environmental degradation which is threatening peoples' economic livelihoods, their social communities, infrastructure, and their natural heritage. However, the actions to date to address the problems fall far short of that required.

Trémont *et al.* (this volume) demonstrate how in the New England region of New South Wales, a rapidly degrading natural resource base (the death of mature trees) provided the stimulus for local action in the form of tree planting. However, the revegetation is not sufficient to replace the dead and dying trees, let alone those that will die in future. Similarly, in New Zealand, despite recognition of massive loss of native forests, significant areas are still being cleared for establishment of exotic plantations (Rosoman and Lucas, this volume). This discrepancy between recognition and action manifest in the Western Australian wheatbelt, the New England Tablelands and New Zealand is also shown in many other areas covered by chapters in this volume. Goldney, Wakefield *et al.* (this volume) put the issue succinctly in their "winning battles and losing the war" analogy. In this case we are winning the battle of raising the conciousness of people about the environmental problems we face, but, we are losing the war of reversing the trends.

What are the impediments to winning the war? The problem is that in many cases scientists and environmentalists have focused on the definition of environmental problems and trying to guage their extent. Neither science nor the environmental movement has produced integrated solutions, nor has either group or governments offered hope to people that there are practical and sustainable solutions. By way of illustration, of the 88 papers submitted for publication in this book, not one addressed all the issues necessary to integrate nature conservation in production environments. The scale of most of the environmental problems we face is such that they can only be dealt with at a catchment or regional level, and this is beyond the scope of an individual or family. Similarly, integrating conservation and production is a landscape-scale issue. In most cases solutions will require the combined action of many like-minded people supported by resources from still others. As pointed out by Hobbs and Saunders (this volume) and Lee and van Hensbergen (in litt.) the impediments to better land management are as much economic, social and political as they are technological. While there is a need to continue development of practical technological solutions to evironmental problems, and integrating nature conservation into management in production environments is included here, there is also a need to effect changes in behaviours, as changes in attitudes are insufficient to effect change on the ground.

From the chapters in this volume, the message that comes through loud and clear is that when it comes to integrating nature conservation in production environments communities, agencies and scientists have a very long road to travel. First, in many cases it is necessary to convince people that conservation is an imperative; i.e., change attitudes. Then it is important to provide practical applications to effect such management and change behaviours to ensure management does take place. Nobody should under-estimate the enormity of the task.

Recently, there have been major changes in the perception of land management and expections of land managers. Governments, society, non-government organizations, and the scientific community are now expecting that all land managers give full consideration to values other than production in the landscape. By way of example, in both Australia and New Zealand since the period of European occupation there has been substantial failure of government policy in relation to natural resource use. Both governments encouraged development of agriculture without any consideration of other values. In addition, conservation was relegated to action through dedicated reserves, usually left after other productive uses had "picked over" the landscape. We are now undergoing a major change in direction with expectations that we maintain production while integrating conservation into production systems.

Unfortunately what is seen and understood about environmental problems do not necessarily translate into behaviour designed to address the problems. For example, Fisher (this volume) notes that in the Central Western Region of New South Wales, most people are able to recognize land cleared of native vegetation or degraded land, but, they fail to make the connection between the actions of clearing the land with the impact on nature or land conservation. In many cases they understand global environmental issues, but they fail to translate this understanding into action at the local scale where they have greater power to act. Too often the economic realities that result from societies' valuation of native biological diversity versus alternative land uses obviates desired behaviour.

Integrating conservation with production requires changes in economics, community sociology, government policies, science and education, and the combined action of many people acting on their patches of ground or pieces of turf. This means that they must agree on what they collectively need to achieve and on the methods needed to address the issues. If all do not act in concert, their approach may fail. In one landcare group in Western Australia, individuals who do not act in concert with the others in their landcare area to address ecological issues are known as "black holes." This is because the conservation actions of the group may fail unless all act to address the problems; what action does take place may appear to be going into a "black hole" without engagement of all land owners in the catchment. Approaches which will reach all natural resource managers, not only those that are in groups that are relatively easy to target, such as landcare groups will need to be developed.

The role of the individual is critical, so it is important to understand individuals and what motivates them. Shrapnel *et al.* (this volume) state that conservation of the environment, as the basis for ecologically sustainable production, is determined by an individual's behaviour towards the environment. Personality characteristics and other psychological factors affect an individual's relationship with the environment. To be effective in managing the environment it is necessary to recognize the quality of this relationship. An individual's personality is an important element in this relationship and will influence how sustainable land management practices are developed and adopted. Each community is a collection of individuals who possess different psychological resources. It is important to recognize that some individuals will be limited in their capacity to recognize, let alone respond to, the present environmental challenges and to change their behaviour. Any integrated plans for landscape management will need to take cognisance of different personalities and find some way of reaching out to the "black holes." This requires a greater range of possible solutions and as Trémont *et al.* (this volume) indicate, this is best served by one-to-one interaction and building relationships.

Briffett *et al.* (this volume) point out that the interaction between humans and wildlife is a complex area. It is influenced by many facets such as attitudes, values, beliefs and behaviours. The relationship between people and nature is highly complex and must be understood within the context of their lives.

Another impediment to winning the war is a lack of vision about how collectively society wants production environments to function and look like in the future. Politicians, government officials, scientists, foresters, farmers, fishers and others blythely write and talk about ecologically sustainable production. However, what is meant? How does the concept translate to a particular landscape? If there is no vision, especially no common vision, we have no idea of our final evironmental destination and, therefore, no idea of how to approach it.

WHO WILL MANAGE THE LANDSCAPE?

The removal of native vegetation and development of land for production is not a random process. The most fertile and productive soils are those selectively cleared for agriculture or set aside for forestry, horticulture or pastoralism. The result in production environments is that the matrix usually occurs on the better soils or more productive areas. The converse is that those areas set aside for nature conservation are often on the poorer areas or are markedly different from the surrounding matrix (Lindenmayer and Franklin; Porteous, both this volume). A consequence of this is that if the aim is to preserve all remaining biota of a region, then all parts of the landscape have some role to play in conservation. Those who own, lease or manage those areas outside the dedicated conservation reserves, need to manage the land under their control in a way that is sympathetic to nature conservation, while still engaged in production. In areas set aside for intensive production such as agriculture, privately-owned land is of critical importance (Davis and Cocklin; Fisher; Porteous, all this volume). The first rule of integrated management is that those who live and work in the landscape should manage that landscape. However, at present local people are not provided with the correct incentives, knowledge and responsibility to make decisions that will conserve biological diversity.

While acknowledging the importance of production lands for nature conservation, it is important to recognize that those involved in production are not in a position to meet all conservation goals of society. Freudenberger and Lefroy and Hobbs (both this volume) argue that conservation of some critical components of biological diversity may be essential to production, but, conservation of all biological diversity that formerly occurred in an area may not be essential for production. The challenge for society, researchers and those involved in production is to recognize and conserve in the production matrix, core elements of biological diversity essential to the ecological sustainability of the production enterprize; for example, the biological diversity responsible for nutrient cycling in the soils that are being mined for agriculture (see below). There is also a need to recognize those elements of biological diversity that are unavoidably destroyed by the production enterprize and to design and implement reserves that reduce the probability of losing these elements.

Individuals working in concert are also important in urban areas. Whisson *et al.* (this volume) state that in urban environments the small size of areas of conservation importance renders them more suited to conservation protection as individual areas of private land. These are more effectively managed by the private land owners, but, conservation security could be effected through some form of incentive such as a covenanting scheme. However, in urban situations, the relative value of this land for other uses is a significant pervese incentive. Any covenant scheme would need substantially more significant incentives attached such as rate relief, tax benefits, payments for services such as carbon sinks, management assistance and follow-up (including enforcement) to be effective. Galbraith (this volume) gives an example of an urban complex that if managed as an integrated unit could provide considerable ecosystem services to a large surrounding urban area.

Marine and aquatic environments may be more problematic, but, individuals involved in production could act as guardians or stewards of those environments (Elder and Cribb, in litt.) in concert with regulatory agencies and the community.

A system of integrated management that involves a range of different landholders as managers needs to take into account the social environment in which it operates (Ebbett, in litt.). In addition, management actions must be practical, developed with active participation of all who have a stake in the process, and locally relevant (Sakulas *et al.*, this volume). They should also be scientifically rigorous, be capable of being communicated and be practical. If necessary, they should be run through the "laugh test", that is, passed through scientific, advisor and manager filters (Burch *et al.*, this volume) to ensure their relevance and practicality.

Jennings and Moore (this volume) point out that the landscape level appears to be the most appropriate for developing plans for ecologically sustainable development. However, the plans should be guided by locally framed community needs and visions. They also make the important point that sustainable development is not solely the maintenance of ecosystem integrity, it is also the preservation of social and cultural systems. These trade-offs and their consequences need to be explicitly considered and addressed as their solutions may be mutually exclusive.

The importance of the individual landholder as part of landscape management draws attention to the role of local government in facilitating successful management. Local governments may play a significant role in the development of integrated approaches to nature conservation in production environments. Jay (this volume) states that local government is the key to facilitating successful management because *"it has legislative power and responsibility for environmental issues at local and regional level; it is accountable to individuals and communities for environmental conditions within their local area, and can harness their energies and commitment for environmental action; and it is potentially the level of government that can provide the ongoing care that is necessary for long-term ecological protection and restoration."* While there is undoubtedly a critical role for local government in environmental management, including managing the environmental effects of tourism, their performance in this role varies enormously (Blaschke *et al.*, this volume). In Australia there is no tradition of major involvement of local government in environment or management of natural resources. The capacity for local government to act in ecological management needs to be supported by higher levels of government so that they are more actively and intimately involved in land use management with conservation of biological diversity being a major objective. Unfortunately this goes against the current trend which is for devolution of responsibility without a corresponding redistribution of financial resources to accompany the increased responsibility.

COMMUNICATIONS: RAISING AWARENESS ABOUT POSSIBLE INTEGRATED SOLUTIONS

The role of effective communications in raising awareness about environmental problems and possible solutions is critical to the successful implementation of sustainable management of conservation in production environments. The importance of communications has been stressed in the last two volumes in this series (Saunders *et al.* 1993b; Craig *et al.* 1995). Unfortunately much communication is a case of preaching to the converted. As an example, at the conference on which this volume is based, 88 papers were submitted for publication, yet only four came from people involved with industries in which there is some attempt to integrate conservation into management for production. The remainder were from people involved in universities (49), government agencies (27), non-government organizations (6), and local government (2) all of whom were directly involved in research into aspects of integration or with development of policy affecting integration. Those from industry were from the viticulture industry (with university co-authors), forestry in native forests (with a co-author from a government agency), plantation forestry, and from the aluminium processing industry (with university co-authors). Ideally, a conference such as this should have been addressed by those managing the production matrices; farmers, pastoralists, miners, professional fishers, etc. The failure to engage these people and their organizations at conferences like these suggests there is little hope that they will pick up the results by reading the proceedings. However, conferences such as these do have important roles to play in allowing the free exchange of ideas so that participants are in a position to learn from other communities. In exchanging ideas, it is as important to know what did not work as it is to know what was successful.

How can environmental awareness and informed debate happen throughout the community? The choices of venue and format are important. The organizers of the conference which gave rise to this volume were praised for running a successful conference. At the plenary discussion session the importance of exchange of information via such conferences was stressed. However, this venue and format may be inappropriate for most of the people we are trying to reach. This point was made forceably at the previous conference in this series by Walsh (1995) in relation to interacting with indigenous people involved in land management. How is it possible to reach "communal farmers" in developing regions who do not have well-developed networks which are relatively easy to tap into (Lee and van Hensbergen, in litt.)? This raises the importance of developing good communication skills, including the ability to be a good listener and incorporate what is heard. It is also essential to target all stakeholders via venues and formats they are most comfortable with. Interacting in venues of the stakeholders' choice is clearly an important option. While this may be time-consuming, it should become an essential component of activites conducted to develop integrated land management.

The importance of involving those managing production matrices in research and development as well as management was stressed many times during the conference (Goldney, Wakefield *et al.*; Lindenmayer and Franklin; Norton; O'Byrne; Quinn *et al.*; Stocker; Stothers; Webber and Fowke, all this volume). It was pointed out that interactive learning and consultation with all those who have a direct stake in the landscape are efficient and essential methods of engaging land managers and others; these are essential elements of adaptive management. Goldney, Wakefield *et al.* (this volume) demonstrate their "Save the Bush Toolkit" designed to allow farmers to assess the conservation status and value of their own patches of native vegetation, individual farm trees and riperian zones, and to devise management to incorporate conservation management into their farm plans. This sort of "bottom-up" approach should prove useful in raising awareness among farmers and gaining their support for active management.

In this context, Norton (this volume) suggests that a regulatory approach to controlling development may be less effective for conservation than a programme of education involving local people in managing their conservation resources better and seeking ways that allow further development and protection of biological diversity assets. This raises the issue of how to decide when development has been taken as far as it should without causing further loss of biota. How do we establish a matrix that includes no-go lines on maps or development and management plans. As mentioned earlier, this will be almost impossible in the face of ever-increasing human populations.

It is also important that research and development are tailored to local conditions and communities.

Many research workers believe that having carried out an important piece of work and published it in the scientific literature, it will be picked up and applied by groups actually managing conservation and/or production. Most managers do not read the technical literature most research workers publish. If they do, they are not in a position to adapt the results to their landscape. To be effective in influencing what happens on the ground, research workers need to go several steps further. They need to tailor their results to local conditions (Trémont *et al.*, this volume) and then work with those local communities to ensure the results are incorporated in management. It is hard to over-estimate the power of personal contact and the value of taking time to share cups of tea or coffee and get to know the people research workers are interacting with. Taylor (this volume) also points out the importance of the personal approach. Working on the properties of individuals allows for the construction of personal relationships, which in turn aids the adoption of results. This may have a flow-on effect to influence other individuals.

Providing financial rewards for sustainable production that includes conservation is an important part of communicating desired goals. The current perverse incentives (e.g., Davis and Cocklin, this volume) are major impediments to achieving conservation across the matrix. As long as costs rest with responsible land owners, fishers etc. who sustain natural assets, and as long as ownership and control remain with governments, resolution of environmental problems will be problematic.

THE NEED FOR EFFECTIVE NETWORKS

There is no doubt that ecological restoration and landscape management require an integrated approach to achieve ecologically sustainable outcomes. This is true of all conservation, production and development environments. This points to the critical importance of involving the local community or production managers in effective networks with ecologists, economists, conservation biologists, landscape architects, resource planners, engineers, hydrologists, social scientists, educators, facilitators, and all tiers of government to be successful in developing ecologically sustainable outcomes. This was the strongest theme emerging from the papers presented at the conference on which this volume is based and was the theme of the previous conference in this series (see Craig *et al.* 1995 for details). There is a major challenge in engaging the community and those seeking to set up networks with the community will need to be innovative, using all avenues available to them and targetting the most effective.

Government agencies should not have a monopoly on nature conservation. Governments have an important role to play, but they need to develop the appropriate blend of support (including resources), policies (including incentives) and regulation (as a last resort to force compliance if necessary). Agencies such as those managing the Golden Gate National Park in California appear to be achieving such approaches (B. O'Neil, pers. comm.). Government agencies also need commitment. Extension work is still an important activity with government agencies involved in natural resouce management, but its importance has been diminishing in this period of economic rationalism, rule of market forces and decrease in government activity. However, it is an arena that non-government organizations such as World Wide Fund for Nature are now entering (Trémont *et al.*, this volume). Trémont *et al.* point out the importance of continuity, commitment and consistency on the part of people involved in extension work. All too often extension workers come from outside the area they are providing advice on and stay for only short periods. As a result, local communities feel let down when the extension workers depart after a relatively short period, often before the work they are advocating has been put into effect. In this regard extension officers recruited from the local community often gain greater acceptance, have greater commitment, and are more effective than the more commonly employed itinerent extension personnel.

To be effective, government agencies need to be explicit about what they are trying to achieve, have no hidden agendas, and work with other agencies co-operatively to achieve ecologically sustainable outcomes. This may mean co-operating across disciplinary boundaries as well as political boundaries. At present there is a tendency to compartmentalize the landscape along political, cadastral, property or some other boundaries of human construct. The administrative boundaries of government agencies often differ. This often reinforces the turf war attitude of government agencies which have overlapping responsibilities for natural resource management (Gunarso and Davie, this volume). In addition, ecosystem boundaries do not follow straight lines or coincide with the boundaries humans impose on the landscape. These mismatches of boundaries have to be taken into account when setting up networks for management.

Trust is also a critical element of effective networks. Lack of trust between any of the links of a network will reduce its efficiency. Moore (1995) discussed the element of trust in detail in the last volume in this series. This was also raised at the conference on which this volume is based, in regard to agency staff, particularly those directly involved with extension and education. Such staff need to establish credibility with the groups they are dealing with before they can be accepted as reliable collaborators or providors of information (Seipen and Dorricott, this volume).

There is a widespread belief that knowledge is power. Playing the knowledge power game can be disasterous for credibility. If there is a feeling among links of a network that they are not trusted with information, they may not collaborate successfully. Effective networks rely on the free and frank exchange of information. In production environments there is a tendency among conservation agencies to withold information from production managers and vice versa. This commonly relates to the location of rare or endangered species on private land (Trémont *et al.*, this volume). In the former case, this is usually because of the belief that production will destroy the species if the locations are known. In the latter case it is because of the belief that more constraints will be placed on the use of the land for production. Trust is essential to developing solutions in these situations although ultimately the problem relates to economics (see below and Davis and Cocklin, this volume).

Politics are an essential part of many networks. Without the political will to act it is impossible to achieve ecological sustainability. There was limited involvement of elected politicians at the conference on which this volume is based. The conference was opened by the New Zealand Minister for Conservation (Smith, see Preface) who presented his opening speech and then left. Two members of local government in New Zealand presented accounts of the activities of the councils in relation to conservation and production (Wilson, this volume; W. Lynch, in litt.). Every network should actively seek to involve politicians in their activities. These should be relevant members of local government, but they may also be representatives of regional and national governments. The involvement of politicians in networks has several objectives; to seek their assistance in the goals of the network and to empower them through increased knowledge of sustainability issues. An alternate approach to gaining the direct involvement of elected politicians, especially those with involvement in relevant policy leverage is to gain the attention and interest of those who are the trusted advisors or "minders" of the elected politicians.

ENVIRONMENTAL ECONOMICS

At present we discount the value of ecosystems and their products (e.g., Costanza *et al.* 1997) because most sit outside the market economy. For example, in agricultural landscapes, native vegetation is usually valued on the basis of the potential of the land if it was in production or, in some cases, on the value the vegetation contributes to the aesthetics of the property and how it contributes to resale value. There is no attempt to value or trade vegetation on its contribution to ecosystem services, e.g., the contribution it makes to water use, its effect on the hydrologic balance, its impact on wind or water erosion, production of oxygen, removal of carbon dioxide, provision of pollination services, disposal of wastes, or the purification of water (see also Galbraith, this volume). As a result, when decisions are made on land use, they are made on the basis of a biased accounting system. The result is that native vegetation is accorded little or no value and is constantly subjected to intentional removal or loss by benign neglect (Curtis *et al.,* this volume). Zoete and Davie (this volume) note that "*In order to make more informed decisions on the use of* M. quinquenervia *wetlands, the goods and services provided by the wetland need to be known. The widespread realization that wetlands perform a range of economically valuable ecological functions is relatively recent* (see Daily 1997), *attesting to the difficulty many people have in observing non-market economic benefits.*"

Hill *et al.* (this volume) discuss environmental care programmes in relation to mineral processing plants. They point out that any money spent on environmental issues beyond that required to comply with operating licences must show a return to shareholders on the investment over and above the minimum required. They state that some environmental care programmes do have demonstrable outcomes and others show their return in a better working environment. This may in turn lead to behavioural change in the workforce and flow through to benefit the company and shareholders.

Industry and individual land managers have to justify environmental programmes financially; but industry often has extra incentives to carry out environmental programmes because of regulatory processes under the polluter-pays approach. Industry also has to take into account the costs of remediation should that be necessary when their industrial activities have ceased. Unfortunately, this is not true of agriculture and pastoralism which have degraded vast areas of the planet. This effectively means that consumers are not paying the true production costs for much primary produce. For example, it has been estimated that to produce a tonne of wheat in some parts of the Australian wheatbelt results in the loss of 7 tonnes of topsoil and other degrading changes to ecosystem processes including massive losses of native biological diversity (Commonwealth of Australia 1996a). The costs of addressing these degrading processes are not included in production or retail costs; they are discounted as externalities and will have to be carried by present and future generations. An economic system aligned toward long-term sustainability rather than just short-term financial gains would include management of both human made and natural assets such as soil, clean water, healthy air, etc. Such a system would build the costs of asset loss into production and retail costs and hence encourage conservation as part of an ecologically sustainable production system.

Current economic approaches are seen as major dis-incentives or impediments to the development of ecological sustainability. It is a widespread belief that it is impossible to "be green when you are financially in the red." Conversely, since environmental degradation is so widespread and pervasive, it may soon become impossible to be "in the black" unless you are green! Economic development is a major key to induce people to develop and support ecologically sustainable production systems. Despite the major differences between them, this is true of both the developed and developing countries. Using China as an example, Zhuge *et al.* (this volume) point out that maintaining the conservation operation of a reserve while the neighbouring people struggle for their survival is the recipe for conflict over the reserve's natural resources, with conservation being the loser.

Environmentally responsible industry codes of practice that include conservation of natural assets such as functioning ecosystem processes can be used as a competitive advantage that enhances profits (Nias *et al.*, this volume). Indeed, Australia and New Zealand both market a "clean green" image to enhance sales to their northern hemisphere markets yet the substance of these claims is declining. The cost to current and future generations in terms of quality of life and financial well being is large. These economic instruments of enlightened self-help, and vertical integration of markets are part of a whole suite of instruments which need to be developed and applied in the approach to managing the matrices of production landscapes.

MATRIX MANAGEMENT: THE HOLISTIC APPROACH

Brock and Jarman (this volume) use wetlands in agricultural landscapes as an example to discuss succinctly the need to focus on the entire matrix, rather than the present approach of focusing on individual elements such as production areas, patches of native vegetation, corridors, conservation reserves, etc. They state that the action of one farmer to conserve wetlands is useful and necessary, but, that by itself will not conserve the biological diversity dependent on wetlands. Successful nature conservation relies on addressing attributes at the landscape-scale or greater. They point out that when wetlands dispersed across the landscape are owned by a range of individuals and/or organizations, management for conservation has to be developed by agreement rather than through purchase and ownership by the leading conservation agency. "*Above all, then, we see nature conservation in wetlands in the agricultural landscape as depending upon interchange of facts, ideas and ideals between farmers, scientists and government conservation managers*."

Approaching ecologically sustainable production needs an acceptance that multiple use will be common in all production environments. For example, in forests this will mean that forest managers will need to accept that timber production should not be the sole consideration in matrix lands; other forest values are important (Gunarso and Davie; Lindenmayer and Franklin, both this volume). To date, short-term economic aspects have overshadowed environmental and social aspects of many production environments. In order to cater for multiple objectives, management will become increasingly complex. This will result in a move away from spatial planning that zones the landscape for particular land uses. Land managers will find themselves managing landscapes with a range of different elements, land tenures, and land uses spanning those with purely conservation objectives to those with purely production objectives (Freudenberger; Lindenmayer and Franklin, both this volume). This represents a fundamental change in responsibility that will need assistance from a range of policy instruments.

Diversity provides resilience and gives more scope for sustainable planning than the more narrow-focused primary production systems commonly employed today. The use of native species for production as with the development of farming game animals in the Northern Province of South Africa (Lee and van Hensbergen, in litt.) or farm forestry in Australia based on plants of local provenance (Lefroy and Hobbs; Curtis *et al.*, both this volume) provides such enterprise diversity. In the former example, use of game species may result in phasing out cattle or goats. This also provides an example of a change in practice that may be difficult to instigate because to some groups social status is indicated by numbers of cattle owned. To change enterprises may involve changing the social system and this is not a trivial matter.

Similarly, changes in practice from production of exotics to local native biological diversity may require a fundemental change in some environmental non-government organizations. Such groups were leaders in the prevention of unsustainable use of native biological diversity. Unthinkingly transferred to a mindset of "zero use" of native biological diversity, they now ensure agricultural and forestry production is solely through exotics to the ongoing detriment of native biological diversity (Grigg *et al.* 1995; Commonwealth of Australia 1998).

The debate over what constitutes ecological sustainability is a vexed one. The word sustainable has been used differently in different contexts. For example, in production environments sustainable may mean the continuation of the production enterprise with no regard for other landscape elements. Thwaites *et al.* (this volume) point out that if production planning does not take a holistic view of the production process, it is unlikely to be sustainable in the long term. While technically the yield in production terms may be potentially sustainable, if other elements of the landscape and socio-economic system are not sustainable, the whole system, including the production industry, may collapse.

ADAPTIVE MANAGEMENT

In the management of Earth's natural resources people are engaged in an enormous uncontrolled experiment. There is little doubt that in many ecosystems we are degrading the natural assets that provide the resource base and learning little to adjust our management. If societies are to be successful in developing ecologically sustainable production which does not compromize native biological diversity further, they must develop a system of management that allows continuous improvement of actions flowing from increased understanding of ecosystem and social processses and human impacts on them. At present there is a major problem of taking ecological theory and translating it into management and policy. It is as an aid to this process that adaptive management becomes critical. Action is needed now. As encapsulated by the precautionary principle, lack of knowledge is not an excuse to prevaricate. Caution can lead to inaction while knowledge is collected. There is a danger that when there is enough knowledge to act, there may be little left to manage.

McIntyre and Hobbs (this volume) point out the experimental nature of current management and advocate gaining knowledge from these experiments. It is essential to combine experimental research with management operations and assessment in an adaptive management system (Hobbs and Saunders; Lindenmayer and Franklin, both this volume). This means much more interaction and successful collaboration between production managers and research workers, with more inter- and multi-disciplinary adaptive management projects. This will involve the breaking down of disciplinary boundaries and active involvement of production managers in research and researchers in management. McIntyre and Hobbs (this volume) suggest the following as part of an adaptive management approach to developing ecologically sustainable production systems: on-ground action with involvement of stakeholders; extensive inter-agency collaboration involving a large inter- and multi-disciplinary team, with a strategic plan, an operational plan, and regular reporting milestones with well-defined performance criteria; a communications plan (with properly identified stakeholders who are appropriately targetted); a technical reference group; a management reference group; and a research steering committee. This provides for technical and some social elements, but not economic and policy requirements.

GUIDELINES FOR INTEGRATING CONSERVATION IN PRODUCTION ENVIRONMENTS

Governments (including local government) and their agencies

Develop policies based on recognition that human society is wholly constrained by the natural ecology of the planet.

Incorporate the economy as a subset of society.

Develop economic systems that value ecosystem supplies and services, rather than the present valuation systems that discount them or treat them as common goods and services that are free to use or abuse.

Foster the development of a vision of ecologically sustainable landscape management and an ethic of stewardship; communicate the vision and the concept of stewardship.

Monitor the environment and report on the change in condition of natural resources; communicate this information widely.

Integrate approaches to environmental adaptive management at the landscape level to break down the traditional compartmentalized approach to management.

Recognize that the goals to developing sustainable management must include ecological, social and economic components.

Develop policies and incentives (e.g., rate relief, tax concessions, management assistance, full cost recovery — including environmental costs of production) that foster the involvement of, and rewards to, individual land managers as those who live and work in the landscape are those who should manage it.

Remove perverse incentives (e.g., tax concessions for clearing native vegetation) that constitute impediments to achieving conservation in production landscapes.

Foster the development of effective networks of all stakeholders necessary to develop integrated landscape management and put it into practice.

Recognize the importance of trust, credibility and sharing knowledge and power in the forging of effective networks and partnerships.

Support environmental extension work, instead of devolving it to members of the community.

Make sure that political rhetoric is matched by action.

Community groups

Develop a vision of ecologically sustainable landscape management and an ethic of stewardship; communicate that vision.

Carry out environmental management at an ecological scale (e.g., catchment) rather than at the scale of some human construct.

Work with other groups to develop integrated solutions to landscape management and help put them into practice.

Implement best management practice approaches to adaptive management.

Recognize that in production landscapes, conservation of biological diversity may be important for sustaining economic production.

Recognize the importance of trust, credibility and sharing knowledge and power in the forging of effective networks and partnerships.

Individuals (including researchers)

Work with others to develop a vision of ecologically sustainable landscape management based at an ecological scale (e.g, catchment); communicate that vision.

Develop an ethic of stewardship.

Use every opportunity to communicate the issue of environmental problems and the need for urgent action.

Tailor the results of research work to local conditions and work with local communities to ensure results are incorporated in management.

Recognize that in production landscapes conservation of biodiversity may be critical for ensuring economic production and develop management strategies accordingly.

Avoid dwelling on the environmental problems being faced; rather focus on potential solutions.

Implement best management practice approaches to adaptive management.

Think globally and act locally.

Recognize the importance of trust, credibility and sharing knowledge and power in the forging of effective networks and partnerships. Adopt a "yes and" rather than a "but" approach to dialogue.

Raise the issue of the need for integrated environmental management with elected politicians.

INCENTIVES FOR CHANGE: THE WAY FORWARD

As mentioned earlier, the fact that a landscape is degrading and losing ecosystem function is not necessarily an incentive to change the behaviours of people dependent on that landscape and its associated ecosystem processes that are being changed. In addition, it takes a considerable period to change people's behaviour and even longer to effect changes in ecosystem function and processes. For example, using revegetation it may take decades to start to lower rising water tables, but it may take centuries to reduce them to the levels they were at prior to the clearing of vegetation that caused them to rise. Nonetheless, the rapid changes in many production landscapes requires immediate action to start the process of change.

Many chapters in this volume stress the need for economic measures to provide incentives and support for developing and maintaining ecologically sustainable production (Burch *et al.*; Crosthwaite and McLeod; Davis and Cocklin; James *et al.*; Lee and Szabo; Mitchell and Craig; Nias *et al.*; Percival and Andrews; Porteous; Rosoman and Lucas; Stothers; Trémont *et al.*; Whisson *et al.*, all this volume). This is particularly important for supporting conservation measures on private land. There have been an array of incentives suggested. These include: covenants; rebates on land rates; deductions on taxable income; direct subsidies; revolving funds; and payments for acting as custodians or stewards of the land and its biota and a re-evaluation of whose problem and who pays.

A rapidly emerging incentive is arising via the power of the consumer (Nias *et al.*, this volume). Quinn *et al.* (this volume) point out that the benefits to farmers of improved environmental performance are likely to become more sharply focused as the concerns of consumers move past the current issue associated with food safety into concerns that production is not degrading the environment and that animals are being handled ethically. There will be an increasing trend towards certification of produce to ensure consumers that production is environmentally sustainable (Nias *et al.*; Rosoman and Lucas, both this volume). Certification is already being carried out for some operations involving harvesting native forests and it is being developed for forestry plantations and fisheries. Nias *et al.* (this volume) state that certification must be based on clear environmental and social performance standards. These standards must be credible and capable of being understood by consumers. They need to be based on objective, measurable and transparent criteria that have been assessed by a party or parties independent of the production process. This provides an important role for non-government organizations like World Wide Fund for Nature and Greenpeace. Certification should be implemented at the production management level (particular forest, biogeographic region, etc.) and linked to a chain of custody and product lifecycle verification. Industry certification is typically driven by industry whereas the potential for environmentally aware consumer groups to drive industry compliance of sustainability is in its infancy.

The chapters in this volume provide eloquent testimony to the considerable amount of work needed to develop and implement ecologically sustainable production. At the final plenary session of the conference that gave rise to this volume someone posed the question "Given the widespread recognition of environmental problems associated with production to support humans, why have we done so little to address the problems?" The chapters in this volume highlight many of the impediments to progress, but, they also provide important ways forward.

There is a need to acknowledge that while people are the core of the problem, they are an essential part of the solution. While there is a considerable way to go before achieving ecologically sustainable production the relevant issues are known. Societies need a collective vision of the future, more effective methods for collaboration and co-ordination of activities, priorities based on a strategy for developing production systems of the future, an economic system that ensures users pay the actual costs of production, and production systems that manage natural assets and harvest resources in a sustainable way. In addition it will be necessary to remove all incentives (such as tax concessions for clearing of native vegetation, the promulgation of iconic images of rural lifestyles regardless of their impacts on the landscape, and the perception of the sanctity of private "land rights" as opposed to the actual responsibilities for sustainable land management) that are impediments to sustainable land management.

In order to be successful in integrating conservation and production it will be necessary to develop a long-term strategy of policy approaches that bring all the elements discussed above together and support them in an integrated manner. While there has been substantial failure in government policy in the past, new policies that are based on shared investment principles in which the community beneficiaries contribute and assist land managers accept new responsibilities in relation to sustainable land management will need to be developed. These should move towards a polluter-pays approach in which land managers become responsible for ensuring that neither land degradation nor pollution continues to occur, and if they do, then those managers acting irresponsibly should be penalized. This will bring natural resource management into line with environmental policy approaches to some industry groups. All of the policy elements discussed in this chapter will be needed in integrated packages in order to develop integrated management approaches to production systems.

The challenges may seem daunting, however, the consequences of not developing ecologically sustainable production systems are disasterous.

ACKNOWLEDGEMENTS

This chapter represents the editors' views of the issues raised at the Nature Conservation in Production Environments conference in Taupo, New Zealand in December 1997. We are grateful to all of the participants at the conference, and especially to the high school students from Auckland and Taupo who took an active part in the conference and shared their views of what they saw as issues of concern for them and the future. We are grateful to Richard Hobbs, V Saunders and Simon Smalley for their constructive criticism of this chapter.

REFERENCES

Commonwealth of Australia, 1996a. Australia State of the Environment 1996. CSIRO Publishing: Collingwood, Victoria.

Commonwealth of Australia, 1996b. The National Strategy for the Conservation of Australia's Biological Diversity. Australian Government Publishing Service: Canberra, Australian Capital Territory.

Commonwealth of Australia, 1998. Commercial Utilization of Australian Native Wildlife. Australian Government Publishing Service: Canberra, Australian Capital Territory.

Costanza, R., d'Arge, R., de Groot, R., Farber, S., Grasso, M., Hannon, B., Limburg, K., Naeem, S., O'Neil, R., Paruelo, J., Raskin, R., Sutton, M. and van den Belt, M., 1997. The value of the world's ecosystem services and natural capital. *Nature* **387:** 253–60.

Craig, J. L., Saunders, D. A. and Mattiske, E. M., 1995. Effective conservation requires people networked for action. Pp. 653–60 *in* Nature Conservation 4: the Role of Networks ed by D. A. Saunders, J. L. Craig and E. M. Mattiske. Surrey Beatty & Sons: Chipping Norton, New South Wales.

Daily, G. C., 1997. Nature's Services: Societal Dependence on Natural Ecosystems. Island Press: Washington, DC.

Department of Conservation and Ministry for the Environment, 1998. New Zealand's Biodiversity Strategy: Our Chance to Turn the Tide. Department of Conservation and Ministry for the Environment: Wellington, New Zealand.

Ehrlich, P. R. and Ehrlich, A. H., 1991. Healing the Planet. Addison-Wesley: Reading, Massachusetts.

Grigg, G., Hale, P. and Lunney, D. (eds), 1995. Conservation through Sustainable Use of Wildlife. University of Queensland: Brisbane, Queensland.

Ministry of the Environment, 1997. The State of New Zealand's Environment. GP Publications: Wellington, New Zealand.

Moore, S. A., 1995. The role of trust in social networks: formation, function and fragility. Pp. 148–54 *in* Nature Conservation 4: the Role of Networks ed by D. A. Saunders, J. L. Craig and E. M. Mattiske. Surrey Beatty & Sons: Chipping Norton, New South Wales.

Saunders, D. A. and Hobbs, R. J. (eds), 1991. Nature Conservation 2: the Role of Corridors. Surrey Beatty & Sons: Chipping Norton, New South Wales.

Saunders, D. A., Hobbs, R. J. and Ehrlich, P. R. (eds), 1993a. Nature Conservation 3: Reconstruction of Fragmented Ecosystems — Global and Regional Perspectives. Surrey Beatty & Sons: Chipping Norton, New South Wales.

Saunders, D. A., Hobbs, R. J. and Ehrlich, P. R., 1993b. Reconstruction of fragmented ecosystems: problems and possibilities. Pp. 305–13 *in* Nature Conservation 3: Reconstruction of Fragmented Ecosystems — Global and Regional Perspectives ed by D. A. Saunders, R. J. Hobbs and P. R. Ehrlich. Surrey Beatty & Sons: Chipping Norton, New South Wales.

Saunders, D. A., Craig, J. L. and Mattiske, E. M. (eds), 1995. Nature Conservation 4: the Role of Networks. Surrey Beatty & Sons: Chipping Norton, New South Wales.

Saunders, D. A., Arnold, G. W., Burbidge, A. A. and Hopkins, A. J. M. (eds), 1987. Nature Conservation: the Role of Remnants of Native Vegetation. Surrey Beatty & Sons: Chipping Norton, New South Wales.

Vitousek, P. M., Mooney, H. A., Lubchenco, J. and Melillo, J., 1997. Human domination of Earth's ecosystems. *Science* **277:** 494–99.

Walsh, F., 1995. Interactions between land management agencies and Australian Aboriginal people: rationale, problems and some lessons. Pp. 88–106 *in* Nature Conservation 4: the Role of Networks ed by D. A. Saunders, J. L. Craig and E. M. Mattiske. Surrey Beatty & Sons: Chipping Norton, New South Wales.

INDEX